NUCLEAR
STRUCTURE 98

NUCLEAR STRUCTURE 98

Gatlinburg, Tennessee   August 1998

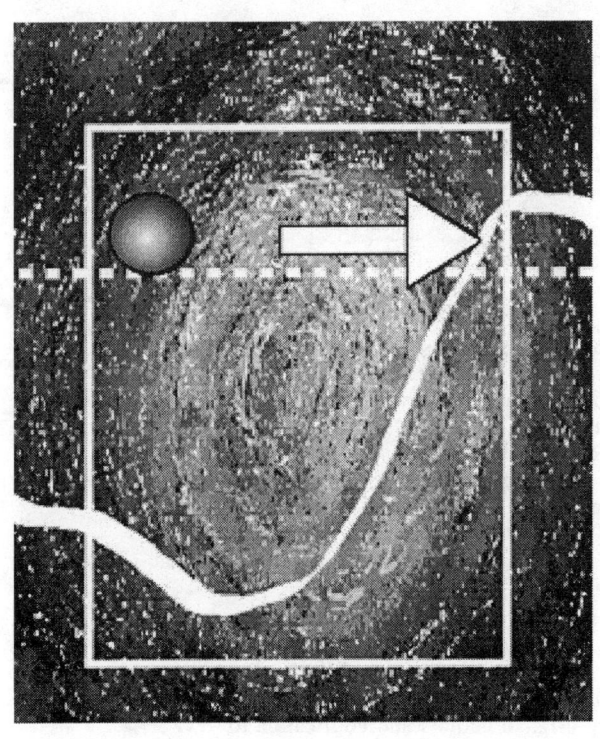

EDITOR
C. Baktash
Physics Division,
Oak Ridge National Laboratory

AIP

American Institute of Physics

**AIP CONFERENCE
PROCEEDINGS 481**

Woodbury, New York

Editor:

C. Baktash
Physics Division
Oak Ridge National Laboratory
P.O. Box 2008
Oak Ridge, TN 37831-6371
U.S.A.

E-mail: baktashc@ornl.gov

The articles on pp. 121–126, 393–406, 464–472, and 527–533 were authored by U.S. Government employees and are not covered by the below mentioned copyright.

L.C. Catalog Card No. 99-63979
ISBN 1-56396-858-4
ISSN 0094-243X
DOE CONF- 990881

Printed in the United States of America

CONTENTS

NUCLEAR MODELS AND EFFECTIVE INTERACTIONS

LIMITS OF ISOSPIN AND DRIP-LINE PHYSICS

PHYSICS NEAR THE PROTON-DRIP LINE

EXOTIC SHAPES AND HIGH-SPIN PHENOMENA

COLLECTIVE MODES OF EXCITATIONS

INTERFACE OF NUCLEAR STRUCTURE AND ASTROPHYSICS

FUTURE DETECTOR SYSTEMS

Preface

The Nuclear Structure 98 conference was held in Gatlinburg, Tennessee, USA, from August 10 to August 15, 1998. This conference was the latest in the series of biennial meetings on nuclear structure that have been sponsored by the low-energy heavy ion laboratories in the U.S. and Canada. The conference was organized by the Physics Division of the Oak Ridge National Laboratory and was sponsored by the Lockheed Martin Energy Research Corporation and the Joint Institute for Heavy Ion Research at Oak Ridge. We also received financial support from a number of industrial sponsors listed on the following pages.

With the advent of powerful gamma-ray detection systems and the first-generation radioactive ion beam facilities, many exciting discoveries have been made in the past few years. Likewise, theoretical activities have continued at a brisk pace to provide interpretation of these new results and to suggest directions for future explorations. The goal of the conference was to provide both a broad perspective of the recent progress in experimental and theoretical nuclear structure physics, and to point out the exciting opportunities that lie ahead in the next few years. A notable feature of the conference was the great emphasis given to theoretical discussions, which represented nearly one third of the presentations. Among the topics addressed at the conference were nuclear models and effective interactions, physics of extreme isospin and nuclear structure with radioactive ion beams, exotic shapes and high-spin phenomena, collective modes of excitation, interface of nuclear structure and astrophysics, and future detector systems and experimental tools.

The program consisted of 31 invited talks that provided an overview of the key subjects addressed at the conference and 32 presentations selected from the contributed abstracts. Texts of 58 of these talks, as well as the conference program and the titles of the poster presentations are included in this book. In response to our call for abstract submission, we received nearly 160 contributions. These abstracts were assembled as Volume I of the Proceedings and distributed among the participants at the time of the conference.

I am most grateful to the members of the Organizing Committee for their invaluable help in the preparation for and smooth running of the conference. Special thanks are due to Carl Gross who managed our Web Home Page, David Radford for setting up and managing the computer network at the conference site, and Michael Smith for preparing the conference signs. I am also indebted to the Program Advisory Committee for their advice in selecting timely topics and excellent speakers who made this conference a great success. I would like to express my deep appreciation to Jackie Smith, the Conference Secretary, for her dedicated and tireless efforts with all organizational aspects of the conference and with the preparation of the two volumes of the Proceedings. I am also grateful to Franda Ervin, Sherry Lamb, Mary-Ruth Lay, Ann McCoy, and Ms. Joy Lee from the ORNL conference office for their invaluable assistance in carrying out the many duties associated with the conference. Finally, I wish to express my appreciation to Edda Reviol for her original painting that appears as the conference logo, and to Suzanne Baktash, Jeanne Bertrand, and Anne Garrett, members of the Hospitality Committee, who organized outings and sightseeing trips for the participants and their companions during the conference.

We look forward to the Nuclear Structure 2000 which will be hosted by the National Superconducting Laboratoy in East Lansing, Michigan, USA.

Cyrus Baktash

INTERNATIONAL PROGRAM ADVISORY COMMITTEE

Bazzacco, D.
Casten, R. F.
Dobaczewski, J.
Donnelly, T. W.
Dracoulis, G. D.
Fossan, D. B.
Friar, J. L.
Gelletly, W.
Grawe, H.
Haas, B.
Hamilton, J. H.
Hansen, P. G.
Hardy, J. C.
Herskind, B.

Janssens, R. V. F.
Julin, R.
Lee, I.-Y.
Matsuyanagi, K.
Mottelson, B.
Mueller, A.
Riley, M. A.
Ring, P.
Sharpey-Schafer, J.
Thielemann, F. K.
Van-Duppen, P.
Waddington, J. C.
Wyss, R.

ORGANIZING COMMITTEE

Akovali, Y. A.
Baktash, C. (Chair)
Beene, J. R.
Bertrand, F. E.
Carter, H. K.
Dean, D. J.
Galindo-Uribarri, A.
Garrett, J. D.
Gross, C. J.
Johnson, N. R.

Nazarewicz, W.
Plasil, F.
Radford, D. C.
Reviol, W.
Rykaczewski, K. P.
Riedinger, L. L.
Strayer, M. R.
Toth, K. S.
Varner, R. L.
Yu, C.-H.

SPONSORS

Physics Division, Oak Ridge National Laboratory
Lockheed Martin Energy Research Corporation
Joint Institute for Heavy Ion Research
EG&G/ORTEC
EURISYS MESURES, Inc.
LeCroy
SCIONIX
WIENER-US

NUCLEAR MODELS AND EFFECTIVE INTERACTIONS

Nuclear Forces and Nuclear Structure

R. Machleidt

Department of Physics, University of Idaho, Moscow, Idaho 83844
Electronic address: machleid@uidaho.edu

Abstract. After a historical review, I present the progress in the field of realistic NN potentials that we have seen in recent years. A new generation of very quantitative (high-quality/high-precision) NN potentials has emerged. These potentials will serve as reliable input for microscopic nuclear structure calculations and will allow for a systematic investigation of off-shell effects. The issue of three-nucleon forces is also discussed.

INTRODUCTION

The goal of nuclear physics is to explain the properties of atomic nuclei in fundamental terms—where the stress is on *fundamental*. The debate then begins with what the fundamental ingredients (or 'first principles') are from which we should start. Nuclear physics is strong interaction physics, so, based upon the Standard Model, it should be quarks exchanging gluons. However, in the spirit of the currently fashionable effective field theories, one may argue that any theory is effective, and what theory is appropriate depends on the energy scale. In fact, Weinberg [1] pointed out 20 years ago that an effective field theory of nucleons and mesons that observes the same symmetries as QCD is equivalent to QCD. This allows us to consider mesons and nucleons as the basic items, on our energy scale. The spin-off from the mesons is the nuclear force.

Thus, it is suggested that the appropriate fundamental ingredients that nuclear structure should be based upon are nucleons and the 'fundamental' nucleon-nucleon (NN) interaction (created by meson exchange). The above discussion also provides clear guideline for how to extend the model if it fails. One may introduce explicit meson degrees of freedom and one may complement the nucleon by meson-nucleon resonances (e.g., the $\Delta(1232)$ isobar); moreover, one may take into account the quark substructure of nucleons. However, it should be stressed that the starting point—besides being fundamental—must also be simple. Nucleons interacting via the two-nucleon force is the simplest of all basic pictures. Only where we have clear evidence that this frame work is insufficient may we extend it. For most problems of conventional nuclear structure, we do not have such evidence at this time.

Starting from nucleons and the bare two-nucleon force, any nuclear many-body problem one may consider is exactly defined. However, due to limitations in computing power, the problem can be solved exactly only up to $A = 8$ [2], and since the needed computing time goes up as factorial of A, there is little hope that we will ever get beyond $A = 12$. Therefore, in most problems of nuclear structure and reactions, the first step is to derive an effective interaction from the bare NN potential. Usually, this involves Brueckner theory or some variation/extension thereof [3,4]. However, it is important to stress that—no matter what many-body theory or model is used—this step has to be done in a *parameter-free* way. Finally, the effective interaction may be applied in the shell model, yielding predictions for nuclear properties.

Following this scheme (which is known as *microscopic* nuclear structure), the predictions depend only on two fundamental items: the bare NN potential and the nuclear many-body theory/model applied. Thus, the comparison of the predictions with the data will allow for conclusions concerning the NN potential (off-shell) and the appropriate nuclear many-body theory. In other words, we will learn something about the foundations of our profession, which is our ultimate goal.

In this conference, we have many contributions that deal with particular effective interactions, shell-model applications, predicted and measured nuclear properties. It is the purpose of this contribution to fill-in on one of the fundamental ingredients that is at the outset of all this, namely, the NN interaction. As it turns out, we

CP481, *Nuclear Structure 98*, edited by C. Baktash

have had substantial progress in recent years and—as a result of this—the starting conditions for *microscopic* nuclear structure calculations are better than ever.

To put the recent advances into perspective, I will start with a brief historical review and then present the new developments. Artificially, I will draw the line between 'history' and recent progress/new developments in the year of 1990.

HISTORICAL PERSPECTIVE

Here, I will give a brief summary of the status in the field of nuclear forces before 1990.[1] This will make it easier for us to appreciate the progress of the past eight years. Historically, one may distinguish between three decades of work on the meson theory of nuclear forces, namely, the 60's, 70's, and 80's. Each decade is characterized by progress in a particular sector of meson theory, namely, the one-boson-exchange model, the two-pion exchange contribution to the NN interaction, and contributions beyond 2π, respectively.

The One-Boson-Exchange Model

The first quantitative meson-theoretic models for the NN interaction were the one-boson-exchange potentials (OBEP). They emerged right after the experimental discovery of heavy mesons in the early 1960's. In general, about six non-strange bosons with masses below 1 GeV are taken into account: the pseudoscalar mesons $\pi(138)$ and $\eta(549)$, the vector mesons $\rho(769)$ and $\omega(783)$, and two scalar bosons $a_0/\delta(983)$ and $\sigma(\approx 550)$. The first particle in each group is isovector (isospin $I = 1$) while the second is isoscalar ($I = 0$). The pion provides the tensor force, which is reduced at short range by the ρ meson. The ω creates the spin-orbit force and the short-range repulsion, and the σ is responsible for the intermediate-range attraction. Thus, it is easy to understand why a model which includes at least four mesons can reproduce the major properties of the nuclear force.[2]

A classic example for an OBEP is the Bryan-Scott potential published in 1969 [7]. Since it is suggestive to think of a potential as a function of r (where r denotes the distance between the centers of the two interacting nucleons), the OBEP's of the 1960's where represented as local r-space potentials. To reduce the original one-meson-exchange Feynman amplitudes to such a simple form, drastic approximations have to be applied. The usual method is to expand the amplitude in terms of p^2/M and keep only terms up to first order (and, in many cases, not even all of them). Commonly, this is called the *nonrelativistic OBEP*. Besides the suggestive character of a local function of r, such potentials are easy to apply in r-space calculations. However, the original potential (Feynman amplitudes) is nonlocal, and, thus, has a very different off-shell behaviour than its local approximation. Though this does not play a great role in two-nucleon scattering, it becomes important when the potential is applied to the nuclear few- and many-body problem. In fact, it turns out that the original nonlocal potential leads to much better predictions in nuclear structure than the local approximation (see below).

Historically, one must understand that after the failure of the pion theories in the 1950's, the one-boson-exchange (OBE) model was considered a great success in the 1960's. However, there are conceptual and quantitative problems with this model.

A questionable point of the OBE model is the fact that it has to introduce a scalar-isoscalar σ-boson in the mass range 500–700 MeV, the existence of which is controversial (for a current status report, see Ref. [8]). Furthermore, the model is restricted to single exchanges of bosons that are 'laddered' in an unitarizing equation. Thus, irreducible multi-meson exchanges, which may be quite sizable (see below), are neglected.

Quantitatively, a major drawback of the *nonrelativistic* OBE model is its failure to describe certain partial waves correctly. The Bryan-Scott nonrelativistic OBE potential predicts $^1P_1$ and $^3D_2$ phases substantially above the data. [9]

An important advance during the 1970's has been the development of the *relativistic OBEP* [10]. In this model, the full, relativistic Feynman amplitudes for the various one-boson-exchanges are used to define the

---

[1] A more comprehensive historical account is given in Refs. [5,6].

[2] The interested reader will find a detailed, pedagogical introduction into the OBE model in sections 3 and 4 of Ref. [5]. In this

4

potential. These nonlocal expressions do not pose any numerical problems when used in momentum space.[3] The quantitative deficiencies of the nonrelativistic OBEP disappear when the non-simplified, relativistic, nonlocal OBE amplitudes are used.

The *Nijmegen potential* [11], published in 1978, is a nonrelativistic r-space OBEP. As a latecomer, it is one of the most sophisticated examples of its kind. It includes all non-strange mesons of the pseudoscalar, vector, and scalar nonet. Thus, besides the six mesons mentioned above, the $\eta'(958)$, $\phi(1020)$, and $S^*(993)$ are taken into account. The model also includes the dominant J=0 parts of the Pomeron and tensor (f, f', A_2) trajectories, which essentially lead to repulsive central Gaussian potentials. For the Pomeron (which from today's point of view may be considered as a multi-gluon exchange) a mass of 308 MeV is assumed. However, this potential is still defined in terms of the nonrelativistic local approximations to the OBE amplitudes. Therefore, it shows exactly the same problems as its ten years older counterpart, the Bryan-Scott potential: the Nijmegen potential fails to predict the $^1P_1$ and $^3D_2$ phase shifts correctly to the same extent as the Bryan-Scott potential. In addition, the Nijmegen potential overpredicts the $^3D_3$ phase shifts by more than a factor of two [9], a problem the models of the 1960's did not have. The inclusion of more bosons does obviously not improve the model and is, thus, unnecessary. It is more important to calculate the one-boson exchange amplitudes correctly and without the local approximations.

The 2π-Exchange Contribution to the NN Interaction

In the 1970's, work on the meson theory of the nuclear force focused on the 2π-exchange contribution to the NN interaction to replace the σ-boson. One way to calculate these contributions is by means of dispersion relations. Around 1970, many groups throughout the world were involved in this approach; we mention here, in particular, the Stony Brook [12] and the Paris [13] groups. These groups could show that the intermediate-range part of the nuclear force is, indeed, decribed correctly by the 2π-exchange as obtained from dispersion integrals.

To construct a complete potential, the Stony Brook as well as the Paris group complemented their 2π-exchange contribution by one-pion-exchange (OPE) and ω exchange. In addition to this, the Paris potential [14] contains a phenomenological short-range potential for $r < 1.5$ fm.

In microscopic nuclear structure calculations, the off-shell behavior of the NN potential is important. The fit of NN potentials to two-nucleon data fixes them only on-shell. The off-shell behavior cannot, in principle, be extracted from two-body data. Theory could determine the off-shell nature of the potential; however, not every theory can do that. Dispersion theory relates observables (equivalent to on-shell T-matrices) to observables; e. g., πN data to NN data. Thus, dispersion theory cannot, in principle, provide any off-shell information. The Paris potential is based upon dispersion theory; thus, the off-shell behavior of this potential is not determined by the underlying theory. On the other hand, every potential does have an off-shell behavior. When undetermined by theory, then the off-shell behavior is a silent by-product of the parametrization chosen to fit the on-shell T-matrix, with which the potential is identified. In summary, due to its basis in dispersion theory, the off-shell behaviour of the Paris potential is not derived on theoretical grounds. This is a serious drawback when it comes to the question of how to interpret nuclear structure results obtained with this potential.

The Field-Theoretic Approach to 2π-Exchange

In a field-theoretic picture, the interaction between mesons and baryons is described by effective Lagrangians. The NN interaction can then be derived in terms of field-theoretic perturbation theory. The lowest order (that is, the second order in terms of meson-baryon interactions) are the one-boson-exchange diagrams, which are easy to calculate.

More difficult (and more numerous) are the irreducible two-meson-exchange (or fourth order) diagrams. It is reasonable to start with the contributions of longest range. These are the graphs that exchange two pions. There are two complications that need to be taken into account: meson-nucleon resonances and meson-meson scattering. The lowest πN resonance, the so-called Δ isobar with a mass of 1232 MeV, gives rise to one set of diagrams. The other set involves $\pi\pi$ interactions. Since it is well established that two pions in relative

---

[3] In fact, in momentum space, the application of a nonlocal potential is numerically as easy as using the momentum-space representation of a local potential.

P-wave form a resonance (the ρ meson), there is no problem with ρ exchange. However, the existence of a proper resonance in $\pi\pi - S$-wave below 900 MeV is controversial [8]. In any case, there are strong correlations between two pions in S-wave at low energies. Durso et al. [15] have shown that these correlations can be described in terms of a broad mass distribution of about 600 ± 260 MeV, which in turn can be approximated by a zero-width scalar-isoscalar (σ) boson of mass 550 MeV [16].

In summary, two approaches are available for calculating the 2π-exchange contribution to the NN interaction: dispersion theory (Paris [13,14]) and field theory (Bonn [16]). One can compare the predictions by the two approaches with each other as well as with the data. For this purpose, one looks into the peripheral partial waves of NN scattering. By and large, one finds satisfactory agreement [16].

$\pi\rho$-Contributions and the Bonn Potential

A model consisting of $\pi + 2\pi + \omega$ describes the peripheral partial waves quite satisfactory. However, when proceeding to lower partial waves (equivalent to shorter internucleonic distances), this model generates too much attraction. This is true for the dispersion-theoretic (Paris) as well as the field-theoretic (Bonn) result. Obviously, further measures have to be taken at short range to arrive at a quantitative model for the NN interaction. It is at this point that the philosophies of the Bonn and Paris groups diverge.

The Paris group decided to give up meson theory at this stage and to describe everything that is still missing by phenomenology. Thus, they added a phenomenological short-range potential for $r < 1.5$ fm which requires many parameters (see Table 1, below) that do not allow for a clear-cut physical interpretation. This short-range potential affects the S-, P-, and D-waves of NN scattering which are, therefore, largely a product of phenomenology in the Paris model.

In contrast, the Bonn group continued to consider further irreducible two-meson exchanges. The next set of diagrams to be considered are the exchanges of π and ρ. As it turns out, these contributions very accurately take care of the discrepancies that remained between theory and experiment [16].

Thus, the $\pi\rho$ contributions provide the short-range repulsion which was still missing. It is important to note that the $\pi\rho$ contributions have only one free parameter, namely the cutoff for the $\rho N\Delta$ vertex. The other parameters involved occur also in other parts of the model and were fixed before (like the πNN and ρNN coupling constants and cutoff parameters). Notice also that the $\pi N\Delta$ and $\rho N\Delta$ coupling constants are not free parameters, since they are related to the corresponding NN coupling constants by $SU(3)$.

In summary, a proper meson-theory for the NN interaction should include the irreducible diagrams of $\pi\rho$ exchange. This contribution has only one free parameter and makes comprehensive short-range phenomenology unnecessary. Thus, meson theory can be truly tested in the low-energy NN system.

In the 1970's and 80's, a field-theoretic model for the NN interaction was developed at the University of Bonn. This model consists of single π, ω, and a_0/δ exchange, the field-theoretic 2π model, and $\pi\rho$ diagrams, as well as a few more irreducible 3π and 4π diagrams (which are not very important, but indicate convergence of the diagrammatic expansion). This quasi-potential has become known as the 'Bonn full model' [16]. It has 12 parameters which are the coupling constants and cutoff masses of the meson-nucleon vertices involved. With a reasonable choice for these parameters, a very satisfactory description of the NN observables up to 300 MeV is achieved (see Table 1). Since the goal of the Bonn model was to put meson theory to a real test, no attempt was ever made to minimize the χ^2 of the fit of the NN data. Nevertheless, the Bonn full model shows the smallest χ^2 for the fit of the NN data among the traditional models.

Summary

In Table 1, we give a summary and an overview of the theoretical input of some representative meson-theoretic NN models of the pre-1990 era. Moreover, this table also lists the χ^2/datum for the fit of the world NN data below 300 MeV laboratory energy, which is 5.12, 3.71, and 1.90 for the Nijmegen [11], Paris [14], and Bonn [16] potentials, respectively. This compact presentation, typical for a table, makes it easy to grasp one important point: The more seriously and consistently meson theory is pursued, the better the results. This table and its trend towards the more comprehensive meson models is the best proof for the validity of meson theory in the low-energy nuclear regime. To show this fact, which is of fundamental importance to our field, was the major achievement of the pre-1990 era.

TABLE 1. Comparison of some typical meson-theoretic nucleon-nucleon models of the pre-1990 era.

	Nijmegen(1978) [11]	Paris Potential(1980) [14]	Bonn full model(1987) [16]
# of free parameters	15	≈ 60	12
Theory includes:			
OBE terms	Yes	Yes	Yes
2π exchange	No	Yes	Yes
$\pi\rho$ diagrams	No	No	Yes
Relativity	No	No	Yes
χ^2/datum for fit of world NN data:			
pp data	2.06	2.31	1.94
np data	6.53	4.35	1.88
pp and np data	5.12	3.71	1.90

While all models considered in Table 1 describe the proton-proton (pp) data well (with χ^2/datum ≈ 2), some models have a problem with the neutron-proton (np) data (with χ^2/datum $\approx 4-6$). For the case of the Paris potential (and, in part, for the Nijmegen potential) this is due to a bad reproduction of the np total cross section data. When the latter data are ignored, the Paris potential fits np as well as pp. The Nijmegen and the Paris potential predict too large np total cross sections because their $^3D_2$ phase shifts are too large [9].

This finishes the review of the developments concerning the NN interaction up to the late 1980's. Next, we turn to the progress made in the 1990's.

RECENT PROGRESS IN NN POTENTIALS

In the 1990's, the focus has been on the quantitative aspect of the NN potentials. Even the best NN models of the past fit the NN data typically with a χ^2/datum ≈ 2 or more. This is still substantially above the perfect χ^2/datum ≈ 1. To put microscopic nuclear structure theory to a reliable test, one needs a perfect NN potential such that discrepancies in the predictions cannot be blamed on a bad fit of the NN data by the potential.

To construct perfect NN potentials one needs, first, a perfect NN analysis. About a decade ago, the Nijmegen group embarked on a program to substantially improve NN phase shift analysis. Finally, in 1993, they could publish a phase-shift analysis of all proton-proton and neutron-proton data below 350 MeV laboratory energy with a χ^2 per datum of 0.99 for 4301 data [17]. Based upon this analysis, new 'high-precision' or 'high-quality' NN potentials have been constructed which, similar to the analysis, fit the NN data with a χ^2/datum ≈ 1.

I will first review the Nijmegen NN analysis and then the new potentials.

The Nijmegen NN Analysis

In spite of the huge NN database available today, conventional phase shift analyses are by no means perfect. For example, the phase shift solutions obtained by Bugg [18] or the VPI group [19] typically have a χ^2/datum of about 1.4, for the energy range 0–425 MeV. This may be due to inconsistencies in the data as well as deficiencies in the constraints applied in the analysis. In any case, it is a matter of fact that within the conventional phase shifts analysis, in which the lower partial waves are essentially unconstrained, a better fit cannot be achieved.

To further improve NN analysis, the Nijmegen group took two decisive measures [17]. First, they 'pruned' the data base; i.e., they scanned very critically the world NN data base (all data in the energy range 0-350 MeV laboratory energy published in a regular physics journal between January 1955 and December 1992) and eliminated all data that had either an improbably high χ^2 (more than three standard deviations off) or an improbably low χ^2; of the 2078 world pp data below 350 MeV 1787 survived the scan, and of the 3446 np data 2514 survived. Second, they introduced sophisticated, semi-phenomenological model assumptions into the analysis. Namely, for each of the lower partial waves ($J \leq 4$) a different energy-dependent potential is adjusted to constrain the energy-dependent analysis. Phase shifts are obtained using these potentials in a Schroedinger equation. From these phase shifts the predictions for the observables are calculated including the χ^2 for the fit of the experimental data. This χ^2 is then minimized as a function of the parameters of the partial-wave potentials. Thus, strictly speaking, the Nijmegen analysis is a *potential analysis*; the final phase shifts are the ones predicted by the 'optimized' partial-wave potentials.

In the Nijmegen analysis, each partial-wave potential consists of a short- and a long-range part, with the separation line at $r = 1.4$ fm. The long-range potential V_L ($r > 1.4$ fm) is made up of an electromagnetic part V_{EM} and a nuclear part V_N:

$$V_L = V_{EM} + V_N \tag{1}$$

The electromagnetic interaction can be written as

$$V_{EM}(pp) = V_C + V_{VP} + V_{MM}(pp) \tag{2}$$

for proton-proton scattering and

$$V_{EM}(np) = V_{MM}(np) \tag{3}$$

for neutron-proton scattering, where V_C denotes an improved Coulomb potential (which takes into account the lowest-order relativistic corrections to the static Coulomb potential and includes contributions of all two-photon exchange diagrams); V_{VP} is the vacuum polarization potential, and V_{MM} the magnetic moment interaction.

The nuclear long-range potential V_N consists of the local one-pion-exchange (OPE) tail V_π (the coupling constant g_π being one of the parameters used to minimize the χ^2) multiplied by a factor M/E and the tail of the heavy-boson-exchange (HBE) contributions of the Nijmegen78 potential [11] V_{HBE}, enhanced by a factor of 1.8 in singlet states; i. e.

$$V_N = \frac{M}{E} \times V_\pi(g_\pi, m_\pi) + f(S) \times V_{HBE} \tag{4}$$

with $f(S = 0) = 1.8$ and $f(S = 1) = 1.0$, where S denotes the total spin of the two-nucleon system. The energy-dependent factor M/E (with $E = \sqrt{M^2 + q_0^2}$, $q_0^2 = MT_{lab}/2$) takes into account relativity in a 'minimal' way, damping the nonrelativistic OPE potential at higher energies.

As indicated, V_π depends on the πNN coupling constant g_π and the pion mass m_π, which gives rise to charge dependence. For pp scattering, the OPE potential is

$$V_\pi^{pp} = V_\pi(g_{\pi^0}, m_{\pi^0}) \tag{5}$$

with m_{π^0} the mass of the neutral pion. In np scattering, we have to distinguish between $T = 1$ and $T = 0$:

$$V_\pi^{np}(T) = -V_\pi(g_{\pi^0}, m_{\pi^0}) + (-1)^{T+1} 2V_\pi(g_{\pi^\pm}, m_{\pi^\pm}) \tag{6}$$

The partial-wave short-range potentials ($r \leq 1.4$ fm) are energy-dependent square-wells (see Figs. 2 and 3 of Ref. [17]). The energy-dependence of the depth of the square-well is parametrized in terms of up to three parameters per partial wave. For the states with $J \leq 4$, there are a total of 39 such parameters (21 for pp and 18 for np) plus the pion-nucleon coupling constants (g_{π^0} and $g_{\pi^\pm}$).

In the Nijmegen np analysis, the $T = 1$ np phase shifts are calculated from the corresponding pp phase shifts (except in $^1S_0$ where an independent analysis is conducted) by applying corrections due to electromagnetic effects and charge dependence of OPE. Thus, the np analysis determines $^1S_0(np)$ and the $T = 0$ states, only.

In the combined pp and np analysis [17], the fit for 1787 pp data and 2514 np data below 350 MeV results in the 'perfect' χ^2/datum = 0.99.

The New High-Precision NN Potentials

Based upon the Nijmegen analysis and the (pruned) Nijmegen data base, new charge-dependent NN potentials were constructed in the early/mid 1990's. The groups involved and the names of their new creations are, in chronological order:

- Nijmegen group [20]: Nijm-I, Nijm-II, and Reid93 potentials.

- Argonne group [21]: V_{18} potential.

- Bonn group [22]: CD-Bonn potential.

All these potentials have in common that they use about 45 parameters and fit the pruned Nijmegen data base with a χ^2/datum ≈ 1 (cf. Table 2, below). The larger number of parameters (as compared to previous OBE models) is necessary to achieve the very accurate fit.

Concerning the theoretical basis of these potential, one could say that they are all—more or less—constructed 'in the spirit of meson theory' (e.g., all potentials include the one-pion-exchange contribution). However, there are considerable differences in the details leading to considerable off-shell differences among the potentials.

To explain these details and differences in a systematic way, let me first sketch the general scheme for the derivation of a meson-theoretic potential.

One starts from field-theoretic Lagrangians for meson-nucleon coupling, which are essentially fixed by symmetries. Typical examples for such Langrangians are:

$$\mathcal{L}_{ps} = -g_{ps}\bar{\psi}i\gamma^5\psi\varphi^{(ps)} \tag{7}$$

$$\mathcal{L}_s = g_s\bar{\psi}\psi\varphi^{(s)} \tag{8}$$

$$\mathcal{L}_v = g_v\bar{\psi}\gamma^\mu\psi\varphi_\mu^{(v)} + \frac{f_v}{4M}\bar{\psi}\sigma^{\mu\nu}\psi(\partial_\mu\varphi_\nu^{(v)} - \partial_\nu\varphi_\mu^{(v)}) \tag{9}$$

where ps, s, and v denote pseudoscalar, scalar, and vector couplings/fields, respectively.

The lowest order contributions to the nuclear force from the above Lagrangians are the second-order Feynman diagrams which, in the center-of-mass frame of the two interacting nucleons, produce the amplitude:

$$A_\alpha(\mathbf{q}',\mathbf{q}) = \frac{\bar{u}_1(\mathbf{q}')\Gamma_1^{(\alpha)}u_1(\mathbf{q})P_\alpha\bar{u}_2(-\mathbf{q}')\Gamma_2^{(\alpha)}u_2(-\mathbf{q})}{(q'-q)^2 - m_\alpha^2} , \tag{10}$$

where $\Gamma_i^{(\alpha)}$ ($i = 1,2$) are vertices derived from the above Lagrangians, u_i are Dirac spinors representing the nucleons, and q and q' are the nucleon relative momenta in the initial and final states, respectively; P_α divided by the denominator is the meson propagator.

The simplest meson-exchange model for the nuclear force is the one-boson-exchange (OBE) potential [5] which sums over several second-order diagrams, each representing the single exchange of a different boson, α:

$$V(\mathbf{q}',\mathbf{q}) = \sqrt{\frac{M}{E'}}\sqrt{\frac{M}{E}}\sum_\alpha iA_\alpha(\mathbf{q}',\mathbf{q})F_\alpha^2(\mathbf{q}',\mathbf{q}) . \tag{11}$$

As is customary, we include form factors, $F_\alpha(\mathbf{q}',\mathbf{q})$, applied to the meson-nucleon vertices, and a square-root factor $M/\sqrt{E'E}$ (with $E = \sqrt{M^2 + \mathbf{q}^2}$ and $E' = \sqrt{M^2 + \mathbf{q}'^2}$; M is the nucleon mass). The form factors regularize the amplitudes for large momenta (short distances) and account for the extended structure of nucleons in a phenomenological way. The square root factors make it possible to cast the unitarizing, relativistic, three-dimensional Blankenbecler-Sugar (BbS) equation for the scattering amplitude [a reduced version of the four-dimensional Bethe-Salpeter (BS) equation] into a form which is identical to the (nonrelativistic) Lippmann-Schwinger equation [5]. Thus, Eq. (11) defines a relativistic potential which can be consistently applied in conventional, nonrelativistic nuclear structure.

Clearly, the Feynman amplitudes, Eq. (10), are in general nonlocal expressions; i. e., Fourier transforming them into configuration space will yield functions of r and r', the relative distances between the two in- and out-going nucleons, respectively. The square root factors create additional nonlocality.

While nonlocality appears quite plausible for heavy vector-meson exchange (corresponding to short distances), we have to stress here that even the one-pion-exchange (OPE) Feynman amplitude is nonlocal. This is important because the pion creates the dominant part of the nuclear tensor force which plays a crucial role in nuclear structure.

Applying $\Gamma^{(\pi)} = g_\pi\gamma_5$ in Eq. (10), yields the Feynman amplitude for neutral pion exchange in pp scattering,

$$iA_\pi(\mathbf{q}',\mathbf{q}) = -\frac{g_\pi^2}{4M^2}\frac{(E'+M)(E+M)}{(\mathbf{q}'-\mathbf{q})^2+m_\pi^2}\left(\frac{\sigma_1\cdot\mathbf{q}'}{E'+M} - \frac{\sigma_1\cdot\mathbf{q}}{E+M}\right)$$
$$\times\left(\frac{\sigma_2\cdot\mathbf{q}'}{E'+M} - \frac{\sigma_2\cdot\mathbf{q}}{E+M}\right) . \tag{12}$$

This is the original and correct result for OPE.

9

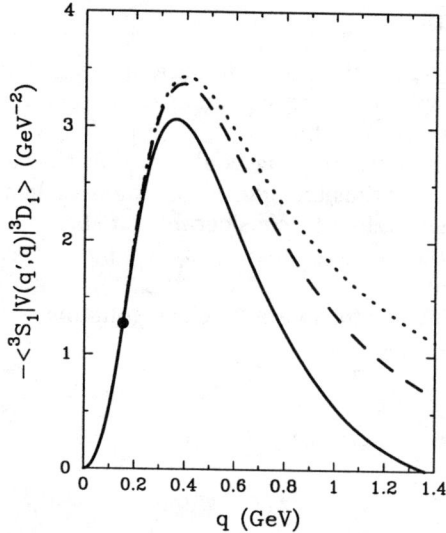

FIGURE 1. Half off-shell $^3S_1$–$^3D_1$ amplitude for the relativistic CD-Bonn potential (solid line), Eq. (11). The dashed curve is obtained when the local approximation, Eq. (13), is used for OPE, and the dotted curve results when this approximation is also used for one-ρ exchange. $q' = 153$ MeV.

If one now introduces the drastic approximation,

$$E' \approx E \approx M ,\tag{13}$$

then one obtains the momentum space representation of the *local* OPE,

$$V_\pi^{(loc)}(\mathbf{k}) = -\frac{g_\pi^2}{4M^2}\frac{(\boldsymbol{\sigma_1}\cdot\mathbf{k})(\boldsymbol{\sigma_2}\cdot\mathbf{k})}{\mathbf{k}^2 + m_\pi^2}\tag{14}$$

with $\mathbf{k} = \mathbf{q}' - \mathbf{q}$. Notice that on-shell, i. e., for $|\mathbf{q}'| = |\mathbf{q}|$, $V_\pi^{(loc)}$ equals $i\mathcal{A}_\pi$. Thus, the nonlocality affects the OPE potential off-shell.

Fourier transform of Eq. (14) yields the well-known local OPE potential in r-space,

$$\begin{aligned}V_\pi^{(loc)}(\mathbf{r}) = \frac{g_\pi^2}{12\pi}\left(\frac{m_\pi}{2M}\right)^2 &\left[\left(\frac{e^{-m_\pi r}}{r} - \frac{4\pi}{m_\pi^2}\delta^{(3)}(\mathbf{r})\right)\boldsymbol{\sigma_1}\cdot\boldsymbol{\sigma_2}\right.\\ &\left.+ \left(1 + \frac{3}{m_\pi r} + \frac{3}{(m_\pi r)^2}\right)\frac{e^{-m_\pi r}}{r}S_{12}\right] ,\end{aligned}\tag{15}$$

where m_π denotes the pion mass. Notice, however, that this 'well-established' local OPE potential is only an approximative representation of the correct OPE Feynman amplitude. A QED analog is the local Coulomb potential *versus* the full field-theoretic one-photon-exchange Feynman amplitude.

It is now of interest to know by how much the local approximation changes the original amplitude. This is demonstrated in Fig. 1, where the half off-shell $^3S_1$–$^3D_1$ potential, which can be produced only by tensor forces, is shown. The on-shell momentum q' is held fixed at 153 MeV (equivalent to 50 MeV laboratory energy), while the off-shell momentum q runs from zero to 1400 MeV. The on-shell point ($q = 153$ MeV) is marked by a solid dot. The solid curve is the CD-Bonn potential which contains the full, nonlocal OPE amplitude Eq. (12). When the static/local approximation, Eq. (13), is made, the dashed curve is obtained. When this approximation is also used for the one-ρ exchange, the dotted curve results. It is clearly seen that the static/local approximation substantially increases the tensor force off-shell. Clearly, we are dealing here not with negligible effects, and the local approximation is obviously not a good one.

Even though the spirit of the new generation of potentials is more sophistication, only the CD-Bonn potential uses the full, original, nonlocal Feynman amplitude for OPE, Eq. (12), while all other potentials still apply the local approximation, Eqs. (14) and (15). As a consequence of this, the CD-Bonn potential has a weaker tensor

TABLE 2. Modern high-precision NN potentials and their predictions for the two- and three-nucleon systems.

	CD-Bonn [22]	Nijm-I [20]	Nijm-II [20]	Reid93 [20]	V_{18} [21]	*Nature*
Character	nonlocal	nonloc. centr. pot. local otherwise	local	local	local	nonlocal
# of parameters	45	41	47	50	40	–
χ^2/datum	1.03	1.03	1.03	1.03	1.09	–
$g_\pi^2/4\pi$	13.6	13.6	13.6	13.6	13.6	14.0(5)
Deuteron properties:						
Quadr. moment (fm^2)	0.270	0.272	0.271	0.270	0.270	0.276(3)[a]
Asymptotic D/S state	0.0255	0.0253	0.0252	0.0251	0.0250	0.0256(4)
D-state probab. (%)	4.83	5.66	5.64	5.70	5.76	–
Triton binding (MeV):						
nonrel. calculation	8.00	7.72	7.62	7.63	7.62	–
relativ. calculation	8.2	–	–	–	–	8.48

[a] Corrected for meson-exchange currents and relativity.

force as compared to all other potentials. This is reflected in the predicted D-state probability of the deuteron, P_D, which is due to the nuclear tensor force. While CD-Bonn predicts $P_D = 4.83\%$, the other potentials yield $P_D = 5.7(1)\%$ (cf. Table 2). These differences in the strength of the tensor force lead to considerable differences in the nuclear structure predictions (see discussion below).

The OPE contribution to the nuclear force essentially takes care of the long-range interaction and the tensor force. In addition to this, all models must describe the intermediate and short range interaction, for which very different approaches are taken. The CD-Bonn includes (besides the pion) the pseudoscalar $\eta(549)$ meson, the vector mesons $\rho(769)$ and $\omega(783)$, and two scalar bosons $a_0/\delta(983)$ and $\sigma(550)$, using the full, nonlocal Feynman amplitudes, Eq. (10), for their exchanges. Thus, all components of the CD-Bonn are nonlocal and the off-shell behavior is the original one that is determined from relativistic field theory.

The models Nijm-I and Nijm-II are based upon the Nijmegen78 potential [11] (discussed in the historical section above) which is constructed from approximate OBE amplitudes. Whereas the Nijm-I uses the totally local approximations for all OBE contributions, the Nijm-I keeps some nonlocal terms in the central force component (but the Nijm-I tensor force is totally local). Nonlocalities in the central force have only a very moderate impact on nuclear structure as compared to nonlocalities in the tensor force. Thus, if for some reason one wants to keep only some of the original nonlocalities in the nuclear force and not all of them, then it would be more important to keep the tensor force nonlocalities.

The Reid93 [20] and Argonne V_{18} [21] potentials do not use meson-exchange for intermediate and short range; instead, a phenomenological parametrization is chosen. The Argonne V_{18} uses local functions of Woods-Saxon type, while Reid93 applies local Yukawas of multiples of the pion mass, similar to the original Reid potential of 1968 [23]. At very short distances, the potentials are regularized either by exponential (V_{18}, Nijm-I, Nijm-II) or by dipole (Reid93) form factors (which are all local functions).

In Fig. 2, the five high-precision potentials (in momentum space) and their phase shift predictions are shown, for the $^1S_0$ and $^3S_1$ states. While the phase shift predictions are indistinguishable, the potentials differ widely—due to the theoretical and mathematical differences discussed. Note that NN potentials differ the most in S-waves and converge with increasing L (where L denotes the total orbital angular momentum of the two-nucleon system).

Before we finish this section, a word about charge dependence is in place. All new potentials are charge-dependent which is essential for obtaining a good χ^2. Thus, each potential comes in three variants: pp, np, and nn. The difference between pp (without electromagnetic effects) and np is essentially given by the charge dependence of OPE, Eqs. (5) and (6), for states with $L > 0$. However, in the $^1S_0$ state, this explains only about 50% of the empirically known difference between the pp and np scattering lengths. The remainder is fitted phenomenologically. Also the pp and nn singlet scattering lengths differ, which breaks charge symmetry. The impact of the mass difference between proton and neutron on the kinetic energy and on the OBE amplitudes, alone, does not explain this difference. Therefore, the CD-Bonn and the Argonne V_{18} nn potentials are phenomenologically adjusted to reproduce the nn scattering length (a_{nn}). The potentials by the Nijmegen group (Nijm-I, Nijm-II, Reid93) do not reproduce a_{nn}.

The above discussion reveals that there is still room for improvement concerning the charge dependence of the new high-precision potentials. It is well-known that there are, besides OPE, other mechanisms which create

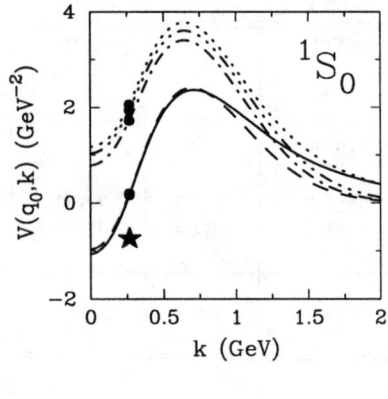
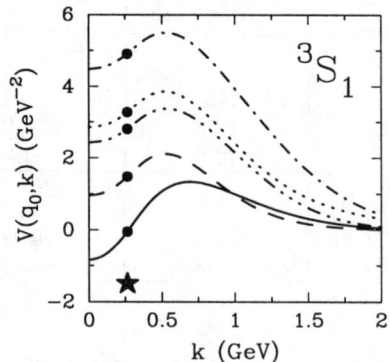
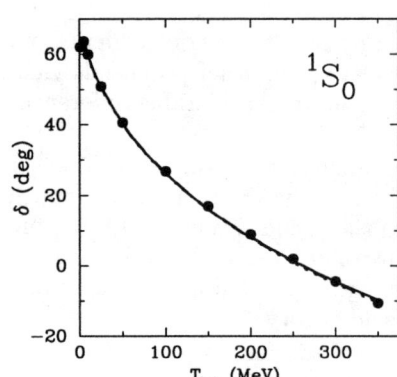
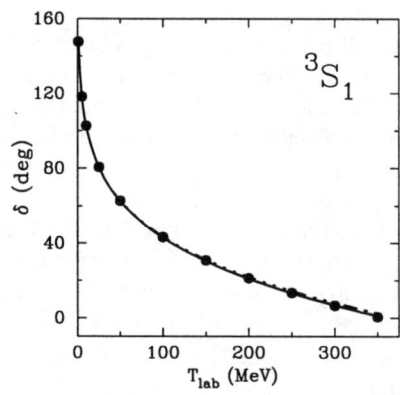

FIGURE 2. **Upper part:** Matrix elements $V(q_0, k)$ of the $^1S_0$ and $^3S_1$ potentials for the CD-Bonn (solid line), Nijm-I (dashed), Nijm-II (dash-dot), Argonne V_{18} (dash-triple-dot) and Reid93 (dotted) potentials. The diagonal matrix elements with $k = q_0 = 265$ MeV/c (equivalent to $T_{lab} = 150$ MeV) are marked by a solid dot. The corresponding matrix element of the full scattering R-matrix is marked by the star. **Lower part:** Predictions for the np phase shifts in the $^1S_0$ and $^3S_1$ state by the five potentials. The five curves are essentially indistinguishable. The solid dots represent the Nijmegen multi-energy np analysis [17].

charge dependence of the nuclear force. Recently, the charge dependence and charge asymmetry as created by a comprehensive microscopic model have been calculated carefully [24,25]. A refined version of the CD-Bonn potential [26] which incorporates the results of this comprehensive study is in preparation. To give an idea of how important additional refinements are, we note that the current charge-dependent, high-precision potentials cannot explain the Nolen-Schiffer anomaly [27]. Taking the charge asymmetry derived in Ref [25] into account, however, explains the anomaly correctly [28].

NN POTENTIALS AND NUCLEAR STRUCTURE

Our goal for this section is to understand how the off-shell NN potential influences nuclear structure predictions. For this, we need to discuss first how the NN t-matrix is calculated, since it is an important quantity in the construction of potentials.

For a given NN potential V, the t-matrix for free-space two-nucleon scattering is obtained from the Lippmann-Schwinger equation, which in the center-of-mass (c.m.) frame reads

$$t(\mathbf{q}', \mathbf{q}; E) = V(\mathbf{q}', \mathbf{q}) - \int d^3k V(\mathbf{q}', \mathbf{k}) \frac{M}{k^2 - ME - i\epsilon} t(\mathbf{k}, \mathbf{q}; E) \tag{16}$$

and in partial-wave decomposition

$$t_{L'L}^{JST}(q', q; E) = V_{L'L}^{JST}(q', q) - \sum_{L''} \int_0^\infty k^2 dk V_{L'L''}^{JST}(q', k) \frac{M}{k^2 - ME - i\epsilon} t_{L''L}^{JST}(k, q; E) , \tag{17}$$

where $\mathbf{q}$, $\mathbf{k}$, and $\mathbf{q}'$ are the relative three-momenta of the two interacting nucleons in the initial, intermediate, and final state; and $q \equiv |\mathbf{q}|$, $k \equiv |\mathbf{k}|$, and $q' \equiv |\mathbf{q}'|$. E denotes the energy of the two interacting nucleons in the c.m. system and is given by

$$E = \frac{q_0^2}{M} \tag{18}$$

with q_0 the magnitude of the initial relative momentum (c.m. on-shell momentum) which is related to the laboratory energy by $T_{lab} = 2q_0^2/M$.

Notice that the integration over the intermediate momenta k in Eqs. (16) and (17) extends from zero to infinity. For intermediate states with $k \neq q_0$, energy is not conserved and the nucleons are off their energy shell ('off-shell'). The off-shell part of the potential (and of the t-matrix) is involved. Thus, in the integral term in Eqs. (16) and (17), the potential (and the t-matrix) contributes off-shell.

Note, however, that in free-space NN scattering, the off-shell potential does not really play a role (it plays a significant role in the few- and many-body problem, see below). The reason for this is simply the procedure by which NN potentials are constructed. The parameters of NN potentials are adjusted so that the resulting on-shell t-matrix fits the empirical NN data. It is important to understand this point, for our later discussion. Therefore, let us consider a case for which the off-shell contributions are particularly large, namely the on-shell t-matrix in the $^3S_1$ state:

$$t_{00}^{110}(q_0, q_0; E) = V_{00}^{110}(q_0, q_0) - \int_0^\infty k^2 dk V_{00}^{110}(q_0, k) \frac{M}{k^2 - ME - i\epsilon} t_{00}^{110}(k, q_0; E)$$
$$- \int_0^\infty k^2 dk V_{02}^{110}(q_0, k) \frac{M}{k^2 - ME - i\epsilon} t_{20}^{110}(k, q_0; E) \tag{19}$$

Up to second order in V, this is

$$t_{00}^{110}(q_0, q_0; E) \approx V_{00}^{110}(q_0, q_0) - \int_0^\infty k^2 dk V_{00}^{110}(q_0, k) \frac{M}{k^2 - ME - i\epsilon} V_{00}^{110}(k, q_0)$$
$$- \int_0^\infty k^2 dk V_{02}^{110}(q_0, k) \frac{M}{k^2 - ME - i\epsilon} V_{20}^{110}(k, q_0) \tag{20}$$

$$\approx V_{00}^{110}(q_0, q_0) - \int_0^\infty k^2 dk V_{02}^{110}(q_0, k) \frac{M}{k^2 - ME - i\epsilon} V_{20}^{110}(k, q_0) \tag{21}$$

where in the last equation, we have neglected the second order term in V_{00}^{110} which is, in general, smaller than the second order in V_{02}^{110}. Without partial-wave decomposition,

$$t(\mathbf{q}_0, \mathbf{q}_0; E) \approx V_C(\mathbf{q}_0, \mathbf{q}_0) - \int d^3k V_T(\mathbf{q}_0, \mathbf{k}) \frac{M}{k^2 - ME - i\epsilon} V_T(\mathbf{k}, \mathbf{q}_0) , \tag{22}$$

where V_C denotes the central force and V_T the tensor force. In words: the most important contributions are the central force in lowest order and the tensor force in second order.

The on-shell t-matrix is related to the observables that are measured in experiments. Thus potentials which fit the same NN scattering data produce the same on-shell t-matrices. However, this does not imply that the potentials are the same. As seen in Eqs. (16) and (17), the t-matrix is the sum of two terms: the Born term and an integral term. When this sum is the same, the individual terms may still be quite different.

As an example, we pick up again the case of the $^3S_1$ state, which is attractive below 300 MeV laboratory energy. Instead of using the t-matrix, it is more convenient to consider the (real) K-matrix (denoted by R below) which, similar to Eq. (19), is given by,

$$R_{00}^{110}(q_0, q_0; E) = V_{00}^{110}(q_0, q_0) - \mathcal{P} \int_0^\infty k^2 dk V_{00}^{110}(q_0, k) \frac{M}{k^2 - ME} R_{00}^{110}(k, q_0; E)$$

$$- \mathcal{P} \int_0^\infty k^2 dk V_{02}^{110}(q_0, k) \frac{M}{k^2 - ME} R_{20}^{110}(k, q_0; E) , \qquad (23)$$

where $\mathcal{P}$ denotes the principal value integral.

In Fig. 2 (upper part), the on-shell R matrix, $R(q_0, q_0; E)$, which corresponds to the empirical phase shift, is denoted by a star and the potential Born term, $V(q_0, q_0)$ is given by a solid dot, for each potential. The difference between star and dot is due to the integral term. Clearly, the size of the integral term (in which the off-shell potential is involved) is very different for different potentials. Hard and strong-tensor-force potentials produce large integral terms, while soft and weak-tensor-force potentials produce small integral terms. Note, however, that by construction the Born and integral terms are balanced such that the same result is obtained for $R(q_0, q_0; E)$ (the star in Fig. 2 is the same for all potentials).

Let us now turn to the many-body problem and consider first the three-nucleon system. The most important point to notice here is that a two-nucleon t-matrix is the input for the three-body Faddeev equations. However, the energy parameter E for this two-body t-matrix is different from the one used in free-space scattering. In the three-body Faddeev equations, one uses

$$E = -B_t - \frac{3}{4} \frac{q^2}{M} \qquad (24)$$

where B_t (≈ 8.5 MeV) is the triton binding energy and q is the magnitude of the momentum of the spectator nucleon. Notice that the energy parameter E is negative here, in contrast to free scattering where E is positive. In a Faddeev calculation, q runs from zero to infinity; thus, the parametric energy for the two-body t-matrix ranges between ≈ -8.5 MeV and $-\infty$. Inspection of Eq. (16) reveals that this negative energy parameter will quench the integral term of the Lippmann-Schwinger equation due to an increase of the energy-denominator. Since this integral term is attractive, the t-matrix will become less attractive as a result of this quenching. This effect will be particularly large for the t-matrix in the $^3S_1$ state, Eq. (19). The consequence is that potentials with large integral terms (i. e., hard potentials, large P_D), will experience more quenching of the attractive integral term and, thus, lose more attraction than potentials with a small integral term (soft potentials).

These arguments can also be stated in more intuitive form. All NN potentials reproduce the deuteron binding energy. In the deuteron, the tensor force couples the $^3S_1$ to the $^3D_1$ state. The triton is a much smaller system than the deuteron. Therefore, in the triton, the D-state is located at much higher energy than in the deutron. This increases the energy gap between the S and D states and makes an S-D transition via the tensor force less likely in the denser system [29]. This 'medium effect' on the second order tensor contribution [cf. Eq. (22)] reduces the binding energy. The stronger the tensor force, the larger the reduction.

This dependence of the triton binding energy predictions on the off-shell potential, particularly the off-shell tensor potential, is confirmed by the results shown in the lower part of Table 2. In a charge-dependent 34-channel momentum-space Faddeev calculation, one obtains 8.00 MeV using the CD-Bonn potential and 7.62(1) MeV applying any of the local potentials. This difference of 0.38 MeV is due to the off-shell differences between the local and nonlocal potentials.

The unacquainted observer may be tempted to believe that this difference of 0.38 MeV is quite small, almost negligible. This is not true. The discrepancy between the predictions of local potentials (7.62 MeV) and experiment (8.48 MeV) is 0.86 MeV. Thus, the problem with the triton binding is that 0.86 MeV cannot be explained in the simplest way, that is all. Any non-trivial contribution must, therefore, be measured against the 0.86 MeV gap between experiment and the simplest theory. On this scale, the nonlocality of the CD-Bonn explains almost 50% of the gap; i. e., it is substantial.

Concerning the remaining difference between theory and experiment, two comments are appropriate. First, besides the relativistic, nonlocal effects that can be absorbed into the two-body potential concept [Eq. (11)], there are further relativistic corrections that come from a relativistic treatment of the three-body system. This increases the triton binding energy by 0.2–0.3 MeV [30–32,22]. Second, notice that the present nonlocal potentials include only the nonlocalities that come from meson-exchange. However, the composite structure (quark substructure) of hadrons should provide additional nonlocalities which may be even larger. It is a challenging topic for future research to derive these additional nonlocalities, and test their impact on nuclear structure predictions.

For heavier systems, the Brueckner G-matrix is the basic quantity for all calculations of nuclear ground and excited states. The G-matrix is the solution of the Bethe-Goldstone equation,

TABLE 3. Predictions by some high-precision NN potentials for the energy per nucleon in nuclear matter (in units of MeV) at $k_F = 1.35$ fm^{-1}. The contributions from some important partial waves and the totals are given.

	CD-Bonn	Nijm-I	Nijm-II
$^1S_0$	−16.75	−16.73	−16.11
$^3S_1$	−19.00	−17.55	−17.05
$^3P_0$	−3.09	−3.11	−3.06
$^3P_1$	9.84	9.73	9.72
$^3P_2$	−7.03	−7.00	−7.01
Total potential energy	−36.35	−35.07	−33.70
Kinetic energy	22.67	22.67	22.67
Total energy	−13.68	−12.40	−11.02

$$G(\mathbf{q'}, \mathbf{q}) = V(\mathbf{q'}, \mathbf{q}) - \int d^3 k V(\mathbf{q'}, \mathbf{k}) \frac{M^\star Q}{k^2 - q^2} G(\mathbf{k}, \mathbf{q}) , \tag{25}$$

which differs from the Lippmann-Schwinger equation, Eq. (16), in two points: the Pauli projector Q and the energy denominator. (For simplicity, I assume in Eq. (25) the so-called continuous choice for the single particle energies in nuclear matter, i. e., the energy of a nucleon is represented by $\epsilon(p) = p^2/(2M^\star) - U_0$ for $p \leq k_F$ as well as $p > k_F$, where $M^\star$ is the effective mass and U_0 a positive constant.) The Pauli projector prevents scattering into occupied states and, thus, cuts out the low momenta in the k integration. The change introduced by the Pauli projector is known as the Pauli effect. The energy denominator gives rise to the so-called dispersive effect. When using the continuous choice, the dispersive effect is given simply by the replacement of M by $M^\star$ ($\approx \frac{2}{3}M$ at nuclear matter density) in the numerator of the integral term, which leads to a reduction of this attractive term by a factor $M^\star/M$. Both effects are in the same direction, namely they quench the integral term. Since the integral term is negative, these effects are repulsive.

In Table 3, nuclear matter results are given for three representative high-precision potentials using the Brueckner pair approximation and the conventional choice for the single particle potential. For the reasons discussed, the largest difference between the predictions occurs in the $^3S_1$ state where the tensor force is involved. In the $^1S_0$ state, only the central force can contribute which is nonlocal for CD-Bonn and Nijm-I. This explains why the binding energy predicted for this state is the same for these two potentials, and is larger than the value predicted by the local potential Nijm-II.

Note that there are further contributions in nuclear matter beyond the Brueckner pair approximation. Three- and four-body clusters contribute about 4 MeV attraction [33] and the medium effect on the Dirac spinors representing the nucleons in nuclear matter creates about 2 MeV repulsion [5] resulting in a net correction of about −2 MeV. This brings the prediction of the nonlocal potential into good agreement with the empirical nuclear matter value of -16 ± 1 MeV.

The trend of the nonlocal Bonn potential to increase binding energies has also a very favorable impact on predictions for spectra of open-shell nuclei [34,35].

THREE-NUCLEON FORCES

Strictly speaking, most many-body forces are an artifact of theory. They are created by 'freezing out' degrees of freedom contained in the full Hamiltonian of the problem. This fact allows us to derive a guideline for dealing with the many-body-force issue in a consistent way: when you introduce a new degree of freedom, do not freeze it out; instead take it into account in the two- *and* many-body problem, consistently. Field-theoretic models for the 2π-exchange contribution to the NN interaction require the $\Delta(1232)$ isobar, which also creates a three-nucleon force (3NF) in the many-body system. Consequently, when introducing the isobar degree of freedom, it should be included in the two- and three-body force simultaneously. When these forces are applied in the many-body problem, then one is faced with two effects: a (repulsive) medium effect on the two-body force [Fig. 3(a)] and an attractive 3NF contribution [Fig 3(b)]. The repulsive medium effect has been known for more than 20 years [39]. *Consistency requires that either both effects are taken into account or none.* Consistent calculations of this kind have first been conducted by the Hannover group [36] for the triton. Recently, these calculations have been improved and extended by Picklesimer and coworkers [37] using the Argonne V_{28} Δ-model [38]. They find an attractive contribution to the triton energy of −0.66 MeV from the 3NF diagrams

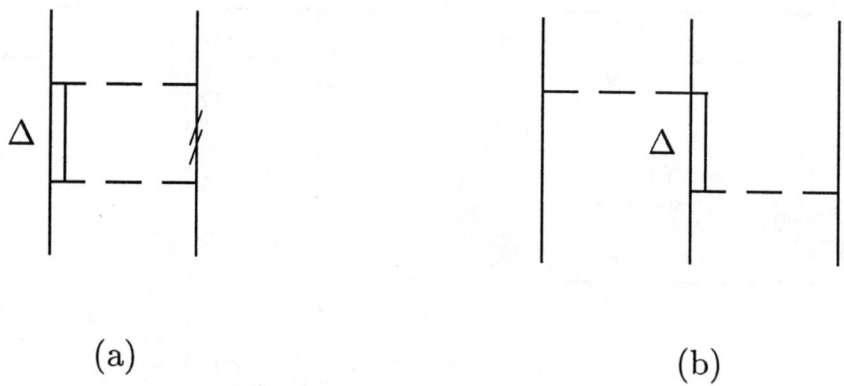

(a)   (b)

FIGURE 3. Two- and three-body forces created by the Δ isobar. Solid lines represent nucleons, double lines Δ isobars, and dashed lines π and ρ exchange. **Part (a)** is a contribution to the two-nucleon force. The double slash on the intermediate nucleon line is to indicate the medium modifications (Pauli blocking and dispersion effects) that occur when this diagram is inserted into a nuclear many-body environment. **Part (b)** is a three-nucleon force.

created by the Δ, and a repulsive contribution of +1.08 MeV from the dispersive effect on the two-nucleon force involving Δ isobars. The total result is *0.42 MeV repulsion* [37].

A contribution that is missing in the Δ 3NF model is the S-wave part of the off-shell πN amplitude. Estimates for this contribution vary between –0.2 and –0.6 MeV [40]. Adding this to the Argonne V_{28} result [i. e., $0.42 - (0.4 \pm 0.2) = 0.0 \pm 0.2$ MeV] yields a vanishing result for the total contribution from three-body forces.

In the more traditional calculations, it was customary to employ the so-called 2π-exchange three-nucleon force. This approach does not take the dispersive effects into account and overbinds the triton, at least if commonly accepted values for the cut-off parameter of the strong pion-nucleon vertex are used. Moreover, this type of 3N force fails in 3N scattering [41] casting additional doubt on its reality.

Besides the Δ-isobar, there are other mechnisms that can give rise to three-nucleon forces. In recent years, it has become fashionable to consider chiral Lagrangians for nuclear interactions. Such Langrangians may create diagrams which represent effective three-nucleon potentials. However, Weinberg has shown that all these diagrams cancel [42].

In conclusion, any *consistent* three-body force calculation—conducted to date—has yielded vanishing results. This implies that the two-body force should provide essentially all binding energies observed in nuclei. The further implication then is that nonlocal/weak-tensor force NN interactions are to be favored over hard/local potentials. It is probably not just an accident that these soft/nonlocal potentials are also more faithful to the underlying theory of nuclear forces.

SUMMARY AND OUTLOOK

Several high-quality/high-precision NN potentials are now available which fit the low-energy NN data with identical perfection. These potentials differ, however, in their off-shell behavior. Thus, the stage is set for a reliable investigation of off-shell effects in microscopic nuclear structure calculations. Such calculations may finally teach us something about the off-shell nature of the nuclear force.

This is *the* time to do *microscopic* nuclear structure calculations!

ACKNOWLEDGEMENT

This work was supported in part by the U.S. National Science Foundation under Grant-No. PHY-9603097.

REFERENCES

1. S. Weinberg, Physica **96A**, 327 (1979).
2. S.C. Pieper, contribution to these Proceedings.
3. T.T.S. Kuo and E. Osnes, *Folded-Diagram Theory of the Effective Interaction in Atomic Nuclei*, Springer Lecture Notes in Physics, Vol. 364 (Springer, Berlin, 1990).
4. M. Hjorth-Jensen, T.T.S. Kuo, and E. Osnes, Phys. Reports **261**, 125 (1995).
5. R. Machleidt, Adv. Nucl. Phys. **19**, 189 (1989).
6. R. Machleidt and G. Q. Li, *Nucleon-nucleon potentials in comparison: Physics or polemics?*, Phys. Reports **242**, 5 (1994).
7. R. Bryan and B. L. Scott, Phys. Rev. **177**, 1435 (1969).
8. Particle Data Group, *Review of Particle Physics*, Phys. Rev. D **54**, 1 (1996), see pp. 329, 355, and 356 therein; Eur. Phys. J. C **3**, 1-794 (1998), see p. 363 and pp. 390-392 therein.
9. A more detailed discussion of this issue and figures illustrating the facts can be found in Ref. [6].
10. A. Gersten, R. Thompson, and A. E. S. Green, Phys. Rev. D **3**, 2076 (1971); G. Schierholz, Nucl. Phys. **B40**, 335 (1972); K. Erkelenz, Phys. Reports **13C**, 191 (1974); K. Holinde and R. Machleidt, Nucl. Phys. **A247**, 495 (1975), *ibid.* **A256**, 479 (1976); J. Fleischer and J. A. Tjon, Nucl. Phys. **B84**, 375 (1975), Phys. Rev. D **24**, 87 (1980).
11. M. M. Nagels, T. A. Rijken, and J. J. de Swart, Phys. Rev. D **17**, 768 (1978).
12. M. Chemtob, J. W. Durso, and D. O. Riska, Nucl. Phys. **B38**, 141 (1972); A. D. Jackson, D. O. Riska. and B. Verwest, Nucl. Phys. **A249**, 397 (1975); G. E. Brown and A. D. Jackson, *The Nucleon-Nucleon Interaction* (North-Holland, Amsterdam, 1976).
13. R. Vinh Mau, J. M. Richard, B. Loiseau, M. Lacombe, and W. N. Cottingham, Phys. Lett. B**44**, 1 (1973); R. Vinh Mau, in *Mesons in Nuclei*, ed. M. Rho and D. H. Wilkinson (North-Holland, Amsterdam, 1979), p. 151.
14. M. Lacombe, B. Loiseau, J. M. Richard, R. Vinh Mau, J. Côté, P. Pirès, and R. de Tourreil, Phys. Rev. C **21**, 861 (1980).
15. J. W. Durso, A. D. Jackson, and B. J. VerWest, Nucl. Phys. **A345**, 471 (1980).
16. R. Machleidt, K. Holinde, and Ch. Elster, Phys. Reports **149**, 1 (1987).
17. V. G. J. Stoks, R. A. M. Klomp, M. C. M. Rentmeester, and J. J. de Swart, Phys. Rev. C **48**, 792 (1993).
18. D. V. Bugg and R. A. Bryan, Nucl. Phys. **A540**, 449 (1992).
19. R. A. Arndt *et al.*, Phys. Rev. D **45**, 3995 (1992).
20. V. G. J. Stoks, R. A. M. Klomp, C. P. F. Terheggen, and J. J. de Swart, Phys. Rev. C **49**, 2950 (1994).
21. R. B. Wiringa, V. G. J. Stoks, and R. Schiavilla, Phys. Rev. C **51**, 38 (1995).
22. R. Machleidt, F. Sammarruca, and Y. Song, Phys. Rev. C **53**, 1483 (1996).
23. R. V. Reid, Ann. Phys. (N.Y.) **50**, 411 (1968).
24. G. Q. Li and R. Machleidt, *Charge-Dependence of the Nucleon-Nucleon Interaction*, Phys. Rev. C, in press; nucl-th/9807080.
25. G. Q. Li and R. Machleidt, *Charge-Asymmetry of the Nucleon-Nucleon Interaction*, Phys. Rev. C, in press; nucl-th/9804023.
26. R. Machleidt, to be published.
27. J. A. Nolen and J. P. Schiffer, Annu. Rev. Nucl. Sci. **19**, 471 (1969).
28. H. Müther, A. Polls, and R. Machleidt, *Isospin symmetry breaking nucleon-nucleon potentials and nuclear structure*, submitted to Phys. Lett. B; nucl-th/9809016.
29. D. R. Inglis, Phys. Rev. **55**, 988 (1939); R. L. Pease and H. Feshbach, Phys. Rev. **88**, 945 (1952).
30. F. Sammarruca, D. P. Xu, and R. Machleidt, Phys. Rev. C **46**, 1636 (1992).
31. G. Rupp and J. A. Tjon, Phys. Rev. C **45**, 2133 (1992).
32. F. Sammarruca and R. Machleidt, Few-Body Systems **24**, 87 (1998).
33. H. Q. Song, M. Baldo, G. Giansiracusa, and U. Lombardo, Phys. Rev. Lett. **81**, 1584 (1998).
34. M. F. Jiang, R. Machleidt, D. B. Stout, and T. T. S. Kuo, Phys. Rev. C **46**, 910 (1992).
35. F. Andreozzi *et al.*, Phys. Rev. C **54**, 1636 (1996); A. Covello, contribution to these Proceedings.
36. C. Hajduk, P. U. Sauer, and W. Strueve, Nucl. Phys. **A405**, 581 (1983).
37. A. Picklesimer, R. A. Rice, and R. Brandenburg, Phys. Rev. C **45**, 2045, 2624 (1992); *ibid.* **46**, 1178 (1992).
38. R. B. Wiringa, R. A. Smith, and T. L. Ainsworth, Phys. Rev. C **29**, 1207 (1984).
39. K. Holinde and R. Machleidt, Nucl. Phys. **A280**, 429 (1977).
40. J. L. Friar, B. F. Gibson, G. L. Payne, and S. A. Coon, Few-Body Systems **5**, 13 (1988).
41. H. Witala, D. Hüber, and W. Glöckle, Phys. Rev. C **49**, 14 (1994).
42. S. Weinberg, Phys. Lett. **B251**, 288 (1990); Nucl. Phys. **B363**, 3 (1991); Phys. Lett. **B295**, 114 (1992).

Large-basis no-core shell-model calculations for p-shell nuclei

B. R. Barrett[a], P. Navrátil[a1], M. J. Thoresen[b] and W. E. Ormand[c]

[a]Department of Physics, University of Arizona, Tucson, AZ 85721 [2]
[b]Physics Education Research Group, Kansas State University, Manhattan, KS 66506
[c]Department of Physics and Astronomy, Louisiana State University, Baton Rouge, LA 70803 [3]

Abstract. Results of large-basis shell-model calculations for nuclei with $A = 7 - 10$ are presented. The effective interactions used in the study were derived microscopically from the Reid93 potential and take into account the Coulomb potential as well as the charge dependence of $T = 1$ partial waves. For $A = 7$, a $6\hbar\Omega$ model space was used, while for the rest of the studied nuclides, the calculations were performed in a $4\hbar\Omega$ model space. It is demonstrated that the shell model combined with microscopic effective interactions derived from modern nucleon-nucleon potentials is capable of providing good agreement with the experimental properties of the ground state as well as with those of the low-lying excited states. In the context of the calculations for $A = 10$ isobars we study in detail the isospin-mixing correction to the $^{10}C \rightarrow ^{10}B$ Fermi matrix element.

INTRODUCTION

Shell-model calculations have been limited by the size of the model space and by the need for an effective interaction within the model space. To simplify the interaction, we propose working in a large model space, in which all A nucleons of a given nucleus are active for a complete $N\hbar\Omega$ basis space and a large value for N. Calculations of this kind have recently been performed [1–8]. By working in a complete $N\hbar\Omega$ space with a single-particle harmonic-oscillator Hamiltonian as our unperturbed Hamiltonian, we can guarantee that all excluded configurations involve an energy of at least $(N + 2)\hbar\Omega$, which should limit any intruder-state difficulties to the less interesting physical states higher in the spectrum [9]. That is, the larger the value of N, the better the guarantee that we have included the major configurations making up the physical low-lying states. At the same time, the no-core basis simplifies the form of the effective interaction, since there are no hole states [4]. In this case, the effective interaction can be interpreted as a generalized A-nucleon G-matrix and associated folded diagrams. For sufficiently large model spaces, the perturbation expansion for the effective interaction may be reasonably expressed in terms of only the two-nucleon G-matrix in the no-core space plus all folded diagrams developed from it [10]. Summing all folded diagrams yields a starting-energy-independent G-matrix. We have developed a new transformation-operator procedure for exactly calculating the energy-independent effective two-nucleon interaction. [5] We can also determine other effective operators, such as effective charges, using this transformation-operator technique [6]. Although this approach limits us to investigations for light nuclei, i.e., $A \leq 16$, it allows us to study how numerical results depend upon the size of the model space as well as upon omitted many-body effects. We have applied this approach to determining various properties of p-shell nuclei [8], including the isospin correction to the superallowed Fermi matrix element [7] in the β decay of $^{10}C \rightarrow ^{10}B$.

The organization for the present contribution is as follows. First, we discuss the shell-model Hamiltonian with a bound center-of-mass and the method used to derive the starting-energy-independent effective interaction. Then we present results of the calculations for $A = 7 - 10$ including the results of the isospin-mixing correction to the Fermi matrix element in $^{10}C \rightarrow ^{10}B$. Finally, we give concluding remarks.

[1)] On leave of absence from the Institute of Nuclear Physics, Academy of Sciences of the Czech Republic, 250 68 Řež near Prague, Czech Republic.
[2)] Supported in part by NSF grant PHY-9605192.
[3)] Supported in part by NSF Cooperative Agreement EPS-9550481, NSF grant 9603006 and DOE Contract DE-FG02-96ER40985.

CP481, *Nuclear Structure 98*, edited by C. Baktash

THE SHELL-MODEL HAMILTONIAN AND EFFECTIVE INTERACTION

In the present contribution we apply the approach discussed in Refs. [5,7]. We start with the one- plus two-body Hamiltonian for the A-nucleon system, i.e.,

$$H^{\Omega} = \sum_{i=1}^{A} \frac{\vec{p}_i^2}{2m} + \sum_{i<j}^{A} V_{\rm N}(\vec{r}_i - \vec{r}_j) + \frac{1}{2} A m \Omega^2 \vec{R}^2 \,, \tag{1}$$

where m is the nucleon mass, $V_{\rm N}(\vec{r}_i - \vec{r}_j)$ the nucleon-nucleon interaction and $\frac{1}{2} A m \Omega^2 \vec{R}^2$ ($\vec{R} = \frac{1}{A} \sum_{i=1}^{A} \vec{r}_i$) is the the center-of-mass harmonic-oscillator potential. The latter potential does not influence intrinsic properties of the many-body system. It provides, however, a mean field felt by each nucleon and allows us to work with a convenient harmonic-oscillator basis. The Hamiltonian (1), depending on the harmonic-oscillator frequency Ω, may be cast into the form

$$H^{\Omega} = \sum_{i=1}^{A} \left[\frac{\vec{p}_i^2}{2m} + \frac{1}{2} m \Omega^2 \vec{r}_i^2 \right] + \sum_{i<j}^{A} \left[V_{\rm N}(\vec{r}_i - \vec{r}_j) - \frac{m \Omega^2}{2A} (\vec{r}_i - \vec{r}_j)^2 \right] \,. \tag{2}$$

The one-body term of the Hamiltonian (2) is then re-written as a sum of the center-of-mass term $H_{\rm cm}^{\Omega} = \frac{\vec{P}_{\rm cm}^2}{2Am} + \frac{1}{2} A m \Omega^2 \vec{R}^2$, $\vec{P}_{\rm cm} = \sum_{i=1}^{A} \vec{p}_i$, and a term depending only on relative coordinates. Shell-model calculations are carried out in a model space defined by a projector P. In the present work, we will always use a complete $N\hbar\Omega$ model space which includes all the configurations up to an energy of $N\hbar\Omega$ relative to the unperturbed ground-state configuration. The complementary space to the model space is defined by the projector $Q = 1 - P$. In addition, from among the eigenstates of the Hamiltonian (2), it is necessary to choose only those corresponding to the same center-of-mass energy. This can be achieved by projecting the center-of-mass eigenstates with energies greater than $\frac{3}{2}\hbar\Omega$ upwards in the energy spectrum. The shell-model Hamiltonian, used in the actual calculations, takes the form

$$H_{P\beta}^{\Omega} = \sum_{i<j=1}^{A} P \left[\frac{(\vec{p}_i - \vec{p}_j)^2}{2Am} + \frac{m \Omega^2}{2A} (\vec{r}_i - \vec{r}_j)^2 \right] P + \sum_{i<j}^{A} P \left[V_{\rm N}(\vec{r}_i - \vec{r}_j) - \frac{m \Omega^2}{2A} (\vec{r}_i - \vec{r}_j)^2 \right]_{\rm eff} P$$
$$+ \beta P (H_{\rm cm}^{\Omega} - \frac{3}{2}\hbar\Omega) P \,, \tag{3}$$

where β is a sufficiently large positive parameter. In Eq. (3), the notation [ ]$_{\rm eff}$ means that the quantity within the square brackets is the residual interaction to be used in the determination of the effective interaction within the model space P (see Ref. [5] for more details).

The effective interaction introduced in Eq. (3) should, in principle, exactly reproduce the full-space results in the model space for some subset of states. In practice, the effective interactions can never be calculated exactly, because, in general, for an A-nucleon system an A-body effective interaction is required. Consequently, large model spaces are desirable when only an approximate effective interaction is used. In that case, the calculation should be less affected by any imprecision of the effective interaction. The same is true for the evaluation of any observable characterized by an operator. In the model space, renormalized effective operators are also required. The larger the model space, the less renormalization is needed.

Usually, the effective interaction is approximated by a two-body effective interaction determined from a two-nucleon system. In this study, we use the procedure, as described in Ref. [5]. To construct the effective interaction we employ the Lee-Suzuki [10] similarity transformation method, which gives an interaction in the form $P_2 V_{\rm eff} P_2 = P_2 V P_2 + P_2 V Q_2 \omega P_2$, with ω the transformation operator satisfying $\omega = Q_2 \omega P_2$. The projection operators $P_2, Q_2 = 1 - P_2$ project on the two-nucleon model and complementary spaces, respectively. Note that we distinguish the two-nucleon system projection operators P_2, Q_2 from the A-nucleon system operators P, Q. Our calculations start with exact solutions of the Hamiltonian

$$H_2^{\Omega} \equiv H_{02}^{\Omega} + V_2^{\Omega} = \frac{\vec{p}_1^2 + \vec{p}_2^2}{2m} + \frac{1}{2} m \Omega^2 (\vec{r}_1^2 + \vec{r}_2^2) + V(\vec{r}_1 - \vec{r}_2) - \frac{m \Omega^2}{2A} (\vec{r}_1 - \vec{r}_2)^2 \,, \tag{4}$$

which is the shell-model Hamiltonian (2) applied to a two-nucleon system. We construct the effective interaction directly from these solutions. Let us denote the two-nucleon harmonic-oscillator states, which form the model space, as $|\alpha_P\rangle$, and those which belong to the Q-space, as $|\alpha_Q\rangle$. Then the Q-space components of an eigenvector $|k\rangle$ of the Hamiltonian (4) can be expressed as a combination of the P-space components with the help of the operator ω

$$\langle \alpha_Q | k \rangle = \sum_{\alpha_P} \langle \alpha_Q | \omega | \alpha_P \rangle \langle \alpha_P | k \rangle . \tag{5}$$

If the dimension of the model space is d_P, we may choose a set $\mathcal{K}$ of d_P eigenvectors, typically the lowest states obtained in each channel, for which the relation (5) will be satisfied. Under the condition that the $d_P \times d_P$ matrix $\langle \alpha_P | k \rangle$ for $|k\rangle \in \mathcal{K}$ is invertible, the operator ω can be determined from (5). Once the operator ω is determined the effective Hamiltonian can be constructed as follows

$$\langle \gamma_P | H_{2\mathrm{eff}} | \alpha_P \rangle = \sum_{k \in \mathcal{K}} \left[\langle \gamma_P | k \rangle E_k \langle k | \alpha_P \rangle + \sum_{\alpha_Q} \langle \gamma_P | k \rangle E_k \langle k | \alpha_Q \rangle \langle \alpha_Q | \omega | \alpha_P \rangle \right] . \tag{6}$$

This Hamiltonian, when diagonalized in a model-space basis, reproduces exactly the set $\mathcal{K}$ of d_P eigenvalues E_k. Note that the effective Hamiltonian is, in general, quasi-Hermitian. It can be hermitized by a similarity transformation determined from the metric operator $P_2(1 + \omega^\dagger \omega) P_2$. The Hermitian Hamiltonian is then given by [11]

$$\bar{H}_{2\mathrm{eff}} = \left[P_2(1 + \omega^\dagger \omega) P_2 \right]^{1/2} H_{2\mathrm{eff}} \left[P_2(1 + \omega^\dagger \omega) P_2 \right]^{-1/2} . \tag{7}$$

Finally, the two-body effective interaction used in the present calculations is determined from the two-nucleon effective Hamiltonian (7) as $V_{\mathrm{eff}} = \bar{H}_{2\mathrm{eff}} - H_{02}^\Omega$.

To at least partially take into account the many-body effects neglected when using only a two-body effective interaction, we employ the recently introduced, so-called, multi-valued effective interaction approach [4]. In that approach, different effective interactions are used for different harmonic-oscillator excitations of the spectators. The effective interactions then carry an additional index indicating the sum of the oscillator quanta for the spectators, N_{sps}, defined by

$$N_{\mathrm{sps}} = N_{\mathrm{sum}} - N_\alpha - N_{\mathrm{spsmin}} = N'_{\mathrm{sum}} - N_\gamma - N_{\mathrm{spsmin}} , \tag{8}$$

where N_{sum} and N'_{sum} are the total oscillator quanta in the initial and final many-body states, respectively, and N_α and N_γ are the total oscillator quanta in the initial and final two-nucleon states $|\alpha\rangle$ and $|\gamma\rangle$, respectively. N_{spsmin} is the minimal value of the spectator harmonic-oscillator quanta for a given system. E.g., for A=7, $N_{\mathrm{spsmin}} = 1$. Different sets of the effective interaction are determined for different model spaces characterized by N_{sps} and defined by projection operators

$$Q_2(N_{\mathrm{sps}}) = \begin{cases} 0 & \text{if } N_1 + N_2 \leq N_{\max} - N_{\mathrm{sps}} , \\ 1 & \text{otherwise} ; \end{cases} \tag{9a}$$

$$P_2(N_{\mathrm{sps}}) = 1 - Q_2(N_{\mathrm{sps}}) . \tag{9b}$$

In Eqs. (9), $N_{\max}$ characterizes the two-nucleon model space. It is an input parameter chosen in relation to the size of the many-nucleon model space. This multi-valued effective-interaction approach is superior to the traditional effective interaction, as confirmed also in a model calculation [12].

CALCULATIONS FOR THE P-SHELL NUCLEI

We apply the formalism outlined in the previous section for selected $0p$-shell nuclei. In the calculations we use the Reid93 nucleon-nucleon potential [13] and consider the following isospin-breaking contributions. First, the Reid93 potential differs in the $T = 1$ channels for proton-neutron (pn) and proton-proton (pp), neutron-neutron (nn) systems, respectively. Second, we add the Coulomb potential to the pp Reid93 potential. Consequently, using the Eqs. (5)-(7), we derive different two-body effective interactions for the pn, pp, and nn systems.

As we derive the effective interaction microscopically from the nucleon-nucleon interaction, the number of freely adjustable parameters in the calculation is limited. First, we have the choice of the model-space size in the shell-model diagonalization. That is, however, constrained by computer capabilities. The largest model space we were able to use was the space allowing all $6\hbar\Omega$ excitations relative to the unperturbed ground state for $A = 7$ nuclei and all $4\hbar\Omega$ excitations relative to the unperturbed ground state for $A > 7$ nuclei, respectively. The calculations were done in the m-scheme using the Many-Fermion-Dynamics Code [14] extended to allow the use of different pn, pp, nn interactions. We note that in this study the same effective interaction is used for each isobaric chain, and thus the same model-space size is employed for all the isobars of given A.

Second, our effective interactions depend on the choice of the two-nucleon model space size. The two-nucleon model space size is related to the many-nucleon model-space size, and, in principle, is determined by that size.

Traditionally, however, the $Q_2 = 0$ space used to determine the G-matrix does not, neccesarily, coincide with the many-particle model space [15,16]. In our calculation, the two-nucleon model space is characterized by a restriction on the number of harmonic-oscillator quanta $N_1 \leq N_{\max}$, $N_2 \leq N_{\max}$, $(N_1 + N_2) \leq N_{\max}$. Here, $N_i = 2n_i + l_i$ is the harmonic-oscillator quantum number for the nucleon $i, i = 1, 2$. This type of restriction guarantees an orthogonal transformation between the two-particle states and the relative- and center-of-mass-coordinate states. With regard to the $4\hbar\Omega$ calculation for the $0p$-shell nuclei, the choice of $N_{\max} = 6$ appears to be appropriate. However, it has been observed in the past [1,3,5] that when the Lee-Suzuki procedure is combined with the G-matrix calculation according to Ref. [15] (which is equivalent to the procedure we are using) and is applied to calculate the two-body effective interaction, the resulting interaction may be too strong. This is, in particular, true, when the multi-valued approach is used. The reason for this is likely the fact that our effective interaction is computed for a two-nucleon system bound in an harmonic-oscillator potential. Therefore, artificial binding from this potential is included in the effective interaction and the many-body effects coming from the large-basis space calculation do not completely compensate for this spurious binding. This effect decreases when the model-space size increases as is demonstrated in our earlier three-nucleon shell-model calculations [17]. Several possible adjustments have been discussed to deal with this problem [1,5] in smaller model spaces and amount to introducing an extra parameter. In the present calculations, we use two methods introduced in the previous papers. For the $4\hbar\Omega$ calculations for the nuclei with $A > 7$ we prefer to treat $N_{\max}$ as a free parameter and use $N_{\max} = 8$ for the $4\hbar\Omega$ calculations. With this choice, which results in an overall weaker interaction than that calculated with $N_{\max} = 6$, we obtain quite reasonable binding energies for all the studied nuclei. On the other hand, in the case of $A = 7$ for the $6\hbar\Omega$ space we follow Ref. [5] and use the parameter k_Q introduced there. While for $k_Q = 1$ there is no modification, the choice of $k_Q < 1$ reduces the contribution of the Q_2-space part of the harmonic-oscillator potential on the two-nucleon effective interaction. For this calculation, $N_{\max} = 8$ appropriate for the $6\hbar\Omega$ many-nucleon space size is employed. In Ref. [5] we used this approach for the $6\hbar\Omega$ calculations for $A = 5$ and 6 nuclei as well as for the $8\hbar\Omega$ calculations for $A = 3$ and 4 nuclei. We note that we are studying ways on how to eliminate these extra parameters by means of renormalizations of the two-body effective interactions utilizing knowledge of the three-body effective interactions. The three-body effective interactions can be calculated following the approach introduced in Ref. [17].

Finally, our results depend on the harmonic-oscillator frequency Ω. For the studied nuclei we choose $\hbar\Omega$ in the range 14-17 MeV and investigate the dependence of the physical quantities on Ω. We keep the same value of $\hbar\Omega$ for each isobaric chain.

Let us note that our calculations do not violate the separation of the center-of-mass and the internal relative motion. In particular, a variation of the parameter β introduced in Eq. (3) does not change the eigenenergies and other characteristics of the physical states. This is so due to the utilization of a complete $N\hbar\Omega$ many-nucleon model space and the triangular two-nucleon model space for deriving the effective interaction as well as due to the procedure used to derive the effective interaction.

Results for A=7-9 nuclei

As examples of our results we present the experimental and calculated excitation spectra of ^{7}Li, ^{8}Be, and ^{9}B, in Figs. 1, 2, and 3, respectively.

The calculations for $A = 7$ were performed in the $6\hbar\Omega$ model space. An harmonic-oscillator frequency of $\hbar\Omega = 17$ MeV was used. As discussed earlier in this section, we employed the additional parameter k_Q, introduced in Ref. [5], and set its value to $k_Q = 0.8$. Note that for ^{7}Li and ^{7}Be, the dimension in the m-scheme reaches 663,527. It is the largest dimension in the present study. The calculations for $A = 8$ and for $A = 9$ were performed in a model space of up to $4\hbar\Omega$ excitations relative to the unperturbed ground-state configuration. An harmonic-oscillator frequency of $\hbar\Omega = 17$ MeV and of $\hbar\Omega = 16$ MeV was used for $A = 8$ and $A = 9$ nuclei, respectively. As explained earlier in this section, the two-body effective interaction was evaluated using $N_{\max} = 8$.

In general, good agreement with experiment is found for both the ground-state characteristics as well as the low-lying excited states. We obtain underbinding in the calculations for the $A = 7$ isobars for ^{7}He and ^{7}B. Note that for an isospin invariant interaction these states would be degenerate with the $T = 3/2$ isospin states of ^{7}Li or ^{7}Be. So this underbinding is equivalent to too much spread in the excitation spectrum. This is a common feature in the no-core shell-model calculations, which diminishes as the model-space size increases. Note in Figs. 1 and 2 the correct ordering of excited states in most cases. In particular, for ^{8}Be we have excellent agreement with experiment for all positive-parity states below an excitation energy of 20 MeV. We note that the $T = 0, 1$ $J = 2^+, 1^+, 3^+$ doublets show significant isospin mixing compared to other calculated states.

In our calculations the magnetic moments are in most cases nicely reproduced. The radii and the quadrupole moments are typically smaller in absolute value, however. We used bare nucleon charges, so there is still need for E2 effective charges despite our large model-space size. For example, in order to reproduce the ^{9}Li and ^{9}Be quadrupole

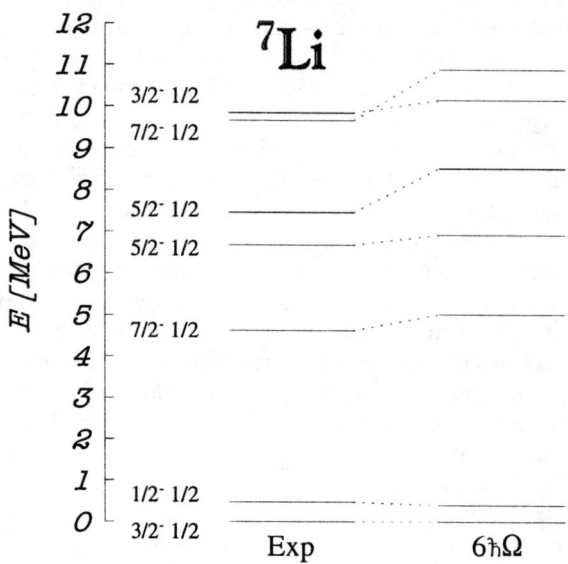

FIGURE 1. The experimental and calculated excitation spectra of ^{7}Li. The results corresponding to the model-space size of $6\hbar\Omega$ relative to the unperturbed ground-state configuration are presented. The harmonic-oscillator energy of $\hbar\Omega = 17$ MeV was used.

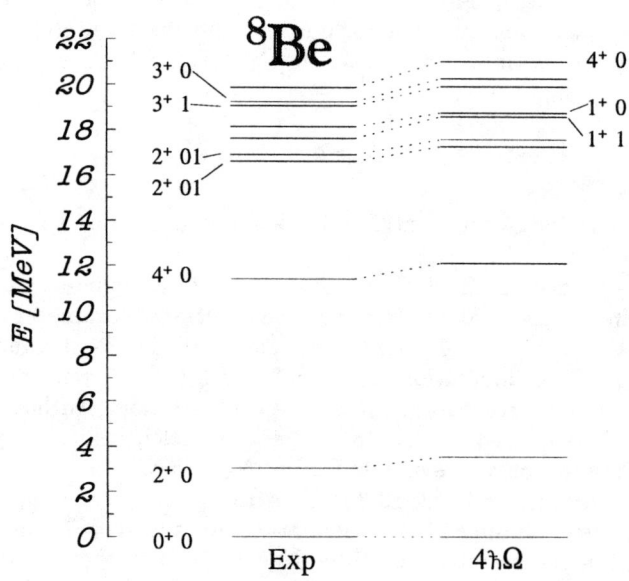

FIGURE 2. The experimental and calculated excitation spectra of ^{8}Be. The results corresponding to the model-space size of $4\hbar\Omega$ relative to the unperturbed ground-state configuration are presented. The harmonic-oscillator energy of $\hbar\Omega = 17$ MeV was used.

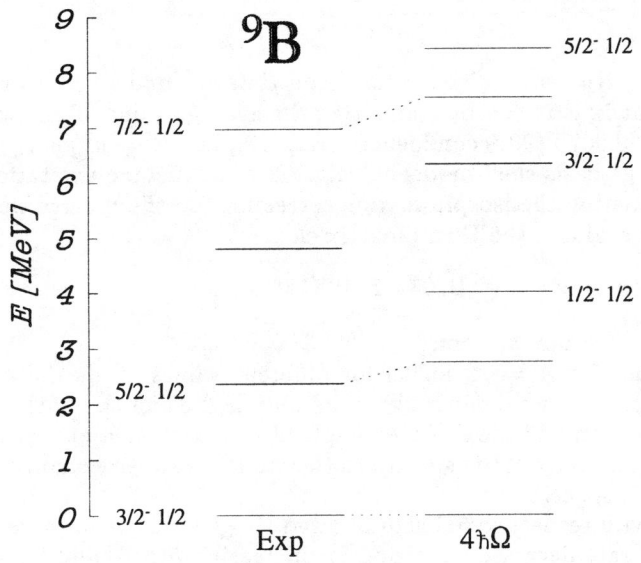

FIGURE 3. The experimental and calculated excitation spectra of ^{9}B. The results corresponding to the model-space size of $4\hbar\Omega$ relative to the unperturbed ground-state configuration are presented. The harmonic-oscillator energy of $\hbar\Omega = 16$ MeV was used.

moments, we would need average effective charges of $e_{\text{eff}}^{\text{p}} = 1.25e$ and $e_{\text{eff}}^{\text{n}} = 0.25e$ for this isobaric chain, with smaller values for ^{9}Li than for ^{9}Be.

Isospin-mixing correction for the ^{10}C$\rightarrow$^{10}B superallowed Fermi transition

Superallowed Fermi β transitions in nuclei, $(J^\pi = 0^+, T = 1) \rightarrow (J^\pi = 0^+, T = 1)$, provide an excellent laboratory for precise tests of the properties of the electroweak interaction. According to the conserved-vector-current (CVC) hypothesis, for pure Fermi transitions the product of the partial half-life, t, and the statistical phase-space factor, f, should be nucleus independent and given by

$$ft = \frac{K}{G_V^2 |M_F|^2},\tag{10}$$

where $K/(\hbar c)^6 = 2\pi^3 \ln 2\hbar/(m_e c^2)^5 = 8.120270(12) \times 10^{-7}$ GeV^{-4}s, G_V is the vector coupling constant for nuclear β decay, and M_F is the Fermi matrix element, $M_F = \langle \psi_f \mid T_\pm \mid \psi_i \rangle$. By comparing the decay rates for muon and nuclear Fermi β decay, the Cabibbo-Kobayashi-Maskawa (CKM) mixing matrix element between u and d quarks (v_{ud}) can be determined and a precise test of the unitarity condition of the CKM matrix under the assumption of the three-generation standard model is possible.

For tests of the standard model, two nucleus-dependent corrections must be applied to experimental ft values. The first is a series of radiative corrections to the statistical phase-space factor embodied in the factors δ_R and Δ_R, The second correction, which is the subject of our contribution, arises because of the presence of isospin-nonconserving (INC) forces (predominantly Coulomb) in nuclei that lead to a renormalization of the Fermi matrix element. This correction is denoted by δ_C [18–20] and modifies the Fermi matrix element by $\mid M_F \mid^2 = \mid M_{F0} \mid^2 (1 - \delta_C)$, where $M_{F0} = [T(T + 1) - T_{z_i} T_{z_f}]^{1/2}$ is the value of the matrix element under the assumption of pure isospin symmetry.

With the corrections δ_R, Δ_R, and δ_C, a "nucleus-independent" $\mathcal{F}t$ can be defined by

$$\mathcal{F}t = ft(1 + \delta_R + \Delta_R)(1 - \delta_C),\tag{11}$$

and the CKM matrix element v_{ud} is given by

$$| v_{ud} |^2 = \frac{\pi^3 \ln 2}{\mathcal{F}t} \frac{\hbar^7}{G_F^2 m_e^5 c^4} = \frac{2984.38(6) \text{ s}}{\mathcal{F}t}, \tag{12}$$

where the Fermi coupling constant, G_F, is obtained from muon β-decay, and includes radiative corrections. The unitarity condition of the CKM matrix is tested by comparing the average value of v_{ud} with the values determined for $v_{us} = 0.2199(17)$ [21] and $v_{ub} < 0.0075$ (90% confidence level) [22], i.e., $v^2 = v_{ud}^2 + v_{us}^2 + v_{ub}^2 = 1$.

We report here the results of large-basis shell-model calculations that include excitations up to $4\hbar\Omega$ for $A = 10$ nuclides, with an emphasis on evaluating the isospin-mixing corrections to the matrix element for the Fermi decay of ^{10}C. Our goal in this study is to evaluate the Fermi matrix element

$$M_F = \langle {}^{10}\text{B}, 0^+1 | T_- | {}^{10}\text{C}, 0^+1 \rangle, \tag{13}$$

which is equal to $\sqrt{2}$ for an isospin-invariant system.

We performed several calculations for $A = 10$ nuclei for different values of the harmonic-oscillator parameter and different model spaces. In particular, we compared results obtained using the $2\hbar\Omega$ and $4\hbar\Omega$ model spaces and three different values of $\hbar\Omega$: 14, 15.5, and 17 MeV. An example of the excitation-energy results obtained for ^{10}B is presented in Fig. 4. We observe a sensitivity of the spectra to the model-space-size change with a clear improvement, compared to experiment, in the $4\hbar\Omega$ space.

In Table 1, the overall behavior with respect to $\hbar\Omega$ is illustrated. In general, we observe a reasonable reproduction of the binding energy, with a moderate decrease occurring for increasing $\hbar\Omega$. Using free-nucleon effective charges, we find that although the quadrupole moment for the 3^+0 state is considerably underestimated, the magnetic dipole moment is well-reproduced. In addition, the point-proton rms radius exhibits a fairly strong dependence and increases with decreasing $\hbar\Omega$. For the rms radius, we find that the best agreement with experiment [24] is achieved for $\hbar\Omega = 14$ MeV.

From the point of view of the beta decay of ^{10}C, a good description of the $T = 1$ states is important. From Fig. 4 we can see that the calculated ^{10}B $T = 1$ states have the right relative positions and are reasonably stable with respect to variations both of the model-space size and $\hbar\Omega$. We have also performed $4\hbar\Omega$ calculations for ^{10}Be to study the splitting of the isospin analog states in the whole isospin-multiplet ^{10}C - ^{10}B - ^{10}Be. The experimental ground-state splitting between ^{10}C and ^{10}Be is 4.66 MeV, while our calculated values are 4.68, 4.83, and 4.94 MeV for $\hbar\Omega = 14, 15.5$, and 17 MeV, respectively. The best agreement with experiment is achieved for $\hbar\Omega = 14$ MeV, where the calculated rms point-proton radius is also in agreement with the experimental value. On the other hand, the splitting between the 0^+1 states of ^{10}C and ^{10}B, which is experimentally 2.69 MeV, is overestimated in our calculations by 8, 11, and 14% for the $\hbar\Omega = 14, 15.5$, and 17 MeV calculations, respectively. Since the correct ^{10}C - ^{10}Be splitting is obtained for $\hbar\Omega = 14$ MeV, the excess in the ^{10}C - ^{10}B splitting suggests that the isospin breaking due the strong $T = 1$ force may be too large. One possible explanation is that our approach for deriving the effective interaction tends to exaggerate the differences between the pn and nn, pp potentials. Such an artificial effect should decrease with increasing model-space size. On the other hand, it is also possible that the Reid93 potential itself overestimates differences between the pn and pp, nn systems in the $T = 1$ channel. For the most part, we find the best overall agreement for the rms point proton radius, binding energy, and Coulomb energy splitting for $\hbar\Omega = 14$ MeV. Given that the isospin mixing is largely driven by the Coulomb interaction, which is then dependent on the size of the nucleus, we feel that the best value for the isospin-mixing correction to the Fermi matrix element will be achieved for $\hbar\Omega = 14$ MeV.

The most important results of our study are also summarized in Table 1 in the last two lines. The calculated isospin-mixing corrections $\delta_C = 1 - \frac{|M_F|^2}{2}$, in %, are presented for all three choices of $\hbar\Omega$ and for both $4\hbar\Omega$ and $2\hbar\Omega$ model spaces. Again, a correlation between the radius and the isospin-mixing correction is clearly observed, as δ_C decreases with increasing radius. This is simply understood in terms of a larger radius implying weaker Coulomb effects. On the other hand, with an increase in the model-space size, a significant increase in the isospin-mixing correction is apparent. This is due to the fact that in the larger model space, the excitation energies of the 1p-1h 0^+1 states decrease, hence leading to greater mixing. For this reason, the more realistic multi-valued effective interaction is important. We have also performed test calculations with the single-valued interaction in the $2\hbar\Omega$ space, and found δ_C to be smaller by approximately 30%.

Our $4\hbar\Omega$ results suggest an isospin-mixing correction $\delta_C \approx 0.08 - 0.1\%$. This is compatible with the previously published value of $\delta_C \approx 0.15(9)\%$ by Ormand and Brown [25]. That value was a sum of two contributions. First, about 0.04% came from the shell-model wave-function renormalization due to the isospin mixing and was obtained in a $0\hbar\Omega$ shell-model calculation using phenomenological effective interactions. Second, the amount 0.09% was due to the deviation from unity of the radial overlap between the converted proton and the corresponding neutron. This effect was attributed to the influence of states lying outside the $0\hbar\Omega$ space. The radial wave-functions were obtained in a Hartree-Fock calculation using Skyrme-type interactions. Because we use a multi-configuration model space in the present calculation, we should have both effects included consistently at the same time.

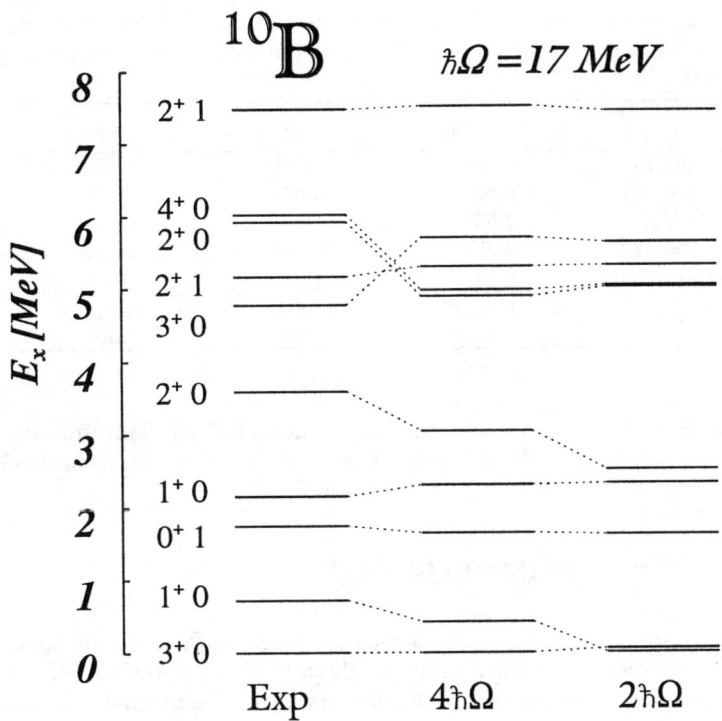

FIGURE 4. The experimental and calculated excitation spectra of ^{10}B. The results corresponding to the model-space sizes of $4\hbar\Omega$ and $2\hbar\Omega$ relative to the ground-state configurations are presented, respectively. The harmonic-oscillator energy of $\hbar\Omega = 17$ MeV was used.

Contrary to previous estimates for the isospin corrections, the present calculation was carried out within a model space that included many $\hbar\Omega$ excitations. As a consequence, the conventional configuration mixing and radial mismatch contributions were evaluated within a unified framework, simultaneously and the usual separation was not necessary. With regard to parameters used within the calculation, we find a correlation between the isospin-mixing correction and the Coulomb splitting between the isotopic multiplets, which, in turn, is governed by the nuclear size through the oscillator parameter. Given that the isospin-mixing correction is primarily a Coulomb effect, the best value for δ_C is taken to coincide with the oscillator parameter that correctly reproduces the Coulomb splittings. With regard to the model-space size, a clear improvement (or an indication towards convergence) in most observables is evident when the size of the model space is increased from $2\hbar\Omega$ to $4\hbar\Omega$, but δ_C is found to increase by only 0.03% (in magnitude) in this case. Hence, our final estimate for δ_C is taken to be 0.12(3)% (where the $4\hbar\Omega$ result has been increased by 0.03% to account for the possible effects of an increased model space). This result also happens to be in excellent agreement with the previous estimates that relied on the conventional separation of the configuration

TABLE 1. Experimental and calculated binding energies, in MeV; magnetic moments, in μ_N; and quadrupole moments, in e fm^2; of ^{10}B. Also the experimental and calculated binding energies, in MeV, and the point proton radius, in fm, of ^{10}C are presented. The results correspond to the $4\hbar\Omega$ calculations. In addition, the isospin-mixing correction δ_C, in %, is shown as obtained both in the $4\hbar\Omega$ calculations and the $2\hbar\Omega$ calculations. Results of three different calculations with the harmonic-oscillator parameter taken to be $\hbar\Omega = 14, 15.5, 17$ MeV, respectively, are presented. The effective interaction used was derived from the Reid 93 nucleon-nucleon potential. The experimental values are taken from Refs. [23,24].

	Exp	$\hbar\Omega = 14$ MeV	$\hbar\Omega = 15.5$ MeV	$\hbar\Omega = 17$ MeV
$E_B(^{10}$B$)$	64.75	63.61	62.78	61.53
$Q(3^+0)$	8.47(6)	5.85	5.64	5.52
$\mu(3^+0)$	1.80	1.86	1.85	1.85
$E_B(^{10}$C$)$	60.32	58.68	58.19	56.83
$\sqrt{\langle r_p^2 \rangle}$	2.31 ± 0.03	2.28	2.21	2.17
$\delta_C(4\hbar\Omega)[\%]$	-	0.084	0.091	0.097
$\delta_C(2\hbar\Omega)[\%]$	-	0.055	0.061	0.067

and radial mismatch contributions. Finally, we note that the magnitude of the isospin-mixing correction obtained in our calculation does not lead to a resolution to the deviation from unitarity for the Cabibbo-Kobayashi-Maskawa matrix.

CONCLUSIONS

We have performed large-basis no-core shell-model calculations for selected $0p$-shell nuclei with $A = 7 - 10$. We used two-body effective interactions derived from the Reid93 nucleon-nucleon potential with the isospin breaking taken into account. We were able to reproduce most of the characteristics of the ground states as well as the correct ordering of the lowest excited states. Our no-core shell-model approach has only a very limited number of freely adjustable parameters, such as the harmonic-oscillator frequency and the size of the model space. The calculations were performed in the $6\hbar\Omega$ and $4\hbar\Omega$ model spaces for the $A = 7$ and $A = 8 - 10$ nuclei, respectively. Our results show, that the multi-configuration shell-model approach combined with the use of microscopic effective interactions is capable of a good qualitative and quantitative description of the $0p$-shell nuclei. We also studied in detail the $A = 10$ nuclei with the emphasis on evaluating the isospin-mixing correction to the ^{10}C$\rightarrow^{10}$B Fermi matrix element.

It is feasible to extend the present $4\hbar\Omega$ calculations to heavier $0p$-shell nuclei as well. We are also presently investigating techniques for including effective three- and four-body interactions in our calculations [17]. In addition, a study related to parity-violating electron scattering is under way.

REFERENCES

1. D.C. Zheng, B.R. Barrett, J.P. Vary and R.J. McCarthy, *Phys. Rev.* **C 49**, 1999 (1994).
2. D.C. Zheng, J.P. Vary and B.R. Barrett, *Phys. Rev.* **C 50**, 2841 (1994).
3. D.C. Zheng, B.R. Barrett, J.P. Vary and H. Müther, *Phys. Rev.* **C 51**, 2471 (1995).
4. D.C. Zheng, B.R. Barrett, J.P. Vary, W.C. Haxton and C.L. Song, *Phys. Rev.* **C 52**, 2488 (1995).
5. P. Navrátil and B. R. Barrett, *Phys. Rev.* **C 54**, 2986 (1996).
6. P. Navrátil, M. Thoresen and B. R. Barrett, *Phys. Rev.* **C 55**, R573 (1997).
7. P. Navrátil, B. R. Barrett and W. E. Ormand, *Phys. Rev.* **C 56**, (1997) 2542.
8. P. Navrátil and B. R. Barrett, *Phys. Rev.* **C 57**, 3119 (1998).
9. T. Schucan and H. A. Wiedenmüller, *Ann. Phys.* **73**, 108 (1972); **76**, 483 (1973).
10. K. Suzuki and S. Y. Lee, *Prog. Theor. Phys.* **64**, 2091 (1980).
11. K. Suzuki, *Prog. Theor. Phys.* **68**, 246 (1982); K. Suzuki and R. Okamoto, *Prog. Theor. Phys.* **70**, 439 (1983).
12. P. Navrátil and B. R. Barrett, *Phys. Lett.* **B 369**, 193 (1996).
13. V. G. J. Stoks, R. A. M. Klomp, C. P. F. Terheggen and J. J. de Swart, *Phys. Rev.* **C 49** 2950 (1994).
14. J. P. Vary and D. C. Zheng, "The Many-Fermion-Dynamics Shell-Model Code", Iowa State University (1994) (unpublished).
15. B. R. Barrett, R. G. L. Hewitt and R. J. McCarthy, *Phys. Rev.* **C 3**, 1137 (1971).

16. M. Hjorth-Jensen, T. T. S. Kuo and E. Osnes, *Phys. Rep.* **261**, 125 (1995).

17. P. Navrátil and B. R. Barrett, *Phys. Rev.* **C 57**, 562 (1998).

18. I. S. Towner, J. C. Hardy, and M. Harvey, *Nucl. Phys.* **A284**, 269 (1977).

19. W. E. Ormand and B. A. Brown, *Phys. Rev. Lett.* **62**, 866 (1989).

20. I. S. Towner, in *Symmetry violation in subatomic physics*, Proc. of the 6th summer institute in theoretical physics, ed. B. Castel and P. J. O'Donnel (World Scientific, Singapore, 1989) p. 211.

21. F.C. Barker, B. A. Brown, W. Jaus, and G. Rasche, *Nucl. Phys.* **A540**, 501 (1992); J. .F. Donoghue, B. R. Holstein, and S. .W. Klint, *Phys. Rev.* **D 35**, 934 (1987).

22. E. D. Thorndike and R. A. Poling, *Phys. Rep.* **157**, 183 (1988).

23. F. Ajzenberg-Selove, *Nucl. Phys.* **A490**, 1 (1988).

24. A. Ozawa, I. Tanihata, T. Kobayashi, Y. Sugahara, O. Yamakawa, K. Omata, K. Sugimoto, D. Olson, W. Christie, H. Wieman, *Nucl. Phys.* **A 608**, 63 (1996).

25. W. E. Ormand and B. A. Brown, *Phys. Rev.* **C 52**, 2455 (1995).

Monte Carlo methods and applications for the nuclear shell model

D.J. Dean[*] and J.A. White[†]

[*]Physics Division, Oak Ridge National Laboratory, Oak Ridge, TN
and Physics Department, University of Tennessee, Knoxville, TN
[†]W. K. Kellogg Radiation Laboratory, 106-38, California Institute of Technology
Pasadena, California 91125 USA

Abstract. The shell-model Monte Carlo (SMMC) technique transforms the traditional nuclear shell-model problem into a path-integral over auxiliary fields. We describe below the method and its applications to four physics issues: calculations of sd-pf- shell nuclei, a discussion of electron-capture rates in pf-shell nuclei, exploration of pairing correlations in unstable nuclei, and level densities in rare earth systems.

I INTRODUCTION

Studies of nuclei far from stability have long been a goal of nuclear science. Nuclei on either side of the stability region, either neutron-rich or deficient, are being produced at new radioactive beam facilities across the world. At these facilities, and with the help of advances in nuclear many-body theory, the community will address many of the key physics issues including: mapping of the neutron and proton drip lines, thus exploring the limits of stability; understanding effects of the continuum on weakly bound nuclear systems; understanding the nature of shell gap modifications in very neutron-rich systems; determining nuclear properties needed for astrophysics; investigating deformation, spin, and pairing properties of systems far from stability; and analyzing microscopically unusual shapes in unstable nuclei.

The range and diversity of nuclear behavior, as indicated in the above list of ongoing and planned experimental investigations, have naturally engendered a host of theoretical models. Short of a complete solution to the many-nucleon problem, the interacting shell model is widely regarded as the most broadly capable description of low-energy nuclear structure, and the one most directly traceable to the fundamental many-body problem. Difficult though it may be, solving the shell-model problem is of fundamental importance to our understanding of the correlations found in nuclei.

One avenue of research during the past few years has been in the area of the nuclear shell model solved not by diagonalization, but by integration. In what follows, we will describe the shell-model Monte Carlo (SMMC) method and discuss several recent and interesting results obtained from theory. These include calculations in sd-pf-shell neutron-rich nuclei, a discussion of electron-capture rates in fp-shell nuclei, pairing correlations in medium-mass nuclei near N=Z, and studies of level densities in rare-earth nuclei.

II SMMC METHODS

In the following we briefly outline the formalism of the SMMC method. We begin with a brief description of statistical mechanics techniques used in our approach, then discuss the Hubbard-Stratonovich transformation, and end with a discussion of Monte Carlo sampling procedures. We refer the reader to previous works [1,3] for a more detailed exposition.

CP481, *Nuclear Structure 98*, edited by C. Baktash
© 1999 American Institute of Physics 1-56396-858-4/99/$15.00

A   Observables

SMMC methods rely on an ability to calculate the imaginary-time many-body evolution operator, $\exp(-\beta H)$, where β is a real c-number. The many-body Hamiltonian can be written schematically as

$$H = \varepsilon \mathcal{O} + \frac{1}{2} V \mathcal{O} \mathcal{O} \ , \tag{1}$$

where $\mathcal{O}$ is a density operator, V is the strength of the two-body interaction, and ε is a single-particle energy. In the full problem, there are many such quantities with various orbital indices that are summed over, but we omit them here for the sake of clarity.

While the SMMC technique does not result in a complete solution to the many-body problem in the sense of giving all eigenvalues and eigenstates of H, it can result in much useful information. For example, the expectation value of some observable, Ω, can be obtained by calculating

$$\langle \Omega \rangle = \frac{\operatorname{Tr} e^{-\beta H} \Omega}{\operatorname{Tr} e^{-\beta H}} \ . \tag{2}$$

Here, $\beta \equiv T^{-1}$ is interpreted as the inverse of the temperature T, and the many-body trace is defined as

$$\operatorname{Tr} X \equiv \sum_i \langle i | X | i \rangle \ , \tag{3}$$

where the sum is over many-body states of the system. In the canonical ensemble, this sum is over all states with a specified number of nucleons (implemented by number projection [2,3]), while the grand canonical ensemble introduces a chemical potential and sums over *all* many-body states.

In the limit of low temperature ($T \to 0$ or $\beta \to \infty$), the canonical trace reduces to a ground-state expectation value. Alternatively, if $|\Phi\rangle$ is a many-body trial state not orthogonal to the exact ground state, $|\Psi\rangle$, then $e^{-\beta H}$ can be used as a filter to refine $|\Phi\rangle$ to $|\Psi\rangle$ as β becomes large. An observable can be calculated in this "zero temperature" method as

$$\frac{\langle \Phi | e^{-\frac{\beta}{2} H} \Omega e^{-\frac{\beta}{2} H} | \Phi \rangle}{\langle \Phi | e^{-\beta H} | \Phi \rangle} \xrightarrow{\beta \to \infty} \frac{\langle \Psi | \Omega | \Psi \rangle}{\langle \Psi | \Psi \rangle} \ . \tag{4}$$

If Ω is the Hamiltonian, then (4) at $\beta = 0$ is the variational estimate of the energy, and improves as β increases. Of course, the efficiency of the refinement for any observable depends upon the degree to which $|\Phi\rangle$ approximates $|\Psi\rangle$.

Beyond such static properties, $e^{-\beta H}$ allows us to obtain some information about the dynamical response of the system. For an operator Ω, the response function, $R_\Omega(\tau)$, in the canonical ensemble is defined as

$$R_\Omega(\tau) \equiv \frac{\operatorname{Tr} e^{-(\beta-\tau)H} \Omega^\dagger e^{-\tau H} \Omega}{\operatorname{Tr} e^{-\beta H}} \equiv \langle \Omega^\dagger(\tau) \Omega(0) \rangle, \tag{5}$$

where $\Omega^\dagger(\tau) \equiv e^{\tau H} \Omega^\dagger e^{-\tau H}$ is the imaginary-time Heisenberg operator. Interesting choices for Ω are the annihilation operators for particular orbitals, the Gamow-Teller, $M1$, or quadrupole moment, etc. Inserting complete sets of A-body eigenstates of H ($\{|i\rangle, |f\rangle\}$ with energies $E_{i,f}$) shows that

$$R_\Omega(\tau) = \frac{1}{Z} \sum_{if} e^{-\beta E_i} |\langle f | \Omega | i \rangle|^2 e^{-\tau(E_f - E_i)}, \tag{6}$$

where $Z = \sum_i e^{-\beta E_i}$ is the partition function. Thus, $R_\Omega(\tau)$ is the Laplace transform of the strength function $S_\Omega(E)$:

$$R_\Omega(\tau) = \int_{-\infty}^{\infty} e^{-\tau E} S_\Omega(E) dE \ ; \tag{7}$$

$$S_\Omega(E) = \frac{1}{Z} \sum_{fi} e^{-\beta E_i} |\langle f | \Omega | i \rangle|^2 \delta(E - E_f + E_i) \ . \tag{8}$$

Hence, if we can calculate $R_\Omega(\tau)$, $S_\Omega(E)$ can be determined. Short of a full inversion of the Laplace transform (which is often numerically difficult), the behavior of $R_\Omega(\tau)$ for small τ gives information about the energy-weighted moments of S_Ω. In particular,

$$R_\Omega(0) = \int_{-\infty}^{\infty} S_\Omega(E) dE = \frac{1}{Z} \sum_i e^{-\beta E_i} |\langle f|\Omega|i\rangle|^2 = \langle \Omega^\dagger \Omega \rangle_A \tag{9}$$

is the total strength,

$$-R'_\Omega(0) = \int_{-\infty}^{\infty} S_\Omega(E) E dE = \frac{1}{Z} \sum_{if} e^{-\beta E_i} |\langle f|\Omega|i\rangle|^2 (E_f - E_i) \tag{10}$$

is the first moment (the prime denotes differentiation with respect to τ).

It is important to note that we usually cannot obtain detailed spectroscopic information from SMMC calculations. Rather, we can calculate expectation values of operators in the thermodynamic ensembles or the ground state. Occasionally, these can indirectly furnish properties of excited states. For example, if there is a collective 2^+ state absorbing most of the $E2$ strength, then the centroid of the quadrupole response function will be a good estimate of its energy. But, in general, we are without the numerous specific excitation energies and wave functions that characterize a direct diagonalization. This is both a blessing and a curse. The former is that for the very large model spaces of interest, there is no way in which we can deal explicitly with all of the wave functions and excitation energies. Indeed, we often don't need to, as experiments only measure average nuclear properties at a given excitation energy. The curse is that comparison with detailed properties of specific levels is difficult. In this sense, the SMMC method is complementary to direct diagonalization for modest model spaces, but is the only method for treating very large problems.

B   The Hubbard-Stratonovich transformation

It remains to describe the Hubbard-Stratonovich "trick" by which $e^{-\beta H}$ is managed. In broad terms, the difficult many-body evolution is replaced by a superposition of an infinity of tractable one-body evolutions, each in a different external field, σ. Integration over the external fields then reduces the many-body problem to quadrature.

To illustrate the approach, let us assume that only one operator $\mathcal{O}$ appears in the Hamiltonian (1). Then all of the difficulty arises from the two-body interaction, that term in H quadratic in $\mathcal{O}$. If H were solely linear in $\mathcal{O}$, we would have a one-body quantum system, which is readily dealt with. To linearize the evolution, we employ the Gaussian identity

$$e^{-\beta H} = \sqrt{\frac{\beta |V|}{2\pi}} \int_{-\infty}^{\infty} d\sigma e^{-\frac{1}{2}\beta |V|\sigma^2} e^{-\beta h}; \quad h = \varepsilon \mathcal{O} + sV\sigma\mathcal{O} . \tag{11}$$

Here, h is a one-body operator associated with a c-number field σ, and the many-body evolution is obtained by integrating the one-body evolution, $U_\sigma \equiv e^{-\beta h}$, over all σ with a Gaussian weight. The phase, s, is 1 if $V < 0$, or i if $V > 0$. Equation (11) is easily verified by completing the square in the exponent of the integrand; since we have assumed that there is only a single operator $\mathcal{O}$, there is no need to worry about non-commutation.

For a realistic Hamiltonian, there will be many non-commuting density operators, $\mathcal{O}_\alpha$, present, but we can always reduce the two-body term to diagonal form. Thus for a general two-body interaction in a general time-reversal invariant form, we write

$$H = \sum_\alpha \left(\epsilon_\alpha^* \bar{\mathcal{O}}_\alpha + \epsilon_\alpha \mathcal{O}_\alpha \right) + \frac{1}{2} \sum_\alpha V_\alpha \left\{ \mathcal{O}_\alpha, \bar{\mathcal{O}}_\alpha \right\} , \tag{12}$$

where $\bar{\mathcal{O}}_\alpha$ is the time reverse of $\mathcal{O}_\alpha$. Since, in general, $[\mathcal{O}_\alpha, \mathcal{O}_\beta] \neq 0$, we must split the interval β into N_t "time slices" of length $\Delta\beta \equiv \beta/N_t$,

$$e^{-\beta H} = [e^{-\Delta\beta H}]^{N_t}, \tag{13}$$

and for each time slice $n = 1, \ldots, N_t$ perform a linearization similar to Eq. 11 using auxiliary fields $\sigma_{\alpha n}$. Note that because the various $\mathcal{O}_\alpha$ need not commute, the representation of $e^{-\Delta\beta h}$ must be accurate through order $(\Delta\beta)^2$ to achieve an overall accuracy of order $\Delta\beta$.

We are now able to write expressions for observables as the ratio of two field integrals. Thus expectations of observables can be written as

$$\langle \Omega \rangle = \frac{\int \mathcal{D}\sigma W_\sigma \Omega_\sigma}{\int \mathcal{D}\sigma W_\sigma}, \tag{14}$$

where

$$W_\sigma = G_\sigma \operatorname{Tr} U_\sigma \; ; \qquad G_\sigma = e^{-\Delta\beta \sum_{\alpha n} |V_\alpha||\sigma_{\alpha n}|^2} \; ;$$

$$\Omega_\sigma = \frac{\operatorname{Tr} U_\sigma \Omega}{\operatorname{Tr} U_\sigma} \; ; \qquad \mathcal{D}\sigma \equiv \prod_{n=1}^{N_t} \prod_\alpha d\sigma_{\alpha n} d\sigma_{\alpha n}^* \left(\frac{\Delta\beta |V_\alpha|}{2\pi} \right), \tag{15}$$

and

$$U_\sigma = U_{N_t} \ldots U_2 U_1 \; ; \qquad U_n = e^{-\Delta\beta h_n} \; ;$$

$$h_n = \sum_\alpha (\varepsilon_\alpha^* + s_\alpha V_\alpha \sigma_{\alpha n}) \bar{\mathcal{O}}_\alpha + (\varepsilon_\alpha + s_\alpha V_\alpha \sigma_{\alpha n}^*) \mathcal{O}_\alpha \; . \tag{16}$$

This is, of course, a discrete version of a path integral over σ. Because there is a field variable for each operator at each time slice, the dimension of the integrals $\mathcal{D}\sigma$ can be very large, often exceeding 10^5. The errors in Eq. 14 are of order $\Delta\beta$, so that high accuracy requires large N_t and perhaps extrapolation to $N_t = \infty$ ($\Delta\beta = 0$).

Thus, the many-body observable is the weighted average (weight W_σ) of the observable Ω_σ calculated in an ensemble involving only the one-body evolution U_σ. Similar expressions involving two σ fields (one each for $e^{-\tau H}$ and $e^{-(\beta-\tau)H}$) can be written down for the response function (5), and all are readily adapted to the canonical or grand canonical ensembles or to the zero-temperature case.

An expression of the form (14) has a number of attractive features. First, the problem has been reduced to quadrature—we need only calculate the ratio of two integrals. Second, all of the quantum mechanics (which appears in Ω_σ) is of the one-body variety, which is simply handled by the algebra of $N_s \times N_s$ matrices. The price to pay is that we must treat the one-body problem for all possible σ fields.

C   Monte Carlo quadrature and the sign problem

We employ the Metropolis, Rosenbluth, Rosenbluth, Teller, and Teller algorithm [4] to generate the field configurations, σ, which requires only the ability to calculate the weight function for a given value of the integration variables. This method requires that the weight function W_σ must be real and non-negative. Unfortunately, many of the Hamiltonians of physical interest suffer from a sign problem, in that W_σ is negative over significant fractions of the integration volume. The fractional variance of a given expectation value becomes unacceptably large as the average sign approaches zero.

It was shown [3] that for even-even and $N = Z$ nuclei there is no sign problem for Hamiltonians if all $V_\alpha \leq 0$. Such forces include reasonable approximations to the realistic Hamiltonian like pairing-plus-multipole interactions. However, for an arbitrary Hamiltonian, we are not guaranteed that all $V_\alpha \leq 0$ (see, for example, Alhassid et al. [6]). However, we may expect that a realistic Hamiltonian will be dominated by terms like those of the schematic pairing-plus-multipole force (which is, after all, why the schematic forces were developed) so that it is, in some sense, close to a Hamiltonian for which the MC is directly applicable. Thus, the "practical solution" to the sign problem presented in Alhassid et al. [6] is based on an extrapolation of observables calculated for a "nearby" family of Hamiltonians whose integrands have a positive sign. Success depends crucially upon the degree of extrapolation required. Empirically, one finds that for all of the many realistic interactions tested in the sd- and pf-shells, the extrapolation required is modest, amounting to a factor-of-two variation in the isovector monopole pairing strength.

Based on the above observation, it is possible to decompose H in Eq. 12 into its "good" and "bad" parts, $H = H_G + H_B$. The "good" Hamiltonian, H_G, includes, in addition to the one-body terms, all the two-body interactions with $V_\alpha \leq 0$, while the "bad" Hamiltonian, H_B, contains all interactions with $V_\alpha > 0$. By construction, calculations with H_G alone have $\Phi_\sigma \equiv 1$ and are thus free of the sign problem.

We define a family of Hamiltonians, H_g, that depend on a continuous real parameter g as $H_g = f(g)H_G + gH_B$, so that $H_{g=1} = H$, and $f(g)$ is a function with $f(1) = 1$ and $f(g < 0) > 0$ that can be chosen to make the extrapolations less severe. (In practical applications, $f(g) = 1 - (1-g)/\chi$ with $\chi \approx 4$, and applied only to the two-body terms in H_G has been found to be a good choice.) If the V_α that are large in magnitude are "good," we expect that $H_{g=0} = H_G$ is a reasonable starting point for the calculation of an observable $\langle \Omega \rangle$. One might then hope to calculate $\langle \Omega \rangle_g = \mathrm{Tr}\left(\Omega e^{-\beta H_g}\right)/\mathrm{Tr}\left(e^{-\beta H_g}\right)$ for small $g > 0$ and then to extrapolate to $g = 1$, but typically $\langle \Phi \rangle$ collapses even for small positive g. However, it is evident from our construction that H_g is characterized by $\Phi_\sigma \equiv 1$ for any $g \leq 0$, since all the "bad" $V_\alpha(> 0)$ are replaced by "good" $gV_\alpha < 0$. We can therefore calculate $\langle \Omega \rangle_g$ for any $g \leq 0$ by a Monte Carlo sampling that is free of the sign problem. If $\langle \Omega \rangle_g$ is a smooth function of g, it should then be possible to extrapolate to $g = 1$ (i.e., to the original Hamiltonian) from $g \leq 0$. We emphasize that $g = 0$ is not expected to be a singular point of $\langle \Omega \rangle_g$; it is special only in the Monte Carlo evaluation. The extrapolation methods we employ have been tested against standard shell-model diagonalizations in many cases, and have, in general, been shown to work very well [3].

III APPLICATIONS

A   *sd-pf* nuclei

Studies of extremely neutron-rich nuclei have revealed a number of intriguing new phenomena. Two sets of these nuclei that have received particular attention are those with neutron number N in the vicinity of the $1s0d$ and $0f_{7/2}$ shell closures ($N \approx 20$ and $N \approx 28$). Experimental studies of neutron-rich Mg and Na isotopes indicate the onset of deformation, as well as the modification of the $N = 20$ shell gap for ^{32}Mg and nearby nuclei [8]. Inspired by the rich set of phenomena occurring near the $N = 20$ shell closure when $N \gg Z$, attention has been directed to nuclei near the $N = 28$ (sub)shell closure for a number of S and Ar isotopes [9,10] where similar, but less dramatic, effects have been seen as well.

In parallel with the experimental efforts, there have been several theoretical studies seeking to understand and, in some cases, predict properties of these unstable nuclei. Both mean-field [11,12] and shell-model calculations [9,10,13–17] have been proposed. The latter require a severe truncation to achieve tractable model spaces, since the successful description of these nuclei involves active nucleons in both the *sd*- and the *pf*-shells. The natural basis for the problem is therefore the full *sd-pf* space, which puts it out of reach of exact diagonalization on current hardware.

SMMC methods offer an alternative to direct diagonalization when the bases become very large. Though SMMC provides limited detailed spectroscopic information, it can predict, with good accuracy, overall nuclear properties such as masses, total strengths, strength distributions, and deformation — precisely those quantities probed by the recent experiments. It thus seems natural to apply SMMC methods to these unstable neutron-rich nuclei. Two questions will arise — center-of-mass motion and choice of the interaction — that are not exactly new, but demand special treatment in very large spaces. These questions were addressed in detail in Ref. [18]. We present a brief selection of results here.

If we fix the sign problem in the same manner as above for $H_{\mathrm{c.m.}}$, we are no longer dealing with a Hamiltonian that pushes *all* spurious components to higher energies — some components might even be lowered for $g < 0$. We will see shortly that this is not a real problem.

We typically choose a minimal extrapolation (linear, quadratic, etc.) in the extrapolation parameter that gives a χ^2 per datum of $\simeq 1$. In much of our work most quantities extrapolate either linearly or quadratically. We measure the center-of-mass contamination by calculating the expectation value of $H_{\mathrm{c.m.}}$. In Fig. 1a, we show the value of $\langle H_{\mathrm{c.m.}} \rangle$ in ^{32}Mg for several different values of $\beta_{\mathrm{c.m.}}$. It is apparent that $\langle H_{\mathrm{c.m.}} \rangle$ decreases as $\beta_{\mathrm{c.m.}}$ increases. We also find that near $\beta_{\mathrm{c.m.}} = 1$, $\langle H_{\mathrm{c.m.}} \rangle \ll 2\hbar\omega \simeq 28$ MeV showing that the center-of-mass contamination is minimal. Note that at approximately $\beta_{\mathrm{c.m.}} = 1.5$ the average of the two different techniques of extrapolation presented in Fig. 1a give $\langle H_{\mathrm{c.m.}} \rangle \simeq 0$ MeV, and the calculations could be fine tuned for each nucleus to obtain this value.

Figure 1a contains two different data sets corresponding to two different methods of extrapolating $\langle H_{\mathrm{c.m.}} \rangle$ to the physical case ($g = 1$). The solid circles show the results of a simple linear extrapolation where for this observable χ^2 per datum is approximately 1. It has been established [3] that $\langle H \rangle$ obeys a variational principle such that the extrapolating curve must have a minimum (slope = 0) at the physical value ($g = 1$). As we sample values of the quantity $\tilde{H}$, it is perhaps reasonable to extrapolate $\langle H_{\mathrm{c.m.}} \rangle$ using this constraint as well (if $\tilde{H}$ were truly separable, this would be an exact procedure). A cubic extrapolation embodying this constraint corresponds to the open circles in Fig. 1a.

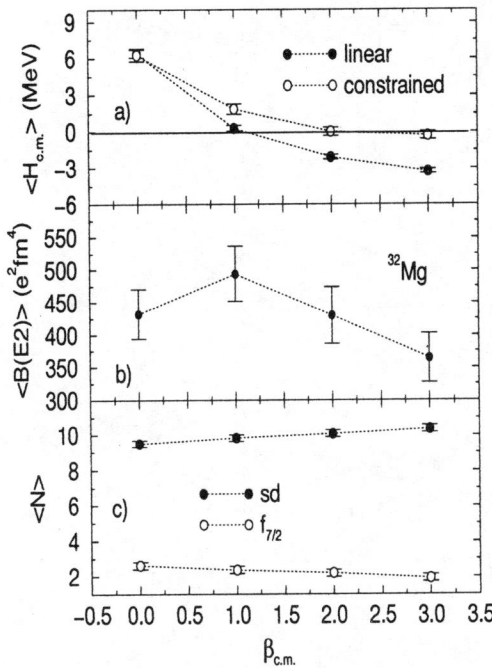

FIGURE 1. (a) The calculated value of $\langle H_{c.m.} \rangle$ as a function of $\beta_{c.m.}$ for ^{32}Mg. Two different extrapolations were performed as described in the text. The center-of-mass contamination is already significantly reduced at $\beta_{c.m.} = 1$. (b) The calculated total $B(E2, 0^+ \rightarrow 2^+)$ as a function of $\beta_{c.m.}$. (c) The sd-shell and $f_{7/2}$ subshell occupations as a function of $\beta_{c.m.}$.

We may further evaluate our extrapolation procedures by comparing SMMC and the standard shell-model results in ^{22}Mg. Shown in Fig. 2a is a detailed comparison for the expectation of the energy $\langle H \rangle$, and in Fig. 2b a comparison for $\langle H_{c.m.} \rangle$. The standard shell-model results were obtained using the code ANTOINE [7]. The SMMC results in Fig. 2a employ a constrained fit, such that $d\langle H \rangle/dg \mid_{g=1} = 0$. The slight deviation from the standard shell model at $g = -0.6, -0.8, -1.0$ is due to increasing interaction matrix elements (with g), while $d\beta$, the imaginary time step, is kept fixed. This deviation is also seen in Fig. 2b. Note that in Fig. 2b neither the constrained fit nor the linear fit (both with χ^2 per datum $\simeq 1$) give a precise description of the standard shell-model results at $g = 1$. An average of the two ways of extrapolation as indicated by the solid line on Fig. 2b, apparently gives the more precise result, and we shall do this for other $H_{c.m.}$ values quoted throughout this paper. The error bar for such an averaged result is given by adding in quadrature the individual errors of both extrapolations.

In Fig. 1b we show the evolution of the total $B(E2)$ and in Fig. 1c we show the occupation of the sd-shell and the $f_{7/2}$-shell as a function of $\beta_{c.m.}$. Note that the occupation of the $f_{7/2}$ orbit decreases as $\beta_{c.m.}$ increases. This is due to a combination of the removal of actual center-of-mass excitations and the "pushing up" in energy of the real states. The $B(E2)$ decreases slowly with $\beta_{c.m.}$, although the uncertainties are consistent with a constant. However, the decrease, particularly at $\beta_{c.m.} = 3$, is likely to be real since we are working in an incomplete $n\hbar\omega$ model space. At extremely large values of $\beta_{c.m.}$ we would remove the pf shell from the calculation and return to the pure sd-shell result, which is substantially smaller than the result shown here. The slow evolution of the $B(E2)$ with $\beta_{c.m.}$ does open the intriguing possibility of studying $B(E2)$s with an interaction that has no sign problem (e.g. Pairing + Quadrature) and no center-of-mass correction with the hope of obtaining reasonable results.

There is limited experimental information about the highly unstable, neutron-rich nuclei under consideration. In many cases only the mass, excitation energy of the first excited state, the $B(E2)$ to that state, and the β-decay rate is known, and not even all of this information is available in some cases. From the measured $B(E2)$, an estimate of the nuclear deformation parameter, β_2, has been obtained via the usual relation

$$\beta_2 = 4\pi\sqrt{B(E2; 0_{gs}^+ \rightarrow 2_1^+)}/3ZR_0^2e \tag{17}$$

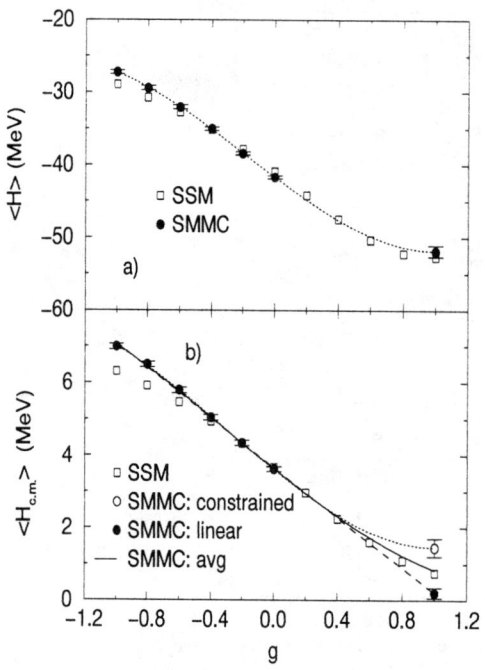

FIGURE 2. (a). The expectation of the Hamiltonian, $\langle H \rangle$ for ^{22}Mg as a function of the extrapolation parameter g. Shown are standard shell-model results and SMMC results. (b). The expectation of the center-of-mass Hamiltonian, $\langle H_{\text{c.m.}} \rangle$ as a function of g. SMMC results are shown for two types of extrapolation procedures, as discussed in the text, and are compared to standard shell-model results.

with $R_0 = 1.2A^{1/3}$ fm and $B(E2)$ given in e^2fm^4.

Much of the interest in the region stems from the unexpectedly large values of the deduced β_2, results which suggest the onset of deformation and have led to speculations about the vanishing of the $N = 20$ and $N = 28$ shell gaps. The lowering in energy of the 2_1^+ state supports this interpretation. The most thoroughly studied case, and the one which most convincingly demonstrates these phenomena, is ^{32}Mg with its extremely large $B(E2) = 454 \pm 78 \, e^2$fm^4 and corresponding $\beta_2 = 0.513$ [8]; however, a word of caution is necessary when deciding on the basis of this limited information that we are in the presence of well-deformed rotors: for ^{22}Mg, we would obtain $\beta_2 = 0.67$, even more spectacular, and for ^{12}C, $\beta_2 = 0.8$, well above the superdeformed bands.

Most of the measured observables can be calculated within the SMMC framework. It is well known that in *deformed* nuclei the total $B(E2)$ strength is almost saturated by the $0_{gs}^+ \rightarrow 2_1^+$ transition (typically 80% to 90% of the strength lies in this transition). Thus the total strength calculated by SMMC should only slightly overestimate the strength of the measured transition. In Fig. 3 the SMMC computed $B(E2, total)$ values are compared to the experimental $B(E2; 0_{gs}^+ \rightarrow 2_1^+)$ values. Reasonable agreement with experimental data across the space is obtained when one chooses effective charges of $e_p = 1.5$ and $e_n = 0.5$. Using these same effective charges, the USD values for the $B(E2, 0_{gs}^+ \rightarrow 2_1^+)$ of the sd-shell nuclei ^{32}Mg and ^{30}Ne are 177.1 and 143.2 e^2fm^4, respectively, far lower than the full sd-pf calculated and experimental values. All of the theoretical calculations require excitations to the pf-shell before reasonable values can be obtained. We note a general agreement among all calculations of the $B(E2)$ for ^{46}Ar, although they are typically larger than experimental data would suggest. We also note a somewhat lower value of the $B(E2)$ in this calculation as compared to experiment and other theoretical calculations in the case of ^{42}S [16]. Also shown in Fig. 3 are the differences between experimental and theoretical binding energies for nuclei in this region. Agreement is quite good overall.

Table 1 gives selected occupation numbers for the nuclei considered. We first note a difficulty in extrapolating some of the occupations where the number of particles is nearly zero. This leads to a systematic error bar that we estimate at ± 0.2 for all occupations shown, while the statistical error bar is quoted in the table. The extrapolations for occupation numbers were principally linear. Table 1 shows that ^{22}Mg remains as an almost pure sd-shell nucleus, as expected. We also see that the protons in ^{30}Ne, ^{32}Mg, and ^{42}Si are almost entirely confined to the sd shell. This latter is a pleasing result in at least two regards. First, it shows that the interaction does not mix the two shells to an unrealistically large extent. Second, if spurious center-of-mass

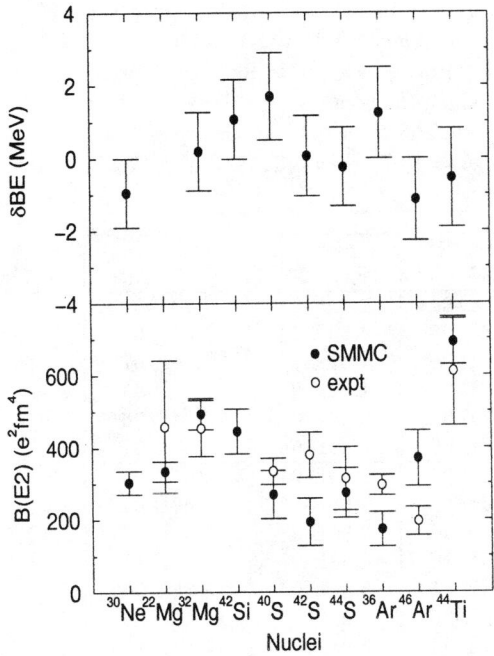

FIGURE 3. Top: Difference between theoretical and experimental binding energies for the sd-pf-shell nuclei studied in this work. Bottom: Experimental and theoretical $B(E2)$ values.

contamination were a severe problem, we would expect to see a larger proton $f_{7/2}$ population for these nuclei due to the $0d_{5/2}$-$0f_{7/2}$ "transition" mediated by the center-of-mass creation operator. The fact that there is little proton $f_{7/2}$ occupation for these nuclei confirms that the center-of-mass contamination is under reasonable control.

An interesting feature of Table 1 lies in the neutron occupations of the $N = 20$ nuclei (^{30}Ne and ^{32}Mg) and the $N = 28$ nuclei (^{42}Si, ^{44}S, and ^{46}Ar). The neutron occupations of the two $N = 20$ nuclei are quite similar, confirming the finding of Fukunishi *et al.* [15] and Poves and Retamosa [14] that the $N = 20$ shell gap is modified. In fact, the neutron $f_{7/2}$ orbital contains approximately two particles before the $N = 20$ closure, thus behaving like an intruder single-particle state. Furthermore, we see that 2p-2h excitations dominate although higher excitations also play some role. We also see that the neutrons occupying the pf-shell in $N = 20$ systems are principally confined to the $f_{7/2}$ sub-shell.

The conclusions that follow from looking at nuclei with $N > 20$, particularly those with $N = 28$, are that the $N = 20$ shell is nearly completely closed at this point, and that the $N = 28$ closure shell is reasonably robust, although approximately one neutron occupies the upper part of the pf shell. Coupling of the protons

TABLE 1. The calculated SMMC neutron (n) and proton (p) occupation numbers for the sd shell, the $0f_{7/2}$ sub-shell, and the remaining orbitals of the pf shell. The statistical errors are given for linear extrapolations. A systematic error of ± 0.2 should also be included.

	N, Z	n-sd	n-$f_{7/2}$	n-$pf_{5/2}$	p-sd	p-$f_{7/2}$	p-$pf_{5/2}$
^{22}Mg	10,12	3.93 ± 0.02	0.1 ± 0.02	-0.05 ± 0.01	2.04 ± 0.02	0.00 ± 0.01	-0.05 ± 0.01
^{30}Ne	20,10	9.95 ± 0.03	2.32 ± 0.03	-0.26 ± 0.02	2.03 ± 0.02	-0.01 ± 0.01	-0.02 ± 0.01
^{32}Mg	20,12	9.84 ± 0.03	2.37 ± 0.03	-0.21 ± 0.02	3.99 ± 0.03	0.05 ± 0.02	-0.05 ± 0.01
^{36}Ar	18,18	9.07 ± 0.03	1.08 ± 0.02	-0.15 ± 0.02	9.07 ± 0.03	1.08 ± 0.02	-0.15 ± 0.02
^{40}S	24,16	11.00 ± 0.03	5.00 ± 0.03	-0.01 ± 0.02	7.57 ± 0.04	0.54 ± 0.02	-0.12 ± 0.02
^{42}Si	28,14	11.77 ± 0.02	7.34 ± 0.02	0.90 ± 0.03	5.79 ± 0.03	0.25 ± 0.02	-0.07 ± 0.01
^{42}S	26,16	11.41 ± 0.02	6.33 ± 0.02	0.25 ± 0.03	7.49 ± 0.03	0.58 ± 0.02	-0.09 ± 0.02
^{44}S	28,16	11.74 ± 0.02	7.18 ± 0.02	1.06 ± 0.03	7.54 ± 0.03	0.56 ± 0.02	-0.12 ± 0.02
^{44}Ti	22,22	10.42 ± 0.03	3.58 ± 0.02	0.00 ± 0.02	10.42 ± 0.03	3.58 ± 0.02	0.00 ± 0.02
^{46}Ar	28,18	11.64 ± 0.02	7.13 ± 0.02	1.23 ± 0.03	8.74 ± 0.03	1.34 ± 0.02	-0.08 ± 0.02

TABLE 2. Comparisons of the SMMC electron capture rates with the total (λ_{ec}) and partial Gamow-Teller (λ_{ec}^{GT}) rates as given in Ref. [20]. Physical conditions at which the comparisons were made are $\rho_7 = 5.86$, $T_9 = 3.40$, and $Y_e = 0.47$ for the upper part of the table, and $\rho_7 = 10.7$, $T_9 = 3.65$, and $Y_e = 0.455$ for the lower part.

Nucleus	λ_{ec} (sec^{-1}) (SMMC)	λ_{ec} (sec^{-1}) (Ref. [20])	λ_{ec}^{GT} (sec^{-1}) (Ref. [20])
^{55}Co	3.89E-04	1.41E-01	1.23E-01
^{57}Co	3.34E-06	3.50E-03	1.31E-04
^{55}Fe	1.20E-08	1.61E-03	1.16E-07
^{56}Ni	3.47E-02	1.60E-02	6.34E-03
^{58}Ni	1.01E-03	6.36E-04	4.04E-06
^{60}Ni	7.39E-05	1.49E-06	4.86E-07
^{59}Co	3.44E-07	2.09E-04	6.37E-05
^{57}Co	2.06E-05	7.65E-03	3.69E-04
^{55}Fe	1.07E-07	3.80E-03	5.51E-07
^{56}Fe	9.80E-06	4.68E-07	6.60E-10
^{54}Fe	3.84E-04	9.50E-04	3.85E-06
^{51}V	1.06E-06	1.24E-05	9.46E-09
^{52}Cr	1.32E-04	2.01E-07	1.59E-10
^{60}Ni	3.61E-04	7.64E-06	2.12E-06

with the low-lying neutron excitations probably accounts for the relatively large $B(E2)$, without the need of invoking rotational behavior.

B   Electron capture rates for Fe-region nuclei

The impact of nuclear structure on astrophysics has become increasingly important, particularly in the fascinating, and presently unsolved, problem of type-II supernovae explosions. One key ingredient of the precollapse scenario is the electron-capture cross section on nuclei [19,20]. An important contribution to electron-capture cross sections in supernovae environments is the Gamow-Teller (GT) strength distribution. This strength distribution, calculated in SMMC using Eqs. (3 and 4) above, is used to find the energy-dependent cross section for electron capture. In order to obtain the electron-capture rates, the cross section is then folded with the flux of a degenerate relativistic electron gas [21]. Note that the Gamow-Teller distribution is calculated at the finite nuclear temperature which, in principle, is the same as the one for the electron gas.

It is important to calculate the GT strength distributions reasonably accurately for both the total strength and the position of the main GT peak in order to have a quantitative estimate for the electron-capture rates. For astrophysical purposes, calculating the rates to within a factor of two is required. We concentrate here on mid-fp-shell results for the electron-capture cross sections [21]. The Kuo-Brown interaction [22], modified in the monopole terms by Zuker and Poves [23], was used throughout these pf-shell calculations. This interaction reproduces quite nicely the ground- and excited-state properties of mid-fp-shell nuclei [24,25], including the total Gamow-Teller strengths and distributions, where the overall agreement between theory and experiment [26] is quite reasonable. The SMMC technique allows one to probe the complete $0\hbar\omega$ fp-shell region without any parameter adjustments to the Hamiltonian, although the Gamow-Teller operator has been renormalized by the standard factor of 0.8.

Do the electron-capture rates presented here indicate potential implications for the precollapse evolution of a type II supernova? To make a judgment on this important question, one should compare in Table I the SMMC rates for selected nuclei with those currently used in collapse calculations [20]. For the comparison, we choose the same physical conditions as assumed in Tables 4–6 in Aufderheide *et al.* [20]. Table I also lists the partial electron-capture rate which has been attributed to Gamow-Teller transitions [20]. Note that for even-parent nuclei, the present rate approximately agrees with the currently recommended *total* rate. A closer inspection,

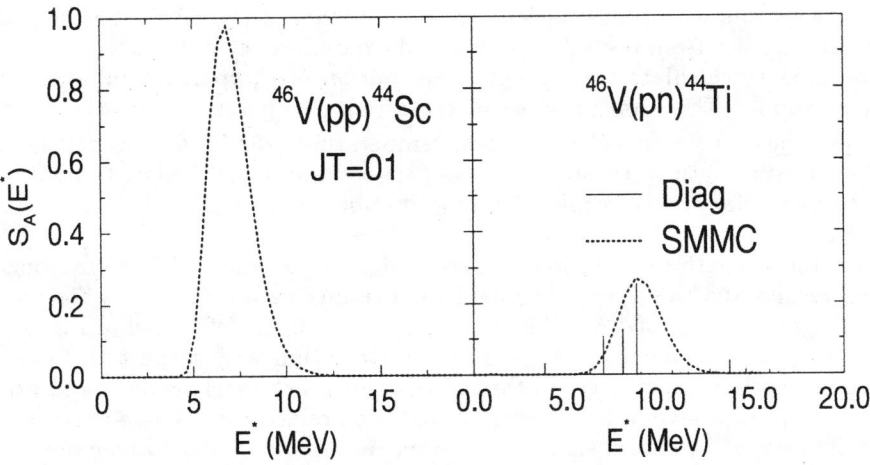

FIGURE 4. Left: proton pair strength distribution for ^{46}V. Right: proton-neutron (T=1) pair strength distribution. SMMC: thick line; direct diagonalization: impulses.

however, shows significant differences between the present rate and the one attributed to the Gamow-Teller transition in Aufderheide *et al.* [20]. The origin of this discrepancy is due to the fact that Fuller *et al.* [27] places the Gamow-Teller resonance for even-even nuclei systematically at too high an excitation energy. This shortcoming has been corrected in Fuller *et al.* [27] and Aufderheide *et al.* [20] by adding an experimentally known low-lying strength in addition to the one attributed to Gamow-Teller transitions. However, the overall good agreement between the SMMC results for even-even nuclei and the recommended rates indicates that the SMMC approach also accounts correctly for this low-lying strength. This has already been deduced from the good agreement between SMMC Gamow-Teller distributions and data including the low-energy regime [26].

Thus, for even-even nuclei, the SMMC approach is able to predict the *total* electron-capture rate rather reliably, even if no experimental data are available. Note that the SMMC rate is somewhat larger than the recommended rate for ^{56}Fe and ^{60}Ni. In both cases, the experimental Gamow-Teller distribution is known and agrees well with the SMMC results [26]. While the proposed increase of the rate for ^{60}Ni is not expected to have noticeable influence on the pre-collapse evolution, the increased rate for ^{56}Fe makes this nucleus an important contributor in the change of Y_e during the collapse (see Table 15 of Aufderheide *et al.* [20]).

For electron capture on odd-A nuclei, observe that the SMMC rates, derived from the Gamow-Teller distributions, are significantly smaller than the recommended total rate. This is due to the fact that for odd-A nuclei the Gamow-Teller transition peaks at rather high excitation energies in the daughter nucleus. The electron-capture rate on odd-A nuclei is therefore carried by weak transitions at low excitation energies. Comparing the SMMC rates to those attributed to Gamow-Teller transitions in Fuller *et al.* [27] and Aufderheide *et al.* [20] reveals that the latter have been, in general, significantly overestimated, which is caused by the fact that the position of the Gamow-Teller resonance is usually put at too low excitation energies in the daughter. The SMMC calculation implies that the Gamow-Teller transitions should not contribute noticeably to the electron-capture rates on odd-A nuclei at the low temperatures studied in Tables 14–16 in Aufderheide *et al.* [20]. Thus, the rates for odd-A nuclei given in these tables should generally be replaced by the non-Gamow-Teller fraction.

C   Pair correlations in nuclei

We would now like to turn to the subject of pair correlations in nuclei, and calculations aimed at their understanding. Nuclei near N=Z offer a unique place to study proton-neutron pairing, particularly in the isospin T=1 channel. In fact, most heavy odd-odd, N=Z nuclei beyond ^{40}Ca have total spin J=0, T=1 ground states. Theoretical studies have shown that many of these nuclei have enhanced T=1 proton-neutron correlations when compared to their even-even counterparts. These correlations are, to a lesser extent, present in even-even systems, but tend to decrease as one moves away from N=Z. In at least one nucleus in the mass 70 region, ^{74}Rb, there is experimental evidence for a ground state $T = 1$ band [28].

Experimentally, pair correlations can best be measured by pair transfer on nuclei. Although total cross sections are typically underpredicted when one employs spectroscopic factors computed from the shell model, relative two-nucleon spectroscopic factors within one nucleus are more reliable. Therefore, it is necessary for one to calculate and measure pair transfer from both the ground and excited states in a nucleus.

The SMMC method may be used to calculate the strength distribution of the pair annihilation operator A_{JTT_z}, as defined in Koonin et al. [3]. The total strength of these pairing operators, i.e. the expectation $\langle A_{JT}^\dagger A_{JT} \rangle$, has been studied previously as a function of mass, temperature [29,30], and rotation [31]. We would like to briefly present here the strength distributions of the pair operators as calculated in SMMC. The strength distribution for the pair transfer spectroscopic factors is proportional to $\langle A - 2 \mid A_{JT} \mid A \rangle$ and is calculated by the inversion of Eq. (4).

In future work, we will discuss the strength distributions in detail. Here we would like to briefly conclude by demonstrating that the SMMC results and the direct diagonalization results agree very nicely for the proton pair strength distributions in the ground state of ^{46}V. This is demonstrated in the left panel of Fig. 2. Shown in the right panel is the isovector proton-neutron pairing strength distribution with respect to the daughter nucleus. Notice that the overall strength is much larger in the proton-neutron channel, as discussed previously in Langanke et al. [30], and that the peak is several MeV lower in excitation relative to the like-particle channel. In both cases the strength distribution in ^{46}V differs significantly from that found in ^{48}Cr, where one finds that the dominant component is a ground-state to ground-state transition involving mainly particles in the $0f_{7/2}$ single-particle state. In both odd-odd N=Z channels, the distribution is fairly highly fragmented.

D Rare earth nuclei

We have recently applied SMMC techniques to survey rare-earth nuclei in the Dy region. This extensive study formed the thesis topic of J.A. White [37], whose goal was to examine how the phenomenologically motivated "pairing-plus-quadrupole" interaction compares in exact shell-model solutions with other methods. We also examined how the shell-model solutions compare with experimental data; static path approximation (SPA) calculations were also performed. There have been efforts recently by others to use SPA calculations, since they are simpler and faster (see [38,39] as examples). However, we found that SPA results are not consistently good. This study was also designed to investigate whether the phenomenological pairing-plus-quadrupole-type interactions can be used in exact solutions for large model spaces, and whether the interaction parameters require significant renormalization when using SPA.

We discuss here one particular aspect of that work, namely level density calculations. Details may be found in [40]. We used the Kumar-Baranger Hamiltonian with parameters appropriate for this region. Our single-particle space included the 50-82 subshell for the protons and the 82-126 shell for the neutrons. While several interesting aspects of these systems were studied in SMMC, we limit our discussion here to the level densities obtained for ^{162}Dy.

SMMC is an excellent way to calculate level densities. $E(\beta)$ is calculated for many values of β which determine the partition function, Z, as

$$\ln[Z(\beta)/Z(0)] = - \int_0^\beta d\beta' E(\beta') \tag{18}$$

$Z(0)$ is the total number of available states in the space. The level density is then computed as an inverse Laplace transform of Z. Here, the last step is performed with a saddle point approximation with $\beta^{-2}C \equiv -dE/d\beta$:

$$S(E) = \beta E + \ln Z(\beta) \tag{19}$$

$$\rho(E) = (2\pi\beta^{-2}C)^{-1/2}\exp(S) \tag{20}$$

SMMC has been used recently to calculate level density in iron region nuclei [41], and here we demonstrate its use in the rare-earth region.

The comparison of SMMC density in ^{162}Dy with the Tveter et al. [42] data is displayed in Fig. 5. The experimental method can reveal fine structure, but does not determine the absolute density magnitude. The SMMC calculation is scaled by a factor to facilitate comparison. In this case, the factor has been chosen to make the curves agree at lower excitation energies. From 1-3 MeV, the agreement is very good. From 3-5 MeV, the SMMC density increases more rapidly than the data. This deviation from the data cannot be accounted for by statistical errors in either the calculation or measurement. Near 6 MeV, the measured density briefly flattens before increasing and this also appears in the calculation, but the measurement errors are larger at that point.

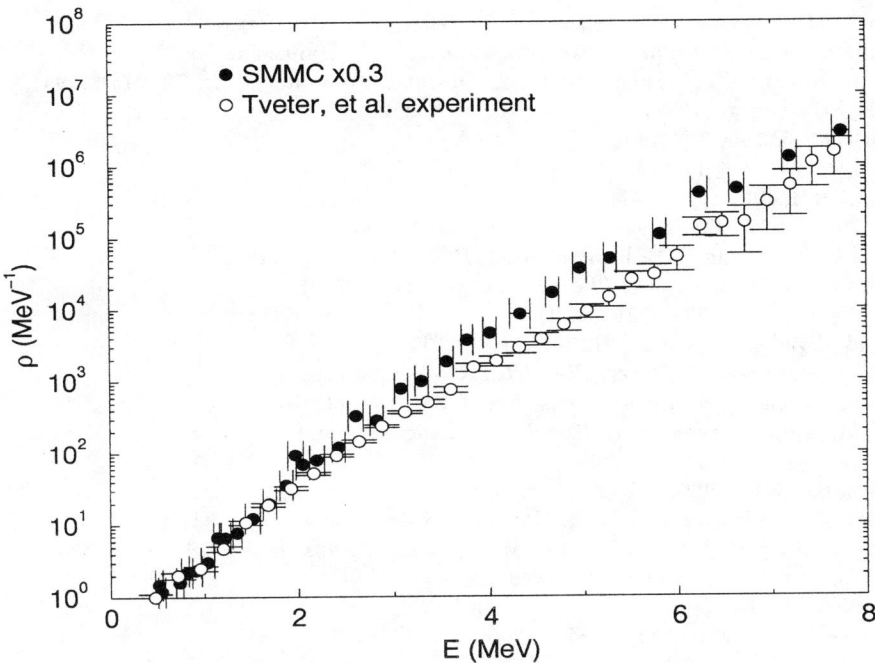

FIGURE 5. SMMC density vs. experimental data in ^{162}Dy.

The measured density includes all states included in the theoretical calculation plus some others, so that one would expect the measured density to be greater than or equal to the calculated density and never smaller. We may have instead chosen our constant to match the densities for moderate excitations and let the measured density be higher than the SMMC density for lower energies (1-3 MeV). Comparing structure between SMMC and data is difficult for the lowest energies due to statistical errors in the calculation and comparison at the upper range of the SMMC calculation, i.e., $E \approx 15$ MeV, is unfortunately impossible since the data only extend to about 8 MeV excitation energy.

IV CONCLUSIONS

In these Proceedings, we have used four specific examples (there are several others) for which the SMMC calculations have proven very useful in understanding the properties of nuclei in systems where the number of valence particles prohibits the use of more traditional approaches. The method has proven to be a valuable tool for furthering our understanding of nuclear structure and astrophysics. Continued developments in both creating useful interactions and shell-model technology should continue to enhance our ability to understand nuclei far from stability in the coming years.

ACKNOWLEDGMENTS

Oak Ridge National Laboratory (ORNL) is managed by Lockheed Martin Energy Research Corp. for the U.S. Department of Energy under contract number DE-AC05-96OR22464. This work was supported in part through grant DE-FG02-96ER40963 from the U.S. Department of Energy. This work was supported in part by the National Science Foundation, Grants No. PHY-9722428, PHY-9420470, and PHY-9412818.

REFERENCES

1. G. H. Lang, C. W. Johnson, S. E. Koonin, and W. E. Ormand, *Phys. Rev.* **C48**, 1518 (1993).
2. W. E. Ormand, D. J. Dean, C. W. Johnson, G. H. Lang, and S. E. Koonin, *Phys. Rev.* **C49**, 1422 (1994).
3. S. E. Koonin, D. J. Dean, and K. Langanke, *Phys. Repts.* **278**, 1 (1997), and references therein.

4. N. Metropolis, A. Rosenbluth, M. Rosenbluth, A. Teller, and E. Teller, *J. Chem. Phys.* **21**, 1087 (1953).

5. E. Y. Loh, Jr. and J. E. Gubernatis, in *Electronic Phase Transitions*, ed. W. Hanke and Yu.V. Kopaev, 177 (1992).

6. Y. Alhassid, D. J. Dean, S. E. Koonin, G. H. Lang, and W. E. Ormand, *Phys. Rev. Lett.* **72**, 613 (1994).

7. E. Caurier, code ANTOINE, Strasbourg, (1989).

8. T. Motobayashi et al., *Phys. Lett.* **B346**, 9 (1995).

9. H. Scheit et al., *Phys. Rev. Lett.* **77**, 3967 (1996).

10. T. Glasmacher et al., *Phys. Lett.* **B395**, 163 (1997).

11. T. R. Werner et al., *Nucl. Phys.* **A597**, 327 (1996).

12. X. Campi, H. Flocard, A. K. Kerman, and S. E. Koonin, *Nucl. Phys.* **A251**, 193 (1975).

13. E. K. Warburton, J. A. Becker, and B. A. Brown, *Phys. Rev.* **C41**, 1147 (1990).

14. A. Poves and J. Retamosa, *Nucl. Phys.* **A571**, 221 (1994).

15. N. Fukunishi, T. Otsuka, and T. Sebe, *Phys. Lett.* **B296**, 279 (1992).

16. J. Retamosa, E. Caurier, F. Nowacki, and A. Poves, *Phys. Rev.* **C55**, 1266 (1997).

17. E. Caurier, F. Nowacki, A. Poves, and J. Retamosa, *Phys. Rev.* **C58**, 2033 (1998).

18. D. J. Dean, M. T. Ressell, M. Hjorth-Jensen, S. E. Koonin, K. Langanke, and A. Zuker, Phys. Rev. **C59**, 2474 (1999).

19. H. A. Bethe, *Rev. Mod. Phys.* **62**, 801 (1990).

20. M. B. Aufderheide, I. Fushiki, S. E. Woosley, and D. H. Hartman, *Astrophys. J. Supp.* **91**, 389 (1994).

21. D. J. Dean, K. Langanke, L. Chatterjee, P. B. Radha, and M. R. Strayer, *Phys. Rev.* **C58**, 536 (1998).

22. T.T.S. Kuo and G. E. Brown, *Nucl. Phys.* **A114**, 241 (1968).

23. A. Poves and A. Zuker, *Phys. Repts.* **70**, 235 (1981).

24. E. Caurier, A. Zuker, A. Poves, and G. Martinez-Pinedo, *Phys. Rev.* **C50**, 225 (1994).

25. K. Langanke, D. J. Dean, P. B. Radha, Y. Alhassid, and S. E. Koonin, *Phys. Rev.* **C52**, 718 (1995).

26. P. B. Radha, D. J. Dean, S. E. Koonin, K. Langanke, and P. Vogel, *Phys. Rev.* **C56**, 3079 (1997).

27. G. M. Fuller, W. A. Fowler, and M. J. Newman, *Astrophys. J.* **48**, 279 (1982).

28. D. Rudolph et al., *Phys. Rev. Lett.* **76**, 376 (1996).

29. K. Langanke, D. J. Dean, P. B. Radha, and S. E. Koonin, *Nucl. Phys.* **A602**, 244 (1996).

30. K. Langanke, D. J. Dean, P. B. Radha, and S. E. Koonin, *Nucl. Phys.* **A613**, 253 (1997).

31. D. J. Dean, S. E. Koonin, K. Langanke, and P. B. Radha, *Phys. Lett.* **B399**, 1 (1997)

32. L. Wilets and M. Jean, *Phys. Rev.* **102**, 788 (1956).

33. O. K. Vorov and V. G. Zelevinsky, *Nucl. Phys.* **A439**, 207 (1985).

34. W. Krips et al., *Nucl. Phys.* **A529**, 485 (1991).

35. T. Otsuka, *Nucl. Phys.* **A557**, 531c (1993).

36. Y. Alhassid, G. Bertsch, D. J. Dean, and S. E. Koonin, *Phys. Rev. Lett.* **77**, 1444 (1996).

37. J. A. White, Ph.D. Thesis, California Institute of Technology, 1998.

38. R. Rossignoli, *Phys. Rev.* **C54**, 3 (1996).

39. R. Rossignoli, N. Canosa, and J. L. Egido, *Nucl. Phys.* **A607**, 3 (1996).

40. J. A. White, S. E. Koonin, and D. J. Dean, submitted to Phys. Rev. C (1998).

41. H. Nakada and Y. Alhassid, *Phys. Rev. Lett.* **79**, 16 (1997).

42. T. Tveter et al., *Phys. Rev. Lett.* **77**, 2404 (1996).

Monte Carlo shell model calculations for medium-mass nuclei

Takaharu Otsuka[1,2], Takahiro Mizusaki[1] and Yutaka Utsuno[1]

[1] *Department of Physics, University of Tokyo, Hongo, Tokyo 113-0033, Japan*
[2] *RIKEN, 2-1 Hirosawa, Wako, Saitama 351-0198, Japan*

Michio Honma

Center for Mathematical Sciences, University of Aizu, Tsuruga, Ikki-machi Aizu-Wakamatsu, Fukushima 965, Japan

Abstract. The formulation and recent applications of the Monte Carlo shell model based upon the Quantum Monte Carlo diagonalization (QMCD) method are reported. The QMCD has been proposed for solving the quantum many-body interacting systems. By the Monte Carlo shell model calculations, the level structure of low-lying states can be studied with realistic interactions, providing a useful tool for nuclear spectroscopy. Some examples of such calculations are presented. We report that the doubly closed shell probability of a proton-rich unstable nucleus ^{56}Ni is shown to be only 49 % in a full pf shell calculation, in contrast to the corresponding probability of ^{48}Ca which reaches 86 %. The Monte Carlo shell model calculation based on the QMCD method is extended so that the structure of non-yrast states can be described as well as yrast states, and it is applied again to ^{56}Ni but on an excited rotational band. This band is nicely described in a good agreement to recent experimental observation. Thus, the Monte Carlo shell model calculation is shown to be quite feasible for the spectroscopic study of nuclei. The Monte Carlo shell model is applied also to the study of unstable nuclei: the level scheme and E2 transition probabilities of neutron-rich nuclei around ^{32}Mg are discussed.

I INTRODUCTION

The nuclear shell model has been successful in the description of various aspects of nuclear structure, partly because it is based upon a minimum number of natural assumptions. Although the direct diagonalization of the Hamiltonian matrix in the full valence-nucleon Hilbert space is desired, the dimension of such a space is too large in many cases, preventing us from performing the full calculations. In order to overcome this difficulty, quantum Monte Carlo approaches have been introduced. As one of them, the Shell Model Monte Carlo (SMMC) method has been proposed successfully [1]. However, the SMMC is basically restricted to ground-state and thermal properties, and is not a proper tool for studying excited states, i.e., level scheme and E2 properties. Moreover, the SMMC suffers from the so-called minus sign problem for realistic interactions. On the other side, the Quantum Monte Carlo Diagonalization (QMCD) method has been proposed [2–5]. The QMCD can describe not only the ground state but also excited states, including their energies and transition matrix elements. The sign problem is irrelevant to the QMCD. Thus, based upon the QMCD method, we introduce the Monte Carlo shell model as a new tool for clarifying the structure of the ground and low-lying states of nuclei.

For the diagonalization in shell model, we usually use the complete bases of a given Hilbert space, and the amplitudes of wave functions distribute over almost all the bases, in general. Consequently, the Hilbert space becomes huge. On the other hand, for the description of nuclear and perhaps some other many-body systems, there may be a small number of important basis vectors, in particular, for low-lying states. Once such appropriate many-body basis states are generated or selected somehow, the Hamiltonian can be diagonalized, to a reasonable approximation, in a subspace spanned by such basis states. Since the number of basis states can then be made much smaller, the practical calculation becomes feasible. In the following we propose a method of choosing the many-body basis states by using, in a way, the auxiliary field Monte Carlo technique, so as to diagonalize the Hamiltonian in a reasonable approximation. This method is referred to as the Quantum Monte Carlo Diagonalization (QMCD) method [2,4,3,5].

The development of the QMCD formulation can be divided into *phases* I, II and III. In the following sections, we describe each phase. In sect. II, the *phase* I is described, after presenting a simple explanation about the quantum Monte Carlo method. In sect. III, the *phase* II is explained very briefly. In sect. IV, we come to the most advanced version, the *phase* III. A test of QMCD in *phase* III is given in sect. V. Applications of the *phase* III will be presented in sects. VI and VII. A summary and remarks will be given in sect. VIII.

II FORMULATION OF QMCD - *PHASE* I -

A Illustrative description of the auxiliary field Monte Carlo method

We first present an illustrative description of the auxiliary field quantum Monte Carlo method, taking a toy Hamiltonian. Although the discussions are quite incomplete, one may find a simple and intuitive picture of the quantum Monte Carlo calculations. We begin with the imaginary-time evolution operator

$$e^{-\beta H},\tag{1}$$

for a given Hamiltonian, H. Here, β is a real parameter. If this operator in eq. (1) is acted on a state $|\Psi^{(0)}\rangle$, one obtains

$$e^{-\beta H}|\Psi^{(0)}\rangle = \sum_i e^{-\beta E_i} c_i |\phi_i\rangle,\tag{2}$$

where E_i is the i-th eigenvalue of H, $|\phi_i\rangle$ is its eigenfunction, c_i stands for amplitude: $|\Psi^{(0)}\rangle = \sum_i c_i |\phi_i\rangle$. For β large enough, amplitudes of the ground and low-lying states become larger in eq. (2) than others. This means that $e^{-\beta H}$ behaves as a projection operator for those states with lower energies. At $\beta \to \infty$, only the ground state survives. Thus, the operator $e^{-\beta H}$ has a nice feature, but its actual handling is very complicated for H containing a two-body (or many-body) interaction.

The Hubbard-Stratonovich transformation [6,7] can be used to ease this difficulty. We shall explain this transformation taking a simple illustrative example. The Hamiltonian is assumed to be of the form,

$$H = \frac{1}{2}VO^2,\tag{3}$$

where V is a coupling constant and O is a one-body operator. For simplicity, we assume $V < 0$. The imaginary-time evolution operator is then written as,

$$e^{-\frac{1}{2}\beta VO^2}.\tag{4}$$

At this point, we note a well-known formula,

$$\int_{-\infty}^{\infty} d\sigma\, e^{-a(\sigma+c)^2} = \sqrt{\pi/a} \quad (a > 0),\tag{5}$$

or

$$e^{-ac^2} = \sqrt{a/\pi} \int_{-\infty}^{\infty} d\sigma\, e^{-a\sigma^2 - 2a\sigma c} \quad (a > 0),\tag{6}$$

where σ means the variable for integration, and a and c are parameters. By substituting $-\frac{1}{2}\beta V^2 \to a$ and $O \to c$, the operator in eq. (4) can be written as,

$$e^{-\frac{1}{2}\beta VO^2} = \int_{-\infty}^{\infty} d\sigma \sqrt{\frac{\beta|V|}{2\pi}} \cdot e^{-\frac{\beta}{2}|V|\sigma^2} \cdot e^{-\beta|V|\sigma O}.\tag{7}$$

This is an exact expression, but is not practical at all. The idea of the quantum Monte Carlo calculation comes in now: the integration in eq. (7) is approximated by a Monte Carlo (MC) sampling as,

$$e^{-\frac{1}{2}\beta V O^2} \approx \sum_{MC:\sigma} \sqrt{\frac{\beta|V|}{2\pi}} \cdot e^{-\beta|V|\sigma O}, \tag{8}$$

where MC stands for Monte Carlo sampling. Here, we introduce a Gaussian,

$$G(\sigma) = e^{-\frac{\beta}{2}|V|\sigma^2}, \tag{9}$$

and the σ variable, called usually *auxiliary field*, is sampled by using the probability weight of eq. (9). Equation (8) becomes exact if the MC sampling is complete.

By combining eqs. (2), (3) and (8), one can see that, apart from the normalization, the ground state can be obtained for $\beta \to \infty$ by

$$|\Phi_g\rangle \sim \sum_{MC:\sigma} e^{-\beta h(\sigma)} |\Phi^{(0)}\rangle, \tag{10}$$

where $h(\sigma)$ is called the *one-body Hamiltonian* and is defined as

$$h(\sigma) = V \sigma O. \tag{11}$$

So far, we assumed $V < 0$. In the case of $V > 0$, the above argument should be modified, and the general definition of $h(\sigma)$ is given by

$$h(\sigma) = s V \sigma O, \tag{12}$$

where s takes the following values,

$$s = 1 \quad for \quad V < O, \tag{13}$$

and

$$s = i \quad for \quad V > O. \tag{14}$$

The ground state energy is given by

$$E_g = \frac{\langle \Phi_g | H | \Phi_g \rangle}{\langle \Phi_g | \Phi_g \rangle}, \tag{15}$$

where Φ_g stands for an unnormalized wave function of the ground state. By combining eq. (10) with this, one obtains

$$E_g \sim \frac{\langle \Phi^{(0)} | H \sum_{MC:\sigma} e^{-\beta h(\sigma)} | \Phi^{(0)} \rangle}{\langle \Phi^{(0)} | \sum_{MC:\sigma} e^{-\beta h(\sigma)} | \Phi^{(0)} \rangle}. \tag{16}$$

The Shell Model Monte Carlo (SMMC) method [1] is based on this equation. Although this expression seems to be appropriate, the calculation does not converge usually, because of variance due to the sampling. Therefore, various techniques have been introduced as described for instance in [1]. We also note that, in the SMMC, the MC sampling is not made for wave functions but for the expectation values in eq. (16). The minus-sign problem occurs generally in most of quantum Monte Carlo calculations. We shall explain briefly what the minus-sign problem is. The denominator of the right-hand side of eq. (16) should be positive definite in principle. This quantity fluctuates in the MC sampling, however. The variance can be much larger than the mean value, producing vanishing or negative values for the denominator. The calculation then becomes unstable. This is called the minus-sign problem in general, and is indeed a common and serious difficulty in quantum Monte Carlo calculations, although it may not occur for the present toy Hamiltonian. We note that there are certain cases without the minus-sign problem [1].

B Auxiliary field Monte Carlo method for general cases

We now generalize the discussions in the previous subsection so that we can solve the general shell model Hamiltonian. Such a Hamiltonian should consists of single particle energies and a two-body interaction:

$$H = \sum_{i,j=1}^{N_{sp}} \epsilon_{ij} c_i^\dagger c_j + \frac{1}{4} \sum_{i,j,k,l=1}^{N_{sp}} v_{ijkl} c_i^\dagger c_j^\dagger c_l c_k, \tag{17}$$

where $c_i^\dagger$ (c_i) denotes the creation (annihilation) operator of a nucleon in a single particle state i. The dimension of the single particle states is denoted as N_{sp}; $N_{sp}=24$ for the sd shell and 40 for the pf shell. This Hamiltonian can be rewritten in the quadratic form of one-body operators O_α:

$$H = \sum_{\alpha=1}^{N_f} (E_\alpha O_\alpha + \frac{1}{2} V_\alpha O_\alpha^2), \tag{18}$$

where the number of the O_α's, called N_f, can be at most N_{sp}^2 and usually appears to be much smaller. We consider the imaginary time evolution operator $e^{-\beta H}$ like the one in eq. (1). In general cases, the imaginary time β is divided into N_t steps,

$$e^{-\beta H} = \prod_{n=1}^{N_t} e^{-\Delta\beta H}, \tag{19}$$

where $\Delta\beta = \beta/N_t$. By applying the Hubbard-Stratonovich transformation [6,7] at each time step, this operator can be expressed as an integral of one body evolution operators with respect to the auxiliary fields $\sigma_{\alpha n}$:

$$e^{-\beta H} \approx \int_{-\infty}^{\infty} \prod_{\alpha,n} d\sigma_{\alpha n} \left(\frac{\Delta\beta |V_\alpha|}{2\pi}\right)^{1/2} \cdot G(\sigma) \cdot \prod_n e^{-\Delta\beta h(\vec{\sigma}_n)}, \tag{20}$$

where $\vec{\sigma}_n$ means a set of auxiliary fields of the n-th time step, $\vec{\sigma}_n = (\sigma_{1n}, \sigma_{2n}, \cdots, \sigma_{N_f n})$, and σ denotes the assembly of the auxiliary fields over all the time steps, $\sigma = \{\vec{\sigma}_1, \vec{\sigma}_2, \cdots, \vec{\sigma}_{N_t}\}$. The Gaussian weight factor $G(\sigma)$ in the general case is defined by

$$G(\sigma) = e^{-\sum_{\alpha,n} \frac{\Delta\beta}{2} |V_\alpha| \sigma_{\alpha n}^2}, \tag{21}$$

and the one-body Hamiltonian $h(\vec{\sigma}_n)$ in the general case is defined, similarly to eq. (12) by

$$h(\vec{\sigma}_n) = \sum_\alpha (E_\alpha + s_\alpha V_\alpha \sigma_{\alpha n}) O_\alpha, \tag{22}$$

where $s_\alpha = \pm 1$ ($= \pm i$) if $V_\alpha < 0$ (> 0).

If the $N_t \times N_f$-dimensional integral is treated with sufficient accuracy, we can obtain the ground state wave function by operating (20) with sufficiently large β on any initial state $|\Psi\rangle$ which is not orthogonal to the ground state. In numerical calculations the integral is evaluated by discretizing the $\sigma_{\alpha n}$ variables, and the integrand is computed for several specific sets σ. In the Monte Carlo integration, each set of σ is generated stochastically according to some weight functions. Then many sets of auxiliary fields are generated and the corresponding wave functions should be added with an equal weight. In many cases this integral does not converge with a tractable number of sets σ, since the variance of the integrand is in general too large. In the SMMC this difficulty can be avoided by considering only the expectation values. In this treatment, the so-called minus-sign problem arises, as mentioned already.

C Basic Process of QMCD

We now move on to the QMCD method which is the major subject of this article. We then return to eq. (10). This equation can be generalized as

$$|\Phi_g\rangle \sim \sum_{MC:\sigma} \prod_{n=1}^{N_t} e^{-\Delta\beta h(\vec{\sigma}_n)} |\Psi^{(0)}\rangle, \tag{23}$$

where σ means $\{\vec{\sigma}_1, \vec{\sigma}_2, \cdots, \vec{\sigma}_{N_t}\}$. This equation implies that the process on the right-hand side should produce all basis vectors needed for the ground state. We can therefore generate basis vectors according to the process in eq. (23), select important ones, and diagonalize the Hamiltonian in a subspace spanned by a small number of those selected basis states. Thus, the ground state and several excited states are expected to be obtained from these basis vectors. Based on this expectation, we have proposed a new method, called QMCD *phase* I. The basis states are obtained in an iterative process outlined as follows (See [2] also).

1. We take an initial intrinsic state $|\Psi^{(0)}\rangle$ which can be determined, for example, by a mean-field method, for instance Hartree-Fock. Then the initial energy is calculated as $E^{(0)} = \langle\Psi^{(0)}|H|\Psi^{(0)}\rangle$.

2. A set of the auxiliary fields σ is given stochastically.

3. We calculate a wave function $|\Phi(\sigma)\rangle$ for the present set σ:

$$|\Phi(\sigma)\rangle \propto \prod_{n=1}^{N_t} e^{-\Delta\beta h(\vec{\sigma}_n)} |\Psi^{(0)}\rangle. \tag{24}$$

4. The state $|\Phi(\sigma)\rangle$ is a candidate for the new basis vector. It is ortho-normalized, by means of the Gram-Schmidt method, with respect to all other basis states obtained previously.

5. By including this state, we diagonalize the Hamiltonian H, and obtain an improved ground state energy E and its eigen wave function $|\Psi\rangle$.

6. The energy eigenvalue E is compared to the eigenvalue obtained in the previous iteration. If the present eigenvalue is lowered sufficiently compared to the eigenvalue in the previous iteration, the present candidate $|\Phi(\sigma)\rangle$ is adopted as the new basis state. Otherwise, this candidate is discarded. Steps from (ii) to (v) are iterated.

The iteration is continued until the the energy E converges reasonably well. We can also confirm the convergence by calculating the expectation value of angular momentum operator.

The adopted basis states are called **QMCD bases**. The number of QMCD bases is referred to as the **QMCD basis dimension**, which is increased as the above steps are repeated. We emphasize that energies and wave functions are determined by the diagonalization, and that we can obtain excited states as well as the ground state, for instance, by monitoring energy eigenvalues of excited states. Since certain basis states are commonly important for low-lying states, several lowest excited states are expected to be obtained with correlated accuracies using the common set of the basis states.

In practical shell model calculations, it is convenient to adopt basis states in the form of Slater determinants:

$$|\Phi\rangle = \prod_{\alpha=1}^{N} a_\alpha^\dagger |-\rangle, \tag{25}$$

where N denotes the number of valence nucleons, $|-\rangle$ is an inert spherical core, and $a_\alpha^\dagger$ represents the nucleon creation operator in a canonical single-particle state α, which is a linear combination of the spherical bases.

$$a_\alpha^\dagger = \sum_{i=1}^{N_{sp}} c_i^\dagger D_{i\alpha}. \tag{26}$$

We can specify the basis state $|\Phi\rangle$ in terms of an $N_{sp} \times N$ complex matrix D. Note that the operation of an exponential of any one-body operator $T = \sum_{ij} T_{ij} c_i^\dagger c_j$ on a state gives a new matrix $D'_{i\alpha} = \sum_j T_{ij} D_{j\alpha}$, while the form of a Slater determinant remains.

The above procedure is called *phase* I of QMCD. Although it works quite well for simple systems [2], it turned out that *phase* I is too naive to carry out realistic large-scale shell-model calculations. So, we had to move to *phase* II as discussed in the next section.

III IMPROVEMENT OF QMCD - *PHASE* II -

In *phase* II, various essential improvements have been made so that the sampling of the basis vectors is made more efficient and symmetries can be restored faster. Although detailed accounts have been published [3,4], we present some discussions in the next two subsections.

A Basis vectors of higher importance

Since the number of manageable bases is finite in practice, we should first select bases of higher importance. We then choose $| \Psi^{(0)} \rangle$ from Hartree-Fock local minima or some states of similar nature, while $| \Psi^{(0)} \rangle$ is not specified in eq.(24). At the same time, $h(\vec{\sigma}_n)$ in eq.(22) is rewritten as

$$h(\vec{\sigma}_n) = h_{\mathrm{MF}} + \sum_{\alpha} V_\alpha s_\alpha \sigma_\alpha O_\alpha, \tag{27}$$

where $h_{\mathrm{MF}} = \sum_\alpha (E_\alpha + V_\alpha c_\alpha) O_\alpha$, with c_α's denoting c-numbers (see eqs.(4-5) of [4]). By choosing appropriate values for the c_α's, h_{MF} can be set a mean-field potential, for instance, Hartree-Fock potential of the given shell-model Hamiltonian. Thus, since h_{MF} is independent of the auxiliary fields σ_α and the sampling is made around $\sigma_\alpha = 0$, the QMCD bases $| \Phi(\sigma) \rangle$ are generated around local minima of the relevant mean fields. The selection of auxiliary fields of higher importance was introduced also in [4].

B Restoration of angular momentum and other symmetries

Since a nucleus has rotational symmetry, the restoration of the total angular momentum, denoted as J, is quite crucial. The basis state in eq. (24) is not an eigenstate of the angular momentum or not even an eigenstate of the z-component of angular momentum. This is because the initial state $| \Psi^{(0)} \rangle$ breaks the rotational symmetry in general, and moreover the operator $e^{-\Delta\beta h(\vec{\sigma}_n)}$ has nothing to do with any symmetry in general.

The z-component of angular momentum is discussed first.

The restoration of the rotational symmetry is very slow if one does it as a result of the present stochastic generation of the basis vectors. Thus we implement several treatment to restore the symmetries. If we have no restriction on the angular momentum of the basis state, we should have degenerate eigenstates for the magnetic quantum number. Here we introduce the projection of z-component of the angular momentum in the QMCD method [3]. In order to extract the component of a given magnetic quantum number, M, from a given QMCD basis, we introduce the following state,

$$| \Phi(\sigma, M) \rangle = P_M | \Phi(\sigma) \rangle = \frac{1}{2\pi} \int_0^{2\pi} d\phi \, e^{-i\phi(J_z - M)} | \Phi(\sigma) \rangle, \tag{28}$$

where P_M is the projector onto the total magnetic quantum number M. Here J_z stands for the z-component of angular momentum operators. Note that J_z is an operator and M is its quantum number. The state $| \Phi(\sigma, M) \rangle$ will be referred to as the M-projected QMCD basis (state). If we consider the eigenstates with angular momentum J, we can use the M-projected QMCD basis states with $M = J$. This M-projection is found to improve the convergence enormously [3].

Next we consider the restoration of the magnitude of the angular momentum, J. The crucial role of J is evident in nuclear spectroscopy. In the QMCD, we diagonalize the Hamiltonian in the laboratory frame by using QMCD bases. If the QMCD bases contain all components (i.e., Slater determinants) required for the coupling to a good angular momentum, the diagonalization restores the rotational symmetry. Such bases, however, cannot be generated in practice by a stochastic way due to unavoidable fluctuation.

We ensure and accelerate this restoration of rotational symmetry, by including rotated states $\exp(-i\theta_y J_y) \exp(-i\theta_z J_z) | \Phi(\sigma) \rangle$ as candidates of new basis states in addition to $| \Phi(\sigma) \rangle$ itself. We refer to this method as J-drive [4]. We have confirmed that the restoration of the angular momentum is remarkably improved by taking several (discrete) values of the angle θ's. The basis vectors thus created do not have good J or M. The J and M quantum numbers are restored by an approximate angular momentum projection. The full angular momentum projector includes $e^{-i\alpha J_z} e^{-i\beta J_y} e^{-i\gamma J_z}$ operators. The α integration corresponds to the

46

M-projection mentioned above, and is actually carried out only for matrix elements, not for wave functions. The $e^{-i\beta J_y} e^{-i\gamma J_z}$ operation is taken care of by the J-drive, and various wave functions are produced as already stated. In other words, we calculate M-projected matrix elements for basis vectors created by J-drive, and then diagonalize the Hamiltonian. The rotational invariance of the Hamiltonian should restore the J quantum number if we have sufficient number of basis vectors. We have found that the projection of the angular momentum can be made quite well by taking only several values of the angles β and γ.

The isospin is important concept for light nuclei. This symmetry can also be restored by the projection in the isospin space. In the QMCD method, the nucleon number conserved basis is used, and hence the isospin projection can be implemented as the one-dimensional integration as [8],

$$P^T_{T_f T_i} | \Phi(\sigma) \rangle = \frac{2T+1}{2} \int_0^\pi d\beta \, \sin \beta \, d^T_{T_f T_i}(\beta) \, e^{-i T_y \beta} | \Phi(\sigma) \rangle, \qquad (29)$$

where $T_i = T_f = (N - Z)/2$. Although the isospin projection is indeed feasible, it is not needed in actual Monte Carlo shell model calculations owing to a special property of the isospin symmetry [1], as stated below and in [5].

Note that any symmetry projections are useful in the QMCD method.

IV QMCD IN *PHASE* III

A Formulation of *Phase* III

We discuss, in this section, the latest crucial improvements [5] of the QMCD method.

We first sketch briefly the QMCD method in *phases* I and II stated in sects. 2 and 3 [2–4]. As already mentioned, the developments of the formulation of the QMCD method can be divided into three phases. In its first version, i.e., phase I, the QMCD bases are generated according directly to eq.(24) [2]. The phase I has been shown to be good for simple systems [2]. It was realized, however, that the phase I is not efficient enough for handling realistic shell model systems [4], even after implementing the projection onto a good quantum number of the z-component of angular momentum, M (i.e., M-projection) [3].

We then introduced several improvements, moving to phase II [4], so as to enhance the efficiency of the QMCD calculations. The basis generation method was improved so that states of higher importance can be generated first.

The other improvement is made on the restoration of symmetries. The nucleus is an isolated system and conserves several symmetries. The restoration of them is practically impossible in stochastic processes except for extremely simple cases as treated in [2], and one has to enforce the restoration. The rotational symmetry is restored by the J-drive [4] and the M-projection [3]. The J-drive means generation of the QMCD bases with the same intrinsic structure but different orientations [4]. The isospin conservation can be kept too.

By combining all the above improvements, we have become able to perform various full one-major-shell calculations with realistic effective interactions. The validity of the calculations have been tested for ^{24}Mg as shown in [4]. Although decent solutions have then been obtained for most cases, it turned out that the calculation cannot be achieved with tractable QMCD dimensions in some cases. We therefore improve the method, resulting in phase III [5].

If one generates basis states by a stochastic method, each basis should contain unnecessary fluctuations which are nothing but noise components. This deficiency is inherent to the stochastic process. We therefore revise the basis-generation method so that each basis contains more relevant components lowering the energy and less irrelevant components to be cancelled by other bases. The Hilbert space used for the Hamiltonian diagonalization can then be much compressed. In fact, this *basis compression* enables us to carry out some QMCD calculations which are otherwise practically infeasible. This compression process is one of the characteristic differences of the QMCD method from other quantum Monte Carlo approaches: In the latter, a much larger number of states in the form of eq. (24) are taken so as to evaluate the effects of their proper superposition, whereas one can vary them and select good ones in phase III of the QMCD.

We now move on to the other major improvement, i.e., the restoration of the angular momentum, J. In phase III, all QMCD basis vectors are projected onto good J and M when their matrix elements are calculated. The projection is carried out by rotating about the three axes of Euler angles. The K-mixing amplitudes are evaluated, for instance, so as to minimize the energy when the basis is added. Thus, the uncertainty about angular momentum is removed completely in practice.

In phase III calculations, the energy monitored in the basis generation is evaluated after projecting onto good J and M or a good M. In the former case, the bases are called J-compressed bases, while the bases are referred to as M-compressed bases in the latter. The J-compressed bases are more refined as the bases, but need more computer time to be calculated. The M-compressed bases are obtained much more easily, and their calculation is much faster. In most practical applications, the M-compressed bases are used, while the J-compressed bases are needed to achieve extrordinarily high accuracy. For more details, please refer to [5]. We emphasize that, at each QMCD dimension, once the last basis is fixed by the above process, the Hamiltonian matrix elements are projected onto good J and M, and are diagonalized. The isospin is treated exactly by utilizing the method in [1].

By combining all the above improvements, phase III has been constructed, and has been proved to be quite efficient in solving large-scale shell model problems [5]. Since the basis vectors are chosen according to their importance, the Monte Carlo shell model is characterized as the (stochastic) **importance truncation** to the full shell model calculation.

V TEST OF QMCD IN *PHASE* III FOR ^{48}CR

The validity of phase III has been confirmed by comparing to the result of the exact diagonalization of the same Hamiltonian. Here, the nucleus ^{48}Cr is taken, and the exact result is obtained from Ref. [9] where an excellent agreement with recent experimental data [10] is presented. The KB3 interaction [11] has been used in the calculation. Figure 1 shows energies of yrast states of ^{48}Cr.

As shown in the top part of Fig. 1, the ground-state energy has been reproduced within 130 keV with the QMCD dimension 40. This result is already $\sim$100 keV below the lower edge of the error bar of the SMMC result with the temperature T=0.5 MeV [12]. One finds in the bottom part of Fig. 1 that the back-bending pattern can be well reproduced by this QMCD calculation.

VI APPLICATION OF QMCD IN *PHASE* III TO ^{56}NI

We apply the QMCD shell model in phase III to an $N=Z$ unstable nucleus, ^{56}Ni, where the $N=Z=28$ closed-shell structure has been expected due to the spin-orbit splitting [13]. Since this closed shell can be destroyed by mixing within the same major shell, the calculation with the full pf-shell configurations is crucial. Such calculations, however, have been limited to lighter pf-shell nuclei. The QMCD calculation presented below [5] is the first full pf-shell calculation for ^{56}Ni.

The single-particle energies and two-body interaction are those called FPD6 [14] and KB3 [11]. The FPD6 is an empirical two-body interaction given in the form of analytic functions with parameters adjusted for A=41$\sim$49 nuclei [14]. The KB3 is based upon the G-matrix in [15] with empirical improvement for the monopole interaction [11]. In both cases, the single-particle energies are obtained from experimental levels of nuclei around ^{40}Ca. We stress that both the FPD6 and KB3 interactions have been designed for full pf-shell calculations, and should be used with the full pf-shell configurations.

Because of better agreement to experimental data as shown later, we discuss calculations with the FPD6 unless otherwise stated, while comparing to experiment.

Figure 2 shows energy levels of ^{56}Ni. One sees a good agreement between calculated levels and experiment [16]. Recently $B(E2; 0_1^+ \rightarrow 2_1^+)$=600$\pm$120 e^2fm^4 has been measured, which is rather large [17]. The FPD6 effective charges [14] produce a somewhat too large value, and the isoscalar charge is re-adjusted by multiplying by a factor of 0.9, resulting in e_p=1.23e and e_n=0.54e. One then obtains $B(E2; 0_1^+ \rightarrow 2_1^+)$=610 e^2fm^4. The $B(E2)$'s are calculated also for the high spin states with these charges, as shown in Fig. 2. Figure 2 also includes results obtained with KB3. Note that the M=0 Hilbert space has a dimension of about 1.1 billion for ^{56}Ni in the full pf shell, precluding a conventional shell model calculation.

One of the salient advantages of the QMCD method over other quantum Monte Carlo approaches including the SMMC method is its capability of direct analysis of the wave function. This is particularly important in the present case in clarifying the $N=Z=28$ closed shell structure: We compute the probability of the $N=Z=28$ closed shell component in the wave function of ^{56}Ni ground state. The result is only 49 %. This is rather small compared to what would be expected for a closed shell nucleus. The occupation probability of $f_{7/2}$ is 0.91 for the ground state. This means that, in the non-doubly-magic part of the wave function, about 3

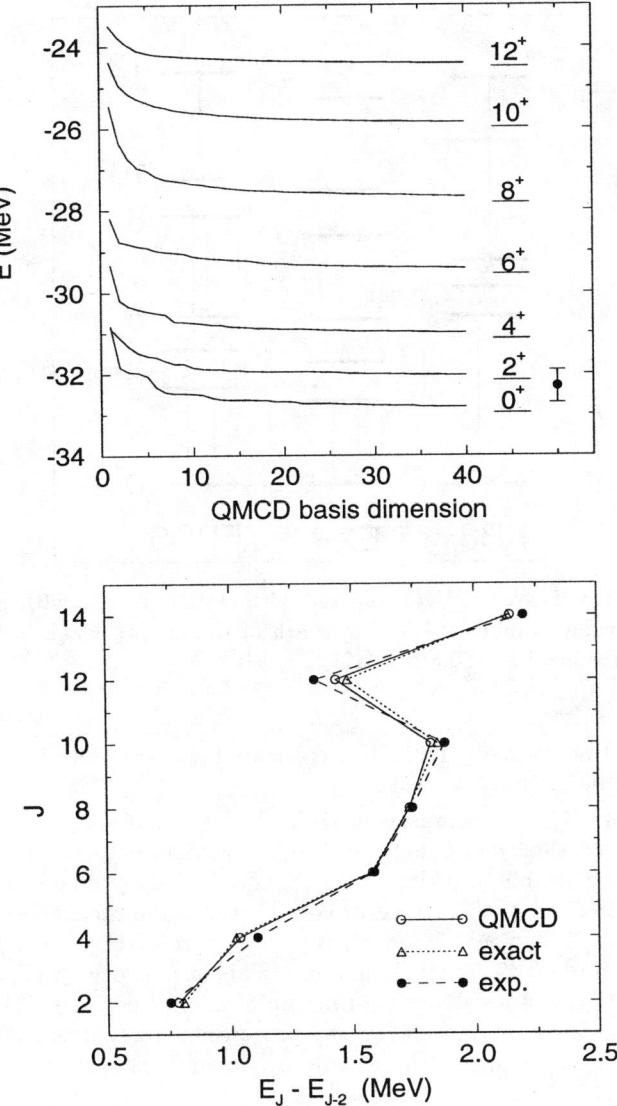

FIGURE 1. Energies of yrast states of ^{48}Cr obtained by QMCD calculation with KB3 Hamiltonian compared with the energies obtained by the exact diagonalization [9]. In the top figure, the energy eigenvalues are shown as functions of the QMCD dimensions. Since the angular momentum projection is made for each basis, the addition of a new basis (i.e., increase of the dimension) implies inclusion of more dynamical degrees of freedom. The point with error bar at far right is the ground state energy of the SMMC calculation with finite temperature T=0.5 MeV [12]. In the bottom figure, a back-bending plot is made for the QMCD (open circle) and exact (triangle) results in comparison with experimental data (closed circle) [10].

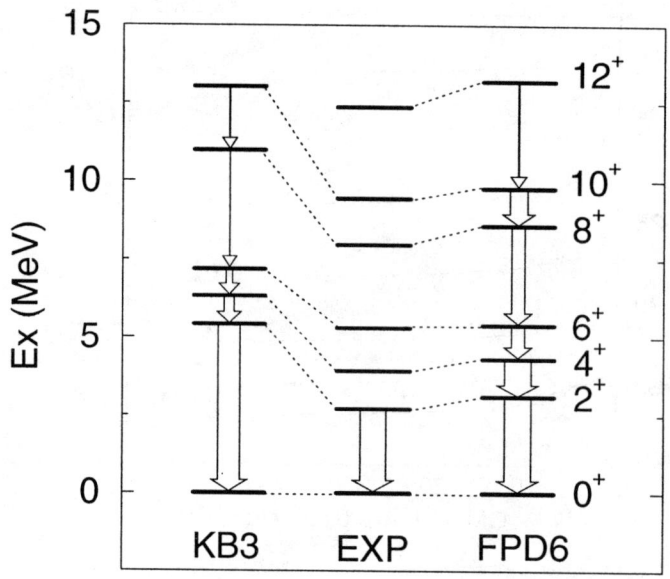

FIGURE 2. Experimental (EXP) yrast levels of ^{56}Ni compared with QMCD results with FPD6 and KB3 Hamiltonians. The $B(E2; (L+2)_1^+ \to L_1^+)$ value is indicated by the width of the arrow, which is so that the experimental $B(E2; 2_1^+ \to 0_1^+)$ value takes its mean value, i.e., 120 e^{2}fm^4 [17].

nucleons are excited from $f_{7/2}$ on the average. Thus, if a truncated shell-model calculation were attempted, 6p-6h excitations from $f_{7/2}$ should be included at least.

We now discuss the structure of ^{48}Ca for comparison [14]. The wave function of the ^{48}Ca ground state contains the $N=28$ and $Z=20$ closed shell component with 86 % probability. This is much larger than the corresponding value for ^{56}Ni. Thus, a sizable breaking of the $N=Z=28$ doubly magic is seen in ^{56}Ni, especially compared to ^{48}Ca. If the $N=Z=28$ shell of ^{56}Ni were broken by the same mechanism as the $N=28$ shell of ^{48}Ca, the closed-shell probability of ^{56}Ni would be given by the square of the corresponding value of ^{48}Ca: $(0.86)^2=0.74$. Clearly, the actual value, 0.49, is much smaller. This is because the $N=Z=28$ shell of ^{56}Ni is broken largely due to interactions between a valence proton and a valence neutron, particularly terms with a quadrupole nature. This seems to be a consequence of strong proton-neutron correlations characterizing $N=Z$ nuclei. On the other hand, the pairing-type interaction should be the major cause of breaking the $N=28$ shell in ^{48}Ca.

We further investigate the structure of non-yrast states of ^{56}Ni [18]. Figure 3 presents the results of QMCD calculations for yrast and additional states up to $J=8^+$. The calculations for non-yrast states will be explained below. Figure 3 includes experimental levels recently observed [19]. Starting around 5 MeV, a well-deformed rotational band appears both in the QMCD calculations and in the experiment [19], although the 0^+ band head has not been seen experimentally. The $B(E2)$ values in this band are quite large, and are consistent with a prolate deformation of $\beta_2 \sim 0.3$.

The structure of this deformed excited band can be seen more transparently in terms of the potential energy surface. For this purpose, the constraint Hartree-Fock calculation is quite useful. We add quadratic constraint for each term of the isoscalar quadrupole operators and also for the x-component of the total angular momentum operator. Here we assume the axially symmetric deformation set by requiring vanished expectation values for $Q_{\pm 1, \pm 2}$. Therefore, the collective rotation occurs about an axis perpendicular to the symmetry axis which is taken to be the z axis. This results in a constraint term proportional to $(<J_x> - \sqrt{J(J+1)})^2$ with a sufficiently large positive coefficient. Figure 4 presents the results of such calculations for the J_x constraints up to $J=12^+$. A broad minimum is seen around the spherical shape for lowest J values, while this minimum is shifted to oblate deformed shape for higher J values. On the other side, a well pronounced minimum is seen at large prolate deformation in all cases. This minimum is nothing but the origin of the rotational band.

In the QMCD calculation for a non-yrast state with J-compressed bases [5], one generally has to monitor the energy eigenvalue of this state, in generating basis vectors to be used mainly for describing this state. Thus, when solving more than one eigenstate of the same spin and parity, one ends up with different sets of basis

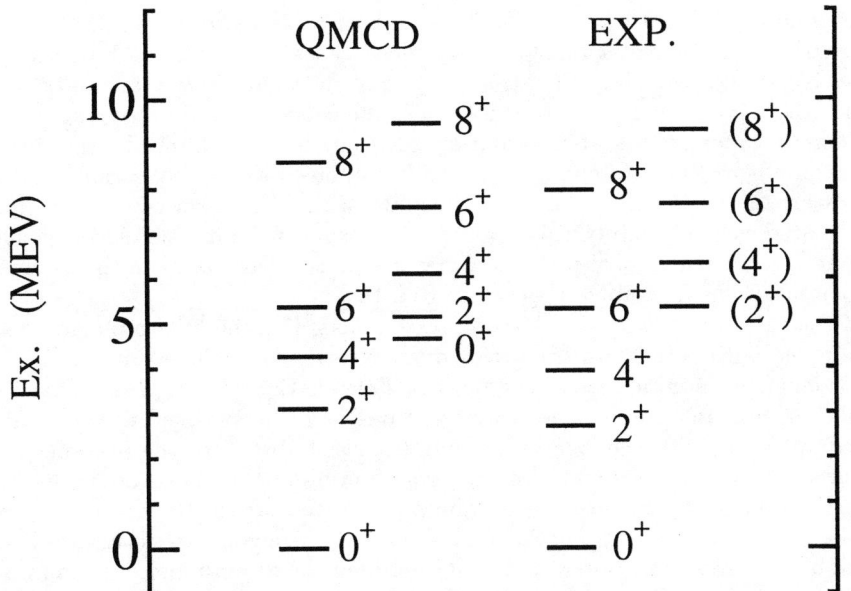

FIGURE 3. Experimental (EXP) energy levels for yrast states and recently observed deformed band [19] of ^{56}Ni compared to QMCD results with FPD6 Hamiltonian.

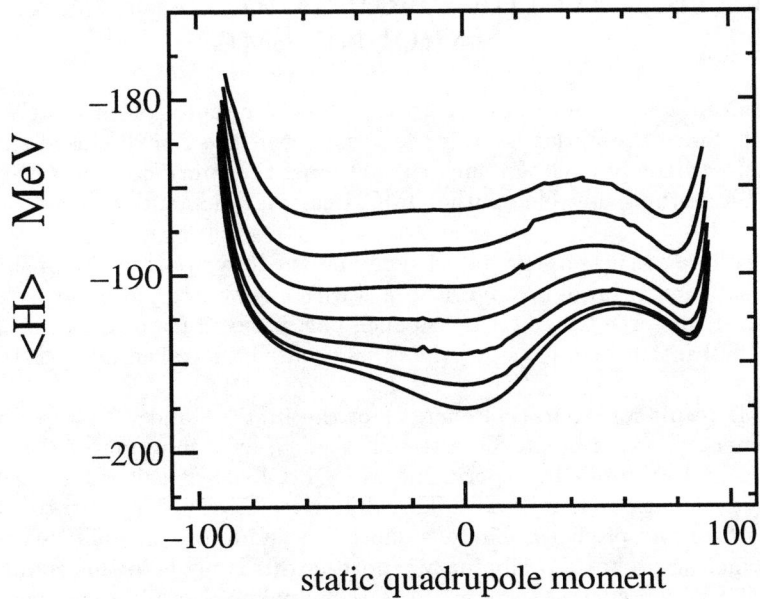

FIGURE 4. Potential energy surface for ^{56}Ni calculated from FPD6 Hamiltonian. Different lines correspond to different constraints for $< J_x >= \sqrt{J(J+1)}$ for J=0, 2, $\cdots$ 12. The axially symmetric deformation is assumed, and the abscissa means the intrinsic quadrupole moment. Tiny fluctuations in the potential curves are only due to numerical complications without any physical significance.

vectors which correspond to different eigenstates. These sets of basis vectors are merged, and the Hamiltonian is diagonalized in a Hilbert subspace spanned by all these bases. Therefore, the orthogonality among the obtained eigenstates is guaranteed. Note that this process is included already in the basis generation for the J-compressed bases. There is another important aspect in the calculation of non-yrast states. The h_{MF} in eq. (27) plays an important role for yrast-state calculations. If there is a pronounced minimum and the eigenstate being obtained corresponds to this minimum, h_{MF} can be replaced by the constrained HF mean-field for the minimum. One can start the basis generation from the state of the local excited minimum, and samples states around this local potential well. Certainly, this is an extended role of the mean field in the QMCD calculations. For general states, however, such a local minimum may not exist, and we have to use another way of fixing h_{MF} in the basis generation, as described in detail in Ref. [18].

If one carries out the calculation with the M-compressed bases [5], one has to include a special care so that proper bases are sampled with respect to the orthogonality to the bases fixed in earlier steps. On the other side, if there is a pronounced minimum like the ones in Fig. 4, the M-compressed basis generation can be done and the calculation remains simpler and faster: the basis can be sampled in the local excited minimum, similarly to the J-compression with the local minimum discussed above. In the present case, the difference in deformation is so large between the ground and the prolate minima that basis states generated in one of the minima are nearly orthogonal to those in the other minimum. Note that the basis states are put together for the diagonalization of the Hamiltonian and the orthogonality among the eigenstates is always exact irrespectively of non-orthogonality of basis states. The potential barrier between the ground and excited minima is high enough to prevent the basis sampling in the prolate well sliding down to the region around the ground minimum in nearly all cases. Thus, quite similarly to the calculations for yrast states, one can obtain basis states by the M-compression for non-yrast states with pronounced local minima. This nice feature, however, cannot be used in a general case without such a local potential minimum. As mentioned above, such a general state is obtained in a different manner [18]. We just note again that such calculations for general states are indeed feasible and more eigenstates of ^{56}Ni have been obtained [18].

VII APPLICATION TO THE STRUCTURE OF UNSTABLE NUCLEI AROUND ^{32}MG

The Monte Carlo shell model has been applied to the structure study of extremely neutron-rich unstable nuclei around ^{32}Mg [20]. Since the major issue is the breaking of the N=20 closed shell [21–25], one has to include both the sd shell and the pf shell. In such calculations, the spurious center-of-mass motion has to be removed. We developed a method suitable for the QMCD calculations, and reduced the spurious component to practically zero.

The model space taken here consists of the full sd shell and the lower part of the pf shell (i.e., $f_{7/2}$ and $p_{3/2}$). This space seems to be sufficiently large up to Si isotopes with the neutron number, N, up to around 24. The effective interaction consists of three parts: (i) the sd shell part is taken from the USD interaction [26], (ii) the pf shell part is from the KB interaction [15], (iii) the cross shell part is taken from [27] which was based upon the MK interaction [28].

Figure 5 shows QMCD result for excitation energies of the first 2^+ and 4^+ states of even-A Mg isotopes, exhibiting a nice agreement to experiment. One sees in Fig. 5, as a function of the neutron number, N, a sudden drop of the first 2^+ level both in experiment and QMCD sd-pf calculation, whereas the calculation within the sd shell shows a continuous increase. This difference clearly demonstrates the importance of the calculation including both sd and pf shells. There is another very important point related to Fig. 5. The first 4^+ energy has not been measured, but a preliminary report on this is made in this conference [29]: the authors report the observation of a higher excited state in ^{32}Mg. The analysis is still in progress in order to prove that this is the 4^+ state. Nevertheless, we would like to emphasize that the ratio $E_x(4_1^+)/E_x(2_1^+)$ is about 2.5 for both experiment and sd-pf QMCD calculation.

Figure 6 shows a similar result for $B(E2; 0_1^+ \rightarrow 2_1^+)$ values. These $B(E2)$ values show a sudden increase at $N = 20$ as observed in [25]. The QMCD calculation reproduces this $B(E2)$ value as well as the one for N=16. This type of systematic calculations are very important and crucial, but were not feasible even in the largest truncated shell model calculation [24], because of exploding dimension. Note that no particle-hole truncation is imposed in QMCD calculations.

Figure 7 shows the result of the constraint Hartree-Fock for ^{32}Mg. The calculation is similar to what has been done for Fig. 4. One sees a quite dramatic pronounced minimum at large deformation. This is clearly

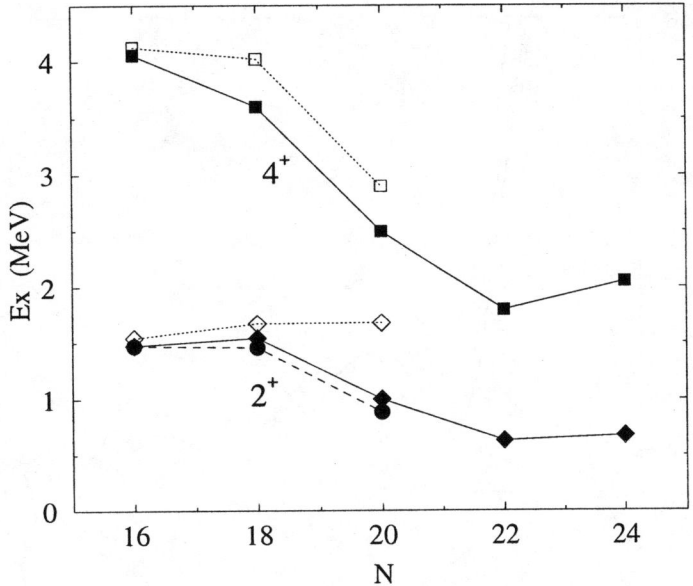

FIGURE 5. Calculated energy levels of the first 2^+ (closed diamond) and 4^+ (closed square) states of even-A Mg isotopes as functions of the neutron number, N. Experimental 2^+ levels are also shown by closed circles. Open symbols indicate results of the calculation within the sd shell.

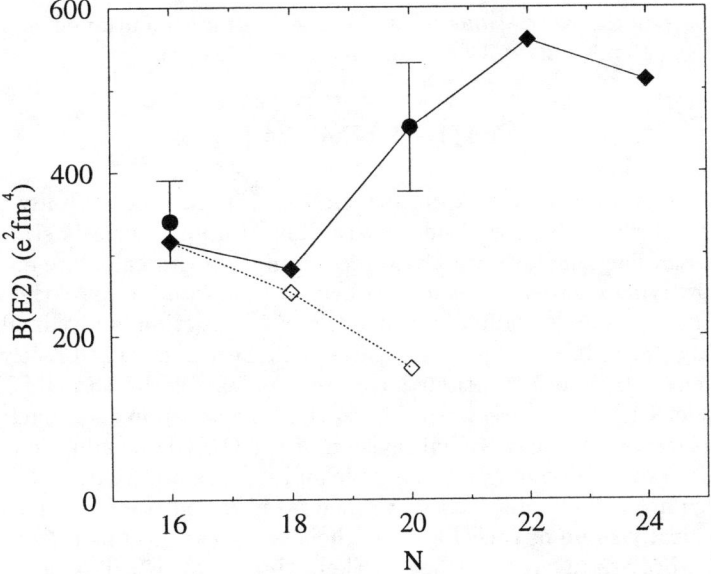

FIGURE 6. $B(E2; 0_1^+ \rightarrow 2_1^+)$ values Mg isotopes. QMCD results are shown by closed diamonds, while open diamonds indicate the results obtained within the sd shell. The experimental value is shown by closed circle [25].

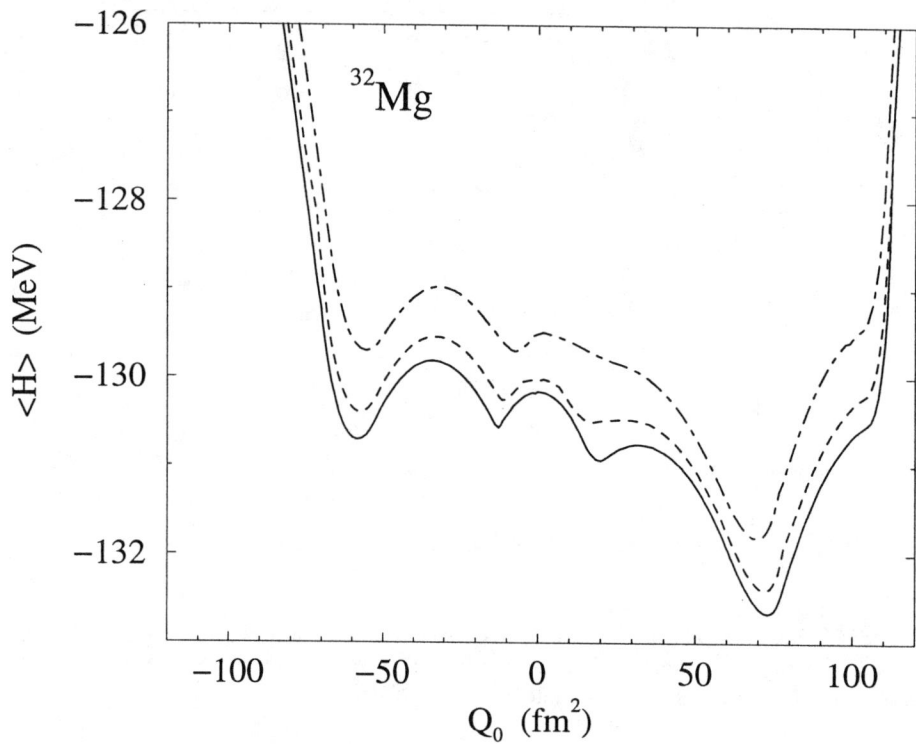

FIGURE 7. Potential energy surface of ^{32}Mg.

due to the excitations of neutrons from the sd shell to the pf shell. One also finds small prolate minima due to the deformation inside the sd shell.

We have carried out systematic calculations with the Hamiltonian mentioned above for O, Ne, Mg and Si isotopes. A report on these calculations will be published elsewhere [20].

VIII SUMMARY

In summary, we have presented a new method for solving quantum many-body problems where particles (fermions or bosons) interact through a two-body interaction. Important basis states can be generated by a method originated in the auxiliary field Monte Carlo technique. We can calculate energies and wave functions of both the ground and low-lying excited states for any two-body interaction. As the system becomes large the QMCD basis dimension increases only gradually. Since the wave function is obtained explicitly, the transition matrix elements, including those between excited states, can be calculated directly. Thus, the shell model calculations with full valence shell configurations have become feasible by the QMCD method, shading light upon the structure of nuclei with more direct relation to the (effective) nucleon-nucleon interaction.

We have presented the latest major and crucial revision of the QMCD formulation. This revision, i.e., phase III, is characterized by the compression of the basis space and the precise treatment of the angular momentum. Thus, in the QMCD calculation, favorable bases are generated based upon their contribution to the energy eigenvalue, and quite naturally some of such bases or their seeds can be taken from mean-field solutions. In other words, the basis vectors are taken according to their "importance". This nice feature can be expressed by (stochastic) **importance truncation** for the shell model.

The **lowest levels of** ^{56}**Ni** are then well described with the FPD6 interaction. It has been shown that the doubly closed shell structure is substantially broken in ^{56}Ni, in contrast to ^{48}Ca. The irregular level structure of higher-spin yrast states of ^{56}Ni is also reproduced, thus ensuring the validity of the present conclusion. The calculation has been developed further so that one can investigate side band structure. An example is shown for recently observed **prolate deformed band in** ^{56}**Ni**. This band has a large deformation, as clearly seen in the potential energy surface analysis which can be incorporated into the QMCD calculation. A large deformation parameter (~ 0.3) has been suggested. This sort of bands seem to overcome the strong spin-orbit splitting owing to a strong quadrupole collectivity, regaining the SU(3) (-like) structure which Elliott suggested for explaining the nuclear rotation [30].

The structure of neutron-rich unstable nuclei around ^{32}Mg is discussed. We reproduce excitation energies, $B(E2)$ values, two-neutron separation energies, *etc.*, within a single unified framework. A large set of the QMCD results for unstable **O, Ne, Mg and Si** isotopes have been obtained, and are being prepared for publication with many stimulating features [20].

We thank Professors B.A. Brown and A. Poves for providing FPD6 and KB3 two-body matrix elements, respectively. This work was supported in part by Grant-in-Aid for Scientific Research (B) (No. 08454058) from the Ministry of Education, Science and Culture.

REFERENCES

1. S. E. Koonin, D. J. Dean, and K. Langanke, Phys. Repts. **278**, 1 (1997); references therein.
2. M. Honma, T. Mizusaki, T. Otsuka (1995): Phys. Rev. Lett. **75**, 1284
3. T. Mizusaki, M. Honma, T. Otsuka (1996): Phys. Rev. **C53**, 2786
4. M. Honma, T. Mizusaki and T. Otsuka (1996): Phys. Rev. Lett. **77**, 3315
5. T. Otsuka, M. Honma and T. Mizusaki (1998): Phys. Rev. Lett. **81**, 1588
6. J. Hubbard J. (1959): Phys. Rev. Lett. **3**, 77
7. R.L. Stratonovich (1957): Dokl. Akad. Nauk. SSSR **115**, 1097 [transl: Soviet Phys. Dokl. **2**, 416 (1957)]
8. T. Otsuka, M. Honma, T. Mizusaki : Proceedings of the Workshop on Contemporary Nuclear Shell model, 29-30 April, 1996, Drexel University, USA, published in Lecture Notes in Physics (Springer-Verlag).
9. E. Caurier, *et al.*, (1994): Phys. Rev. **C50**, 225
10. S.M. Lenzi, *et al.*, (1996): Z. Phys. **A354**, 117
11. A. Poves, A. Zuker (1981): Phys. Rep. **70**, 235
12. K. Langanke, *et al.*, (1995): Phys. Rev. **C52**, 718
13. For instance, A. Bohr and B. R. Mottelson, *Nuclear Structure* Vol.1, (Benjamin, New York, 1969).
14. W.A. Richter, *et al.*, (1991): Nucl. Phys. **A523**, 325
15. T. T. S. Kuo and G. E. Brown, Nucl. Phys. **A114**, 241 (1968).
16. R. B. Firestone, *et al.*, (ed.), Table of Isotopes, (Wiley, New York, 1996).
17. G. Kraus, *et al.*, Phys. Rev. Lett. **73**, 1773 (1994).
18. T. Mizusaki, T. Otsuka, Y. Utsuno, M. Honma and T. Sebe, to be submitted.
19. D. Rudolph, private communications; D. Rudolph, *et al.*, submitted to Phys. Rev. Lett.
20. Y. Utsuno, T. Otsuka, T. Mizusaki and M. Honma, to be submitted.
21. D. Guillemaud-Mueller, *et al.*, (1984): Nucl. Phys. **A426**, 37
22. A. Poves and J. Retamosa, Phys. Lett. **B184**, 311 (1987); A. Poves and J. Retamosa, Nucl. Phys. **A571**, 221 (1994).
23. E.K. Warburton, J.A. Becker and B.A. Brown, Phys. Rev. **C41**, 1147 (1990)
24. N. Fukunishi, T. Otsuka and T. Sebe, (1992): Phys. Lett. **B296**, 279
25. T. Motobayashi, *et al.*, (1995): Phys. Lett. **B346**, 9
26. B.A. Brown and B.H. Wildenthal, Ann. Rev. Nucl. Part. Sci. **38**, 29 (1988)
27. D.E. Alburger, J.A. Becker, B.A. Brown and S. Raman, Phys. Rev. **C34**, 1031 (1996)
28. D.J. Millener and D. Kurath, Nucl. Phys. **A255**, 315 (1975)
29. F. Azaiez, *et al.*, presentation in this conference.
30. J.P. Elliott, Proc. Roy. Soc., **A245**, 128, 562 (1968)

Realistic Effective Interactions and Nuclear Structure Calculations

A. Covello,[1] L. Coraggio,[1] A. Gargano,[1] N. Itaco,[1] and T. T. S. Kuo[2]

[1]*Dipartimento di Scienze Fisiche, Università di Napoli Federico II, and Istituto Nazionale di Fisica Nucleare, Complesso Universitario di Monte S. Angelo, Via Cintia, 80126 Napoli, Italy*
[2]*Department of Physics, SUNY, Stony Brook, New York 11794*

Abstract. We address the two main questions relevant to microscopic nuclear structure calculations starting from a free NN potential. These concern the accuracy of these kinds of calculations and the extent to which they depend on the potential used as input. Regarding the first question, we present some results obtained for nuclei around doubly magic ^{132}Sn and ^{208}Pb by making use of an effective interaction derived from the Bonn A potential. Comparison shows that our results are in very good agreement with the experimental data. As for the second question, we present the results obtained for the nucleus ^{134}Te by making use of four different NN potentials. They indicate that nuclear structure calculations may help in understanding the off-shell nature of the NN potential.

INTRODUCTION

The shell model is the basic framework for nuclear structure calculations in terms of nucleons. Since the early 1950s many hundreds of shell-model calculations have been carried out, most of them being very successful in describing a variety of nuclear structure phenomena. In any standard shell-model calculation one has to start by defining a model space, namely by specifying a set of active single-particle (s.p.) orbits. The choice of the model space is of course conditioned by the size of the matrices to be set up and diagonalized. The rapid increase in computer power and the development of high-quality codes in the last decade has greatly extended the feasibility of large-scale calculations [1]. While these technical improvements add to the practical value of the shell model, much uncertainty still exists for what concerns the model-space effective interaction V_{eff}. In most of the existing calculations to date either empirical effective interaction containing several adjustable parameters have been used or the two-body matrix elements have been treated as free parameters, this latter approach being limited to small model spaces.

This uncertainty in shell model work can only be removed by taking a more fundamental approach, namely by deriving the effective interaction from the free nucleon-nucleon (NN) potential. As is well known, the first step in this direction was taken in the mid 1960s by Kuo and Brown [2] who derived an s-d shell effective interaction from the Hamada-Johnston potential [3]. Since that time there has been substantial progress towards a microscopic approach to nuclear structure calculations starting from a free NN potential. On the one hand, high-quality NN potentials have been constructed which reproduce quite accurately all the known NN data. On the other hand, the many-body methods for calculating the matrix elements of the effective interaction have been largely improved. A review of modern NN potentials is given in Ref. [4] while the main aspects of the derivation of V_{eff} are discussed in Ref. [5]. These improvements have brought about renewed interest in shell-model calculations with realistic effective interactions. In this context, the two crucial questions are: i) how accurate is an effective interaction derived from the NN potential? ii) to which extent can nuclear structure calculations distinguish between different NN potentials?

Recent calculations for nuclei in the ^{100}Sn and ^{132}Sn regions [6–11] have achieved very good agreement with experiment indicating the ability of realistic effective interactions to provide a description of nuclear structure properties at least as accurate as that provided by traditional, empirical interactions. To our knowledge, no systematic investigation concerning the second question has been carried out thus far. The main interest in trying to answer this question stems from the fact that two potentials which fit equally well the NN data up to the inelastic threshold may differ substantially in their off-shell behavior. Thus, from microscopic nuclear structure calculations we may learn something about the off-shell properties of the nuclear potential.

The main aim of this paper is to report on some achievements of our current work relevant to both the above questions. We shall first present some results of realistic shell-model calculations for nuclei having either few protons outside doubly magic ^{132}Sn or few neutron holes in doubly magic ^{208}Pb. They are ^{135}I, ^{136}Xe and ^{206,205,204}Pb. In all of these calculations we have made use of a realistic effective interaction derived from the Bonn A free NN potential [12]. Then we shall present the results obtained for the two proton-nucleus ^{134}Te by making use of four different potentials, Paris [13], Nijmegen93 [14], Bonn A and CD Bonn [15], which are all based on the meson theory of nuclear force.

We shall see that while the former study confirms what was learned from our previous calculations with the Bonn potential, the latter indicates a dependence of nuclear structure results on the kind of potential used as input.

OUTLINE OF CALCULATIONS

As already mentioned in the Introduction, for all the six nuclei considered in this paper we have employed an effective interaction derived from the Bonn A potential. For ^{134}Te we have also performed calculations employing three other effective interactions derived from the Paris, Nijmegen93 and CD Bonn potential, respectively. These effective interactions were all obtained using a G-matrix folded-diagram formalism, including renormalizations from both core polarization and folded diagrams. For the $N = 82$ isotones ^{134}Te, ^{135}I and ^{136}Xe we have considered ^{132}Sn as an inert core and let the valence protons occupy the five single-particle (s.p.) orbits $0g_{7/2}$, $1d_{5/2}$, $2s_{1/2}$, $1d_{3/2}$, and $0h_{11/2}$. For the Pb isotopes, we have treated neutrons as valence holes with respect to the ^{208}Pb closed core and included in the model space the six single-hole (s.h.) orbits $2p_{1/2}$, $1f_{5/2}$, $2p_{3/2}$, $0i_{13/2}$, $1f_{7/2}$, and $0h_{9/2}$. A description of the derivation of our V_{eff} for the $N = 82$ isotones and for the Pb isotopes can be found in Refs. [16] and [17], respectively. For the shell-model oscillator parameter $\hbar\omega$ we have used the value 7.88 MeV for the $N = 82$ isotones and 6.88 MeV for the Pb isotopes, as obtained from the relationship $\hbar\omega = 45A^{-1/3} - 25A^{-2/3}$ for A= 132 and A= 208, respectively.

As regards the s.p. energies, for the $N = 82$ isotones we have taken three s.p. spacings from the experimental spectrum of ^{133}Sb [18,19]. In fact, the $g_{7/2}$, $d_{5/2}$, $d_{3/2}$, and $h_{11/2}$ states can be associated with the ground state and the 0.962, 2.439 and 2.793 MeV excited levels, respectively. As for the $s_{1/2}$ state, its position has been determined by reproducing the experimental energy of the $\frac{1}{2}^+$ level at 2.15 MeV in ^{137}Cs. This yields the value $\epsilon_{s_{1/2}} = 2.8$ MeV. Regarding the Pb isotopes, the s.h. energies have all been taken from the experimental spectrum of ^{207}Pb [20]. They are (in MeV) $\epsilon_{p_{1/2}} = 0$, $\epsilon_{f_{5/2}} = 0.570$, $\epsilon_{p_{3/2}} = 0.898$, $\epsilon_{i_{13/2}} = 1.633$, $\epsilon_{f_{7/2}} = 2.340$, and $\epsilon_{h_{9/2}} = 3.414$.

RESULTS

In Fig. 1 we report the experimental [21,22] and theoretical spectra of the three-proton nucleus ^{135}I. As regards the two-proton nucleus ^{134}Te, the results obtained by using the Bonn A potential are to be found in Fig. 6. The spectra of both these nuclei have already been presented in a previous paper [16], where a detailed comparison between theory and experiment is made. In that paper we also reported the calculated $E2$ and $E3$ transition rates in ^{134}Te and compared them with the available experimental data. While we refer the reader for details to the above paper, we emphasize here the very good agreement between the calculated spectra and the experimental ones, as is shown by the value of the rms deviation σ [23], which are 106 keV and 58 keV for ^{134}Te and ^{135}I, respectively.

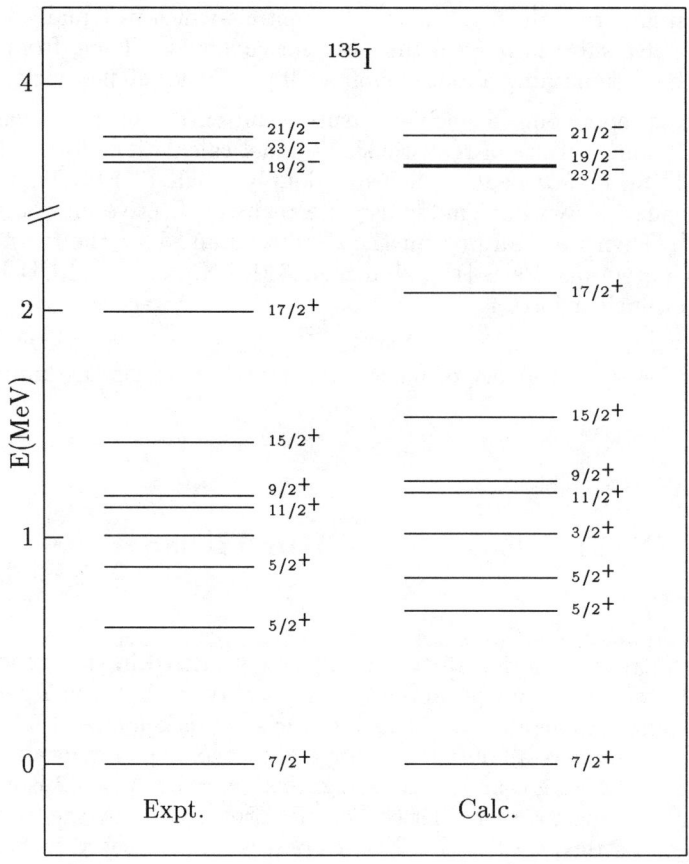

FIGURE 1. Experimental and calculated spectrum of ^{135}I.

The experimental [24] and calculated spectra of the four-proton nucleus ^{136}Xe are compared in Fig. 2, where all the calculated and experimental levels up to about 2.5 MeV excitation energy are reported. Excluding the three states with $J^\pi = 3^+, 4^+$, $J^\pi = (4^+)$, and $J^\pi = 4^+$ at 2.13 , 2.46, and 2.56 MeV, respectively, each level in the observed spectrum can be unambiguously identified with a level predicted by the theory. The quantitative agreement between theory and experiment is quite satisfactory. In fact, a rather large discrepancy (248 keV) occurs only for the 0_2^+ state, the calculated excitation energies of all other states differing by less than 110 keV from the experimental values. The rms deviation relative to the eight identified excited states is 107 keV. As regards the three above mentioned states, for which we have not attempted to establish a correspondence between theory and experiment, a firm spin assignment to the levels at 2.13 and 2.46 MeV is needed to clarify the situation.

Let us now come to the Pb isotopes. The experimental [25–27] and theoretical spectra of ^{206}Pb, ^{205}Pb and ^{204}Pb are compared in Figs. 3, 4 and 5, where we report all the calculated and experimental levels up to 2.5, 1.5 and 2.0 MeV, respectively. In the high-energy regions we only compare the calculated high-spin states with the observed ones. From Figs. 3-5 we see that a very good agreement with experiment is obtained for the low-energy spectra. In particular, in each of the three nuclei the theoretical level density reproduces remarkably well the experimental one.

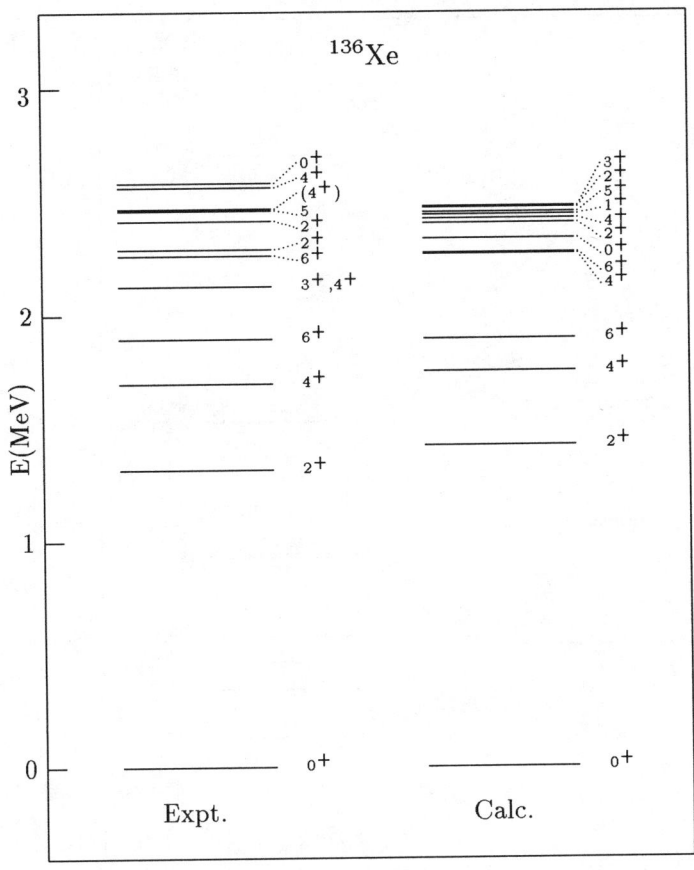

FIGURE 2. Experimental and calculated spectrum of ^{136}Xe.

Note too that each state of a given J^π in any of three calculated spectra has its experimental counterpart, with a few exceptions. In fact, as may be seen in Fig. 4, the $\frac{5}{2}^-$, $(\frac{3}{2}, \frac{1}{2})^-$, and $(\frac{9}{2}, \frac{7}{2})^-$ states observed at 1.265, 1.374 and 1.499 MeV in ^{205}Pb cannot be safely identified with levels predicted by the theory. As regards ^{204}Pb, we find the 0_4^+ state at 1.954 MeV while the experimental one, which is not reported in Fig. 5, lies at 2.433 MeV. It should be mentioned, however, that the theory predicts four more 0^+ states in the energy interval 2.2–2.6 MeV. Aside from these uncertainties, the agreement between calculated and experimental spectra is such as to allow us to identify experimental states with no firm or without spin-parity assignment. For ^{206}Pb our results suggest that the observed levels at 2.197 and 2.236 MeV have $J^\pi = 3^+$ and 1^+, respectively. As for ^{205}Pb, we predict $J^\pi = \frac{1}{2}^-$ and $\frac{3}{2}^-$ for the experimental levels at 0.803 an 0.998 MeV.

Regarding the quantitative agreement between our results and experiment, the rms deviation σ is 207 and 216 keV for ^{206}Pb and ^{204}Pb, respectively. For ^{205}Pb the σ value is 74 keV, excluding the three above mentioned states, for which we have not attempted any identification. Concerning the high-spin states in ^{206}Pb and ^{205}Pb, from Figs. 3 and 4 we see that they are also well described by the theory. In ^{204}Pb the agreement between theory and experiment is rather worse for the states lying above 4.3 MeV excitation energy, the largest discrepancy being about 400 keV for the 16_2^+ state.

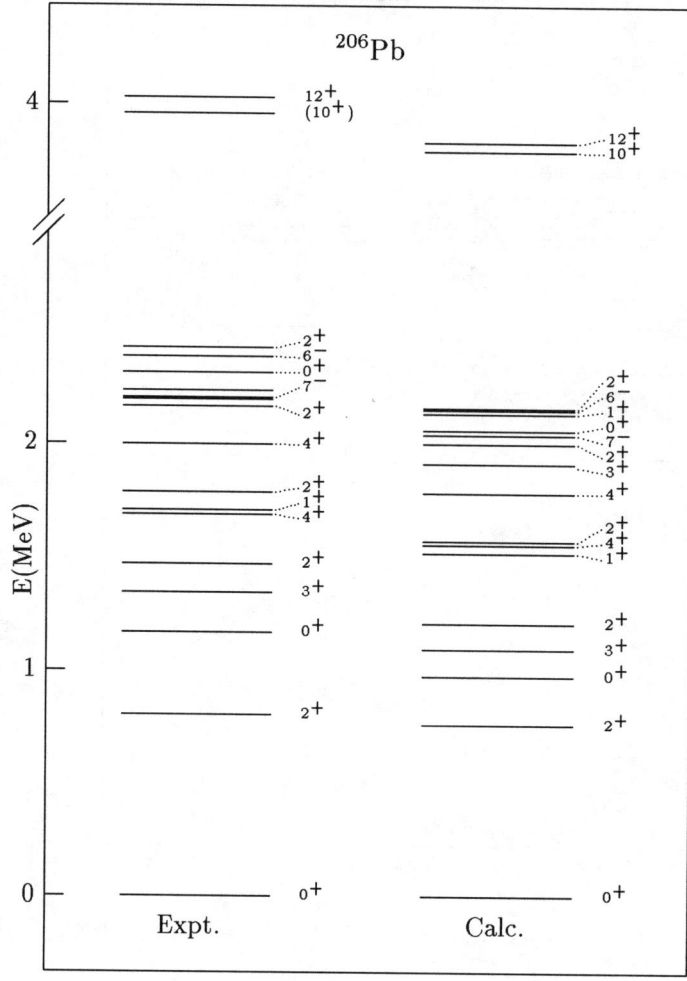

FIGURE 3. Experimental and calculated spectrum of ^{206}Pb.

We have also calculated the electromagnetic properties for each of the three isotopes. For the sake of brevity we do not report these results in the present paper, but refer the reader to Ref. [17], where a detailed comparison with the available experimental data is also made. We only mention here that a very good overall agreement is obtained.

As already mentioned in the Introduction, we are currently investigating the dependence of nuclear structure results on the NN potential used to derive the model space effective interaction. Here we present the results obtained for the nucleus ^{134}Te which provides a very good testing ground for this investigation. In fact, this nucleus has only two valence protons and thus offers the opportunity to test directly the matrix elements of the calculated effective interactions. Furthermore, all the s.p. energies, except $\epsilon_{s_{1/2}}$, are known from experiment. The uncertainty in the position of the $s_{\frac{1}{2}}$ level, however, has practically no influence on the low-energy spectrum.

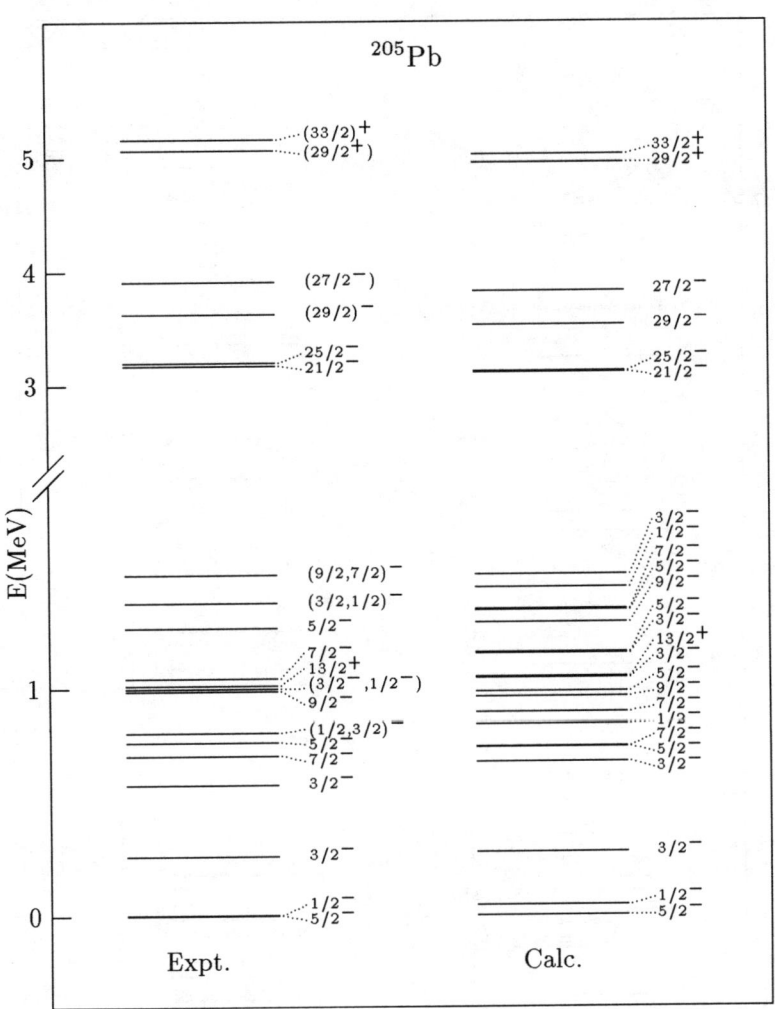

FIGURE 4. Experimental and calculated spectrum of ^{205}Pb.

CP481, *Nuclear Structure 98*, edited by C. Baktash
© 1999 American Institute of Physics 1-56396-858-4/99/$15.00

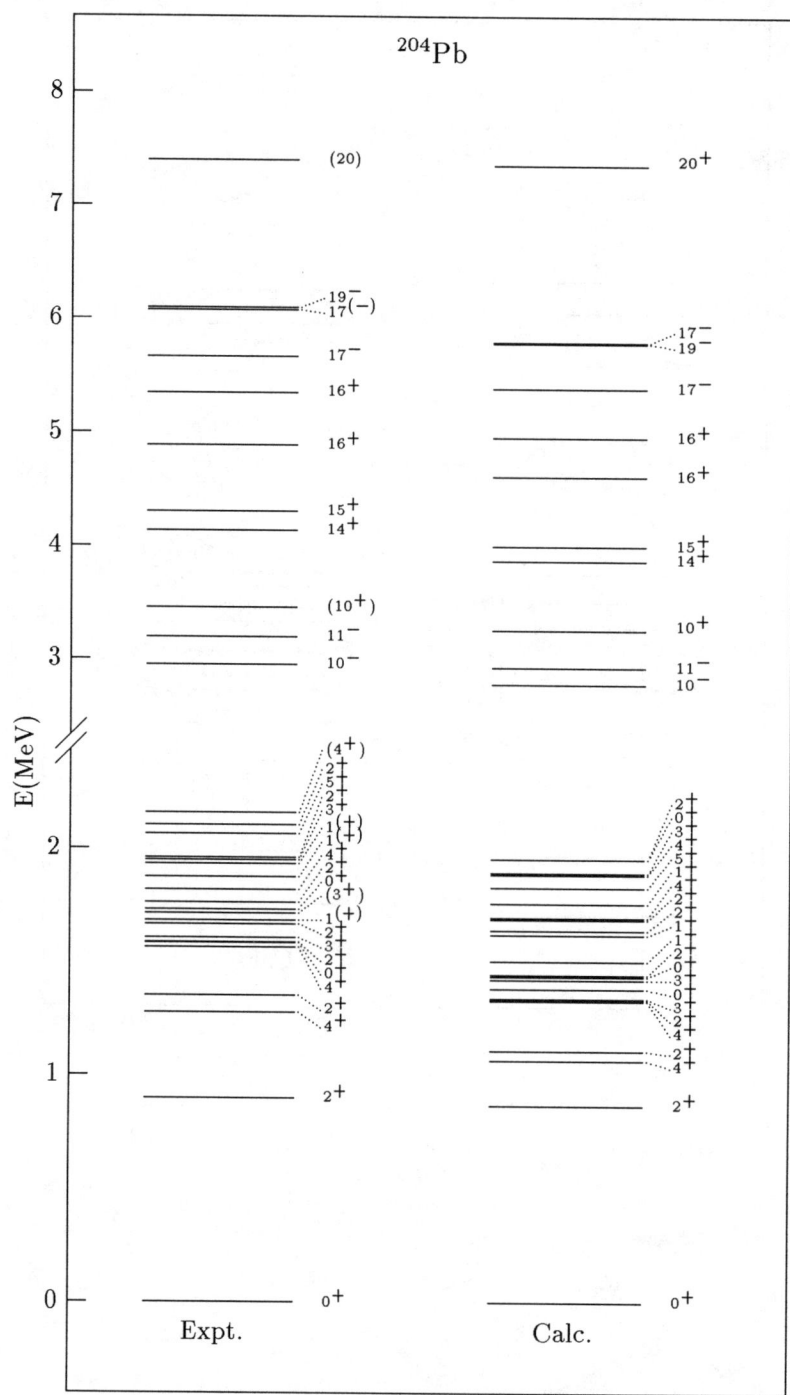

FIGURE 5. Experimental and calculated spectrum of ^{204}Pb.

In Fig. 6 we report the four theoretical spectra obtained by using the Paris, Nijmegen93, CD Bonn and Bonn A potentials together with the experimental one [28,29]. We see that the best agreement with experiment is produced by the Bonn A effective interaction. The rms deviation σ is 106, 160, 211, and 346 keV for Bonn A, CD Bonn, Nijm93 and Paris, respectively. These results show that different NN potentials produce somewhat

62

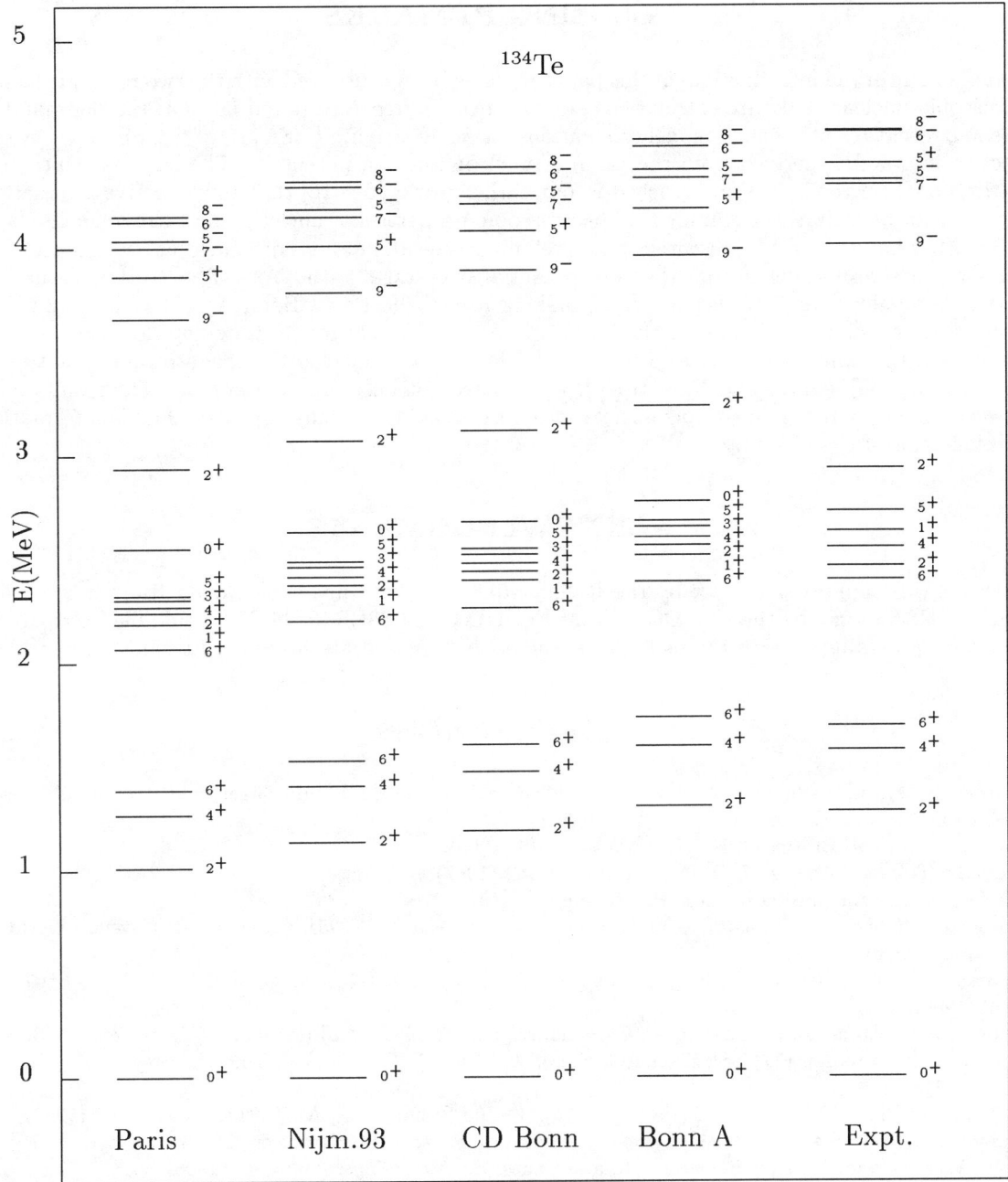

FIGURE 6. Spectrum of ^{134}Te. Predictions by various NN potentials are compared with experiment.

different nuclear structure results. Of course, the significance of these differences, which do not exceed 100 keV if we exclude the Paris potential, may be a matter of discussion. The following remark is, however, in order. The four potentials considered differ in the strength of the tensor-force component as measured by the predicted D-state probability of the deuteron P_D. This is 4.4% for Bonn A, 4.83% for CD Bonn, 5.76% for Nijm93, and 5.8% for Paris. Our results suggest that potentials with a weak tensor force may lead to a better description of nuclear structure properties. This is quite an interesting point since differences in P_D may in turn be traced to off-shell differences.

CLOSING REMARKS

As pointed out in the Introduction, in this paper we have been concerned with the two main problems related to microscopic nuclear structure calculations starting from a free NN potential. On the one hand, we have shown some recent results of shell-model calculations for nuclei around ^{132}Sn and ^{208}Pb obtained by employing an effective interaction derived from the Bonn A nucleon-nucleon potential. The success achieved by these calculations confirm the conclusion reached in our earlier works [6–8,16] that this effective interaction is able to describe with quantitative accuracy the spectroscopic properties of nuclei around closed shells.

On the other hand, we have presented some preliminary results of a study aimed at ascertaining how much nuclear structure results depend on the NN potential one starts with. As a first testing ground, we have chosen the two valence-proton nucleus ^{134}Te, making use of the Paris, Bonn A, CD Bonn and Nijm93 NN potentials. Since this study is still in the initial stage, it is premature to draw any definite conclusion. We plan to extend this kind of calculations to several other nuclei and to other modern high-quality potentials, like Argonne v_{18} [30], Nijm I and Njim II [14], which have not been considered here. The results obtained so far, however, indicate that microscopic nuclear structure calculations may provide valuable information on the off-shell behavior of the NN potential.

ACKNOWLEDGMENTS

This work was supported in part by the Italian Ministero dell'Università e della Ricerca Scientifica e Tecnologica (MURST) and by the U.S. DOE Grant No. DE-FG02-88ER40388. We would like to thank Ruprecht Machleidt for providing us with the matrix elements of NN potentials and for valuable comments.

REFERENCES

1. Caurier, E., Martínez-Pinedo, G., Nowacki, F., Poves, A., Retamosa, I., and Zuker, A. P., to be published in *Phys. Rev.* C, and references therein.
2. Kuo, T. T. S., and Brown, G. E., *Nucl. Phys.* **85** 40 (1966).
3. Hamada, T., and Johnston, I. D., *Nucl. Phys.* **34** 382 (1962).
4. Machleidt, R., contribution to these Proceedings
5. Kuo, T. T. S., in *New Perspectives in Nuclear Structure (Ravello 1995)*, edited by A. Covello (World Scientific, Singapore, 1996), p. 159.
6. Andreozzi, F., Coraggio, L., Covello, A., Gargano, A., Kuo, T. T. S., Li, Z. B., and Porrino, A., *Phys. Rev.* C **54**, 1636 (1996).
7. Andreozzi, F., Coraggio, L., Covello, A., Gargano, A., Kuo, T. T. S., and Porrino, A., *Phys. Rev.* C **56**, R16 (1997).
8. Covello, A., Andreozzi, F., Coraggio, L., Gargano, A., Kuo, T. T. S., and Porrino, A. *Prog. Part. Nucl. Phys.* **38**, 165 (1997).
9. Holt, A., Engeland, T., Osnes, E., Hjorth-Jensen, M., and Suhonen, J., *Nucl. Phys.* **A618**, 107 (1997).
10. Suhonen, J., Toivanen, J., Holt, A., Engeland, T., Osnes, E., and Hjorth-Jensen, M., *Nucl. Phys.* **A628**, 41 (1998).
11. Holt, A., Engeland, T., Hjorth-Jensen, M., and Osnes, E., *Nucl. Phys.* **A634**, 41 (1998).
12. Machleidt, R., Holinde, K., and Elster, Ch., *Phys. Rep.* **149**, 1 (1987).
13. Lacombe, M., Loiseau, B., Richard, J. M., Vinh Mau, R., Côté, J., Pires, P., and de Torreil, R., *Phys. Rev.* C **21**, 861 (1980).
14. Stoks, V. G. J., Klopmp, R. A. M., Terheggen, C. P. F., and de Swart, J. J., *Phys. Rev.* C **49**, 2950 (1994).
15. Machleidt, R., Sammarruca, F., and Song, Y., *Phys. Rev.* C **53**, R1483 (1996).
16. Covello, A., Coraggio, L., and A. Gargano, in *Proceedings of the SNEC98 Intern. Conference (Padova 1997)*, *Nuovo Cimento* A, in press.
17. Coraggio, L., Covello, A., Gargano, A., Itaco, N., and Kuo, T. T. S., *Phys. Rev.* C, **58** (1998).
18. Sergeenkov Yu., V., and Sigalov, V. M., *Nucl. Data Sheets* **49**, 39 (1986).
19. Sanchez Vega, M., Fogelberg, B., Mach, H., Taylor, R. B. E., Lindroth, A., and Blomqvist, J., *Phys. Rev. Lett.* **80**, 5504, (1998).
20. Martin, M. J., *Nucl. Data Sheets* **70**, 313 (1993).
21. Zhang, C. T., *et al. Phys. Rev. Lett.* **77**, 3743 (1996).
22. Sergeenkov, Yu. V., *Nucl. Data Sheets* **52**, 205 (1987).
23. We define $\sigma = \{\frac{1}{N_d} \sum_i [E_{\exp}(i) - E_{\text{calc}}(i)]^2\}^{1/2}$, where N_d is the number of data.

24. Tuli, J. K., *Nucl. Data Sheets* **71**, 1 (1994).
25. Helmer, R. G., and Lee, M. A., *Nucl. Data Sheets* **61**, 93 (1990).
26. Rab, S., *Nucl. Data Sheets* **69**, 679 (1993).
27. Schmorak, M. R., *Nucl. Data Sheets* **72**, 409 (1994).
28. Omtvedt, J. P., Mach, H., Fogelberg, B., Jerrestam, D., Hellström, M., Spanier, L., Erokhina, K. I., Isakov, V. I., *Phys. Rev. Lett.* **75**, 3090 (1995).
29. Sergeenkov, Yu. V., *Nucl. Data Sheets* **71**, 557 (1994).
30. Wiringa, R. B., Stoks, V. G. J., and Schiavilla, R., *Phys. Rev.* C **51**, 38 (1995).

LIMITS OF ISOSPIN AND DRIP-LINE PHYSICS

Nuclear Structure Near the Drip Lines

Witold Nazarewicz

Department of Physics, University of Tennessee, Knoxville, Tennessee 37996[1]
Physics Division, Oak Ridge National Laboratory, Oak Ridge, Tennessee 37831[2]
Institute of Theoretical Physics, University of Warsaw, ul. Hoża 69, PL-00-681 Warsaw, Poland

Abstract. Experiments with beams of unstable nuclei will make it possible to look closely into many aspects of the nuclear many-body problem. Theoretically, exotic nuclei represent a formidable challenge for the nuclear many-body theories and their power to predict nuclear properties in nuclear terra incognita.

INTRODUCTION

One of the main frontiers of nuclear structure today is the physics of radioactive nuclear beams (RNB). One of the indications of the potential of this field is the large international interest in it [1–5]. At present there are only a few laboratories with radioactive ion beam capabilities. However, the prospects for new experiments and the success of the current programs have led to a number of RNB facilities under development and a number of further proposals worldwide, including the construction of the next-generation facilities in Europe, U.S., and Japan. There are several major key themes behind the RNB program [2]. They include: (i) nuclear structure (the nature of nucleonic matter), (ii) nuclear astrophysics (the origin of the Universe), and (iii) physics of fundamental symmetries (tests of the Standard Model).

The nuclear landscape, the territory of the RNB physics, is shown in Fig. 1. Black squares indicate stable nuclei; there are less than 300 stable nuclei, or those long-lived, with half-lives comparable to or longer than the age of Earth. Some of the unstable nuclei can be found on Earth, some are man-made, and several thousand nuclei are the yet-unexplored exotic species belonging to nuclear "terra incognita". Moving away from stable nuclei by adding either protons or neutrons, one finally reaches the particle drip lines where the nuclear binding ends. The nuclei beyond the drip lines are unbound to nucleon emission; that is, for those systems the strong interaction is unable to cluster A nucleons as one nucleus. (An exciting question is whether there can possibly exist *islands* of stability beyond the neutron drip line. One of such islands, indicated in Fig. 1, is a neutron star which exists thanks to gravitation. So far, calculations for light neutron drops have not produced permanent binding [6,7].)

The uncharted regions of the (N,Z) plane contain information that can answer many questions of fundamental importance for science: What are the limits of nuclear existence? What are the properties of nuclei with an extreme N/Z ratio? What is the effective nucleon-nucleon interaction in the nucleus having a very large neutron excess? There are also related questions in the field of nuclear astrophysics. Since radioactive nuclei are produced in many astrophysical sites, knowledge of their properties is crucial for our understanding of the underlying processes [8,9].

Nuclear life far from stability is different from that around the stability line; the promised access to completely new combinations of proton and neutron numbers offers prospects for new structural phenomena. The main objective of this talk is to discuss some of the challenges and opportunities of research with radioactive nuclear beams.

---

[1] Research supported by the U.S. Department of Energy under Contract DE-FG02-96ER40963.

[2] Research supported by the U.S. Department of Energy under Contract DE-AC05-96OR22464 with Lockheed Martin Energy Research Corp.

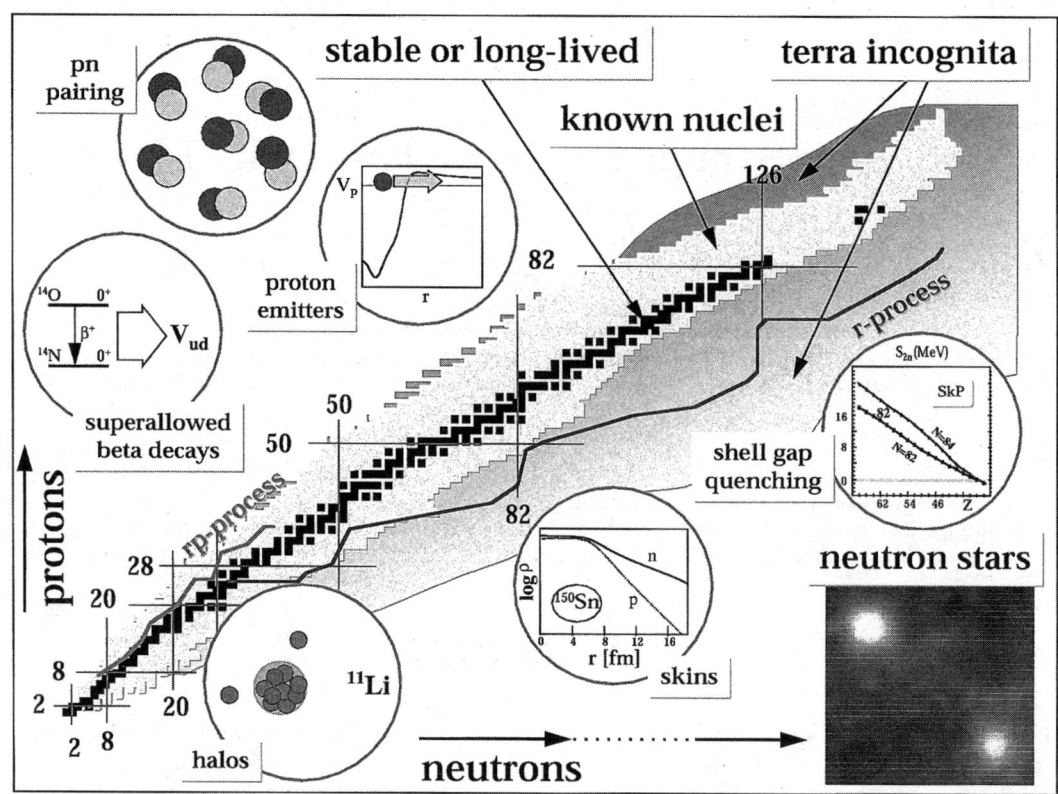

FIGURE 1. The nuclear landscape: territory of radioactive nuclear beams. Some of the important physics themes are indicated schematically. They include: weakly bound nuclear halos, neutron skins and their excitation modes, modifications of shell structure at large N/Z ratios, proton emitters, proton-neutron superconductivity, nuclear tests of the Standard Model, and nucleosynthesis.

NUCLEAR MANY-BODY PROBLEM

The main goal of nuclear structure is to understand how does the nucleus, a fascinating many-body system bound by strong interaction, work. The answer to this question is of fundamental importance. After all, nucleonic matter makes up 99.9 percent of the Universe we live in.

The common theme for the field of nuclear structure is that of the nucleon-nucleon (NN) interaction which clusters nucleons together into one composite system. Figure 2 shows, schematically, our main strategy in our quest for understanding the nuclear force in the context of the hadronic and nucleonic many-body problem. The free NN force can be viewed as a residual interaction of the underlying quark-gluon dynamics of QCD, similar to the intermolecular forces that stem from QED. Some of the challenges related to the nature of the bare NN interaction and the structure of light nuclei have been addressed in the talk by Machleidt [11]; see also the review by Friar [12]. Due to in-medium effects, the NN force in heavy nuclei differs considerably from the free NN interaction. A challenging task is to relate this effective force to that between free nucleons (see Refs. [13–16] for recent developments). Figure 2 illustrates the intellectual connection between the hadronic many-body problem (quark-gluon description of a nucleon) and the nucleonic many-body problem (nucleus as a system of Z protons and N neutrons). It probably would be very naive to think of the behavior of a heavy nucleus directly in terms of the underlying quark-gluon dynamics, but undoubtedly the understanding of the bridges in Fig. 2 will make this goal qualitatively possible.

From a theoretical point of view, spectroscopy of exotic nuclei offers a unique test of those components of effective interactions that depend on the isospin degrees of freedom. Here, the important questions asked are [17]: What is the density dependence of the two-body force? What is the density and radial dependence of the one-body spin-orbit force? What is the form of the pairing interaction in weakly bound nuclei? What is the

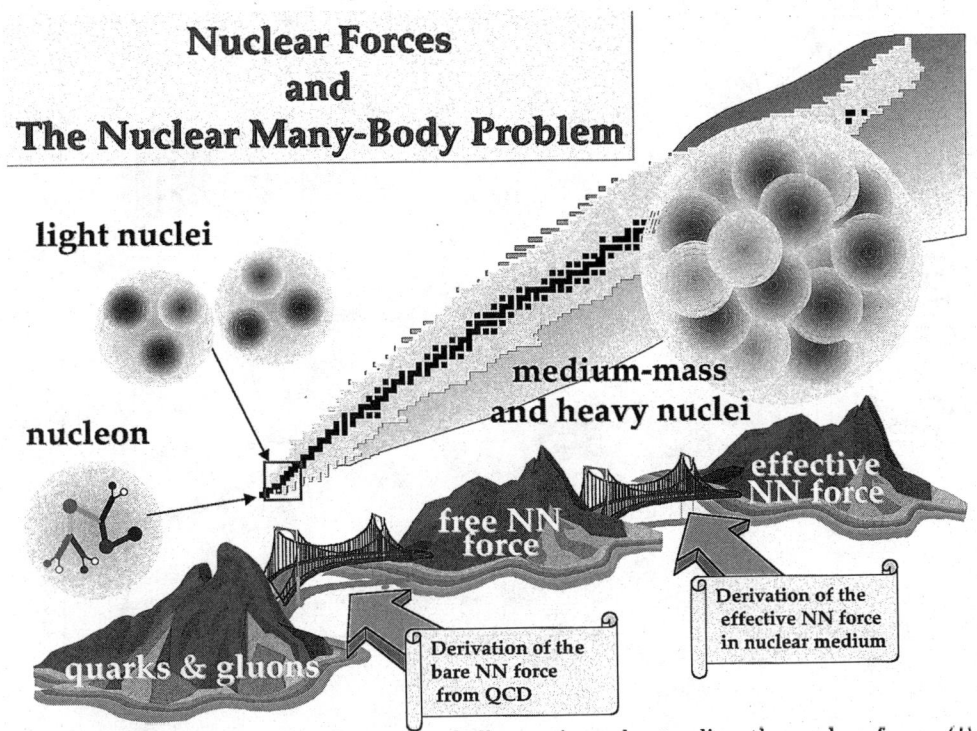

FIGURE 2. From hadrons to heavy nuclei: main challenges in understanding the nuclear force. (From Ref. [10].)

role of the medium effects (renormalization) and of the core polarization in the nuclear exterior (halo or skin region) where the nucleonic density is small?

From the perspective of low-energy nuclear physics, the *effective* building blocks of a nucleus are protons and neutrons. In spite of this simplifying assumption, the dimension of the nuclear many-body problem is overwhelming. Nuclei are exceedingly difficult to describe; they contain too many nucleons to allow for an exact treatment and far too few to disregard finite-size effects. We have learned a great deal about nuclear modes using phenomenological models, often based on ingenious intuition and symmetry considerations. These models and approximations have been extremely successful in interpreting nuclear states and classifying nuclear excitations and decays. Based on this experience, and thanks to the developments in theoretical modeling and computer technology, we are now on the edge of the *microscopic description*. By taking advantage of modern many-body algorithms, one can shorten the cycle theory↔experiment↔theory.

While the very light nuclei can nowadays be described as A-body clusters bound by a free NN force, the conceptual framework of larger nuclei is still that of the nuclear shell model. There has been substantial progress in the area of the conventional shell model. The tractable configuration spaces are growing steadily, no-core calculations are coming, and microscopic effective interactions are being developed and employed. Traditional shell-model techniques make it possible to approach the collective nuclei from the pf shell; the calculations involve model spaces with dimensions of several millions. However, progress in this area is going to be very slow due to exploding dimensions when increasing the number of valence nucleons.

For nuclei close to the magic ones, the low-energy properties depend primarily on the behavior of a few valence nucleons. However, for nuclei with many valence particles, as well as for intruder configurations involving many particle-hole excitations, the concept of valence nucleons is less useful, and the valence and inner-shell nucleons have to be treated on an equal footing. Here, a useful starting point is the self-consistent mean-field theory with the density-dependent effective NN interaction. Thanks to developments in computational techniques, the Hartree-Fock-Bogolyubov (HFB) and relativistic mean-field (RMF) approaches employing microscopic effective interactions are now widely used and – in terms of their predictive power – favorably compare with results of more phenomenological macroscopic-microscopic models.

Figure 3 displays two-neutron separation energies for the Sn isotopes calculated in several state-of-the-art models based on the self-consistent mean-field theory: HFB-D1S (based on finite-range Gogny interaction D1S [18]), HFB-SkP and HFB-SLy4 (based on zero-range Skyrme parametrizations SkP [19] and SLy4 [20]), LEDF

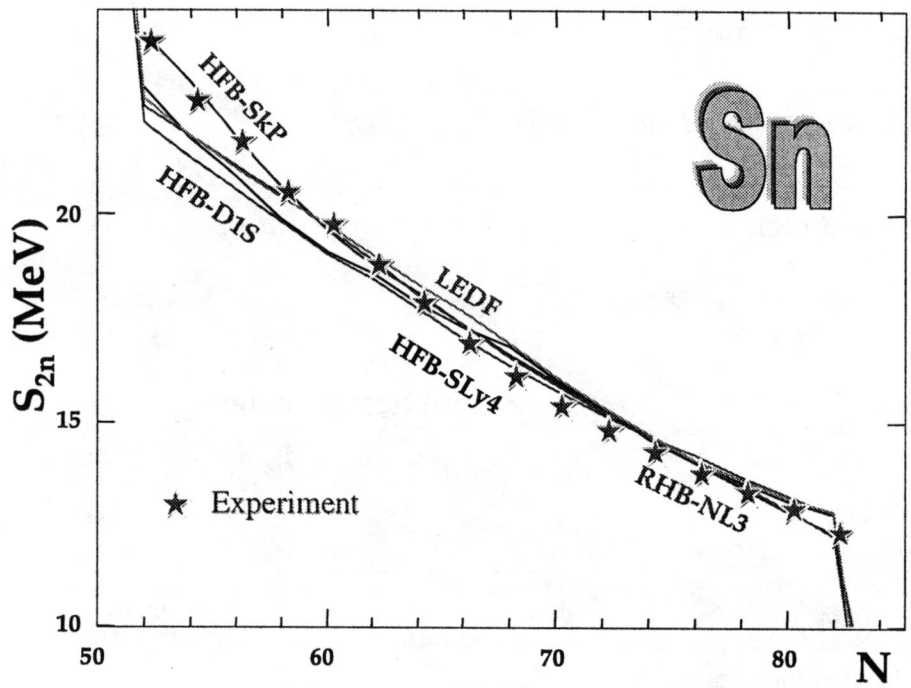

FIGURE 3. Two-neutron separation energies, S_{2n}, for the Sn isotopes calculated in five microscopic models: HFB-D1 (courtesy of J. Dechargé), HFB-SkP and HFB-SLy4 (courtesy of J. Dobaczewski), LEDF (courtesy of S. Fayans), and RHB-NL3 (courtesy of G. Lalazissis). The experimental data are indicated by stars.

(local energy-density functional model with parametrization FaNDF^0 [21]), and RHB-NL3 (relativistic Hartree-Bogolyubov model with NL3 parametrization [22]). All models nicely describe the existing experimental data; some interesting deviations are seen when approaching the proton drip line.

THEORETICAL CHALLENGES FAR FROM STABILITY

Nuclear life at extreme N/Z ratios is different from that around the stability line. The unique structural factor is the weak binding; hence the closeness to the particle continuum. For weakly bound nuclei, the Fermi energy lies very close to zero, and the decay channels must be taken into account explicitly. As a result, many cherished approaches of nuclear theory must be modified. (For an extensive discussion of the theoretical perspectives far from stability, see the recent review [17].) But there is also a splendid opportunity: the explicit coupling between bound states and continuum, and the presence of low-lying scattering states invite strong interplay and cross-fertilization between nuclear structure and reaction theory.

How can one extend traditional tools of nuclear theory to account for the scattering of nucleons from bound single-particle orbitals to unbound states? The closeness of the particle continuum reverberates in two aspects of the theoretical description. Firstly, the particles forming a bound nuclear state can virtually scatter back and forth into the particle continuum phase space. This process must conserve the localization of the nuclear wave function which remains bound even with such a virtual scattering taken into account. A theoretical description of this kind of effect still remains virgin territory, although some progress has been made in the analysis of the virtual pair scattering [19,23]. Secondly, nucleons can very easily leave the nucleus altogether and enter the particle continuum through the real scattering. This is an old problem which, in the context of excited states near or above the particle threshold, has been addressed by the continuum shell model (CSM) [24–26]. In the CSM, the continuum states (decay channels) and bound states are treated on an equal footing. Consequently, correlations due to the coupling to resonances, the spatial extension effects in weakly bound states, the structure of resonances, and the structure of particle transfer form factors are properly described by the CSM. Unfortunately, in many shell-model calculations for weakly bound nuclei (including those presented at this meeting!) the continuum aspects are completely disregarded; hence their conclusions should be taken with a grain of salt (see, however, the recent study [27]).

Often, particle continuum is approximated by the quasibound states, i.e., the states resulting from the diagonalization of a finite potential in a large basis [28,29] or by enclosing the finite nuclear potential within an infinite well with walls positioned at a large distance from the nuclear surface [30,31]. More sophisticated methods of discretizing continuum include the Sturmian function expansions [32–35] and resonant (Gamow) state expansions.

How do well-established microscopic models of nuclear structure perform when extrapolated to exotic nuclei? Due to many uncertainties, we do not know a simple answer to this question. As an example, Fig. 4 shows

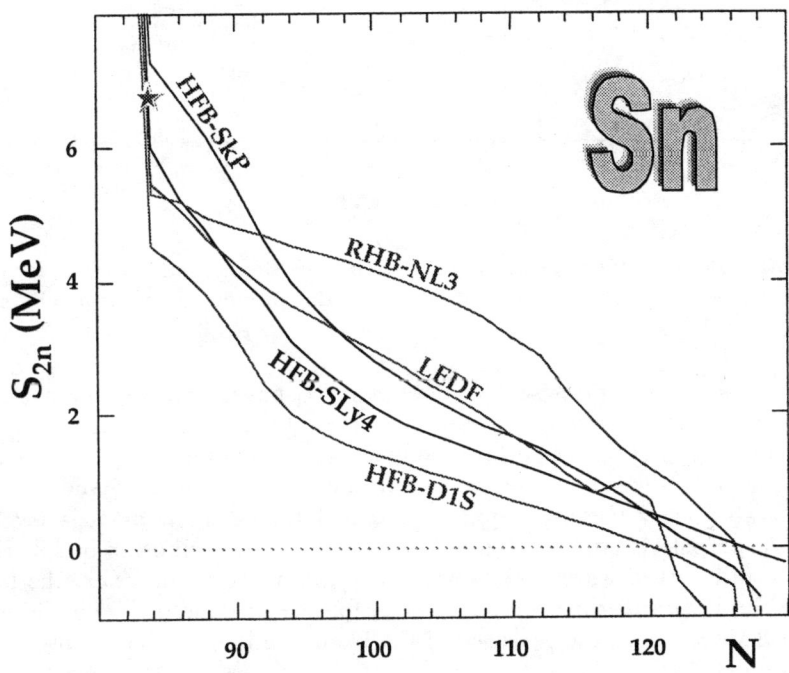

FIGURE 4. Similar to Fig. 3 except for very neutron-rich Sn isotopes.

the predicted two-neutron separation energies for the tin isotopes using the same models as in Fig. 3. Clearly, the differences between forces are greater in the region of "terra incognita" than in the region where masses are known. As seen in Fig. 4, the position of the neutron drip line for the Sn isotopes slightly depends on the effective interaction used; it varies between $N=120$ (HFB-D1S) and $N=126$ (RHB-NL3). Therefore, the uncertainty due to the largely unknown isospin dependence of the effective force gives an appreciable theoretical "error bar" for the position of the drip line. Unfortunately, the results presented in Fig. 4 do not tell us much about which of the forces discussed should be the preferred one since one is dealing with dramatic extrapolations far beyond the region known experimentally. However, a detailed analysis of the force dependence of results may give us valuable information on the relative importance of various force parameters.

CONTINUUM SHELL MODEL AND GAMOW STATES

The Gamow states are eigenstates of the time-independent Schrödinger equation with complex eigenvalues [36–38]. They are regular at $r=0$ and satisfy a purely outgoing wave type of asymptotics with the complex energy eigenvalue. The real part of the complex energy eigenvalue is the expectation value of the one-body Hamiltonian, while the imaginary part is related to the total decay width of the quasi-stationary state. The Gamow states are the poles of the S-matrix on the complex energy plane lying below the positive real axis. The closer they lie to the real axis, the more they resemble the bound states, and they can be associated with narrow resonances.

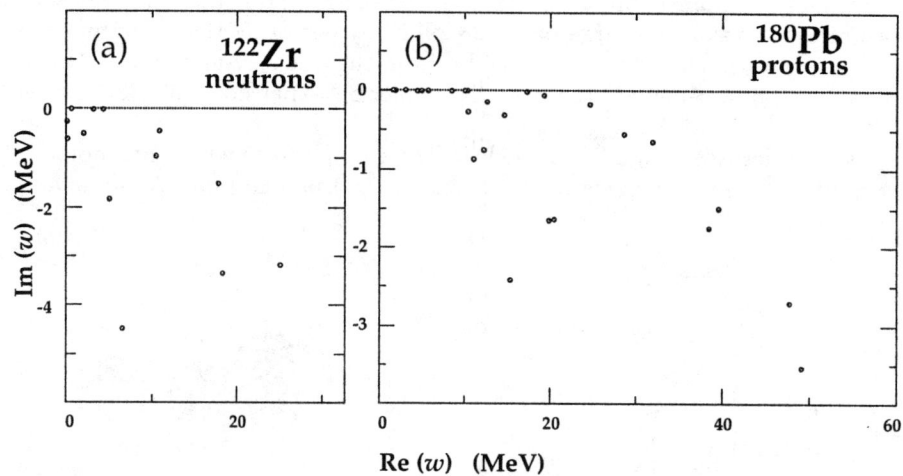

FIGURE 5. The distribution of Gamow energy eigenvalues w_i in the (Re(w), Im(w)) plane for (a) neutron drip-line nucleus ^{122}Zr (neutron eigenvalues), and (b) proton-rich nucleus ^{180}Pb (proton eigenvalues). (From Ref. [39].)

A generalized completeness relation proposed by Berggren [37] paved the way for using Gamow states as basis states in a similar way as the ordinary bound states are used. Using the generalized completeness relation, one can treat a selected set of resonant states on the same footing as bound states. The remaining part of the continuum is treated by means of the integral along a path in a complex energy plane.

Recently, the Gamow-state theory has been applied to calculate the single-particle level density and shell corrections for finite depth potentials [40,39]. Figure 5 illustrates the distribution of spherical Gamow eigen-values for ^{122}Z and ^{180}Pb (all partial waves are shown). For the neutrons in ^{122}Zr, the large-width Gamow states appear just above the Re(w)=0 threshold. However, for the protons in ^{180}Pb, the particle continuum is shifted effectively by ~8 MeV due to the presence of the Coulomb barrier. The proton Gamow states that appear at low energies are extremely narrow resonances, usually discussed in the context of proton emitters. Other applications of Gamow states to the description of resonances and drip-line nuclei can be found in Refs. [38,41,42].

SHELL STRUCTURE OF DRIP-LINE NUCLEI

A significant new theme concerns shell structure near the particle drip lines. Since the isospin dependence of the effective nucleon-nucleon interaction is largely unknown, the structure of single-particle states, collective modes, and the behavior of global nuclear properties is very uncertain in nuclei with extreme N/Z ratios. For instance, some calculations predict [17,43,44] that the shell structure of neutron drip-line nuclei is different from what is known around the beta-stability valley. According to other calculations [45], a reduction of the spin-orbit splitting in neutron-rich nuclei is expected. The gradual change of shell structure with neutron number is believed to give rise to new sorts of collective phenomena [29,46]. It is to be noted that the experimentally observed collapse of magic gaps seen in some neutron-rich light nuclei (the so-called islands of inversion [47–50]) might also be related to the predicted quenching of magic gaps [51].

The effect of the weakening of known shell effects in drip-line nuclei has significant consequences for the description of the astrophysical r-process. Although many of the astrophysically important neutron-rich nuclei, belonging to the terra incognita of Fig. 1 will not be experimentally accessible in the foreseeble future, their part of the (Z, N) chart has been already visited by Mother Nature in cataclysmic events such as supernovae. Consequently, valuable information on nuclear structure in terra incognita can be offered by astrophysical data. In particular, the analysis of the r-process abundances can provide an indirect experimental confirmation of the shell-quenching effect [52,53].

Nuclear Halos

Halo nuclei are symbols of RNB physics. The very weak binding of the outermost neutrons leading to a rather good decoupling of halo from the core simplifies many aspects of unrelying nuclear structure and reaction mechanism. Much has been learned about halo nuclei. Few-body calculations, especially those based on the method of hyperspherical harmonics and the adiabatic hyperspherical method, have reached a high level of sophistication [54,55]. They give the basic understanding of many observed phenomena (momentum distributions, electromagnetic strength distributions) but often neglect the important structure aspects. Some of the questions asked in the context of halo nuclei are: What are the main dynamical degrees of freedom which affect the core-halo coupling? What is the degree of clusterization of the cores? Is the clusterization mechanism enhanced close to the neutron drip line? If so, what are its manifestations in heavier systems? What are modifications of the effective interaction in the halo region? Through halos, one hopes to learn more about the microscopic mechanism of clusterization found in "normal" nuclei (see the Ikeda diagram shown in Fig. 6), such as in the hypothetical three-alpha state of ^{12}C.

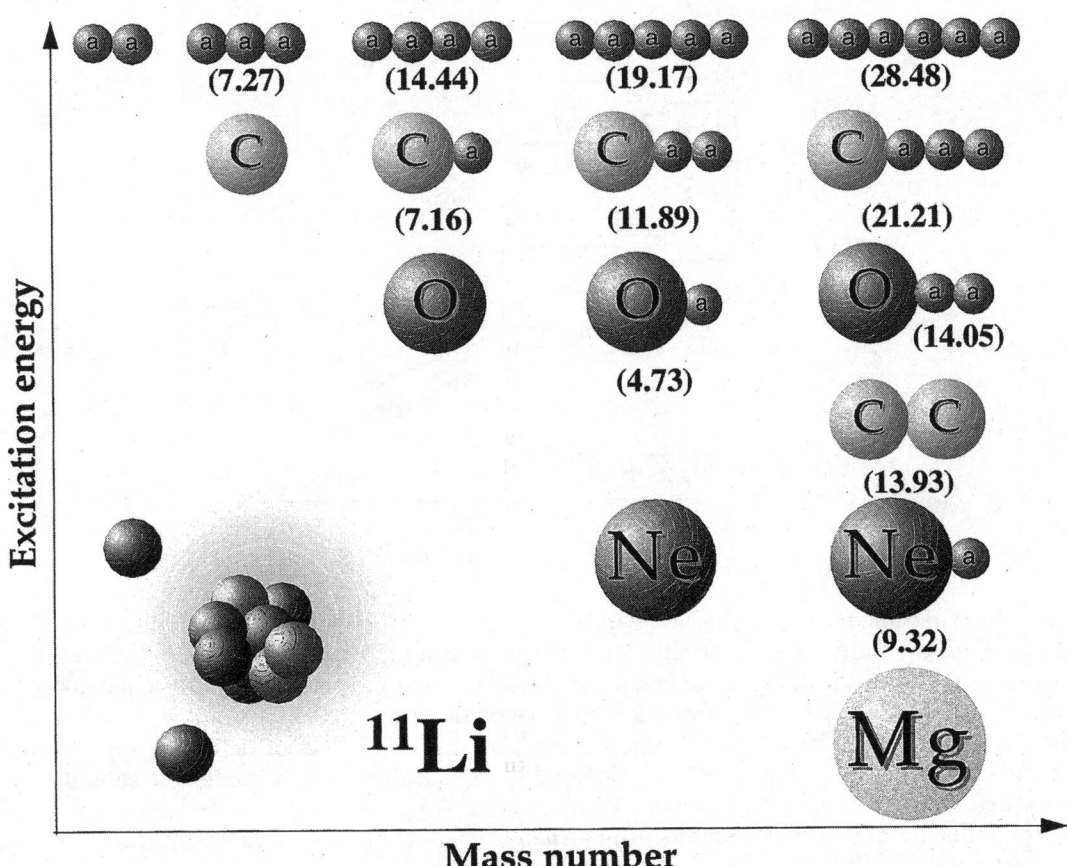

FIGURE 6. Ikeda diagram for light nuclei [56]. The threshold energy for each decay mode (in MeV) is indicated. The halo nucleus ^{11}Li, viewed as a three-body borromean system, is also shown.

SPECTROSCOPY OF PROTON EMITTERS

Proton radioactivity is an excellent example of the elementary three-dimensional quantum-mechanical tunneling. Lifetimes of proton emitters directly provide an indication of the angular momentum content of the narrow proton resonance [57,58]. Experimental and theoretical investigations of proton emitters (or theoretically predicted ground-state di-proton emitters) are just opening up a wealth of exciting physics associated with the residual interaction coupling between bound states and extremely narrow resonances in the region of very low density of single-particle levels.

In general, proton emission half-lives depend mainly on the proton separation energy and orbital angular momentum, but depend rather weakly on the details of the intrinsic structure of proton emitters, e.g., on the parameters of the proton-core potential. This suggests that the lifetimes of deformed proton emitters will

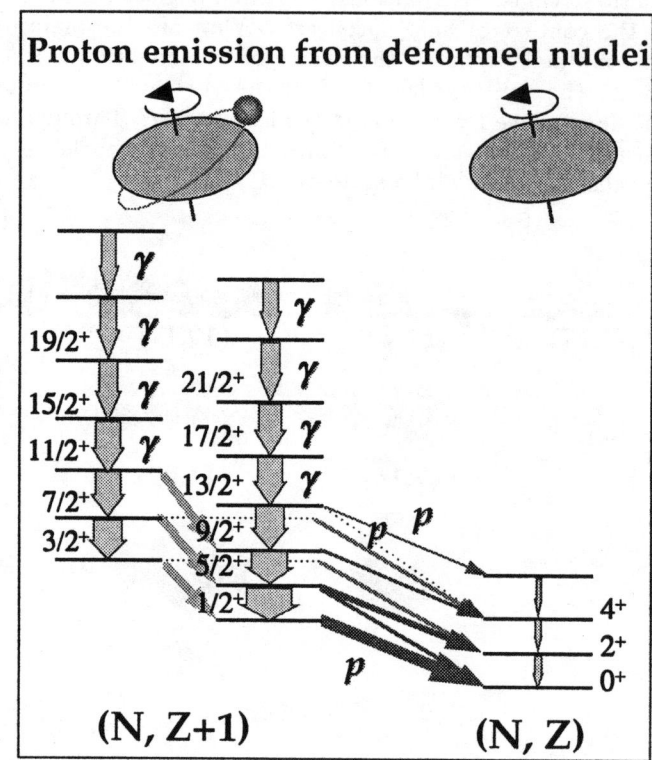

FIGURE 7. Competition between gamma and proton decays in a deformed proton emitter.

provide direct information on the angular momentum content of the associated Nilsson state, and hence on the nuclear shape. Figure 7 shows, schematically, a possible scenario of proton-gamma competition in a deformed proton emitter. Of course, the energy window for such a process is expected to be fairly narrow.

Recently, a method of calculating deformed proton resonances by means of the coupled-channel technique with Gamow states has been proposed [42,59]. In another work, half-lives of deformed proton emitters were analyzed by means of the time-dependent Schrödenger equation [60]. In this context it should be noted that while in very deformed nuclei proton resonances can be treated by means of the strong coupling approach, to investigate the influence of the angular momentum dependence of the proton decay width, the Coriolis coupling should be considered.

Another exciting avenue is the competition between gamma-radiation and the emission of prompt protons. Here, spectacular examples are proton-emitting intruder bands in ^{58}Cu [61] and ^{56}Ni [62]. In ^{56}Ni, where two intruder bands have been observed, the lower rotational band can be explained by large-scale shell-model calculations in the pf shell. Also the results of cranked mean-field calculations indicate that this band is built upon a 4p-4h excitation within the pf shell (see Fig. 8, 4^04^0 band). The second band, however, is expected to involve particles in the $1g_{9/2}$ orbit, which is supported by its nearly identical behavior to the band in ^{58}Cu. However, the best scenario for this band [62] is based on *one* proton promoted to the $1g_{9/2}$ orbit (4^04^1 band in Fig. 8), while a neutron *and* a proton occupies this orbit in ^{58}Cu. Radioactive medium-mass nuclei such as ^{56}Ni are fantastic territories where various theoretical approaches can be confronted: state-of-the-art shell model, mean-field models, and cluster models. The spherical structures in ^{56}Ni are well described by the large-scale shell model, the collective 4^04^1 band can be understood in terms of the self-consistent mean-field theory, and

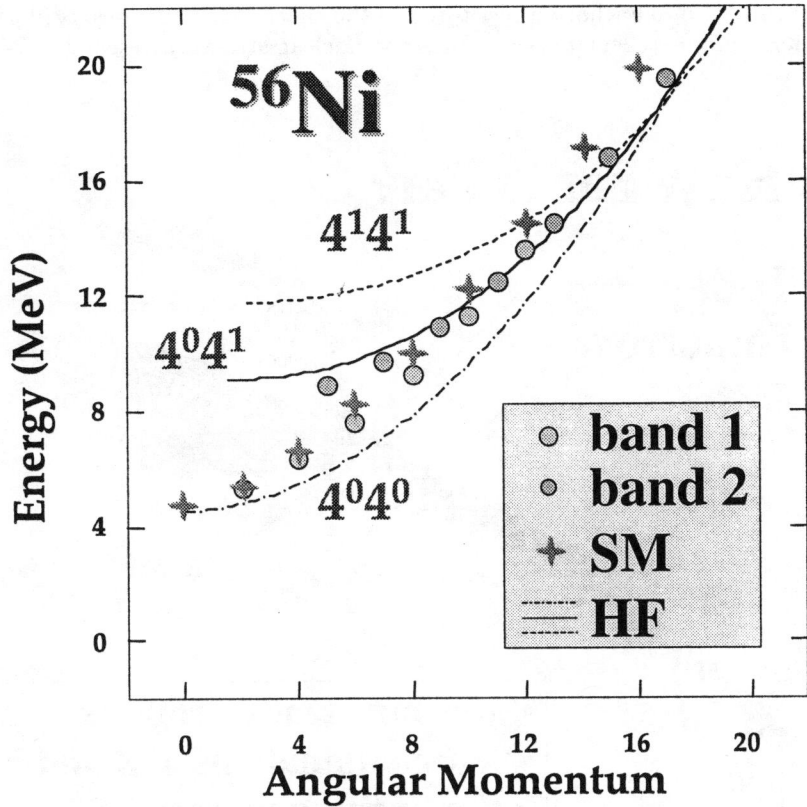

FIGURE 8. Excitation energy versus angular momentum for experimental and calculated intruder bands in ^{56}Ni. Lines indicate the $4^0 4^0$, $4^0 4^1$, and $4^1 4^1$ bands calculated in the cranked HF+SLy4 model, while the KBF shell-model calculations (SM) are indicated by crosses.

the $4^0 4^0$ band can be described by both methods (see also Refs. [63–65]).J

CONCLUSIONS

In trying to see the phenomena of a "new physics", one should ask the fundamental question of "how far is far" (see Fig. 9)? There is very little doubt that progress in the RNB research is not going to be rapid. Especially for the neutron-rich heavy elements, it will be a real struggle to obtain basic spectroscopic information. To put things in perspective, it is instructive to recall the discovery of polonium by Maria and Pierre Curie at the end of June 1898. The longest-lived isotope of polonium is ^{209}Po with a half-life of 102 y. The isotope ^{218}Po was studied by Rutherford already in 1904 [66]; its half-life is 3.1 min [67]. Amazingly, it took ninety-four years to add one more neutron to ^{218}Po; only recently have the neutron-rich isotopes, ^{219,220}Po, been produced at GSI by the international collaboration led by M. Pfützner [68]. It has been much easier to approach the neutron-poor side. The lightest polonium known, ^{190}Po, was found in 1988 at GSI [69]; its half-life, 2.4 ms [70], is long enough to enable detailed spectroscopic studies.

The doubly magic $N=Z=50$ nucleus ^{100}Sn is a paradigm of RNB physics at the proton-rich side. Although it was found experimentally three years ago [71,72], it took more than two years to roughly determine its mass [73], and it will still take quite a few years to find its first excited state. On an optimistic note, many nuclei very close to the $N=Z=50$ corner have already been approached spectroscopically in recent years [74], so the prospects for more spectroscopic news on ^{100}Sn are good. It will be much harder to approach another doubly magic nucleus, the neutron-rich ^{78}Ni. It was produced in 1995 [75], but our knowledge about this system and its neighbors is very scarce. Indeed, the heaviest nickel isotope known spectroscopically is ^{70}Ni [76], which is as far as eight neutrons away!

An experimental excursion into uncharted territories of the chart of the nuclides, exploring new combinations of Z and N, will offer many excellent opportunities for nuclear structure research. What is most exciting,

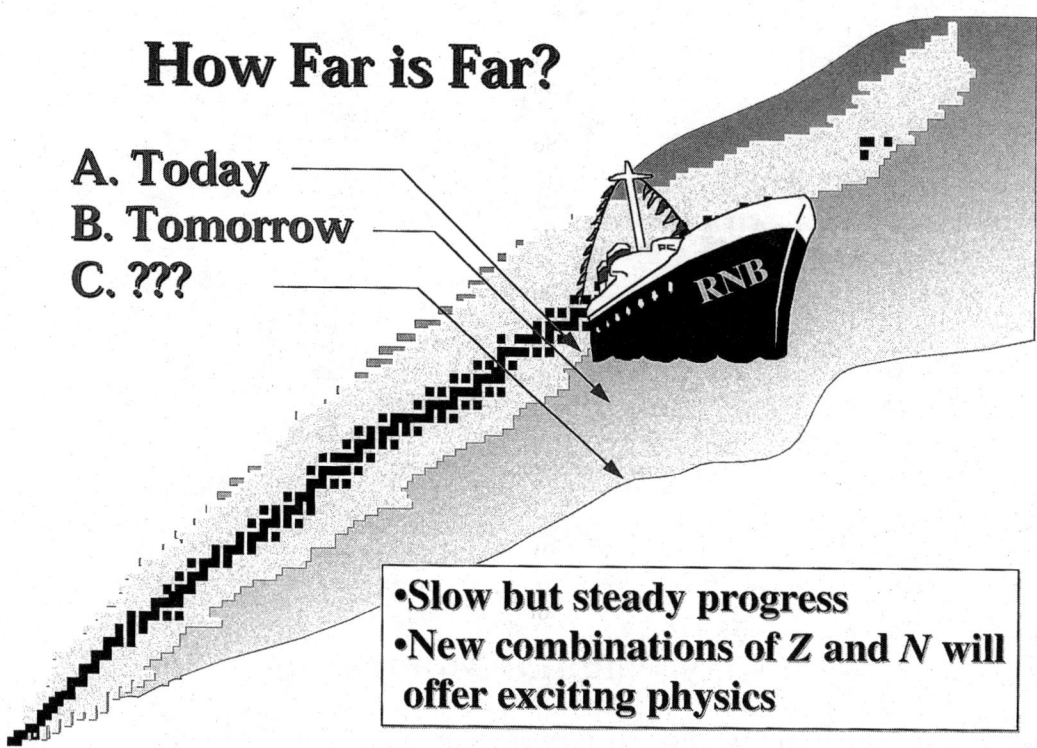

FIGURE 9. "Fishing expedition" to terra incognita far from stability.

however, is that there are many unique features of exotic nuclei that give prospects for entirely new phenomena likely to be different from anything we have observed to date. We are only at the beginning of a most exciting journey.

ACKNOWLEDGEMENTS

It is a great pleasure to acknowledge a very close collaboration with Jacek Dobaczewski, which resulted in many ideas and results presented in this talk.

REFERENCES

1. W. Nazarewicz, B. Sherril, I. Tanihata, and P. Van Duppen, Nuclear Physics News **6**, 17 (1996).
2. *Scientific Opportunities With an Advanced ISOL Facility*, Report, November 1997; http://www.er.doe.gov/production/henp/isolpaper.pdf.
3. The K500+K1200 Project, NSCL, MSU, http://www.nscl.msu.edu/facility/ccp/home.html.
4. SPIRAL project at GANIL, http://ganinfo.in2p3.fr/spiral/.
5. RI Beam Factory, RIKEN, http://www.rarf.riken.go.jp/ribf/doc/ribf94e/.
6. B.S. Pudliner, A. Smerzi, J. Carlson, V.R. Pandharipande, S.C. Pieper, and D.G. Ravenhall, Phys. Rev. Lett. **76**, 2416 (1996).
7. A. Smerzi, D.G. Ravenhall, and V.R. Pandharipande, Phys. Rev. **C56**, 2549 (1997).
8. M. Wiescher, this volume.
9. M. Wiescher, H. Schatz, and A.E. Champagne, Phil. Trans. R. Soc. Lond. A **356**, 2105 (1998).

10. W. Nazarewicz, Nucl. Phys. **A630**, 239c (1998).
11. R. Machleidt, this volume.
12. J.L. Friar, in: "Nuclear Physics with Effective Field Theory", ed. by R. Seki, U. van Kolck, and M.J. Savage, (World Scientific, Singapore, 1998), p. 145.
13. M. Hjorth-Jensen, T.T.S. Kuo, and E.Osnes, Phys. Rep. **261**, 125 (1995).
14. P. Navratil, M.Thoresen, and B.R. Barrett, Phys. Rev. **C55**, R573 (1997).
15. L. Engvik, E. Osnes, M. Hjorth-Jensen, and T.T.S. Kuo, Nucl. Phys. **A622**, 553 (1997).
16. B.R. Barrett, this volume.
17. J. Dobaczewski and W. Nazarewicz, Phil. Trans. R. Soc. Lond. A **356**, 2007 (1998).
18. J. Dechargé and D. Gogny, Phys. Rev. **C21**, 1568 (1980).
19. J. Dobaczewski, H. Flocard and J. Treiner, Nucl. Phys. **A422**, 103 (1984).
20. E. Chabanat, *Interactions effectives pour des conditions extrêmes d'isospin*, Université Claude Bernard Lyon-1, Thesis 1995, LYCEN T 9501, unpublished.
21. S.A. Fayans, ENAM 98, ed. by B.M. Sherrill, D.J. Morrissey, and C.N. Davids, AIP Conference Proceedings **455**, 310 (1998).
22. G.A. Lalazissis, J. König, and P. Ring, Phys. Rev. **C55**, 540 (1997).
23. J. Dobaczewski, W. Nazarewicz, T.R. Werner, J.-F. Berger, C.R. Chinn, and J. Dechargé, Phys. Rev. **C53**, 2809 (1996).
24. U. Fano, Phys. Rev. **124**, 1866 (1961).
25. C. Mahaux and H. Weidenmüller, *Shell Model Approaches to Nuclear Reactions* (North-Holland, Amsterdam, 1969).
26. R.J. Philpott, Fizika **9**, suppl. 3, 21 (1977).
27. K. Bennaceur, F. Nowacki, J. Okołowicz, and M. Płoszajczak, J. Phys. **G24**, 1631 (1998).
28. M. Bolsterli, E.O. Fiset, J.R. Nix, and J.L. Norton, Phys. Rev. **C5**, 1050 (1972).
29. W. Nazarewicz, T.R. Werner, and J. Dobaczewski, Phys. Rev. **C50**, 2860 (1994).
30. J.R. Bennett, J. Engel, and S. Pittel, Phys. Lett. **B368**, 7 (1996).
31. F. Ghielmetti, G. Colo, E. Vigezzi, P.F. Bortignon, and R.A. Broglia, Phys. Rev. **C54**, R2143 (1996).
32. W. Glöckle, J. Hufner, and H.A. Weidenmueller, Nucl. Phys. **A90**, 481 (1967).
33. J.S. Vaagen, B.S. Nilsson, J. Bang, and R.M. Ibarra, Nucl. Phys. **A319**, 143 (1979).
34. G. Rawitscher, Phys. Rev. **C25**, 2196 (1982).
35. M. Buballa, S. Dróżdż, S. Krewald, and J. Speth, Ann. Phys. **208**, 346 (1991).
36. J. Humblet and L. Rosenfeld, Nucl. Phys. **26**, 529 (1961).
37. T. Berggren, Nucl. Phys. **A109**, 265 (1968).
38. T. Vertse, P. Curutchet, and R.J. Liotta, Lecture Notes in Physics **325** (Springer Verlag, Berlin 1987), p. 179.
39. T. Vertse, R.J. Liotta, W. Nazarewicz, N. Sandulescu, and A.T. Kruppa, Phys. Rev. **C57**, 3089 (1998).
40. N. Sandulescu, O. Civitarese, R.J. Liotta, and T. Vertse, Phys. Rev. **C55**, 1250 (1997).
41. L.S. Ferreira, E. Maglione, and R.J. Liotta, Phys. Rev. Lett. **78**, 1640 (1997).
42. E. Maglione, L.S. Ferreira, and R.J. Liotta, Phys. Rev. Lett. **81**, 538 (1998).
43. F. Tondeur, Z. Phys. **A288**, 97 (1978).
44. J. Dobaczewski, I. Hamamoto, W. Nazarewicz, and J.A. Sheikh, Phys. Rev. Lett. **72**, 981 (1994).
45. G.A. Lalazissis, D. Vretanar, W. Poschl, and P. Ring, Phys. Lett. **418B**, 7 (1998).
46. W.-T. Chou, R.F. Casten, and N.V. Zamfir, Phys. Rev. **C51**, 2444 (1995).
47. C. Détraz *et al.*, Phys. Rev. **C19**, 164 (1979).
48. F. Touchard *et al.*, Phys. Rev. **C25**, 2756 (1982).
49. H. Scheit *et al.*, Phys. Rev. Lett. **77**, 3967 (1996).
50. T. Glasmacher *et al.*, Phys. Lett. B **395**, 163 (1997).
51. P.-G. Reinhard *et al.*, in preparation.
52. B. Chen, J. Dobaczewski, K.-L. Kratz, K. Langanke, B. Pfeiffer, F.-K. Thielemann, and P. Vogel, Phys. Lett. **B355**, 37 (1995).
53. B. Pfeiffer, K.-L. Kratz, and F.-K. Thielemann, Z. Phys. **A357**, 235 (1997).
54. B.V. Danilin, I.J. Thompson, J.S. Vaagen, and M.V. Zhukov, Nucl. Phys. **A632**, 383 (1998).
55. E. Garrido, A. Cobis, D.V. Fedorov, and A.S. Jensen, Nucl. Phys. **A630**, 409c (1998).
56. H. Horiuchi, K. Ikeda, and Y. Suzuki, Suppl. Prog. Theor. Phys. **52**, 89 (1972).
57. P.J. Woods and C.N. Davids, Ann. Rev. Nucl. Part. Sci. **47**, 541 (1997).
58. P.J. Woods, this volume.
59. A. Kruppa *et al.*, in preparation.
60. P. Talou, N. Carjan, and D. Strottman, Phys. Rev. **C58**, 3280 (1998).
61. D. Rudolph, C. Baktash, J. Dobaczewski, W. Nazarewicz, W. Satuła, M.J. Brinkman, M. Devlin, H.-Q. Jin, D.R. LaFosse, L.L. Riedinger, D.G. Sarantites, and C.-H. Yu, Phys. Rev. Lett. **80**, 3018 (1998).
62. D. Rudolph *et al.*, submitted to Phys. Rev. Lett.

63. J. Dobaczewski, this volume.

64. D. Rudolph, this volume.

65. T. Otsuka, this volume.

66. E. Rutherford, Phil. Trans. Roy. Soc. **204**, 169 (1904).

67. M. Curie, A.S. Eve, H. Gerger, O. Hahn, S.C. Lind, S. Meyer, E. Rutherford, and E. Schweidler, Rev. Mod. Phys. **3**, 427 (1931).

68. M. Pfützner et al., GSI 98-1, p. 32 (1998); submitted to Phys. Lett.

69. A.B. Quint et al., GSI-88-1, p.16 (1988).

70. J.C. Batchelder *et al.*, Phys. Rev. **C55**, R2142 (1997).

71. R. Schneider *et al.*, Z. Phys. A **348**, 241 (1994).

72. M. Lewitowicz *et al.*, Phys. Lett. **B332**, 20 (1994).

73. M. Chartier *et al.*, Phys. Rev. Lett. **77**, 2400 (1996).

74. H. Grawe *et al.*, Z. Phys. **A358**, 185 (1997).

75. Ch. Engelmann *et al.*, Z. Phys. **A352**, 351 (1995).

76. R. Grzywacz *et al.*, Phys. Rev. Lett. **81**, 766 (1998).

Exotic nuclei in self–consistent mean–field models

M. Bender[a], K. Rutz[b,c], T. Bürvenich[b],
P.-G. Reinhard[d,e], J. A. Maruhn[b,e], W. Greiner[b,e]

[a] *Department of Physics and Astronomy, University of North Carolina at Chapel Hill, U.S.A.*
[b] *Institut für Theoretische Physik, Universität Frankfurt am Main, Germany*
[c] *Gesellschaft für Schwerionenforschung mbH, Darmstadt, Germany*
[d] *Institut für Theoretische Physik II, Universität Erlangen–Nürnberg, Germany*
[e] *Joint Institute for Heavy–Ion Research, Oak Ridge National Laboratory, Tennessee, U.S.A.*

Abstract. We discuss two widely used nuclear mean–field models, the relativistic mean–field model and the (nonrelativistic) Skyrme–Hartree–Fock model, and their capability to describe exotic nuclei with emphasis on neutron–rich tin isotopes and superheavy nuclei.

INTRODUCTION

This contribution discusses nuclear mean–field models and their capabilities to describe exotic nuclei. The two most widely used mean–field models are considered, the relativistic mean–field model (RMF) [1,2] and the nonrelativistic Skyrme–Hartree–Fock approach (SHF), see [3] for an early review. Both models aim at a fully self–consistent description of nuclei employing effective interactions developed for the purpose of effective Hartree or Hartree–Fock calculations. Thus the models stay in between the more elaborate many–body theories which try a description in terms of a basic nucleon–nucleon interaction [4,5] and the more phenomenological approach on the basis of macroscopic–microscopic models [6]. We ought to mention that there are two additional self–consistent mean–field models in the literature, the nonrelativistic finite–range Gogny force [7] and the relativistic point–coupling model, see e.g. [8], which we omit here.

The actual effective interactions of RMF and SHF can be restricted by very general arguments to a form with only few free (6–10) parameters which are to be adjusted to a few key data of nuclear structure. The models give a very good description of almost all stable nuclei from ^{16}O on up to the heaviest elements [2,9]. Stable nuclei, however, represent only a very narrow valley in the N–Z–plane and thus there remain several loosely fixed aspects in the models, particularly concerning isovector properties. Exotic nuclei which are now becoming more and more accessible are a challenge for mean–field models and will provide at the same time useful key data for further development. It is the aim of this contribution to explore the predictions of SHF and RMF when proceeding into the exotic regions of the nuclear chart, here in particular to neutron–rich isotopes and to superheavy nuclei.

In view of the uncertainties in extrapolations, we consider a broad selection of parameterizations with about comparable quality concerning normal nuclear properties but differences in less well determined aspects, e.g. spin–orbit interaction and isovector properties.

The Skyrme interaction was originally designed as an effective two–body force for nuclear structure calculations. It has the technical advantage that the exchange terms in the Hartree–Fock equations have the same structure as the direct terms and therefore the numerical solution of the Hartree–Fock equations is as simple as in case of the Hartree approach. The total binding energy can be formulated in terms of an energy functional which depends on local densities and currents only. This links the Skyrme–Hartree–Fock model to the effective energy functional theory in the Kohn–Sham approach. The Hohenberg–Kohn theorem states that the non–degenerated ground–state energy of a many–Fermion system with local two-body interactions is a unique functional of the local density only. The Kohn–Sham scheme relies on the Hohenberg–Kohn theorem but keeps for the kinetic energy the full dependence on the single–particle wavefunctions which allows to maintain the full shell structure while employing for the rest rather simple functionals in local–density approximation. This

CP481, *Nuclear Structure 98*, edited by C. Baktash

point-of-view can be carried over to the case of nuclei where, however, the non-local two–body interaction requires en extension of the energy functional by a dependence on other densities and currents (e.g. spin–orbit). In any case, there is no need for a fundamental two–body force in an effective many–body theory, but one can start from an effective energy functional which is formulated directly at the level of one–body densities and currents, see e.g. [10] and references therein. The energy functional used here is given by

$$\mathcal{E} = \mathcal{E}_{kin} + \mathcal{E}_{Sk} + \mathcal{E}_{C} + \mathcal{E}_{pair} \quad , \tag{1}$$

where $\mathcal{E}_{kin}$ is the kinetic energy, $\mathcal{E}_{C}$ the Coulomb energy including the exchange term in local density approximation and $\mathcal{E}_{pair}$ the pairing energy functional described below. For the Skyrme functional we use the form

$$\mathcal{E}_{Sk} = \int d^3 r \left[\frac{b_0}{2} \rho^2 + b_1 \rho \tau - \frac{b_2}{2} \rho \Delta \rho + \frac{b_3}{3} \rho^{\alpha+2} - b_4 \rho \nabla \cdot \mathbf{J} \right.$$
$$\left. - \sum_q \left(\frac{b_0'}{2} \rho_q^2 + b_1' \rho_q \tau_q - \frac{b_2'}{2} \rho_q \Delta \rho_q + \frac{b_3'}{3} \rho^\alpha \rho_q^2 + b_4' \rho_q \nabla \cdot \mathbf{J}_q \right) \right] \tag{2}$$

with $q \in \{p, n\}$, see e.g. [11]. ρ_q, τ_q, and $\mathbf{J}_q$ denote the local density, kinetic density, and spin–orbit current of protons and neutrons respectively, while densities without index denote total densities, e.g. $\rho = \rho_p + \rho_n$. The equations–of–motion of the single–particle states are derived from a variational principle. The b_i, b_i', and α are the adjustable parameters of the Skyrme functional which have to be fitted to experimental data.

We consider the parameterizations SkM* [12], SkP [13], SkI1, SkI3, SkI4 [9], and SLy6 [14] which all employ the standard form but differ in bias. All these parameterizations have been obtained from fits to nuclear ground–state data, binding energies, radii, and a few selected spin–orbit splittings. The fit of SkI1 includes even the whole electromagnetic formfactor at low q (i.e. up to the first maximum). The force SkP uses effective mass $m^*/m = 1$ and is designed to describe both the particle–hole and particle–particle channel of the effective interaction. The other forces all have smaller effective masses around $0.7 \leq m^*/m \leq 0.8$. The force SkM* was first to deliver acceptable incompressibility and fission properties. The force SLy6 stems from an attempt to cover properties of pure neutron matter together with normal nuclear ground–state properties. The force SkI1 stems from a recent systematic fit already embracing data from exotic nuclei.

The forces SkI3 and SkI4 are fitted exactly as SkI1 but use a variant of the Skyrme parameterization where the spin–orbit force is complemented by an explicit isovector degree–of–freedom [9]. They are designed to overcome the different isovector trends of spin–orbit coupling between conventional Skyrme forces and the RMF. For standard Skyrme forces holds $b_4 = b_4'$ in (2). SkI3 contains a fixed isovector part exactly analogous to the RMF with $b_4' = 0$, whereas SkI4 is adjusted allowing free variation of both parameters b_4 and b_4' of the spin–orbit force. All forces provide a good description of the gross properties of stable nuclei.

The relativistic mean–field model is formulated in terms of a Lagrangian of the form [2]

$$\mathcal{L}_{RMF} = \mathcal{L}_N + \mathcal{L}_M + \mathcal{L}_{NM} + \mathcal{L}_{em} \quad , \tag{3}$$

where $\mathcal{L}_N$ is the free Lagrangian of the nucleons and $\mathcal{L}_{em}$ the Lagrangian of the electromagnetic field. The Lagrangians of the fields $\mathcal{L}_M$ and their coupling to the nucleons $\mathcal{L}_{NM}$ are given by

$$\mathcal{L}_M = \frac{1}{2}(\partial_\mu \Phi_\sigma \partial^\mu \Phi_\sigma + m^2 \Phi_\sigma^2) + \mathcal{L}_{nonl}$$
$$- \frac{1}{2}\left[\frac{1}{2}(\partial_\mu \Phi_{\omega,\nu} - \partial_\nu \Phi_{\omega,\mu})\partial^\mu \Phi_\omega^\nu - m_\omega^2 \Phi_{\omega,\mu} \Phi_\omega^\mu \right]$$
$$- \frac{1}{2}\left[\frac{1}{2}(\partial_\mu \Phi_{\rho,\nu} - \partial_\nu \Phi_{\rho,\mu}) \cdot \partial^\mu \Phi_\rho^\nu - m_\rho^2 \Phi_{\rho,\mu} \cdot \Phi_\rho^\mu \right],$$
$$\mathcal{L}_{NM} = -g_\sigma \Phi_\sigma \rho^s - g_\omega \Phi_{\omega,\mu} \rho^\mu - g_\rho \Phi_{\rho,\mu} \cdot \rho^\mu \quad . \tag{4}$$

The model includes couplings of the scalar–isoscalar (Φ_σ), vector–isoscalar ($\Phi_{\omega,\mu}$), and vector–isovector ($\Phi_{\rho,\mu}$) field to the corresponding scalar–isoscalar (ρ^s), vector–isoscalar (ρ^μ) and vector–isovector (ρ^μ) densities of the nucleons. $\mathcal{L}_{nonl}$ is a self–interaction of the scalar–isoscalar field that is needed to give the model a proper density dependence and to reproduce the compressibility of nuclear matter. The coupling constants g_i, the masses of the fields m_i with $i \in \{\sigma, \omega, \rho\}$ and the parameters of $\mathcal{L}_{nonl}$ are the adjustable parameters of the Lagrangian. It turns out that the quality of an actual fit is quite insensitive to the masses of the vector fields

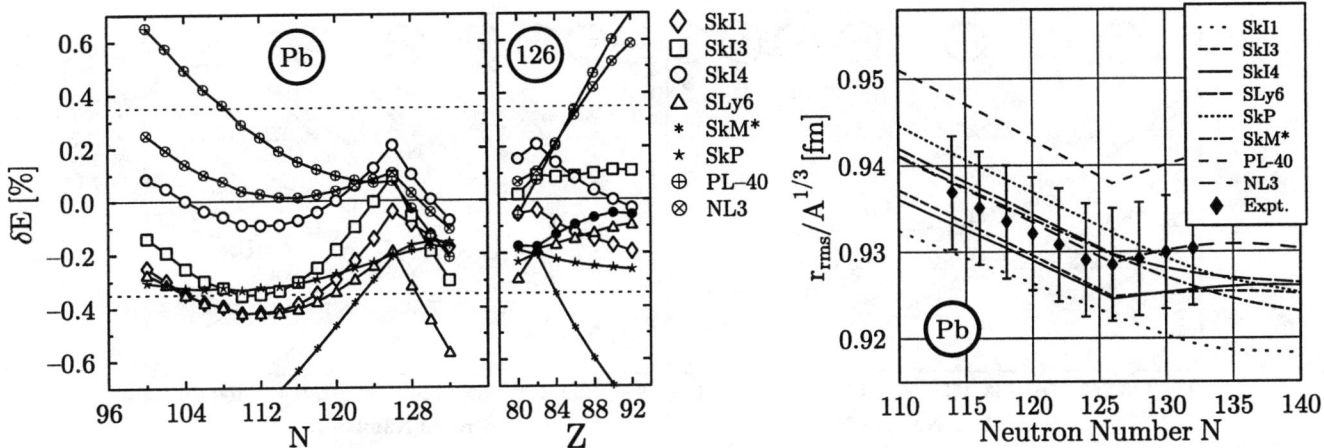

FIGURE 1. Left panel: Relative error (in %) of the binding energies for lead isotopes and $N = 126$ isotones for a selection of typical mean–field parameterizations as indicated. Negative values stand for under-bound nuclei, positive values for over-bound nuclei. Right panel: Rescaled r.m.s. radii of the charge distribution for Pb isotopes.

m_ω, m_ρ, therefore for these the experimental values of the corresponding mesons are used reducing the number of free parameters of the RMF Lagrangian [2].

We consider here the parameterizations NL3 [15] and PL–40 [16]. The force NL3 results from a recent fit including nuclear matter data within the standard *ansatz*. It gives a good description of both nuclear ground states and systematics of giant resonances. The force PL–40 aims at a best fit to nuclear ground–state properties with a stabilized form of the scalar nonlinear self–coupling, that behaves like the standard *ansatz* used for NL3 at typical nuclear scalar densities, but with an overall positive–definite curvature to avoid instabilities at high scalar densities, see [2,16] for details. Like in the case of SHF, the equations–of–motion of the nucleons and fields are derived variationally.

Both SHF and RMF models for the mean field of the nucleons have to be complemented by an appropriate treatment of pair correlations. The residual pairing interaction can be formulated in terms of an effective energy functional as well, unless otherwise noted we use the *ansatz*

$$\mathcal{E}_{\text{pair}} = \frac{1}{4} \sum_q \int \mathrm{d}^3 r \, G_q(r) \, \chi_q^2 \quad , \tag{5}$$

where χ_q is the pairing density including state–dependent cutoff factors to restrict the pairing interaction to the vicinity of the Fermi surface [17]. We consider pairing between like particles only. G_q is the strength of the pairing functional, that is allowed to have a density dependence. For most calculations we employ a simple functional where $G_q(r) = V_q$ is constant and which corresponds to a delta pairing force (DF). In some cases a density–dependent delta interaction (DDDI) is used with $G_q(r) = V_q(1 - \rho(r)/\rho_{\text{n.m.}})$, where $\rho_{\text{n.m.}} = 0.16\,\text{fm}^{-3}$ is the saturation density of nuclear matter. The strengths V_p for protons and V_n for neutrons depend on the actual mean–field parameterization. They are optimized by fitting for each parameterization separately the pairing gaps in isotopic and isotonic chains of semi–magic nuclei throughout the chart of nuclei.

The numerical procedure solves the coupled SHF and RMF equations on a grid in coordinate space with the damped gradient iteration method [18]. The codes for the solution of both SHF and RMF models have been implemented in a common programming environment sharing all the crucial basic routines.

RESULTS AND DISCUSSION

Quality of the Parameterizations

As pointed out above, nowadays mean–field models provide a very good description of the gross properties of stable nuclei. We will demonstrate that here, pars pro toto, for the example of nuclei around doubly magic ^{208}Pb. The left panel in fig. 1 shows the relative error $\delta E = (E_{\text{calc}} - E_{\text{expt}})/E_{\text{expt}}$ in binding energies of the

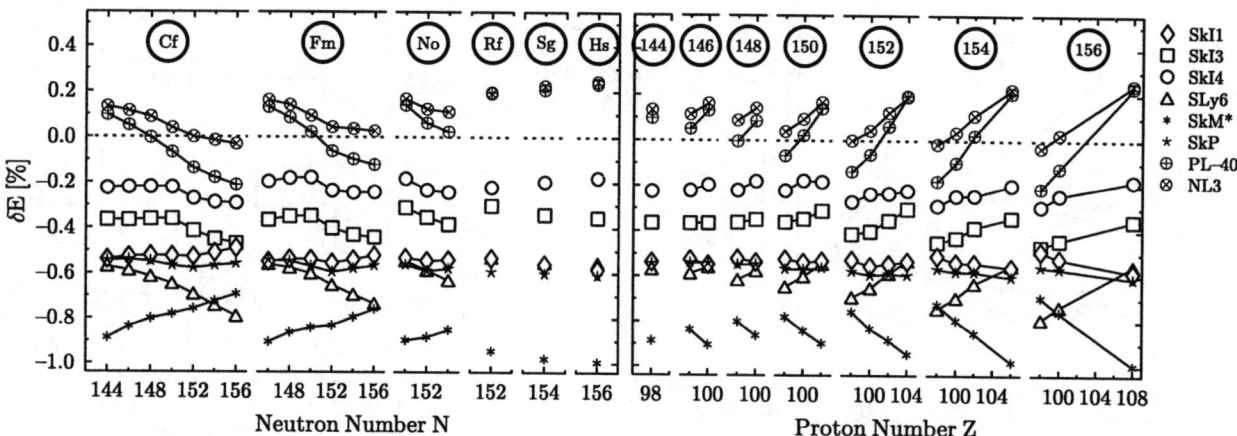

FIGURE 2. Relative error (in %) of binding energies, for a variety of superheavy elements and a selection of typical mean–field parameterizations as indicated. In the left panel, the data are drawn for isotopic chains, while the right panel shows the same data drawn for isotonic chains.

chain of lead isotopes and the chain of $N = 126$ isotones. The data for the neutron–poor lead isotopes and ^{218}U were measured just recently [19], while the other data are taken from [20]. The two horizontal lines at 0.35% indicate the average error in binding energies allowed for good fits [2,9]. One sees that all results remain essentially within these bounds (note that extreme isotopes represent already an extrapolation of the model to short–living nuclei). At second glance, however, one realizes unresolved trends. There are pronounced kinks at the magic numbers $N = 126$ and $Z = 82$. This indicates that the mean–field models produce too large two–nucleon shell gaps for both neutrons and protons in ^{208}Pb. But there are chances to overcome this problem. Research in that direction is underway. Another problem are the large slopes of the errors, which correspond to an error in the two–nucleon separation energies. This problem occurs especially in the RMF parameterizations and the old Skyrme force SkM*. While for SkM* the total binding energy decreases too fast when going to neutron–poor lead isotopes and increases to slow when going to proton–rich $N = 126$ isotones from the fit nucleus ^{208}Pb, this is the other way round for the RMF force PL–40. For NL3 only a slope in the chain of $N = 126$ isotones is visible. This hints that the isovector channel of the RMF is too rigid.

The right panel in Fig. 1 shows rescaled charge r.m.s. radii in comparison with available experimental data (diamonds) taken from [21]. The error bars are to indicate the average error of good fits which amounts to about 0.7% deviation, the error of the experimental values is smaller than the markers used to plot them. Again, most forces remain within these bounds except for SkI1 and PL–40. Again, we see deviations in trends. The data have a pronounced kink at the magic $N = 82$. This kink is not reproduced by all SHF forces in conventional parameterization but emerges correctly in the RMF. The problem in the SHF comes from a too restricted form of the spin–orbit force. An appropriate generalization (inspired be the RMF) solves the problem [9,22] and yields the SHF forces SkI3 and SkI4 which follow the given trend [9]. Note that the kink became apparent only after data for the neutron–rich isotopes were available. ^{214}Pb can already be considered as an exotic nucleus, and the example demonstrates how new data deepen our understanding of mean–field models.

Extrapolation to Superheavy Nuclei

The search for new elements was a strong motivation for heavy–ion physics, and with great experimental effort a large amount of new superheavy isotopes has been accessed [23,24]. Superheavy nuclei are a critical probe for nuclear structure models because the results are very sensitive to details of the models due to the high density of single–particle levels in this region. It is the question how all these parameterizations which are comparable for stable nuclei perform when extrapolating to the region of superheavy nuclei. A first answer comes from comparison of the ground states from deformed mean–field calculations with the recently measured, heaviest known even–even nuclei which are located between the deformed neutron shell closure at $N = 152$ and a region of enhanced stability in the vicinity of $^{270}_{162}$Hs [29]. For this nucleus the occurrence of a doubly deformed shell closure is predicted within macroscopic–microscopic models [25,26]. Fig. 2 shows the relative error in binding energies of these nuclei, the left panel for isotopic chains, the right panel for isotonic chains. It is obvious that

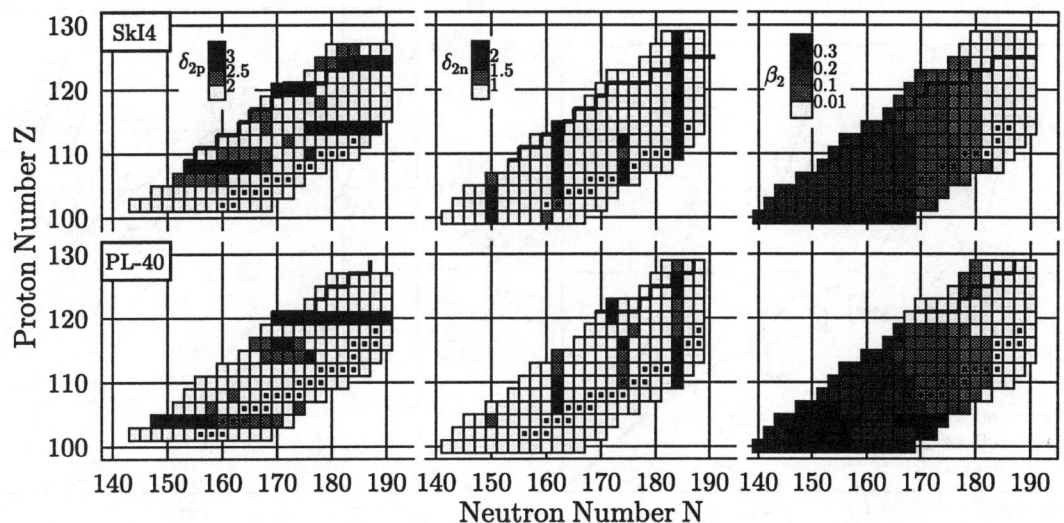

FIGURE 3. Two–proton gaps δ_{2p} (left column), two–neutron gaps δ_{2n} (middle column) and the corresponding relative quadrupole momenta β_2 (right column) in the N–Z plane, resulting from axially deformed calculations with the forces SkI4 and PL–40. Nuclei that are stable with respect to β decay and the two-proton drip-line are emphasized.

in this extrapolation the errors of the various forces spread more than in the case of the lead isotopes shown before. It is gratifying, however, to see that there remain three forces which stay well within the bounds of 0.35% error. It is noteworthy that from the SHF models only SkI4 succeeds which employs the extended form of spin–orbit coupling. As in the case of nuclei around ^{208}Pb, there remain nonetheless unresolved trends, here mainly in the form of non–zero slopes. A slope in δE means that the two–neutron separation energies S_{2n} deviate from the experimental values. Most SHF forces produce flat curves which means that the error has no trend and remains about the same throughout the whole region. All RMF parameterizations, on the other hand, show the same strong isotopic and isotonic trends (in different directions) in the errors of the binding energies that already showed up around ^{208}Pb. From the three well performing forces, SkI4 gives the best two–neutron separation energies S_{2n} [29].

Having preselected three successful forces in the regime of existing superheavy nuclei, we can now try to extrapolate further to even heavier systems. The most interesting question is the location of the next spherical doubly magic system. These can be found by scanning the two–neutron shell gaps $\delta_{2n} = E(N + 2, Z) - 2E(N, Z) + E(N - 2, Z)$ and two–proton shell gaps $\delta_{2p} = E(N, Z + 2) - 2E(N, Z) + E(N, Z - 2)$. Large values indicate a shell closure. We have explored systematically the superheavy nuclei in the range up to $Z = 128$ and $N = 190$ with axially symmetric deformed mean–field calculations with the two preselected forces SkI4 and PL–40 [29]. Fig. 3 shows the shell gaps of protons (left panel) and neutrons (middle panel) as well as the relative quadrupole moment β_2 for the RMF force PL–40 (lower row) and the SHF force SkI4 (upper row). The dimensionless relative quadrupole deformation is defined as $\beta_2 = 4\pi \langle r^2 Y_{20}\rangle/(3Ar_0^2)$ with $r_0 = 1.2\, A^{1/3}$ fm and provides a more immediate geometrical understanding than the quadrupole moment as such. Dark squares indicate large shell gaps, and doubly magic nuclei are to be expected where large neutron and proton shell gaps coincide. For the lower Z and N we have the regime of deformed ground states, and there we find that SkI4 predicts a doubly magic (deformed) ^{270}Hs$^{162}_{108}$ whereas PL–40 has a magic $Z = 104$ proton shell but no convincing accompagning neutron shell. The two forces thus differ significantly in their predictions. There are now first experimental data available on the ground–state deformation of a nucleus in the transfermium region, i.e. for ^{254}No [30]. The predictions of both SkI4 ($\beta_2 = 0.30$) and PL–40 ($\beta_2 = 0.31$) are only slightly larger than the experimental value $\beta_2 = 0.26$.

The difference between the forces persists into the regime of the spherical doubly magic systems where SkI4 points at $Z = 114, N = 184$ but PL–40 at $Z = 120, N = 172$. Other forces like SLy6 or SkI1 predict no doubly spherical magic system in this region at all, while SkM* and SkP predict the even heavier $Z = 126, N = 184$ [28]. This shows that faint differences in the spectral distribution carry forth to substantially different extrapolations for superheavy nuclei. These are thus a critical probe for nuclear structure models.

The full potential energy surface (PES) is also of interest as it allows to estimate the stability against

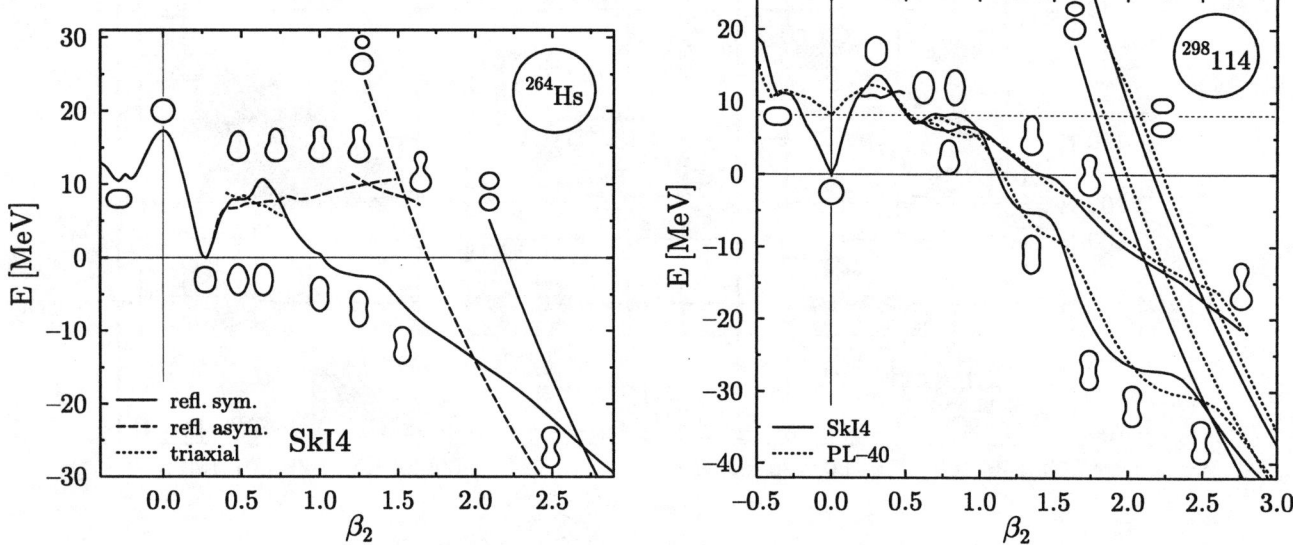

FIGURE 4. Left panel: Valleys in the PES of ^{264}Hs, from calculations in axial symmetry with ("refl. sym.") and without ("refl. asym.") reflection symmetry. In the vicinity of the first barrier also the result from a non–axial calculation ("triaxial") is shown. To give an impression of the nuclear shapes along the path, mass density contours at $\rho_0 = 0.07$ fm^{-3} are drawn near the corresponding curves. Right panel: Valleys in the PES of $^{298}_{184}$114 for SkI4 and PL–40. In the vicinity of the first barrier for SkI4 also the result from a non–axial calculation is shown, that lowers the barrier.

spontaneous fission and the fusion path for the synthesis of these nuclei. As an example, Figure 4 shows the paths of minimum potential energy in the PES of ^{264}Hs, computed with SkI4 and a quadrupole constraint. The higher multipole deformations have been left free to adjust themselves to the minimum configuration, see e.g. [27]. The PES of transfermium nuclei shows significant deviations from the fission barriers of actinides. There is no fission isomer and the second barrier vanishes. The fission path will follow the reflection symmetric solution. Although the first barrier has similar width and height as the first barrier of typical actinide nuclei, the absence of the second barrier will lower the lifetime against spontaneous fission dramatically. The first barrier is a bit lowered if one allows for triaxial configurations. The reflection asymmetric solution does not lower the overall barrier, but it coexists far inside the barriers. It connects the asymptotically separated combination ^{210}Po + ^{54}Cr with the ground state and corresponds to the fusion path. This combination of projectile and target differs only slightly from the experimentally successful choice ^{207}Pb(^{58}Fe, n)^{264}Hs [32]. A calculation in the RMF using PL–40 gives qualitatively the same results [31].

The right panel of Fig.4 shows the paths of minimum potential energy in the PES of $^{298}_{184}$114, calculated with SkI4 (solid line) and PL–40 (dotted line). The PES from calculations with PL–40 is shifted with respect to SkI4 in such a way that the (spurious) shallow symmetric minimum at $\beta_2 \approx 0.6$ has the same energy in both models. For deformations larger than $\beta_2 \approx 0.5$ both forces coincide in their prediction for the PES. The second barrier vanishes if asymmetric shapes are taken into account, but at large deformations the symmetric path is energetically favored. The significant difference between the forces occurs at small deformations $\beta_2 < 0.5$. The binding energy of the spherical configuration, measured from the reference point, is lowered by 7 MeV for SkI4, but raised by approximately 1.3 MeV for PL–40. This is caused by very different shell effects in ^{298}114. SkI4 predicts this nucleus to be doubly magic with strong shell effects in both protons and neutrons, while PL–40 predicts a weak neutron shell closure only. It remains to be noted that around $\beta_2 \approx 0.3$ the barrier is slightly lowered for triaxial shapes, for SkI4 by approximately 3 MeV, while for PL–40 the gain in energy is only a few hundred keV.

Neutron–Rich Tin Isotopes

Another direction is the extrapolation towards neutron–rich isotopes which is particularly important for astrophysical purposes, i.e. the description of nucleosynthesis in stars and the explanation of abundances of

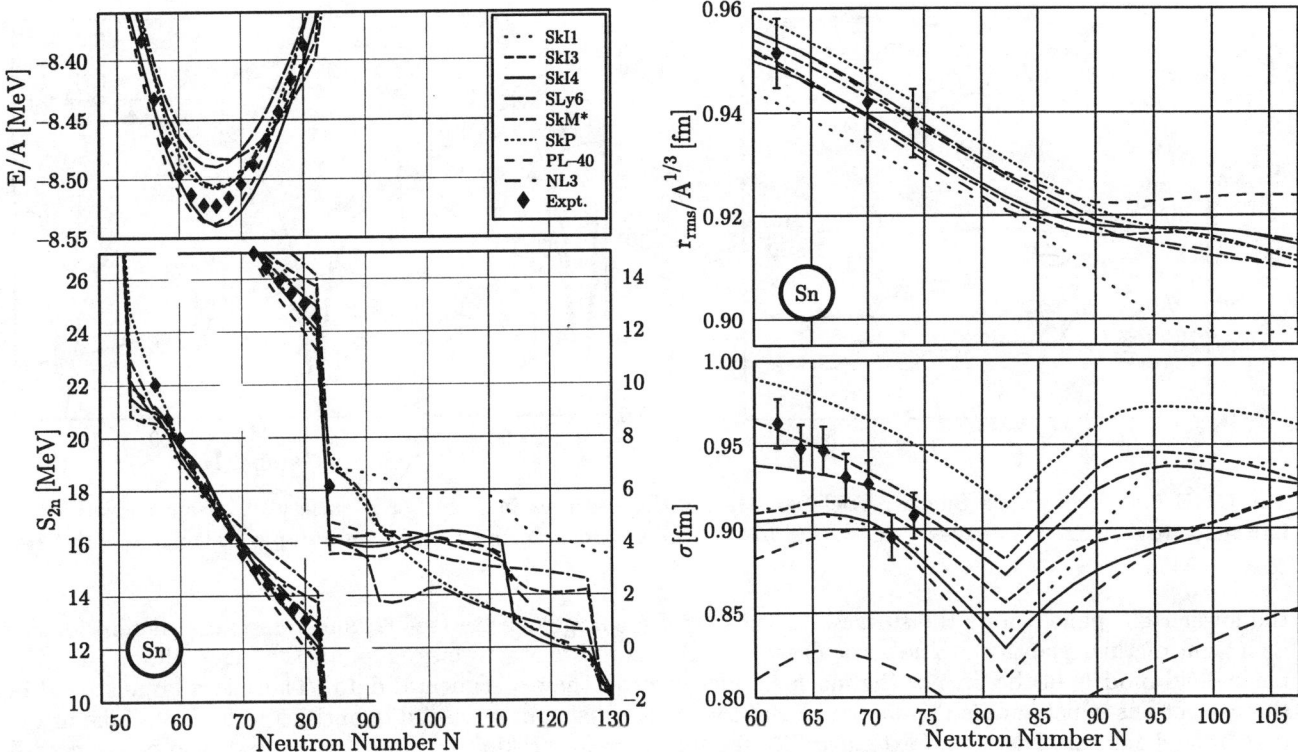

FIGURE 5. Left panel: Binding energy per nucleon E/A (upper panel) and two–neutron separation energy S_{2n} for the chain of tin isotopes calculated with the forces as indicated. Right panel: Root–mean–square radii (upper panel) and surface thickness (lower panel) of the charge density for the chain of tin isotopes calculated with the forces as indicated. Note that the plot of the S_{2n} is splitted.

heavy elements in the solar system or other astrophysical objects.

The left panels of Fig. 5 show the binding energy per nucleon E/A (upper panel) and two–neutron separation energy S_{2n} (lower panel) for the tin isotopes. Note that the plot of the separation energies is splitted at the neutron shell closure at $N = 82$ to enlarge the differences between the various parameterizations. The labels on the left side are for the isotopes at small neutron numbers, the labels on the right side for the neutron–rich isotopes. The large step comes at the magic number $N = 82$ as expected. Left to that we have the regime of stable nuclei where most forces differ only little. But some forces, however show already here deviations from the experimental trend. Like in the case of lead isotopes, SkM*shows the largest deviations from the experimental values of all forces under investigation, predicting too small an increase of the binding energy going from stability towards neutron–rich nuclei, while for PL–40 the increase is somewhat to small. The Skyrme forces SkP, SkI4, SLy6 and the RMF force NL3 give the best reproduction for isotopes with $N < 82$. There is so far only one experimental value for a two–neutron separation energy beyond $N = 82$, and none of the parameterizations is able to reproduce the amplitude of the jump in S_{2n} at the closed neutron shell. For SkP it comes out a little too small, while for all other forces it comes out too large. For larger neutron numbers, we see substantial differences in the predictions, and many of these differences become manifest right after the shell closure. The forces differ in their predictions for the next shell closure and the slope of the S_{2n}. SLy6 predicts a subshell closure at $N = 90$, all RMF forces and Skyrme forces SkI3 and SkI4 with extended spin–orbit coupling predict $N = 112$ as the next neutron shell closure, while most standard Skyrme forces predict $N = 126$. For SkP there appears no significant shell structure at all beyond $N = 82$. This means that we will have useful key data soon when the next exotic Sn isotopes are measured. Note that these differences are hard to see in the plot of the binding energy per nucleon E/A. But this plot gives other useful information: most forces fail to reproduce the exact location of the valley of stability and the binding energy comes out somewhat too small for most of the parameterizations, although the relative errors stay within the 0.35% allowed in the fit.

The upper right panel of Fig. 5 shows the charge r.m.s. radii for a variety of SHF and RMF forces, while

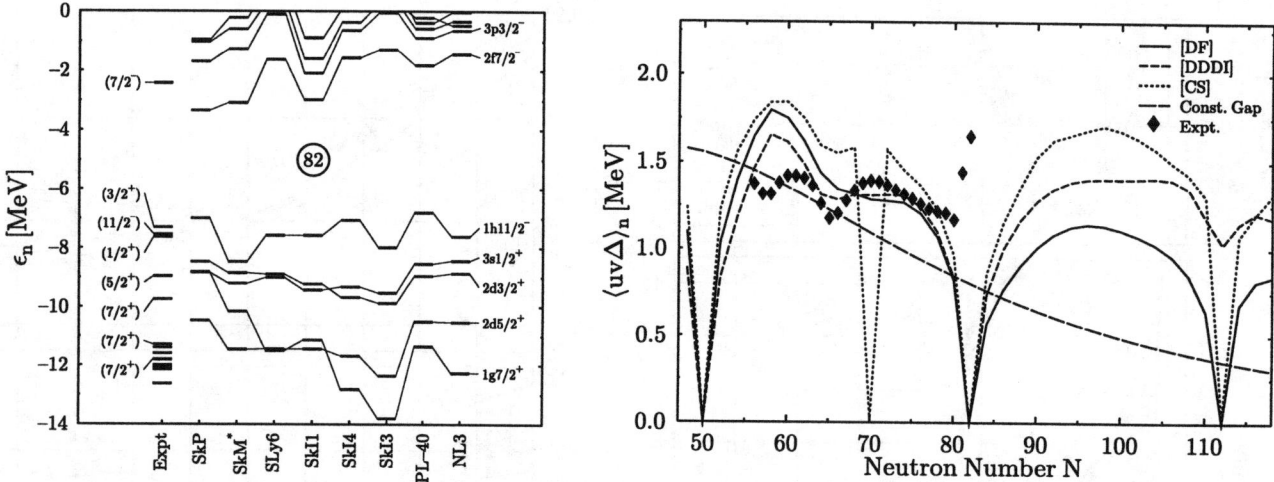

FIGURE 6. Left panel: Single–particle spectra for the neutrons in ^{132}Sn for a variety of parameterizations as indicated. Right panel: Average pairing gap $\langle uv\Delta \rangle_n$ of the neutrons for the chain of tin isotopes, calculated with SkI4.

the lower right panel shows the surface thickness of the charge density (see [2] and references therein for the definition of this quantity). The error drawn for the experimental values is again the error allowed for the mean–field models in the fit, not the much smaller error of the experimental data. There is a large spread in the predictions which makes the observables of the charge distribution useful quantities to look at. Like in the case of lead the relativistic and extended Skyrme forces predict a kink in the r.m.s. radii at the neutron shell closure. While all forces reproduce the experimental data for r.m.s. radii, most of the forces fail to reproduce the surface thickness, especially with NL3 it comes out by far too small. This has yet to be further explored.

The left panel of Fig. 6 compares the single–neutron spectra in ^{132}Sn for a variety of forces with experimental information. The critical detail is the sequence of hole states next to the Fermi surface at $N = 82$. All mean–field models place the $1h_{11/2}$ as first hole state (often with large distance to the next) whereas data show it only as second state. We have tried to refit parameterizations which maintain the same overall quality and fix that problem, but did not succeed. There are indications that this hints to a deeper defect in the energy functional concerning the density dependence of the effective nucleon mass m^*/m.

Pairing

So far we have discussed the quality and extrapolation properties of the mean–field models only. But all mean–field models have to be complemented by the treatment of pairing correlations. For those an effective interaction has to be used as well. The right panel of of Fig. 6 shows the average pairing gaps of the neutrons

$$\langle uv\Delta \rangle_n = \frac{\sum_{k \in \Omega_n} f_k v_k u_k \Delta_k}{\sum_{k \in \Omega_n} f_k u_k v_k} \tag{6}$$

where k is the single–particle expectation value of the pairing potential, see [17] for details, for four different commonly used pairing forces within the BCS approach. "DF" is a pairing energy functional corresponding to a delta force, "DDDI" is a density–dependent delta interaction, and "CS" is the widely used constant strength approach. "Const. gap" is a different approach, where the pairing gap rather than a pairing strength is used as input, which is parameterized in dependence on mass number and relative neutron excess. The experimental values shown are calculated with the five–point formula given in [33]. For CS the level–density dependent parameterization of the pairing strength described in [34] is used, while the pairing strength of the DF and DDDI functionals are fitted to the experimental pairing gaps in several chains of spherical semi–magic nuclei. All forces (DF, DDDI, CS) agree in the region of known nuclei, but cannot reproduce exactly the experimental values. The deviation around $N \approx 60$ is connected to deficiencies in the single–particle spectra in the underlying mean–field force SkI4. The BCS pairing scheme breaks down in nuclei with small level densities, e.g. at magic numbers leading to vanishing $\langle uv\Delta \rangle_n$. The breakdown with the CS force at $N = 70$ corresponds to a subshell closure predicted by SkI4 in this nucleus. Differences between the pairing forces appear immedeatly beyond

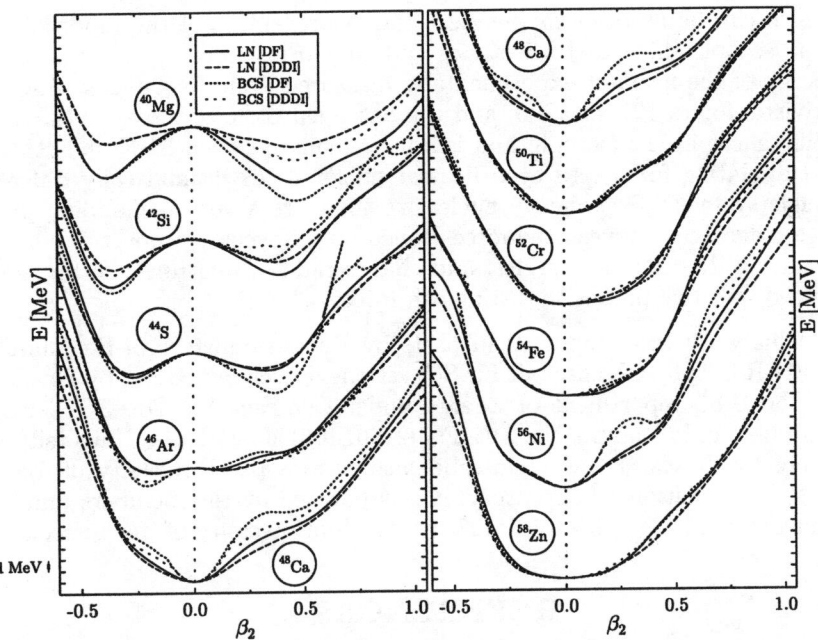

FIGURE 7. Potential energy surfaces for the chain of $N = 28$ isotones, calculated with SkI4 and the pairing models as indicated.

$N = 82$. CS predicts the largest average gaps, DF the smallest ones while DDDI stays in between. This influences both the odd–even mass staggering and the deformation properties of these nuclei and shows the importance of fixing the pairing functionals with the same precision as the mean–field forces. The constant gap approach gives a fair parameterization of the experimental gaps in nuclei close to the valley of stability but deviates already when going to the already known neutron–rich tin isotopes and gives highly questionable results when extrapolated to very neutron–rich systems.

Finally, Fig. 7 shows the potential energy surfaces of the $N = 28$ isotones, calculated with SkI4 and several pairing models. "BCS" indicates the standard BCS treatment of pairing correlations, while "LN" denotes an improved pairing model with approximate particle–number projection before variation within the Lipkin–Nogami scheme, see [34] and references therein. For the pairing interaction, we restrict ourselves here to the DF and DDDI pairing functionals. All heavy $N = 28$ isotones with $Z \geq 20$ have a spherical minimum in the PES, but for the nuclei away from the closed proton shells at $Z = 20$ and $Z = 28$ the PES are rather soft. For the doubly magic nuclei ^{48}Ca and ^{56}Ni there occur significant differences between the pairing models. While the BCS model predicts rather stiff minima, the PES calculated with the LN scheme are much more smeared out. This will have a large effect on excited states. For proton numbers smaller than the magic $Z = 20$ appears a transition from spherical to deformed ground states, but the deformed minima are differently developed. Like in the case of heavier $N = 28$ isotones, the LN scheme smears out the minima. This shows the large influence of the corrections for broken symmetries (in this case uncertainty of the particle number) on observables of small nuclei. Other corrections of the same order are corrections for spurious center–of–mass, rotational and vibrational modes, see e.g. [35] and references therein. Note that the pairing force leads to uncertainties of the same order; comparing the DF and DDDI forces within the same approach, the DF force gives more distinct minima than the DDDI functional. In small nuclei, the differences arising from the pairing force have a similar size as the differences from the mean–field functionals. Therefore the pairing energy functional has to be parameterized with the same care as the mean–field functionals.

CONCLUSIONS

We have investigated the description of nuclei in the framework of relativistic and non–relativistic mean–field models. There is a large selection of SHF and RMF forces which perform very well for the bulk properties of stable nuclei with, e.g., an average error in the energy of 0.35%. There are, however, unresolved isotopic and isotonic trends which can be used for further test and improvement of the models, as demonstrated on the connection between isotopic shifts in Pb isotopes and spin–orbit force. A comparison with the binding and the separation energy of the heaviest experimentally measured even-even nuclei shows a clear preference for the standard relativistic forces PL–40, NL3, and the extended Skyrme forces SkI4 and SkI3. There are, however, significant differences in the isotopic and isotonic trends between SHF and RMF. Taking the pre-selected forces and extrapolating further to even heavier nuclei yields dramatically different predictions for superheavy doubly–magic systems. Superheavy nuclei are therefore a very demanding probe for mean–field models. There are huge differences between the forces concerning isovector properties which become manifest for very neutron–rich nuclei. Results on energies, and charge radii of neutron–rich Sn isotopes demonstrate that these data are indeed sensitive probes to distinguish forces.

Acknowledgments. This work was supported in parts by Bundesministerium für Bildung und Forschung (BMBF), project No. 06 ER 808, by Gesellschaft für Schwerionenforschung (GSI), by Graduiertenkolleg Schwerionenphysik, and by the U.S. Department of Energy under Contract No. DE–FG02–97ER41019 with the University of North Carolina and Contract No. DE–FG02–96ER40963 with the University of Tennessee. The Joint Institute for Heavy Ion Research has as member institutions the University of Tennessee, Vanderbilt University, and the Oak Ridge National Laboratory; it is supported by the members and by the Department of Energy through Contract No. DE–FG05–87ER40361 with the University of Tennessee.

REFERENCES

1. Serot, B. D., and Walecka, J. D., Adv. Nucl. Phys. **16**, 1 (1986).
2. Reinhard, P.–G., *Rep. Prog. Phys.* **52**, 439 (1989).
3. Quentin, P., and Flocard, H., *Ann. Rev. Nucl. Part. Sci.* **28**, 523 (1978).
4. Covello, A., contribution to this volume.
5. Machleidt, R., contribution to this volume.
6. Möller, P., Nix, J. R., Myers, W. D., and Swiatecki, W. R., *At. Data and Nucl. Data Tables* **59**, 185 (1995).
7. Dechargé, J., Gogny, D., *Phys. Rev. C* **21**, 1568 (1980).
8. Nikolaus, B. A., Hoch, T., and Madland, D. G., *Phys. Rev. C* **46**, 1757 (1992).
9. Reinhard, P.–G., and Flocard, H., *Nucl. Phys.* **A584**, 467 (1995).
10. Reinhard, P.–G., and Toepffer, C., *Int. J. of Mod. Phys.* **E3**, 435 (1994).
11. Reinhard, P.–G., *Ann. Phys. (Leipzig)* **1**, 632 (1992).
12. Bartel, J., Quentin, P., Brack, M., Guet, C., and Håkansson, H.–B., *Nucl. Phys.* **A386**, 79 (1982).
13. Dobaczewski, J., Flocard, H., and Treiner, J., *Nucl. Phys.* **A422**, 103 (1984).
14. Chabanat, E., Bonche, P., Haensel, P., Meyer, J., and Schaeffer, R., *Nucl. Phys.* **A635**, 231 (1998).
15. G. A. Lalazissis, J. König, and P. Ring, *Phys. Rev. C* **55**, 540 (1997).
16. Reinhard, P.–G., *Z. Phys.* **A 329**, 257 (1988).
17. Bender, M., Reinhard, P.–G., Rutz, K., and Maruhn, J. A., submitted to *Phys. Rev. C* (1998).
18. Blum, V., Lauritsch, G., Maruhn, J. A., and Reinhard, P.–G., *J. Comp. Phys.* **100**, 364 (1992).
19. Kerscher, T. F., Ph. D. Thesis, Ludwig–Maximilians–Universität, München, 1996.
20. Audi, G., and Wapstra, A. H., *Nucl. Phys.* **A595**, 409 (1995).
21. Fricke, G., et al., *Atomic Data Nucl. Data Tables* **60**, 177 (1995).
22. Sharma, M. M., Lalazissis, G. A., König, J., and Ring, P., *Phys. Rev. Lett.* **74**, 3744 (1995).
23. Hofmann, S., *et al.*, *Z. Phys.* **A350**, 277 (1995), *Z. Phys.* **A350**, 281 (1995), and *Z. Phys.* **A354**, 229 (1996).
24. Lazarev, Yu. A., *et al.*, *Phys. Rev. C* **54**, 620 (1996).
25. Patyk, Z., and Sobiczewski, A., *Nucl. Phys.* **A533**, 132 (1991).
26. Möller, P., and Nix, J. R., *Nucl. Phys.* **A549**, 84, (1992); *J. Phys. G* **20**, 1681, (1994).
27. Rutz, K., Maruhn, J. A., Reinhard, P.–G., and Greiner, W., *Nucl. Phys.* **A590**, 680 (1995).
28. Rutz, K., Bender, M., Bürvenich, T., Schilling, T., Reinhard, P.–G., Maruhn, J. A., and Greiner, W., *Phys. Rev.* **C56**, 238 (1997).
29. Bürvenich, T., Rutz, K., Bender, M., Reinhard, P.–G., Maruhn, J. A., Greiner, W., *EPJ* **A** (1998), in print.
30. Reiter, P., *et al.*, contribution to this volume.
31. Bender, M., Rutz, K., Reinhard, P.–G., Maruhn, J. A., and Greiner, W., *Phys. Rev. C* (1998), in print.
32. Münzenberg, G., *et al.*, *Z. Phys.* **A317**, 235 (1984).
33. Madland, D. G., and Nix, J. R., Nucl. Phys. **A476**, 1 (1988).
34. Reinhard, P.–G., Nazarewicz, W., Bender, M., and Maruhn, J. A., *Phys. Rev. C* **53**, 2776 (1996).
35. Urbano, J. N., Goeke, K., and Reinhard, P.–G., *Nucl. Phys.* **A370**, 329 (1981).

Relativistic mean-field description
of nuclei at the drip-lines

D. Vretenar[1,2], P. Ring[2], and G.A. Lalazissis[2]

[1] *Physics Department, Faculty of Science, University of Zagreb, Croatia*
[2] *Physik-Department der Technischen Universität München, D-85748 Garching, Germany*

Abstract. We present a review of recent applications of the relativistic mean-field theory to the structure of nuclei close to the drip-lines. For systems with extreme isospin values, the relativistic Hartree-Bogoliubov model provides a unified and self-consistent description of mean-field and pairing correlations. The model has been applied in studies of structure phenomena that include: formation of neutron skin and of neutron halos in light nuclei in the mass region above the s-d neutron shell, the strong isospin dependence of the effective spin-orbit interaction and the resulting modification of surface properties, the suppression of the spherical $N = 28$ shell gap for neutron-rich nuclei and the related phenomenon of deformation and shape coexistence, the proton drip-line in the spherical nuclei $14 \leq Z \leq 28$, and ground-state proton radioactivity in the region of deformed nuclei $59 \leq Z \leq 69$.

INTRODUCTION

Experimental and theoretical studies of exotic nuclei with extreme isospin values present one of the most active areas of research in nuclear physics. Experiments with radioactive nuclear beams provide the opportunity to study very short-lived nuclei with very large neutron to proton ratios N/Z. New theoretical models and techniques are being developed in order to describe unique phenomena in nuclei very different from those usually encountered along the line of stability. On the neutron-rich side in particular, exotic phenomena include the weak binding of the outermost neutrons, pronounced effects of the coupling between bound states and the particle continuum, regions of nuclei with very diffuse neutron densities, formation of the neutron skin and halo structures. The modification of the effective nuclear potential produces a suppression of shell effects, the disappearance of spherical magic numbers, and the onset of deformation and shape coexistence. Isovector quadrupole deformations are expected at the neutron drip-lines, and possible low-energy collective isovector modes have been predicted. Because of their relevance to the r-process in nucleosynthesis, nuclei close to the neutron drip-line are also very important in nuclear astrophysics. Their properties determine the astrophysical conditions for the formation of neutron-rich stable isotopes.

Proton-rich nuclei also display many interesting structure phenomena which are important both for nuclear physics and astrophysics. These nuclei are characterized by exotic ground-state decay modes such as direct emission of charged particles and β-decays with large Q-values. The properties of many proton-rich nuclei should also play an important role in the process of nucleosynthesis by rapid-proton capture. In addition to decay properties (particle emission, β-decay), of fundamental importance are studies of atomic masses and separation energies, and especially the precise location of proton drip-lines.

Particularly important are nuclear systems with an equal number of proton and neutrons. Due to the high symmetry between the proton and neutron degrees of freedom, it is possible to study the details of the effective off-shell $n-p$ interaction. Protons and neutrons occupy the same shell-model orbitals and therefore these nuclei present unique systems in which $(S = 0, T = 1)$ and $(S = 1, T = 0)$ pairing can be studied. While for $A \leq 40$ the $N = Z$ nuclei are β-stable, in heavier systems the Coulomb interaction drives the beta-stability line towards neutron-rich isotopes. Heavier $N = Z$ nuclei approach the proton drip-line and display a variety of structure phenomena: shape coexistence, superdeformation, alignment of proton-neutron pairs, proton radioactivity from highly excited states.

CP481, *Nuclear Structure 98*, edited by C. Baktash

Some of the most interesting examples of nuclear systems with large isospin values are provided by the recent experimental results on the synthesis and stability of the heaviest elements. The periodic system has been extended with three new elements $Z = 110, 111, 112$. These elements are found beyond the macroscopic limit of nuclear stability and are stabilized only by quantal shell effects. All the heaviest elements found recently are believed to be well deformed. Still heavier and more neutron-rich elements are expected to be spherical and stabilized by shell effects, which strongly influence the spontaneous fission and alpha-decay half-lives.

While an impressive amount of experimental data on properties of exotic nuclei has been published in the last decade, the development of theoretical models has proceeded at a somewhat slower pace. Significant progress has been reported in the development of mean-field theories and theoretical models which use effective interactions to describe low energy nuclear states. Calculations based on modern Monte Carlo shell-model techniques provide a relatively accurate description of the structure of light and medium-mass exotic nuclei. For medium-heavy and heavy systems the only viable approach at present are large scale self-consistent mean-field calculations (Hartree-Fock, Hartree-Fock-Bogoliubov, relativistic mean-field). However, the present accuracy of model predictions, in general, is not comparable with the precision of modern experimental research facilities, and there are many serious technical difficulties in the calculation of properties of complex nuclear systems at the drip-lines.

Models based on quantum hadrodynamics [1] provide a framework in which the nuclear system is described by interacting baryons and mesons. In comparison with conventional non relativistic descriptions, relativistic models explicitly include mesonic degrees of freedom and consider the nucleons as Dirac particles. A variety of nuclear phenomena have been described in the relativistic framework: nuclear matter, properties of finite spherical and deformed nuclei, hypernuclei, neutron stars, nucleon-nucleus and electron-nucleus scattering, relativistic heavy-ion collisions. In particular, relativistic models based on the mean-field approximation have been successfully applied in the description of properties of spherical and deformed β-stable nuclei, and more recently in studies of exotic nuclei far from the valley of beta stability.

THE RELATIVISTIC HARTREE-BOGOLIUBOV MODEL

Detailed properties of nuclear matter and of finite nuclei along the β-stability line have been very successfully described with relativistic mean-field models [2]. The nucleus is described as a system of Dirac nucleons, which interact in a relativistic covariant manner through the exchange of virtual mesons: the isoscalar scalar σ-meson, the isoscalar vector ω-meson and the isovector vector ρ-meson. The effective Lagrangian density of quantum hadrodynamics reads

$$
\begin{aligned}
\mathcal{L} = &\ \bar{\psi}\left(i\gamma \cdot \partial - m\right)\psi \\
&+ \frac{1}{2}(\partial\sigma)^2 - U(\sigma) - \frac{1}{4}\Omega_{\mu\nu}\Omega^{\mu\nu} + \frac{1}{2}m_\omega^2\omega^2 - \frac{1}{4}\vec{R}_{\mu\nu}\vec{R}^{\mu\nu} + \frac{1}{2}m_\rho^2\vec{\rho}^2 - \frac{1}{4}F_{\mu\nu}F^{\mu\nu} \\
&- g_\sigma\bar{\psi}\sigma\psi - g_\omega\bar{\psi}\gamma \cdot \omega\psi - g_\rho\bar{\psi}\gamma \cdot \vec{\rho}\vec{\tau}\psi - e\bar{\psi}\gamma \cdot A\frac{(1-\tau_3)}{2}\psi .
\end{aligned}
\tag{1}
$$

Vectors in isospin space are denoted by arrows, and bold-faced symbols will indicate vectors in ordinary three-dimensional space. The Dirac spinor ψ denotes the nucleon with mass m. m_σ, m_ω, and m_ρ are the masses of the σ-meson, the ω-meson, and the ρ-meson. g_σ, g_ω, and g_ρ are the corresponding coupling constants for the mesons to the nucleon. $e^2/4\pi = 1/137.036$. The coupling constants and unknown meson masses are parameters, adjusted to fit data of nuclear matter and of finite nuclei. $U(\sigma)$ denotes the non-linear σ self-interaction

$$
U(\sigma) = \frac{1}{2}m_\sigma^2\sigma^2 + \frac{1}{3}g_2\sigma^3 + \frac{1}{4}g_3\sigma^4,
\tag{2}
$$

and $\Omega^{\mu\nu}$, $\vec{R}^{\mu\nu}$, and $F^{\mu\nu}$ are field tensors

$$
\Omega^{\mu\nu} = \partial^\mu\omega^\nu - \partial^\nu\omega^\mu
\tag{3}
$$

$$
\vec{R}^{\mu\nu} = \partial^\mu\vec{\rho}^\nu - \partial^\nu\vec{\rho}^\mu
\tag{4}
$$

$$
F^{\mu\nu} = \partial^\mu A^\nu - \partial^\nu A^\mu.
\tag{5}
$$

The lowest order of the quantum field theory is the *mean-field* approximation: the meson field operators are replaced by their expectation values. The A nucleons, described by a Slater determinant $|\Phi\rangle$ of single-particle spinors ψ_i, $(i = 1, 2, ..., A)$, move independently in the classical meson fields. The sources of the meson fields are defined by the nucleon densities and currents. The ground state of a nucleus is described by the stationary self-consistent solution of the coupled system of Dirac and Klein-Gordon equations. Due to time reversal invariance, there are no currents in the static solution for an even-even system, and therefore the spatial vector components ω, ρ_3 and $\mathbf{A}$ of the vector meson fields vanish. The Dirac equation reads

$$\left\{ -i\boldsymbol{\alpha} \cdot \boldsymbol{\nabla} + \beta(m + g_\sigma \sigma) + g_\omega \omega^0 + g_\rho \tau_3 \rho_3^0 + e\frac{(1 - \tau_3)}{2} A^0 \right\} \psi_i = \varepsilon_i \psi_i \tag{6}$$

The effective mass $m^*(\mathbf{r})$ is defined as

$$m^*(\mathbf{r}) = m + g_\sigma \, \sigma(\mathbf{r}), \tag{7}$$

and the potential $V(\mathbf{r})$ as

$$V(\mathbf{r}) = g_\omega \, \omega^0(\mathbf{r}) + g_\rho \, \tau_3 \, \rho_3^0(\mathbf{r}) + e\frac{(1 - \tau_3)}{2} A^0(\mathbf{r}). \tag{8}$$

In nuclei with odd numbers of protons or neutrons time reversal symmetry is broken. The odd particle induces polarization currents and the time-odd components in the meson fields. These fields play an essential role in the description of magnetic moments and of moments of inertia in rotating nuclei. However, their effect on deformations and binding energies is very small and can be neglected to a good approximation.

In order to describe ground-state properties of open-shell nuclei, pairing correlations have to be taken into account. For nuclei close to the β stability line, pairing has been included in the relativistic mean-field model in the form of a simple BCS approximation [3]. However, for nuclei far from stability the BCS model presents only a poor approximation. In particular, in drip-line nuclei the Fermi level is found close to the particle continuum. The lowest particle-hole or particle-particle modes are often embedded in the continuum, and the coupling between bound and continuum states has to be taken into account explicitly. The BCS model does not provide a correct description of the scattering of nucleonic pairs from bound states to the positive energy continuum [4]; levels high in the continuum become partially occupied. Including the system in a box of finite size leads to unreliable predictions for nuclear radii depending on the size of this box. In the non-relativistic case, a unified description of mean-field and pairing correlations is obtained in the Hartree-Fock-Bogoliubov (HFB) theory [4]. The ground state of a nucleus $|\Phi>$ is represented as the vacuum with respect to independent quasi-particles. The quasi-particle operators are defined by a unitary Bogoliubov transformation of the single-nucleon creation and annihilation operators. The generalized single-particle Hamiltonian of HFB theory contains two average potentials: the self-consistent field $\hat{\Gamma}$ which encloses all the long range ph correlations, and a pairing field $\hat{\Delta}$ which sums up the pp-correlations. The expectation value of the nuclear Hamiltonian $< \Phi|\hat{H}|\Phi >$ can be expressed as a function of the hermitian density matrix ρ and the antisymmetric pairing tensor κ. The variation of the energy functional with respect to ρ and κ produces the single quasi-particle Hartree-Fock-Bogoliubov equations

$$\begin{pmatrix} \hat{h} - \lambda & \hat{\Delta} \\ -\hat{\Delta}^* & -\hat{h} + \lambda \end{pmatrix} \begin{pmatrix} U_k \\ V_k \end{pmatrix} = E_k \begin{pmatrix} U_k \\ V_k \end{pmatrix}. \tag{9}$$

HFB-theory, being a variational approximation, results in a violation of basic symmetries of the nuclear system. Among the most important is the non conservation of the number of particles. In order that the expectation value of the particle number operator in the ground state equals the number of nucleons, equations (9) contain a chemical potential λ which has to be determined by the particle number subsidiary condition. The column vectors denote the quasi-particle wave functions, and E_k are the quasi-particle energies.

If the pairing field $\hat{\Delta}$ is diagonal and constant, HFB reduces to the BCS-approximation. The lower and upper components $U_k(\mathbf{r})$ and $V_k(\mathbf{r})$ are proportional, with the BCS-occupation amplitudes u_k and v_k as proportionality constants. For a more general pairing interaction this will no longer be the case. As opposed to the functions $U_k(\mathbf{r})$, the lower components $V_k(\mathbf{r})$ are localized functions of $\mathbf{r}$, as long as the chemical potential λ is below the continuum limit. Since the densities are bilinear products of $V_k(\mathbf{r})$ (see Eqs.(15)-(18)), the system is always localized. The HFB wave function can be written either in the quasiparticle basis as a product of independent quasi-particle states, or in the *canonical basis* as a highly correlated BCS-state. In the *canonical basis* nucleons occupy single-particle states. If the chemical potential is close to the continuum, the pairing interaction scatters pairs of nucleons into continuum states. Because of additional pairing correlations, particles which occupy those levels cannot evaporate. Mathematically this is expressed by an additional non-trivial transformation which connects the particle operators in the canonical basis to the quasi-particles of the HFB wave function.

The relativistic extension of the HFB theory was introduced in Ref. [5]. In the Hartree approximation for the self-consistent mean field, the Relativistic Hartree-Bogoliubov (RHB) equations read

$$\begin{pmatrix} \hat{h}_D - m - \lambda & \hat{\Delta} \\ -\hat{\Delta}^* & -\hat{h}_D + m + \lambda \end{pmatrix} \begin{pmatrix} U_k(\mathbf{r}) \\ V_k(\mathbf{r}) \end{pmatrix} = E_k \begin{pmatrix} U_k(\mathbf{r}) \\ V_k(\mathbf{r}) \end{pmatrix}. \tag{10}$$

where $\hat{h}_D$ is the single-nucleon Dirac Hamiltonian (6), and m is the nucleon mass. The RHB equations have to be solved self-consistently, with potentials determined in the mean-field approximation from solutions of Klein-Gordon equations

$$\left[-\Delta + m_\sigma^2 \right] \sigma(\mathbf{r}) = -g_\sigma \, \rho_s(\mathbf{r}) - g_2 \, \sigma^2(\mathbf{r}) - g_3 \, \sigma^3(\mathbf{r}) \tag{11}$$

$$\left[-\Delta + m_\omega^2 \right] \omega^0(\mathbf{r}) = g_\omega \, \rho_v(\mathbf{r}) \tag{12}$$

$$\left[-\Delta + m_\rho^2 \right] \rho^0(\mathbf{r}) = g_\rho \, \rho_3(\mathbf{r}) \tag{13}$$

$$-\Delta \, A^0(\mathbf{r}) = e \, \rho_p(\mathbf{r}). \tag{14}$$

for the sigma meson, omega meson, rho meson and photon field, respectively. Due to charge conservation, only the 3rd-component of the isovector rho meson contributes. The source terms in equations (11) to (14) are sums of bilinear products of baryon amplitudes

$$\rho_s(\mathbf{r}) = \sum_{E_k > 0} V_k^\dagger(\mathbf{r}) \gamma^0 V_k(\mathbf{r}), \tag{15}$$

$$\rho_v(\mathbf{r}) = \sum_{E_k > 0} V_k^\dagger(\mathbf{r}) V_k(\mathbf{r}), \tag{16}$$

$$\rho_3(\mathbf{r}) = \sum_{E_k > 0} V_k^\dagger(\mathbf{r}) \tau_3 V_k(\mathbf{r}), \tag{17}$$

$$\rho_{em}(\mathbf{r}) = \sum_{E_k > 0} V_k^\dagger(\mathbf{r}) \frac{1 - \tau_3}{2} V_k(\mathbf{r}), \tag{18}$$

where the sums run over all positive energy states. For M degrees of freedom, for example number of nodes on a radial mesh, the HB equations are $2M$-dimensional and have $2M$ eigenvalues and eigenvectors. To each eigenvector (U_k, V_k) with eigenvalue E_k, there corresponds an eigenvector (V_k^*, U_k^*) with eigenvalue $-E_k$. Since baryon quasi-particle operators satisfy fermion commutation relations, it is forbidden to occupy the levels E_k and $-E_k$ simultaneously. Usually one chooses the M positive eigenvalues E_k for the solution that corresponds to a ground state of a nucleus with even particle number.

The pairing field $\hat{\Delta}$ in (10) is an integral operator with the kernel

$$\Delta_{ab}(\mathbf{r}, \mathbf{r}') = \frac{1}{2} \sum_{c,d} V_{abcd}(\mathbf{r}, \mathbf{r}') \kappa_{cd}(\mathbf{r}, \mathbf{r}'), \tag{19}$$

where a, b, c, d denote quantum numbers that specify the Dirac indices of the spinors, $V_{abcd}(\mathbf{r}, \mathbf{r}')$ are matrix elements of a general two-body pairing interaction, and the pairing tensor is defined

$$\kappa_{cd}(\mathbf{r}, \mathbf{r}') = \sum_{E_k > 0} U_{ck}^*(\mathbf{r}) V_{dk}(\mathbf{r}'). \tag{20}$$

The integral operator $\hat{\Delta}$ acts on the wave function $V_k(\mathbf{r})$:

$$(\hat{\Delta} V_k)(\mathbf{r}) = \sum_b \int d^3 r' \Delta_{ab}(\mathbf{r}, \mathbf{r}') V_{bk}(\mathbf{r}'). \tag{21}$$

The self-consistent solution of the Dirac-Hartree-Bogoliubov integro-differential eigenvalue equations and Klein-Gordon equations for the meson fields determines the nuclear ground state. For systems with spherical symmetry, i.e. single closed-shell nuclei, the coupled system of equations has been solved using finite element methods in coordinate space [6–10], and by expansion in a basis of spherical harmonic oscillator [11–13]. For deformed nuclei the present version of the model does not include solutions in coordinate space. The Dirac-Hartree-Bogoliubov equations and the equations for the meson fields are solved by expanding the nucleon spinors $U_k(\mathbf{r})$ and $V_k(\mathbf{r})$, and the meson fields in terms of the eigenfunctions of a deformed axially symmetric oscillator potential [3]. Of course for nuclei at the drip-lines, solutions in configurational representation might not provide an accurate description of properties that crucially depend on the spatial extension of nucleon densities, as for example nuclear radii.

The eigensolutions of Eq. (10) form a set of orthogonal (normalized) single quasi-particle states. The corresponding eigenvalues are the single quasi-particle energies. The self-consistent iteration procedure is performed in the basis of quasi-particle states. The self-consistent quasi-particle eigenspectrum is then transformed into the canonical basis of single-particle states. The canonical basis is defined to be the one in which the matrix $R_{kk'} = \langle V_k(\mathbf{r}) | V_{k'}(\mathbf{r}) \rangle$ is diagonal. The transformation to the canonical basis determines the energies and occupation probabilities of single-particle states, which correspond to the self-consistent solution for the ground state of a nucleus.

The details of ground-state properties of nuclei at the drip-lines will depend on the coupling constants and masses of the effective Lagrangian, on the pairing interaction and coupling between bound and continuum states. Several parameter sets of the mean-field Lagrangian have been derived that provide a satisfactory description of nuclear properties along the β-stability line. In particular, NL1 [14], NL-SH [15], and NL3 [16]. The effective forces NL1 and NL-SH have been frequently used to calculate properties of nuclear matter and of finite nuclei, and have become standard parameterizations for relativistic mean-field calculations. The parameter set NL3 has been derived more recently [16], by adjusting to ground state properties of a large number of spherical nuclei. Properties calculated with NL3 indicate that this is probably the best RMF effective interaction so far, both for nuclei at and away from the line of β-stability.

In many applications of the relativistic mean-field model pairing correlations have been taken into account in a very phenomenological way in the BCS-theory with monopole pairing force adjusted to the experimental odd-even mass differences. This framework obviously cannot be applied to the description of the coupling to the particle continuum in nuclei close to the drip-line. The question therefore arises, which pairing interaction $V_{abcd}(\mathbf{r}, \mathbf{r}')$ should be used in Eq. (19). In principle, in Ref. [5] a fully relativistic derivation of the pairing force has been developed, starting from the Lagrangian (1). Using the Gorkov factorization technique, it has been possible to demonstrate that the pairing interaction results from the one-meson exchange (σ-, ω- and ρ-mesons). In practice, however, it turns out that the pairing correlations calculated in this way, with coupling constants taken from standard parameter sets of the RMF model, are much too strong. The repulsion produced by the exchange of vector mesons at short distances results in a pairing gap at the Fermi surface that is by a factor three too large. However, as has been argued in many applications of the Hartree-Fock-Bogoliubov theory, there is no real reason to use the same effective forces in both the particle-hole and particle-particle channel. In a first-order approximation, the effective interaction contained in the mean-field $\hat{\Gamma}$ is a G-matrix, the sum over all ladder diagrams. The effective force in the pp channel, i.e. in the pairing potential $\hat{\Delta}$, should be the K matrix, the sum of all diagrams irreducible in pp-direction. Since very little is known about this matrix in the relativistic approach, in most applications of the RHB model a phenomenological pairing interaction has been used, the pairing part of the Gogny force [17],

$$V^{pp}(1, 2) = \sum_{i=1,2} e^{-((\mathbf{r}_1 - \mathbf{r}_2)/\mu_i)^2} (W_i + B_i P^\sigma - H_i P^\tau - M_i P^\sigma P^\tau), \tag{22}$$

with the set D1S [17] for the parameters μ_i, W_i, B_i, H_i and M_i ($i = 1, 2$). This force has been very carefully adjusted to the pairing properties of finite nuclei all over the periodic table. In particular, the basic advantage of the Gogny force is the finite range, which automatically guarantees a proper cut-off in momentum space. The fact that it is a non-relativistic interaction has little influence on the results of RHB calculations. A relativistic pairing force should include both a Lorentz-scalar and a Lorentz-vector part, as for example the interactions derived in Refs. [5]. However, as opposed to the mean-field Γ, where the contribution of the small components in the Dirac spinors is reflected in a very large spin-orbit splitting and where the difference between the scalar density (15) and the vector density (16) leads to saturation, the contribution of these components in the pairing channel can be neglected to a very good approximation. In the pairing channel only levels in the vicinity of the Fermi surface are really important, or in other words, pairing is a completely non-relativistic effect.

In order to test the combination of the NL3 effective interaction for the mean-field Lagrangian, and the Gogny interaction with the parameter D1S in the pairing channel, ground-state properties of a chain of Ni and Sn isotopes have been calculated in Ref. [12]. The analysis has shown that the NL3 effective force provides an accurate description of systems with very different number of neutrons. The correct isospin dependence of NL3 implies that this interaction can be used to make reliable predictions about drip-line nuclei. In Fig. 1 the one- and two-neutron separation energies are shown for the Sn ($50 \leq N \leq 88$) isotopes. The values that correspond to the self-consistent RHB ground-states are compared with experimental data and extrapolated values from Ref. [18]. The theoretical values reproduce in detail the experimental separation energies. The model describes not only the empirical values within one major neutron shell, but it also reproduces the transitions between major shells. The agreement with experimental data is also very good for the Ni isotopes. For the total binding energies, except for the region around ^{60}Ni and for $^{100-102}$Sn, the absolute differences between the calculated and experimental masses are less than 2 MeV. For Ni the model predicts weaker binding for $N \leq 40$. Compared to experimental values, the theoretical binding energies are ≈ 1 MeV larger for neutrons in the $1g_{9/2}$ orbital ($40 \leq N \leq 50$). For Sn isotopes the results of model calculation in general display stronger binding. The differences are somewhat larger for $^{100-102}$Sn, but for these nuclei it is expected that the masses might be strongly affected by proton-neutron residual short-range correlations. Excellent results have also been obtained in the comparison between calculated and experimental neutron and proton rms radii for the Ni and Sn isotopes.

In Ref. [19] it has been shown that constrained RMF calculations with the NL3 effective force reproduce the excitation energies of superdeformed minima relative to the ground-state in ^{194}Hg and ^{194}Pb. In the same work the NL3 interaction was also used for calculations of binding energies and deformation parameters of rare-earth nuclei. All theoretical analyses have shown that the combination NL3 + D1S could be expected to predict relevant results also in the regions of drip-line nuclei. In all the applications of the RHB model which are described in the present review, the NL3 effective force has been used in the mean-field Lagrangian, and the Gogny D1S interaction in the pairing channel.

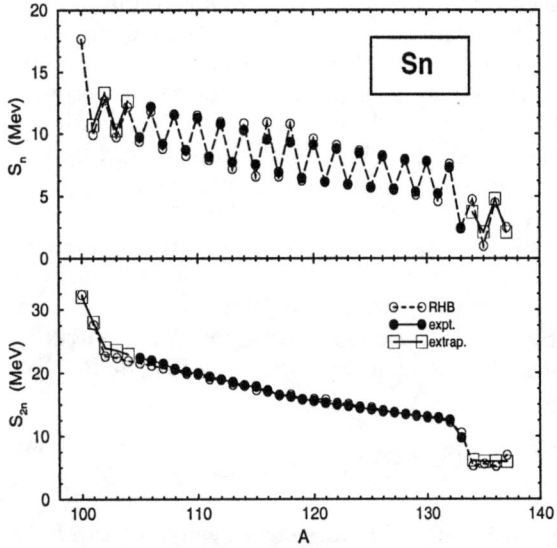

FIGURE 1. One and two-neutron separation energies for Sn isotopes calculated in the RHB model and compared with experimental (filled circles) and extrapolated (squares) data from the compilation of G. Audi and A. H. Wapstra.

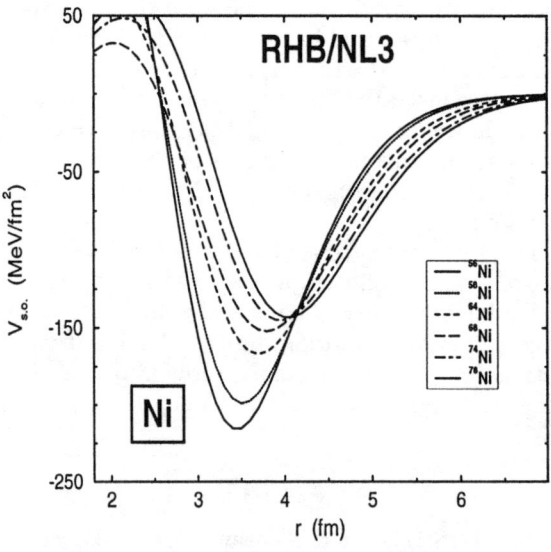

FIGURE 2. Radial dependence of the spin-orbit term of the potential in self-consistent solutions for the ground-states of Ni ($28 \leq N \leq 50$) nuclei.

The spin-orbit interaction plays a central role in nuclear structure. It is rooted in the basis of the nuclear shell model, where its inclusion is essential in order to reproduce the experimentally established magic numbers.

In non-relativistic models based on the mean-field approximation, the spin-orbit potential is included in a phenomenological way. Of course such an ansatz introduces an additional parameter: the strength of the spin-orbit interaction. The value of this parameter is usually adjusted to the experimental spin-orbit splittings in spherical nuclei, for example ^{16}O. On the other hand, in the relativistic framework the nucleons are described as Dirac spinors. This means that in the relativistic description of the nuclear many-body problem, the spin-orbit interaction arises naturally from the Dirac-Lorenz structure of the effective Lagrangian. No additional strength parameter is necessary, and relativistic models reproduce the empirical spin-orbit splittings. It is then of special interest to study the predictions of the relativistic model for the isovector properties of the spin-orbit term of the effective interaction [9]. In the first order approximation, and assuming spherical symmetry, the spin-orbit term can be written as

$$V_{s.o.} = \frac{1}{r}\frac{\partial}{\partial r}V_{ls}(r),$$ (23)

where V_{ls} is the spin-orbit potential [2].

$$V_{ls} = \frac{m}{m_{eff}}(V - S).$$ (24)

V and S denote the repulsive vector and the attractive scalar potentials, respectively. m_{eff} is the effective mass

$$m_{eff} = m - \frac{1}{2}(V - S).$$ (25)

Using the vector and scalar potentials from the self-consistent ground-state solutions, from (23) - (25) the corresponding spin-orbit terms have been computed for the Ni isotopes. They are displayed in Fig. 2 as function of the radial distance from the center of the nucleus. The magnitude of the spin-orbit term $V_{s.o.}$ decreases with the number of neutrons. If ^{56}Ni with ^{78}Ni are compared, the reduction is $\approx 35\%$ in the surface region. This implies a significant weakening of the spin-orbit interaction. The minimum of $V_{s.o.}$ is also shifted outwards, and this reflects the larger spatial extension of the scalar and vector densities, which become very diffuse on the surface. The effect is stronger in light nuclei [9], while it has been shown in Ref. [12] that the reduction of the spin-orbit term seems to be less pronounced in the Sn isotopes. These results indicate that the weakening of the spin-orbit interaction might not be that important in heavy nuclei. The effect is reflected in the calculated energy splittings between spin-orbit partners. The gradual decrease of the energy gap between single-particle levels is consistent with the weakening of the spin-orbit term of the effective interaction.

NEUTRON HALO IN LIGHT NUCLEI

In some loosely bound systems at the drip-lines, the neutron density distribution displays an extremely long tail: the neutron halo. The resulting large interaction cross sections have provided the first experimental evidence for halo nuclei [20]. The neutron halo phenomenon has been studied with a variety of theoretical models [21,22]. For very light nuclei in particular, models based on the separation into core plus valence space nucleons (three-body Borromean systems) have been employed. In heavier neutron-rich nuclei one expects that mean-field models should provide a better description of ground-state properties. In a mean-field description, the neutron halo and the stability against nucleon emission can only be explained with the inclusion of pairing correlations. Both the properties of single-particle states near the neutron Fermi level, and the pairing interaction, are important in the formation of the neutron halo.

The details of the formation of the neutron halo in Ne isotopes have been studied in Ref. [7]. In Fig. 3a the rms radii for Ne isotopes are plotted as functions of neutron number. Neutron and proton rms radii are shown, and the $N^{1/3}$ curve normalized so that it coincides with the neutron radius in ^{20}Ne. The neutron radii follow the mean-field $N^{1/3}$ curve up to $N \approx 22$. For larger values of N the neutron radii display a sharp increase, while the proton radii stay practically constant. This sudden increase in neutron rms radii has been interpreted as evidence for the formation of a multi-particle halo. The phenomenon is also observed in the plot of proton and neutron density distributions 4. The proton density profiles do not change with the number of neutrons, while the neutron density distributions display an abrupt change between ^{30}Ne and ^{32}Ne. The microscopic origin of the neutron halo has been found in a delicate balance of the self-consistent mean-field and the pairing field. This is shown in Fig. 3b, where the neutron single-particle states $1f_{7/2}$, $2p_{3/2}$ and $2p_{1/2}$ in the canonical basis, and the Fermi energy are plotted as function of the neutron number. For $N \leq 22$ the triplet of states is high in the continuum, and the Fermi level uniformly increases toward zero. The triplet approaches zero energy, and a gap is formed between these states and all other states in the continuum. The shell structure dramatically changes at $N \geq 22$. Between $N = 22$ and $N = 32$ the Fermi level is practically constant and very close to the continuum. The addition of neutrons in this region of the drip does not increase the binding. Only the spatial extension of neutron distribution displays an increase. The formation of the neutron halo is related to the quasi-degeneracy of the triplet of states $1f_{7/2}$, $2p_{3/2}$ and $2p_{1/2}$. The pairing interaction promotes neutrons from the $1f_{7/2}$ orbital to the $2p$ levels. Since these levels are so close in energy, the total binding energy does not change significantly. Due to their small centrifugal barrier, the $2p_{3/2}$ and $2p_{1/2}$ orbitals form the halo. A similar mechanism has been suggested in Ref. [23] for the experimentally observed halo in the nucleus ^{11}Li. There the formation of the halo is determined by the pair of neutron levels $1p_{1/2}$ and $2s_{1/2}$. A giant halo has been also predicted for Zirconium isotopes [24]. In that case the halo originates from the neutron orbitals $2f_{7/2}$, $3p_{3/2}$ and $3p_{1/2}$.

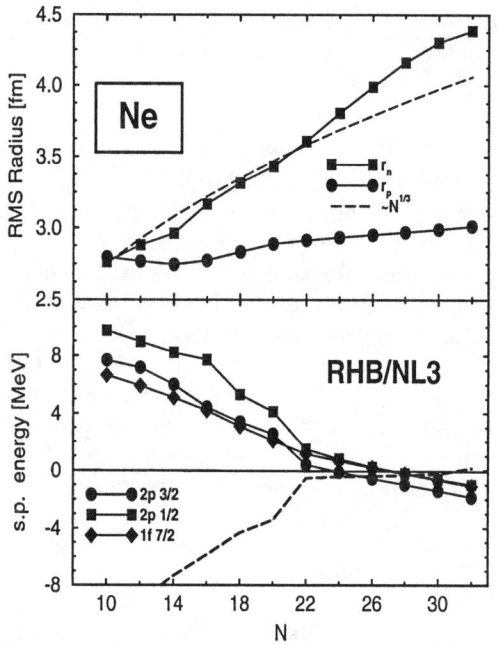

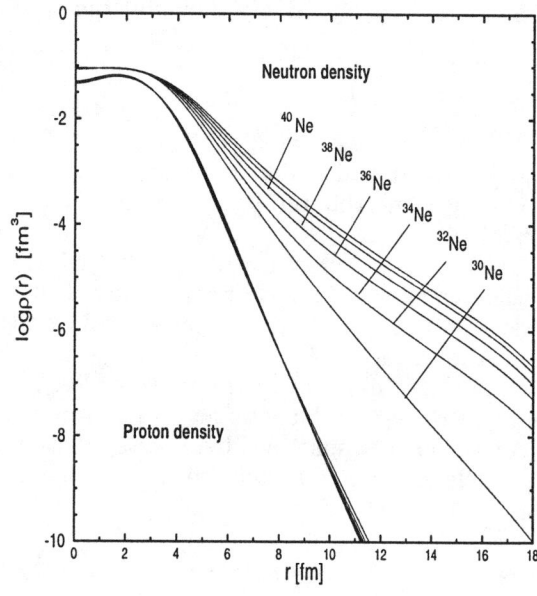

FIGURE 3. Calculated proton and neutron *rms* radii for Ne isotopes (top), and the 1f-2p single-particle neutron levels in the canonical basis (bottom).

FIGURE 4. Proton and neutron density distribution for Ne isotopes.

RHB calculations reported in Ref. [8] have shown that the triplet of single-particle states near the neutron Fermi level: $1f_{7/2}$, $2p_{3/2}$ and $2p_{1/2}$, and the neutron pairing interaction determine the location of the neutron drip-line, the formation of the neutron skin, or eventually of the neutron halo in light nuclei. For C, N, O and F the triplet is still high in the continuum at $N = 20$, and the pairing interaction is to weak to promote pairs of neutrons into these levels. All mean-field effective interactions predict similar results, and the neutron drip is found at $N = 18$ or $N = 20$. For Ne, Na, and Mg the states $1f_{7/2}$, $2p_{3/2}$ and $2p_{1/2}$ are much lower in energy, and for $N \geq 20$ the neutrons populate these levels. The neutron drip can change by as much as twelve neutrons. The model predicts the formation of neutron skin, and eventually neutron halo in Ne and Na. This is due to the fact that the triplet of states is almost degenerate in energy for $N \geq 20$. For Mg the $1f_{7/2}$ lies deeper and neutrons above the $s - d$ shell will exclusively populate this level, resulting in a deformation of the mean field. When the neutron single-particle levels are displayed as function of the number of protons, it turns out that the binding of neutrons is determined by the $1f_{7/2}$ orbital which is found to be parallel to the slope of the Fermi level. The reduction of the spin-orbit splitting $1f_{7/2}$ - $1f_{5/2}$ close to the neutron drip-line has been related to the isospin dependence of the spin-orbit interaction. Density distributions have been analyzed, and the resulting surface thickness and diffuseness parameter reflect the reduction of the spin-orbit term of the effective potential in the surface region of neutron-rich nuclei.

In Ref. [10] an interesting study of Λ-hypernuclei with a large neutron excess has been reported. In particular, the effects of the Λ hyperon on Ne isotopes with neutron halo have been studied. Although the inclusion of the Λ does not produce excessive changes in bulk properties of these nuclei, it can shift the neutron drip by stabilizing an otherwise unbound core nucleus at the drip-line. The microscopic mechanism through which additional neutrons are bound to the core originates from the increase in magnitude of the spin-orbit term in presence of the Λ particle. The Λ in its ground state produces only a fractional change in the central mean-field potential. On the other hand, through a purely relativistic effect, it notably changes the spin-orbit term in the surface region, providing additional binding for the outermost neutrons. The effect can be illustrated on on the example of ^{30}Ne and the corresponding hypernucleus $^{31}_{\Lambda}$Ne. The mean field potential, in which the nucleons move, results from the cancelation of two large meson potentials: the attractive scalar potential S and the repulsive vector potential V: V+S. The spin-orbit potential, on the other hand, arises from the very strong anti-nucleon potential V-S. Therefore, while in the presence of the Λ the changes in V and S cancel out in the mean-field potential, they are amplified in V_{ls}. For the core ^{30}Ne the values of the scalar (S) and vector (V) potential in the center of the nucleus are -380 MeV and 308 MeV, respectively. For $^{31}_{\Lambda}$Ne the corresponding values are: -412 MeV and 336 MeV. The addition of the Λ particle changes the value of the mean-field potential in the center of the nucleus by 4 MeV, but it changes the anti-nucleon potential by 60 MeV. This is reflected in the corresponding spin-orbit term of the effective potential, which provides more binding for states close to the Fermi surface. The additional binding stabilizes the hypernuclear core.

SHAPE COEXISTENCE IN THE DEFORMED N = 28 REGION

The region of neutron-rich $N \approx 28$ nuclei displays many interesting phenomena: the average nucleonic potential is modified, shell effects are suppressed, large quadrupole deformations are observed as well as shape coexistence, isovector quadrupole deformations are predicted at drip-lines. The detailed knowledge of the microscopic structure of these nuclei is also essential for a correct description of the nucleosynthesis of the heavy Ca, Ti and Cr isotopes. In the the RHB model the structure of exotic neutron rich-nuclei with $12 \leq Z \leq 20$ has been studied, and in particular the light $N = 28$ nuclei. Especially interesting is the influence of the spherical shell $N = 28$ on the structure of nuclei below ^{48}Ca, the deformation effects that result from the $1f7/2 \rightarrow fp$ core breaking, and the shape coexistence phenomena predicted for these γ-soft nuclei.

The results of Skyrme Hartree-Fock + BCS and RMF + BCS calculations of Ref. [25] have shown that neutron-rich Si, S and Ar isotopes can be considered as γ-soft, with deformations depending on subtle interplay between the deformed gaps $Z = 16$ and 18, and the spherical gap at $N = 28$. Because of cross-shell excitations to the $2p_{3/2}$, $2p_{1/2}$ and $1f_{5/2}$ shells, the $N = 28$ gap appears to be broken in most cases. In the RMF + BCS analysis of Ref.l [26], a careful study of the phenomenon of shape coexistence was performed for nuclei in this region. It was shown that several Si and S isotopes exhibit shape coexistence: two minima with different deformations occur in the binding energy. The energy difference between the two minima is of the order of few hundred keV. For ^{44}S the calculated difference was only 30 keV. The results obtained with the two models, HF and RMF, were found to be similar, although also important differences were calculated, as for example, the equilibrium shape of the $N = 28$ nucleus ^{44}S. The comparison with results of FRDM and ETFSI calculations, has also shown the importance of a unified and self-consistent description of mean-field and pairing correlations in this region of transitional nuclei.

In Fig. 5 the two-neutron separation energies are plotted for the even-even nuclei $12 \leq Z \leq 24$ and $24 \leq Z \leq 32$. The values that correspond to the self-consistent RHB ground-states (symbols connected by lines) are compared with experimental data and extrapolated values from Ref. [18] (filled symbols). Except for Mg and Si, these isotopes are not at the drip-lines. The theoretical values reproduce in detail the experimental separation energies, except for ^{48}Cr. In general, it has been found that RHB model binding energies are in very good agreement with experimental data when one of the shells (proton or neutron) is closed, or when valence protons and neutrons occupy different major shells (i.e. below and above N and/or $Z = 20$). The differences are more pronounced when both protons and neutrons occupy the same major shell, and especially for the $N = Z$ nuclei. For these nuclei additional correlations should be taken into account, and in particular proton-neutron pairing could have a strong influence on the masses.

The predicted mass quadrupole deformations for the ground states of $N = 28$ nuclei are shown in the upper

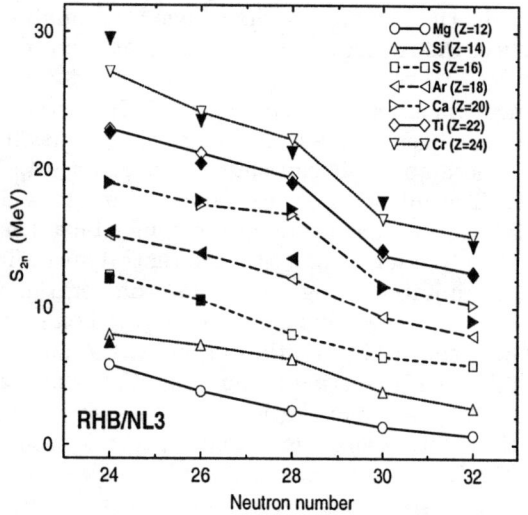

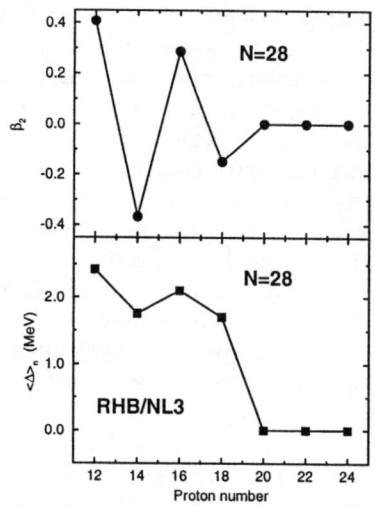

FIGURE 5. Two-neutron separation energies in the $N \approx 28$ region calculated in the RHB model and compared with experimental data (filled symbols) from the compilation of G. Audi and A. H. Wapstra.

FIGURE 6. Self-consistent RHB quadrupole deformations for ground-states of the $N = 28$ isotones (top). Average neutron pairing gaps $< \Delta_N >$ as function of proton number (bottom).

panel of Fig. 6. The staggering between prolate and oblate configurations indicates that the potential is γ-soft. The absolute values of the deformation decrease towards the $Z = 20$ closed shell. Starting with Ca, the $N = 28$ nuclei are spherical in the ground state. The calculated quadrupole deformations are in agreement with previously reported theoretical results [25] (prolate for $Z = 16$, oblate for $Z = 18$), and with available experimental data: $|\beta_2| = 0.258(36)$ for ^{44}S [27,28], and $|\beta_2| = 0.176(17)$ for ^{46}Ar [29]. Experimental data (energies of 2_1^+ states and $B(E2; 0_{g.s.}^+ \to 2_1^+)$ values) do not determine the sign of deformation, i.e. do not differentiate between prolate and oblate shapes. In the lower panel of Fig. 6 the average values of the neutron pairing gaps for occupied canonical states are displayed. $< \Delta_N >$ provides an excellent quantitative measure of pairing correlations. The calculated values of $< \Delta_N > \approx 2$ MeV correspond to those found in open-shell Ni and Sn isotopes [12]. The spherical shell closure $N = 28$ is strongly suppressed for nuclei with $Z \leq 18$, and only for $Z \geq 20$ neutron pairing correlations vanish.

In the fully microscopic and self-consistent RHB model, the details of the single-neutron levels can be analyzed, and the formation of minima in the binding energy can be studied. Figs. 7-9 display the single-neutron levels in the canonical basis for the $N = 28$ nuclei ^{42}Si, ^{44}S, and ^{46}Ar, respectively. The single-neutron eigenstates of the density matrix result from constrained RHB calculations performed by imposing a quadratic constraint on the quadrupole moment. The canonical states are plotted as function of the quadrupole deformation, and the dotted curve denotes the position of the Fermi level. In the insert the corresponding total binding energy curve as function of the quadrupole moment is shown. For ^{42}Si the binding energy displays a deep oblate minimum ($\beta_2 \approx -0.4$). The second, prolate minimum is found at an excitation energy of ≈ 1.5 MeV. Shape coexistence is more pronounced for ^{44}S. The ground state is prolate deformed, the calculated deformation in excellent agreement with experimental data [27,28]. The oblate minimum is found only ≈ 200 keV above the ground state. Finally, for the nucleus ^{46}Ar a very flat energy surface is found on the oblate side. The deformation of the ground-state oblate minimum agrees with experimental data [29], the spherical state is only few keV higher. It is also interesting to observe how the spherical gap between the $1f_{7/2}$ orbital and the $2p_{3/2}$, $2p_{1/2}$ orbitals varies with proton number. While the gap is really strongly reduced for ^{42}Si and ^{44}S, in the $Z = 18$ isotone ^{46}Ar the spherical gap is ≈ 4 MeV. Of course from ^{48}Ca the $N = 28$ nuclei become spherical. Therefore the single-neutron canonical states in Figs. 7-9 clearly display the disappearance of the spherical $N = 28$ shell closure for neutron-rich nuclei below $Z = 18$.

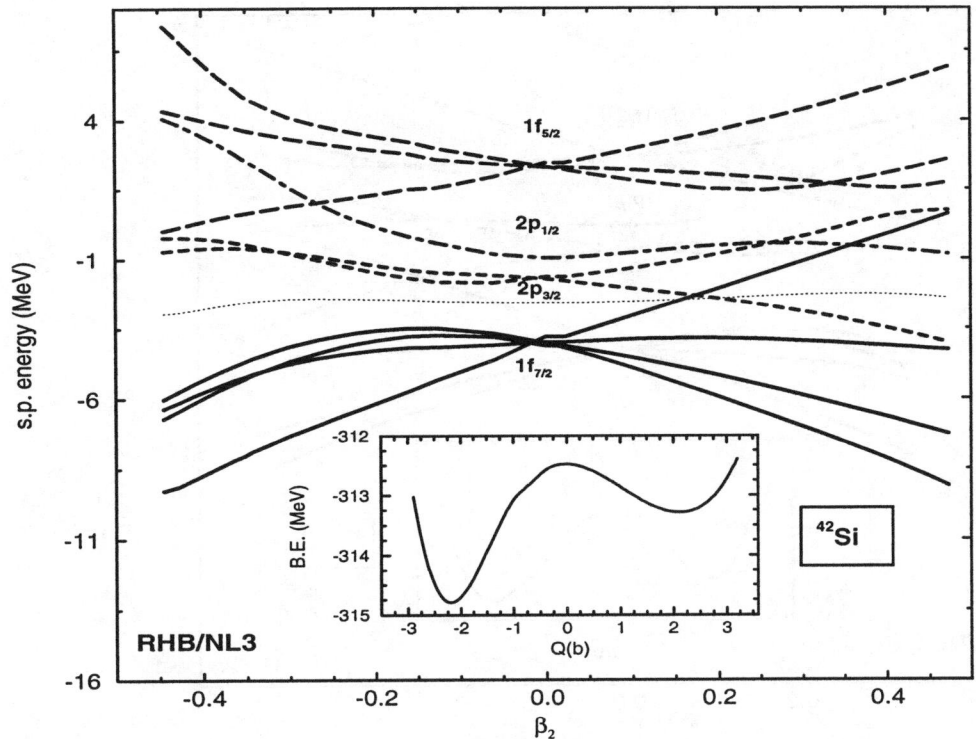

FIGURE 7. The neutron single-particle levels for ^{42}Si as function of the quadrupole deformation. The energies in the canonical basis correspond to ground-state RHB solutions with constrained quadrupole deformation. The dotted line denotes the neutron Fermi level. In the insert the corresponding total binding energy curve is shown.

NUCLEI AT THE PROTON DRIP-LINES

The decay by direct proton emission provides the opportunity to study the structure of systems beyond the drip-line. The phenomenon of ground-state proton radioactivity is determined by a delicate interplay between the Coulomb and centrifugal terms of the effective potential. While low-Z nuclei lying beyond the proton drip-line exist only as short lived resonances, the relatively high potential energy barrier enables the observation of ground-state proton emission from medium-heavy and heavy nuclei. At the drip-lines proton emission competes with β^+ decay; for heavy nuclei also fission or α decay can be favored. The proton drip-line has been fully mapped up to $Z = 21$, and possibly for odd-Z nuclei up to In [30]. No examples of ground-state proton emission have been discovered below $Z = 50$.

The RHB model with finite range-pairing has been applied in the study of ground-state properties of spherical even-even nuclei $14 \leq Z \leq 28$ and $N = 18, 20, 22$ [13]. While for these neutron numbers the nuclei with $14 \leq Z \leq 20$ are not really very proton-rich, nevertheless they are useful for a comparison of the model calculations with experimental data. Of particular interest are the predictions of the model for the proton-rich nuclei in the $1f_{7/2}$ region. These nuclei have recently been extensively investigated in experiments involving fragmentation of ^{58}Ni. The principal motivation of many experimental studies in this region is the possible occurrence of the two-proton ground-state radioactivity. In particular, the region around ^{48}Ni is expected to contain nuclei which are two-proton emitters. On the other hand, because of the Coulomb barrier at the proton drip-line, the emission of a pair of protons may be strongly delayed for nuclei with small negative two-proton separation energies.

The two-proton separation energies that correspond to the self-consistent RHB ground-states for the even-even nuclei $14 \leq Z \leq 28$ and $N = 18, 20, 22$, have been compared with experimental data and extrapolated values from Ref. [18]. The theoretical values reproduce in detail the experimental separation energies, except for ^{38}Ca and ^{44}Ti. In Table 1 the calculated total binding energies for the $N = 18, 20, 22$ isotones are displayed in comparison with empirical values. As has been already shown in the previous section for the neutron rich

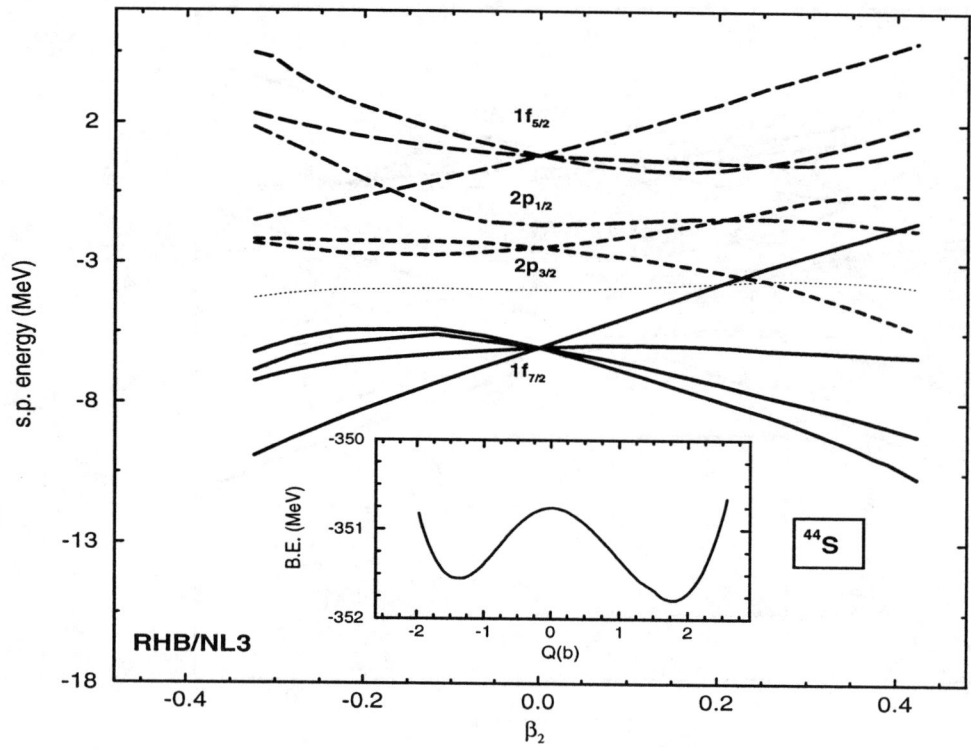

FIGURE 8. The same as in Fig. 7, but for ^{44}S.

$N = 28$ nuclei, the RHB model results are in very good agreement with experimental data when one of the shells (proton or neutron) is closed, or when valence protons and neutrons occupy different major shells (i.e. bellow and above N and/or $Z = 20$). The absolute differences between the calculated and experimental masses are less than 2 MeV. The differences are larger when both proton and neutron valence particles (holes) occupy the same major shell, and especially for the $N = Z$ nuclei ^{36}Ar and ^{44}Ti.

TABLE 1. Comparison between calculated and empirical binding energies. All values are in units of MeV; empirical values are displayed in parentheses.

^{32}Si	269.02 (271.41)	^{40}Ar	343.97 (343.81)	^{44}Cr	351.65 (349.99)
^{34}Si	284.42 (283.43)	^{38}Ca	313.11 (313.04)	^{46}Cr	380.19 (381.98)
^{36}Si	293.08 (292.02)	^{40}Ca	341.99 (342.05)	^{44}Fe	312.07 (-)
^{34}S	288.10 (291.84)	^{42}Ca	362.95 (361.90)	^{46}Fe	352.25 (350.20)
^{36}S	307.98 (308.71)	^{40}Ti	315.39 (314.49)	^{48}Fe	384.42 (385.19)
^{38}S	320.77 (321.05)	^{42}Ti	348.35 (346.91)	^{46}Ni	306.72 (-)
^{36}Ar	302.52 (306.71)	^{44}Ti	373.15 (375.47)	^{48}Ni	349.92 (-)
^{38}Ar	327.34 (327.06)	^{42}Cr	314.94 (314.20)	^{50}Ni	385.52 (385.50)

The RHB results should be also compared with recently reported self-consistent mean-field calculations of Ref. [31], and with properties of proton-rich nuclei calculated with the shell model [32]. The calculations of Ref. [31] have been performed for several mean-field models (Hartree-Fock, Hartree-Fock-Bogoliubov, and relativistic mean-field), and for a number of effective interactions. The results systematically predict the two-proton drip-line to lie between ^{42}Cr and ^{44}Cr, ^{44}Fe and ^{46}Fe, and ^{48}Ni and ^{50}Ni. Recent studies of proton drip-line nuclei in this region have been performed in experiments based on ^{58}Ni fragmentation on a beryllium target, and evidence has been reported for particle stability of ^{50}Ni [33]. In the shell-model calculations of

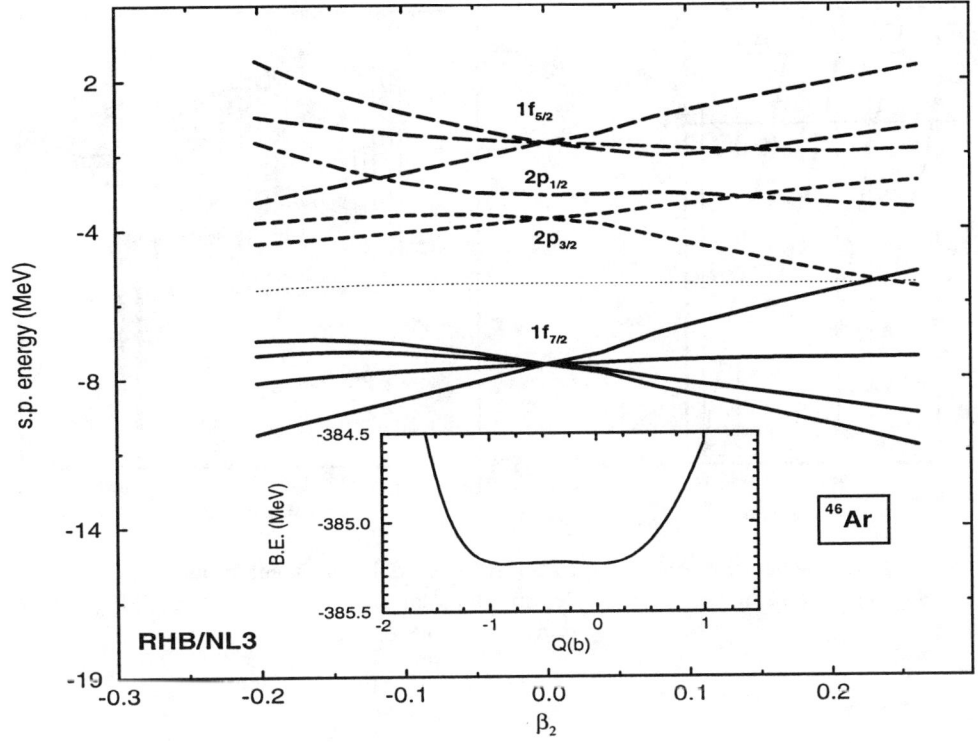

FIGURE 9. The same as in Fig. 7, but for ^{46}Ar.

Ref. [32] absolute binding energies were evaluated by computing the Coulomb energy shifts between mirror nuclei, and adding this shift to the experimentally determined binding energy of the neutron rich isotope. The calculated two-proton separation energies predicted a proton drip-line in agreement with experimental data and with the mean-field results [31]. Compared to the RHB results, the shell-model total binding energies are in somewhat better agreement with experimental data. However, the two models give almost identical values for the extracted two-proton separation energies of the drip-line nuclei. The self-consistent RHB NL3+D1S two-proton separation energies at the drip-line are also very close to the values that result from non-relativistic HFB+Gogny (D1S) calculation of Ref. [31].

Proton radioactivity in odd-Z nuclei has been investigated in the two spherical regions from $51 \leq Z \leq 55$ and $69 \leq Z \leq 83$. The systematics of proton decay spectroscopic factors is consistent with half-lives calculated in the spherical WKB or DWBA approximations. Recently reported proton decay rates [34] indicate that the missing region of light rare-earth nuclei contains strongly deformed nuclei at the drip-lines. The lifetimes of deformed proton emitters provide direct information on the last occupied Nilsson configuration, and therefore on the shape of the nucleus. Modern models for proton decay rates from deformed nuclei have only recently been developed. However, even the most realistic calculations are not based on a fully microscopic and self-consistent description of proton unstable nuclei. In particular, such a description should also include important pairing correlations.

Fig. 10 displays the one-proton separation energies for the odd-Z nuclei $59 \leq Z \leq 69$, as function of the number of neutrons. The model predicts the drip-line nuclei: ^{124}Pr, ^{129}Pm, ^{134}Eu, ^{139}Tb, ^{146}Ho, and ^{152}Tm. In heavy proton drip-line nuclei the potential energy barrier, which results from the superposition of the Coulomb and centrifugal potentials, is relatively high. For the proton decay to occur the odd valence proton must penetrate the potential barrier, and this process competes with β^+ decay. Since the half-life for proton decay is inversely proportional to the energy of the odd proton, in many nuclei the decay will not be observed immediately after the drip-line. Proton radioactivity is expected to dominate over β^+ decay only when the energy of the odd proton becomes relatively high. This is also a crucial point for the relativistic description of proton emitters, since the precise values of the separation energies depend on the isovector properties of the spin-orbit interaction.

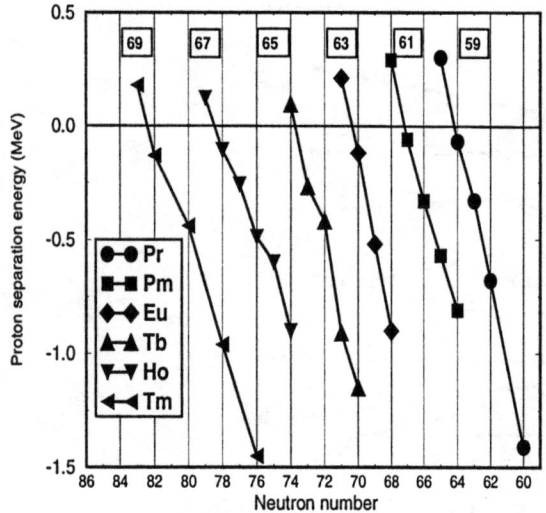

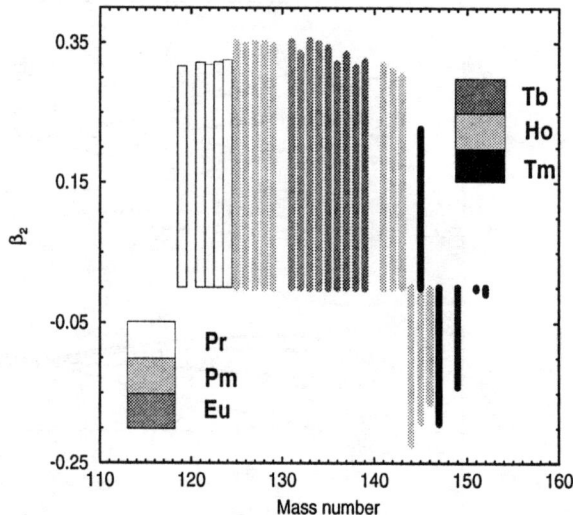

FIGURE 10. Calculated one-proton separation energies for odd-Z nuclei $59 \leq Z \leq 69$ at and beyond the drip-line.

FIGURE 11. Self-consistent ground-state quadrupole deformations for the odd-Z nuclei $59 \leq Z \leq 69$ at the proton drip-line.

The calculated separation energies should be compared with recently reported experimental data on proton radioactivity from ^{131}Eu, ^{141}Ho [34], ^{145}Tm [35], and ^{147}Tm [36]. The ^{131}Eu transition has an energy $E_p = 0.950(8)$ MeV and a half-life 26(6) ms, consistent with decay from either $3/2^+$[411] or $5/2^+$[413] Nilsson orbital. For ^{141}Ho the transition energy is $E_p = 1.169(8)$ MeV, and the half-life 4.2(4) ms is assigned to the decay of the $7/2^-$[523] orbital. The calculated proton separation energy, both for ^{131}Eu and ^{141}Ho, is of -0.9 MeV. In the RHB calculation for ^{131}Eu the odd proton occupies the $5/2^+$[413] orbital, while the ground state of ^{141}Ho corresponds to the $7/2^-$[523] proton orbital. This orbital is also occupied by the odd proton in the calculated ground states of ^{145}Tm and ^{147}Tm. The calculated proton separation energies: -1.46 MeV in ^{145}Tm, and -0.96 MeV in ^{147}Tm, are compared with the experimental values for transition energies: $E_p = 1.728(10)$ MeV in ^{145}Tm, and $E_p = 1.054(19)$ MeV in ^{147}Tm. When compared with spherical WKB or DWBA calculations [37], the experimental half-lives for the two Tm isotopes are consistent with spectroscopic factors for decays from the $h_{11/2}$ proton orbital. Though the predicted RHB ground-state configuration $7/2^-$[523] indeed originates from the spherical $h_{11/2}$ orbital, the two nuclei are deformed. ^{145}Tm has a prolate quadrupole deformation $\beta_2 = 0.23$, and ^{147}Tm is oblate in the ground-state with $\beta_2 = -0.19$. Calculations also predict possible proton emitters ^{136}Tb and ^{135}Tb with separation energies -0.90 MeV and -1.15 MeV, respectively. In both isotopes the predicted ground-state proton configuration is $3/2^+$[411].

The calculated mass quadrupole deformation parameters for the odd-Z nuclei $59 \leq Z \leq 69$ at and beyond the drip-line are shown in Fig. 11. Pr, Pm, Eu and Tb isotopes are strongly prolate deformed ($\beta_2 \approx 0.30-0.35$). By increasing the number of neutrons, Ho and Tm display a transition from prolate to oblate shapes. The absolute values of β_2 decrease towards the spherical solutions at $N = 82$. The quadrupole deformations calculated in the RHB model with the NL3 effective interaction, are found in excellent agreement with the predictions of the macroscopic-microscopic mass model [38].

A detailed analysis of single proton levels, including spectroscopic factors, can be performed in the canonical basis which results from the fully microscopic and self-consistent RHB calculations. For the Eu isotopes this is illustrated in Fig. 12, where the proton single-particle energies in the canonical basis are shown as function of the neutron number. The thick dashed line denotes the position of the Fermi level. In particular, for the proton emitter ^{131}Eu, the ground-state corresponds to the odd valence proton in the $5/2^+$[413] orbital.

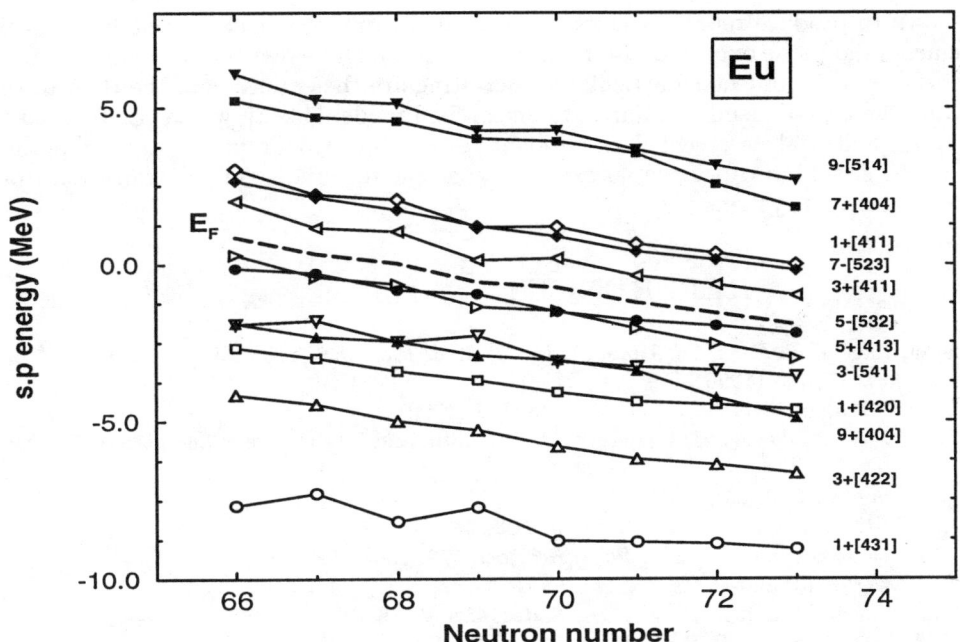

FIGURE 12. The proton single-particle levels for the Eu isotopes. The dashed line denotes the position of the Fermi level. The energies in the canonical basis correspond to ground-state solutions calculated with the NL3 effective force of the mean-field Lagrangian. The parameter set D1S is used for the finite range Gogny-type interaction in the pairing channel.

CONCLUSIONS

Models based on the relativistic mean-field approximation provide a microscopically consistent, and yet simple and economical treatment of the nuclear many-body problem. By adjusting just a few model parameters, coupling constants and effective masses, to global properties of simple, spherical and stable nuclei, it has been possible to describe in detail a variety of nuclear structure phenomena over the whole periodic table, from light nuclei to superheavy elements. When also pairing correlations are included in the self-consistent Hartree-Bogoliubov framework, the relativistic mean-field theory can be applied to the physics of exotic nuclei at the drip-lines. In the present review recent applications of the relativistic Hartree-Bogoliubov model have been described: description of the neutron drip-line in light nuclei, and of the microscopic mechanism for the formation of neutron halos; Λ-hypernuclei with large neutron excess; the study of the reduction of the effective spin-orbit interaction and the resulting modification of surface properties in drip-line nuclei; the suppression of shell effects and the onset of deformation and shape coexistence; the proton drip-line, and ground-state proton emitters in the region of deformed rare-earth nuclei.

The excellent results reported in these studies clearly reflect the basic advantages of the relativistic models over the more traditional non-relativistic approaches to nuclear structure: (i) models based on quantum hadrodynamics are more fundamental and they explicitly include mesonic degrees of freedom; (ii) they incorporate important relativistic effects, for instance the strong scalar and vector potentials and the resulting spin-orbit interaction. Of course, at the present stage, the most successful relativistic models are phenomenological. Although important results have been reported in the microscopic derivation of the low-energy hadronic Lagrangian, it is still not possible to bridge the gap between QCD as the underlying theory of the strong interaction, and the effective theories that have to be used in the nuclear many-body problem. In applications to nuclear structure phenomena, one would also like to see a more consistent treatment of pairing correlations. The relativistic theory of pairing presents a very active area of research. However, only phenomenological effective forces have been shown to produce reliable results when applied to finite nuclei, especially in exotic regions. Important improvements should be expected also from models that include density dependent interactions and describe nucleons as composite objects. Particularly interesting are the extension of the RPA to the relativistic framework [39], and the time-dependent relativistic mean-field model [40,41], which can be used to describe excited collective states and analyze recent data on giant resonances at drip-lines. Large amplitude collective motion in exotic nuclei should also provide an excellent example for studies of the transition from regular to chaotic dynamics in quantum systems [42].

REFERENCES

1. B.D. Serot and J.D. Walecka, *Adv. Nucl. Phys.* **16**, 1 (1986); *Int. J. Mod. Phys.* **E6**, 515 (1997).
2. P. Ring, *Progr. Part. Nucl. Phys.* **37**, 193 (1996).
3. Y.K. Gambhir, P. Ring, and A. Thimet, *Ann. Phys. (N.Y.)* **198**, 132 (1990).
4. J. Dobaczewski, W. Nazarewicz, T. R. Werner, J. F. Berger, C. R. Chinn, and J.Decharge, *Phys. Rev. C* **53**, 2809 (1996).
5. H. Kucharek and P. Ring, *Z. Phys. A* **339**, 23 (1991).
6. W. Pöschl, D. Vretenar and P. Ring, *Comput. Phys. Commun.* **103**, 217 (1997).
7. W. Pöschl, D. Vretenar, G.A. Lalazissis, and P. Ring, *Phys. Rev. Lett.* **79**, 3841 (1997).
8. G.A. Lalazissis, D. Vretenar, W. Pöschl, and P. Ring, *Nucl. Phys.* **A632**, 363 (1998).
9. G.A. Lalazissis, D. Vretenar, W. Pöschl, and P. Ring, *Phys. Lett.* **B418**, 7 (1998).
10. D. Vretenar, W. Pöschl, G.A. Lalazissis, and P. Ring, *Phys. Rev. C* **57**, R1060 (1998).
11. T. Gonzalez-Llarena, J.L. Egido, G.A. Lalazissis, and P. *Ring, Phys. Lett.* **B379**, 13 (1996).
12. G.A. Lalazissis, D. Vretenar, and P. Ring, Phys. Rev. C **57**, 2294 (1998).
13. D. Vretenar, G.A. Lalazissis, and P. Ring, *Phys. Rev. C* **57**, 3071 (1998).
14. P.G. Reinhard, M.Rufa, J. Maruhn, W. Greiner and J. Friedrich, *Z. Phys. A* **323**, 13 (1986).
15. M.M. Sharma, M.A. Nagarajan, and P. Ring, *Phys. Lett.* **B312**, 377 (1993).
16. G.A. Lalazissis, J. König, and P. Ring, *Phys. Rev. C* **55**, 540 (1997).
17. J. F. Berger, M. Girod and D. Gogny, *Nucl. Phys.* **A428**, 32 (1984).
18. G. Audi and A. H. Wapstra, *Nucl. Phys.* **A595**, 409 (1995).
19. G.A. Lalazissis and P. Ring, *Phys. Lett.* **B427**, 225 (1998).
20. I. Tanihata et al., *Phys. Rev. Lett.* **55**, 2676 (1985); *Phys. Lett.* **B206**, 592 (1988).
21. I. Tanihata, *Prog. Part. Nucl. Phys.* **35**, 505 (1995).
22. P. Hansen, A.S. Jensen, and B. Jonson, *Annu. Rev. Nucl. Part. Phys.* **45**, 591 (1995).
23. J. Meng and P. Ring, *Phys. Rev. Lett.* **77**, 3963 (1996).
24. J. Meng and P. Ring, *Phys. Rev. Lett.* **80**, 460 (1998).
25. T. R. Werner, J. A. Sheikh, M.. Misu, W. Nazarewicz, J. Rikovska, K. Heeger, A. S. Umar and M. R. Strayer, *Nucl. Phys.* **A597**, 327 (1996).
26. G. A. Lalazissis, A. R. Farhan and M. M. Sharma, *Nucl. Phys.* **A628**, 221 (1998).
27. T. Glasmacher et al., *Phys. Lett.* **B395**, 163 (1997).
28. T. Glasmacher, it Nucl. Phys. **A630**, 278c (1998).
29. H. Scheit et al., *Phys. Rev. Lett.* **77**, 3967 (1996).
30. P.J. Woods and C.N. Davids, *Annu. Rev. Nucl. Part. Sci.* **47**, 541 (1997).
31. W. Nazarewicz, J. Dobaczewski, T. R. Werner, J. A. Maruhn, P.-G. Reinhard, K. Rutz, C. R. Chinn, A. S. Umar and M. R. Strayer, *Phys. Rev. C* **53**, 740 (1996).
32. W. E. Ormand, *Phys. Rev. C* **53**, 214 (1996).
33. B. Blank et al., *Phys. Rev. C* **50**, 2398 (1994).
34. C.N. Davids et al., *Phys. Rev. Lett.* **80**, 1849 (1998).
35. J.C. Batchelder et al., *Phys. Rev. C* **57**, R1042 (1998).
36. P.J. Sellin et al., *Phys. Rev. C* **47**, 1933 (1993).
37. S. Åberg, P.B. Semmes, and W. Nazarewicz, *Phys. Rev. C* **56**, 1762 (1997).
38. P. Möller, J.R. Nix, W.D. Myers, and W.J. Swiatecki, *At. Data Nucl. Data Tables* **59**, 185 (1995).
39. Z. Ma, N. Van Giai, and H. Toki, *Phys. Rev. C* **55**, 2385 (1997).
40. D. Vretenar, H. Berghammer, and P. Ring, *Nucl. Phys.* **A581**, 679 (1995).
41. D. Vretenar, G. A. Lalazissis, R. Behnsch, W. Pöschl and P. Ring, *Nucl. Phys.* **A621**, 853 (1997).
42. D. Vretenar, P. Ring, G. A. Lalazissis and W. Pöschl, *Phys. Rev. E* **56**, 6418 (1997).

Mass Determination in Storage Rings

H. Wollnik

2. Physikalisches Institut, Universität Giessen, Germany

Abstract. The precise mass determination of short lived nuclei is discussed and here especially the mass determination from their frequency of rotation in storage rings. So far such measurements have been done in accelerator storage rings with highly stripped high energy ions. It is possible, however, also to use for this purpose also much smaller low energy storage rings in which the manipulation of ions becomes much simpler. Such rings can not only be used for the measurement of the masses of ions but also to separate ions of different masses and thus could be used to prepare beams of ions of metastable nuclei.

Introduction

One of the basic characteristics of nuclei are their masses. For stable nuclei these masses are known, while they are widely unknown for unstable nuclei. This knowledge is crucial, however, for the understanding of the formation of elements in stars, where such short lived nuclei exist abundantly.

When selecting one of the different mass measuring techniques it is necessary to choose one that finishes the measurement in a time that is comparable to the shortest half life time of the nuclei under consideration and that achieves a high mass accuracy but requires only very few of the difficult to produce ions.

So far used Mass Measuring Techniques

For short lived nuclei mass measurements have been performed:

1. by determining their Qβ-values using plastic- [1] or Ge-detectors [2]. These Qβ-values relate the masses neighboring nuclei within one isobaric chain, provided sufficient information about the daughter's decay scheme is available.

2. by directly determining the masses of ions in laterally dispersive mass spectrometers [3]. In these systems the required high mass resolving power m/(m is achieved by narrow entrance and exit slits which inevitably cause a small ion transmission.

    In such systems actually the ions' momentum/charge ratio is determined in magnetic sector fields which directly reveals the masses only for monoenergetic ions. Providing, however, also for some energy dispersion by additional electrostatic sector fields,, the system can be made to b energy-achromatic while still being laterally mass dispersive [4].

3. by directly determining ion masses in longitudinally dispersive mass analyzers in which the ion transmission -

CP481, *Nuclear Structure 98*, edited by C. Baktash

at least to first order - is decoupled from the mass resolving power of the mass analyzer [5,6,7,8].

In such systems actually the ion velocity is measured which directly reveals the masses only for ions whose energies K are all equal or individually known [9]. There is a solution also for ions whose energies K_0 by $\pm\Delta K$ in which case one must design the ion optics for such a system such that the more energetic and thus faster ions are sent on a detour of proper length. In this case the over all ion flight time becomes energy-isochronous with the system still being longitudinally mass dispersive [5,6,7,8].

Though longitudinally dispersive mass analyses have been performed with high precision in single pass systems [10] their real power has become apparent only in multiturn storage devices like:

1. the Penning trap in which the ions of mass/charge ratio m/q move in a homogeneous magnetic flux density B with a velocity v along circles of radius $\rho = mv/Bq$. Thus the flight time per turn is in μsec: $t \approx 2\pi\rho/v \approx 0.0651212(m/B)$ with m in mass units u and B in Tesla. This t is independent of the ion energy K so that the ion motion is always energy-isochronous. The ion motion in such devices is also influenced by electrostatic fields which by their dc-component keep the ions from escaping in the direction of the magnetic field lines and by their ac-component modulate the ion motion in a mass selective fashion.

2. This technique works only for ions that live for around one second since for a precise mass analysis times of about one second are necessary. The thus performed mass measurements, however, are very accurate and usually have only errors of a few keV or even less [11,12].

3. accelerator storage rings in which the ions move in a ring structure that consists of magnetic sector fields and quadrupole lenses. For mass measurements such systems can be used

    3.1 either in their standard setting in which case the ion energy spread (K must be reduced by beam cooling [13] or
    3.2 after proper tuning the quadrupoles in the ring to achieve energy isochronicity [8].
    In the second case, however, usually only a reduced ion transmission can be achieved.

Mass determination in accelerator storage rings

So far mass measurements in accelerator storage rings have been performed only in the ESR of the GSI in Germany. This storage ring is rather large and has a circumference of (100m. In these measurements very energetic ions have been used which were cooled by their interaction with monoenergetic electrons in some section of the ESR ring [13-17].

In this experiment a large number of ions of different masses can be investigated simultaneously (see Fig.4 in ref [14]) since the ion revolution times are determined by Fourier analyzing the signal that is obtained from electrodes close to the ion beam. In this way one can resolve mass peaks that differ by only (1 MeV (see Fig.2 in ref [15]) as long as there are at least ≈10 highly charged energetic ions in the ring. Detrimental in these measurements is that the ion beam cooling requires some time. Thus this type of mass measurement is only applicable for ions that live at least several seconds and preferably 10 times more. Also care must be taken in order that neighboring nuclei do not influence each other during the cooling process.

With this technique recently the masses of more than 100 neutron deficient heavy nuclei have been determined newly [15,16]. In many cases these measurements also allowed to determine the masses of a specific nucleus in its ground-state and in an excited metastable state (see for instance Fig.1 and Fig.4 in [17] for the cases of "Mn" and "Fe") pro-vided the life time of this excited state was long enough. Observing the intensities of these two peaks over some time also revealed the half lives of the corresponding states.

Though this is not a common procedure yet, one also could eject ions of a specific nucleus from the ESR in two ways for nuclear reaction work. Once one could do this in the standard fashion in which case the nuclei of some of the ions are in their ground state and some are in their metastable state. Then one could repeat the experiment after a time during which one type of these nuclei has decayed. This would allow to populate the different levels of the finally produced nucleus with different intensities in the two experiments.

Instead of cooling the ion beam in the storage ring so that the revolution frequencies of ions reveal the ion masses, there is also the possibility to tune the ion optics of an accelerator storage ring like the ESR so as to make the ion motion energy isochronous [8]. Also in this case the revolution frequencies of ions reveal their masses directly allowing mass measurements with accuracies of ≈ 100 keV as has been shown experimentally (see Fig.1 in [18]).

This energy-isochronous technique is applicable not only to long lived but also to short lived nuclei since no beam cooling is required. Though different ion detection methods are feasible, so far only internal targets were used from which secondary electrons were extracted whenever a single ion had passed. By this technique individual ions were tracked for more than 100 revolutions so far.

Mass determinations in
energy-isochronous table-top storage rings

In the experiments in the ESR of the GSI the ions moved with about half of the velocity of light with ≈ 500 nsec per turn. Since the ion passage through a thin foil could be determined with an accuracy of ≈ 100 psec (see Fig.2 in [18]) thus mass resolving powers $m/\Delta m$ of a few 1000 per turn were realized which resulted in overall mass resolving powers of slightly more than 10^5.

Actually also other storage rings could feasibly be used as mass analyzers as long as they guarantee an energy-isochronous ion motion [19]. In case of a ring of 3m circumference, for instance, 10 keV ions of mass 100 would make one revolution in $\approx 30 \mu$sec and for ion pulses of 1nsec length achieve[1] a mass resolving power $m/\Delta m$ of possibly 20'000 per turn. Such an energy-isochronous ring could be built from magnetic or from electrostatic sector fields and quadrupoles. A feasible realization of such a ring is shown in Fig.1. Ion energies of 10 keV are not high enough to pass them through a thin foil without drastically changing the individual ion flight paths. However, since low energy ions can easily be bunched or deflected there are very promising alternative ways:

1. one can control the ions by beam bunchers and deflectors before they enter and after they leave the storage ring . For this purpose one could:

    1.1 bunch the ions into the ring such that the ions of different masses all start their energy-isochronous flight in the ring simultaneously. (In Fig.1 they would start at point "B").

    1.2 transport them - after a certain number of revolutions in the ring - energy-isochronously to some external position. (In Fig.1 this would be to point "C"). At this position the arrival times of the different ions can be recorded or ions of a specific arrival time be pulsed out and transported to some experimental site either directly or after additional acceleration.

3. one can use a very high frequency device to accelerate or decelerate the ions during every passage around the ring at some intermediate time focus that advantageously coincides with a beam waist. (In Fig.1 such a time focus and beam waist would be at point "B"). In this case only ions of a precise rotation frequency can stay in the ring while all others are eliminated by slits at intermediate beam waists at which the lateral energy dispersion does not vanish.

---

[1] For a single pass time-of-flight mass analyzer that uses an ion reflector instead of sector fields to deflect the ions we have experimentally achieved mass resolving powers $m/\Delta m$ of more than 10'000 (limited by the available electronic circuitry) for 6keV ions [19] and overall flight times of $\approx 30 \mu$sec.

4. one can use a very high frequency device to deflect the ions during every passage around the ring at some intermediate time focus that advantageously is located between lateral beam waists. Also in this case only ions of a precise rotation frequency can stay in the ring while all others are eliminated by slits at intermediate beam waists.

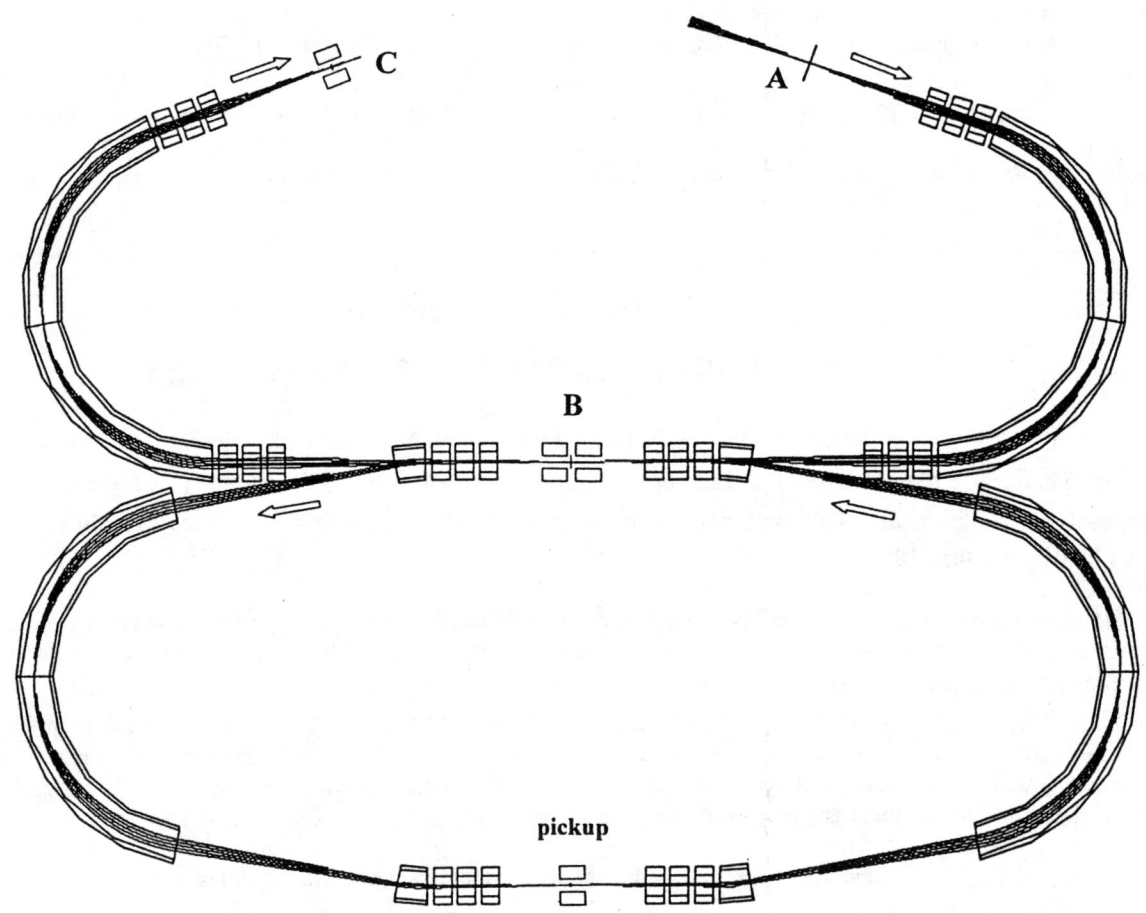

Fig.1 Shown is an energy-isochronous table-top storage ring preceded and followed by energy-isochronous beam guidance systems that ensure that an ion beam can be brought into the ring (from "A" to "B") and extracted out of the ring (from "B" to "C") without destroying its bunch structure.The storage ring is designed to consist of electrostatic sector fields and quadrupoles only. All ion optical elements are considered to be powered in a dc fashion except two of the 10° deflectors whose voltage must be switched off when ions must enter or leave the ring. Providing ion beams of $\approx$10keV to the ring that are bunched to $\approx$1nsec, mass resolving powers m/Δm$\geq$100'000 can be expected. This mass resolving power could be used for mass determinations by Fourier transforming the signal from some pick-off electrodes, or by recording the fluorescense of the ion cloud caused by a properly timed psec laser [20], or by placing a good time detector at point "C". Using a pulsed beam deflector (a TOF mass selector) at point "C" the ions of a specific mass also could be separated out for further experiments. The finally achieved mass resolving power could also be improved by using high frequency deflectors or accelerators gaps at energy-isochronous foci (like point "B") in the ring.

A feasible solution for such an energy isochronous storage ring is shown in Fig.1. In this ring the ion beam has a beam waist after every 90° of rotation and is energy isochronous after 180° of rotation. In Fig.1 one can also see an entrance beam guidance system (from "A" to "B") and an exit beam guidance system (from "B" to "C") both of which are energy isochronous and have beam waists at their entrance and exit points. In order that the ion beam physically can be entered into the ring, the electrostatic sector fields in this storage ring are designed to consist of two 160° sectors each of which is preceded and followed by 10° sectors. The voltages applied to the electrodes of these 10° sectors would be the same as those applied to the 160° sectors. However, for two of the 10° sectors these voltages would be switched off during times when ions must enter or leave the ring.

An energy-isochronous storage ring like the one shown in Fig.1 could be used for the precise mass analysis of short lived nuclei and should achieve mass resolving powers $m/\Delta m \geq 100'000$ after ≈ 5 or more turns. Such a ring, however, could be used also to boost the mass resolving power of the large sector field mass separators [21] used for radioactive ion beam facilities. While such separators usually are expected to provide a mass separated dc ion beam, a table-top storage ring like the one of Fig.1 will only accept ions for a few μsec and require ≈ 1 msec for a sufficiently good mass analysis. Thus only less than 1% of the dc ion beam delivered by a sector field mass analyzer can be used - at this point - and at the end be highly mass resolved. In detail a storage ring as shown in Fig.1 would require that the ions, that are entered into the ring are bunched such as to arrive within perhaps 1nsec at point "A" from where they are transported to point "B" - energy-isochronously - and after a longitudinal mass analysis brought over many turns energy- isochronically from point "B" to point "C". There a pulsed beam deflector could separate out ions of a specific mass for further acceleration.

Acknowledgement

This work has been supported in part by the German "Minister für Forschung und Technologie".

References

1. M.Gross et al., Nucl. Instr. and Methods A, **311**(1992)512
2. R.Decker et al., Z.f.Physik A, **294**(1980)35
3. M.Epherre et al., Nucl. Phys. A, **340**(1980)1
4. F.W.Aston, Philosoph. Mag., **38**(1919)709
5. W.Poschenrieder, Int. J. Mass Spectr. and Ion Phys., **6**(1971)413; **9**(1972)357
6. H.Wollnik, Nucl. Instr. and Methods, **186**(1981)441
7. D.Vieira et al., Phys. Rev. Lett., **57**(1986)3253
8. Wollnik, "Optics of Charged Particles", 1987, Acad. Press,Orlando
9. W.Mittig et al., Ann. Rev. Nucl. Part. Science, **47**(1997)27
10. M.D.Lunney et al., Hyperfine Int., **99**(1996)105
11. D.Beck et al., Nucl. Phys. A, **626**(1997)343c
12. G.Gabrielse et al., Phys. Rev. Lett., **74**(1995)3544
13. B.Franzke, Nucl. Instr. and Meth. B, **24/25**(1987)18
14. B.Schlitt et al., Hyperfine Interactions, **99**(1996)117
15. H.Wollnik et al., Nuclear Physics A, **616**(1997)346
16. T.Radon et al., Phys. Rev. Lett., **78**(1997)4071
17. H.Irnich et al., Phys. Rev. Lett., **75**(1995)4182
18. H.Wollnik et al., Nucl. Phys., **626**(1997)327c
19. H.Wollnik in Biological Mass Spectr., ed T.Matsuo, San-El Publ., Kyoto (1992) 204
20. D.Slaugther, private communication
21. H.Wollnik, Part. Acc., **47**(1994)145

Isoscalar and Isovector Dipole Mode in β−stable and Drip Line Nuclei

H.Sagawa[a,1] , I.Hamamoto[b] and X.Z.Zhang[b,c]

[a] Center for Mathematical Sciences, University of Aizu
Ikki-machi, Aizu-Wakamatsu, Fukushima 965-8580, Japan.
[b] Department of Mathematical Physics
Lund Institute of Technology at University of Lund, Lund, Sweden.
and
[c] Institute of Atomic Energy, Beijing
The People's Republic of China.

Abstract. The isoscalar and the isovector dipole mode in nuclei are investigated, using the self-consistent Hartree-Fock plus the random-phase approximation with Skyrme interactions. Including simultaneously both the isoscalar and the isovector correlation, the RPA response function is estimated in the coordinate space so as to take into account properly the continuum effect. In β-stable nuclei such as ^{208}Pb the frequency of the isovector giant dipole resonance (IVGDR) is lower than that of the isoscalar giant dipole resonance (ISGDR, compression mode). In contrast, in lighter drip line nuclei a major part of the isoscalar compression dipole strength lies at an energy much lower than the energy of the IVGDR, due to the threshold effect of loosely-bound nucleons. The transition densities and displacement fields of ISGDR and IVGDR are discussed in comparison with collective model predictions.

INTRODUCTION

The structure of nuclei far from β−stability is an exciting research field since many new phenomena have been observed and are expected in connection with the small binding energies of the nucleons close to the Fermi surface. Some of the new experimental observations are the halo and the skin effects on the density distributions, and the soft dipole mode in halo nuclei. The dynamical response of drip line nuclei to various external fields is expected to show an interesting exotic structure, due to the presence of the low-lying threshold strength unique in those nuclei. We have studied the response functions of drip line nuclei [1] for various multipole operators, performing the Hartree-Fock (HF) calculation with Skyrme interactions and then using the random-phase-approximation (RPA). In the present work we study the IS dipole mode (compression mode) in comparison with the IV dipole mode, taking into account both the isoscalar (IS) and isovector (IV) correlation in the RPA [2]. The RPA response function is solved in coordinate space using Green's functions, to evaluate properly the coupling to the continuum which becomes extremely important near the drip lines. Taking away carefully the IS spurious (center of mass) component from the calculated dipole spectra, we compare the obtained dipole response of drip line nuclei with that of β-stable nuclei.

Various types of giant resonances (GR) are listed in Table 1 together with experimental systematics. The IV giant dipole resonance (IVGDR) is well established and the oldest one among various giant resonances in nuclei [3]. Systematic data are also available for IS giant monopole resonances (GMR) and IS giant quadrupole resonances (GQR). The IS dipole resonance was theoretically studied already more than 20 years ago [3,4]. Numerical calculations were also made for some doubly-closed β-stable nuclei [5,6], for which hadron inelastic scattering experiments could be easily done. Recently, the observation of an IS giant dipole resonance (ISGDR), the IS dipole compression mode, in ^{208}Pb was reported in ref. [7], using the (α, α') cross sections at forward angles. In β-stable nuclei the ISGDR may well be expected at such a high

---

[1] presented by H. Sagawa

TABLE 1. Giant resonances in β-stable nuclei. The transition operator is denoted by $\hat{O}$. The Coulomb energy is given by V_c

| | isospin operator | 1 or τ_z | |
Name	λ^π	$\hat{O}^{\lambda,\tau}$	Empirical excitation energy (MeV)
IS GMR	0^+	r^2	$80/A^{1/3}$
IV GDR	1^-	$rY_{1\mu}\tau_z$	$79/A^{1/3}$
IS GQR	2^+	$r^2 Y_{2\mu}$	$58/A^{1/3}$
IV GQR	2^+	$r^2 Y_{2\mu}\tau_z$	
IV spin GR	1^+	$\sigma_\mu \tau_z$	
IS GDR	1^-	$r^3 Y_{1\mu}$	

| | isospin operator | $\tau_\pm$ | |
Name	λ^π	$\hat{O}^{\lambda,\tau}$	Empirical excitation energy
IAS	0^+	$t_\pm$	$V_c(\text{daughter}) - V_c(\text{mother}) \pm (m_n - m_p)c^2$
Gamow-Teller GR	1^+	$\sigma_\mu t_\pm$	
spin-dipole GR	$0^-, 1^-, 2^-$	$[\sigma \times Y_1]^\lambda t_\pm$	

energy. In contrast, it is an interesting open question whether a considerable amount of the IS compression dipole strength will appear in the low-energy threshold region, since it was pointed out [1,2] that the response functions for various non-compression multipoles show in general a large transition strength in the low-energy region just above the threshold.

RPA RESPONSE FUNCTION

The RPA theory is based on the time dependent HF theory and, in the small amplitude limit, the time dependent vibration is described by the the RPA response function G^{RPA} which is expressed as

$$G_{\text{RPA}} = G^{(0)} + G^{(0)} v_{ph} G_{\text{RPA}}$$
$$= (1 - G^{(0)} v_{ph})^{-1} G^{(0)} \tag{1}$$

where the unperturbed response $G^{(0)}$ is given by

$$G^{(0)}(\mathbf{r}, \mathbf{r}' : \omega) = \sum_{h,p} \varphi_h^*(\mathbf{r}) \varphi_p^*(\mathbf{r}) \left\{ \frac{1}{\epsilon_p - \epsilon_h - \omega - i\eta} + \frac{1}{\epsilon_p - \epsilon_h + \omega - i\eta} \right\} \varphi_p(\mathbf{r}') \varphi_h(\mathbf{r}') \tag{2}$$

The particle-hole (p-h) interaction v_{ph} is obtained from the second functional derivatives of the hamiltonian density with respect to the density for each isospin-spin channel. The wave functions of the particle and hole states are denoted by φ_p and φ_h, respectively. Because of the cancellation between the first term and the second term in Eq.(2), we can extend the sum of particle states (denoted by p) to all the particle and hole states (denoted by $i \in all$) as

$$G^{(0)}(\mathbf{r}, \mathbf{r}' : \omega) = \sum_{h, i \in all} \varphi_h^*(\mathbf{r}) \varphi_i^*(\mathbf{r}) \left\{ \frac{1}{\epsilon_i - \epsilon_h - \omega - i\eta} + \frac{1}{\epsilon_i - \epsilon_h + \omega - i\eta} \right\} \varphi_i(\mathbf{r}') \varphi_h(\mathbf{r}')$$
$$= \sum_h \varphi_h^*(\mathbf{r}) \left\langle \mathbf{r} \left| \frac{1}{H_0 - \epsilon_h - \omega - i\eta} + \frac{1}{H_0 - \epsilon_h + \omega - i\eta} \right| \mathbf{r}' \right\rangle \varphi_h(\mathbf{r}') \tag{3}$$

In the last equation, the sum i on all states is expressed in closed form and the energy ϵ_i is replaced by the HF hamiltonian H_0. The RPA response (1) is calculated by using the p-h interaction v_{ph} derived from the hamiltonian density of Skyrme interaction. Since the Skyrme interaction has the momentum dependent terms, we have to propagate several different operators like $f^\lambda(r) Y_{\lambda\mu}(\hat{r}) \times \left(1, \nabla^2, (\vec{\nabla}_p \pm \vec{\nabla}_h)\right)$ in the coordinate space to calculate the RPA response (1) for the λ-pole transition. Here we denote the transition operator as $\hat{O}^\lambda = f^\lambda Y_{\lambda\mu}$. The response to the operator $\hat{O}^\lambda$ is used to calculate the transition density, while $f^\lambda Y_{\lambda\mu} \times \overrightarrow{\nabla}$ is used for the transition current calculations. Explicit expressions for $f^\lambda Y_{\lambda\mu}$ are given in Table 1 and in Eqs. (5) and (6).

We study the RPA strength function

$$S(E) \equiv \sum_n |<n\,|\,\hat{O}\,|\,0>|^2\, \delta(E - E_n) = \frac{1}{\pi} Im\, Tr(\hat{O}^\dagger\, G_{RPA}(E)\, \hat{O}) \tag{4}$$

In Eq.(4) $\hat{O}$ represents the one-body operators

$$\hat{O}_\mu^{\lambda=1,\,\tau=1} = \sum_i \tau_z(i)\, r_i\, Y_{1\mu}(\hat{r}_i) \qquad \text{for isovector dipole strength} \tag{5}$$

and

$$\hat{O}_\mu^{\lambda=1,\,\tau=0} = \sum_i r_i^3\, Y_{1\mu}(\hat{r}_i) \qquad \text{for isoscalar dipole strength.} \tag{6}$$

The transition density for an excited state $|n>$,

$$\rho_{n0}^{tr}(\vec{r}) \equiv <n\,|\, \sum_{i=1}^A \delta(\vec{r} - \vec{r}_i)\,|\,0> \qquad , \tag{7}$$

can be calculated from the RPA response, and the radial transition density $\rho_n^{tr}(r)$ is defined by

$$\rho_{n0}^{tr}(\vec{r}) \equiv \rho_n^{tr}(r)\, Y_{\lambda\,\mu}(\hat{r}) \qquad . \tag{8}$$

The transition current, so called the convection current, for the excited state is also evaluated as

$$\vec{J}_{n0}^{tr}(\vec{r}) = <n\,|\, \sum_{i=1}^A \frac{1}{2mi}\{\delta(\vec{r} - \vec{r}_i)\overrightarrow{\nabla}_i - \overleftarrow{\nabla}_i\delta(\vec{r} - \vec{r}_i)\}\,|\,0> . \tag{9}$$

The transition current (9) is expanded in the complete set of the vector spherical harmonics $\overrightarrow{Y}_{\lambda l,\mu}^*(\hat{r})$,

$$\vec{J}_{n0}^{tr}(\vec{r}) = (-i) \sum_{l=\lambda\pm1} J_{\lambda l}(r)\overrightarrow{Y}_{\lambda l,\mu}^*(\hat{r}). \tag{10}$$

The radial part of the transition current is defined as

$$J_{\lambda l}(r) = i \int \overrightarrow{Y}_{\lambda l,\mu}(\hat{r})\vec{J}_{n0}^{tr}(\vec{r})d\hat{r}$$

$$= <n\,|\, \sum_{i=1}^A \frac{1}{2m}\{\delta(r - r_i)[Y_l^*(\hat{r}_i) \times (\overrightarrow{\nabla}_i^+ - \overleftarrow{\nabla}_i^+)]_{\lambda\mu}\}\,|\,0> \tag{11}$$

and the velocity field (the displacement field) is given by

$$\vec{v}(\vec{r}) = \vec{J}_{n0}^{tr}(\vec{r})/\rho_o(r) \tag{12}$$

which has a similarity to the classical expression and allows us to compare it with that in macroscopic models of nuclear vibrations.

Since our calculated RPA strength function does not contain the spreading width, namely the coupling to 2p-2h (or more complicated) configurations relative to the ground states, the calculated width of each peak cannot be directly compared with experiments. In order to simulate the coupling to those configurations, in Figs.1 and 2, we show also the averaged strength functions,

$$\bar{S}(E) = \int S(E_0)\, \rho(E - E_0)\, dE_0 \tag{13}$$

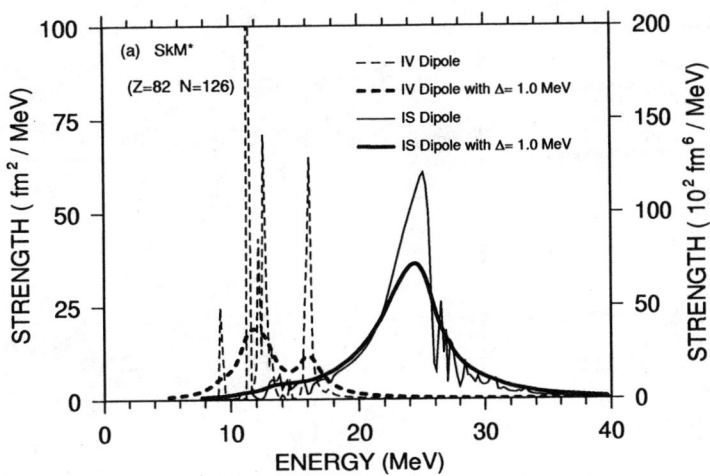

FIGURE 1. The RPA IS and IV dipole strength in the β-stable nucleus $^{208}_{82}\text{Pb}_{126}$ as a function of excitation energy. The scale of the IS dipole strength is shown on the right-hand-side, while that of the IV dipole strength is denoted on the left-hand-side. The solid line expresses the IS dipole strength, while the dashed line denotes the IV dipole strength. The thick lines are obtained by averaging the calculated RPA strength (denoted by the respective thin lines) using Eq.(13) with Δ =1 MeV. The strength appearing below the threshold due to the averaging procedure has no meaning. The SkM* interaction is used. The figure is taken from ref. [8]

with the weight function

$$\rho(E - E_0) = \frac{1}{\pi} \frac{\Delta}{(E - E_0)^2 + \Delta^2} \tag{14}$$

In plotting the figures we have used the value of Δ=1 MeV, which is somewhat arbitrarily chosen. The strength, which appears in the energy region below the threshold due to the averaging procedure, has no meaning.

IS AND IV DIPOLE RESPONSE

We discuss in this section the IS and IV dipole response of several nuclei. Most of the calculated results are taken from refs. [8,10]. In Fig.1 the calculated RPA strength functions corresponding to the IV dipole operator $\hat{O}_\mu^{\lambda=1,\tau=1}$ in (5) and the IS dipole operator $\hat{O}_\mu^{\lambda=1,\tau=0}$ in (6) are shown for the β-stable nucleus, $^{208}_{82}\text{Pb}_{126}$. Both the RPA strength functions (4) and the averaged ones with Δ =1 MeV in Eq.(13) are shown. The SkM* interaction is used both in the HF and the RPA calculation. In $^{208}_{82}\text{Pb}_{126}$ the peak energy of the ISGDR in the RPA strength function (4) is 25.0 MeV, while the energy defined by the formula

$$\bar{E} = \frac{m_1}{m_0} \tag{15}$$

is equal to 23.4 MeV. In Eq.(15) the energy-weighted moments, m_k , are defined by

$$m_k = \int S(E) E^k \, dE . \tag{16}$$

115

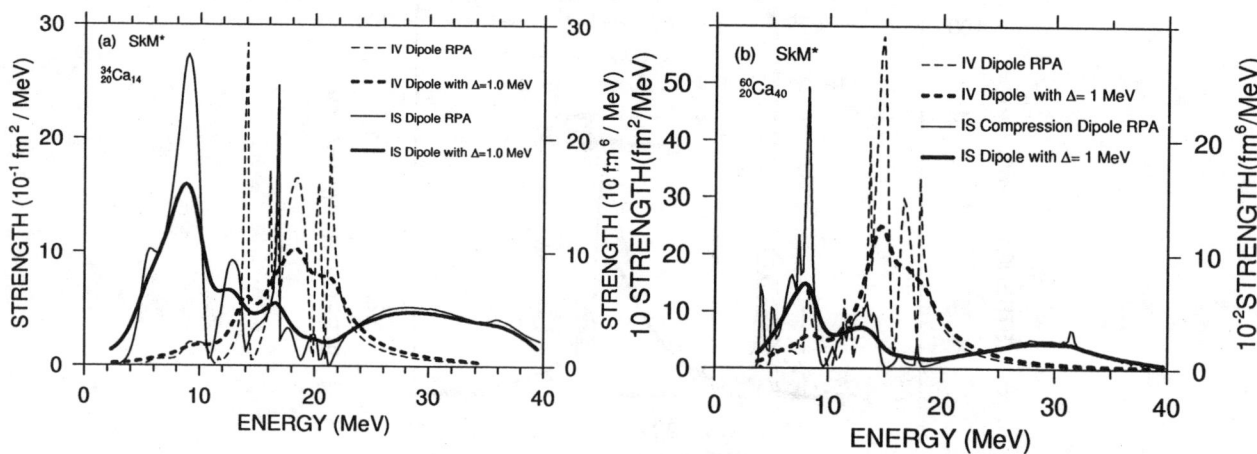

FIGURE 2. The RPA IS and IV dipole strength as a function of excitation energy in ; (a) the proton drip line nucleus $^{34}_{20}Ca_{14}$, and (b) the neutron drip line nucleus $^{60}_{20}Ca_{40}$. See the caption to Fig.1 for details.

The ISGDR in the energy region of 18-30 MeV of Fig.1, which is expressed by the thin solid line, consumes about 85 percent of the EWSR coming from the calculated intrinsic excitations without the spurious components. It is seen from Fig.1 that in $^{208}_{82}Pb_{126}$ the calculated IVGDR lies energetically about 10 MeV lower than the estimated ISGDR. The measured peak energy of the IVGDR is 13.4 MeV, which is almost equal to our averaged calculated value, $\bar{E}$=13.3 MeV, while the ISGDR is reported [7] to be found around 22.5 MeV.

In Figs.2a and 2b we show the IV and the IS dipole strength functions for the proton drip line nucleus $^{34}_{20}Ca_{14}$ and the neutron drip line nucleus $^{60}_{20}Ca_{40}$ respectively. It is well known that in light nuclei it is difficult to interpret the observed IV dipole strength in terms of a single resonance ("IVGDR") frequency, since the difference between the energies of relevant p-h excitations may be comparable with or even larger than the width of the possible giant resonance. The multiple peak structure can be seen in the IV dipole strength function of $^{34}_{20}Ca_{14}$ and $^{60}_{20}Ca_{40}$ in Fig. 2. In those light drip line nuclei it is seen that the considerable part of the IS dipole transition strength is consumed by the threshold strength and lies clearly below the "IVGDR", while the higher-lying IS dipole strength is observed as a very broad peak with an extremely large tail. The IVGDR lies energetically between the low-energy IS dipole peaks and the very broad "ISGDR".

In Figs.3a and 3b the RPA dipole response of protons and that of neutrons to the operator (6) are shown for $^{34}_{20}Ca_{14}$ and $^{60}_{20}Ca_{40}$, respectively, in comparison with the total RPA IS dipole response. The low-energy threshold strength consists predominantly of proton excitations in the proton drip line nucleus ^{34}Ca , while it comes exclusively from neutron excitations in the neutron drip line nuclei such as ^{60}Ca. The unique structure of the threshold strength, which comes essentially from the uncorrelated excitations of protons or neutrons with small binding energies, is very similar to that for other non-compression multipoles studied previously. However, in Figs.3a and 3b it is very interesting to observe that in the high-energy "ISGDR" region the neutron contribution interferes constructively with the proton contribution, as expected for the IS collective mode. In contrast, in the lower energy region the neutron and proton contributions almost always interfere destructively.

We have performed numerical calculations with several Skyrme interactions and confirmed that the conclusions drawn by using the SkM* interaction remain the same for all Skyrme interactions, except for numerical details. For example, the calculated frequency of the IVGDR is sensitive to the value of the symmetry energy coefficient of the Skyrme interactions used. In contrast, since the ISGDR is a compression mode, the calculated frequency is sensitive to the incompressibility of the Skyrme interaction employed. A higher frequency of the ISGDR is obtained for Skyrme interactions with a higher incompressibility. For example, the peak

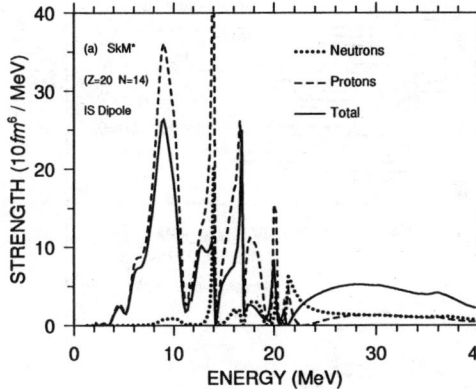
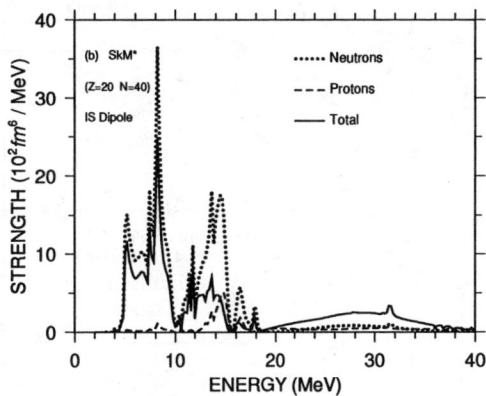

FIGURE 3. The RPA dipole strength of protons and neutrons to the IS operator (6); (a) the proton drip line nucleus $^{34}_{20}\text{Ca}_{14}$, and (b) the neutron drip line nucleus $^{60}_{20}\text{Ca}_{40}$. The spurious components are eliminated in the estimate of the strength function. See ref. [8] and the caption to Fig.1 for details.

energy of the ISGDR calculated by using the SIII interaction is considerably higher than that estimated by employing the SkM* interaction. See, for example, ref. [6].

The transition densities of IV GDR and IS GDR at three energies in ^{208}Pb are shown on the LHS of Fig. 4. Because of large fragmentation of the IV dipole strength, the transition density of two peaks at $E_x=12.6$ and 16.3MeV are different as seen in Figs. 4(a) and 4(b). A collective model for IVGDR, the Goldhaber-Teller (G-T) model, gives the transition density

$$\rho^{tr}_{\lambda=1\tau=1}(r) \propto \frac{d\rho_0(r)}{dr} . \tag{17}$$

The transition density at $E = 12.6\text{MeV}$ in Fig. 4(a) is somewhat different to the G-T model prediction for the peak position, while that at $E = 16.3\text{MeV}$ in Fig. 4(b) has a similar peak position to that of G-T model. On the other hand, the IS GDR is a single peak and the transition density has a similar radial dependence to that of the collective model [5,6],

$$\rho^{tr}_{\lambda=1\tau=0}(r) \propto 10r\rho_0(r) + (3r^2 - \frac{3}{5} <r^2>_m)\frac{d\rho_0(r)}{dr} . \tag{18}$$

assuming that the IS GDR state exhausts 100% of the EWSR.

2D plots of displacement fields on the RHS of Figs. 4(a) and 4(b) looks very similar to a simple translation in z-direction inside the core denoted by the solid quarter circle. However the peak at $E_x=16.3\text{MeV}$ shows a node of the displacement around $r = 9$ fm in Fig. 4(b). This characteristic feature for the displacement field of ISGDR is also predicted by a microscopic model [11]. Because the two current components $J_{\lambda=1,l=\lambda\pm1}(r)$ are competing in magnitude with different phases, the 2D plot of the displacement field looks complex in Fig. 4(c) with two nodes.

The transition densities (LHS) and the displacement fields (RHS) of IV GDR and IS GDR in ^{60}Ca are shown in Fig. 5. The proton and neutron contributions for the transition density are very much energy dependent in the case of IV GDR. The transition density at $E_x=14.8\text{MeV}$ has a very large proton contribution in the core of the nucleus , while the neutron contribution is very much concentrated in the neutron skin region. This asymmetry is due to the effect of loosely-bound neutron excitation from $1f_{5/2}$ state in ^{60}Ca. On the other hand, the neutron and proton contributions are almost the same magnitude with opposite signs at

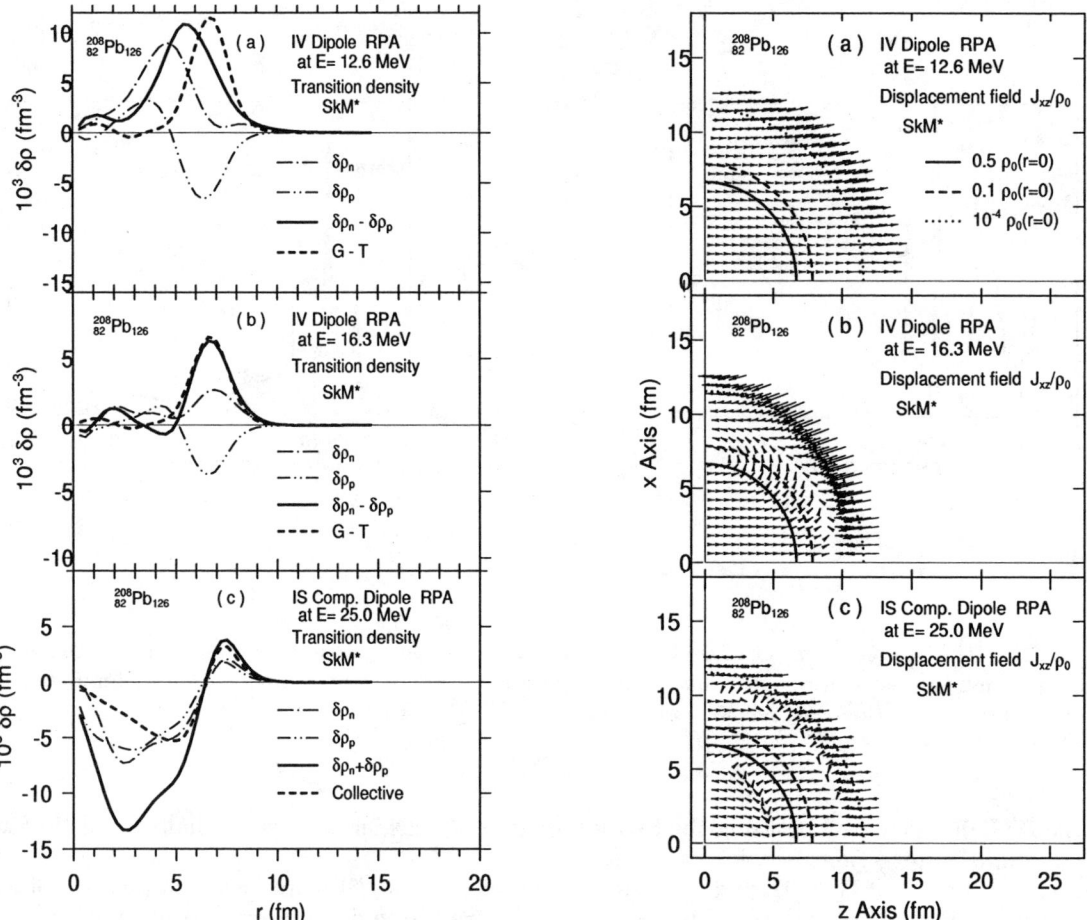

FIGURE 4. Transition densities (LHS) and displacement fields (RHS) of RPA excitation modes of $^{208}_{82}\text{Pb}_{126}$, (a) the lower peak of IV GDR at 12.60 MeV, (b) the higher peak of IV GDR at 16.30 MeV, and (c) IS compression GDR at 25.00 MeV in Fig. 1. The results of RPA excitations are compared with the collective model predictions. The absolute magnitude of the transition density of collective model is normalized to have the same B(Eλ) value as that of RPA excitation.

E_x=19.0MeV, which is a typical feature of the IV mode. The transition density of IS GDR has also almost equal contributions of protons and neutrons with the same phase as seen in Fig. 5(c). Nevertheless, we can see a slightly enhanced neutron contribution outside of the core region $r > 5$fm, compared with those densities in ^{208}Pb on the LHS of Fig. 4(c).

Inside of the dashed circle where the density is 10% of the central density, the 2D plot of the displacement field of IVGDR looks almost identical to the classical translational flow pattern . However, the nodes in the 2D plot of the displacement fields appears sometimes outside of the nuclear surface as can be seen in Figs. 4(b), 5(a) and 5(b). The ISGDR in Fig. 5(c) shows a flow pattern similar to that of the collective model and also to that of ^{208}Pb. However, the innermost node of the displacement field lies close to the nuclear surface in ^{60}Ca, while that of ^{208}Pb is seen well inside of the surface in the case of the ISGDR.

CONCLUSION

We have studied the IS and the IV dipole strength function of drip line nuclei in comparison with those of β-stable nuclei, using the self-consistent HF plus the RPA with Skyrme interactions. The spurious (c.m.)

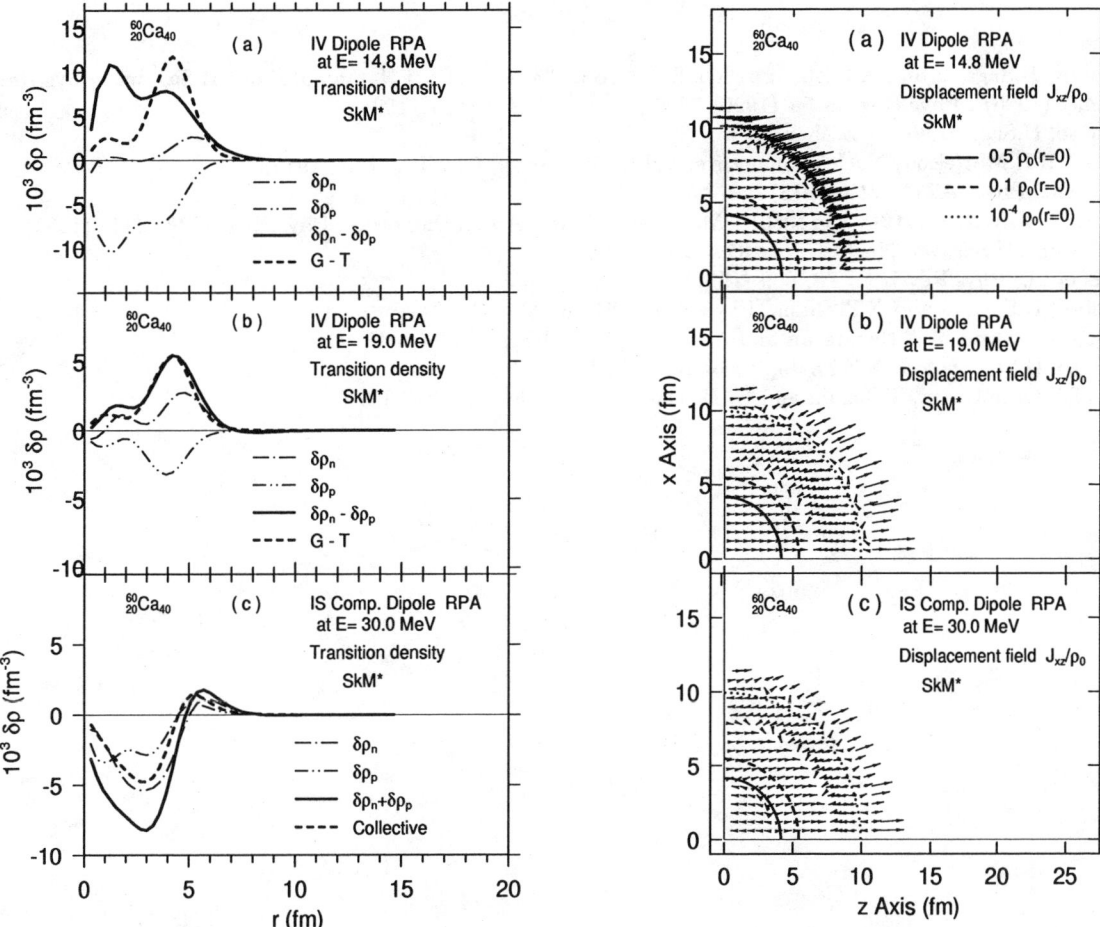

FIGURE 5. Transition densities (LHS) and displacement fields (RHS) of RPA excitation modes of $^{60}_{20}\text{Ca}_{40}$, (a) the lower peak of IV GDR at 14.80 MeV, (b) the higher peak of IV GDR at 19.00 MeV, and (c) IS compression GDR at 30.00 MeV. See the caption to Fig. 4 for details.

component is carefully subtracted from the calculated spectra. In lighter drip line nuclei the low-energy threshold strength consumes a considerable part of the IS dipole strength, while the high-energy ISGDR (compression mode) becomes much broader than that of the stable nuclei. It is shown that the low-energy IS dipole peaks in the drip line nuclei lie clearly lower than the IVGDR, while the frequency of the ISGDR is definitely much higher than that of the IVGDR in β-stable nuclei such as ^{208}Pb and ^{40}Ca. The presence of the low-lying IS dipole peaks in drip line nuclei may play a very important role in electron or hadron scattering experiments, though it may not have so much effect on photon scattering.

The transition densities and displacement fields of IS and IV GDR in ^{60}Ca are discussed in comparison with those of ^{208}Pb. We found that the transition density and displacement field of IVGDR are very much energy dependent. Namely, the transition densities of high energy peaks of IVGDR in the two nuclei are well described by that of the Goldhaber-Teller model, while the low energy peaks show substantial difference from the collective model. The displacement field of IVGDR shows a typical translational pattern inside the surface in general. On the other hand, there appears irregular flow pattern outside of the dashed circle where the density is 10% of the central density. The transition densities of ISGDR in the two nuclei have qualitatively similar radial dependence. The displacement field of ^{60}Ca has a node close to the surface, while the node appears well inside of the surface in the case of ^{208}Pb.

One of the authors (X.Z.Z.) acknowledges the financial support provided by the Wenner-Gren Foundation, which makes it possible for him to work at the Lund Institute of Technology.

REFERENCES

1. I.Hamamoto, H.Sagawa and X.Z.Zhang, Phys.Rev. **C53**, 765 (1996) : I.Hamamoto and H.Sagawa, Phys.Rev. **C53**, R1492 (1996) ; Phys.Rev. **C 54** (1996) 2369 ; Phys.Lett.**B394**, 1 (1997).
2. I.Hamamoto, H.Sagawa and X.Z.Zhang, Phys.Rev. **C55**, 2361 (1997).
3. A.Bohr and B.R.Mottelson, Nuclear Structure, Vol.II (Benjamin, Reading. MA, 1975).
4. T.J.Deal, Nucl.Phys. **A217**, 210 (1973).
5. M.N.Harakeh, Phys.Lett. **B90**, 13 (1980) : M.N.Harakeh and A.E.L.Dieperink, Phys.Rev. **C23**, 2329 (1981).
6. N.Van Giai and H.Sagawa, Nucl.Phys. **A371**, 1 (1981).
7. B.F.Davis et al., Phys.Rev.Lett. **79**, 609 (1997).
8. I.Hamamoto, H.Sagawa and X.Z.Zhang, Phys.Rev. **C57**, R1064 (1998)
9. F.Catara, E.G.Lanza, M.A.Nagarajan and A.Vitturi, Nucl. Phys. **A624**, 449 (1997).
10. I.Hamamoto, H.Sagawa and X.Z.Zhang, to be published.
11. F.E.Serr, T.S.Dumitrescu, T.Suzuki and C. H. Dasso, Nucl. Phys. **A404**, 359 (1983).

Structure and Formation Mechanism of the Transfermium Isotope ^{254}No

P. Reiter[1], T.L. Khoo[1], C.J. Lister[1], D. Seweryniak[1], I. Ahmad[1], M. Alcorta[1],
M.P. Carpenter[1], J.A. Cizewski[1,3], C.N. Davids[1], G. Gervais[1], J.P. Greene[1], W.F. Henning[1],
R.V.F. Janssens[1], T. Lauritsen[1], S. Siem[1,8], A.A. Sonzogni[1], D. Sullivan[1], J. Uusitalo[1],
I. Wiedenhöver[1], N. Amzal[2], P.A. Butler[2], A.J. Chewter[2], K.Y. Ding[3], N. Fotiades[3],
J.D. Fox[4], P.T. Greenlees[2], R.-D. Herzberg[2], G.D. Jones[2], W. Korten[5], M. Leino[6], K. Vetter[7]

[1] *Argonne National Laboratory, Argonne, Illinois 60439.*
[2] *University of Liverpool, Liverpool L69 7ZE, England.*
[3] *Rutgers University, New Brunswick, New Jersey 08903.*
[4] *Florida State University, Tallahassee, Florida 32306.*
[5] *DAPNIA/SPhN, CEA Saclay, F-91191 Gif-sur-Yvette Cedex, France.*
[6] *University of Jyväskylä, Jyväskylä, Finland.*
[7] *Lawrence Berkeley National Laboratory, Berkeley, California 94720.*
[8] *University of Oslo, Oslo, Norway.*

Abstract. The ground-state band of the Z=102 isotope ^{254}No has been identified up to spin 14, indicating that the nucleus is deformed. The deduced quadrupole deformation, $\beta = 0.27$, is in agreement with theoretical predictions. These observations confirm that the shell-correction energy responsible for the stability of transfermium nuclei is partly derived from deformation. The survival of ^{254}No up to spin 14 means that its fission barrier persists at least up to that spin.

INTRODUCTION

Experimental information about the formation and structure of heavy elements is of pivotal interest in the search for the long-predicted island of spherical superheavy nuclei. Despite a vanishing liquid drop fission barrier, the elements with Z~100 are stable against fission. A large shell correction energy creates a fission barrier of up to 8 MeV [1,2] and a 'peninsula' of stability has been found from Z=102 to 112, where nuclei decay preferentially by α emission [3]. For the predicted 'island' of superheavy nuclei at Z=114, N=184 the stability is based on doubly-closed spherical proton and neutron shells [4]. The shell correction energy of lighter transfermium nuclei (TFN) is maximized for deformed shapes, with not only quadrupole, but also higher multipole moments being important around ^{256}No [5]. However, hitherto there is no direct experimental confirmation of the deformation. The observation of the ground state band in at least one nucleus will provide important confirmation of the models, and provide a benchmark for extrapolation to heavier systems.

Observation of high spin states provides important information about the fission barrier at high angular momentum, which is intimately related to the formation mechanism of heavy elements via fusion-evaporation reactions. Theoretical predictions of the barrier and the shell correction energies are calculated for only the ground state of TFN. The barrier may decrease with increasing angular momentum and it is not obvious that high spin states will survive due to a high fissility.

EXPERIMENTAL PROCEDURE

The unprecedented sensitivity achieved by combining Gammasphere and the Argonne Fragment Mass Analyzer (FMA) makes it possible to study the nuclear structure of TFN with (HI,xn) reactions. The interesting

CP481, *Nuclear Structure 98*, edited by C. Baktash
1999 American Institute of Physics 1-56396-858-4

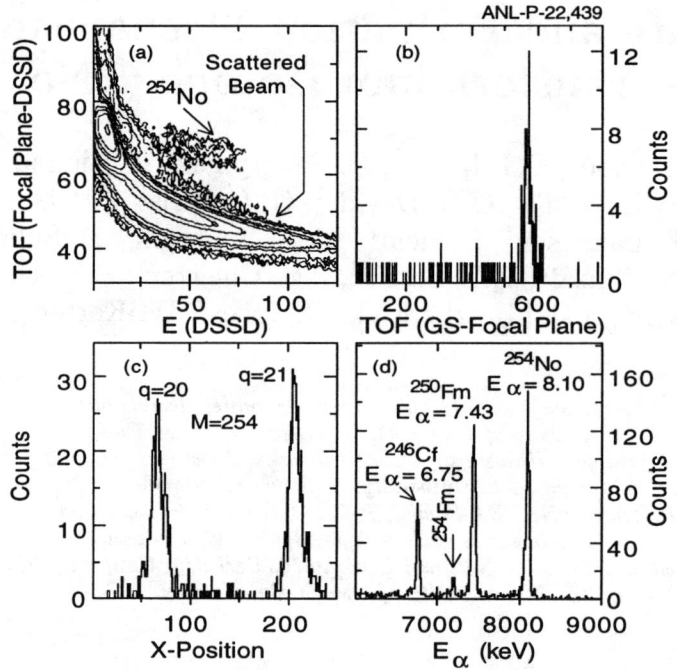

FIGURE 1. (a) Two-dimensional spectrum of the flight time from the focal-plane detector to the DSSD *vs.* the evaporation-residue implant energy in the DSSD. (b) Time-of-flight spectrum from Gammasphere to the focal-plane detector. (c) Mass/charge spectrum at the focal-plane detector. (d) Alpha spectrum from the DSSD detector, showing the peaks from three generations of α decay, starting with ^{254}No. (The ^{254}Fm α peak follows two successive electron-capture decays from ^{254}No). In (a-c), the axes are in channel numbers.

reaction products are formed with cross sections in the μb region. Prompt γ rays are required to be in coincidence with the very weakly produced evaporation residues, which have to be identified from a $> 10^4$ more intense fission background in the residue separator. Moreover, Gammasphere is an efficient detector array with high granularity, which is suitable for determining the two-dimensional multiplicity (fold)/ sum-energy distribution of evaporation residues. From these observables, the entry spin and the excitation energy of the residues can be obtained.

A first experiment was performed to produce ^{254}No via the ^{208}Pb(^{48}Ca,2n) reaction at a bombarding energy of 215 MeV. Excitation function measurements [6] determined a production cross section of $\approx 3\mu$b for the cold-fusion reaction at this beam energy. The target and projectile combination of two doubly-closed shell nuclei results in a large negative Q-value and, hence, a low compound nucleus excitation energy (19.3 MeV). As a consequence, the 2n evaporation channel is essentially the only open one, and charged particle evaporation is not observed [6].

The Argonne superconducting linear accelerator ATLAS provided ^{48}Ca beams of up to 9 pnA. The ^{208}Pb targets (500μg/cm^2) were mounted on a rotating target wheel to prevent destruction by the beam. In addition, the beam was dispersed vertically by $\pm$2.5mm through wobbling with a magnetic steerer.

Prompt γ rays from ^{254}No were detected with Gammasphere, consisting of 101 Compton suppressed Ge detectors. At the FMA focal plane, charged particles were detected using either a position-sensitive multi-channel plate detector or a parallel-plate avalanche counter. After transmission through the focal-plane detector, the residues were implanted in a double-sided Si strip detector (DSSD), consisting of 1600 1x1 mm pixels.

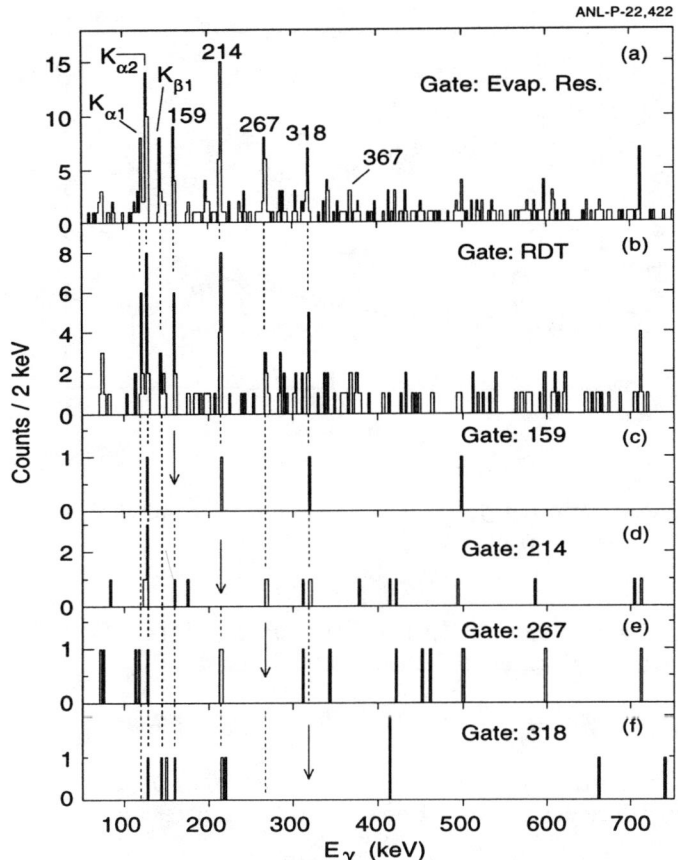

FIGURE 2. (a) Gamma spectrum obtained using coincidence gates on the 'peaks' in Fig. 1 (a-c). The background level is 0.4 counts/channel. Peaks labelled by energy are assigned as transitions within the ground-state band of ^{254}No. (b) Spectrum with an additional requirement on the ^{254}No α peak shown in Fig. 1d. (c-f) Coincidence spectra from gates set on the transitions indicated by arrows. Vertical dashed lines help visual alignment of peaks in the different panels.

DATA ANALYSIS AND RESULTS

Ground State Band and Moments of Inertia

To separate the nobelium residues from scattered beam particles and fission fragments, gates were set on the two-dimensional histogram of flight time between the focal-plane detector and the DSSD vs. DSSD implant energy (Fig. 1a). Coincidence gates were also required on: (i) the time of flight of the residues from the the target to the focal plane (Fig. 1b) and (ii) the focal-plane detector positions corresponding to two charge states (q = 20 and 21) of ^{254}No (Fig. 1c). After implantation of a ^{254}No evaporation residue in a specific pixel, subsequent α-decay in that pixel is used to identify ^{254}No, with additional requirements of energy (8.093 MeV) and decay time (Fig. 1d). The measured ^{254}No half-life $t_{1/2} = 51 \pm 6$ s agrees with the known 55 s half-life.

Correlations of γ rays with a focal-plane signal corresponding to a properly identified No residue constitutes the recoil decay tagging (RDT) technique, whereby unambiguous identification of the γ-ray parentage is achieved in the spectrum shown in Fig. 2b. This spectrum affirms that the spectrum in Fig. 2a, which is generated without the additional RDT requirements, contains only ^{254}No γ rays. Coincidence gates on individual transitions, are given in Fig. 2 c-f. They support the assignment that the transitions labeled with energies constitute a cascade corresponding to the ground-state band of ^{254}No.

The identification of a rotational band in ^{254}No immediately establishes that the nucleus is deformed and confirms predictions of theories calculating shell-correction energies of TFN. Fig. 3 shows the moments of inertia,

123

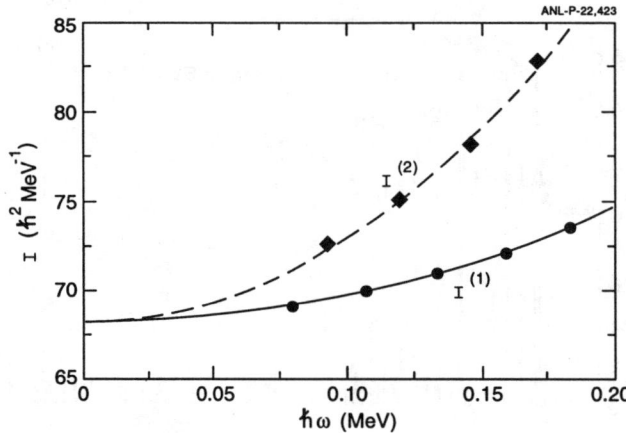

FIGURE 3. Moments of inertia, $\mathcal{J}^{(1)}$ and $\mathcal{J}^{(2)}$, for the ground state band of ^{254}No. The lines are fits to the data, using $\mathcal{J}_0 = 68.2\ \hbar^2 MeV^{-1}$, $\mathcal{J}_1 = 164.9\ \hbar^2 MeV^{-1}$.

$\mathcal{J}^{(1)}$ and $\mathcal{J}^{(2)}$, for the ground-state band of ^{254}No. $(\mathcal{J}^{(1)} = \hbar^2(2I-1)/E_\gamma(I)$; $\mathcal{J}^{(2)} = 4\hbar^2/(E_\gamma(I) - E_\gamma(I-2))$; $\hbar\omega = E_\gamma/2$.) The Harris parameterization of the moments of inertia

$$\mathcal{J}^{(1)} = \mathcal{J}_0 + \mathcal{J}_1\omega^2, \qquad \mathcal{J}^{(2)} = \mathcal{J}_0 + 3\mathcal{J}_1\omega^2 \qquad (1)$$

provides excellent fits of $\mathcal{J}^{(2)}(\omega)$ and $\mathcal{J}^{(1)}(\omega)$. From the parameters $\mathcal{J}_0$ and $\mathcal{J}_1$, we deduced the spins of the emitting states, using a procedure described in Ref. [7] and the expression

$$I = \mathcal{J}_0\omega + \mathcal{J}_1\omega^3 + 1/2. \qquad (2)$$

The spins have even integer values between 6 and 14 (within .01), providing support for assigning the transitions to the ground-state band.

The proposed level scheme of ^{254}No is shown in Fig. 4. We estimate the energies of transitions from the 2^+ and 4^+ states as 44(1) and 102(1) keV, respectively. These γ rays were not detected because the states decay almost entirely by internal electron conversion. The deduced transition energies also conform to those extrapolated from neighboring lighter nuclei, providing additional support for the assigned spins.

Sum-energy/Multiplicity Distribution

The analysis of the multiplicity vs. sum-energy distribution (Fig. 5) suggests that residues are formed with spins up to $\sim 18\hbar$. At E_{Lab}=215MeV the ^{208}Pb(^{48}Ca,2n) cold-fusion reaction can populate excited states in a limited range above the (measured) yrast line, up to the maximum excitation energy $E^{max} = E_{CN} - S_{n1} - S_{n2}$, after emission of two neutrons. The phase space is further limited by the kinetic energy of the neutrons. The populated states after neutron emission would lie below the yrast line for spins larger than 22. A new measurement at higher beam energies, leading to higher excitation energies, would bring larger spins into the No residue, so that the angular momentum dependence of the fission barrier can be studied and higher-spin states of the ground-state band perhaps identified.

DISCUSSION AND CONCLUSION

The B(E2) values of rotors are related to the 2^+ level energies by empirical formulae [8–10]. By using Eq. 4 of Ref. [9] and equations from Ref. [10] relating the B(E2), quadrupole moment and deformation, we deduce a quadrupole deformation parameter of $\beta = 0.27(2)$ for ^{254}No. (The uncertainty is given by the systematic deviations between measured B(E2) values in heavy nuclei and those deduced from the empirical relationship of Ref. [9].)

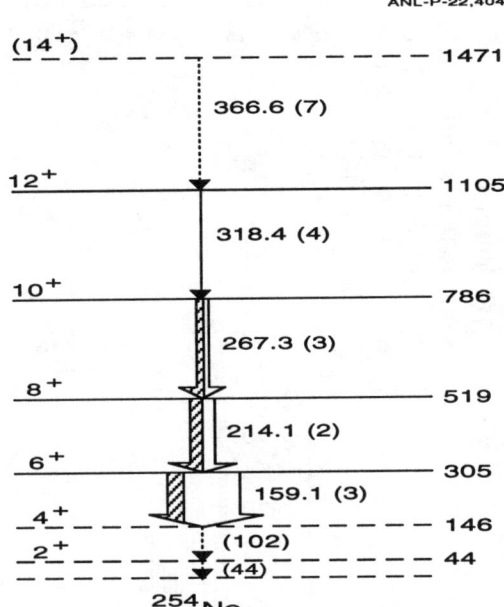

FIGURE 4. Proposed level scheme for the ground-state band of ^{254}No. Spins are deduced using Eq. 3; the parity is assumed to be positive. The energies of the lowest two transitions, which were not detected decay because they decay by internal conversion electrons, were deduced by extrapolation. The $14^+ \rightarrow 12^+$ assignment is tentative. The widths of the filled and open arrows are proportional to the γ and electron intensities, respectively; the latter were computed by assuming that the transitions have E2 multipolarity. Note the large conversion coefficients in nobelium. The transition intensities decrease as spin grows, as expected for a (HI,xn) reaction.

This value is in agreement with a value of 0.25 given by different macroscopic-microscopic model calculations [5,11,12], and with respective values of 0.27 and 0.26 from a Hartree-Fock-Bogoliubov (HFB) calculation with the SLy4 force [13] and from a relativistic Hartree-Bogoliubov calculation with the NL3 Lagrangian parameterization [14]. Other HFB and relativistic mean-field calculations with other force parameterizations [15] predict values between 0.28 and 0.31. Hence, the properties of TFN can, in principle, test the predictive power of the different interactions used in HFB and relativistic mean-field calculations for nuclei far from stability.

The measured moments of inertia increase with spin, as seen in Fig. 3, probably due to the gradual alignment of quasiparticles, specifically those occupying the high-j proton $i_{13/2}$ and neutron $j_{15/2}$ orbitals. Hence, the increase of $\mathcal{J}^{(2)}$ and $\mathcal{J}^{(1)}$ with frequency can provide a stringent test of theory. However, no calculations of finite-spin properties have been published so far.

The observation of states with spin up to 14 implies that neutron evaporation can compete against fission up to at least that spin. A fission barrier must still exist up to that angular momentum in ^{254}No. Preliminary analysis of the multiplicity distribution from our experiment suggests that residues are formed with spin up to $\sim$18. Our data further imply that the shell-correction energy, which creates the barrier, is reasonably robust with spin. They also demonstrate that high-spin states of TFN can be studied by means of (HI,xn) reactions. For future work, it will be interesting to investigate the fission barrier and its dependence on angular momentum to provide better insight into the production mechanism of the heaviest elements.

We acknowledge the efforts of the ATLAS crew in providing the required currents of ^{48}Ca and of B. Nardi in the operation of the target wheel. We are grateful for discussions with and private communications from P-H. Heenen, G. Lalazissis, A. Macchiavelli, W. Nazarewicz, K. Rutz, R. Smolanczuk and A. Sobiczewski. The work is supported by the U.S. Dept. of Energy, under Contract Nos. W-31-109-ENG-38 and DE-AC03-76SF00098, the U.S. National Science Foundation, the U.K. Engineering and Sciences Research Council and a NATO grant (for S.S.) through the Research Council of Norway.

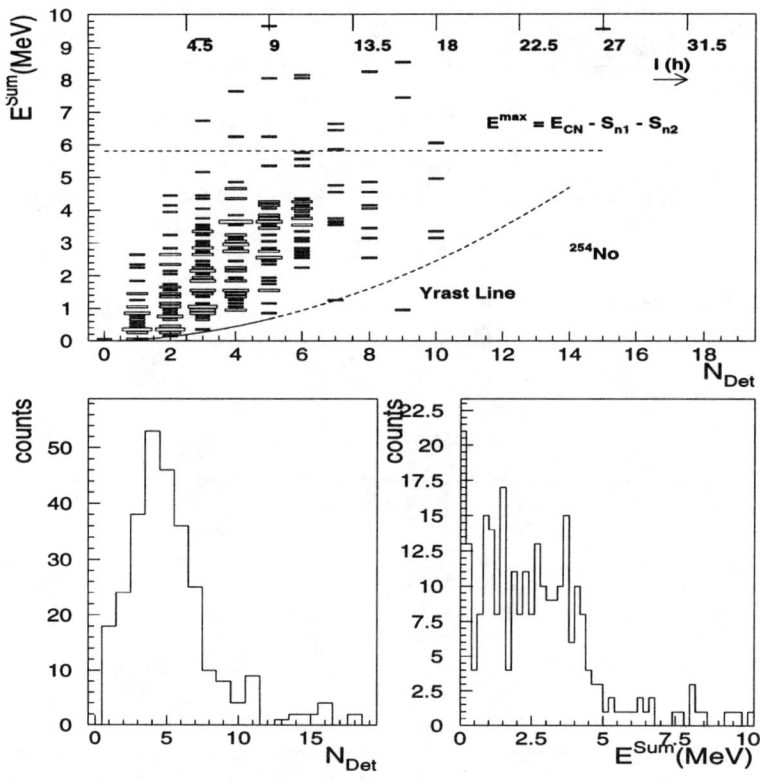

FIGURE 5. Two-dimensional spectrum of the Gammasphere module multiplicity vs. the sum-energy and the total projections. (A Gammasphere module is a combined detector unit consisting of a Ge-detector and the surrounding BGO shield.) The measured distribution is concentrated between the measured yrast line and the maximum excitation energy after neutron emission $E^{max} = E_{CN} - S_{n1} - S_{n2}$.

REFERENCES

1. P. Möller and J.R. Nix, J. Phys. **G20**, 1681 (1994).
2. R. Smolanczuk *et al.*, Phys. Rev. **C52**, 1871 (1995).
3. S. Hofmann, Rep. Prog. Phys. **61**, 639 (1998).
4. S. G. Nilsson *et al.*, Nucl. Phys. **A115**, 545 (1968).
5. Z. Patyk and A. Sobiczewski, Nucl. Phys. **A533**, 132 (1991).
6. H.W. Gäggeler *et al.*, Nucl. Phys. **A502**, 561c (1989).
7. J.E. Draper *et al.*, Phys. Rev. **C42**, R1791 (1990); J. Becker *et al.*, *ibid.* **C46**, 889 (1992); G. Hackman *et al.*, Phys. Rev. Lett. **79**, 4100 (1997).
8. L. Grodzins, Phys. Lett. **2**, 88 (1962).
9. S. Raman *et al.*, At. Data Nucl. Data Tables **42**, 1 (1989).
10. W. Nazarewicz and I. Ragnarsson, *Handbook Of Nuclear Properties*, Clarendon Press, Oxford, 1996, ed. D. Poenaru and W. Greiner, p. 97.
11. S. Cwiok *et al.*, Nucl. Phys. **A573**, 356 (1994).
12. P. Möller *et al.*, At. Data Nucl. Data Tables **59**, 185 (1995).
13. S. Cwiok *et al.*, Nucl. Phys. **A611**, 211 (1996); W. Nazarewicz and P-H. Heenen, priv. comm. (1998).
14. G. Lalazissis *et al.*, priv. comm. (1998); Nucl. Phys. **A608**, 202 (1996).
15. T. Bürvenich *et al.*, Eur. Phys., in press; J. K. Rutz *et al.*, Phys. Rev. C **56**, 238 (1997).

PHYSICS NEAR THE PROTON-DRIP LINE

Superallowed Fermi Beta Decay and Coulomb Mixing in Nuclei

J.C. Hardy[a] and I. S. Towner[b]

[a] Cyclotron Institute, Texas A & M University,
College Station, TX 77843
[b] Physics department, Queen's University,
Kingston, Ontario K7L 3N6, Canada

Abstract. Superallowed $0^+ \to 0^+$ nuclear beta decay provides a direct measure of the weak vector coupling constant, G_V. We survey current world data on the nine accurately determined transitions of this type, which range from the decay of ^{10}C to that of ^{54}Co, and demonstrate that the results confirm conservation of the weak vector current (CVC) but differ at the 98% confidence level from the unitarity condition for the Cabibbo-Kobayashi-Maskawa (CKM) matrix. We examine the reliability of the small calculated corrections that have been applied to the data, and conclude that there are no evident defects although the Coulomb correction, δ_C, depends sensitively on nuclear structure and thus needs to be constrained independently. The potential importance of a result in disagreement with unitarity, clearly indicates the need for further work to confirm or deny the discrepancy. We examine the options and recommend priorities for new experiments and improved calculations. Some of the required experiments depend upon the availability of intense radioactive beams. Others are possible with existing facilities.

INTRODUCTION

In probing the properties of the weak interaction, superallowed $0^+ \to 0^+$ nuclear beta decay has become a singularly valuable tool, principally because of its relative independence from the effects of nuclear structure, which are notoriously difficult to account for with high accuracy. Since the measured $0^+ \to 0^+$ transitions are between $T = 1$ analog states, nuclear-structure effects only enter at the level of the differences between the parent and daughter wave functions. These differences, which are caused by Coulomb and charge-dependent nuclear forces, turn out to be very small, and introduce a correction of order 1% when the experimental ft-values are used to extract a value for the effective weak vector coupling constant, G'_V. Even a conservative estimate of the uncertainties in this correction indicate that structure-dependent uncertainties should not afflict the experimental determination of G'_V above the level of approximately $\pm 0.1\%$. As a result, considerable effort has gone into making ft-value measurements that achieve this level of experimental precision or better.

In this paper, we begin by summarizing the current status of world data on superallowed Fermi beta decays and demonstrating the extent to which these data test the Conserved Vector Current (CVC) hypothesis and the unitarity of the Cabibbo-Kobayashi-Maskawa (CKM) matrix. In fact, we show that the success of the measurements has been such that experiment has outstripped theory. The theoretical uncertainties in calculated corrections – including the nuclear-structure-dependent charge-corrections – are now the limiting factor when superallowed beta decay is used to test the validity of the Standard Model. On the one hand, this can be viewed as a serious limitation to the continued usefulness of such measurements in future or, at least, as an indication that any improvements in their precision will have to be used to probe charge-dependence in nuclear structure rather than the fundamental properties of weak interactions. On the other hand, it can also be taken as a challenge to nuclear-structure theorists and experimenters alike to establish the effects of charge-dependence by independent means and thus to improve our ability overall to calculate them precisely.

Following our survey of world data, we outline the current approach to calculating charge-dependent effects, describe measurements that have been used to test these calculations independently, and consider future

CP481, *Nuclear Structure 98*, edited by C. Baktash

prospects for improved results. In particular, we shall consider the prospects for enlarging the sample of well-measured $0^+, T = 1$ superallowed emitters, and examine what needs to be done before such measurements can usefully contribute to our fundamental understanding of the weak interaction.

CURRENT STATUS OF WORLD DATA

Superallowed Fermi $0^+ \rightarrow 0^+$ nuclear beta decays [1,2] provide both the best test of the CVC hypothesis in weak interactions and, together with the muon lifetime, the most accurate value for the up-down quark-mixing matrix element of the CKM matrix, V_{ud}. Because the axial current cannot contribute in lowest order to transitions between spin-0 states, the experimental ft-value is related directly to the vector coupling constant. Specifically, for an isospin-1 multiplet,

$$ft(1 + \delta_R) = \frac{K}{G_V'^2 \langle M_V \rangle^2}, \tag{1}$$

with

$$
\begin{aligned}
G_V' &= G_V(1 + \Delta_R^V)^{1/2}, \\
\langle M_V \rangle^2 &= 2(1 - \delta_C), \\
K/(\hbar c)^6 &= 2\pi^3 \hbar \ln 2/(m_e c^2)^5 = (8120.271 \pm 0.012) \times 10^{-10} GeV^{-4} s,
\end{aligned} \tag{2}
$$

where f is the statistical rate function, t is the partial half-life for the transition, $\langle M_V \rangle$ is the Fermi matrix element and G_V is the primitive vector coupling constant. The physical constants used to evaluate K were taken from the most recent Particle Data Group publication [3]. These equations also include three calculated correction terms – all of order 1%. We write δ_R as the nucleus-dependent part of the radiative correction, Δ_R^V as the nucleus-independent part of the radiative correction, and δ_C as the isospin symmetry-breaking correction. A general description of these three correction terms and the methods used in their calculation has appeared elsewhere [1,2,4]. In the present context, it is sufficient to note that nuclear structure plays a small role in the determination of δ_R, but it is predominant for that of δ_C.

Equations (1) and (2) can now be combined into a form that is convenient for the analysis of experimental results:

$$\mathcal{F}t \equiv ft(1 + \delta_R)(1 - \delta_C) = \frac{K}{2G_V^2(1 + \Delta_R^V)}. \tag{3}$$

Here we have defined $\mathcal{F}t$ as the "corrected" ft-value. From this equation, it is evident that the $\mathcal{F}t$-values obtained from $0^+ \rightarrow 0^+$ transitions in different nuclei can constitute a stringent test of CVC, which requires them all to be equal.

To date, superallowed $0^+ \rightarrow 0^+$ transitions have been measured to $\pm 0.1\%$ precision or better in the decays of nine nuclei ranging from ^{10}C to ^{54}Co. World data on Q-values, lifetimes and branching ratios – the results of over 100 independent measurements – were thoroughly surveyed [1] in 1989 and then updated several times since, most recently for the WEIN98 conference [4]. The resulting weighted averages are given in the first three columns of Table 1. Using the calculated electron-capture probabilities [1] given in the next column, we obtain the "uncorrected" ft-values listed in column 5 with partial half-lives determined from the formula $t = t_{1/2}(1 + P_{EC})/R$.

To convert these results for ft into $\mathcal{F}t$-values, we apply the δ_C and δ_R corrections given in columns 6 and 7. We describe the δ_C calculations and their dependence on nuclear structure in a later section of this paper; the δ_C values used in Table 1 were taken from the last column of Table 3. The δ_R values come from our previous recent analyses [2,4] and arise from a variety of primary sources [1,5–8]. It is important to appreciate that the values of δ_C and δ_R result from more than one independent calculation. In the case of δ_R, the calculations are in complete accord with one another; for δ_C, we have used an average of two independent calculations with assigned uncertainties that reflect the (small) scatter between them. Thus, in a real sense, both experimentally and theoretically, the $\mathcal{F}t$-values given in Table 1 and plotted in Fig. 1 represent the totality of current world knowledge. The uncertainties reflect the experimental uncertainties and an estimate of the *relative* theoretical uncertainties in δ_C There is no statistically significant evidence of inconsistencies in the data ($\chi^2/\nu = 1.1$), thus verifying the expectation of CVC at the level of 3×10^{-4}, the fractional uncertainty quoted on the average $\mathcal{F}t$-value.

TABLE 1. Experimental results (Q_{EC}, $t_{1/2}$ and branching ratio, R), electron-capture probabilities (P_{EC}) and calculated corrections (δ_R and δ_C) for $0^+ \to 0^+$ transitions.

	Q_{EC} (keV)	$t_{1/2}$ (ms)	R (%)	P_{EC} (%)	ft (s)	δ_C (%)	δ_R (%)	$\mathcal{F}t$ (s)
^{10}C	1907.77(9)	19290(12)	1.4645(19)	0.296	3038.7(45)	0.16(3)	1.30(4)	3072.9(48)
^{14}O	2830.51(22)	70603(18)	99.336(10)	0.087	3038.1(18)	0.22(3)	1.26(5)	3069.7(26)
^{26m}Al	4232.42(35)	6344.9(19)	$\geq$ 99.97	0.083	3035.8(17)	0.31(3)	1.45(2)	3070.0(21)
^{34}Cl	5491.71(22)	1525.76(88)	$\geq$ 99.988	0.078	3048.4(19)	0.61(3)	1.33(3)	3070.1(24)
^{38m}K	6044.34(12)	923.95(64)	$\geq$ 99.998	0.082	3049.5(21)	0.62(3)	1.33(4)	3071.1(27)
^{42}Sc	6425.58(28)	680.72(26)	99.9941(14)	0.095	3045.1(14)	0.41(3)	1.47(5)	3077.3(23)
^{46}V	7050.63(69)	422.51(11)	99.9848(13)	0.096	3044.6(18)	0.41(3)	1.40(6)	3074.4(27)
^{50}Mn	7632.39(28)	283.25(14)	99.942(3)	0.100	3043.7(16)	0.41(3)	1.40(7)	3073.8(27)
^{54}Co	8242.56(28)	193.270(63)	99.9955(6)	0.104	3045.8(11)	0.52(3)	1.40(7)	3072.2(27)
						Average, $\overline{\mathcal{F}t}$		3072.3(9)
						χ^2/ν		1.10

In using this average $\mathcal{F}t$-value to determine V_{ud} and test CKM unitarity, we must account for additional uncertainty: *viz*

$$\overline{\mathcal{F}t} = 3072.3 \pm 0.9 \pm 1.1, \tag{4}$$

where the first error is the statistical error of the fit (as illustrated in Fig. 1), and the second is an error related to the systematic difference between the two calculations of δ_C by Towner, Hardy and Harvey [9,10] and by Ormand and Brown [11] that we have combined in reaching this result. (For a more complete discussion of how we treat these theoretical uncertainties, see reference [1].) We now add the two errors linearly to obtain the value we use in subsequent analysis:

$$\overline{\mathcal{F}t} = 3072.3 \pm 2.0. \tag{5}$$

The value of V_{ud} is obtained by relating the vector constant, G_V, determined from this $\overline{\mathcal{F}t}$ value, to the weak coupling constant from muon decay, $G_F/(\hbar c)^3 = (1.16639 \pm 0.00001) \times 10^{-5}$ GeV^{-2}, according to:

$$V_{ud}^2 = \frac{K}{2G_F^2(1 + \Delta_R^V)\overline{\mathcal{F}t}}. \tag{6}$$

With the nucleus-independent radiative correction adopted from Sirlin [12], $\Delta_R^V = (2.40 \pm 0.08)\%$, we obtain the result

$$|V_{ud}| = 0.9740 \pm 0.0005, \tag{7}$$

We are now in a position to test the unitarity of the CKM matrix by evaluating the sum of squares of the elements in its first row. With the value just obtained for V_{ud}, combined with the values of V_{us} and V_{ub} quoted by the Particle Data Group [3], the unitarity sum becomes

$$|V_{ud}|^2 + |V_{us}|^2 + |V_{ub}|^2 = 0.9968 \pm 0.0014, \tag{8}$$

which differs from unity at the 98% confidence level.

IS NON-UNITARITY REAL?

The result in equation (8) is a very provocative one. If it is taken at face value, it indicates the need for some extension to the electroweak Standard Model, possibly indicating the presence of right-hand currents or of a

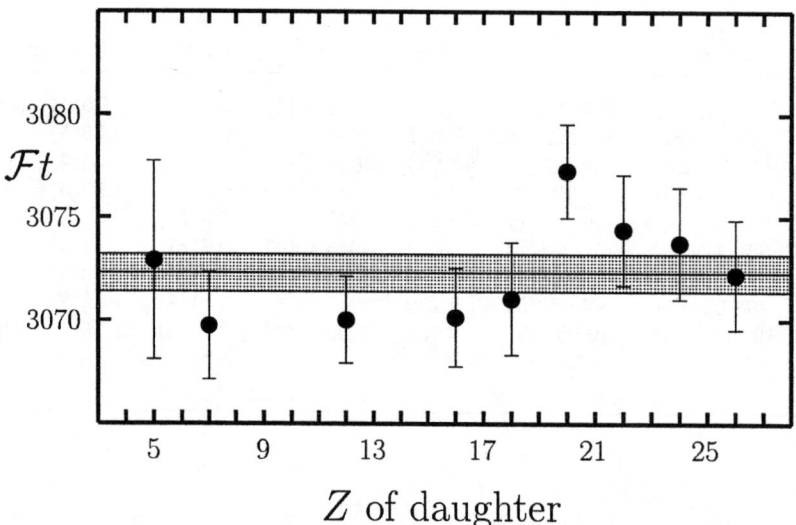

FIGURE 1. $\mathcal{F}t$-values for the nine precision data, and the best least-squares one-parameter fit.

scalar interaction [4]. This would have profound implications. However, the result could have a more trivial explanation. It could instead reflect some undiagnosed inadequacy in the calculated radiative or Coulomb corrections used to evaluate V_{ud} – or possibly a comparable inadequacy in the evaluation of V_{us}. What can be stated with some certainty is that the experimental results for the nine nuclei listed in Table 1 cannot be at fault. Not only do they originate from a large number of independent measurements, but also the error bar associated with $|V_{ud}|$ is *not* predominantly experimental in origin. In fact, if experiment were the sole contributor, the uncertainty would be only ±0.0001. The largest contributions to the $|V_{ud}|$ error bar come from $\Delta_{\text{R}}^{\text{V}}$ (±0.0004) and δ_C (±0.0003).

Thus, if we are to determine whether the minimal Standard Model has failed, we must eliminate all possible "trivial" explanations for the apparent non-unitarity. To do so, nuclear physicists focus on the reliability of the calculated corrections in V_{ud}. (Others are re-evaluating V_{us} – see reference [4].) But, if there is a fault in the corrections, what size effect are we seeking? To restore unitarity, the calculated radiative corrections (δ_R or $\Delta_{\text{R}}^{\text{V}}$) for all nine superallowed transitions would all have to be shifted downwards by 0.3%, or the calculated Coulomb corrections, δ_C, all shifted upwards by 0.3%, or some combination of the two. Such changes would constitute a substantial fraction of the total values of these small quantities. We have recently re-examined [4] the calculation of the various correction terms to see whether such large changes are plausible. Our conclusion is largely negative, based on arguments that are now briefly presented.

The radiative correction has been conveniently divided into terms that are nucleus-dependent, δ_R, and terms that are not, $\Delta_{\text{R}}^{\text{V}}$. These are written

$$\delta_R = \frac{\alpha}{2\pi}\left[\bar{g}(E_m) + \delta_2 + \delta_3 + 2C_{NS}\right]$$

$$\Delta_{\text{R}}^{\text{V}} = \frac{\alpha}{2\pi}\left[4\ln(m_z/m_p) + \ln(m_p/m_{\text{A}}) + 2C_{\text{Born}}\right] + \cdots, \tag{9}$$

where the ellipses represent further small terms of order 0.1%. In these equations, E_m is the maximum electron energy in beta decay, m_z the Z-boson mass, m_{A} the a_1-meson mass, and δ_2 and δ_3 the order $Z\alpha^2$ and $Z^2\alpha^3$ contributions. The electron-energy dependent function, $g(E_e, E_m)$, was derived by Sirlin [5]; it is here averaged over the electron spectrum to give $\bar{g}(E_m)$.

Typical values are

$$\delta_R \simeq 0.95 + 0.43 + 0.05 + (\alpha/\pi)C_{NS}\%, \tag{10}$$

where $(\alpha/\pi)C_{NS}$ is of order -0.3% for $T_z = -1$ beta emitters, ^{10}C and ^{14}O, and of order five times smaller for the $T_z = 0$ emitters, ranging from -0.09% to $+0.03\%$. Thus for $T_z = 0$ emitters $\delta_R \simeq 1.4\%$. If the failure to obtain unitarity in the CKM matrix with V_{ud} from nuclear beta decay is due to the value of δ_R, then δ_R must be reduced to 1.1%. This is not likely. The leading term, 0.95%, involves standard QED and is well verified. The order-$Z\alpha^2$ term, 0.43%, while less secure has been calculated twice [6,7] independently, with results in accord.

For the nucleus-independent term

$$\Delta_R^V = 2.12 - 0.03 + 0.20 + 0.1\% \simeq 2.4\%, \tag{11}$$

of which the first term, the leading logarithm, is unambiguous. Again, to achieve unitarity of the CKM matrix, Δ_R^V would have to be reduced to 2.1%, i.e. all terms other than the leading logarithm summing to zero. This also seems unlikely.

Because the leading terms in the radiative corrections are so well founded, attention has focused more on possible weaknesses in the Coulomb correction. Although smaller than the radiative correction, the Coulomb correction is clearly sensitive to nuclear-structure issues. It comes about because Coulomb and charge-dependent nuclear forces destroy isospin symmetry between the initial and final states in superallowed beta-decay. The consequences are twofold: there are different degrees of configuration mixing in the two states, and, because their binding energies are not identical, their radial wave functions differ. Thus, we accommodate both effects by writing $\delta_C = \delta_{C1} + \delta_{C2}$.

There have been several independent calculations of δ_C. The first followed methods developed by Towner, Hardy and Harvey [9] with refinements presented in more recent publications [10,15]. They use shell-model calculations to determine δ_{C1}, and full-parentage expansions in terms of Woods-Saxon radial wave functions to obtain δ_{C2}. A second calculation, by Ormand and Brown [11], also employed the shell model to obtain δ_{C1} but derived δ_{C2} from a self-consistent Hartree-Fock calculation. The results of these two calculations agree remarkably well with one another, the Towner-Hardy-Harvey values being systematically only 0.07% higher than Ormand-Brown ones. It is an average of these two sets of values that we have used for δ_C in our analysis as given in Table 1.

Two more recent calculations provide a valuable check that these δ_C values are not suffering from severe systematic effects. Sagawa, van Giai and Suzuki [13] have added RPA correlations to a Hartree-Fock calculation that incorporates charge-symmetry and charge-independence breaking forces in the mean-field potential to take account of isospin impurity in the core; the correlations, in essence, introduce a coupling to the isovector monopole giant resonance. The calculation is not constrained, however, to reproduce known separation energies as were the two calculations already described. Finally, a large-basis shell-model calculation has been mounted for the $A = 10$ case by Navrátil, Barrett and Ormand [14]. Both of these two new works have produced values of δ_C very similar to, but actually *smaller* than those used in our analysis, i.e. worsening rather than helping the unitarity problem.

The typical value of δ_C is of order 0.4%. If the unitarity problem is to be solved by improvements in δ_C, then δ_C has to be raised to around 0.7%. There is no evidence whatsoever for such a shift from recent works. Even so, considerable effort has already gone into making independent experimental checks on the accuracy of the δ_C calculations. These are described in the next section.

TESTS OF THE COULOMB CORRECTION

As shown in equation (2) the Coulomb correction, δ_C, modifies the square of the nuclear matrix element: $|M_V|^2 \to |M_V|^2(1 - \delta_C)$. Here M_V is the Fermi matrix element, the expectation value of the isospin ladder operator, which for isospin $T = 1$ states has the value $M_V = \sqrt{2}$. The value of δ_C clearly depends on the detailed structure of the nuclei involved. In the preceding section we have written δ_C as the sum of two components, δ_{C1} and δ_{C2}, the first reflecting the *difference* between configuration mixing in the initial and final states, while the second reflects the differences in their radial wave functions. Our calculations for δ_{C2} have been documented in a paper by Towner, Hardy and Harvey [9] and so will not be repeated here. It is sufficient to stress that constraints are placed on the δ_{C2}-calculations by insisting that the asymptotic forms of the proton and neutron radial functions match the known separation energies. Our calculations for δ_{C1} are documented here.

It is instructive to consider a simplified two-state mixing case as it will illustrate the issues involved. As a specific example, take the case of ^{42}Sc decaying to ^{42}Ca in a superallowed Fermi transition. In the calculation we admit two 0^+ states, the ground state and one excited state. One of these states might be a two-particle state, $|2p\rangle$, relative to a closed ^{40}Ca core, the other might be a four-particle two-hole state, $|4p\text{-}2h\rangle$. The strong interaction heavily mixes the two states, so the ground state, ψ_0, and excited state, ψ_1, will have wave functions

$$\psi_0 = A|2p\rangle + B|4p\text{-}2h\rangle$$
$$\psi_1 = B|2p\rangle - A|4p\text{-}2h\rangle, \tag{12}$$

where the mixing amplitudes A and B will depend on the details of the strong interaction. The strong interaction, however, is isospin invariant so the same wave functions describe the states in ^{42}Sc and the isospin mirror states in ^{42}Ca. Thus the Fermi matrix element between ground states is

$$\begin{aligned}\langle\psi_0|\tau_+|\psi_0\rangle &= A^2\langle 2p|\tau_+|2p\rangle + B^2\langle 4p\text{-}2h|\tau_+|4p\text{-}2h\rangle \\ &= \sqrt{2}(A^2 + B^2) \\ &= \sqrt{2}, \end{aligned} \tag{13}$$

while the Fermi matrix element between the ^{42}Sc ground state and the ^{42}Ca excited state is

$$\begin{aligned}\langle\psi_1|\tau_+|\psi_0\rangle &= AB\langle 2p|\tau_+|2p\rangle - AB\langle 4p\text{-}2h|\tau_+|4p\text{-}2h\rangle \\ &= \sqrt{2}(AB - AB) \\ &= 0. \end{aligned} \tag{14}$$

This latter result is just a reminder that the operator for superallowed Fermi decay, being the isospin ladder operator, only connects with states of the same isospin multiplet, *i.e.* analogue states.

Now, suppose we add to the Hamiltonian Coulomb and other charge-dependent forces. These terms, no doubt, will be much weaker than the strong-interaction forces, but their impact is to modify slightly the wave functions ψ_0 and ψ_1 and by differing amounts in ^{42}Sc and ^{42}Ca. Thus these wavefunctions are now written

$$\begin{aligned}\psi_i(^{42}Sc) &= b_0\psi_0 + b_1\psi_1 \\ \psi_{f0}(^{42}Ca) &= a_0\psi_0 + a_1\psi_1 \\ \psi_{f1}(^{42}Ca) &= a_1\psi_0 - a_0\psi_1, \end{aligned} \tag{15}$$

with a_0 and b_0 both being close to unity. The Fermi matrix element between ground states becomes

$$\begin{aligned}\langle\psi_{f0}(^{42}Ca)|\tau_+|\psi_i(^{42}Sc)\rangle &= a_0b_0\langle\psi_0|\tau_+|\psi_0\rangle + a_1b_1\langle\psi_1|\tau_+|\psi_1\rangle \\ &= \sqrt{2}(a_0b_0 + a_1b_1), \end{aligned}$$
$$|\langle\psi_{f0}(^{42}Ca)|\tau_+|\psi_i(^{42}Sc)\rangle|^2 \simeq 2\left(1 - (a_1 - b_1)^2\right), \tag{16}$$

using $a_0^2 + a_1^2 = 1$, $b_0^2 + b_1^2 = 1$. Further the matrix element between ^{42}Sc ground state and the ^{42}Ca excited state is

$$\begin{aligned}\langle\psi_{f1}(^{42}Ca)|\tau_+|\psi_i(^{42}Sc)\rangle &= a_1b_0\langle\psi_0|\tau_+|\psi_0\rangle - a_0b_1\langle\psi_1|\tau_+|\psi_1\rangle \\ &= \sqrt{2}(a_1b_0 - a_0b_1), \end{aligned}$$
$$|\langle\psi_{f1}(^{42}Ca)|\tau_+|\psi_i(^{42}Sc)\rangle|^2 \simeq 2(a_1 - b_1)^2, \tag{17}$$

and is no longer zero. If we write the ground state to ground state matrix element squared as: $|M_V^0|^2 = 2(1 - \delta_{C1}^0)$, and the ground state to excited state matrix element squared as: $|M_V^1|^2 = 2\delta_{C1}^1$, then for two-state mixing the corrections are equal:

$$\delta_{C1}^0 = \delta_{C1}^1 = (a_1 - b_1)^2. \tag{18}$$

This is a specific result for two-state mixing, in general they would not be equal. Further the correction δ_{C1} depends on the *difference* in the degree of isospin mixing in ^{42}Sc relative to ^{42}Ca. This is clearly evident if we use perturbation theory to estimate the small amplitudes a_1 and b_1. Let $V_C(^{42}Ca)$ and $V_C(^{42}Sc)$ be the Coulomb and other charge-dependent forces operative in ^{42}Ca and ^{42}Sc respectively, then

$$\delta_{C1}^0 = \delta_{C1}^1 = \frac{\left[\langle\psi_1|V_C(^{42}Ca)|\psi_0\rangle - \langle\psi_1|V_C(^{42}Sc)|\psi_0\rangle\right]^2}{(E_1 - E_0)^2}, \tag{19}$$

TABLE 2. Experimental branching ratios, R, for non-analogue Fermi transitions, and values of δ_{C1} from experiment and theory.

Nuclide	Expt[a]		Theory	
	R(ppm)	δ_{C1}^1(%)	δ_{C1}^1(%)	δ_{C1}^0(%)
^{38m}K	< 19	< 0.28	0.096	0.100
^{42}Sc	59(14)[b]	0.040(9)	0.041	0.049
^{46}V	39(4)	0.053(5)	0.046	0.087
^{50}Mn	< 3	< 0.016	0.051	0.068
^{54}Co	45(6)	0.035(5)	0.037	0.045

[a] From Hagberg *et al.* [15]
[b] Daehnick *et al.* [19] averaged with earlier results [20–22]

where $E_1 - E_0$ is the energy separation between the excited- and ground- 0^+ states from the charge-independent Hamiltonian. Thus δ_{C1} is inversely proportional to the square of this energy difference.

In a shell-model calculation it is notoriously difficult to obtain correctly the experimental $E_1 - E_0$ energy separation. This is because the $|4p\text{-}2h\rangle$ is a deformed state, while the $|2p\rangle$ is a spherical state and these two distinct aspects are difficult to realise in a truncated calculation in a spherical basis. Thus in the calculations we are about to describe, the values of δ_{C1}^0 and δ_{C1}^1 obtained are both corrected using

$$\delta_{C1} = \delta_{C1}^{\text{calc}} \times \frac{(E_1 - E_0)^2_{\text{calc}}}{(E_1 - E_0)^2_{\text{expt}}}. \tag{20}$$

In this way, we believe we are adjusting the δ_{C1} value approximately for the imperfections of the underlying strong-interaction Hamiltonian and the necessity of using model-space truncations.

Here we discuss calculations for ^{38m}K, ^{42}Sc, ^{46}V, ^{50}Mn and ^{54}Co, which were recomputed recently to compare with the excited state non-analogue Fermi transitions measured by Hagberg *et al.* [15]. In each case we use the largest model space practicable in a proton-neutron (pn) basis. For ^{38m}K and ^{42}Sc this involved orbitals in both the (s, d) and (p, f) shells and the effective interaction constructed by Warburton, Becker, Millener and Brown (WBMB) [16] was used for the underlying strong interaction. For ^{46}V, ^{50}Mn and ^{54}Co, the orbitals span the (p, f) shell and the strong interaction was taken to be FPMI3 from Richter *et al.* [17]. The single-particle energies were fixed from experimental values at the closed-shell-plus-one configuration.

The Coulomb and charge-dependent interaction terms to be added were constrained so that the ground-state masses of the isotriplet of states, $T_z = -1, 0, +1$, were fitted. The isobaric multiplet mass equation (IMME) writes these masses as

$$M(T_z) = a + bT_z + cT_z^2, \tag{21}$$

where the coefficient a represents the result from a charge-independent Hamiltonian, and b and c are charge-dependent corrections. Specifically

$$\begin{aligned} b &= \left(M(T_z = +1) - M(T_z = -1) \right)/2 \\ c &= \left(M(T_z = +1) + M(T_z = -1) - 2M(T_z = 0) \right)/2, \end{aligned} \tag{22}$$

so b is governed by the difference between pp and nn forces, and c by the difference between the pn and nn forces. Our strategy was to multiply the two-body Coulomb matrix elements by a constant factor so that the b-coefficient of the IMME is reproduced, and likewise the pn matrix elements are multiplied by a constant factor to reproduce the c-coefficient. This strategy of adjusting the strength of the Coulomb and charge-dependent nuclear forces to reproduce the IMME equation was pioneered by Ormand and Brown [18].

The results of these calculations for δ_{C1}^0 and δ_{C1}^1 for nuclei ^{38m}K, ^{42}Sc, ^{46}V, ^{50}Mn and ^{54}Co are given in Table 2. Only the δ_{C1}^0 value is needed for the analysis of the superallowed Fermi data. However, the companion δ_{C1}^1 value can be subjected to an experimental test. If the Fermi transition to an excited 0^+, non-analogue, state can be determined then the measured branching ratio, R, is proportional to δ_{C1}^1. These branching ratios are very small, parts per million (ppm), so their measurement requires a dedicated effort, and the work of

TABLE 3. Calculated Coulomb correction, δ_C in percent units.

| Nuclide | Towner-Hardy[a] | | Ormand-Brown[b] | | Adopted Value[c] |
	$\delta_{C1}^0(\%)$	$\delta_{C2}(\%)$	$\delta_{C1}^0(\%)$	$\delta_{C2}(\%)$	$\delta_C = \delta_{C1} + \delta_{C2}(\%)$
^{10}C	0.006	0.17	0.04	0.11	0.16(3)
^{14}O	0.004	0.28	0.01	0.14	0.22(3)
^{26m}Al	0.057	0.27	0.01	0.29	0.31(3)
^{34}Cl	0.024	0.62	0.06	0.51	0.61(3)
^{38m}K	0.100	0.54	0.11	0.48	0.62(3)
^{42}Sc	0.049	0.35	0.11	0.31	0.41(3)
^{46}V	0.087	0.36	0.09	0.29	0.41(3)
^{50}Mn	0.068	0.40	0.02	0.33	0.41(3)
^{54}Co	0.045	0.56	0.04	0.40	0.52(3)

[a] Refs. [9,10]
[b] Ref. [11]
[c] Average of Towner-Hardy and Ormand-Brown values; assigned uncertainties reflect the *relative* scatter between these calculations.

the Chalk River group in this regard is described in Hagberg *et al.* [15]. If t_0 is the partial half-life for the ground-state decay and t_1 the the partial half-life to the excited 0^+ state, then

$$R = \frac{t_0}{t_1} = \frac{f_1}{f_0}\frac{f_0 t_0}{f_1 t_1} = \frac{f_1}{f_0}\frac{2\delta_{C1}^1}{2(1 - \delta_{C1}^0)} \simeq \frac{f_1}{f_0}\delta_{C1}^1, \tag{23}$$

where f_0 and f_1 are phase space integrals for the ground state and excited state respectively. Table 2 lists the experimental value of R and δ_{C1}^1. The comparison between theory and experiment is exceptional in all cases except ^{50}Mn and so provides a lot of confidence that the companion δ_{C1}^0 values used in the superallowed Fermi data analysis are reasonable. The ^{50}Mn case is interesting in that the experiment was unable to locate a branch to an excited 0^+ state in the expected energy region and so deduced that any such branching ratio would be less than 3 ppm.

The δ_{C1}^0 values for lighter superallowed Fermi emitters, ^{10}C, ^{14}O, ^{26m}Al and ^{34}Cl were obtained in a similar manner, as described in Towner [10]. The $\delta_{C1} = \delta_{C1}^0$ values are quite small, of the order of 0.1%, but seem under reasonable control. The larger δ_{C2} values associated with radial overlap integrals, potentially can have more uncertainty. We have found that our results based on using Saxon-Woods radial functions lead to systematically larger δ_{C2} values than the calculations of Ormand and Brown [11] with Hartree-Fock functions. We therefore allow for this systematic difference in our data analysis by using average values and introducing a systematic uncertainty to $\overline{\mathcal{F}t}$ (see equation (4)). In Table 3 we give our δ_{C1} and δ_{C2} values, together with those of Ormand and Brown, and our adopted final numbers, $\delta_C = \delta_{C1} + \delta_{C2}$. These are the values that appear in Table 1 and are used to analyze the world data for superallowed $0^+ \rightarrow 0^+$ decays.

FUTURE DIRECTIONS

With the experimental evidence so far completely in support of the calculated values for δ_C, the current world data on superallowed $0^+ \rightarrow 0^+$ beta decay are tantalizingly close to a result in definitive disagreement with CKM unitarity. Naively one might expect such a situation would prompt an urgent new round of experiments with the goal of further reducing the quoted uncertainty in $|V_{ud}|$ but, unfortunately, the next step cannot be so straightforward. As we have already noted, the error bar associated with $|V_{ud}|$ in equation (7) is now dominated by uncertainties in the calculated correction terms. Any improvements in precision made within the existing body of experimental data will be effectively lost once the results are applied to the unitarity test so long as there are no improvements in the calculations.

Clearly of highest priority in future must be to increase the precision of the correction terms, particularly Δ_R^V, which is the largest contributor to the uncertainty of $|V_{ud}|$. Its importance is manifest not only in the unitarity test based on superallowed $0^+ \rightarrow 0^+$ beta decay but also in any tests based on neutron or pion decays. To date, the experimental data on these non-nuclear decays are considerably less precise [4] than those on the $0^+ \rightarrow 0^+$ transitions but, year by year, improvements are being made in the neutron-decay measurements largely motivated by the prospect of a unitarity test unfettered with the structure-dependent

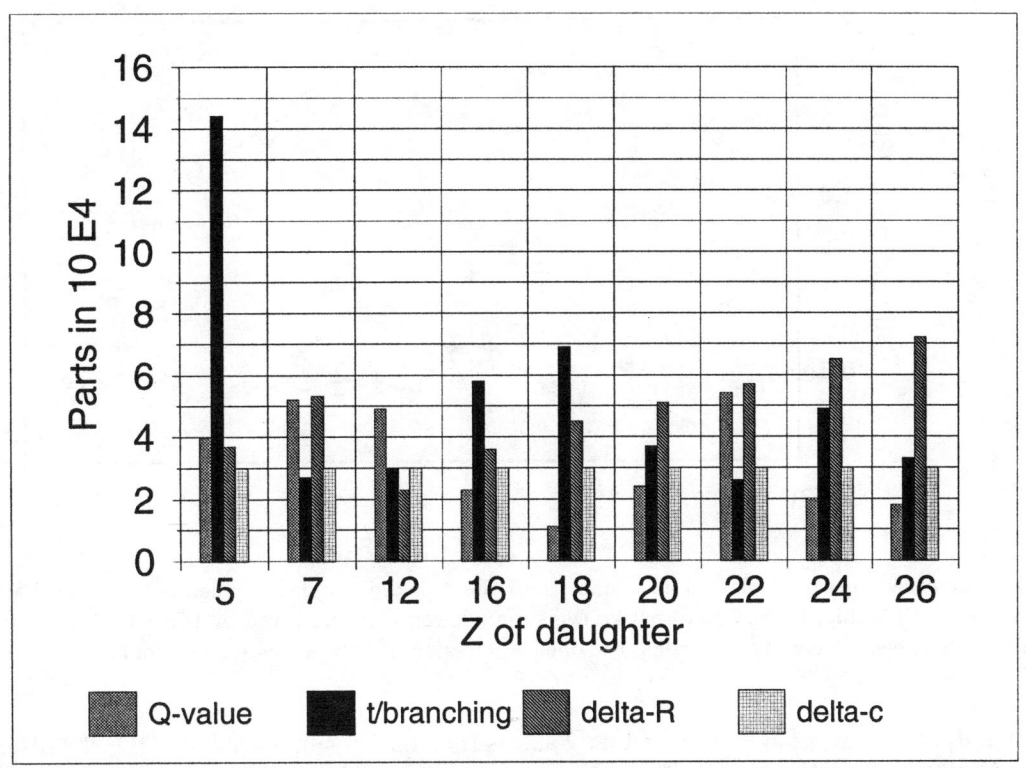

FIGURE 2. Contributions from experiment and theory to the overall $\mathcal{F}t$-value uncertainty for each superallowed transition listed in Table 1.

Coulomb correction, δ_C, which vanishes for the neutron. In spite of this simplification in the neutron decay, however, any potential advantage to the unitarity test is lost unless Δ_R^V can be calculated with greater precision, since *all* determinations of $|V_{ud}|$ depend directly on Δ_R^V.

Until such time as the Δ_R^V calculation is refined and the neutron decay measurements rival the precision of the nuclear results, the best hope for improvements to the unitarity test lies in increasing our confidence in the calculated values of δ_C. Though a reasonable estimate of the uncertainty in δ_C has been incorporated into the derivation of $|V_{ud}|$ (see Tables 1 and 3), there is no doubt that the dominant role of nuclear structure in the calculation of δ_C leaves some people questioning whether the real uncertainty is larger than the one actually quoted. Under the circumstances, any experiment that can probe the veracity of the δ_C calculations will make a valuable contribution to the whole problem. There are at least three different approaches that can be taken in devising such experiments.

First, improvements can be sought in results for the nine superallowed transitions whose ft-values are already known to within a fraction of a percent. This would not be a fruitless endeavour. It is certainly true that, given the large quantity of careful measurements now contributing to the content of Table 1, there is little chance that the central value of $\overline{\mathcal{F}t}$ will be changed significantly by a few more. And, it is also true that improvements in the experimental uncertainties will not be directly reflected in a reduced uncertainty for $|V_{ud}|$ unless the Δ_R^V correction has been improved too. But, the test of CVC can be made more demanding as the experimental precision is increased and, to the extent that the $\mathcal{F}t$-values continue to agree with one another, this would demonstrate at the same time the reliability of the δ_C calculations, which compensate for the transition-to-transition variations evident in the uncorrected ft-values. Of course, it is only the *relative* values of δ_C that can be tested by this method, but it would be a pathological fault indeed that could calculate in detail the required variations in δ_C while failing to obtain their *absolute* values to comparable precision.

The various experimental and theoretical contributions to the $\mathcal{F}t$-value uncertainties are shown in Fig. 2 for the nine superallowed transitions whose ft values are known to within a fraction of a percent. If we accept that it is valuable for experiment to be at least a factor of two more precise than the calculations for δ_R and δ_C (*relative*

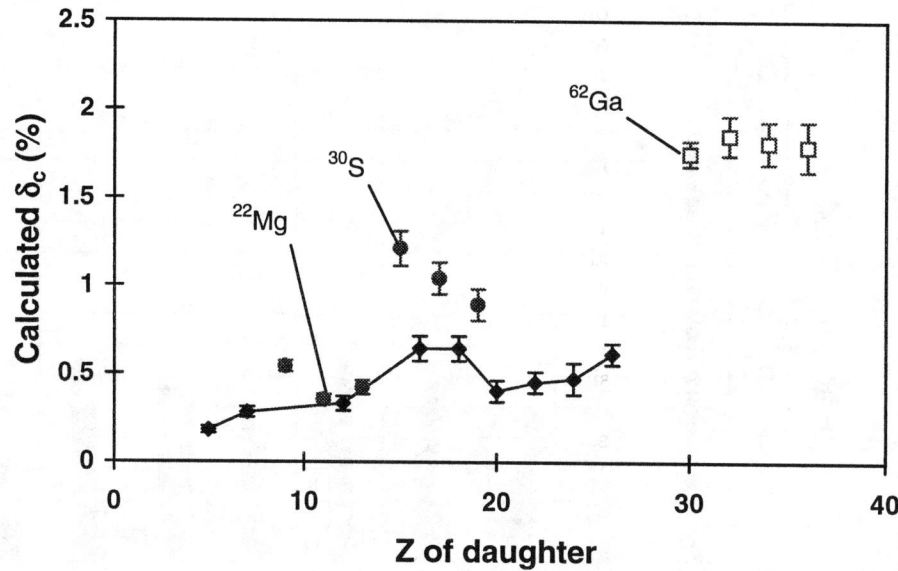

FIGURE 3. Calculated δ_C values plotted as a function of the Z of the daughter nucleus. The solid diamonds joined by the line represent the values for the nine well known superallowed emitters listed in Table 1. The circles are for the $T_z = -1$ emitters between ^{18}Ne and ^{38}Ca; while the open squares are for the $T_z = 0$ cases from ^{62}Ga to ^{74}Rb.

uncertainties only), then an examination of Fig. 2 shows that the Q-values for ^{10}C, ^{14}O, ^{26m}Al and ^{46}V, the half-lives of ^{10}C, ^{34}Cl and ^{38m}K, and the branching ratio for ^{10}C can all bear improvement. Such improvements will soon be feasible. The Q-values will reach the required level (and more) as mass measurements with new on-line Penning traps become possible; half-lives will likely yield to measurements with higher statistics as high-intensity beams of separated isotopes are developed for radioactive-beam facilities; and, finally, an improved branching-ratio measurement on ^{10}C has already been made with Gammasphere and simply awaits analysis [23].

Another experimental approach to testing δ_C is offered by the possibility of increasing the number of superallowed emitters accessible to precision studies. The greatest attention recently has been paid to the $T_z = 0$ (odd-odd) emitters with $A \geq 62$, since these nuclei are expected to be produced at new radioactive-beam facilities, and their calculated Coulomb corrections, δ_C, are predicted to be large [11,13,24], as is illustrated in Fig. 3. In principle, then, they could provide a valuable test of the accuracy of δ_C calculations. It is likely, though, that these heavy emitters will not provide ft-values with sufficient precision to be useful directly in extracting competitive $\mathcal{F}t$-values in the near future. All of the well known emitters listed in Table 1, with the exception of ^{10}C, have the special advantage that the superallowed branch from each is by far the dominant transition in its decay ($> 99\%$). This means that the branching ratio for the superallowed transitions can be determined to high precision from relatively imprecise measurements of the other weak transitions, which can simply be subtracted from 100%. In contrast, the decays of the heavier $T_z = 0$ emitters – nuclides such as ^{62}Ga, ^{66}As, ^{70}Br and ^{74}Rb – will be of considerably higher energy and each will therefore involve several allowed transitions of significant intensity in addition to the superallowed transition. Branching-ratio measurements will thus be very demanding, particularly with the limited radioactive-beam intensities likely to be available initially for these rather exotic nuclei. Lifetime measurements will be similarly constrained by statistics. As to the Q_{EC} values, even Penning traps will be hard pressed to produce the required precision of a few parts in 10^9 for the masses of these short-lived ($t_{1/2} \leq 100$ ms) nuclides.

More immediately achievable among these heavier superallowed emitters are measurements of the type described in the preceding section, in which non-analogue Fermi transitions were observed [15] and compared with model calculations as a test of the techniques used in calculating δ_C. Such measurements would yield important information on Coulomb mixing and they would be an important prerequisite for any serious attempt to obtain $\mathcal{F}t$-values in this region of nuclei. Another important prerequisite would be the experimental determination of the coefficients of the IMME for all relevant 0^+ $T = 1$ states. These demanding experiments are interesting in their own right but, as preliminaries to future $\mathcal{F}t$-value determinations, they are especially important because model calculations in this region of rapid shape changes are likely to be far less reliable than they are in the (s, d) and (p, f) shells. Without the constraints provided by these experiments – and possibly even with them – tests of δ_C from the measured $\mathcal{F}t$-values could end up reflecting more about the limitations of the nuclear models used than about the underlying physics of the weak interaction. As a first probe of these issues, an investigation of non-analogue Fermi transitions from ^{62}Ga is already underway [25]. More will undoubtedly follow.

In the near future, the most promising experimental approach that can actually increase the number of precisely measured $\mathcal{F}t$-values is to study the $T_z = -1$ superallowed emitters with $18 \leq A \leq 38$. There is good reason to explore them. For example, as shown in Fig. 3, the calculated value [9] of δ_C for ^{30}S decay, though smaller than the δ_C's expected for the heavier nuclei, is actually 1.2% – about a factor of two larger than for any other case currently known – while ^{22}Mg has a very low value of 0.35%. If the ft-values for these two nuclei can be determined to a precision of a few tenths of a percent or better, and the large predicted difference is confirmed, then it will do much to increase our confidence in the calculated Coulomb corrections. This would be especially convincing since the calculation involves the same model space as was used for the presently known cases. To be sure, these decays will provide an experimental challenge, particularly in the measurement of their branching ratios, but the required precision should be achievable with isotope-separated beams that are currently available. In fact, such experiments are also in their early stages at the Texas A&M cyclotron [26].

CONCLUSIONS

The current world data on superallowed $0^+ \to 0^+$ beta decays lead to a self-consistent set of $\mathcal{F}t$-values that agree with CVC but differ provocatively, though not yet definitively, from the expectation of CKM unitarity. There are no evident defects in the calculated radiative and Coulomb corrections that could remove the problem, but suspicion continues to fall on the calculations of Coulomb mixing, which depend sensitively on the details of nuclear structure. If any progress is to be made in firmly establishing (or eliminating) the discrepancy with unitarity, additional experiments are required that focus on this issue. We have indicated what some relevant nuclear experiments might be, and have particularly emphasized that experiments to measure the ft-values of heavy $T_z = 0$ odd-odd superallowed emitters with $A \geq 62$, which have been proposed for new radioactive-beam facilities, are very difficult and should be preceded by measurements in the same mass region of non-analogue Fermi decays and IMME coefficients.

On the theoretical side, the most important requirement for all tests of CKM universality that depend upon V_{ud} is an improved determination of the nucleus-independent radiative correction, Δ_R^V. It is not only the results from nuclear superallowed decays that must be subjected to this correction term, but also the results from the neutron and pion decays if they also are to be used to extract V_{ud}. Though the latter decays are currently known with less precision than the nuclear decays, one can reasonably expect them to improve significantly over the next decade. Thus, it is of highest priority to reduce the uncertainty currently attached to the calculation of Δ_R^V. That having been stated, it must be noted that nuclear decays will also require more reliable δ_C calculations to remain competitive, especially if the uncertainties are reduced on Δ_R^V. In any case, improved model calculations of nuclear structure and Coulomb mixing in nuclei with $N \simeq Z$ and $A \geq 62$ are an important requirement for the future if ft-value measurements are to be attempted in this region.

The work of JCH was supported by the U.S. Department of Energy under Grant number DE-FG05-93ER40773 and by the Robert A. Welch Foundation.

REFERENCES

1. Hardy, J.C., Towner, I.S., Koslowsky, V.T., Hagberg, E., and Schmeing, H., *Nucl. Phys.* A**509**, 429 (1990).
2. Towner, I.S., and Hardy, J.C., in *Symmetries and Fundamental Interactions in Nuclei*, eds. Henley, E.M., and Haxton, W.C., Singapore: World-Scientific, 1995, pp.183-249.
3. Caso, C., *et al.*, *Eur. Phys. J.* C**3**, 1 (1998).
4. Towner, I.S., and Hardy, J.C., in *Proceedings of the V Int. Symp. on Weak and Electromagnetic Interactions* Singapore: World-Scientific, 1998, to be published.
5. Sirlin, A., *Phys. Rev.* **164**, 1767 (1967).
6. Sirlin, A., *Phys. Rev.* D**35**, 3423 (1987); Sirlin, A., and Zucchini, R., *Phys. Rev. Lett.* **57**, 1994 (1986).
7. Jaus, W., and Rasche, G., *Phys. Rev.* D**35**, 3420 (1987).

8. Towner, I.S., *Phys. Lett.* B**333**, 13 (1994).

9. Towner, I.S., Hardy, J.C., and Harvey, M., *Nucl. Phys.* A**284**, 269 (1977).

10. Towner, I.S., in *Symmetry Violations in Subatomic Physics*, eds. Castel, B., and O'Donnel, P.J., Singapore: World-Scientific, 1989, p.211.

11. Ormand, W.E., and Brown, B.A., *Phys. Rev.* C**52**, 2455 (1995).

12. Sirlin, A., in *Precision Tests of the Standard Electroweak Model*, ed. Langacker, P., Singapore: World-Scientific, Singapore, 1994.

13. Sagawa, H., van Giai, N., and Suzuki, T., *Phys. Rev.* C**53**, 2163 (1996).

14. Navrátil, P., Barrett, B.R., and Ormand, W.E., *Phys. Rev.* C**56**, 2542 (1997).

15. Hagberg, E., Koslowsky, V.T., Hardy, J.C., Towner, I.S., Hykawy, J.G., Savard, G., and Shinozuka, T., *Phys. Rev. Lett.* **73**, 396 (1994).

16. Warburton, E.K., Becker, J.A., Brown, B.A., and Millener, D.J., *Ann. Phys. (N.Y.)* **187**, 471 (1988).

17. Richter, W.A., van dar Merwe, M.G., Julies, R.E., and Brown, B.A., *Nucl. Phys.* A**523**, 325 (1991).

18. Ormand, W.E., and Brown, B.A., *Nucl. Phys.* A**440**, 274 (1985).

19. Daehnick, W.W., and Rosa, R.D., *Phys. Rev.* C**5**, 1499 (1985).

20. Sandorfi, A.M., Lister, C.J., Alburger, D.E., and Warburton, E.K., *Phys. Rev.* C**22**, 2213 (1980).

21. DelVecchio, R.M., and Daehnick, W.W., *Phys. Rev.* C**17**, 1809 (1978).

22. Ingalls, R.D., Overley, J.C., and Wilson, H.S., *Nucl. Phys.* A**293**, 117 (1977).

23. Freedman, S.J., private communication.

24. Hardy, J.C., in *Proceedings of the Workshop on the Production and Use of Intense Radioactive Beams at the Isospin Laboratory, 1992*, ed. J.D. Garrett (Joint Institute for Heavy Ion Research, Oak Ridge, Tennessee: Conf. 9210121) pg. 51.

25. Hyman, B.C., Azhari, A., Gagliardi, C.A., Hardy, J.C., Sandu, T., Tang, X.D., Trache, L. and Tribble, R.E., in *Progress in Research, 1997/1998, Cyclotron Institute, Texas A& M University*, pg. I-31.

26. Hardy, J.C., Azhari, R. Burch, A., Gagliardi, C.A., Lipnik, P., Mayes, E., Trache, L. and Tribble, R.E., in *Progress in Research, 1997/1998, Cyclotron Institute, Texas A& M University*, pg. I-34.

Wigner energy, odd-even mass staggering and the time-odd mean-fields

Wojciech Satuła

Institute of Theoretical Physics, University of Warsaw, ul. Hoża 69, 00-681 Warszawa, Poland
Joint Institute for Heavy Ion Research, Oak Ridge, TN 37831-6374, USA
Department of Physics, University of Tennessee, Knoxville, TN 37996, USA

Abstract. Various properties of single-particle Hartree-Fock ground-state solutions in $N \sim Z$ nuclei are investigated. The emphasis is on a role of single-particle mean-field in odd-even mass staggering. It is shown that, unlike in traditional scenario originating from the Fermi gas or macroscopic models, the symmetry energy contribution to odd-even mass staggering is nearly cancelled by the contribution coming from the average level density. It allows to construct indicators probing both pairing as well as mean-field components to the odd-even mass staggering. The impact of the single-particle Hartree-Fock field on Wigner energy and residual pn interaction in odd-odd nuclei is also discussed.

INTRODUCTION

Mean-field is considered as a standard nuclear model to describe medium mass and heavy nuclei. Indeed, various applications demonstrated ability of this relatively simple concept to describe wide range of nuclear phenomena. However, the accuracy of mean-field results or predictions is far from satisfactory when confronted with the present day, high precision experimental data. The predictions in many cases differ substantially for various parametrizations of the same effective interaction and the situation calls for systematic program to optimize the forces. Recent work of Chabanat *et al.* [1] can be considered as the first step towards finding comprehensive parametrization of the Skyrme force [2]. Indeed, the so called Lyon (SLy) forces developed in Ref. [1] are very widely used and are considered to be one of the best among the Skyrme forces available nowadays.

The aim of this work is to search for certain generic features of the ground-state Skyrme-Hartree-Fock (SHF) solutions which may help in better understanding of basic properties of this force through the nature of self-consistent solutions. First section discusses a role of single-particle SHF mean-field in odd-even mass staggering (OES) which, in atomic nuclei, is usually attributed to pairing [3,4]. Proper understanding of mean-field and pairing contributions to OES is crucial to better understand and parametrize both channels. The second section investigates the possibility to obtain an additional binding energy in the $N = Z$ nuclei within single-particle self-consistent SHF model or conventional, i.e. including only $T = 1, |T_z| = 1$ pairing correlations, self-consistent Hartree-Fock-Bogolyubov model. Finally, the third section discusses properties of self-consistent SHF solutions in time-odd channel.

ODD-EVEN MASS STAGGERING

The odd-even mass staggering in atomic nuclei is usually attributed to the presence of pairing correlations [3]. Indeed, in the simplest scenario based on standard BCS theory of superfluidity [5], the ground state energies of three adjacent isotopes (isotones) can be approximately interrelated by a simple readjustment of their chemical potentials λ:

$$B(N), \quad B(N \pm 1) \approx B(N) + E_k \pm \lambda, \tag{1}$$

141

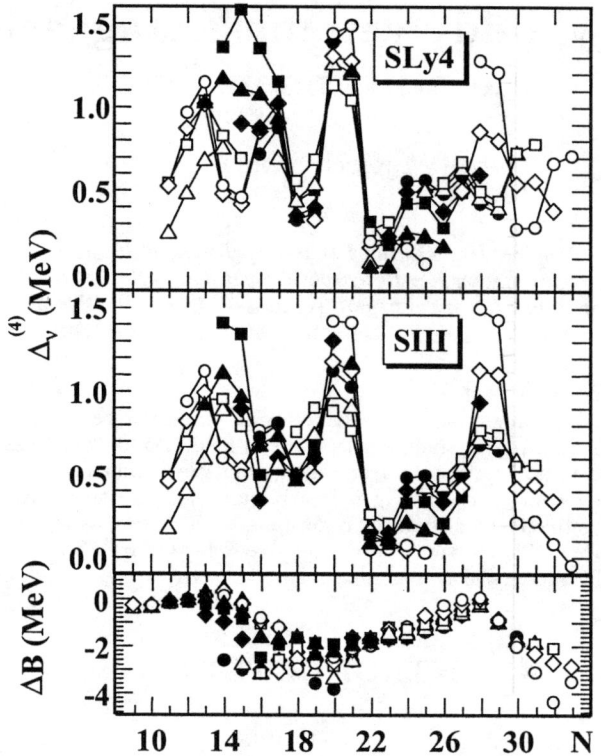

FIGURE 1. The OES $\Delta_\nu^{(4)}$ as a function of neutron number N computed from theoretical masses calculated using SHF theory with SLy4 (upper part) and SIII (middle part) Skyrme interaction. Note that the pattern of OES is almost independent on the interaction. The lowest panel shows the difference of nuclear masses calculated in both cases. Unlike the OES (local mass correlations) calculated masses are strongly interaction dependent.

where $E_k \approx \Delta$ denotes energy of the lowest quasiparticle in odd-N nucleus. Consequently, the quantity

$$\Delta_\nu^{(3)}(N) = \frac{\pi_N}{2}(B(N-1) - 2B(N) + B(N+1)) \approx E_k \approx \Delta, \tag{2}$$

is often interpreted as a measure of empirical pairing gap, Δ [6]. In Eq. (2), $\pi_N = (-1)^N$ is the number parity and $B(N)$ is the (negative) binding energy of the system with N particles. It has been noticed long time ago based on the Fermi gas model that the quantity (2) contains strong contribution from the symmetry energy [7]. Because $\Delta_\nu^{(3)}(N)$ is a finite step approximation of derivative of the second order:

$$\Delta_\nu^{(3)}(N) \approx \frac{\partial^2 B(N)}{\partial N^2}, \tag{3}$$

the symmetry energy $[\sim a_I(A)(N-Z)^2]$ will indeed strongly contribute to OES according to this criterion. However, this contribution is independent on nucleon-number parity and, therefore, the symmetry energy effect can be removed simply by replacing the three-point indicator (2) by the higher order expression like the one containing four masses [8]:

$$\Delta^{(4)}(N) \equiv \frac{\pi_N}{4}\left[3B(N-1) - 3B(N) - B(N-2)\right.$$

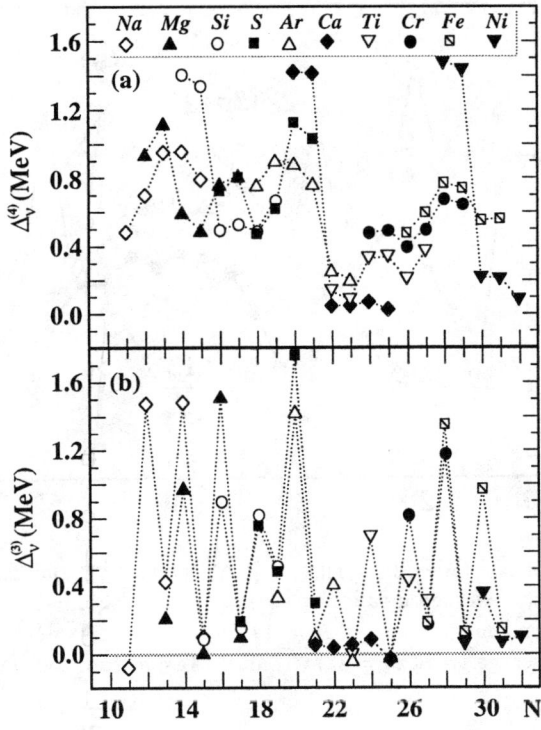

FIGURE 2. The OES $\Delta_\nu^{(4)}$ (upper part) and $\Delta_\nu^{(3)}$ (lower part) computed from theoretical masses calculated using SHF theory with SIII Skyrme interaction. Note, that rather complicated pattern of OES $\Delta_\nu^{(4)}$ simplifies for $\Delta_\nu^{(3)}$. The $\Delta_\nu^{(3)}$ shows clear alternating pattern with large values for even-N and small for odd-N. Taken from [4].

$$+ B(N+1)] = \frac{1}{2}[\Delta^{(3)}(N) + \Delta^{(3)}(N-1)]. \tag{4}$$

Consequently, in the traditional picture originating from the Fermi gas or liquid drop (macroscopic) models the mean-field contribution to the odd-even mass staggering is predominantly due to symmetry energy and can essentially be removed by means of high-order indicators like (4), see [9] and refs. quoted therein.

Contrary to the above so far commonly accepted scenario, the indicator (4) generates sizable OES when applied to the mass table calculated using single-particle Skyrme-Hartree-Fock model, see [4] and Fig. 1. These mass calculations has been performed for nuclei with $9 \leq Z \leq 28$ and $-2 \leq N-Z \leq 6$, using HFODD code (v1.75) of [10] and two parametrizations of the Skyrme force: SIII [11] and SLy4 [1]. The pattern of OES, which reaches from 30% to 50% of experimental value, appears to be generic feature of the SHF solutions i.e. is almost independent on parametrization although calculated masses differ substantially for the two parametrizations employed, see Fig. 1.

A rather complicated pattern of OES in Fig. 1 simplifies when criterion $\Delta^{(3)}$ is used instead of $\Delta^{(4)}$, see Fig. 2. Indeed, values of $\Delta^{(3)}(N)$ are large for even-N and small (close to zero) for odd-N. Furthermore, it appears possible to explain this alternating $\Delta^{(3)}(N)$ behavior obtained directly in fully microscopic SHF calculations by using simple arguments based on Strutinsky energy theorem [12]. According to this theorem the results of self-consistent calculations can be well approximated by the microscopic-macroscopic shell-correction method i.e. the total binding energy can be written as $B = E_{\rm sp} - \tilde{E}_{\rm sp} + E_{\rm macro}$, where

$$E_{\rm sp} = \sum_{k=1}^{A} e_k \tag{5}$$

is the shell-model energy (sum of single-particle energies of occupied states), $\tilde{E}_{\rm sp}$ is the Strutinsky-averaged shell-model energy, and $E_{\rm macro}$ stands for the macroscopic liquid-drop energy.

The contribution to OES from shell-model energy is:

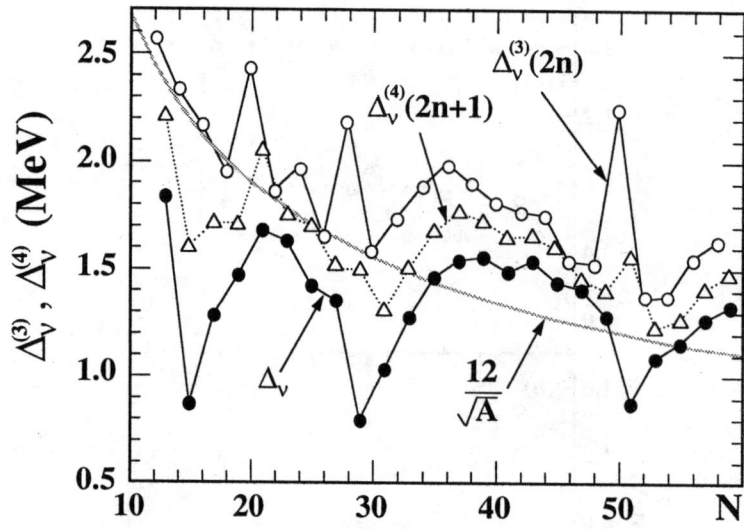

FIGURE 3. The empirical values of $\Delta_\nu^{(3)}(N = 2n + 1)[\equiv \Delta_\nu]$, $\Delta_\nu^{(3)}(N = 2n)$, and $\Delta_\nu^{(4)}(N = 2n + 1)$ as a function of neutron number. Each point represents average mean over several even-Z isotones. Note, that only $\Delta_\nu^{(3)}(N = 2n + 1)$ shows characteristic quenching at (semi)magic gaps at $N = 14, 28, 50$ (no effect at $N = 20$ is seen) as anticipated for pairing. Thick gray line shows the average trend $\Delta = 12/\sqrt{A}$ MeV. Taken from Ref. [4].

$$\delta e \equiv \Delta_{\rm sp}^{(3)}(N) \approx \frac{1}{4}(1 + \pi_N)(e_{n+1} - e_n) = \left\{ \begin{array}{cc} 0 & \text{if} \quad N = 2n + 1 \\ (e_{n+1} - e_n) & \text{if} \quad N = 2n \end{array} \right. \tag{6}$$

This single-particle mechanism behind OES was first noticed in Ref. [13] and applied to OES in metal clusters [14–16]. The contribution coming from smooth Strutinsky energy can be expressed through average level density at the Fermi energy $g(\lambda)$:

$$\Delta^{(3)}(N) \approx \frac{\pi_N}{2g(\lambda)}. \tag{7}$$

The average single-particle level density is $g(\lambda) = 3a/\pi^2(m^*/m)$ [13]. The empirical value of the level density parameter for light nuclei is $a \approx A/8$ MeV what agrees well with the estimate based on realistic potentials [17]. A correction due to effective mass is $m^*/m = 0.76(0.70)$ for SIII and SLy4 forces, respectively. Consequently, the contribution to OES coming from smooth Strutinsky energy equals $\delta\Delta^{(3)} \approx -18/A$ MeV. Contribution from macroscopic energy is essentially due to symmetry energy $[\frac{a_I}{2A}(N - Z)^2]$. The empirical value of the symmetry energy strength in light nuclei is $a_I = 38$ MeV [23]. Hence, the liquid-drop contribution to OES is $\delta\Delta^{(3)} \approx 19/A$ MeV what indeed nearly cancels out the contribution coming from the smooth Strutinsky energy.

The alternating pattern of $\Delta^{(3)}(N)$ allows to extract both pairing and mean-field component of OES. Indeed, at odd-N mean-field component to $\Delta^{(3)}(N)$ is small and its value can be associated mainly with pairing:

$$\Delta_\nu(N) \equiv \Delta_\nu^{(3)}(N = 2n + 1), \tag{8}$$

while the differences:

$$e_{n+1} - e_n = 2\left[\Delta_\nu^{(3)}(N = 2n) - \Delta_\nu^{(3)}(N = 2n + 1)\right], \tag{9}$$

provide information about *effective Nilsson* single-particle spectra [for more details concerning this aspect see discussion and Fig. 4 in Ref. [4]].

Empirical values of $\Delta_\nu^{(3)}$ and $\Delta_\nu^{(4)}$ are shown in Fig. 3. Note, that only values of $\Delta_\nu^{(3)}(N = 2n + 1)$ show oscillatory behavior reflecting expected quenching of neutron pairing correlations at the magic (semi-magic) gaps at $N = 14, 28$, and 50. Rather surprisingly no quenching effect is seen at $N = 20$. For $\Delta_\nu^{(3)}(2n)$ and $\Delta_\nu^{(4)}$

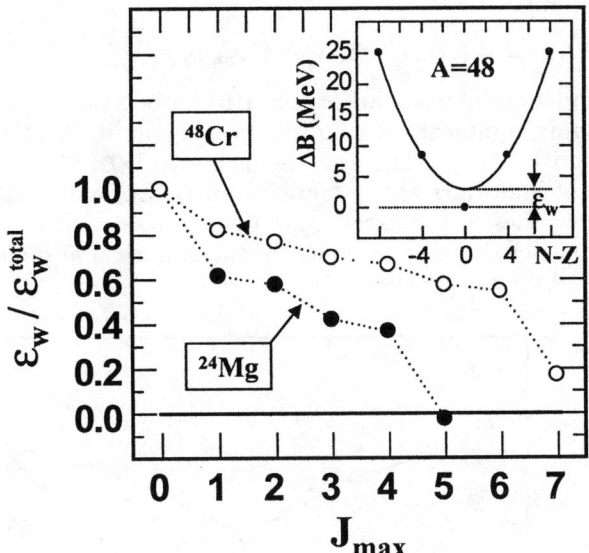

FIGURE 4. The contributions to the energy ε_W coming from isoscalar pairs of different angular momenta in ^{24}Mg (solid circles) and ^{48}Cr (open circles). The energy ε_W defines additional binding energy in $N = Z$ nucleus as a displacement of its binding energy from the average parabolic $\sim (N - Z)^2$ behavior in the $N \neq Z$ isostones, see inset. The details of this $0\hbar\omega$ shell-model calculation can be found in Ref. [22].

the expected quenching of pairing correlations is counterbalanced by single-particle effects and no minima are seen in these cases. These empirical observations neatly support the above scenario.

Similar interpretation of quantities (8) and (9) can be drown also based on seniority, equidistant-level [Richardson], and pairing-plus-quadrupole models [18]. Let's consider here as an example a seniority (or pairing quasispin) model. The ground state energy in this model [see e.g. [6] p. 222] equals:

$$B(N,s) = -\frac{G}{4}(N - s)(2\Omega - s - N + 2) \quad \text{where} \quad \begin{cases} s = 0 & \text{for} \quad N = 2n \\ s = 1 & \text{for} \quad N = 2n + 1 \end{cases}, \tag{10}$$

where s is a seniority quantum number, Ω stands for degeneracy of the shell and G denotes pairing strength. After simple algebra one gets:

$$\Delta^{(3)}(N) = \begin{cases} \frac{1}{2}G\Omega + \frac{1}{2}G & \text{for} \quad N = 2n \\ \\ \frac{1}{2}G\Omega & \text{for} \quad N = 2n + 1 \end{cases}. \tag{11}$$

Note that in accordance with (8) indicator $\Delta^{(3)}(N = 2n + 1)$ probes only collective pairing energy while $\Delta^{(3)}(N = 2n)$ contains also weak, $\delta e = G$ according to (9), *mean-field* contribution.

In the BCS approximation single-particle potential contributes both to $\Delta^{(3)}(N = 2n)$ and $\Delta^{(3)}(N = 2n+1)$. In this case one obtains Eq. (11) only in open shell regime where number of pairs is $n \sim \Omega/2 \gg 1$. In this regime, where BCS is considered to be very good approximation, also $\Delta_{BCS} \approx G\Omega/2 [\equiv \Delta^{(3)}(2n + 1)_{exact}]$.

WIGNER ENERGY

It has been known since the early work of [19] that macroscopic-microscopic approaches systematically underbinds $N \approx Z$ nuclei. Similar situation holds in semi-classical Thomas-Fermi models [20,21] as well as in *spherical* Skyrme-Hartree-Fock calculations irrespectively on parametrization of the Skyrme force [22,23]. This extra binding energy, which is characterized by $\sim |N - Z|$ type singularity at the $N = Z$ line, is dubbed Wigner energy and is usually parametrized as:

$$E_W = W(A)|N - Z| + d(A)\pi_{pn}\delta_{NZ}, \qquad (12)$$

where $W(A)$ and (A) are smooth functions of mass, and $\pi_{pn} = 1(0)$ for odd-odd(other) nuclei. Parametrization (12) can be justified based on simple arguments of counting of np-pairs in identical spatial orbitals and on elementary properties of np interaction [24,9]. This estimate gives also $W/d=1$. In Ref. [22] (for more details see [23]) we constructed a family of indicators able to probe both $W(A)$ and $d(A)$. It appears that indeed $W(A) \approx d(A)$ provided that the energies of the lowest $T = 0$ states are used in these indicators instead of masses of heavy $N = Z$ odd-odd nuclei. It provides independent empirical argument that Wigner energy is indeed predominantly related to $T = 0$ np-interaction.

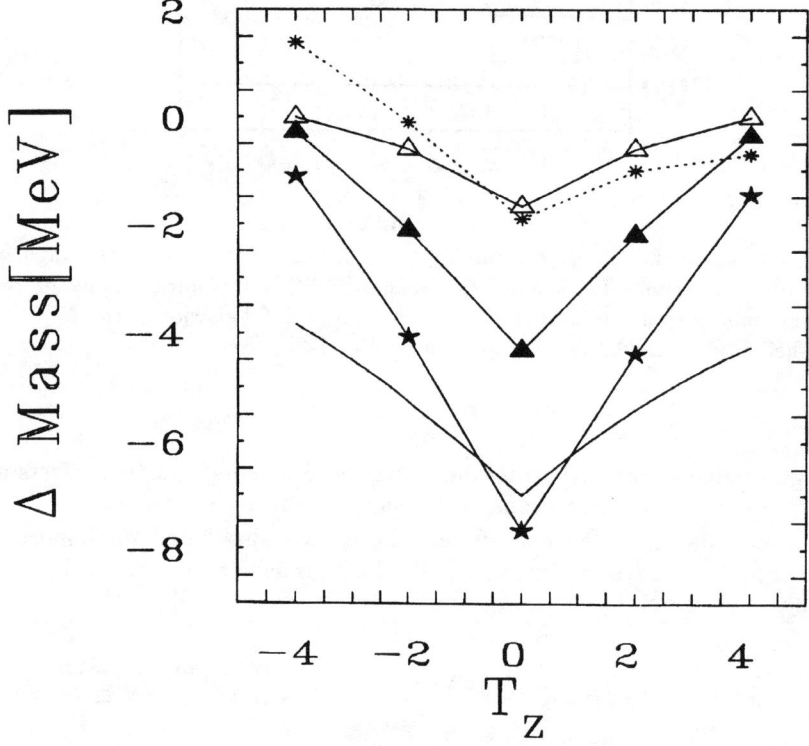

FIGURE 5. The mass excess $\Delta M = B(x^T) - B(x^T = 1)$ calculated using generalized BCS plus Lipkin-Nogami theory as a function of T_z. Different curves correspond to different values of x^T being a ratio of the isoscalar to the isovector coupling constants. Dotted line marks the results of extended Thomas-Fermi calculations [20] while solid line without symbols Wigner energy according to formula from Ref. [19].

Unlike mean-field, the state-of-the-art shell-model calculations reproduce empirical Wigner energy very well [22]. One can, therefore, use this model to gain some knowledge about microscopic structure of the Wigner energy. The shell-model indeed relates the Wigner energy to $T = 0$ interaction. It has been demonstrated first by Brenner et al. [25] in sd shell. The detailed microscopic analysis in sd and fp shells [22] revealed, however, rather complex structure of the Wigner energy in terms of nucleonic isoscalar pairs of different angular momenta. Although the largest contributions come from pairs coupled to $J = 1$ and $J = J_{max} = 5(7)$ for sd and fp shells, respectively, all matrix elements of intermediate angular momenta seem to be rather important, see Fig. 4.

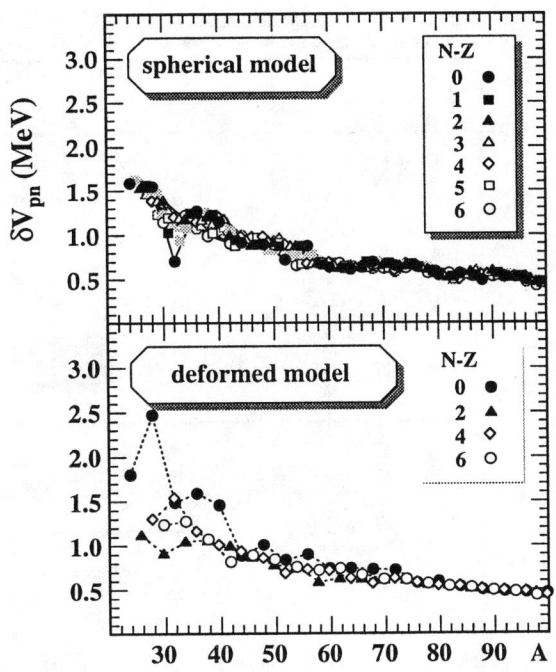

FIGURE 6. The values of δV_{pn} (see Eq. (13)) calculated using theoretical masses computed by means of *spherical* SHF model with SLy4 Skyrme force and density dependent, surface active delta interaction in pairing channel (upper part) and theoretical masses computed by Tajima et al. [30] using *deformed* SHF-plus-BCS theory with SIII Skyrme interaction (lower part). Figure taken from [23].

The deficiencies of conventional mean-field which allows only for $T = 1$, $|T_z|=1$ pairing can be remedied in generalized mean-field theory which takes into account also np-pairing correlations. It has been demonstrated in Ref. [26] that isoscalar np-pairing correlations can provide missing binding energy at $N = Z$ line and its closest vicinity as shown in Fig. 5. The model Hamiltonian employed in [26] was very simple, based on similar pair counting mechanism like discussed above in connection with the Wigner energy parametrization. For such a simple Hamiltonian generalized BCS solutions do not mix $T = 0$ and $T = 1$ pair correlations (see also [32]) and the energy gain is possible only when the isoscalar correlations are on the average stronger than the isovector. The latter cannot give an extra binding energy (in even-even nuclei) as long as isospin is (approximately) conserved. Simple estimates of isoscalar and isovector pair correlations in *simple nuclei* like ^{42}Sc suggest indeed that isoscalar correlations are on the average stronger than isovector, see discussion in [27,28]. However, connection between empirical interactions deduced from the spectra of simple nuclei and the mean-field effective pairing interaction is not obvious. Fast disappearance of isoparing with increasing $T_z = (N - Z)/2$ shows up naturally in generalized BCS model and can be easily understood in terms of blocking of np-pairing due to proton or neutron excess [26,29]. One should mention here, that recent generalized BCS calculations with G-matrix interaction [31] in fpg shell ($A \sim 80$) do allow for mixing of $T = 0$ and $T = 1$ pair correlations similar to the shell-model solutions [32].

The np-pairing is not necessarily the only contribution to the Wigner energy. In principle also self-consistent mean-field itself can contribute to the extra binding in $N = Z$ nuclei because of congruent nodal structure of a neutron and proton wave functions which can lead to stronger np-interaction as suggested by [21]. The congruence energy can manifest itself only in fully microscopic models. Spherical, conventional Hartree-Fock-Bogolyubov calculations with $T = 1, |T_z| = 1$ pairing only do not show any congruence effects as shown in Fig. 6, see also Refs. [22,23]. This result neither depends on Skyrme interaction nor pairing force. However, deformed SHF-plus-BCS calculations do show around $\sim 30\%$ of the empirical Wigner energy. Both spherical and deformed calculations are illustrated in Fig. 6 which shows values of the so called double-difference indicator [35]

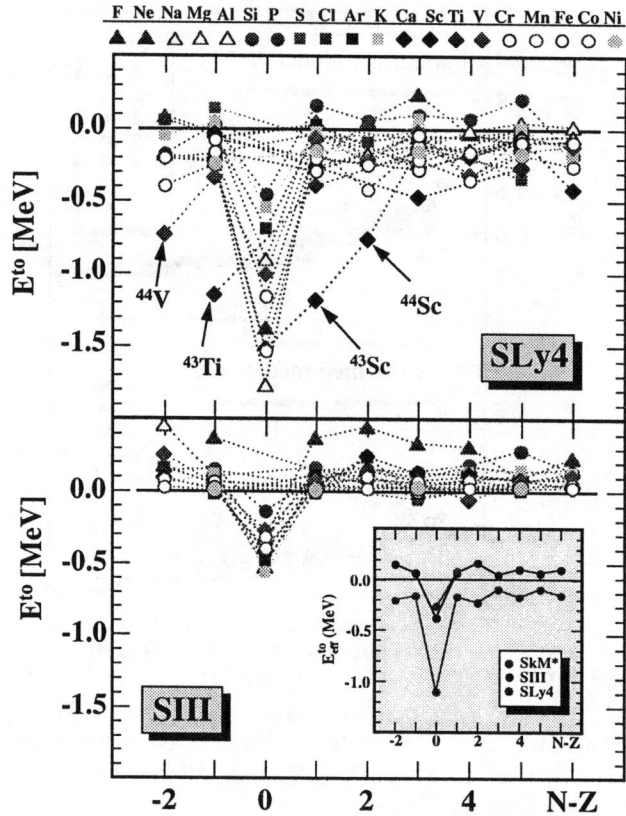

FIGURE 7. The effective energetical contribution to nuclear mass due to time-odd mean-field as a function of $N - Z$. The calculations were performed using SHF model with Sly4 (upper part) and SIII (lower part) Skyrme forces, respectively. An inset represents configuration independent contributions i.e. arithmetic averages over T_z=const nuclei.

$$\delta V_{np} = [B(N,Z) - B(N-2,Z) - B(N,Z-2) + B(N-2,Z-2)]/4 \approx \frac{\partial^2 B(N,Z)}{\partial N \partial Z}. \tag{13}$$

This quantity is particularly convenient to probe Wigner energy [22]. In deformed calculations (only even-even nuclei) masses computed by Tajima *et al.* [30] were used to calculate δV_{np}. Very similar effect (about 30% of experimental value) was obtained also in our single-particle deformed SHF calculations [23]. In these calculations also odd-odd nuclei has been included. Rather surprisingly, an enhancement of δV_{np} in odd-odd $N = Z$ nuclei comes entirely from the *time-odd isoscalar field*, see next section for greater detail.

THE TIME-ODD MEAN-FIELDS

Investigation of odd and odd-odd nuclei require the time-odd mean-fields to be systematically taken into account in the calculations. The time-odd part of the energy density $\mathcal{H}^{odd}$ reads:

$$\mathcal{H}_t^{odd} = C_t^s s_t^2 + C_t^{\Delta s} s_t \cdot \Delta s_t + C_t^T s_t \cdot T_t + C_t^j j_t^2 C_t^{\nabla j} s_t \cdot (\nabla \times j_t) \tag{14}$$

where s, j, T are time-odd spin, momentum, and vector kinetic densities, respectively, and subscript $t[=0,1]$ denotes isospin [33,10].

The effective contribution of the time-odd mean-fields to the total energy E^{to} can be investigated by comparing full SHF calculations with the calculations restricted to time-even fields only. The result is shown in Fig. 7. As seen from the figure everywhere beyond $N = Z$ line E^{to} is small but force dependent. Indeed, the time-odd fields are effectively attractive for SLy4 but repulsive for SIII forces, respectively. Also beyond $N = Z$ line simple additivity for the averages [see inset in Fig. 7]:

$$\overline{E}^{to}(N - Z = 2n) \approx \overline{E}^{to}(N - Z = 2n - 1) + \overline{E}^{to}(N - Z = 2n + 1), \tag{15}$$

is rather well fulfilled, reflecting single-particle nature of the effect. The most remarkable observation is an enhancement of the time-odd effects in $N = Z$ nuclei. The effect is essentially due to cancellation of two large components in (14), namely a repulsive $C_0^s s_0^2$ field is overbalanced by an attractive $C_0^{\Delta s} s_0 \cdot \Delta s_0$ field. The effect reflects most likely spontaneous spin polarization at the nuclear surface but its nature and consequences are not yet fully recognized and are under study [36]. It is noteworthy, that any observable consequences of this *spin collectivity* may help to establish certain constraints on time-odd coupling constants C_0^s and $C_0^{\Delta s}$. Note that C_t^s and $C_t^{\Delta s}$ are the only *free* time-odd coupling constants for Skyrme force. The remaining coupling constants (14) are related to time-even coupling constants because of local gauge invariance of the Skyrme force [34].

It is interesting to observe also strong instabilities in time-odd fields for $N \neq Z$ ^{43}Sc, ^{44}Sc nuclei and in their isobaric analogs ^{43}Ti, ^{44}V. These instabilities show up only for SLy4 force and only at the very bottom of the $f_{7/2}$ shell. This is, most likely, due to the extended spatial dimensions of these nuclei and due to strong attractiveness of C_0^s component of the SLy4 force at low densities where $C_0^s(\rho \to 0) = -207.8\,\text{MeV fm}^3$. For most of the commonly used Skyrme forces $C_0^s(\rho \to 0) > 0$, see table I in Ref. [33].

Let's finally consider briefly an indicator

$$
\begin{aligned}
\epsilon_{pn} = (-1)^{(A+1)}[&-B(N - 1, Z - 1) + 2B(N, Z - 1) - B(N + 1, Z - 1) + \\
&2B(N - 1, Z) - 4B(N, Z) + 2B(N + 1, Z) \\
&-B(N - 1, Z + 1) + 2B(N, Z + 1) - B(N + 1, Z + 1)]/4,
\end{aligned} \tag{16}
$$

which probes residual pn-interaction in odd-odd nuclei, see [9]. The data indicate, that for light $N \sim Z$ nuclei ϵ_{pn} is weakly dependent on nuclear mass and $\epsilon_{pn} \approx 500\,\text{keV}$. Only in the closest vicinity of $N = Z$ line ϵ_{pn} is enhanced due to the Wigner energy, see [22,23]. Applying the indicator (16) to single-particle SHF mass table gives *no effect* i.e. $\epsilon_{pn} \approx 0$ in odd-odd $N \neq Z$ nuclei. The cancellation of time-odd effects in (16) is a consequence of the above mentioned, see Eq. (15), energetical additivity of time-odd contributions in odd and odd-odd nuclei. Only in $N = Z$ odd-odd cases $\epsilon_{pn} \neq 0$ as a result of an enhancement of the time-odd fields in these nuclei. It seems therefore justified to conclude that the residual neutron-proton interaction in odd-odd nuclei as seen through the indicator (16) goes entirely beyond the mean-field.

SUMMARY

Selected properties of single-particle self-consistent Skyrme mean-field have been analyzed. It has been shown by direct microscopic calculations that the mean-field component to OES according to the criterion (2) shows strong nucleon number parity $\pi_{N(Z)}$ dependence with $\Delta_{\nu(\pi)}^{(3)}$ being large (small) for $\pi_{N(Z)} = +1(-1)$, respectively. This alternating pattern is due to (i) cancellation of the contributions from symmetry energy and average level density, and (ii) due to nuclear Jahn-Teller effect. It allows to interpret $\Delta_{\nu(\pi)}^{(3)}(N = 2n + 1)$ directly as due to (almost) pure pairing and construct higher order indicator (9) to probe also the mean-field component to OES. This scenario is entirely different than the commonly accepted scenario derived from the Fermi gas model where mean-field component to OES is due to symmetry energy and thus is independent on the number parity.

It has been demonstrated that *spherical* HFB calculations including only $T = 1, |T_z| = 1$ pairing generate no congruence energy i.e. $W(A) \approx 0$. It appears also, that *deformed* single-particle SHF (or SHF-plus-BCS) mean-field can generate only relatively small fraction $\sim 30\%$ of the empirical Wigner energy strength $W(A)$ in even-even $N = Z$ nuclei. Therefore, most of the Wigner energy strength in $N = Z$ nuclei seems to be beyond the Skyrme mean-field and, most likely, is due to neutron-proton pairing.

One of the most spectacular observations is strong enhancement of the time-odd effects in $N = Z$ nuclei. Closer examination shows that this effect depends essentially on the first two components (more precisely is due to their isoscalar parts) in (14). Because C_0^s and $C_0^{\Delta s}$ are independent (free) time-odd coupling constants for Skyrme force any observable consequences of this effect can help to establish empirical constraints for their values. In contrast to the $N = Z$ nuclei, in $N \neq Z$ nuclei effective contributions to the total binding energy coming from the time-odd fields reflect simple additivity pattern between odd and odd-odd nuclei typical for a single-particle model. As a consequence, binding energy between valence proton and neutron in odd-odd nuclei meassured by means of the indicator (16) is *zero*. In other words ϵ_{pn} is entirely beyond SHF field.

ACKNOWLEDGMENTS

The material presented here was obtained in collaboration with D.J. Dean, J. Dobaczewski, W. Nazarewicz, and R. Wyss. This research was supported in part by the U.S. Department of Energy under Contract Nos. DE-FG02-96ER40963 (University of Tennessee), DE-FG05-87ER40361 (Joint Institute for Heavy Ion Research), DE-AC05-96OR22464 with Lockheed Martin Energy Research Corp. (Oak Ridge National Laboratory), and by the Polish Committee for Scientific Research (KBN) under Contract No. 2 P03B 040 14.

REFERENCES

1. E. Chabanat E., Bonche P., Haensel P., Meyer J., and Schaeffer R., *Nucl. Phys* **A627**, 710 (1997).
2. Skyrme T.H.R., *Phil. Mag.* **1**, 1043 (1956); *Nucl. Phys.* **9**, 615 (1959).
3. Bohr A., Mottelson B.R., and Pines D., *Phys. Rev.* **110**, 936 (1958).
4. Satuła W., Dobaczewski J., and Nazarewicz W., *Phys. Rev. Lett.* in print, and nucl-th/9804060
5. Bardeen J., Cooper L.N., and Schrieffer J.R., *Phys. Rev.* **110**, 1175 (1957)
6. Ring P., and Schuck P., The Nuclear Many-Body Problem, Berlin: Springer-Verlag, (1980)
7. Bohr A., and Mottelson B.R., *Nuclear Structure*, vol. 2, New York: W.A. Benjamin, 1975.
8. Nilsson S.G., and O. Prior, *Mat. Fys. Medd. Dan. Vid. Selsk.* **32**, No. 16 (1961).
9. Jensen A.S., Hansen P.G., and Jonson B., *Nucl. Phys.* **A431**, 393 (1984).
10. Dobaczewski J., and Dudek J., *Comp. Phys. Commun.* **102**, 166, 183 (1997), and to be published.
11. Beiner M., Flocard H., Nguyen Van Giai, and Quentin P., *Nucl. Phys.* **A238**, 29 (1975).
12. Strutinsky V.M., *Nucl. Phys.* **A218**, 169 (1974).
13. Bohr A. and Mottelson B.R., *Nuclear Structure*, vol. 1, New York: W.A. Benjamin, 1969.
14. Clemenger K., *Phys. Rev.* **B32**, 1359 (1985).
15. Manninen M., Mansikka-aho J., Nishioka H., and Takahashi Y., *Z. Phys.* **D31**, 259 (1994).
16. Yannouleas C., and Landman U., *Phys. Rev.* **B51**, 1902 (1995).
17. Shlomo S., *Nucl. Phys.* **A539**, 17 (1992).
18. Dobaczewski J., Nazarewicz W., and Satuła W., to be published.
19. Myers W., and Swiatecki W., *Nucl. Phys.* **A81**, 1 (1966).
20. Abboussir Y., Pearson J.M., Dutta A.K., and Tondeur F., *Atomic Data and Nuclear Data Tables* **61**, 127 (1995)
21. Myers W., and Swiatecki W., *Nucl. Phys.* **A612**, 249 (1997).
22. Satuła W., Dean D.J., Gary J., Mizutori S., and Nazarewicz W., *Phys. Lett.* **B407**, 103 (1997).
23. Satuła W., Dean D.J., Dobaczewski J., Garrett J., and Nazarewicz W., to be published.
24. Myers W., *Droplet Model of Atomic Nuclei*, New York: Plenum, 1977.
25. Brenner D.S, Wesselborg C., Casten R.F., Warner D.D., and Zhang J.-Y., *Phys. Lett.* **B243**, 1 (1990).
26. Satuła W., and Wyss R.A., *Phys. Lett.* **B393**, 1 (1997).
27. Rosenfeld L., *Nuclear Forces*, Amsterdam: North-Holland, 1948.
28. N. Anantaraman N., and Schiffer J.P., *Phys. Lett.* **B37**, 229 (1971), and Schiffer J.P., *Ann. Phys.* **66**, 78 (1971).
29. Satuła W., and Wyss R., to be published.
30. Tajima N., Takahara S., and Onishi N., *Nucl. Phys.* **A603**, 23 (1996).
31. Goodman A.L, *these proceedings*.
32. Engel J., Langanke K., and Vogel P., *Phys. Lett.* **B389**, 211 (1996).
33. Dobaczewski J., and Dudek J., *Phys. Rev.* **C52**, 1827 (1995).
34. Engel Y.M., Brink D.M., Goeke K., Krieger S.J., and D. Vautherin, *Nucl. Phys.* **A249**, 215 (1975).
35. Zhang J.-Y., Casten R.F., and Brenner D.S., *Phys. Lett.* **B227**, 1 (1989).
36. Satuła W., Dobaczewski J., and Nazarewicz W., to be published.

T=0 Neutron-Proton Correlations at high angular momenta

R. Wyss

Royal Institute of Technology (KTH), Physics Department Frescati, Frescativ. 24, S-104 05 Stockholm

Abstract. The properties of $T = 0$ neutron-proton correlations are discussed within the framework of different model calculations. Single-j shell calculations reveal that the $T = 0$ correlations remain up to the highest frequencies. They are more complex than the $T = 1$ corrlations and cannot be restricted to $L = 0$ pairs only. Whereas it may be difficult to find clear evidence for $T = 0$ pairing at low spins, $T = 0$ correlations are found to induce a new excitation scheme at high angular momenta.

INTRODUCTION

Pairing correlations have always played a decisive role for the low energy structure of atomic nuclei. In close analogy to the BCS-theory of superconductivity, it was suggested early on that the ground state of most nuclei is formed by a coherent superposition of pairs of nucleons, moving in time reversed orbits. [1] Although the BCS-theory of the nuclear pairing interaction is charge independent, one in general considers only scattering between pairs of like particles (neutron-neutron and proton-proton).

Atomic nuclei are composed of two different kind of fermions which can occupy identical states. This is an unique situation, that is not found in other fermionic systems. Since the short range nuclear force is attractive, nuclei in close vicinity of the $N = Z$ line may form a condensate of neutron-proton (np) pairs that have totally symmetric wave-functions in spin and ordinary space. Different approaches to generalize the pairing interaction to fully take into account the iso-spin degree of freedom have been discussed in the literature in the 60ties and 70ties, see e.g. Ref. [2] and references therein. Following the experimental observation of ^{100}Sn and the spectroscopic study of a large number of heavy $N = Z$ nuclei in recent years, the quest for neutron-proton pairing has gained renewed interest. [3–5] Indeed, the massdefect at the $N = Z$-line, the so called Wigner energy finds a microscopic explanation in a generalized pairing theory [4]. In the following we use the standard notations for isospin (t, T), intrinsic spin (s, S), orbital angular momentum (l, L) and total angular momentum (j, J), where small letters denote the single particle content and capital letters the vector added quantities.

Pairs of particles moving in time-reversed orbits have the largest momentum exchange when their total spin S and angular momentum L are coupled to zero. Such pairs form a triplet in iso-space $T = 1$, being decomposed of either proton (neutron-) pairs with $T_z = \pm 1$ or a neutron-proton pair with $T_z = 0$. Strong short range correlations of particle-particle (pp) type, are in general treated by means of the standard 'monopole' pairing force ($T = 1$) with $T_z = \pm 1$ pairs only. In order to treat protons and neutrons on the same footing, one may extend the formalism to include $T_z = 0$ pairs. [6] For even-even nuclei, such extensions may be considered less interesting, since the $T_z = 0$ pairs are simply related to $T_z = \pm 1$ pairs via rotation in iso-space. The binding energy of even-even nuclei e.g. is not affected by the inclusion of $T_z = 0$ pairing [4]. However, for odd-odd $N = Z$ nuclei, many interesting properties may emerge, that will be addressed in a forthcoming paper [7].

In nuclear physics it is well known that the Deuteron is bound only in the spin triplet, $S = 1$ iso-spin singlet, $T = 0$ state. Also nucleon-nucleon scattering data show stronger attraction for the spin triplet than spin singlet state. One may therefore expect that neutron-proton pairing is more important at the $N = Z$ line than neutron-neutron and proton-proton pairing.

In the context of the shell model, one defines a 'monopole' pairing interaction for the $L = 0$, $T = 1$ singlet state [8], and one may proceed in an analoguous manner for the $T = 0$, $L = 0$ and $S = 1$ triplet state. This

CP481, *Nuclear Structure 98*, edited by C. Baktash

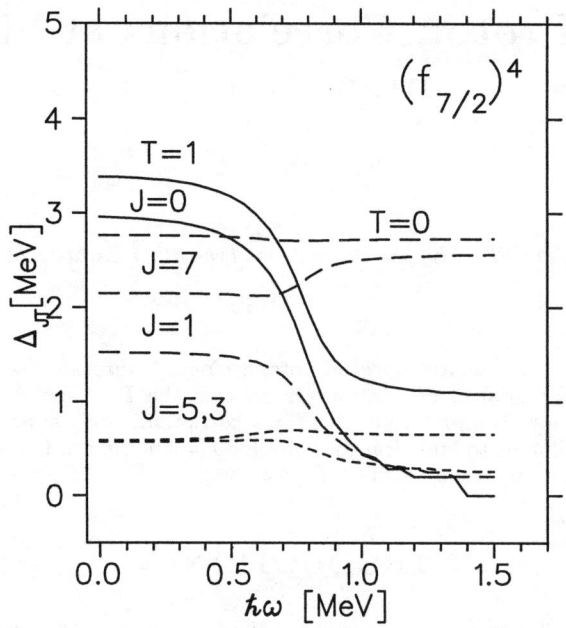

FIGURE 1. Decomposed (J, T) pairing energy for 2 protons and 2 neutrons in the $f_{7/2}$ shell. The results for 4 protons and 4 neutrons is similar.

restricted definition is of advantage for algebraic models [9–11] since it limits the model-space when analyzing properties of np-correlations. It may be appealing, to consider $L = 0$ pairs only for the pairing interaction. However, this neglects several important aspects. Whereas the $L = 0$ part of a short range $T = 1$ interaction like the δ-force is dominant, it plays only a minor role for the $T = 0$ interaction [8]. This is nicely seen in the experimental spectra of odd-odd nuclei, where the low-lying 0^+-state is in close vicinity to states with odd spins and even parity corresponding to the $(j^2)_{J=1}$ and $(j^2)_{J=2j}$ configuration.

For mean field calculations, it does not make much sense to restrict to a $L = 0$ pairing force, especially not for the $T = 0$ channel. Already the $T = 0$ pair with lowest J, $J = 1$ has a strong $L = 2$ component that needs to be taken into account. Not even the $T = 1$ pairing channel can be restricted to $L = 0$-pairing, when one aims at a quantitative description of the rotational motion and pair-breaking mechanism in deformed nuclei. One has to include at least $L = 2$ ('quadrupole'-)pairing [14,15]. The contribution to the binding energy steming from the $L = 2$-pairing may be negligible small, of the order of a tens of keV, but it is certainly coherent [15].

In the present paper, we investigate the influence of the $T = 0$ correlations on properties of rotational bands at high angular momenta. Single-j shell calculations are presented in section 1 and mean-field calculations based on the HFB-method in section 2.

NEUTRON-PROTON CORRELATIONS IN A SINGLE-J SHELL

In a single-j shell, one can diagonalize exactly the two-body interaction and also investigate the dependence of different correlations on angular frequency [16]. The pair-gap, is not defined in such a model. Nevertheless, one may decompose the expectation value of the two-body interaction, E_{exp}, with respect to the contributions coming from the different angular momenta J. We have performed a single-j shell calculation for different number of particles in the $f_{7/2}$ shell and relate a 'gap'-parameter Δ_{JT} to the expectation energy via the following equation

$$E_{exp} = \frac{1}{2} \sum_{JT} \frac{\Delta_{JT}\Delta_{JT}^*}{E_{JT}}, \tag{1}$$

where E_{JT} is the value of the normalized, anti-symmetric two-body matrix elements of the δ-function interaction [17]. As expected, the single-j shell calculations indeed result in a strong binding for maximum and minimum J in the $T=0$ channel, but also shows that states with intermediate J are important, see fig. 1. Whereas the binding for the $T=0$-pairs is rather complex, the $T=1$ part of the interaction is dominated by a single J. It is the dominant role of the $J=0$ pairing in the $T=1$ channel that may justify the restriction to $L=0$ in that channel. These properties of the $T=1$ and $T=0$ correlations have been known since long [8,18], see also the recent investigation in [19]. When we start to rotate the nucleus, the Coriolis force tends to align the quasi-particles along the rotational axis, resulting in the well known pair breaking mechanism. The $T=1$ pairing becomes reduced and the $L=0$ part totally quenched. Since the $J=1$ part of the $T=0$ correlations also involve particles in which the intrinsic j's are mainly coupled antiparallel, this part is reduced in a similar fashion as the $J=0$ correlations. At the same time, the $J=7$ correlation pick up in strength and compensate the loss from the $J=1$. Although there exist only one $J=7$ state in the $f_{7/2}$ shell, this state has the highest degeneracy, $2J+1$, implying that many pairs may contribute coherently to this state. The total $T=0$ correlations remain unchanged as a function of frequency. Apparently, the $T=0$ energy is not affected by rotation and indeed, we do not observe any effect on the alignment by switching on or off that part of the interaction.

Recent shell model calculations question the importance of $T=0$ correlations for the spectrum of ^{48}Cr [12]. However, the only matrix elements considered in this analysis are those of $L=0$. As shown above, in contrast to the $T=1$ channel, the $T=0$ interaction is most attractive for parallel $J=2j$ and antiparallel $J=1$ coupling of the individual angular momenta j, Fig. 1. Even the $T=0$, $J=1$ matrix element contains contribution from both $L=2$ and $L=0$. The requirement of total $L=0$ and $S=1$ for a l^2-pair of $T=0$ implies that the individual projection of the orbital angular momentum, l, (l_m) have to be antiparallel and the individual spins parallel. For a given l^2 configuration, one may couple the individual orbital angular momenta and intrinsic spins s, to a total angular momentum j, $j=l\pm1/2$. When decomposing the contribution to the $L=0$ multiplet in terms the 'spin-orbit' partner pairs $(j=l+\frac{1}{2}, j=l-\frac{1}{2})$, pairs with $(j=l+1/2)^2$ and $(j=l-1/2)^2$, respectively, we find from the appropriate $6j$ symbols a ratio of 8:7:2 (for $l=2$). This implies of course, that the $L=0$, $T=0$ pairing becomes strongly reduced by the nuclear mean-field in the presence of the spin-orbit interaction. The concept of a restricted $L=0$, $T=0$ pairing may be meaningful up to the sd-shell, where the LS-coupling still has validity. When entering the $f_{7/2}$ shell, we do not expect the $L=0$-pairing in the $T=0$ channel to play a very important role. Therefore, it is not at all surprising that this interaction turns out to have little influence on the spectrum of ^{48}Cr.

One should remember, that the definition of the pairing gap originates from the mean-field approximation where it defines the average gap or potential felt by a pair of particles at the Fermi-surface [1,20,21]. The pairing interaction is thus intimitely linked to the symmetry-breaking of the mean-field approach, which strictly separates between a particle-particle (pp) and particle-hole (ph) field. In contrast, the two-body 'pairing' matrix element of the shell-model is a different entity containing both particle-hole and particle-particle part and one has to be carefule when comparing results of the two different approaches. The question of a static pair-gap in the $T=0$ channel can be addressed meaningfully in the mean-field: if the correlations are strong enough, it will show up in a 'deformed' solution, otherwise not. This is of course linked to the strength of the interaction, which still needs to be determined. Generalized BCS- and HFB- calculations clearly indicate that $T=0$ correlations are coherent and necessary to understand the experimental data [2,4,13].

We are thus lead to the following conclusions: i)the spin-orbit force in the nuclear potential quenches part of the $L=0$ coupling and ii) the restriction to $L=0$ pairs only takes into account a minor fraction of the real correlations. To proper investigate the full spectrum of $T=0$, np-correlations, one needs to take into account all J-values (implying all possible L). The $T=0$ pairing as we define in the following, will have contributions from all J's. In an even-even nucleus e.g. the spin $I=0^+$ state has a contribution from the standard monopole-pairing with pairs of $J=0$ but of course also the contribution from all odd J-values. Note, however, that even the δ-force strongly simplifies the nucleon short-range interaction, since it allows only the coupling of even L's to take part. The interaction related to space antisymmetric states is not taken into account in any model involving a δ-force like interaction.

BAND TERMINATION AT HIGH ANGULAR MOMENTA

As discussed in the previous section, one may not see obvious fingerprints of the $T=0$ pairing in the low spin regime, since it is dominated by the breaking of pairs with low J and with respect to that, there is little difference in $J=0$ and $J=1$. In order to investigate the importance of the $T=0$ correlations at high angular momenta, we have performed a case study for the nucleus ^{48}Cr [13], that has been investigated quite extensively in experiment and theory [22,23]. The valence particles of ^{48}Cr are placed in the middle of the $f_{7/2}$ shell and the evolution of collective motion and band termination is well described by shell-model calculations [24]. The shell model faces difficulties to address states beyond the fp-shell. Hence, mean-field calculations may shed some light on the properties of states beyond the aligned $16\hbar^+$ state.

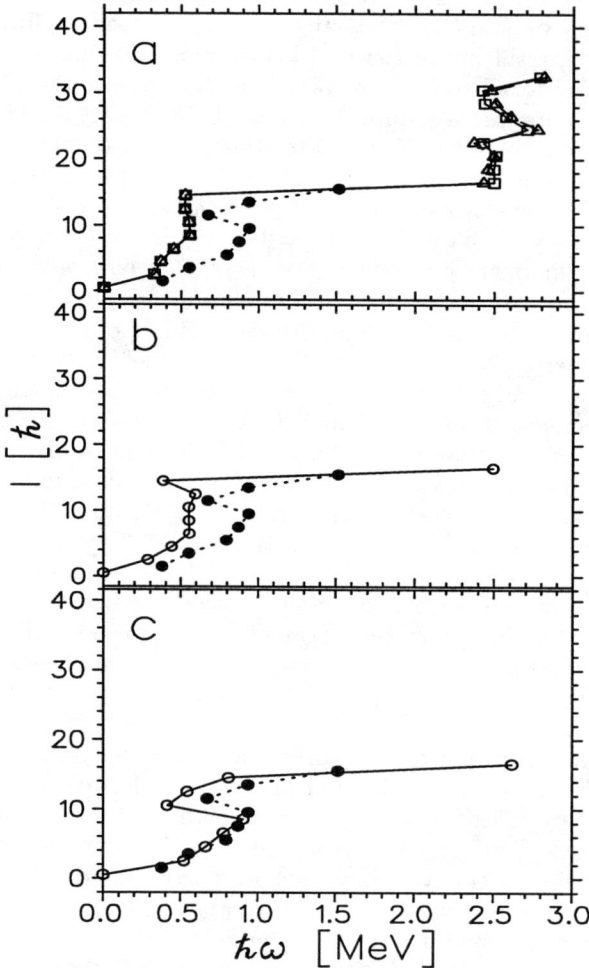

FIGURE 2. Comparison of calculated and experimentally deduced I_x as a funciton of rotational frequency, $\hbar\omega$. In a) we show two sets of calculations, with $G_{T0} = 1.1G_{T1}$ and $1.3\ G_{T1}$. The low spin part, $I \leq 16\hbar$ has only $T = 1$ pairing, wheras above that spin value, a transition to $T = 0$ pairing occurs. In b) the HFB-solution is dominated by $T = 0$ pairing correlations, although a small part of $T = 1$ pairing is present simultanously. c) The contribution of the $T = 0$ pairing energy is added to the solution of a), according to $d\omega = \frac{dE_{pair}}{dI}$.

We have performed cranked HFB-calculations in R-space based on the Skyrme-force [25,26] where the two-body pp interaction has been extended to incorporate both the $T = 0$ and $T = 1$ channel [13]. The only symmetry restriction used in the calculations is parity [13]. In the low spin regime, we found two HFB-solutions, dominated by either the $T = 0$ or $T = 1$ correlations, that were almost degenerate in energy, see a) and b) in Fig.2. Note, that for the case of the $T = 0$ solution, both pairing modes are present, showing the capability of our approach to incorporate the two different pairing channels simultanously. As shown in the previous section, the exact solution will always mix both pairing modes whereas the HFB approximation usually gives preference to either one. As discussed in ref. [4], particle number projection results in mixed solutions and one may expect that also iso-spin projection will become necessary. These steps will be considered in a further development of our model.

The results from the low-spin solutions appear very interesting: As seen in Fig. 2, the gross features of the alignment of the solution that is dominated by $T = 0$ pairing (b) does not differ very much from the solution with $T=1$ (a). This underlines the previous discussion - the low spin regime will always be characterized by the pair-breaking mechanism - be it $J=1$ or $J=0$ pairs. Nevertheless, the slope of I_x at low spins of the $T=0$ solution somewhat better reproduces the experimental values.

The small value of the moments of inertia of atomic nuclei have been taken as a fingerprint for the $T=1$ pairing correlations, where the size of the reduction with respect to the value of the rigid body indicates the strength of the correlations. Apparently, in the low spin regime, we obtain a similar reduction for the $T=0$-pairing, where again the size of the reduction is a measure of the correlations. Unfortunately, one cannot disentagle what is the contribution due to $T=0$ and $T=1$ pairing. Since the two solution are exptected to mix, we may artificially add the contribution of the pairing energy from the $T=0$-pairing to the $T=1$-solution. The result of such a mixing is shown in Fig. 2 c), resulting in a much improved agreement with experiment. To really investigate the effect of $T=0$ and $T=1$ pairing at low spins, one needs to go beyond the mean-field description.

In fig.3 the values of the pairing energy for the two solutions are compared. Indeed, the $T=0$ pairing is considerably more resistant to rotation than the $T=1$. The $T=0$ pairing energy reveal a similar drop as the $T=1$, but it does not approach zero and it increases again at the $12\hbar$ aligned state, which has a shape close to spherical. This dip in the pairing energy is related to the calculated backbend. Also the intrinsic quadrupole moments of the two solutions differ. For transitional nuclei, the difference can become quite large. The nucleus ^{44}Ti e.g. is calculated to have spherical shape in the presence of $T=1$ pairing whereas it has a calculated quadrupole moment of $\approx 50eb$ for the $T=0$ solution.

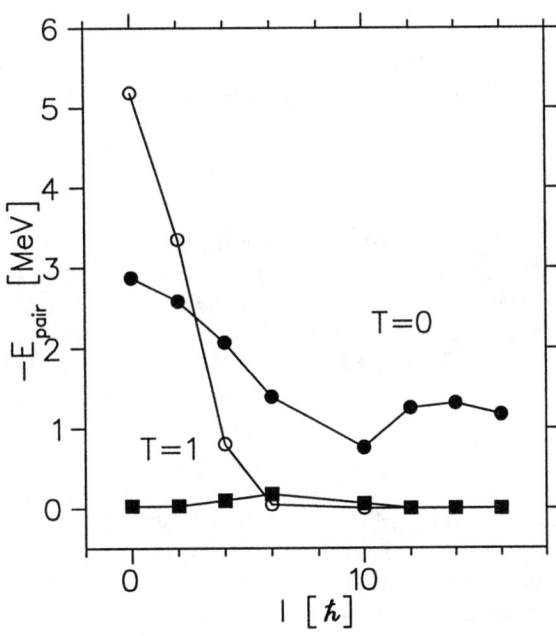

FIGURE 3. The pairing energy of the $T=0$ (solid symbols) and $T=1$ (open symbols) Skyrme-HFB solutions at low spins. Note the small value of the $T=1$ pairing for the $T=0$ solution.

The latter point is rather important and we will discuss it shortly. The pair-breaking mechanism is similar for $T=0$ and $T=1$, and one may not expect strong influence on the band-crossing frequencies. Nevertheless, since the $T=0$ pairing field affects the shape of the nucleus, and band-crossing frequencies quite sensitively depend on deformation [27], the two effects are linked together. For the case of ^{72}Kr e.g., a large delay of the band crossing has been observed [28]. A possible mechanism emerging from our result would be that the nucleus stays at a more deformed shape, thus delaying the crossing frequency. In addition, since in real nuclei more states than a single j-shell contributes to the pairing energy, one expects more binding from all different $J=1$ states, resulting in additional ground-state correlations.

For the high spin regime, we find only a solution with finite $T=0$ gap. This result clearly demonstrates the onset of $T=0$ pairing at high angular momenta, since the static $T=1$ correlations are quenched at these

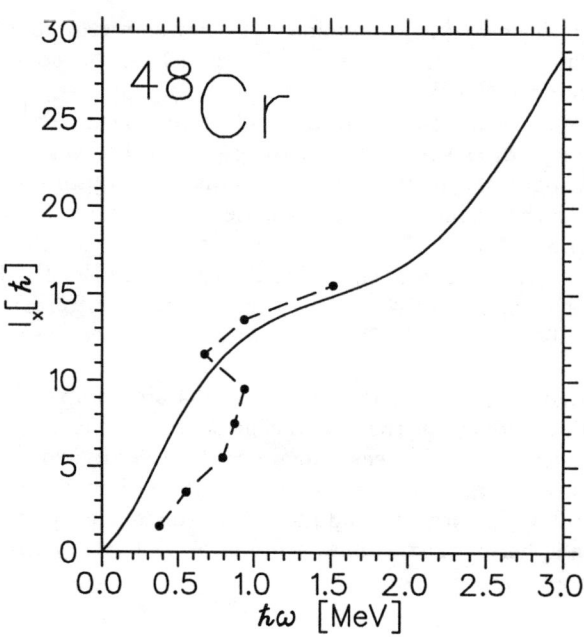

FIGURE 4. Cranked Woods-Saxon calculations for ^{48}Cr. The calculations are done at spherical shape and allow for both $T=0$ and $T=1$ pairing. The $T=0$ pairing field allows for a smooth occupation of the aligned deformation driving intruder orbitals

spin-values. The $T=0$ pairing energy rises after the terminating $I=16\hbar$-state and maximizes at $I=24\hbar$, where it has approximatively the same value as the nn (pp) $T=1$ pairing at $I=0\hbar$ (2.2 MeV). The size of the $T=0$ correlation implies the presence of a static $T=0$ pairing field. The increase in angular momenta beyond the terminating $I=16\hbar$ state is due to excitation from $d_{3/2}$ and $f_{7/2}$ into $g_{9/2}$ and $f_{5/2}$ orbits. Note that these particle-hole excitations are caused by the short range $T=0$ pairing force. Without the $T=0$ correlations, no cranked HFB- or HF-solutions are found in the region between $I=16$ and $I=32\hbar$ [13].

The effect of the $T=0$ pairing is to allow for a smooth occupation of the aligned $[(d_{3/2})^{-2}_{3+}(f_{7/2})^{-2}_{+1} - (f_{5/2})^2_{+5}(g_{9/2})^2_{9+}]_{18+}$ configuration (the $(f_{7/2})^{-2}$ holes are with respect to the $(f_{7/2})^8_{16+}$ configuration). The occupation probability of this configurations is increasing smoothly, resulting in an almost constant increase in angular momenta and deformation as a function of frequency. Since the configuration is exactly half filled at $I=24\hbar$, we obtain a maximum for the pairing correlations at that spin value [13].

Note that this 'vertical' excitation is completely different to the common ph-excitations, that are associated with the shape coexisting band structures. One may e.g. promote 2 particles from the $d_{3/2}$ into the $g_{9/2}$. This 2p-2h excitation couples to angular momentum $I=0\hbar$ and may result in a new rotational band. In this case, the $g_{9/2}$ is fully occupied and the $d_{3/2}$ is empty and the angular momentum gain arises from the alignment of the valence particles. This kind of 'horizontal' ph-excitation form the well-know intruder bands or highly- and super-deformed band structures. In the case of the $T=0$ pairing, one instead smoothly occupies the aligned configuration, resulting in a smooth deformation change. This presents a new kind of collective excitations, where the $T=0$ pairing field induce an increase in deformation and angular momenta.

The same effect is seen in cranked Woods-Saxon calculations [7]. Here $T=0$ and $T=1$ pairing are present simultanously and contribute to the total energy. Approximate particle number projection is performed by means of the Lipkin Nogami procedure. More details can be found in [4,7]. An important difference of the cranked Woods-Saxon calculations is the lack of feed back from the ocupation of intruder orbitals to the single particle potential. Although the same configurations are occupied as in the cranked Skyrme-HFB calculations, the shape stays spherical, since the single particle potential does not feel the intrinsic deformation of the ph-excitations. Remarkably, the presence of $T=0$ pairing allows for a smooth increase of the angular momenta at spherical shape. Since we are cranking around a symmetry axis, this is in contrast to the step wise increase which is obtained for normal cranking with or without $T=1$, where each step in angular momentum is

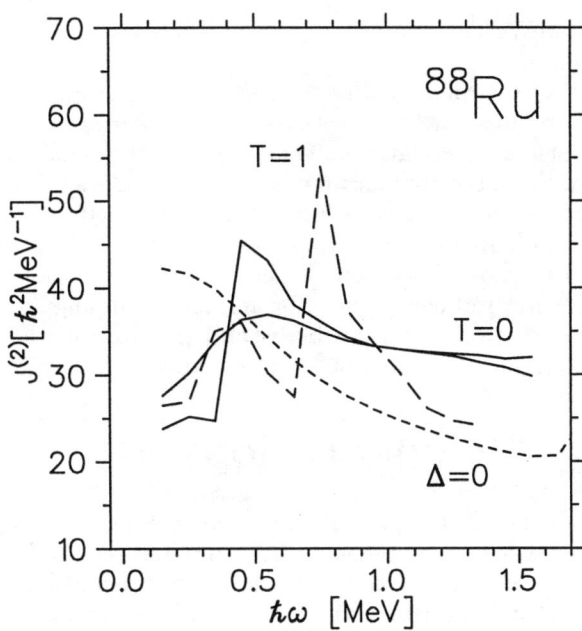

FIGURE 5. The J^2 moment of inertia resulting from the minimum of the TRS-calculations with different parameters of the pairing correlations. For further information, see text.

associated to the $< j_z >$ expectation value of each particular pair. The presence of $T = 0$ pairing allows instead a smooth 'filling' of the aligned orbital. Again, we encounter a new mechanism to create angular momentum and deformation.

The feature discussed in this context is a general property of nuclei in the close vicinity of $N = Z$. Cranked HFB calculations for ^{44}Ti e.g. show the same mechanism. After the terminating state at $I = 12\hbar$, $T = 0$ correlations start to show up, resulting in a further increase of the angular momenta. For the case of ^{44}Ti, the $T = 0$-solution is considerably more deformed than the spherical $T = 1$-solution. Again, we expect only one solution to be physical, but in the presence of $T = 0$ correlations, there is a clear tendency to more deformed shapes, that shows up especially in transitional nuclei. In general, there will always be the mechanism to go beyond the terminating state by means of $T = 0$ induced ph excitations. These 'vertical' excitations will of course have to compete with the 'horizontal' excitations, where by deforming the nucleus a particular set of favoured ph-excitations can lead to a more deformed shape and result in a new collective rotational band.

An interesting situation may arise at super-deformed shape in $N = Z$ nuclei. Cranked Woods-Saxon calculations predict very deep minima at super- and hyperdeformed shapes in ^{88}Ru and ^{92}Pd. The very stable shapes of these doubly magic super- and hyper-deformed nuclei have similar configuration as the harmonic oscillator at 2:1 shape for $N = Z = 40$. The experimental investigation of this shell structure is still in progress, but we may expect with the event of radio-active beam facilities to probe such exotic nuclei. Since the valence shell is particular large for super-deformed nuclei, the fingerprints of $T = 0$ type pairing may become enhanced, especially at high angular momenta.

Results of our total routhian surface calculations are shown in Fig.5. We compare four different sets of calculations: The standard solution where we allow for $T = 1$ pairing only (dashed line), shows a sharp crossing related to the alignment of $h_{11/2}$ protons and neutrons. After that bandcrossing, the moment of inertia drops and approaches the curve, where no pairing correlations are present ($\Delta = 0$, short-dashed line). The curves shown with solid lines correspond to a calculation, where we allow both $T = 0$ and $T = 1$ pairing to be present. The strength of the $T = 0$ pairing field is scaled with respect to $T = 1$, see the discussion in [4]. The curve that shows only a smooth hump in the frequency range of $\hbar\omega \approx 0.5$ MeV has a strenght of $G_{T=0} = 1.3G_{T=1}$, implying that the $T = 0$ pairing field is present already at $\hbar\omega = 0.0$ MeV. No sharp band crossing is observed for that case. When we choose a strength that is undercritical, $G_{T=0} = 1.1G_{T=1}$, implying that the pairing is of $T = 1$ type at $\hbar\omega = 0.0$ MeV, we experience a pairing phase transition from $T = 1$ to $T = 0$, which is related to the sharp crossing at $\hbar\omega = 0.5$ MeV. This kind of transition has been discussed previously in Ref. [29]. and reflects the mean-field approximation.

Note that the moment of inertia for a rotational band, where $T = 0$ pairing correlations are present, exceeds by a far amount the value that is obtained for 'rigid' rotation (i.e. no pairing correlations). The nucleus appears thus more 'rigid' in the presence of $T = 0$ correlations. This is exactly the same phenomenon as discussed above concerning the band-termination. Since the angular momentum in a nucleus is built up from the contribution of individual nucleons, the angular momentum space is limited for each specific configuration. This is in contrast to a rigid body, where there is in principal no limitation, beyond fission. In the presence of $T = 0$ pairing, the configuration space is not limited anymore, since now the short range correlations will always scatter pairs into orbitals with larger angular momenta, and hence allow for an increase in angular momentum. Hence, rotation in that case really resembles the rotation of a rigid body, and the drop in the moment of inertia, which is obtained in all nuclei at a certain frequency, may not be observed.

CONCLUSIONS.

We have presented different model studies that explore the iso-spin degree of freedom in the particle-particle channel. The $T = 0$ pairing correlations are expeted to influence the structure of $N = Z$ nuclei. In order to discuss these correlations, one certainly cannot restrict to the narrow definition of $L = 0$ pairing. As shown for the case of a single j-shell, all L-values contribute to the correlations. In the context of the mean-field, it is clear that a $T = 0$ pairing gap is obtained, i.e. the correlations are coherent. The $T = 0$ pairing is resistant to rotation and modifies the rotational spectrum at high angular velocities. A new kind of collective excitation mode appears, allowing for smooth and continous occupation of high-j orbits.

ACKNOWLEDGEMENTS.

This work is done in collaboration with P.-H. Heenen, U.L.B Brussels, W. Satuła, Univ. Warsaw and KTH, J. Sheikh, KTH and Tata Institute of Fundamental Research, and J. Terasaki, KTH. Discussions with J. Blomquist are acknowledged. We are thankful to the support given by the Göran Gustaffsson foundation, the Swedish Institute, the Swedish Natural Research Council (NFR) and the Axel och Margaret Ax:son Johnson Stiftelse.

REFERENCES

1. A. Bohr, B.R. Mottelson, D. Pines, Phys. Rev. **110** 936 (1958).
2. A.L. Goodman, Adv. Nucl. Phys. **11** 263 (1979).
3. J. Engel, K. Langanke and P. Vogel, Phys. Lett. **B389** 211 (1996).
4. W. Satuła and R. Wyss, Phys. Lett. **B393** 1 (1997).
5. A. Goodman, this issue
6. A. Goswami and L.S. Kisslinger, Phys. Rev. **140** B26 (1965).
7. W. Satuła and R. Wyss, in preparation
8. I. Talmi, Rev. Mod. Phys., **34**, 704 (1962).
9. G.G. Dussel, E.E. Maqueda, R.P.J. Perazzo and J.A. Evans, Nucl. Phys. **A450** 164 (1986).
10. J.A. Evans, G.G. Dussel, E.E. Maqueda and R.P.J. Perazzo, Nucl. Phys. **A367** 77 (1981).
11. P. Van Isacker and D. D. Warner, Phys. Rev. Lett. **78**, 3266 (1997).
12. A. Poves and G. Martinez-Pinedo, Phys. Lett. **B430** 203 (1998).
13. J. Terasaki, R. Wyss P.-H. Heenen, Phys. Lett. **B437** 1 (1998).
14. M. Diebel, Nucl. Phys. **A419** 221 (1984).
15. W. Satuła and R. Wyss, Phys. Rev. **C50** 2888 (1994).
16. J. A. Sheikh, N. Rowley, M. A. Nagarajan and H. G. Price Phys. Rev. Lett **64** 376 (1990).
17. J. Sheikh and R. Wyss, to be published
18. J.P. Schiffer, and W.W. True, Rev. Mod. Phys **48** 191 (1976).
19. W. Satuła, D. Dean, J. Gary, S. Mizutori and W. Nazarewicz, Phys. Lett. **B407** 103 (1997).
20. N.N. Bogoliubov, Sov. Phys. JETP, **7** 41 (1958).
21. S.T. Belyaev, Kg. Danske Videnskab. Selskab, Mat.–Fys. Medd. **31**, No. 11 (1959).

22. S. M. Lenzi et al. Z. Phys. A **354** 117 (1996).

23. J. A. Cameron et al. Phys. Lett. B **387** 266 (1996).

24. E. Caurier, J.L. Egido, G. Martinez-Pinedo, A. Poves, J. Retamosa, L.M. Robledo and A.P. Zuker, Phys. Rev. Lett. **75** 2466 (1995).

25. P. Bonche, H. Flocard and P.-H. Heenen, Nucl. Phys. A **467** 115 (1987).

26. J. Terasaki, P.-H. Heenen, P. Bonche, J. Dobaczewski and H. Flocard, Nucl. Phys. A **593** 1 (1995).

27. R.Wyss et. al. Nucl. Phys. **A505** 337 (1989).

28. G. de Angelis et. al. Phys. Lett **B415** 217 (1997).

29. E.M. Müller, K. Mühlhans, K. Neergård and U. Mosel, Phys. Lett. **B105** (1981) 329, and Nucl. Phys. **A383** 233 (1982).

Neutron-Proton Pairing In N = Z Nuclei

Alan L. Goodman

Physics Department, Tulane University, New Orleans, Louisiana 70118, USA

Abstract. The isospin generalized pairing theory is used to determine the ground states of even $N = Z$ nuclei with mass number $A = 76 - 96$. The calculations include isospin $T = 1$ (pp, nn, and np) pair correlations as well as isospin $T = 0$ (np) pair correlations. There is a transition from $T = 1$ pairing at the beginning of this isotope sequence to $T = 0$ pairing at the end of the sequence, with a mixed phase containing both $T = 0$ and $T = 1$ pairing near the middle of the sequence.

INTRODUCTION

By the early 1960's it was recognized that the original theory of nuclear pair correlations was incomplete [1]. In this theory each Cooper pair contained two protons or two neutrons. However it did not include the possibility that a Cooper pair might contain one neutron and one proton. During the period 1964-72, it was shown how the pairing theory could be generalized to include np Cooper pairs [2]. The ideal condition for finding np pair correlations occurs in $N = Z$ nuclei, where the neutrons and protons occupy the same spatial orbitals and have the maximum spatial overlap. The new radioactive beam facilities have produced exotic $N = Z$ nuclei up to ^{100}Sn. Consequently the mass region $A = 80 - 100$ provides new terrain for investigating Cooper pairs consisting of one neutron and one proton.

In the ground state of an even $N = Z$ nucleus, the nucleon orbitals occur in quartets $|\alpha p\rangle$, $|\alpha n\rangle$, $|\overline{\alpha} p\rangle$, and $|\overline{\alpha} n\rangle$, where $|\alpha\rangle$ denotes a space-spin orbital and $|\overline{\alpha}\rangle$ is the time-reverse of $|\alpha\rangle$. The isospin generalized theory of nucleon pair correlations [3] includes the following Cooper pairs: $|\alpha p, \overline{\alpha} p, T = 1\rangle$, $|\alpha n, \overline{\alpha} n, T = 1\rangle$, $|\alpha n, \overline{\alpha} p, T = 1\rangle$, $|\alpha n, \overline{\alpha} p, T = 0\rangle$, and $|\alpha n, \alpha p, T = 0\rangle$, as well as the time-reverse of these np pairs. Observe that for np pairs, the two particles may occupy identical space-spin orbitals, as well as time-reversed space-spin orbitals.

T = 0 AND T = 1 PAIRING THEORY

The conventional BCS theory (no np pairing) defines the quasiparticles by a two-dimensional transformation, in which each quasiparticle operator $a_\alpha^\dagger$ is defined as a linear combination of the particle operators $C_\alpha^\dagger$ and $C_{\overline{\alpha}}$.

The isospin generalized BCS theory [3] includes all Cooper pairs mentioned in the Introduction. This is accomplished by introducing the eight-dimensional transformation

$$
\begin{pmatrix} \mathbf{a}^\dagger(\alpha) \\ \mathbf{a}(\alpha) \end{pmatrix} = \begin{pmatrix} u(\alpha) & -v(\alpha) \\ -v^*(\alpha) & u^*(\alpha) \end{pmatrix} \begin{pmatrix} \mathbf{C}^\dagger(\alpha) \\ \mathbf{C}(\alpha) \end{pmatrix} \tag{1}
$$

where $\mathbf{a}^\dagger(\alpha)$ and $\mathbf{C}^\dagger(\alpha)$ are the four-component vectors

$$
\mathbf{a}^\dagger(\alpha) = \begin{pmatrix} a_{\alpha 1}^\dagger \\ a_{\alpha 2}^\dagger \\ a_{\overline{\alpha} 1}^\dagger \\ a_{\overline{\alpha} 2}^\dagger \end{pmatrix}, \qquad \mathbf{C}^\dagger(\alpha) = \begin{pmatrix} C_{\alpha p}^\dagger \\ C_{\alpha n}^\dagger \\ C_{\overline{\alpha} p}^\dagger \\ C_{\overline{\alpha} n}^\dagger \end{pmatrix} \tag{2}
$$

CP481, *Nuclear Structure 98*, edited by C. Baktash

Observe that each quasiparticle contains both neutron and proton components. In the ground state of a nucleus with $N = Z =$ even, time-reversal symmetry and isospin symmetry each generate a degeneracy factor of two. Then the four-dimensional matrices $u(\alpha)$ and $v(\alpha)$ acquire the simplified forms

$$u(\alpha) = u_\alpha I \tag{3}$$

where I is the four-dimensional unit matrix, and

$$v(\alpha) = \begin{pmatrix} 0 & v_{\alpha 1} & v_{\alpha 2} & v_{\alpha 3} \\ -v_{\alpha 1} & 0 & v_{\alpha 3}^* & -v_{\alpha 2} \\ -v_{\alpha 2} & -v_{\alpha 3}^* & 0 & v_{\alpha 1}^* \\ -v_{\alpha 3} & v_{\alpha 2} & -v_{\alpha 1}^* & 0 \end{pmatrix} \tag{4}$$

where u_α, $v_{\alpha 2}$ are real and $v_{\alpha 1}$, $v_{\alpha 3}$ are complex. The isospin generalized pairing wave-function for the ground state of even $N = Z$ nuclei has the form

$$|\Phi_0\rangle = \prod_{\alpha > 0} (u_\alpha + v_{\alpha 1}^* C_{\alpha p}^\dagger C_{\alpha n}^\dagger + v_{\alpha 2} C_{\alpha p}^\dagger C_{\overline{\alpha} p}^\dagger + v_{\alpha 3}^* C_{\alpha p}^\dagger C_{\overline{\alpha} n}^\dagger)$$

$$\times (u_\alpha + v_{\alpha 1} C_{\overline{\alpha} p}^\dagger C_{\overline{\alpha} n}^\dagger - v_{\alpha 2} C_{\alpha n}^\dagger C_{\overline{\alpha} n}^\dagger + v_{\alpha 3} C_{\alpha n}^\dagger C_{\overline{\alpha} p}^\dagger)|0\rangle \tag{5}$$

This wave-function contains all of the Cooper pairs listed in the Introduction. The wave-function is the vacuum for the quasiparticles defined in Eq. (1). Each one of the orbitals $|\alpha p\rangle$, $|\alpha n\rangle$, $|\overline{\alpha} p\rangle$, and $|\overline{\alpha} n\rangle$ has an occupation probability

$$v_\alpha^2 = |v_{\alpha 1}|^2 + |v_{\alpha 2}|^2 + |v_{\alpha 3}|^2 \tag{6}$$

Since the coefficients $v_{\alpha 1}$ and $v_{\alpha 3}$ are complex, the pn pair potential is also complex. The pn pair potential decomposes into $T = 1$ and $T = 0$ components, as follows:

$$\mathrm{Re}\,\Delta_{\alpha p, \overline{\alpha} n} = \Delta_{\alpha p, \overline{\alpha} n}^{T=1} \tag{7}$$

$$\mathrm{Im}\,\Delta_{\alpha p, \overline{\alpha} n} = \Delta_{\alpha p, \overline{\alpha} n}^{T=0} \tag{8}$$

$$\Delta_{\alpha p, \alpha n} = \Delta_{\alpha p, \alpha n}^{T=0} \tag{9}$$

Two procedures are used to determine the ground state wave-function. The first method uses the Hartree-Fock (HF) theory to calculate the orbitals $|\alpha\rangle$. Then the isospin generalized BCS equations [3] are used to determine the coefficients $v_{\alpha 1}$, $v_{\alpha 2}$, and $v_{\alpha 3}$. In this BCS theory there are separate gap equations for each pair mode, and the equations are coupled through the quasiparticle energy. The BCS energy matrix is eight-dimensional. It has eigenvalues (i.e., the quasiparticle energies) which are four-fold degenerate, with the form

$$E_\alpha = [(\epsilon_\alpha - \lambda)^2 + \Delta_\alpha^2]^{1/2} \tag{10}$$

where the pair potential Δ_α is the coherent sum of the contributions from each pair mode

$$\Delta_\alpha^2 = |\Delta_{\alpha p, \overline{\alpha} p}|^2 + |\Delta_{\alpha p, \overline{\alpha} n}|^2 + |\Delta_{\alpha p, \alpha n}|^2 \tag{11}$$

This method has the virtue of simplicity, in that the isospin generalized coupled gap equations are not more difficult to solve than the conventional BCS gap equation. However it has the defect that the occupation probabilities v_α^2 generated by the pair correlations are not permitted to renormalize the orbitals $|\alpha\rangle$ through the Hartree-Fock field. (A caveat to the reader: Do not be tempted into considering iteration between the HF equation and the BCS equations in the vain hope of achieving self-consistency in both the orbitals and their occupations. In general this will not bring you any closer to the fully self-consistent wave-function.)

The second method for determining the ground state wave-function is the Hartree-Fock-Bogoliubov (HFB) theory, which provides full self-consistency in both the Hartree-Fock and pairing degrees of freedom. The isospin generalized BCS wave-function [Eq. (5)] is used as the input trial wave-function for the HFB calculation. To

161

include both $T = 0$ and $T = 1$ pair correlations in the same wave-function, it is necessary to use complex quasiparticle coefficients and complex pair potentials. The usual definition of the HFB pair potential is

$$\Delta_{ij} = \frac{1}{2} \sum_{kl} \langle ij|v_a|kl\rangle \langle C_l C_k\rangle \tag{12}$$

where $|i\rangle$ denotes $|nljm\tau\rangle$, and τ is p or n. The expectation value of the particle operators $\langle C_l C_k\rangle$ is determined with respect to the HFB quasiparticle vacuum. Each quasiparticle contains both neutron and proton components. The HFB pair potential [Eq. (12)] includes all of the different Cooper pairs which have been discussed above.

Even though HF + BCS is not equivalent to HFB, it has been demonstrated [3] that the HFB ground state wave-function for even $N = Z$ nuclei can be given in the simple form of Eq. (5). This is provided by the quasi-canonical basis: The HFB density matrix $\rho_{ij} = \langle C_j^\dagger C_i\rangle$ is four-fold degenerate. The quartet of eigenvectors $|\alpha p\rangle$, $|\alpha n\rangle$, $|\overline{\alpha}p\rangle$, and $|\overline{\alpha}n\rangle$ correspond to the same eigenvalue v_α^2. This basis also provides a convenient way to characterize the HFB pair potential Δ_{ij}, which is a matrix with complex components. Average pair gaps $\overline{\Delta}$ can be defined for each of the different pair modes by representing Δ in the quasi-canonical basis

$$\overline{\Delta}_{p\overline{p}} = -\overline{\Delta}_{n\overline{n}} = \frac{1}{m} \sum_{\alpha=1}^{m} |\Delta_{\alpha p,\overline{\alpha}p}| \tag{13}$$

$$\overline{\Delta}_{p\overline{n}} = \frac{1}{m} \sum_{\alpha=1}^{m} |\Delta_{\alpha p,\overline{\alpha}n}| \tag{14}$$

$$\overline{\Delta}_{pn} = \frac{1}{m} \sum_{\alpha=1}^{m} |\Delta_{\alpha p,\alpha n}| \tag{15}$$

where $\overline{p}$ or $\overline{n}$ means that the nucleon occupies one of the time-reversed space-spin orbitals $|\overline{\alpha}\rangle$. (This does not imply that the pair potential is block diagonal in the quasi-canonical quartets. In general there will be non-zero components $\Delta_{\alpha_1\tau_1,\alpha_2\tau_2}$ and $\Delta_{\alpha_1\tau_1,\overline{\alpha}_2\tau_2}$, where $\alpha_1 \neq \alpha_2$.)

ISOSPIN GENERALIZED PAIRING CALCULATIONS

The calculations use a model space which includes the $2p_{1/2}$, $2p_{3/2}$, $1f_{5/2}$, and $1g_{9/2}$ shells. This corresponds to a closed core of $^{56}_{28}$Ni. The effective interaction was calculated by T. Kuo from the Paris potential. The monopole terms were modified by A. Nowacki. The same model space and interaction were used by Dean et. al. [4] in their study of ^{74}Rb.

The pair potential and the Hartree-Fock potential are calculated with this effective interaction. The pair potential includes the multipoles $J = 0, 1, 2, 3, 4, 5, 6, 7, 8, 9$. For nuclei around ^{80}Zr the BCS pair potential is smaller than the experimental odd-even mass differences. To obtain better agreement between calculated and experimental pair gaps, all nucleon-nucleon pairing matrix elements are scaled by the factor $S_p = 1.45$.

It is well known that in the mass region around ^{80}Zr, the structure properties are extremely sensitive to the choice of the parameters in the calculation, such as the core single-nucleon energies e_j in the Hamiltonian. For this reason, all calculations have been repeated for three choices of the core single-nucleon energies. For the first choice, the energies e_j are taken from the experimental energies of the ^{57}Ni spectrum. Then the values of e_j for the $p_{1/2}$, $p_{3/2}$, $f_{5/2}$, and $g_{9/2}$ shells are 1.113, 0, 0.769, and 3.0 MeV, respectively. For the second choice, the energies e_j are adjusted so that the spherical Hartree-Fock single-nucleon energies ϵ_j for ^{80}Zr are equal to the spherical Nilsson energies (taken from a Nilsson diagram [5] specifically constructed to fit ^{80}Zr), i.e.

$$\epsilon_j(\text{HF}, {}^{80}\text{Zr}, \beta = 0) = e_j + U_j(\text{HF}, {}^{80}\text{Zr}, \beta = 0)$$
$$= \epsilon_j(\text{Nilsson}, \beta = 0) \tag{16}$$

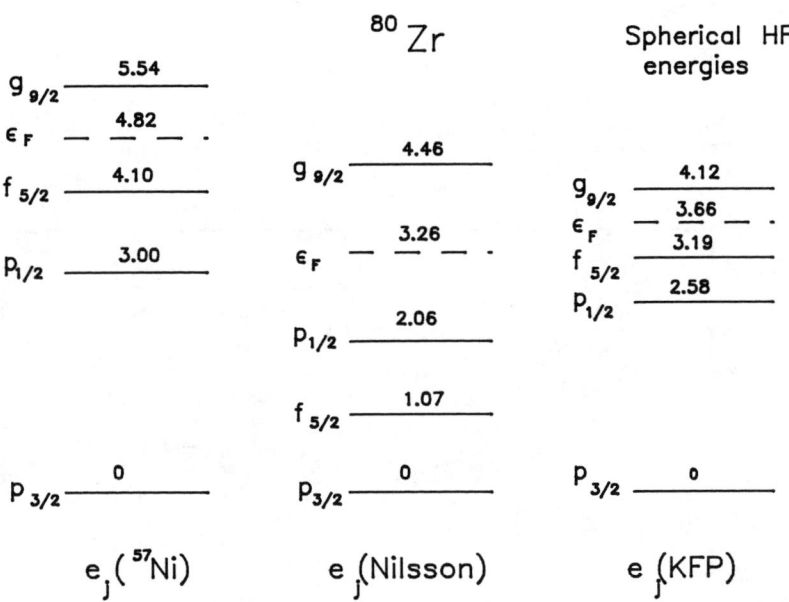

FIGURE 1. Spherical Hartree-Fock single particle energies for ^{80}Zr for three different choices of the energy e_j.

where β is the quadrupole deformation and U_j is the Hartree-Fock potential. These values of e_j for the $p_{1/2}$, $p_{3/2}$, $f_{5/2}$, and $g_{9/2}$ shells are 12.446, 12.275, 10.014, and 14.192 MeV, respectively. For the third choice, the energies e_j are taken from Table II in Kirchuk, Federman, and Pittel (KFP) [6], who determine e_j from experimental spectra in the $A \approx 90$ mass region. Their values of e_j are 0.69, 0, -0.14, and 1.58 MeV, respectively.

The difference between these three choices of e_j is illustrated in Fig. 1, which shows the spherical Hartree-Fock energies ϵ_j for ^{80}Zr. The energies ϵ_j are shifted so that the lowest energy is 0. The $e_j(^{57}$Ni) spectrum is compressed relative to the $e_j(^{57}$Ni) spectrum, and the two spectra have the levels in the same order with about the same relative spacings. The e_j(Nilsson) spectrum is also compressed relative to the $e_j(^{57}$Ni) spectrum, and it has the $f_{5/2}$ shell below the $p_{1/2}$ shell.

First the Hartree-Fock equation is solved to obtain all possible minima, maxima, and saddle points. Triaxial shapes are considered, as well as axially symmetric shapes. Then the isospin generalized BCS equations are solved for each Hartree-Fock solution. The BCS equations are used in several different ways: (1) Solutions are obtained when all pairing modes are activated simultaneously. (2) Solutions are obtained when each pairing mode is activated while the others are turned off. All of these provide self-consistent BCS solutions.

Now we discuss the lowest energy BCS state for each isotope. Each nucleon orbital $|\alpha\rangle$ has a different pair gap Δ_α. The average pair gap $\overline{\Delta}$ is shown in Fig. 2. For the ^{57}Ni choice of the single-nucleon energy e_j, the isotopes ^{76}Sr, ^{80}Zr, and ^{84}Mo have $T = 1$ pair correlations, whereas ^{88}Ru and ^{92}Pd have a mixed phase with $T = 0$ and $T = 1$ pair correlations, and ^{96}Cd has $T = 0$ pair correlations. For the Nilsson choice of the energy e_j, ^{76}Sr has $T = 1$ pairing, ^{92}Pd has $T = 0$ pairing, while ^{80}Zr and ^{96}Cd have the mixed phase with $T = 0$ and $T = 1$ pairing. For the KFP choice of the energy e_j, ^{76}Sr, ^{80}Zr and ^{96}Cd have $T = 1$ pairing, ^{92}Pd has $T = 0$ pairing, and ^{84}Mo has the mixed phase with $T = 0$ and $T = 1$ pairing. It is sometimes stated that $T = 0$ pairing and $T = 1$ pairing are mutually exclusive in the isospin generalized BCS theory, and that a mixed phase containing both $T = 0$ and $T = 1$ pairing is never obtained. This calculation clearly demonstrates that the mixed phase can be obtained in the BCS ground state.

Finally, each BCS state is used as an input trial wave-function for the HFB calculations. The HFB equation is used in two manners: (1) Solutions are obtained when all pairing modes are activated simultaneously. (2) Solutions are obtained when each pairing mode is activated while the others are turned off. All of these are self-consistent HFB solutions. Now we consider the lowest energy HFB state for each isotope. The average pair gap $\overline{\Delta}$ is shown in Fig. 3. For the ^{57}Ni choice of the single-nucleon energy e_j, the isotopes ^{76}Sr, ^{80}Zr, and ^{84}Mo have $T = 1$ pairing, while ^{88}Ru has a mixed phase with $T = 0$ pairing and $T = 1$ pairing, and ^{92}Pd and ^{96}Cd have $T = 0$ pairing. For the Nilsson choice of the energy e_j, ^{76}Sr and ^{80}Zr have $T = 1$ pairing, ^{84}Mo and ^{88}Ru

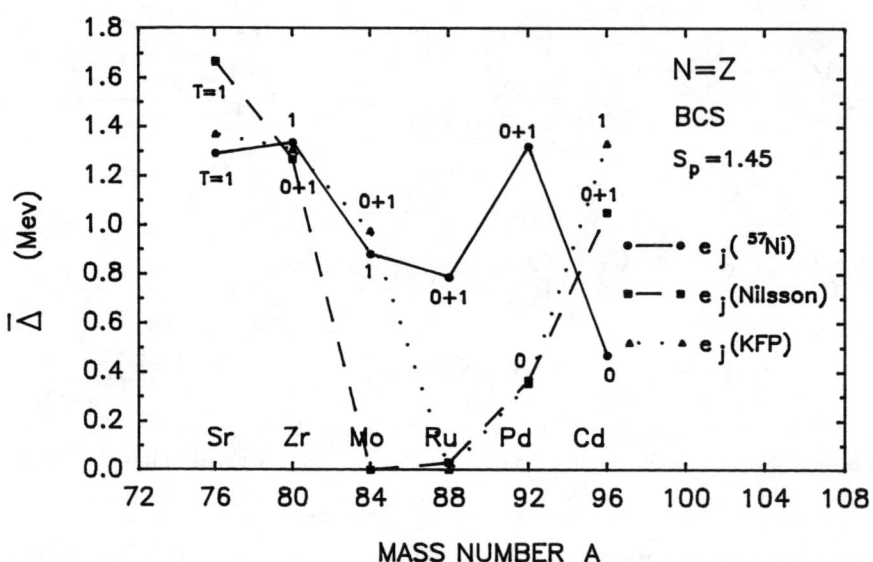

FIGURE 2. Average pair gap $\overline{\Delta}$ versus mass number A for isospin generalized BCS calculations.

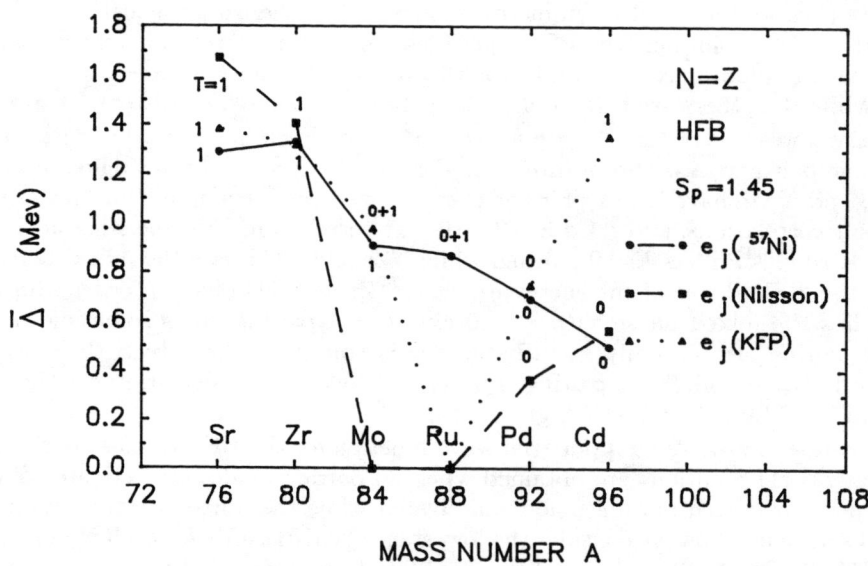

FIGURE 3. Average pair gap $\overline{\Delta}$ versus mass number A for HFB calculations.

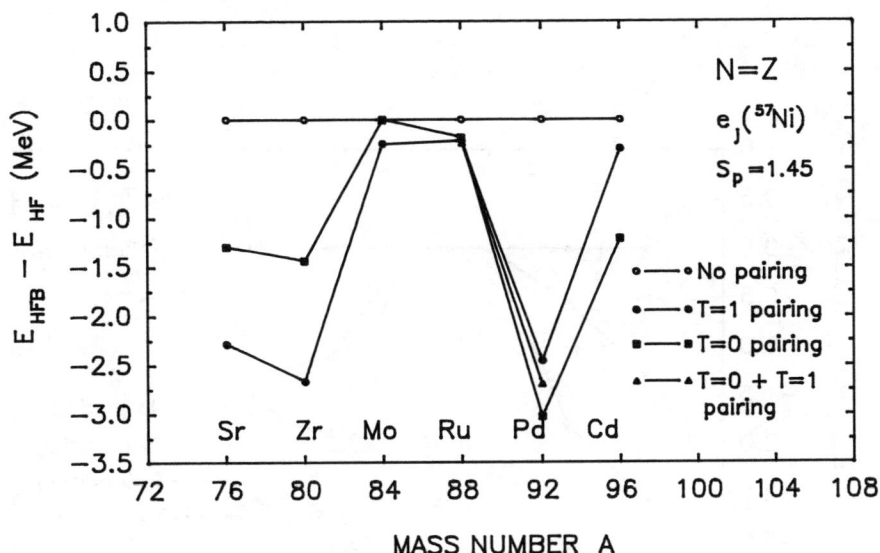

FIGURE 4. Difference between HFB energy and Hartree-Fock energy for each pair mode. The ^{57}Ni single nucleon energies e_j are used.

have no pairing, and ^{92}Pd and ^{96}Cd have $T = 0$ pairing. For the KFP choice of the energy e_j, ^{76}Sr, ^{80}Zr and ^{96}Cd have $T = 1$ pairing, ^{84}Mo has the mixed phase with $T = 0$ and $T = 1$ pairing, ^{88}Ru has no pairing, and ^{92}Pd has $T = 0$ pairing. This demonstrates that $T = 0$ pairing and $T = 1$ pairing are not mutually exclusive in HFB calculations, and that the mixed phase can occur in the HFB ground state.

The HFB energies may be compared for each of the different pair modes. This is achieved as follows: The lowest energy HFB state originates from a particular Hartree-Fock state. Using this same Hartree-Fock state, each of the different pair modes is then activated to find the self-consistent HFB state corresponding to each type of pair correlation. The relative energies of these various HFB states are shown in Figs. 4 - 6 . Fig. 4 shows the relative energies for the ^{57}Ni choice of e_j . For ^{76}Sr and ^{80}Zr, $T = 1$ pairing provides significantly more binding energy than $T = 0$ pairing. In ^{88}Ru the mixed phase ground state with $T = 0$ and $T = 1$ pairing is slightly below the state with only $T = 1$ pairing. For ^{92}Pd the lowest energy state has $|\alpha n, \alpha p, T = 0\rangle$ Cooper pairs, while there is a higher energy state with the mixed phase containing $|\alpha n, \overline{\alpha} p, T = 0\rangle$ Cooper pairs and $T = 1$ Cooper pairs. For ^{96}Cd $T = 0$ pairing gives substantially more binding energy than $T = 1$ pairing. Fig. 5 shows the relative energies for the Nilsson choice of e_j. For ^{76}Sr and ^{80}Zr, $T = 1$ pairing provides substantially more binding energy than $T = 0$ pairing. For ^{92}Pd and ^{96}Cd, $T = 0$ pairing provides significantly more binding than $T = 1$ pairing. For ^{96}Cd the lowest energy state has $|\alpha n, \alpha p, T = 0\rangle$ Cooper pairs, while there is a higher energy state with the mixed phase containing $|\alpha n, \overline{\alpha} p, T = 0\rangle$ Cooper pairs and $T = 1$ Cooper pairs. Fig. 6 shows the relative energies for the KFP choice for e_j. For ^{76}Sr and ^{80}Zr, $T = 1$ pairing provides significantly more binding energy than $T = 0$ pairing. For ^{84}Mo the mixed phase ground state with $T = 0$ and $T = 1$ pairing is slightly below the state with only $T = 1$ pairing. For ^{92}Pd the $T = 0$ pairing state is 0.241 MeV below the $T = 1$ pairing state. For ^{96}Cd the $T = 1$ pairing state is only 0.131 MeV below the $T = 0$ pairing state.

CONCLUSIONS

The isospin generalized BCS equations and the HFB equation have been used to study pairing correlations in the ground states of even $N = Z$ nuclei with $A = 76 - 96$. The following Cooper pairs were considered simultaneously, as well as individually: $|\alpha p, \overline{\alpha} p, T = 1\rangle$, $|\alpha n, \overline{\alpha} n, T = 1\rangle$, $|\alpha n, \overline{\alpha} p, T = 1\rangle$, $|\alpha n, \overline{\alpha} p, T = 0\rangle$, and $|\alpha n, \alpha p, T = 0\rangle$, as well as the time-reverse of these np pairs. The lowest energy HFB state has the following

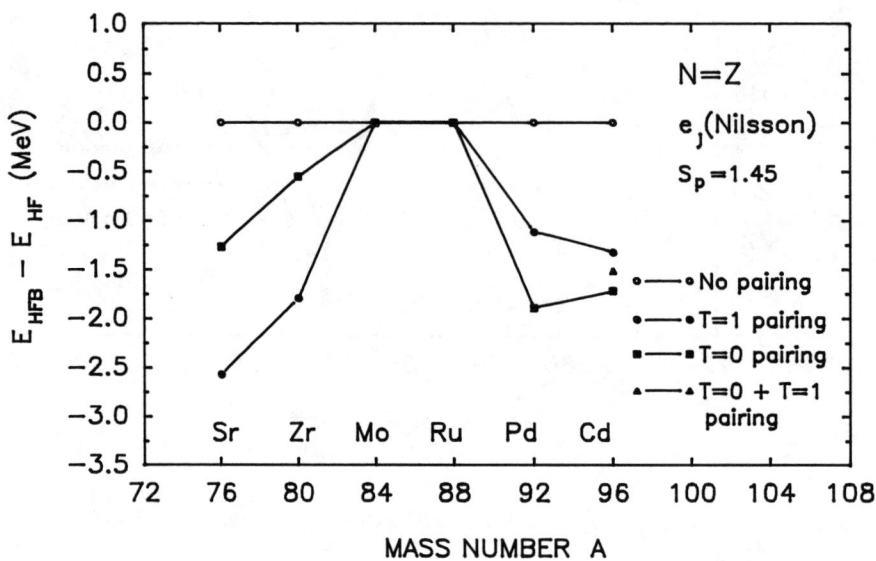

FIGURE 5. Difference between HFB energy and Hartree-Fock energy for each pair mode. The Nilsson single nucleon energies e_j are used.

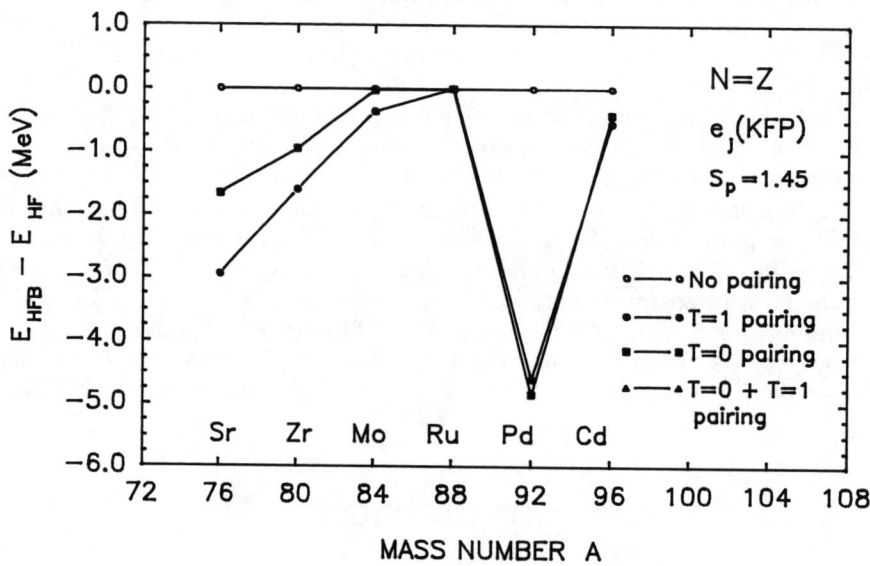

FIGURE 6. Difference between HFB energy and Hartree-Fock energy for each pair mode. The KFP single nucleon energies e_j are used.

pairing properties: For all three choices of the single-nucleon energy e_j, there is a transition from $T = 1$ pairing at the beginning of this isotope sequence to $T = 0$ pairing at the end of the sequence. (The only exception is ^{96}Cd with the KFP choice of e_j, where the $T = 1$ pairing state is slightly below the $T = 0$ pairing state.) For the ^{57}Ni and KFP choices of e_j, there is a mixed phase with both $T = 0$ and $T = 1$ pairing near the middle of the isotope sequence. It should be emphasized that in this mixed phase two different types of Cooper pairs ($T = 0$ and $T = 1$) co-exist in the same wave-function, as indicated by Eq. (5). (The mixed phase does not correspond to a $T = 0$ pairing state and a $T = 1$ pairing state which are degenerate.) This demonstrates that $T = 0$ pairing and $T = 1$ pairing are not mutually exclusive in isospin generalized BCS and HFB calculations.

ACKNOWLEDGMENTS

This work was supported in part by the National Science Foundation. The author is grateful to D. Dean for providing matrix elements of the nucleon-nucleon interaction.

REFERENCES

1. A.M. Lane, *Nuclear Theory* (Benjamin, New York, 1964) Ch. 5.1.
2. A.L. Goodman, in *Advances In Nuclear Physics*, edited by J. Negele and E. Vogt (Plenum, New York, 1979), Vol. 11, p. 263, Ch. 4.11 and 5.2.
3. A.L. Goodman, *Nucl. Phys.* A **186**, 475 (1972).
4. D. Dean, S. Koonin, K. Langanke, and P. Radha, *Phys. Lett.* B **399**, 1 (1997).
5. K. Heyde, J. Moreau, and M. Waroquier, *Phys. Rev.* C **29**, 1859 (1984).
6. E. Kirchuk, P. Federman, and S. Pittel, *Phys. Rev.* C **47**, 567 (1993).

Band structure in ^{79}Y and the question of $T = 0$ pairing

<u>S.D. Paul</u>[1,2], C. Baktash[1], W. Satuła[3,4,5], C. J. Gross[1,2], I. Birriel[6], R.M. Clark[7],
R.A. Cunningham [8], M. Devlin[9], P. Fallon[7], A. Galindo-Uribarri[1], T. Ginter[10],
D.R. Lafosse[#9], J. Kay[8], F. Lerma[9], I.Y. Lee[7], C. Leyland[8], A.O. Macchiavelli[7],
B.D. MacDonald[11], S.J. Metcalfe[8], A. Piechaczek[12], D.C. Radford[1], W. Reviol[4],
L.L. Riedinger[4], D. Rudolph[*1], K. Rykaczewski[1], D.G. Sarantites[9], J.X. Saladin[6],
D. Shapira[1], G.N. Sylvan[13], S.L. Tabor[13], K.S. Toth[1], W. Weintraub[4], D.F. Winchell[6],
V.Q. Wood[6], R. Wyss[14], and C.H. Yu[1]

[1] *Physics Division, Oak Ridge National Laboratory, Oak Ridge, Tennessee 37831*
[2] *Oak Ridge Institute for Science and Education, Oak Ridge, Tennessee 37831*
[3] *Joint Institute of Heavy Ion Research, Oak Ridge, Tennessee 37831*
[4] *Department of Physics, University of Tennessee, Knoxville, Tennessee 37996*
[5] *Institute of Theoretical Physics, Warsaw University, PL-00681 Warsaw, Poland*
[6] *Department of Physics, University of Pittsburgh, Pittsburgh, Philadelphia 15260*
[7] *Nuclear Science Division, Lawrence Berkeley National Laboratory, Berkeley, California 94720*
[8] *CCLRC, Daresbury Laboratory, Warrington WA4 4AD, United Kingdom*
[9] *Chemistry Department, Washington University, St. Louis, Missouri 63130*
[10] *Department of Physics and Astronomy, Vanderbilt University, Nashville, Tennessee 37235*
[11] *School of Physics, Georgia Institute of Technology, Atlanta, Georgia 30332*
[12] *Department of Physics, Louisiana State University, Baton Rouge, Louisiana 70803*
[13] *Florida State University, Tallahassee, Florida 32306*
[14] *The Royal Institute of Technology, Physics Department Frescati, S-104 05 Stockholm, Sweden*

Abstract.
Excited states in the $N=Z+1$ nucleus ^{79}Y were identified using the reaction ^{28}Si(^{54}Fe, $p2n$)^{79}Y at a 200 MeV beam energy and an experimental set up consisting of an array of Ge detectors and the Recoil Mass Spectrometer at Oak Ridge National Laboratory. With the help of additional γ-γ coincidence data obtained with Gammasphere, these γ-rays were found to form a strongly-coupled rotational band with rigid-rotor-like behavior. Results of conventional Nilsson-Strutinsky cranked shell model calculations, which predict a deformation of $\beta_2 \sim 0.4$, are in excellent agreement with the properties of this band. Similar calculations for the neighboring $N=Z$ and $N=Z+1$ nuclei are also in good agreement with experimental data. This suggests that the presence of the putative $T=0$ neutron-proton pairing does not significantly affect such simple observables as the moments of inertia of these bands at low spins.

INTRODUCTION

The structure of highly neutron defficient nuclei have recently become the subject of very active experimental and theoretical studies [1–4]. This is partly because of the richness of physical phenomena appearing in these nuclei such as shape coexistence, prolate-oblate shape mixing or shape transitions to name a few. Another important reason for such activities is that they can be understood with a variety of theoretical approaches, including the Monte Carlo shell model [5], symmetry-conserving models [6], as well as mean-field or algebraic methods [7,8]. Therefore, they provide excellent oppertunity to study effective nuclear forces and methods, approximations, and coupling schemes. However, the reason which makes them very attractive to study is the possibility of observing new effects due to the neutron proton intereaction namely the emergence of collective T=0 pairing phase. Due to particles occupying strongly overlapping orbital in nuclei closer to the $N=Z$ line, one expects an enhancement of neutron-proton (np) pairing, including the exciting possibility of observing

CP481, *Nuclear Structure 98*, edited by C. Baktash

np superconductivity. However, despite vigorous investigation of this problem, we still face many questions and conceptual difficulties regarding the existence of a np-pairing phase, its fundamental building blocks, and its experimental signatures. For example, the shell-model adapts rigorous definition of the Cooper pairs in terms of isospin-spin (T, J) quantum numbers, but lacks a natural definition of the order parameter. In contrast, the mean-field approach offers a natural definition of the order parameter Δ, but is less rigorous concerning parametrization of the pairing interaction. One of the possible manifestations of the np pairing is the extra binding energy in $N=Z$ nuclei, known as the Wigner energy (see, e.g., [9,10] and references therein). Indeed, conventional mean-field models, which only allow for $T=1$, $|T_z|=1$ pairing irrespective of its form, systematically underbind $N=Z$ nuclei [11]. But a generalized mean-field approach, which also allows for the $T=0$ np pairing, can naturally account for this extra binding energy which is characterized by an $\sim|N-Z|$ behavior [12]. Similarly, detailed microscopic shell-model calculations that correctly reproduce the Wigner energy show that the Wigner energy is indeed due to $T=0$ interaction [9,10]. However, its structure is very complex when expressed in terms of isoscalar nucleonic pairs of various angular momenta [10].

Medium mass $N \sim Z$ nuclei perhaps offer the most favorable conditions for the manifestation of np pairing phase in nuclear matter mainly because of the large number of valence protons and neutrons. A recent study of ^{74}Rb [1] has already provided some evidence for the presence of (collective) $T=1$, $T_z=0$ pairing in the even-spin band based on its similarity to the ground-state band in ^{74}Kr, its isobaric analogue. It seems, however, that the odd-spin $T=0$ band in this nucleus reflects mostly the (non-collective) coupling of a pair of [431]3/2 neutron and proton orbitals. Although shell model Monte Carlo calculations [5] seem to support this interpretation, it is not entirely clear whether the structure of ^{74}Rb reflects collective or non-collective components of np pairing. Therefore, systematic experimental studies of heavy $N\approx Z$ nuclei are needed to provide more clear clues concerning the question of isoscalar np pairing. The present study of ^{79}Y is part of our systematic studies of $T_z=1/2$ nuclei [2–4]. Earlier reports of this work has been presented in refs. [13–15].

EXPERIMENTAL PROCEDURE

The results presented in this work have been obtained in two separate experiments. In the first experiment, γ-rays associated with ^{79}Y were identified at the Holifield Radioactive Ion Beam Facility (HRIBF) using the reaction ^{28}Si(^{54}Fe, $p2n$)^{79}Y at 200 MeV and a beam intensity of $\sim$10 pnA. The target consisted of a layer of 0.5 mg/cm^2 ^{28}Si evaporated onto a 1 mg/cm^2 Ta foil that faced the beam. The emitted γ rays were detected by an array of six segmented-Clover and four Compton-Suppressed HPGe detectors. The recoils were separated in different groups according to the mass to charge ratio using the the Recoil Mass Spectrometer (RMS) [16] at HRIBF and detected at the focal plane of RMS by Position Sensitive Avalanch Counter (PSAC).

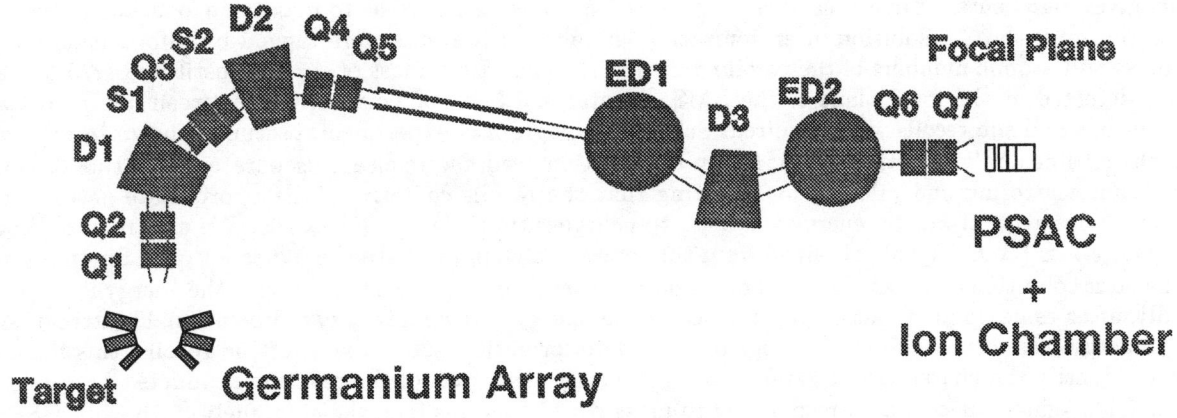

FIGURE 1. Schematic of experimental setup used to identify gamma-rays associated with ^{79}Y.

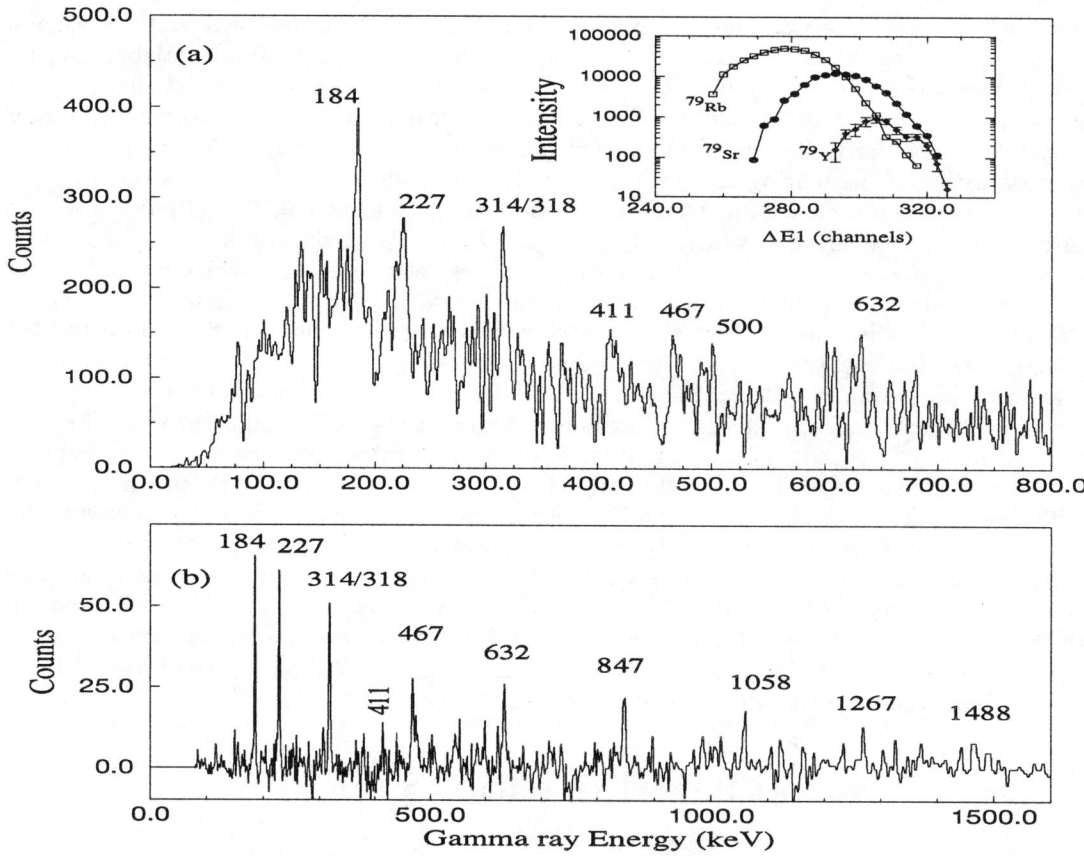

FIGURE 2. (a) A spectrum of characteristic γ rays in ^{79}Y ($N=Z+1$) gated with the RMS and ionization chamber at HRIBF. Gamma rays assigned to ^{79}Y have been marked by their energies in keV. Note that the 500 keV γ-ray is not placed in the level scheme. The inset shows the γ-ray intensity as a function of the energy-loss signal (ΔE) for ^{79}Rb, ^{79}Sr and ^{79}Y. (b) A spectrum obtained by summing several double gates on transitions belonging to the favored signature of the ground-state band in ^{79}Y.

The information regarding atomic number and total kinetic energy of recoiling nucleus is obtained by using Ionization Chamber [17]. Figure 1 shows the schematics of experimental setup. An inverse kinematic reaction which gives high center of mass recoil velocity, $\frac{v}{c} \sim 5.5\%$ was chosen so as to maximize focusing of recoils near 0° and improve the Z resolution in an ionization chamber. All events were tagged by information regarding the mass and atomic numbers of the recoiling nuclei. Recoils with a mass of $A=79$ constituted $\approx 70\%$ of all the recoils detected at the focal plane of the RMS. A total of 1.5×10^8 coincidences between one- and two-fold gamma-rays and the recoils were acquired. Schematic diagram of experimental setup is shown in Figure 1.

In the off-line analysis of these data, fusion-evaporation events were cleanly separated from those associated with beam scattering and pile up, by requiring that the recoils conform to the appropriate gates in a two-dimensional matrix of kinetic energy vs. mass-to-charge ratio (A/q) of the recoils. We also required that the two energy-loss (ΔE) signals obtained from the ionization chamber have the expected ratio for the recoils.

The later condition eliminated some contaminated events and improved the shape of the energy loss spectrum. Finally, after removing the energy dependence of the energy-loss signals, a two-dimensional matrix of ΔE vs. γ-ray energy was formed. Since ΔE signals provide information about the Z of the recoils, this matrix was used to identify the characteristic gamma rays associated with each of the reaction products.

The (A/q)-gated spectrum corresponding to mass $A=79$ contains four nuclei, namely ^{79}Rb ($3p$), ^{79}Sr ($2pn$), ^{79}Y ($p2n$) and ^{76}Kr ($\alpha 2p$). (The last nucleus appears in this gate due to the mass-to-charge ratio ambiguity.) With the help of a two-dimensional gate on the Total Energy vs. (A/q) matrix, a large fraction of the ^{76}Kr events was removed. The relative intensities of ^{79}Rb, ^{79}Sr, and ^{76}Kr in the mass-79 spectrum were 67%, 26%, and 6%, respectively. To identify the characteristic γ rays associated with the weakly populated nucleus ^{79}Y, we followed the following iterative procedure. First, using the known gamma rays in the strongly populated

170

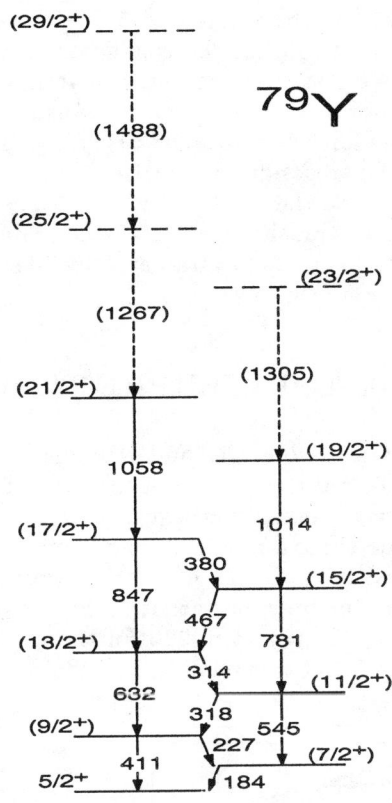

FIGURE 3. A partial level scheme for ^{79}Y obtained in the present work.

^{79}Rb and ^{79}Sr nuclei, we projected out their corresponding ΔE spectra. From both the shapes and centroids of these so-called Z spectra, we could determine the optimal ΔE gates for ^{79}Rb, ^{79}Sr, and ^{79}Y. In the second step, using these Z gates, we obtained total gamma-ray spectra for each of these three channels. The resulting spectrum for ^{79}Rb was free of contaminants, and was used to subtract out any contributions from this channel to the ^{79}Sr spectrum. Finally, a fraction of each of these two "purified" spectra were subtracted from the ^{79}Y spectrum to identify the characteristic gamma rays associated with this nucleus. The resulting γ-ray spectrum is shown in Fig. 2(a). In all, 6 γ-rays, 184 keV, 227 keV, 318 keV, 411 keV, 467 keV and 632 keV were assigned to ^{79}Y. Gamma-ray intensities as a function of the energy-loss signal in the ionization chamber confirmed that all these γ-rays belong to ^{79}Y. One such spectrum for the 184 keV transition is compared with those associated with γ rays in ^{79}Rb and ^{79}Sr in the inset of Fig. 2(a). We may define the quality factor for Z-resolution as $(P1 - P2)/FWHM$, where P1 and P2 are the centroids of the ΔE spectra for two isobaric nuclei with $\Delta Z = 1$ and FWHM is the full width at half maximum of these spectra. We obtained a quality factor of 0.7 for the present experiment. The partial cross section for ^{79}Y was estimated to be less than 200 μb.

In order to establish the coincidence relationship between the identified gamma rays in ^{79}Y, a γ-γ matrix and a γ-γ-γ cube were created from the data obtained in a second experiment using the reaction ^{58}Ni(^{28}Si, $\alpha p2n)^{79}$Y. The 130 MeV ^{28}Si beam was provided by the 88-Inch Cyclotron at the Lawrence Berkeley National Laboratory. Reaction γ-rays were detected by 57 Ge detectors of the Gammasphere Phase-I array [18], while charged particles were detected by 95 CsI detectors of Microball [19]. The target consisted of an enriched ^{58}Ni foil with a thickness of $\sim$0.4 mg/cm^2. A total of 1.5×10^9 events with a γ-ray coincidence fold of three or higher were collected. The level structure obtained from the analysis of these data is shown in Fig. 3. Lenz et al. [20] have previously reported a level structure for ^{79}Y. But, except for the pair of 184 and 227 keV transitions, our analysis did not find these gamma-rays to be in coincidence with each other. In accordance with the beta-decay results given in Ref. [21], we have adopted $5/2^+$ for the ground-state spin and parity. This is consistent with theoretical assignment of $g_{9/2}$ for the configuration of the ground-state band, as will be discussed below. Although lack of adequate statistics prevented us from confirming the tentative spin and parity assignments shown in Fig. 3, the presence of several interband transitions that connect the two signature partners of the band support our assignments for levels up to $I^\pi=(17/2^+)$. The 1488 keV $(29/2 \to 25/2)$ and 1267 keV $(25/2 \to 21/2)$ transitions in the favored signature of the band are shown as dotted because their placement could not be confirmed with respect to the 1058 keV $(21/2 \to 17/2)$ transition. However, our data indicate that they are in coincidence with other transitions in this band. Similar arguments hold for the 1305 keV $(23/2 \to 19/2)$ transition. Fig. 2(b) shows sum of spectra gated by 184, 227, 314, 318, 467, 632, and 847 keV transitions.

THEORETICAL INTERPRETATION

Figure 4(a) shows the experimental kinematic, $J^{(1)}[\equiv I/\omega]$ and dynamical, $J^{(2)}[\equiv dI/d\omega]$ moments of inertia (MoI) for the positive parity, positive signature $[(\pi,\alpha)=(+,+1/2)]$ band in ^{79}Y. As can be seen from this Figure, these two moment of inertias are nearly constant and equal $(J^{(1)} \sim J^{(2)} \sim 19~\hbar^2 \text{MeV}^{-1})$ over the entire observed frequency range. The Moment of Inertia of a rigid spheroidal nucleus of mass $A=79$ and deformation of $\beta_2=0.4$ (as predicted by TRS calculations) is $J_{rig} \sim 22~\hbar^2~\text{MeV}^{-1}$ which is slightly ($\sim 10\%$) larger than what is observed in ^{79}Y. It has been pointed out previously that such an equality of kinematic and dynamic moments of inertia is a signature of rigid body like rotation [22]. Furthermore, Peker $et~al.$ [23] in their study of ^{98}Y have noticed $J^{(2)} \sim J_{rig}$ and have argued that this relationship signifies quenching of pairing correlations

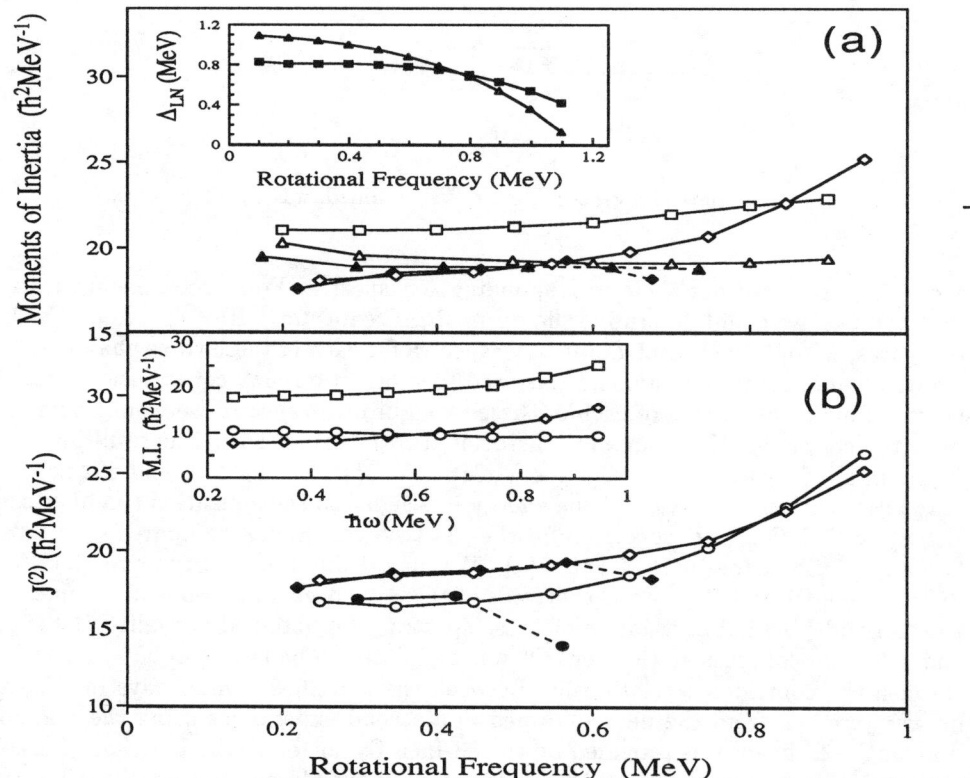

FIGURE 4. (a) Comparison of experimental (full symbols) and theoretical (open symbols) Moments of Inertia for the favored signature band in ^{79}Y as a function of $\hbar\omega$. The $J^{(1)}$ and $J^{(2)}$ moments of inertia are marked by triangles and diamonds, respectively. The open squares show the $J^{(1)}$ values from calculations with no pairing. The inset shows the pairing order parameters Δ_{LN} for protons (squares) and neutrons (triangles), respectively. (b) Comparison between the experimental (full symbols) and theoretical (open symbols) $J^{(2)}$ values for both signatures. The inset shows the $J^{(2)}$ contribution by neutrons (open diamonds) and protons (open circles) as predicted by theoretical calculations. The total $J^{(2)}$ contribution is shown by open squares.

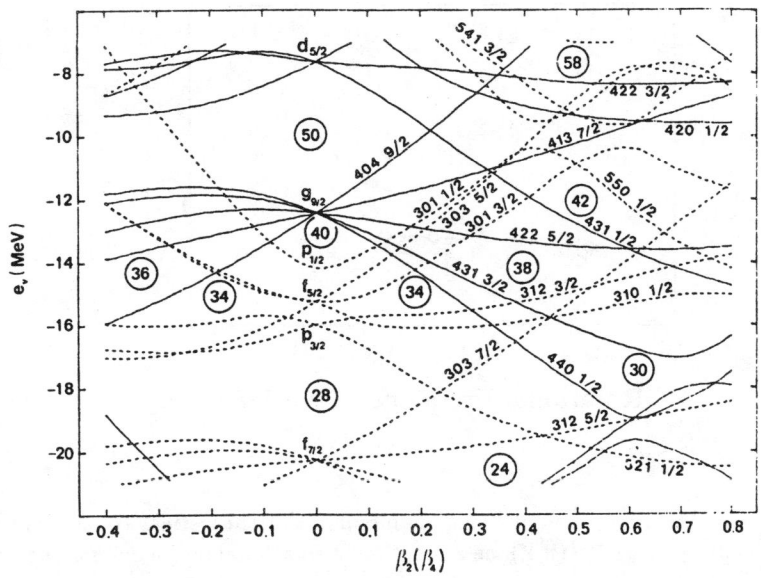

FIGURE 5. Single particle energy levels (Nilsson diagram) for nuclei in Mass-80 region

in ^{98}Y. As we shall see below, despite the fact that $J^{(2)} \sim J_{rig}$, our detailed theoretical calculations points out the importance of pairing correlations in ^{79}Y.

To better understand the structure of the observed bands, we have performed deformation and pairing self-consistent Total Routhian Surface (TRS) calculations using a Saxon-Woods potential. The pairing channel includes seniority and doubly-stretched quadrupole pairing interactions to avoid spurious shape dependence. To avoid a superfluid-to-normal phase transition due to the mean field approximation, we employed an approximate particle-number projection known as Lipkin-Nogami method [25,26]. These calculations reveal that correct treatment of pairing is crucial for a quantitative understanding of the MoI despite the presence of large shell gaps that weaken pairing correlations. The role of pairing is illustrated in Fig. 4 where we have compared the calculated $J^{(1)}$ values for the paired (open triangles) and unpaired (open squares) systems. The calculated unpaired $J^{(1)}$ overestimates the experimental MoI by 2-3 $\hbar^2$MeV^{-1}; i.e. by more than 10 % . The Lipkin-Nogami order parameters Δ_{LN} (i.e., the seniority type correlations) are shown in the inset of Fig. 4(a). They are weakly dependent on the rotational frequency below the point where $\nu g_{9/2}$ aligns and are $\Delta_{LN}^{(\pi)} \approx 0.8$ MeV and $\Delta_{LN}^{(\nu)} \approx 1.1$ MeV for protons and neutrons, respectively. These values may be compared with an estimate of the static pairing gap for this mass-region, namely $\Delta \approx 12/\sqrt{A} \approx 1.3$ MeV. Since Lipkin-Nogami order parameters take into account also the pairing fluctuations, indeed the calculated values of Δ_{LN} indicate weakened pairing correlations.

The fact of constancy of MoI in ^{79}Y is not unexpected and may be anticipated from the Nilsson diagram.(Please see Fig.5 for Nilsson diagram of the single-particle energies in this mass region.) The structure of the nucleus ^{79}Y (Z=39, N=40) is governed by the large shell gaps that appear at particle numbers N=38 and 40 at a deformation of $\beta_2 \sim 0.4$. These two gaps are bound by [431]3/2 on lower side and by [431]1/2 orbital on higher side and separated by Nilsson orbital [422]5/2 which is occupied by the unpaired proton in ^{79}Y at deformation of $\beta_2 \sim 0.4$. As a result of the [422]5/2 orbital occupancy of last proton, this orbital gets blocked and an effective *super*-gap at Z=38-40 gets formed. The formation of such a super gap thus facilitate the reduction of proton contribution to the pairing correlations in ^{79}Y. These observations are supported on quantitative basis by detailed theoretical calculations described above. Figure 4(c) shows calculated contribution of Proton and neutron to the dynamic moment of inertia $J^{(2)}$. As can be seen from Figure 4(c) and may be anticipated based on our discussion above, proton contribution remains nearly constant because of formation of an effective *super*-gap at Z=38-40 while the neutron contribution to $J^{(2)}$ increases with increasing rotational frequency. Results of the paired calculations for MoI are in excellent agreement with the data for both signatures as seen in Fig. 4. The detailed TRS calculations fully confirm all the anticipated trends. The $\sim$10% difference between the unpaired value of the MoI and the data may be attributed to the presence of (weak) pairing correlations. Another important fact that emerges from these calculation is the stability of prolate deformed minima over observed frequency range. The calculated deformation ($\beta_2 \sim 0.4$) of the strongly-coupled yrast band built on

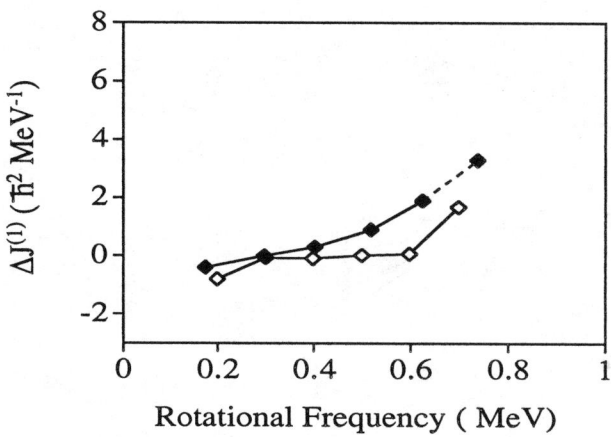

FIGURE 6. Comparison of the experimental (solid diamonds) and theoretical (open diamonds) differences of the moments of inertia, $\Delta J^{(1)} = J^{(1)}(^{77}\text{Sr}) - J^{(1)}(^{79}\text{Y})$, as a function of rotational frequency for the positive-signature bands..

the $g_{9/2}$ proton orbital remains almost constant up to $\hbar\omega \approx 0.7\,\text{MeV}$. The stability of such a minima is an important consideration in present analysis since it eliminates the possible disagreement between theory and experiment which can be produced by effect such as shape variation or shape coexistence as these cannot be treated completely in mean field framework of our theoretical calculations.

The agreement between theory and experiment in ^{79}Y is remarkable. Our earlier studies in this mass region [2,3,24] have shown that the agreement is not accidental: Very good agreement between theory and experiment has been obtained also in the lighter $T_z = 1/2$ nuclei ^{75}Rb [2] and ^{77}Sr [3]. As pointed out by Gross et.al. [2], for $T_z = \frac{1}{2}$ nucleus ^{75}Rb these calculations were able to reproduce observed experimental trends rather well. Although the calculated crossing is somewhat delayed compared to observed one, this can be attributed to the deficiency of cranking model in bandcrosssing region. It is also noteworthy that the ground-state band in ^{77}Sr ($Z=38$, $N=39$) shares many similarities with that in ^{79}Y: Occupation of the [422]5/2 orbital by the unpaired neutron creates a supergap at $N=38$-40 and deformation of $\beta_2 \sim 0.40$. Furthermore, the gap at $Z=38$ in ^{77}Sr closely resembles that at $N=40$ in ^{79}Y. Indeed, these two bands were found to be nearly identical both experimentally and theoretically. Figure 6 shows the differences in the values of $J^{(1)}$ for the ground-state bands in these two nuclei calculated from data and the theory. The agreement is again excellent.

Recently Satula et.al. have carried out Nilsson Strutinsky cranked shell model calculationsfor T=0 band observed in N=Z nucleus, ^{74}Rb. In these calculations, they could reproduce the observed feature of T=0 band (e.g routhians and moment of inertia) to a good accuracy.

It is rather unexpected that our conventional TRS calculations, which do not explicitly include the $T=0$ np interactions, can explain the experimental data in the $N \sim Z$ nuclei so well. (Some np interaction, however, enters into these calculations indirectly and through the assumption of common deformation for protons and neutrons. This interaction is of particle-hole type and can be well represented as a separable interaction in proton and neutron indices as discussed by Dobaczewski et. al. [27].) Two reasons may be suggested. First, such simple observables as the moments of inertia may not be sensitive to the presence of $T=0$ np interactions. Alternatively, effects due to the $T=0$ np interaction may manifest themselves more clearly at very high spins where Coriolis antipairing nearly quenches the $T=1$ interaction. Therefore, high-spin states in the $N=Z$ nuclei may provide the best data set to look for $T=0$ pairing correlation.

FUTURE OUTLOOK

Encouraged by the success of mean field approach in this region in explaining the observed band structure, we hereby give prediction for N=Z nuclei ^{78}Y and ^{76}Sr [28]. As can be seen from Figure 7, for ^{78}Y the dynamic moment of inertia, $J^{(2)}$, remains constant for large frequency range again signifying the importance of blocking of orbitals $[422]\frac{5}{2}$ (see Figure 5) by last odd proton and odd neutron thus creating *super*gaps. In comparison, even-even, N=Z nuclei ^{76}Sr shows quite different behavior. The $J^{(2)}$ initially remains constant but after $\hbar\omega \sim 0.5$ MeV shows increase in its value indicating the possible neutron alignment. These results are important because traditionally, it is believed that odd-odd N=Z nuclei are the most amenable to the np pairing thus forming important case for the test of the np pairing.

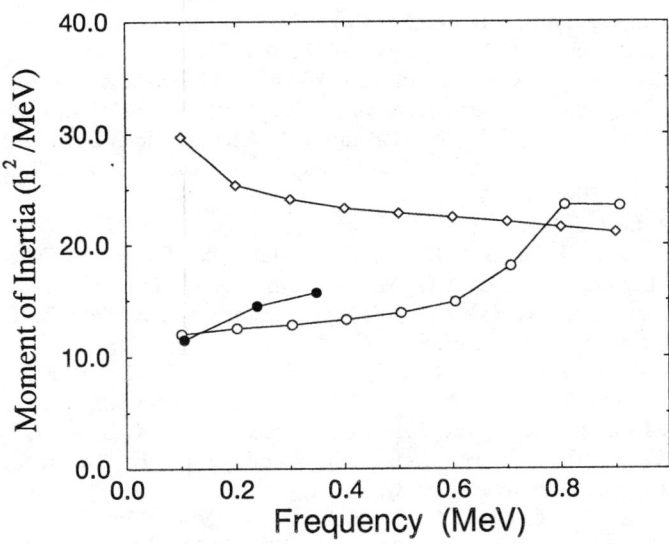

FIGURE 7. Predictions for heavy N=Z nuclei ^{78}Y (open diamonds) and ^{76}Sr (open circles). Experimental points for ^{76}Sr are shown in filled circles.

SUMMARY

To summarize, by combining data from two separate experiments we have identified a strongly-coupled band in the $T_z=1/2$ nucleus ^{79}Y which shows a rigid-rotor-like behavior. The favored and unfavored members of this band extend up to spins of $(29/2)\hbar$ and $(23/2)\hbar$, respectively. Conventional TRS calculations, which do not invoke any explicit $T=0$ proton-neutron correlations, are in excellent agreement with the experimental data for this nucleus, as well as its neighboring $T_z=1/2$ nuclei ^{75}Rb and ^{77}Sr. This suggests that the presence of the putative $T=0$ neutron-proton pairing does not significantly affect such simple observables as the moments of inertia of these bands. However, high-spin states in $N=Z$ nuclei may provide some sensitivity to the effects of $T=0$ np interaction.

We wish to thank the staff of the HRIBF for excellent help during the experiment. Oak Ridge National Laboratory is managed by Lockheed Martin Energy Research Corp. for the U.S. DOE under contract DE-AC05-96OR22464. This work is supported in part by the U.S. DOE under contract numbers DE-FG02-88ER-40406 (Washington University), DE-AC03-76SF00098 (LBNL), DE-FG02-96ER40963 (University of Tennessee), and DE-FG05-87ER40361 (JIHIR), and by National Science Foundation under grant numbers PHY9319934 (Pittsburgh) and PHY9523974 (FSU), and by Polish Committee for Scientific Research (KBN) under Contract No. 2 P03B 040 14. This research was also supported by DOE through contract Nos. DE-AC05-760R00033 (Oak Ridge Institute for Science and Education). The Oak Ridge National Laboratory Postdoctoral research program is administered jointly by Oak Ridge National Laboratory and the Oak Ridge Institute for Science and Education. We also acknowledge travel support from NATO collaborative research grant.

REFERENCES

* Present address: Department of Physics, Lund University, S-22100 Lund, Sweden.

\# Present address: State University of New York, Stony Brook, NY 11794.

1. D. Rudolph , C.J. Gross, J.A. Sheikh, D.D. Warner, I.G. Bearden, R.A. Cunningham, D. Foltescu, W. Gelletly, F. Hannachi, A. Harder, T.D. Johnson, A. Jungclaus, M.K. Kabadiyski, K.P. Lieb, H.A. Roth, J. Simpson, Ö. Skeppstedt, B.J. Varley, and M. Weiszflog, Phys. Rev. Lett. **76**, 376 (1996).

2. C.J.Gross *et al.*, D. Rudolph, W. Satula, J. Alexander, C. Baktash, P.J. Coleman-Smith, D.M. Cullen, R.A. Cunningham, J.D. Garrett, W. Gelletly, A. Harder, M.K. Kabadiyski, I. Lazarus, K.P. Lieb, H.A. Roth, D.G. Sarantites, J.A. Sheikh, J. Simpson, Ö. Skeppstedt, B.J. Varley, and D.D. Warner, Phys. Rev. C **56**, R591 (1997).

3. C.J. Gross, C. Baktash, D.M. Cullen, R.A. Cunningham, J.D. Garrett, W. Gelletly, F. Hannachi, A. Harder, M.K. Kabadiyski, K.P. Lieb, C.J. Lister, W. Nazarewicz, H.A. Roth, D. Rudolph, D. G. Sarantites, J.A. Sheikh, J. Simpson, Ö. Skeppstedt, B.J. Varley, and D.D. Warner, Phys. Rev. C **49**, R580 (1994).

4. C.J. Gross, W. Gelletly, M.A. Bentley, H.G. Price, J. Simpson, B.J. Varley, J.L. Durell, Ö. Skeppstedt, and S. Rastikerdar, Phys. Rev. C **43**, R5 (1991).

5. D.J. Dean, S.E. Koonin, K. Langanke, and P.B. Radha, Phys. Lett. B **399**, 1 (1997).

6. A. Petrovici, K.W. Schmid, and A. Faessler, Nucl. Phys. **A605**, 290 (1996)

7. P. Bonche, H. Flocard, P.H. Heenen, S.J. Krieger, and M.S. Weiss, Nucl. Phys. **A443**, 39 (1985).

8. W. Nazarewicz, J. Dudek, R. Bengtsson, T. Bengtsson, and I. Ragnarsson, Nucl. Phys. **A435**, 397 (1985).

9. D.S. Brenner, C. Wesselborg, R.F. Casten, D.D. Warner and J.-Y. Zhang, Phys. Lett. B **243**, 1 (1990).

10. W. Satuła, D.J. Dean, J. Gary, S. Mizutori and W. Nazarewicz, Phys. Lett. B **407**, 103 (1997).

11. W. Myers and W. Swiatecki, Nucl. Phys. **81**, 1 (1966).

12. W. Satuła and R. Wyss, Phys. Lett. B **393**, 1 (1997).

13. S.D. Paul, C. Baktash, W. Satuła, C. J. Gross, I. Birriel, R.M. Clark , M. Devlin, P. Fallon , A. Galindo-Uribarri, T . Ginter, D.R. Lafosse, F. Lerma, I.Y. Lee, A.O. Macchiavelli , B. Mcdonald, A. Piechaczek, D.C. Radford, W. Reviol, L.L. Riedinger, D. Rudolph, K. Rykaczewski, D.G. Sarantites, J.X. Saladin, D. Shapira, G.N. Sylvan, S.L. Tabor, K.S. Toth, W. Weintraub, D.F. Winchell, V.Q. Wood, and C.H. Yu, BAPS **43**, 1134 (1998).

14. S.D. Paul, C. Baktash, W. Satuła, C. J. Gross, I. Birriel, R.M. Clark , M. Devlin, P. Fallon , A. Galindo-Uribarri, T . Ginter, D.R. Lafosse, F. Lerma, I.Y. Lee, A.O. Macchiavelli , B. Mcdonald, A. Piechaczek, D.C. Radford, W. Reviol, L.L. Riedinger, D. Rudolph, K. Rykaczewski, D.G. Sarantites, J.X. Saladin, D. Shapira, G.N. Sylvan, S.L. Tabor, K.S. Toth, W. Weintraub, D.F. Winchell, V.Q. Wood, and C.H. Yu, in *Book of abstract, Nuclear Structure '98 conference*, (Gatlinburg, Tennessee, 1998), Vol. I, p. 101.

15. S.D. Paul, C. Baktash, W. Satuła, C. J. Gross, I. Birriel, R.M. Clark , M. Devlin, P. Fallon , A. Galindo-Uribarri, T . Ginter, D.R. Lafosse, F. Lerma, I.Y. Lee, A.O. Macchiavelli , B. Mcdonald, A. Piechaczek, D.C. Radford, W. Reviol, L.L. Riedinger, D. Rudolph, K. Rykaczewski, D.G. Sarantites, J.X. Saladin, D. Shapira, G.N. Sylvan, S.L. Tabor, K.S. Toth, W. Weintraub, D.F. Winchell, V.Q. Wood, and C.H. Yu, Physical Review C **58**, R3037 (1998).

16. J.D. Cole, T.M. Cormier, J.H. Hamilton, and A.V. Ramayya, Nucl. Instr. and Meth. **B70**, 343 (1992).

17. C. j. Gross *et al* To be published; A. N. James, P. A. Butler, T.P. Morrison, J. Simpson and K.A. Connell Nucl. Inst. and Meth. **212**, 545 (1983).

18. I.-Y. Lee, Nucl. Phys. **A520**, 641c (1990).

19. D.G. Sarantites *et al.*, Nucl. Instrum. Meth. **A381**, 418 (1996).

20. U. Lenz, K.E.G. Löbner, U. Quade, K. Rudolph, W. Schomburg, S.J. Skorka, and M. Steinmayer, in *Nuclei Far From Stability*, ed. I.S. Towner (AIP, Rosseau Lake, canada, 1987), p. 389.

21. J. Mukai, A. Odohara, R. Nakatani, Y. Haruta, H. Tomura, B.J. Min, K. Heiguchi, S. Suematsu, S. Mitarai, and T. Kuroyanagi, Z. Phys. **342**, 393 (1992).

22. J. D. Garret, J. Phys. Soc. Jpn. **54** (1985) Suppl.II,456.

23. L. K. Peker, F. K. Wohn, J.C. Hill, and R.F. Petry Phys. Lett. **B169**, 323 (1986).

24. D. Rudolph, C. Baktash, C.J. Gross, W. Satula, R. Wyss, I. Birriel, M. Devlin, H.-Q. Jin, D.R. LaFosse, F. Lerma, J.X. Saladin, D. G. Sarantites, G.N. Sylvan, S. L. Tabor, D.F. Winchell, V.Q. Wood, and C.H. Yu, Phys. Rev. C **56**, 98 (1997).

25. W. Satuła, R. Wyss, and P. Magierski, Nucl. Phys. **A578**, 45 (1994).

26. W. Satuła and R. Wyss, Phys. Rev. C **50**, 2888 (1994).

27. J. Dobaczewski, W. Nazarewicz, J. Skalski and T. Werner, Phys. Rev. Lett. **60**, 2254 (1988).

28. W. Satuła, private communications.

Nuclear structure near the doubly-magic ^{100}Sn

H. Grawe[1] in collaboration with

M. Górska[1,2] [†], M. Lipoglavšek[3,4], C. Fahlander[3,5], J. Nyberg[6], A. Gadea[5], G. de Angelis[5], Z. Hu[1], E. Roeckl[1], the EBGSI[1], GASP[5], PEX-NORDBALL[7] collaborations

[1] GSI, Darmstadt, Germany; [2] IEP, University of Warsaw, Poland; [3] Dept. of Physics, Lund University, Sweden; [4] J. Stefan Institute, Ljubljana, Slovenia; [5] INFN LNL Legnaro, Italy; [6] TSL, Uppsala, Sweden; [7] NBI, Roskilde, Denmark.

Abstract. The single particle (hole) energies in ^{100}Sn, as extrapolated by a shell model analysis of the neighbouring nuclei, show a remarkable similarity to those in ^{56}Ni, one major shell lower. This is borne out in nearly identical $I^\pi = 2^+$ excitation energies, implying $E(2^+) \simeq 3$ MeV in ^{100}Sn, and a large neutron effective E2 charge $e \geq 1.6e$. In contrast a small proton polarisation charge $\delta e \leq 0.3e$ is found, pointing to a large isovector charge. Mean field predictions for single particle energies show substantial deviations from the experimental extrapolation. ¿From the experimental two-proton hole spectrum in ^{98}Cd an improved empirical interaction is extracted for the $\pi(p_{1/2}, g_{9/2})$ model space yielding a good description of the N=50 isotones ^{95}Rh to ^{98}Cd. In ^{104}Sn, for the first time in this region, strong E3 transitions with B(E3)$\geq$ 17 W.u. were identified, indicating $E(3^-) \simeq 3$ MeV in ^{100}Sn. New experimental devices, as the Ge-cluster cube and total absorption spectrometers, applied in a pioneering experiment to the β^+/EC decay of ^{97}Ag, have led to a consistent picture of the Gamow-Teller quenching around ^{100}Sn. The experimental results are discussed in the framework of various shell model approaches by using both empirical and realistic interactions.

I INTRODUCTION

With the present progress in the instrumentation of nuclear structure experiments and the development of radioactive ion beams a wealth of new experimental data have become and will be available on exotic nuclei at N$\simeq$Z and the proton dripline. The development of efficient γ- arrays, recoil separators and selective ancillary detectors for evaporation neutrons and charged particles enable in-beam spectroscopy of proton-rich nuclei up to ^{100}Sn with fusion-evaporation reactions and stable beam/target combinations [1–4]. Decay spectroscopy of mass separated fusion residues and of fragmentation products have mapped the N=Z and proton dripline to the N=Z=50 closed shells [3–7]. Efficient γ-detection is a mandatory prerequisite for decay studies following mass separation and in-flight identification of fusion, fission or fragmentation products.

The main topics to be studied are shell structure, the mean field and the residual NN interaction at extreme isospin $T_z \simeq 0$. At N$\simeq$Z the proton-neutron interaction, its influence on shell closures and pairing, M1, E2 and E3 core excitations, Fermi- and Gamow-Teller decay, isospin mixing, proton decay and exotic shapes are the challenging study objects. The experimental approach to ^{100}Sn, as summarized recently [8], is landmarked by the T_z=1 nuclei [1,2] ^{98}Cd and ^{102}Sn in in-beam studies, and by ^{94}Ag [3] and ^{100,101}Sn in β^+/EC decay [4–6].

The single particle structure of ^{100}Sn as extracted from a shell model analysis of neighbouring nuclei [9] is shown in fig. 1. It shows close resemblance to ^{56}Ni, one major shell lower and likewise ls-open, which is corroborated by nearly identical I^π=2$^+$ excitation energies in the neighbouring isotopes and isotones of ^{100}Sn and ^{56}Ni [8]. From this striking similarity, $E_x(2^+) \simeq 3 MeV$ can be expected in ^{100}Sn [8], in agreement with a recent RPA calculation [10] predicting $E_x(2^+)$=2.86 MeV. The uncertainties in the single particle (hole) energies of fig. 1 are 100 keV for $\pi p_{1/2}, g_{9/2}$, 300 keV for $\nu d_{5/2}, g_{7/2}, h_{11/2}$ and 500 keV for $\pi d_{5/2}, g_{7/2}, \nu s_{1/2}, d_{3/2}$. In fig. 1 the extrapolated single particle energies are compared to mean field predictions with Skyrme SkX [11] and relativistic PL40 [12] forces. Clearly there are substantial deviations both in relative positions and

---

[†] The author is a scholar of the Foundation for Polish Science

CP481, *Nuclear Structure 98*, edited by C. Baktash

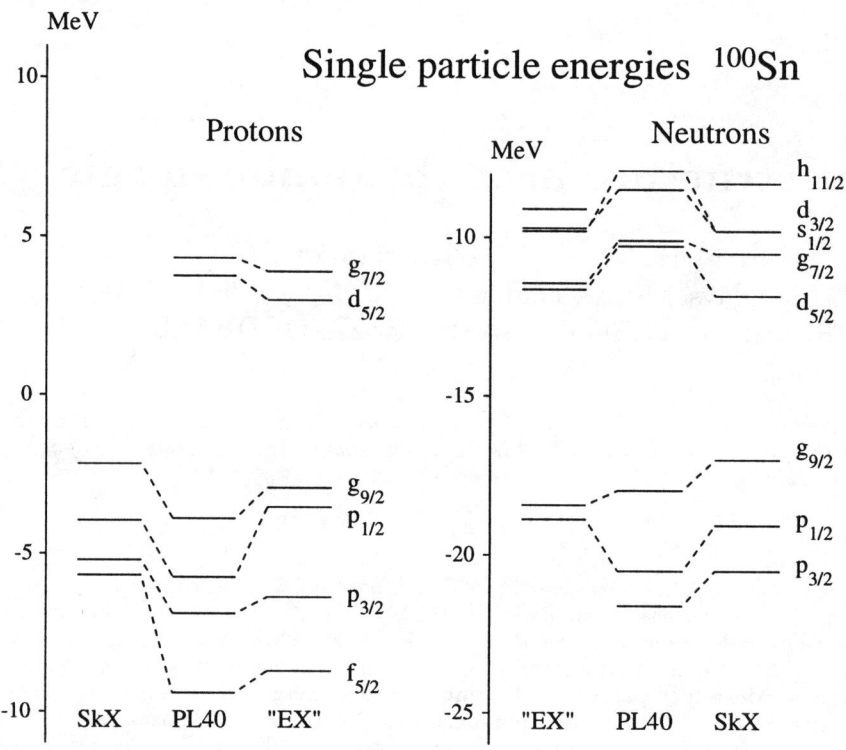

FIGURE 1. Single particle energies in ^{100}Sn and recent predictions from a new Skyrme parametrization [11] and a relativistic mean field approach [12].

the shell gaps. Due to the non ls-closed shell model core one expects strong $g_{9/2} - d_{5/2}$ (E2), $g_{9/2} - g_{7/2}$ ($\sigma\tau$), and $p_{1/2} - d_{5/2}, g_{7/2}$ $l_1 \pm l_2 = 3$ (E3) particle-hole excitations.

In this overview we will discuss recent experimental developments, empirical and realistic residual interactions, E2 and E3 core excitation (polarisation), and the Gamow-Teller strength in ^{100}Sn.

II EXPERIMENTAL TOOLS

The γ-ray spectroscopy of exotic nuclei requires highly efficient, selective and versatile detector arrays to discriminate the interesting residues of fusion-evaporation reactions against the abundantly produced less exotic species. For γ-detection the large multidetector-arrays EUROBALL [13] and GAMMASPHERE [14] and their precursors, respective early implementations, EUROGAM [15], GASP [16], PEX-NORDBALL [8,17] and the EUROBALL-cluster arrays [18–20] represent most powerful tools. Besides recoil mass separators, enabling recoil decay tagging techniques [21] to access extremely proton rich and particle emitting nuclei, ancillary detector arrays for neutrons and charged particles are indispensible prerequisits for selective filtering, exit channnel identification and particle spectroscopy close to the proton dripline [8]. Detailed spectroscopy is possible down to the sub-μbarn level [1,2], corresponding to production rates of less than 0.1 atoms/s.

In fig. 2a the $\Omega = 1\pi$ forward neutron wall [22] as attached to the central clover Ge-detector section of EUROBALL is shown. It consists of 15 3-fold segmented pseudohexagons in the geometry of the backward EUROBALL cluster cap, and a central 5-fold segmented pentagon (fig. 2b). The total efficiency for neutrons evaporated in a symmetric heavy-ion reaction was measured as 30 %. Fig. 3 shows the performance of nγ-discrimination by combining neutron time-of-flight (TOF) and pulse shape discrimination by the zero-crossing (Z/C) method. The spectrum was taken with neutrons from a ^{248}Cf fission source, which exhibit a spectrum

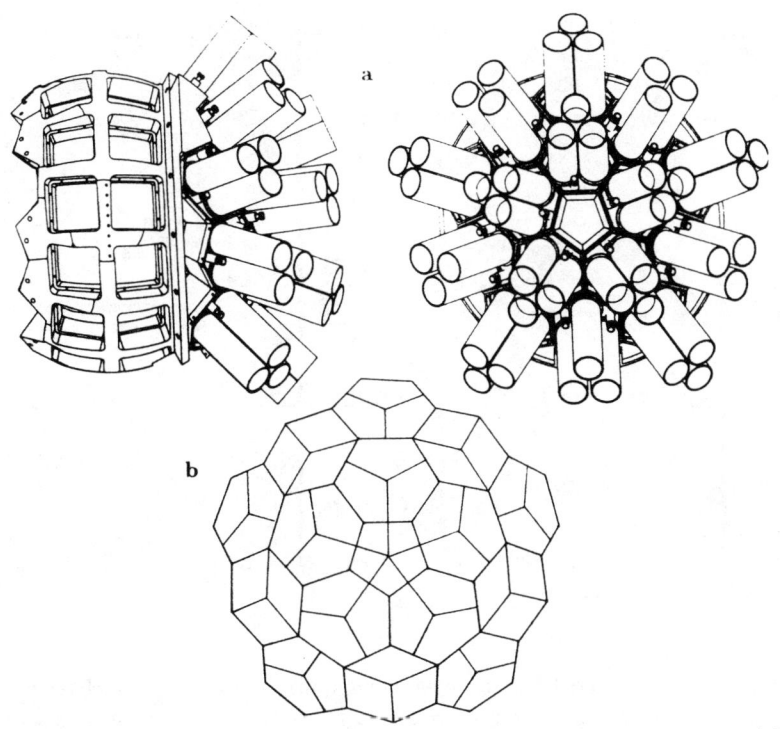

FIGURE 2. The EUROBALL neutron wall attached to the central section clover frame (a) and segmentation of the array (b).

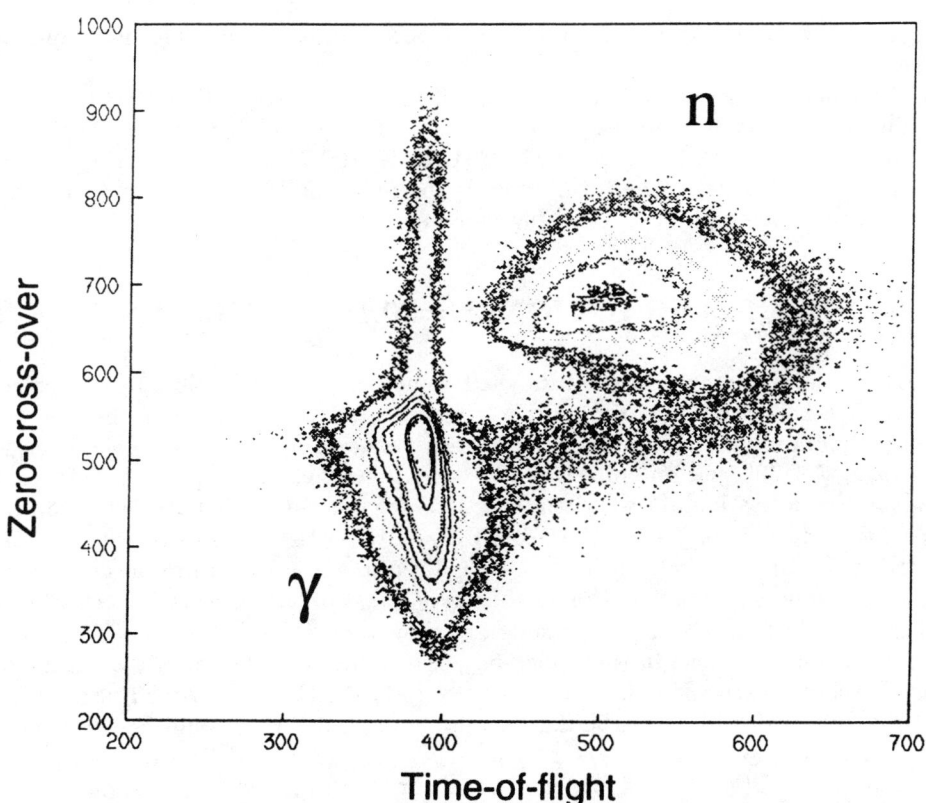

FIGURE 3. nγ - discrimination; neutron TOF vs. Z/C time.

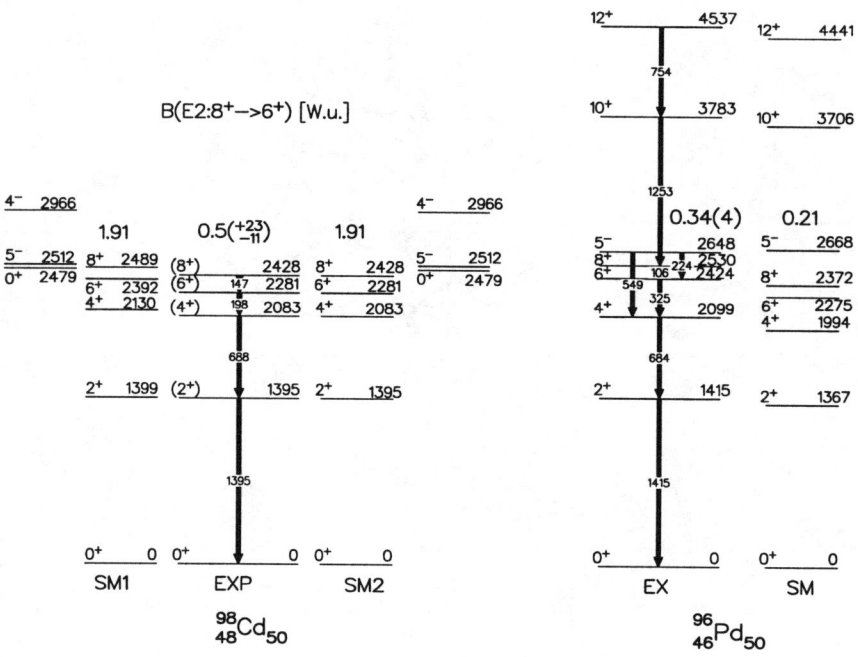

FIGURE 4. Experimental ^{98}Cd and ^{96}Pd level schemes in comparison to shell model results for level energies and B(E2;$8^+ \to 6^+$) obtained by using a new parametrisation of the $\pi(p_{1/2}, g_{9/2})$ empirical interaction.

identical to those from a fusion reaction. The quality of separation at a lower threshold equivalent to 50 keV electrons is excellent.

For light charged particle (LCP) detection the Si-ball ISIS [23] is used in EUROBALL, which is based on ΔE-E telescopes and is presently upgraded by segmentation and by exploiting pulse shape discrimination of LCP [24]. The scintillator Microball [25] used in GAMMASPHERE is complemented in forward direction by a highly segmented Si strip detector. The future goal is to enable LCP detection in spectroscopy quality to study discrete LCP emission from highly excited high spin states.

III   T=1 RESIDUAL INTERACTION: N=50 ISOTONES AND SN ISOTOPES

A doubly magic nucleus is characterized by its shell structure and the residual interaction in the single particle states at the Fermi surface. Due to the hitherto scarce spectroscopic data on the one- and two-particle (hole) neighbours of ^{100}Sn, only indirect information is available on single particle (hole) energies, as outlined in the introduction and Ref. [9], and on empirical two-body matrix elements [9,26–30]. Increasing effort has recently been made in the derivation of realistic interactions from various NN potentials [28,31–33].

In a recent experiment during the PEX-NORDBALL Pre-EUROBALL campaign at NBI, the ^{100}Sn $T_z=1$ neighbours ^{98}Cd [1] and ^{102}Sn [2] were identified. As ^{98}Cd represents the two-proton hole spectrum in ^{100}Sn, the N=50 empirical shell model parametrisation in the $\pi p_{1/2}, g_{9/2}$ model space [26,27] can now be improved to fit the experimental data. In figs. 4 and 5 the results are shown for ^{98}Cd, ^{96}Pd and ^{97}Ag, ^{95}Rh, respectively. The agreement obtained for the nuclei in the upper $\pi g_{9/2}$ shell starts to deteriorate towards midshell, as the limited model space excludes excitations from the lower π (p,f) shell and the N=50 neutron core. A revised fit in the spirit of ref. [30], which includes the data available beyond ^{96}Pd, would be the next step towards a reliable empirical T=1 interaction in the $\pi(f_{5/2}, p_{3/2}, p_{1/2}, g_{9/2})$ model space. On the other hand only marginal agreement is observed for the B(E2) values for transitions from the $\pi(g_{9/2})^n_{8+,21/2+}$ isomers calculated with an effective proton charge $e_\pi = 1.72\,e$ [9]. For the midshell nucleus ^{95}Rh this is not surprising, as the leading

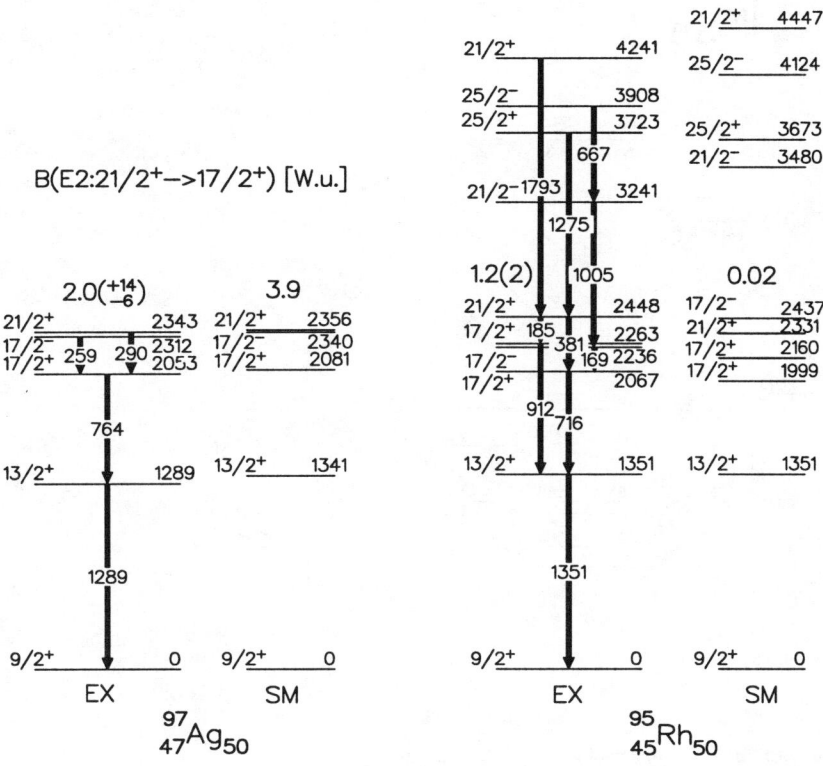

FIGURE 5. Same as Fig.4 for ^{95}Rh and ^{97}Ag and B(E2;21/2$^+$ → 17/2$^+$).

configuration $\pi(g_{9/2})^5$ contribution vanishes, which opens the field for neutron ph excitation across the N=50 shell gap to become effective. The discrepancy in the E2 strength for ^{98}Cd in contrast to the fairly well reproduced ^{96}Pd cannot be explained at all (see sect. IV). The large experimental uncertainty for ^{97}Ag will be reduced in the ongoing analysis of a recent experiment [34].

In fig. 6 the experimental level scheme below the I^π=6$^+$ isomer in ^{102}Sn is compared to various shell model predictions that are based on empirical and realistic interactions [9,28,31,32]. Good agreement is observed especially for the small 6$^+$ - 4$^+$ splitting, measured recently to be 48 keV [35]. The deviations among the predictions of the experimentally unknown higher lying states are mainly due to different assumptions for the single particle energies within the uncertainties discussed in the introduction.

IV   CORE EXCITATION OF ^{100}SN: E2 AND E3 POLARISATION

Particle-hole excitations of the type Δl=2 of an ls-open magic core give rise to low lying E2 strength and large E2 polarisation charges. The N=Z magic nuclei ^{56}Ni and ^{100}Sn provide a unique study ground, as the constructive isoscalar interaction of protons and neutrons occupying identical shell model orbits is expected to cause large polarisation effects. Likewise the shell structure of ^{100}Sn (see fig. 1) allows for E3 excitation of the $l_1 \pm l_2 = 3$ type, where l_2 corresponds to the $p_{1/2}$ hole and l_1 to the $d_{5/2}$ and $g_{7/2}$ particles.

The B(E2) values deduced for the $\pi(g_{9/2})^{-2}_{8^+}$ and $\nu(d_{5/2},g_{7/2})_{6^+}$ isomers in ^{98}Cd and ^{102}Sn correspond to $e_\pi = 0.94\binom{14}{10}\ e$ [1] and $e_\nu = 2.0\binom{3}{4}\ e$ [35]. The neutron effective charge supports results [36] deduced from ^{104}Sn and expectations from ^{56}Ni, based on the identical shell structure and I^π=2$^+$ energies. The small proton polarisation charge $\delta e_\pi \leq 0.1\ e$ seems to contradict a general E2 softness, as invoked for ^{56}Ni [37] and indicates a large isovector charge $e_{IV} = (e_\nu - e_\pi + 1)/2$. This is at variance with theoretical predictions based on Skyrme RPA calculations [39]. However, even a polarisation charge $\delta e_\pi = 0.3\ e$, as deduced for ^{97}Ag [38],

^{102}Sn

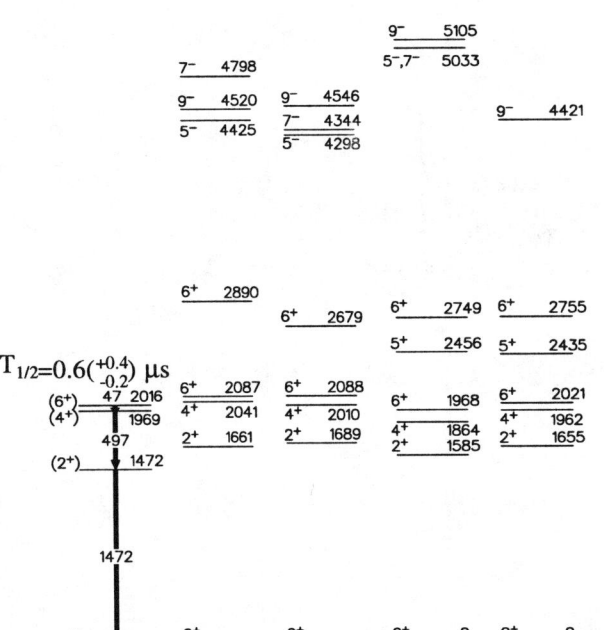

FIGURE 6. Experimental and shell model level schemes for ^{102}Sn for various approaches [9,28,32,31] (left to right).

which would correspond to a ^{98}Cd halflife shorter by a factor of two, would yield an isovector charge $e_{IV} \geq 0.6$, whereas $e_{IV} = 0.08$ is predicted by theory [39]. It should be emphasized, that the uncertainty in the neutron charge is mainly due to the uncertain position of the $s_{1/2}$ single particle energy (see sect. I), whereas for ^{98}Cd the existence of a higher lying core excited isomer, which escaped observation, could strongly affect the experimental halflife.

In fig. 7 results of a recent high spin study [40] of ^{104}Sn at GASP (LNL) are shown in a partial level scheme. States up to 7 MeV are well reproduced for all interactions shown for ^{102}Sn in fig. 6. For illustration only one of them [9] is included in fig. 7. The sequence of γ-rays above 8 MeV has M1 character and is most likely due to magnetic rotation based on proton ph excitation beyond the present pure neutron space. A similar M1 band was observed recently in the neighbouring ^{105}Sn [41]. Lifetimes were measured employing the recoil distance Doppler shift (RDDS) method. When taking the uncertainties in the single particle energies of the low spin orbitals into account, the B(E2) values for the $6^+ \rightarrow 4^+$ and $10^+ \rightarrow 8^+$ transitions are consistent with the neutron effective charge extracted for ^{102}Sn. For the first time in the ^{100}Sn region enhanced E3 transitions are observed. They cannot be reproduced by the shell model calculation without invoking admixture of a $I^\pi = 3^-$ phonon state. The observed enhancement of 12(4) and 27(10) W.u. for the $13^-_{1,2} \rightarrow 10^+$ transitions, repectively, compares well with the expected strength of an E3 phonon in ^{100}Sn, indicating $E_x \simeq 3$ MeV for this state, which in a recent RPA calculation was predicted at 3.56 MeV [10]. It should be noted, that these values are based on effective RDDS lifetimes, i.e. they represent lower limits as feeding times were not accounted for.

V ^{100}SN β^+/EC DECAY: GT QUENCHING IN ^{97}AG

The β^+/EC decay of ^{100}Sn is governed by the $\pi g_{9/2} \rightarrow \nu g_{7/2}$ Gamow-Teller (GT) transformation. In fig. 8 the decay scheme is shown as predicted in the simplest shell model approach [9] ignoring ph admixtures in the ^{100}Sn ground state. The dominant β-feeding populates the $(\pi g_{9/2}^{-1} \nu g_{7/2})_{1^+}$ state in ^{100}In. While the

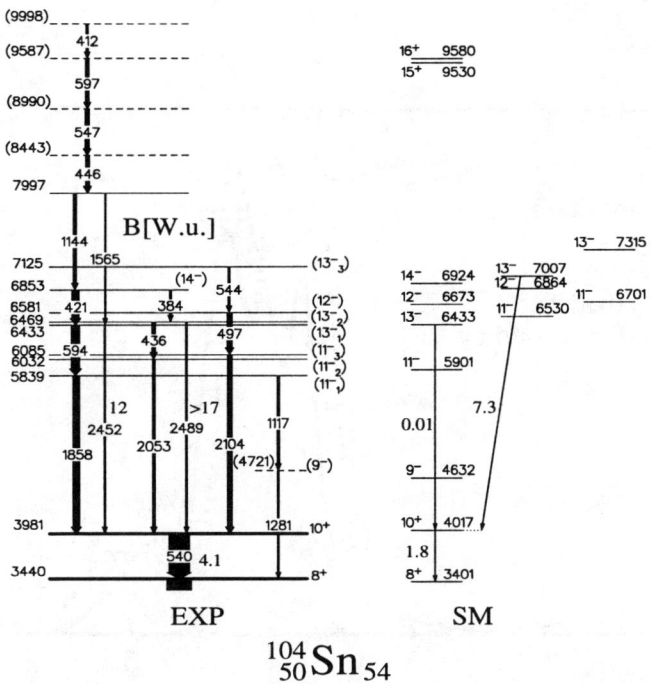

FIGURE 7. ^{104}Sn partial level scheme and comparison to shell model predictions. B(σ L) values are given for the $10^+ \to 8^+$ E2 and $13^-_{1,2} \to 10^+$ E3 transitions.

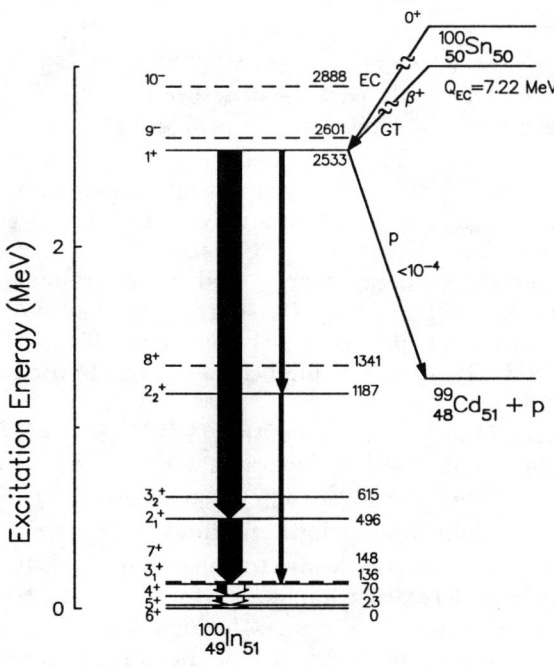

FIGURE 8. Shell model prediction for the ^{100}Sn β^+/EC decay.

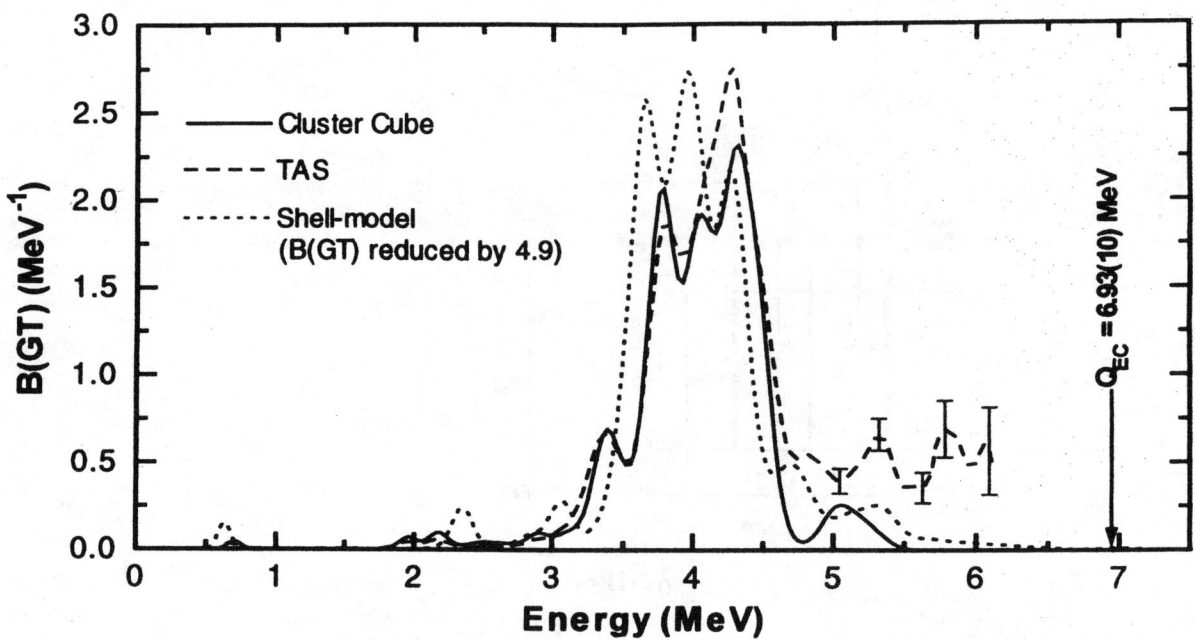

FIGURE 9. ^{97}Ag B(GT) distribution; cluster cube and TAS experiment vs. shell model. For comparison the high resolution cluster cube and shell model data are folded with the TAS resolution.

feeding pattern can be reliably predicted, the reduced GT strength B(GT) is less well determined in shell model calculations disregarding core polarisation. The single particle structure of ^{100}Sn (fig. 1) implies strong $\sigma\tau$, i.e. GT (and M1) matrix elements between the spin orbit partners $g_{9/2}$ and $g_{7/2}$.

The general quenching of the GT strength observed in β^+/EC decay [28,42–44] posed a theoretical as well as an experimental challenge that has been overcome only recently. In standard decay studies of nuclei with large Q_{EC} windows, a large portion of primary γ-rays, depopulating highly excited states embedded in a region of high level density, is not detected due to low efficient γ-detection. Consequently an apparent feeding is assigned to low-lying states, which due to an inappropriate decay energy used in determining the phase space factor f, translates to a too small reduced strength B(GT)= $6160/[ft_{1/2}(g_A/g_V)^2]$. The problem has been attacked in two ways, by the use of high-efficient γ-arrays like the Ge-cluster cube [45] and/or by total absorption spectrometers (TAS), consisting of large NaI(Tl) crystals of high efficiency but limited energy and multiplicity resolution [46].

In fig. 9 the B(GT) distribution as extracted from the analysis of the β^+/EC decay of ^{97}Ag with both methods is shown [45]. A well developed GT resonance is observed, which originates from the decay of the paired protons in the $\pi(g_{9/2})^n_{0+}g_{9/2}$ 1qp ground state (n=6,8) of ^{97}Ag to the 3qp daughter states in ^{97}Pd with configurations $[\pi g_{9/2}(\pi g_{9/2}^{-1}\nu g_{7/2})_{1+}]_{7/2+,9/2+,11/2+}$. The experimental strength distributions obtained with the two methods agree well, though the high resolution experiment still seems to miss some high-lying GT strength. The experimental results are compared to a shell model calculation in the $\pi(p_{1/2}, g_{9/2})\,\nu(d_{5/2}, g_{7/2}, s_{1/2}, d_{3/2}, h_{11/2})$ model space [28,45] . For better comparison with the TAS results, the high resolution and shell model data are folded with the TAS resolution function. The position and width of the GT resonance is well reproduced in the shell model, though a detailed comparison to the remaining fine structure originating from the $\pi\nu$ residual interaction in the daughter nucleus cannot be made presently, due to missing firm spin assignments. The overall strength of the shell model results was reduced by a factor 4.9, which compares well with quenching factors deduced from the decay of even-even N=50 and Z=50 nuclei close to ^{100}Sn, which are less affected by experimental deficiencies [42]. The established general quenching factor after considering configuration mixing in a $0\hbar\omega$ space h=$h_{cp} \times h_{ho} \simeq 5$ is explained by factors $h_{cp} \simeq 2-3$ from core polarisation and $h_{ho} \simeq 1.6$

from higher order corrections including subnucleonic degrees of freedom, which can be taken into account by a renormalized coupling constant $(g_A/g_V)^2$ [42,43]. Finally it should be emphasized that the high resolution study, resulting in 575 new γ-rays and 147 new states assigned to the decay of ^{97}Ag, is unprecedented in β-decay studies far from stability to date.

CONCLUSION

The recent developments in detector technology and in-beam and online identification techniques have enabled detailed spectroscopy of exotic nuclei at $N \simeq Z$ and at the proton dripline up to ^{100}Sn. The single particle structure of the N=Z=50 closed shell can be inferred from the new data, and first evidence is obtained on empirical two-body matrix elements and the adequacy of realistic interactions in the T=1 channel. The knowledge of the T=0 interaction, beyond N=50 is rather limited. First hints on the E2 and E3 polarizability of the closed core, which allow conclusions on the position of low-lying $I^\pi = 2^+$ and 3^- states, were obtained. However, a consistent explanation of these data is still missing. Experimentally, the data on the Gamow-Teller quenching in the ^{100}Sn region seem to converge, which may stimulate refined theoretical work.

At present, even with the new technical progress, we are still far from a complete understanding of the unexpected phenomena observed so far in this region, nor are we able to study subtle details in Fermi decay, isospin mixing, proton-neutron pairing, shape evolution with spin, and dripline effects thus far from stability. In the future, with intense radioactive ion beams and a detector technology, which is standard nowadays for stable-isotope beam experiments, considerable progress in detailed spectroscopy at and beyond the proton dripline will become possible, especially in connection with particle emission from ground states, high spin levels and from shape isomers.

ACKNOWLEDGEMENTS

The authors greatly appreciate the support and discussions with their collegues in the EBGSI, GASP, PEX-NORDBALL and EUROBALL collaborations. The invaluable help of the technical staff in the various laboratories involved is gratefully acknowledged. The detectors for the EUROBALL cluster campaigns were financially supported by the German BMBF and the Swedish Research Council.

REFERENCES

1. M. Górska *et al.*, *Phys. Rev. Lett.* **79**, 2415 (1997).
2. M. Lipoglavšek *et al.*, *Z. Phys.* **A356**, 239 (1996).
3. K. Schmidt *et al.*, *Z. Phys.* **A350**, 99 (1994).
4. Z. Janas *et al.*, *Phys. Scr.* **T56**, 262 (1995).
5. R. Schneider *et al.*, *Z. Phys.* **A348**, 241 (1994).
6. M. Lewitowicz *et al.*, *Phys. Lett.* **B332**, 20 (1994).
7. R. Grzywacz *et al.*, *Phys. Rev.* **C55**, 1126 (1997).
8. H. Grawe *et al.*, *Z. Phys.* **A358**, 185 (1997).
9. H. Grawe *et al.*, *Phys. Scr.* **T56**, 71 (1995).
10. S. Kamerdzhiev *et al.*, *Z. Phys.* **A346**, 253 (1993).
11. B.A. Brown, *Phys. Rev.* **C58**, 220 (1998).
12. K. Rutz *et al.*, *Nucl. Phys.* **A634**, 67 (1998).
13. C. Rossi Alvarez, *Proc. 6th Int. Spring Seminar on Nuclear Physics "Highlights of Modern Nuclear Structure", S. Agata sui due Golfi, Italy, May 18-22, 1998, in print.*
14. I.Y. Lee, *Progr. Part. Nucl. Phys.* **38**, 65 (1997).
15. F.A. Beck, *Progr. Part. Nucl. Phys.* **28**, 443 (1992).
16. C. Rossi Alvarez, *Nucl. Phys. News* **3**, 10 (1993).
17. C.D. O'Leary *et al.* *Phys. Rev. Lett.* **79**, 4349 (1997).
18. A. Jungclaus *et al.*, *Nucl. Phys.* **A637**, 346 (1998).
19. M. Rejmund *et al.*, *Z. Phys.* **A359**, 243 (1997).

20. Z. Hu *et al.*, *Nucl. Instr. and Meth.* **A**, in print.
21. E.S. Paul *et al.*, *Phys. Rev.* **C51**, 78 (1995).
22. Ö. Skeppstedt *et al.*, *Nucl. Instr. and Meth.* **A**, in print.
23. E. Farnea *et al.*, *Nucl. Instr. and Meth.* **A400**, 87 (1997).
24. G. Pausch *et al.*, *IEEE Transactions on Nuclear Science* **43**, 1097 (1996).
25. D.G. Sarantites *et al.*, *Nucl. Instr. and Meth.* **A381**, 418 (1996).
26. R. Gross, A. Frenkel, *Nucl. Phys.* **A267**, 85 (1976).
27. J. Blomqvist, L. Rydström, *Phys. Scr.* **31**, 31 (1985).
28. B. A. Brown, K. Rykaczewski *Phys. Rev.* **C50**, R2270 (1994).
29. D.H. Gloeckner, F.J.D. Serduke, *Nucl. Phys.* **A220**, 477 (1974).
30. X. Ji, B.H. Wildenthal, *Phys. Rev.* **C37**, 1256 (1988).
31. F. Andreozzi *et al.*, *Phys. Rev.* **C54**, 1663 (1996).
32. M. Hjorth-Jensen *et al.*, *Phys. Rep.* **261**, 25 (1995).
33. J. Sinatkas *et al.*, *J. Phys.* **G18**, 1377 (1992).
34. A. Axelsson *et al.*, *private communication.*
35. M. Lipoglavšek, D. Seweryniak *et al.*, *Phys. Lett. B*, in print.
36. R. Schubart *et al.*, *Z. Phys.* **A352**, 373 (1995).
37. G. Kraus *et al.*, *Phys. Rev. Lett.* **73**, 1173 (1994).
38. D. Alber *et al.*, *Z. Phys.* **A335**, 265 (1990).
39. I. Hamamoto, H. Sagawa, *Phys. Lett.* **B394**, 1 (1997).
40. M. Górska *et al.*, *Phys. Rev.* **C58**, 108 (1998).
41. A. Gadea *et al.*, *Phys. Rev.* **C55**, R1 (1998).
42. K. Rykaczewski, *Inst. Phys. Conf. Ser.* **No. 132**, 517 (1992).
43. I.S. Towner, *Nucl. Phys.* **A444**, 402 (1985) and *AIP Conf. Proc.*, **164**, 593 (1988).
44. I.P. Johnston *et al.*, *Phys. Rev.* **C44**, 1476 (1991).
45. Z. Hu *et al.*, *Proc. 2nd Int. Conf. on Exotic Nuclei and Atomic Masses (ENAM 98), June 23-27, 1998, Bellaire, Michigan, USA)*, in print and to be published.
46. J. Agramunt *et al.*, *contribution to this conference.*

Two-Proton Decay of the First Excited State of ^{17}Ne

M.J. Chromik[1,2], P.G. Thirolf[1,2], M. Thoennessen[1],
M. Fauerbach[1], T. Glasmacher[1], R. Ibbotson[1], R.A. Kryger[1],
H. Scheit[1], and P.J. Woods[3]

[1] *National Superconducting Cyclotron Laboratory and Department of Physics & Astronomy,
Michigan State University, East Lansing MI48824, USA*
[2] *Sektion Physik, Ludwig-Maximilians Universität München, D-85748 Garching, Germany*
[3] *Department of Physics, University of Edinburgh, Edinburgh EH9 3JZ, United Kingdom*

Abstract. The first excited state of ^{17}Ne has been populated via intermediate energy Coulomb excitation with a radioactive beam of ^{17}Ne on a ^{197}Au target to search for two-proton decay, which is in competition with the γ-decay back to the ground state of ^{17}Ne. The reconstructed invariant mass spectrum of the outgoing ^{15}O in coincidence with two protons shows evidence for a two-proton transition from the first excited state in ^{17}Ne as well as for transitions from higher excited states in ^{17}Ne, which will decay via the emission of two sequential protons.

INTRODUCTION

So far all experimental attempts to identify two-proton radioactivity at or near the proton dripline have been unsuccessful (e.g. [1]). A promising candidate is ^{17}Ne, where the first excited state ($J^\pi = 3/2^-$, $E^* = 1.288$ MeV) is bound by 168 keV with respect to one-proton emission but unbound with respect to two-proton emission by 344 keV (for details see [2,3]). Therefore this state can decay via a simultaneous emission of protons to ^{15}O, because the widths of the low lying states in ^{16}F are too small ($\simeq 40$keV) for a sequential decay through the tails. The two-proton decay is in competition with the γ-decay to the ground state of ^{17}Ne. In a recent intermediate energy Coulomb excitation experiment the γ-decay from the first excited state to the ground state ($J^\pi = 1/2^-$) has been measured and the experimental yield has been compared to the theoretically expected cross section. The measured γ-ray yield accounts for only 43% of the predicted yield from an excitation cross section of 18.7 mbarn, thus encouraging the investigation of a potential two-proton decay branch [2].

EXPERIMENTAL PROCEDURE

The experiment to search for the two-proton decay of ^{17}Ne was performed at the National Superconducting Cyclotron Laboratory at Michigan State University. A 60MeV/u radioactive ^{17}Ne beam was produced from a primary ^{20}Ne beam using the A1200 fragment separator. A Wien filter was used to further purify the secondary beam and 90% pure beam with an intensity of $\simeq 5000$ ^{17}Ne particles/s was achieved. In order to identify the two-proton decay from the first excited state in ^{17}Ne a complete reconstruction of the decay kinematics in the center-of-mass system (CM) was necessary. Thus the interaction point on the target as well as the energies and directions of all outgoing decay particles had to be measured.

Fig. 1 shows the experimental setup. Two position sensitive parallel plate avalanche counters (PPAC) in front of the target served to determine the interaction point of the ^{17}Ne in the target and a thin plastic scintillator 40m upstream was used for time-of-flight (ToF) measurements to identify the incoming particle. The target was surrounded by the MSU-NaI-array [4] to simultaneously measure the γ-ray decay. The particle fragments were analyzed in a multiple stage particle telescope, which was positioned at 0° relative to the

CP481, *Nuclear Structure 98*, edited by C. Baktash

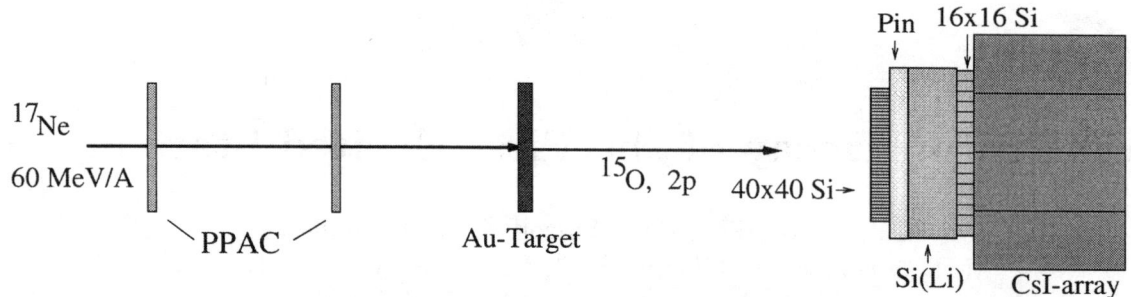

FIGURE 1. Experimental Setup

beam axis. The position of the heavy fragment was measured in a 500μm thick 40×40 double-sided silicon strip detector and the ΔE in a Si-Pin Diode (500μm), which also delivered a time signal. The heavy fragments were then stopped in a 5mm thick Si(Li)-Detector. The light fragments penetrated these detectors and were then detected in a 300 μm thick 16×16 double-sided silicon strip detector, which served as a ΔE-detector as well as for position measurements. Finally the protons were stopped in a 4×4 CsI array, (consisting of 16 crystals 1.7cm×1.7cm x 5cm, read out by photo diodes). These detectors were packed in a close geometry and placed 16.3cm behind the target, thus covering an opening angle from 0° to 7°. A 6.5cm thick lead collimator (with a central opening adapted to the size of the 40×40 strip detector) was placed in front of the detectors in order to shield the surrounding NaI-array from γ-rays originating from interactions with the detector material.

DATA ANALYSIS AND RESULTS

The incoming ^{17}Ne particles had to be identified and separated from the 10% of contamination which primarily consisted of ^{15}O. Fig. 2(a) shows the clear separation in the ToF spectrum between the plastic scintillator (40m upstream) and the Si-PIN. In addition the ^{17}Ne was spatially separated from the ^{15}O after a bending magnet at the end of the Wien-filter so it also could be separated in the 40×40 strip detector (Fig. 2(b).

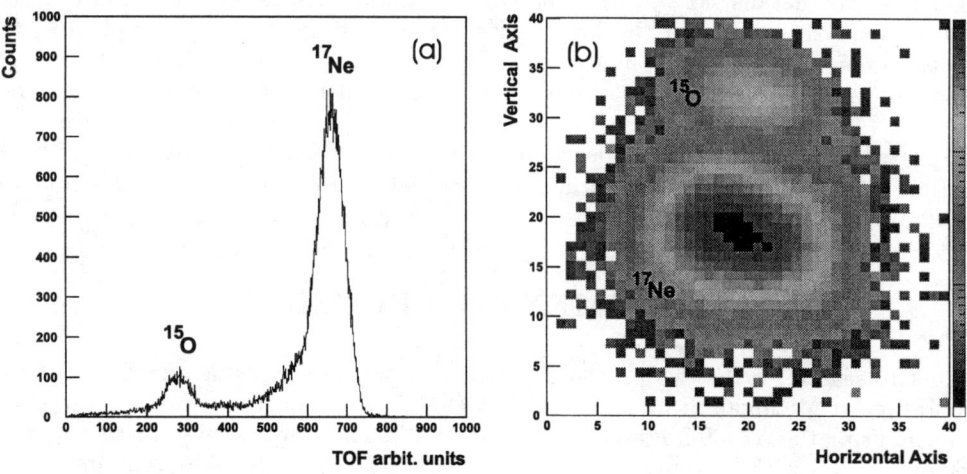

FIGURE 2. Particle Identification of the incoming beam. The incoming ^{17}Ne was identified with TOF-measurements (a) and the position in the 40×40 strip detector (b).

In the two-dimensional ΔE(PIN)-E(Si5) spectrum (not shown) it is possible to identify ^{15}O fragments from the (^{17}Ne,^{15}Opp) reaction and separate them clearly from the ^{17}Ne beam.

After identifying the ^{15}O from the reaction it is necessary to determine two-proton events. In Fig. 2 the sum of the ΔE in all the strips of the 16×16 strip detector is plotted against the sum of the deposited energy in all CsI detectors. In a spectrum gated on incoming ^{17}Ne only (a), one can clearly see the proton- and the deuteron-band, and barely the triton-band. An intensive band, consisting of events with twice the ΔE and E-values of the proton band can be identified with the two-proton band. At even larger energy losses and energies one can also see the ^{3}He and ^{4}He bands. An additional condition requiring single (Fig. 2(b)) or double hits (Fig. 2(c)) in the 16×16 verifies the identification of the two-proton band.

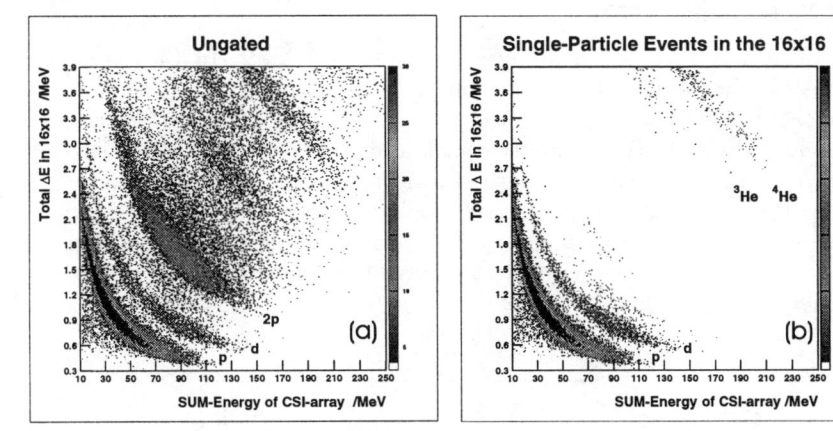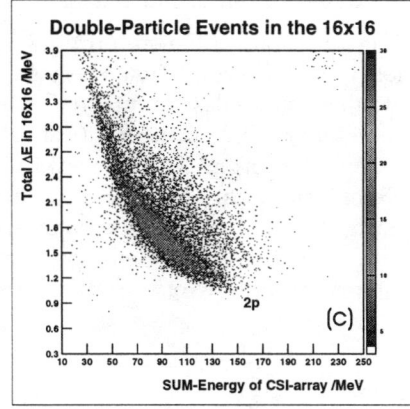

FIGURE 3. Sum of the ΔE in all the strips of the 16×16 strip detector versus the sum of the deposited energy in all CsI detectors with no gates (a), one particle gate (b) and two-particle gate (c).

Using the geometric correlation between events in the 16×16 strip detector and the CsI-array one can then extract the energies and directions for each proton. With the information of the energies of the outgoing fragments and their trajectories one can perform the transformation into the CM-System in order to obtain the invariant mass spectrum which is shown in Fig. 4(a). The Gaussian fit curves indicate the preliminary assigned peaks, the lowest one at an energy of $295 \pm 40^{stat} \pm 50^{syst}$ keV corresponding to the simultaneous two-proton decay of the first excited state in ^{17}Ne.

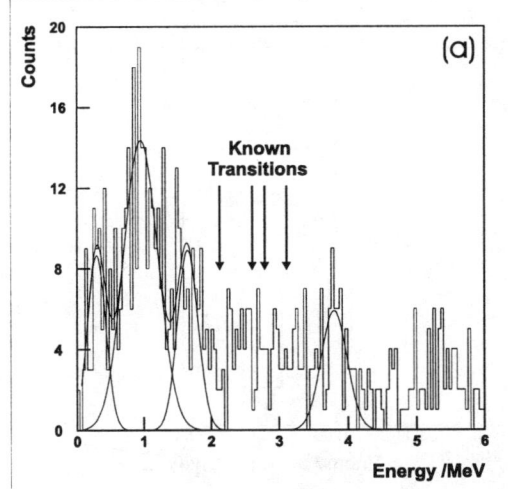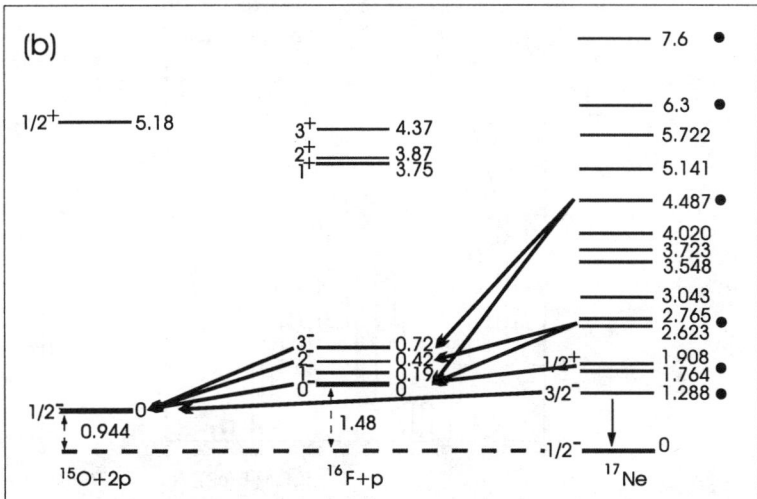

FIGURE 4. Decay energy spectrum (a) and decay scheme (b) of ^{17}Ne.

The measured decay energies agree with the known values [5]. The arrows between 2 and 3 MeV indicate positions of known transitions, which could not be resolved. The energy resolution is on the order of 150-200keV, mainly dominated by the error in the determination of the interaction point on the target.

Fig. 4(b) shows the decay scheme of ^{17}Ne. The levels of ^{17}Ne up to 6 MeV are from [5]. In order to transform the decay energies from Fig. 4(a) to the excitation energies one has to add the two-proton separation energy of 944keV. The full circles indicate states that were identified in the present experiment. Two new states at 6.3 MeV and 7.6 MeV were observed. In the meantime the 6.3 MeV state has been also identified and published in a transfer reaction [6].

Before the lowest energy peak can cleanly be identified as a two-proton decay it has to be shown that the decay of the upper states can be understood as sequential decays via intermediate states in ^{16}F. As an example we discuss the decay structure of the third peak at a decay energy of $\sim$ 1.7MeV corresponding to the decay of the doublet at an excitation energy of 2.623MeV and 2.765MeV.

Fig. 5 shows different proton spectra of events gated on a decay energy of ~1.7 MeV. In Fig. 5(a) the energy of the second (low energy) proton (The protons within one event were ordered by energy). It shows a double humped structure indicating two different distinguishable decay path. This decay pattern can be verified by gating on the left (light shaded) and the right (dark shaded) peak and displaying the energy spectra of the first proton. These spectra are shown in Fig. 5(b) and Fig. 5(d) resulting in two peaks with distinctly different energies. Another way to display the sequential decay scheme is the difference spectrum shown in Fig. 5(c). The low energy second proton together with the high energy first proton result in a large difference peak, whereas the high energy second proton and the low energy first proton have almost the same energy, thus resulting in a very small difference energy.

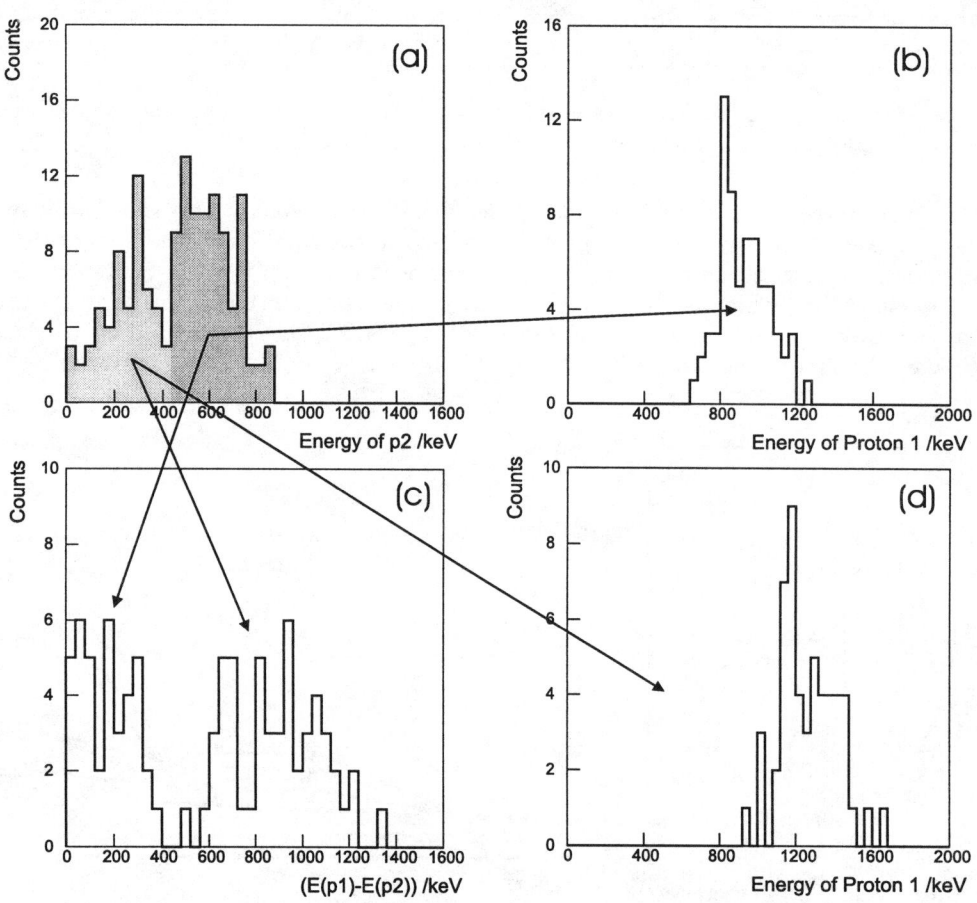

FIGURE 5. Decay structure of the peak at ~1.7 MeV. The energy spectra of the second proton (a) and the first proton gated by the low (light area, (b)) and high (dark area, (d)) second proton. The difference energy spectrum is shown in part (c).

A similar analysis has to be performed for the lowest energy peak of Fig. 4(a) in order to determine the decay structure of this two-proton decay. However, in a preliminary analysis of the opening angle of the protons in the CM-system indicates an isotropic distribution, which would be consistent with a simultaneous uncorrelated two-proton emission. A preliminary analysis of the lifetime indicates a two-proton lifetime in the order of picoseconds, compared to a lifetime of $\simeq$ 0.1 ps for the γ-decay from the first excited state in ^{17}Ne.

DISCUSSION AND CONCLUSION

In conclusion, we observed several sequential two-proton transitions at their expected energies as well as evidence for two-proton radioactivity of the first excited state in ^{17}Ne. The preliminary lifetime of picoseconds is too fast for the emission of two protons from the $1d_{5/2}$ shell, which is calculated to be a factor of $\simeq 300$ slower [2]. Using an improved experimental setup with optimized efficiency and energy resolution will help to clarify the remaining uncertainties in the context of the reported first evidence for two proton radioactivity.

One of us (MJC) acknowledges the support and hospitality of the NSCL and the support of the "Studienstiftung des Deutschen Volkes".
We acknowledge the help of A. Azhari and S. Yokoyama during the experiment and thank J. Brown, D.J. Morrissey and M. Steiner for producing the radioactive ^{17}Ne beam. This work was supported by the National Science Foundation under grant PHY95-28844.

REFERENCES

1. R. A. Kryger *et al.*, Phys. Rev. Lett. **74**, 860 (1995)
2. M. J. Chromik *et al.*, Phys. Rev. C **55**, 1676 (1997)
3. P. Woods *et al.* Proposal for CYCLONE, Louvain-la-Neuve 1995, unpublished.
4. H. Scheit *et al.*, submitted to Nucl. Instr. Meth. A (1998)
5. V. Gũimaraes *et al.*, Z. Phys. A **353**, 117 (1995)
6. V. Gũimaraes *et al.*, Phys. Rev. C **58**, 116 (1998)

Prompt Particle Decays of Deformed Bands and Nuclear Structure near ^{56}Ni

D. Rudolph*

*Department of Physics, Lund University, S-22100 Lund, Sweden

Abstract. Excited states in neutron-deficient mass $A \sim 60$ nuclei were studied by means of several heavy-ion fusion-evaporation reactions. The GAMMASPHERE array in conjunction with ancillary detector systems such as MICROBALL and neutron detectors allowed for the identification of well deformed rotational structures in the second well of nuclei in the direct vicinity of the doubly-magic isotope ^{56}Ni. Most interestingly, some of these bands were found to decay by prompt particle emission in competition to the expected γ decay-out. Experimental details of this decay mode are presented. The rotational bands are interpreted with mean-field and large-scale shell-model calculations.

INTRODUCTION

An interesting facet of nuclei close to the proton drip line is that proton (or even alpha-particle) emission might compete with γ decay of excited states due to the very low Coulomb barrier. In fact, proton radioactivity was first discovered in the decay of the $I^\pi = 19/2^-$ isomeric state in ^{53}Co [1]. Lateron, β-delayed protons [2] and proton emission from highly excited Gamow-Teller resonance states [3] have also been observed in nuclei around ^{56}Ni. However, the process of particle emission can take place only in a narrow energy window [4]. This might be the main reason that no ground-state proton emitters have been identified in the $A \sim 60$ region [5] though the phenomenon is by and large well established in heavier nuclei ($A \geq 120$) along the proton drip line (cf. contributions by J.C. Batchelder and P.J. Woods and Refs. [5,6]).

The theoretical description of (spherical) proton emission accompanied and projected the experimental work because "as studies of exotic nuclei progress, it will be very important to determine the empirical ordering of different single-particle states near and beyond the drip-line, to classify excited states, and, possibly, to investigate the competition between the proton radioactivity and other decay modes (such as gamma decay)" [7]. Moving from spherical to deformed proton emitters [8], a very recent theoretical study points out that "proton decay can be used to probe small components of the deformed wave function in the mother nucleus, which would otherwise be very difficult, if at all possible, to measure" [9]. The competition between prompt γ radiation and particle (proton or neutron) emission has already been studied theoretically more than twenty years ago. T. Døssing and co-workers conclude that "future experiments on high angular momenta might reveal yrast traps or ... states ... from which the γ decay is strongly hindered. If such states exist above the critical angular momentum for particle emission then the nucleon will be emitted with just one energy, or a few possible energies, and this nucleon plus the following γ-decay cascade would tell a lot about the structure of the high spin state" [10]. However, experimental evidence for these exotic decays has been lacking until recently.

Indeed, experiments have by now revealed a plethora of states in superdeformed or second minima of the nuclear potential from which the γ decay back *is* strongly hindered. At latest, superdeformation was established in the mass $A \sim 60$ regime (cf. contribution by C.E. Svensson and Ref. [11]), and very recently we succeeded to observe for the first time a prompt ($\tau < 3$ ns) decay of a *well-deformed* excited rotational band in the $N=Z$ nucleus ^{58}Cu via emission of monoenergetic protons into a *spherical* excited state in ^{57}Ni [12]. In the following, a second case was established in the decay-out of a rotational band in the doubly magic nucleus ^{56}Ni [13], and there is evidence for a third example in the neighboring isotope ^{59}Cu [14]. Moreover, a weak ($\sim 5\%$) decay-out branch from the second minimum in ^{58}Ni constitutes the first observation of a prompt monoenergetic α decay

CP481, *Nuclear Structure 98*, edited by C. Baktash

into the spherical 2949 keV 6^+ yrast state in ^{54}Fe [15–17]. It is noted that long-lived particle decays (half-lifes in the order of hours), possibly associated with the decay of states in a second minimum of the potential, have been reported earlier for both α particles [18] and protons [19]. However, the experimental basis shown by Marinov *et al.* is considered weak if not inconclusive.

This contribution presents more details of the results in this new field of nuclear structure studies in the $A \sim 60$ region, and summarizes other compelling results on high-spin spectroscopy near ^{56}Ni:

- Spectroscopic data from doubly magic and neighboring nuclei provide essential information for the parameter sets of (here: large-scale) spherical shell-model calculations, i.e., single-particle energies and two-body matrix-elements and, simultaneously, put the most stringent constraints on these calculations. Consequently, they define and relate the effective nuclear forces. So far, large-scale shell-model calculations in the full pf configuration space have been successfully used to describe the collective structures of $1f_{7/2}$ mid-shell nuclei [20–22].

- The deformed states near ^{56}Ni involve multi-particle multi-hole excitations across the $N,Z = 28$ shell gap from the $1f_{7/2}$ shell to the $2p_{3/2}$, $1f_{5/2}$, and $2p_{1/2}$ orbits thus providing an excellent testing ground to confront the nuclear shell-model with cluster models [23,24] and the approaches based on the mean-field theory [11,12,25,26]. Excitations to the higher lying $1g_{9/2}$ intruder orbit explain the recently observed well-deformed or superdeformed rotational bands in the second minima of, e.g., ^{58}Cu [12] and ^{62}Zn [11,27]. At high excitation energies, reaction studies have revealed evidence for hyperdeformed resonances in the ^{56}Ni compound system [28].

- The investigated nuclei are lying at or even beyond the $N = Z$ line. This implies the coherent occurence of proton and neutron shell effects which may lead to pronounced shapes but also shape coexistence effects. In addition, the well characterized (super)deformed rotational bands in the $A \sim 60$ $N = Z$ nuclei might shed some light on the influence of isospin $T = 0$ neutron-proton correlations in the rotational behavior.

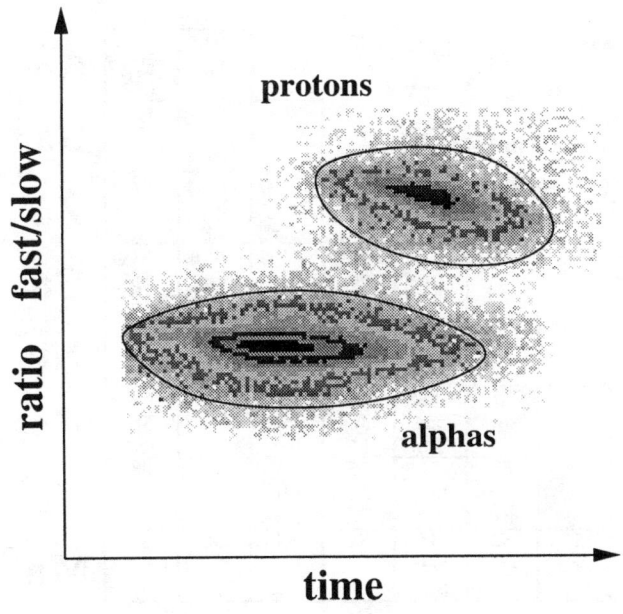

FIGURE 1. Two-dimensional spectrum from the MICROBALL detector element 35 which is situated in the fourth ring at 52° with respect to the beam axis. The leading edge discriminator time is plotted versus the charge ratio of the fast and the slow part ('particle identification') of the signal, both in arbitrary units. The spectrum is gated by two two-dimensional gates in similar matrices 'energy vs. time' and 'energy vs. ratio' for protons as well as α-particles. The third two-dimensional gate for each species, indicated by the solid lines in the plot, finally determines the unambigous particle identification which is then used for further analysis.

- The overall shell structure in the vicinity of ^{56}Ni is very similar to the one at the next (and last) doubly magic $N = Z$ nucleus, namely ^{100}Sn (cf. contribution by H. Grawe). However, the $A \sim 60$ regime is considerably easier to access not only with heavy-ion fusion evaporation reactions but also light-ion induced reactions and fragmentation processes. Likewise, in-beam studies of ^{100}Sn and its closest neighbors demand the presence of radioactive ion beams.

EXPERIMENTS

Mainly light ion induced reactions have been previously used to study excited states in ^{56}Ni [29,30] and neighboring nuclei. The yrast line was followed typically up to less than 10 MeV excitation energy and spins $I \leq 10\ \hbar$. The experiment pioneering the high-spin regime of nuclei near ^{56}Ni was performed at the 88-Inch Cyclotron at the Lawrence Berkeley National Laboratory. The fusion-evaporation reaction ^{28}Si(^{36}Ar,$xpynz\alpha$) at 143 MeV beam energy was used. A 0.42 mg/cm^2 thin target layer of 99.1 % enriched ^{28}Si was evaporated onto a 0.9 mg/cm^2 Tantalum support foil which faced the beam leading to a reduction of some 7 MeV in the beam energy. The GAMMASPHERE array [31] comprised 82 Germanium detectors and the γ rays were measured in prompt coincidence with the evaporated light particles to provide reaction channel selection. The charged particles were detected in the 4π CsI ball MICROBALL [32] while neutrons were measured in fifteen liquid scintillator neutron detectors.

Protons and α particles were identified in the MICROBALL by pulse-shape discrimination techniques [32]. For the present experiment we employed a triple two-dimensional gating procedure for each particle species. They are set in the three two-dimensional spectra which can be created using the three signals from each

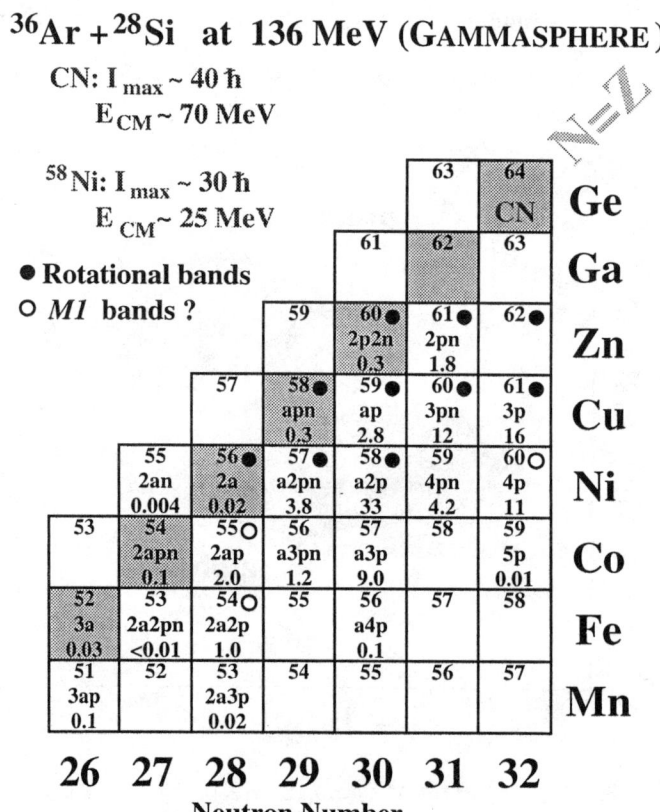

FIGURE 2. Nuclei in the vicinity of ^{56}Ni populated in the reaction ^{36}Ar+^{28}Si at 136 MeV beam energy. For each isotope the mass number, the reaction channel, and the experimental relative fusion-evaporation cross section (%) are given. Filled circles indicate the observation of deformed or superdeformed bands. Open circles mark the presence of sequences which might form so-called $M1$ bands.

MICROBALL element: time, energy, and the charge ratio between the early and late part of the pulse ('particle identification'). Figure 1 illustrates the final gates (solid lines) in a two-dimensional spectrum time vs. ratio. of MICROBALL detector element 35, situated in the fourth ring around 52° with respect to the beam axis. The 52° laboratory angle correspond to nearly 90° in the center-of-mass frame. The spectrum shown in Fig. 1 is already gated by two two-dimensional gates set in the matrices 'energy vs. time' and 'energy vs. ratio' for protons as well as α-particles. Clearly, the third gate unambiguously identifies the kind of the charged particle detected, and the total of by and large 600 two-dimensional gates are optimized in the course of three to four iteration loops. The inverse kinematics and restrictive particle gating (aiming at the most effective channel selection, cf. Ref. [33]) resulted in a proton detection *and* identification efficiency of nearly 80 % while that of α particles amounted to some 65 %.

Neutrons and γ rays were very well discrimated by the combination of a pulse-shape analysis of the neutron detector signals and time-of-flight measurements relative to the RF-frequency of the 88-Inch Cyclotron. The energies of the charged particles detected in the MICROBALL were used to evaluate the momenta of the individual recoiling nuclei. This kinematic correction results in a more precise Doppler-shift determination of the γ-ray energies and, hence, significantly improves the energy resolution of the peaks in the spectra, particularly for these light systems.

The events were sorted off-line into various E_γ projections, E_γ-E_γ matrices, and E_γ-E_γ-E_γ cubes subject to appropriate evaporated particle and analysis conditions. Careful successive subtractions of contaminations from higher fold charged-particle channels, which may leak through when one or more proton or α escaped detection or identification, resulted in purified singles projections, $\gamma\gamma$ matrices, and γ-gated spectra for a given isotope. Based on the yields of (known) ground-state transitions, corrected for γ- and particle detection efficiency, experimental relative cross sections were deduced. Altogether some 25 nuclei were produced with measurable cross section. Among them, the strongest channel, ^{58}Ni+1α2p, comprises nearly one third of the total fusion cross section. In contrast, ^{55}Ni+2α1n represents the weakest channel identified, having about 0.004 % relative fusion-evaporation cross section [33]. Figure 2 provides the relevant part of the chart of nuclides, indicating the compound nucleus (CN) ^{64}Ge of the reaction ^{36}Ar+^{28}Si. For each isotope the mass number, the reaction channel, and the experimental relative fusion-evaporation cross section in percent are given. Filled circles indicate the observation of deformed or superdeformed bands. Open circles mark the presence of sequences which might form so-called $M1$ bands.

THE NATURE OF THE ROTATIONAL BANDS NEAR ^{56}NI

The rotational bands in the vicinity of ^{56}Ni are based on four-particle four-hole (4p-4h) excitations across the magic $N = Z = 28$ shell gap between the $1f_{7/2}$ orbit and the so-called upper pf shell ($2p_{3/2}$, $1f_{5/2}$, and $2p_{1/2}$ orbits). In the Nilsson scheme, illustrated at the top of Fig. 3, the crossing between the upsloping [303]7/2 and the downsloping [321]1/2 orbits opens a second distinct energy gap at quadruple deformations $\beta_2 \sim 0.3$–0.5. Taking ^{56}Ni as an example, this configuration is reflected by the sequence of Routhians ($\beta_2 = 0.4$) in the upper right part of Fig. 3 labelled 4^04^0: The levels below the *deformed* shell gap at $N = Z = 28$ are filled, and none of the (quasi)particles has been excited into the $\mathcal{N} = 4$ $1g_{9/2}$ intruder orbit [440]1/2.

The Routhians presented in the lower left part of Fig. 3 originate from a theoretical description obtained within the Skyrme-Hartree-Fock (HF) method and the cranking approximation [34] with the Skyrme interaction SLy4 [35] at a quadrupole deformation $\beta_2 \sim 0.4$. Due to the large Coriolis force, the downsloping high-j low-Ω [440]1/2 neutron (and proton) orbit creates an energy gap of several MeV at (for the mass $A \sim 60$ region) medium high rotational frequencies of $\hbar\omega \approx 1.0$ MeV. For ^{56}Ni the excitation of one neutron and one proton (4^14^1 in Fig. 3) costs about 5 MeV excitation energy at low spins, but it is predicted to become the yrast configuration beyond spin $I = 20$ $\hbar$. Due to the expected shape driving effect of the $1g_{9/2}$ particles the calculated quadrupole deformation increases from $Q_t = 1.65$ eb for the 4^04^0 band to 1.89 eb for the 4^14^1 band (see the right hand side of Fig. 3).

The odd-odd $N = Z = 29$ nucleus ^{58}Cu has one proton and one neutron in addition to ^{56}Ni. Consequently, it forms a deformed "doubly-magic" core at $\beta_2 \sim 0.4$ — the yrast 4^14^1 band is very well separated from other collective structures by a gap of some 4 MeV [12]. Moving to even larger deformations, the second [440]1/2 Routhian also becomes more quickly favored, leading to a superdeformed "doubly magic" core at $\beta_2 \sim 0.5$ for the $N = Z = 30$ system, i.e., ^{60}Zn. In fact, this band has recently been identified, and it is discussed in the contribution by C.E. Svensson and Ref. [36] in more detail.

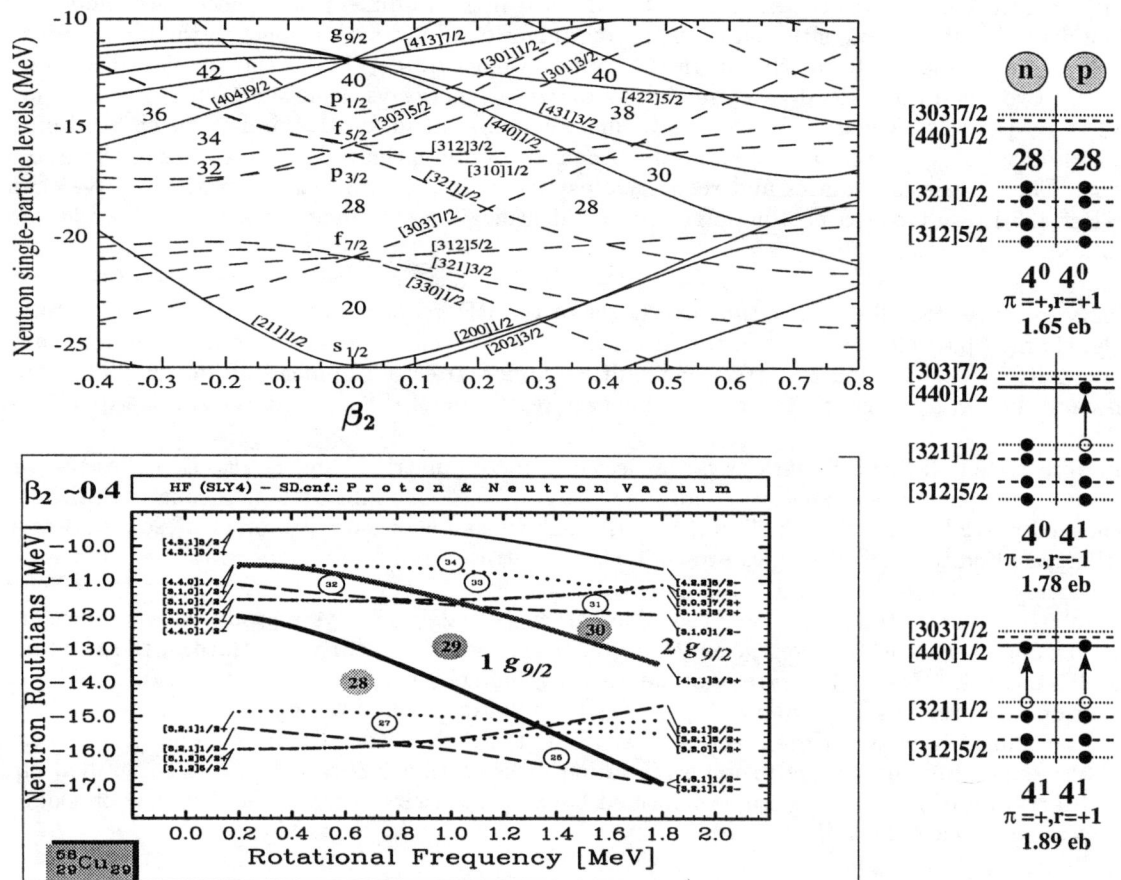

FIGURE 3. The relevant neutron single-particle levels for the mass $A \sim 60$ region from Woods-Saxon calculations are presented in the upper left part. The Routhians, originating from cranked Hartree-Fock calculations, are shown underneath for a quadrupole deformation of $\beta_2 \sim 0.4$. On the right hand side, three possible single-particle configurations for deformed bands in ^{56}Ni are sketched with zero neutrons and zero protons ($4^0 4^0$), one proton ($4^0 4^1$), and one neutron and one proton ($4^1 4^1$) excited from the pf shell into the $\mathcal{N} = 4$ [440]1/2 $1g_{9/2}$ intruder level.

PROMPT PROTON DECAY IN ^{58}CU

The first example of a prompt ($\tau < 3$ ns) proton decay of a state associated with a deformed secondary minimum in the potential has been observed in ^{58}Cu [12]. A schematic drawing of the irregularly spaced states in the first, spherical minimum, and the rotational band in the second minimum of this nucleus is shown in Fig. 4. The spin values of the states in the first minimum were measured while those in the second minimum were inferred from the most probable configuration of the band, i.e., $\pi 1g_{9/2} \times \nu 1g_{9/2}$, as well as the best estimates originating from partly measured spins of initial and final states of both γ and proton decay out of the band. Within the band angular distribution and correlation measurements clearly indicate quadrupole character for the γ-ray transitions [17].

From the γ-ray spectrum at the bottom of Fig. 4 we can infer that one final state at 3701 keV excitation energy in the daughter nucleus ^{57}Ni is dominantly ($> 90\%$ of the *proton* decay intensity) populated by the main branch observed at 2.3(1) MeV proton center-of-mass energy. The γ-ray spectrum at the bottom was gated by one detected α particle, proton, and neutron, and measured in coincidence with the second and third transition in the rotational band (1197 and 1576 keV). It shows the other transitions in the band (830, 1955, 2342, 2748, and 3181 keV), the discrete linking transition at 4.2 MeV, and γ rays between states in the

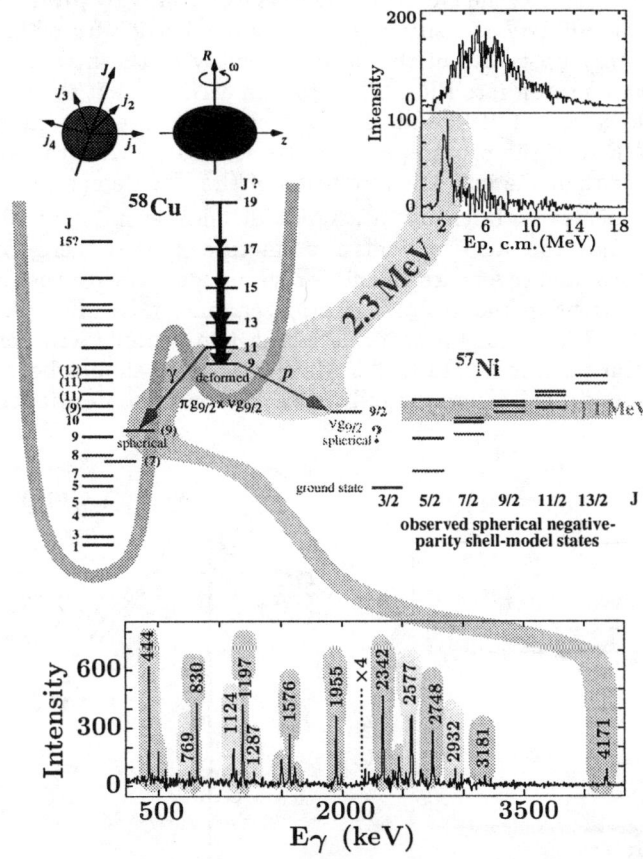

FIGURE 4. Sketch of the prompt proton decay from the deformed band in the second potential well of ^{58}Cu into a spherical daughter state in ^{57}Ni. The γ-ray spectrum at the bottom was gated by one detected alpha, proton, and neutron, and measured in coincidence with the second and third transition in the rotational band (1197 and 1576 keV). It shows the other transitions in the band (830, 1955, 2342, 2748, and 3181 keV), the discrete linking transition at 4.2 MeV, and γ rays between states in the spherical minima of both ^{58}Cu (e.g., 444 keV) and ^{57}Ni (e.g., 1124 and 2577 keV). In the upper right part of the figure proton center-of-mass energy spectra are shown gated by one α particle, one or two protons, and one neutron, in prompt coincidence with the 444 keV ground-state transition in ^{58}Cu (top) and the 830 keV transition which feeds the proton decaying ^{58}Cu band head.

spherical minima of both ^{58}Cu (e.g., 444 keV) and ^{57}Ni. The 1124 and 2577 keV transitions mark the main γ-decay branch of the 3701 keV level in ^{57}Ni [37,17]. A weaker branch from the same level proceeds via the 2932 and 769 keV lines while another proton branch might be inferred from the weak presence of the 1287 keV $11/2^- \to 7/2^-$ yrast transition of ^{57}Ni in the spectrum. The known levels in ^{57}Ni with observed *negative* parity and spins $I = (9/2 \pm 2)\,\hbar$ are also indicated in the figure [37,17]. Though several of them have very similar excitation energies as compared to the 3701 keV 9/2 [17] state, only the latter is significantly populated through the proton decay. In fact, a $9/2^-$ state is well established only 12 keV apart. Since shell-model calculations indicate a high degree of mixing within the negative-parity states in ^{57}Ni [17], this fact plus the apparent selectivity of the proton decay clearly hint towards positive parity of the 3701 keV state. However, it yet remains to be determined experimentally.

In the upper right part of Fig. 4 center-of-mass energy spectra of protons are shown. The spectra are preselected by gating on one detected α particle, one or two protons, and one neutron. The upper spectrum is in prompt coincidence with the 444 keV ground-state transition in ^{58}Cu and reveals a shape expected for an energy distribution following a fusion-evaporation reaction. Contrary, the lower spectrum provides a distinct peak at 2.3(1) MeV and a FWHM of 0.7(1) MeV. It is in coincidence with the 830 keV transition which directly feeds the proton decaying ^{58}Cu band head. The yield of the peak amounts to 51(2)% of the total number of counts in this spectrum, indicating that the band-head decays essentially by proton emission. In fact, analysing a γ-ray spectrum gated by the 830 keV transition, a γ decay-out branch from this state can be estimated to be less than 3%. Contrary, only up to 8% of the decay intensity of the first excited state in the rotational band may proceed via a 3 MeV proton into the $11/2^-$ state in ^{57}Ni — next to the 830 keV in-band transition this state decays through two γ rays at 1519 and 4171 keV back into the first minimum of ^{58}Cu. It should be noted that the measured FWHM of the proton peak can be associated with the kinematic broadening due to the finite opening angle rather than the intrinsic resolution of the CsI elements.

Next to the proton energy spectra $\gamma\gamma$ intensity ratios provide another unique tool to identify prompt particle decays. Figures. 5(a) and (b) show the ratios of yields as a function of γ-ray energy for transitions in coincidence with one α particle, two protons, and one neutron, and γ rays associated with the proton decay ("^{58}Cu gated") as well as normal yrast transitions in the $1\alpha 2p1n$ reaction channel ^{57}Ni. For part (a) an additional proton energy restriction of $E_{p,c.m.} < 3$ MeV for at least one of the two protons was demanded for the numerator. Since the energy of the proton peak amounts to 2.3 MeV, this ratio should be close to unity for the "^{58}Cu gated" transitions as it was confirmed experimentally. On the contrary, normal transitions within ^{57}Ni must

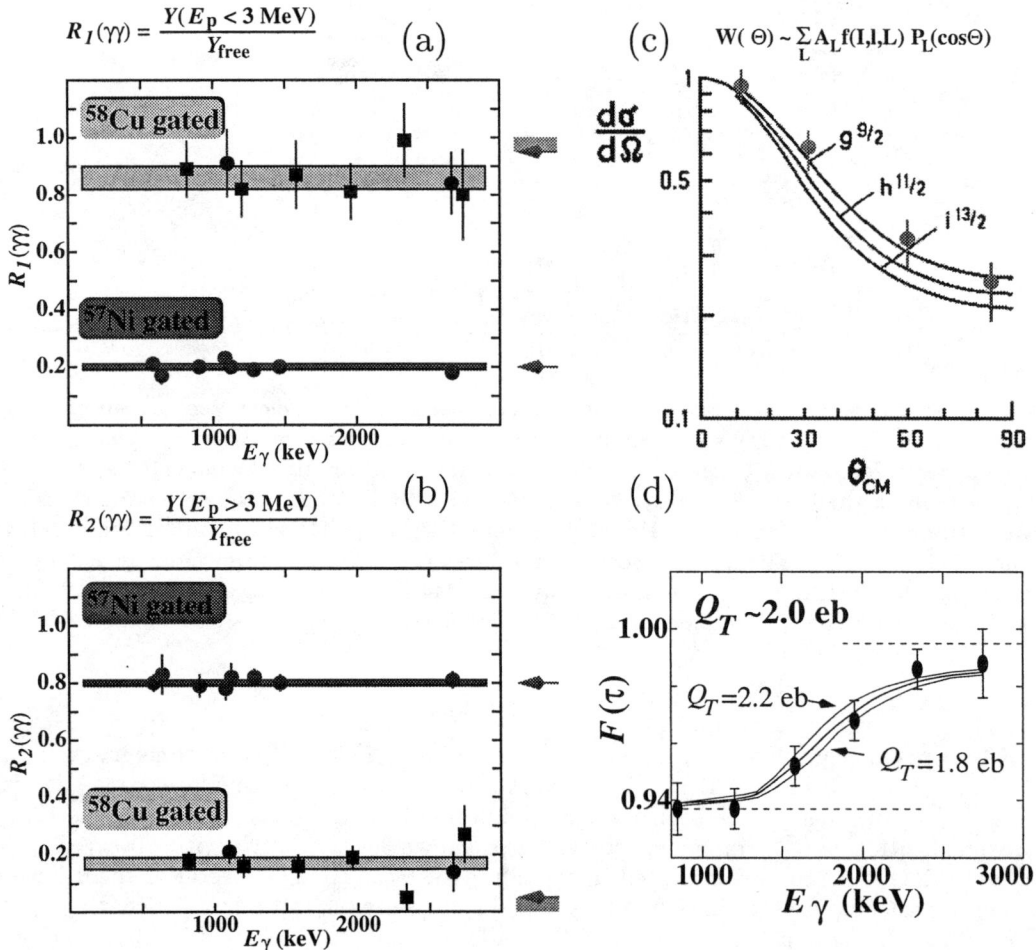

FIGURE 5. Part (a) and (b) provide ratios of yields of transitions associated with the prompt proton decay ("^{58}Cu gated") and between normal spherical states in ^{57}Ni. An overall particle gating of one α particle, two protons, and one neutron was applied. Part (c) illustrates the angular distribution of the protons detected in rings one through four in MICROBALL. They were normalized to calculations taken from Ref. [10]. The fractional Doppler shifts $F(\tau)$ of transitions in the deformed band of ^{58}Cu are shown in part (d). They are consistent with an average quadrupole deformation of $Q_T = 2.0(2)$ eb. The dashed horizontal lines represent the full shift close to $F = 1.0$ and the shift for decays outside the thin target foil, respectively.

be reduced significantly because the vast majority of protons from the fusion reaction have energies in excess of 3 MeV (cf. Fig. 4). This can be clearly seen in Fig. 5(a). Figure 5(b) provides the opposite energy restriction in the nominator: Transitions from ^{57}Ni are barely reduced while those associated with the proton decay nearly disappeared from the spectra.

For the first four rings of MICROBALL the proton energies of the 2.3 MeV peak in the laboratory frame were sufficiently high to safely cross the absorbers in front of the individual detector elements and their energy thresholds. This allowed the analysis of the proton angular distribution up to 90° in the center-of-mass frame. By using the proton spectra in coincidence with the 830, 1197, and 1576 keV γ-ray transitions in appropriately gated E_γ vs. $E_{p,c.m.}$ matrices, summing over the all detectors in one ring of MICROBALL, and taking into account the solid angle in the center-of-mass frame the four data points in Fig. 5(c) were deduced. They were then normalized to calculations taken from Ref. [10] which were performed for an angular momentum $I = 50\ \hbar$ and full alignment. As the spin values are smaller and since the alignment is less pronounced for the ^{58}Cu case, i.e., corresponding to somewhat more flattened theoretical curves, we associate an angular momentum of $l = 3$-5 with the observed proton decay.

To confirm the strong deformation of the rotational band an analysis of so-called residual or fractional Doppler shifts $F(\tau)$ [38] was performed. The data points are shown in Fig. 5(d) and compared to three curves representing simulations of the slowing down process of the recoils in the thin target foil assuming different average but constant quadrupole moments Q_T of the band. The best fit was achieved for $Q_T = 2.0$ eb. The dashed horizontal lines represent the full shift close to $F = 1.0$ and the shift for decays outside the thin target foil, respectively. The quadrupole moment corresponds to an axial deformation of $\beta_2 = 0.37$ or an average collectivity of some 100 W.u. It is in good agreement with the HF cranking calculations which predict a decreasing (proton) quadrupole moment of $Q_0 = 2.5$ eb at a rotational frequency $\omega = 0$ to $Q_0 = 1.5$ eb at $\hbar\omega - 2$ MeV. This variation can be attributed to a smooth band termination [39] based on gradually aligning nucleons which form the intrinsic configuration. The resulting decrease of the dynamical moment of inertia is also in accordance with the observations.

PROMPT PROTON DECAY IN ^{56}NI

As mentioned earlier, the $4^1 4^1$ assignment to the band in ^{58}Cu is straightforward not only based on the HF calculations presented in Ref. [12] but also on results from other types of cranking calculations. Supported by the proton angular distribution measurement, the observed proton branch could be associated with the emission of the $1g_{9/2}$ proton from the bottom of the $4^1 4^1$ band. In addition, the $1g_{9/2}$ proton is one of the two shape driving particles, i.e., with a high probability it is moving close to the surface of the deformed ^{58}Cu nucleus. After the emission of this proton, the remaining object would be left with a neutron in the $1g_{9/2}$ orbit and, apparently, rearranges its shape to sphericity. So far, this proton decay scenario and the other experimental knowledge is consistent with the fact that the 3701 keV 9/2 daughter state represents the neutron $1g_{9/2}$ single-particle state in ^{57}Ni [17].

Coincidence, intensity balance, and summed energy relations were used to significantly extend the known level scheme of ^{56}Ni. It is shown in Fig. 6. The previously reported spherical yrast sequence [30] was confirmed. In addition, yrare (8$^+$) and (10$^+$) states were identified. The yrast sequence was extended beyond $I = 12\ \hbar$ which is the maximum possible spin for simple 2p-2h excitations in the pf shell-model space. Spin and parity assignments are based on directional correlations of oriented states (DCO ratios), angular distributions from the purified singles projections at different Ge-detector angles, and on the regular increase of γ-ray energies in the rotational bands [17]. The highlights of the results [13] include the observation of two rotational bands, labelled 1 (ND) and 2 (SD) in Fig. 6. In addition, band 2 was found to have a 50% prompt proton decay branch into the ground state of ^{55}Co. This manifests the second example of a prompt proton decay from states in the second minimum of the potential.

The rotational band 1 decays back via several high-energy γ rays into the spherical minimum. A 349 keV $E2$ transition connecting the second 2$^+$ with the second 0$^+$ state (which presumably forms the deformed band head) has not been observed. Band 2 is firmly linked into band 1 via the 1627 keV line. Moreover, the angular distribution of the 1627 keV interband transition is consistent with that of a stretched dipole, i.e., band 2 consists of odd-spin states. Gamma-ray spectra gated with two α particles and the 1572, 1945, and 2318 keV transitions in band 2 reveal the presence of transitions at 1201 and probably 846 keV. They are likely candidates for lower-spin band members due to the regular decrease in transitional energy. Most interestingly, however, in the spectrum gated with the 1201 keV transition, γ-decay-out intensity is missing because summed yields

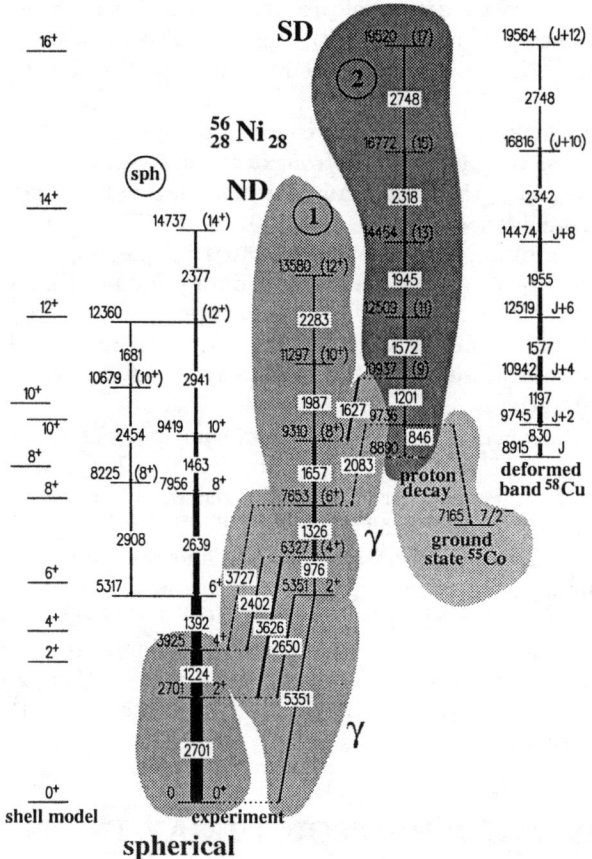

FIGURE 6. Proposed partial level scheme ^{56}Ni. The energy labels are given in keV. The widths of the arrows are proportional to the relative intensities of the γ rays. Tentative transitions and levels are dashed. On the left hand side, the energy levels arising from 6p-6h large-scale shell model calculations [40] in the pf model space are included. The superdeformed band 2 (dark shaded) was found to decay via prompt proton emission into the ground state of ^{55}Co. In addition, it reveals transition energies nearly identical to those in the rotational band identified for ^{58}Cu. This band is included on the right hand side of the for comparison.

of the (tentative) 846 and 2083 keV transitions or the 2701 and 5351 keV ground-state transitions account for only 52(9)% of the yield of the 1572 keV transition. Possible explanations are that the level at 9736 keV is an isomer with a half-life of $\tau > 10$ ns. This is not very likely as the γ-ray of 846 keV indicates the continuation of the rotational band 2 with an estimated partial half-life of about 1 ps. Secondly, the γ decay might proceed through several unobserved weak branches. This option can be ruled out because decays from other low-spin states [29] have not been seen, and the main decay flux from them should be collected by the 2701 keV ground-state transition in any case. The third possibility would be a prompt proton emission into the ground state of ^{55}Co.

As inferred from the well known masses [41], the binding-energy difference between ^{56}Ni and ^{55}Co amounts to 7.165(11) MeV. This implies $Q_p = 2.571(12)$ MeV for the proton branch. However, the verification is more complicated than in the previous case. Firstly, the ^{56}Ni+2α channel was populated more than ten times weaker than ^{58}Cu (cf. Fig. 2). Secondly, the lack of especially the neutron particle gate leads to a worse signal to noise ratio in the γ-ray and proton spectra due to a considerably higher background arising from the much more intense ^{55}Co+2α1p and ^{54}Fe+2α2p reactions. Finally and most importantly, no potential coincidence with γ rays in the daughter nucleus is observable because only the ground state of ^{55}Co can be expected to be populated by the proton decay according to Q-value considerations.

The direct evidence, i.e., the proton center-of-mass energy spectra based on a 2α1p-gated E_γ vs. $E_{p,c.m.}$

matrix is presented in Fig. 7(c). It is the sum of spectra gated by the 1201, 1572, and 1945 keV transitions in band 2 of ^{56}Ni — a peak-like structure at a proton energy of 2.5(1) MeV appears. The full "curve" in that spectrum represents the (normalized) total projection of the E_γ vs. $E_{p,c.m.}$ matrix. The shape of the total projection reflects the expected distribution of proton energies following fusion-evaporation reactions. The other two panels in Fig. 7 show the proton spectra in coincidence with a low-lying [1657 keV, part (a)] and a high-lying [1333 keV, part(b)] transition in the excitation scheme of ^{55}Co [17]. These spectra reveal the typical evaporation-type distributuion which peaks at $E_p \sim$ 4-6 MeV. It should be noted that for the sake of cleanliness and safety only rings one through four of MICROBALL (cf. Fig. 1) were used for demanding the proton coincidence.

Further evidence for the proton decay arises from the investigation of a $\gamma\gamma$ matrix gated by two α-particles and one proton, with and without an additional energy restriction for the proton of $E_{p,c.m.} < 3$ MeV [cf. Fig. 7(d)]. This energy limit reduced the yields of transitions in γ-gated spectra belonging to the $2\alpha 1p$ channel ^{55}Co to 4(1)% as compared to the yields deduced from the unrestricted $2\alpha 1p$-gated spectra. However, the yields of transitions of the ^{56}Ni band 2 remained more or less unchanged: 83(15)% are still observed. This number is very similar to the average number obtained for the ^{58}Cu case when performing a similar exercise [cf. Figs. 5(a)

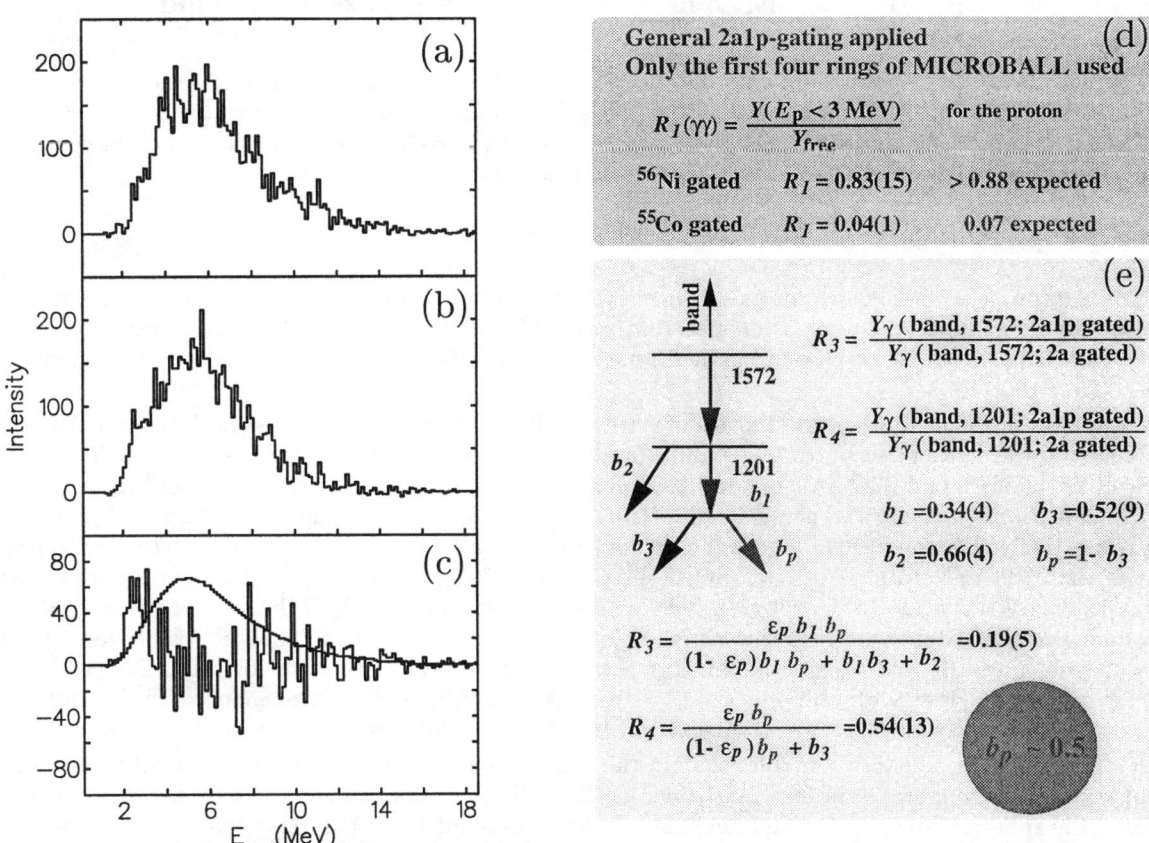

FIGURE 7. Proton center-of-mass energy spectra based on a $2\alpha 1p$-gated E_γ vs. $E_{p,c.m.}$ matrix are shown on the left hand side. Parts (a) and (b) are in coincidence with the $31/2 \rightarrow 29/2$ 1333 and $17/2^- \rightarrow 15/2^-$ 1657 keV transition in the $2\alpha 1p$ reaction channel ^{55}Co [17], respectively. Part (c) is the sum of spectra gated by the 1201, 1572, and 1945 keV transitions in band 2 of the 2α channel ^{56}Ni — a peak-like structure at low proton energies appears. The full "curve" represents the (normalized) total projection of the E_γ vs. $E_{p,c.m.}$ matrix. On the right hand side part (d) recalls the mean value for the ratio $R_1(\gamma\gamma)$ [cf. Fig. 5(a) and (b)]. Part (e) summarizes the branching ratios and ratios of yields of transitions from spectra gated by the 1201 and 1572 keV γ rays in band 2 of ^{56}Ni in specific particle-gated $\gamma\gamma$ matrices. Using the proton detection efficiency of $\epsilon_p = 0.78(1)$ a proton branch of 49(14)% out of the 9736 keV level can be determined. See text for details.

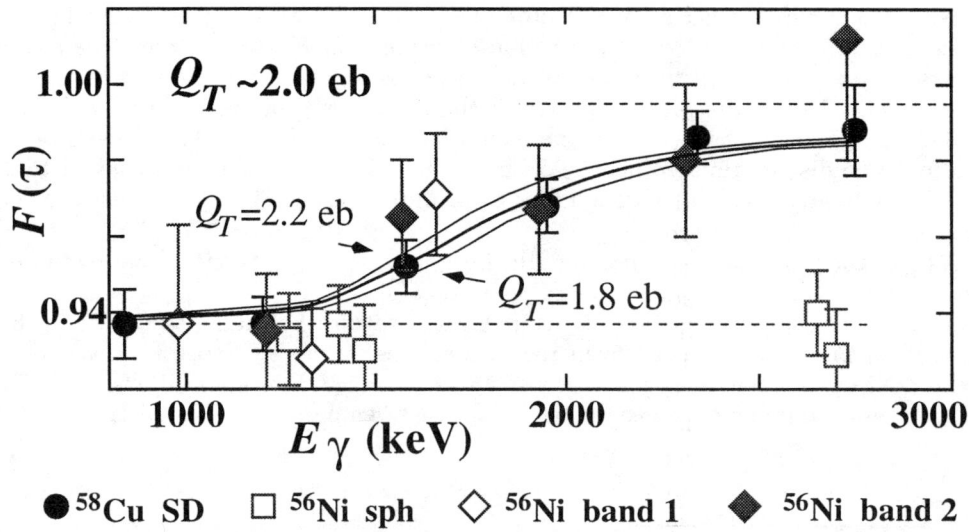

FIGURE 8. The fractional Doppler shifts $F(\tau)$ of transitions in ^{56}Ni are compared to those of the deformed band in ^{58}Cu [filled circles, cf. Fig. 5(d)]. Open squares reflect values for transitions from spherical states in ^{56}Ni while open and filled diamonds correspond to those from band 1 and band 2, respectively. The dashed horizontal lines represent the full shift close to $F = 1.0$ and the shift for decays outside the thin target foil.

and (b)]. By summing the γ-ray spectra gated by the 1201, 1572, and 1945 keV transitions in the restricted $2\alpha1p$-gated matrix one obtains a very clean spectrum providing the peaks from rotational band 2 of ^{56}Ni but none of the transitions from ^{55}Co. Due to the mentioned intensity relations, the band cannot have an identical partner in ^{55}Co either.

Finally, Fig. 7(e) presents another set of ratios of γ-ray yields which are used to measure the size of the proton branch. It shows a sketch of the bottom of band 2 and its proposed decay. "band" reflects the higher-lying transitions of 1945, 2318, and 2748 keV in band 2. The ratios R_3 and R_4 are the ratios of yields averaged over the "band" transitions in spectra in coincidence with the 1572 keV and 1201 keV line in 2α and $2\alpha1p$ gated matrices, respectively. For the 1572 keV gate the nominator of R_3 reflects the flux of band 2 *observed* to proton decay out of the 9736 keV state. It is given by the proton detection efficiency ϵ_p multiplied by the branching ratio of the 1201 keV line b_1 and the proton branch b_p. The denominator of R_3 is the sum of the *non-observed* proton branch and the γ decay-outs via the 1627 keV line (b_2) and the 846 and 2083 keV transitions ($b_1 \cdot b_3$). Since $b_p = 1 - b_3$ both R_3 and R_4 provide an independent measure of the proton branch. Plugging in the γ branching ratios [cf. Fig. 7(e)] one obtains $b_p = 0.49(14)$ which is in nice agreement with the previously mentioned missing γ-decay strength $1 - b_3 = 0.48(9)$ in the decay out process of the 9736 keV level.

Though statistics are considerably smaller than in the ^{58}Cu case it was possible to perform only a qualitative analysis of fractional Doppler shifts for transitions in ^{56}Ni. The results are shown in Fig. 8 for transitions depopulating the spherical states (open squares) and the levels of band 1 and band 2 (open and filled diamonds, respectively). They are compared to the values deduced for the band in ^{58}Cu [cf. Fig. 5(d)]. Clearly, band 2 is of collective nature with about the same quadrupole moment as the rotational band in ^{58}Cu. Due to lack of statistics it was not possible to follow band 1 to higher spins, but already the data point for the 1657 keV transition is distinctively different from the average value expected for the slower transitions in the first minimum.

A theoretical description of excited states in ^{56}Ni requires state-of-the-art shell-model and mean-field methods. The spherical states in the vicinity of a doubly magic nucleus are a clear case for the spherical shell model. The orbits involved near ^{56}Ni are the $1f_{7/2}$ orbit below and the $2p_{3/2}$, $1f_{5/2}$, and $2p_{1/2}$ (upper pf shell) orbits above the $N,Z = 28$ shell gap. For this so-called full pf model space two common parameter sets exist, namely the FPD6 interaction by Richter, van der Merwe, Julies, and Brown [42], and the KB3 interaction introduced by the Madrid-Strasbourg group [43]. Using more sophisticated diagonalisation algorithms [44], the deformed

$N \approx Z$ nuclei in the mid $1f_{7/2}$ shell (up to mass $A = 50$) could recently be well explained by large-scale full pf shell-model calculations [22,25,45–47]. However, the original KB3 interaction yields a too large energy gap at the $N,Z = 28$ shell closure. This has been cured by making the diagonal matrix elements which connect the $1f_{7/2}$ orbit with the others 100 keV more attractive [48] for the calculations around ^{56}Ni.

The current computational limit for calculations in the full pf configuration space are mass $A = 52$ nuclei [49]. There the basis dimensions exceed 100 million states. Therefore, the configuration space for the shell model in the direct vicinity of ^{56}Ni has to be truncated. For the study of ^{56}Ni the excitation of up to six particles from the $1f_{7/2}$ into the upper pf shell was allowed, and the results proved very successful in explaining not only the spherical states but also the rotational band 1 (see below).

Nevertheless, particles excited into the high-j $1g_{9/2}$ intruder orbit are necessary to allow for the strongly or superdeformed bands in the mass region [11,12]. Moreover, the spherical $1g_{9/2}$ states have been identified in ^{59}Cu and ^{59}Ni [50] and likewise in ^{57}Ni [17]. Therefore, attempts are under way to incorporate this shell in the present pf model space. Of course, this would imply an additional dramatic increase in the dimensions for the conventional shell-model studies, such that a considerable truncation in the number of particles either crossing the $N,Z = 28$ gap and/or being put into the $1g_{9/2}$ orbit will be unavoidable.

More recent developments try to make a detour round the problems associated with huge dimensions in the model space. The Shell-Model Monte-Carlo (SMMC) [51] and Quantum Monte-Carlo Diagonalisation (QMCD) [52,53] methods are tracing only the most important components of the wave functions (cf. contribution by T. Otsuka). These calculations were very successful in describing, e.g., ground state properties of the pf shell nuclei [54] or the spherical yrast sequence in ^{56}Ni [55]. The average number of particles occupying the $1f_{7/2}$ shell was found to be ~ 14 for the spherical shell-model states and ~ 10 for the first rotational band in ^{56}Ni [56]. The latter number also indicates that the conventional shell-model calculations with up to 6p-6h excitations shall be sufficient to describe the spherical states and band 1 in ^{56}Ni. Finally, the SMMC and QMCD models may provide the favorable tools for the (full) implementation of the $1g_{9/2}$ shell.

On the left hand side of Fig. 9 the experimental results for the spherical yrast states of ^{56}Ni are compared to the results of the QMCD model [55,56] and those based on the (modified) 6p-6h KB3 interaction [40]. The thickness of the vertical lines reflect the $B(E2)$ values. The corresponding numbers are also presented. At the top right of each state the deduced half-lifes are given. Experimentally, only the $B(E2; 2^+ \rightarrow 0^+) = 120(24)$ e^2fm^4 has been measured [57]. The fact that the observed γ-ray transitions in the yrast cascade are (mainly) stopped in the spectra of the backed target experiment indicates that at least the life time of the 9419 keV 10^+ state must be larger than several picoseconds. In addition to the QMCD and SM KBF calculations, results from rather simple shell-model calculations performed with the code RITSSCHIL [58] are included in Fig. 9, too. Only up to two particles are allowed to be excited across the $N,Z = 28$ shell gap. This keeps the dimension of basis states below 210 for ^{56}Ni (and below 30000 for other nuclei near ^{56}Ni [17]). The FPD6 parameter set [42] was used for the two-body residual interaction, and the single-particle energies are based on Ref. [59]. However, the simplicity of the 2p-2h excitations demanded a modification to them. Firstly, the single-particle energies of the upper pf shell were slightly changed such that the $3/2^-$ ($2p_{3/2}$ orbit) ground states and first excited $5/2^-$ ($1f_{5/2}$ orbit) and $1/2^-$ ($2p_{1/2}$ orbit) states in the mass $A = 57$ mirror pair ^{57}Cu [60] and ^{57}Ni were reproduced within some 10 keV. Secondly, the gap between the upper pf shell and the $1f_{7/2}$ orbit was decreased by 2.4 MeV. This accounts for the softness of the core, and reproduces the excitation energies of the 10^+ and 12^+ yrast states in ^{56}Ni. Apart from the too high 2^+ energy the results are comparable to those obtained with the more sophisticated models. It should be noted that the predictions for the electromagnetic decay propertiesat intermediate spins, namely the branching ratios, crucially depend on the exact locations of the single-particle levels while the calculated level energies remain essentially unchanged.

To describe the deformed bands, the cranked Hartree-Fock (HF) [34] and Hartree-Fock-Bogolyubov (HFB) [61] methods were employed with the Skyrme interaction SLy4 [35]. In the HFB calculations, a zero-range density-dependent pairing force has been used, and the particle-number projection have was done using the Lipkin-Nogami prescription. As illustrated on the right hand side of Fig. 9, the deformed band 1 can be described both by the shell-model and mean-field calculations. In the shell-model it has a dominant 4p-4h character relative to the *spherical* closed core of ^{56}Ni. The predicted intrinsic quadrupole moment decreases from 1.3 eb at low spins to $\sim$0.8 eb at $I = 12$ $\hbar$. Though the calculated energies agree well with data (cf. Fig. 9) a closer look shows that rotational law $E_x \sim I(I + 1)$ is less closely followed by the calculations than observed due to the necessary truncation of the shell-model space.

In the mean-field calculations, band 1 can be ascribed to the 4^04^0 configuration with respect to a closed *deformed* core of ^{56}Ni. The band-head of the 4^04^0 configuration agrees with the excitation energy of band 1.

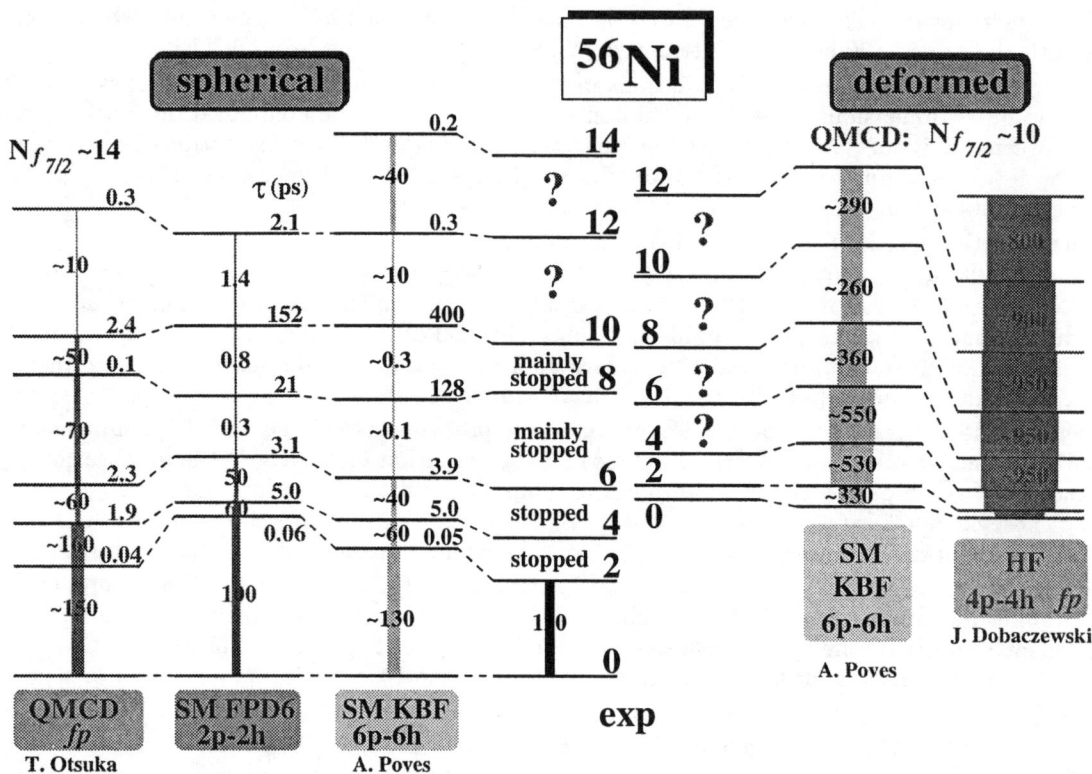

FIGURE 9. Comparison of the observed spherical states and band 1 in ^{56}Ni with different calculations.

However, as visible in Fig. 9, its alignment is predicted too large because pairing correlations were not considered. In fact, the HFB calculations show that pairing is large for band 1 [13]. By promoting one particle across the deformed gap from the [321]1/2 to the [440]1/2 Routhian, one obtains the $4^1 4^0$ and $4^0 4^1$ bands which have negative parity and odd spins. Both bands behave extremely similar with the $4^0 4^1$ structure is lower in energy by some one- to twohundred keV. Their energies agree well with band 2 and the dynamic moment of inertia is consistent up to $\hbar\omega \approx 1.1$ MeV. However, a predicted band crossing between the [321]1/2 and [312]5/2 Routhians around $\hbar\omega \approx 1.5$ MeV has not been observed yet.

To illustrate the similarity of band 2 in ^{56}Ni with the deformed band in ^{58}Cu, the latter is also displayed in Fig. 6. Both the γ-ray energies and the excitation energies relative to the respective ground states are nearly identical. Assuming a band-head spin $J = 9$ for the band in ^{58}Cu the $4^0 4^1$ configuration of ^{56}Ni has a relative alignment close to -4. This is consistent with the angular momentum proposed for band 2. The calculated $4^1 4^1$ band in ^{56}Ni carries $i_{\text{rel}} \approx -0.5$, which is difficult to accommodate experimentally. Therefore, we associate a probably mixed $4^1 4^0$ and $4^0 4^1$ structure with band 2. Unfortunately, the predicted dynamic moments of inertia of bands $4^0 4^1$ in ^{56}Ni and $4^1 4^1$ in ^{58}Cu differ both at low spins and at high spins which makes it difficult to understand their near identicality.

SUMMARY AND OUTLOOK

To conclude, extensive new experimental data in the mass $A \sim 60$ region around the doubly-magic nucleus ^{56}Ni was obtained. The most compelling results include the first observation of *prompt* particle decays from states in the secondary minimum of the potential. Different from β-delayed proton emission or proton emitters along the drip line they compete with fast electromagnetic radiation. By now, two cases of proton decay have

TABLE 1. Summary of particle decays from deformed states in the mass $A = 60$ region.

Isotope	Band			Decay (%)				Reference
	$E_{x,max}$	I_{max}	total γ	obs.disc.γ	Particle	E_{part}		
^{56}Ni	20 MeV	$\sim 17\ \hbar$	~ 50	~ 50	~ 50	2.5(1) MeV		D. Rudolph $et\ al.$, PRL, submitted.
^{58}Ni	30 MeV	$\sim 24\ \hbar$	~ 96	~ 35	~ 4	6.7(3) MeV		D. Rudolph $et\ al.$, to be published.
^{58}Cu	23 MeV	$\sim 23\ \hbar$	~ 33	~ 33	~ 67	2.3(1) MeV		D. Rudolph $et\ al.$, PRL80, 3018 (1998).
^{59}Cu	32 MeV	$\sim 28\ \hbar$	> 95	~ 50	< 5	< 3 MeV		C. Andreoiu $et\ al.$, to be published.

been identified in ^{58}Cu and ^{56}Ni. A third decay in the neighboring isotope ^{59}Cu seems likely but awaits final experimental proof. In addition, an α decay branch from the superdeformed band in ^{58}Ni was observed. The present information concerning the particle decays is summarized in Table 1. Future experiments dedicated to perform particle spectroscopy are scheduled both at EUROBALL and GAMMASPHERE. While results from the EUROBALL experiment shall provide the level lifetimes, spins and parities, the GAMMASPHERE run is supposed to deliver more precisely the energies and branching ratios for presumably weak particle decay branches. Therefore, the first three rings of MICROBALL shall be replaced by highly segmented Silicon strip detectors.

Next to the particle decays the well- and superdeformed rotational bands in the region provide a challenging area of research themselves for both theory (cf. contributions by J. Dobaczewski and T. Otsuka) and experiment (cf. contribution by C.E. Svensson). Finally, the comprehensive data in the first minimum of nuclei near ^{56}Ni thoroughly test the predictions from (large-scale) spherical shell-model calculations. Recent progress for the conventional type of calculations and those based on Monte-Carlo methods allow also the description of the moderately deformed rotational bands which in turn makes the nuclei close to ^{56}Ni unique to compare and relate those models with the ones based on mean-field theory.

ACKNOWLEDGEMENT

First of all, I would like to thank C. Baktash for many invaluable discussions and the opportunity to present our latest results on prompt particle decays at this enjoyable conference. I am also grateful for the grant received from the Joint Institute of Heavy Ion Research to visit Tennessee for a few weeks in August 1998. Furthermore I am indebted to E. Caurier, D.J. Dean, J. Dobaczewski, P.-H. Heenen, W. Nazarewicz, F. Nowacki, A. Poves, and W. Satula for the effort they put into the theoretical description and understanding of the spherical and deformed states in the mass region in general and ^{56}Ni in particular and letting me use their results. Without the perfect and persistent work of the collaborators from Washington University, namely M. Devlin, D.R. LaFosse, and last but not least D.G. Sarantites, this work would not have been possible at all. Finally, I want to thank M.J. Brinkman, R.M. Clark, P. Fallon, H.-Q. Jin, R. Krücken, I.-Y. Lee, R. MacLeod, A.O. Macchiavelli, L.L. Riedinger, and C.-H. Yu and the operating crew of the 88-Inch Cyclotron for their assistance during this experiment, and D. Balamuth, S. Freeman, M. Leddy, and C.J. Lister for both providing and setting up the neutron detector array at GAMMASPHERE. Oak Ridge National Laboratory is managed by Lockheed Martin Energy Research Corp. for the U.S. Department of Energy under contract DE-AC05-96OR22464. This research was supported in part by the Swedish Natural Science Research Councils, the U.S. Department of Energy under grant No. DE-FG02-96ER40963 (UT) and No. DE-FG05-88ER40406 (WU).

REFERENCES

1. K.P. Jackson $et\ al.$, Phys. Lett. **33B**, 281 (1970).
2. J. Cerny and J.C. Hardy, Annu. Rev. Nucl. Sci **27**, 333 (1977).
3. H. Akimune $et\ al.$, Ann. Rep. RCNP (Osaka) 1995, p.20.
4. V.I. Goldansky, Nucl. Phys. **19**, 482 (1960).
5. S. Hofmann, Radiochim. Acta **70/71**, 93 (1995).
6. P.J. Woods and C.N. Davids, Annu. Rev. Nucl. Part. Sci. **47**, 541 (1997).
7. S. Åberg, P.B. Semmes, and W. Nazarewicz, Phys. Rev. **C56**, 1762 (1997).
8. C.N. Davids $et\ al.$, Phys. Rev. Lett. **80**, 1849 (1998).
9. E. Maglione, L.S. Ferreira, and R.J. Liotta, Phys. Rev. Lett. **81**, 538 (1998).
10. T. Døssing, S. Frauendorf, and H. Schulz, Nucl. Phys. **A287**, 137 (1977).

11. C.E. Svensson *et al.*, Phys. Rev. Lett. **79**, 1223 (1997).

12. D. Rudolph *et al.*, Phys. Rev. Lett. **80**, 3018 (1998).

13. D. Rudolph *et al.*, submitted to Phys. Rev. Lett.

14. C. Andreoiu *et al.*, to be published.

15. J. Styczen *et al.*, Nucl. Phys. **A327**, (1979) 295.

16. J. Huo, H. Sun, W. Zhao, and Q. Zhou, Nucl. Data Sheets **68**, 887 (1993).

17. D. Rudolph *et al.*, to be published.

18. A. Marinov, S. Gelberg, and D. Kolb, Mod. Phys. Lett. **A11**, 861 (1996).

19. A. Marinov, S. Gelberg, and D. Kolb, Mod. Phys. Lett. **A11**, 949 (1996).

20. H. Nakada, T. Sebe, and T. Otsuka, Nucl. Phys. **A571**, 467 (1994).

21. K. Langanke, D.J. Dean, P.B. Radha, Y. Alhassid and S.E. Koonin, Phys. Rev. **C52** 718 (1995).

22. S. Lenzi *et al.*, Phys. Rev. **C56**, 1313 (1997).

23. J. Zhang, A.C. Merchant, and W.D.M. Rae, Phys. Rev. C **49**, 562 (1994).

24. M. Freer, R.R. Betts, and A.H. Wuosmaa, Nucl. Phys. **A587**, 36 (1995).

25. E. Caurier *et al.*, Phys. Rev. Lett. **75**, 2466 (1995).

26. J. Terasaki *et al.*, Nucl. Phys. **A600**, 371 (1996).

27. C.E. Svensson *et al.*, Phys. Rev. Lett. **80**, 2558 (1998).

28. R.R. Betts, B.B. Back, and B.G. Glagola, Phys. Rev. Lett. **47**, 23 (1981).

29. J. Huo, Nucl. Data Sheets **67**, 523 (1992).

30. J. Blomqvist *et al.*, Z. Phys. A **322**, 169 (1985).

31. I.-Y. Lee, Nucl.Phys. **A520**, 641c (1990).

32. D.G. Sarantites *et al.*, Nucl. Instrum. Meth. **A381**, 418 (1996).

33. D. Rudolph *et al.*, Z. Phys. **A358**, 379 (1997).

34. J. Dobaczewski and J. Dudek, Comp. Phys. Commun. **102**, 166 (1997); **102**, 183 (1997); to be published.

35. E. Chabanat *et al.*, Nucl. Phys. **A627**, 710 (1997).

36. C.E. Svensson *et al.*, to be published.

37. K. Spohr *et al.*, Acta Phys. Pol. **B26**, 297 (1995), and K. Spohr, PhD thesis, FZ Jülich, Jül-3171, ISSN 0944-2952, 1996.

38. B. Cederwall *et al.*, Nucl. Intrum. Meth. A **354**, 591 (1995).

39. I. Ragnarsson, Acta Phys. Pol. B **27**, 33 (1996).

40. A. Poves, E. Caurier, and F. Nowacki, priv. comm.

41. G. Audi and A.H. Wapstra, Nucl. Phys. **A565**, 1 (1993); Nucl. Phys. **A595**, 409 (1995).

42. W. A. Richter, M. G. van der Merwe, R. E. Julies, and B. A. Brown, Nucl. Phys. **A523** 325 (1991).

43. A. Poves and A. Zuker, Phys. Rep. **70**, 235 (1981).

44. E. Caurier, shell-model code, Strasbourg (1990).

45. E. Caurier, A.P. Zuker, A. Poves, and G. Martinez-Pinedo, Phys. Rev. **C50**, 225 (1994).

46. G. Martinez-Pinedo *et al.*, Phys. Rev. **C54**, R2150 (1996).

47. S. Lenzi *et al.*, Z. Phys. **A354**, 117 (1996).

48. G. Martinez-Pinedo *et al.*, Phys. Rev. **C55**, 187 (1997).

49. C.A. Ur *et al.*, Phys. Rev. **C**, in press.

50. S. Juutinen, J. Hattula, M. Jääskeläinen, A. Virtanen, and T. Lönnroth, Nucl. Phys. **A504**, 205 (1989).

51. S.E. Koonin, D.J. Dean, and K. Langanke, Phys. Rep. **278**, 1 (1996), and references therein.

52. M. Honma, T. Mizusaki, and T. Otsuka, Phys. Rev. Lett. **75**, 1284 (1995).

53. M. Honma, T. Mizusaki, and T. Otsuka, Phys. Rev. Lett. **77**, 3315 (1996).

54. K. Langanke *et al.*, Nucl. Phys. **A613**, 253 (1997).

55. T. Otsuka, M. Honma, and T. Mizusaki, Phys. Rev. Lett. **81**, 1588 (1998).

56. T. Otsuka, priv. comm.

57. G. Kraus *et al.*, Phys. Rev. Lett. **73**, 1773 (1994).

58. D. Zwarts, Phys. Comm. **38**, 365 (1985).

59. L. Trache *et al.*, Phys. Rev. **C54**, 2361 (1996).

60. M.R. Bhat, Nucl. Data Sheets **67**, 195 (1992).

61. P. Bonche, H. Flocard, and P.-H. Heenen, Nucl. Phys. **A598**, 169 (1996).

Proton Radioactivity

Philip J. Woods

Department of Physics and Astronomy, Edinburgh University, EH9 3JZ UK

Abstract. This paper reviews the recent developments in the study of proton-radioactivity. A large and wide ranging data base now exists to study this phenomenon. Theoretical models of proton emission from spherical nuclei can predict the detailed trends of proton decay spectroscopic factors very well. The first examples of proton decay from highly deformed nuclei have been discovered and are well reproduced by a theoretical approach based on Nilsson states. Such measurements provide an insight into the fragmentation of single particle strength in deformed nuclei. In the case of ^{131}Eu proton decay fine structure has been identified for the first time, thereby providing direct information on the degree of deformation. The technique of Recoil Decay Tagging and its particular application to the study of the structure of deformed proton radioactive nuclei is discussed.

INTRODUCTION

The proton drip-line represents the dividing line between isotopes that are either bound or unbound to the emission of a proton from their ground-states. For light elements, nuclei lying beyond the proton drip-line only exist in the form of resonances, ^{39}Sc being the heaviest system to be studied to date [1]. The Coulomb barrier experienced by an unbound proton increases progressively with element number, Z, until it becomes likely that systems that are proton unbound can be directly detected. For example the isotope ^{77}Y has recently been identified [2] using the LISE3 separator at GANIL although mass formulae suggest the odd proton is energetically unbound. Progressing beyond Z=50 it becomes more probable than not that odd Z nuclei have at least one proton-radioactive isotope. This is due to a combination of the large height of the Coulomb barrier and the decrease in the rate of change of proton decay Q-value with neutron number $\triangle Q_p/\triangle N$ which varies with an approximate inverse dependence on the mass number, A [3]. For proton radioactivity to be observed experimentally it must have a significant decay branch which means that in practise the drip-line must be crossed by several isotopes. For example, ^{171}Ir is the heaviest proton unbound Ir isotope but proton emission is first observed for the isotope ^{167}Ir [4]. A continuous chain of odd-Z proton emitting elements has been identified from Z=67-83 [5], these results along with other examples now constitute a large and wide ranging data base of proton transitions from which to explore the phenomenon of proton radioactivity.

SPHERICAL PROTON EMITTERS

In its simplest form the proton decay transition probability can be calculated using a semi-classical WKB approach [6]. A recent theoretical review of proton emission from spherical nuclei by Aberg et al. [7] has shown that such calculations agree surprising well with more exact DWBA and two-potential treatments when the same proton potential parameter set is used. The choice of a realistic potential introduces an uncertainty ~ 2 into the calculation [8]. A typical potential set is the global scattering potential of Becchetti and Greenlees [9] used in Figure 1 for the cases of ground and isomeric proton emission from ^{167}Ir which occur from $s_{1/2}$ and $h_{11/2}$ orbitals, respectively. As can be seen from the table inset into Figure 1 the proton decay rates are extremely sensitive to the orbital angular momentum of the unbound proton, l, and shell model assignments can be confidently made despite the uncertainty in the choice of potential.

A very attractive aspect of proton decay is that the proton can be considered to be preformed inside the nucleus thereby avoiding some of the uncertainty associated with alpha-decay transition rate calculations. Nonetheless, in order to calculate proton decay rates correctly a spectroscopic factor must be introduced which is defined theoretically as [10]

$$S = \frac{1}{2J_i+1}| < J_i||a^\dagger_{nlj}||J_f > |^2$$

CP481, *Nuclear Structure 98*, edited by C. Baktash

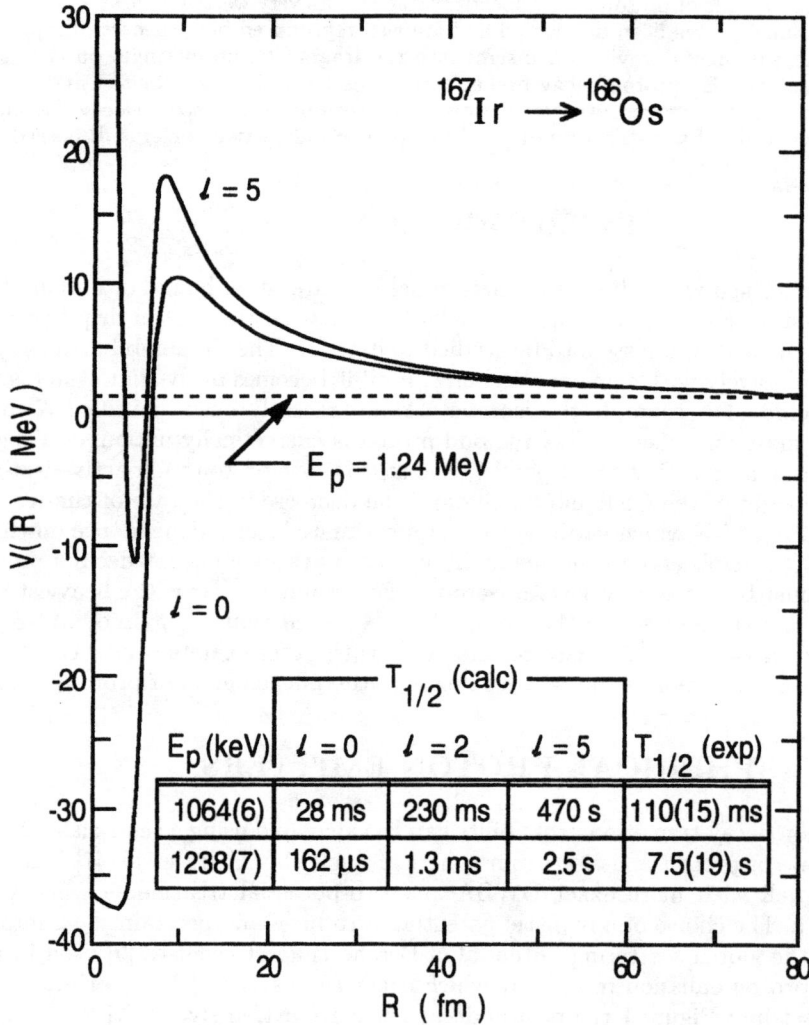

Figure 1. Proton-nucleus potentials calculated for the proton emitter ^{167}Ir. The inset shows proton decay partial half-lives calculated using the WKB approximation for three values of the orbital angular momentum l, compared to the experimental values for the ground and isomeric state configurations.

where (nlj) represents a single particle state , J_i and J_f represent the total angular momentum of the nucleus in the parent and daughter systems and j represents the angular momentum of the emitted proton.

Davids et al. [4] used a low seniority shell model calculation to predict proton radioactivity spectroscopic factors for spherical nuclei with 64<Z<82. The calculation assumed degenerate $s_{\frac{1}{2}}$, $d_{\frac{3}{2}}$ and $h_{\frac{11}{2}}$ protons and spectator neutrons and predicted

$$S = \frac{P}{9}$$

where P represents the number of pairs of proton holes in the daughter nucleus wrt the Z=82 closed shell. In Figure 2 this calculation is compared with experimental values for spectroscopic factors defined as the ratio of the theoretical and experimental proton partial half-lives

$$S^{exp} = \frac{t_{\frac{1}{2}}^{wkb}}{t_{\frac{1}{2}}^{exp}}$$

In general the spectroscopic factors are reproduced remarkably well although the experimental values for the $d_{3/2}$ transitions are consistently lower than the model predictions. Recent calculations of spectroscopic factors using a more sophisticated BCS approach [7] are also able to reproduce the trends of spectroscopic factors well although the discrepancies for $d_{3/2}$ transitions remain present.

DEFORMED PROTON EMITTERS

It has been known for some time that the decay rates of the proton emitters [109]I and [113]Cs cannot be reproduced using calculations of the type described above [11]. One possible explanation for this behaviour was the onset of modest prolate deformations in the region of the proton drip-line above Z=50. Bugrov and Kadmensky [12] developed a model for proton emission from deformed nuclei using Nilsson wavefunctions with quadrupole deformation treated as a free parameter. They were able to reproduce the anomalous half-lives of [109]I and [113]Cs using relatively modest deformations $\beta{\sim}0.1$ consistent with values expected for this transitional region. The macroscopic-microscopic mass model of Moller et al. [13] predicted the onset of much higher prolate deformations ($\beta{\sim}0.3$) immediately below Z=69 along the region of the proton drip-line. Experiments using the Fragment Mass Analyzer (FMA) [14] at Argonne have recently identified ground-state proton radioactivity from [141]Ho (Z=67) lying just inside this region, and [131]Eu (Z= 63) [15] lying at the heart of the region, with predicted quadrupole deformation parameters of 0.29 and 0.33 [13], respectively. Spherical proton decay calculations signally fail to reproduce the half-lives. Davids et al. [15] demonstrated that the decay rates could be well reproduced using the calculational approach of Bugrov and Kadmensky obtaining Nilsson configurations and quadrupole deformations consistent with the predictions of Moller et al. [13] (see Figure 3). This successful extension of the theory to highly deformed nuclei represents a significant deepening of our understanding of the proton decay phenomenon. These transitions provide a unique insight into the fragmentation of single particle strength in deformed nuclei. In the case of [141]Ho a weakly produced isomeric proton transition has been identified in experiments at Oak Ridge which can also be well understood in terms of a highly deformed Nilsson configuration [16].

DISCOVERY OF PROTON DECAY FINE STRUCTURE

Following the identification of proton radioactivity from the highly deformed nucleus [131]Eu it was decided to revisit this nucleus in an experiment at Argonne in order to search for the previously unobserved phenomenon of proton decay fine structure on the basis that the first excited 2^+ level in the daughter nucleus [130]Sm should be low enough for a significant decay branch [17]. Figure 4 shows the energy spectrum for decays occurring within 100ms of an A = 131 ion implanting into the same quasi-pixel of a Double-sided Silicon Strip Detector situated behind the focal plane of the Argonne FMA. The more intensely produced peak at higher energy corresponds to the previously identified ground-state proton transition from [131]Eu. The second peak produced with approximately one tenth of the intensity is at an energy $\sim$120 keV lower. The peak has the same half-life within errors as the previously idntified transition and is assigned to the proton decay fine structure of [131]Eu. Using the Grodzin's formula [18,19] this implies a value of $\beta{\sim}0.34$ for the daughter nucleus [130]Sm in excellent agreement with the value of 0.33 predicted by Moller et al. [13]. This also provides a consistency check on the high deformation necessary to reproduce the partial half-life of the main ground-state proton transition from [131]Eu using the deformed DWBA calculational approach [15]. This approach will be applied to reproduce the proton branching ratio from the ground-state. The spectroscopic factor

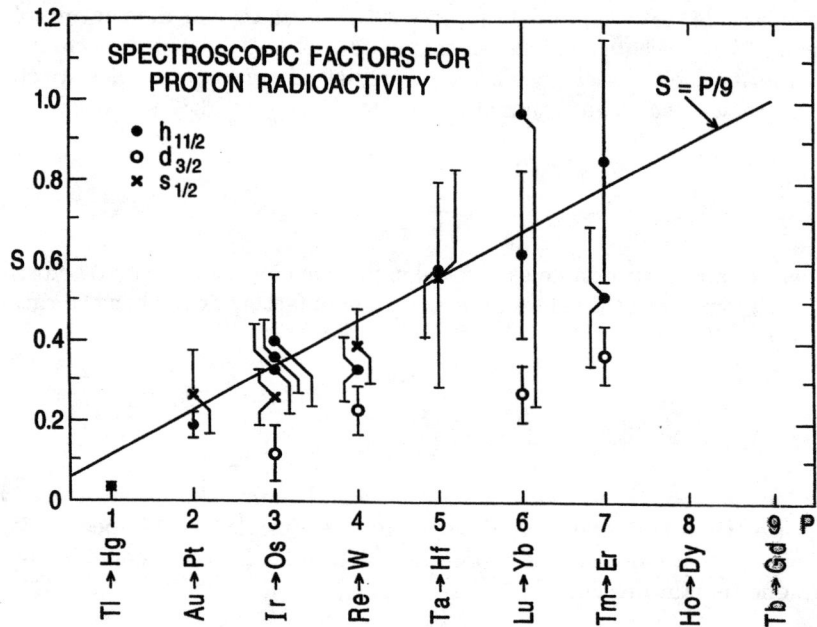

Figure 2. Experimental proton decay spectroscopic factors compared with the low seniority shell model calculation prediction of P/9 for the region 64<Z<82, where P represents the number of proton hole pairs wrt the Z = 82 shell closure in the daughter nucleus.

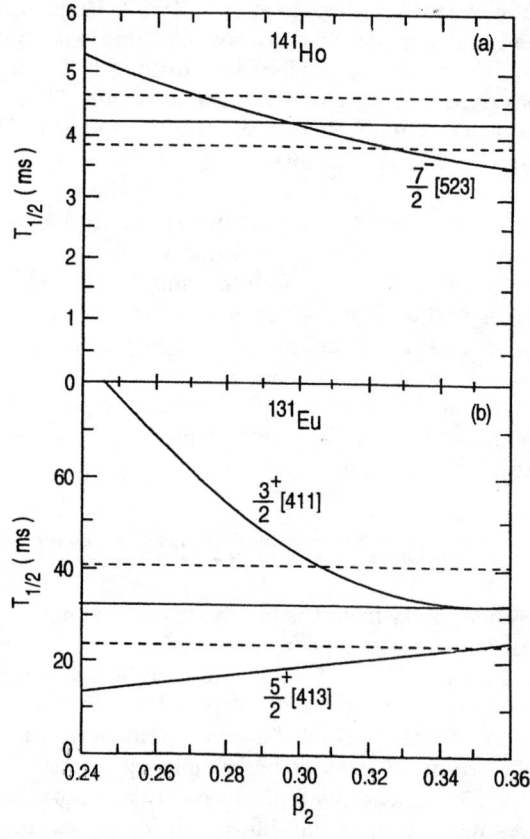

Figure 3. Calculated proton decay half-lives for Nilsson configurations of ^{141}Ho and ^{131}Eu as a function of deformation. Experimental values are shown as a horizontal solid line with a dashed line representing the uncertainty.

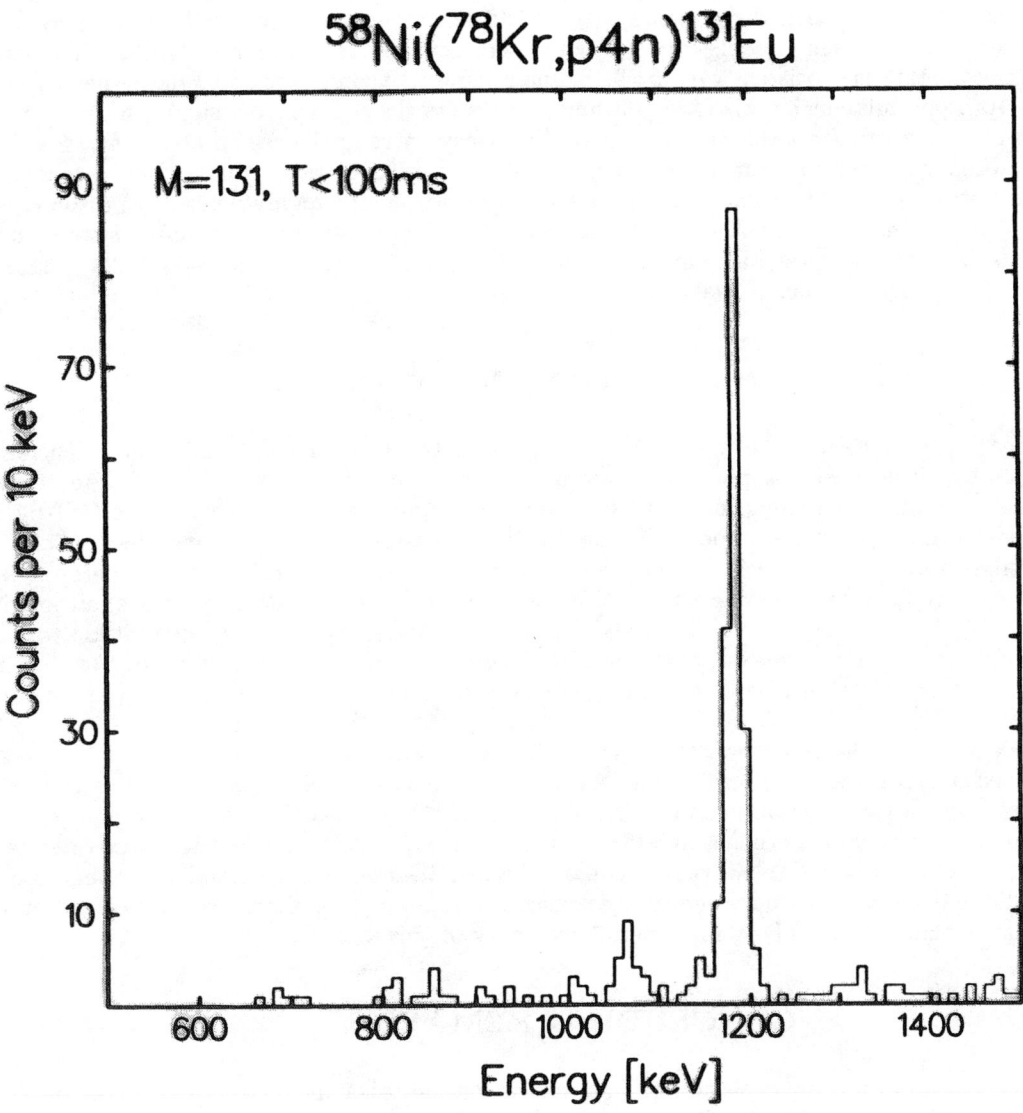

Figure 4. The discovery of proton decay fine structure in ^{131}Eu using the Argonne Fragment Mass Analyzer. The lower energy peak corresponds to a proton decay branch to the first excited 2$^+$ state in the daughter nucleus ^{130}Sm.

for the fine structure transition to the 2^+ state should provide an insight into the relationship between single particle states and the microscopic structure of rotational collective excitations.

GAMMA-RAY SPECTROSCOPY OF PROTON EMITTERS USING RECOIL DECAY TAGGING

Although theories of spherical and deformed proton emitters are now being tested over a wide range of nuclei, including the new phenomenon of proton decay fine structure, it is desirable to have independent information on the structure of these nuclei. In particular high resolution in-beam gamma-ray studies can provide insights into the nuclear deformation. The technique of Recoil Decay Tagging (RDT) (see Figure 5) [20,21] is an ideal tool that was developed [20] with this particular goal in mind and is now used extensively in the study of neutron-deficient and heavy nuclei exhibiting charged particle decay modes. The RDT technique was successfully applied to ground and isomeric proton emission from ^{147}Tm (see figure 6) in an experiment on the Argonne FMA [22]. The gamma-ray band built on the ground-state is consistent with $\beta=0.13$. Interestingly, Kadmensky and Bugrov have applied their model for deformed proton emission to this case [23] and the results do not disagree significantly from spherical calculations, unlike the effect of such a deformation on proton decay rates in the region above Z = 50. It appears that Tm and Ho proton emitters lie right at the interface of the region of rapid shape change to high prolate deformations. Clearly it is desirable to identify in-beam gamma-rays from the highly deformed proton emitters. In a very recent RDT experiment using the Argonne FMA coupled to Gammasphere, gamma-rays were successfully identified from the ground and isomeric states in ^{141}Ho, the latter having a cross-section~50nb [24], these data are shown in Figure 7 and are currently being analysed.

TWO PROTON EMISSION

Two proton radioactivity has yet to be observed, it is expected to occur in highly neutron - deficient even Z nuclei when the proton pairing energy suppresses single proton emission. The most promising candidate isotope discovered to date is ^{45}Fe identified using high energy heavy ion fragmentation on the FRS at GSI by Blank et al. [25], with the yet to be discovered isotope ^{48}Ni being another possible candidate if it indeed exists. As with one proton radioactivity these relatively low Z nuclei will have decay rates very sensitive to the decay Q-value and an alternative approach would be too look in heavier regions using fusion-evaporation reactions, although present techniques will probably have to improve in sensitivity by approximately two orders of magnitude to make such searches feasible. One should also not rule out two proton radioactivity from an isomeric state in a nucleus not lying beyond the proton drip-line, this after all was how the first example of one proton radioactivity was serendipitously discovered [26].

The two proton decay mechanism can be studied from resonant states. Pioneering studies of the beta-delayed two proton decay mechanism showed a predominance of sequential emission [27]. More recent experiments using radioactive beams to populate two proton-unbound resonances [28,29] have also yielded similar results. In order to suppress the sequential process an experiment at MSU has used inelastic scattering of a radioactive ^{17}Ne beam to populate an excited state that is bound to one proton emission. The initial results reported here at this conference by Thoenesson et al. [30] indicate the existence of a decay branch corresponding to simultaneous two proton emission. A further experiment is planned to study in more detail the decay of this state.

CONCLUDING REMARKS

The recent years have produced an explosion of information on the phenomenon of proton radioactivity with 38 transitions now being known. Theoretical models are able to reproduce in detail the systematic variation of proton decay spectroscopic factors for a wide range of spherical nuclei [4,7] thereby sensitively testing the nuclear shell model at the extreme edge of stability. The first examples of proton emission from highly deformed nuclei have been discovered [15]. These transitions have provided a direct insight into the fragmentation of single particle strength within highly deformed nuclei. Decay rates from these nuclei are found to agree well with theoretical calculations assuming Nilsson states [15,12]. Proton decay fine structure is reported here for the first time providing independent confirmation of the high deformations involved. In-beam studies of the gamma-rays using the RDT technique are providing complementary nuclear structure information on proton-radioactive nuclei [20,22] that will assist in constraining theoretical calculations of proton decay rates. Furthermore, such studies will provide new insights into the behaviour of proton unbound nuclei at high spin and excitation energy. These aspects of nuclear

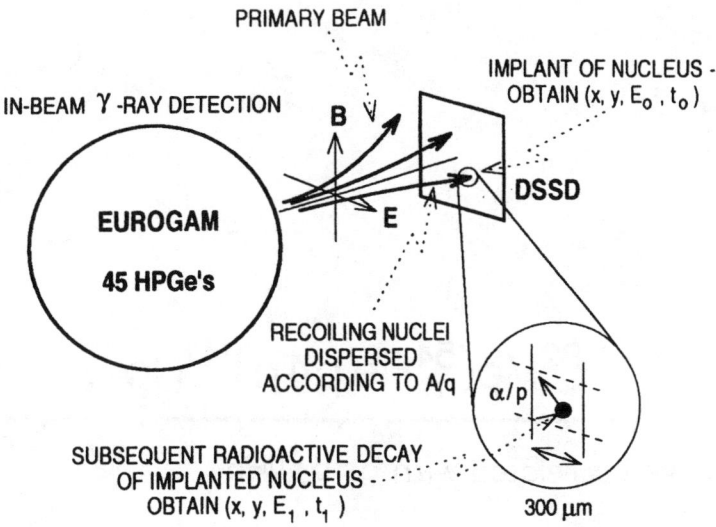

Figure 5. Schematic diagram of the Recoil Decay Tagging (RDT) technique applied using the Daresbury RMS to identify in-beam γ-rays from the ground-state proton emitter [109]I

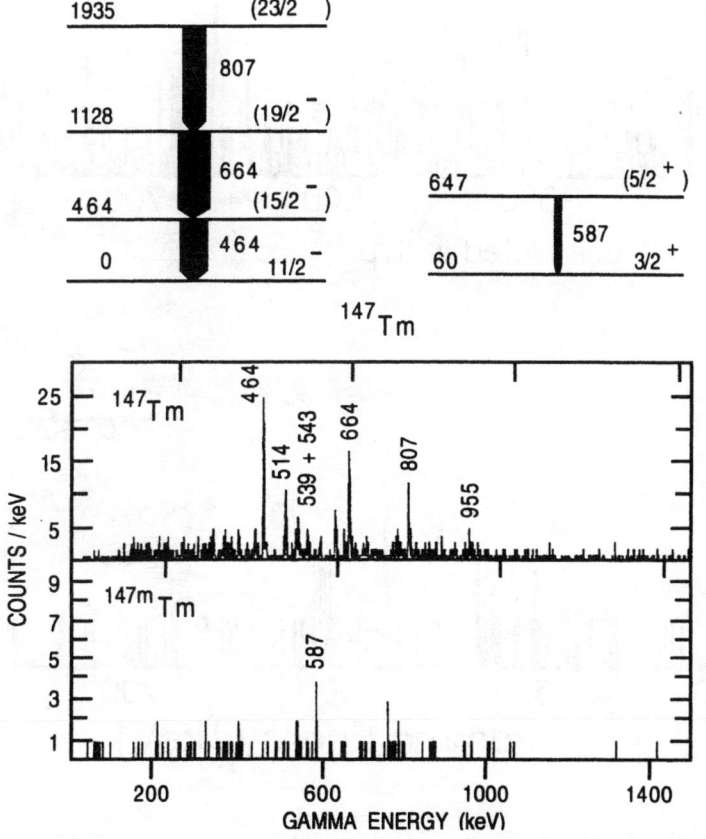

Figure 6. Energy level scheme and gamma spectra for [147]Tm obtained using the RDT technique on the Argonne FMA. The upper and lower spectra show gamma-rays preceding the proton-decays of the $h_{11/2}$ ground-state and $d_{3/2}$ isomeric state, respectively.

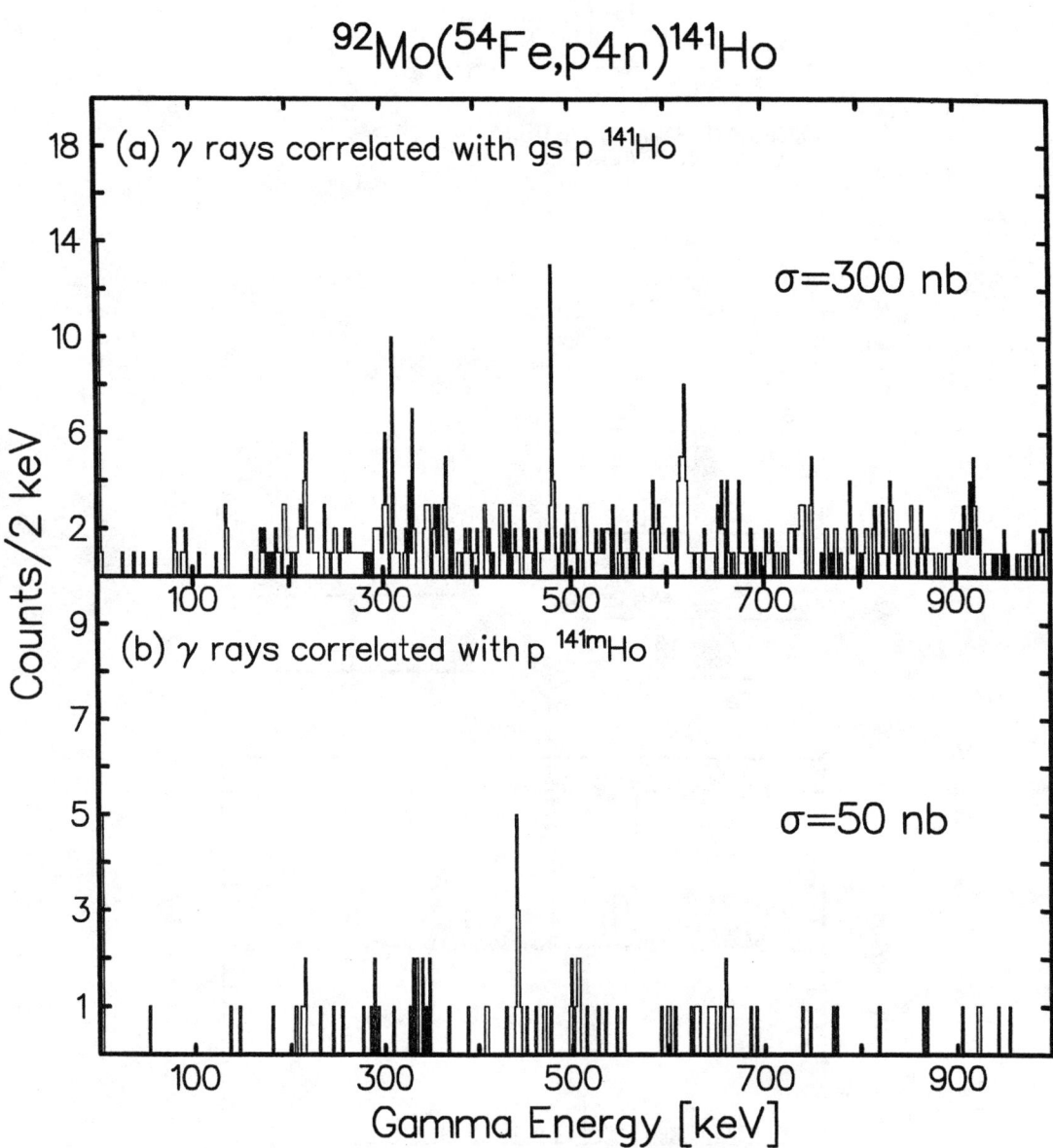

Figure 7. Recoil Decay Tagging gamma-ray spectra for the ground and isomeric states of the highly deformed proton emitter ^{141}Ho obtained using the
Gammasphere array on the Argonne Fragment Mass Analyzer.

structure are now also being studied using particle detection at the target position where much shorter lifetime proton decay transitions can be studied in competition with gamma-decays [31].

In summary there has been a great advance in our detailed understanding of the phenomenon of proton radioactivity. This has derived from the large increase in known transitions and the varied nuclear landscape in which this process is now found.

ACKNOWLEDGEMENTS

I would like to thank all my colleagues involved in experiments conducted at Argonne and Oak Ridge.

REFERENCES

1. M.F. Mohar et al., *Phys. Rev.* **C38**, 747 (1988).
2. P.H. Regan et al., *Acta Polonika* in Press.
3. P.J. Woods, Proceedings of the 4th International School of Heavy Ion Physics, Erice, 315 (1997).
4. C. N. Davids, P. J. Woods, J. C. Batchelder, C. R. Bingham, D. J. Blumenthal, L. T. Brown, B. C. Busse, L. F. Conticchio, T. Davinson, S. J. Freeman, D. J. Henderson, R. J. Irvine, R. D. Page, H. T. Pentilla, D. Seweryniak, K. S. Toth, W. B. Walters and B. E. Zimmerman, *Phys. Rev.* **C55**, 2255 (1997).
5. P.J. Woods and C.N. Davids, *Ann. Rev. Nucl. Part. Sci.* **47**, 541 (1997).
6. S. Hofmann, W. Reisdorf, G. Munzenberg, F.P. Hessberger, J.R.H. Schneider, P. Armbruster, *Z. Phys.* **A305**, 111 (1982).
7. S. Aberg, P.B. Semmes, W. Nazarewicz, *Phys. Rev.* **C56**, 1762 (1997).
8. P.J. Sellin, P.J. Woods, T. Davinson, N.J. Davies, A.N. James, K. Livingston, R.D. Page, A.C. Shotter, *Phys. Rev.* **C47**, 1933 (1993).
9. F.D. Becchetti and G.W. Greenlees, *Phys. Rev.* **182**, 1190 (1969).
10. A. Bohr and B.R. Mottelson, *Nuclear Structure 1* (W.A. Benjamin , New York 1969).
11. A. Gillitzer, T. Faestermann, K. Hartel, P. Kienle and E. Nolte, *Z. Phys.* **A326**, 107 (1987).
12. V.P. Bugrov and S.G. Kadmensky, *Sov. J. Nucl. Phys.* **49**, 967 (1989).
13. P. Moller, J.R. Nix, K.-L. Kratz, *At. Data Nucl. Data Tables* **66**, 131 (1997)
14. C. N. Davids et al., *Nucl. Instrum. Methods* **B70**, 358 (1992).
15. C.N. Davids, P.J. Woods, D. Seweryniak, A.A. Sonzogni, J.C. Batchelder, C.R. Bingham, T. Davinson, D.J. Henderson, R.J. Irvine, G.L. Poli, J.Uusitalo, W.B. Walters, *Phys. Rev. Lett.* **80**, 1849 (1998).
16. J.C. Batchelder et al., Proceedings of this Conference.
17. C.N. Davids. P.J. Woods, D. Seweryniak, A.A. Sonzogni, M. Carpenter, J. Ressler, J. Schwarz, J. Uusitalo and W.B. Walters, paper in preparation.
18. L. Grodzins, Phys. Lett. 2, 88 (1962).
19. F. Stephens et al., Phys. Rev. Lett. 29, 438 (1972).
20. E.S. Paul, P. J. Woods, T. Davinson, R. D. Page, P. J. Sellin, C. W. Beausang, R. M. Clark, R. A. Cunningham, S. A. Forbes, D. B. Fossan, A. Gizon, J. Gizon, K. Hauschild, I. M. Hibbert, A. N. James, D. R. LaFosse, I. Lazarus, H. Schnare, J. Simpson, R. Wadsorth and M. D. Waring, *Phys. Rev.* **C51**, 78 (1995).
21. R. S. Simon, K-H. Schmidt, F. P. Hessberger, S. Hlavae, M. Honusek, G. Munzenberg, H. G. Clerc, U. Gollerthan, W. Schwab, *Z. Phys.* **A325**, 197 (1986).
22. D. Seweryniak, C. N.Davids, W. B. Walters, P. J. Woods, I. Ahmad, H. Amro, D. J. Blumental, L.T. Brown, M. P. Carpenter, T. Davinson, S.M. Fischer, D.J. Henderson, R.V.F. Janssens, T. L. Khoo, I. Hibbert, R.J. Irvine, R. J. Irvine, C.J. Lister, J. A. Mckenzie, D. Nisius, C. Parry, and R. Wadsworth *Phys. Rev.* **C55**, R2137 (1997).
23. S.G. Kadmensky and V.P. Bugrov, *Phys. At. Nucl.* **59**, 399 (1996).
24. P.J. Woods, private communication (1998).
25. B. Blank et al., *Phys. Rev. Lett.* **77**, 2893 (1996).
26. K.P. Jackson et al., *Phys. Lett.* **B33**, 281 (1970).
27. R. Jahn et al., *Phys. Rev.* **C31**, 1576 (1985).
28. R. A. Kryger, A. Azhari, M. Hellstrom, J. H. Kelley, T. Kubo, R. P. Pfaff, E. Ramakrishnan, B. M. Sherril, M. Thoenesson, S. Yokoyama, R. J. Charity, J. Dempsey, A. Kirov, N. Robertson, D. G. Sarantites, L. G. Sobotka, J. A. Winger, *Phys. Rev. Lett.* **74**, 861 (1995).
29. C. R. Bain, P. J. Woods, R. Coszach, T. Davinson, P. Decrock, M. Gaelens, W. Galster, M. Huyse, R. J. Irvine, P. Leleux, M. Loiselet, C. Michotte, R. Neal, A. Ninane, G. Ryckewaert, A. C. Shotter, G. Vancraeynest, J. Vervier and J. Wauters, *Phys. Lett.* **B373**, 35 (1996).
30. M. Thoenesson, M. Chromik, P.G. Thirolf, , M. Fauerbach, T. Glasmacher, R. Ibbotson, R.A. Kryger, H. Scheit, P.J. Woods, Conference Proceedings.
31. D. Rudolph, Proceedings of this Conference.

Proton Decay Studies of the Light Lu, Tm and Ho Isotopes*

J. C. Batchelder[1], C. R. Bingham[2,3], C. J. Gross[2,7], R. Grzywacz[3],
K. Rykaczewski[2,4], K. S. Toth[2], E. F. Zganjar[5], Y. Akovali[2], T. Davinson[6],
T. N. Ginter[7], J. H. Hamilton[7], Z. Janas[4], M. Karny[4], S. H. Kim[3],
B. D. MacDonald[9], J. F. Mas[2], J. W. McConnell[2], A. Piechaczek[5], J. J. Ressler[10],
R. C. Slinger[6], J. Szerypo[11], W. Weintraub[3], P. J. Woods[6], C.-H. Yu[2]

1 Oak Ridge Associated Universities, Oak Ridge TN, 37831 USA
2 Physics Division, Oak Ridge National Laboratory, Oak Ridge, 37831 TN USA
3 University of Tennessee, Knoxville TN 37996 USA
4 IEP, Warsaw University, 00681 Warsaw, Hoza 69, Poland
5 Louisiana State University, Baton Rouge, LA 70803 USA
6 University of Edinburgh, Edinburgh, EH9 3JZ, United Kingdom
7 Vanderbilt University, Nashville TN 37235 USA
8 Oak Ridge Institute for Science and Education, Oak Ridge, TN 37831 USA
9 Georgia Institute of Technology, Atlanta GA 30332 USA
10 University of Maryland, College Park, MD 20742 USA
11 Joint Institute for Heavy Ion Research, Oak Ridge TN, 37831 USA

Abstract.

A double-sided Si-strip detector system has been installed and commissioned at the focal plane of the Recoil Mass Spectrometer at the Holifield Radioactive Ion Beam Facility. The system can be used for heavy charged particle emission studies with half-lives as low as a few μsec. In this paper we present identification and study of the decay properties of the five new proton emitters: ^{140}Ho, ^{141m}Ho, ^{145}Tm, ^{150m}Lu and ^{151m}Lu.

INTRODUCTION

Nuclei which are energetically unbound to the emission of a proton are located beyond the proton drip line. Observation of protons emitted from these isotopes allows us not only to establish the limits of stability for a given element, but also gives information on the structure and mass of the parent nucleus. Study of the decay of proton-emitting isotopes allows one to study nuclear structure effects in nuclei that are inaccessible by means of in-beam experiments. The emitted proton tunnels through the Coulomb and centrifugal barriers, and the decay probability depends strongly on the energy of the proton and on its angular momentum ℓ. Because of this, the ℓ value of the emitted proton can often be determined through the use of a simple spherical WKB calculation of the expected rate of the tunneling process. However, this calculation does not take into account the details of the nuclear structure effects. Therefore, the difference between the calculated and experimental half-lives is due to nuclear structure effects such as an occupation of a respective proton orbital, and is known as the spectroscopic factor. The ratio of the calculated half-life to the measured value is defined as the experimental spectroscopic factor [1]:

CP481, *Nuclear Structure 98*, edited by C. Baktash

$$S_p^{exp} = \frac{T_{1/2}^{calc}}{T_{1/2}^{exp}} . \tag{1}$$

The theoretical spectroscopic factor (S_p^{th}) can be calculated via an independent quasi-particle BCS approximation [2]. The spectroscopic factor is given by:

$$S_p^{th} = u_j^2 , \tag{2}$$

where u_j^2 is the probability that the proton orbital (n ℓ j) is empty in the daughter nucleus.

The calculations [3] predicting the ground-state deformation of the isotopes reveal that there is a region of high prolate deformation between Z = 50 and 67, and then a region of relatively low deformation between Z = 69 to 82. The effect of deformation on the rate of quantum tunneling is a subject of great interest. To gain insight, however, one needs to understand both spherical and deformed proton emission rates. We can therefore divide proton emission from nuclei with Z > 50 into two categories: deformed and spherical proton emission.

While calculated values of S_p agree quite well for $75 \leq Z \leq 81$, the comparison is rather inconclusive for nuclei with $69 \leq Z \leq 71$, because the known proton emitters in this region have either large uncertainties in their measured branching ratios (as is the case for ^{147}Tm: 15(5)% [4]), or the β-decay branch is unknown (^{146}Tm [5] and ^{150,151}Lu [6]) and is estimated from the gross β-decay theory of Takahashi et al [7]. If the estimate is in error by a factor of 2 or more, its use would result in a corresponding uncertainty in the spectroscopic factor. These nuclei lie to the neutron-deficient side of the N=82 closed shell, so that the major competition with proton emission is β$^+$ rather than α decay. Because β–decay branching ratios are difficult to measure accurately, large uncertainties exist in the corresponding partial proton half-lives. With this in mind we have concentrated on those proton emitters in this region that were predicted to have half-lives far too short for beta decay to compete with the expected proton emission.

Until recently, the only known ground state proton emitters considered to be deformed were ^{109}I [8] and ^{112,113}Cs [8,9]. These isotopes exhibited longer half-lives than what would have been expected from a simple spherical model. Calculations by Bugrov and Kadmenskii [10] were able to reproduce the experimental data by using a deformation of β$_2$~ 0.15 in the parent nucleus ^{109}I and β$_2$ ~ 0.21 for ^{113}Cs. More recently, Davids et al. [11] have observed proton emission from the ground state of ^{141}Ho, with a proton energy of 1169(8) keV and half-life of 4.2(4) ms. They showed that the data do not agree with spherical WKB predictions but rather with calculations that take deformation into account. In this work, we present new data on ^{141}Ho confirming the earlier measurement, and report for the first time proton emission from an isomeric state in ^{141}Ho, along with proton emission from ^{140}Ho.

Experimental Setup

The experiments presented here were performed at The Oak Ridge Holifield Radioactive Ion Beam Facility (HRIBF), using the 25 MV Tandem Accelerator. After bombardment on the appropriate target, recoil nuclei of interest were separated spatially according to their mass/charge (A/Q) values through the HRIBF Recoil Mass Spectrometer (RMS) [12,13,14]. A gas-filled position sensitive avalanche counter (PSAC) at the focal plane was used to identify the A/Q of the recoils. In reactions similar to those presented herein, the RMS transmission efficiency for the central ion has been determined to be 3-4% for a typical production target of 0.5 - 1.0 mg/cm^2 [15]. Following the PSAC, the central ions were implanted into a ~ 60-μm thick double-sided silicon strip detector (DSSD) [16] with 40 horizontal and 40 vertical strips. This strip arrangement results in a total of 1600 pixels, each acting as an individual detector. Depending on the settings of the RMS, either one or two masses were directed onto the DSSD. When two masses are put on the detector, mass identification is made by the relative x-position of the recoils on the DSSD.

For each event in the DSSD, the time (from a continuously running clock), energy, and event type (recoil or decay, depending on whether it is in coincidence with the PSAC or not) were recorded. By using this time information, the half-life of the decaying nuclide could be determined. The shortest decay observable with our setup is determined by the flight time through the RMS (typically ~ 2 μs) and the recovery time of the amplifiers after overload due to the implantation of the high-energy recoil. The overload effect causes decay events which occurs shortly (< ~30 μsec) after the implantation to have a slightly worse energy resolution because each amplifier's response is somewhat different. Individual strips were gain-matched in software through the use of external alpha sources, and internally from ^{147}Tm [17], ^{113}Cs [8] and ^{109}I [8] protons produced in reactions of ^{54}Fe and ^{58}Ni beams on ^{58}Ni and ^{92}Mo targets.

New Spherical Proton Emitters: ^{145}Tm, ^{151m}Lu and ^{150m}Lu

In the region of $64 < Z < 82$, the orbitals $0h_{11/2}$, $1d_{3/2}$, and $2s_{1/2}$ are expected to lie close together near the Fermi surface. Of these orbitals, the $d_{3/2}$ state can mix with the $s_{1/2}$ state coupled to the 2^+ phonon of the even-even core excitation. The $h_{11/2}$ state is expected, however, to have rather small admixtures. Therefore to rigorously test the predictive power of the BCS-based description of spectroscopic factors, precise S_p^{exp} values for $\pi h_{11/2}$ emitters are essential. With this in mind, we investigated the proton decay of ^{145}Tm.

Thulium-145 [18] was produced via the ^{92}Mo(^{58}Ni,p4n) reaction. A 0.91-mg/cm^2 thick target of ^{92}Mo (97 % enrichment) was bombarded with 315-MeV ^{58}Ni ions (307 MeV at the target mid-point) extracted from the HRIBF Tandem Accelerator, with an average beam current on target of ~ 15 pnA over a period of 50 hours. Figure 1(a) shows the total decay spectrum accumulated in the ^{58}Ni + ^{92}Mo irradiation. One observes α-decay peaks of nuclei produced in reactions on target isotopic impurities whose A/Q value is similar to that of the central ion. The low energy part of the spectrum is dominated by a broad distribution of "escape" events resulting from α decays in the backward direction, i.e., only part of the energy is deposited in the DSSD. Also visible is a peak at 1.12 MeV which we attribute to protons from ^{146}Tm [5,19], and which provided an internal energy calibration. Figure 1(b) shows only those events that had a time between implantation and decay of ≤ 50 μs. In this spectrum, a peak with an energy of 1.728(10) MeV is clearly seen. We assign it to the proton decay of ^{145}Tm as no other nucleus with A = 145 and Z < 69 is proton unbound [20]. Since all the events in this peak came at very short times after recoil implantation, the energy resolution is poorer than what is normally observed for decays with longer half-lives. By correcting for the overload effect of the amplifiers, (based on the measured proton spectra of ^{113}Cs; $T_{1/2} = 18$ μs) the resolution of this peak was improved as shown in Fig. 1(c). If the overall detection efficiency of the RMS is taken to be 3% (for this reaction), the production cross section for ^{145}Tm is estimated to be 500 nb.

As mentioned above, half-life information can be obtained by correlating the times of a recoil and the next decay event in a given pixel. The resulting half-life for the ^{145}Tm proton peak is 3.5(10) μs. A WKB approximation calculation for protons emitted from the $0h_{11/2}$ ($\Delta \ell = 5$), $1d_{3/2}$ ($\Delta \ell = 2$), and $2s_{1/2}$ ($\Delta \ell = 0$) orbitals of ^{145}Tm was performed with the experimental E_p as input. The optical potential was taken from the real part of the optical potential of Becchetti and Greenlees [21]. The calculated half-lives are $1.8^{+0.3}_{-0.2}$ μs, 0.7(1) ns and 80(12) ps, respectively for $\Delta \ell = 5$, 2, and 0. Table 1 compares these values to the experimental value of 3.5(10) μs and clearly indicates a $\Delta \ell = 5$ transfer and thus an $0h_{11/2}$ assignment for the parent proton state which, based on level systematics in this mass region, is probably the ground state in ^{145}Tm. Taking the WKB value of 1.8 μs for the theoretical half-life results in a spectroscopic factor of 0.51(16) for this $0h_{11/2}$ emitter. The error should be considered as a lower limit since uncertainties due to the optical potential are not included. This value is consistent with an overall spherical description for this nucleus, and agrees well with the value of 0.64 for the S_p^{th} obtained by Åberg et al [2] using the BCS approximation. This work represents the first S_p^{exp} value with error bars small enough to be meaningful for the $0h_{11/2}$ proton emission from odd-mass Tm isotopes.

The lutetium isotopes were among the first nuclei where proton emission was observed. In fact ^{151}Lu was the first ground state proton emitter discovered [22]. However, for both the cases of ^{151}Lu and ^{150}Lu, only one high-spin proton-emitting isomer has been observed. Neighboring nuclei however, have been observed to emit a

TABLE 1. Comparison of the observed proton half-lives with values calculated using the spherical WKB approximation.

species	E_p (MeV)	Exp $T_{1/2}$	calculated $T_{1/2}$ (WKB)		
			$0h_{11/2}$ ($\Delta \ell = 5$)	$1d_{3/2}$ ($\Delta \ell = 2$)	$2s_{1/2}$ ($\Delta \ell = 0$)
^{145}Tm	1.728(10)	3.5(10) μs	$1.8^{+0.3}_{-0.2}$ μs	0.7(1) ns	80(12) ps
^{151m}Lu	1.310(15)	16(1) μs	13^{+6}_{-4} ms	$4.9^{+2.0}_{-1.4}$ μs	$0.55^{+0.28}_{-0.14}$ μs
^{150}Lu	1.263(4)*	49(5) ms	36(3) ms	13(1) μs	1.5(1) μs
^{150m}Lu	1.290^{+20}_{-30}	30^{+40}_{-10} μs	19^{+20}_{-7} ms	7^{+7}_{-3} μs	$0.7^{+1.5}_{-0.2}$ μs
^{141m}Ho	1.230(20)	$8.4^{+3.1}_{-1.8}$ μs	7^{+4}_{-3} ms	2(1) μs	0.2(1) μs
^{141}Ho	1.169(8)†	3.9(5) ms	32(7) ms	$8.6^{+1.9}_{-1.6}$ μs	$0.89^{+0.20}_{-0.16}$ μs
^{140}Ho	1.086(10)	$6.1^{+3.1}_{-1.5}$ ms	320^{+100}_{-80} ms	81^{+26}_{-20} μs	8^{+3}_{-2} μs

* Energy from Ref 6.
† Energy from Ref. 11.

proton from both a high- and low-spin isomer. As with the Tm isotopes, it is expected that the $0h_{11/2}$, $1d_{3/2}$, and $2s_{1/2}$ orbitals lie close to the ground state. With this in mind, we reinvestigated the proton decays of ^{151}Lu and ^{150}Lu.

Lutetium-151 was produced via the ^{96}Ru(^{58}Ni,p2n) reaction with a beam energy of 266 MeV. Figure 2(a) shows the spectrum obtained from the above reaction with a constraint on the time between recoil and decay of $\leq$ 500 ms. A further constraint on the time of $\leq$ 50 μs is shown in Fig. 2(b). It clearly shows a peak at 1310(15) keV which we attribute to the proton decay of an isomer located at an energy of 80(15) keV [23] above the previously known $h_{11/2}$ ground state. The half-life associated with this new radioactivity is 16(1) μs. A WKB approximation calculation for protons emitted from the $0h_{11/2}$ ($\Delta \ell = 5$), $1d_{3/2}$ ($\Delta \ell = 2$), and $2s_{1/2}$ ($\Delta \ell = 0$) orbitals of ^{151}Lu was performed with the experimental E_p as input (see table 1). As can be seen from the table, the calculated value that matches most closely is the value of $4.9^{+2.0}_{-1.4}$ μs μs for emission from the $d_{3/2}$ state, with resulting spectroscopic factor of $0.31^{+0.12}_{-0.09}$. This value is low when compared to the calculated u_j^2 of 0.74 [24], which is typical of the $d_{3/2}$ proton emitters (see Fig 4).

For the case of the odd-odd nucleus ^{150}Lu, one might expect that it also exhibits two proton emitting isomers, similar to the situation for ^{146}Tm. If one considers coupling the neutron quasi-particles to the proton orbitals, the resulting lowest proton unbound states are $\pi h_{11/2} \otimes \nu h_{11/2}$, $\pi h_{11/2} \otimes \nu d_{3/2}$, $\pi d_{3/2} \otimes \nu d_{3/2}$ and $\pi d_{3/2} \otimes \nu h_{11/2}$. The known [6] isomer is attributed to arise from decay out of the 5^- or 6^- state to the $1/2^+$ or $3/2^-$ proton decaying state in the ^{149}Yb daughter. During our reinvestigation of this nucleus, we discovered a second short-lived proton emitting state in ^{150}Lu. We produced ^{150}Lu via the ^{96}Ru(^{58}Ni,p3n) reaction at a beam energy of 292 MeV (at the target midpoint). Figure 2(c) shows the total decay spectrum with a time constraint between decay and recoil of $\leq$ 100 ms. The known ^{150}Lu proton peak is clearly shown at 1261 keV. Figure 2(d) shows the same spectrum with a further constraint on the time of $\leq$ 100 μs. In this figure a higher energy peak of 1290(30) keV is observed, which we attribute to the decay of a short-lived isomer of ^{150}Lu. The half-lives of these two transitions resulting from this measurement are 49(5) ms (compared to 35(10) ms from Ref. [6]), and 30^{+40}_{-10} µs for the new isomer. Ytterbium-149 is expected to have a ground-state spin of $1/2^+$ or $3/2^+$ and an isomer of $11/2^-$ similar to the structure observed for ^{151}Yb [25]. We therefore propose that the new proton transition from ^{150}Lu is the decay of the isomeric $\pi d_{3/2} \otimes \nu d_{3/2}$ state to a $\nu d_{3/2}$ state (probably the ground state) in ^{149}Yb.

The known experimental (including this work) values for the spherical S_p are compared with the theoretical ones in Fig. 4 (a) for $h_{11/2}$ and 4(b) for $d_{3/2}$. Comparing the two figures, one can see that for $h_{11/2}$ values, the agreement between experimental and theoretical S_p values is fairly good (except for the large error bars), while the experimental $d_{3/2}$ values are consistently too low compared to the theoretical values. This probably indicates that the $d_{3/2}$ proton emitting states have a significant amount of mixing with other configurations such as $2^+ \otimes \pi s_{1/2}$.

New Deformed Proton Emitters: ^{140}Ho and ^{141m}Ho

Deformation has been predicted [26] to increase substantially along the proton drip-line between the Tm ($\beta_2 \lesssim 0.15$) and the Ho isotopes ($\beta_2 \gtrsim 0.25$). Therefore, the study of the proton decay of the Ho isotopes will give us some insight into the effects of large deformation on the tunneling process. Previous to this study, the only highly deformed proton emitters known were ^{141}Ho and ^{131}Eu [11]. Our investigations have led to the discovery of proton decay from a short-lived isomer of ^{141}Ho and from a new isotope ^{140}Ho [27]. Holmium-140 and ^{141}Ho were produced via the ^{92}Mo(^{54}Fe,p5n) and ^{92}Mo(^{54}Fe,p4n) reactions, respectively, using 315-MeV ^{54}Fe ions extracted from the HRIBF Tandem Accelerator. In this experiment, ions of mass 140 were deposited on one side (strips # 20-40) of the DSSD, and mass 141 were deposited on the other (# 1-19).

Figure 3(a) shows the decay spectrum obtained from the above reaction with A=141, and a time constraint between decay and recoil of T $\leq$ 16 ms. In this spectrum a peak is observed at ~1170 keV with a rather broad shoulder on the high energy side. Further constraints on the time reveals that it consists of two peaks: the known proton peak from ^{141}Ho, and a new proton transition from a short-lived isomer of ^{141}Ho. In figure 3(b) we constrain the time to 200 µs $\leq$ T $\leq$ 16 ms, showing only the known ^{141}Ho proton peak. The time is constrained in Fig. 3(c) to those events of T $\leq$ 50 µs. In this spectrum, a higher energy peak is observed. The energies and half-lives of these two peaks are E_p = 1230(20) keV, $T_{1/2}$= $8.4^{+3.1}_{-1.8}$ µs for the new isomer of ^{141}Ho and E_p = 1169(8) keV (this peak was used as a calibration point), $T_{1/2}$= 3.9(5) ms for the previously known proton arising from the ground state of ^{141}Ho, (which compares well with the literature value 4.2(4) ms [11]). Figure 3(d) shows the decay spectrum from mass 140 with the time constraint of 200 µs $\leq$ T $\leq$ 16 ms. Here, two lines are observed; one from the ^{141}Ho ground state (due to overlap of the tail of mass 141), and one from the new isotope ^{140}Ho, with values of E_p = 1086(10) keV, and $T_{1/2}$= $6.1^{+3.1}_{-1.5}$ ms.

As can be seen from Table 1, the spherical WKB approach gives values which can not easily explain the experimental half-lives of the Ho isotopes. In Ref. 11 Davids *et al.*, demonstrate that the ^{141}Ho ground-state decay can be explained using the formalism developed by Bugrov and Kadmenskii [28], resulting in a good fit for $\beta_2 \sim 0.3$ for proton emission from the $\pi 7/2^-$ [523] state originating from the $h_{11/2}$ orbital. Our calculations [29], which minimize the energies of the ^{141}Ho states in deformation space, indicate that the $\pi 1/2^+$[411] and the $\pi 7/2^-$ [523] have nearly the same energy and deformation ($\beta_2 \sim 0.27$ and $\beta_4 \sim -0.07$, $\beta_6 \sim 0.01$). The next excited state, $\pi 5/2^-$ [532], is predicted to be 250 keV higher. Proton emission from the $\pi 1/2^+$[411] and $\pi 7/2^-$ [523] states to the ground state of ^{140}Dy would have $\Delta \ell$ equal to 0 and 3 respectively. The spherical WKB approach predicts the proton $T_{1/2}$ values to be 0.2 µs and 1.7 µs, respectively, overestimating the proton emission probability. The proton decay rate from the deformed states is clearly reduced due to the complex structure of deformed wave function and tunneling through the deformed Coulomb barrier [11,28,29]. Based on the measured half-lives, we assign $\pi 7/2^-$ [523] to the ground state, and $\pi 1/2^+$[411] of the 60(20) keV isomer of ^{141}Ho.

Proton emission from the odd-odd nucleus ^{140}Ho is more complicated as one has to consider both the proton and neutron quasi-particles. The aforementioned calculations predict that the $\pi 7/2^-$[523] $\otimes \nu 9/2^-$[514], $\pi 7/2$[523] $\otimes \nu 5/2^+$[402], $\pi 1/2^+$[411] $\otimes \nu 9/2^-$[514] and $\pi 1/2^+$[411] $\otimes \nu 5/2^+$[402] levels all lie within 20 keV of the ground state. Based on information from the ^{141}Ho proton decay, the $\pi 7/2^-$[523] orbital coupled with either the $\nu 9/2^-$[514] or $\nu 5/2^+$[402] orbital can be assigned to the observed proton-emitting state in ^{140}Ho.

ACKNOWLEDGEMENTS

* UNIRIB is a consortium of universities, Oak Ridge Associated Universities, and is supported by them and the U.S. Department of Energy under contract No. DE-AC05-76OR00033 with the Oak Ridge Associated Universities. ORNL is managed by Lockheed Martin Energy Research under contract number DE-AC05-96OR22464 with the U.S. Department of Energy. ZJ and MK are partially supported by KBN, Poland, contract KBN 2 P03B 039 13.

REFERENCES

1. M. H. McFarlane, and J. B. French, Rev. Mod. Phys. **32**, 567 (1960).
2. S. Åberg, P. B. Semmes, and W. Nazarewicz, Phys. Rev. C **56**, 1762, (1997).
3. P. Möller, et al., At. Data Nucl. Data Tables **59**, 185 (1995).
4. K. S. Toth, D. C. Sousa, P. A. Wilmarth, J. M. Nitschke, and K. S. Vierinen, Phys. Rev. C. **47** 1804 (1993).
5. K. Livingston, P. J. Woods, T. Davinson, N. J. Davis, S. Hoffman, A. N. James, R. D. Page, P. J. Sellin, and A. C. Shotter, Phys. Lett. **312B** 46, (1993).
6. P. J. Sellin, P. J. Woods, T. Davinson, N. J. Davis, K. Livingston, R. D. Page, A. C. Shotter, S. Hoffman, and A. N. James, Phys Rev C **47**, 193 (1993).
7. K. Takahashi, M. Yamada and T. Kondoh, At. Data Nucl. Data Tables **12**, 101 (1973).
8. A. Gillitzer, T. Faestermann, K. Hartel, P. Kienle, and E. Nolte, Z. Phys. **A326**, 107 (1987).
9. R. D. Page, P. J. Woods, R. A. Cunningham, T. Davinson, N. J. Davis, S. Hoffman, A. N. James, K. Livingston, P. J. Sellin, and A. C. Shotter, Phys. Rev. Lett. **68**, 1287 (1992).
10. V. P. Bugrov and S. G. Kadmenskii, Sov. J. Nucl. Phys. **49**, 967 (1989)..
11. C. N. Davids, P.J. Woods, D. Seweryniak, A. A. Sonzogni, J. C. Batchelder, C. R. Bingham, T. Davinson, D.J. Henderson, R. J. Irvine, G. L. Poli, J. Uusitalo, and W. B. Walters, Phys. Rev. Lett. **80**, 1849 (1998)
12. T. M. Comier, J. D. Cole, J. H. Hamilton and A. V. Ramayya, Nucl. Instrum. Methods Phys. Res. **A297**, 199, (1990).
13. J. D. Cole, T. M. Cormier, J. H. Hamilton and A. V. Ramayya, Nucl. Instrum. Methods Phys. Res. **B70**, 343 (1992).
14. C. J. Gross, Y. A. Akovali, M. J. Brinkman, J. W. Johnson, J. Mas, J. W. McConnell, W. T. Milner, D. Shapira, and A. N. James, Application of Accelerators in Research and Industry, AIP conference proceedings 392, Woodbury, NY, Vol 1, (1997) p 401.
15. C. J. Gross, private communication.
16. P. J. Sellin, P. J. Woods, D. Branford, T. Davinson, N. J. Davis, D. G. Ireland, K. Livingston, R. D. Page, A. C. Shotter, S. Hofmann, R. A. Hunt, A. N. James, M. A. C. Hotchkis, M. A. Freer, S. L. Thomas, Nucl. Instrum. Methods Phys. Res. **A311**, 217 (1992).
17. P. O. Larsson, T. Batsch, R. Kirchner, O. Klepper, W. Kurcewicz, E. Roeckl, D. Schart, W. F. Feix, G. Nyman, P. Tidemand-Petersson, Z. Phys. **A314**, 9 (1983).
18. J. C. Batchelder, C. R. Bingham, K. Rykaczewski, K. S. Toth, T. Davinson, J. A. McKenzie, P. J. Woods, J. H. Hamilton, W. B. Walters, E. F. Zganjar, C. Baktash, T. N. Ginter, J. Greene, C. J. Gross, J. Mas, J. McConnell, S. D. Paul, D. Shapira, X. Xu, and C. -H. Yu, Phys Rev C 57, **R1042** (1998).
19. P. J. Woods, T. Davinson, N. J. Davis, S. Hoffman, A. N. James, K. Livingston, R. D. Page, P. J. Sellin, and A. C. Shotter, Nucl. Phys. **A553**, (1993).
20. G. Audi and H. Wapstra, Nucl. Phys. **A565**, 66 (1993).
21. F. D. Becchetti, Jr. and G. W. Greenlees, Phys. Rev. **182**, 1190 (1969).
22. S. Hoffman, W. Reisdorf, G. Münzenberg, F. P. Hessberger, J. R. H. Schneider, and P. Armbruster, Z. Phys. **A305**, 111 (1982).
23. C. R. Bingham, J. C. Batchelder, K. Rykaczewski, K. S. Toth, C. -H. Yu, T. N. Ginter, C. J. Gross, R. Grzywacz, M. Karney, S. H. Kim, B. D. MacDonald, J. Mas, J. W. McConnell, J. Szerypo, and W. Weintraub, to be published.
24. W. Nazarewicz, private communication.
25. E. Nolte, G.Korschinek, Ch. Setzensack Z. Phys. A309, 33 (1982)
26. P. Möller, J. R. Nix, W. D. Myers and j. Swiatecki, At. Data Nucl. Data Tables **59**, 185 (1995).
27. K. Rykaczewski, J. C. Batchelder, C. R. Bingham, K. S. Toth, T. Davinson, T. N. Ginter, C. J. Gross, R. Grzywacz, M. Karny, B. D. MacDonald, J. F. Mas, J. W. McConnell, A. Piechaczek, R. C. Slinger, P. J. Woods, E. F. Zganjar, to be published.
28. V. P. Bugrov and S. G. Kadmenskii, Sov. J. Nucl. Phys. **49**, 967 (1989).
29. W. Nazarewicz, M. A. Riley, and J. D. Garrett, Nucl. Phys. **A512**, 61 (1990).

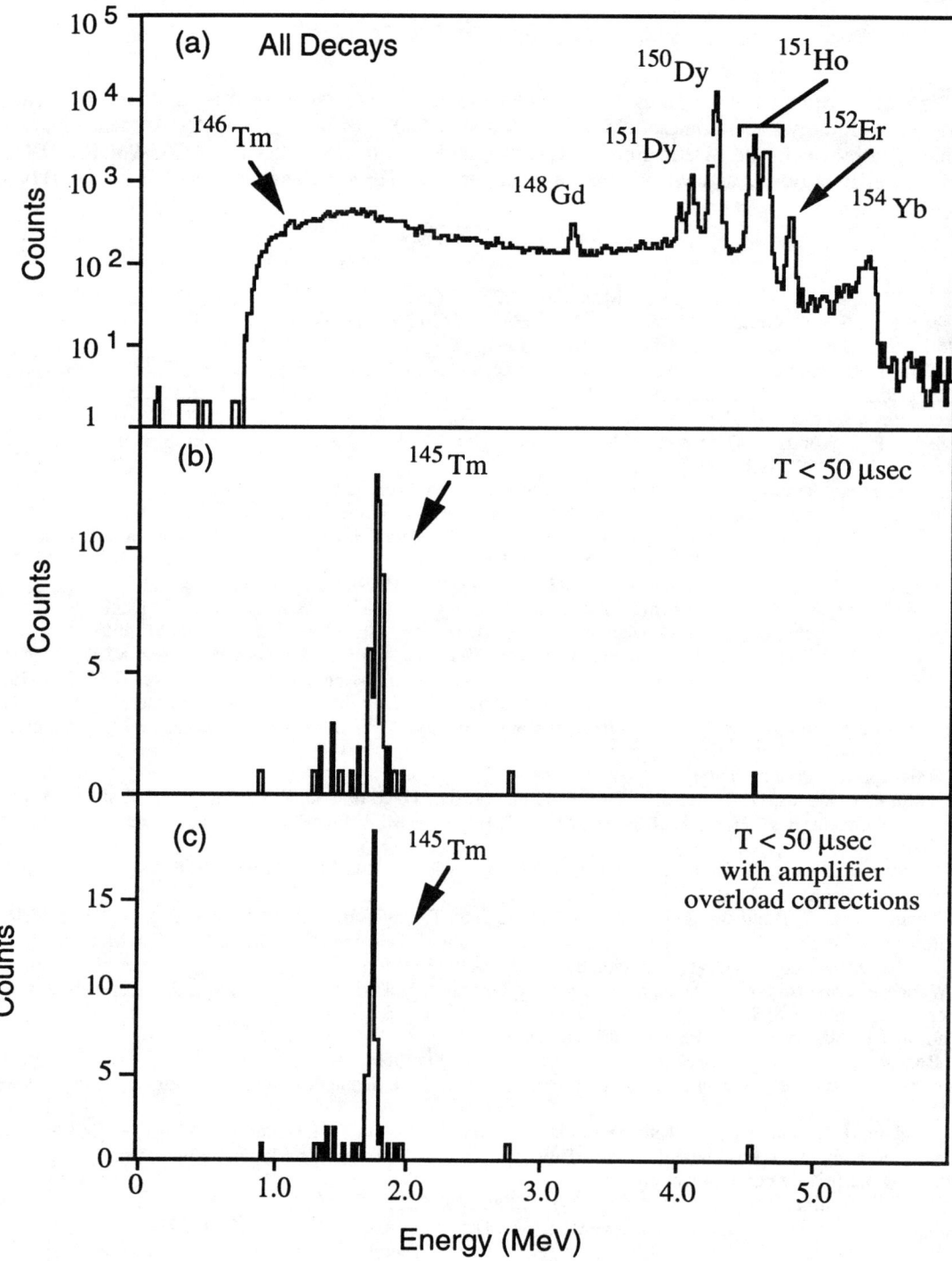

FIGURE 1. (a) Total decay spectrum observed in the DSSD during 307-MeV ^{58}Ni bombardments of ^{92}Mo. The α peaks above 3 MeV arise from nuclei formed in reactions on isotopic impurities present in the ^{92}Mo target. (b) The same spectrum gated on A = 145 recoils, with a time between decay and recoil implantation of ≤ 50 μs. (c) The spectrum in (b) with corrections to the resolution resulting from the amplifier overload of each strip.

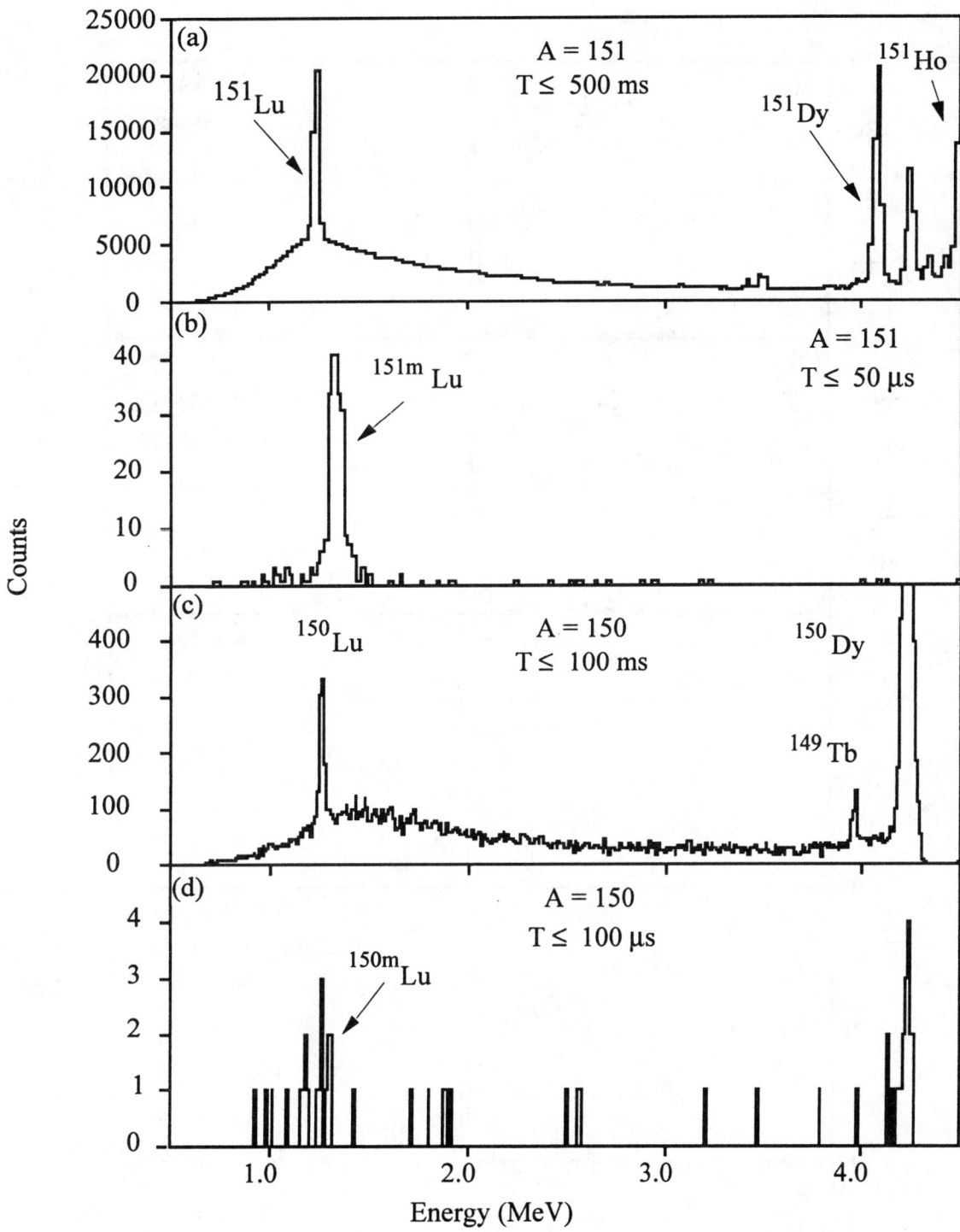

FIGURE 2. (a) Decay spectrum observed during 270-MeV ^{58}Ni bombardments of ^{96}Ru, with a time between decay and recoil implantation of ≤ 500 ms. (b) The decay spectrum from (a) with a further time constraint of ≤ 50 μs.(c) Decay spectrum observed during 292 MeV ^{58}Ni bombardments on ^{96}Ru, with a time constraint of T ≤ 100 ms between decay and recoil. (d) The same spectrum as (c) with a time constraint of T ≤ 100 μs

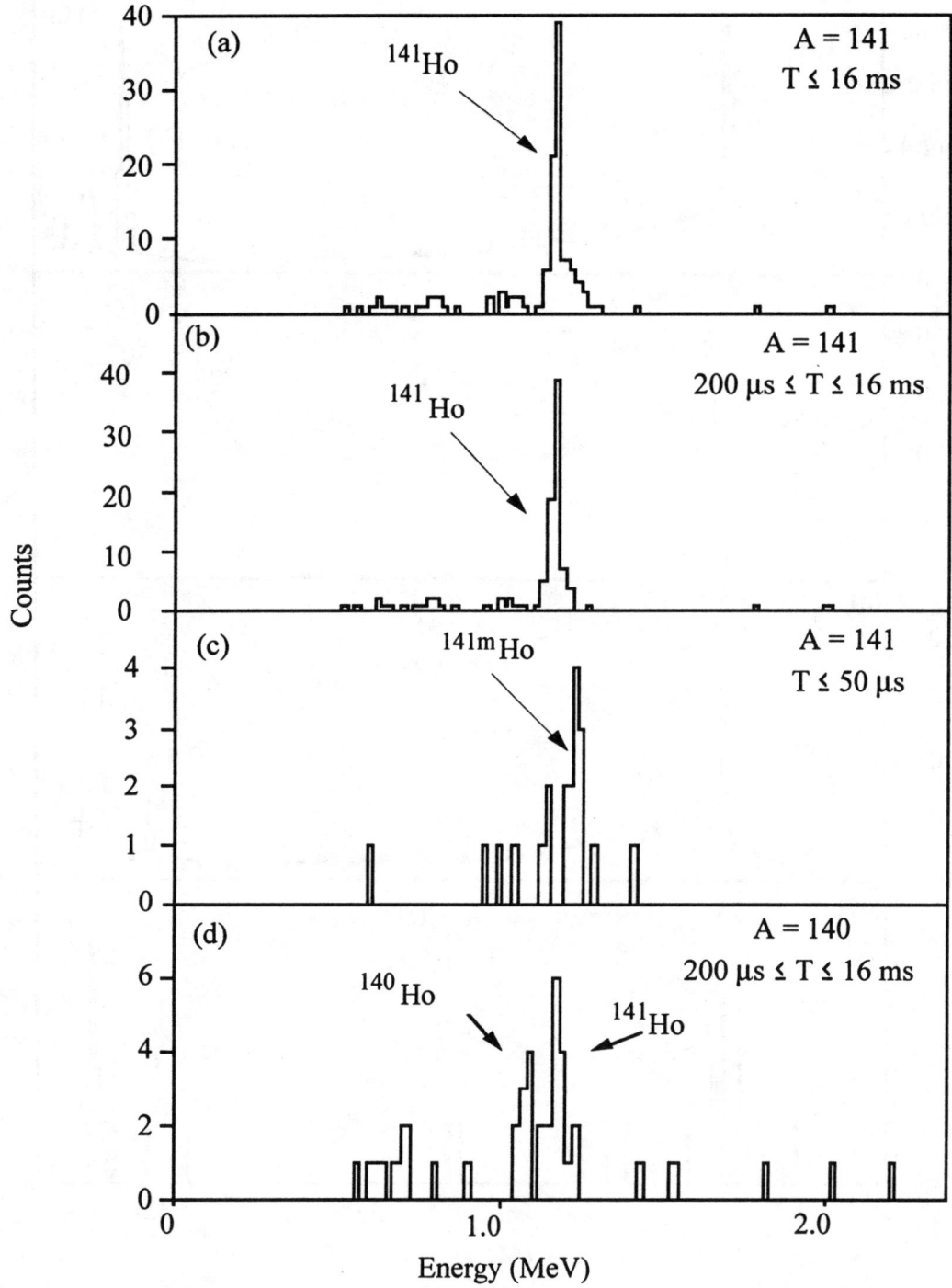

FIGURE 3. (a) Decay spectrum from 315-MeV ^{54}Fe bombardments of ^{92}Mo, with A = 141, and a time between decay and recoil implantation of ≤ 160 ms. The same spectrum is shown with a time constraint of (b) 200 μs ≤ T ≤160 ms, and (c) T ≤ 50 μs. The decay spectrum with A =140 with a time constraint of 200 μs ≤ T ≤ 160 ms is shown in (d).

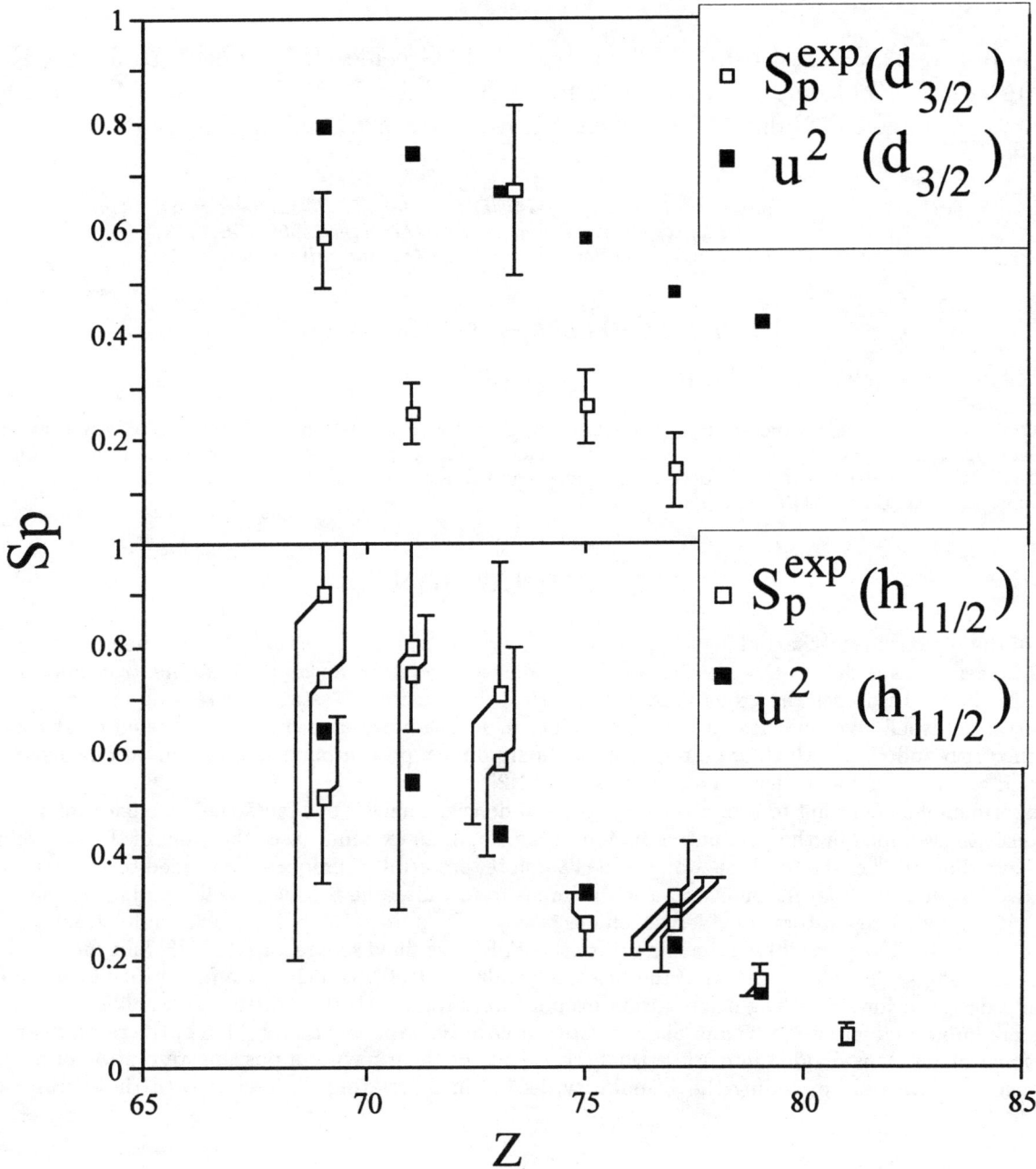

FIGURE 4. Comparison of S_p^{exp} and u^2 for known proton emitters with $69 \leq Z \leq 81$. Part (a) compares the values for protons emitted form the $\pi d_{3/2}$ orbital, while (b) contains the values for the $\pi h_{11/2}$ orbitals. The values for u^2 are taken from reference 2.

Recoil Decay Tagging Studies of Onset of Collectivity Near the Z = 82 Shell Gap

S. Juutinen*, J.F.C. Cocks*, O. Dorvaux*, P.T. Greenlees[†], P. Jones*, R. Julin*, K. Helariutta*, H. Kankaanpää*, H. Kettunen*, P. Kuusiniemi*, M. Leino*, M. Muikku*, R.D. Page[†], P. Rahkila*, A. Savelius*, W.H. Trzaska* and J. Uusitalo*[‡]

* *Department of Physics, University of Jyväskylä, P.O.Box 35, 40351 Jyväskylä, Finland*
† *Department of Physics, University of Liverpool, Liverpool L69 7ZE, U.K.*
‡ *Present address: Argonne National Laboratory, Argonne, Illinois 60439, USA*

and the JUROSPHERE and SARI collaborations

Abstract. The nuclear structure of neutron deficient Z $\approx$ 82 nuclei has been studied in fusion-evaporation reactions using in-beam γ-ray spectroscopic methods. Prompt γ rays were observed using the JUROSPHERE and SARI arrays of Ge detectors. The assignment of γ rays to specific nuclei was based on the recoil-decay tagging (RDT) technique.

INTRODUCTION

One of the regions in the chart of nuclei presently receiving much interest is that close to the neutron deficient Z $\approx$ 82 nuclei. This is due to the presence of shape coexistence in the nuclei near the neutron midshell at N = 104. In the lead isotopes the ground state is spherical, but excited 0^+ states decreasing in energy towards the neutron midshell give evidence for competing shapes. In-beam experiments have revealed yrast rotational bands in ^{186}Pb and ^{188}Pb which are interpreted as being due to proton particle-hole excitations across the Z = 82 shell gap driving the nucleus to a prolate shape [1,2].

In experiments attempting to produce heavy neutron deficient nuclei, one faces the problem that the γ rays of interest are swamped by the dominating fission background. In extreme cases the yield of the desired nuclei may be less than 10^{-6} of the total reaction yield. Therefore, powerful techniques are needed to separate wanted rare γ rays originating from fusion-evaporation residues from those due to the huge fission background.

The use of recoil separators provides an efficient way to suppress the γ-ray background from unwanted reaction channels. The present experiments utilise the JYFL gas-filled separator RITU [3] and the recoil-decay tagging method [4,5] for identification of the reaction products. RITU is a charge and velocity focusing device especially designed for separating heavy fusion evaporation products with high (up to $\approx$ 40 %) transmission. Recoils are implanted into a 80×35 mm Si strip detector covering typically about 70 % of the recoil distribution at the focal plane. It is divided into 16 vertical strips each of them having a position resolution of about 300 μm in vertical direction, providing the granularity needed in correlating the recoils with their subsequent α decays.

THE JUROSPHERE AND SARI COLLABORATIONS

During the years 1997 and 1998, a series of in-beam γ-ray experiments using the JUROSPHERE and SARI Ge detector arrays were carried out at the Department of Physics, University of Jyväskylä (JYFL). In 1997 during a 7 month campaign, some 3500 h of beam time were devoted to 25 experiments using the JUROSPHERE array. The SARI array was used for a period of about 5 months in 1998. During this time about 1840 h

CP481, *Nuclear Structure 98*, edited by C. Baktash

of beam time were allocated to 10 experiments. Most of these experiments were RDT measurements. In all experiments, beams provided by the JYFL ECR ion source were accelerated by the K = 130 MeV cyclotron. Typical beam intensities were 5-15 pnA. Both Ge detector arrays were provided by the French/UK Loan Pool.

The JUROSPHERE array consisted of 10 TESSA-type [6] and 15 EUROGAM Phase I [7] Compton-suppressed Ge detectors. The TESSA detectors were positioned at angles of 78° and 101° and the Eurogam detectors at angles of 134° and 158° with respect to the beam direction. The absolute photo peak efficiency of the array for 1.3 MeV γ rays was $\approx$ 1.5 %. Additionally, one TESSA type Ge detector was used to observe the γ rays at the RITU focal plane.

The SARI array consisted of three or four segmented clover detectors positioned around the target position. The clover detectors were operated without Compton suppression, but were shielded by a thick layer of lead against the radiation from the beam tube and the surroundings. The absolute efficiency of the array was 2-4 % at 1.3 MeV γ-ray energy depending on the number of the detectors and the target to detector distance. The SARI set-up also involved Ge detectors at the RITU focal plane to detect the γ rays from long-lived isomers and the γ rays emitted following the α decays. A close geometry set-up of four TESSA detectors was used in several experiments. The efficiency of this system was 0.8 %. In some of the experiments, a large volume clover detector (from GSI) was used to detect the γ rays at the focal plane. The clover detector is twice as efficient as the set-up of four TESSA detectors.

During the JUROSPHERE and SARI campaigns, excited states have been identified for the first time in many heavy nuclei near the proton drip line, with production cross sections reaching as low as about 100 nb. Examples of the results from the recoil-decay tagging experiments will be presented below. Other highlights from the RDT experiments performed at JYFL include structure studies of following nuclides:

- o ^{226}U from the reaction ^{208}Pb(^{22}Ne,4n) [8]
- o ^{198}Rn from the rection ^{166}Er(^{36}Ar,4n) [9]
- o ^{168,170}Pt from the reaction ^{112}Sn(^{58,60}Ni,2n) [10]
- o ^{171}Ir from the reaction ^{116}Sn(^{58}Ni,p2n) [11]
- o ^{184}Pb from the reaction ^{148}Sm(^{40}Ca,4n) [12]
- o ^{164}Os from the reaction ^{106}Cd(^{60}Ni,2n) [13]
- o ^{254}No from the reaction ^{208}Pb(^{48}Ca,2n) [14]

PROBING THE SHAPE OF ^{176}Hg ALONG THE YRAST LINE

In the neutron-deficient even-mass Hg isotopes the weakly deformed oblate band remains yrast down to ^{190}Hg. In the lighter isotopes, the oblate band is crossed by an intruding deformed band associated with a prolate-deformed energy minimum [15–17]. As expected, the prolate states minimize their energies in ^{182}Hg [18] but they still lie above the ground state [19] which is predicted to evolve from the oblate shape towards a spherical shape [17].

Recently, yrast levels in ^{178}Hg were identified [20] using the recoil-decay tagging (RDT) technique. An increase in the excitation energy of the prolate band was observed, in accordance with the theoretical predictions [17]. In the same experiment three relatively high-energy γ rays were assigned to ^{176}Hg. They were tentatively associated with an E2 cascade de-exciting the lowest 2^+, 4^+ and 6^+ states in ^{176}Hg. However, on the basis of this experimental information the question of a possible appearance of a prolate structure in ^{176}Hg remained unresolved.

In the present work [21] we have carried out an improved in-beam γ-ray spectroscopy study of ^{176}Hg to confirm the tentative assignments of ref. [20] and to probe further its yrast line towards higher spin. Excited states of ^{176}Hg were populated via the ^{144}Sm(^{36}Ar,4n) fusion-evaporation channel at a bombarding energy of 190 MeV. The target consisted of a single 500 μg/cm^2 self-supporting metallic ^{144}Sm foil of 92.4% enrichment. Prompt γ rays from ^{176}Hg were resolved from those arising from the dominant background of fission and other reaction products using the characteristic properties of the α decay of ^{176}Hg (E$_\alpha$ = 6750 keV, t$_{1/2}$ = (21 $\pm$ 3) ms) in a recoil-decay tagging measurement. Altogether, 90000 ^{176}Hg alpha decay events were recorded and the cross section for the reaction used to produce the nucleus ^{176}Hg was estimated to be about 5μb.

The energy spectrum of γ rays in coincidence with recoils is shown in Fig. 1(a). It is dominated by γ rays from ^{176}Pt produced in the (^{36}Ar,2p2n) fusion-evaporation channel. These γ rays are absent in Fig. 1(b), which shows a recoil-gated γ-ray spectrum obtained by correlating with the ^{176}Hg α decay. In this spectrum there are seven strong lines (400.9, 453.2, 500.5, 529.9, 551.0, 613.3 and 756.4 keV) which we firmly assign

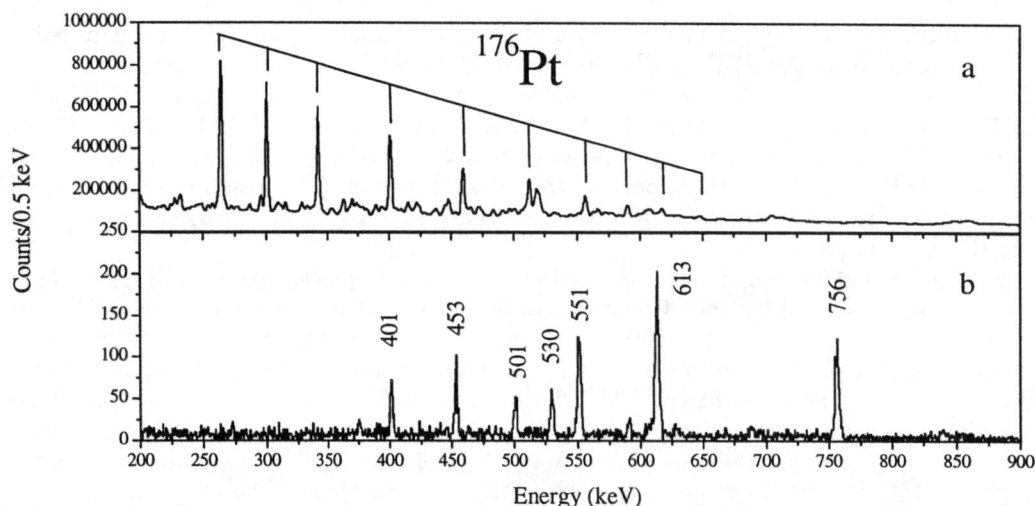

FIGURE 1. a) Energy spectrum of γ rays in coincidence with fusion-evaporation recidues detected in the RITU focal plane detector. b) γ ray spectrum in coincidence with fusion-evaporation recidues and tagged with ^{176}Hg α decays.

to originate from ^{176}Hg. The three strongest lines were seen by Carpenter *et al.* [20]. Despite the relatively low production cross section, a significant amount of recoil-gated α-tagged γ-γ coincidence data were obtained. Analysis of such data revealed that the 453.2, 500.5, 551.0, 613.3 and 756.4 keV γ rays are emitted as a cascade. The 400.9 and 529.9 keV transitions are in coincidence with each other and the three strongest lines in Fig. 1(b).

The intensity ratio of γ rays observed by the Ge detectors at 134° and 158° to those observed by the 79° and 101° Ge detectors was found to be consistent with the stretched E2 character for the 551.0, 613.3 and 756.4 keV transitions. For the weaker 453.2 and 500.5 keV transitions only tentative E2 assignments were possible. These arguments together with γ-ray coincidence and intensity information were used to generate the decay scheme of Fig. 2.

The level-energy systematics of even-mass Hg isotopes down to ^{176}Hg is displayed in Fig. 3. The energies of the first excited 2^+ and 4^+ states in ^{176}Hg lie higher than in any other Hg isotope except the closed-shell

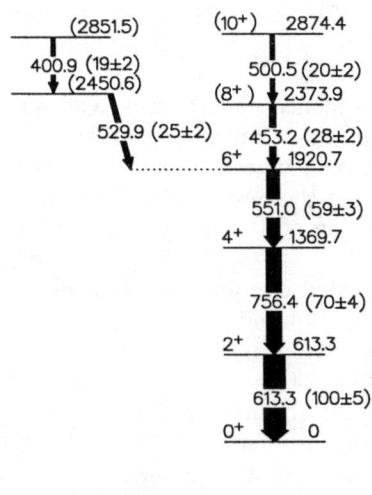

FIGURE 2. Level scheme of ^{176}Hg deduced from the present data. Intensities are given in parentheses.

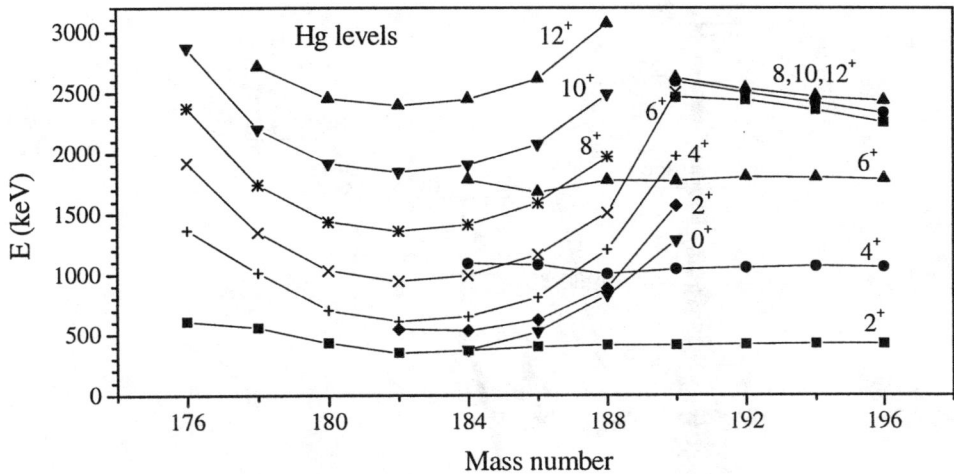

FIGURE 3. Energy level systematics in Hg nuclei.

nucleus ^{206}Hg$_{126}$. In accordance with the theoretical predictions [17], the rise in the 2$^+$ and 4$^+$ level energies suggests a transition towards a spherical ground state as already discussed in ref. [20]. The lowering of the transition energies in the yrast band above the 6$^+$ state is interpreted to be the result of the crossing of the prolate band.

A two-band mixing model similar to that in ref. [22] was used to extract the energy difference between the assumed prolate and weakly-oblate band heads. A value of about 1300 keV was extracted using the VMI parameters and the prolate-oblate interaction strength (about 100 keV) which reproduce the ^{178}Hg level scheme [20]. The extracted value is about 600 keV higher than in ^{178}Hg revealing a rapid increase in the excitation energy of the prolate intruder structure with decreasing neutron number.

ONSET OF DEFORMATION IN Po NUCLEI

The Po nuclei with two protons outside the Z = 82 shell closure form an interesting series of isotopes. Alpha- and beta-decay studies strongly support the view that the drop of level energies in light even-mass Po isotopes is due to the proton 4p-2h intruder configurations [23,24]. However, yrast and near yrast states in ^{194}Po and ^{196}Po have also been described within the vibrator picture [25,26]. Recently, we observed for the first time excited states in ^{192}Po and showed that the level energies in the ground state band continue the decreasing trend seen in the heavier isotopes [27].

In order to obtain a better insight into the structure of the ^{194}Po nucleus, experiments have been carried out using the JUROSHERE and SARI arrays and the ^{171}Yb(^{28}Si,5n) reaction at a bombarding energy of 155 MeV [28]. The target was an about 500 μg/cm^2 thick enriched self-supporting foil. In the reaction used the production cross section of ^{194}Po was estimated to be about 500 μb.

In the experiments, a total of about 8×10^5 ^{194}Po α decays were recorded. Recoil gated γ-γ coincidence data with and without α-decay tagging were analysed. The preliminary level scheme of ^{194}Po is shown in Fig. 4. The multipolarity of the transitions was deduced from the angular distribution ratios, as in ^{176}Hg. For the yrast states, the level scheme by Younes et al [25] is confirmed up to I$^\pi$ = 10$^+$. The 329 keV transition was shown to be of E2 type, thereby extending the yrast band up to I$^\pi$ = 12$^+$. Several non-yrast levels were identified for the first time. Furthermore, the α tagged γ-ray spectrum for the delayed transitions recorded at the focal plane reveals the existence of a long-lived isomer, which feeds the yrast 6$^+$ and 8$^+$ states. Isomeric I$^\pi$ = 11$^-$ and I$^\pi$ = 12$^+$ states have been observed in heavier Po isotopes and assigned to the πh$_{9/2}$i$_{13/2}$ and νi$^2_{13/2}$ configurations, respectively [29,30]. The isomeric state in ^{194}Po may be of similar origin. However, the half-lives in ^{196}Po and ^{198}Po are less than 1 μs, but in ^{194}Po it appears to be much longer.

The systematics of the positive parity states in the even Po nuclei is plotted in Fig. 5. A remarkable drop of energy occurs for low-lying states, particularly for the second 0$^+$ state. This behaviour is similar to that

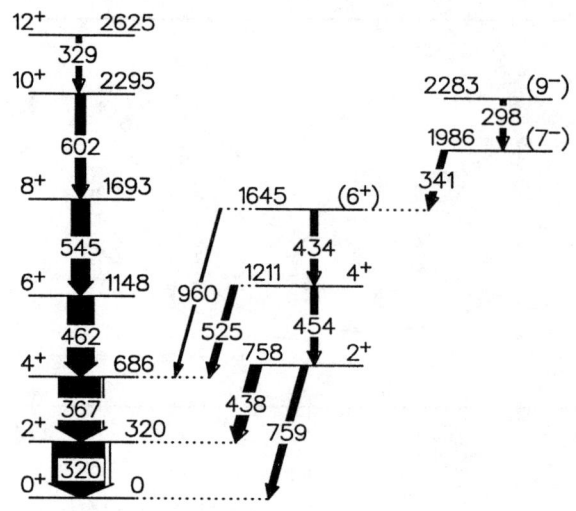

FIGURE 4. Tentative level scheme of ^{194}Po.

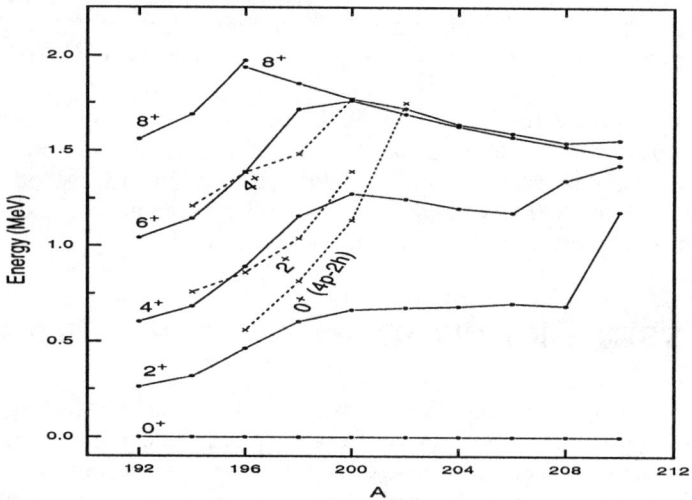

FIGURE 5. Energy level systematics for even-mass Po isotopes.

in the Pb isotopes and suggests shape coexistence. According to the Nilsson-Strutinsky type calculations, the deformed intruder states in Po nuclei with A $\geq$ 190 have an oblate shape with $\beta_2 \approx$ -0.15 [31].

Two-band mixing calculations similar to those for ^{176}Hg and ^{178}Hg were also performed for the Po isotopes, using the excited 0$^+$ state in ^{196}Po [23] to deduce the interaction strength. These calculations indicate that the spherical and intruder 0$^+$ states get strongly mixed and that while the unmixed spherical 0$^+$ state is still below the deformed one, the ordering is reversed in ^{192}Po. This picture is in accordance with that deduced from the α decay hindrance factors [23,24].

New experiments are being planned to find the first excited 0$^+$ state, as well as the long-lived isomeric state in ^{194}Po.

ACKNOWLEDGEMENTS

This work was supported by the Academy of Finland and by the Access to Large Scale Facility program under the Training and Mobility of Researchers program of the European Union. PTG and RDP acknowledge funding by the Engineering and Physical Sciences Research Council of the UK.

REFERENCES

1. Heese J. *et al., Phys. Lett.* **B 302**, 390 (1993)
2. Baxter A.M. *et al., Phys. Rev.* **C 48**, R 2140 (1993)
3. Leino M. *et al., Nucl. Instr. Meth.* **B 99**, 653 (1995)
4. Simon R.S. *et al, Z. Phys.* **A 325**, 197 (1986)
5. Paul E.S. *et al., Phys. Rev.* **C 51**, 78 (1995)
6. Nolan P.J.,Gifford D.W. and Twin P.J., *Nucl. Intr. Meth.* **A 236**, 95 (1985)
7. Nolan P.J., *Nucl. Phys.* **A 520**, 657c (1990)
8. Greenlees P.T. *et al., J. Phys. G: Nucl. Phys.* **24**, L63 (1998)
9. Taylor R.B.E. *et al.*, submitted to Phys. Rev. **C**
10. King S.L. *et al.*, Phys. Lett. **B**, in press
11. Bark R.A. *et al.*, to be published
12. Cocks J.F.C. *et al., Eur. Phys. J.* **A 3**, 17 (1998)
13. Simpson J. *et al.,* to be published
14. Leino M. *et al.,* to be published
15. Wood J.L., Heyde K., Nazarewicz W., Huyse M. and Van Duppen P., *Phys. Rep.* **215**, 101 (1992)
16. Frauendorf S., Pashkevich V.V., *Phys. Lett.* **B 55**, 365 (1975)
17. Nazarewicz W., *Phys. Lett.* **B 305**, 195 (1993)
18. Dracoulis G.D. et al., *Phys. Lett.* **B 208**, 365 (1988)
19. Bindra K.S. *et al., Phys. Rev.* **C 51**, 401 (1995)
20. Carpenter M.P. *et al., Phys. Rev. Lett.,* **78**, 3650 (1997)
21. Muikku M. *et al.*, Phys. Rev. **C**, in press
22. Dracoulis G.D., *Phys. Rev.* **C 49**, 3324 (1994)
23. Bijnens N. *et al., Phys. Rev. Lett.* **75**, 4571 (1995)
24. Allatt R.G. *et al., Phys. Lett.* **B**, in press
25. Younes W. *et al., Phys. Rev.* **C 52**, 1723 (1995)
26. Bernstein L.A. *et al., Phys. Rev.* **C 52**, 621 (1995)
27. Helariutta *et al., Phys. Rev.* **C 54**, R2799 (1996)
28. Helariutta *et al.*, to be published
29. Alber D. *et al., Z. Phys.* **A 339**, 225 (1991)
30. Maj A. *et al., Nucl. Phys.* **A 509**, 413 (1990)
31. May F.R., Pashkevich V.V. and Frauendorf S., *Phys. Lett.* **B 68**, 113 (1977)

The Spectroscopy of ^{183}Tl: an Extreme Case of Prolate-Oblate Shape-Competition

W. Reviol[*], L. L. Riedinger[*], C. R. Bingham[*], W. Weintraub[*], D. Jenkins[†], R. Wadsworth[†],
A. N. Wilson[†], S. Juutinen[‡], K. Helariutta[‡], M. P. Carpenter[§], R. V. F. Janssens[°],
D. Seweryniak[°], J. Uusitalo[°], I. Wiedenhöver[°], C. J. Gross[‖], K. S. Toth[‖], J. C. Batchelder[¶],
J. A. Cizewski[§]

[*]*Department of Physics, University of Tennessee, Knoxville, TN 37996*
[†]*Department of Physics, University of York, Heslington, York YO1 5DD, UK*
[‡]*Department of Physics, University of Jyväskylä, 40351 Jyväskylä, Finland*
[°]*Physics Division, Argonne National Laboratory, Argonne, IL 60439*
[‖]*Physics Division, Oak Ridge National Laboratory, Oak Ridge, TN 37831*
[¶]*UNIRIB, Oak Ridge Associated Universities, Oak Ridge, TN 37831*
[§]*Department of Physics, Rutgers University, New Brunswick, NJ 08903*

Abstract. The yrast sequence in ^{183}Tl has been studied for the first time in recoil-mass and decay tagged γ-ray spectroscopic measurements. A rotational-like cascade of seven transitions is observed down to the bandhead with spin $13/2^+$. In contrast to adjacent nuclei, links from the yrast band to a lower lying weakly deformed (oblate) structure are not observed. It appears that the prolate energy minimum in ^{183}Tl drops significantly compared to ^{185}Tl and minimizes below the neutron $i_{13/2}$ midshell ($N \leq 102$). Possibilities for the decay out of the band in ^{183}Tl are discussed.

INTRODUCTION

The mercury-lead region, particularly at $A \leq 190$, has provided textbook examples for the shape coexistence phenomenon [1]. Shape transitions from weakly deformed ($\beta_2 \sim 0.15$) oblate (Hg nuclei) or spherical states (Pb nuclei) to excited well deformed ($\beta_2 \sim 0.25$) prolate minima have been observed at very low spin ($I > 2\hbar$), see Refs. [1] and [2–4]. Coexistence between these oblate and prolate shapes is also seen in light Tl nuclei ($Z = 81$) [5]. The proton intruder orbitals such as the $i_{13/2}$ states are low-K and downsloping as a function of deformation for Tl nuclei and thus are expected to stabilize the prolate minimum that is analogous to that found in adjacent Hg and Pb nuclei. Prolate bands built on $\pi i_{13/2}$, $\pi h_{9/2}$, and $\pi f_{7/2}$ orbitals have been observed in ^{187}Tl [6,7] and in ^{185}Tl [7]. Whereas the prolate band in even-mass Hg and Pb nuclei minimizes in excitation energy at $N = 103$ [8] or close to this neutron number [4], the prolate bands in Tl nuclei *viz*. the yrast bands based on the $i_{13/2}$ intruder are predicted [7] to drop further in the next lighter isotopes ($N = 102$ and 100). This trend seems to highlight the shape-driving nature of the high-j proton intruders involved in the formation of the prolate minimum in the Hg, Tl, and Pb isotopes. Clearly, the exploration of the yrast structure of the Tl isotopes towards lower neutron numbers is of importance.

The results of a γ-ray spectroscopic study of the next lighter isotope, ^{183}Tl, are discussed in the present paper. This nucleus and other very neutron-deficient nuclei in this mass region are difficult to study, since the fusion-evaporation cross sections are small compared to the fission yield, which typically accounts for $> 90\%$ of the decay of the compound system and introduces a large background. Therefore, to be sensitive for excited states of a particular evaporation residue, in-beam γ-ray spectroscopic methods need to be combined with mass separation of the recoiling nuclei or/and recoil decay tagging [9]. However, the strongest known line in the α decay of ^{183}Tl ($E_\alpha(9/2^- \rightarrow 9/2^-) = 6.38$ MeV) [10] represents a branch of $\sim 1.5\%$ only. Thus, in the present case the RDT method is less suitable than *e.g.* for the neighboring Pb nuclei and the results presented hereafter

CP481, *Nuclear Structure 98*, edited by C. Baktash

are mainly based on the spectroscopy of ^{183}Tl with mass identification and identification of the atomic number by the x-ray yield.

EXPERIMENTAL TECHNIQUES AND RESULTS

In a recent experiment with GAMMASPHERE [11] at the Argonne Fragment Mass Analyzer (FMA) [12], the ^{108}Pd(^{78}Kr,p2n) reaction at 340 MeV was used to study ^{183}Tl. Prompt γ radiation from the target was detected with 93 Ge counters of coaxial and four of planar (LEPS) type, all surrounded by BGO Compton suppressors. The FMA was used to separate the evaporation residues from fission products and primary beam. The mass/charge ratio of the evaporation residues was determined at the FMA focal plane using a position-sensitive parallel-grid avalanche counter (PGAC). These recoils were subsequently implanted in a double-sided Si strip detector (DSSD) located 40 cm downstream, which was used to detect the energies of recoils (E_r) and decay particles (E_α) as well as their positions. The time-of-flight (TOF) of the recoils between GAMMASPHERE and PGAC and between PGAC and DSSD was also measured. A total of $\sim 10^8$ events were recorded under the trigger conditions, recoil-γ^n ($n \geq 1$) or DSSD-decay.

The choice of a nearly symmetric reaction, which is in favor of a high FMA efficiency, causes a rather large amount of scattered beam at the FMA focal plane. In order to reduce the contamination from the scattered beam particles in the mass spectrum, a two-dimensional gate on the E_r versus TOF(PGAC-DSSD) matrix is required for the data analysis. The resulting mass spectrum obtained for the PGAC x-position is shown in Fig. 1 (top). Peaks corresponding to $A = 180$ to 184 recoils are observed for three charge states ($Q = 30, 31, 32$). The bottom part of Fig. 1 shows the γ-ray spectrum enhanced in $A = 183$ by appropriate gating on the x- and y-position of recoils at the focal plane and on the TOF between GAMMASPHERE and the PGAC. Most of the strong γ-ray lines in this mass 183 gate can be identified as known transitions in ^{183}Au [13] and ^{183}Hg [14]. The remaining strong peaks in this mass-gated spectrum, labeled by their energies (in keV), are assigned to ^{183}Tl. This assignment is based on (i) the absence of concidence relationships between these new γ rays and the known transitions in ^{183}Au or ^{183}Hg, and (ii) coincident Tl x-rays, when gating on the 160-, 260-, 355-keV lines. It is further supported by statistical model calculations, which for the given reaction predict that the p2n channel leading to ^{183}Tl is stronger than the competing fusion-evaporation channels.

Fig. 2 shows γ-ray coincidence spectra attributed to ^{183}Tl, obtained from the mass-gated E_γ-E_γ matrix (unpacked γ-γ-γ and higher fold events). All peaks labeled are in concidence with each other. The top spectrum shows the sum of gates placed on the 160-keV to 514-keV transitions. At the bottom, the spectrum gated by the 160-keV γ ray is shown for comparison. Striking features of these spectra are (i) the regular energy spacings of the peaks, (ii) the absence of additional peaks that would disturb this spacing, and (iii) an intensity pattern that allows us to arrange these γ rays in a cascade with increasing E_γ from the bottom to the top of the sequence (see also next section). Obviously, the spectra of Fig. 2 display the yrast sequence in ^{183}Tl, which exhibits a rotational behavior.

Spectroscopy of the ^{182,183}Tl nuclei has been also a byproduct of a study of ^{182}Pb at RITU (Jyväskylä) [15], where a similar experimental setup and a ^{42}Ca induced reaction was used. At RITU, a 6.40 MeV α line (^{183m}Tl, $t_{1/2} = 35$ ms) was found [15], thought to be unobserved in the α spectra of Refs. [10,16] and tentatively attributed to ^{183}Tl, however, with some ambiguity on the mass assignment. The present data confirm this α line and allow for a firm assignment of it to ^{183}Tl [15]. The γ-ray spectrum correlated with this α line obtained at RITU [15] and a preliminary RDT analysis of the present data is consistent with the proposed assignment of the γ rays in Fig. 2 to ^{183}Tl. Furthermore, there is some indication for a weak 421 keV γ ray in both the RDT and mass-gated spectra (not seen in Fig. 2), being the best candidate for the expected $11/2^- \rightarrow 9/2^-$ transition ^{183}Tl which is seen in the heavier odd-mass isotopes and associated there with an oblate shape. However, the association of this γ ray with ^{183}Tl should be viewed as a tentative assignment.

DISCUSSION

While the 260-, 355-, 439-, 514-, 581-keV sequence in ^{183}Tl resembles the well-deformed yrast bands in the heavier Tl isotopes and neighboring Hg and Pb isotopes, the bottom part of this band and its decay-out are significantly different from those cases in that several strong transitions other than the 160-keV line are absent in the spectra. This difference is probably best seen in the spectrum gated by the 160-keV line (Fig. 2, bottom). If the 160-keV line in ^{183}Tl were the analog to the 207-keV $17/2^+ \rightarrow 13/2^+$ "decay-out" transition in ^{185}Tl (see

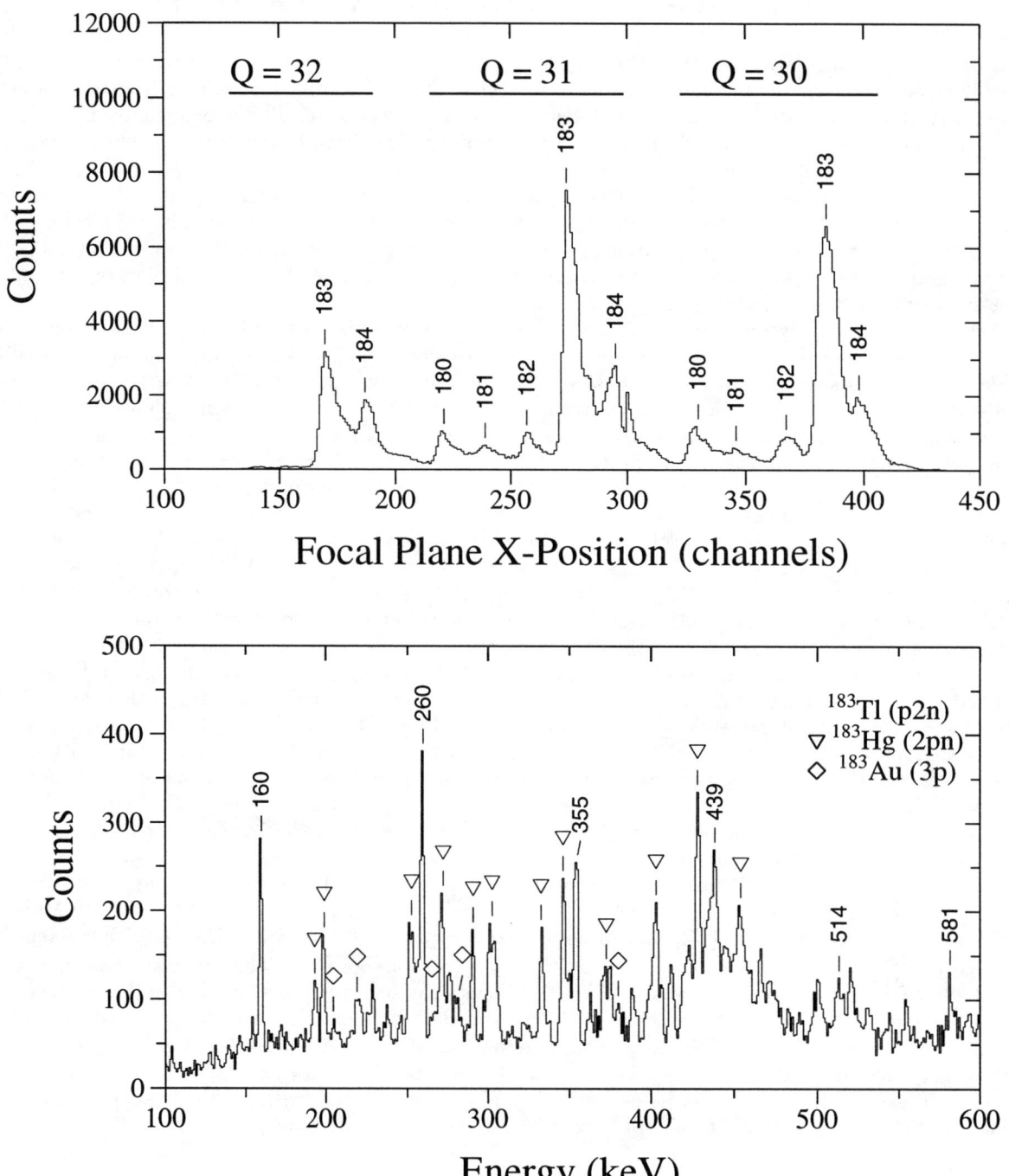

FIGURE 1. Top: A/Q mass spectrum for ^{78}Kr on ^{108}Pd obtained at the horizontal position of the FMA focal plane. Peaks are labeled by the mass number, A, of the recoiling ions and fall into three groups corresponding to charge states $Q = 30$, 31, and 32. Bottom: Spectrum of prompt γ rays from the target in coincidence with $A = 183$ events in the focal plane x-y position spectrum. Both spectra are generated under appropriate time-of-flight conditions (see text).

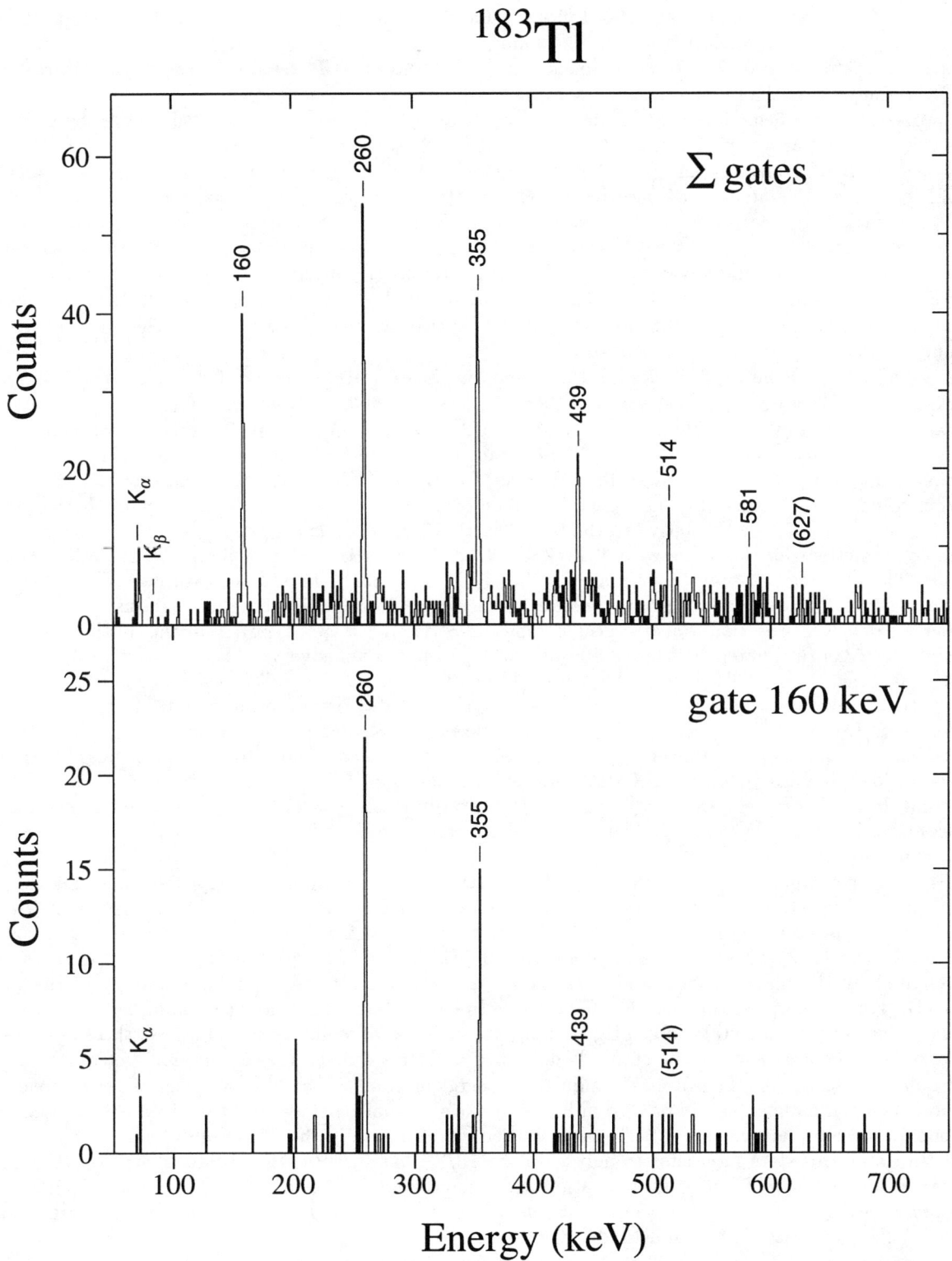

FIGURE 2. Sum of projections (top) from the mass 183 selected E_γ-E_γ coincidence data with gates on the 160-, 260-, 355-, 439-, 514-keV lines and an "individual" gate (bottom) for comparison. Members of the rotational band (see text) are labeled by their energies in keV, x-rays of thallium are identified as well.

Ref. [7], Fig. 8), one would expect to see in the gated γ-ray spectrum additional significant peaks at 400 - 450 keV ($11/2^- \rightarrow 9/2^-$) and probably at 100 - 150 keV. This is evidently not the case in the yrast spectrum of ^{183}Tl, that is, there is no indication for strong linking transitions from the yrast sequence to the known [10] $9/2^-$ isomeric state (oblate). By the same token, the yrast band in ^{183}Tl must be lower lying with respect to the $9/2^-$ isomeric state than in the heavier isotopes.

The proposed level scheme for ^{183}Tl is shown in Fig. 3. Spins and parities of the levels above the $9/2^-$ state are assigned by comparison with the yrast sequences in neighboring odd-mass Tl nuclei. The 160-keV γ ray is identified as an $E2$ transition and placed at the bottom of the band based on intensity considerations. The inset of Fig. 3 shows the intensity after correction for detection efficiency and internal conversion as function of initial spin for the $\pi i_{13/2}$ bands in ^{185,187}Tl (labeled by the Nilsson quantum numbers [660]) [7] and the present data. A common normalization is used to ease the comparison of these intensity patterns. An $E2$ assignment for the strongly converted 160-keV γ ray in ^{183}Tl ensures intensity conservation for the whole cascade, while $M1$ or $E1$ assignments are ruled out for the same reason. It appears that the $\pi i_{13/2}$ band in ^{183}Tl is populated all the way down to the $13/2^+$ bandhead due to its low excitation energy, even though the final decay out of this band is not clear yet.

How can the $\pi i_{13/2}$ band in ^{183}Tl decay? Depending on the excitation energy of the bandhead with respect to the $9/2^-$ isomeric state, E_{rel}, there are two possible decay-out scenarios which define at the same time the upper and lower limit of E_{rel}. *Upper limit: decay by M2 transition.* Accepting the presence of a $11/2^-$ level 421 keV above the $9/2^-$ state, a $13/2^+ \rightarrow 9/2^-$ $M2$ transition would be competitive with a $13/2^+ \rightarrow 11/2^-$ $E1$ transition if $E_\gamma(E1) \leq 3$ keV according to a Weisskopf estimate ($E_{rel} \leq 424$ keV). In this scenario, the $E1$ transition could not be observed. *Lower limit: α-decay to ^{179}Au.* By knowing the location of the $13/2^+$ level in the nucleus ^{179}Au (390 keV relative to the $9/2^-$ level) [17] and the energy $E_\alpha(9/2^- \rightarrow 9/2^-) = 6.38$ MeV, reasonable predictions for the α energy of a $13/2^+ \rightarrow 13/2^+$ decay or the relative energy of the initial state can be made. The previous upper-limit estimate for the $13/2^+$ state in ^{183}Tl and the estimate ($E_{rel} = 410$ keV) obtained for an 6.40 MeV α decay from this state to the $13/2^+$ level in the daughter nucleus ^{179}Au would be compatible. However, one would also expect competing γ decay (of $M2$ type) from this initial state to the $9/2^-$ state in ^{183}Tl (estimated half-life < 100 ns), which is not conclusive yet. For a lower-limit estimate of E_{rel}, it is assumed that a 0.3% α-decay branch could be observed and the following consideration is made. A decay to $13/2^+$ level in the daughter nucleus ^{179}Au is calculated to be competitive with a $13/2^+ \rightarrow 9/2^-$ $M2$ transition if $E_\gamma(M2) \leq 95$ keV ($E_{rel} \geq 95$ keV). The energy of the corresponding α line is $E_\alpha \leq 6.10$ MeV. So far, a search for a new α-line in this energy range has lead to a negative result. However, it is possible that this line is embedded in a strong $E_\alpha = 6.13$ MeV line from the decay of ^{180}Hg, a contaminant reaction product in the present data. To proceed, an average value of the two limits, $E_{rel} = 256$ keV, is assumed in the following discussion. Notice that a 256 keV $M2$ transition with an estimated half-life of 800 ns could not be observed in prompt γ-ray spectroscopy.

Fig. 4 compares experimentally obtained prolate-oblate and oblate-oblate energy differences for the near-yrast coexisting level structures in Tl isotopes. These data are given relative to the $9/2^-$ isomeric state (labeled [505]), where the prolate bandheads for $^{185-189}$Tl have been derived from extrapolations of the levels observed at and above the $17/2^+$ using a variable-moment-of-inertia fit [7]. The new information obtained for ^{183}Tl is represented by the open symbols and bars (upper and lower limit for the relative energy of the prolate bandhead). For comparison, the energy differences between prolate bandheads (extrapolated 0_2^+ states) and the oblate deformed ground state in the Hg isotones are included as well. As stated above, these oblate states in Hg and Tl nuclei are viewed to be similarly deformed and thus can serve as a common "reference".

The oblate states in the Tl isotopes minimize in energy around $N = 108$ and then slightly move up as function of neutron number, as can be seen in Fig. 4 by the systematic behavior of the $11/2^- \rightarrow 9/2^-$ transitions (diamonds). For completeness, the $13/2_1^+$ state (labeled [606]), found in the Tl isotopes for $A \geq 185$, is plotted as well. In ^{185}Tl, this state interacts strongly with the $13/2^+$ bandhead of the prolate structure (leading to a ~ 30 keV displacement). There is no indication yet for another $13/2^+$ state in ^{183}Tl nor for a perturbation of the proposed $13/2^+$ bandhead of the prolate structure (judging *e.g.* from a systematic comparison of the aligned angular momenta in the bands under discussion).

The prolate-oblate differences in Tl nuclei as a function of neutron number follow approximately the trend seen in the Hg isotones for $N \geq 104$ (dashed lines). However, near $N = 103$ (neutron $i_{13/2}$ midshell) the prolate band in Hg nuclei reaches its minimum energy, while by the upper-limit estimate for ^{183}Tl the band in Tl nuclei is further downsloping with decreasing N. Interestingly, the ^{179,181,183}Au nuclei [17,13] show a trend in prolate-prolate ([660]-[532]) differences (minimum at $N \leq 100$) that is perhaps comparable with E_{rel} versus N for the [660] bands in the Tl isotopes. On the other hand, for the even-mass Pb isotopes the trend

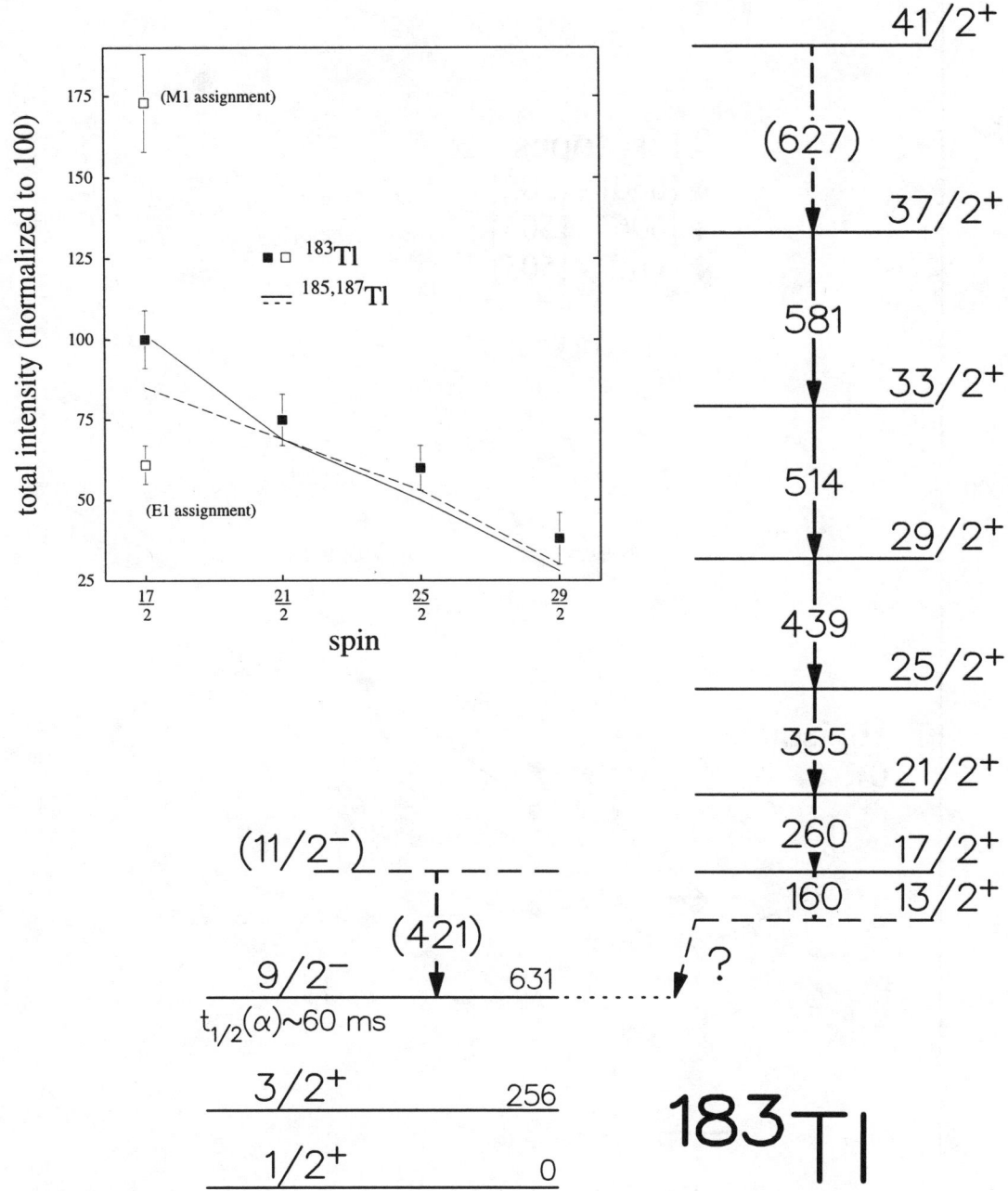

FIGURE 3. Proposed level scheme for ^{183}Tl from mass-gated γ ray and α spectroscopy. Transitions placed with less confidence are given in parenthesis. The low-lying $3/2^+$ and $9/2^-$ levels and the quoted half-life are adopted from independent α-decay work (Refs. [10,16]). The assignment of spins and positive parity to the yrast band (this work) is based on analogies with structures in neighboring Tl isotopes and the measured intensity pattern, as shown in the inset of this figure. Symbols for the inset: ^{183}Tl squares, ^{185}Tl full line, and ^{187}Tl dashed line.

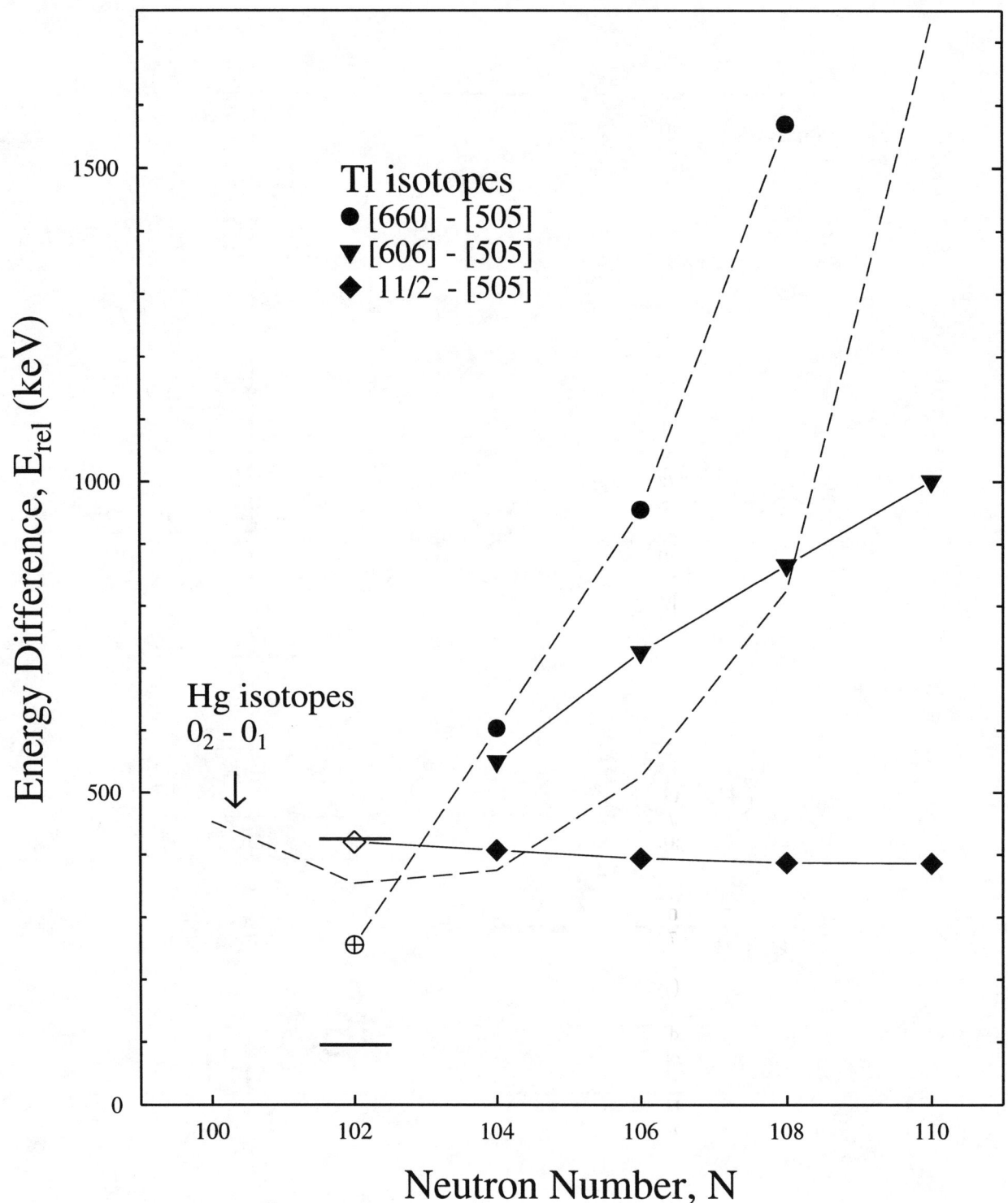

FIGURE 4. Energies of coexisting level structures in Tl isotopes relative to the corresponding [505]9/2⁻ isomeric states versus neutron number as obtained from experiment. The data point at $N = 102$ (^{183}Tl) is an average of upper and lower limit estimates (see text). For comparison, the trend of the prolate-oblate energy difference in the Hg isotopes is indicated as well.

of prolate-spherical $(0_3^+$-$0_1^+)$ energy differences found near $N = 103$ [4] is similar to that for the Hg isotopes shown in Fig. 4. These findings indicate that the odd proton has considerable impact on the formation of the prolate minimum and gives rise to speculations such as polarization of the quadrupole core when coupling a deformation driving $i_{13/2}$ proton to it. Recent calculations [7] predict that the prolate states in odd-mass Tl nuclei continue to drop in energy past midshell, in agreement with the current results. However, as stated in Ref. [7], further theoretical investigations are necessary to better understand this behavior of the prolate states.

CONCLUSIONS

The yrast sequence in ^{183}Tl has been observed for the first time in recoil-mass and decay tagged γ-ray spectroscopic measurements, recently performed with GAMMASPHERE at the FMA and JUROSPHERE at RITU. While the level spacings of this new sequence resemble the well-deformed (prolate) excited bands in adjacent nuclei of Hg, Tl, and Pb, its decay-out properties are different from those cases in two respects. (*i*) The rotational-like sequence is observed from medium spin down to the $I^\pi = 13/2^+$ bandhead. (*ii*) A strong γ-decay branch from the prolate band to a slightly-oblate structure, like in heavier Tl nuclei, is not observed. These features indicate that the prolate energy minimum in ^{183}Tl has dropped significantly compared to ^{185}Tl and minimizes below the neutron $i_{13/2}$ midshell, as predicted by theory. The low excitation energy of the band in ^{183}Tl possibly gives rise to a rare mode of α-decay, from the $13/2^+$ bandhead to a deformed $13/2^+$ state in the daughter nucleus ^{179}Au, presently under investigation. These findings, made possible by recent instrumental developments, represent an extreme case of prolate-oblate shape competition.

This work was supported by the US Department of Energy under contracts nos. DE-FG05-87ER40361 (U. Tenn.), W-31-109-ENG-38 (ANL), DE-AC05-96OR22464 (ORNL), and in part by the US National Science Foundation.

REFERENCES

1. Wood J. L., *et al.., Phys. Rep.* **215**, 211 (1992).
2. Heese J., *et al., Phys. Lett.* **B 302**, 390 (1993).
3. Baxter A. M., *et al., Phys. Rev.* **C 48**, R 2140 (1993).
4. Cocks J. F. C., *et al., Europ. Phys. J.* **A 3**, 17 (1998).
5. Kreiner, A. J., *et al., Phys. Rev.* **C 38**, 2674 (1988).
6. Reviol W., et al., *Phys. Rev.* **C 49**, R 587 (1994).
7. Lane G. J., et al., *Nucl. Phys.* **A 586**, 316 (1995).
8. Dracoulis G., et al., *Phys. Lett.* **B 208**, 365 (1988).
9. Paul E. S., *et al., Phys. Rev.* **C 51**, 78 (1995).
10. Batchelder, J. C., *et al., Europ. Phys. J.* **A**, submittted (1998).
11. Lee I. Y., *Nucl. Phys.* **A 520**, 641c (1990).
12. Davids, C. N., *et al., Nucl. Instrum. Methods Phys. Res.* **B 70**, 358 (1992).
13. Mueller, W. F. *et al.*, to be published (1998).
14. *Table of Isotopes*, Eighth Edition, Vol. II.
15. Jenkins, D., to be published.
16. Schrewe U. J. et al., *Phys. Lett.* **B 91**, 46 (1980).
17. Mueller, W. F., et al., *Univ. of Tennessee Progress Report*, March 1, 1997.

A GLIMPSE OF NEUTRON-RICH NUCLEI

IN-BEAM GAMMA SPECTROSCOPY OF VERY NEUTRON-RICH NUCLEI AT GANIL

F. Azaiez, M. Belleguic, O. Sorlin, S. Leenhardt, C. Bourgeois, C. Donzaud, J. Duprat,
D. Guillemaud-Mueller, A.C.Mueller, F. Pougheon.
IPN Orsay (France)

M.G.Saint-Laurent, M. J. Lopez, N. L. Achouri, J. M. Daugas, M. Lewitowicz, W. Mittig,
F. De Oliveira, P. Roussel-Chomaz, H. Savajols.
GANIL Caen (France)

Yu-E. Penionzhkevich, Yu. Sobolev.
FLNR-KINR Dubna (Russia)

C. Borcea, M. Stanoiu.
IAP Bucarest (Roumania)

J. C. Angelique, S. Grevy, N. Orr.
LPC Caen (France)

I. Deloncle, J. Kiener, M. G. Porquet.
CSNSM Orsay (France)

F. Marie, A. Gillibert.
DAPNIA-CEA (France)
J. E. Sauvestre.
DAM-CEA (France)

Z. Dlouhy.
RPI-REZ (Czech Republic)

W. Shuying.
GSI Darmstadt (Germany)

Abstract. *The structure of nuclei far from stability can be investigated by means of in-beam gamma spectroscopy using both stable and radioactive beams. A γ-array consisting of BaF2 and Ge detectors have been used at GANIL in order to perform two of such experiments :*
i) Secondary beams of neutron-rich Ge, Zn and Ni isotopes around N=40 were produced from the fragmentation of a primary ^{86}Kr beam, and analyzed by means of the LISE3 spectrometer. For the transitions between the 0^+ ground state and the first 2^+ state, B(E2) values have been extracted from Coulomb excitation cross-section measurement..Some of the obtained results related to the effectiveness of the N=40 spherical sub-shell closure as well as future improvements of the quality of the data are discussed.
ii) In-beam γ-spectroscopy of neutron-rich nuclei around ^{32}Mg produced by fragmentation of ^{36}S have been recently performed. Gamma decay of relatively higher excited states have been measured in a large number of exotic light nuclei. Preliminary results obtained in a number of even-even nuclei around N=20, such as ^{20,22}O, ^{26,28}Ne and ^{30,32}Mg are presented.

INTRODUCTION

One of the most challenging goals of experiments with radioactive beams is to determine how the structure of atomic nuclei changes near the drip lines as the binding energies of single particles approach zero. Because of the relatively low intensity of the existing radioactive beams, the structure of very exotic nuclei has been investigated

CP481, *Nuclear Structure 98*, edited by C. Baktash
© 1999 American Institute of Physics 1-56396-858-4/99/$15.00

mainly through beta-decay and isomer-decay studies. Only recently, elastic and inelastic secondary reactions, induced by radioactive beams on stable targets, has been used in order to extract nuclear structure information. Among them, inelastic scattering in inverse kinematics and Coulomb excitation are known to provide valuable nuclear structure information such as the energy and the collectivity of the first excited states of nuclei. Inelastic proton scattering to the first excited 2^+ state of the doubly magic ^{56}Ni nucleus is a nice example of such experiment with radioactive secondary beams [1]. On the other hand, Coulomb excitation has been applied for a long time to studies of low lying states of stable nuclei. The incident energy is usually set below the Coulomb barrier in order to avoid any contribution from the nuclear interaction. At higher incident energy (a few tens of MeV/A) and for scattering angles below the grazing angle, the nuclear contribution has been found to be negligible with respect to a dominant Coulomb interaction. The large cross section of the Coulomb excitation process at intermediate incident energies is a key factor to overcome the difficulties imposed by the weak intensity of presently available radioactive beams. This has been recently demonstrated in Coulomb excitation experiments of unstable beams of ^{11}Be [2,3], ^{32}Mg [4] and neutron rich Ar and S isotopes [5,6]. Coulomb excitation of secondary radioactive beams as well as a novel method based on in-beam γ-spectroscopy of exotic nuclei produced by projectile fragmentation have been recently used at GANIL. The description of the two experiments together with the obtained results on the spectroscopy of neutron rich nuclei around N=40 and N=20 will be presented.

NEUTRON-RICH NUCLEI AROUND N=40

The neutron-rich side of stability is expected to exhibit new magic numbers. Near the edge of the valley of stability, the surface of the nuclei would be essentially composed of a diffuse neutron-matter. This diffuseness should already be felt before the drip-line, for nuclei with large N/Z ratios. Calculations in the framework of Hartree Fock Bogoliubov (HFB) and Relativistic Mean Field (RMF) theories suggest that for such neutron-rich nuclei, a better description would be obtained with a rounded potential that can be schematically simulated by the harmonic-oscillator potential [7]. The increase of the N=40 gap arises naturally from this approach. As a consequence the well-pronounced shell-gap at N=50 should be reduced. The ^{68}Ni$_{40}$ nucleus exhibits a high 2^+ energy of 2.03 MeV [8] in contrast to its neighboring isotopes ^{66}Ni [8] and ^{70}Ni [9] which have lower 2^+ energies of 1425 keV and 1259 keV, respectively. The sudden increase of the 2^+ energy at N=40 (see figure1) suggests that ^{68}Ni is spherical and can be considered as a good core to modelize more neutron-rich Ni isotopes up to ^{78}Ni [10]. The unknown reduced transition probability B(E2: $0^+ \rightarrow 2^+$) of ^{68}Ni should then be measured in order to confirm this shell effect. In this respect, it would be interesting to compare this B(E2) value to that of the doubly magic ^{56}Ni, which was found to be 600 (120) e^2fm^4 [1]. For heavier N=40 isotones, proton excitations in the fp-shell increase the collectivity to B(E2)=1600(140) e^2fm^4 for ^{70}Zn [11] and 2130(60) e^2fm^4 for ^{72}Ge[11] (see figure2).

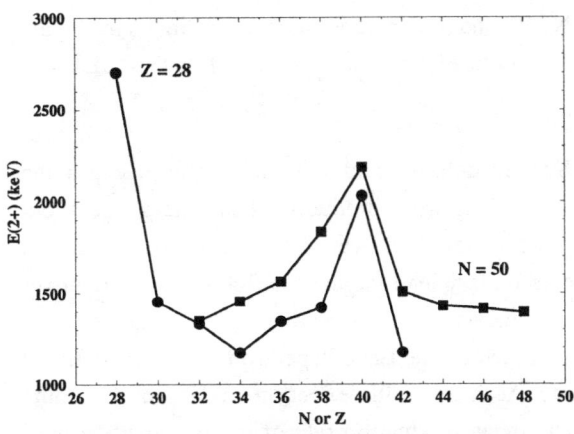

Figure 1 : *Systematic of the excitation energies of the first 2^+ states in Z=28 isotopes and N=50 isotones*

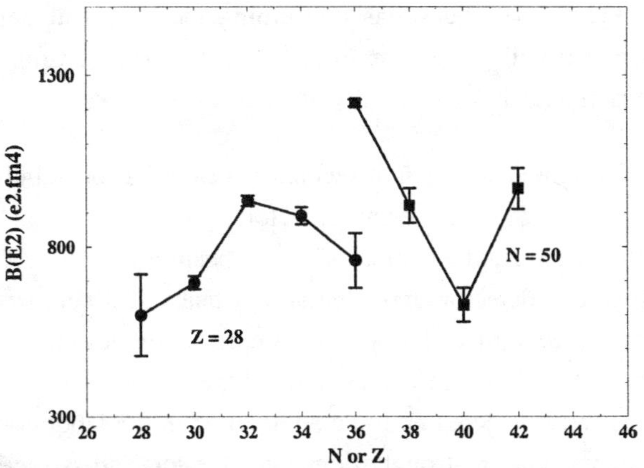

Figure 2 : *Systematic of the B(E2) values of the 2^+ to 0^+ transitions in Z=28 isotopes and N=50 isotones*

Another feature of the N=40 nuclei is that they exhibit a 0^+_2 state at relatively low excitation energy. For ^{68}Ni and ^{72}Ge, this state is below the 2^+ state, and is therefore an E0 isomer since it cannot decay to the 0^+_1 g.s. by γ-emission. The structure of the 0^+_2 isomer in ^{68}Ni is not yet understood and needs further experimental investigations. It could be viewed, on one hand, as a two quasi-particle excitation of the core [8]. On the other hand, the self-consistent HFB calculations using Gogny [12] and Skyrme [13] effective interactions interpret it as a shape isomer with a large quadrupole deformation of $\beta_2 \approx 0.4$ [12]. The examination of a Nilsson diagram, where a shell gap is clearly visible at such deformation parameter, also supports this interpretation. Whether the predicted shape-isomer corresponds to the known 0^+_2 level or to a more excited 0^+_3 state is still debated. In addition to the 2^+ level, information about excited states of ^{68}Ni have been obtained by Broda et al. [8] from multi-nucleon transfer reactions. In this study, a 5^- isomer at 2.847 MeV was discovered and interpreted as due to a neutron particle-hole ($g_{9/2} \otimes p_{1/2}^{-1}$) excitation. Since this isomer has a long half life of $T_{1/2}$=0.86ms, it can be used as a secondary beam in order to induce coulomb excitation to study its configuration. This holds true for the 0^+_2 isomer with $T_{1/2}$=220ns [14], even if a fraction of it can be lost during the flight time in the spectrometer. Recently, an experimental program has started at GANIL in order to

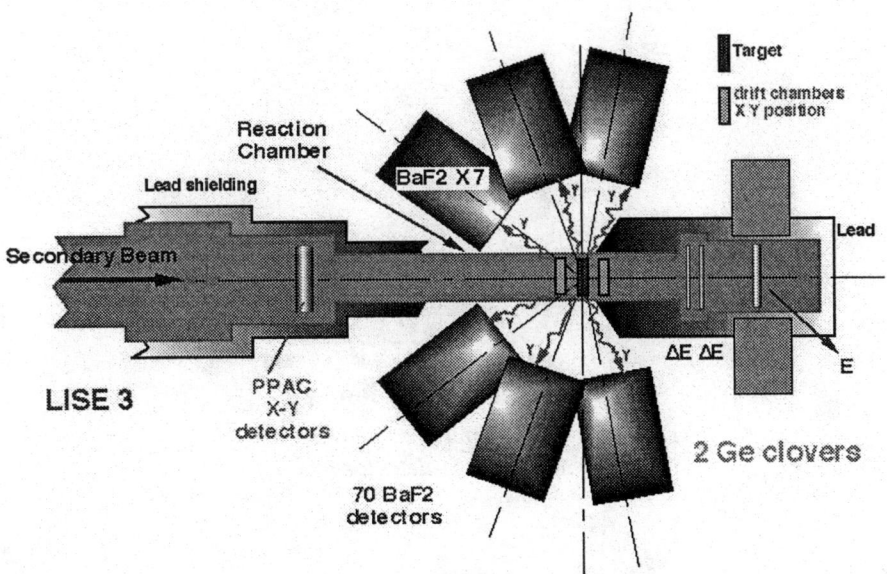

Figure 3 : *Schematic view of the experiment*

address the opened questions concerning the spherical gap at N=40 and the nature of isomers in ^{68}Ni. In the following, we will present preliminary results from a Coulomb excitation experiment, performed at GANIL, with neutron-rich nuclei in the vicinity of ^{68}Ni .

The Coulomb excitation of secondary beams has been induced by the Coulomb field of a thick target placed in the center of a large gamma-array detector. The nuclei ^{76}Ge, ^{70,72}Zn and ^{68}Ni were produced at an energy of about 50MeV/u by the fragmentation of a ^{86}Kr beam at 65 MeV/A. They were identified event-by-event in two large-area (25cm^2) silicon detectors mounted at a distance of 50 cm from the secondary lead target (see figure. 3). Two clover Ge-detectors were placed around the implantation detector in order to measure the γ's originating from the decay of isomers produced in the fragmentation of the primary beam and transmitted by the LISE3 spectrometer. The scattered fragments were detected up to an angle of 3° in the laboratory frame. At these small deflection angles, the Coulomb inelastic contribution strongly dominates the total cross section. The mean production rate of the fragments was of about 100 particles/second for ^{76}Ge, ^{72}Zn and 20 particles/second for ^{68}Ni and ^{70}Zn. The lead target of 220 mg.cm^{-2} thickness was surrounded by the γ-array of 70 BaF$_2$ detectors of 'Chateau de Cristal', mounted in the 4π geometry of the first-generation TAPS-detector (figure. 3). The diameter of each crystal is about 9 cm with a length of 14 cm. The γ-rays of interest, subsequent to the coulomb excitation, are emitted in flight with a velocity of v/c=0.3. Consequently, the Doppler effect induces a broadening of the γ-lines. By placing the BaF$_2$ detectors at a distance of 35 cm from the lead target the opening angle viewed from the target is reduced, keeping the broadening to a reasonable value of 15% at 1 MeV. A time-spectrum between scattered ^{76}Ge fragments detected in the large-area silicon detector and a γ-ray in the BaF$_2$ array is shown in figure. 4. The narrow peak of 3ns width in this time-spectrum is due to prompt γ's following the coulomb excitation of the projectile. This width is a convolution between the time-resolution of the silicon detector and all 70 BaF2 crystals. The flat component of figure. 4 originates from random coincidences with γ-background from the experimental room and from the BaF$_2$ crystals (these crystals have an internal α radioactivity). In the spectra of figure 5, background subtraction and Doppler shift corrections as a function of the emission angle of the γ-rays have been applied.

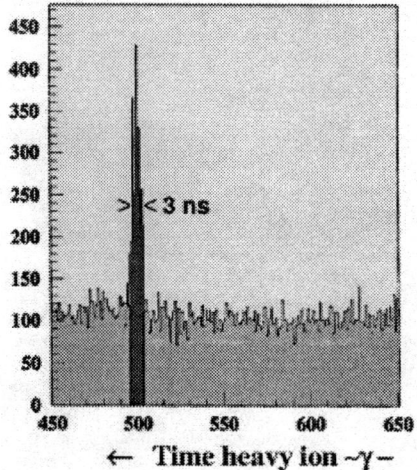

Figure 4: *Time difference between an incoming ion in the Si-detector and a γ in the BaF$_2$ array.*

The ^{76}Ge nucleus is a well-known case of deformed nucleus, exhibiting a low 2^+ state (563 keV) and a strong excitation probability, B(E2)↑ = 2680 e^{2}fm^4. It has been used as a calibration measurement in order to determine B(E2) values of ^{70,72}Zn and ^{68}Ni. Even if the spectrometer was not optimized for ^{70}Zn, an excitation probability B(E2) = 1500(400) e^{2}fm^4 has been determined for the 2^+ state at 885 keV. The agreement between the present value and the one measured at low energy [11], where the coulomb excitation is the dominant fraction of the 2^+ excitation, is a clear indication of the validity of other extracted B(E2) values from our data. The experimental results [15] are summarized in figure. 6 which includes also values for neighboring isotopes. The B(E2) values of the Zn isotopic chain indicate that the collectivity is increasing at N=40, the maximum of collectivity occurring at N=42. Using the prescription of Raman et al. [16] for deformed nuclei, a deformation parameter of β_2=0.23 is found. But it seems that the collectivity for N=40 isotones is more of a vibrational type. This is reflected by the low-energy level schemes of ^{70}Zn, ^{72}Ge, and ^{74}Se which feature E(4^+)/E(2^+) ratios very close to 2.0. This shows that protons are quickly washing out the N=40 effect due to the large number of 2qp-excitations in the fp shell above Z=28. For the coulomb excitation of ^{68}Ni, very few counts are found around the expected energy E(2^+) = 2.03 MeV. Therefore, only an upper limit of B(E2) = 800 e^{2}fm^4 can be derived. This limit does not fully establish the sphericity of ^{68}Ni. The only structure which could be visible in the spectrum of ^{68}Ni is the one around 490keV. The nature of this structure is not yet clear. One explanation could be that the 490 keV γ-line is originating from the Coulomb excitation of either the 5^- or the 0_2^+ isomer states in ^{68}Ni. These isomers were populated with a fairly high rate of the total ^{68}Ni intensity [9].

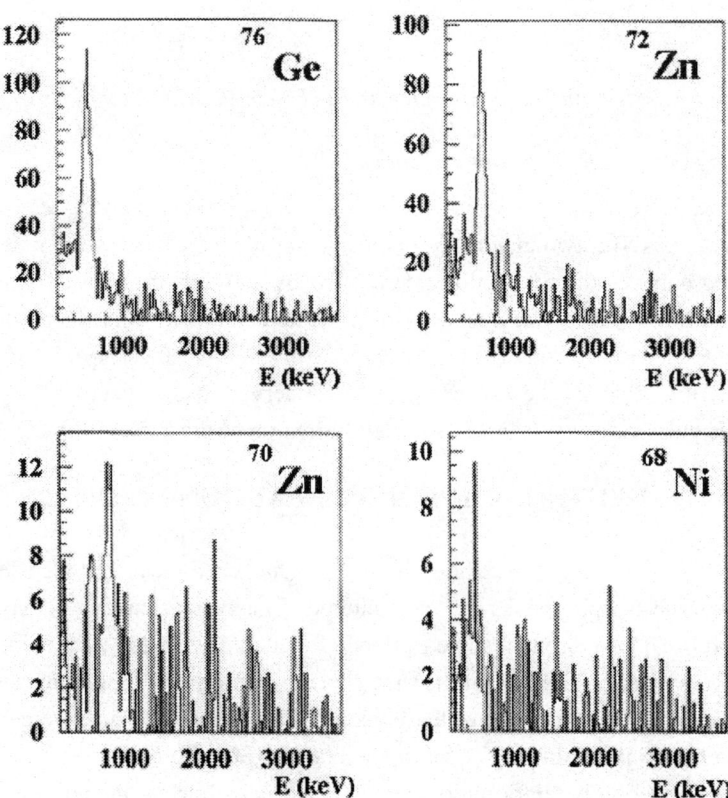

Figure 5: *Background subtracted and Doppler corrected energy-spectra obtained from the Coulomb excitation of ^{76}Ge, ^{70,72}Zn, and ^{68}Ni. The x-axis is in keV*

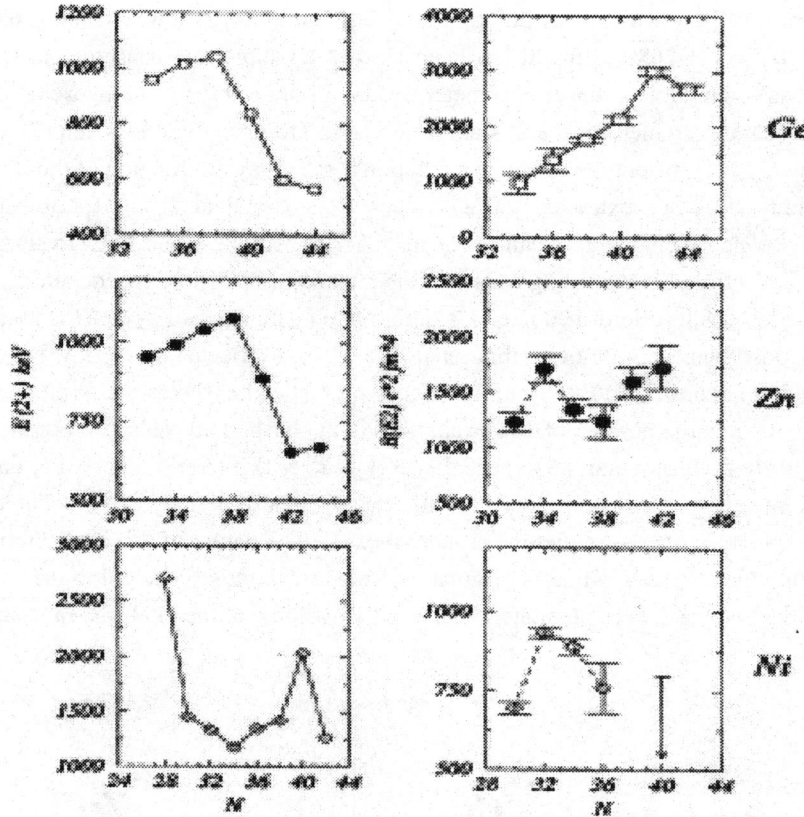

Figure 6 : *Systematics of 2⁺ energies and B(E2) in the Ge, Zn and Ni isotopes [8]. New experimental results for the B(E2) of ⁶⁸Ni[15], ⁷²Zn[15] and for the 2⁺ energy of ⁷⁰Ni[9] have been included*

For future experiments at GANIL, two major improvements are planned. The use of a neutron-rich primary beam ⁷⁰Zn, closer to ⁶⁸Ni, should enhance the production rate of ⁶⁸Ni by a factor of at least 100. In addition, the use of a higher resolution γ-array consisting of four segmented clover Ge detectors will offer the possibility of detecting γ-rays with energies as low as 100 keV (to be compared to 400keV with the present set-up). The improved γ-detection set-up should help elucidating the origin of the low energy γ-ray observed in the spectrum of ⁶⁸Ni. The increase of the secondary beam intensities will permit an unambiguous determination of the B(E2: $0^+ \rightarrow 2^+$) value in ⁶⁸Ni.

NEUTRON-RICH NUCLEI AROUND N=20

From the study of the structure of light neutron-rich nuclei, it has been recently suggested that some major shell-gaps are weakened when large isospin values are encountered. The typical cases of ³²Mg and ⁴⁴S, where a large collectivity has been found [4-6] have brought some evidence for such a shell-gap weakening at large neutron excess. Though, information on the excitation energies of the first 2⁺ states and on the B(E2) values of the 2⁺ to 0⁺ transitions is not sufficient to fully understand the structure of these nuclei. For instance the measurement of the E(4⁺)/E(2⁺) ratio should shed some light on the origin of the large quadrupole collectivety observed.

In order to bring more spectroscopic information on ³²Mg and neighboring nuclei, a novel experimental method has been used. This method is based on the production of very neutron-rich nuclei in relatively higher excited states, through the projectile fragmentation process, and on the detection of their in-beam gamma-decay. Such experiment aiming for the measurement of the E(4⁺)/E(2⁺) ratio in ³⁰,³²Mg, ²⁶⁻²⁸Ne and ²²O has been recently performed at GANIL. A ³⁶S beam, at 77MeV/u was used on a 2.77 mg/cm² Be target. The target was located at the entrance of the

248

SPEG spectrometer which was used to analyze the different fragments produced in the reaction. At a proper $B\rho = 3.4$ T.m setting of the spectrometer (for this setting the transmission of the spectrometer was optimized for the ^{32}Mg fragments mainly with respect to elastic scattered ^{36}S beam), many neutron rich exotic nuclei have been produced and identified. The produced nuclei are identified in a time of flight versus energy loss plot. The energy loss is measured in an ionization chamber at the focal plane of the SPEG spectrometer, whereas the time of flight of the fragments is given by the time difference between the RF pulse from the accelerator and the signal from a plastic scintillator located at the SPEG focal plane. It is worth pointing out that most of the produced nuclei are TERRA INCOGNITA for nuclear spectroscopy and thus gamma spectroscopy of these nuclei (such as $^{22-23}$O, $^{27-28}$Ne, $^{32-33}$Mg) is completely unknown. Gamma-spectroscopy for all the produced exotic nuclei is obtained by performing coincidences between the analyzed fragments and γ-rays emitted in fight during their decay to the ground state. For that purpose a highly efficient (25 % at 1.33MeV) gamma array consisting of 74 BaF2 crystals (the same used for the Coulomb excitation experiment) was used around the target covering symmetrically the upper and lower hemispheres (roughly 80% of the solid angle around the target is covered). This array is supposed to provide fragment-γ as well as fragment-γ-γ coincidences. The latter is needed to build-up a level scheme for each fragment. In addition to the BaF2 array, four 70% high resolution Ge detectors were used at the most backward angles (in between the two hemispheres) in order to help identifying some more complex BaF2 spectra (see spectra below). Even though the analysis of the data is in progress, enough results are obtained today to prove the feasibility and the power of the method. In the following, part of the results obtained in even-even nuclei are presented. After gating on the proper fragment and on the true fragment-γ coincidences (subtracting the random coincidences contribution), some Doppler corrected γ-spectra are presented and commented. No result has been obtained yet on the gamma angular distribution and correlation for angular moment assignment. Therefore our discussion will be based on the assumption that the strongest γ-line in a spectrum of an even-even nucleus represents the 2^+ to 0^+ transition. This is a fairly valid assumption for light nuclei where no low-lying octupole states are expected. Nevertheless this has to be proven yet.

i) The γ-spectra obtained by gating on the ^{18}C fragment (see figure 7), reveals for the first time γ-spectroscopy information on this neutron rich nucleus. A γ-line is clearly visible at 1.6 MeV in the BaF2 spectrum, probably the 2^+ to 0^+ transition. This 2^+ state has been already seen by transfert reactions [17]. Furthermore The same spectrum shows a shoulder at gamma energy around 2.0 MeV, indicating the decay of unknown higher excited state. Despite the low statistics in the Ge spectrum, one can barely see the same features.

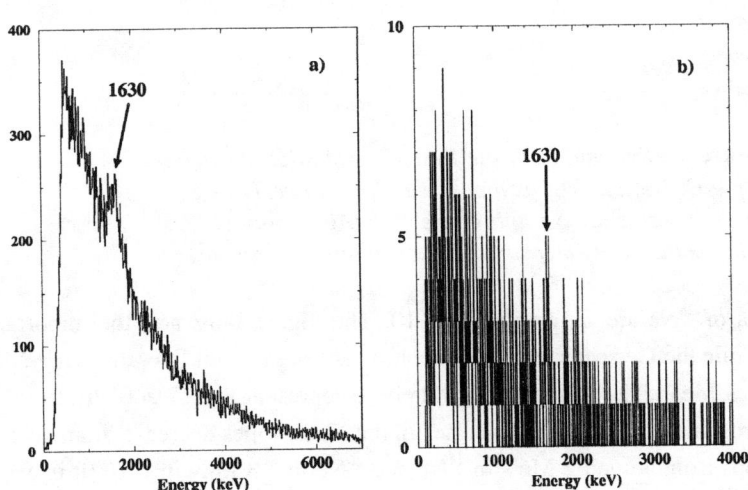

Figure 7 : *BaF2(a) and Ge (b) spectra obtained from the in-beam γ decay of ^{18}C*

ii) The obtained γ-spectra of ^{22}O from both the BaF2 and Ge detectors is presented in figure. 8. If we assume, from intensity argument, that the γ-line observed for the first time at 3.1 MeV represents the 2^+ to 0^+ transition, this extends the systematic of the 2^+ transition energies of Oxygen isotopes up to N= 14. One can see in figure 9, that Oxygen isotopes exhibit the lowest 2^+ energy at half-occupancy of the $d_{5/2}$ state (N=12) just like the Ne and Mg isotopes do (the energy scale of figure 9 is to be multiplied by a factor 4 for the Oxygen isotopes). Whether it will continue to follow the same trend up to N=16 or not is a key point to understand why the last bound Oxygen isotope seems to be ^{24}O [18-20].

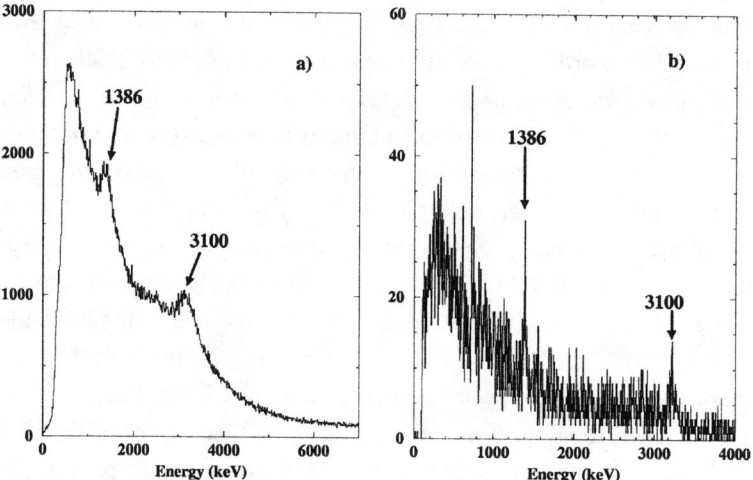

Figure 8 : *BaF2 (a) and Ge (b) spectra of the in-beam γ-decay of ^{22}O*

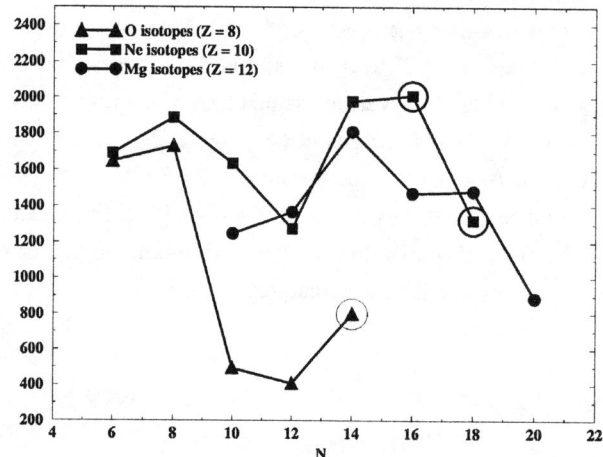

Figure 9 : *Systematic of the first 2^+ excitation energies in the O, Ne and Mg isotopes The labels are Triangles for O, squares for Ne and circles for Mg (empty circles indicate the newly measured values in our experiment).*

iii) The BaF2 and Ge spectra of ^{28}Ne are shown in figure 10. This figure illustrates the importance of the high efficiency for this experiment. While the Ge spectrum does not show any significant line, the BaF2 spectrum exhibits a quite convincing structure at 1.3 MeV. This γ-line is very likely to represent the 2^+ to 0^+ transition in ^{28}Ne which shows for the first time that, approaching N=20, the 2^+ energies in the Ne isotopes decrease dramatically. One can see in figure 9 that the 2^+ energy drops from around 2 MeV in ^{24}Ne and ^{26}Ne to 1.3 Mev in ^{28}Ne (it is worth pointing out that the 2^+ excitation energy of ^{26}Ne has been already measured in a β-decay experiment at GANIL[21]). This behavior is presumably a sign of shell structure change for neutron rich Ne isotopes similar to the one observed long time ago in the Mg isotopes [22].

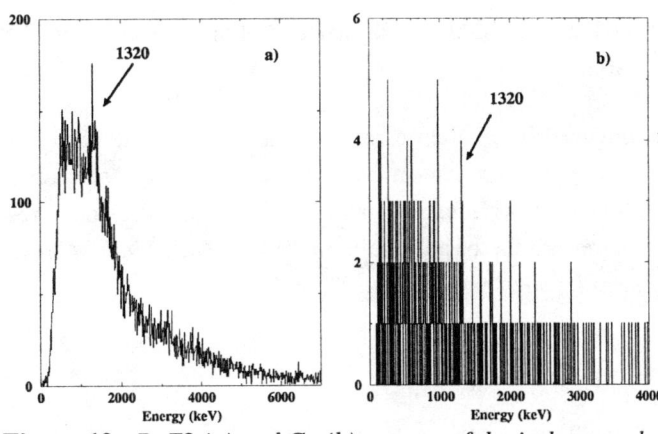

Figure 10 : *BaF2 (a) and Ge (b) spectra of the in-beam γ-decay of* ^{28}Ne

iv) In figure 11 and 12, BaF2 spectra and Ge spectra of both ^{30}Mg and ^{32} Mg are shown. The spectra of ^{30}Mg are good illustration of the importance of the high resolution Ge detectors for the experiment. Only from the Ge spectrum one can tell that the shoulder (around 1.8MeV) observed in the Baf2 spectrum is a not a single line but instead a

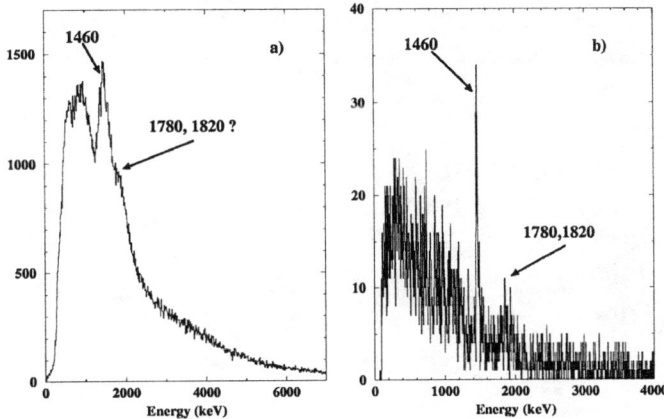

Figure 11 : *BaF2 (a) and Ge (b) spectra of the in-beam γ-decay of* ^{30}Mg.

complex structure made of two maybe three lines. Like for many other produced fragment, the gamma spectra (Ge and BaF2 spectra) of ^{32}Mg (see fig. 12) exhibit more than one line. For all these fragments, γ-angular distribution and γ-γ coincidence between BaF2 detectors has to be analyzed in order to deduce a level scheme. This type of analysis is quite in progress for ^{32}Mg and reveals that the two line: the 885Kev (the well known 2^+ to 0^+ transition in ^{32}Mg [22]) and the 1.4 Mev newly observed line, are in coincidence. The nature (multipolarity) of the 1.4Mev γ-ray is not yet extracted from the data. Though, it is likely to be either the 4^+ to 2^+ transition or a transition from a second 2_2^+ to a

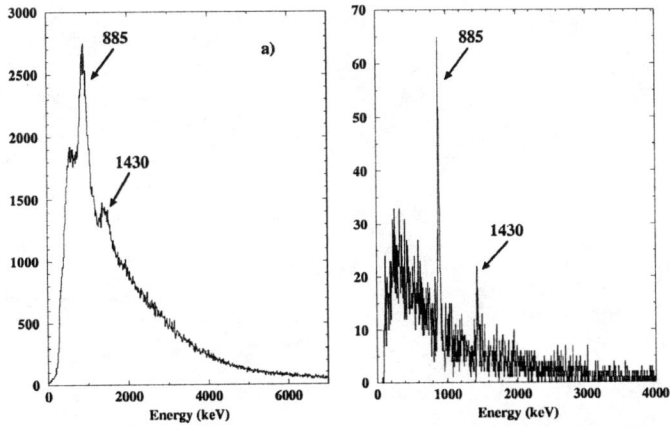

Figure 12 : *BaF2 (a) and Ge (b) spectra of the in-beam γ-decay of* ^{32}Mg

first 2_1^+ state. In both cases this will shed more light on the physics underlying the so-called shell-effect quenching at the neutron rich side of the valley of stability.

Beside the importance of the obtained preliminary results, this experiment shows that the **'in-beam-γ-fragmentation'** method is very promising for exploring nuclear structure far from stability. It also highlights the important role of the gamma detection system from the point of view of efficiency, resolution and Doppler Broadening reduction. These three features are the basic requirements for which EXOGAM [23] was built and thus make it the ideal gamma detection system for such experiment.

REFERENCES

1. G. Kraus et al., *Phys. Rev. Lett.* 73, 1773 (1994)

2. R. Anne et al., *Z. Phys. A* 352 (1995) 397.

3. T. Nakamura et al., *Phys. lett. B* 394 (1997) 11.

4. T. Motobayashi et al., *Phys. Lett. B* **346**, 9 (1995)

5. H. Scheit et al., *Phys. Rev. Lett.* **77**, 3967 (1996)

6. T. Glasmacher et al., *Phys. Lett. B* **395**, 163 (1997)

7. J. Dobaczewski et al., *Phys. Rev. Lett.* **72**, 981 (1994)

8. R. Broda et al., *Phys. Rev. Lett.* **74**, 868 (1995)

9. R. Grzywacz, *Phys. Rev. Lett.* **81**, 766 (1998)

10. H. Grawe et al., *Prog. in Part. and Nucl. Phys.* **38**, 15 (1997)

11. P. H. Stelson and F. K. McGowan, *Nucl. Phys.A* **32**, 652 (1962)

12. M. Girod et al., *Phys. Rev. C* **37**, 2600 (1988)

13. P. Bonche et al., *Nucl. Phys. A* **500**, 308 (1989)

14. M. Bernas et al., *Phys. Lett. B* **113**, 279 (1982)

15. S. Leenhardt, thesis work, I.P.N Orsay

16. S. Raman et al., *Phys. Rev. C* **43**, 556 (1991)

17. K. Fifield et al., *Nucl. Phys. A***385**, 505 (1982)

18. D. Guillemaud-Mueller et al., *Phys.Rev* **C41**, 937 (1990)

19. M. Fauerbach et al., *Phys. Rev. C***53**, 647 (1996)

20. O. Tarasov et al, *Phys. Lett.* **B409**, 64 (1997)

21. A. T. Reed et al. private communication

22. D. Guillemaud-Mueller et al., *Nucl. Phys. A***426**, 37 (1984)

23. F. Azaiez and W. Korten., *Nucl. Phys. News,* Vol. **7**, No. **4**, (1997)

Proton Scattering on Radioactive Proton *sd*-Shell Nuclei and Comparison to Coulomb Excitation

T. Glasmacher[*], Y. Blumenfeld[†], P.D. Cottle[‡], K.W. Kemper[‡], R.W. Ibbotson[*], J.H. Kelley[†], K. Jewell[‡], F. Maréchal[†‡], L. Riley[‡], H. Scheit[*], T. Suomijärvi[†]

[*]*Department of Physics and Astronomy and National Superconducting Cyclotron Laboratory, Michigan State University, East Lansing, Michigan, 48824*
[†]*Institut de Physique Nucléaire, IN2P3-CNRS, 91406 Orsay, France*
[‡]*Department of Physics, Florida State University, Tallahassee, Florida 32306*

Abstract. Systematic measurements of the first excited states in neutron-rich nuclei in the $\pi(sd)$-shell have been performed in the last two years using the technique of intermediate energy Coulomb excitation. These experiments were designed to study the evolution of nuclear shell structure as the neutron-dripline is approached and to provide experimental data which can be easily compared to nuclear model predictions. These measurements indicated a breakdown of the N=20 shell closure for ^{32}Mg and a weakening of the N=28 shell closure below ^{48}Ca, as well as a region of moderate collectivity around ^{40,42}S. However, this experimental technique is sensitive only to the electromagnetic response of the nucleus.

To probe the neutron distributions we performed a series of proton scattering experiments in inverse kinematics on neutron-rich sulfur and argon isotopes. Proton scattering at energies of about 30 MeV/nucleon is more sensitive to neutrons than to protons by about a factor of three. The combination of the proton scattering results with the Coulomb excitation measurements allows a determination of the ratio of the neutron transition matrix element M_n to the proton transition matrix element M_p. Deviations from the homogeneous collective model expectation $M_n/M_p=N/Z$ can be indicative of isovector deformation predicted to occur for neutron-rich sulfur isotopes.

Proton scattering experiments and Coulomb excitation experiments on the same isotopes have been performed at the NSCL with radioactive beams produced in the A1200 fragment separator and protons detected in an array of Si strip–Si-CsI telescopes.

INTRODUCTION

For the structure of nuclei close to the valley of β-stability, a sophisticated level of understanding has been achieved in the last 40 years. This is indicated by the good agreement between experimental data and theoretical nuclear structure predictions. Further away from stability, in particular towards the neutron-dripline, experimental data become sparse and model predictions are less easily compared to measurements. We do know that the binding energy per nucleon changes from about 8-9 MeV in the valley of β-stability to zero at the dripline. Thus the nuclear shells, which are strong close to stability and are the origin of many nuclear structure effects, must weaken as the nuclear binding energy decreases.

Exotic ion-beam facilities can provide beams of many different β-unstable isotopes for experimental study. At projectile fragmentation facilities these beams have velocities of about 30%-50% of the speed of light ($\beta \approx 0.3$–0.5). Their intensities vary greatly with isospin, and the further the secondary beam is removed from stability, the less intense it will be. The secondary beam energies are determined by the production mechanism (in-flight fragmentation) and are on the order of $E \approx 50 - 200$ MeV/nucleon.

Experiments to study the structure of such "fast" beams must be performed in inverse kinematics, i.e. the exotic beam impinges on a stable target. In addition, the larger the cross section and the more efficient the detector, the fewer particles are needed to perform a meaningful measurement. An experimental technique that probes the electromagnetic response of the nucleus is the technique of intermediate-energy Coulomb excitation. A complementary technique that probes the distribution of hadrons in the nucleus is hadron scattering. In particular, proton scattering at energies around 30-50 MeV/nucleon is about three times more sensitive to the neutron distribution

CP481, *Nuclear Structure 98*, edited by C. Baktash

in the nucleus than the proton distribution (1). Proton targets can be realized in the laboratory by using $(CH_2)_n$ foils. Stable neutron or pion targets cannot be made in the laboratory.

The combination of the proton scattering results with the Coulomb excitation measurements allows a determination of the ratio of the neutron transition matrix element M_n to the proton transition matrix element M_p. Deviations from the homogeneous collective model expectation $M_n/M_p = N/Z$ can be indicative of isovector deformation predicted to occur for neutron-rich sulfur isotopes (2-4).

INTERMEDIATE ENERGY COULOMB EXCITATION

The experimental technique of intermediate energy Coulomb excitation (5) takes advantage of the high exotic beam energies available and allows for the measurement of the energy of the first excited state and the transition matrix element (B(E2) value) between the first excited state and the ground state with low exotic beam intensities.

The experimental concept is illustrated in Fig. 1. The secondary beam particles are identified on an event-by-event basis and their directions are determined before interaction with a heavy secondary target. There, the beam is excited and emits a photon in flight. The scattered beam is detected in a forward detector to ensure that the scattering angle is small and thus the minimum impact parameter large enough to have no nuclear contribution to the excitation mechanism. The energies and interaction points of the emitted photons are detected around the secondary target. In order to reconstruct the energy of the transition in the projectile, the photon energy spectrum is Doppler-shifted on an event-by-event basis into the frame of the moving projectile. The intrinsic energy resolution of the NaI-based γ-ray detector and the finite opening angle of the detector are the major contributors to the broadening of the photopeak in the energy spectrum. With the availability of position sensitive segmented high-purity Germanium (HPGe) detectors (6-9) it will be practically possible to eliminate the contribution due to the intrinsic detector resolution, and thus to drastically reduce the broadening. High resolution gamma ray spectroscopy ($\Delta E/E \leq 0.5\%$) will then be possible with fast exotic beams. An array of eighteen 32-fold segmented HPGe detectors for use in fast-beam experiments is under construction at the NSCL.

Intermediate energy Coulomb excitation experiments with photon detection have been performed at GANIL (10), RIKEN (11,12) and at the NSCL (13-16). A detailed description of the experimental method can be found in (17).

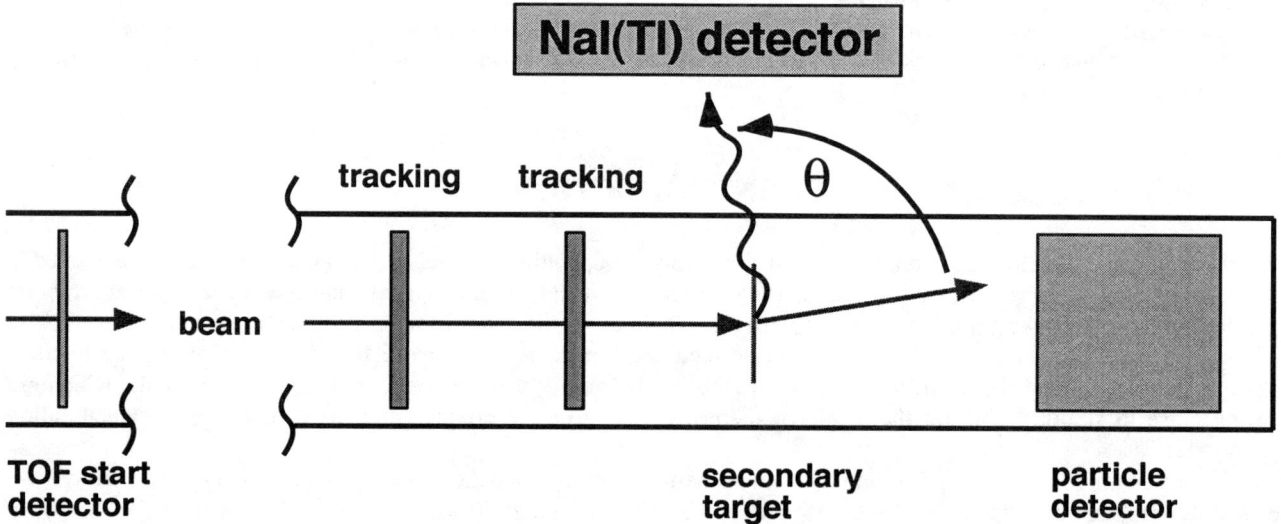

FIGURE 1. Schematic setup of an intermediate energy Coulomb excitation experiment. The photon detectors (here NaI(Tl) detectors) have to measure the energy and the interaction point of the γ-ray so that the energy of the photon in the frame of the beam particle can be deduced from the measured Doppler-shifted energy.

PROTON SCATTERING IN INVERSE KINEMATICS

The experimental setup of the proton scattering experiments in inverse kinematics is similar to the Coulomb excitation experiments, however the photon detectors are replaced by position sensitive proton detectors. Telescopes (see Fig. 2) of Si-strip detectors (300 μm thick, 5cm x 5cm in size), Si diode (500 μm thick, 5cm x 5cm in size), followed by 1 cm of CsI were used to identify protons and to measure their energies in a range of 0.5 MeV – 25 MeV. In addition to the energy measurement, the strip detector allows the simultaneous determination of the proton's emission angle with respect to the incident beam with a precision of about 0.6°.

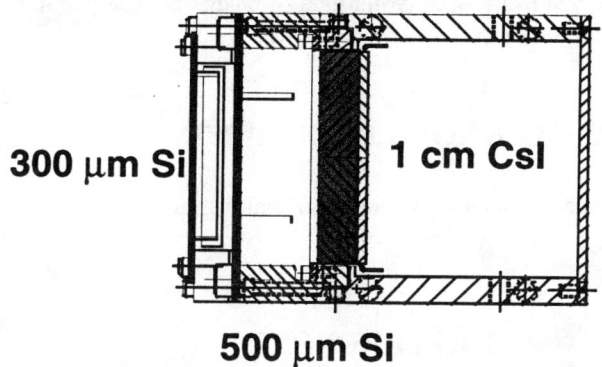

FIGURE 2. Schematic of the particle telescopes used in the proton scattering experiments.

The eight strip detector telescopes are arranged close to 90° with respect to the incoming beam. Protons emitted forward to these angles correspond to protons emitted close to zero degrees in the center of mass. The detectors covered an angular range of about 0°-40° in the center of mass. Figure 3 illustrates the experimental setup. From the flux of the incoming particles and the target thickness, along with the measured yields of protons at different scattering angles, angular distribution in the inelastic and elastic proton scattering cross section can be determined as long as the excited states are separated from the ground state and each other by more than approximately 700 keV.

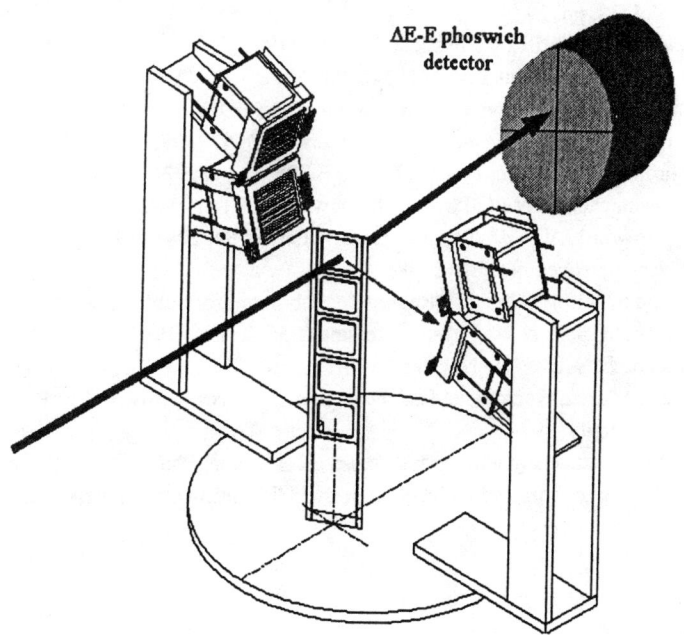

FIGURE 3. Arrangement of the particle telescopes around the $(CH_2)_n$ target. The scattered protons are detected close to 90° in the laboratory while the scattered beam particle is identified in telescope located at 0° in the laboratory.

RESULTS

Measurements of Coulomb excitation cross sections have been performed on neutron rich argon, sulfur and silicon isotopes (13,16,18). Proton scattering experiments have been performed on ^{38}S (19,20), ^{40}S (21), ^{36,42,44}Ar (22) and ^{43}Ar (21). In addition, we also performed proton scattering experiments on ^{18}Ne (23) and ^{20}O (24). Unlike in Coulomb excitation experiments where the extraction of deformation parameters can be performed in a model-independent fashion, the extraction of deformation parameters from proton scattering experiments has to rely on nuclear models to calculate the structure of the nuclei of interest. At Florida State University, we are currently developing the tools necessary to analyze proton scattering data of chains of isotopes in a consistent way. This will allow the extraction of neutron and proton multipole matrix elements.

ACKNOWLEDGEMENTS

This work was supported by the National Science Foundation under grants PHY-9523974, PHY-9528844, PHY-9602927 and PHY-9724299.

REFERENCES

1. Bernstein, A. M., Brown, V. R., and Madsen, V. A., Comments on Nucl. Part. Phys. **11**, 203 (1983).

2. Werner, T. R., Sheikh, J. A., Nazarewicz, W. *et al.*, Physics Letters B **335**, 259 (1994).

3. Werner, T. R., Sheikh, J. A., Misu, M. *et al.*, Nuclear Physics A **597**, 327 (1996).

4. Lalazissis, G. A., Vretenar, D., Poschl, W. *et al.*, Nuclear Physics A **632**, 363 (1998).

5. Winther, A. and Alder, K., Nuclear Physics A **319**, 518 (1979).

6. Goulding, F. S. and Landis, D. A., IEEE Trans. Nucl. Sci. **41** (4), 1145 (1994).

7. Kröll, T., Peter, I., Elze, T. W. *et al.*, Nuclear Instruments and Methods in Physics Research A **371**, 489 (1996).

8. Eberth, J., Thomas, H. G., von Brentano, P. *et al.*, Nuclear Instruments and Methods in Physics Research A **369**, 135 (1996).

9. Eberth, J., Thomas, H. G., Weisshaar, D. *et al.*, "Development of segmented Ge detectors for future γ-ray arrays," in *Progress in Particle and Nuclear Physics* (1997), Vol. 38, pp. 29.

10. Anne, R., Bazin, D., Bimbot, R. *et al.*, Zeitschrift für Physik A **352**, 397 (1995).

11. Motobayashi, T., Ikeda, Y., Ando, Y. *et al.*, Physics Letters B **346**, 9 (1995).

12. Nakamura, T., Motobayashi, T., Ando, Y. *et al.*, Physics Letters B , in print (1997).

13. Scheit, H., Glasmacher, T., Brown, B. A. *et al.*, Physical Review Letters **77**, 3967 (1996).

14. Glasmacher, T., Brown, B. A., Chromik, M. J. *et al.*, Physics Letters B **395**, 163 (1997).

15. Fauerbach, M., Chromik, M. J., Glasmacher, T. *et al.*, Physical Review C **56**, R1 (1997).

16. Ibbotson, R. W., Glasmacher, T., Brown, B. A. *et al.*, Physical Review Letters **80**, 2081 (1998).

17. Glasmacher, T., Ann. Rev. Nucl. Part. Science , in print (1998).

18. Ibbotson, R. W., Glasmacher, T., and Scheit, H., Physical Review C , submitted for publication (1998).

19. Kelley, J. H., Suomijärvi, T., Hirzebruch, S. E. *et al.*, Physical Review C **56**, R1206 (1997).

20. Suomijärvi, T., Kelley, J. H., Hirzebruch, S. E. *et al.*, Nuclear Physics A **616**, 295c (1997).

21. Maréchal, F., "Diffusion de protons par des noyaux instables," Docteur en sciences, Université de Paris-Sud, 1998.

22. Scheit, H., Ph.D., Michigan State University, 1998.

23. Riley, L. A., "Inverse kinematics proton scattering with a radioactive ^{18}Ne beam," Ph.D., Florida State University, 1997.

24. Jewel, J. K., "Fragmentation of the low energy octupole phonon in ^{196}Pt and inverse kinematics proton scattering from the radioactive beam ^{20}O," Ph.D., Florida State University, 1997.

Microsecond Isomers beyond ^{68}Ni

R. Grzywacz[a,b], J.M. Daugas[c], L. Achouri[c], J.C. Angelique[d],
D. Baiborodin[e], R. Béraud[f], R. Bentida[f], C. Bingham[a],
C. Borcea[g], W.N. Catford[h], A. Emsallem[f], G. de France[c],
H. Grawe[i], K.L. Jones[h], M. Lewitowicz[c], R. Lemmon[h,j], M.J.
Lopez[c], F. de Oliveira,[c], M. Pfützner[b], P.H. Regan[h],
K. Rykaczewski[k,b], M. Sawicka[b], J.E. Sauvestre[l], G. Sletten[m],
M. Stanoiu[g]

[a] *University of Tennessee, Knoxville, Knoxville, TN 37996, USA*
[b] *IFD, Warsaw University, Pl-00681 Warsaw, Hoża 69, Poland*
[c] *GANIL, BP 5027, 14021 Caen Cedex, France*
[d] *LPC-ISMRA, Bld du Marechal Juin,14050 Caen, France*
[e] *NPI,250 68 Rez, Czech Republic*
[f] *IPN Lyon, 69622 Villeurbane Cedex, France*
[g] *IAP, Bucharest-Magurele P.O.Box MG6, Rumania*
[h] *Department of Physics, University of Surrey, Guilford, Surrey, GU2 5XH, United Kingdom*
[i] *GSI, Postfach 110552, D-64220, Darmstadt, Germany*
[j] *CLRC, Daresbury Laboratory, UK*
[k] *ORNL, Physics Division, Oak Ridge, TN 37830, USA*
[l] *CE Bruyères-le-Châtel, BP 12, F-91680 Bruyères-le-Châtel, France*
[m] *Niels Bohr Institute, University of Copenhagen, Copenhagen, Denmark*

Abstract. Excited states in the nuclei with Z$\sim$28 and N$\sim$ 40 have been studied. These neutron-rich nuclei have been produced in the fragmentation reaction of a ^{86}Kr beam at 60.3 MeV/nucl. The decays of isomers populated in the fragmentation are detected at the final focus of the spectrometer in event-by-event microsecond correlation with identified recoils. Several new isomers with half-lives $\sim$ 300 ns have been discovered including the very exotic ^{78m}Zn. A puzzling result of the performed experiment, the non-observation of expected I$^{\pi}$=8^{+} isomers in ^{72}Ni and ^{74}Ni nuclei is also reported.

INTRODUCTION

The region of neutron-rich nuclei around ^{68m}Ni is particulary interesting for tracking the evolution of magic shell gaps towards the neutron-drip line [1]. However, since the production of these nuclei is difficult, information on the excited states is rather scarce. This suggested the studies of μs-isomers in the heavy Ni region

CP481, *Nuclear Structure 98*, edited by C. Baktash

produced via the fragmentation of relativistic heavy-ions [2]. Interesting results obtained in the first experiment with a ^{86}Kr beam, like the identification of the 8^+ isomer in ^{70}Ni, interpreted as a "valence mirror" of ^{92}Mo, triggered the continuation of the programme, in particular a search for 8^+ isomers in ^{72}Ni and ^{74}Ni. The successful identification of the 8^+ state in ^{78}Zn was made, but the preliminary analysis does not show any evidence for the expected metastable states in heavy nickel isotopes.

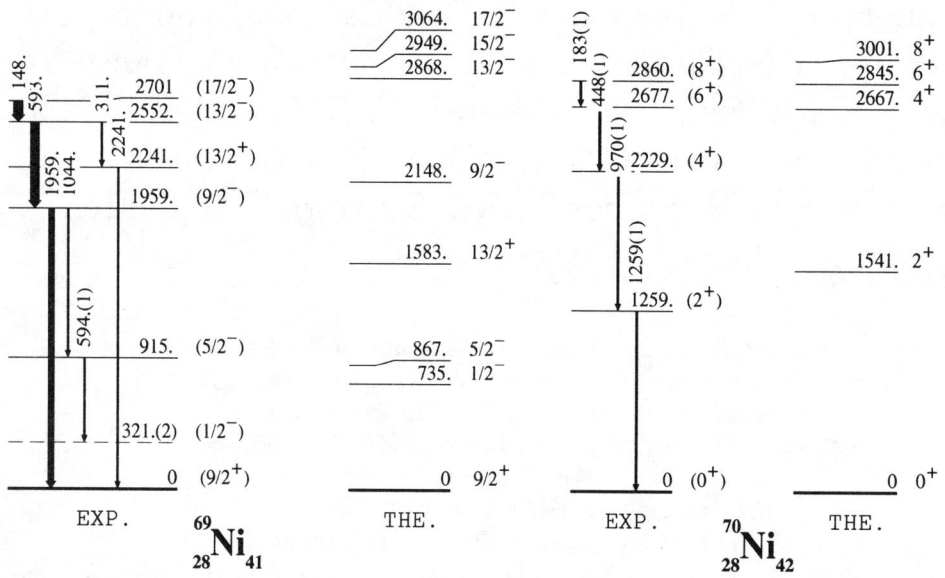

FIGURE 1. The decay schemes of ^{69}Ni an ^{70}Ni as deduced from the observed γ-radiation following the decay of isomeric states. The theoretical predictions are obtained using the TBME obtained from realistic interaction by M. Hjorth-Jensen [3,2].

EXPERIMENTAL METHOD AND PREVIOUS RESULTS

The spectroscopy of μs-isomers after the heavy-ion fragmentation requires an efficient gamma ray detector system and a recoil spectrometer with in flight identification capability. So far, most of these experiments have been performed at GANIL with intermediate energy heavy-ion beams [4,5,7,6,2,8] with the exception of a study performed at GSI with relativistic ^{238}U ions [9]. The straightforward way of producing exotic nuclei, by impinging intense beam ($\sim$ 30 pnA) of heavy projectiles on a thick ^{nat}Ni ($\sim$100mg/cm^2) rotating target and separating them electromagnetically, proved to be quite efficient, even though the production cross sections in this type of reactions are low.

The direct identification of recoil properties like mass, atomic number and charge with a 100 % efficiency allows for low-rate experiments. Such a setup with germanium detectors at the final focus (where the ions are stopped) and the event-by-event ion-gamma correlation electronics (as demonstrated in the pioneering experiments on fission fragments [10] and fusion-evaporation residues [11]) can be very selective in measuring gamma radiation occuring within microseconds of the recoil stopping. The fragmentation reaction is characterized by wide mass and charge distribution. Thus gamma decaying, **exotic** isomers are the objects to be searched and studied with this method. The isomeric states can be studied simultaneously over a large fraction of the nuclidic chart. These experiments provide the unique information on excited states very far from stability, which can serve as a landmark for future more complete spectroscopic studies with intense radioactive beams.

There have been two attempts to find and study isomers beyond $^{68}_{28}Ni_{40}$ using fragmentation of a ^{86}Kr beam. The first, reported in [2], resulted in a number of new isomers, like ^{54m}Sc, ^{54m}V, ^{59m}Cr, ^{65m}Fe, ^{66m1,m2}Co, ^{67m}Fe, ^{69m1,m2}Ni, ^{70m}Ni, ^{71m}Cu, ^{72m}Cu, ^{78m}Zn and ^{79m}As. Information on the excited states of the produced isomeric nuclei was obtained for the first time. The primary goal of this experiment, namely the discovery of the 8^+ isomer in ^{70}Ni was achieved. A sequence of four gamma ray transitions have been identified in this nucleus with energies consistant with those expected from the valence mirror concept [2], (see fig. 1). The relatively short half-live of 210(50) ns for ^{70}Ni compared to the long flight time (TOF $\sim$ 1300 ns) resulted in considerable in-flight losses nevertheless the statistics obtained were sufficient to give conclusive information on the proposed decay scheme.

Among others, two isomers have been identified in the neighboring nickel isotope – ^{69}Ni. In addition to the $I^\pi=17/2^-$ ($E^*=2701$, $T_{1/2}=439\pm3$ ns) isomer we were able to identify a metastable state at $E^*=321$ keV basing on the $\gamma-\gamma$ coincidences and intensities for the observed transitions. The former of these structures has a $(\nu g_{9/2})^2$ confiuration of 8^+ isomer in ^{70}Ni with a coupled $p_{1/2}$ neutron. The second isomer is a single particle excitation of one of the $p_{1/2}$ neutrons to the $g_{9/2}$ orbital, thus creating a $I^\pi=1/2^-$ state very close to the $I^\pi=9/2^+$ ground state. Due to the large spin difference the $1/2^-$ state should predominantly decaying via β^- emission. The β^- decay properties of this state has been recently reported by the Leuven group [12].

NEW RESULTS

In late July 1998 the second experiment with a ^{86}Kr beam was performed at GANIL. Tha aim was to identify, among others, 8^+ isomers in more exotic even-even nuclei. The improvements in this experiment were related to the shorter time-of-flight (reduced to about 200 ns), gamma-ray efficiency, and primary beam intensity. The combined result of these improvements is clearly demonstrated by the collected ^{70m}Ni spectrum, see fig. 2 [13]. Several new short lived isomers have been identified on-line : ^{78m}Ga, ^{78m}Zn, ^{74m}Cu, ^{68m}Co, ^{70m}Co, ^{62m}Mn, ^{60m}V. The

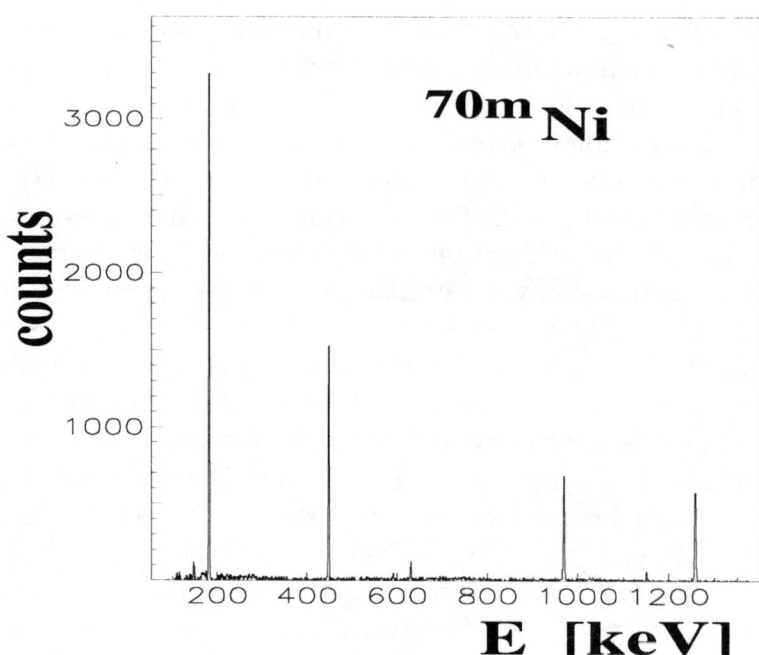

FIGURE 2. Spectrum of the decay of the 8$^+$ isomer in ^{70}Ni [13]. Statistics is improved by factor of ~200 in comparison to earlier work [2].

isomers observed in the previous experiment are also seen clearly on the isomer identification plot (fig. 3) as pronounced groups of gamma-correlated counts (right panel). The left panel gives an idea of the overall number of produced and identified isotopes. Among other, the exciting results of this experiment are the discovery of 8$^+$ isomer in ^{78}Zn and the absence of isomers in ^{72}Ni and ^{74}Ni.

The known excited states in even zinc isotopes with N>40 did not reveal the existence of 8$^+$ isomer [14]. This may be due in part to the limitation of beta decay studies, which do not favour population of high spin states. The excitation predicted in [14] of the lowest 8$^+$ in ^{76}Zn is very high (3.163 MeV), with high energy spacing between 8$^+$ and 6$^+$. This implies very short lifetime for such a state, below our observation limit. ^{78}Zn has a ground state configuration of two neutron holes in $g_{9/2}$ shell and 2 proton particles in the fp shell. The observed sequence of four transitions (fig. 4) together with measured half-life of 319(9) ns suggests a I$^\pi$=8$^+$ spin and parity assignment for the isomeric state. This state would be purely a two $g_{9/2}$ neutron hole state of the same origin as expected for ^{76}Ni or observed for $g_{9/2}$ protons in ^{98}Cd [15] and ^{100}Cd [16]. The 2$^+$ and 4$^+$ states which we tentatively place at 730 keV and 1620 keV are the excitations of mixed character of $p_{3/2}$ and $f_{5/2}$ protons and $g_{9/2}$, $p_{1/2}$ neutrons. The spin 6$^+$ and 8$^+$ states at 2528 keV and 2673 keV, respectively, should not have any multiparticle proton admixture at these excitation energies. Thus, the isomerism is of the same character as for ^{70}Ni case. There is not a clear explanation for the non-observation of other I$^\pi$=8$^+$ isomers in zinc isotopes but the even more intriguing experimental result is the absence of the

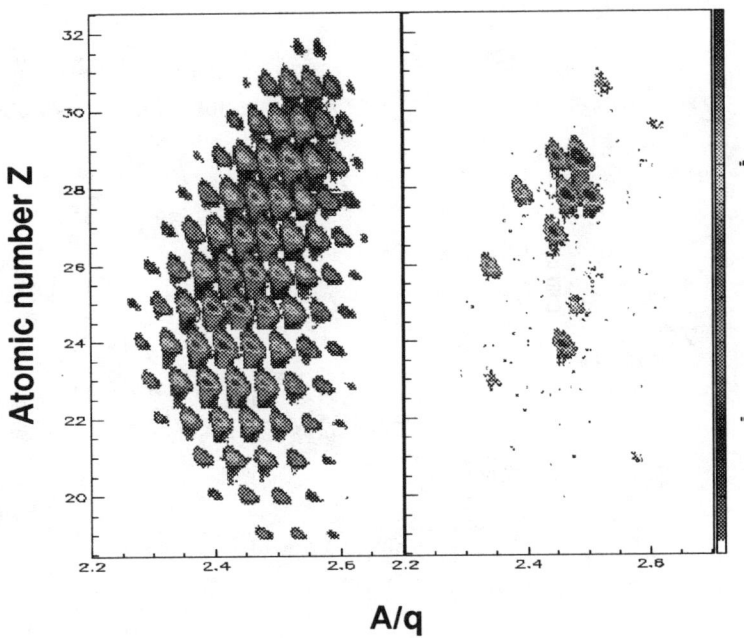

FIGURE 3. The heavy ion identification plot showing the isotopes (left panel) and most abundant microsecond isomers (right panel) produced in fragmentation of ^{86}Kr beam. The spectrum in the right panel is obtained from by selecting the events in delayed coincidence with gamma radiation, the random background subtracted [13].

^{72}Ni and ^{74}Ni isomers. Following the valence mirror concept, which worked for the ^{70m}Ni case, we would expect that the existence of isomers persists when filling the $g_{9/2}$ shell with pairs of neutrons in the presence of the Z=28 magic proton shell. This is a symmetrical situation as for the Z=50 and 40<N<50 nuclei. Similarly we were expecting longer half-lives for the nickel isomers. The surprising result is that within our observation limits, which are compatible with expected half-life values, we did not observe the gamma cascade for either of these cases. In the case of ^{72}Ni from the number of observed counts, knowing the time of flight (about 200 ns) and correlation interval (25 μs), we can deduce upper and lower limits for the half-life of the isomer. The preliminary estimate, basing on assumptions that isomeric ratio and gamma efficiency are the same for ^{70}Ni and ^{72}Ni, gives 40 ns > $T_{1/2}$ > 4000μs. Let us consider these limits for the yrast 8^+ isomeric state. The lower half-life limit requires either the large increase of B(E2) value from ^{70}Ni to ^{72}Ni (5 to 10 times), or increase of the isomeric transition energy by about 30 %. The upper half-life limit would require very dramatic changes in B(E2), a factor of about a few thousand, or the reduction of transition energy down to tens of keV. The other option would be an intruder 7^- state below 8^+ state which would destroy the isomerism. There is, however, no experimental evidence for the existence of such an yrast state. These somewhat speculative elaborations need a more careful analysis from a full set of collected data to prove or disprove the existence of ^{72m}Ni and ^{74m}Ni. The other

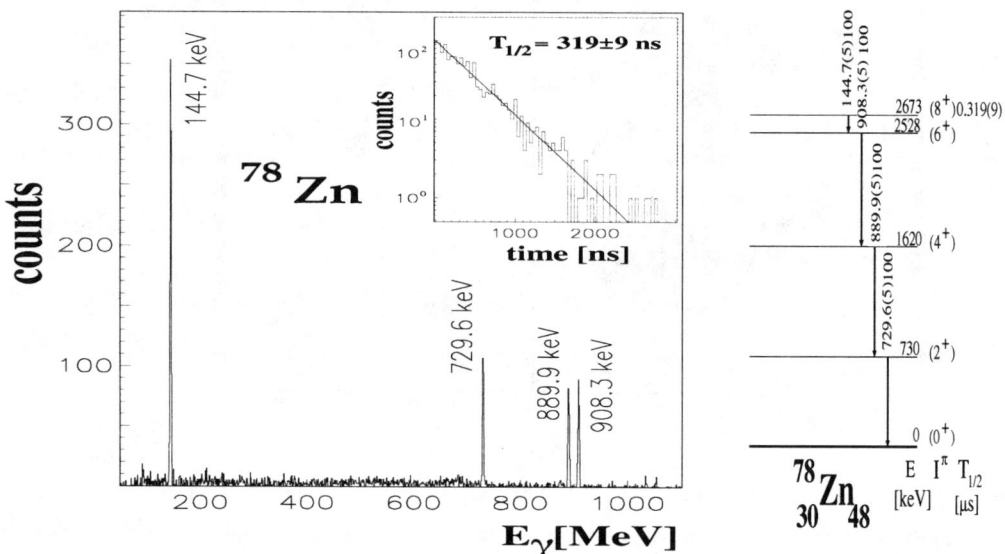

FIGURE 4. Decay γ and time spectrum (inset) of the isomer observed in ^{78}Zn. The decay scheme as suggested by 8^+ assignment of the isomeric level [13].

option to be considered is that the isomer is not produced in the reaction, which in view of previous results [4,17] obtained for yrast isomers, is very unlikely.

SUMMARY AND OUTLOOK

In the experiments performed with the ^{86}Kr beam at GANIL about 20 new isomers in neutron-rich nuclei have been identified and studied. New spectroscopic information has been obtained for the first time on the excited states of these nuclei. Among several new isomers observed, the ^{78}Zn represents the closest experimental approach to doubly magic ^{78}Ni in terms of excited states study. This rich nuclear structure information is the subject of systematical interpretation studies. The spherical shell-model with empirical TBME [18] (or TBME values deduced from realistic interaction [3,2]) which correctly described the ^{69,70}Ni isomerism, is confronted now with the non-observation of the predicted ^{72m}Ni and ^{74m}Ni.

Nuclear physics at The University of Tennessee is supported by the U.S. Department of Energy through Contract No. DE-FG02-96ER4098. Oak Ridge National Laboratory is managed by Lockheed Martin Energy Research Corporation under contract DE-AC05-96O22464 with U.S. Department of Energy. This work has been also supported by IN2P3, the Polish Committee for Scientific Research (KBN), and the EPSRC.

REFERENCES

1. J. Dobaczewski *et al*, Phys. Rev. Lett. **72** (94) 981
2. R. Grzywacz *et al*, Phys.Rev.Lett. **81** (1998) 766
3. M. Hjorth-Jensen, T.T.S. Kuo, E. Osnes, Phys. Rep. **261** (1995) 125
4. R. Grzywacz *et al*, Phys. Lett. **B355** (95) 439
5. M. Robinson *et al*, Phys. Rev. **C53** (96) R1465
6. C. Chandler *et al*, Phys. Rev. **C56** (97) R2924
7. R. Grzywacz *et al*, Phys. Rev. **C55** (97) 1126
8. R. Grzywacz *et al*, Phys.Lett. **429B** (1998) 247
9. M. Pfützner *et al*, submitted to Phys. Lett. B
10. J.W. Grüter *et al*, Phys.Lett. **33B** (1970) 474
11. S.Hofmann *et al*, Z.Phys. **A325** (1986) 37
12. W. Mueller *et al*, this proceedings
13. J.M. Daugas *et al*, to be published
14. J.A. Winger *et al*, Phys. Rev. **C42** (90) 954
15. M. Górska *et al.*, Phys. Rev. Lett. **79** (1997) 2415
16. M. Górska *et al*, Z.Phys. **A350** (1994) 181
17. R. Grzywacz, Proc. of the Int. School of Heavy Ion Physics, Erice, Italy, June 1997, R. Broglia and P.G. Hansen, World Scientific 1998, p. 399
18. H. Grawe, Prog. Part. Nucl. Phys. **38** (97) 15

Beta decay of neutron-rich Co: Probing single-particle states at and above the $N = 40$ subshell closure

W.F. Mueller*, B. Bruyneel*, S. Franchoo*, M. Huyse*, U. Köster†, K.-L. Kratz‡,
K. Kruglov*, Y. Kudryavtsev*, B. Pfeiffer‡, R. Raabe*, I. Reusen*, P. Thirolf‖,
P. Van Duppen*, J. Van Roosbroeck*, L. Vermeeren*, W.B. Walters¶, L. Weissman*, and
A. Wöhr*[1]

*Instituut voor Kern- en Stralingsfysica, University of Leuven, B-3001 Leuven, Belgium
†Physik-Department, Technical University of Munich, D-85748 Garching, Germany
‡Institut für Kernchemie, University of Mainz, D-55099 Mainz, Germany
‖Sektion Physik, University of Munich, D-85748 Garching, Germany
¶Department of Chemistry, University of Maryland, College Park, Maryland 20742

Abstract. Neutron-rich Co nuclei with $A = 66-70$ were produced by the laser-ionization isotope-separation on-line method. The β decay from these nuclei has been studied. A case example is given by reporting on the observed decay scheme of ^{68}Co. The half life of the ground-state decay of this nucleus was measured to be 0.21(3) seconds. In addition, a new β decaying isomer half life of 1.16(25) seconds was discovered. The level scheme of ^{68}Ni has been significantly extended, and an interpretation of the observed levels is made by assuming that the $N = 40$ gap has the characteristics of a shell closure.

INTRODUCTION

In recent years, the development of new experimental techniques for producing neutron-rich nuclei in the region of Ni isotopes has inspired many research projects. For example, identification and half lives for Ni nuclei up to $A = 74$ were first measured by Bernas *et al.* [1]. Nuclear fragmentation production methods were used to measure half lives up to ^{75}Ni [2], and also resulted in the first identification of ^{78}Ni [3]. Other results using nuclear fragmentation include the identification of many new microsecond isomers around neutron-rich Ni [4] and measurement of $B(E2)$ matrix elements in many of these nuclei [5]. Excited states in ^{68}Ni were identified for the first time in multi-particle transfer reactions [6], and studies of ^{68}Ni were continued with deep-inelastic scattering experiments by Broda *et al.* [7]. Another technique to produce neutron-rich nuclei is to induce fission of a heavy actinide nucleus and subsequently separate the product of interest from the other isotopes (the ISOL method). The success of the ISOL method was demonstrated in experiments where $^{68-74}$Ni were produced and their β decay studied [8]. Results from the ISOL experiments provide some of the first detailed spectroscopy studies of these nuclei. Such experiments have been continued to include β-decay studies of $^{66-70}$Co.

The study of these neutron-rich nuclei will provide experimental evidence to compare with predictions of the trends in the astrophysical r process [9]. In addition to this, these nuclei present an interesting challenge to understand the evolution of single-particle structure as one proceeds to extreme isospin values. For example, one such feature is the so-called shell quenching effect, also known as the monopole correction to shell-model single-particle energies [10,11]. Such an effect was discovered in the even-even Cu isotope chain from $A = 68$ to 74 [8].

In this article, we concentrate on the results of the decay of ^{68}Co. Information about other aspects of this experiment have been presented in other sources. For example, production yields for $^{66-70}$Co and a partial decay scheme for ^{69}Co have been presented in Ref. [12].

---

[1] Present address: Physics Department, University of Oxford, Oxford OX1 3PU, United Kingdom

CP481, *Nuclear Structure 98*, edited by C. Baktash
© 1999 American Institute of Physics 1-56396-858-4/99/$15.00

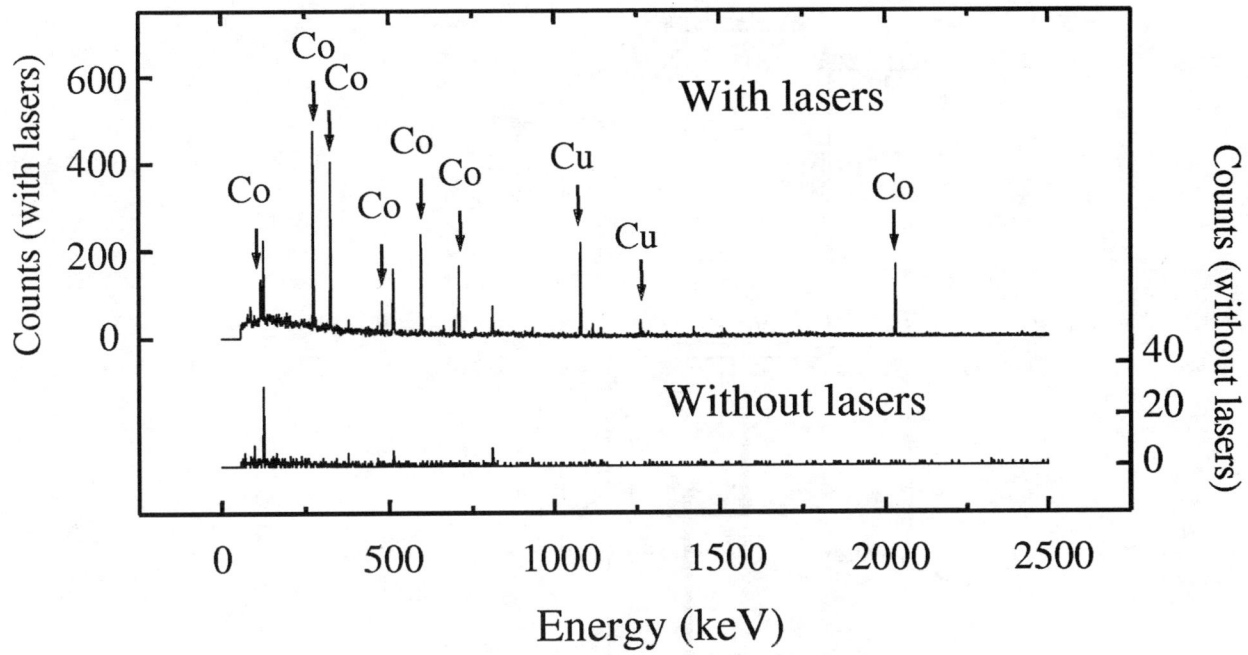

FIGURE 1. Representative spectra for $A = 68$.

EXPERIMENTAL SETUP

A pulsed 30-MeV proton beam from the Cyclone accelerator at the Louvain-la-neuve cyclotron facility was used to induce fission of ^{238}U to produce the Co nuclei. The thin ^{238}U target foils were contained within the cell of the IGLIS ion-guide laser ion source where the fission by-products were captured in a gas of 500 mb Ar. The recoils become neutralized in the buffer gas and transported by the gas flow to the extraction point of the IGLIS cell where the Co isotopes were selectively charged in a two-step laser-ionization process. The Co ions were then extracted from the gas cell through a Sextupole Ion Guide (SPIG) and injected into the LISOL mass separator. The Co nuclei of a specific mass were then deposited in a tape at the detection point. The details of the IGLIS-LISOL setup can be found in Refs. [13–15]. The mass separator was run with a 0.6-sec implantation cycle followed by a 1.0-sec "beam-off" period such that half-life information of the deposited nuclei could be extracted from the time behavior of their decay.

Surrounding the detection point were two high-purity Germanium detectors (70% and 75% relative efficiency) and three ΔE plastic detectors arranged in a compact configuration. The trigger for the data acquisition system was set to accept β-γ and γ-γ coincidence events. In addition Ge singles were collected by a separate system. Long-lived daughter products as well as other background contamination at the detection point were removed by periodically moving the implantation tape.

RESULTS

The β decay of Co isotopes with $A = 66 - 70$ were studied and excited states for a given nucleus were positively identified by comparing the β-delayed γ spectra with laser irradiation to corresponding spectra without lasers. Representative spectra illustrating this is shown in Fig. 1. In this figure, the lower histogram is the Ge energy spectrum for $A = 68$ when the lasers are turned off. The events in this spectrum are largely the result of neutron-induced reactions in the Ge detector and other minor external contaminants. When the lasers are tuned to Co resonance one observes the upper spectrum in Fig. 1. The contaminating peaks are still present, however, all other lines can be associated with the γ rays resulting from the β decay of ^{68}Co, with the exception of ^{68}Cu lines which are the result of the ^{68}Ni-daughter decay.

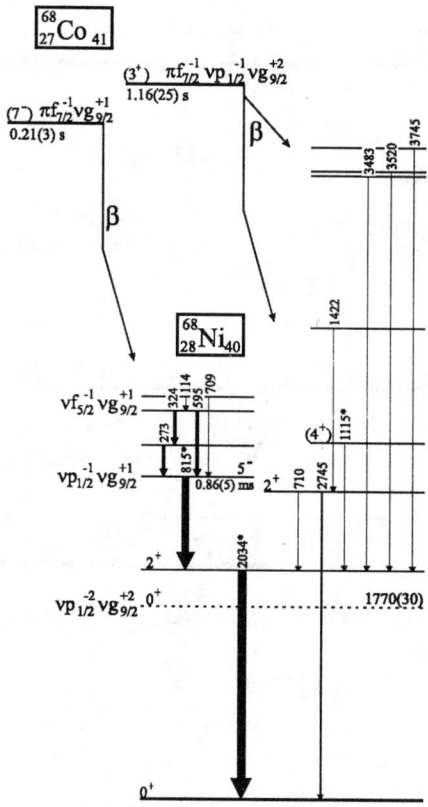

FIGURE 2. Preliminary decay scheme for ^{68}Co.

A preliminary decay scheme for ^{68}Co is shown in Fig. 2. The γ rays that were previously identified by Broda *et al.* [7] are indicated with asterisks. In addition to the first excited 2^+ state, Broda *et al.* were able to identify a 5^- state, which is interpreted as the $\nu p^{-1} \nu g_{9/2}^{+1}$ configuration, as well as a state with a tentative spin and parity assignment of 4^+ or 3^-. The first excited 0^+ state at ≈ 1770 keV as well as other states were also observed by Bernas *et al.* [6]. We do not have evidence that we feed this 0^+ level, but include it here for completeness. The half-life analysis of the γ rays provides clear evidence that the decay of ^{68}Co proceeds through two different states with half-lives of 0.21(3) and 1.16(25) seconds respectively. The ^{68}Ni level scheme in Fig. 2 is drawn to indicate the levels fed by these two states in ^{68}Co. The levels in the left part of the scheme are fed by the short half-life state, while those one the right can be associated with the longer half-life state.

Analysis of the γ rays characterized by the short half life determined that they are not in coincidence with the 2034-keV transition. Some options for the placement of these γ rays could be feeding either the ground state of ^{68}Ni, the 0.86(5)-ms 5^- isomer, or alternatively the 1770-keV 0^+ state. Ground-state feeding can be ruled out because the relatively low energies of the γ rays (e.g. 273, 324, etc.) would place several levels lower in energy than the previously identified 0^+ and 2^+ states. If such states existed, they would most probably have been detected in the previous experiments. Another option is to have these transitions feed into the 5^- isomer. Strong support for this placement, rather than feeding the 1770-keV excited state, arises from the result that the 815-keV γ ray is observed to have the same half life behavior as the other short lived transitions.

As can be seen in Fig. 2, β decay is also observed from a second state in ^{68}Co. The decay pattern from this longer lived state is somewhat more complex. Greatest proportion of the β decay intensity of this state is observed to directly feed the first 2^+ state in ^{68}Ni. There are several other higher-energy states in ^{68}Ni that are also directly fed by the longer lived β decay. Most of these states are observed to then decay by one or two-step γ transitions to the 2034-keV level.

DISCUSSION

^{68}Ni can be treated as a doubly-magic nucleus with $Z = 28$ and $N = 40$, where the neutron gap occurs between the f-p shell and the $1g_{9/2}$ orbital. Using the extreme single-particle approximation, one can make reasonable interpretations for the structure of both ^{68}Co and ^{68}Ni. In consideration of this approximation the ground state of $^{68}_{27}$Co$_{41}$ is formed by adding a hole to the proton shell and a particle to the neutron shell, and the resulting configuration is $\pi f_{7/2}^{-1} \nu g_{9/2}^{+1}$. Based on studies for coupling the angular momentum of particle and hole states [16,17], the spin of the lowest state for this configuration is $7\hbar$ with parity of -1. Likewise, one can consider the configuration for the isomeric state. In this case, the first excited state would involve an excitation of a pair of neutrons across the $N = 40$ gap into the $\nu g_{9/2}$ orbital. The resulting configuration would then be $\pi f_{7/2}^{-1} \nu p_{1/2}^{-1} \nu g_{9/2}^{+2}$ with a spin and parity of 3^+.

For both of these states in ^{68}Co, the energetically favored β-decay mode would be the $\nu f_{5/2} \rightarrow \pi f_{7/2}$ Gamow-Teller transition. For the case of the ground-state decay, feeding would be observed primarily to the 6^- and 7^- $\nu f_{5/2}^{-1} \nu g_{9/2}^{+1}$ broken-pair states in ^{68}Ni. As can be seen in Fig. 2, β decay to such states is observed and followed by γ decay to the 5^- isomer.

The β decay of the 3^+ isomer is more complex. As in the ground-state decay, the decay of the isomer would still involve the conversion of an $f_{5/2}$ neutron to fill the $f_{7/2}$ proton hole. The resulting configuration in ^{68}Ni would be $\nu f_{5/2}^{-1} \nu p_{1/2}^{-1} \nu g_{9/2}^{+2}$. Such a configuration is not energetically favored since a pair of neutrons must be excited across the $N = 40$ gap. As a consequence, states of lower energy which mix with this configuration will compete for the β feeding and the decay intensity is spread over several levels as a result.

CONCLUSION

In conclusion, the laser ion-guide setup has been used to perform measurements for the decay of very neutron-rich cobalt nuclei. The study of the β-decay of ^{68}Co has resulted in the discovery of a long-lived β-decaying isomer in this nucleus. This isomer can be interpreted as a $\pi f_{7/2}^{-1} \nu p_{1/2}^{-1} \nu g_{9/2}^{+2}$ excitation. The ground-state β decay of ^{68}Co decay is found to proceed to states in ^{68}Ni that subsequently γ decay to the 5^- level. The β decay of the ^{68}Co isomer, however, is observed to decay to several different states in ^{68}Ni. This is due to the fact that the favored configuration ($\nu f_{5/2}^{-1} \nu p_{1/2}^{-1} \nu g_{9/2}^{+2}$) is much higher in excitation energy than competing levels.

We gratefully thank J. Gentens and P. Van den Bergh for running the LISOL separator. This work is supported by the Inter-University Attraction Poles (IUAP) Research Programme and by the Fund for Scientific Research – Flanders (FWO-Belgium). M.H. is Research Director, L.V. Postdoctoral Researcher, and S.F. Research Assistant of the FWO, Belgium.

REFERENCES

1. M. Bernas et al., Z. Phys. **A336**, 41 (1990).
2. F. Ameil et al., Eur. Phys. J. **A1**, 275 (1998).
3. Ch. Engelmann et al., Z. Phys. **A352**, 351 (1995).
4. R. Grzywacz et al., Phys. Rev. Lett. **81**, 766 (1998) and these proceedings.
5. F. Azaiez, these proceedings.
6. M. Bernas et al., J. Physique Lett. **45**, L851 (1984); M. Girod et al., Phys. Rev. C **37**, 2600 (1988).
7. R. Broda et al., Phys. Rev. Lett. **74**, 868 (1995).
8. S. Franchoo et al., Phys. Rev. Lett. accepted for publication.
9. K.-L. Kratz et al., Z. Phys. **A332**, 419 (1989).
10. W. B. Walters, in proceedings *Int. Symp. on Nucl. Phys. of our Times*, Sanibel 1992, p. 457.
11. K. Heyde et al., Nucl. Phys. **A466**, 189 (1987).
12. S. Franchoo et al., in proceedings *ENAM '98*, Michigan 1998.
13. Y. Kudryavtsev et al., Nucl. Instr. and Meth. B **114**, 350 (1996).
14. L. Vermeeren et al., Nucl. Inst. and Meth. **B126**, 81 (1997).
15. P. Van den Bergh et al., Nucl. Instr. and Meth. B **126**, 194 (1997).
16. M. Moinester, et al., Phys. Rev. **179**, 984 (1969).
17. V. Paar, Nucl. Phys. **A331**, 16 (1979).

Spectroscopy near ^{132}Sn

Henryk Mach

Department of Neutron Research, Uppsala University,
S 611-82 Nyköping, Sweden

Abstract. Over the last few years one observes an accelerated pace of new developments related to the experimental structure investigations of nuclei at double shell closures. In the following, the current status of experimental information on the nuclei in a close vicinity of ^{132}Sn and the prospects of further developments in the nearest future are discussed. A special attention is given to the selected, key developments in four areas of approach to this region: β-decay at the ISOL facilities, laser-induced ion-sources, prompt γ-rays in spontaneous fission using multi-detector arrays, and the fragmentation of energetic Uranium beams.

1. INTRODUCTION

Nuclei at double shell closures remain in the focus of experimental and theoretical studies, as they provide key understanding of the multi-nucleon systems, which the nucleus is, by offering the simplest multi-nucleon laboratory possible – the one with up to a few particles or holes coupled to a doubly magic core. Recently, a special attention was given to the doubly closed shell regions far-off stability line, where thanks to the novel techniques, new experimental information is rapidly emerging. These nuclei provide an opportunity to elaborate on the exotic systems placed even further away from the regions currently explored, where new phenomena are expected, particularly for the loosely bound systems with a large neutron excess. In this respect the doubly magic region at ^{132}Sn is of special interest – it is the only such region, for all of the medium to heavy neutron-rich nuclei, currently accessible for detailed studies. The following presentation will constitute perhaps the first attempt to review the experimental situation on nuclei in a very close vicinity of ^{132}Sn. This area of research expands very rapidly, and although an attempt was made by the author to provide a comprehensive and representative review, nevertheless almost by definition the enclosed material is somewhat selective.

As of now, the far-off stability neutron-rich ^{132}Sn region is accessible via fission of heavy targets. In recent years, the range of available techniques for accessing this region has greatly increased as well as the quality of data coming from established methods has strongly improved. One can identify four areas of research at ^{132}Sn that have a strong impact and can be classified via a distinctive method of production of exotic nuclei. These include β decay studies via the traditional route of fission product isotope separation as utilized at ISOLDE at CERN or OSIRIS at Studsvik using "standard" ion sources; these studies are discussed in Section 2. Section 3 provides an overview of a new approach offered by the laser techniques, which assist both the ionization and mass separation processes, and although still under development, can be already credited with exciting results. An alternative approach is offered by studies of prompt γ-rays in spontaneous fission. These studies, discussed in Section 4, have been strongly invigorated by a new generation of multi-detector arrays, EUROGAM or GAMMASPHERE, which provide the necessary level of sensitivity to allow for the first time a detailed investigation of nuclei at ^{132}Sn. Finally, Section 5 includes a brief overview of an important technique that has recently emerged – the fragmentation of energetic Uranium beams. This new approach was already shown at GSI to provide means of production of a host of new nuclei near ^{132}Sn.

This presentation is meant to review the current status of the experimental data at ^{132}Sn. A few comments on the status of the techniques and prospects for the nearest future are enclosed in Section 6. No discussion is included, however, on the major new RNB facilities under development, which will certainly have a dramatic impact on the experimental data at ^{132}Sn in a longer perspective.

CP481, *Nuclear Structure 98*, edited by C. Baktash

2. β-DECAY STUDIES

The first information on the excited states in ^{132}Sn has been obtained more than 25 years ago by Kerek *et al.* (1) at the OSIRIS fission product mass separator at Studsvik. These authors have investigated the β-decay of ^{132}In and identified a transition at 4041(2) keV, which was assigned to de-excite the first excited state in ^{132}Sn. Importantly, this pioneering work has established ^{132}Sn as the strongest bound nuclear system (2) after correction is taken for the $A^{-1/3}$ mass dependence.

Despite the time factor, the β-decay studies at the ISOL facilities remain in the front-line of research on nuclei in the ^{132}Sn region providing the bulk of spectroscopic information. More efficient methods of beam production and a new generation of multi-coincidence experimental techniques provide motivation for a vigorous research in this area. Currently the most active is the Uppsala group of Fogelberg *et al.* with the research conducted mainly at the OSIRIS facility (3). The ANUBIS ion-source developed (4) and implemented a few years ago at Studsvik, provides a strong production enhancement of exotic beams. The basic production process is a thermal neutron induced fission of a small ^{235}U target. Although the beam intensities available at OSIRIS are lower than at ISOLDE at CERN, the beam is basically free from strong impurities (like Cs), while operation of the separator as an in-house facility allows for longer experimental times. There are three lines of the experimental techniques developed or maintained at Studsvik: high precision Q_β spectroscopy, level lifetime measurements in the picosecond range via advanced time-delayed method, and nuclear orientation studies. These techniques will be discussed later in this section. An essential weakness of the facility is a lack of a modern detector array (e.g. a set of Compton-suppressed or Clover-type detectors even in a basic cube configuration), that would allow for very high sensitivity γ spectroscopy including the angular correlation and linear-polarization measurements. That issue, including comments on the experimental possibilities at OSIRIS will be discussed in Section 6.

In contrast to OSIRIS, the ISOLDE separator at CERN is a large-scale facility, which naturally attracts development of sophisticated equipment. The development of laser ion sources is discussed further in Section 3, while some comments on the application of High Resolution Separator, MISTRAL RF spectrometer, ISOLTRAP and REX-ISOLDE for the future studies at ^{132}Sn are enclosed in Section 6. ISOLDE provides much higher beam intensities at ^{132}Sn due to a different production process of spallation, fission or fragmentation of heavy targets by 1 GeV protons. In particular, the beam intensities of heavy In cannot be matched at OSIRIS and thus the β decays of ^{133,134}In are investigated at ISOLDE. The essential progress in studies of the heavy In decays is achieved by the application of multi-coincidences involving β-delayed neutrons. One should note, that the studies discussed further in this section involve the use of "standard" ion-sources with limited chemical selectivity. The range of nuclei in a close vicinity of ^{132}Sn, which structure can be studied that way, is limited to the β decays of In, Sn and Sb.

2.1 Detailed Structure Studies at ^{132}Sn

In the last few years, there has been a series of detailed γ spectroscopic studies at ^{132}Sn of essentially new quality using the new capability of the ion-source and novel experimental techniques at OSIRIS. Detailed spectroscopic information has been obtained on ^{132}Sn (5-7), ^{132}Sb (8), ^{134}Te (9) and ^{133}Sb (10). Studies of the excited states in ^{130}Sn, ^{131}Sn and a new investigation of ^{132}Sb are in progress. The work at ISOLDE includes the investigation of single neutron states in ^{133}Sn (11) via multi-coincidences with β-delayed neutrons. The latter study will be continued in the end of 1998.

^{132}**Sn:** This nucleus represents the doubly magic *core*. The recent study on the excited states in ^{132}Sn from the decay of ^{132}In performed by Fogelberg *et al.* (5–7) provided a dramatic improvement in the knowledge of this key nucleus. In particular, 13 new excited states and 12 new β-transition rates have been established, while the angular momenta and parities of the low-lying states have been firmly determined from the conversion electron measurements. Furthermore, 6 new level lifetimes have been determined via the fast timing βγγ(t) method yielding 20 new electromagnetic transition rates. The first excited negative parity state, the 3⁻ level at 4351.9-keV (see Fig. 1), was shown to have a collective octupole character with B(E3) ≥ 7 W.u. Several particle-hole multiplets have been identified. The three positive parity states near 5.5 MeV were identified as members of the $\pi g_{7/2} g^{-1}_{9/2}$ multiplet, thus representing the first observation of proton p-h states in ^{132}Sn. More than about 20% of the theoretically estimated bound states of ^{132}Sn have been identified. Figure 1, which schematically illustrates a portion of the new data, clearly shows that the first excited negative parity state at 4351.9 keV of spin/parity 3⁻ is of collective character and in particular, cannot be simply assigned as the 3⁻ member of the $\nu f_{7/2} d_{3/2}^{-1}$ multiplet. The broken lines indicate positions of levels not observed experimentally but predicted from a comparison of the interaction strength with the equivalent multiples in ^{208}Pb.

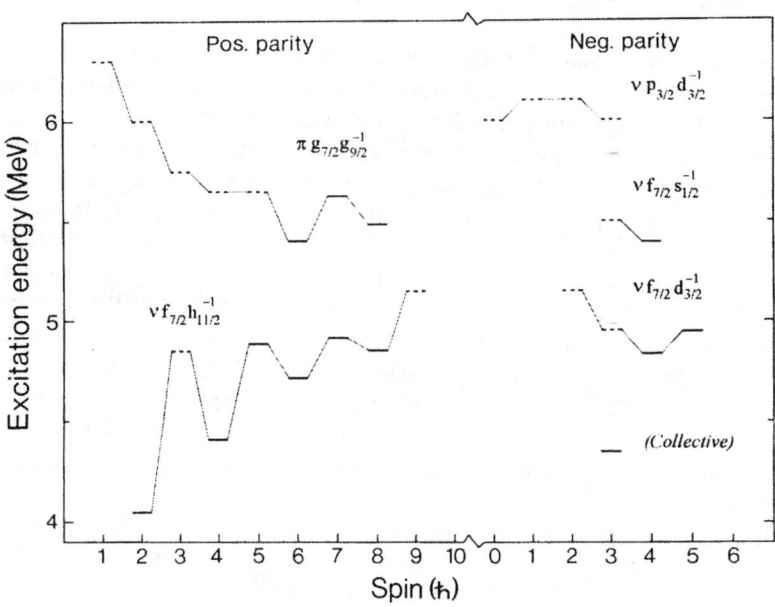

Figure 1 Schematic representation of the particle-hole multiplets in ^{132}Sn identified in the γ spectroscopic measurements by Fogelberg et al. (5,7).

^{130}**Sn:** *core–2n.* The excited states in ^{130}Sn, which can be represented as two-neutron holes coupled to the doubly-magic core, can be populated via β decays of the low-spin, $1^{(-)}$, medium spin, (5^+), and a high-spin, (10^-), decays of ^{130}In. This spin combination, large Q_β energy of more than 10 MeV and strong beam intensities of all isomers of ^{130}In available at OSIRIS and ISOLDE, make this nucleus a good candidate in which a complete set of low-lying multiplets can be established. Studies of this nucleus are in progress at OSIRIS.

^{131}**Sn:** *core–1n.* There has been no detailed study yet of the excited states in ^{131}Sn that would have a comparable sensitivity to the other recent studies at OSIRIS. A new investigation has been initiated (12) that should verify the location of the single neutron-hole states and measure the branching ratios for weak transitions connecting these levels.

^{132}**Sb:** *core+ 1p–1n.* The structure of ^{132}Sb (8) provides a unique opportunity to study the coupling of a valence proton particle to a neutron hole outside the doubly magic ^{132}Sn core. In fact, it represents the best case in the Sn region from A=100 to 132, to study the T=0 matrix elements of the effective interaction. The previous study of ^{132}Sb by Mach *et al.* (8), which included conversion electron, γγ(θ) angular correlation and fast timing βγγ(t) measurements, has established firm spin/parity assignments to the states of low-spin that are strongly populated in the β decay of ^{132}Sn, see Fig. 2. The B(M1) and B(E2) transition rates were obtained for a number of transitions from the measured level lifetimes and δ(E2/M1) ratios. However, no information was obtained on several low-energy transitions reported by Clark *et al.* (13) and Stone *et al.* (14) in the previous study of ^{132}Sb. These transitions may reveal the properties of the important low-lying states of higher spin formed by coupling of the $g_{7/2}$ proton to the $d_{5/2}$ and to the $h_{11/2}$ neutron holes. Consequently, ^{132}Sb is reinvestigated (15) with about two orders of magnitude higher sensitivity than the previous study. The preliminary results confirm the presence of a set of low-energy transitions observed before and the level-lifetime of 100 ns reported by Clark et al. (13). Strong coincidence relations that are observed should reveal several new levels in this nucleus as well as clarify the placements of the low-energy transitions.

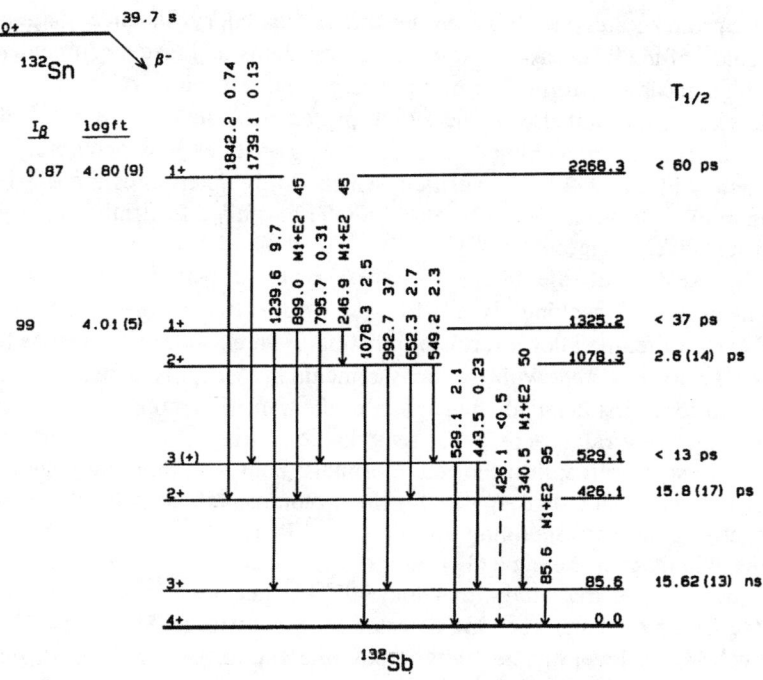

Figure 2 Partial decay scheme of ^{132}Sn to ^{132}Sb summarizing the results of study by Mach *et al.* (8).

^{133}**Sn**: *core+1n.* The properties of the single neutron states in ^{133}Sn have been a subject of many unsuccessful investigations prior to the seminal work by Hoff *et al.* (11). The "traditional" β-decay spectroscopy of ^{133}In was severely handicapped by very strong delayed neutron emission probability of 85(10) %. However, by the same token the levels of interest were successfully investigated at ISOLDE as populated via the delayed neutron emission in the decay of ^{134}In. The single neutron states of $p_{3/2}$, $h_{9/2}$ and $f_{5/2}$ were found at the excitation energy of 853.7, 1560.9, and 2004.6 keV. Crucial to the new investigation were the multi-coincidences involving the β delayed neutrons.

Further work is planned (16), however, to clarify the nature of a few weak γ transitions which show possible association with ^{133}Sn but have eluded a firm identification. It is expected that one of them is de-exciting the $p_{1/2}$ single neutron state, which has not been identified yet. Moreover, an attempt will be made to measure the lifetime of the single neutron $p_{3/2}$ state at 853.7 keV.

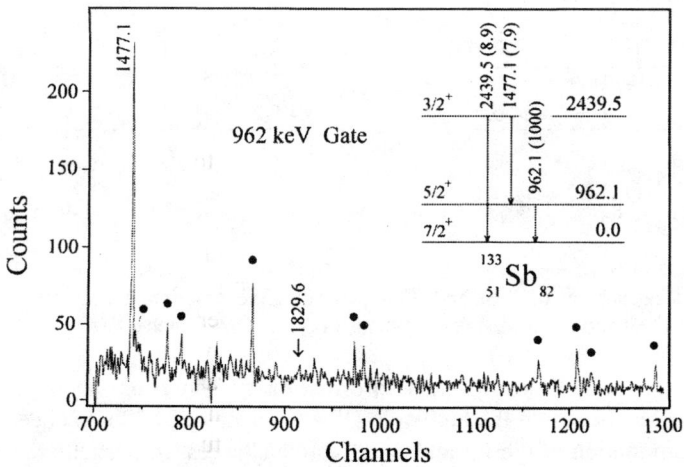

Figure 3 Coincidence spectrum with the 962-keV transition firmly defining the single proton $d_{3/2}$ level at 2439.5 keV in ^{133}Sb; from the study by Sanchez-Vega *et al.* (10).

271

^{133}Sb: *core+1p.* The single proton states in ^{133}Sb have been investigated by Sanchez-Vega *et al.* (10) via a high-sensitivity γ spectroscopic study of the β$^-$ decay of ^{133}Sn. The experiments included γγ coincidences and a series of γ-ray multi-spectra that provided accurate information on the energy and intensities of the transitions. The fast timing βγγ(t) method was used to extract the half-life of the single proton $h_{11/2}$ state. As a result the $d_{3/2}$ state has been identified at 2439.5 keV and its γ-ray branching to the $d_{5/2}$ and $g_{7/2}$ states was accurately determined. Figure 3 illustrates the coincidence gate with the 962-keV transition, which firmly determines the $d_{3/2}$ level. One should note that the previous investigation of ^{133}Sb from the decay of ^{133}Sn (17) has failed to firmly identify this state, although a possible candidate was very tentatively suggested.

For the $h_{11/2}$ state at 2791.3 keV, a half-life of $T_{1/2}$ = 11.4(4.5) ps and a γ-ray branching to the $d_{5/2}$ were measured for the first time (10). The measured γ branching gives B(M1; $d_{3/2} \rightarrow d_{5/2}$)/B(E2; $d_{3/2} \rightarrow g_{7/2}$) = 0.037(3) if the rates are expressed in W.u., while the level lifetime allows for determination of an exceptionally fast M2 rate of B(M2; $h_{11/2} \rightarrow g_{7/2}$) = 0.55(22) W.u. and a E3 rate consistent with the measurements in ^{134}Te, discussed below, B(E3; $h_{11/2} \rightarrow d_{5/2}$) = 22(13) W.u. This work allowed for a first detailed comparison of the transition strengths in the single proton nuclei of ^{133}Sb and ^{209}Bi, and the corresponding rates were found remarkable similar. The firm identification of the $d_{3/2}$ state allowed for a comparison of the spin-orbit splitting of the *d* orbitals, both as proton particles in ^{133}Sb and as neutron holes in ^{131}Sn. The splitting is found about 10% smaller for the proton orbitals than for the neutron ones, which is in contrast to the situation regarding the corresponding *f* orbitals in ^{209}Bi and ^{207}Pb, where it is the neutron holes that show smaller splitting but the magnitude is the same (10).

The level of sensitivity achieved in that study, of about 10^{-5} per decay of ^{133}Sn, was possible due to the first application at OSIRIS of the Compton-suppressed Ge detector for γ spectroscopic studies. This study demonstrates quite clearly that even a much higher level of sensitivity is possible due to the low level of radiation background at OSIRIS, if one would use a compact array of modern detectors. That would open the way for studies of very weak γ branches in the nuclei at ^{132}Sn. Figure 4 provides a summary of the new results in ^{133}Sb.

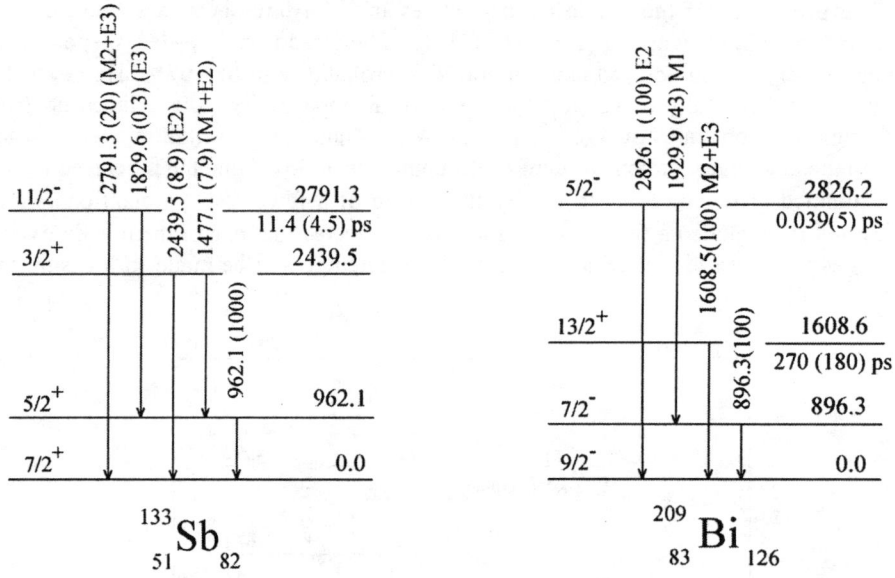

Figure 4 A partial decay scheme of ^{133}Sb and ^{209}Bi illustrating the experimental results on the single proton states in ^{133}Sb and the corresponding states in ^{209}Bi; from Sanchez-Vega *et al.* (10).

^{134}Te: *core+2p.* This nucleus yields (9) an important information on the coupling of a pair of valence proton particles. A detailed spectroscopy of ^{134}Te from the high-spin (J$^\pi$ = 7$^-$) decay of ^{134}Sb has been performed by Omtvedt *et al.* (9) and resulted in the determination of the E2 and E3 effective charges. In particular, the lowest-lying member of the $g_{7/2}h_{11/2}$ proton multiplet was found to be the 9$^-$ state, which was identified as an Yrast isomer with $T_{1/2}$ = 703(26) ps. This state is de-excited by two E3 transitions to the 6_1^+ and 6_2^+ levels. The latter transition involves the single particle transformation $h_{11/2} \rightarrow d_{5/2}$, which is very suitable for defining the octupole effective charge.

272

The E3 transition, that represents the single particle transformation $h_{11/2} \rightarrow d_{5/2}$, has a B(E3) rate of 8.0(1.3) W.u., which is enhanced over the single particle rate due to the coupling of the octupole phonon of the core. For the ^{132}Sn region the effective charges were found (9) to be $e_{eff}(E2) \approx 1.9\ |e|$ and $e_{eff}(E3) \approx 1.9\ |e|$, which lead to the estimate of the B(E2) and B(E3) strength in ^{132}Sn as 11 W.u. and 15 W.u., respectively. The latter value is very consistent with the B(E3) rate of 22(13) W.u. found (10) in ^{133}Sb, as discussed above.

2.2 Highly Accurate Q_β Measurements

The accurate nuclear mass data carry important information on the nuclear structure. At or near closed shells, the mass values give information on the single particle or few particle binding energies and on the interaction energies in few nucleon systems. A few years ago a program of precise Q_β measurements has been initiated at OSIRIS with the aim to provide highly reliable and accurate mass data in the vicinity of closed shell nuclei. The technique of $\beta\gamma$-coincidences is not very reliable for complex and strongly fragmented β decays, which are typical for nuclei in the deformed regions. However, at a closed shell, and particularly at doubly closed shells, the level density is low and the β decay is strongly concentrated in a few intense transitions of the Gamov-Teller type. In this case, the reliability and accuracy of the $\beta\gamma$-coincidence measurements is strongly enhanced. One has to note, that the overall accuracy in the mass data for exotic nuclei remains subject to the accuracy in the mass determinations closer to the line of stability and to the reliability of the decay scheme information relevant to the measurement.

The accepted masses of the neutron rich N=82 isotones in the vicinity of ^{132}Sn have been recently questioned as seriously inconsistent (18) with predictions based on shell model technique (the issue is also discussed in Section 4). In particular, the mass "window" W, constructed from a specific combination of the N = 82 ground state masses, was found to be significantly different from the value derived using the experimental level energies for selected Yrast states in ^{134}Te and ^{135}I. A difference of about 500 keV between W = −3570 keV from spectroscopy and W = −3080(150) keV from the N = 82 masses, strongly contrasts with a difference of less than 5 keV observed for the corresponding N = 126 isotopes at ^{208}Pb. This discrepancy has sparked a wider discussion on the applicability of the shell model reduction techniques, and has generated a keen interest in the method, but these issues go beyond the scope of the review.

In order to address this controversy Fogelberg et al. (19) have re-investigated 14 total β decay energies, Q_β, in the isobaric chains involving ^{132}Sn and the nearest N = 82 isotones. This comprehensive investigation has covered the decays of isotopes of Sn, Sb, Te, I and Xe in the mass range A = 130–135. In particular, since the atomic masses of the far-from-stability nuclides are obtained by adding the Q_β energies to the known mass values of nuclei closer to the stability line, the studies included reinvestigations of the Q_β energies and decay schemes for the decays placed near the line of stability. Figure 5 illustrates two selected end-point energy plots: one for ^{135}Xe and one for ^{134}I.

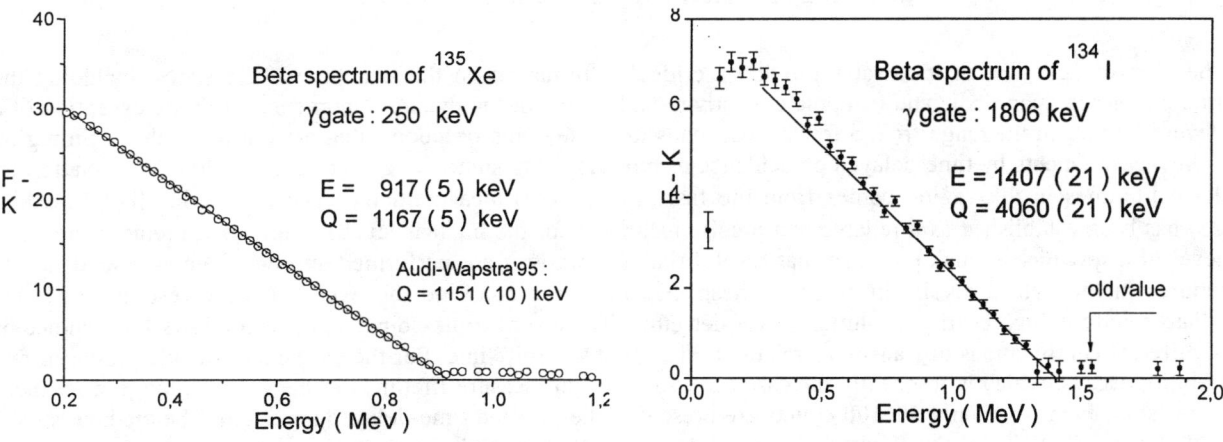

Figure 5. The β energy spectra measured with a Si(Li) detector and gated by γ rays in ^{135}Xe and ^{134}I. A solid line shows a Fermi-Kurie fit to the data. Results from Fogelberg et al. (19).

The new results derived for the atomic masses at ^{132}Sn are significantly more precise that the previously accepted values (20) and differ significantly from them in a few cases. A strong discrepancy of 123 keV was reported (19) for the decay of ^{134}I, which affected the mass determinations also for the more exotic A = 134 nuclei. The error in the previously measured value (21) has been mostly attributed to the fact that a key β energy spectrum, which strongly influenced the previous result, was in fact gated by a γ ray found (19) to be a doublet (with the weaker component of only a few percent of the intensity of the dominant one). A summing effect in the β energy spectrum due to a long series of cascading, and thus coincident, γ transitions could have played a role as well. These effects were corrected by gating on several γ transitions, which were found in the γγ-coincidences to be highly pure. Furthermore, a thin Si(Li) spectrometer was used in the β energy measurements placed at a fair distance from the source to minimize summing.

Table 1. New Mass Excess (ME) values at ^{132}Sn measured by Fogelberg et al. (19) compared to the recent ME compilation by Audi and Wapstra (20).

Nuclide	*ME* (MeV) Fogelberg *et al.*	*ME* (MeV) Audi-Wapstra'95
	N = 82	
^{135}I	-83.793 (8)	-83.788 (23)
^{134}Te	-82.559 (11)	-82.400 (30)
^{133}Sb	-78.951 (28)	-78.960 (80)
^{132}Sn	-76.577 (24)	-76.621 (26)

Table 1 provides a comparison of the new mass data for the N = 82 nuclei to the Mass Excess (ME) values from the most recent compilation by Audi and Wapstra (20). The results selected for Table 1 are those which are involved in the mass window determination discussed above. By taking the new results given in Table 1, one obtains a new value for the mass window W = −3608 (94) keV, which is in an agreement within the experimental uncertainty, with the value of W = −3570 keV from the spectroscopy of excited states (18). Quite clearly the new data resolve the Q_β Puzzle at N = 82 in favor of the shell model technique, but the technique is not expected (19) to be applicable to all cases at ^{132}Sn.

2.3 Measurements of Dynamic Moments

The electromagnetic transition rates provide a critical information on the structure of the states, including the information on the quadrupole and octupole collectivity and details of the dynamic moments. With the exception of a few level lifetimes in the range from a few nanoseconds to the few microseconds, that are known in the ^{132}Sn region from the measurements by time-delayed coincidence techniques using small Ge detectors, the bulk of information on the level lifetimes in this region comes from the fast timing βγγ(t) measurements. The fast timing βγγ(t) method (22,23) has been established as a reliable and precise technique for the nuclear level lifetime measurements down to the level of a few picoseconds provided that careful time calibrations are performed and one follows a sequence of procedures in the data analysis. In essence, these measurements combine the excellent time resolution of fast scintillators and the high energy resolution of Ge detectors. The use of triple-coincidence data allows for a choice of many different combinations of transitions related to the same level lifetime. For the exotic and weakly produced far-off-stability decays, which is frequently the case at ^{132}Sn, one can measure lifetimes in the ~100 ps range with about 20% precision, even when only 50–100 counts are present in the selected time-delayed spectrum. The method is used at ISOLDE and OSIRIS by the Studsvik group of Mach et al. This technique can be also applied to the lifetime measurements of the intermediate states populated in the decay of isomers – as a follow-up experiment to the proposed search for microsecond isomers in the ^{132}Sn region using the projectile fragment separator (FRS) at GSI and discussed in Section 5.

All the spectroscopic studies performed at OSIRIS discussed in section 2.1, have included level lifetime measurements via the fast timing method. Selected measurements were also performed in the ~100 ns range using small Ge crystals to clarify the discrepancies in the nanosecond-range measurements done many years ago when the time reference could have been the main source of uncertainty in the result (see for example results quoted in Ref. (9)).

The measurements in ^{134}Te (9) allowed for determination of the E2 and E3 effective charges and elaborate on the quadrupole and octupole collectivity of the core. In particular, these results led to the estimated (9) B(E2) rate of 11 W.u. for the transition de-exciting the 2_1^+ state in ^{132}Sn, a value comparable to the E2 strength in ^{208}Pb, while the estimated B(E3) rate of 15 W.u. for transition from the 3_1^- state is about 2 times smaller than in ^{208}Pb. A strong similarity for the M1, E2, and M2 transition rates between the corresponding single proton states has been established (10) for ^{133}Sb and ^{209}Bi. This similarity is also extended to include ^{132}Sb (8), where strong configuration mixing has been established for the low-lying multiplets.

In the case of nuclei considered here, the other types of measurements discussed in sections 4–6 currently do not master sufficient statistics to allow for level lifetime determinations. However, an important exception must be noted: Zhang *et al.* (24) has measured the half-life of 80(15) ns for the 6^+ state in the two neutron particle nucleus of ^{134}Sn giving first measure of the E2 effective charge for the $f_{7/2}$ neutrons. This important measurement has been done using the EUROGAM II array and the ^{248}Cm source.

2.4 Static Magnetic Moment Measurements via Nuclear Orientation

A very interesting program of measurements of the static magnetic moments for nuclei in vicinity of ^{132}Sn is run by the Studsvik-Oxford-Kyoto-Niigata collaboration led by Stone *et al.*, using the OSIRIS dilution refrigerator for the Low Temperature Nuclear Orientation (LTNO) studies. The unit was brought in 1994 from Daresbury and installed on-line to the OSIRIS ISOL facility in 1995. Precision values of magnetic moments are obtained by nuclear magnetic resonance on oriented nuclei. The experiments at OSIRIS have shown that the conditions for the on-line studies of short-lived nuclides differ strongly from traditional LTNO work using long-lived sources. The traditional work is made using samples that have been annealed after the implantation of radioactive atoms. As this is not possible in on-line work, it is important to prepare the Fe implantation foils in such a way that a high fraction of the radioactive atoms goes into good sites as a result of the implantation process itself. However, the precision of magnetic resonance results is as high as in a traditional work, provided that a reference isotope of the same species is implanted in the foil being used during actual measurements of unknown magnetic moments.

The recent on-line studies were focussed on the odd Sb (25) and Te (26) nuclei in the vicinity of doubly magic ^{132}Sn. A summary of the magnetic moment measurements on the odd-Sb nuclei, including the single proton particle system ^{133}Sb, has been provided by Stone *et al.* (25). In particular, the magnetic moment of $\mu = 3.00(1)$ μ_N has been measured for the ground state of ^{133}Sb, which represents one proton coupled to the ^{132}Sn core. This result provided the first sensitive test in medium-heavy nuclei of mesonic exchange current and core polarization effects (25). On the other hand, the data on the $7/2^+$ ground state moments in $^{123-133}$Sb revealed a systematic variation of the moments as a function of the neutron number. This has been interpreted (25) as evidence for core contributions with negative collective g factors supporting microscopic calculations.

3. LASER ISOTOPE AND ISOMER SEPARATION OF HEAVY Ag

In order to search for very exotic nuclei one needs exceptionally high sensitivity, which can be only achieved with isotopically clean beam. This goal can be reached either via high-resolution mass separation, multi-fold coincidences, or combined time-of-flight and $\Delta E/E$ identification. Recently, a concentrated effort has been made to extract spectroscopic information on the heavy Cadmium nuclei at N = 82 from the decay of Silver. The collaboration of Mainz-Maryland-Troitzk-CERN led by Kratz *et al.* (27) has made a remarkable progress by using a chemically selective ion source. The aim of the group is to extract the spectroscopic information of the N = 82 nuclei below ^{132}Sn that is strongly relevant to the r-process nucleosynthesis. The particular aim is to directly verify the N = 82 shell quenching (a profound weakening of the gap at N = 82 for Z < 50) that has been stipulated by the properties of nearby nuclei. The new results obtained at ISOLDE have been recently summarized by Kratz *et al.* (28). The data indicate quite convincingly that indeed quenching is taking place at N = 82. Although, the Collaboration did not study any of the nuclei under review here (their next effort is directed towards two proton hole ^{130}Cd), yet the crucial evidence they present has a profound implication on the structure of nuclei at ^{132}Sn.

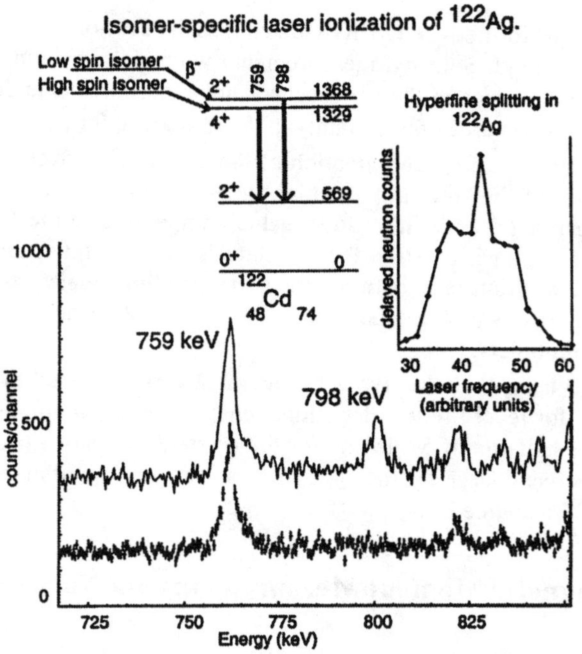

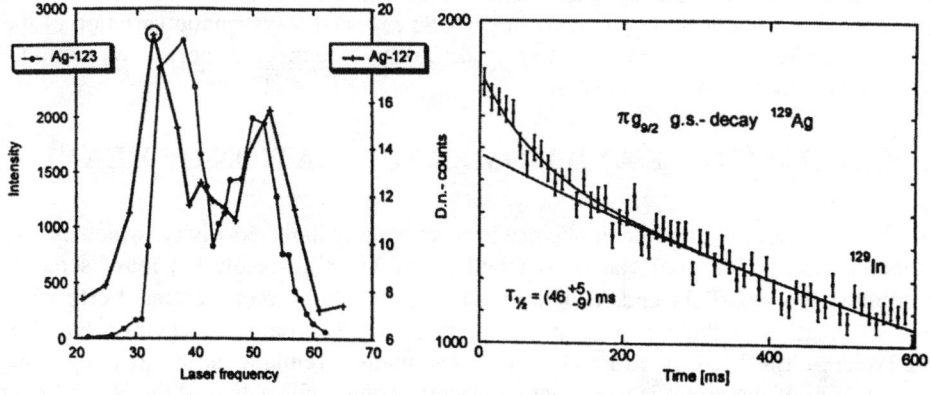

Figure 6. Fragments of γ spectra from the decay of ^{122}Ag illustrating the effect of isomeric separation by a judicious selection of laser frequency; from Kratz *et al.*, (28). See text for discussion.

The production of exotic neutron-rich Silver nuclei is achieved via a chemically selective laser ion source at the ISOLDE facility. The system is supported by a host of special precautions including micro-gating procedures to suppress the background arising from the surface-ionized In and Cs isobars. Since the production rate is quite small, an additional selectivity must be provided. Recent progress in these studies has been achieved by the ability to select, via laser frequency, the high-spin decay while suppressing production of the low-spin isomer. Figure 6 illustrates the effect of isomeric separation in the decay of ^{122}Ag, which proceeds via low- and high-spin isomers. The insert to the figure illustrates a composite peak: a strongly split peak associated with the high-spin 9$^-$ configuration and a narrow peak in the center arising due to the low-spin, 1$^-$. Indeed when the laser frequency is shifted away from the center (to 35 units), ionization of the low-spin isomer is suppressed and the γ-spectrum (the lower spectrum in Fig. 6) indicates only decay of the high-spin isomer.

Figure 7. The left panel illustrates broad-band laser frequency scans for ^{123}Ag and ^{127}Ag using delayed neutron counting. The right panel shows the delayed neutron decay spectrum at A = 129 that was measured with the laser frequency set at the maximum of the high-spin yield (open circle at 30 units). This additional selection has permitting the first identification of the ^{129}Ag decay, which is now clearly visible on top of the activities from two isomers in ^{129}In; from Kratz *et al.*, (28).

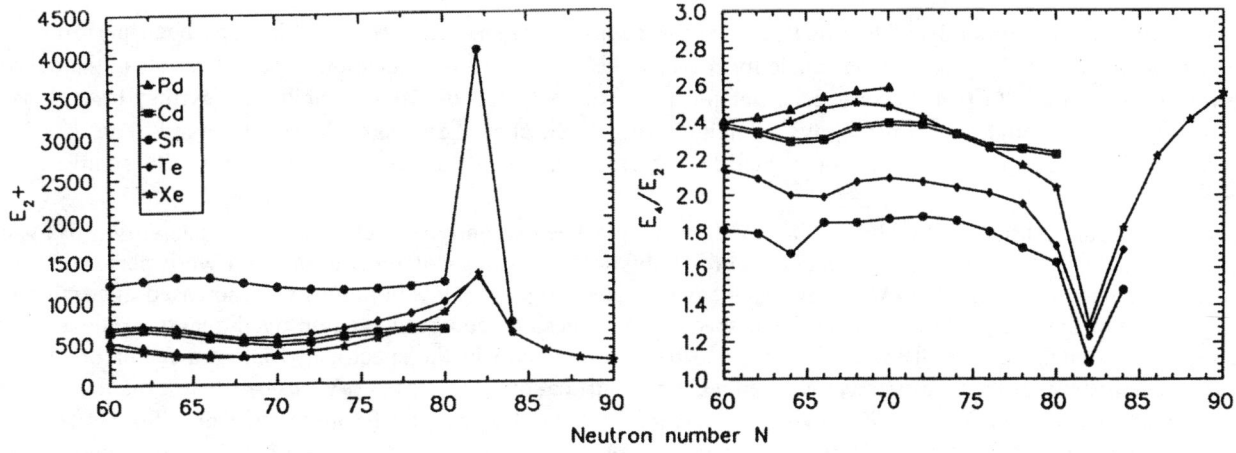

Figure 8. Left panel illustrates systematics of first excited 2+ states, while the right panel shows ratios of energies of the first excited 4+ and 2+ states in neutron-rich Pd (Z=46) to Xe (Z=54) isotopes; from Kratz *et al.*, (28).

With the recently obtained excitation energies of the first excited 2^+ and 4^+ states in heavy Cd nuclei (28), the systematics of these states is now complete up to and including ^{128}Cd at N = 80, see figure 8. It is clear from these systematics, that when approaching N = 82, the Cd nuclei show a remarkable deviation from the general pattern. This includes lowering of the excitation energy for the 2_1^+ states from 657 keV in ^{126}Cd (N = 78) to 645 keV in ^{128}Cd (N = 80), and a constant ratio of the $4^+/2^+$ energies of about 2.2 for the three heaviest Cd . Furthermore, in a recent study of excited states in ^{134}Sn by Zhang et al. (24), a strong lowering in energy of the 2_1^+ state was found in ^{134}Sn (N = 84) in contrast to almost uniform values for the even Sn isotopes in the range N = 52–80. These data imply weakening of the N = 82 shell gap below ^{132}Sn for Z < 50 and perhaps a similar effect for heavy Sn nuclei at N > 82. A search for the first excited state in ^{130}Cd at N = 82 is planned for the next Ag run at ISOLDE expected (27) in the late part of 1998.

4. HIGH SPIN SPECTROSCOPY AT ^{132}Sn

β decay studies, which were described in Section 2 and 3, frequently yield information on the long-lived high-spin isomeric traps often observed at doubly closed shells. Yet the preferred technique to study high-spin states in this region are prompt γ rays from the spontaneous fission of heavy nuclei. The drawback is low yield for the exotic nuclei that must be compensated by high sensitivity of a multi-detector array. Only recently, full-scale research programs on prompt fission have been initiated coincidental with the arrival of the EUROGAM and GAMMASHPERE arrays. The work on prompt-fission at GAMMASPHERE has been strongly concentrated on the octupole soft nuclei at ^{146}Ba and the γ-soft region at ^{108}Ru (see for example a review of these studies by Hamilton *et al.* (29)). However, two groups working with the EUROGAM data on the fission of ^{248}Cm, Daly *et al.* (30) and Urban *et al.* (31), have started a highly successful program focussed on the nuclei at ^{132}Sn. Daly et al. have summarized (30) recent studies on a series of N = 82, 83 and 84 isotones at ^{132}Sn. Our attention here is given to the A = 134 nuclei.

The main tool of investigation are high-fold γ coincidences. However, the prompt fission studies are truly complex, due to a tedious process of identification of the required channel. A given data set is quite massive by including coincidences due to a few hundred different nuclei. The main process of identification involves complementary light and heavy fission fragments with a well defined proton value Z, while the mass number A is somewhat less defined due to a statistical nature of the neutron release process in fission. If the fragments complementary to the exotic ones, happen to be placed closer to the stability line, then they are usually somewhat known from other studies. The latter could include β-decay work. However, the results on β decay may become critical (as the only source of information), if the nucleus of interest is exotic and neutron-rich. In particular, β-decay work may identify one or two Yrast transitions and may provide the starting blocks for reconstruction of the high-spin decay. Study of ^{134}Te (18) represents such a case. For ^{134}Sn, there was no information on the excited states prior to the study (24), while in ^{134}Sb (30,31), only a few states of very low-spin have been identified in β decay. However, population of these states by an indirect feeding from high-spin levels, seems to be too weak.

^{133}**Sn:** *core+1n.* Initial study of this nucleus, via the fission product γ rays from ^{248}Cm, has been performed by Bhattacharyya *et al.* (32). A search was made for the 1561-keV $\nu h_{9/2} \rightarrow \nu f_{7/2}$ transition, reported earlier in the delayed neutron β-decay of ^{134}In (see Section 2.1), but no trace was detected in cross coincidences with γ transitions in ^{110,111,112}Pd, even though the Yrast γ rays of neighboring ^{132}Sn and ^{134}Sn were clearly observed. This has been attributed (32) to the anticipated drastic reduction in γ-ray cascade intensities due to the unusually small neutron separation energy.

The same data set obtained with EUROGAM II array, has been recently reinvestigated by Urban *et al.*, (31) using improved analysis techniques. In particular, more sophisticated sortings of the fission data were performed with selected add-back in the EUROGAM Clover detectors, higher order energy calibrations and increased dispersion and energy range of the analyzed three-dimensional spectra (31). These procedures have improved quality of the analyzed data giving an enhancement of the critical peak to background ratio in the spectra by almost a factor of two. An increased sensitivity of the measurement allowed a clear identification of the 1561-keV $\nu h_{9/2} \rightarrow \nu f_{7/2}$ transition in triple coincidence with the 534–349 keV $4^+ \rightarrow 2^+ \rightarrow 0^+$ cascade in ^{112}Pd, the strongest fission partner of ^{133}Sn. However, no candidates were found for γ decay of the $\nu i_{13/2}$ level. From the measured yield for the 1561-keV transition, authors concluded (31) that prompt γ decay of excited states in ^{133}Sn is strongly suppressed, which can be explained by small neutron separation energy and low density of the high-spin states in this single valence neutron nucleus.

^{134}**Sn:** *core+2n.* The first information on properties of excited states in the two neutron particle system ^{134}Sn, has come from the work of Zhang *et al.* (24) who have used EUROGAM II to study spontaneous fission of ^{248}Cm. Three transitions of energy 725.4, 347.3, and 173.7 keV were identified in coincidence with known γ rays from the fission partner nuclei of ^{110,111,112}Pd. The observed sequence 174–347–725 was classified as de-exciting the neutron $f_{7/2}^2$ multiplet. A rather low excitation energy of the 2_1^+ state in ^{134}Sn, 725.4 keV, indicates weakening of the Z = 50 shell gap for N > 82, quite in accord with theoretical predictions (see discussion in Section 3). A half-life measurement of 80(15) ns for the 1246.4-keV state, which spin is expected to be 6^+, gives the B(E2; $6^+ \rightarrow 4^+$) rate of 36(7) $e^2 fm^2$ and allows an estimate of E2 effective charge for the $f_{7/2}$ neutron configuration.

^{134}Sn has been reinvestigatied by Urban *et al.* (31). The latter study has revealed an addition state at 2507.9 keV de-excited by a 1261.5-keV line to the 6^+ level at 1246.4 keV. The new state has been interpreted (31) as 8^+ member of the $\nu f_{7/2}\nu h_{9/2}$ multiplet based on very close agreement with the prediction (measured 1261.5 keV vs predicted 1270 keV), that was obtained from scaling interaction from the corresponding multiplet in ^{210}Pb, nuclear masses at ^{134}Sn and energy of the $\nu h_{9/2}$ state in ^{133}Sn.

^{134}**Te:** *core+2p.* High spin states in ^{134}Te have been investigated by Butler-Moore *et al.*, (33) using a 20 Compton suppressed Ge detector Close-Packed Ball at the Holifield Heavy Ion Research Facility. Three series of measurements have been performed with ^{252}Cf and one with ^{242}Pu. In the first run, a gas-filled ionization chamber was used to allow measurements of fission-fragment–γ–γ coincidences, with the purpose to suppress γ rays emitted in radioactive decays. The analysis included triple γ–γ–γ coincidences, where two gates were set on transitions in the two partner-fragments to enhance selection process. In both studies (with ^{252}Cf and ^{242}Pu, thus with different partner-fragments), two strong γ rays of energy 2322 and 2865 keV have been identified in ^{134}Te and assigned as transitions from the 4015- and 4558-keV levels feeding the 6_1^+ state at 1693 keV. (A note of interest: such strong and high-energy transitions are rare in spontaneous fission, thus the case has a practical application for detection the presence of fissionable material in weapons (29).) The first of these transitions, at the energy of 2321.8 keV, has been earlier identified in the decay of ^{134}Sb $\rightarrow$ ^{134}Te by Meyer *et al.* (34), while the level it de-excites, at 4013.2 keV, has been interpreted by Heyde and collaborators (35) as a likely candidate for the 9^- member of the $\pi g_{7/2}\pi h_{11/2}$ multiplet. Both transitions have been seen in the detailed β-decay study by Omtvedt *et al.* (9) who identified the 4013.3-keV state as an isomeric trap and indeed, established its identity as 9^- member of the $\pi g_{7/2}\pi h_{11/2}$ multiplet (see a discussion in section 2.1). The second state, at 4557.5 keV, has been interpreted (9) as 6^- member of the same multiplet.

A subsequent fission study by Zhang *et al.* (18) using EUROGAM II (with 52 escape-suppressed spectrometers incorporating 124 Ge detector elements and supported further by four additional LEPS detectors), has substantially extended the earlier work (33). About 12 additional γ rays have been incorporated to the decay scheme in prompt fission. Here, the levels at 4014, 4299 and 4563 keV, observed earlier in β decay (9), have been also interpreted as members of the $\pi g_{7/2}\pi h_{11/2}$ multiplet. However, six new states build on top of the 4557-keV level, as well as the 4557 keV level itself, have been interpreted as positive-parity core-excited states, with the configuration $\pi g_{7/2}^2 \nu f_{7/2} h_{11/2}^{-1}$.

The study by Zhang *et al.* (18) included two N = 82 cases: the two proton nucleus ^{134}Te, and ^{135}I – a system with three valence protons outside of the ^{132}Sn core. The spin-parity assignments and interpretation of the excited states have been obtained via shell model calculations with the nucleon-nucleon interaction taken either from the know data at ^{132}Sn or estimated from the multiplets in ^{208}Bi and ^{210}Bi with scaling as $A^{-1/3}$ to take account of the nuclear size variation (2). The excellent overall agreement with experiment has been noted for relative energies of the excited states but not when the ground state masses were included. This led to a rather provocative conclusion (18) "that one or more of the accepted N = 82 masses is inaccurate by considerably more than the estimated errors", which can be seen as a rare example of a case when disagreement with theoretical model interpretation is taken as a proof not of the deficiency of the model, but of the inaccuracy in experiment. A subsequent experimental reinvestigation of the mass values at ^{132}Sn by Fogelberg *et al.* (19), already discussed in Section 2.2, confirmed the model interpretation to be correct. This is, however, a very significant issue for the interpretation of the high-spin fission data at ^{132}Sn, where due to weak intensities and lack of more powerful experimental possibilities (angular correlations, linear polarization data, etc...), it is the shell model estimate that remains the most important tool in the spin/parity assignments and interpretation of the levels. The situation is reminiscent of the first β-decay studies in this region. A steady progress in the experimental techniques and analysis is clearly visible, as seen in the case of ^{134}Sb discussed below.

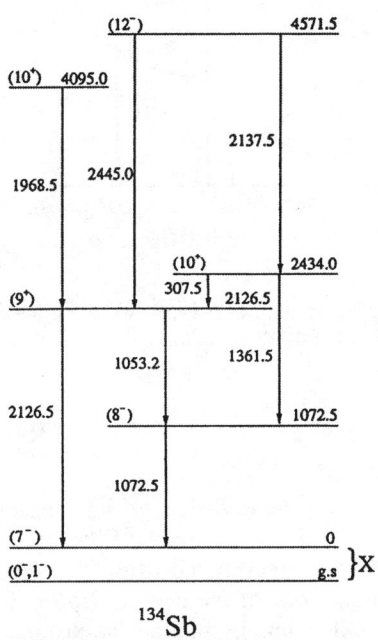

Figure 9. A partial decay scheme of ^{134}Sb obtained in the work of Urban *et al.* (31). The excited states at 1072 and 2126 keV have been first reported by Bhattacharyya *et al.* (32). The excitation energy of the 7$^-$ level remains unknown, which is symbolically indicated by the excitation energy X. Figure from Ref. (31).

^{134}**Sb:** *core+p+n*. Three γ rays of energy 1053, 1073 and 2126 keV, have been assigned to ^{134}Sb in the study of high spin states by Bhattacharyya *et al.*, (32). In the analysis, first the Yrast cascades were identified in $^{110-113}$Rh using the known Yrast transitions in ^{131}Sb and ^{133}Sb. Then the newly identified lines in Rh were selected as gates to identify transitions in ^{134}Sb. The three γ rays in ^{134}Sb define two levels at the excitation energy of 1073 and 2126 keV relative to the 7$^-$ isomeric state (see Figure 9). The observed states are interpreted as due to $(\pi g_{7/2} \nu f_{7/2})7^-$, $(\pi g_{7/2} \nu h_{9/2})8^-$, and $(\pi h_{11/2} \nu f_{7/2})9^+$, respectively. The approximate excitation energies for these configurations have been calculated using the single particle energies and the proton-neutron interaction energies estimated from the corresponding configurations in ^{210}Bi. The resultant values are 1120 and 2140 keV, which provide a convincing agreement with the experimental energies of 1073 and 2126, respectively. (A different interpretation of the 2126-keV states has been proposed by Oros *et al.* (36). According to authors of Ref. (36) these states are due to the $(\pi g_{7/2} \nu f_{7/2})7^-$, $(\pi g_{7/2} \nu h_{9/2})8^-$, and $(\pi g_{7/2} \nu i_{13/2})9^+$ configurations, with the calculated relative excitation energies of 1071 and 2021 keV. Here, the unknown single particle energy for $\nu i_{13/2}$ neutron is taken as $\varepsilon(i_{13/2}) = 1.90$ MeV. This requires a considerable reinterpretation of the results not only for ^{134}Sb, but also for ^{133}Sn and ^{135}Te; for details see Ref. (36).)

A recent reinvestigation of the same data set by Urban *et al.* (31), see a discussion on [133]Sn earlier in this Section, has allowed for a considerable extension of the level scheme. Of particular interest were the 2434.0- and 4571.5-keV states, see Fig. 9, which were interpreted (31) as the $(\pi g_{7/2} \nu i_{13/2})10^+$ and $(\pi h_{11/2} \nu i_{13/2})12^-$ configurations, respectively. The analysis within the shell model, which treated the $\nu i_{13/2}$ single particle energy as unknown, provided an empirical estimate for it as 2694 keV (see also a discussion by Daly *et al.* in (30)). Another important aspect of this investigation was the good quality of data obtained in fission that allowed for the angular correlation measurements. Some of these results are illustrated in Fig. 10.

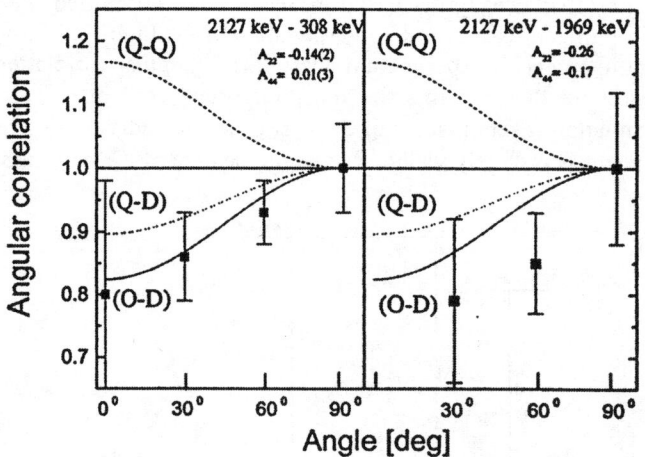

Figure 10. Angular correlations for the 2127–308 and 2127–1969 keV cascades in [134]Sb plotted against the expected quadrupole–quadrupole (Q-Q), quadrupole–dipole (Q-D) and octupole–dipole (O-D) correlations; from Urban *et al.* (31).

5. PROJECTILE FISSION OF ENERGETIC [238]U

In the first pioneering experiment with fission of relativistic [238]U ions on a Lead target performed at the fragment separator FRS at GSI-Darmstadt, Bernas *et al.* (37) have discovered fifty-three new neutron-rich fission-fragments. In a subsequent study using relativistic [238]U on Be target additional fifty-eight new fission products in the lighter neutron rich region have been discovered (38). Although, none of the new neutron-rich nuclei were in the very close vicinity of [132]Sn considered here, nevertheless this novel technology will have a profound influence on the study of this region in the near future. The strength of the method is a nonselective extraction of the fission fragments and full A and Z separation. Furthermore, study of the fission mechanism itself becomes possible in these experiments. The most neutron rich nuclei are produced via low-excitation energy process dominated by the Coulomb fission of the Uranium projectile (39). For low energy fission, the velocities of forward-emitted fragments are higher by about 5% as compared to the projectile velocity, which allows for their selection.

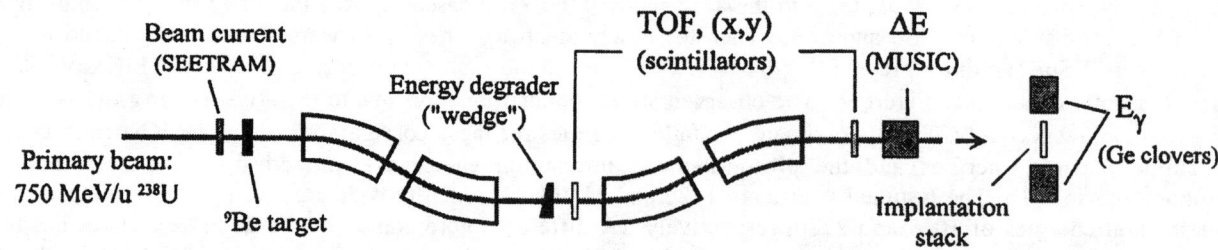

Figure 11. Schematic drawing of the FRS operating in the achromatic mode with a beam catcher and set of Ge detectors – a setup for the delayed γ-ray spectroscopy of fission fragments. Fragments are identified in the second stage of the spectrometer by a ΔE–ToF–Bρ method using two plastic scintillators (for Time of Flight) located in the intermediate and final focal planes, and with the ionization chamber MUSIC (for ΔE); by courtesy of M. Hellström (40).

Recent upgrades of the beam intensities at the GSI facility allows for a more detailed spectroscopy of the neutron rich nuclei obtained from projectile fission. The Lund-Darmstadt-Orsay-Warsaw-Grenoble-Caen collaboration headed by Hellström et al. (40) will explore in early 1999 the time-delayed γ-ray spectroscopy in the regions around doubly magic ^{78}Ni and ^{132}Sn, see Fig. 11. Many high-spin isomers are expected near ^{132}Sn, including the case of special interest: ^{130}Cd. The challenging question for ^{130}Cd is whether any weakening of the N = 82 shell gap, discussed in Section 3, will washout the expected 8^+ isomer for the $g_{9/2}^{-2}$ proton-hole configuration.

6. PROSPECTS FOR THE NEAR FUTURE

The prospects for future studies at the exotic region of ^{132}Sn, depend very much on a single factor: the intensity of the exotic beams. To meet this end, there is a long-term vigorous program of source development at ISOLDE, and one notes a steady improvement in the beam quality at GSI. For OSIRIS, there is a concept of a new type of ion source (3) based on ^{238}U target and utilization of a sizable flux of fast neutrons at the R-2-0 reactor. For the studies of prompt fission via multi-detector arrays, further improvement is in the more efficient detector systems and in the technology of data analysis as shown by Urban et al. (31). However, there is still a choice of different sources, like the use of thermal neutron induced fission of ^{235}U, which has shown in the test studies (41) a strong yield enhancement at ^{132}Sn.

The ISOLDE facility has enormous potential for the spectroscopy at ^{132}Sn. The most interesting prospect is to use excellent beam intensities in the region of ^{132}Sn for Coulomb excitation in the inverse kinematics. That is certainly on the agenda for REX-ISOLDE, but in a later stage of development. More imminent are the prospects of direct mass measurements in the ISOLTRAP, where very high precision could be expected, and with the MISTRAL RF-spectrometer. A steady progress in the performance of laser ion sources and related measurement techniques gives a well-deserved optimism that in the very near future some details will be learned from the β spectroscopy of ^{130}Ag and ^{130}Cd. On the other hand the β spectroscopy from multicoincidences with neutrons, championed by Hoff et al. (11), is still waiting for its first full-length run on the highly interesting cases of ^{133,134}Sn. That run is scheduled for late 1998.

The β spectroscopy of the ^{132}Sn region at OSIRIS is still a long way before exhausting its experimental possibilities. The highly precise Q_β measurements should be extended to include the A = 129 chain and below, which is an interesting region from the point of view of astrophysics and weakening of the N = 82 shell gap, discussed in Section 3. Recent measurements of the static magnetic moments included highly interesting case of ^{132}Sb, with the first results soon to be released. Further such measurements are planned for nuclei in this region. The detailed γ spectroscopy of ^{130,131}Sn and ^{132}Sb, including highly important dynamic moment measurements, is in progress. However, as a necessary and logical step for the future, one has to utilize at OSIRIS a compact array of modern detectors, which will allow for a high precision determination of weak branches and presence of weakly populated states, like for example the missing single proton $s_{1/2}$ state in ^{133}Sb. Such measurement would require about 1–2 orders of magnitude higher sensitivity than obtained in the recent work of Sanchez-Vega et al. (10). Furthermore, strongly needed good quality angular correlation and linear polarization measurements would then be also possible. OSIRIS with clean, good intensity beams and flexible beam time schedule provides an ideal place for such measurements.

The prompt fission studies show a steady progress. The most interesting will be the first results on the excited states in the heavy In isotopes at N = 81–83. However, new and challenging results from fission studies are expected basically on all nuclei in this region. Measurement of the isomeric states at ^{132}Sn at the FRS spectrometer at GSI-Darmstadt, will open a series of further experiments, like for example direct mass measurements in the storage ring.

ACKNOWLEDGMENTS

It is a great pleasure to acknowledge contribution, advice and comments from many co-workers in this exciting field of study. Particular thanks to M. Bernas, J. Blomqvist, P. Daly, B. Fogelberg, M. Hellström, K.-L. Kratz, W. Kurcewicz, K. Mezilev, M. Sanchez-Vega, and W. Urban who contributed their materials or information. Appreciation is also expressed to G. Bollen, J. Hamilton, M. Pfützner, H. Ravn and P. Van Duppen for many useful comments on the subject. This work was supported in part by the Swedish Natural Science Research Council.

REFERENCES

1. Kerek, A., Holm, G. B., De Geer, L.-E., and Borg, S., Phys. Lett. **B 44**, 252 (1973).

2. Blomqvist, J., In *Proceedings to the 4ᵗʰ International Conference on Nuclei Far From Stability*, Helsingör, 1981 (CERN Report No. 81-09, CERN, Geneva, 1981) p. 536.
3. Fogelberg, B., Mach, H., and Jacobsson, L., in *Research with Fission Fragments*, eds.: T. von Egidy et al., (Singapore, World Scientific, 1997) p. 69.
4. Jacobsson, L., *et al.*, Nucl. Instrum Methods **B 26**, 223 (1987), see also: Fogelberg, B., *et al.*, Nucl. Instrum. Methods **B 70**, 137 (1992).
5. Fogelberg, B., *et al.*, Phys. Rev. Lett. **73**, 2413 (1994).
6. Fogelberg, B., *et al.*, In *Proceedings of the 6ᵗʰ International Conference on Nuclei Far From Stability and the 9ᵗʰ International Conference on Atomic Masses and Fundamental Constants*, eds.: R. Neugart and A. Wöhr, Institute of Physics Conference Series No. 132 (IOP Publishing Ltd, Bristol, 1993) p. 569.
7. Fogelberg, B., *et al.*, Physica Scripta **T 56**, 79 (1995).
8. Mach, H., *et al.*, Phys. Rev. **C 51**, 500 (1995).
9. Omtvedt, J. P., *et al.*, Phys. Rev. Lett. **75**, 3090 (1995).
10. Sanchez-Vega, M., *et al.*, Phys. Rev. Lett. **80**, 5402 (1998).
11. Hoff, P., *et al.*, Phys. Rev. Lett. **77**, 1020 (1996).
12. Sanchez-Vega, M., *et al.*, to be published.
13. Clark, R. G., Glendenin, L. E., and Talbert W. L., in *Proccedings of the Symposium on the Physics and Chemistry of Fission*, Rochester, 1973 (IAEA, Vienna, 1974), Vol. 2, p. 221.
14. Stone, A. C., Faller, S. H., and Walters, W. B., Phys. Rev. **C 39**, 1963 (1989).
15. Mach, H., *et al.*, to be published
16. ISOLDE IS338 Collaboration, spokesman P. Hoff, Department of Chemistry, University of Oslo, Norway.
17. Blomqvist, J., *et al.*, Z. Phys. **A 314**, 199 (1983).
18. Zhang, C. T., *et al.*, Phys. Rev. Lett. **77**, 3743 (1996).
19. Fogelberg, B., *et al.*, to be published; see also: Fogelberg, B., *et al.*, in *Proceedings of the 2ⁿᵈ International Conference on Exotic Nuclei and Atomic Masses, ENAM98*, 23–27 June 1998, Shanty Creek, Michigan, USA.
20. Audi, G., and Wapstra, A.H., Nucl. Phys. **A 595**, 409 (1995).
21. Mezilev, K.A., *et al.*, Physica Scripta **T 56**, 272 (1995).
22. Mach, H., *et al.*, Nucl. Instr. Methods **A 280**, 49 (1989), and references therein.
23. Mach, H., *et al.*, Nucl. Phys. **A 523**, 197 (1991).
24. Zhang, C. T., *et al.*, Z. Phys. **A 358**, 9 (1997).
25. Stone, N. J., *et al.*, Phys. Rev. Lett. **78**, 820 (1997).
26. White, G., *et al.*, to be published.
27. ISOLDE IS333 Collaboration, spokesman K.-L.Kratz, Department of Chemistry, Mainz University, Germany.
28. Kratz, K.-L., in *ENAM 98, Exotic Nuclei and Atomic Masses*, 23-27 June 1998, Bellaire, Michigan, USA, AIP Conference Proceedings 455, eds. B.M. Sherrill, D. J. Morrissey, and C.N. Davids (AIP, New York, 1998), pp. 827-836.
29. Hamilton, J. H., *et al.*, Prog. Part. Nucl. Phys. **35**, 635 (1995).
30. Daly, P. J., *et al.*, in *Proceedings of the 6ᵗʰ International Spring Seminar on Nuclear Physics: Highlights of Modern Nuclear Structure*, 18–22 May 1998, S. Agata Sui Due Golfi, Italy.
31. Urban, W., Kurcewicz, W., Nowak, A., *et al.*, to be published.
32. Bhattacharyya, P., *et al.*, Phys. Rev. **C 56**, R2363 (1997).
33. Butler-Moore, K., *et al.*, in *Proceedings of the 6ᵗʰ International Conference on Nuclei Far From Stability and the 9ᵗʰ International Conference on Atomic Masses and Fundamental Constants*, eds.: R. Neugart and A. Wöhr, Institute of Physics Conference Series No. 132 (IOP Publishing Ltd, Bristol, 1993) p. 551.
34. Meyer, R. A., and Henry, E. A., in *Proceedings of the International Conference on Nuclear Spectroscopy of Fission Products*, ed. T. von Egidy (Institute of Physics, Bristol, 1980) Vol. 51, p. 59.
35. Heyde, K., *et al.*, Phys. Rev. **C 25**, 3193 (1982).
36. Oros, A. -M., *et al.*, in *Proceedings of the 6ᵗʰ International Spring Seminar on Nuclear Physics: Highlights of Modern Nuclear Structure*, 18–22 May 1998, S. Agata Sui Due Golfi, Italy.
37. Bernas, M., *et al.*, Phys. Lett. **B 331**, 19 (1994).
38. Bernas, M., *et al.*, Phys. Lett. **B 415**, 111 (1997).
39. Donzaud, D., *et al.*, Eur. Phys. J. **A 1**, 407 (1998).
40. Proposal S210, GSI-Darmstadt, spokesman, M. Hellström, Physics Department, Lund University, Sweden.
41. Hamilton, J., private communication.

Spectroscopy of Neutron-rich Nuclei Populated in the Spontaneous Fission of ^{252}Cf and ^{248}Cm

A.G.Smith, G.S.Simpson, J.Billowes, J.L.Durell, W.R.Phillips, P.J.Dagnall, S.J.Freeman, M.Leddy, A.A.Roach, J.F.Smith

Department of Physics and Astronomy, University of Manchester, M13 9PL, U.K

A.Jungclaus, K.P.Lieb, C.Teich

Physikalisches Institut, Universität Göttingen, D-37073 Göttingen, Germany

N.Schulz, B.J.P.Gall, F.Hoellinger

Centre de Recherches Nucléaires, IN2P3-CNRS, Université Louis Pasteur, 67037 Strasbourg, France

I.Ahmad, J.Greene

Argonne National Laboratory, Argonne, IL 60439, U.S.A

A.Algora

Laboratoi Nazionali di Legnaro, 35020 Legnaro (PD), Italy

INTRODUCTION

Spontaneous fission provides a very natural production mechanism for neutron-rich nuclei. The large excesses of neutrons over protons in heavy systems near the valley of stability, coupled with the tendency for the charge-to-mass ratio to be preserved in the fission process, results in fission fragments that lie well to the neutron-rich side of stability. The excitation energy of the primary fragments produced at scission results in the evaporation of neutrons (Fig 1) to produce secondary fragments whose excitation energy is dissipated through γ-ray emission. In the last few years the study of such γ rays has produced a dramatic increase in the available information on the structure of neutron-rich nuclei. This renaissance in fission-fragment spectroscopy has been due primarily to the much increased resolving power of large arrays of germanium γ-ray detectors such as Euroball and Gammasphere. Through the improved detection efficiency and high granularity, these arrays have made it possible to obtain reasonable rates for the detection of three coincident γ rays out of the cascades of multiplicity ten that are typically produced in spontaneous fission. In addition to the determination of the energies of excited levels, developments in spectroscopic techniques have allowed for information to be deduced regarding the spins and parities [1,2] and lifetimes [3,4] of such states of extreme isospin.

In this paper we present research that has been carried out using the Euroball and Eurogam arrays to detect γ rays emitted from spontaneously fissioning ^{248}Cm and ^{252}Cf. The paper focuses on three sub-areas of current activity, namely, the measurement of yields of secondary fragment pairs, the measurement of state lifetimes at around spin 10, and recent measurements of g-factors of excited states in fission fragments.

CP481, *Nuclear Structure 98*, edited by C. Baktash
© 1999 American Institute of Physics 1-56396-858-4/99/$15.00

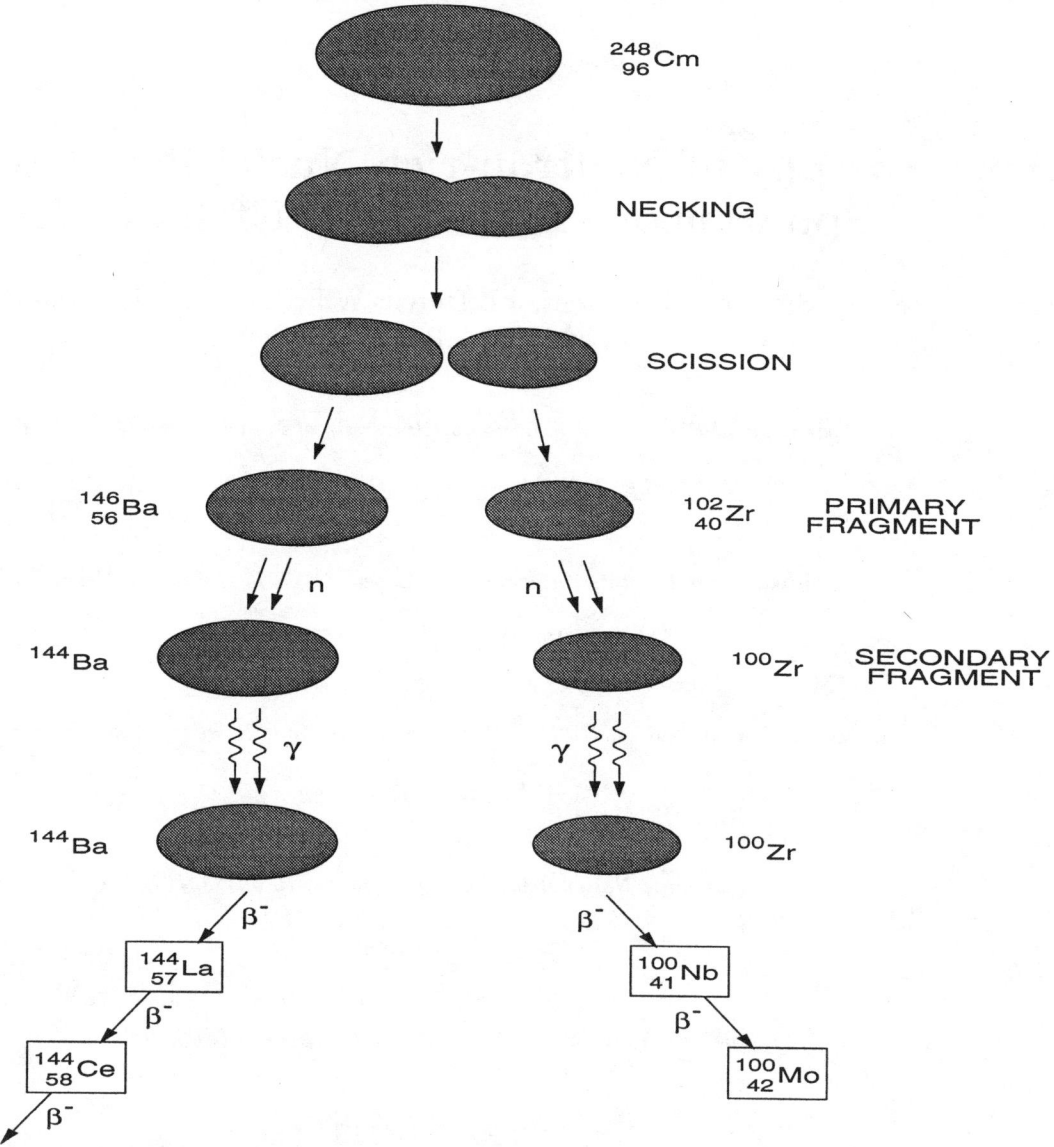

FIGURE 1. A schematic illustration of the process of spontaneous fission, with ^{248}Cm as the fissioning system.

I YIELD MEASUREMENTS

Lately there has been significant interest in the secondary yield measurements for ^{252}Cf fission generated by the observation [5] of larger-than-expected yields for fragment pairs corresponding to the emission of high numbers (8-10) of neutrons in the Ba/Mo split. The yield distributions for Ba's with Mo's have been interpreted in terms of a bimodal fission process, of which the low TKE (Total Kinetic Energy) mode has been associated with a highly deformed configuration of the nascent Ba fragments. We have undertaken a measurement of the yields of complimentary Ba and Mo fragments using the ^{252}Cf data set (comprising 10^{10} events of at least three coincident γ rays). The method requires the accurate determination of the intensities of the often-weak coincidences occurring between γ rays emitted from fragment pairs corresponding to different numbers of emitted neutrons. To fit the required coincidence peaks in an even-handed way, we opted to sort the data into a γ-γ-γ cube which was then searched and fit using the AMPSAF [6] software using a local projection method for the determination of background. The result of this fit was a database of photopeak centroid and intensity information which was then interrogated for the coincidences of interest. In Fig 2 the AMPSAF database was gated on the the $2 \rightarrow 0$ transition in ^{144}Ba and pairs of γ rays were used to determine the intensity of each

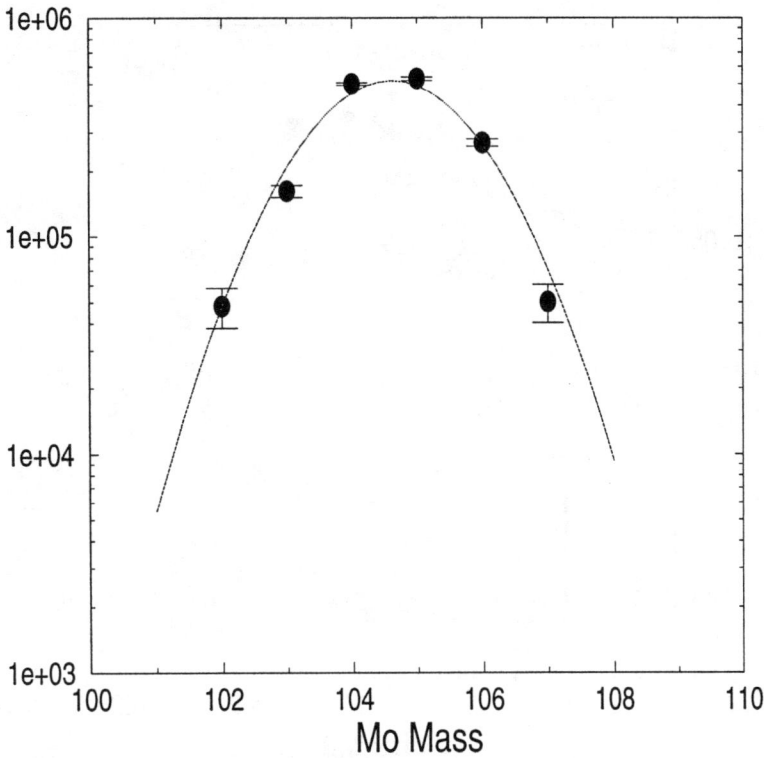

FIGURE 2. Yields of Mo isotopes observed in coincidence with [144]Ba

Mo isotope. Fig 3 shows results for which [104]Mo was isolated in a spectrum composed of a double gate on $2 \to 0$ and $6 \to 4$, added to a double gate on the $4 \to 2$ and $6 \to 4$ transitions. In this combined spectrum, the intensities of low-lying transitions in the Ba isotopes were used to give the secondary-fragment yields of Ba isotopes in coincidence with [104]Mo. Special care was required to isolate [104]Mo from [108]Mo, which has similar energies for the $2 \to 0$ and $4 \to 2$, hence our use of double gates involving the $6 \to 4$ transition. Superimposed on the measured distributions is a discretised Gaussian function with width $\sigma = 1.08$ neutrons. This curve provides a good representation of the distribution of $(\nu - \bar{\nu})$, where ν is the number of emitted neutrons, and $\bar{\nu}$ is the mean, for a variety of low energy fission processes [7]. It can be seen that the measurements of secondary-fragment yields presented here are in reasonable agreement with the trend of this distribution. The results of our preliminary analysis for these and other Ba/Mo combinations have so far failed to provide substantiating evidence of the low TKE mode.

II LIFETIME MEASUREMENTS

In this work the spontaneously fissioning isotope [248]Cm was used as a source of neutron-rich fission fragments whose electromagnetic decay properties were to be studied. The source consisted of about 5 mg of curium oxide (giving a fission rate of about 7×10^4 per second) embedded in a pellet of potassium chloride. The pellet was large enough (0.5mm thick and 5mm in diameter) that the fission fragments were stopped within it. The Eurogam phase 2 array of Compton-suppressed germanium detectors was used to detect the γ rays emitted in the decay of the excited fission fragments. Around 2.5 billion coincidence events with more than three γ rays were recorded on magnetic tape. From these data a $\gamma\gamma\gamma$ cube and a $\gamma\gamma\gamma\gamma$ hypercube [8] were constructed. To obtain spectra that emphasised a particular decay sequence, for the most part triple-gated 1D spectra from the hypercube were used, though in a few cases where the statistics were poor, double-gated 1D spectra from

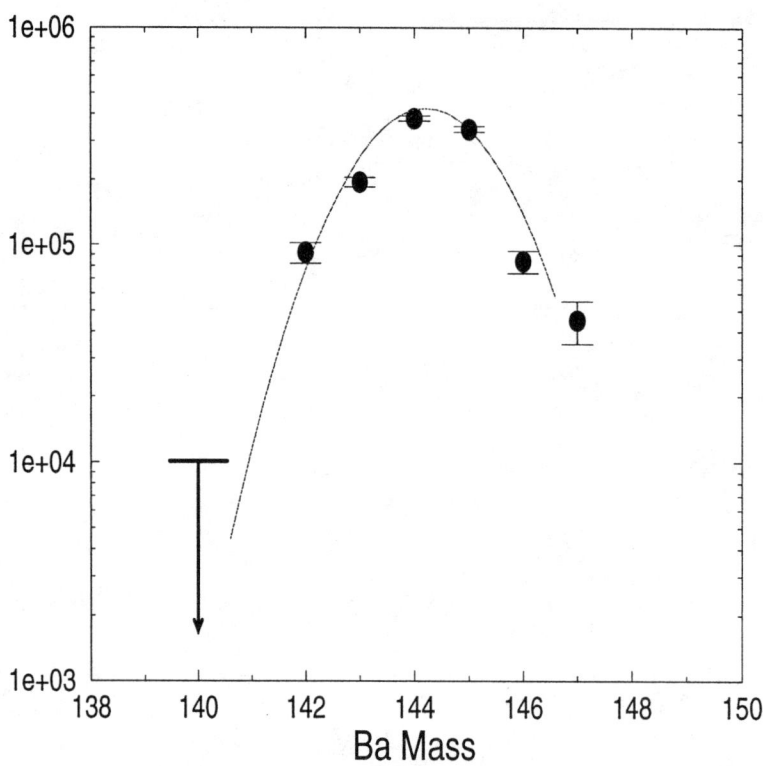

FIGURE 3. Yields of Ba isotopes observed in coincidence with[104]Mo

the cube were used.

At spins of $I \approx 10$, symmetrically Doppler–broadened lineshapes are observed in the γ–ray energy spectra of many of the light fission fragments. The broad lineshapes correspond to decays from states that have lifetimes comparable to (or faster than) the stopping time (1-2 ps) of the fission fragments in the source pellet, the broadening being due to the variable Doppler-shift that is observed for the time distribution of decays from such states.

The Doppler–profile method (DPM) [3] combines a simulation of the stopping of the isotropically–directed fission fragment with a simulation of the electromagnetic decay to generate a lineshape that can be compared directly with the data and thereby extract state lifetimes. The stopping of the fragment in the KCl pellet is simulated with a Monte-Carlo computer code using electronic and nuclear stopping powers given by the computer code ZBL [9]. The initial fragment kinetic energy distribution was assumed to be Gaussian with its centroid taken from Ref. [10] (with a Coulomb correction for the particular charge split) and width taken to be the same as that measured [11] for the light fragments from [252]Cf spontaneous fission. The simulation produced a two–dimensional data matrix (Fractional Doppler Shift, or FDS matrix) whose contents gave the probability of observing a particular fractional Doppler shift ($\beta \cos\theta$, where β is the velocity of the fragment and θ is the angle between the fragment velocity and the γ–ray detection axis) in the emitted radiation at any given time.

In modelling the electromagnetic decay of the excited fission fragment, consideration must be given to the time distribution of the intensity feeding the states whose lifetimes are to be determined. For the highest–spin state whose decay produces an observed lineshape it is not possible to extract any information directly from the γ–ray spectrum about the feeding distribution, so the feeding of this state is simulated by a two-state feeding model in which the lifetimes of both feeding states are varied in the fit. For all other states there is some information regarding the feeding distribution contained in the lineshape of the decay into the state, although a large fraction (40-60%) of the population of the state is provided by unobserved "side–feeding"

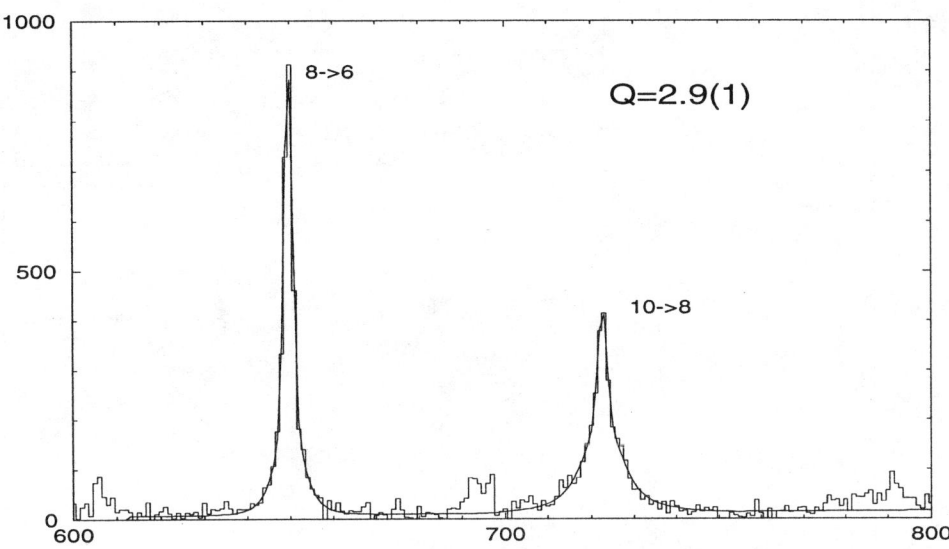

FIGURE 4. Observed lineshapes for the 8 → 6 and 10 → 8 transitions in ^{112}Ru. The spectrum is produced from a hypercube, gated on the 2 → 0, 4 → 2 and 12 → 10 transitions, thereby reducing the sensitivity of the resulting Q to uncertainties in the time distribution of the unobserved side feeding.

transitions. Measurements of γ-ray multiplicities following spontaneous fission suggest that a yrast state at intermediate spin is fed predominantly by a combination of fast statistical γ rays, and the transition from the next-highest yrast state. Although slow side-feeding components have been observed they usually constitute only a few percent of the decay intensity. The time distribution of the side-feeding was simulated in the fitting procedure by a two-state model with a variable feeding time. To reduce the effect of possible contamination of the lineshapes with sharp lines from the heavy fragments the rotational band (between I=6 and I=12) was assumed to correspond to the rotation of a nucleus of constant intrinsic quadrupole moment (Q). Within this model, the mean lifetime (τ in ps) of a state of spin I, decaying within a rotational band by a stretched E2 γ ray of energy E_γ (in MeV) is given by

$$\frac{1}{\tau} \approx 1.217 < I, K, 2, 0 | I - 2, K >^2 Q^2 E_\gamma^5,$$

where $< I, K, 2, 0 | I - 2, K >$ is a Clebsch–Gordan coefficient and Q is measured in e barns (eb). Q was used to vary the state lifetimes in a manner consistent with this assumption, the best fit to the data resulting in a solution for Q. The quantum number K represents the projection of the intrinsic angular momentum on the symmetry axis. It is well defined only in the case of axially-symmetric nuclei at low rotational frequencies. In all even-nuclei examined here, K was taken as zero, whereas for the odd nuclei the bandhead value of K was used.

Fig 5 shows the quadrupole moments deduced from lifetime measurements for the light fission-fragments. The onset of deformation in the Sr and Zr isotopes is clearly seen, as is the trend to lower quadrupole moments with increasing Z. The quadrupole moments of the even Mo isotopes are lower than expected, when compared to either predicted ground-state quadrupole moments, or to those deduced from measurements of lifetimes

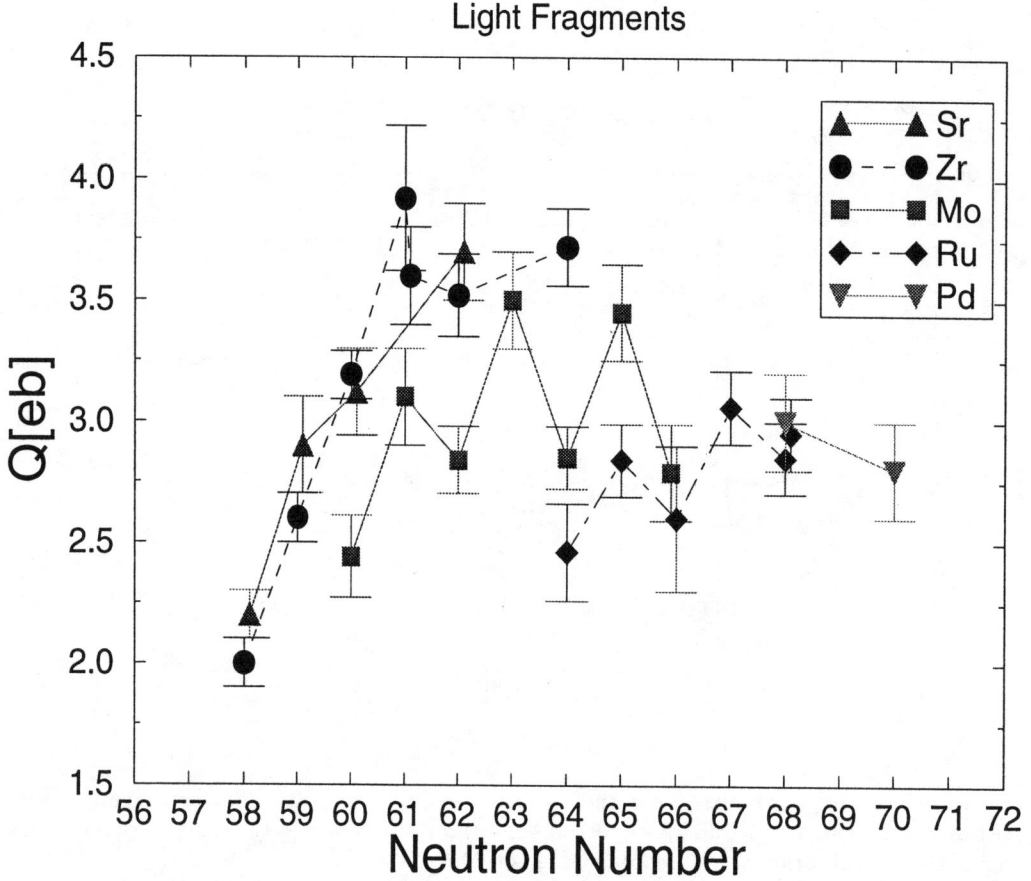

FIGURE 5. A compilation of quadrupole moments deduced from our DPM lifetime measurements for rotational bands in A~100 fission fragments

for the first 2^+ states. This difference between the quadrupole moments at low and intermediate spin has previously been interpreted in terms of a change in the triaxiality due to $h_{11/2}$ netron alignment [4]. In the data presented here, it appears that the quadrupole moments of the odd Mo isotopes show greater stability with increasing spin, having values similar to the Zr isotopes at $I \sim 10$. All of the measurements made in the odd Mo and Ru nuclei correspond to bands of $h_{11/2}$ character and it may be that the occupation of the moderately prolate-driving $h_{11/2}$ orbitals by the odd neutron is sufficient to stabilise the axial symmetry. However, detailed calculations are required to determine whether such an effect would reproduce the observed phenomenon.

MAGNETIC MOMENT MEASUREMENTS OF EXCITED STATES

To date there have been no measurements of the g-factors of excited states in fission fragments using prompt γ rays, although there have long been experiments that measure g-factors of states populated in the beta-decay of mass-separated fission fragments. The advantage of being able to use the prompt γ rays as opposed to those emitted following β decay is two fold. Using the prompt γ rays allows access to yrast states up to intermediate spin (8-16$\hbar$) whereas post-β-decay γ rays are generally emitted from states whose spin is close to that of the parent nucleus. In addition, the use prompt γ rays may allow access to more neutron-rich species with shorter β-decay lifetimes. In this section, we describe an experiment to measure for the first time, the g-factors of excited states in neutron-rich fission fragments through the time-integral perturbed angular correlation functions between pairs of prompt γ rays.

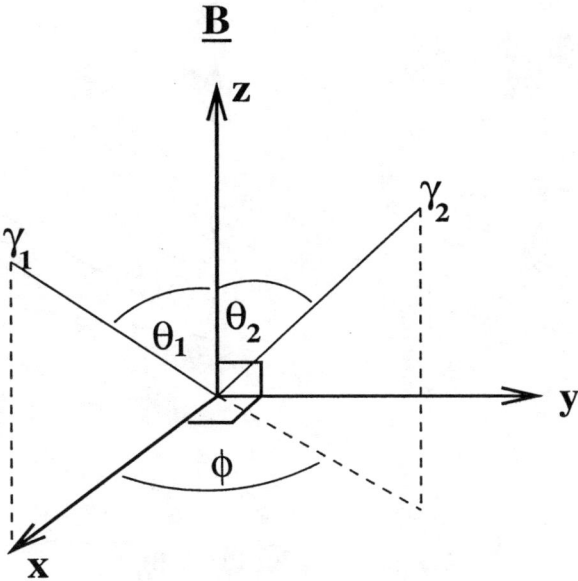

FIGURE 6. Geometry used in the discussion of the measurement of Larmor precessions caused by the hyperfine magnetic field B

The experiment involved the use of a ^{252}Cf source of total activity $120\mu Ci$ sandwiched between two layers of Gd metal. Prior to the deposition of the Cf, the Gd metal foils (each $20 mgcm^{-2}$ thick) were annealed in an oven at $650°C$ for ten minutes. The magnetisation of these foils as a function of temperature and applied field was measured using a magnotometer. The results of this measurement showed that the magnetic moment of the Gd reached 87 percent of its calculated maximum value with the temperature held at 80K. The Cf was then electro plated onto the surface of one of the Gd foils and a layer of indium metal ($200\mu gcm^{-2}$ thick) evaporated over a second Gd foil. The layer of indium acted as an aid to adhesion between the active foil and the second foil which was rolled on top to produce a closed source in which the fission fragments stop in Gd. The source was placed in a specially designed chamber at the centre of the Euroball array. In this chamber the source was clamped between a Cu strip and an Al plate, the Cu strip being attached to a cold Cu block that was maintained at around 86K by a constant flow of liquid-nitrogen. A pair of small permanent magnets applying a field of 0.2 Tesla were placed either side of the source in the direction normally reserved for the beam to Euroball. The direction of the applied field could be reversed by rotating the magnet assembly through an angle of 180°. A system of colour-coded LED's with optical-fibre links to the outside of the Euroball array was used to confirm the orientation of the magnet assembly. The experiment ran for 14 days with the field direction being changed every two hours and the data written in event-by-event mode to magnetic tape. After presorting, which involved adding together the energies of Compton events in the composite cluster and clover detectors, as well as software-gainmatching the energy signals for the whole array, 9.9×10^9 events of fold 3 or greater were recorded.

In determining the magnitude and sign of the Larmor precession for a complex array such as Euroball, it is important to combine the data from many different detector pairs. In a coordinate system Fig 6 where the applied field is along the z-axis, θ_1 and θ_2 represent the polar angles of the two detection axes respectively and ϕ is the azimuthal angle between the detection axes, we take the Larmor precession to be of an angle $\delta\phi_r$ about the z-axis. The effect of the rotation on W (the known angular correlation function) is then given by,

$$\frac{\partial W}{\partial \phi}\delta\phi_r = -\frac{\partial W}{\partial cos\theta}sin\theta_1 sin\theta_2 sin\phi \delta\phi_r.$$

From this expression it is clear that there are two geometric parameters that have to be recorded in any event if the data from different detector pairs are to be combined, namely, $c_{12} = cos\theta$ and $s_{12} = sin\theta_1 sin\theta_2 sin\phi$. The procedure adopted in the analysis described here was to divide Euroball into 256 bins in the space of c_{12} versus s_{12} and to label each bin with an integer index, $j(1,2)$. Since the effect of a precession on the count rate in a detector pair goes to zero as $s_{12} \to 0$, data were only used for detector combinations with $s_{12} > 0.1$

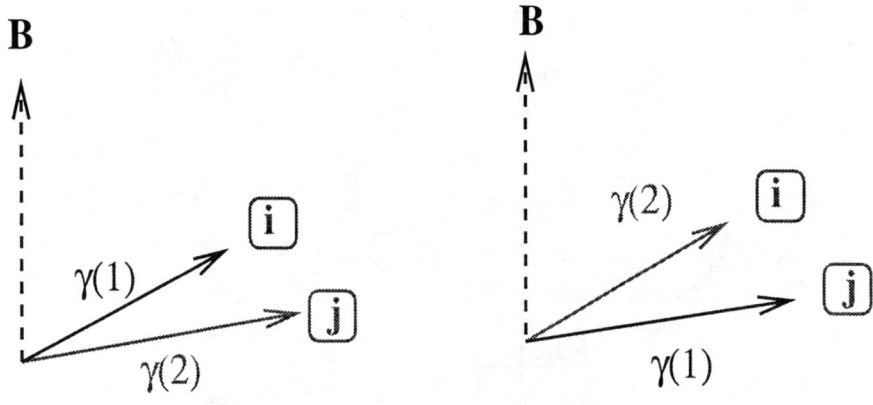

$$\rho_{ij} = \left[\frac{\uparrow N(c_{ij}, s_{ij}) \quad \downarrow N(c_{ij}, -s_{ij})}{\downarrow N(c_{ij}, s_{ij}) \quad \uparrow N(c_{ij}, -s_{ij})} \right]^{1/2}$$

$$X = \frac{(\rho_{ij} - 1)}{(\rho_{ij} + 1)} \frac{1}{s_{ij}} = \frac{W'(c_{ij})}{W(c_{ij})} \delta\phi$$

FIGURE 7. An illustration showing the symmetry used in producing the double ratio ρ_{ij}. An interchange of the γ rays going into detectors i and j produces and equal, but opposite, change in the coincidence count rate and corresponds to a change in sign of the parameter s.

The large number of different nuclei populated in fission means that in an experiment such as this (where the fission-fragments are not directly detected), it is necessary to use at least three specified γ-ray energies to gain adequate selectivity of the decay path. Since the angular correlation measurement concerns only two γ rays, a third γ ray is used as a so-called 'isotropic' selector. To allow for flexibility in the application of gates on various transitions, it was decided to sort the data into a 4-dimensional, high-resolution hypercube [8], whose axes corresponded to, e_0, e_1, e_2, $j(1,2)$ (e_0 is the energy of the isotropic selector, and the axis $j(1,2)$ is the angle index for the detection of the γ rays with energies e_1 and e_2). Data for both field directions were stored in the same hypercube, differing only in the allocated range on the 4th axis, thus ensuring that the applied gates were identical for the field-up and field-down data. Uncertainties caused by normalising the field-up to the field-down data, and concerns about possible changes in the detector efficiencies between runs are dealt with through the use of double ratios,

$$\rho(c_{12}, s_{12}) = \left[\frac{\uparrow N(c_{12}, s_{12}) \downarrow N(c_{12}, -s_{12})}{\downarrow N(c_{12}, s_{12}) \uparrow N(c_{12}, -s_{12})} \right]^{1/2},$$

where $\uparrow N(c_{12}, s_{12})$ represents, for example, the number of counts with field up recorded in detector combinations with c_{12} and s_{12} values identified through the index $j(1,2)$. The double ratio as defined above makes use of the symmetry that for small precessions, the fractional change in the number of counts in any detector pair is equal and opposite if the γ rays are interchanged, ie. if $(c_{12}, s_{1,2}) \to (c_{12}, -s_{1,2})$. The double ratio can then be used to calculate the quantity,

$$X(cos\theta) = \left[\frac{\rho(c_{12}, s_{12}) - 1}{\rho(c_{12}, s_{12}) + 1} \right] \frac{1}{s_{12}} = \frac{1}{W} \frac{dW}{d(cos\theta)} \delta\phi_r,$$

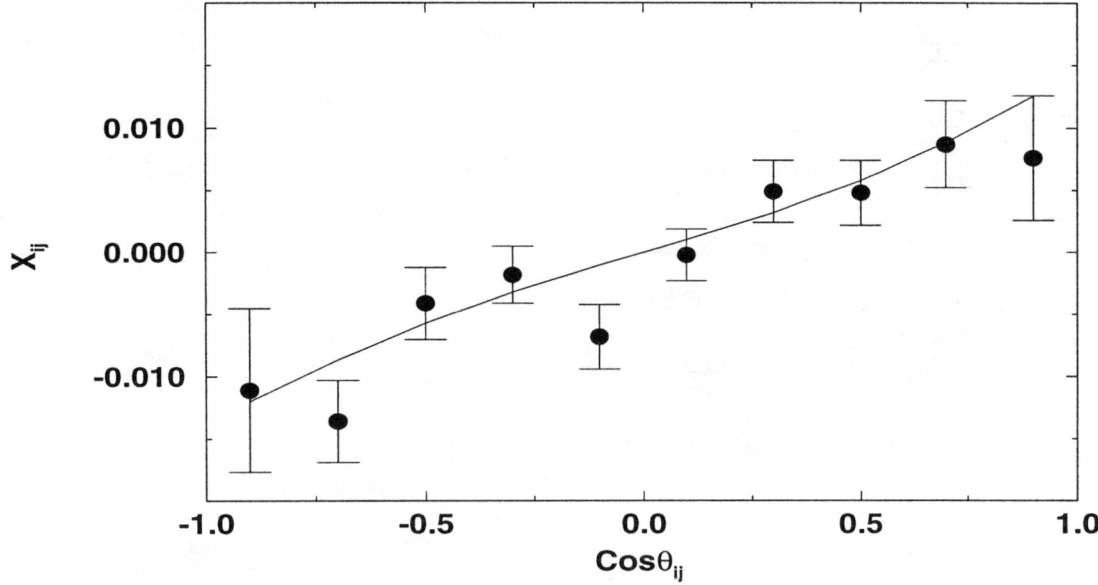

FIGURE 8. Plot of the mean value of X against the cosine of the angle between detectors, for the first 2^+ state in 144Ba. Superimposed on the data is a fit proportional to the logarithmic derivative of the measured angular correlation function between the $4 \rightarrow 2$ and $2 \rightarrow 0$ transitions.

a product of the logarithmic derivative of the angular correlation function and the precession angle. In the data analysed here, a weighted mean of $X(cos\theta)$ is taken over all the detector combinations corresponding to a given $cos\theta$, and the results fitted with the known logarithmic derivative of W to extract the precession angle.

Figure 8 shows a measured precession in ^{144}Ba, using the correlation between the $2^+ \rightarrow 0^+$ and $4^+ \rightarrow 2^+$ yrast transitions. The isotropic gate is placed on the $6^+ \rightarrow 4^+$. Plotted in the figure is $X(cos\theta)$ against $cos\theta$, from which the magnitude and sign of the precession are obtained. Several such measurements are possible for the $I^\pi = 2^+$ state, just by choosing a different isotropic selector, or by using, for example, the correlation between the $2^+ \rightarrow 0^+$ and the $6^+ \rightarrow 4^+$ γ rays. Fifteen different gates were analysed to obtain the resulting precession of 47(7) mrad. Using the known g-factor [12] and lifetime [13] for the ^{144}Ba 2^+ state, it is then possible to determine the static hyperfine field experienced by the implanted ions, using an approximate formula valid for small precessions,

$$\delta\phi_r \sim 47.8gB\tau$$

where the angle is in mrad, the field (B) in Tesla and the state lifetime (τ) is in nanoseconds. This gives the hyperfine field for implanted Ba nuclei as 2.8(4) Tesla. It should be noted that the measured field is much lower than expected, typical static hyperfine fields in Gd vary with the Z of the implanted ion, and from sample to sample, but are often above 10 Tesla. Analysis of the 2^+ states in the even Zr isotopes suggests that the field strengths may be as much as 5 times smaller than have previously been measured in in-beam experiments. The cause of the low fields is not known with any certainty, but may be due to damage of the Gd lattice caused by the fission fragments in the 8 months between the source being manufactured and the experiment being performed.

In spite of the disappointing field strengths, it has proved possible to measure precessions for several states with nanosecond lifetimes. Measurements have been made for the first $I^\pi = 2^+$ states in ^{144,146}Ba and ^{146,148,150}Ce as well as for the $9/2^-$ state at 117 keV in ^{143}Ba and the $7/2^-$ state at 114 keV in ^{145}Ba. The deduced g-factors are presented in Table 1. Also shown in figure 9 is a comparison between the results

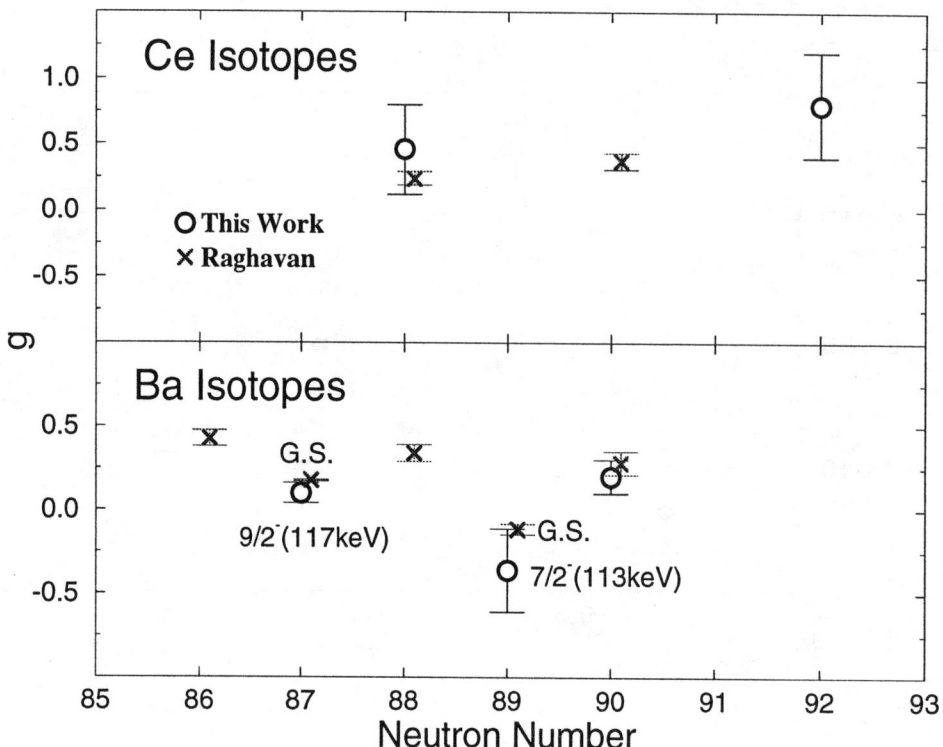

FIGURE 9. The g-factor results from this work are compared with those compiled in Raghavan. The first 2^+ states in ^{144}Ba and ^{148}Ce are used to calibrate the field strengths.

obtained here and those compiled in reference [12]. It can be seen from this comparison that in general the results obtained here are consistent with previously known g-factors, for those cases where measurements have been made. The measurement of the 2^+ state in ^{146}Ba is consistent with the previous result and provides supporting evidence for a downward trend in the g-factors of the 2^+ states of even-even Ba isotopes. It has been suggested [14,15] that this decrease may be explained within the framework of the IBA as due to the increasing number of neutron valence bosons that occurs in the first half of a neutron shell and the effective g-factor associated with the neutrons.

The measurements of the $9/2^+$ and $7/2^+$ states in ^{143}Ba and ^{145}Ba, respectively are the first measurements for these states. In figure 9 these data are compared with the known g-factors for the ground states of these nuclei. In both cases, there is no measurable change in the g-factor between the ground states and first excited states, within the rather large experimental uncertainties. The ^{150}Ce result is the first g-factor measurement in this nucleus, but due to the large experimental uncertainty, it is difficult to say whether the trend of decreasing g-factors for the 2^+ states appears also in the Ce isotopes.

III   CONCLUDING REMARKS

This paper has discussed three areas of research related to fission-fragment spectroscopy. The measurement of coincidence yields of fission fragments, particularly for the Ba/Mo split, is an area of topical interest and efforts will continue in search of supporting evidence for the low TKE fission mode. The lifetime measurements for the A~ 100 nuclei presented here now provide the first comprehensive survey of the evolution of nuclear shape in this region at intermediate spin, from the well deformed Sr and Zr isotopes through to the much softer Pd nuclei. Also presented in this paper has been the first attempt to measure g-factors of excited states in fission-fragments using prompt γ rays. There is considerable scope for the refinement of the techniques presented here, both for the measurement of lifetimes and for g-factors. The dividends of such developments being access to information on the electromagnetic properties of excited states in nuclei which are of the most

Nucleus	State	Lifetime (ns)	Precession (mrad)	g-factor
^{143}Ba	9/2⁻ (117keV)	3.8 ±1.2	47 ±28	0.10 ±0.06
^{144}Ba	2⁺ (199keV)	1.06 ±0.06	47 ±7	0.34 ±0.05 †
^{145}Ba	7/2⁻ (113keV)	0.35 ±0.1 ‡	-16 ±11	-0.36 ±0.25
^{146}Ba	2⁺ (181keV)	1.26 ±0.09	32 ±16	0.20 ±0.10
^{146}Ce	2⁺ (259keV)	0.36 ±0.06	-27 ±19	0.46 ±0.34
^{148}Ce	2⁺ (159keV)	1.5 ±0.1	-89 ±17	0.37 ±0.06 §
^{150}Ce	4⁺ (306keV)	0.26 ±0.10 ‡	-47 ±21	0.8 ±0.4

TABLE 1. g-factor results. †g-factor from Raghavan, At. Data Nucl. Data Tables 42,189(1989) used to calibrate static hyperfine field for Ba (B=+2.8(4)Tesla) §g-factor from Raghavan, At. Data Nucl. Data Tables 42,189(1989) used to calibrate static hyperfine field for Ce (B=-3.4(8)Tesla) ‡Estimated lifetime as no available experimental measurements.

neutron-rich that can be produced using available techniques.

This work was supported by the US Department of Energy under contract No. W-31-109-ENG-38. The authors are also indebted for the use of the ^{252}Cf to the Office of Basic Energy Sciences, US Department of Energy, through the Transplutonium element production facilities at the Oak Ridge National Laboratory.

REFERENCES

1. W.Urban, *et al.* Nucl. Instr. Meth. **A365**,596 (1995).
2. M.A.Jones, *et al.* Nucl. Phys. **A605**,133 (1996).
3. A.G.Smith, *et al.* Phys. Rev. Lett. **73**,1540 (1994).
4. A.G.Smith, *et al.* Phys. Rev. Lett. **77**,1711 (1996).
5. G.M.Ter-Akopian, *et al.* Phys. Rev. C **55**,1146 (1997).
6. A.G.Smith and W.J.Vermeer, Nucl. Instr. Meth. **A350**,314 (1994).
7. R.Vandenbosch and J.R.Huizenga Nuclear Fission, Academic Press (1973).
8. A.G.Smith, Nucl. Instr. Meth. **A381**,517 (1996).
9. J.F.Ziegler, J.P.Biersack and U.Littmark, The Stopping and Range of Heavy Ions in Solids, Permagon Press (1985).
10. V.E.Viola, K.Kwiatkowski and M.Walker, Phys. Rev. **C31**, 1550 (1985).
11. S.L Whetstone Jr., Phys. Rev. **131**, 1232 (1963).
12. P.Raghavan, At. Data Nucl. Data Tables **42**, 189 (1989).
13. S.Raman, At. Data Nucl. Data Tables **36**, 1 (1987).
14. A.Wolf, 2nd International Workshop on Fisson-Product Spectroscopy, Seyssins (1998), Priv. Com.
15. M.Sambataro *et al.*, Nucl. Phys. **A423**,333 (1984)

Fission studies of secondary beams from relativistic uranium projectiles: The proton even-odd effect in fission fragment charge yields

A.R. Junghans[a], J. Benlliure[a], K.-H. Schmidt[a], B. Voss[a], C. Böckstiegel[b], H.-G. Clerc[b], A. Grewe[b], A. Heinz[b], M. de Jong[b], J. Müller[b], S. Steinhäuser[b], and M. Pfützner[c]

[a] GSI Darmstadt, Planckstraße 1, D-64291 Darmstadt, Germany
[b] Institut für Kernphysik, Technische Universität Darmstadt, Schloßgartenstraße 9, D-64289 Darmstadt, Germany
[c] Institute of Experimental Physics, University of Warsaw, Ul Hoza 69, 00-381 Warszawa, Poland

Abstract. Nuclear-charge yields of fragments produced by fission of neutron-deficient isotopes of uranium, protactinium, actinium, and radium have been measured. These radioactive isotopes were produced as secondary beams, and electromagnetic fission was induced in a lead target with an average excitation energy around 11 MeV. The local even-odd effect in symmetric and in asymmetric fission of thorium isotopes is found to be independent of Z^2/A. The charge yields of the fission fragments of the odd-Z fissioning protactinium and actinium show a pronounced even-odd effect. In asymmetric fission the unpaired proton predominantly sticks to the heavy fragment. A statistical model based on the single-particle level density at the Fermi energy is able to reproduce the overall trend of the local even-odd effects both in even-Z and odd-Z fissioning systems.

INTRODUCTION

The even-odd effect in nuclear fission is a very pronounced experimental fact. In the low-energy fission of a nucleus with even proton number an enhanced production of fission fragments with even proton number is observed. As an example, the charge yields from ^{229}Th(n_{th}, f) [1] are shown in Fig. 1.

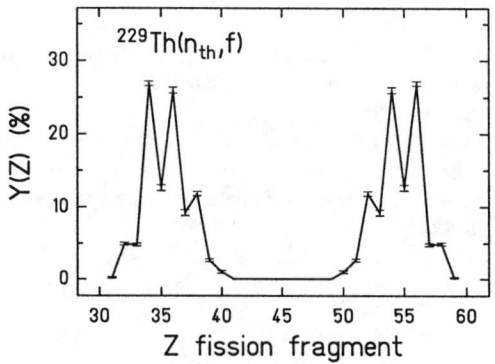

FIGURE 1. Fission-fragment charge yields from thermal-neutron induced fission of ^{229}Th(n_{th}, f) [1]. Please note that only the light fission-fragment group $Z = 31 - 40$ was observed experimentally.

CP481, *Nuclear Structure 98*, edited by C. Baktash

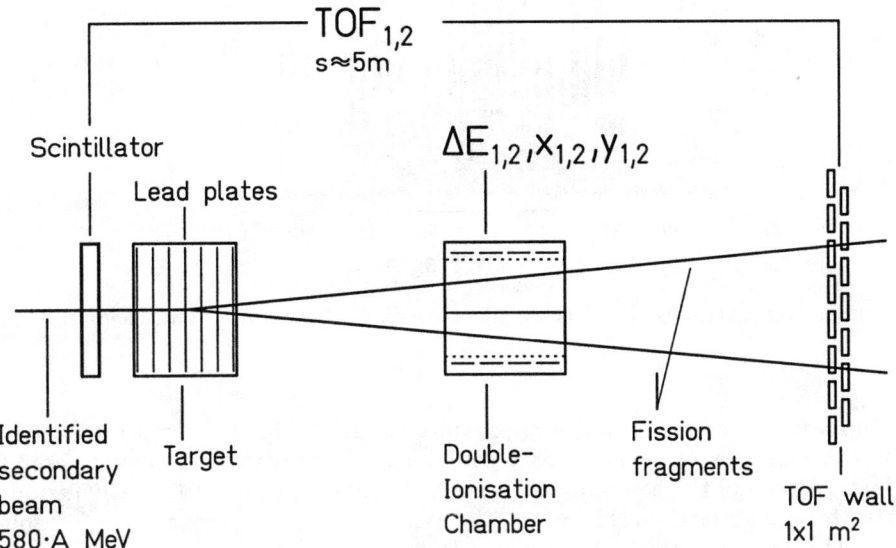

FIGURE 2. Schematic drawing of the experimental setup for the secondary-beam fission experiment behind the fragment separator.

From the even-odd effect, conclusions concerning the dynamics of fission may be drawn. The enhanced production of fission fragments with even proton numbers has been interpreted as a direct measure of the quasi-particle excitation at scission [2]. Once a proton pair is broken, it is assumed that the unpaired protons are distributed with equal probabilities to the two nascent fragments and the even-odd effect vanishes. In this interpretation, the even-odd effect is due to the partial survival of the fully paired superfluid phase during the fission process and is destroyed by the onset of dissipation in the motion from saddle towards scission. In contrast to the neutron number of the fission fragments which is changed by neutron evaporation following the fission process – the proton number is a direct information on the charge split at scission.

Experimental information on nuclear charges of fission fragments is not at all easy to obtain. The best results were obtained at the Institute Laue-Langevin (ILL), Grenoble, e.g. [1,3]. Here, we present a comprehensive study on the even-odd structure in the fission-fragment charge yields of neutron-deficient isotopes of uranium, protactinium, thorium, and actinium performed at Gesellschaft für Schwerionenforschung (GSI), Darmstadt [4]. By use of secondary beams produced from relativistic ^{238}U projectiles fission from short-lived nuclei was investigated in inverse kinematics.

EXPERIMENT

The secondary-beam facility of GSI offers unique possibilities to provide beams of neutron-deficient actinides and preactinides produced by fragmentation of relativistic ^{238}U projectiles. Nuclear charge and mass number can be selected in a wide range within the limits given by the primary-beam intensity and the production cross section [5,6] by accordingly tuning the fragment separator FRS [7].

Behind the FRS, fission in flight of neutron-deficient actinides and preactinides was studied with an experimental setup, see Fig. 2, which was optimized on the secondary-beam characteristics: With an average beam energy of 530 A MeV, the secondary projectiles, identified in mass and nuclear charge, hit a 3 g/cm^2 lead target. By the strong electromagnetic field of the heavy target nucleus the giant resonances (mainly the giant dipole resonance) are excited in the secondary projectiles.

This reaction mechanism allows to study fission at the available secondary-beam intensities of typically 100/s because the high cross section of the electromagnetic excitation leads to sufficient fission rates from the decay of the giant resonances in the secondary projectiles. As typical values we calculate for the electromagnetically induced fission of ^{234}U(430 A MeV) in the lead target an average excitation energy at fission of 11 MeV, a width

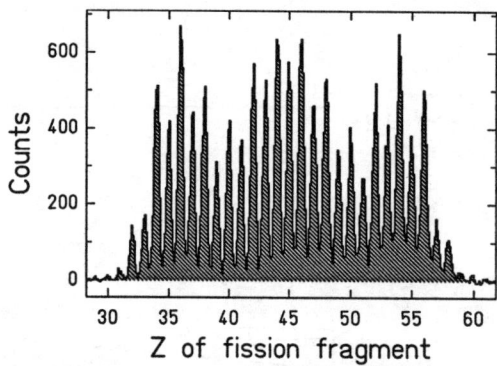

FIGURE 3. The measured charge response for electromagnetic-induced fission of ^{226}Th. The overall charge resolution is $Z/\Delta Z \approx 125$.

of 7 MeV (FWHM) and a fission cross section of approximately 2.5 barn. This results in an electromagnetic fission rate of the order of 1%. The excitation-energy distribution is similar for all investigated nuclei.

Due to the high kinetic energy of the fissioning nuclei of 430 A MeV on the average the fission fragments are strongly focussed in the forward direction. Therefore, the experimental setup covers the full solid angle in the center-of-mass system. The detection efficiency amounts to 81%.

The energy loss of both fission fragments is measured independently in a double ionisation chamber. To achieve a good charge resolution, in addition the time of flight of both fission fragments is measured with a plastic-scintillator wall. In Fig. 3, the charge response for electromagnetic fission of ^{226}Th is shown. With our setup we reach a typical charge resolution of $Z/\Delta Z \approx 125$ for both light and heavy fission fragments, which cannot be reached with conventional techniques.

Fission after nuclear reactions can very efficiently be separated from electromagnetic-induced fission in the lead plates of the target by setting a condition on the charge sum of both fission fragments. This eliminates most of the nuclear reactions which change the nuclear charge sum. Then, the remaining part of nuclear-induced fission is removed by substraction of the suitably scaled nuclear-induced fission measured in the plastic scintillator. This procedure is described in detail in Ref. [8].

CHARGE YIELDS

In the experiment, the charge yields of fission fragments from $^{209,...,220}$Ra, $^{214,...,223}$Ac, $^{220,...,229}$Th, $^{224,...,232}$Pa, and $^{231,...,234}$U have been measured. In Fig. 4 the charge yields of ^{233}U, ^{228,226,224}Th, and ^{214}Ra are shown as examples. The measured yields document the transition from asymmetric fission typical for the heavy actinides like ^{233}U to symmetric fission typical for the lighter elements. Especially, the uranium and thorium isotopes show a pronounced even-odd structure. As the measured data cover long isotopic chains of several elements a systematic study of the even-odd effect is feasible.

For the first time, even-odd effects could be observed in *symmetric* fission. Up to this experiment even-odd effects could only be studied in asymmetrically fissioning systems, e. g. see Fig. 1. With secondary beams, also symmetrically fissioning systems like ^{224}Th, see Fig. 4 are accessible for low-energy fission.

AVERAGE AND LOCAL EVEN-ODD EFFECT

The average even-odd effect is defined by the ratio

$$\delta_p = \frac{Y_e - Y_o}{Y_e + Y_o},$$

where $Y_{e,o}$ are the cumulative yields of all even (odd) elements. In the light of the present understanding, one expects a higher even-odd effect for fissioning systems with a low fissility Z^2/A as the saddle-point deformation

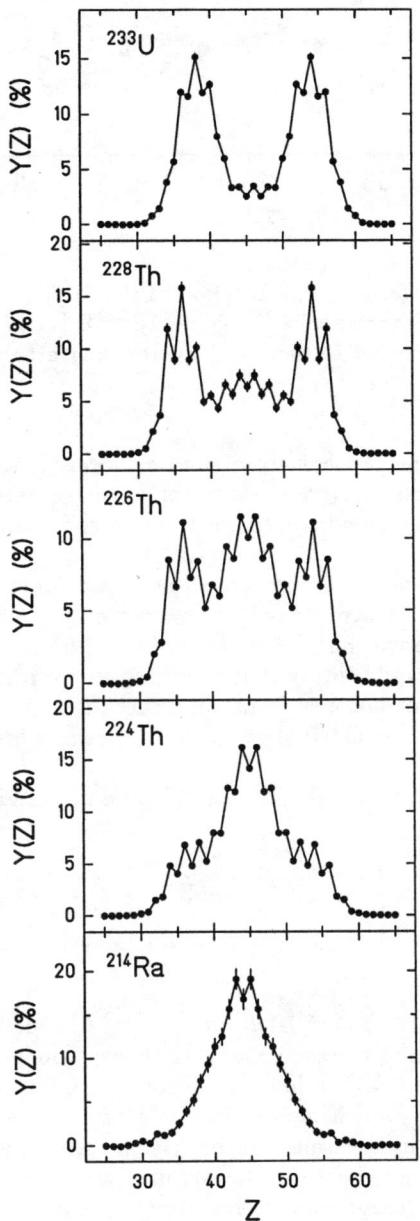

FIGURE 4. Measured charge yields for [233]U, [228,226,224]Th, and [214]Ra.

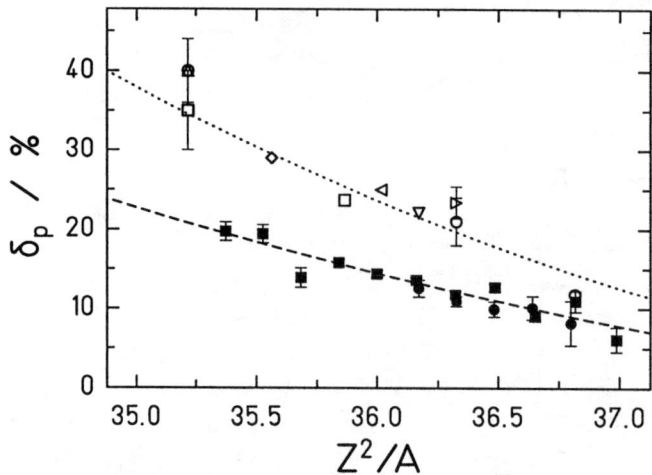

FIGURE 5. Measured average even-odd effects δ_p as a function of the fissility parameter Z^2/A. The open symbols correspond to data from low-energy fission for the nuclei ^{230}Th [9–11], ^{233}U [11,12], ^{234}U [13], ^{235}U [14], ^{236}U [3], ^{238}U [15], and ^{240}Pu [16]. The full symbols correspond to thorium (squares) and uranium (circles) isotopes investigated in this work.

in these systems is comparably large, the descent from saddle to scission is short and not much energy is dissipated into intrinsic excitation. Hence, the even-odd effect is rather large. Whereas for a system with a higher fissility the saddle point is less deformed, therefore the descend from saddle to scission is longer and consequently the even-odd effect is lower.

In Fig. 5 the experimental average even-odd effects are shown as a function of fissility. At first, one notes that up to now the experimental data were scarce; only a few systems have been investigated. These values nevertheless show the trend discussed above that for a low fissility the even-odd effect is high and for a high fissility the even-odd effect is low. This would imply that the intrinsic excitation energy at scission grows with Z^2/A. The data for thorium and uranium isotopes from our experiment follow the same trend although they are generally lower ($\approx 40\%$) which could be attributed to the somewhat higher excitation energies in fission after electromagnetic excitation.

To draw more specific conclusions, the even-odd effect can also be investigated locally from four consecutive yields by a third-difference method [17]

$$\delta_p\left(Z + \frac{3}{2}\right) = \frac{1}{8} \cdot (-1)^{Z+1} \cdot \left(\ln Y(Z+3) - \ln Y(Z)\right.$$
$$\left. - 3 \cdot \left[\ln Y(Z+2) - \ln Y(Z+1)\right]\right), \tag{1}$$

where $Y(Z)$ is the yield of the element Z.

The local even-odd effect $\delta_p(Z)$ is plotted for ^{235}U(n_{th}, f) in Fig. 6. One notes a strong increase from 20% up to 40% for the most asymmetric charge splits. From this it was concluded that very asymmetric fission is a weakly dissipative process leading to rather low excitation energy at scission [18]. Our data for ^{234}U also shown in Fig. 6 are generally lower, which again may be attributed to the higher excitation energy in electromagnetically induced fission than in neutron-induced fission. Although they do not extend to the very asymmetric charge splits they show an increase towards asymmetry. The structures of $\delta_p(Z)$ for both nuclei are similar but not identical because the fissioning systems are two neutrons apart.

With our new data, the local even-odd effect $\delta_p(Z)$ can be investigated both for symmetric and asymmetric fission. In Fig. 7, $\delta_p(Z)$ is shown separately for symmetric and asymmetric fission of the thorium isotopes. The even-odd effect for asymmetric fission amounts roughly to 20% and is higher than the even-odd effect for symmetric fission with approximately 9%. But surprisingly the local even-odd effect for symmetric and asymmetric fission is rather independent of fissility Z^2/A for the row of thorium isotopes. This implies that the dependence of the average even-odd effect δ_p with Z^2/A is not genuine, but is merely simulated by the different admixtures of symmetric and asymmetric fission. The neutron-rich thorium isotopes fission predominantly asymmetrically, therefore the average even-odd effect δ_p (again shown in Fig. 7) is close to the value of the local

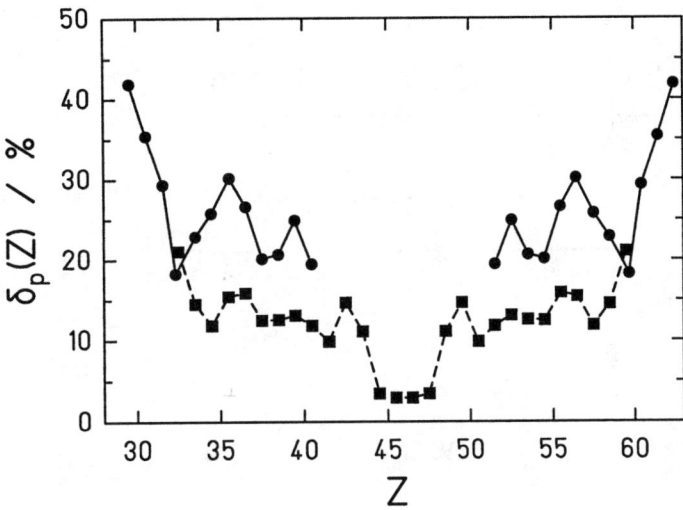

FIGURE 6. Measured local even-odd effects $\delta_p(Z)$ of $^{235}U(n_{th}, f)$(circles) [3,18] and from electromagnetic induced fission of ^{234}U (squares, this work). Please note that in thermal-neutron induced fission only the light fission fragments where observed experimentally.

even-odd effect $\delta_p(Z)$ for asymmetric fission. With decreasing neutron number symmetric fission becomes more dominant, and the even-odd effect decreases accordingly. For the most neutron-deficient isotopes measured, fission is mainly symmetric and the average even-odd effect corresponds to the local even-odd effect in symmetric fission. So the conclusion from the average even-odd effect on the dissipated energy during the descend from saddle to scission was rash. It seems more appropriate to first investigate how the local even-odd effect $\delta_p(Z)$ is influenced by pair breaking or unpaired protons before jumping to conclusions about dissipation in the fission process. An interesting object to study in this context is the local even-odd effect in fissioning systems with an odd atomic number, where there is always at least one unpaired proton.

LOCAL EVEN-ODD EFFECT FROM UNPAIRED PROTONS

The fission-fragment charge yields and the local even-odd effects for ^{220}Ac and ^{228}Pa are shown in Fig. 8. The local even-odd effect shows a striking result for these odd-Z fissioning nuclei: The values of $\delta_p(Z)$ tend to be positive for the light fragments and negative for the heavy fragments. This means that the unpaired proton predominantly sticks to the heavy fragment. For the extremely asymmetric charge splits $\delta_p(Z)$ is bigger than 20%. This behaviour is found systematically for all measured protactinium and actinium isotopes.

The local even-odd effect in odd-Z fissioning nuclei cannot be explained by the current models of the even-odd effect [2], which predict that the even-odd effect vanishes in the presence of unpaired protons.

In view of these results one should consider any relation between the pair breaking and the enhanced production of even-charge fission fragments with caution. Moreover, it would be challenging to find a general explanation for the even-odd effect in odd-Z and even-Z fissioning nuclei.

The even-odd effect stemming from unpaired protons can be explained by a statistical-model consideration [4]. In the following, only the basic idea and the results will be discussed shortly.

The basic assumption is, that the even-odd effect from unpaired protons is governed by the single-particle level density at the Fermi-energy. As in the experiment the charge yields have been measured we need to consider the single-particle level density $g(Z)$ of the protons in the nucleus. It can be shown that $g(Z)$ is proportional to the nuclear charge under the assumption of a nearly unchanged charge density in the fission process [4]. The probability $p(Z)$ for the unpaired proton to stick to either fragment is then determined by the available phase space

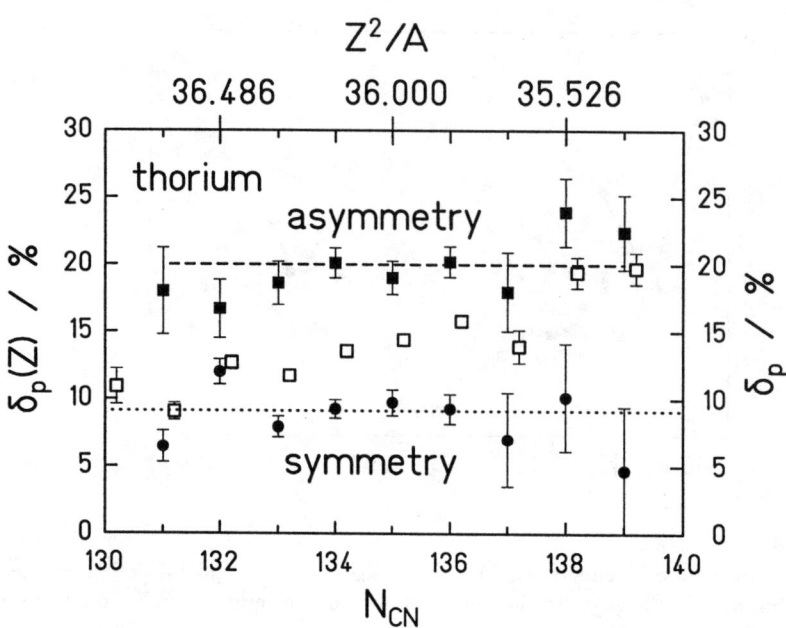

FIGURE 7. The local even-odd effect $\delta_p(Z)$ as defined by Eqn. (1) shown as a function of neutron number and Z^2/A for symmetric and asymmetric fission of thorium isotopes. The lines are shown to guide the eye. The open symbols show the average even-odd-effect δ_p for comparison.

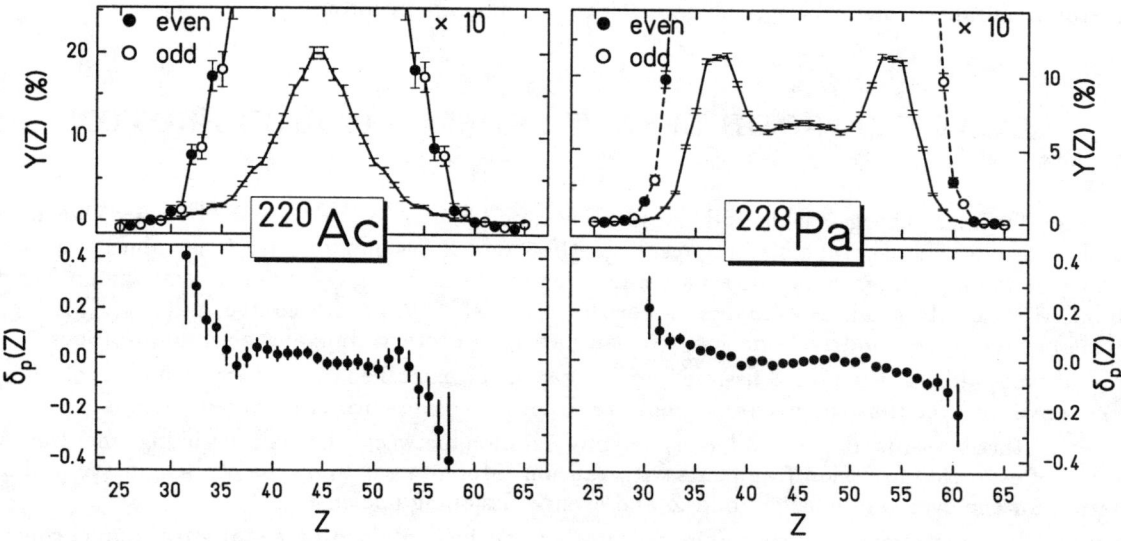

FIGURE 8. The upper panel shows the measured fission-fragment charge yields for ^{220}Ac and ^{228}Pa. To make the even-odd staggering visible, the asymmetric yields are plotted with a magnification factor of 10, and the even and odd charges are marked. The lower panel shows the deduced local even-odd effects for the same isotopes.

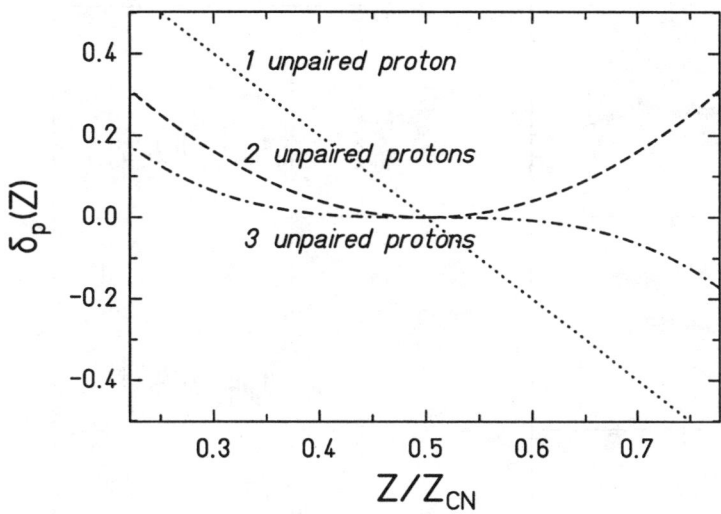

FIGURE 9. The theoretical even-odd effect arising from unpaired protons as given by Eqn.(2) for 1,2,3 unpaired protons.

$$p(Z) = \frac{g(Z)}{g(Z) + g(Z_{cn} - Z)} = \frac{Z}{Z_{cn}},$$

where Z_{cn} is the charge of the fissioning compound nucleus. From a short straightforward combinatorical calculation one gets for the even-odd effect from N unpaired protons:

$$\delta_p(Z) = (1 - 2p(Z))^N \qquad (2)$$

In Fig. 9 the theoretical even-odd effect according to Eqn.(2) is plotted as a function of Z/Z_{cn} for $N = 1, 2, 3$. In symmetric fission there arises no contribution from unpaired protons to the even-odd effect because $p(Z) = 1/2$ and the probability to stick to either fragment is the same. This implies, that conclusions on the survival of the superfluid fully paired phase during the motion from saddle to scission may be drawn from the magnitude of the local even-odd effect in symmetric fission only, while in asymmetric fission the even-odd effect arising from our statistical consideration has to be taken into account as well.

For one unpaired proton one observes a linear dependence of the theoretical even-odd effect on Z/Z_{cn}, for one pair broken a parabolic dependence, while for three unpaired protons it is a polynomial of third order. Furthermore, the magnitude of the even-odd effect decreases with increasing number of unpaired protons. The fundamental difference is, that the even-odd effects of even-Z fissioning nuclei stay positive even in higher orders, while for odd-Z fissioning nuclei they change their sign at symmetry.

To explain the observed local even-odd effects in our data we need to take into account the contribution from both the fully paired phase and from unpaired particles as well. For even-Z fissioning nuclei the observed even-odd effect can be explained by a linear combination of the fully paired phase $\delta_p \equiv 1$ and by the contribution with one pair broken ($N = 2$ in Eqn.(2)). The higher orders are generally smaller and are neglected for the sake of simplicity. The weighting factor for the survival of the superfluid phase can be determined from the local even-odd effect in symmetry as stated above. A value of 10% can be derived from the symmetric charge splits from ^{233}U and ^{226}Th. This means, the probability of one pair broken is 90% for the electromagnetically induced fission studied in this work.

For odd-Z fissioning nuclei, the observed even-odd effect can be explained in the same way, but naturally here is always one unpaired proton present. Thus, the linear combination will be of one unpaired proton and three unpaired protons ($N = 1, 3$ in Eqn.(2)). Higher orders are again neglected.

By employing the same weighting factors in the even-Z fissioning systems (0, 2 unpaired protons) and in the odd-Z fissioning systems (1, 3 unpaired protons) we can already describe the systematic behaviour of the measured data. In Fig. 10 a selection of uranium, thorium, protactinium, and actinium isotopes is shown

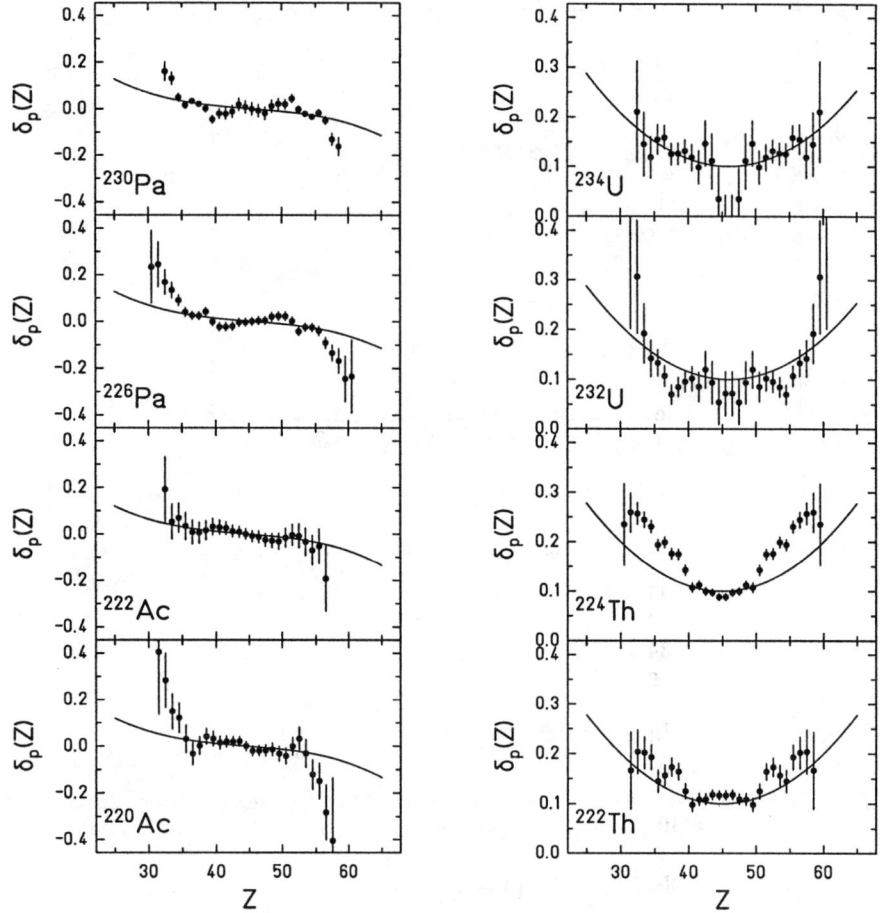

FIGURE 10. The measured local even-odd effects (data points) for isotopes of uranium, thorium, protactinium, and actinium are shown in comparison with the model prediction (full lines) as described in the text.

together with the model prediction. The typical behaviour as discussed above is also found in the experimental data. The model can account for a non-vanishing local even-odd effect in odd-Z fissioning nuclei and the variation of the local even-odd effect in even-Z fissioning nuclei. It follows that a big part of the increasing even-odd effect with asymmetry is just due to the phase space of the unpaired protons and is not related to the dissipated energy from saddle to scission. But one also sees discrepancies which may be partly due to structural effects not included in the very schematic single-particle level density g. The remaining discrepancies between the model and the data as far as they do not relate to structural effects in the level density may then indeed be due to a reduced dissipation in asymmetric charge splits.

One might suspect that also the Q-value of even-even charge splits favors the production of even elements in fission. By comparing the total number of levels below a given excitation energy in even- and odd-charge nuclei it can be concluded that this is true only for the very low excitation energies below 2Δ, where Δ is the pairing gap, while at higher excitation energies the differences in the total number of levels are to small to have an effect [4]. Thus, the Q-value even-odd staggering generally does not favor the even-odd effect, when proton pairs are broken.

In our model it is assumed, that the unpaired protons are distributed according to the available phase space. An additional contribution to the pair-breaking by a local necking-in process at scission [19] is not included which would inevitably lead to a separation of the two protons. In this scenario the phase-space argument of our model would not be valid. The separation of the two protons into different fragments would thus diminish the observed increase of the local even-odd effect with charge asymmetry and deteriorate the agreement of the model with the experimental data. Therefore, one may infer that the observed local even-odd effect can not result from a pair-breaking at scission (random neck rupture [20]).

CONCLUSION

We have shown that the interpretation of the even-odd effect in terms of energy dissipation in fission is more complex than thought up to now. The strong variations of the observed even-odd effects in the elemental yields of neutron-deficient isotopes from uranium to actinium as a function of the fissioning system and as a function of the asymmetry of the charge split are attributed to a great part to a statistical contribution rather than to a variation of the intrinsic excitation energy at scission. More detailed studies are required to gain a fully quanitative understanding of the contributions of these two totally different mechanisms leading to the observed even-odd effect.

ACKNOWLEDGEMENT

This work has been supported by the GSI Hochschulprogramm and by the Bundesminister für Bildung und Forschung (BMBF) under contract 06 DA 473. The responsibility for the contents rests with the authors.

REFERENCES

1. J.P. Bocquet et al., *Z.Phys.* **A 335**, 41 (1990)
2. H. Nifenecker et al., *Z. Phys.* **A 308**, 39 (1982)
3. W. Lang et al., *Nucl. Phys.* **A 345**, 34 (1980)
4. St. Steinhäuser et al., *Nucl. Phys.* **A 634**, 89 (1998)
5. H.G. Clerc et al., *Nucl. Phys.* **A 590**, 785 (1995)
6. A.R. Junghans et al., *Nucl. Phys.* **A 629**, 635 (1998)
7. H. Geissel et al., *Nucl. Instr. Meth.* **B 70**, 286 (1992)
8. A. Grewe et al., *Nucl. Phys.* **A 614**, 400 (1997)
9. J.P. Bocquet et al., *Nucl. Phys.* **A 502**, 213c (1989)
10. G. Mariolopoulos et al., Nucl. Phys. **A 361**, 213 (1989)
11. M. Djebara et al., *Nucl. Phys.* **A 425**, 120 (1984)
12. J. Kaufmann et al., *Z. Phys.* **A 341**, 319 (1992)
13. U. Quade et al., *Nucl. Phys.* **A 487**, 1 (1988)
14. K. Persyn et al., *Proc. Int. Workshop Dynamical Aspects of Nuclear Fission, Smolenice, Czechoslovakia 1991*
15. S. Pommé et al. *Nucl. Phys.* **A 560**, 689 (1993)
16. C. Schmitt et al., *Nucl. Phys.* **A 430**, 21 (1984)
17. B.L. Tracy et al., *Phys. Rev.* **C 5**, 222 (1972)
18. J.L. Sida et al., *Nucl. Phys.* **A 502**, 233c (1989)
19. J.F. Berger et al., *Nucl. Phys.* **A 502**, 85c (1989)
20. U. Brosa et al., *Phys. Rep.* **197**, 167 (1990)

Incomplete-fusion reactions for γ-ray spectroscopy: application to the study of high-spin states in ^{234}U

G.J. Lane*, G.D. Dracoulis, A.P. Byrne†, T.R. McGoram and A.R. Poletti‡

Dept. of Nuclear Physics, Australian National University, Canberra ACT 0200, AUSTRALIA

* *Present Address: Lawrence Berkeley National Laboratory, 1 Cyclotron Rd, MS-88, Berkeley CA 94720, USA*
† *Joint Appointment: Dept. of Physics, Australian National University, Canberra ACT 0200, AUSTRALIA*
‡ *Perm. Address: Dept. of Physics, University of Auckland, Private Bag 92019, Auckland, NEW ZEALAND*

Abstract. Incomplete-fusion reactions occur when breakup of the projectile results in only part of the beam particle fusing with the target, the remnant being emitted with an energy equivalent to the beam velocity. Such reactions have been demonstrated to populate slightly neutron-rich nuclei compared to conventional fusion-evaporation reactions, opening possibilities for the study of nuclei along the neutron-rich side of the line of stability. Results from a study of ^{211}Po are presented to illustrate the use of incomplete-fusion reactions for γ-ray spectroscopy. New results from a test-run which populated high-spin states in ^{234}U via the ^{232}Th(^{9}Be,$\alpha 3n$) reaction are also presented. An interesting feature of the latter reaction is that the high fission probabilities for the compound nuclei which follow complete fusion, results in the residues from incomplete fusion forming the dominant residue channels.

INTRODUCTION

This paper will give a brief overview of some recent efforts to apply incomplete-fusion reactions to the study of nuclei along the neutron-rich side of the line of stability. These studies are timely in light of the strong interest in nuclei far from stability on both the proton and neutron-rich sides, and the efforts which are being put into the development of radioactive ion beams to populate these exotic nuclei. One method of radioactive ion beam production is the fragmentation of nuclei at relativistic energies, however, this method results in products which are too energetic to immediately induce heavy-ion fusion-evaporation reactions. Nevertheless, in certain circumstances it is possible to obtain "radioactive beams" with appropriate beam energies for heavy-ion fusion by the breakup of a stable beam, e.g. ^{9}Be $\rightarrow \alpha +$ "^{5}He". This breakup process occurs with appreciable cross-section at energies near the Coulomb barrier and it is then possible for the "^{5}He" fragment to fuse with the target, while the the α particle is emitted directly. In this way, neutron-rich nuclei can be produced. Furthermore, the dynamics of the incomplete-fusion reaction mean that even for a beam which breaks up into two stable isotopes, with one subsequently fusing with the target, it is possible to populate slightly more neutron-rich products than would occur with conventional complete fusion with the stable fragment as a beam.

This paper focusses on the experimental techniques and methods. We first discuss those features of incomplete-fusion reactions which make them amenable to γ-ray spectroscopy studies of neutron-rich nuclei. A demonstration is provided by the ^{208}Pb(^{9}Be,$\alpha 2n$)^{211}Po reaction. The results of a test-run which populated high-spin states in ^{234}U using the ^{232}Th(^{9}Be,$\alpha 3n$) reaction are also presented, and the outlook for further work is discussed. A previous report on the progress of these studies can be found in Ref. [1,2].

INCOMPLETE-FUSION REACTIONS FOR γ-RAY SPECTROSCOPY

Early studies of incomplete-fusion reactions for γ-ray spectroscopy, focussed on the possibility of obtaining a narrow spin distribution in the compound system corresponding to collisions near grazing [3,4]. Since then

CP481, *Nuclear Structure 98*, edited by C. Baktash

the use of incomplete-fusion reactions to populate nuclei for the purpose of conventional γ-ray spectroscopy has not been wide-spread.

There is a tendency for incomplete-fusion reactions to populate neutron-rich nuclei [1,2]. The reasons for this are immediately obvious when one of the breakup fragments corresponds to a radioactive product, e.g. ^{9}Be $\to \alpha +$ "^{5}He" or ^{7}Li $\to \alpha + t$. However, it also occurs when the beam breaks up into stable products; this is because of the separation energy associated with the breakup [1,2]. For example, if the $(^7$Li,$2n)$ reaction is sub-barrier and would have a peak cross-section at an energy E_{peak}, then the $(^{11}$B,$\alpha 2n)$ reaction which occurs via incomplete-fusion would have a peak cross-section at $E = \frac{11}{7}(E_{peak} + E_{sep}(^{11}B \to \alpha + ^7Li))$. Since the separation energy E_{sep} is large and positive ($+8.7$ MeV), this reaction can occur, *when it proceeds via breakup*. On the other hand, if complete fusion occurs, it is unlikely that an α-particle and two neutrons will be evaporated.

While the possible population of neutron-rich isotopes is itself interesting, there are other differences in the dynamics of complete and incomplete fusion which can be exploited for γ-ray spectroscopy. The differences in the spin input distributions have been touched upon already. More important in the current work is the different character of the emitted charged particles. Those from the equilibrated compound system following complete fusion are emitted isotropically in the centre of mass and have a distribution of energies. In contrast, the left-over fragment after incomplete fusion is emitted with essentially the beam velocity in a direction peaked around the grazing angle.

It has been realised only recently that there is also a correlation between the angle at which the charged-particles are emitted and the type of residual nucleus which results; typically the heavier nuclei with fewer neutrons evaporated are associated with more strongly forward-focussed charged particles [1,2]. This is likely related to the relative cross-sections for population of an isotope via incomplete fusion and complete fusion. More neutron-rich isotopes are populated by incomplete fusion and consequently are associated with a higher proportion of forward-going charged particles. The resultant differences can be exploited for the assignment of γ-rays to particular residue channels. This has been especially important in the studies of extremely long-lived, high-spin isomers in the $A \approx 180$ and trans-lead nuclei, since there the long lifetimes decouple the high-spin states from those below the isomer. One of the best known high-spin isomers is the $t_{1/2} = 31$ year, 16^+ isomer in ^{178}Hf, above which no excited states were known until recently. The γ-rays comprising the rotational band built upon this isomer were observed and assigned to ^{178}Hf using the characteristics of the α-particle angular distributions in a ^{9}Be-induced incomplete-fusion reaction [5]. Similar methods were used to assign excited states to bands built upon a number of long-lived states in other hafnium isotopes and these results have been discussed by Simon Mullins in these proceedings [6], while results for bands in tantalum [7] isotopes have been presented elsewhere.

To illustrate some of the above mentioned features we will look at the example of population and identification of high-spin states in ^{211}Po.

^{211}PO CASE STUDY

The nuclei ^{211,212}Po both have high-spin α-decaying isomers with $J^\pi = \frac{25}{2}^+, (18^+)$ and lifetimes $\tau_m = 36.4, 65$ s, respectively. Until recently, no high-spin states above these isomeric states were known. The physics of these high-spin states is rather interesting from a shell-model perspective, so that the isomeric states have been the subject of considerable experimental and theoretical effort. (Note that ^{212}Po is formed by adding an α-particle to the ^{208}Pb core.)

The population of high-spin states in ^{211,212}Po via conventional heavy-ion fusion reactions is problematic, as they are relatively neutron-rich. The only available xn reactions which use stable beams and targets are light-ion induced reactions, and, of these, only the ^{208}Pb$(\alpha,n)^{211}$Po reaction is close to or above the Coulomb barrier. Alternatives might be the ^{208}Pb$(^9$Be,$\alpha xn)$ reactions with $x = 1$ or 2 for ^{212}Po and ^{211}Po, respectively. However, statistical model calculations for this system predict that the evaporation of α-particles is unlikely and the residue channels are almost entirely due to neutron evaporation, with ^{210}Po being the heaviest polonium nucleus predicted to have any yield above a couple of millibarns. Nevertheless, if break-up and incomplete fusion were to play a role then one might expect the ("^{5}He",xn) reaction to populate states in ^{211,212}Po.

An experiment using the ^{9}Be $+ ^{208}$Pb reaction was performed at the Australian National University with the CAESAR [8] array of six Compton-suppressed, 25% efficient HPGe detectors and the Particle Detector Ball (PDB) [9] with 14 plastic phoswich detectors in a close-packed geometry covering 85% of 4π. The PDB has three rings of essentially equivalent detectors, i.e. four "forward", six "middle" and four "backward" detectors,

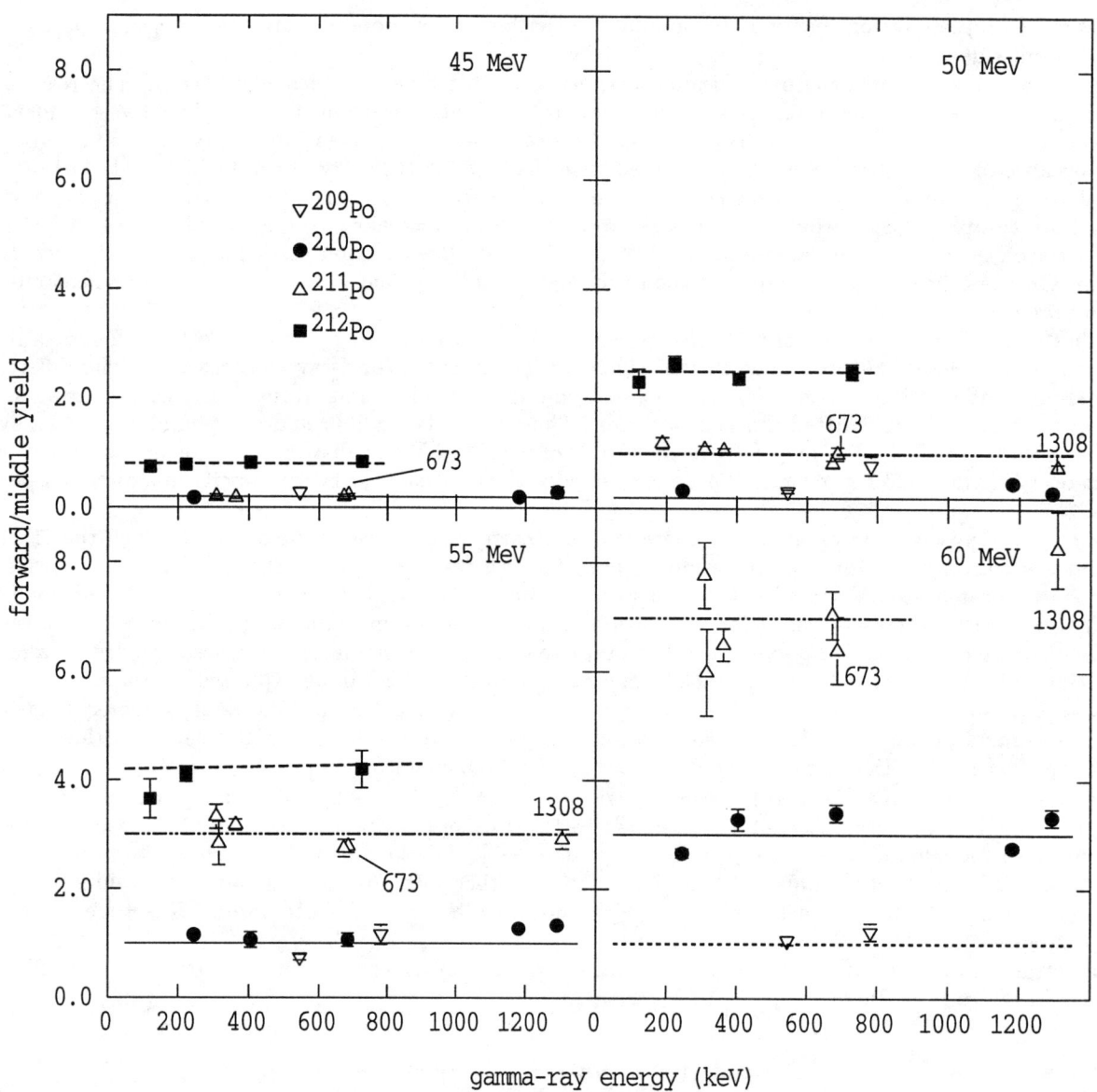

FIGURE 1. Ratios of yields of γ-rays measured in coincidence with particles in the forward and middle rings of detectors in the PDB. Results are shown for four different beam energies. The 673 and 1308 keV γ-rays were previously unknown, but can be assigned to ^{211}Po based on the yield ratios. Figure taken from Ref. [10]

subtending angles of approximately $20°$ to $60°$, $60°$ to $120°$ and $120°$ to $165°$, respectively. The forward detectors were covered with aluminium shielding foils of 45 mg/cm^2 thickness to prevent rate limitations from the scattering of beam particles. These foils were sufficient to stop ^{9}Be ions with energies up to ~ 57 MeV, but also stopped α-particles of less than ~ 19 MeV energy, thus reducing the α-detection efficiency at forward angles. Beams of ^{9}Be ions with energies ranging from 45 to 60 MeV were provided by the 14UD tandem accelerator, incident on 2.3 and 3.3 mg/cm^2 ^{208}Pb foils, isotopically enriched to > 98%.

High-spin states above the $\frac{25}{2}^+$ isomer in ^{211}Po were identified using the correlation between the number of neutrons emitted and the angle of emission of the α-particles. This was done by measuring the yields of γ-rays in coincidence with α-particles in the three rings of the PDB. The ratio of yield in coincidence with the forward ring compared to the yield in coincidence with the middle ring is the same for γ-rays from the same nucleus, and varies from nucleus to nucleus. This is shown in Fig. 1 where the strongest new γ-rays which feed above the α-decaying isomer in ^{211}Po (673 and 1308 keV) show a ratio which clearly follows the value measured for

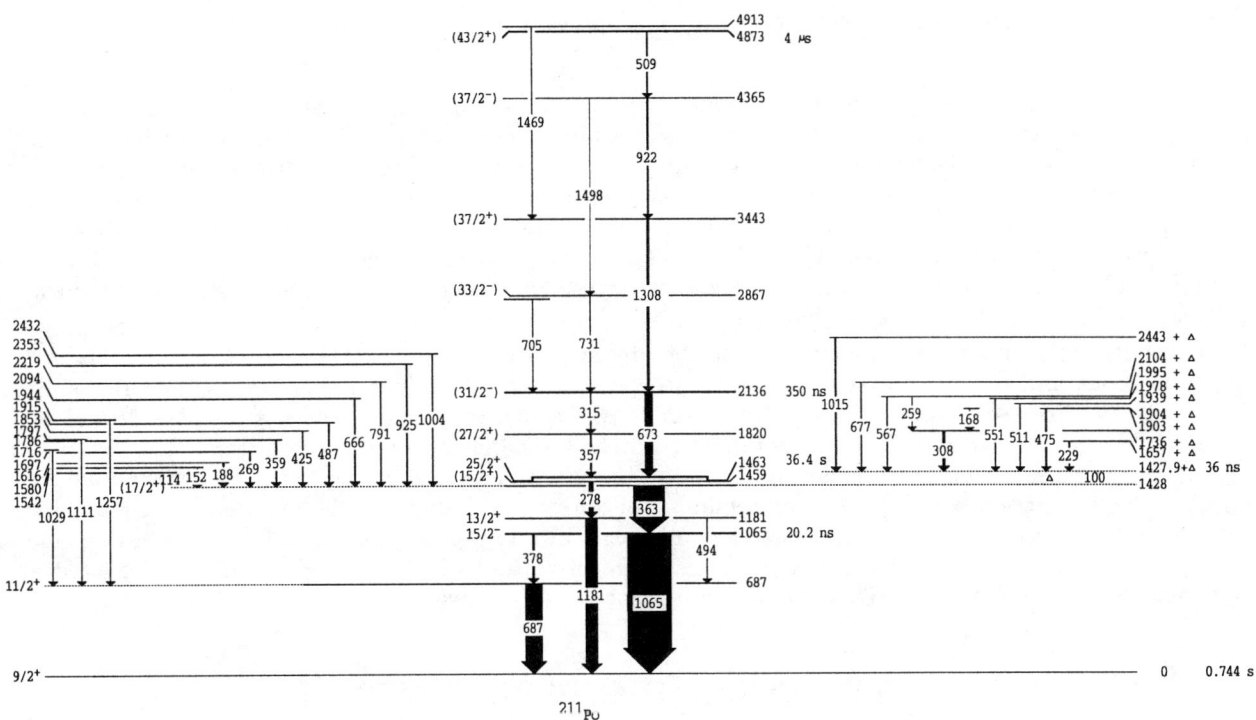

FIGURE 2. Partial level scheme for ^{211}Po showing the new high-spin states above the $\frac{25}{2}^+$ isomer (from Ref. [10]).

the γ-rays known previously [11] below the isomer.

The figure also shows the dependence of the angular yield ratio as a function of the beam energy. At the lowest measured energy the yield ratios for the different channels are similar, but with a measurable enhancement of α-particles in the forward direction for γ-rays from ^{212}Po. As the beam energy increases, the yield ratios for the channels populated through incomplete fusion all increase, corresponding to the shifting of the grazing angle to the forward direction. Furthermore, there is a clear tendency for more neutron-rich nuclei to be associated with a more forward peaked angular distribution, indicating that a larger component of their total cross-section comes from incomplete fusion.

The main γ-γ coincidence measurements with the ^{9}Be beam were made at a beam energy of 50 MeV. After γ-rays were assigned to ^{211}Po, the level scheme (see Fig. 2) was built up further from data collected using the ^{208}Pb(^{7}Li,$p3n$) reaction at 56 MeV. Since this bombardment was higher above the Coulomb barrier than the ^{9}Be-induced reaction, it populated higher-spin states in ^{211}Po. These measurements were made without the PDB, so it could not be verified whether the breakup of ^{7}Li into $t + \alpha$ played a significant role. Previous work suggests that breakup is important for the ^{7}Li + ^{208}Pb reaction [12].

The γ-ray measurements show that the polonium isotopes are populated more strongly than the statistical model predictions. The absolute residue cross-sections have been measured directly by Dasgupta *et al.* to try and understand the origin of this increased polonium cross-section [13]. For the ^{9}Be + ^{208}Pb reaction they find the polonium nuclei make up approximately 30% of the total residue yield, with ^{211}Po having the largest polonium cross-section (as large as the ^{213}Rn ($3n$) channel), while ^{212}Po has a peak cross-section of 15 mb. Furthermore, the shapes of the excitation functions for the polonium nuclei are distinctly different from those for the radon nuclei, indicative of a different reaction mechanism than fusion-evaporation. Dasgupta *et al.* also formed the same compound system using the ^{13}C + ^{204}Hg reaction and again measured the residue yields. For this system, negligible amounts of polonium nuclei are produced, the vast majority of the fusion cross-section instead going into xn channels populating radon nuclei. Modelling the cross-sections they found good agreement with theory for the ^{13}C + ^{204}Hg system, but the measured absolute xn yields for the ^{9}Be + ^{208}Pb system were only approximately 70% of the theoretical values. Their conclusion is that the missing cross-section corresponds to break-up of the beam prior to fusion, and that it is the subsequent fusion of a fragment of the beam (either ^{4}He, ^{5}He or ^{4}He+n) which leads to population of the polonium nuclei.

OTHER METHODS OF POPULATING ^{211}PO

It should be noted that in parallel to the present γ-ray measurements, Fornal *et al.* [14] also identified high-spin states above the isomer in ^{211}Po. They used a pulsed beam of 450 MeV ^{76}Ge ions from the ALPI accelerator at Legnaro incident on a thick ^{208}Pb target. The γ-rays were observed using the GASP array of 40 Compton-suppressed HPGe detectors. Deep-inelastic processes populated a large range of isotopes in this reaction, with the polonium isotopes populated mainly in coincidence with zinc isotopes due to the transfer of $2p + xn$ from the beam to the target. Thus, Fornal *et al.* could first assign the 673 keV γ-ray to ^{211}Po by its delayed coincidence with known prompt lines in zinc isotopes. Following this assignment, the rest of their level scheme for states above the $\frac{25}{2}^+$ isomer (see Fig. 1 of Ref. [14]) could be built from γ-γ coincidences.

These results make an interesting contrast with the present method. Deep-inelastic reactions populate many (many!) isotopes and certainly have the potential to populate more neutron-rich isotopes than incomplete-fusion reactions. However, even with the 40 detector GASP array, the resulting level scheme of Fornal *et al.* for ^{211}Po is less comprehensive than that obtained with the CAESAR array of six detectors. The reasons are two-fold: (i) the cross-section for populating ^{211}Po via the incomplete fusion reaction is far higher than using the deep-inelastic reaction, and (ii) the smaller number of competing channels in the incomplete-fusion reaction results in less γ-ray background. This shows that for nuclei in a limited region along the neutron-rich side of the line of stability, incomplete fusion reactions are an effective tool which can and should be exploited as a complement to deep-inelastic reactions and as a precursor to radioactive ion beam experiments.

VERY HEAVY NUCLEI: HIGH-SPIN STATES IN ^{234}U

Recognising that break-up may be even more prevalent with heavier targets, and in an effort to further probe the usefullness of incomplete-fusion reactions, we performed a test run to try and populate excited states in ^{234}U via the ^{232}Th(^{9}Be,$\alpha 3n$) reaction. This approach contrasts with the obvious choice of xn fusion-evaporation reaction, ^{232}Th(α,$2n$), which has been used previously [15–17] to populate high-spin states in ^{234}U up to $I^\pi = 16^+$.

We performed a γ-ray spectroscopy experiment using a beam of 52 MeV ^{9}Be ions, supplied by the ANU 14UD Pelletron accelerator, incident upon a 1.0 mg/cm^2 self-supporting target of ^{232}Th. The target was chosen to be thick enough that most of the recoiling fusion products are stopped in the target, but thin enough that the majority of the fission fragments escape. The experimental set-up was essentially the same as that used for the ^{211}Po experiments described above.

The measured γ-ray singles spectrum is shown in Fig. 3a. The strongest lines present are the x-rays and Coulomb excitation γ-rays arising from the ^{232}Th target [17], together with light-ion contaminants formed from reactions on oxygen (present in the form of oxide in the target). An unresolved continuum of γ-rays due to fission fragments is present, while the γ rays from the ground state band in ^{234}U are very weak.

To enhance the ^{234}U component of the reaction, the γ-γ coincidence data were sorted offline into E_γ-E_γ matrices with various time and particle conditions. Figure 3b shows the projection of a γ-γ matrix created requiring two γ rays to be observed within ± 170 ns of each other and also in prompt coincidence with the detection of an α particle. A total of 4×10^6 α-γ-γ coincidence events were sorted into the matrix. An unresolved continuum of γ-rays due to fission is the dominant feature of the matrix projection, with prominent γ rays from Coulomb excitation of the ^{232}Th target also present. The γ-rays from the ground-state band in ^{234}U are strongly enhanced compared to the singles spectrum.

Figure 4 presents a sum of five background-subtracted coincidence spectra obtained from the γ-γ coincidence matrix by setting gates on the $E2$ transitions in the cascade connecting the 4^+ and 14^+ excited states in ^{234}U. The ground-state band is observed clearly up to the 374 keV, $18^+ \rightarrow 16^+$ transition. Note that the states in the ground-state band of ^{234}U above $16\hbar$ were identified previously using Couloumb excitation [18] and heavy-ion induced transfer [19] reactions, not using the ^{232}Th(α,$2n$) reaction, which does not populate ^{234}U to such high angular momenta.

With the limited statistics collected, the current work did not observe the states from the known octupole, β or γ bands in ^{234}U. The inset to Fig. 4 shows the energy region in which the γ rays from the decay of these bands to the ground-state band are expected [17]. The current work was also sensitive to the identification of isomeric states, although no delayed population of the ground state band was observed at the level of $\sim 5\%$ of the total ^{234}U channel intensity.

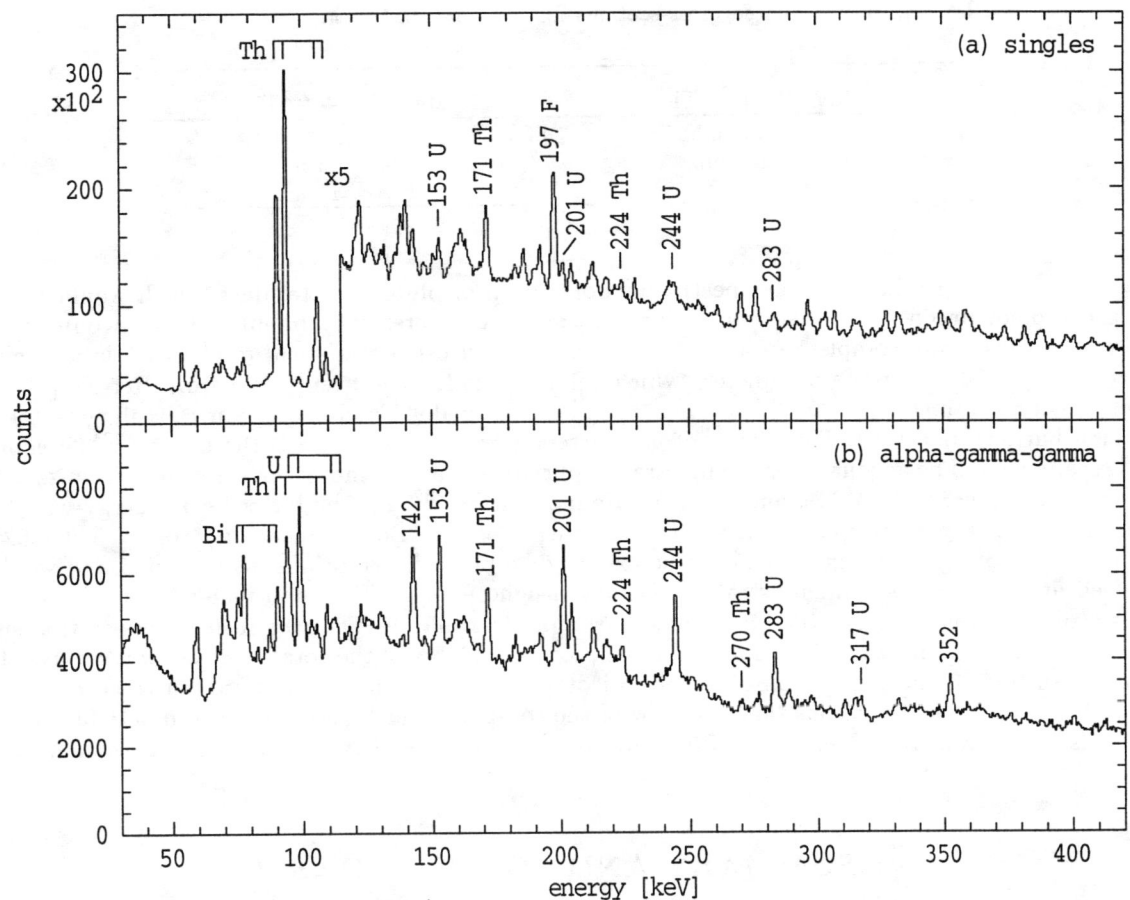

FIGURE 3. (a) Singles spectrum for the ^{9}Be + ^{232}Th reaction at 52 MeV. (b) Projection of the α-gated γ-γ matrix.

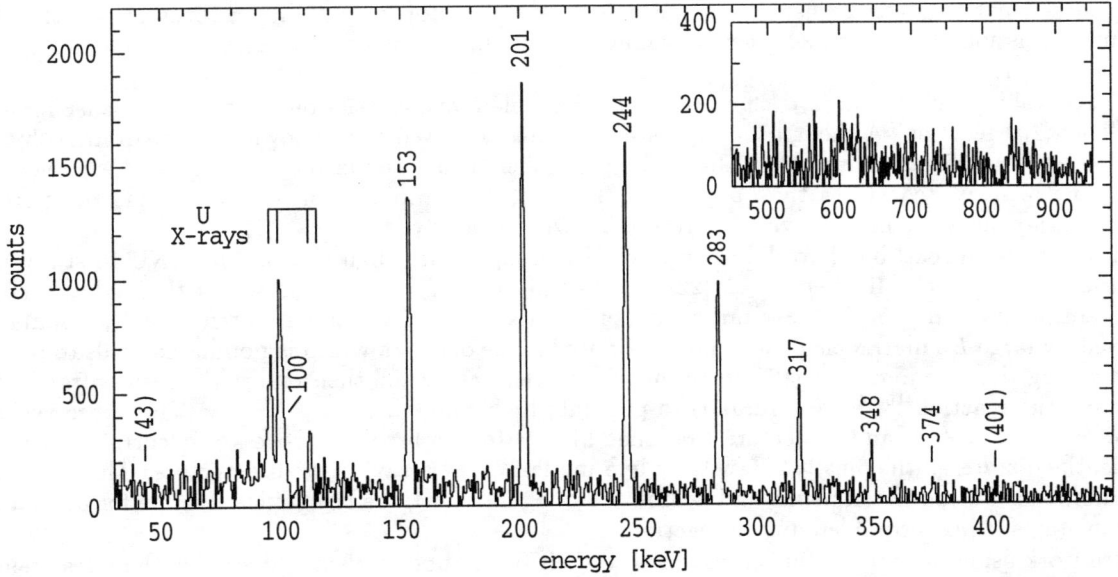

FIGURE 4. Sum of spectra with gates on the $E2$ transitions between the 4^+ and 14^+ excited states in the ground-state band of ^{234}U.

309

TABLE 1. Calculated cross-sections for reactions on ^{232}Th.

52 MeV ^{9}Be + ^{232}Th		29 MeV ^{5}He + ^{232}Th	
Fission	437 mb	Fission	662 mb
^{236}Pu	0.4 mb	^{234}U	13 mb
		^{233}U	11 mb

There was no evidence in the γ-ray spectra for population of plutonium residues which would be formed following complete fusion of ^{9}Be + ^{232}Th. The reason why the residues populated by incomplete fusion dominate over those from complete fusion can be demonstrated using statistical model calculations with the PACE code [20]. Values for the parameters which enter the code have been taken from Ref. [21], including the level density parameters, $a_n = A/8.8$ MeV^{-1} and $a_f/a_n = 0.90$, and the Kramers scaling factor of the Sierk fission barrier, $k_f = 0.88$. This choice of parameters describes the decay of the nearby ^{224}Th compound system [21] and should be adequate for our illustrative purposes. Table 1 shows the calculated cross-sections for two different reactions, 52 MeV ^{9}Be and 29 MeV ^{5}He incident on ^{232}Th. The latter beam energy was deduced by assuming an energy sharing between the two fragments of the ^{9}Be beam, i.e. $E(^5\text{He}) = 5/9 \times 52$ MeV. (Note that fusion of the ^{5}He fragment would be expected to occur for collisions near grazing and would thus be localised at high l-waves. Our illustrative calculation ignores this effect, and instead assumes a standard triangular spin distribution.) For the ^{9}Be-induced reaction, less than 0.1% of the fusion cross-section survives fission. This compares with the ^{5}He-induced reaction, for which 3.5% of the total cross-section survives fission, eventually resulting in the population of uranium isotopes via neutron evaporation. In essence, the higher fissility of the $Z = 94$ system means that very few of the compound nuclei formed by complete fusion, survive fission. This means that the observation of uranium products implies they were populated by an incomplete-fusion process.

SUMMARY AND CONCLUSIONS

With a modest γ-ray / particle-detector array combination, it is possible to perform extensive spectroscopy on a range of neutron-rich nuclei using incomplete-fusion reactions. The characteristic angular distributions of the charged particles can be used to assign new γ-rays to isotopes, while it has been shown that for very heavy nuclei, fission of the complete fusion products results in the incomplete fusion products dominating the residue population.

The encouraging results for ^{234}U show that the incomplete-fusion reaction does probe higher in spin than the ^{232}Th(α,2n) reaction. Although the highest-spin states observed were known already from both Coulomb excitation and transfer experiments, it should be remembered that the latter two processes are selective with respect to which excited states they populate. Thus, the incomplete-fusion reaction has the potential to populate high-spin states in ^{234}U which were not known previously.

Incomplete-fusion reactions have been employed by groups other than that at the ANU to study neutron-rich nuclei. For example, Hartlein et al. [22] discuss the cases of ^{160,162}Dy, in which the highest known spin states were observed in Coulomb excitation measurements. These experiments preferentially populate states connected by large $E2$ matrix elements, and the limited range of states which is populated leads to problems in understanding the s-band crossings in these nuclei. Hartlein et al. [22] then report the results from a test run which used the reaction ^{158,160}Gd(^{7}Li,pxn) to populate high-spin states in ^{160,162}Dy. They observed strongly forward-focussed protons and conjecture that the ^{7}Li $\rightarrow$ ^{6}He + p breakup process is occurring. Furthermore, they identified states with spins $\geq 20\hbar$ with only a modest γ-ray array of seven detectors. This is higher than the Coulomb excitation experiments, and demonstrates again that there is significant scope for populating new high-spin states using incomplete-fusion reactions.

Future work using incomplete-fusion reactions to their fullest potential should employ the latest generations of charged particle and γ-ray detector arrays; this will give considerable improvement over the experiments described here which used modest arrays. Also, detailed investigations of the reaction mechanism need to be performed utilising a particle-detector array with high detection efficiency and good discrimination between p, d, t, α, ^{5}He,^{7}Li etc. These sorts of experiments would be a useful precursor to studies with radioactive ion beams and may provide a sneak preview of some of the interesting new physics which radioactive ion beams promise to provide.

Acknowledgements: The authors wish to thank N. Dasgupta *et al.* for making their fusion cross-section measurements available prior to publication. GJL is grateful to H. Grawe for bringing the ^{160,162}Dy results to his attention. The ^{211}Po results comprise part of the MSc thesis of T.R. McGoram and have also been published elsewhere [10]. Figures 1 and 2 are reprinted from Nuclear Physics A, Vol 637, T.R. McGoram *et al.*, "High-spin isomers in ^{211}Po and related structures in ^{210}Po and ^{212}Po", pp 469, Copyright 1998, with permission from Elsevier Science.

REFERENCES

1. G.D. Dracoulis, *Proc. of the Conf. Nuclear Structure at the Limits*, Argonne 1996, ANL/PHY-97/1, p148.
2. G.D. Dracoulis, A.P. Byrne, T. Kibédi, T.R. McGoram and S.M. Mullins, *J. Phys. G* **23**, 1191 (1997).
3. T. Inamura, M. Ishihara, T. Fukuda, T. Shimoda and H. Hiruta, Phys. Lett. B **68**, 51 (1977).
4. D.R. Zolnowski *et al.*, Phys. Rev. Lett. **41**, 92 (1978).
5. S.M. Mullins, G.D. Dracoulis, A.P. Byrne, T.R. McGoram, S. Bayer, W.A. Seale, and F.G. Kondev, Phys. Lett. B **393**, 279 (1997).
6. S.M. Mullins, *these proceedings*.
7. G.D. Dracoulis *et al.*, Phys. Rev. C **58**, 1444 (1998); G.D. Dracoulis *et al.*, Phys. Rev. C **58**, 1837 (1998).
8. G.D. Dracoulis and A.P. Byrne, Dept. of Nucl. Physics Annual Report 1989 ANU-P/1052, *unpublished*.
9. G.J. Lane, A.P. Byrne, and G.D. Dracoulis, Dept. of Nucl. Physics Annual Report 1992 ANU-P/1118, *unpublished*.
10. T.R. McGoram, G.D. Dracoulis, A.P. Byrne, A.R. Poletti, and S. Bayer, Nucl. Phys. **A637**, 469 (1998).
11. B. Fant, T. Lönnroth and V. Rahkonen, Nucl. Phys. **A355**, 171 (1981).
12. O. Hausser, A.B. McDonald, T.K. Alexander, A.J. Ferguson, and R.E. Warner, Phys. Lett. B **38**, 75 (1972).
13. M. Dasgupta *et al.*, Phys. Rev. Lett. **82**, 1395 (1999).
14. B. Fornal *et al.*, Eur. Phys. J. A **1**, 355 (1998).
15. P. Zeyen *et al.*, Z. Phys. A **328**, 399 (1987).
16. W.Z. Venema, J.F.W. Jansen, R.V.F. Janssens, and J. Van Klinken, Phys. Lett. B **156**, 163 (1985).
17. R.B. Firestone, *Table of Isotopes*, 8th ed., edited by V.S. Shirley (Wiley, New York, 1996).
18. H. Ower, Th. W. Elze, J. Idzko, K. Stelzer, E. Grosse, H. Emling, P. Fuchs, D. Schwalm, H.J. Wollersheim, N. Kaffrell, and N. Trautmann, Nucl. Phys. **A388**, 421 (1982).
19. K.G. Helmer *et al.*, Phys. Rev. C **44**, 2598 (1991).
20. A.F. Gavron, Phys. Rev. C **21**, 230 (1980).
21. C.R. Morton *et al.*, Phys. Rev. C **52**, 243 (1995).
22. T. Hartlein *et al.*, Prog. Part. Nucl. Phys. **38**, 309 (1997).

EXOTIC SHAPES AND HIGH-SPIN PHENOMENA

Superdeformation: Perspectives and prospects

Jacek Dobaczewski

Institute of Theoretical Physics, Warsaw University, Hoża 69, PL-00681, Warsaw, Poland
Joint Institute for Heavy Ion Research, Oak Ridge National Laboratory, P.O. Box 2008, Oak Ridge, TN 37831, U.S.A.
Department of Physics, University of Tennessee, Knoxville, TN 37996, U.S.A.
Institut de Recherches Subatomiques, Université Louis Pasteur, Strasbourg I, F-67037 Strasbourg Cedex 2, France

Abstract. We present a review of the mean-field approaches describing superdeformed states, which are currently used and/or being developed. As an example, we discuss in more details the properties of superdeformed $A \sim 60$ nuclei, and present results of calculations for the rotational band in the doubly magic superdeformed nucleus ^{32}S.

INTRODUCTION

Physics of superdeformed nuclear shapes is already a mature field, and a number of excellent review articles exist [1–4]. For the recent developments in experiment we refer the reader to other contributions to this conference. In theory, after the early calculations, which have predicted the existence of the superdeformed shape, see Ref. [2] for the review, and its experimental discovery in 1985 [5], numerous groups have developed sophisticated techniques to perform calculations allowing the description of rotational bands for very elongated shapes. At present, one witnesses a very impressive activity in this field. Almost all these theoretical approaches use the mean-field concept of the nucleus. One can say that the phenomenon of superdeformation constitutes a spectacular manifestation of the mean-field properties of nuclei.

In the present contribution, we review the mean-field methods presently used to describe the rotating superdeformed states, and discuss results obtained in recent years. As an example, we present in more details the calculations for nuclei in the $A \sim 60$ region, and for ^{32}S.

WHAT HAS BEEN DONE IN THEORY AFTER 1995

In this section we attempt at describing the present status of the mean-field theory of superdeformed states. In Table 1 we enumerate groups working in this domain in 1998. They are listed according to types of the mean field used, which are: the Nilsson (NS) [107,108] or Woods-Saxon (WS) [109,110,108] phenomenological mean fields, the non-relativistic self-consistent Skyrme [111–113,108] and Gogny [114,115,108] mean fields, or the self-consistent Relativistic Mean Field (RMF) [116]. As shown in Table 1, calculations are performed either by neglecting the pairing correlations (none), or by including them at various degrees of sophistication, ranging from the simplest $\Delta(\omega)$ method [117,118], or the Bardeen-Cooper-Schrieffer (BCS) or Bogolyubov (HFB) approximations [108], to the Lipkin-Nogami (LN) [119–124] or the particle-number-projection (PNP) [108] methods. Approaches which do include pairing correlations use various interactions in the particle-particle channel. These can be: the seniority (P^+P) or quadrupole pairing (Q^+Q) forces, the zero-range or zero-range and density-dependent forces, which induce the volume (Vol.) or surface (Surf.) pairing correlations, respectively, or the particle-particle force given by the finite-range Gogny interaction. Aiming at covering only the most recent studies, and assuming the review in Ref. [3] as the baseline, we quote in Table 1 references to papers published since 1995 till the present day.

Ingemar Ragnarsson and collaborators [6–22] have published numerous analyses of superdeformed bands, providing theoretical interpretation at the same time as many initial experimental discoveries have been made. In their calculations they have been minimizing the total energy (obtained within the Strutinsky prescription)

CP481, *Nuclear Structure 98*, edited by C. Baktash

TABLE 1. 1998 Who Is Who in the mean-field theory of superdeformed states.

	Refs.	Mean Field					Pairing						Pairing Force				
		NS	WS	Skyrme	Gogny	RMF	none	$\Delta(\omega)$	BCS	HFB	LN	PNP	P^+P	Q^+Q	Vol.	Surf.	Gogny
I.Ragnarsson et al.	[6–22]	X					X			X			X				
T.Nakatsukasa[a] et al.	[23–31]	X					X	X					X				
Y.Sun[b] et al.	[32–35]	X							X				X	X			
R.Wyss et al.	[36–57]		X							X	X	X	X	X	X		
T.R.Werner et al.	[58–65]		X				X										
R.R.Chasman et al.	[66–68]		X				X										
P-H.Heenen[c] et al.	[69–79]			X					X	X	X			X		X	X	
J.Dobaczewski et al.	[80–89]			X			X										
K.Matsuyanagi et al.	[90]			X			X										
L.Egido et al.	[91–93]				X		X			X	X	X					X
M.Girod[d] et al.	[94–96]				X					X	X						X
P.Ring et al.	[97–103]					X	X		X				X				
H.Madokoro et al.	[104–106]					X	X										

[a] Plus collective RPA correlations.
[b] Plus the angular momentum projection and coupling to selected multi-qp states. No cranking.
[c] Plus collective GCM correlations.
[d] Plus collective equation of motion.

with respect to deformation. Depending on the physical situation, pairing correlations have been included within the NS+HFB approach (self-consistent treatment of pairing within a fixed set of NS single-particle energies). Some studies required a given configuration to be followed across the ϵ-γ plane, and then the pairing correlations have been neglected, while the level crossings have been removed by the diabatic procedure [125]. In this way, in the $A{\sim}60$ region of superdeformed nuclei, several bands could be interpreted in terms of the band-termination phenomenon [9,15,19,20]. Following the statistical analysis of experimental data [126], which aimed at quantifying the frequency of occurrence of the identical bands, Karlsson, Ragnarsson et al. presented an analogous analysis of the calculated moments of inertia [16,18]. The same group has also performed the analysis of relative quadrupole moments [21] and confirmed their additivity (obtained originally within the HF method [84]), as well as studied the quadrupole polarization charges in terms of the harmonic oscillator (HO) single-particle quantum numbers. Recently, Afanasjev, Ragnarsson, and Ring [22] have presented a joint Nilsson and RMF study of several superdeformed nuclides in the $A{\sim}60$ region.

Takashi Nakatsukasa and collaborators [23–31] have studied superdeformed bands in the $A{\sim}190$ (and $A{\sim}150$ [23]) region of nuclei, with the particular emphasis on the collective excitations. They have been solving the RPA equations in the rotating frame and have used separable multipole interactions in doubly-stretched coordinates [127]. Pairing correlations have either been neglected [23], or included within the $\Delta(\omega)$ method. They have proposed an interpretation of several excited superdeformed bands in terms of collective quadrupole or octupole excitations.

Yang Sun and collaborators [32–35] have used the Projected Shell Model (PSM) [128,32] to describe the superdeformed bands in several nuclei. In two aspects this model goes beyond the mean-field approximation. Firstly, it uses the BCS wave functions projected on good angular momentum. (A restriction to axial shapes allows to perform the projection by doing one-dimensional integrals [108].) Secondly, the PSM allows for a mixing of the mean-field ground state with low-lying two- and four-quasiparticle states (in case of an even-even system). The model does not use the cranking approximation, i.e., the high-angular momentum states are obtained by projecting appropriate components from non-rotating states. A diagonalization in such a basis allows to some extent to take into account the configuration changes with increasing spin. Conservation of the angular momentum removes known disadvantages of the cranking approximation which fails when two configurations cross each other. On the other hand, large spaces are probably required to accommodate rotation-induced modifications of states when they are treated in a non-rotating basis (especially in the pairing channel). By conserving the angular momentum, the model also allows to calculate the transition matrix elements in a consistent way, although for large deformations the standard intrinsic-frame treatment is probably precise enough. The model certainly deserves further attention, especially when it will be enriched by the particle-number projection, which can by crucial in a weak-pairing regime typical for the rotating superdeformed states. The model is very successful in describing properties of superdeformed bands in $A\sim190$ nuclei [33].

Ramon Wyss and collaborators [36–57] have provided theoretical interpretation of a large number of experimentally observed superdeformed bands in the $A\sim130$, 150, and 190 regions nuclei. They have used the WS+HFB approach (self-consistent treatment of pairing within a fixed set of WS single-particle energies) together with the LN method to restore the numbers of particles. The total routhians, obtained by the Strutinsky procedure, have been minimized with respect to the β_2, γ, and β_4 deformations. The method constitutes by now a standard approach to describe rotational states, and proved to be extremely successful in reproducing properties of superdeformed bands, see for example Ref. [36].

Tomek Werner and collaborators [58–65] have used a similar method, based on the WS mean field and with no pairing correlations, to describe numerous superdeformed bands in the $A\sim80$ region.

Richard Chasman and collaborators [66–68] have employed another version of the Strutinsky approach, based on the WS mean field, with emphasis on the octupole degree of freedom for nuclei in the $A\sim150$ and 190 regions.

Paul-Henri Heenen and collaborators [69–79] have been using the HFB+Skyrme code in space coordinates, with pairing correlations treated within the LN method and various interactions used in the particle-particle channel. By analyzing superdeformed bands in $A\sim150$ and 190 nuclei they have pointed out the necessity of using the pairing forces having surface character [72]. They have also studied collective correlations at superdeformed shapes by employing the Generator Coordinate Method (GCM) in the quadrupole-octupole [70] and hexadecapole [71] channels. Excitation energies and particle separation energies in secondary minima have also been recently studied [75]. Satuła, Heenen, and collaborators have also performed the joint WS and HF+Skyrme analysis of single-particle alignments in $A\sim190$ nuclei [79].

The present author and collaborators [80–89] have used the HF+Skyrme code constructed in the Cartesian HO basis [85] to describe superdeformed bands in $A\sim60$ and 150 nuclei. The code is fast enough to allow massive calculations for multiple nuclei and configurations. Such results have been obtained for bands around ^{152}Dy, which allowed to formulate the additivity principle for relative quadrupole moments [84], and to study influence of various time-odd components on properties of superdeformed bands [80,83]. Results for nuclei in the $A\sim60$ region [89] are discussed below.

Masayuki Yamagami and Kenichi Matsuyanagi [90] have recently constructed a new HF+Skyrme code in space coordinates, in which they break all space symmetries. As the first application they present at this conference the calculations for rotational bands in ^{32}S.

Luis Egido and collaborators [91–93] have studied the superdeformed bands in the $A\sim190$ region by using the HFB method, the Gogny interaction, and the LN and PNP treatments of pairing correlations. They have pointed out the importance of a consistent treatment of the complete interaction in the LN method, and studied this method in the presence of density-dependent terms. In a recent study [93] they have shown that the dynamic moment of inertia of ^{152}Dy, which in the self-consistent calculations was always slightly larger than in the experiment, cf. Refs. [73,82], comes out just right when the exact PNP calculation is performed.

Michel Girod and collaborators [94–96] have also used the HFB theory and the Gogny interaction, as well as the LN method, and studied the rotational spectra either in the cranking approximation [94,95] or by diagonalizing the collective Bohr Hamiltonian with the inertial functions calculated at zero spin via the Gaussian Overlap Approximation to the GCM [95,96].

Peter Ring and collaborators [97–103] have used the RMF approach without pairing correlations to describe the superdeformed bands in $A\sim140$-150 nuclei [99,102]. They have performed systematic calculations of moments of inertia, quadrupole moments and alignments, and attributed specific configurations to experimentally discovered bands. Similar method has also been used in $A\sim80$ nuclei [97]. By using the BCS pairing correlations (at no spin) they have also analyzed excitation energies of superdeformed minima in $A\sim190$ nuclei [103].

Hideki Madokoro and Masayuki Matsuzaki [104–106] have very recently constructed a cranking code based on the RMF mean field. Their first applications aim at studying the newly discovered superdeformed states in the $A\sim60$ region.

The review presented above is by no means exhaustive, because it covers only the recent theoretical investigations of rotational bands at superdeformation, performed within the mean-field methods. Although these methods constitute at present the mainstream activity, there are also several others which have not been

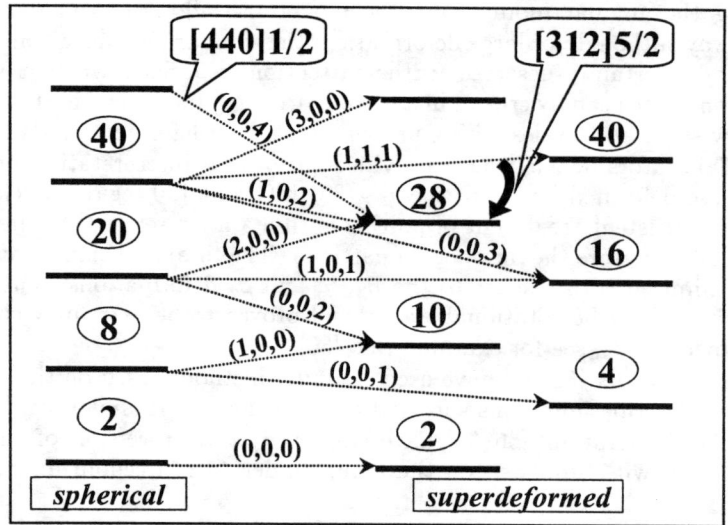

FIGURE 1. Spectra of the spherical (axis ratio 1:1, left) and superdeformed (axis ratio 2:1, right) harmonic oscillator (HO). Magic numbers, including two-fold spin degeneracy, are shown in ovals. Dotted arrows show how the states with given Cartesian quantum numbers $(n_x n_y n_z)$ move when the HO potential is deformed from the spherical to superdeformed shape. At every arrow only one state is indicated, although several other ones may move in the same way, for example, states (2,0,1) and (0,2,1) move together with (1,1,1). Nilsson quantum numbers of two important orbitals are shown explicitly, namely, [440]1/2 indicates the positive-parity intruder $N_0=4$ orbital, while [312]5/2 shows the only orbital which is pushed down by the spin-orbit interaction below the HO magic gap for 28 particles. From [89].

discussed here.

Judging just from the sheer number of studies performed in recent years, one can say that the domain is rich and very active. Within it there are several varieties of approaches, and various groups collaborate and/or compete in achieving the best results.

Close collaboration between the experimental and theoretical groups is a definite asset of the performed studies. In this domain, experiments, interpretations, and predictions are strongly interwoven and mutually supporting, for the benefit of our understanding of the underlying physics.

SUPERDEFORMATION IN $A{\sim}60$ NUCLEI

The recent discovery of the next region where the superdeformed bands exist, the region of nuclei with $A{\sim}60$, see contributions to this conference in Refs. [129,130], has prompted a number of theoretical investigations [9,15,19,20,22,86,89,105,106]. Structure of these nuclei can easily be understood by analyzing the simple HO mean field. In Fig. 1 we present the spherical (1:1) and superdeformed (2:1) HO spectra in light nuclei. Energies of the spherical states read

$$\epsilon_{n_x n_y n_z} = \hbar\omega_0(n_x + \tfrac{1}{2}) + \hbar\omega_0(n_y + \tfrac{1}{2}) + \hbar\omega_0(n_z + \tfrac{1}{2}) = \hbar\omega_0(N_0 + \tfrac{3}{2}), \tag{1}$$

where $(n_x n_y n_z)$ are the standard Cartesian HO quantum numbers, and $N_0 = n_x + n_y + n_z$ is the total number of quanta, i.e., the principal HO quantum number. Energies of the superdeformed HO oscillator can be grouped in two families having even and odd numbers of quanta n_z in the z direction, respectively, see Refs. [131,132] and references cited therein, i.e.,

$$\begin{aligned}
\epsilon_{n_x n_y n_z} &= \hbar\omega_0'(n_x + \tfrac{1}{2}) + \hbar\omega_0'(n_y + \tfrac{1}{2}) + \tfrac{1}{2}\hbar\omega_0'(n_z + \tfrac{1}{2}) \\
&= \begin{cases} \hbar\omega_0'(N_0' + \tfrac{3}{2}); & \text{for} \quad n_z = 2n_z' \\ \hbar\omega_0'(N_0' + \tfrac{3}{2}) + \tfrac{1}{2}\hbar\omega_0'; & \text{for} \quad n_z = 2n_z' + 1 \end{cases}
\end{aligned} \tag{2}$$

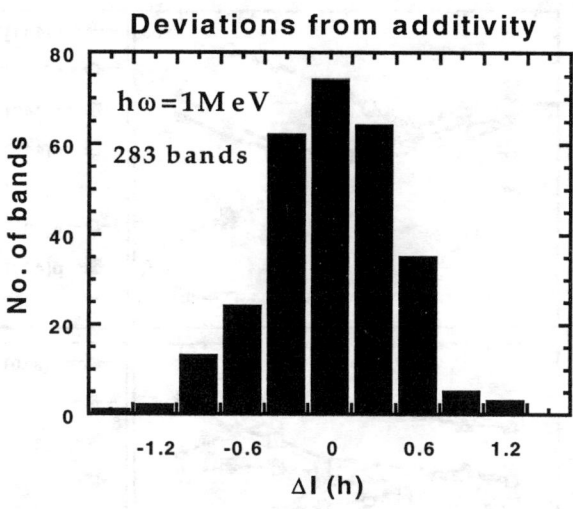

FIGURE 2. Histogram of deviations between the HF values of angular momenta calculated at $\hbar\omega=1\,\text{MeV}$ and the values given by the additivity formula (3).

Here, $N_0' = n_x + n_y + n_z'$ is the principal HO quantum number of each of the two spherical-like HO spectra, and ω_0' is the corresponding new frequency which is simply related to ω by the volume-conservation condition.

Dotted arrows in Fig. 1 indicate the correspondence between spherical and superdeformed HO states. For example, the three degenerate $N_0=1$ spherical states, $(n_x n_y n_z)=(1,0,0)$, $(0,1,0)$, and $(0,0,1)$, split and move to either one of the two families. Namely, states $(1,0,0)$ and $(0,1,0)$ join the $N_0'=1$ shell of the lower family, while state $(0,0,1)$ forms the $N_0'=0$ shell of the higher family.

One can see that the standard sequence of the spherical magic HO numbers, 2, 8, 20, 40,..., is at the superdeformed shape replaced by the sequence 2, 4, 10, 16, 28, 40.... The numbers in this sequence are either doubled spherical magic numbers or sums of two consecutive spherical magic numbers. Had the HO mean-field been exactly realized in nature, one would therefore observe doubly-magic superdeformed states in ^{8}Be, ^{20}Ne, ^{32}S, ^{56}Ni, ^{80}Zr, and so on.

The simple HO picture can very well be associated with the cluster states in light nuclei [131], for example, with the α-α cluster ground state of ^{8}Be or the ^{16}O-α cluster state in ^{20}Ne. The next candidate in the sequence is, as yet unobserved, the ^{16}O-^{16}O cluster (or superdeformed) state in ^{32}S (see the next section). Still higher in the sequence one has to take into account the spin-orbit effects which perturb the simple HO mean field. In $A\sim 60$ nuclei, the influence of the spin-orbit force is still relatively weak, and amounts to shifting *only one* Nilsson orbital, [312]5/2, below the superdeformed gap. As a result, we observe the superdeformed structures mostly in the nuclei slightly heavier than ^{56}Ni [129], and we expect the nucleus ^{60}Zn to be the doubly magic superdeformed core. In Fig. 1, the Nilsson orbital [312]5/2 is marked by a thick arrow.

Another very important orbital in this region is the $[440]1/2=(0,0,4)$ intruder state, also explicitly indicated in Fig. 1. In fact, together with the next higher intruder state, [431]3/2, these are the only positive-parity states near the Fermi level of $A\sim 60$ superdeformed nuclei. Moreover, since these two orbitals are strongly deformation-driving, one obtains significant changes of deformation between states differing in numbers of occupied $N_0=4$ routhians [86]. Therefore, it is useful to label the superdeformed configurations in this region as $4^n 4^p$, where n and p are the numbers of the occupied $N_0=4$ routhians [88]. For example, the doubly magic superdeformed configuration in ^{60}Zn can be labelled as $4^2 4^2$. Interestingly, the best theoretical scenario to describe the pair of identical bands in ^{56}Ni and ^{58}Cu involves configurations with *different* numbers of occupied $N_0=4$ states [130].

In view of a rapid increase in the amount and quality of experimental data on superdeformed states in $A\sim 60$ nuclei, one can aim at systematic studies of their properties. Theoretically, such systematic investigations indicate [89] that the quadrupole moments of nuclei in this region, similarly as those in $A\sim 150$ nuclei [84], very well obey [86] the additivity principle. In the present paper we analyze the similar hypothesis for the relative angular momenta, namely, we suppose that at a given angular frequency ω, the angular momentum of a given $A\sim 60$ superdeformed configuration can be described through the additive contributions,

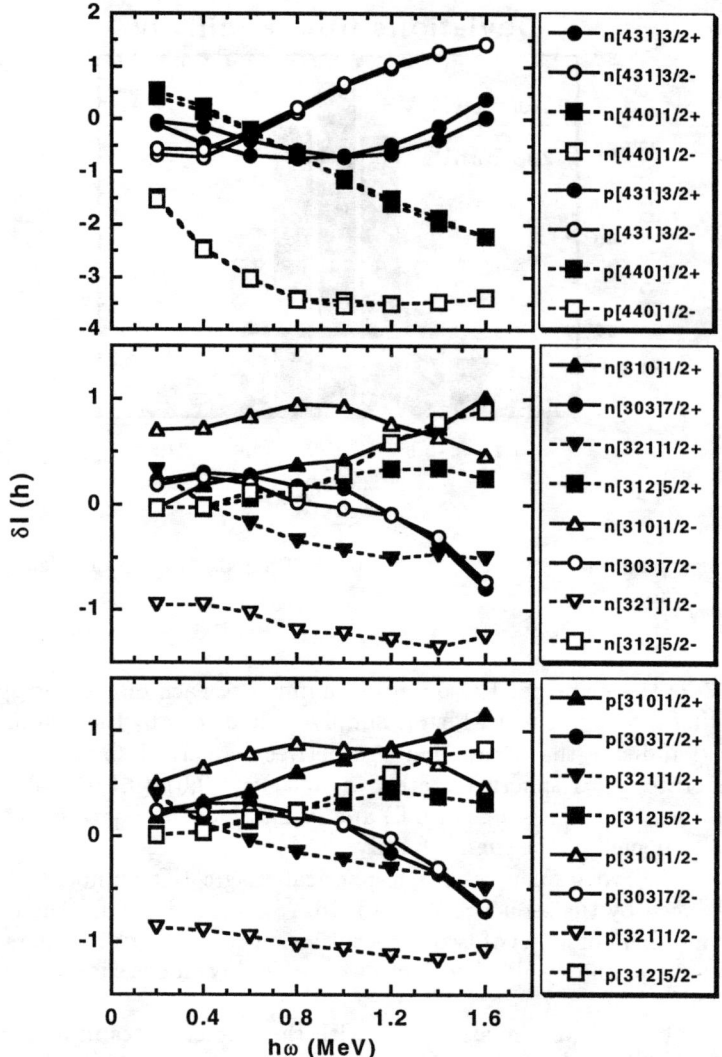

FIGURE 3. Contributions of particle and hole orbitals (shown with solid and dashed lines, respectively) to angular momenta in $A{\sim}60$ superdeformed nuclei. Closed and open symbols correspond to the signature quantum numbers $r{=}{+}i$ and $r{=}{-}i$, respectively.

$$I_{\mathrm{add}}(\omega) = I\left[^{60}\mathrm{Zn}\right](\omega) + \sum_p \delta I_p(\omega) n_p + \sum_h \delta I_h(\omega) n_h, \qquad (3)$$

with respect to the $^{60}\mathrm{Zn}$ core. In Eq. (3), $I\left[^{60}\mathrm{Zn}\right]$ denotes the angular momentum of the doubly magic $4^2 4^2$ configuration in $^{60}\mathrm{Zn}$, while n_p and n_h denote respectively the occupation numbers (equal 0 or 1) of particles and holes which have to be created in the $^{60}\mathrm{Zn}$ core in order to obtain the given configuration in an adjacent nucleus.

Individual contributions of particles (δI_p) and holes (δI_h) can be obtained by fitting formula (3) (separately at each value of ω) to the calculated HF angular momenta of a large set of configurations. Details of this procedure will be given elsewhere [89]. As an example, in Fig. 2 we give the histogram of deviations,

$$\Delta I(\omega) = I_{\mathrm{HF}}(\omega) - I_{\mathrm{add}}(\omega), \qquad (4)$$

between the HF results and the additivity formula (3), obtained at $\hbar\omega{=}1\,\mathrm{MeV}$. One can see that the additivity hypothesis for the angular momenta does not work so precisely as it does for the quadrupole moments. Typical

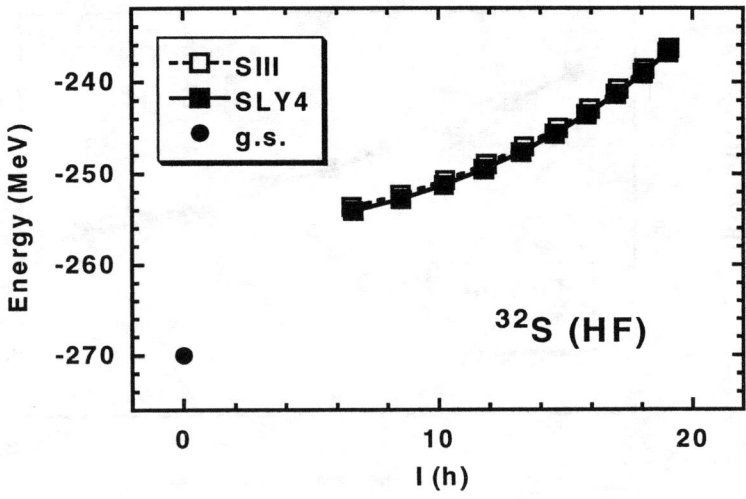

FIGURE 4. Energies of the superdeformed band in ^{32}S as functions of spin, calculated within the HF approximation for the SIII (open squares) and SLy4 (full squares) Skyrme interactions. The dot indicates the calculated HF energy of the ground-state configuration.

deviations turn out to be of about $0.5\,\hbar$, which is large compared to the precision of relative alignments we want to describe. Therefore, in all particular cases, measured relative alignments should be compared with calculated values for a given particular pair of configurations.

However, the obtained contributions of individual routhians, shown in Fig. 3, may serve as indicators of which orbital is most important in providing the angular momentum to the nucleus. As seen in the top panel of Fig. 3, the intruder $N_0=4$ orbitals carry much larger angular momenta than the negative-parity states shown in the lower two panels (note the difference of scale). Contributions of proton and neutron intruder orbitals are almost equal. One should also note that beyond $\hbar\omega\simeq0.6\,\mathrm{MeV}$, the proton and neutron Nilsson routhians [440]1/2− carry almost constant alignments. Therefore, they may lead to identical bands in nuclei which *differ* by the occupation of this particular orbital. Similarly, the negative-parity orbitals [321]1/2−, which carry constant alignments of about $-1\hbar$ may also lead to identical bands.

Results shown in Fig. 3 may also be useful in giving the changes of dynamic moments of inertia $\mathcal{J}^{(2)}(\omega)=dI(\omega)/d\omega$, which arise when a given orbital becomes occupied. Indeed, slopes of curves presented in Fig. 3 are equal to relative changes of the dynamic moments of inertia. From these results it is clear that the values of $\mathcal{J}^{(2)}$ should not depend on the occupation of the first intruder routhian [440]1/2−, and should increase with occupying the following three intruder routhians [440]1/2+, [431]3/2−, and [431]3/2+ (note that the *hole* contribution of [440]1/2+ is shown in Fig. 3).

THE LAST CHALLENGE: ^{32}S

As illustrated in Fig. 1, the nucleus ^{32}S (16 protons and 16 neutrons) is the missing link between the known region of superdeformed nuclei around ^{60}Zn and the cluster-like structures in lighter nuclei, like in ^{20}Ne and ^{8}Be. In fact, the first indication that the cluster states in ^{32}S may exist is provided by the measurements in the ^{16}O-^{16}O breakup channel [133]. On the other hand, numerous mean-field calculations, both non-self-consistent [134] and self-consistent [135], as well as the α-cluster calculations [136], predict in this nucleus the existence of the 2:1 deformed structures. Such states should coexist with numerous low-deformation states already known in this nucleus [137], which are very well described by the sd-shell model [138].

In Figs. 4 and 5 we show results of the HF calculations performed with the Skyrme SIII [139] and SLy4 [140,141] interactions. Both interactions give very similar predictions for the superdeformed band which is based on the 3^23^2 configuration, i.e., with altogether four $N_0=3$ intruder states (originating from the $1f_{7/2}$ spherical orbital) occupied. The band should begin around $I=6\hbar$ and continue beyond the spin of $20\hbar$, where it

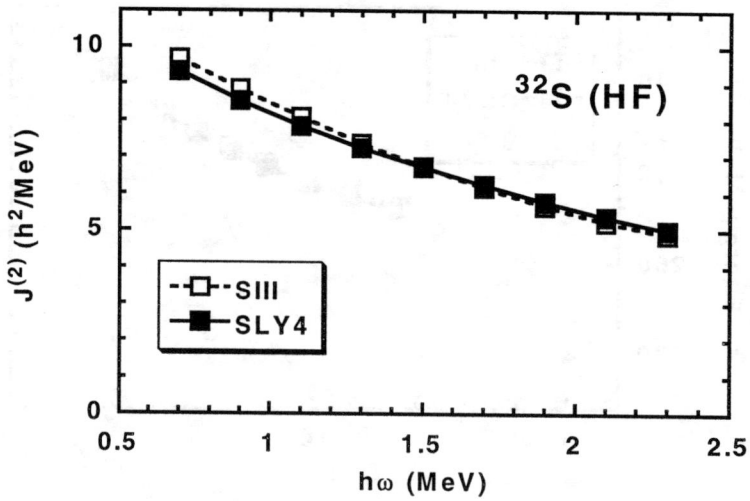

FIGURE 5. Same as in Fig. 4 for the dynamic moments of inertia as functions of the rotational frequency.

may be crossed by a rapidly descending configuration based on the [440]1/2 orbital, cf. Ref. [90]. The existence of the superdeformed band in ^{32}S is predicted by these same models and approaches as those successfully used in describing the superdeformed structures in heavier nuclei. Its discovery would therefore constitute a spectacular confirmation of the predictive power of mean-field methods.

ACKNOWLEDGMENTS

Collaboration, discussions, help, and friendship of my colleagues Witek Nazarewicz and Wojtek Satuła are gratefully acknowledged. This research was supported in part by the Polish Committee for Scientific Research (KBN) under Contract No. 2 P03B 040 14, and by the computational grants from the Interdisciplinary Centre for Mathematical and Computational Modeling (ICM) of the Warsaw University, from the *Institut du Développement et de Ressources en Informatique Scientifique* (IDRIS) of CNRS, France, and from the *Regionales Hochschulrechenzentrum Kaiserslautern*, Germany.

REFERENCES

1. Nolan P. and Twin P., Annu. Rev. Nucl. Part. Sci. **38**, 533 (1988).
2. Åberg S., Flocard H., and Nazarewicz W., Annu. Rev. Nucl. Part. Sci. **40**, 439 (1990).
3. Baktash C., Haas B., and Nazarewicz W., Annu. Rev. Nucl. Part. Sci. **45**, 485 (1995).
4. Baktash C., Prog. Part. Nucl. Phys., **38**, 291 (1997).
5. Twin P.J., Nyakó B.M., Nelson A.H., Simpson J., Bentley M.A., Cranmer-Gordon H.W., Forsyth P.D., Howe D., Mokhtar A.R., Morrison J.D., Sharpey-Schafer J.F., Sletten G., Phys. Rev. Lett. **57**, 811 (1986).
6. Semmes P.B., Ragnarsson I., and Åberg S., Phys. Lett. B, **345**, 185 (1995).
7. Flibotte S., Hackman G., Ragnarsson I., Theisen C., Andrews H.R., Ball G.C., Beausang C.W., Beck F.A., Belier G., Bentley M.A., Byrski T., Curien D., de France G., Disdier D., Duchene G., Haas B., Haslip D.S., Janzen V.P., Jones P.M., Kharraja B., Kuehner J.A., Lisle J.C., Merdinger J.C., Mullins S.M., Paul E.S., Prevost D., Radford D.C., Rauch V., Smith J.F., Styczen J., Twin P.J., Vivien J.P., Waddington J.C., Ward D., and Zuber K., Nucl. Phys. **584**, 373 (1995).
8. Mullins S.M., Flibotte S., Hackman G., Rodriguez J.L., Waddington J.C., Afanasjev A.V., Ragnarsson I., Andrews H.R., Galindo-Uribarri A., Janzen V.P., Radford D.C., Ward D., Cromaz M., DeGraff J., Drake T.E., and Pilotte S., Phys. Rev. **C52**, 99 (1995).
9. Ragnarsson I., Acta Phys. Pol. **27**, 33 (1996).

10. Savajols H., Korichi A., Ward D., Appelbe D., Ball G.C., Beausang C., Beck F.A., Byrski T., Curien D., Dagnall P., de France G., Disdier D., Duchene G., Erturk S., Finck C., Flibotte S., Gall B., Galindo-Uribarri A., Haas B., Hackman G., Janzen V.P., Kharraja B., Lisle J.C., Merdinger J.C., Mullins S.M., Pilotte S., Prevost D., Radford D.C., Rauch V., Rigollet C., Smalley D., Smith M.B., Stezowski O., Styczen J., Theisen C., Twin P.J., Vivien J.P., Waddington J.C., Zuber K., and Ragnarsson I., Phys. Rev. Lett. **76**, 4480 (1996).

11. Galindo-Uribarri A., Mullins S.M., Ward D., Cromaz M., DeGraaf J., Drake T.E., Flibotte S., Janzen V.P., Radford D.C., and Ragnarsson I., Phys. Rev. **C54**, R454 (1996).

12. Afanasjev A.V., and Ragnarsson I., Nucl. Phys. **A608**, 176 (1996).

13. Theisen Ch., Khadiri N., Vivien J.P., Ragnarsson I., Beausang C.W., Beck F.A., Belier G., Byrski T., Curien D., de France G., Disdier D., Duchene G., Finck Ch., Flibotte S., Gall B., Haas B., Hanine H., Herskind B., Kharraja B., Merdinger J.C., Nourreddine A., Myako B.M., Perez G.E., Prevost D., Stezowski O., Rauch V., Rigollet C., Savajols H., Sharpey-Schafer J., Twin P.J., Wei L., and Zuber K., Phys. Rev. **C54**, 2910 (1996).

14. Nisius D., Janssens R.V.F., Moore E.F., Fallon P., Crowell B., Lauritsen T., Hackman G., Ahmad I., Amro H., Asztalos S., Carpenter M.P., Chowdhury P., Clark R.M., Daly P.J., Deleplanque M.A., Diamond R.M., Fischer S.M., Grabowski Z.W., Khoo T.L., Lee I.Y., Macchiavelli A.O., Mayer R.H., Stephens F.S., Afanasjev A.V., and Ragnarsson I., Phys. Lett. B, **392**, 18 (1997).

15. Svensson C.E., Baktash C., Cameron J.A., Devlin M., Eberth J., Flibotte S., Haslip D.S., LaFosse D.R., Lee I.Y., Macchiavelli A.O., MacLeod R.W., Nieminen J.M., Paul S.D., Riedinger L.L., Rudolph D., Sarantites D.G., Thomas H.G., Waddington J.C., Weintraub W., Wilson J.N., Afanasjev A.V., and Ragnarsson I., Phys. Rev. Lett. **79**, 1233 (1997).

16. Åberg S., Joensson L.O., Karlsson L.B., and Ragnarsson I., Z. Phys. **A358**, 269 (1997).

17. Sarantites D.G., LaFosse D.R., Devlin M., Lerma F., Wood V.Q., Saladin J.X., Winchell D.F., Baktash C., Yu C-H., Fallon P., Lee I.Y., Macchiavelli A.O., MacLeod R.W., Afanasjev A.V., and Ragnarsson I., Phys. Rev. **C57**, R1 (1998).

18. Karlsson L.B., Ragnarsson I., and Åberg S., Phys. Lett. B, **416**, 16 (1998).

19. Svensson C.E., Baktash C., Ball G.C., Cameron J.A., Devlin M., Eberth J., Flibotte S., Galindo-Uribarri A., Haslip D.S., Janzen V.P., LaFosse D.R., Lee I.Y., Macchiavelli A.O., Macleod R.W., Nieminen J.M., Paul S.D., Radford D.C., Riedinger L.L., Rudolph D., Sarantites D.G., Thomas H.G., Waddington J.C., Ward D., Weintraub W., Wilson J.N., Afanasjev A.V., and Ragnarsson I., Phys. Rev. Lett. **80**, 2558 (1998).

20. Galindo-Uribarri A., Ward D., Ball G.C., Janzen V.P., Radford D.C., Ragnarsson I., and Headly D., Phys. Lett. B **422**, 45 (1998).

21. Karlsson L.B., Ragnarsson I., and Åberg S., Nucl.Phys. **A639**, 654 (1998).

22. Afanasjev A.V., Ragnarsson I., and Ring P., Report nucl-th/9809074.

23. Nakatsukasa T., Matsuyanagi K., Mizutori S., and Nazarewicz W., Phys. Lett. B **343,**, 19 (1995).

24. Crowell B., Carpenter M.P., Janssens R.V.F., Blumenthal D.J., Timar J., Wilson A.N., Sharpey-Schafer J.F., Nakatsukasa T., Ahmad I., Astier A., Azaiez F., du Croux L., Gall B.J.P., Hannachi F., Khoo T.L., Korichi A., Lauritsen T., Lopez-Martens A., Meyer M., Nisius D., Paul E.S., Porquet M.G., and Redon N., Phys. Rev. **C51**, R1599 (1995).

25. Nakatsukasa T., Acta Phys. Pol. **B27**, 59 (1996).

26. Nakatsukasa T., Matsuyanagi K., Mizutori S., and Shimizu Y.R., Phys. Rev. **C53**, 2213 (1996).

27. Wilson A.N., Timar J., Sharpey-Schafer J.F., Crowell B., Carpenter M.P., Janssens R.V.F., Blumenthal D.J., Ahmad I., Astier A., Azaiez F., Bergstrom M., Ducroux L., Gall B.J.P., Hannachi F., Khoo T.L., Korichi A., Lauritsen T., Lopez-Martens A., Meyer M., Nisius D., Paul E.S., Porquet M.G., Redon N., Wilson J.N., and Nakatsukasa T., Phys. Rev. **C54**, 559 (1996).

28. Fallon P., Stephens F.S., Asztalos S., Busse B., Clark R.M., Deleplanque M.A., Diamond R.M., Krucken R., Lee I.Y., Macchiavelli A.O., MacLeod R.W., Schmid G., Vetter K., and Nakatsukasa T., Phys. Rev. **C55**, R999 (1997).

29. Bouneau S., Azaiez F., Duprat J., Deloncle I., Porquet M.G., van Severen U.J., Nakatsukasa T., Aleonard M.M., Astier A., Baldsiefen G., Beausang C.W., Beck F.A., Bourgeois C., Curien D., Dozie N., Ducroux L., Gall B.J.P., Huebel H., Kaci M., Korten W., Meyer M., Redon N., Sergolle H., and Sharpey-Schafer J.F., Z. Phys. **358**, 179 (1997).

30. Hackman G., Khoo T.L., Carpenter M.P., Lauritsen T., Lopez-Martens A., Calderin I.J., Janssens R.V.F., Ackermann D., Ahmad I., Agarwala S., Blumenthal D.J., Fischer S.M., Nisius D., Reiter P., Young J., Amro H., Moore E.F., Hannachi F., Korichi A., Lee I.Y., Macchiavelli A.O., Dossing T., and Nakatsukasa T., Phys. Rev. Lett. **79**, 4100 (1997).

31. Azaiez F., Bouneau S., Duprat J., Deloncle I., Porquet M-G., van Severen U.J., Nakatsukasa T., Aleonard M.M., Astier A., Baldsiefen S., Beausang C.W., Beck F.A., Bourgeois C., Curien D., Dozie N., Ducroux L., Gall B., Hubel H., Kaci M., Korten W., Meyer M., Redon N., Sergolle H., and Sharpey-Shafer J.F., Acta Phys. Hung. **6**, 289 (1997).

32. Sun Y. and Guidry M., Phys. Rev. **C52**, R2844 (1995).

33. Sun Y., Zhang J-Y., and Guidry M., Phys. Rev. Lett. **78**, 2321 (1997).

34. Brown T.B., Pfohl J., Riley M.A., Hartley D.J., Sarantites D.G., Devlin M., LaFosse D.R., Lerma F., Archer D.E., Clark R.M., Fallon P., Hibbert I.M., Joss D.T., Nolan P.J., O'Brien N.J., Paul E.S., Sheline R.K., Simpson J., Wadsworth R., and Sun Y., Phys. Rev. **C56**, R1210 (1997).

35. Zhang J-Y., Sun Y., Riedinger L.L., and Guidry M.,, Phys. Rev. **C58**, 868 (1998).

36. Satuła W. and Wyss R., Phys. Rev. **C50**, 2888 (1994).

37. Satuła W. and Wyss R., Phys. Scr. **T56**, 159 (1995).

38. Hauschild K., Wadsworth R., Clark R.M., Fallon P., Fossan DB., Hibbert I.M., Macchiavelli A.O., Nolan P.J., Schnare H., Semple A.T., Thorslund I., Walker L., Satuła W., and Wyss R., Phys. Lett. B **353**, 438 (1995).

39. Wyss R. and Satuła W., Phys. Lett. B **351**, 393 (1995).

40. Rzaca-Urban T., Lieder R.M., Utzelmann S., Gast W., Georgiev A., Jaeger H.M., Bazzacco D., Lunardi S., Menegazzo R., Rossi-Alvarez C., de Angelis G., Napoli D.R., Vedovato G., and Wyss R., Phys. Lett. B **356**, 456 (1995).

41. Bouneau S., Wilson A.N., Azaiez F., Sharpey-Schafer J.F., Korichi A., Deloncle I., Porquet M.G., Timar J., Astier A., Bergstrom M., Bourgeois C., Ducroux L., Duprat J., Gall B.J.P., Hannachi F., Kaci M., Le Coz Y., Lopez-Martens A., Meyer M., Paul E.S., Perrin N., Pilotte S., Redon N., Riley M.A., Schuck C., Sergolle H., and Wyss R., Phys. Rev. **C53**, R9 (1996).

42. de Angelis G., Wyss R., Bazzacco D., De Poli M., Gadea A., Lunardi S., Napoli D.R., Petrache C.M., Alvarez C.R., Sferrazza M., and Rubio B., Phys. Rev. **C53**, 679 (1996).

43. Satuła W. and Wyss R., Acta Phys. Pol. **B27**, 121 (1996).

44. Bouneau S., Wilson A.N., Azaiez F., Shrapey-Schafer J.F., Korichi A., Deloncle I., Porquet M.G., Timar J., Astier A., Bergstrom M., Bourgeois C., Ducroux L., Duprat J., Gall B.J.P., Hannachi F., Kaci M., Le Coz Y., Lopez-Martens A., Meyer M., Paul E.S., Perrin N., Pilotte S., Redon N., Riley M.A., Schuck C., Sergolle H., and Wyss R., Acta Phys. Pol. **B27**, 197 (1996).

45. Petrache C.M., Bazzacco D., Lunardi S., Rossi Alvarez C., Venturelli R., Bucurescu D., Ur C.A., De Acuna D., Maron G., Napoli D.R., Medina N.H., Oliveira J.R.B., and Wyss R., Phys. Lett. B **373**, 275 (1996).

46. Petrache C.M., Bazzacco D., Lunardi S., Alvarez C.R., Venturelli R., Pavan P., Medina N.H., Rao M.N., Burch R., De Angelis G., Gadea A., Maron G., Napoli D.R., Zhu L., and Wyss R., Phys. Rev. Lett. **77**, 239 (1996).

47. Piiparinen M., Atac A., Blomqvist J., Hagemann G.B., Herskind B., Julin R., Juutinen S., Lampinen A., Nyberg J., Sletten G., Tikkanen P., Toermaenen S., Virtanen A., and Wyss R., Nucl. Phys. **A605**, 191 (1996).

48. Petrache C.M., Rao M.N., Medina N.H., Ribas R.V., Bazzacco D., Lunardi S., Rossi Alvarez C., Venturelli R., Burch R., Pavan P., De Angelis G., De Poli M., Vedovato G., Zhu L.H., and Wyss R., Phys. Lett. B **383**, 145 (1996).

49. Petrache C.M., Bazzacco D., Lunardi S., Rossi Alvarez C., Venturelli R., Burch R., Pavan P., Maron G., Napoli D.R., Zhu L.H., and Wyss R., Phys. Lett. B **387**, 31 (1996).

50. Hibbert I.M., Wadsworth R., Hauschild K., Hubel H., Korten W., Van Severen U.J., Paul E.S., Wilson A.N., Wilson J.N., Byrne A.P., Satuła W., and Wyss R., Phys. Rev. **C54**, 2253 (1996).

51. Hackman G., Krucken R., Janssens R.V.F., Deleplanque M.A., Carpenter M.P., Ackermann D., Ahmad I., Amro H., Asztalos S., Blumenthal D.J., Clark R.M., Diamond R.M., Fallon P., Fischer S.M., Herskind B., Khoo T.L., Lauritsen T., Lee I-Y., MacLeod R.W., Macchiavelli A.O., Nisius D., Schmid G.J., Stephens F.S., Vetter K., and Wyss R., Phys. Rev. **C55**, 148 (1997).

52. Lieder R.M., Rzaca-Urban T., Jensen H.J., Gast W., Georgiev A., Jager H., Utzelmann S., Ur C.A., Bolzonella G.P., Bazzacco D., Lunardi S., Medina N.H., Menegazzo R., Petrache C.M., Rizzuto M.A., Rossi-Alvarez C., De Angelis G., De Acuna D., Maron G., Napoli D.R., Vedovato G., Zhu L.H., and Wyss R., Prog. Part. Nucl. Phys. **38**, 41 (1997).

53. Lunardi S., Zhu L.H., Petrache C.M., Bazzacco D., Medina N.H., Rizzuto M.A., Rossi Alvarez C., De Angelis G., Maron G., Napoli DR., Utzelmann S., Gast W., Lieder R.M., Georgiev A., Xu F., and Wyss R., Nucl. Phys. **A618**, 238 (1997).

54. Hibbert I.M., Wadsworth R., Hauschild K., Huebel H., Korten W., van Severen U.J., Paul E.S., Wilson A.N., Wilson J.N., Byrne A.P., Satuła W., and Wyss R., Z. Phys. **358**, 199 (1997).

55. Lunardi S., Bazzacco D., Fabris D., Lunardon M., Medina N.H., Nebbia G., Petrache C.M., Rizzuto M.A., Rossi Alvarez C., Viesti G., de Angelis G., Cinausero M., De Acuna D., De Poli M., Farnea E., Fioretto E., Maron G., Napoli D.R., Prete G., Zhu L.H., Utzelmann S., Gast W., Lieder R.M., Georgiev A., Xu F., and Wyss R., Acta Phys. Hung. **6**, 241 (1997).

56. Semple A.T., Paul E.S., Boston A.J., Hibbert I.M., Joss D.T., Nolan P.J., O'Brien N.J., Parry C.M., Shepherd S.L., Wadsworth R., and Wyss R., Jour. Phys. **G24**, 1125 (1998).

57. Bouneau S., Azaiez F., Duprat J., Deloncle I., Porquet M.G., Astier A., Bergstroem M., Bourgeois C., Ducroux L., Gall B.J.P., Kaci M., Le Coz Y., Meyer M., Paul E.S., Redon N., Riley M.A., Sergolle H., Sharpey-Schafer J.F., Timar J., Wilson A.N., and Wyss R., Eur. Phys. Jour. **A2**, 245 (1998).

58. Baktash C., Cullen D.M., Garrett J.D., Gross C.J., Johnson N.R., Nazarewicz W., Sarantites D.G., Simpson J., and Werner T.R., Phys. Rev. Lett. **74**, 1946 (1995).

59. LaFosse D.R., Hua P-F., Sarantites D.G., Baktash C., Akovali Y.A., Brinkman M., Cederwall B., Cristancho F., Doring J., Gross CJ., Jin H-Q., Korolija M., Landulfo E., Lee I.Y., Macchiavelli A.O., Maier M.R., Rathbun W., Saladin J.X., Stracener D.W., Tabor S.L., Vander Mollen A., and Werner T.R., Phys. Lett. B **354**, 34 (1995).

60. Smith A.G., Dagnall P.J., Lisle J.C., Smalley D.H., Werner T.R., Chapman R., Finck C., Haas B., Leddy M., Nazarewicz W., Prevost D., Rowley N., and Savajols H., Phys. Lett. B **355**, 32 (1995).

61. Jin H-Q., Baktash C., Brinkman M.J., Gross C.J., Sarantites D.G., Lee I.Y., Cederwall B., Cristancho F., Doring J., Durham F.E., Hua P-F., Johns G.D., Korolija M., LaFosse D.R., Landulfo E., Macchiavelli A.O., Rathbun W., Saladin J.X., Stracener D.W., Tabor S.L., and Werner T.R., Phys. Rev. Lett. **75**, 1471 (1995).

62. Cristancho F., LaFosse D.R., Baktash C., Winchell D.F., Cederwall B., Doering J., Gross C.J., Hua P.F., Jin H.Q., Korolija M., Landulfo E., Lee I.Y., Macchiavelli A.O., Maier M.R., Rathbun W., Saladin J.X., Sarantites D., Stracener D.W., Tabor S.L., Vander Mollen A., and Werner T.R., Phys. Lett. B **357**, 281 (1995).

63. Werner T.R. and Dudek J., At. Data Nucl. Data Tab. **59**, 1 (1995).

64. Dagnall P.J., Smith A.G., Lisle J.C., Smalley D.H., Chapman R., Finck C., Haas B., Leddy M.J., Prevost D., Rowley N., Savajols H., Werner T.R., and Nazarewicz W., Acta Phys. Pol. **B27**, 155 (1996).

65. Saladin J.X., Winchell D.F., Cristancho F., LaFosse D.R., Baktash C., Cederwall B., Doring J., Gross C.J., Hua P-F., Jin H-Q., Korolija M., Landulfo E., Lee I.Y., Macchiavelli A.O., Maier M.R., Rathbun W., Sarantites D., Stracener D.W., Tabor S.L., Mollen A.V., and Werner T.R., Proc. of the *Workshop on Gammasphere Physics*, eds. Deleplanque M.A., Lee I.Y., and Macchiavelli A.O., (World Scientific,1996) p. 89.

66. Chasman R.R., Phys. Lett. B **364**, 137 (1995).

67. Chasman R.R. and Robledo L.M., Phys. Lett. B **351**, 18 (1995).

68. Fischer S.M., Janssens R.V.F., Riley M.A., Chasman R.R., Ahmad I., Blumenthal D.J., Brown T.B., Carpenter M.P., Hackman G., Hartley D.J., Khoo T.L., Lauritsen T., Ma W.C., Nisius D., Simpson J., and Varmette P.G., Phys. Rev. **C54**, R2806 (1996).

69. Heenen P-H., Bonche P., and Flocard H., Nucl. Phys. **A588**, 490 (1995).

70. Meyer J., Bonche P., Weiss M.S., Dobaczewski J., Flocard H., and Heenen PH., Nucl. Phys. **A588**, 597 (1995).

71. Magierski P., Heenen P-H., and Nazarewicz W., Phys. Rev. **C51**, R2880 (1995).

72. Terasaki J., Heenen P.H., Bonche P., Dobaczewski J., and Flocard H., Nucl. Phys. **A593**, 1 (1995).

73. Bonche P., Flocard H., and Heenen P.H., Nucl. Phys. **A598**, 169 (1996).

74. Terasaki J., Flocard H., Heenen P-H., and Bonche P., Phys. Rev. **C55**, 1231 (1997).

75. Heenen P-H., Dobaczewski J., Nazarewicz W., Bonche P., and Khoo T.L., Phys. Rev. **C57**, 1719 (1998).

76. Heenen P-H. and Janssens R.V.F., Phys. Rev. **C57**, 159 (1998).

77. Dancer H., Bonche P., Flocard H., Heenen P-H., Meyer J., and Meyer M., Phys. Rev. **C58**, 2068 (1998).

78. Rigollet C., Heenen P-H., Bonche P., and Flocard H., to be published.

79. Fallon P., Heenen P-H., Satuła W., Clark R.M., Stephens F.S., Deleplanque M.A., Diamond R.M., Lee I.Y., Macchiavelli A.O., and Vetter K., abstract to this conference, to be published.

80. Dobaczewski J. and Dudek J., Phys. Rev. **C52**, 1827 (1995).

81. Hackman G., Wadsworth R., Haslip D.S., Clark R.M., Dobaczewski J., Dudek J., Flibotte S., Hauschild K., Hibbert I.M., Lee I-Y., Mullins S.M., Macchiavelli A.O., Pilotte S., Semple A.T., Thorslund I., Timar J., Vaska P., Waddington J.C., and Walker L., Phys. Rev. **C52**, R2293 (1995).

82. El Aouad N., Dobaczewski J., Dudek J., Li X., Luo W.D., Molique H., Bouguettoucha A., Byrski Th., Beck F., Finck C., Kharraja B., preprint CRN 96-38, Report nucl-th/9612048.

83. Dobaczewski J. and Dudek J., Acta Phys. Pol. **B27**, 45 (1996).

84. Satuła W., Dobaczewski J., Dudek J., and Nazarewicz W., Phys. Rev. Lett. **77**, 5182 (1996).

85. Dobaczewski J. and Dudek J., Comp. Phys. Comm. **102**, 166 (1997); **102**, 183 (1997).

86. Dobaczewski J., Invited talk at the *XVII RCNP International Symposium on Innovative Computational Methods in Nuclear Many-Body Problems*, in press, preprint JIHIR 98-01, Report nucl-th/9801056.

87. Rudolph D., Baktash C., Dobaczewski J., Nazarewicz W., Satufa W., Brinkman M.J., Devlin M., Jin H-Q., LaFosse D.R., Riedinger L.L., Sarantites D.G., and Yu C-H., Phys. Rev. Lett. **80**, 3018 (1998).

88. Rudolph D., Baktash C., Satuła W., Dobaczewski J., Nazarewicz W., Brinkman M.J., Devlin M., Jin H-Q., LaFosse D.R., Riedinger L.L., Sarantites D.G., and Yu C-H., Nucl. Phys. **A630**, 417 (1998).

89. Dobaczewski J., Satuła W., Nazarewicz W., and Baktash C., to be published.

90. Yamagami M. and Matsuyanagi K., Report nucl-th/9809038; contribution to this conference.

91. Valor A., Egido J.L., and Robledo L.M., Phys. Lett. B **392**, 249 (1997).

92. Villafranca A. and Egido J.L., Phys. Lett. B **408**, 35 (1997).

93. Auguiano M., Egido J.L., and Robledo L.M., to be published.

94. Girod M., Delaroche J.P., Berger J.F., and Libert J., Phys.Lett. **325B**, 1 (1994).

95. Girod M., Delaroche J.P., Berger J.F., Peru S., and Libert J., Z. Phys. **A358**, 177 (1997).

325

96. Girod M. and Berger J.F., to be published.

97. Afanasjev A.V., König J., and Ring P., Phys. Lett. B **367**, 11 (1996).

98. Zhu Zhiyuan, König J., and Ring P., High En. Phys. Nucl. Phys. **20**, 448 (1996).

99. Afanasjev A.V., König J., and Ring P., Nucl. Phys. **A608**, 107 (1996).

100. Ring P. and Afanasjev A.V., Prog. Part. Nucl. Phys. **38**, 137 (1997).

101. Afanasjev A.V., Lalazissis G.A., and Ring P., Acta Phys. Hung. **6**, 299 (1997).

102. Afanasjev A.V., Lalazissis G.A., and Ring P., Nucl. Phys. **A634**, 395 (1998).

103. Lalazissis G.A. and Ring P., Phys. Lett. B **427**, 225 (1998).

104. Madokoro H. and Matsuzaki M., Phys. Rev. **C56**, R2934 (1997).

105. Madokoro H. and Matsuzaki M., Report nucl-th/9712063.

106. Madokoro H. and Matsuzaki M., Report nucl-th/9803058.

107. Nilsson S.G., Mat. Fys. Medd. Dan. Vid. Selsk. **29**, No. 16 (1955).

108. Ring P. and Schuck P., *The Nuclear Many-Body Problem* (Springer-Verlag, Berlin, 1980).

109. Woods R.D. and Saxon D.S., Phys. Rev. **95**, 577 (1954).

110. Ćwiok S., Dudek J., Nazarewicz W., Skalski J. and Werner T., Comp. Phys. Comm. **46**, 379 (1987).

111. Skyrme T.H.R., Phil. Mag. **1**, 1043 (1956).

112. Skyrme T.H.R., Nucl. Phys. **9**, 615 (1959).

113. Vautherin D. and Brink D.M., Phys. Rev. **C5**, 626 (1972).

114. Gogny D., in: *Proc. Int. Conf. on Nuclear Physics*, eds. De Boer J. and Mang H. (North-Holland, Amsterdam, 1973).

115. Dechargé J. and Gogny D., Phys. Rev. **C21**, 1568 (1980).

116. Ring P., Prog. Part. Nucl. Phys. **37**, 193 (1996).

117. Nazarewicz W., Wyss R., and Johnson A., Nucl. Phys. **A503**, 285 (1989).

118. Wyss R., Satuła W., Nazarewicz W. and Johnson A., Nucl. Phys. **A511**, 324 (1990).

119. Lipkin H.J., Ann. of Phys., **9**, 272 (1960).

120. Nogami Y., Phys. Rev. **134**, 313 (1964).

121. Nogami Y., Zucker I.J., Nucl. Phys., **60**, 203 (1964).

122. Nogami Y., Phys. Lett. **15**, 4 (1965).

123. Goodfellow J.F. and Nogami Y., Can. J. Phys., **44**, 1321 (1966).

124. Pradhan H.C., Nogami Y., and Law J., Nucl. Phys. **A201**, 357 (1973).

125. Bengtsson T., Nucl. Phys. **A496**, 56 (1989).

126. de France G., Baktash C., Haas B., and Nazarewicz W., Phys. Rev. **C53**, R1070 (1996).

127. Sakamoto H. and Kishimoto T., Nucl. Phys. **A501**, 205 (1989).

128. Hara K. and Sun Y., Int. Jour. Mod. Phys. E **4**, 637 (1995).

129. Svensson C.E., contribution to this conference.

130. Rudolph D., contribution to this conference.

131. Nazarewicz W. and Dobaczewski J., Phys. Rev. Lett. **68**, 154 (1992).

132. Nazarewicz W., Ćwiok S., Dobaczewski J., Saladin J.X., Acta Phys. Pol. **B26**, 189 (1995).

133. Curtis N., Murphy A.St.J., Clarke N.M., Freer M., Fulton B.R., Hall S.J., Leddy M.J., Pople J.S., Tungate G., and Ward R.P., Phys. Rev. **C53**, 1804 (1996).

134. Leander G.A. and Larsson S.E., Nucl. Phys. **A239**, 93 (1975).

135. Flocard H., Heenen P.H., Krieger S.J., and Weiss M., Prog. Theor. Phys. **72**, 1000 (1984).

136. Zhang J., Rae W.D.M., and Merchant A.C., Nucl. Phys. **A575**, 61 (1994).

137. Brenneisen J., Erhardt B., Glatz F., Kern Th., Ott R., Röpke H., Schmälzlin J., Siedle P., and Wildenthal B.H., Z. Phys. **A357**, 157 (1997).

138. Brenneisen J., Erhardt B., Glatz F., Kern Th., Ott R., Röpke H., Schmälzlin J., Siedle P., and Wildenthal B.H., Z. Phys. **A357**, 377 (1997).

139. Beiner M., Flocard H., Van Giai N., and Quentin P., Nucl. Phys. **A238**, 29 (1975).

140. Chabanat E., Bonche P., Haensel P., Meyer J., and Schaeffer F., Phys. Scr. **T56**, 231 (1995).

141. Chabanat E., *Interactions effectives pour des conditions extrêmes d'isospin*, Université Claude Bernard Lyon-1, Thesis 1995, LYCEN T 9501, unpublished.

Exotic Shapes in ^{32}S suggested by the Symmetry-Unrestricted Cranked Hartree-Fock Calculations

Masayuki Yamagami and Kenichi Matsuyanagi

Department of Physics, Graduate School of Science, Kyoto University, Kitashirakawa, Kyoto 606-8502, Japan

Abstract.
High-spin structure of ^{32}S is investigated by means of the cranked Skyrme-Hartree-Fock method in the three-dimensional Cartesian-mesh representation. Some interesting suggestions are obtained: 1) An internal structure change (toward hyperdeformation) may occur at $I > 20$ in the superdeformed band, 2) A non-axial Y_{31} deformed band may appear in the yrast line with $5 \leq I \leq 13$.

Introduction

Since the discovery of the superdeformed(SD) band in ^{152}Dy, about two hundreds SD bands have been found in various mass (A=60, 80, 130, 150, 190) regions [1]. Yet, the doubly magic SD band in ^{32}S, which has been expected quite a long time [2,3] remains unexplored, and will become a great challenge in the coming years.

Quite recently, we have constructed a new computer code for the cranked Skyrme Hartree-Fock (HF) calculation based on the three-dimensional (3D) Cartesian-mesh representation, which provides a powerful tool for exploring exotic shapes (breaking both axial and reflection symmetries in the intrinsic states) at high spin in unstable nuclei as well as in stable nuclei. As a first application of this new code, we have investigated high-spin structure of ^{32}S and obtained some interesting results on which we are going to discuss below.

Cranked Skyrme HF Calculation

We solve the cranked HF equation

$$\delta < H - \omega_{rot} J_x >= 0 \tag{1}$$

in the 3D Cartesian-mesh representation. We adopt the standard algorithm [4–7] but completely remove various restrictions on spatial symmetries. When we allow for the simultaneous breaking of both reflection and axial symmetries, it is crucial to fulfill the center-of-mass condition

$$< \sum_{i=1}^{A} x_i >=< \sum_{i=1}^{A} y_i >=< \sum_{i=1}^{A} z_i >= 0, \tag{2}$$

and the principal-axis condition

$$< \sum_{i=1}^{A} x_i y_i >=< \sum_{i=1}^{A} y_i z_i >=< \sum_{i=1}^{A} z_i x_i >= 0. \tag{3}$$

CP481, *Nuclear Structure 98*, edited by C. Baktash

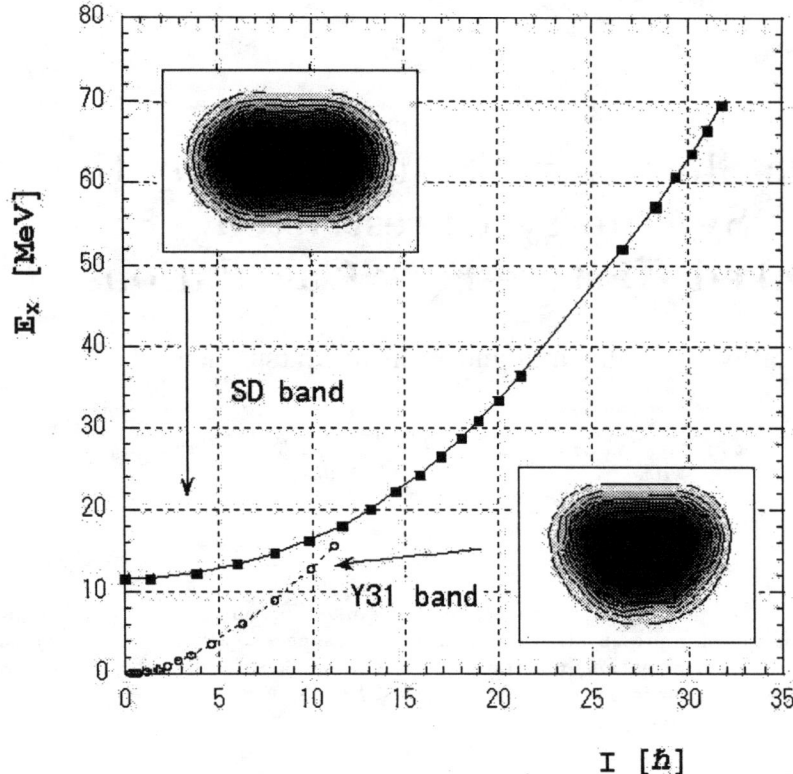

FIGURE 1. Excitation energy versus angular-momentum plot of the yrast structure of ^{32}S calculated with the Skyrme III interaction. Density distributions projected on the plane perpendicular to the rotation axis are shown, as insets, for the SD band (solid line with filled squares). and the Y_{31} band (broken line with open circles).

Special care is taken to accurately fulfill the above conditions during the iteration procedure. We solve these equations inside the sphere with radius $R=8$[fm] and mesh size $h=1$[fm], starting with various initial configurations. We use the Skyrme III interaction which has been successful in describing systematically the ground-state quadrupole deformations in a wide area of nuclear chart [7]. Results of the calculation are presented in Figs. 1-3. Figure 1 shows the structure of the yrast line. The expected superdeformed(SD) band becomes the yrast for $I \geq 14$. In addition to the SD band, we obtained an interesting band possessing the Y_{31} deformation, which appears in the yrast line with $5 \leq I \leq 13$. Let us call this band "Y_{31} band." The calculated angular momentum I and deformation δ for the SD band and the Y_{31} band are shown in Figs. 2 and 3 as functions of the rotational frequency ω_{rot}. Below we shall first discuss the SD band and then about the Y_{31} band.

High-Spin Limit of the Superdeformed Band

The SD band is obtained from $I = 0$ to about $I = 20\hbar$. The potential energy surface for the SD state at $I = 0$ is shown in Fig. 4. We see that the excitation energy of the SD state at $I = 0$ is about 12 MeV. [1] It becomes the yrast above $I = 14\hbar$.

A particularly interesting point is the behavior of the SD band in the high-spin limit: It is clearly seen in

---

[1] The rotational zero-point energy corrections are evaluated to be -4.3 MeV and -1.1 MeV for the SD and the ground-state configurations, respectively. If we take these corrections into account, the excitation energy becomes about 9 MeV.

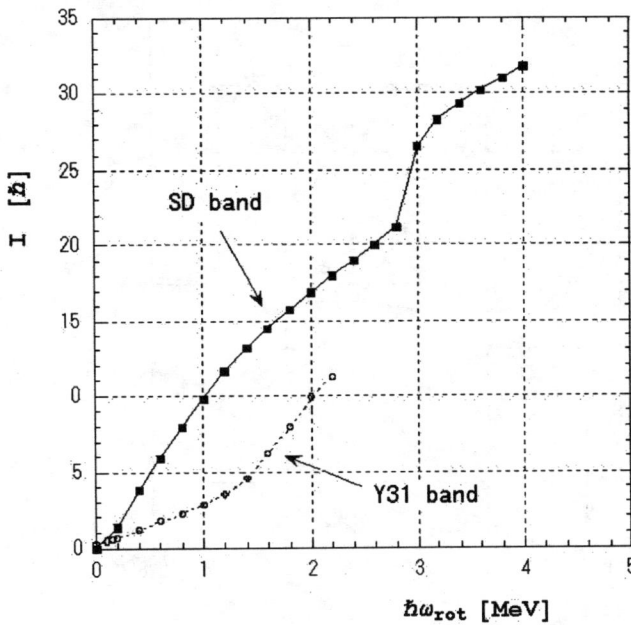

FIGURE 2. Angular momenta I plotted as a function of the rotation frequency ω_{rot} for the SD band (solid line with filled squares) and the Y_{31} band (broken line with open circles).

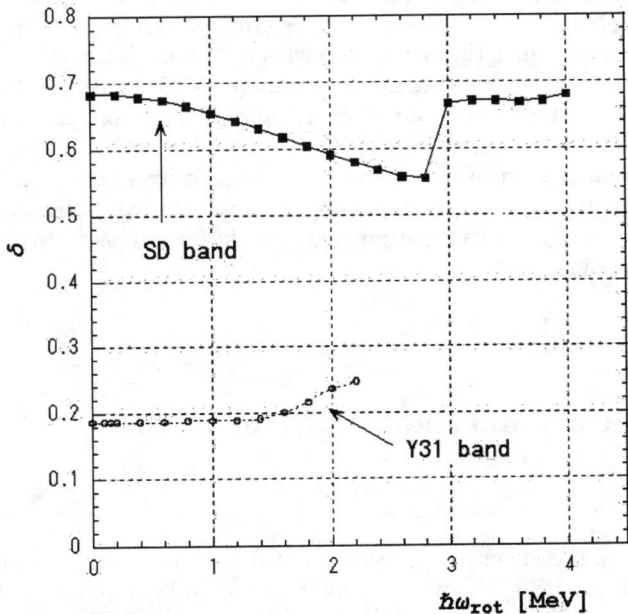

FIGURE 3. Deformation δ plotted as a function of the rotation frequency ω_{rot} for for the SD band (solid line with filled squares) and the Y_{31} band (broken line with open circles). δ is defined as $\delta = \frac{3}{4} < \sum_{i=1}^{A}(2x_i^2 - y_i^2 - z_i^2) > / < \sum_{i=1}^{A}(x_i^2 + y_i^2 + z_i^2) >$.

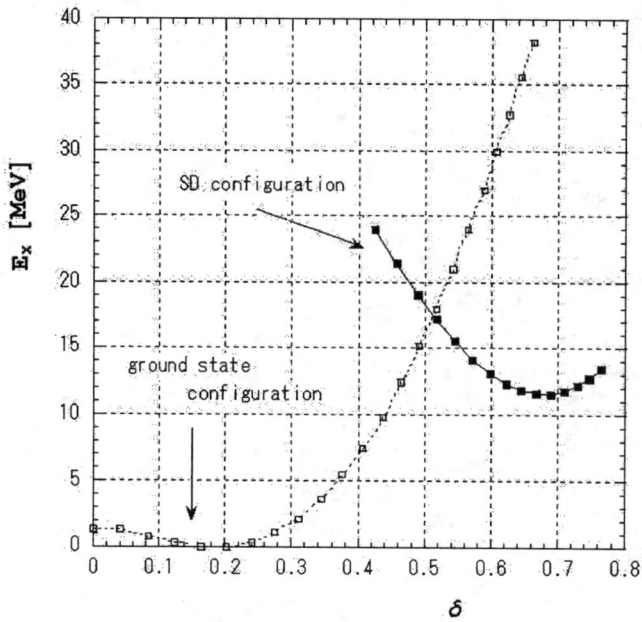

FIGURE 4. Potential energy surface for the SD configuration at $I = 0$ (solid line with filled squares) relative to that for the ground state configuration (dotted line with open squares). This calculation was done by means of the constrained HF procedure [8].

Figs. 2 and 3 that a jump occurs both in the angular momentum I and the deformation δ at $\omega_{rot} \simeq 3$ MeV/$\hbar$. At this point, I jumps from about 22 to $26\hbar$, and δ increases from about 0.56 to 0.66. This is due to the level crossing with the rotation-aligned $[440]\frac{1}{2}$ orbit. Thus the states above $I \simeq 24\hbar$ may be better characterized as the hyperdeformed configuration rather than the SD configuration. Such a singular behavior of the SD band can be noticed also in the previous cranked HF calculation with the BKN force [9], but no explanation of its microscopic origin was given there. Let us note that if we regard the SD configuration as to correspond to the j-j-coupling shell model $4p$-$12h$ configuration $\pi[(f_{7/2})^2(sd)^{-6}] \otimes \nu[(f_{7/2})^2(sd)^{-6}]$ (relative to ^{40}Ca) in the spherical limit, the maximum angular momentum that can be generated by aligning the single-particle angular momenta toward the direction of the rotation axis is $24\hbar$, and thus "the SD band termination" may be expected at this angular momentum. Interestingly, our calculation suggests that a crossover to the hyperdeformed band takes place just at this region of the yrast line.

Effects of Time-Odd Components

It would be interesting to examine the effect of rotation-induced time-odd components in the mean field. In Fig. 5 we compare the results of calculation with and without the time-odd components. From this figure we can easily confirm that the dynamical moment of inertia $J^{(2)} = \partial I/\partial \omega_{rot}$ of the SD band increases about 30% due to the time-odd components. This increase is well compared with the effective-mass ratio $m/m^* = 1/0.76 \simeq 1.3$ for the Skyrme III interaction, and seems to be consistent with what expected from the restoration of the local Galilean invariance [10] (more generally speaking, local gauge invariance [11]) of the Skyrme force; namely, the major effect of the time-odd components is to restore the decrease of the moment of inertia due to the effective mass m^* and bring it back to the value for the nucleon mass m.

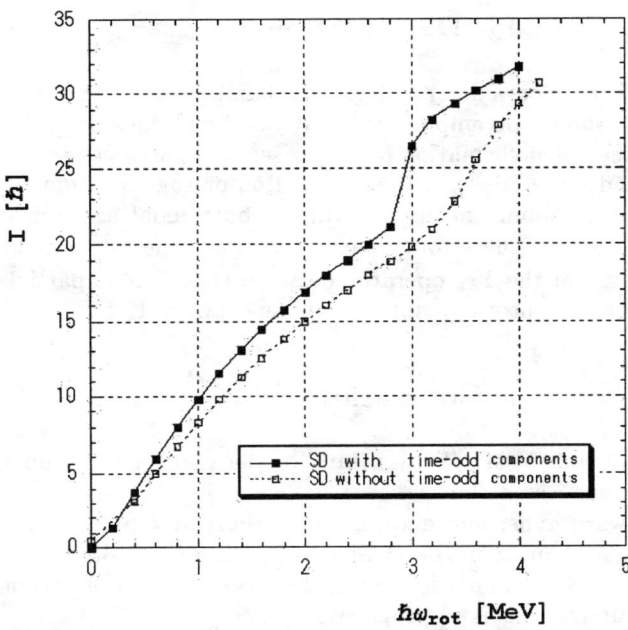

FIGURE 5. Angular momenta I of the SD band plotted as a function of the rotation frequency ω_{rot}. The solid line with filled squares (dotted line with open squares) indicates the result with(without) the time-odd components.

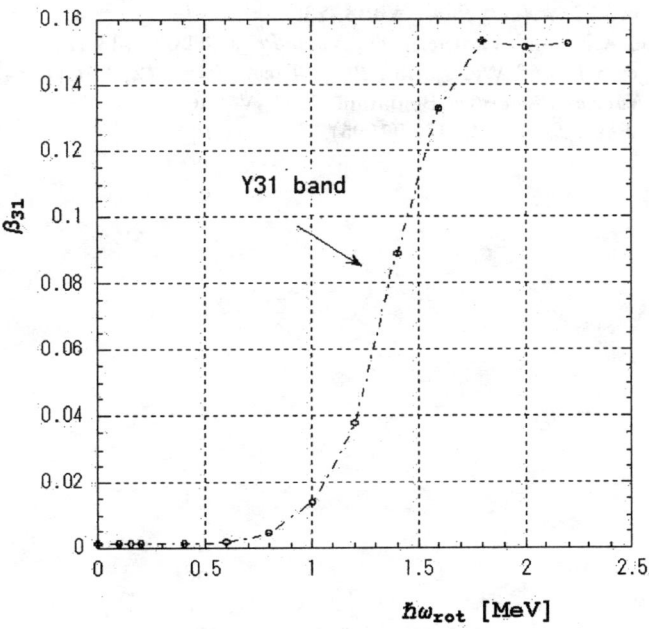

FIGURE 6. Non-axial octupole deformation β_{31} of the Y_{31} band, plotted as a function of the rotation frequency ω_{rot}. β_{31} is defined through the mass-octupole moments in the usual manner .

331

Y_{31} Deformation

As noticed in Fig. 1, we found that a non-axial Y_{31} deformed band ($\delta \simeq 0.2$ and $\beta_{31} = 0.1 \sim 0.15$) appears in the yrast line with $5 \le I \le 13$. It should be emphasized that this band does not exist at $I = 0$ but emerges at high spin: As shown in Fig. 6, the Y_{31} deformation quickly rises when ω_{rot} exceeds 1 MeV/$\hbar$.

Formation mechanism of this band is well described as a function of angular momentum I by means of the new cranked HF code allowing for the simultaneous breaking of both axial and reflection symmetries. We found that this band emerges as a result of the strong coupling between the rotation-aligned $[330]\frac{1}{2}$ orbit and the $[211]\frac{1}{2}$ orbit. The matrix element of the Y_{31} operator between these single-particle states is large, since they satisfy the selection rule for the asymptotic quantum numbers ($\Delta\Lambda = 1, \Delta n_z = 2$).

Conclusions

We have investigated high-spin structure of ^{32}S by means of the cranked Skyrme HF method in the 3D Cartesian-mesh representation, and suggested that
1) an internal structure change (toward hyperdeformation) may occurs at $I > 20$ in the superdeformed band,
2) a non-axial Y_{31} deformed band may appear in the yrast line with $5 \le I \le 13$.

We have obtained similar results also in calculations with the Skyrme M* interaction. More detailed study including dependence on effective interactions is in progress.

REFERENCES

1. Dobaczewski J., contribution to this conference.
2. Sheline R.K., Ragnarsson L., and Nilsson S.G., *Phys. Lett.* **41B**, 115(1972).
3. Leander G., and Larsson S.E., *Nucl. Phys.* **A239**, 93 (1975).
4. Davies K.T.R., Flocard H., Krieger S.J., and Weiss M.S., *Nucl. Phys.* **A342**, 111(1980).
5. Bonche P., Flocard H., Heenen P.H., Krieger S.J., and Weiss M.S., *Nucl. Phys.* **A443**, 39(1985).
6. Bonche P., Flocard H., Heenen P.H., *Nucl. Phys.* **A467**, 115(1987).
7. Tajima N., Takahara S., and Onishi N., *Nucl. Phys.* **A603**, 23(1996).
8. Flocard H., Quentin P., Kerman A.K., and Vautherin D., *Nucl. Phys.* **A203**, 433(1973).
9. Flocard H., Heenen P.H., Krieger S.J., and Weiss M.S., *Prog. Theor. Phys.* **72**, 1000(1984).
10. Bohr A., and Mottelson B.R., *Nuclear Structure*: Benjamin, 1975, Vol. 2.
11. Dobaczewski J., and Dudek J., *Phys. Rev.* **C52**, 1827(1995).

Superdeformation and Band Termination in $A \sim 60$ Nuclei

C. E. Svensson

Department of Physics and Astronomy, McMaster University, Hamilton, Ontario, Canada L8S 4M1

Abstract. High-spin states in proton-rich $A \sim 60$ nuclei on and near the $N = Z$ line have been studied in a series of experiments performed with Gammasphere and the Microball charged-particle detector array. These experiments have led to the identification of more than 40 deformed and superdeformed rotational bands in more than a dozen different Ni, Cu, and Zn isotopes. Because of the limited spin content of the single-particle configurations in these relatively light nuclei, many of these bands are observed to, or close to, their terminating states. Lifetime measurements for two sets of strongly coupled rotational bands in ^{62}Zn have confirmed the predicted loss of collectivity associated with the phenomenon of smooth band termination. Superdeformed bands have been observed in ^{60,61,62,65}Zn, firmly establishing an island of superdeformation for nuclei with particle numbers $N, Z \approx 30$. Linking transitions connecting the doubly-magic SD band in the $N = Z$ nucleus ^{60}Zn to the yrast line have been identified. These linking transitions not only provide the first definite spin, parity, and excitation energy measurements for SD states in the $A \sim 60$ mass region, but their stretched $E2$ character and relatively large $B(E2)$ values suggest that the decay-out process in this region differs substantially from that observed in heavier systems.

INTRODUCTION

Recent high-spin studies of proton-rich nuclei in the $A \sim 60$ mass region have revealed a rich pattern of shape transformations and noncollective-collective transitions with increasing angular momentum. With a limited number of valence particles outside doubly-magic ^{56}Ni, the low-spin decay schemes of these nuclei are dominated by spherical shell model states. At intermediate spins, the presence of high-j orbitals both below ($f_{7/2}$) and above ($g_{9/2}$) the spherical shell gaps at $N, Z = 28$ lead to well-deformed ($\beta_2 \sim 0.3$) collective rotational bands in these nuclei built on particle-hole excitations across these shell gaps. However, with only a small number of particles and holes occupying these high-j orbitals, the maximum spin available in these rotational bands is often limited to ~ 20–30 $\hbar$. Band termination is thus a common phenomenon in this mass region, as evidenced by the rapidly decreasing dynamic moments of inertia of many of the bands shown in Fig. 1. In a few cases, the fully-aligned terminating states of these bands have been identified, and recent lifetime measurements for the terminating bands in ^{62}Zn [1] have confirmed the predicted loss of collectivity associated with smooth band termination [2].

At still higher spins, the large superdeformed (SD) shell gaps for particle numbers $N, Z = 30$ [3,4] lead to the prediction [5,6] of highly collective superdeformed bands in this mass region. Such superdeformed bands have now been identified in ^{62}Zn [7], ^{61}Zn and ^{65}Zn [8], and the $N = Z$ nucleus ^{60}Zn [9]. The predicted island of superdeformation around $N, Z \approx 30$ has thus been firmly established. The observation of linking transitions connecting the doubly-magic ^{60}Zn SD band to the yrast line has also provided the first definite spin, parity, and excitation measurements for SD states in the $A \sim 60$ mass region and, as discussed below, the stretched $E2$ character and relatively large $B(E2)$'s of these transitions indicate that the decay-out process differs substantially from that which has been observed for SD bands in the mercury region.

Another fascinating result of these investigations has been the observation of well-deformed rotational bands which decay to spherical states not only by γ-ray emission, but also by discrete-line charged-particle emission [10]. A number of the rotational bands in Ni and Cu isotopes identified in these studies and the most recent results on charged-particle decay in this mass region are discussed in D. Rudolph's contribution

CP481, *Nuclear Structure 98*, edited by C. Baktash

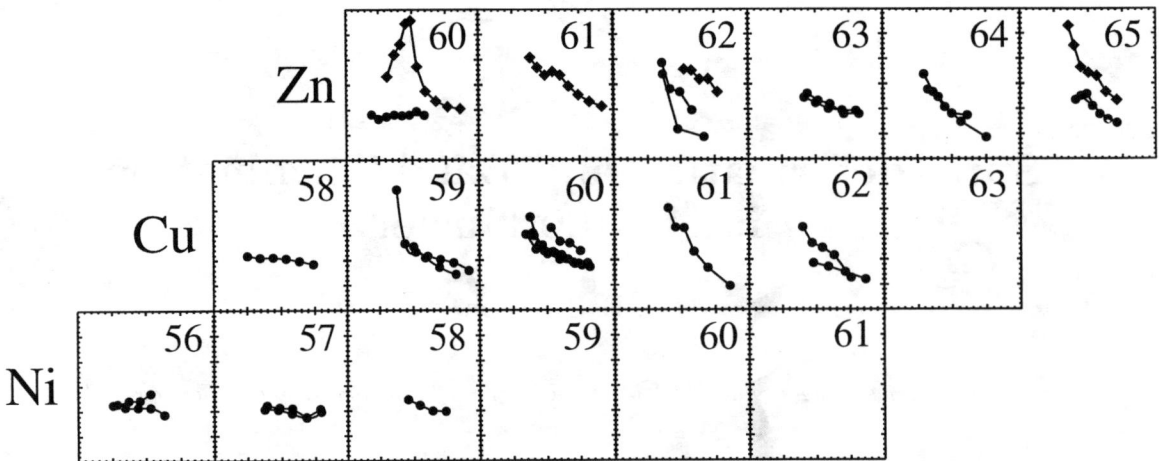

FIGURE 1. Dynamic moment of inertia $\mathcal{J}^{(2)}$ versus rotational frequency $\hbar\omega$ for a representative sampling of the 41 rotational bands identified in $A \sim 60$ nuclei in the experiments discussed here. The superdeformed bands in ^{60,61,62,65}Zn are indicated by diamonds. The $\mathcal{J}^{(2)}$ scale ranges from 0 to 30 $\hbar^2$MeV^{-1} and the $\hbar\omega$ scale covers the range from 0 to 2 MeV. Many of these bands are, as yet, unpublished and will be the subject of a series of papers by colleagues whose institutions are listed in the acknowledgements.

to these proceedings. Here we focus on studies of band termination and superdeformation in the proton-rich Zn isotopes.

EXPERIMENTS

A systematic study of high-spin states in proton-rich $A \sim 60$ nuclei with the Gammasphere array [11] has been carried out in four experiments; GS54, GS90, GS112, and GSFMA9. The reactions used were ^{28}Si(^{36}Ar,$x\alpha ypzn$), ^{40}Ca(^{28}Si,$x\alpha ypzn$), ^{40}Ca(^{29}Si,$x\alpha ypzn$), and ^{40}Ca(^{32}S,$x\alpha ypzn$), with 143-MeV ^{36}Ar, 125-MeV ^{28}Si, and 130-MeV ^{29}Si beams provided by the 88-Inch Cyclotron at Lawrence Berkeley National Laboratory, and a 134-MeV ^{32}S beam provided by the ATLAS facility at Argonne National Laboratory. In the first and second experiments, Gammasphere comprised 82 and 83 HPGe detectors and approximately 2×10^9, γ^3 and higher-fold coincidence events were collected, while in the third and fourth experiments, 100 and 101 HPGe detectors were used and approximately 1.5×10^9, γ^4 and higher-fold events were recorded.

Because of the relatively low Coulomb barrier in this mass region and statistical pressures for the proton-rich compound systems to decay toward stability by charged-particle emission, the evaporation process in these reactions is characterized by a general competition between proton, alpha, and neutron evaporation. As shown in Fig. 2, this leads to a fragmentation of the total fusion cross section into a large number of exit channels. It can be noted that, in a typical reaction with 3 to 5 evaporated particles, there are only 3 channels open in a (HI,xn) reaction, while for the more general case of a (HI,$x\alpha ypzn$) reaction there are 46 different channels with $x + y + z = 3$–5. Although not all of these channels are populated with significant cross section, it is standard in high-spin $A \sim 60$ experiments to populate 20 or more different nuclei with measurable yields (see Fig. 2). A highly efficient and selective means of identifying reaction channels is thus crucial to the success of these experiments, particularly in studies of the weak reaction channels leading to nuclei on or beyond the $N = Z$ line.

A variety of channel selection techniques have been employed in the experiments discussed here. In all four experiments, protons and alpha particles were detected and identified with the Microball [12], a 4π array of 95 CsI(Tl) scintillators. Neutrons were detected in the GS54 and GS90 experiments with 15 liquid scintillator detectors located at forward angles relative to the beam axis, while recoil ions were identified with the Argonne Fragment Mass Analyzer [13] in the GSFMA9 experiment. In the GS90, GS112, and GSFMA9 experiments the Hevimet collimators were also removed from the Gammasphere detectors in order to enable event-by-event γ-ray multiplicity and sum-energy measurements [14] and additional channel selectivity based on energy conservation requirements [15].

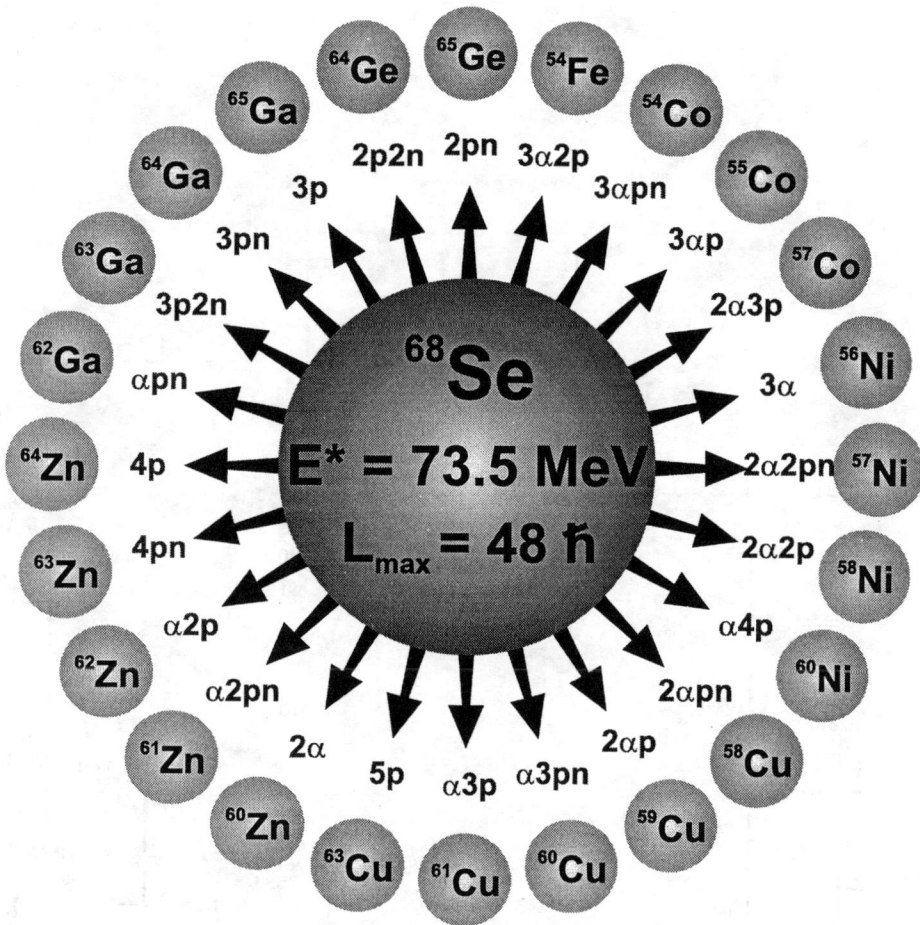

FIGURE 2. An illustration of the fragmentation of the total fusion cross section typical in the reactions used to populated high-spin states in proton-rich $A \sim 60$ nuclei. In this case, the compound system ^{68}Se is formed in the fusion of 125-MeV ^{28}Si on ^{40}Ca.

$A \sim 60$ BAND TERMINATION

The first collective rotational bands identified in the $A \sim 60$ mass region were a pair of strongly coupled bands in ^{64}Zn studied by Galindo-Uribarri *et al.* [16,17] in an experiment performed with the 8π Spectrometer [18] and the Miniball [19] charged-particle detector array at the former TASCC facility of the Chalk River Laboratories. The dynamic moments of inertia of these bands (see Fig. 1) show the rapid decrease with rotational frequency characteristic of the smoothly terminating rotational bands first identified [20] in the $A \sim 110$ mass region. Although linking transitions for the bands in ^{64}Zn were not established and definite spin and parity assignments were not possible in this experiment, comparisons with configuration-dependent cranked Nilsson-Strutinsky calculations favor a single-particle configuration based on a proton excitation across the $Z = 28$ spherical shell gap from the $f_{7/2}$ to the $g_{9/2}$ orbital and the excitation of two valence neutrons to $g_{9/2}$. This assignment suggests that both signatures of the ^{64}Zn band were observed to within one transition of their respective terminating states at $I^{\pi} = 25^-$ and 26^- [16].

Two sets of strongly coupled rotational bands in ^{62}Zn [1] were also identified in the 8π experiment mentioned above. These bands were further studied in the GS90 experiment, where the greater statistics and improved photopeak efficiency for high-energy γ rays enabled extensions of the bands to higher spin and observations of the linking transitions connecting them to the remainder of the decay scheme. A partial level scheme for ^{62}Zn from these studies is shown in Fig. 3. Many of the spherical states up to the yrast 11^- state were previously known [21,22] and a number of the higher-spin noncollective states were also identified recently by Furutaka *et al.* [23]. Here we focus on the properties of the strongly coupled rotational

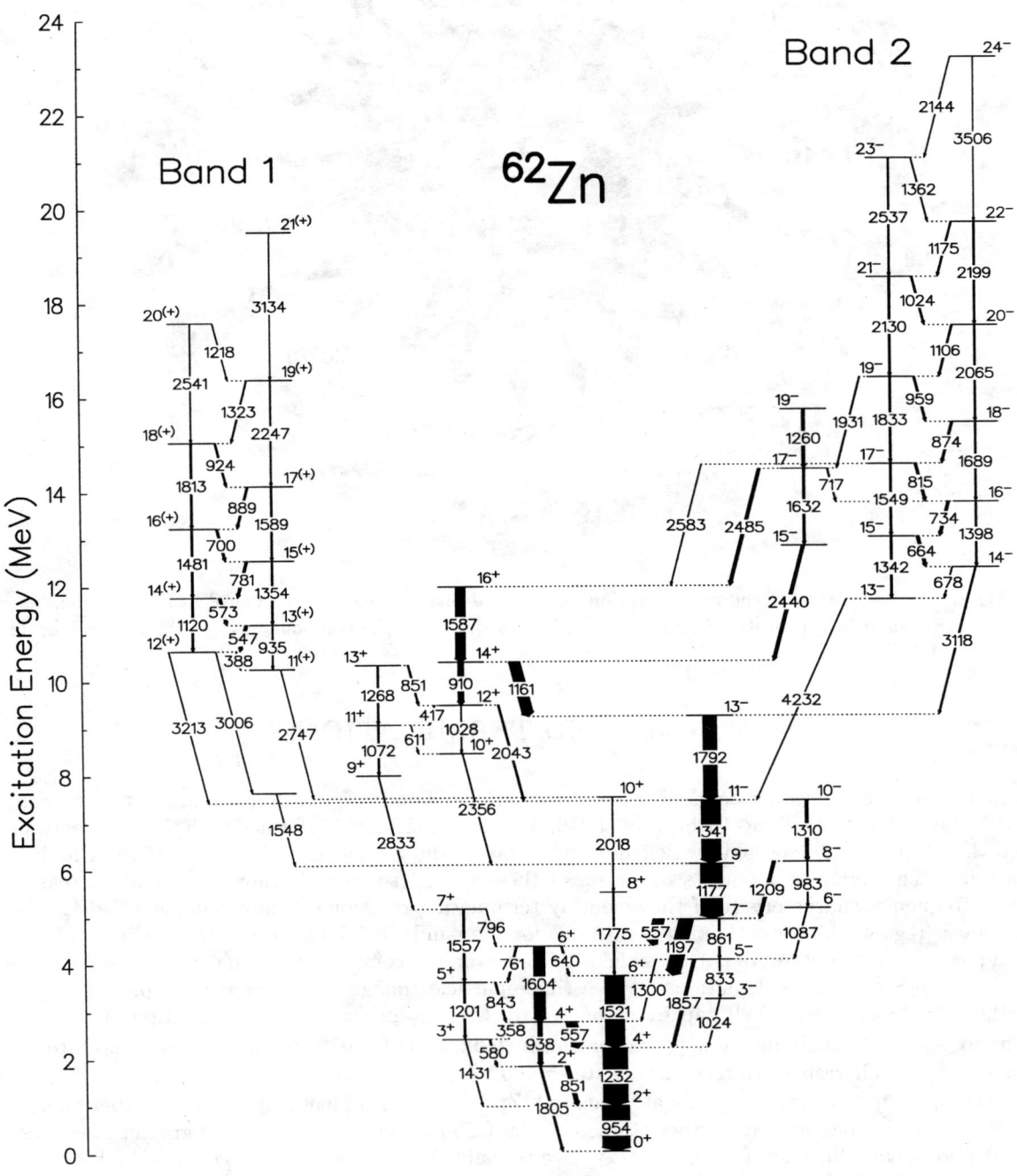

FIGURE 3. Partial decay scheme for ^{62}Zn. Transition energies are given to the nearest keV. The unlinked superdeformed band observed in ^{62}Zn is not shown.

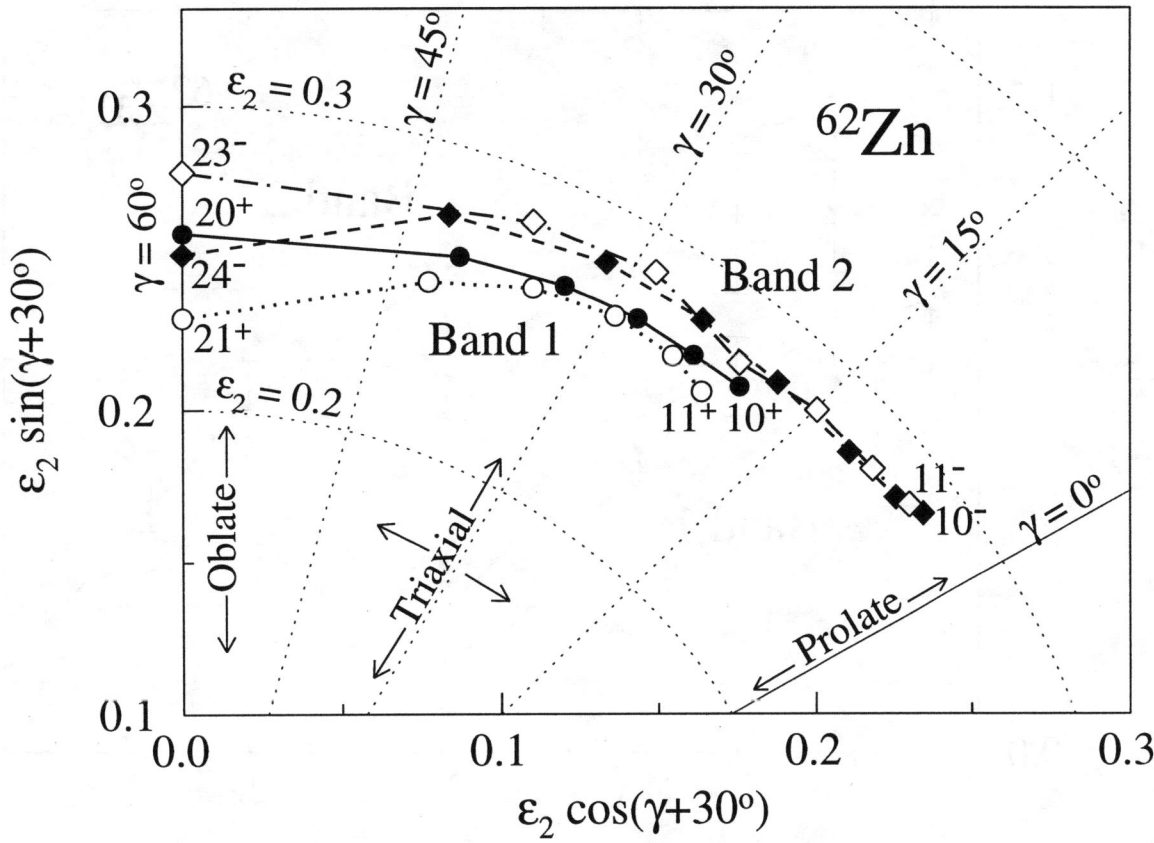

FIGURE 4. Calculated shape trajectories in the (ε_2, γ) plane for band 1 (circles) and band 2 (diamonds) in ^{62}Zn.

bands, labeled bands 1 and 2 in Fig. 3. To assign configurations to these bands, calculations for high-spin states in ^{62}Zn were carried out employing the configuration-dependent shell-correction approach [2,24] with the cranked Nilsson potential (parameters from Ref. [24]). Based on the excellent agreement between these calculations and the experimental observations [1], configurations involving one $f_{7/2}$ proton hole, one $g_{9/2}$ proton, and one (two) $g_{9/2}$ neutrons were assigned to bands 1 (2) (the [11,01] and [11,02] configurations in shorthand notation[1]). These assignments indicate that bands 1 and 2 have been observed up to their respective terminating states in which all of the angular momentum available from their single-particle configurations has been exhausted.

The strongly coupled bands in ^{64}Zn and bands 1 and 2 in ^{62}Zn are examples of smoothly terminating rotational bands [25] which are observed over an extended spin range prior to termination. The calculated shape trajectories for the ^{62}Zn bands are shown in Fig. 4. As this figure indicates, the nucleus is calculated to be triaxial for the intermediate-spin states in these bands and to change shape gradually, terminating in noncollective oblate ($\gamma = 60°$) states in which all of the angular momenta of the valence particles and holes are maximally aligned along the "rotation" axis. This predicted transition to an oblate terminating state implies a loss of collectivity [2], and hence transition quadrupole moments in these bands which should decrease as termination is approached. It is also interesting to note that, unlike the bands in the $A \sim 110$ mass region which are predicted to lose quadrupole deformation as they approach termination [2], the bands in ^{62}Zn and ^{64}Zn are calculated to remain well deformed ($\varepsilon_2 \sim 0.23$–0.29) over their entire spin range. The predicted loss of collectivity in these $A \sim 60$ bands is thus entirely associated with the γ degree of freedom, whereas in the $A \sim 110$ region it results from a combination of decreasing ε_2 and increasing γ.

---

[1] A useful shorthand configuration notation in this region is to label the high-j particles and holes by $[p1p2, n1n2]$ where $p1$ ($n1$) is the number of proton (neutron) $f_{7/2}$ holes and $p2$ ($n2$) is the number of proton (neutron) $g_{9/2}$ particles. The remaining (unlabeled) valence particles are distributed over the (mixed) $f_{5/2}$, $p_{3/2}$, and $p_{1/2}$ orbitals.

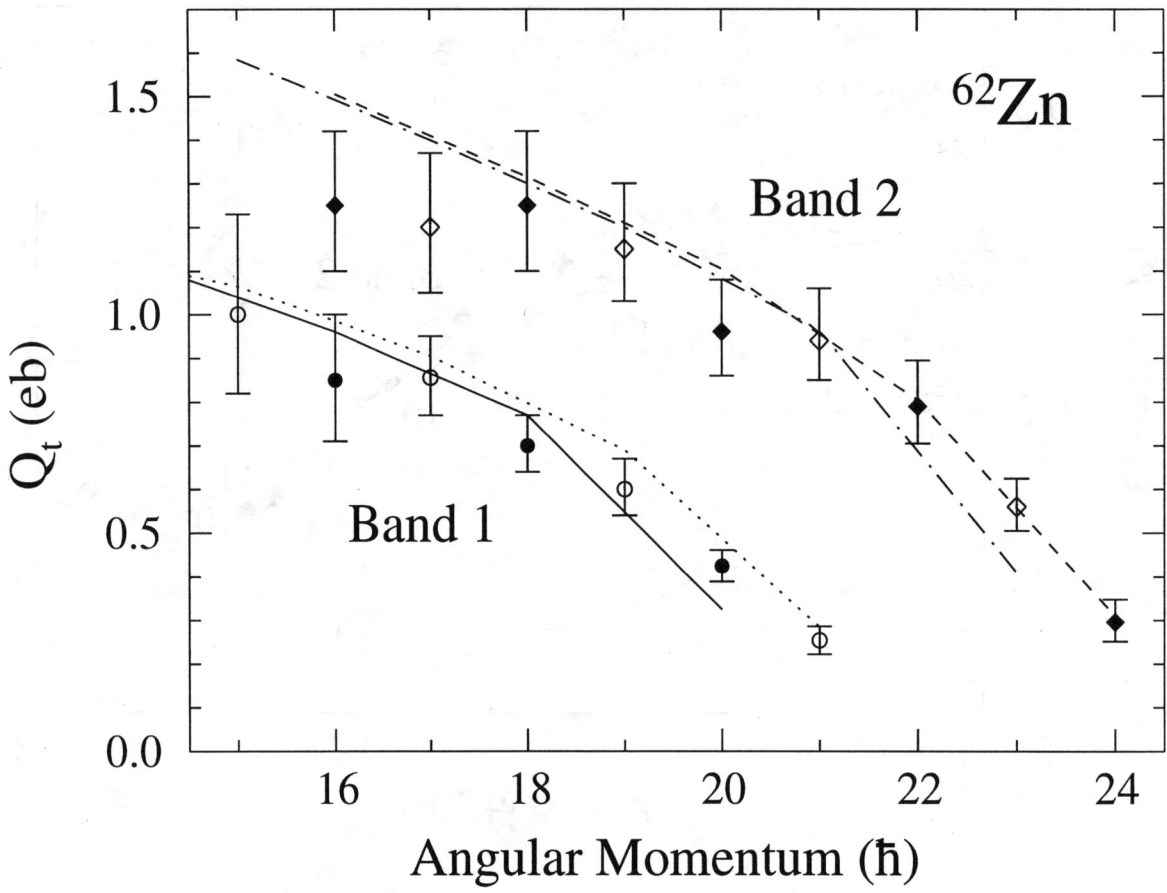

FIGURE 5. Transition quadrupole moments for band 1 (circles) and 2 (diamonds) in ^{62}Zn plotted versus the spin of the initial state. The lines show the calculated Q_t values for these bands.

An experimental test of the loss of collectivity inherent in the present theoretical interpretation of smoothly terminating bands requires lifetime measurements close to the terminating states. In fact, the relatively low spins at which bands 1 and 2 in ^{62}Zn terminate enabled such measurements to be performed all the way to termination. Despite the predicted loss of collectivity in these bands, the high γ-ray transition energies lead to state lifetimes which are short compared to the average time (of order 100 fs) taken by the recoiling nuclei to escape from the thin target. Lifetime measurements for high-spin bands in this mass region are thus performed by the thin-target Doppler-shift attenuation method [26] in the same experiments used for spectroscopic studies. The details of such lifetime measurements for the terminating bands in ^{62}Zn, including a discussion of the uncertainties associated with stopping powers and side feeding, are given in Ref. [1]. The results of these measurements are shown in Fig. 5, where the deduced transition quadrupole moment Q_t for each transition is plotted versus the spin of the initial state. These Q_t values do indeed decrease with increasing angular momentum and reach values corresponding to noncollective transitions strengths of ~ 1 Weisskopf unit (W.u.) ($Q_t \sim 0.2$ eb) at the highest spins. The lines in Fig. 5 represent the theoretical predictions for the ^{62}Zn bands. The calculated Q_t for the $I \rightarrow I-2$ transition has been taken as the average of the Q_t values calculated from the equilibrium deformations of the I and $I-2$ states (as described in Ref. [27]). Although these calculations do not account for the imperfect overlap of the shape wavefunctions for the initial and final states, and should thus be treated with caution close to termination where the shape change per transition is large [2] (see Fig. 4), the decreases in Q_t which they predict are in excellent agreement with experiment. These results, together with recent lifetime measurements approaching the terminating states of rotational bands in ^{108}Sn and ^{109}Sb [28], provide the first experimental confirmation of the predicted loss of collectivity associated with the smooth termination of rotational bands.

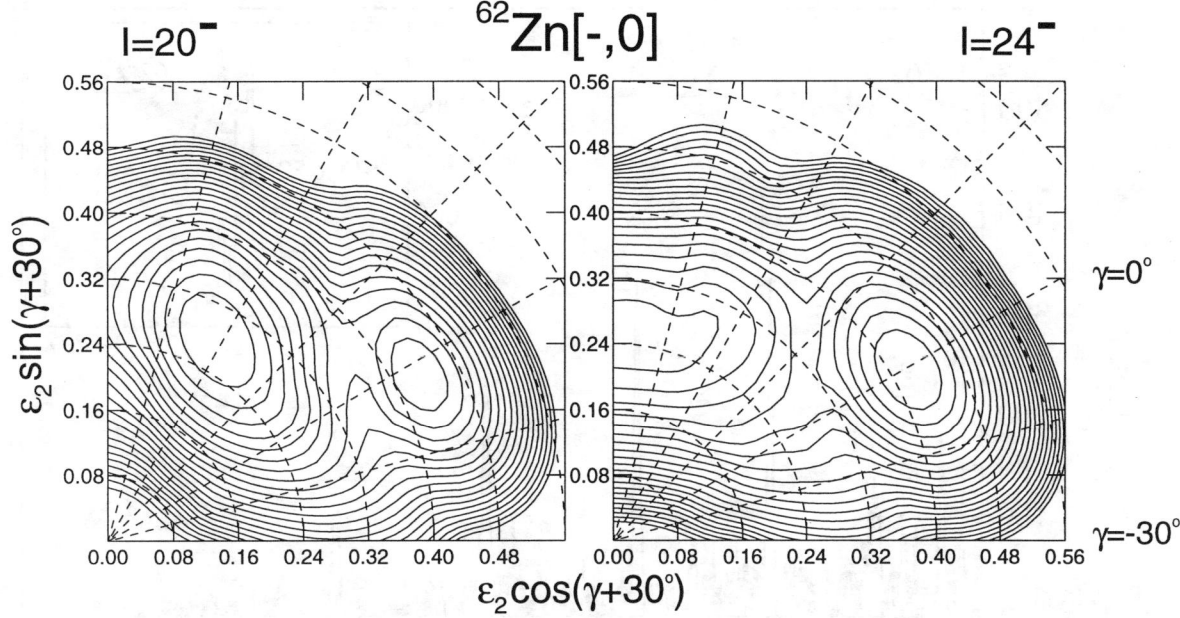

^{62}Zn[-,0]

I=20$^-$ I=24$^-$

γ=0°

γ=-30°

$\varepsilon_2\sin(\gamma+30°)$

$\varepsilon_2\cos(\gamma+30°)$

FIGURE 6. Calculated potential energy surfaces for ^{62}Zn at $I^\pi = 20^-$ (left) and $I^\pi = 24^-$ (right). Equipotential lines separated by 0.25 MeV are shown up to 5.5 MeV above the minima.

$A \sim 60$ SUPERDEFORMATION

As noted in the introduction, the large superdeformed shell gaps for particle numbers $N, Z = 30$ [3,4] lead to the prediction [5,6] of superdeformed bands in this mass region. Representative high-spin potential energy surfaces for ^{62}Zn at $I^\pi = 20^-$ and 24^- are shown in Fig. 6. At spin 20^-, the absolute minimum at $\varepsilon_2 \approx 0.28, \gamma \approx 30°$ corresponds to band 2 discussed above. However, a well-defined superdeformed second minimum is clearly visible at $\varepsilon_2 \approx 0.44, \gamma \approx 0°$. At spin 24^-, band 2 terminates in an oblate ($\gamma = 60°$) shape and the prolate superdeformed minimum becomes yrast.

The first SD band in the $A \sim 60$ mass region was identified in ^{62}Zn [7] in the GS90 experiment. The γ-ray spectrum obtained by summing coincidence gates set on the members of this band, which was populated with $\sim 1\%$ of the ^{62}Zn channel intensity, is shown in Fig. 7. Although an additional γ ray at 3986 keV was observed in coincidence with the band and is a candidate for a decay-out transition, this band could not be linked conclusively to the remainder of the ^{62}Zn decay scheme. The parity, spins, and excitation energies of the SD states are thus uncertain. However, coincidences between the members of the SD band and high-spin ND transitions in the ^{62}Zn level scheme suggest that the observed SD transitions cover the spin range from approximately $18\hbar$ to $30\hbar$. In addition, the excitation energy of the entry region which feeds this band could be accurately determined by measuring the total center of mass energy T_p of the evaporated charged particles in coincidence with γ rays in the SD band. As shown in the inset of Fig. 7, the mean T_p of 28.8 MeV for events feeding the SD band was 6.2 MeV below the average for the $\alpha 2p$ channel. The corresponding mean excitation energy of 32.0 MeV for events feeding the SD band was thus 6.2 MeV higher than the ^{62}Zn average entry excitation energy of 25.8 MeV. These measurements clearly demonstrate that the SD band was populated by only the highest excitation energy, and hence highest spin, components of the ^{62}Zn entry distribution. The dynamic moment of inertia of this band, shown in Fig. 1, is considerably larger than those of the normal deformed bands in this mass region at the same rotational frequency. The large deformation and high collectivity of this band were also established by a thin-target Doppler shift attenuation measurement of its transition quadrupole moment. Assuming an axially symmetric shape, the measured Q_t of $2.7^{+0.7}_{-0.5}$ eb corresponds to a quadrupole deformation $\beta_2 = 0.45^{+0.10}_{-0.07}$.

The properties of this band are in excellent agreement [7] with theoretical calculations which indicate that SD bands with deformations $\beta_2 = 0.41$–0.49 become yrast in ^{62}Zn for spins $I \geq 24$ $\hbar$. However, the lack of exact spin, parity, and excitation energy measurements for this band prevented a definite configuration assignment. All of the favored SD bands in ^{62}Zn have the neutron and proton orbitals filled up to the

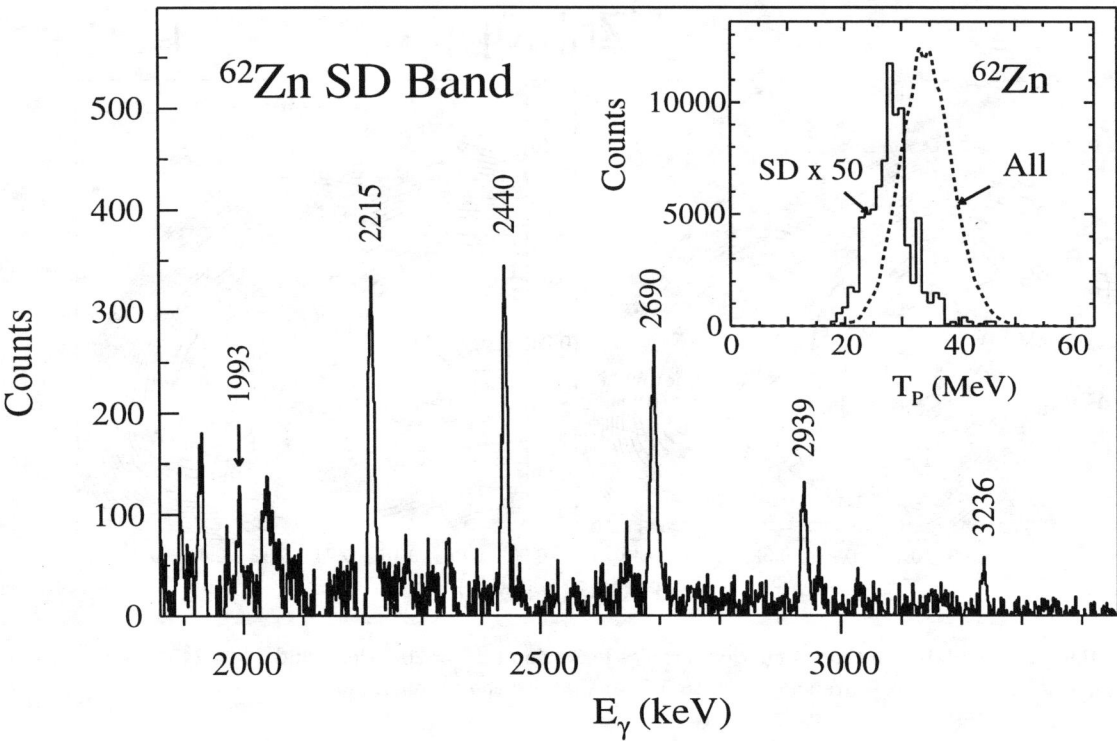

FIGURE 7. Gamma-ray spectrum of the ^{62}Zn superdeformed band obtained by summing coincidence gates set on all of the band members. The inset shows spectra of total center of mass charged-particle energy T_p in coincidence with triple gates set on members of the SD band (solid line) and low-spin transitions in ^{62}Zn (dashed line).

$N, Z = 30$ SD shell gaps corresponding to the [22,22] configuration, i.e. $[f_{7/2}]^{-2}[g_{9/2}]^2$ for both the proton and neutron subsystems. Different SD bands are formed depending on the orbitals occupied by the last two neutrons. The Q_t measurement, which has a relatively large uncertainty because of the very short lifetimes (of order 1 fs) of the states in this highly collective band (see Ref. [7]), is consistent with a [22,2n] configuration with n, the number of $g_{9/2}$ neutrons, being 2–4. Comparison of the $\mathcal{J}^{(2)}$ for this band with the results of cranked Hartree-Fock calculations [29,30] does, however, favor one of the [22,23] configurations.

As shown in Fig. 1, superdeformed bands based on the $N, Z = 30$ SD shell gaps have also been identified in ^{61}Zn and ^{65}Zn [8], and the doubly-magic superdeformed nucleus ^{60}Zn [9]. Although the deformation decreases rapidly when particles are removed from the ^{60}Zn SD core (e.g. $\beta_2 \sim 0.37$ for the [21,21] configuration in ^{58}Cu [10] compared to $\beta_2 \sim 0.47$ for the [22,22] configuration in ^{60}Zn [9]), the strongly deformed bands in Cu and Ni isotopes with two proton and two neutron holes in the $f_{7/2}$ orbital (see D. Rudolph's contribution to these proceedings) may also be regarded as part of the island of super/strong-deformation for nuclei with particle numbers $N, Z \approx 30$. It should also be noted that a superdeformed band has been identified in ^{68}Zn by M. Devlin *et al.* [31] in an experiment aimed at studying neutron-rich nuclei in this mass region. This band is interpreted as arising from a combination of the proton configuration involved in the $A \sim 60$ SD bands and a neutron configuration similar to those involved in the $A \sim 80$ SD bands, and represents the first observation of superdeformation in a new neutron-rich island around $A \sim 70$.

THE DOUBLY-MAGIC SUPERDEFORMED BAND IN ^{60}ZN

The doubly-magic superdeformed configuration in the $A \sim 60$ mass region, corresponding to filling the single-particle energy levels up to the $N, Z = 30$ SD gaps, should be observed in ^{60}Zn. The study of high-spin states in $N = Z$ nuclei in this mass region is, however, hindered by the small cross sections for populating these nuclei in fusion-evaporation reactions with stable beams and targets. The $N = Z$ nucleus ^{60}Zn was studied in the GS90 and GSFMA9 experiments, via the ^{40}Ca(^{28}Si,2α)^{60}Zn and ^{40}Ca(^{32}S,3α)^{60}Zn reactions,

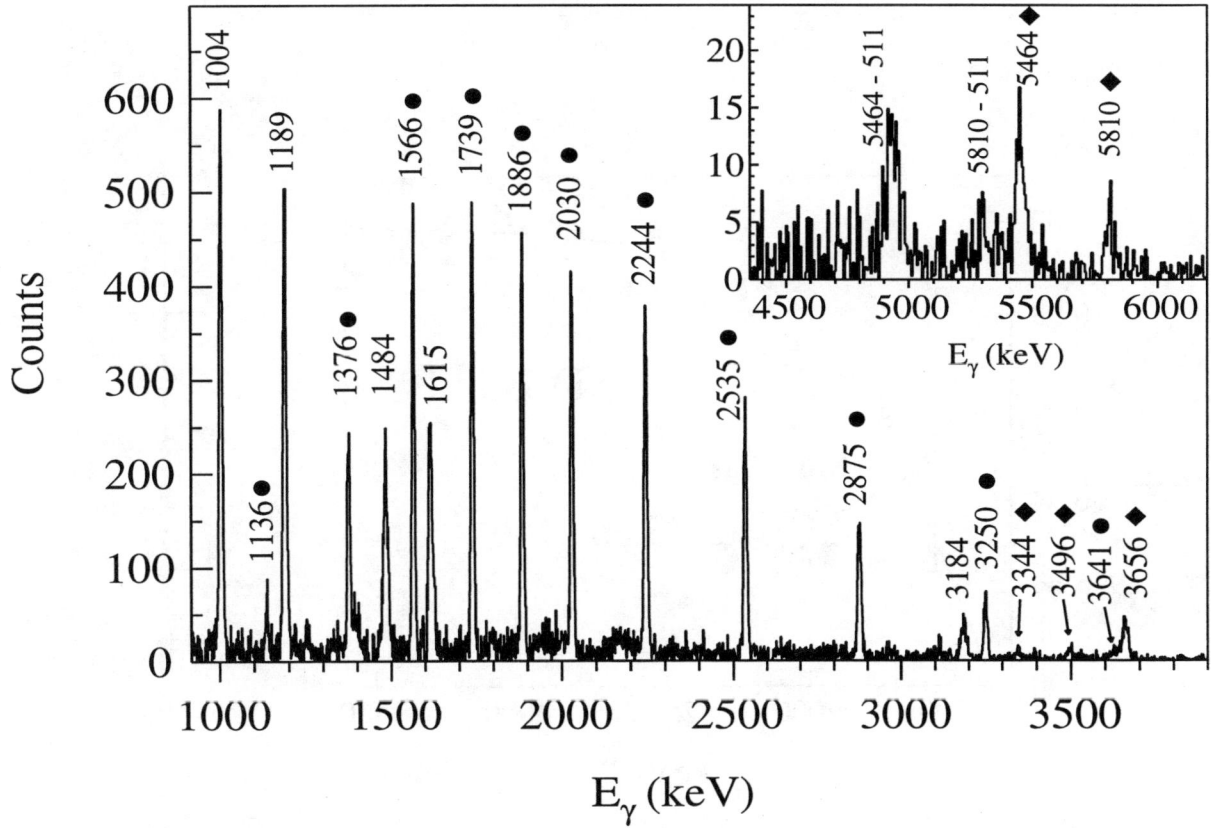

FIGURE 8. Gamma-ray spectrum of the ^{60}Zn superdeformed band obtained by summing coincidence gates set on all of the band members (circles). Diamonds label transitions linking the band to the remainder of the decay scheme.

and was populated with only $\sim 0.1\%$ of the total fusion cross section in each case. However, cranked Nilsson and cranked relativistic mean field calculations for ^{60}Zn [32], indicate that the doubly-magic SD configuration in this nucleus becomes yrast at relatively low spin ($I \sim 16\hbar$) and by $I \sim 24\hbar$ is separated from excited states by a large (~ 2 MeV) energy gap. If ^{60}Zn were populated at high-spin, a substantial fraction of the channel intensity would thus be expected to feed the doubly-magic SD band. In fact, this band was populated in these experiments with a much larger fraction of the channel intensity (60(4)% and 34(3)% in the GS90 and GSFMA9 experiments, respectively) than any previously observed SD band. The combination of this very large relative intensity and the clean selection of the weak ^{60}Zn channel by applying the total energy method [15] to events in which all of the evaporated alpha particles were detected in the Microball, permitted the identification and detailed analysis of the yrast SD band in this $N = Z$ nucleus.

The γ-ray spectrum obtained by summing coincidence gates set on the members of the ^{60}Zn SD band is shown in Fig. 8 and a partial decay scheme for ^{60}Zn is shown in Fig. 9. In addition to observing the doubly-magic SD band, linking transitions connecting this band to the yrast line were identified. An angular distribution measurement for the 3656 keV linking transition yields $a_2 = 0.38 \pm 0.09$ and $a_4 = -0.20 \pm 0.11$ and strongly favors a stretched $E2$ assignment for this transition. This assignment, which is consistent with the expectation that the yrast SD band in ^{60}Zn should have positive parity and even spins, is also supported by the absence of the 6.84 MeV stretched $E2$ transition to the yrast 8^+ state that would be energetically favored if the 3656 keV link were of $\Delta I = 0$ mixed $E2/M1$ character. Stretched $E2$ character is therefore assigned to this transition and, hence, to all of the observed linking transitions. These linking transitions thus establish the spins, parity, and excitation energies of the yrast SD states in ^{60}Zn, as indicated in Fig. 9.

The fractional Doppler shifts measured for the transitions in the ^{60}Zn SD band are shown in the inset of Fig. 9. Combining the statistical uncertainties in these $F(\tau)$ measurements with estimated uncertainties associated with the stopping powers and side feeding, the best fit to these data gives a transition quadrupole moment $Q_t = 2.75 \pm 0.45$ eb, corresponding to a deformation $\beta_2 = 0.47 \pm 0.07$ (for $\gamma = 0°$). If this Q_t is

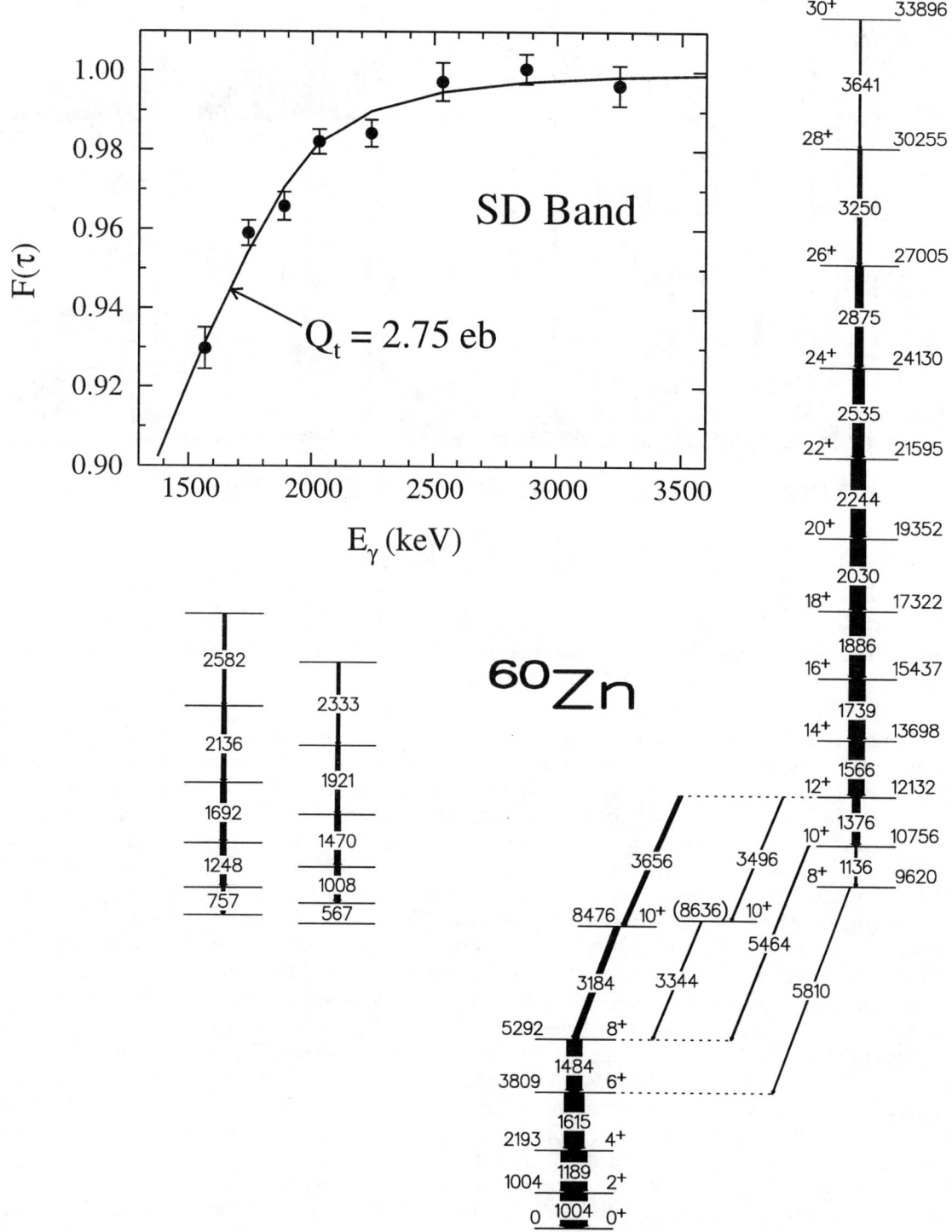

FIGURE 9. Partial decay scheme for ^{60}Zn. The order of the 3496 and 3344 keV linking transitions is uncertain, as are the spins and excitation energies of the unlinked normal deformed bands. The inset shows fractional Doppler shifts for members of the SD band and a fit giving $Q_t = 2.75$ *eb*.

assumed to remain approximately constant during the decay-out process, the measured in-band to decay-out branching ratios give $B(E2)$'s of 0.2(1), 0.8(2), and 0.04(2) W.u. for the 3496, 3656, and 5464 keV decay-out transitions. Further assuming a γ-ray energy of 900 keV and the upper limit of $\sim 4\%$ of the SD band intensity for the unobserved $8^+ \rightarrow 6^+$ SD transition gives a lower limit of 0.01 W.u. for the 5810 keV transition.

Stretched $E2$ decay out of highly deformed ($\beta_2 \sim 0.30$–0.35) bands has been observed in a number of the Nd isotopes [33–35]. However, in these nuclei the deformations of the highly deformed bands and the normal deformed ($\beta_2 \sim 0.20$–25) bands to which they decay are similar and no well-defined barrier exists between the two minima at the decay-out point. In ^{60}Zn, the deformation change between the SD band and the near-spherical states to which it decays is larger and the observation of the SD band to the point where it is 4.33 MeV above the yrast line suggests a substantial barrier between the two minima. In this sense, the decay out of the ^{60}Zn SD band is similar to that observed for the SD bands in ^{194}Hg [36,37] and ^{194}Pb [38,39]. The stretched $E2$ character and relatively large $B(E2)$'s of the decay-out transitions in ^{60}Zn are, however, difficult to reconcile with the weak-mixing statistical model [40] which successfully explains the dominance of $E1$ linking transitions and the strongly hindered[2] decay-out transition strengths observed in the mercury region.

A number of authors have pointed out the importance of pairing correlations in the decay-out process for SD bands in heavier nuclei [42–44]. In ^{60}Zn, the doubly-magic [22,22] SD configuration differs from the ground-band configuration only by the movement of a pair of protons and a pair of neutrons from the $f_{7/2}$ orbital to the $g_{9/2}$ orbital. The mixing of SD and ND states in ^{60}Zn resulting from pairing interactions is thus expected to be more significant than in the $A \sim 150$ and $A \sim 190$ regions where a substantial number (on the order of 10) pairs of nucleons must be rearranged. Furthermore, the upbend in the dynamic moment of inertia of the ^{60}Zn SD band (see Fig. 1), interpreted as a manifestation of the alignment of $g_{9/2}$ protons and neutrons, provides a clear indication of the importance of pairing in the low-spin states of this SD band. A detailed theoretical study of the role of pairing correlations in the decay-out process for $A \sim 60$ SD bands remains to be performed. It is hoped that these first results on the decay out of the doubly-magic SD band in ^{60}Zn will motivate such studies in the near future.

SUMMARY

High-spin states in proton-rich $A \sim 60$ nuclei have been studied in a series of experiments with the Gammasphere array and the Microball charged-particle detector. These experiments have, to date, led to the identification of more than 40 high-spin rotational bands in more than a dozen Ni, Cu, and Zn isotopes. Some of the highlights of these studies presented here include (i) transition quadrupole moment measurements for terminating bands in ^{62}Zn which confirm the predicted loss of collectivity associated with the smooth termination of rotational bands, (ii) the observation of superdeformed bands in ^{60,61,62,65}Zn, and (iii) the identification of linking transitions in ^{60}Zn which establish the spins, parity, and excitation energies of the yrast SD states in this $N = Z$ nucleus and indicate that the decay-out process in ^{60}Zn differs substantially from that observed in heavier systems.

ACKNOWLEDGEMENTS

The results presented here have involved the collaboration of a large number of researchers from many institutions. Special thanks go to all of my collaborators at Argonne National Laboratory, Lawrence Berkeley National Laboratory, Lund University, McMaster University, Oak Ridge National Laboratory, Staffordshire University, Universität zu Köln, Universität München, University of Tennessee, and Washington University. This work has been partially funded by NSERC (Canada), the DOE under Contracts Nos. DE-AC05-96OR22464, DE-AC03-76SF00098, W-31-109-ENG-38, DE-FG05-88ER40406, and DE-FG05-93ER40770, EPSRC (U.K.), BMBF (Germany) under Contract Nos. 06-OK-668 and 06-LM-868, the Swedish Natural Science Research Council, and the Royal Swedish Academy of Sciences.

---

[2] It can be noted that the decay-out $B(E2)$'s in ^{60}Zn are 2–4 orders of magnitude larger than the *upper limit* of $\sim 10^{-4}$ W.u. (before correcting for the weak SD/ND mixing amplitudes) set on the decay-out $B(E2)$'s in ^{194}Pb [41].

REFERENCES

1. C. E. Svensson *et al.*, Phys. Rev. Lett. **80**, 2558 (1998).
2. A. V. Afanasjev and I. Ragnarsson, Nucl. Phys. **A591**, 387 (1995).
3. R. K. Sheline, I. Ragnarsson, and S. G. Nilsson, Phys. Lett. B **41**, 115 (1972).
4. J. Dudek *et al.*, Phys. Rev. Lett. **59**, 1405 (1987).
5. I. Ragnarsson, in *Proceedings of the Workshop on the Science of Intense Radioactive Ion Beams,* Los Alamos National Laboratory Report LA-11964-C, p. 199 (1990).
6. R. K. Sheline, P. C. Sood, and I. Ragnarsson, Int. J. Mod. Phys. **A6**, 5057 (1991).
7. C. E. Svensson *et al.*, Phys. Rev. Lett. **79**, 1233 (1997).
8. C.-H. Yu *el al.*, abstracts to this conference p.154,155, and to be published.
9. C. E. Svensson *et al.*, Phys. Rev. Lett. (submitted).
10. D. Rudolph *et al.*, Phys. Rev. Lett. **80**, 3018 (1998).
11. I.-Y. Lee, Nucl. Phys. **A520**, 641c (1990).
12. D. G. Sarantites *et al.*, Nucl. Instrum. Methods Phys. Res., Sect. A **381**, 418 (1996).
13. C. N. Davids *et al.* Nucl. Instrum. Methods Phys. Res., Sect. B **70**, 358 (1992).
14. M. Devlin *et al.*, Nucl. Instrum. Methods Phys. Res., Sect. A **383**, 506 (1996).
15. C. E. Svensson *et al.*, Nucl. Instrum. Methods Phys. Res., Sect. A **396**, 228 (1997).
16. A. Galindo-Uribarri *et al.*, Phys. Lett. B **422**, 45 (1998).
17. V. P. Janzen *et al.*, in *Proceedings of the Conference on Nuclear Structure at the Limits,* Argonne National Laboratory Report No. ANL/PHY-91/1, p. 171, (1997).
18. H. R. Andrews *et al.*, Atomic Energy of Canada Limited, Report No. AECL-8329 (1984).
19. A. Galindo-Uribarri, Prog. Part. Nucl. Phys. **28**, 463 (1992).
20. V. P. Janzen *et al.*, Phys. Rev. Lett. **72**, 1160 (1995).
21. L. Mulligan, R. W. Zurmühle, and D. P. Balamuth, Phys. Rev. C **19**, 1295 (1979).
22. M. M. King, Nucl. Data Sheets **60**, 337 (1990).
23. K. Furutaka *et al.*, Z. Phys. A **358**, 279 (1997).
24. T. Bengtsson and I. Ragnarsson, Nucl. Phys. **A436**, 14 (1985).
25. I. Ragnarsson *et al.*, Phys. Rev. Lett. **74**, 3935 (1995).
26. B. Cederwall *et al.*, Nucl. Instrum. Methods Phys. Res., Sect. A **354**, 591 (1995).
27. A. V. Afanasjev and I. Ragnarsson, Nucl. Phys. **A608**, 176 (1996).
28. R. Wadsworth *et al.*, Phys. Rev. Lett. **80**, 1174 (1998).
29. J. Dobaczewski, W. Nazarewicz, W. Satuła, and C. Baktash, (to be published), J. Dobaczewski, private communication (1998).
30. D. Rudolph, *et al.* Nucl. Phys. **A630**, 417c (1998).
31. M. Devlin *et al.*, abstract to this conference p.26, and to be published.
32. A. V. Afanasjev, I. Ragnarsson, and P. Ring, (to be published).
33. D. Bazzacco *et al.*, Phys. Rev. C **49**, R2281 (1994).
34. S. Lunardi *et al.*, Phys. Rev. C **52**, R6 (1995).
35. M. A. Deleplanque *et al.*, Phys. Rev. C **52**, R2302 (1995).
36. T. L. Khoo *et al.*, Phys. Rev. Lett. **76**, 1583 (1996).
37. G. Hackman *et al.*, Phys. Rev. Lett. **79**, 4100 (1997).
38. A. Lopez-Martens *et al.*, Phys. Lett. B **380**, 18 (1996).
39. K. Hauschild *et al.*, Phys. Rev. C **55**, 2819 (1997).
40. E. Vigezzi, R. A. Broglia, and T. Døssing, Phys. Lett. B **249**, 163 (1990).
41. R. Krücken *et al.*, Phys. Rev. C **55**, R1625 (1997).
42. P. Bonche *et al.*, Nucl. Phys. **A519**, 509 (1990).
43. Y. R. Shimizu *et al.*, Phys. Lett. B **274**, 253 (1992).
44. G. F. Bertsch, Nucl. Phys. **A574**, 169c (1994).

Non-axial Octupole Deformations of $N = Z$ Nuclei in $A \sim 60 - 80$ Mass Region

M. Matsuo*, S. Takami[†], and K. Yabana[†]

*Yukawa Institute for Theoretical Physics, Kyoto University, Kyoto 606-8502, Japan
[†] Graduate School of Science and Technology, Niigata University, Niigata 950-2101, Japan

Abstract. By performing a fully three dimensional Hartree-Fock calculation with use of the Skyrm forces, we demonstrate possibility of exotic deformations violating both the reflection and the axial symmetries of $N = Z$ nuclei in $A \sim 60 - 80$ mass region. The Y_{32} tetrahedral shape predicted in excited ^{80}Zr arises from a shell gap at $N, Z = 40$ which is enhanced for the tetrahedron deformation. Softness toward the Y_{33} triangular deformation of the oblate state in ^{68}Se is also predicted.

INTRODUCTION

Octupole deformations that violate the reflection symmetry have attracted many attentions recently in studies of nuclear structure [1]. Although axial octupole deformation is well established in actinides and in neutron rich Xe and Ba region, there seems no experimental evidence of more exotic shapes that violate both the reflection and the axial symmetries, except for light alpha-clustering nuclei such as ^{12}C [2].

Non-axial octupole deformations have been predicted theoretically for actinides [3] and in superdeformed nuclei [4,5]. Instead we point out possibility of the non-axial octupole shapes in $N = Z$ $A \sim 60 - 80$ region of the nuclear chart [6,7]. For $N = Z$ nuclei the deformation driving shell effects which arise from both protons and neutrons cooperate coherently. The shell effect is indeed large in this mass region as is illustrated by the prolate-oblate shape coexistence in Kr and Se isotopes [8] and the sudden onset of large prolate deformation for $N, Z \approx 38 - 40$ [9]. Furthermore octupole instability due to the single-particle structure in nuclear mean-field is predicted for $N, Z \sim 34$ (in addition to $56, 90, 134$) [10,11] for which the coupling between $g_{9/2}$ and $p_{3/2}$ orbits are mostly responsible.

It is found recently that the shell structures plays important roles for aggregates of metallic atoms (metal clusters). In this system the mean-field for the valence electrons resembles with the nuclear mean-field except for its effectively zero ls potential [12]. Non-axial octupole deformatios seem to be one of the dominant deformations in this system as demonstrated by means of schematic models [13,14] and more recent mean-field calculations [15,16]. In particular, the tetrahedron shape or the Y_{32} deformation is predicted to be stable because of the shell gaps generated at $N_{electron} = 40, 70, 112, 156$ [13,15,16]. In this paper, we will show that the tetrahedron shape is expected in the nucleus ^{80}Zr with $N = Z = 40$, being in parallel with the metal cluster system in spite of the difference in the ls potential and the pairing correlation present in nuclei. Furthermore, non-axial octupole deformations besides Y_{32} are important deformation modes in $N = Z$ nuclei in $A \sim 60 - 80$ mass region. See also the previous publication [6].

A FULLY THREE DIMENSIONAL HARTREE-FOCK CALCULATION

For investigating non-axial reflection asymmetric deformations by means of the mean-field theory, we have to exclude any symmetry assumptions on nuclear deformations. Furthermore, it is important to use a description which allows arbitrary deformation when we search unknown shapes. From this view point, we adopt the Hartree-Fock method and perform a fully three dimensional calculation using the Cartesian mesh representation without any requirements on nuclear shapes and its symmetries. The Skyrm force is used as the effective

CP481, *Nuclear Structure 98*, edited by C. Baktash

nucleon-nucleon force. The pairing correlation is treated by means of the BCS method for the seniority force with a cutoff of the single-particle orbits, which is chosen the same as Ref. [17,18]. The imaginary time method is used for the iteration procedure [19]. Thus the basic formulation is in parallel with that of Ref. [17], except for the symmetry treatment (they assumed the reflection symmetries) and calculational details. A new code has been developed independently so that it can also include the parity projection and full variation after projection [20]. The parity projected Hartree-Fock with Skyrm force, but without BCS treatment, has been applied to light nuclei [20] (but not to the present investigation).

For the results presented below, we use the 3D mesh which are enclosed by a sphere of radius $R = 13$ fm, and the mesh width of 1 fm. The parameter set of SIII [21] is used for the Skyrm force in most calculations while SkM* [22] is also employed for comparison. Concerning the pairing force, we use $G_p = 16.5/(11 + Z)$ MeV for protons as given in Ref. [17]. For neutrons, we use $G_n = 16.5/(11 + N)$ MeV [23] since it is natural to take the same value as protons for $N = Z$ nuclei. In order to search not only the ground state but also local minimum states, we performed different runs of iterations starting with different initial conditions. To characterize deformation of the obtained solutions, we use the mass multipole moments,

$$\alpha_{lm} \equiv \frac{4\pi \langle \Phi | \sum_i^A r_i^l X_{lm}(i) | \Phi \rangle}{3AR^l}, (m = -l, \cdots, l), \tag{1}$$

where A is the mass number and $R = 1.2A^{1/3}$ fm. Here X_{lm} is a real basis of the spherical harmonics,

$$X_{l0} = Y_{l0},$$
$$X_{l|m|} = \frac{1}{\sqrt{2}}(Y_{l-|m|} + Y_{l-|m|}^*),$$
$$X_{l-|m|} = \frac{-i}{\sqrt{2}}(Y_{l|m|} - Y_{l|m|}^*), \tag{2}$$

where the quantization axis is chosen as the largest and smallest principal inertia axes for prolate and oblate solutions, respectively. We put the constraints $\alpha_{1m} = 0 (m = -1, 0, 1)$ for the center of mass, and $\alpha_{2m} = 0 (m = -2, -1, 1)$ for the principal axes. This is done by adding the constraining fields $-\lambda_{lm} r^l X_{lm}$ to the mean field, where the constraining Lagrange multipliers are determined so that the constraints are satisfied in each iteration step [20].

For the quadrupole moment, we use ordinary (β, γ) notation, i.e., $\alpha_{20} = \beta \cos \gamma, \alpha_{22} = \beta \sin \gamma$, mapped in the $\beta > 0, 0 < \gamma < \pi/3$ section. To represent magnitude of the octupole deformation, we define

$$\beta_3 \equiv (\sum_{m=-3}^3 \alpha_{3m}^2)^{\frac{1}{2}}, \quad \beta_{3m} \equiv (\alpha_{3m}^2 + \alpha_{3-m}^2)^{\frac{1}{2}} \quad (m = 0, 1, 2, 3). \tag{3}$$

TABLE 1. The ground and local minimum HF solutions obtained with the SIII force. The number in the first line is the excitation energy in MeV measured from the ground state solution. The second and the third line list the quadrupole and octupole deformations of the solutions.

	Oblate $\beta > 0$, $30° < \gamma \leq 60°$	$\beta = 0$	Prolate $\beta > 0$, $0° \leq \gamma \leq 30°$
^{64}Ge			g.s. $\beta, \gamma = 0.28, 25°$ (triaxial) $\beta_3 = 0.00$
^{68}Se	g.s. $\beta, \gamma = 0.28, 60°$ $\beta_3 = \beta_{33} = 0.14$		0.25 $\beta, \gamma = 0.25, 0°$ $\beta_3 = 0.00$
^{72}Kr	g.s. $\beta, \gamma = 0.35, 60°$ $\beta_3 = 0.00$		1.47 $\beta, \gamma = 0.40, 0°$ $\beta_3 = 0.00$
^{76}Sr	2.72 $\beta, \gamma = 0.14, 60°$ $\beta_3 = 0.13, \beta_{32} = 0.12$		g.s. $\beta, \gamma = 0.50, 0°$ $\beta_3 = 0.00$
^{80}Zr	1.56 $\beta, \gamma = 0.19, 60°$ $\beta_3 = 0.00$	0.96 $\beta, \gamma = 0.00, 0°$ $\beta_3 = \beta_{32} = 0.23$	g.s. $\beta, \gamma = 0.51, 0°$ $\beta_3 = 0.00$

SHAPES OF GROUND AND LOCAL MINIMUM STATES

The HF ground states and local minimum solutions for the even-even nuclei ^{64}Ge, ^{68}Se, ^{72}Kr, ^{76}Sr and ^{80}Zr are listed in Table 1. The result is essentially the same as the previous calculation [6] while the convergence of the HF iteration is taken care of more carefully in the present calculation. It is noted that many of these nuclei have local minimum solutions with low excitation energy, suggesting presence of shape coexisting excited states. The ground state shape changes from the triaxial (^{64}Ge), the oblate (^{68}Se,^{72}Kr), to the strong prolate shape (^{76}Sr,^{80}Zr) as N, Z increases. This is consistent with the experimental trends [9,24,25].

The reflection asymmetric solutions with non-zero octupole deformation ($\beta_3 > 0$) are obtained as the ground state of ^{68}Se, and as the shape coexisting excited states in ^{76}Sr and ^{80}Zr. These solutions *violate not only the reflection symmetry but also the axial one*. The ^{68}Se ground state has finte Y_{33} deformation ($\beta_{33} = 0.14$) together with oblate quadrupole deformation. The density profile shows a triangular shape, as seen in Fig.1. In ^{80}Zr, the first local minimum solution with the excitation energy of 0.96 MeV has large Y_{32} deformation ($\beta_{32} = 0.23$) whereas the solution has no quadrupole deformation. It has a tetrahedral shape as shown in Fig. 1. Although this solution violates both the reflection and the axial symmetry, the tetrahedral symmetry (T_d) of the point group emerges as a spontaneous symmetry. Indeed the calculated single-particle energy spectrum shows a four-fold degeneracy which is a characteristic feature of the Fermion mean-field with the T_d symmetry [3] (See also Fig.3(c)).

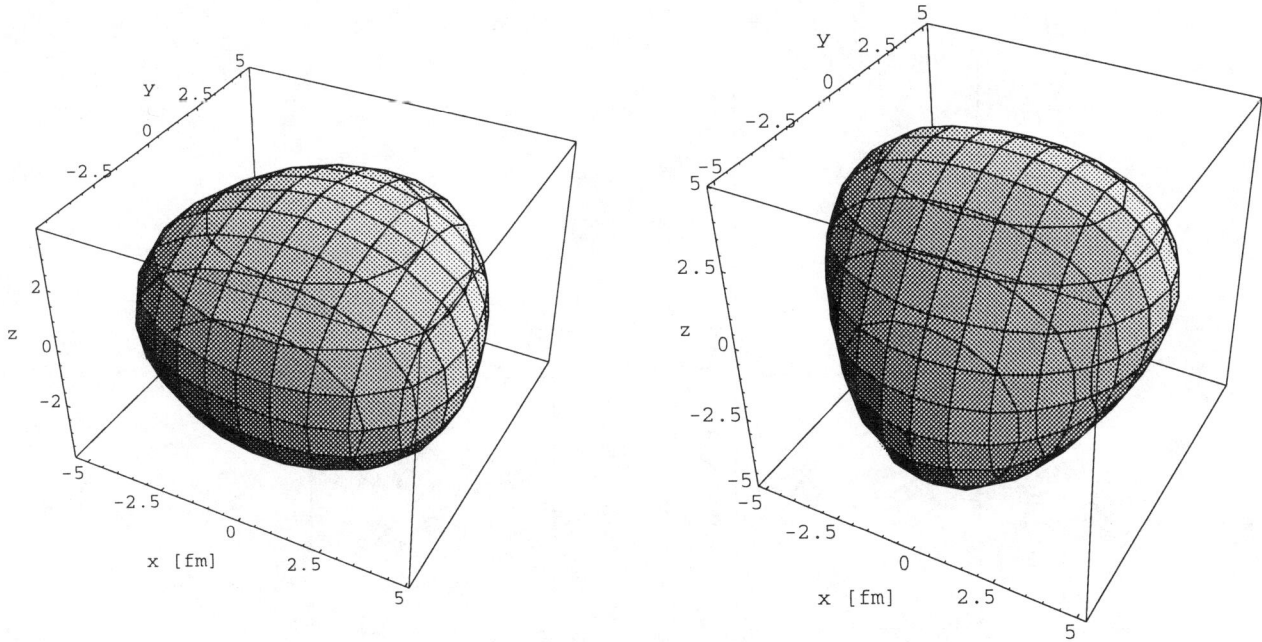

FIGURE 1. The density contour surface at the half central density for the triangular ground state solution in ^{68}Se (left), and the tetrahedral first local minmum solution in ^{80}Zr (right), listed in Table 1 obtained with the SIII force.

NON-AXIAL OCTUPOLE DEFORMATIONS

Potential energy curves

To evaluate softness of the obtained solutions toward the octupole deformations, we calculate the potential energy curve as a function of the deformation parameters α_{3m} for all the independent components $m = 0, 1, 2, 3$ of octupole deformations by means of the constraint Hartree-Fock calculation. For this calculation, we introduce additional nine constraints for the quadrupole deformation α_{20}, α_{22} (equivalent to β, γ) and the octupole deformations $\alpha_{3m}(m = -3, .., 3)$. The values of β, γ is fixed to the ones of a HF solution under consideration,

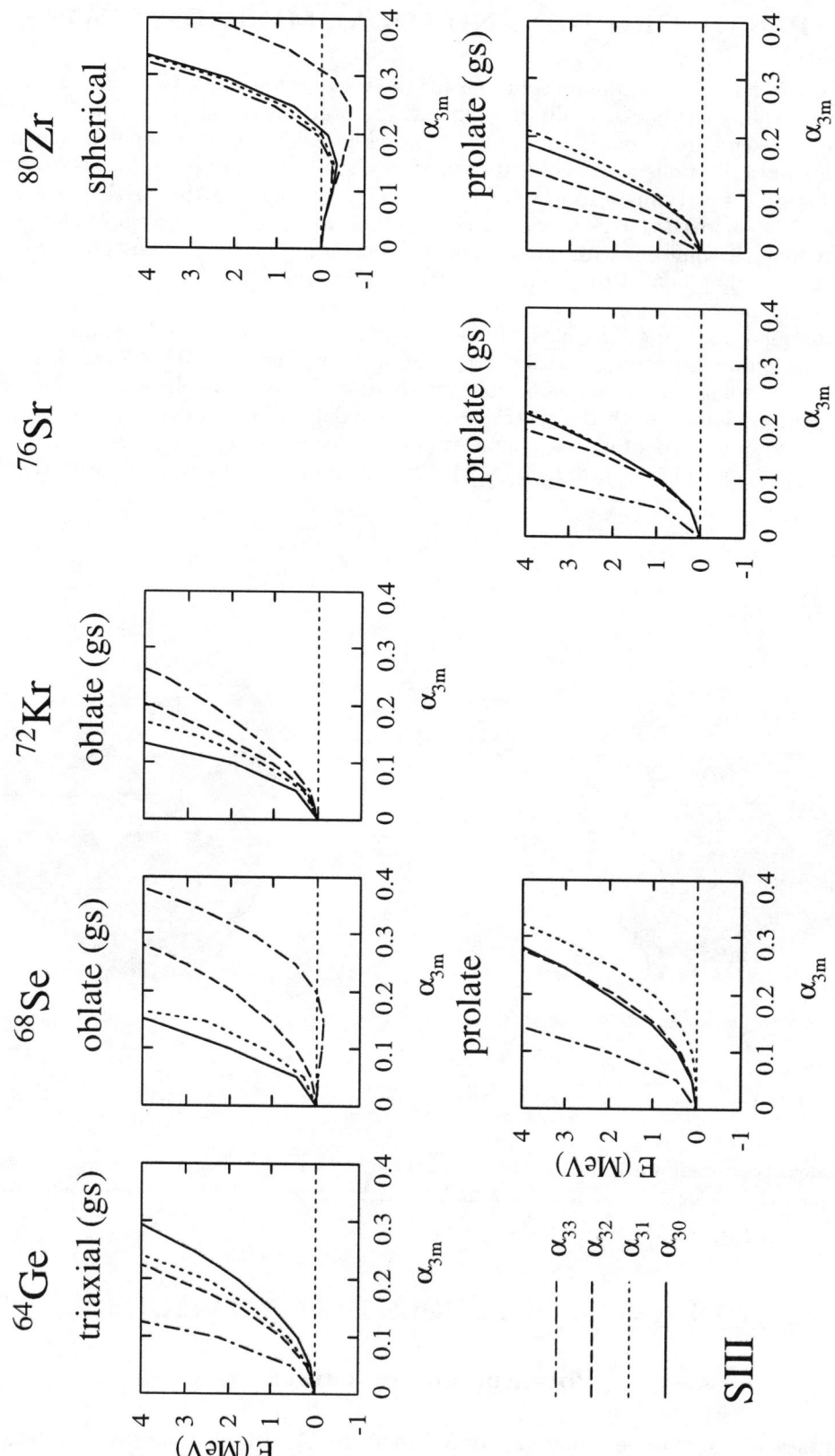

FIGURE 2. Potential energy curve as a function of the octupole deformation α_{3m} ($m = 0, 1, 2, 3$) associated with the ground and local minimum solutions listed in Table 1.

and one of $\alpha_{3m}(m = 0, 1, 2, 3)$ is varied while other α_{3m}'s are fixed to zero. The calculated potential energy curve is plotted in Fig.2.

Tetrahedron shape and associated shell gap in ^{80}Zr

The tetrahedral solution in ^{80}Zr corresponds to the minimum at $\alpha_{32} = 0.23$ of the α_{32} potential energy curve with the spherical constraint ($\beta = 0$). The energy gain due to the tetrahedron deformation or equivalently the energy difference between the minimum at $\alpha_{32} = 0.23$ relative to $\alpha_{32} = 0$ is as large as 0.70 MeV. It is noted that the potential energy curves for $\alpha_{3m}(m = 0, 1, 3)$ do not show such deep minimum. The reason of this can be seen in the single-particle energy diagram, Fig.3, plotted as a function of the octupole deformations α_{3m}. Figure 3 indicates that the single-particle energy spectrum accompanies a large gap at $N, Z = 40$ which increases as increasing the tetrahedral deformation α_{32} while the other octupole deformations $m = 0, 1, 3$ do not have this feature. This is the gap specific to the tetrahedron deformation, and presence of such gap is due to the high symmetry (the T_d symmetry of the point group) [13]. Note that the tetrahedron shell gap at particle number 40 is found also in the metal cluster systems [16] for which there is no ls potential. Figure 3 indicates that the tetrahedron shell gap emerges even under influence of the large ls term in nuclear potential. It is also noted that the tetrahedron minima is obtained with including the pairing. The realistic Hartree-Fock

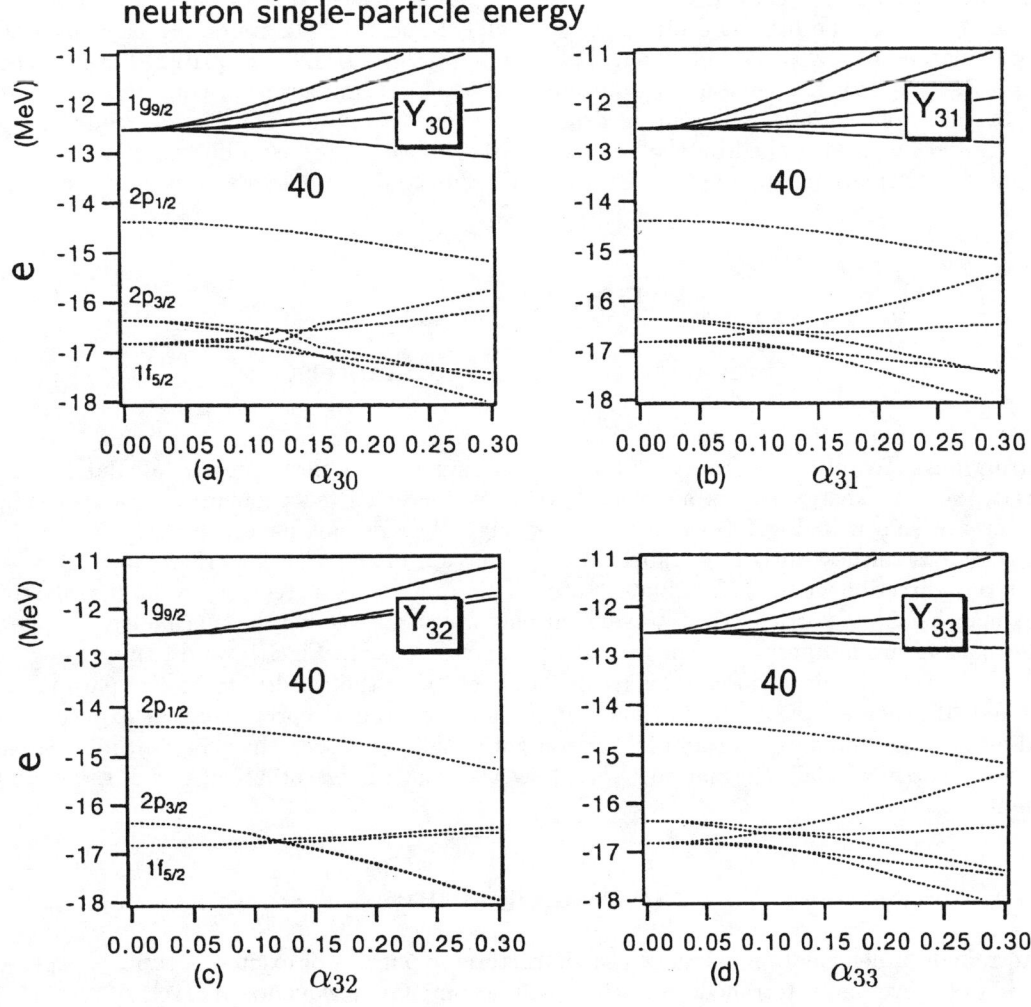

FIGURE 3. Neutron single-particle energies as a function of octupole deformation α_{3m} for $m = 0, 1, 2, 3$ ((a),(b),(c),(d), respectively) calculated for ^{80}Zr with $\beta = 0$. The quadrupole constraint is fixed to $\beta = 0$. The SIII force is used. The proton spectrum is almost the same as neutrons.

calculation thus supports the correspondence between nuclei and clusters suggested in Ref. [16] on the presence of the tetrahedron shell gap.

Triangular Softness in ^{68}Se

Concerning the oblate triangular solution of ^{68}Se, the minima is quite shallow with respect to the α_{33} direction. The energy gain due to the Y_{33} deformation (the energy difference between the minimum and the solution with $\beta_3, \alpha_{33} = 0$) is just 0.15 MeV. This indicates that the oblate state is extremely soft toward the α_{33} deformation rather than a rigid triangular deformation. It is noted that the potential energy curves for $\alpha_{3m} = 0 (m = 0, 1, 2)$ are not as soft as the α_{33} curve.

Systematics

In Fig.2 and Table 1 are seen systematic trends of the non-axial octupole deformations in $N = Z, A \sim 60 - 80$ nuclei. The octupole softness is enhanced at $N, Z \sim 34$, (representative is oblate ground state of ^{68}Se which is soft for triangular deformation), and at $N, Z \sim 40$ (the tetrahedral deformation). It is also seen that the Y_{33} mode is the softest among the four octupole deformations for the oblate states whereas Y_{31} and Y_{30} is favored for the prolate states. This is explained in terms of the single-particle shell structures near the Fermi surface [6]. For the oblate states, the high Ω orbits such as [404]9/2 and [413]7/2 stemming from $g_{9/2}$ are located near the Fermi surface of $N, Z \sim 34$ and strong Y_{33} coupling with [301]3/2 and [310]1/2 orbits emerges. On the other hand the $g_{9/2}$ low Ω orbits are situated far above the Fermi surface. Thus Y_{30} Y_{31} couplings are disfavored. For the prolate states Y_{31} is favored in a reversed reasoning. The Y_{32} deformations favored at $N, Z \sim 40$ is caused by the tetrahedron shell gap discussed above. The instability and the softness for the non-axial octupole deformations are quite contrasting with the axial octupole instability known in other mass regions.

DISCUSSION

Sensitivity to force parameters

Besides SIII, other parameter sets such as SkM* [22] are known to provide a reasonable description of nuclear deformations. To check sensitivity to the force parameters we performed a calculation using SkM*. The calculated potential energy curves associated with an oblate HF solution in ^{68}Se and a spherical HF solution in ^{80}Zr are shown in Fig.4 (a) and (c). Although they do not have a minima at finite α_{3m}, the potential curve shows softness for the tetrahedral α_{32} direction of the ^{80}Zr spherical state ($\beta = 0$) and for the triangular α_{33} direction of the ^{68}Se oblate state. This qualitative agreement with the SIII results can be expected since the single-particle structures which play a central role for the mechanism for the non-axial octupole deformations are not very different for both parameter sets. In Fig.4(b) and (d), we show the results of SIII but with increased pairing force strength in order to show the sensitivity to the pairing correlation. Here we assumed 10% increase of $G_{n,p}$ which is within a reasonable range reproducing the experimental pairing gap (the odd-even mass difference). From comparison with Fig.2, it is seen that the pairing correlation have an effect to reduce the non-axial deformation. Note, however, that the essential features of the potential curve remain the same.

Spectral signatures

If a rigid tetrahedron deformation is realized, a characteristic pattern of excited spectra is expected. Rotational excitation of a even-even tetrahedron nucleus will accompany a sequence of levels $0^+, 3^-, 4^+, 6^+, 7^-, ...$ following the rotational energy relation $E(I) - E(0) = I(I + 1)/2\mathcal{J}$ in parallel with the tetrahedron molecule [26]. Here the spectrum is labeled by the irreducible representation of the T_d group. The rotational levels are connected by strong E3 transitions instead of E2. Viewing the potential energy curve for the tetrahedral solution in ^{80}Zr, we can expect not a rigid deformation but rather a transitional situation between the tetrahedron

350

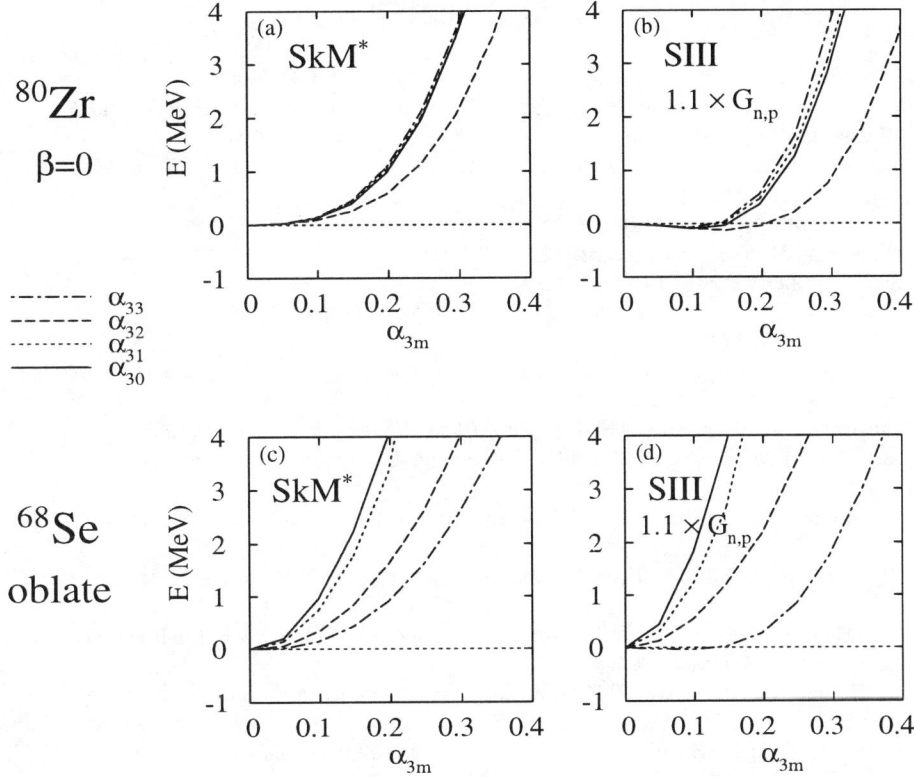

FIGURE 4. Potential energy curve calculated with the SkM* force (a,c), and with SIII and an increased pairing force $G_{n,p} = 18.15/(11 + N, Z)$ MeV (b,c), for the excited spherical state in ^{80}Zr (a,b) and the oblate state in ^{68}Se (c,d).

rotation and the octupole vibration around the spherical state. From the experimental systematics of the first 3^- excitation energy in the neighboring mass region [27], a typical octupole vibrational energy is about a few MeV. Using this estimate and the calculated potential energy curve for the α_{32} direction, amplitude associated with the transitional tetrahedral excitation is estimated to be about $\alpha_{32}(\beta_3) \sim 0.3$. This corresponds to $B(E3) \sim 60$ W.u. on the basis of $B(E3) = (3ZeR^3\beta_3/4\pi)^2$, which is much larger than the systematic values [28] in neighboring mass region.

For the oblate state in ^{68}Se, the potential energy curve suggests soft vibrational excitation for the Y_{33} direction which couples with the rotational excitation. We can expect for such case that the excitation spectra showing a $K^\pi = 0^+$ rotational band ($0^+, 2^+, 4^+, ..$) associated with the ground state, and a $K^\pi = 3^-$ rotational band ($3^-, 4^-, 5^-, ..$) associated with the vibrational excitation. From the calculated potential energy curve, the vibrational amplitude is estimated as $\alpha_{33}(\beta_3) \sim 0.2 - 0.3$, corresponding to $B(E3) \sim 20 - 40$ W.u. for interband transitions.

The nuclei under discussion are close to the proton drip line, and experimental information on the excited spectra is not very rich at present. Candidates of negative parity states are found in ^{64}Ge, which suggests presence of some but not very strong octupole collectivity [24]. Quite recently candidate negative parity states are found in ^{68}Se, and the observed spectra looks similar to that in ^{64}Ge although the experimental information is not enough to see the octupole collectivity [25].

CONCLUSIONS

The fully three dimensional Hatree-Fock calculation suggests that the non-axial octupole deformations are important in $N = Z, A \sim 60 - 80$ nuclei. Prominent examples are the tetrahedral deformation of an shape coexisting excited state in ^{80}Zr, and the triangular softness of the oblate state in ^{68}Se. The tetrahedron deformation causes a shell gap at $N, Z = 40$, which founds the microscopic origin of the exotic shape.

REFERENCES

1. As a review, P. A. Butler and W. Nazarewicz, *Rev. Mod. Phys.* **68**, 349 (1996).
2. As a review, Y. Fujiwara, H. Horiuchi, K. Ikeda, M. Kamimura, K. Katō, Y. Suzuki and E. Uegaki, *Prog. Theor. Phys. Suppl.* **68**, 29 (1980);
   H. Horiuchi and K. Ikeda, in *International Review of Nuclear Physics*, ed. T. T. S. Kuo and E. Osnes (World Scientific, Singapore, 1986), Vol. 4, p. 1.
3. X. Li and J. Dudek, *Phys. Rev.* **C94**, R1250 (1994).
4. X. Li, J. Dudek, and P. Romain, *Phys. Lett.***B271**, 281 (1991).
5. R. R. Chasman, *Phys. Lett.* **B266**, 243 (1991).
6. S. Takami, K. Yabana and M. Matsuo, *Phys. Lett.* **B431**, 242 (1998).
7. J. Skalski, *Phys. Rev.* **C43**, 140 (1991).
8. As a review, J.L. Wood, K. Heyde, W. Nazarewicz, M. Huyse and P. Van Duppen *Phys. Rep.* **215**, 101 (1992).
9. C.J.Lister, M.Campbell, A.A.Chishti, W.Gelletly, L.Goettig, R.Moscrop, B.J.Vary, A.N.James, T.Morrison, H.G.Price, J.Simpson, K.Connel, and O.Skeppstedt, *Phys. Rev. Lett.* **59**, 1270 (1987);
   B.J.Vary, M.Campbell, A.A.Chishti, W.Gelletly, L.Goettig, C.J.Lister, A.N.James, and O.Skeppstedt, *Phys. Lett.* **B194**, 463 (1987);
   H.Dejbakhsh, T.M.Cormier, X.Zhao, A.V.Ramayya, L.Chaturvedi, S.Zhu, J. Kormicki, *Phys. Lett.* **B249**, 195 (1990);
   Kr C.J.Lister, P.J. Ennis, A.A. Chishti, B.J.Varley, W.Gelletly, H.G. Price, A.N. James, *Phys. Rev.* **C42**, R1191 (1990);
   W. Gelletly, M.A. Bentley, H.G. Price, J.Simpson, C.J. Gross, J.L. Durell, B.J. Varley, O.Skeppstedt and S. Rastikerdar, *Phys. Lett.* **B253**, 287 (1991);
   G.de Angelis, C.Fahlander, A.Gadea, E.Farnea, W.Gelletly, A.Aprahamian, D.Bazacco, F.Becker, P.G.Bizzeti, A.Bizzeti-Sona, F.Brandolini, D.de Acuña, M. De Poli, J. Eberth, D. Foltescu, S.M.Lenzi, S.Lunardi, T.Martinez, D.R.Napoli, P.Pavan, C.M.Petrache, C.Rossi Alvarez, D.Rudolph, B.Rubio, W.Satuła, S.Skoda, P.Spolaore, H.G.Thomas, C.A. Ur, and R.Wyss, *Phys. Lett.* **B415**, 217 (1997).
10. W. Nazarewicz, P. Olanders, I. Ragnarsson, J. Dudek, G. A. Leander, P. Mőller and E. Ruchowska, *Nucl. Phys.* **A429**, 269 (1984).
11. G. A. Leander, R. K. Sheline, P. Mőller, P. Olanders, I. Ragnarsson and A. J. Sierk, *Nucl. Phys.* **A388**, 452 (1982).
12. W.A. de Heer, *Rev. Mod. Phys.* **65**, 611 (1993);
    M. Brack, *Rev. Mod. Phys.* **65**, 677 (1993).
13. I. Hamamoto, B. Mottelson, H. Xie and X. Z. Zhang, *Z. Phys.* **D21**, 163 (1991).
14. F. Frisk, I. Hamamoto and F. R. May, *Phys. Scr.* **50**, 628 (1994).
15. S.M. Reimann, M. Koskinen, H. Häkkinen, P.E. Lindelof, and M. Manninen, *Phys. Rev.* **B56**, 12147 (1997).
16. J. Kolehmainen, M. Koskinen, H. Häkkinen, M. Manninen, and S. Reimann, *Czech. J. Phys.* **48**, 679 (1998).
17. P. Bonche, H. Flocard, P.-H. Heenen, S. J. Krieger and M. S. Weiss, *Nucl. Phys.* **A443**, 39 (1985).
18. N. Tajima, S. Takahara and N. Onishi, *Nucl. Phys.* **A603**, 23 (1996).
19. K. T. R. Davies, H. Flocard, S. Krieger and M. S. Weiss, *Nucl. Phys.* **A342**, 111 (1980).
20. S. Takami, K. Yabana and K. Ikeda, *Prog. Theor. Phys.* **96**, 407 (1996);
    S. Takami, K. Yabana and K. Ikeda, *Proc. XVII RCNP Int. Symp. on Innovative Computational Methods in Nuclear Many-Body Problems*, ed. H.Horiuchi, M.Kamimura, H.Toki, Y.Fujiwara, M.Matsuo and Y.Sakuragi (World Scientific, Singapore, 1998) pp.112-116.
21. M. Beiner, H.Flocard, Nguen Van Giai and P. Quentin, *Nucl. Phys.* **A238**, 29 (1975).
22. J. Bartel, P. Quentin, M. Brack, C. Guet and H.-B. Håkansson, *Nucl. Phys.* **A386**, 79 (1982).
23. P.-H.Heenen, J.Skalski, P. Bonche, and H. Flocard, *Phys. Rev.* **C50**, 802 (1994).
24. P. J. Ennis, C. J. Lister, W. Gelletly, H. G. Price, B. J. Varley, P. A. Butler, T. Hoare, S. Cwiok and W. Nazarewicz, *Nucl. Phys.* **A535**, 392 (1991).
25. S. Skoda, B.Fielder, F.Becker, J. Eberth, S.Freund, T.Steinhardt, O.Stuch, O. Thelen, H.G. Thomas, L. Käubler, J. Reif, H.Shnare, R.Schwengner, T.Servene, G.Winter,V.Fischer, A.Jungclaus, D.Kast, K.P.Lieb, C.Teich, C,Ender, T.Hä rtlein, F.Köck, D.Schwarlm and P.Baumann, *Phys. Rev.* **C58**, R5 (1998).
26. G. Herzberg, *Molecular Spectra and Molecular Structure* Vol. II (D. Van Nostrand Company, New York, 1945).
27. P.D. Cottle, *Phys. Rev.* **C42**, 1264 (1990).
28. R.H. Spear and W.N. Catford, *Phys. Rev.* **C41**, R1351 (1990).

Comparative Lifetimes of Superdeformed Bands in A $\sim$ 80 Nuclei

F. Lerma,[1] M. Devlin,[1] D. R. LaFosse,[1,5] D. G. Sarantites,[1] R. M. Clark,[2] I. Y. Lee,[2] A. O. Macchiavelli,[2] R. W. MacLeod,[2] S. L. Tabor,[3] D. Soltysik,[3] C. Baktash[4]

[1] *Chemistry Department, Washington University, St. Louis, MO 63130*
[2] *Nuclear Science Division, Lawrence Berkeley National Laboratory, Berkeley, CA 94720*
[3] *Department of Physics, Florida State University, Tallahassee, FL 32306*
[4] *Physics Division, Oak Ridge National Laboratory, Oak Ridge TN 37830*
[5] *Current address: Department of Physics, SUNY at Stony Brook, Stony Brook, NY 11794*

Abstract. A comparative measurement of the transition quadrupole moments of the yrast superdeformed bands in $^{80-83}$Sr, ^{83}Y, ^{84}Zr has been performed using the Doppler-shift attenuation method. Thus, we have accurately measured the relative deformations of these structures, establishing for the first time, clear trends in the deformations of these bands. The yrast SD bands in the $^{80-83}$Sr isotopes are shown to possess similar transition quadrupole moments, while significantly larger values are obtained for the ^{83}Y and ^{84}Zr cases. These results provide a stringent evaluation of the intruder orbital assignments of these structures, suggesting new assignments in some of these cases.

INTRODUCTION

Since the discovery of a discrete superdeformed (SD) band in ^{83}Sr [1,2], highly improved detection capabilities have allowed the study of a multitude of SD bands in several nuclei of mass $\sim$ 80. Superdeformed bands in these nuclei have been characterized as largely deformed prolate structures ($\beta_2 \sim 0.5$) due to their high spin nature, large dynamical moments of inertia ($\mathcal{J}^{(2)}$), and in various cases, from measured transition quadrupole moments (see for example, [3–7]). However, large experimental uncertainties have limited the accuracy of previous lifetime measurements, hampering clear estimates of the deformations of the SD bands in those nuclei. Consequently, well defined trends in the deformation of SD bands across the $A \sim 80$ nuclei have not been established.

In this work, we present the results from an experiment aimed to measure comparative transition quadrupole moments, Q_t, of the SD bands in $^{80-83}$Sr, ^{83}Y, and ^{84}Zr via the Doppler-shift attenuation method [8]. Recent lifetime measurements [3–7,9] have established that these bands are highly deformed ($\beta_2 \gtrsim 0.4$) prolate rotors. However, substantial variations in the systematic errors in those values hampered a clear understanding of the relative deformations of those bands. As a result, we have attempted to measure the deformations of those structures to establish clear comparisons with present intruder orbital assignments and to obtain trends in the evolution of superdeformation in these nuclei.

EXPERIMENTAL

The experiment was performed at the 88– inch Cyclotron at the Lawrence Berkeley National Laboratory. A single 570 μg/cm^2 ^{58}Ni target foil backed with 3.2 mg/cm^2 of gold was used in two separate reactions. In the first reaction, the target was bombarded with a ^{28}Si beam at 130 MeV, populating high spin states in ^{80,82}Sr, and ^{83}Y, via the $\alpha 2p$, $4p$ and $3p$ fusion-evaporation exit channels, respectively. In the second reaction, a ^{29}Si beam at 130 MeV was used to populate high-spin states in ^{81,83}Sr, and ^{84}Zr via the $\alpha 2p$, $4p$ and $2pn$ exit channels, respectively. The γ rays emitted from the reactions were detected by the GAMMASPHERE array,

CP481, *Nuclear Structure 98*, edited by C. Baktash

which consisted of 100 hyper-pure Ge detectors fitted with BGO Compton suppressors, and the residual charged particles were detected by the MICROBALL charged-particle detector array [10]. Each reaction was run for 2 days, accumulating $\sim 9 \times 10^8$ events per reaction. The hevimet absorber shields were removed during the experiment to obtain total γ-ray energy (H_γ), and γ fold (k_γ) information per event [11], and the event trigger was set to accept events with clean γ fold equal to 4 and higher. The MICROBALL provided good charged particle detection and identification, which were used to sort the data into individual exit channels. Residual contaminants in the particle gates resulting from unidentified protons, alpha particles and neutrons were sharply reduced by placing two-dimensional gates on H_γ and excitation energy, E^* (where, $E^* = const.$ $- \Sigma\ particle\ energies$) [12].

The energies of the charged particles detected with the MICROBALL were used to reconstruct the momenta of the recoiling nuclei on an event-by-event basis. These momenta were used to perform a precise Doppler-shift correction to the γ-ray energies which was varied as a function of E_γ to compensate for the slowing of the nuclei in the target. Events from individual channels were then sorted into seven $E_\gamma - E_\gamma$ matrices. Each matrix contained coincidences of detectors in a narrow angular segment (relative to the beam axis) and any other detector. The individual angular segments consist of groups of rings of detectors in the GAMMASPHERE array at average angles $\overline{\theta} = 29.9°$, $52.9°$, $74.3°$, $90.0°$, $105.1°$, $127.1°$, $150.1°$. The matrices were gated on the energies of in-band transitions of the SD bands, and one-dimensional spectra were projected onto the axis containing counts from the angular segment, $\overline{\theta}$. The centroids of the SD band transitions (E_γ) were measured in the resultant spectra. Subsequently, a residual Doppler shift (β_{res}) was fitted to the seven (E_γ, $\overline{\theta}$) pairs (for each transition in the SD bands), using the expression $E_\gamma = E_\gamma^0 \sqrt{1 - \beta_{res}^2}/(1 - \beta_{res}\cos(\overline{\theta}))$, by a least-squares procedure. The residual and applied Doppler shifts of individual transitions add relativistically to obtain an average recoil velocity β, which is expressed as a fraction of the initial recoil velocity, $\beta\ /\ \beta_0 \equiv F(\tau)$. A one dimensional spectrum resulting from the sums of single gates on the SD band in ^{84}Zr is shown in Fig. 1(a). Inserts 1(b) and 1(c) contain spectra gated on two separate angular segments of detectors.

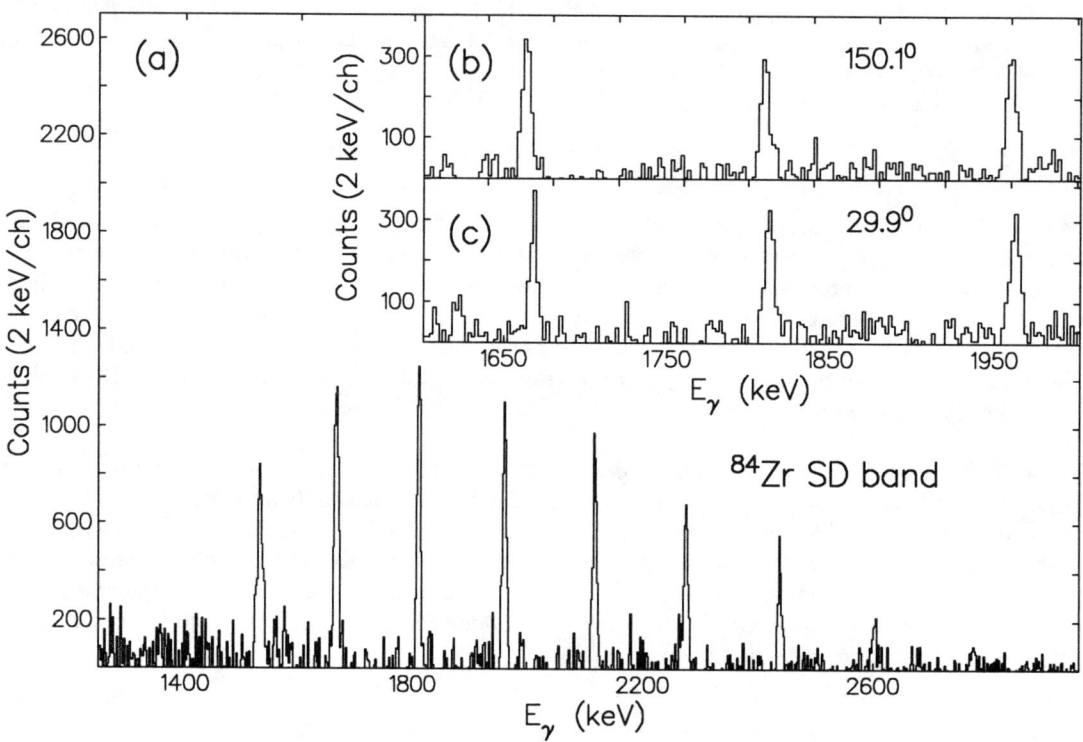

FIGURE 1. A spectrum resulting from the sums of clean single gates on in-band transitions of the SD band in ^{84}Zr is shown in panel (a). Spectra single gated on the ^{84}Zr SD band on all coincidences with angular segments of 15 detectors at 150.1° and 29.9° (relative to the beam axis) are shown in panels (b) and (c) respectively.

The average lifetimes of the SD bands were measured by fitting calculated fractional Doppler shift curves to the measured $F(\tau)$ curves for each SD band, using the Doppler-shift attenuation method described in Ref.

[5,8]. In this method, the decay times of the states in the SD bands are related to the velocity of the recoiling nuclei as they slow down in the target. Side feeding was modeled based on measured intensity patterns, and a constant Q_t was fitted to all levels in-band and those in the side-feeding cascades. Thus, side-feeding times were assumed to be equal to those of preceding in-band transitions.

Absolute values obtained from this method contain systematic errors which result from the electronic and nuclear stopping power estimates that are used to calculate the slowing of the recoiling nuclei in the target foil, and from the measured thickness of the target foil and backing, which are known approximately. For the time scales and therefore the velocities involved in the decay of the SD states the electronic stopping dominates. This reduces large systematic errors due to nuclear stopping when the recoiling species slow down in the same media of the target and the Au backing. Thus, accurate and meaningful comparisons of the extracted values could be made.

RESULTS AND DISCUSSION

The $F(\tau)$ curves extracted from the backed target experiment displayed a significant improvement in sensitivity to the lifetimes of the SD bands compared to those values extracted from thin target experiments. An example of this is shown in Fig. 2, where $F(\tau)$ curves are compared for the yrast SD band in [81]Sr. The fitted curve to the extracted $F(\tau)$ values of the SD band in [81]Sr represents the best fit performed in this analysis, in contrast, the fit to the SD band in [84]Zr was the least favorable fit obtained. It is likely that the poorer fit results from the reduced sensitivity to the shorter decay times of the states in the SD band in [84]Zr.

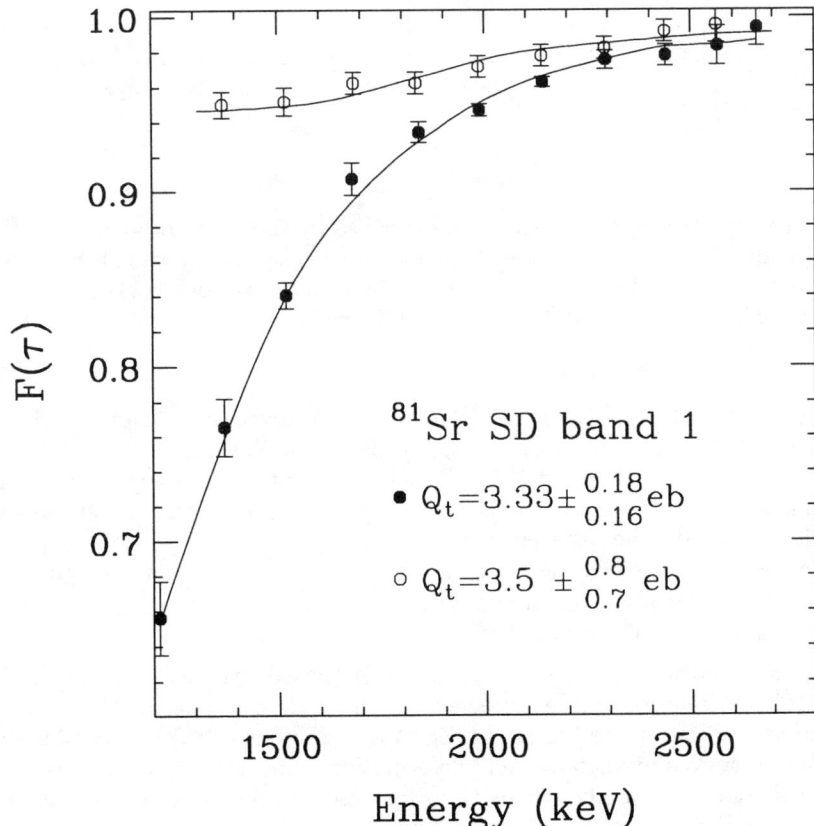

FIGURE 2. Fractional Doppler-shifts $F(\tau)$ of the yrast SD band in [81]Sr obtained in the backed target experiment (solid circles) are compared to those from an earlier work [5] using a thin target. The fractional shift of the last transition in the band from the backed target experiment is reduced to 65%, versus 95% obtained from the thin target experiment.

The Q_t of the SD bands in the [80−83]Sr isotope chain are shown in Fig. 3(a) and those of the SD bands in ^{82}Sr, ^{83}Y, and ^{84}Zr in panel (b). Previously measured values, and values deduced from previous intruder orbital assignments are shown for comparison.

Intruder orbital assignments indicate a rise in deformation for the SD bands in the $^{80-83}$Sr isotopes upon the addition of neutrons to the ^{80}Sr core, and previous experimental evidence displays a similar pattern in the $^{80-82}$Sr cases, suggesting partial confirmation of that trend. In the case of the SD bands in the ^{82}Sr, ^{83}Y and ^{84}Zr isotone chain, large values are predicted in all cases from intruder orbital calculations, and those values are satisfactorily reproduced by previous experimental values (within the large experimental uncertainties), possibly suggesting the same intruder orbital configurations for these structures.

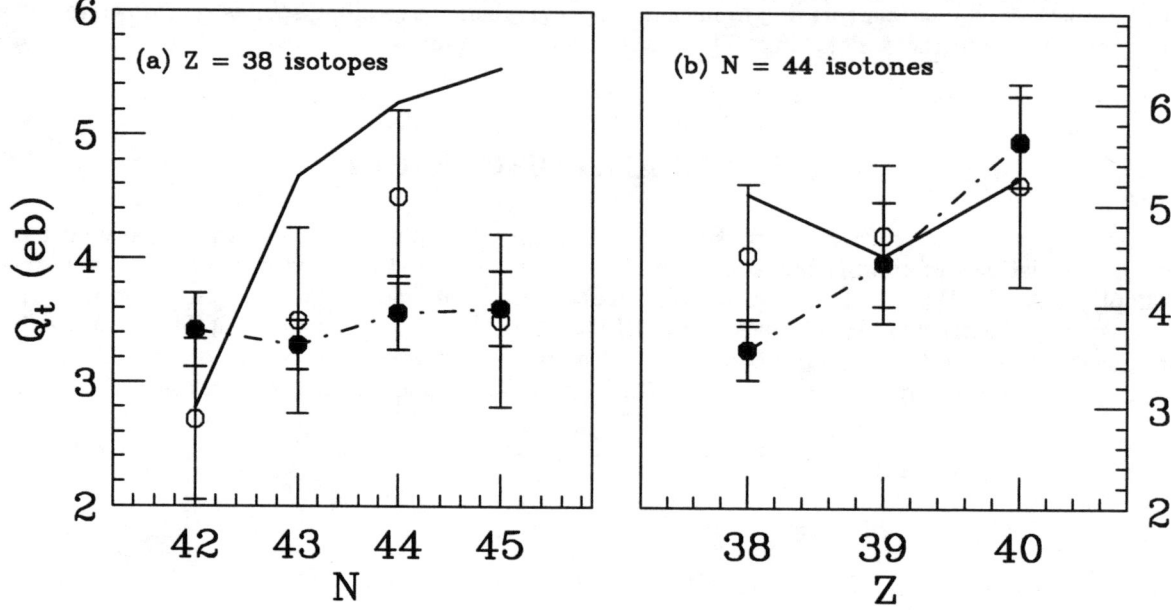

FIGURE 3. The average transition quadrupole moments obtained for the yrast SD bands in $^{80-83}$Sr are shown in panel (a), and those values obtained for the ^{82}Sr, ^{83}Y, and ^{84}Zr cases are displayed in panel (b). Solid circles represent results obtained in this work. Open circles which represent results obtained in previous works and a solid line respresenting values estimated from intruder orbital calculations are shown for comparison.

However, the present results allow accurate comparisons which reveal significantly different trends. A nearly constant transition quadrupole moment is observed for the yrast SD bands in $^{80-83}$Sr, and a rise in Q_t is observed for the SD bands in ^{82}Sr, ^{83}Y, and ^{84}Zr. These new results suggest similar intruder-orbital configurations for the SD bands in the $^{80-83}$Sr isotope chain, and the significant rise in deformation of the SD bands in the $N = 44$ isotones possibly indicates that the additional protons in the ^{83}Y and ^{84}Zr structures occupy deformation-driving $h_{11/2}$ orbitals which the SD band in ^{82}Sr does not. This is in contrast to a previous work [6], in which the SD bands in ^{82}Sr and ^{84}Zr were reported to have similar deformations, and consequently were interpreted to possess the same configuration.

Theoretical predictions from intruder-orbital assignments (based primarily on the $\mathcal{J}^{(2)}$ character of these bands) predict large deformations for all SD bands except the yrast SD band in ^{80}Sr, showing good agreement with the experimental Q_t values for the ^{83}Y and ^{84}Zr cases. However, the yrast SD bands in $^{81-83}$Sr possess moderate Q_t values, contrary to a predicted rise in deformation from intruder orbital assignments, and a larger deformation is observed than that predicted for the ^{80}Sr case. Earlier theoretical calculations indicate that various bands may be populated at the spins considered, and that there are not unique solutions that can be selected based on the extracted $\mathcal{J}^{(2)}$ values only. Therefore, the transition quadrupole moment is a more robust quantity which can lead to a better interpretation of the intruder orbitals involved. In Ref. [13], a class of solutions are presented for the ^{80}Sr case, covering a wide range of deformations and spins. Similarly, various solutions are mentioned for the ^{81,82,83}Sr cases in previous works. However, those solutions are not presented in sufficient detail to make new assignments to accomodate the newly measured Q_t values of the SD bands in $^{80-83}$Sr.

SUMMARY

We have presented an experiment in which the relative transition quadrupole moments of the SD bands in $^{80-83}$Sr, $^{80-83}$Y, and $^{80-83}$Zr have been accurately measured. The use of a backed target has demonstrated a significant improvement in the sensitivity to the lifetimes of these structures, and the use of the same experimental conditions (primarily the use of the same target and backing) have allowed accurate comparison of these results. Previous trends in the deformation of the SD bands in these nuclei have been discussed, and a new well-defined trend has been presented. Large discrepancies have been shown between the present results and those derived from previously assigned intruder orbital configurations for the SD bands in $^{80-83}$Sr suggesting new theoretical calculations are necessary to interpret the configuration of these structures.

AKNOWLEDGEMENTS

The work at Washington University is supported in part by the U.S. Department of Energy (DOE), Division of Nuclear Physics under grant No. DE-FG02-88ER-40406. Oak Ridge National Laboratory is managed by Lockheed Martin Energy Research Corp. for the US DOE under contract DE-AC05-96OR22464. The work at Lawrence Berkeley National Laboratory is supported in part by the U.S. DOE, Division of Nuclear Physics under grant No. DE-AC03-76SF00098. The U.S. National Science Foundation provides support for the work at Florida State University under grant No. PHY-9210082. The authors are grateful to A. Lipski for making the target for this experiment.

REFERENCES

1. C. Baktash *et al.*, Phys. Rev. Lett. **74**, 1946 (1995).
2. D. R. LaFosse *et al.*, Phys. Lett. B **354**, 34 (1995).
3. H.-Q. Jin *et al.*, Phys. Rev. Lett. **75**, 1471 (1995).
4. D. G. Sarantites *et al.*, Phys. Rev. C **57**, 1 (1998).
5. M. Devlin *et al.*, Phys. Lett. B **415**, 328 (1997).
6. C.-H. Yu *et al.*, Phys. Rev. C **57**, 113 (1998).
7. D. Rudolph *et al.*, Phys. Lett. B **389**, 463 (1996).
8. B. Cederwall *et al.*, Nucl. Instr. and Meth. A **354**, 591 (1995).
9. H.-Q. Jin *et al.*, to be published, private communication.
10. D. G. Sarantites *et al.*, Nucl. Instr. and Meth. A **381**, 418 (1996).
11. M. Devlin *et al.*, Nucl. Instr. and Meth. A **383**, 506 (1996).
12. C. E. Svensson *et al.*, Nucl. Instr. and Meth. A **396**,228 (1996).
13. W. Nazarewicz *et al.*, Nucl. Phys. A **435**, 397 (1985).

Smooth Band Termination in the Mass A=110 Region

D.R. LaFosse,* A. Boston,† C.J. Chiara,* R.M. Clark,‖ M. Devlin,‡ D.B. Fossan,*
G.J. Lane,‖ I.-Y. Lee,‖ F.A. Lerma,‡ A.O. Macchiavelli,‖ E.S. Paul,† D.G. Sarantites,‡
J.M. Sears,* A.T. Semple,† J.F. Smith,¶ K. Starosta,* R. Wadsworth,§ A.V. Afanasjev,** and
I. Ragnarsson**

* *Department of Physics and Astronomy, State University of New York at Stony Brook, Stony Brook, NY 11794*
† *Oliver Lodge Laboratory, University of Liverpool, Liverpool L69 3BX, United Kingdom*
‡ *Department of Chemistry, Washington University, St. Louis, MO 63130*
‖ *Nuclear Science Division, Lawrence Berkeley National Laboratory, Berkeley, CA 94720*
¶ *Department of Physics, University of Manchester, Manchester M13 9PL, United Kingdom*
§ *Department of Physics, University of York, Heslington, York, YO1 5DD, United Kingdom*
** *Department of Mathematical Physics, Lund Institute of Technology, Box 118, S-22100 Lund, Sweden*

Abstract.

The systematics of smoothly terminating rotational bands based on proton $2p-2h$ excitations in the $A = 110$ mass region are presented. Terminating bands (or nearly so) based on this proton excitation have been found in nuclei ranging from ^{107}In, up to ^{114}Te, and possibly extending to $_{54}$Xe nuclei. The impressive agreement between experimental data and theoretical calculations is also presented. However, recently discovered structures based on proton $1p-1h$ excitations begin to show disagreement with theoretical calculations. These new bands are also discussed. The current and future directions of research into smooth band termination will be presented.

INTRODUCTION

As far back as 1975, Bohr and Mottelson described a process in which the deformation of a rotational band decreases as the valence nucleons align their angular momenta with the rotation axis [1]. Once all the available valence nucleons have aligned, the rotational band terminates leaving a nucleus with a noncollective oblate shape. For some time the textbook case was ^{20}Ne, in which the ground state sequence terminates in a fully aligned state at a relatively low spin of $I = 8\hbar$.

In order to observe this phenomenon in heavier nuclei, where a single deformed configuration may be followed over a large range of spin, one should look in nuclei near doubly-closed shells. This is due to several factors. First of all, such nuclei have small numbers of valence nucleons, resulting in terminating spins attainable in heavy-ion fusion-evaporation reactions. A second consideration is the number of available deformed configurations near the yrast line. A nucleus having a large number of valence nucleons will also have a large number of deformed configurations. This makes it unlikely that a single configuration can be observed over a large range of spin; rather, one configuration will be crossed by another and the non-yrast terminating states will not be observed. Near closed shells, however, particle-hole excitations across the shell gap are typically necessary to provide deformation. This tends to limit the number of near-yrast deformed configurations. Finally, nuclei having too few valence nucleons are not likely to possess stable deformed configurations. Calculations suggest that in the $A = 110$ mass region nuclei having 10-15 valence nucleons outside of the ^{100}Sn doubly-closed shell are ideal candidates.

The classic doubly-magic nucleus is ^{100}Sn, and third-generation γ-ray detector arrays have made it possible to study nuclei near $Z = N = 50$ to high spin. As a result, a number of spectacular examples of smooth band termination have been found in the $A = 110$ mass region. The first bands described as smoothly terminating were found in ^{109}Sb [2] using data from the 8π array at Chalk River Laboratories; these results were later confirmed and improved [3] employing data from the Early Implementation of Gammasphere. Since then,

CP481, *Nuclear Structure 98*, edited by C. Baktash

many examples of smooth band termination have been documented in the mass $A = 110$ region, and we are now able to study systematic properties of such structures. In what follows we will present some of these systematics, and compare the results to theoretical calculations obtained with the Configuration Dependent Shell Correction method [4]. Some new results will also be presented, and some ideas for directions of future research as well.

SMOOTH BAND TERMINATION: GENERAL CONSIDERATIONS

The experimental manifestation of smooth band termination can be seen in Fig. 1. In this figure the dynamic moments of inertia $(\mathcal{J}^{(2)})$ for three rotational bands in ^{109}Sb are shown, as well as the energies of the levels versus spin, where the energies are plotted relative to a rotating liquid drop reference (henceforth this will be referred to as an $E - E_{LD}$ curve). Smooth band termination is evident as a decreasing $\mathcal{J}^{(2)}$ at the highest rotational frequencies seen in the top part of the figure, or as a characteristic parabolic shape of the $E - E_{LD}$ curve in the bottom half. Both demonstrate a loss of collectivity as the rotational bands reach the highest observed spins.

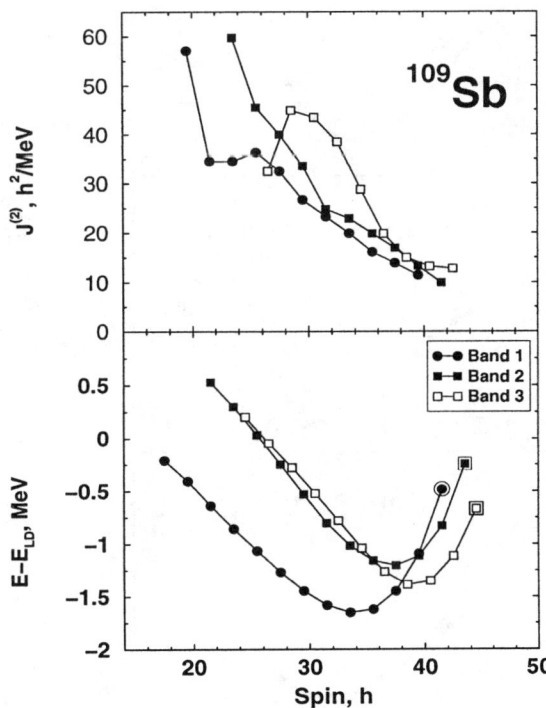

FIGURE 1. Top: The dynamic moment of inertia for three of the known terminating bands in ^{109}Sb. Bottom: Level energies minus a rigid rotor reference for the same bands in ^{109}Sb. In the bottom panel, the terminating states of the rotational bands are depicted by open symbols.

The theoretical interpretation of this phenomenon is provided by the Configuration Dependent Shell Correction (CDSC) method [4]. Calculations based on this method show that the band initially undergoes a collective (macroscopic) rotation, but microscopic considerations become important as the the valence nucleons gradually align their spins with the rotation axis. The shape of the nucleus changes from prolate to triaxial (and less deformed) as a result. Increasingly more energy is required to gain an additional $2\hbar$ alignment, resulting in the decrease in the $\mathcal{J}^{(2)}$, and the characteristic minimum in the $E - E_{LD}$ curve. Eventually all available valence particles outside the ^{100}Sn core have aligned their spins. At this point the nucleus attains a noncollective oblate shape, and the rotational band must end, further rotation and higher spin being impossible without a change in configuration.

In general, the calculations do a remarkable job of reproducing the experimental data. An example is shown in Fig. 2. Here the $E - E_{LD}$ curves are shown for 3 different structures, the yrast bands in ^{108}Sn, ^{109}Sb, and ^{110}Sb. The configurations shown are the [21,2] configurations in ^{108}Sn, ^{109}Sb, and the [21,3] configuration for ^{110}Sb which have been assigned to these bands. (The notation $[p_1 p_2, n]$ can be understood as follows: p_1 refers to the number of $g_{9/2}$ holes in the configuration, and p_2, n refer to the number of occupied $h_{11/2}$ protons and neutrons, respectively. If one allows that the $d_{5/2}$ and $g_{7/2}$ orbitals are indistinguishable in these calculations, this notation is sufficient to completely describe the configuration of the system.) Note that these calculations do not predict the absolute energies of states, and pairing is not included in the calculations. It is therefore the shape of the curves above spin $I \sim 20\hbar$ which is important. In all cases, the calculations accurately reproduce the shapes and positions of the minima in the $E - E_{LD}$ curves. The calculations also reproduce the terminating spins, shown as open circles in the figure.

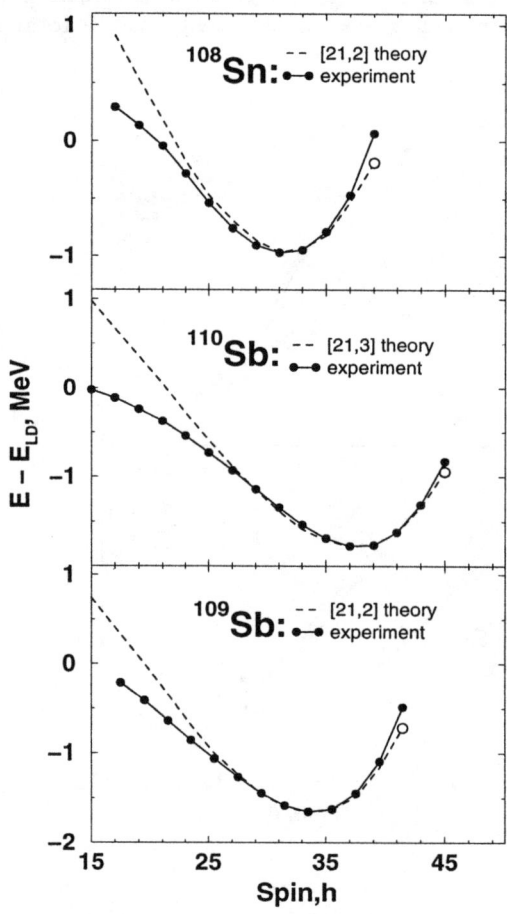

FIGURE 2. $E - E_{LD}$ curves for the yrast rotational bands in ^{108}Sn, ^{109}Sb, and ^{110}Sb.

It is important to recognize that the decision that a band has reached termination can usually be made without recourse to the CDSC calculations. This is due to the fact that nuclei near the ^{100}Sn doubly-closed shell have a limited number of valence nucleons. The configuration of a band can be determined by its properties at low spin, for example by the presence or absence of pair alignments, the spin sequence, and parity of the band. However, once the configuration of a band is known it is a simple application of the Pauli Principle and shell model to determine the terminating spin. For example, the $\mathcal{J}^{(2)}$ for Band 1 in ^{109}Sb is shown in Fig. 1. The high value of the $\mathcal{J}^{(2)}$ at the lowest frequencies is suggestive of an $h_{11/2}$ neutron crossing; the peak at $I \approx 26\hbar$ can be ascribed to a $g_{7/2}$ proton crossing. Other structure information comes from the need for a 2p–2h proton excitation to supply a modest deformation of $\beta \approx 0.25$. Finally, the band is known from the

DCO ratios of the linking transitions to have negative parity and spins from $35/2\hbar$ to $83/2\hbar$. This information suggests the following complete configuration for the band: $\pi[(g_{9/2})^{-2}(g_{7/2},d_{5/2})^2h_{11/2}]\,\nu[(g_{7/2},d_{5/2})^6(h_{11/2})^2]$, or in our notation, [21,2]. Then, the terminating spin is determined through the Pauli Principle to be $83/2\hbar$: $\pi[(g_{9/2})_8^{-2}(g_{7/2},d_{5/2})_6^2(h_{11/2})_{5.5}]_{19.5}\,\nu[(g_{7/2},d_{5/2})_{12}^6(h_{11/2})_{10}^2]_{22}$. It can be seen from the bottom panel of Figs. 1 and 2 that this band extends to $83/2\hbar$ and has therefore reached the terminating state for this configuration.

SYSTEMATICS OF 2p-2h BANDS

A series of experiments has uncovered numerous rotational bands which show characteristics of termination. The majority of these structures are based on $2p-2h$ proton excitations across the $Z = 50$ shell gap. Terminating bands with such $[2p_1,n]$ configurations have been identified in nuclei as light as $Z = 49$ [107]In [5], up to $Z = 53$ [113]I [6] and $Z = 52$ [114]Te [7] and possibly beyond. In this section some systematics are presented, as well as a discussion of lifetime measurements which convincingly demonstrate the validity of the termination interpretation.

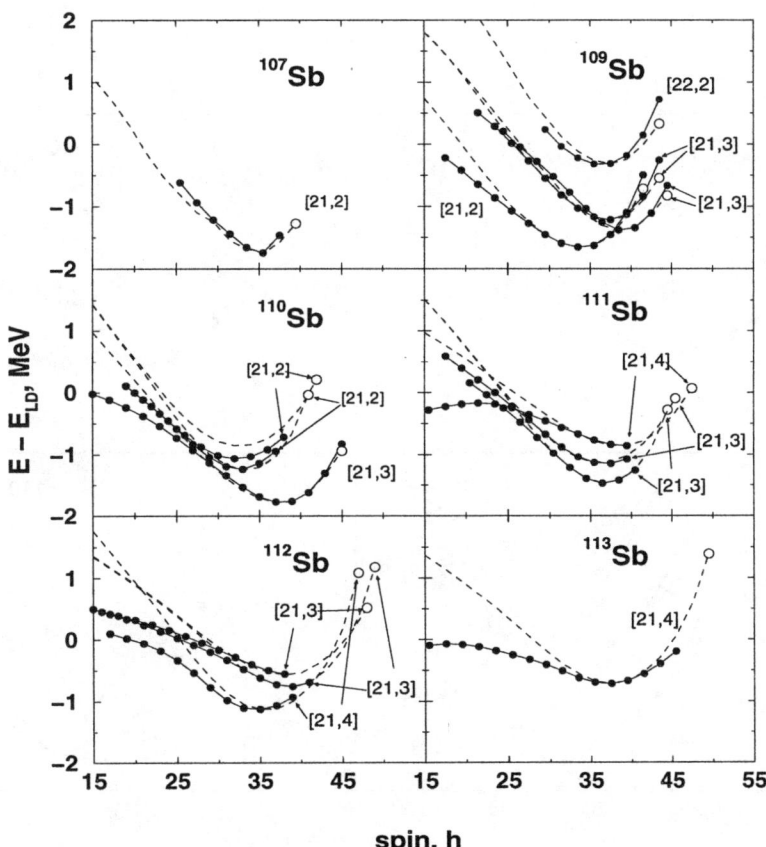

FIGURE 3. $E - E_{LD}$ curves extracted from several terminating bands in the $Z = 51$ Sb isotope chain. Solid lines connecting data points indicate the experimental measurements, and dashed lines indicate the theoretical calculations. The predicted terminating states are depicted as open circles at the ends of the dashed lines.

A   The Z=51 isotopes

Some of the best examples of smooth band termination can be found in the $_{51}$Sb nuclei. Rotational bands based on $3p2h$ proton configurations are known from ^{107}Sb up to ^{119}Sb. (Only in ^{108}Sb have such structures not yet been identified.) In the lighter isotopes, several bands are observed to termination, or observed only 1 or 2 transitions shy of the expected terminating spin. In Fig. 3, several of the $E - E_{LD}$ curves for these bands are shown for nuclei from ^{107}Sb to ^{113}Sb. In addition, CDSC calculations for the relevant configurations are shown.

In all cases, the theoretical curves match the experimental measurements extremely well. The shapes and positions of the minima are well reproduced, and one can see that the bands in ^{109}Sb and a single band in ^{110}Sb are observed to their predicted terminating states. It is worth pointing out that a wide range of configurations are covered by these structures. Some, such as the band shown from ^{107}Sb possess only a single $h_{11/2}$ proton in their configurations ([21,2]). Others possess two $h_{11/2}$ protons, for example the [22,2] band in ^{109}Sb. A wide range of neutron configurations where two to four $h_{11/2}$ neutrons are occupied are also well reproduced by the calculations.

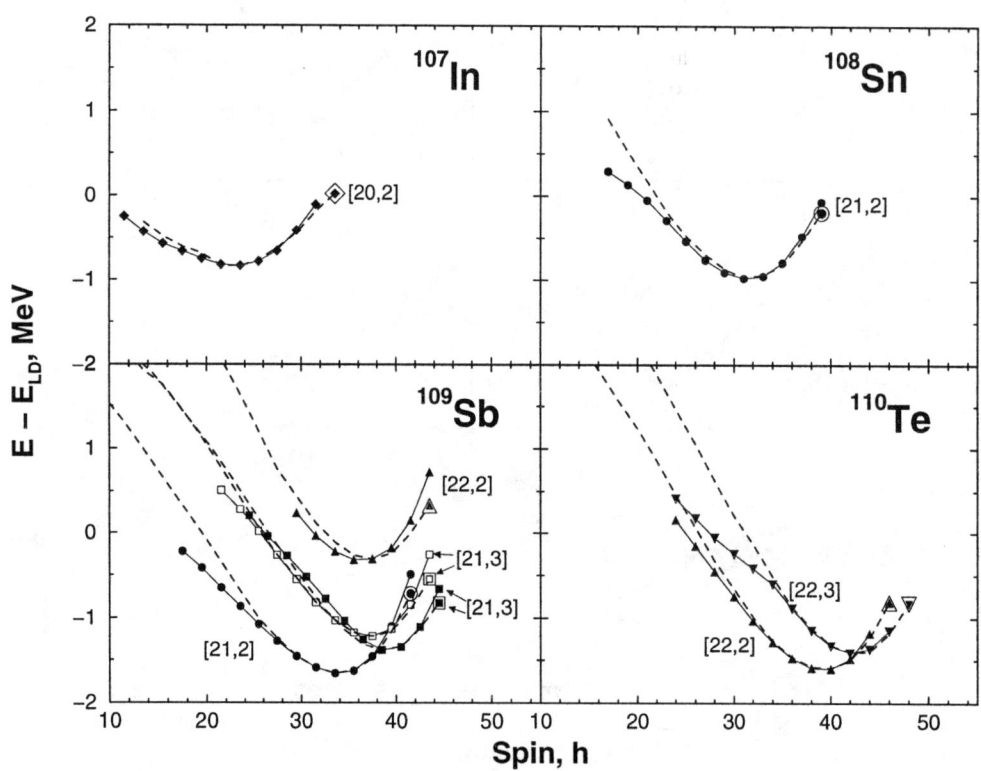

FIGURE 4. $E - E_{LD}$ curves extracted from several terminating bands in the $N = 58$ isotone chain. Solid lines connecting data points indicate the experimental measurements, and dashed lines indicate the theoretical calculations. The predicted terminating states are depicted as open circles at the ends of the dashed lines.

B The N=58 isotones

The systematics of terminating structures as a function of Z can also be studied. For this we choose the $N = 58$ isotones, where a number of nuclei have been found to possess terminating bands. The lightest of these is [107]In, the heaviest [110]Te. Both theoretical calculations and experimental data can be seen in Fig. 4.

As was the case for the $Z = 51$ isotopes, the calculations are very successful in reproducing the experimental results. In this case, a wide variety of proton configurations are accurately explained by the theory. In [107]In, there are no $h_{11/2}$ proton orbitals occupied, implying a fairly modest deformation. In [109]Sb, there are configurations involving one or two occupied $h_{11/2}$ proton orbitals. The results presented in this and the previous section demonstrate that the calculations are exceedingly successful in explaining the experimental results. This is true even though a large range of both proton and neutron configurations is represented. It is worth noting, however, that all the above results involve bands having a $2p-2h$ proton excitation. This will be discussed further in a subsequent section of this report.

C Lifetime measurements in [108]Sn and [109]Sb

The agreement between theory and experiment presented in the previous sections is very impressive. But a definitive demonstration of band termination is provided by lifetime measurements performed on the nuclei [108]Sn and [109]Sb. If the band termination picture is in fact correct, the quadrupole moments of the bands should decrease steadily with increasing spin. The results of the lifetime measurements for the [21,2] band in [109]Sb are shown in Fig. 5, taken from Ref. [8]. The transition quadrupole moments do in fact decrease with increasing spin. In addition, the theory nearly perfectly agrees with the experimental measurements. Measurements performed for two bands in [108]Sn show similar agreement with theory [8].

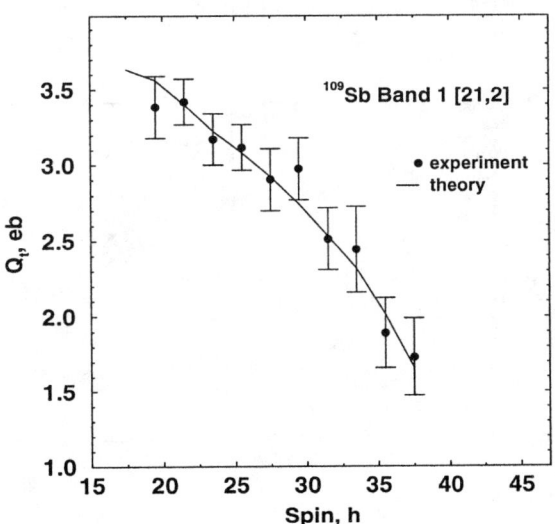

FIGURE 5. The measured and calculated transition quadrupole moments versus spin for the [21,2] band in [109]Sb.

NEW STRUCTURES BASED ON [11,n] CONFIGURATIONS

The terminating structures discussed so far all have a $2p-2h$ excitation across the $Z = 50$ shell gap in common. Since these nuclei are heavily influenced by the $Z = 50$ major shell gap, this excitation is necessary to supply deformation. This is accomplished by the β-driving influence of the $\pi g_{9/2}$ hole orbitals. However, it would be a good test of the theoretical calculations to compare them to rotational structures based on a different number of $g_{9/2}$ proton holes.

For some time it has been known that $1p-1h$ excitations can also provide deformation in the $_{51}$Sb nuclei [9]. For example, the odd-mass $^{107-119}_{51}$Sb nuclei are all known to possess strongly coupled rotational bands based on a proton $(g_{9/2})^{-1}(g_{7/2}, d_{5/2})^2$ high-K configuration. In our notation, these bands would be designated as [10,n], having a single $g_{9/2}$ proton hole, zero $h_{11/2}$ protons, and some number n of $h_{11/2}$ neutrons. These bands are only modestly deformed however, and are typically not observed beyond $I = 20\hbar$, and hence not observed to termination. Also, since the CDSC calculations do not include pairing, they are not expected to be able to adequately describe the bands at these low spins.

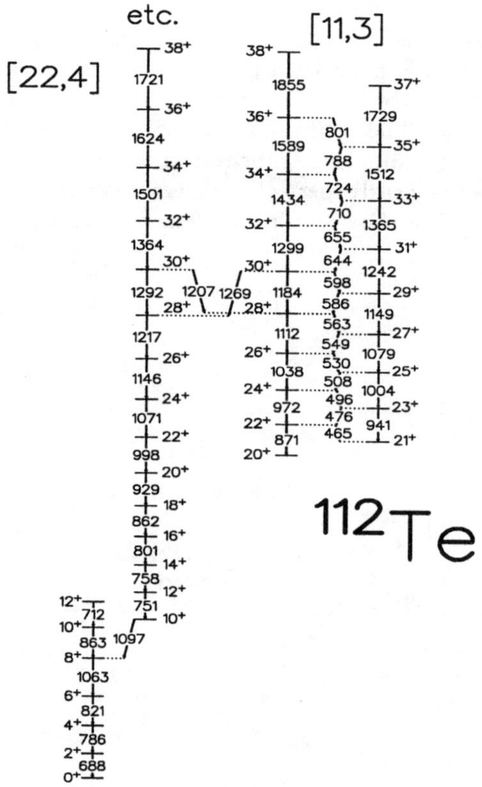

FIGURE 6. A partial level scheme of ^{112}Te, showing the [22,4] band and the [11,3] signature partner bands. The [22,4] band extends to spins higher than shown in the figure.

Recently however, new structures have been observed in ^{110,112}Te. These nuclei were produced in a Gammasphere/Microball experiment, using the ^{58}Ni(^{58}Ni,α2p;4p) reactions, respectively. In ^{112}Te, a pair of signature partner bands with strong $M1$ connecting transitions was found. This structure has been linked to low-spin states, and thus the spin sequences of these bands are known. Figure 6 shows the pair of bands, and the rather unconventional manner in which they have been linked to the low-spin states. As can be seen in the figure the two bands show a very small signature splitting, and strong $M1$ transitions linking the two signatures. The observed properties of these bands are consistent with the presence of a single high-K proton $g_{9/2}$ hole in their configuration, and in fact the CDSC calculations predict a low-lying [11,3] configuration.

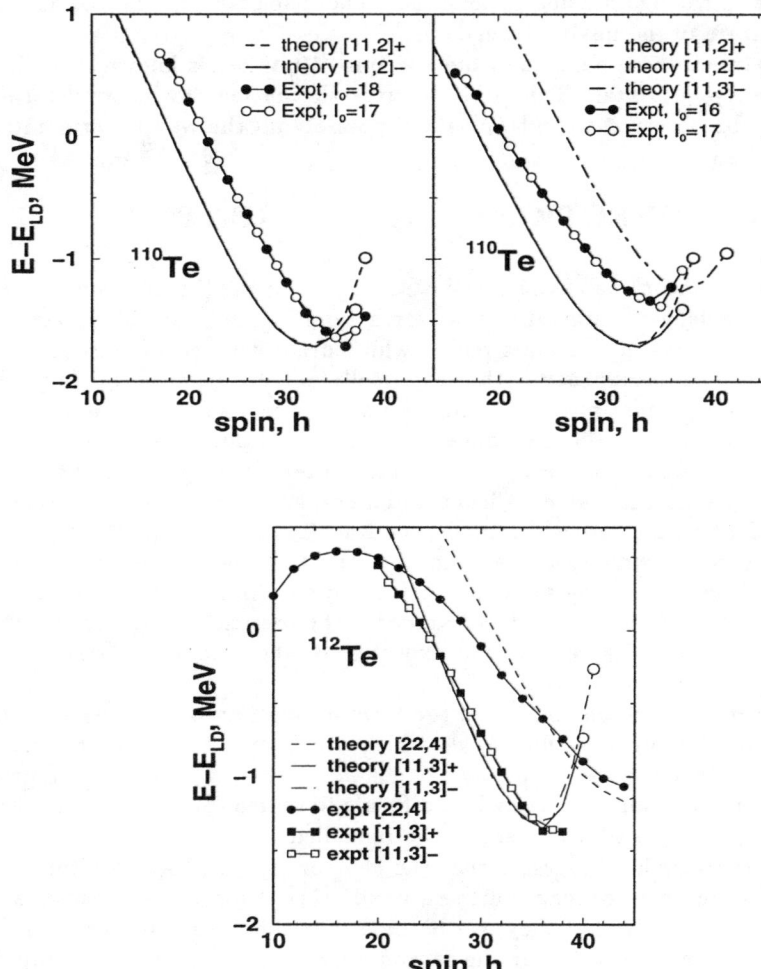

FIGURE 7. Plots of the energy minus a rigid rotor reference for the linked coupled bands in ^{110,112}Te. Both possible spin assignments are shown for ^{110}Te, as well as the possible configurations for each assignment.

In ^{110}Te, two such pairs of signature partner bands were identified. One pair has been linked to the low-spin states, but the spins are uncertain by $1\hbar$ since the DCO ratios of the two linking transitions could not be reliably measured. Therefore the parity of the bands is uncertain as well. Interestingly, the CDSC calculations predict a single low-lying [11,2] and two [11,3] configurations in ^{110}Te, all of which show a small signature splitting due to the single hole in the proton $g_{9/2}$ orbital.

The $E - E_{LD}$ plots for the linked coupled bands are shown in Fig. 7. For ^{110}Te, both possible spin sequences are shown together with possible configuration assignments. The top left panel shows the band assuming the two linking transitions have $E2$ character. This fixes the parity of the band, and only the [11,2] configuration is possible. The top right panel of the figure shows the band assuming the linking transitions are dipoles. In this case positive or negative parity is possible, depending on whether the linking transitions are $M1$ or $E1$. The [11,2] and one of the [11,3] configurations are shown in the figure. Finally, the bottom panel shows the pair of coupled bands in ^{112}Te along with calculations for the [11,3] configuration, which is the only plausible assignment. It should be pointed out that there is an additional [11,3] configuration in ^{110}Te which is not shown in the figure. Also, the analysis and interpretation of ^{110}Te is not finished at this date and should be considered preliminary.

In Fig. 7 one can see that the agreement between theory and experiment is not as good as it is for the bands having two holes in the proton $g_{9/2}$ orbital. In ^{112}Te the calculations predict the minimum in the $E - E_{LD}$ plot to occur at least $2\hbar$ lower than is experimentally observed. The situation in ^{110}Te is similar, the disparity between theory and experiment again being at least $2\hbar$. The disparity may be considerably higher however, depending on the true nature of the linking transitions.

The main difference between these bands and the low spin [10,n] bands known in $_{51}$Sb and $_{53}$I nuclei is the occupation of the proton $h_{11/2}$ orbital. This orbital drives the nucleus to larger deformations, allowing the [11,n] rotational bands to be observed to high spins and possibly all the way to termination.

FUTURE DIRECTIONS AND CURRENT ACTIVITIES

The disagreement between theory and experiment with regard to the [11,n] bands known in ^{110,112}Te points to one area where our knowledge is limited. It is worth pointing out that these are the only terminating (or nearly terminating) structures in this mass region which are not based on a $2p-2h$ excitation. Thus it is worthwhile to study further this discrepancy both theoretically and experimentally. In general, studies of bands having configurations [p_1p_2,n], where $p_1 \neq 2$ should be undertaken. At present a search for predicted [31,n] bands in ^{109}Sb is currently underway [10] using data from a recent Gammasphere/Microball/FMA experiment.

Preliminary results in ^{117,119}Xe [11] suggest the possibility of terminating bands having [02,n] configurations. However, the current evidence does not allow a firm assignment, since there are competing [22,n] configurations which must be considered as well. Planned lifetime experiments will be necessary to distinguish between the [22,n] and [02,n] configurations. Finally, searches continue for additional [$2p_1,n$] bands, mainly at the extremes of the mass region. Recent data are being searched for the presence of such bands in ^{116,118}Xe, where particle-hole excitations across the $Z = 50$ shell are not necessary for the presence of significant deformations. Searches closer to ^{100}Sn continue as well, with recent data expected to show deformed structures in ^{108}Sb and $_{49}$In isotopes, among others.

Another area of future research is the linking of the terminating bands to states of known spin. The three bands shown in Fig. 2 are actually the only bands which are known to terminate *and* which are linked to the lower-spin states. (The linked signature partner bands in ^{110}Te may represent a fourth, depending on the multipolarity of the linking transitions.) Clearly it is difficult to search for discrepancies with theory when spin assignments are uncertain, and this situation should be remedied.

Finally, recent theoretical work has suggested that the neutron $d_{3/2}$ and $s_{1/2}$ orbitals may also be important in determining the terminating spin of many of these bands [12]. However, the energies of these orbitals are not known very precisely. It is likely that they will become more important in nuclei having a relatively large number of neutrons (^{113}Sb, for example) near the terminating spin. Thus it is also important to observe the rotational bands to (near) termination in nuclei closer to the neutron midshell.

REFERENCES

1. Aage Bohr, and Ben R. Mottelson, *Nuclear Structure, Vol 2*, Reading, Ma: W. A. Benjamin, Inc., 1975, p.43.
2. V.P. Janzen, D.R. LaFosse, H. Schnare, D.B. Fossan, A. Galindo-Uribarri, J.R. Hughes, S.M. Mullins, E.S. Paul, L. Persson, S. Pilotte, D.C. Radford, I. Ragnarsson, P. Vaska, J.C. Waddington, R. Wadsworth, D. Ward, J. Wilson, and R. Wyss, *Phys. Rev. Lett.* **72**, 1160 (1994).
3. H. Schnare, D.R. LaFosse, D.B. Fossan, J.R. Hughes, P. Vaska, K. Hauschild, I.M. Hibbert, R. Wadsworth, V.P. Janzen, D.C. Radford, S.M. Mullins, C.W. Beausang, E.S. Paul, J. DeGraaf, I.-Y. Lee, A.O. Macchiavelli, A.V. Afanasjev, and I. Ragnarsson, *Phys. Rev.* **C54**, 1598 (1996).
4. I. Ragnarsson and A. Afanasjev, *Nucl. Phys.* **A591**, 387 (1995).
5. P. Vaska. Ph.D. Thesis, State University of New York at Stony Brook.

6. M.P. Waring, E.S. Paul, C.W. Beausang, R.M. Clark, R.A. Cunningham, T. Davinson, S.A. Forbes, D.B. Fossan, S.J. Gale, A. Gizon, K. Hauschild, I.M. Hibbert, A.N. James, P.M. Jones, M.J. Joyce, D.R. LaFosse, R.D. Page, I. Ragnarsson, H. Schnare, P.J. Sellin, J. Simpson, P. Vaska, R. Wadsworth, and P.J. Woods, *Phys. Rev.* **C51**, (1995) 2427; K. Starosta, private communication.

7. I. Thorslund, D.B. Fossan, D.R. LaFosse, H. Schnare, K. Hauschild, I.M. Hibbert, S.M. Mullins, E.S. Paul, I. Ragnarsson, J.M. Sears, P. Vaska, and R. Wadsworth, *Phys. Rev.* **C52**, (1995) R2839.

8. R. Wadsworth, R.M. Clark, J.A. Cameron, D.B. Fossan, I.M. Hibbert, V.P. Janzen, R. Krücken, G.J. Lane, I.-Y. Lee, A.O. Macchiavelli, C.M. Parry, J.M. Sears, J.F. Smith, A.V. Afanasjev, and I. Ragnarsson, *Phys. Rev. Lett.* **80**, (1998) 1174.

9. A.K. Gaigalas, R.E. Shroy, G. Schatz, and D.B. Fossan, *Phys. Rev. Lett.* **35**, 555 (1975); *Phys. Rev. C* **19**, 1324 (1975).

10. G.J. Lane, private communication.

11. E.S. Paul, private communication.

12. A.V. Afanasjev and I. Ragnarsson, *Nucl. Phys.* **A628**, 580 (1998).

Decay out of the highly-deformed bands in the A=130 mass region and experimental Δ_ν for Nd nuclei

C.M. Petrache*[1], D. Bazzacco*, G. Falconi*, S. Lunardi*, E. Maglione*, R. Menegazzo*, C. Rossi Alvarez*, R. Venturelli*, G. Viesti*, G. de Angelis†, M. De Poli†, C. Fahlander†, E. Farnea†, A. Gadea†, G. Maron†, D.R. Napoli†, Zs. Podolyàk†, S. Perriès•, A. Astier•, L. Ducroux•, M. Meyer•, N. Redon•, A. Bracco°, S. Frattini°, S. Leoni°, I. Deloncle°, M.G. Porquet°, N. Marginean‡, C.A. Ur‡, B. Cederwall×, A. Johnson×, R. Wyss×

*Dipartimento di Fisica and INFN, Sezione di Padova, Padova, Italy

†INFN, Laboratori Nazionali di Legnaro, Legnaro, Italy

•Institut de Physique Nucléaires de Lyon, IN2P3/CNRS, Université C. Bernard, Lyon, France

°Dipartimento di Fisica and INFN, Sezione di Milano, Milano, Italy

°Centre de Spectrométrie Nucléaire et de Spectrométrie de Masse, IN2P3/CNRS, Orsay, France

‡Institute of Physics and Nuclear Engineering "Horia Hulubei", Experimental Fundamental Physics Division, Bucharest, Romania

×Royal Institute of Technology, Physics Department Frescati, Stockholm, Sweden

Abstract. A review of the results on the decay out of highly-deformed bands in the A=130 mass region is presented, including new experimental information on the decay out of two highly-deformed bands in ^{132}Nd and ^{134}Nd. The previously observed E2 linking transitions are explained by the crossing with observed non-yrast bands of normal deformation. Large interaction matrix elements and mixing amplitudes between the interacting highly-deformed and normal-deformed bands are deduced. The large normal-deformed admixture into the lowest observed highly-deformed states explains why the decay out occurs at a certain spin along the band and shows that the barrier between the highly-deformed and normal-deformed minima disappears at low spins. Results on the decay out of the highly-deformed band of ^{136}Nd are reported and the extraction of Δ_ν for the highly-deformed Nd nuclei is discussed.

INTRODUCTION

One of the most exciting questions about the superdeformation phenomenon concerns the feeding out of the superdeformed bands and the understanding of the main features of this mechanism: discrete γ-rays

---

[1] Permanent address: Institute of Physics and Nuclear Engineering, Bucharest, Romania.

CP481, *Nuclear Structure 98*, edited by C. Baktash

against statistical paths, properties of the preferentially populated normal-deformed states, pairing effects in the superdeformed bands. The study of the decay out is the only safe way to determine the excitation energy and to assign the spin and parity to superdeformed states. At the point of decay towards normal-deformed states, the superdeformed rotational sequence disappears suddenly and the decay intensity fragments into so many weak branches that a direct identification of the linking transitions has, in general, not been possible. From the theoretical point of view this feature of the decay-out mode has been explained in statistical terms, as a process triggered by small admixtures of normal-deformed states into the superdeformed wavefunctions [1]. From the experimental point of view, an attempt to identify the linking pattern by summing the energy of two consecutive transitions has lead to a tentative location of the superdeformed band of ^{143}Eu [2]; however, the same method was not successful in other cases. In the heavier A=190 mass region the statistical nature of the process has been confirmed by Henry et al. [3], who isolated a quasi-continuous spectrum of the decay-out transitions in ^{192}Hg; this spectrum was qualitatively reproduced by the calculations of Døssing et al. [4]. The first discrete linking transitions in the A=190 mass region have been discovered by Khoo et al. in ^{194}Hg [5]. They account for ~5% of the decay strength, with the remaining strength assigned to unresolved (statistical) gamma rays. Two more superdeformed bands have been linked through discrete transitions to low-lying states, an excited band in ^{194}Hg [6] and one band in ^{194}Pb [7,8].

The highly-deformed bands of nuclei in the A=130 region are based on ellipsoidal shapes with a quadrupole deformation β~$0.3 - 0.4$, which is not very different from that of the normal-deformed shapes ($\beta_2 = 0.15 - 0.25$) [9]. The observation of discrete linking transitions has been favoured here by the lower excitation energy of the highly-deformed states with respect to the normal-deformed yrast line in the decay-out region (~1 MeV, in contrast with energies of ~ 4.5 MeV for superdeformed bands in the Hg, Pb nuclei), as well as by the relatively higher population intensity of the bands in odd-even nuclei. The transition between the two deformations requires only one or two level crossings on the path from high to low deformation and therefore induces less fragmentation of the decay-out flux than in the other regions of superdeformation [10].

The elongated highly-deformed shapes are stabilized by large shell gaps which develop in the *static* level diagram for $N = 72$ and $Z = 58$. The corresponding "doubly-magic" nucleus is ^{130}Ce. However, in rotating nuclei the two signature partners of the $\Omega = 1/2$ $\nu i_{13/2}$ intruder orbital cross the $N = 72$ shell gap and give rise to *dynamic* single-particle shell gaps at $N = 73$ and $N = 74$. The very high spin states in the A=130 mass region are most efficiently generated by the occupation of the $\nu i_{13/2}$ intruder orbitals lying above the $N = 72$ gap and carrying large intrinsic angular momentum [9,11]. The proton shell gap appears in Nd nuclei ($Z = 60$) at a smaller deformation ($\beta_2 \sim 0.3$) compared to the shell gap at $Z = 58$ corresponding to Ce nuclei ($\beta_2 \sim 0.4$) [11]. In a simplified picture, the proton shell gap at $Z = 58$ has two holes in the $g_{9/2}$ orbitals with respect to the Nd isotopes at $Z = 60$. However, due to pairing correlations, the $g_{9/2}$ states are partly occupied also in the highly-deformed bands of the Nd isotopes at high rotational frequencies. In contrast to the neutron shell gaps, the proton shell gaps stay stable over the entire frequency range of interest. Hence, favoured particle-hole excitations involve neutrons.

Many new results have been recently obtained at GASP in nuclei of the A=130 mass region ranging from Pr to Gd. In particular, the decay-out of the highly-deformed bands has been established in the nuclei ^{132}Nd [12], ^{133}Nd [13,14], ^{134}Nd [15], ^{135}Nd [16,17], ^{136}Nd [18] and ^{137}Nd [19,20]. On the other side, the decay out of the highly-deformed bands in Ce isotopes [21–26] could not be identified, owing to the large deformation of the bands ($\beta_2 \approx 0.4$) and to the higher excitation energy of their lowest levels with respect to the normal-deformed yrast line. In the present contribution I will discuss mainly the decay out in the even-even ^{132}Nd, ^{134}Nd and ^{136}Nd nuclei.

REVIEW OF THE DECAY-OUT RESULTS IN THE A=130 MASS REGION

The decay-out pattern of bands based on configurations involving neutron intruder orbitals ($i_{13/2}$, $f_{7/2}$, $h_{9/2}$), which, depending on their quadrupole deformation, have been called superdeformed, highly-deformed, well deformed, enhanced deformation bands or simply intruder bands, is related to their excitation energy and to the difference in deformation with respect to normal-deformed bands, as well as to the underlying configurations. The most extended decay-out studies have been performed for Nd nuclei, which led to the observation of linking transitions over a series of six nuclei. In none of the neighboring isotopic chains, from La (Z=57) to Gd (Z=64), have the decay out been established over such an extended series of nuclei. In order to establish some common characteristics of the various decay-out patterns, we grouped together several nuclei in four groups: the first one involves even-Z isotopes which lie on the nuclear chart above the Nd nuclei (^{137,139}Sm and ^{139}Gd), the second one involves the Ce nuclei and the odd-even Pr and Pm isotopes, the third one includes the odd-odd isotopes of Pr and Pm, and the fourth one is formed by the Nd nuclei. In the following I will shortly review the results on the decay out for each group of nuclei.

Nuclei above Nd (^{137,139}Sm, ^{139}Gd)

In the odd-even Sm and Gd isotopes the highly-deformed bands are populated down to much lower excitation energy and spin than in the Nd nuclei, and the decay out is less fragmented [27,28]. Nearly 100% of the band intensity has been identified in discrete linking transitions.

In ^{137}Sm and ^{139}Gd the decay out takes place mainly via enhanced E1 transitions. The highly-deformed band in ^{137}Sm decays also via an E2 transition of 586 keV from the $21/2^+$ state to the $17/2^+$ state of band 3. This transition can be explained by an accidental mixing of the two $21/2^+$ states. By means of a simple two-band mixing calculation, from the strengths of the transitions de-exciting the $21/2^+$ highly-deformed state we determined a normal-deformed admixture in the highly-deformed state of $\beta^2 = 0.17$. This large admixture indicates that no barrier is present at the transition from high to low deformation in ^{137}Sm, and extrapolating, in the similar nucleus ^{139}Gd.

Information about the strength of the E1 linking transitions has been deduced from the intensity ratio between the E1 and the competing in-band E2 transitions. The transitional quadrupole moment for the highly-deformed band of ^{139}Gd has been measured to be $Q_0 \sim 7eb$. Assuming a constant Q_0 down to the bottom of the band, a B(E1) strength of $7 \times 10^{-4} e^2 fm^2$ has been obtained for the 616 keV $17/2^+ \rightarrow 15/2^-$ transition. As a tentative explanation of this enhanced E1 strength, the coupling through the 3^- phonon with states of the $\nu f_{7/2}$ configuration has been suggested.

The decay out of the highly-deformed band of ^{139}Sm has been observed at the level of $\sim$100%. An accidental level mixing at spin $33/2^+$ with a non-yrast state was observed, which takes away $\sim$20% of the band intensity. The highly-deformed band ends at spin $21/2^+$, where the decay out proceeds via E2 and M1 transitions towards low-lying states. This again indicates that there is no barrier and that the lowest states of the highly-deformed band are very mixed with normal-deformed states.

Summarizing, the observation of nearly 100% of the decay-out flux in the odd-even Sm and Gd nuclei can be explained by the low excitation of the highly-deformed band and the relatively small difference in deformation between the normal-deformed and the $\nu i_{13/2}$ highly-deformed configurations. The smaller difference in deformation between the intruder and normal-deformed configurations in the Sm and Gd (Z=62,64) isotopes with respect to the corresponding Nd (Z=60) isotones is due to the increased collectivity (quadrupole deformation of the normal-deformed states) with increasing proton number from Z=60 to Z=62,64, and to the nearly constant deformation of the intruder bands as a function of Z.

The Ce and the odd-even Pr and Pm nuclei

The observed highly-deformed bands in Ce and odd-even Pr and Pm nuclei can be separated in two categories: (i) semi-decoupled bands based on the $\pi g_{9/2}^{-1}$ configuration and (ii) bands based on the $\nu i_{13/2}$ and/or $\pi g_{9/2}^{-1}$ configurations. Bands based on the $\pi g_{9/2}^{-1}$ configuration have been observed in the odd-even ^{129,131}Pr [29,30] and ^{133,135}Pm [31,32] nuclei. The deformation of these bands is larger than the deformation of the normal-deformed states. They have a low excitation energy with respect to the yrast states, which facilitated the observation of $\sim$100% of the decay-out flux.

The bands discovered in the Ce [21–26] and odd-even ^{133}Pr [33] nuclei involve one or two neutron particles in the $i_{13/2}$ intruder orbital, as well as one or two proton holes in the $g_{9/2}$ extruder orbital. Their measured deformation of $\beta_2 \sim 0.4$ is larger than in other A=130 nuclei. The observed bands are built on a second minimum in the potential energy surface which coexists over a wide energy range with a normal-deformed minimum from which is separated, at the point of decay, by a barrier which is especially pronounced in Ce nuclei. No discrete linking transitions to the normal-deformed states around spin $\approx 20\hbar$ has been established yet. The situation resembles the one of superdeformed nuclei, although the difference in deformation between the two minima is less pronounced: the decay-out flux is fragmented into several pathways which share the already weak intensity of the bands. The difficulty to observe discrete linking transitions in these cases can be explained by both the higher excitation energy of the bands and the larger difference in deformation between the normal-deformed and highly-deformed potential energy minima.

Odd-odd La, Pr and Pm nuclei

In the odd-odd nuclei of the A=130 mass region the prevalent structures are semi-decoupled deformed rotational bands built on $\pi h_{11/2} \otimes \nu h_{11/2}$, $\pi g_{7/2}/d_{5/2} \otimes \nu h_{11/2}$ and $\pi h_{11/2} \otimes \nu g_{7/2}$ configurations [34–36]. Above N = 72, the occupation of the deformation driving $\nu(f_{7/2}, h_{9/2})$ and $\nu i_{13/2}$ intruder orbitals becomes possible. Bands tentatively assigned to configurations built on these orbitals have indeed been observed in ^{130}La [37], ^{132}Pr [38–40], ^{134}Pr [34,38], ^{134}Pm [41], and ^{136}Pm [42]. Total Routhian surface calculations predict an enhanced deformation for bands built on the $\nu i_{13/2}$ intruder orbital, whereas for the bands built on the $h_{9/2}$ and $f_{7/2}$ intruder orbitals the predicted deformation is only slightly larger than that of the normal-deformed bands. Until recently, lifetime measurements to experimentally demonstrate the predicted difference in deformation between the doubly-decoupled bands based on $\nu i_{13/2}$ or $h_{9/2}/f_{7/2}$ orbitals were very scarce. The only results were from a fractional Doppler shift analysis of a doubly-decoupled band in ^{134}Pm [41], which led to the conclusion that the odd neutron occupies the $[541]1/2^-$ orbit of mixed $h_{9/2}$ and $f_{7/2}$ parentage. Very recently, a centroid shift and lineshape analysis of the intruder bands of ^{134}Pr has been performed [43], which gives the first experimental proof that the deformation of the band based on the $\nu h_{9/2}/f_{7/2}$ orbital is smaller than the deformation of the band based on the $\nu i_{13/2}$ orbital, being instead similar to the deformation of the bands based on non-intruder configurations. Very recent lifetime results for ^{130}Pr and ^{132}Pr have also been presented at this conference, giving further evidence about the relatively low deformation of the bands built on the $h_{9/2}$ and $f_{7/2}$ neutron orbitals. In this case, it is completely inappropriate to speak about minima with different deformation associated to the normal-deformed and intruder bands. The small deformation of these bands and their low excitation energy favours the mixing and explain the observation of $\sim$100% of the decay-out flux.

The bands built on the $\nu i_{13/2}$ intruder orbital have a larger deformation ($\beta_2 \sim 0.3 - 0.35$) and a higher excitation with respect to the normal-deformed yrast states ($\beta_2 \sim 0.2$). For none of these bands has the decay out been found.

The Nd nuclei

The highly-deformed bands in Nd nuclei have different decay-out patterns depending (i) on the type of nucleus (even-even or odd-even) and (ii) on the parity of the highly-deformed band.

The bands in the odd-even Nd nuclei are strongly populated ($\sim$5-10% of the reaction channel) and this certainly facilitated the early observation of discrete linking transitions in the ^{133,135,137}Nd nuclei [13,14,16,19]. The excitation energy of the highly-deformed bands with respect to the yrast line is lower than in the even-even nuclei and decreases with decreasing neutron number. This is related to the decreasing difference in deformation between the highly-deformed and normal-deformed bands when going towards lighter, more collective Nd nuclei. The fraction of the band intensity observed as discrete linking transitions changes from 100% for ^{133}Nd, to 75% for ^{135}Nd and 30% for ^{137}Nd. As can be seen in fig. 1, depending of the nucleus and its specific level structure, the decay out towards low-lying normal-deformed states proceeds via E2 transitions (^{133}Nd), E2 plus E1 transitions (^{135}Nd) and cascades of E2 plus M1 transitions (^{137}Nd).

In ^{133}Nd the decay of the highly-deformed band has been quantitatively explained in terms of level mixing which involves orbitals that differ by two major oscillator quantum numbers [13,14]. A different interpretation has been given for the decay out in ^{135}Nd, where the highly-deformed band ends at spin $25/2^+$ and no significant mixing with normal-deformed states has been observed [16]. This "band termination" at low spin has been explained by the change of the highly-deformed minimum into a triaxial one with lower β_2 deformation. "Ultimate Cranking" calculations [16] with the $\nu i_{13/2}$ orbital kept occupied, have shown that this change can be related to the shift of one proton pair and one neutron pair out of the deformation-driving Nilsson orbitals. In ^{137}Nd, the sudden depopulation of the highly-deformed band has been explained by total Routhian surface calculations through a change of the nuclear shape which is microscopically related to the transfer of the valence neutron from a N = 6 to a N = 4 Nilsson orbital [19]. However, as can bee seen in fig. 1, the decay out in ^{135,137}Nd occurs when a state of the highly-deformed band has an energy close to that of an non-yrast intermediate state, with which can strongly mix and induce the observed fragmentation of the decay-out flux. The hot non-yrast states most probably have normal deformation and can indeed involve in their configuration the predicted N=4 Nilsson orbital which crosses the N=6 intruder orbital at low frequency.

The highly-deformed bands in even-even Nd nuclei, which are based on two quasi-particle neutron excitations, are much weaker than in the odd-even ones ($\sim$0.5-1.5%). The identification of discrete linking transitions becomes as difficult as in the other regions of superdeformation. Such transitions have been only observed in

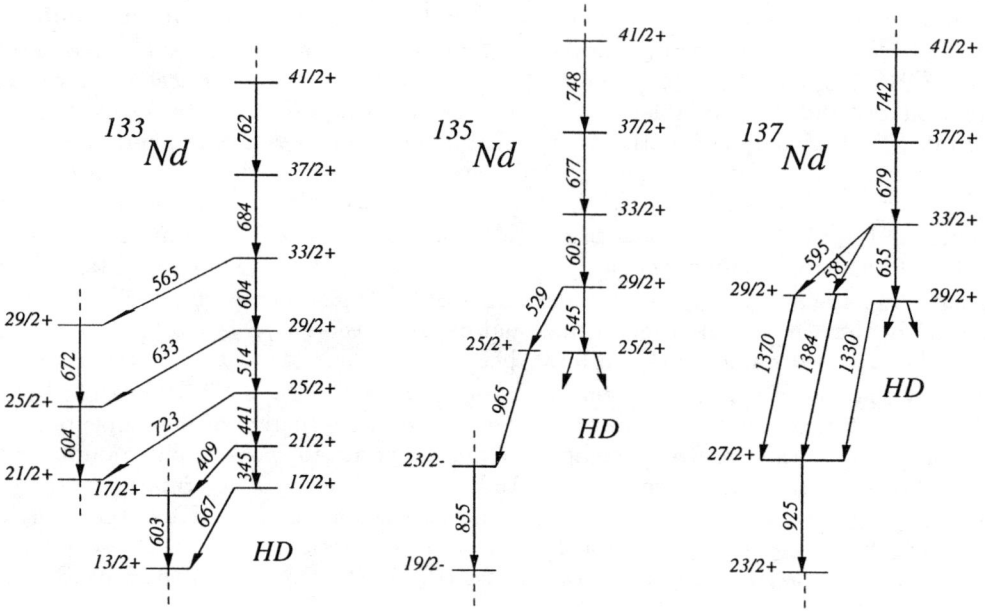

FIGURE 1. Decay out of the highly-deformed bands in odd-even Nd nuclei.

^{132,134}Nd [12,15] and very recently in ^{136}Nd [18]. The total intensity of the decay-out transitions decreases with increasing neutron number, from 65% for ^{132}Nd, to 50% for ^{134}Nd and 20% for ^{136}Nd. The observed highly-deformed bands in even-even Nd nuclei are based on two types of configurations, which lead to either negative parity (the yrast bands observed in ^{132,134,136}Nd shown in fig. 2) or positive parity (the excited bands observed in ^{132,134}Nd).

Different decay-out patterns have been identified: the negative-parity bands decay in two steps via E2 + enhanced E1 transitions with strengths of $\sim 10^{-3}$ W.u., whereas the positive-parity bands decay in one step via E2 transitions with strengths of ~ 1 W.u. This difference in the decay-out pattern has been observed first in ^{134}Nd [15], where the yrast highly-deformed band decays towards intermediate states via E2 transitions and then via enhanced E1 transitions to normal-deformed states, whereas the excited highly-deformed band decays via E2 transitions directly to the same normal-deformed bands. At that time it was not clear why the excited band decays at a certain spin and why the E2 linking transitions were enhanced. The suggestion was made that the mixing with normal-deformed states could be large. Later on, the study of the lighter ^{132}Nd nucleus [12] revealed a similar situation for the positive-parity band identified in this nucleus (band HD-2). A definite conclusion could not be drawn, since the lowest in-band transitions were chosen on the basis of their higher intensity, which, in the presence of crossing bands is not always a correct assumption (see below). It can happen that for certain positions of the interacting levels the inter-band transitions are stronger than the in-band transitions. As we will see in the following, this is the case for band HD-2 of ^{132}Nd.

The decay-out pattern of the negative-parity highly-deformed bands in ^{134,136}Nd is different from that in ^{132}Nd, where the band built on the intruder orbitals has a deformation similar to that of the low-lying states, its excitation is smaller and the mixing with normal-deformed states is strong. The 21$^-$ states of the highly-deformed bands in ^{134,136}Nd decay towards intermediate 19$^-$ states, with energies close to the energy of the 19$^-$ in-band state. These intermediate states have most probably normal deformations and, due to the closeness in energy with the 19$^-$ highly-deformed state, can induce strong normal-deformed admixtures in the highly-deformed state. The nature of the negative-parity intermediate states can be different from that of the intermediate states observed in odd-even Nd nuclei, having a certain octupole collectivity, which would give the most straightforward explanation of the observed enhanced E1 transitions from the 19$^-$ states to the 18$^+$

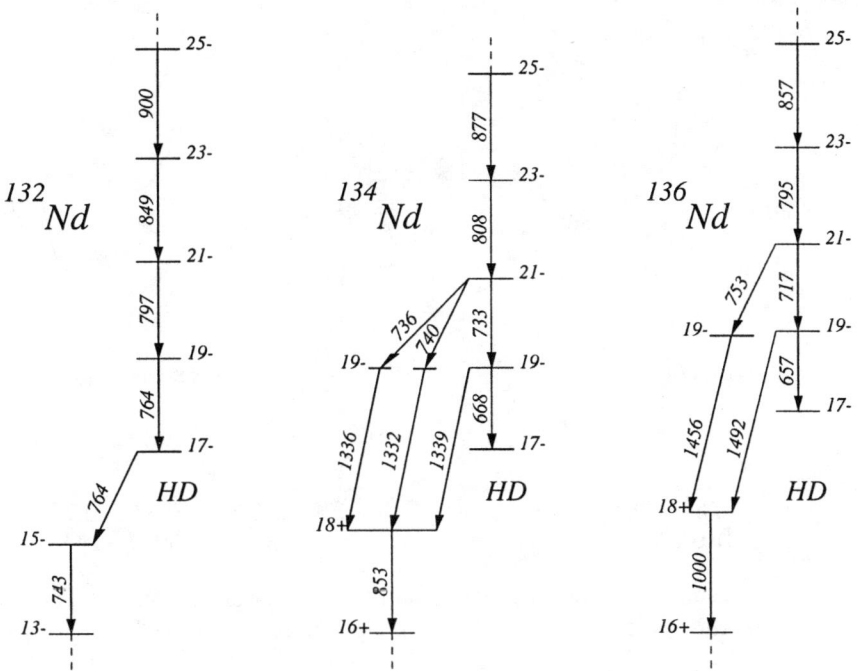

FIGURE 2. Decay out of the negative-parity highly-deformed bands in even-even Nd nuclei.

normal-deformed states (see also the section on the decay out in ^{136}Nd).

In the following I will review the decay-out results obtained on ^{132}Nd and ^{134}Nd, and discuss new results obtained from a revisited analysis of the decay out of the positive-parity bands in the two nuclei.

DECAY OUT IN ^{132}ND AND ^{134}ND

The ^{132}Nd nucleus

The bands built on intruder orbitals observed in ^{132}Nd have lower excitation energies and higher intensities than the highly-deformed bands of ^{134}Nd [12]. This is a clear evidence of the rotational quenching of the $N = 72$ shell gap at intermediate spins. From the recently published results on ^{132}Nd [12], two features related to band HD-2 are relevant for the present discussion: the interaction with the S-band at high spins and the decay out via E2 transitions towards both the S-band and to states of another non-yrast band.

The interaction at high spin between band HD-2 and the S-band is clear from the dynamical moment of inertia of the bands (see fig. 3), which have perturbations of opposite sign at $\hbar\omega \approx 0.65$ keV and $\hbar\omega \approx 0.7$ keV, respectively. These perturbations are induced by an accidental degeneracy of the 34$^+$ levels of the two bands. The search for inter-band transitions led to the identification of a 1313 keV transition between the 36^+_{HD-2} and 34^+_{S-band} states (see fig. 2 of ref. [12]), which fixes the relative positions of the bands and the even spins and positive parity of band HD-2. This helped in the search for the discrete linking transitions of band HD-2. The observed decay paths account for ~65% of the band intensity.

A consistent configuration assignment for band HD-2 involves a hole in the $\nu[411]1/2$ ($\alpha = -1/2$) orbital coupled to the $\nu i_{13/2}$ ($\alpha = +1/2$) intruder orbital, which leads to a $(\pi,\alpha) = (+,0)$ two-quasiparticle configuration (see fig. 3b of ref. [12]). The ground-state band is calculated to have large softness in the γ and β-direction. After the $h_{11/2}$ proton alignment, the deformation of the S-band is calculated to increase from $\beta_2 \sim 0.27$ to $\beta_2 \sim 0.33$, the deformation associated to the $N = 72$ shell gap (see fig. 4). The calculations show that the

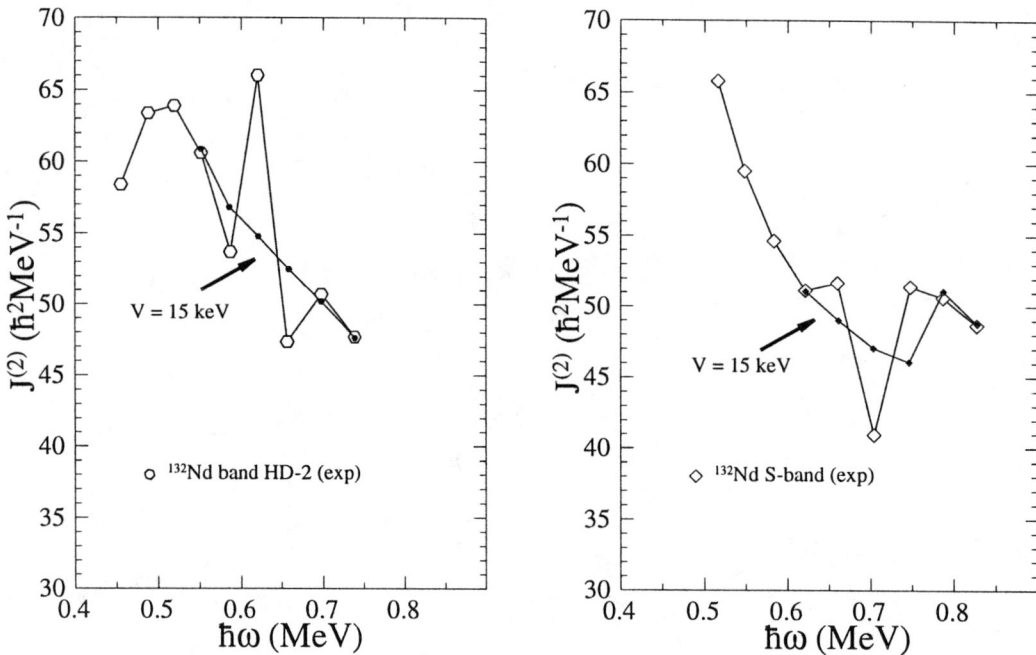

FIGURE 3. Dynamical moments of inertia of band HD-2 and of the S-band of ^{132}Nd. The smooth J$^{(2)}$ curves (filled symbols) are obtained from the calculated unperturbed energies, which are deduced from the experimental values assuming an interaction V=15 keV.

S-band is crossed at intermediate spins by bands based on particle-hole configurations involving excitations across the $N = 72$ gap, like the observed band HD-2. This clearly shows that the $N = 72$ gap changes with increasing rotational frequency. It becomes quenched due to the lowering in energy of the $i_{13/2}$ intruder orbital from above the shell gap (see fig. 3 of ref. [11]).

The ^{134}Nd nucleus

In the ^{134}Nd nucleus, the decay out of both the yrast and excited highly-deformed bands has been established [15]. The excitation energy and spins for all observed bands in ^{134}Nd could be thus determined, allowing for a detailed comparison of the different structures observed at high spin. The excited highly-deformed band involves the $\nu i_{13/2} - [402]5/2^+$ configuration that is crossed at higher frequencies by the $\nu(i_{13/2})^2$ one, whereas the yrast highly-deformed band involves the $\nu i_{13/2} - \nu h_{9/2}/f_{7/2}$ configuration [11]. A crossing at high spins has been observed also in band 3 of this nucleus, which marks the occupation of the first $\nu i_{13/2}$ intruder orbital [44]. All normal-deformed and highly-deformed bands observed in ^{134}Nd exhibit a step-wise alignment pattern at very high spins, indicating the successive occupation of the first and second $N = 6$ $\nu i_{13/2}$ intruder orbitals through adiabatic configuration changes. Lifetime measurements have been performed to determine the deformation as a function of the occupied $\nu i_{13/2}$ intruder orbitals [45]. For the yrast highly-deformed band involving one $\nu i_{13/2}$ intruder orbital, a quadrupole moment of 6.8(3) eb has been extracted, which is slightly larger than the quadrupole moment of the excited highly-deformed band, which is Q_t=6.4(4) eb. This difference in deformation is induced by the [541]1/2 orbital of mixed $h_{9/2}/f_{7/2}$ parentage present in the configuration of the yrast highly-deformed band. Band 3 of ^{134}Nd exhibits a bandcrossing at spin $\sim 30\hbar$. The fit of the upper part, which involves one $\nu i_{13/2}$ intruder orbital, gives a quadrupole moment of 6.5 eb, whereas the fit of the lower part gives a quadrupole moment of 4.9(3) eb. The quadrupole moment of band 3 below the crossing is confirmed by recent results from a plunger measurement [46], which gives for the 16$^+$ state a quadrupole moment of Q_t=4.23$^{+0.50}_{-0.37}eb$.

The analysis of the decay-out pattern in ^{134}Nd [15] led to the following results. The main decay-out flux of the yrast highly-deformed band occurs from the 21$^-$ state towards two intermediate 19$^-$ levels, lying within

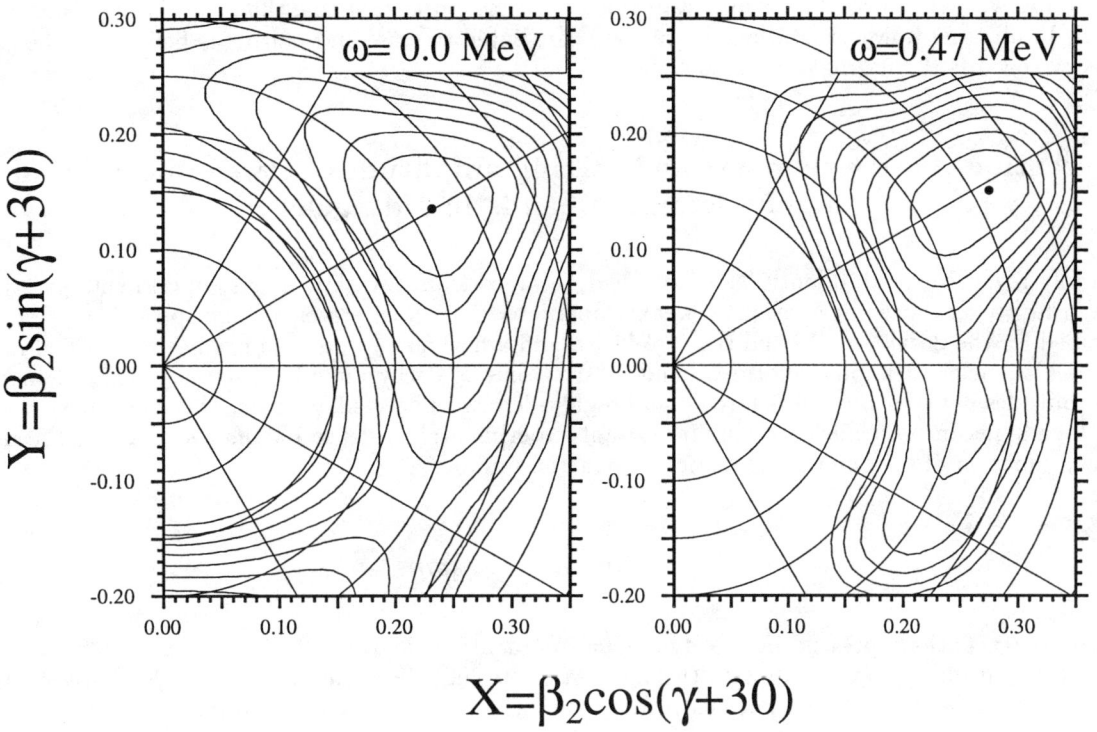

FIGURE 4. Total routhian surfaces for the $(\pi = +, \alpha = 0)$ configuration in ^{132}Nd at two rotational frequencies.

7 keV of excitation energy with respect to the highly-deformed 19^- state. The two intermediate 19^- states most probably involves states based on normal-deformed shapes. The presence of even a very small interaction will lead to a sizable mixing of the corresponding wavefunctions, which would induce the observed similar de-excitation pattern via enhanced E1 transitions of the three 19^- states. The observed linking transitions account for only $\sim$40% of the band intensity. From the B(E1)/B(E2) ratio of the last in-band 668 keV transition (E2) and the decay-out 1339 keV transition (E1), we extracted a B(E1)-strength of $\sim 10^{-3}$ W.u., comparable to those observed in the heavy Ba-Sm region, for which stable octupole deformations have been predicted. Since the neutron Fermi-surface at high deformation is almost between the $f_{7/2}$ and $i_{13/2}$ orbitals and the highly-deformed yrast band of ^{134}Nd is assumed to involve the $\nu i_{13/2} - h_{9/2}/f_{7/2}$ configuration, one can indeed expect octupole couplings to be present.

Three linking transitions of 1780, 1889 and 1928 keV have been identified for the excited highly-deformed band (fig. 5), which carry $\sim$40% of the band intensity. Based on the DCO ratios which do not show significant anisotropy between 90° and 34° when gating on E2 transitions, a $\Delta I = 2$, E2 character has been assigned to these transition, which leads to even spins and positive parity for the band. The B(E2)-strengths of the 1780 and 1928 keV transitions has been estimated from the ratio between the decay-out and the in-band 726 keV transition intensities by a simple energy scaling of the transition operators. From the measured in-band B(E2)-strength of 360 W.u., B(E2)-strengths of 1.5 W.u. and 0.7 W.u. are obtained for the 1780 and 1928 keV transitions, respectively. Such sizable out-of-band strengths clearly indicate a spread of the normal-deformed states into the highly-deformed minimum, which leads to a significant mixing between the highly-deformed and normal-deformed states. When no pairing is present, the transition between the states at different deformations are severely hindered, since the related operators act only on a single particle. The enhanced strength of the E2 linking transitions indicate the presence of pairing correlations, which helps the nucleus to slide between different deformations. However, as the pairing correlations are smaller in the two-quasiparticle highly-deformed bands of ^{134}Nd than in the one-quasiparticle highly-deformed band of ^{133}Nd, and the decay out occurs at similar excitation energies above yrast in both nuclei, the conclusion was drawn that the mixing with the normal-deformed states plays a more important role than the sliding via the pairing interaction. In all cases however, static pair gaps are calculated to exist and pairing is believed to be essential for the decay towards the normal-deformed states.

Interaction matrix elements and mixing amplitudes between normal- and highly-deformed bands in Nd nuclei

We recently revisited the data on ^{132}Nd and ^{134}Nd obtained from two thin-target experiments performed at GASP, in order to get more information on weakly populated non-yrast states at high spins. New results have been obtained for the decay out of band HD-2 of ^{132}Nd and a new non-yrast band has been identified in ^{134}Nd, which crosses the excited highly-deformed band at the point of decay out. Interaction matrix elements and mixing amplitudes between normal-deformed and highly-deformed bands have been obtained in the decay-out region at high spins in both nuclei, giving important experimental evidence on the evolution of the different minima as a function of spin and as a function of neutron number.

Interaction at high spins

^{132}Nd. The perturbations present in the dynamical moments of inertia of the S-band and band HD-2 of ^{132}Nd (see fig. 3) allows one to estimate the interaction strength and mixing amplitudes between the two bands. As in the case of the ^{133}Nd nucleus [14], by smoothing out the dynamical moments of inertia in the interaction region one can deduce the positions of the unperturbed levels, and by performing a two-band mixing calculation with constant interaction one can estimate the mixing amplitudes. As can be seen in fig. 3, we could not obtain a regular behaviour for the S-band at the highest spins, as in the case of band HD-2. This reveals the presence of a crossing between the S-band and another unobserved band. An upper limit of $\sim$15 keV has been obtained for the interaction strength at spin 34^+, while the amplitudes of the highly-deformed and normal-deformed configurations in the highly-deformed 34^+ state could only be estimated to be $\alpha^2=0.84$ and $\beta^2=0.16$, respectively. The relatively large normal-deformed admixture, which, as we will see in the following, is of the same order of magnitude as the values obtained at the bottom of the band, indicates a similar deformation for the two bands and a substantial mixing between the normal and intruder configurations. Another consequence of these results is that in the light Nd nuclei the minima associated with the two configurations are not separated by any barrier, being in excellent agreement with the total Routhian surface calculations which predict only a sliding of the minimum from $\beta_2 \sim 0.27$ to $\beta_2 \sim 0.33$ with increasing frequency (see fig. 4).

^{134}Nd. A perturbation at high spins has been observed in the dynamical moment of inertia of band 3 in ^{134}Nd, which marks the crossing between configurations based on one and two $\nu i_{13/2}$ intruder orbitals [44]. However, the excitation energy of the state with spin 40^+ in band 3 ($E_x = 18.108$ MeV) is close to the excitation energy of the state with similar spin in the highly-deformed band ($E_x = 18.151$ MeV). The two states can easily interact, due to the small difference in energy (43 keV) and to the similar configuration, which is based on a highly-deformation shape induced by the occupied $\nu i_{13/2}$ intruder orbitals. A search for possible inter-band transitions at high spin has been undertaken. We have been able to find evidence for a transition of 1384 keV linking the 42^+ state of band 3 to the 40^+ state of the highly-deformed band. The 1384 keV transition appears in the sum-spectrum obtained by double-gating on the transitions of the highly-deformed band, but due to the limited statistics at very high spins, we could not get clear evidence about the mutual coincidence with individual transitions from the highly-deformed band and could not extract the branching ratio of the 42^+ state of band 3. A maximum interaction matrix element of ~ 20 keV is expected for degenerate unperturbed states. We could not obtain a precise estimation of the interaction strength because of the crossing exhibited by band 3 in the same spin range. Anyhow, the observation of the 1384 keV inter-band transition indicates an appreciable mixing between the 40^+ states, which suggests an interaction strength of at least several keV. In ^{134}Nd, at difference with ^{132}Nd where the S-band is observed up to high spins and interact with the band built on the $\nu i_{13/2}$ orbital, the S-band ends much earlier, since, due to its smaller deformation with respect to the highly-deformed bands, becomes rapidly non-yrast; the interaction at high spins observed in ^{134}Nd occurs between bands built on $\nu i_{13/2}$ intruder orbitals with similar large deformation.

^{132}Nd. The two lowest states of band HD-2 in ^{132}Nd have been chosen in ref. [12] on the basis of their intensity and on yrastness criteria. From the present analysis, and after a comparison with the heavier ^{134}Nd nucleus (see below), it results that the lowest two states of band HD-2 are the previously assigned intermediate 18$^+$ and 20$^+$ states decaying through the 1956 and 1884 keV transitions (see fig. 5).

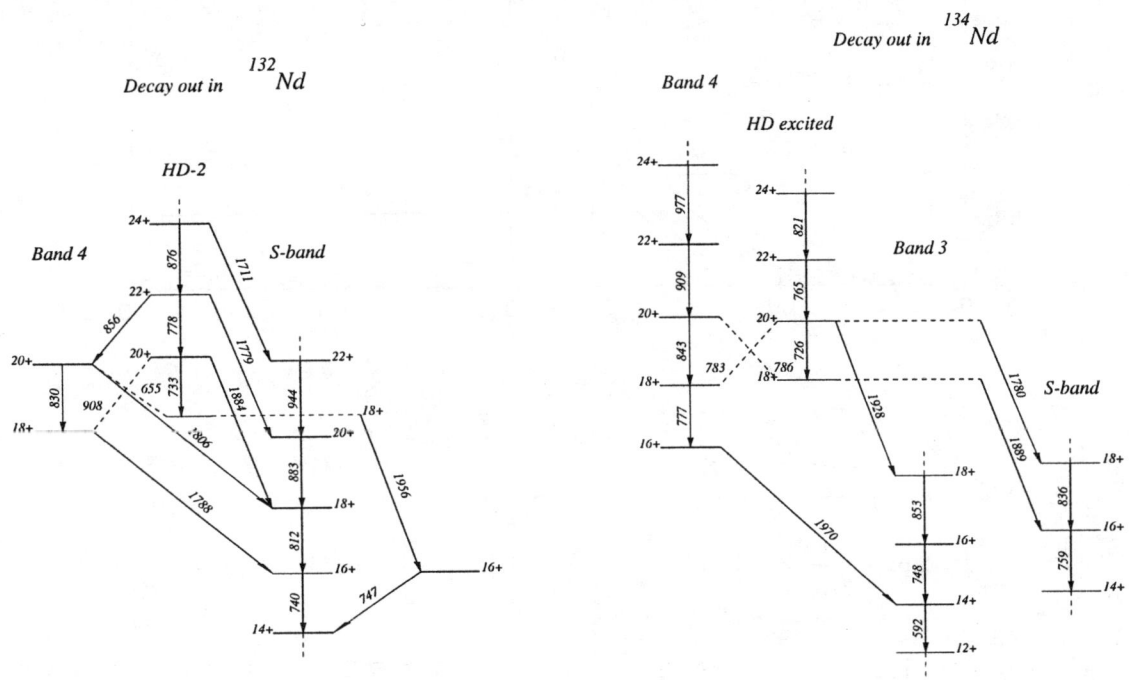

FIGURE 5. Decay out of the band HD-2 of ^{132}Nd and of the excited highly-deformed band of ^{134}Nd.

The two lowest transitions of band HD-2 are therefore the 778 and 733 keV transitions. The 856 keV transition becomes an inter-band transition and the 830 keV transition becomes a transition between the 20$^+$ and 18$^+$ states of a band (that we will call band 4) which interacts with band HD-2 and decays through the 1806 and 1788 keV transitions towards the 18$^+$ and 16$^+$ states of the S-band.

In order to estimate the interaction matrix element and the mixing amplitudes, a careful search for inter-band transitions has been performed. Excepting the already identified 856 keV ($22^+_{HD} \rightarrow 20^+_{ND}$) transition, due to lack of statistics and the presence of several contaminating transitions in the region of interest, no definite evidence could be obtained on the existence of other inter-band transitions. However, upper limits have been established for the intensities of the 655 keV ($20^+_{ND} \rightarrow 18^+_{HD}$) and 908 keV ($20^+_{HD} \rightarrow 18^+_{ND}$) transitions, which lead to upper limits for the intensity ratios I(655)/I(830)<0.20 and I(908)/I(733)<0.3. A similar accurate search has been performed for an inter-band transition of energy ~870 keV, which would connect an hypothetical 22$^+$ state of band 4 decaying to the 20$^+$ state of band 4 through a ~950 keV transition, but again without success. The only intensity ratio of inter-band to in-band transitions could be extracted for the 22^+_{HD} level, which is I(778)/I(856)= 0.58±0.05. The results on the intensity ratios have been used to estimate the matrix elements between the crossing bands, from a comparison with the calculated I_{in}/I_{out} ratios as a function of the interaction. Calculations of I_{in}/I_{out} have been performed assuming different ratios of the quadrupole moments Q_{ND}/Q_{HD}=0.7,0.8,0.9,1. The three experimentally extracted intensity ratios are shown in fig. 6, together with the calculated values. One can see that the range of the interaction 30 keV<V<34 keV extracted from the branching of the 22^+_{HD} level is in good agreement with the upper limits extracted from the 20^+_{HD} and 20^+_{ND} states. The same calculations have been performed assuming the previously assigned two lowest transitions in band HD-2 [12]; they are in complete disagreement with the experimental branching ratio for the 22^+_{HD} state (see lower-right panel of fig. 6), supporting therefore the present assignment of the lowest in-band transitions.

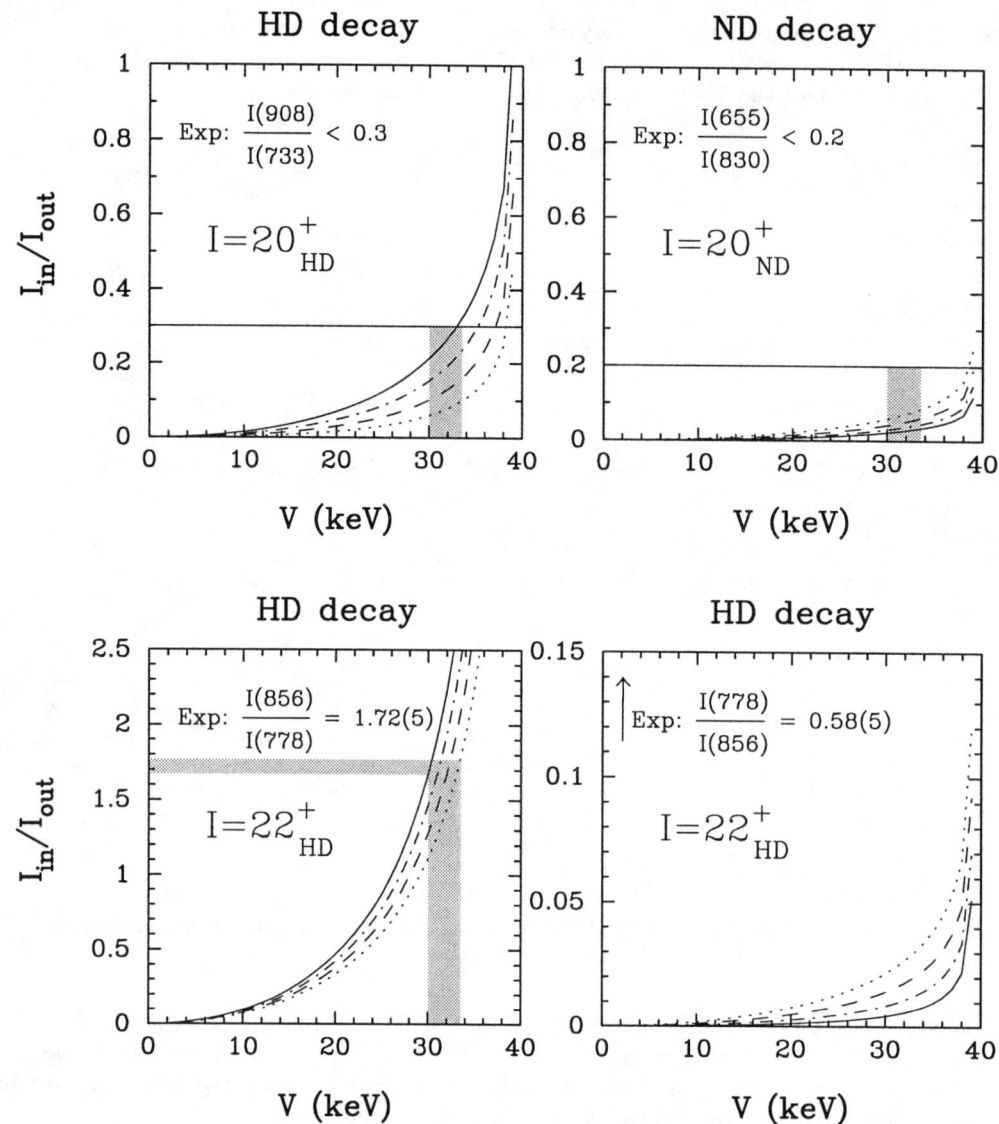

FIGURE 6. Calculated ratios of in-band (I_{in}) and out-of-band (I_{out}) intensities as a function of the interaction strength for ^{132}Nd. The different curves correspond to Q_{ND}/Q_{HD} ratios of 0.7 (dotted line), 0.8 (dashed line), 0.9 (dotted-dashed line) and 1.0 (solid line).

By using the deduced interaction matrix elements we could calculate the composition of the 20^+_{HD} and 22^+_{HD} wavefunctions, which, depending on the quadrupole moments, results to have normal-deformed admixtures of $\beta^2 = 0.18 \div 0.28$ and $\beta^2 = 0.12 \div 0.18$, respectively. These values are of the same order of magnitude as the values deduced from the interaction at high spin between the S-band and band HD-2, indicating that band 4 has a deformation not much different from the deformation of the bands built on intruder orbitals. It is therefore inappropriate to speak about bands built on minima with different deformation in the light Nd nuclei, since the deformation of the different bands seems to be very similar and no barrier can develop between the points corresponding to different deformations in the potential energy surface. The wavefunctions of the different bands appears to be mixed.

The two interacting bands have similar decay-out patterns towards states of the S-band, consisting of high energy E2 transitions. If one assumes a configuration involving $h_{11/2}$ protons for band 4, e. g. $(\pi h_{11/2})^4$, and a $(\pi h_{11/2})^2$ configuration for the S-band, one can explain the decay of band 4 to the S-band by the closely related configurations. The mixing between band 4 and band HD-2 leads to admixtures of the assumed $(\pi h_{11/2})^4$ configuration in the lowest states of band HD-2, which would explain the existence of the decay-out transitions towards the S-band. Finally, the decay of the lowest state of band HD-2 through the 1956 keV transition to the 16^+ non-yrast state can be explained by the same mechanism, if one assume a mixing between the non-yrast 16^+ state with the 16^+ state of the S-band.

^{134}Nd. A new non-yrast band, most probably of normal deformation, has been identified in ^{134}Nd (fig. 5). It is based on a 16^+ state, which decays to the 14^+ state of band 3 via an E2 transition of 1970 keV. Six transitions with energies of 777, 843, 909, 977 1062 and 1068 keV have been observed in this new band, which we will call band 4 (as in the case of ^{132}Nd). The 18^+ and 20^+ states have energies close to the energies of the 18^+ and 20^+ states of the excited highly-deformed band: the 18^+_{ND} state is 57 keV below the 18^+_{HD} state and the 20^+_{ND} state is 60 keV above the 20^+_{HD} state. Due to this small difference in energy, the states with the same spin interact. A careful search for the inter-band transitions 783 keV ($20^+_{HD} \rightarrow 18^+_{ND}$) and 786 keV ($20^+_{ND} \rightarrow 18^+_{HD}$) has been performed. However, only upper limits for the transition intensity ratios could be obtained. As in the case of ^{132}Nd discussed before, the intensity ratios have been used to estimate the matrix elements V between the interacting states, which results to be $V \leq 10 \div 15$ keV, depending on the assumed ratio of the quadrupole moments of the bands. Even if only an upper limit has been obtained for the interaction, one can safely suppose that the real value is in the range of several keV. Such a value is of the same order of magnitude as the interaction at high spin between the highly-deformed band and band 3, which, being built on the same minimum at large deformation are not separated by any barrier. One can therefore conclude that also between the minima of the excited highly-deformed band and of band 4 there is no barrier, even if in this case the calculated minima have different deformations.

As in the case of ^{132}Nd, the crossing between band 4 and the excited highly-deformed band generates normal-deformed admixtures in the lowest states of the excited highly-deformed band. Assuming again a $(\pi h_{11/2})^4$ configuration for band 4 and $(\pi h_{11/2})^2$ for band 3 (as assigned in ref. [12]), one could explain the existence of the 1928 keV ($20^+_{HD} \rightarrow 18^+_{band-3}$) linking transition. It is not easy to account for the other linking transitions towards the 16^+ and 18^+ states of the S-band, with an assigned $(\nu h_{11/2})^2$ configuration [44]. However, the connecting transitions between band 3 and band HD-2 (see ref. [44]), as well as the calculated softness of the ^{134}Nd nucleus at high spin, suggest that band 3 and the S-band are rather mixed. In such a case, at least a fraction of the strength of the 1780 keV ($20^+_{HD} \rightarrow 18^+_{S-band}$) and 1889 keV ($18^+_{HD} \rightarrow 16^+_{S-band}$) transitions can be explained by the wavefunctions of normal-deformed admixtures in the lowest highly-deformed states.

Conclusions. The present analysis shows that the decay out of the highly-deformed positive-parity bands of ^{132}Nd and ^{134}Nd occurs at a certain spin along the band due to the crossing with a non-yrast normal-deformed band. No barrier exists at the point of decay out and the lowest highly-deformed states have large normal-deformed admixtures, which can account for some of the observed E2 linking transitions. Interactions have been observed at high spins in both nuclei, which are important pieces of evidence about the configurations of the bands and their deformation.

DECAY OUT IN ^{136}ND: EXPERIMENTAL Δ_ν FOR THE HIGHLY-DEFORMED CONFIGURATIONS IN ND NUCLEI

The highly-deformed bands of the even-even Nd nuclei, which are based on two-quasiparticle excitations, are much weaker than in the odd-even ones. The band intensity is about 1% of the total population of the nucleus, and therefore the identification of discrete linking transitions becomes as difficult as in the other regions of superdeformation. Such transitions have been only recently observed in the ^{132,134}Nd nuclei [12,15].

The search for linking transitions in the ^{136}Nd nucleus was one of the aims of an experiment performed recently with EUROBALL [18]. Indications about the mixing of one of the lowest highly-deformed states with a close lying state existed already from a previous experiment performed with the GASP array, which led to the observation of several high-spin bands in ^{136}Nd [47].

The much higher statistics obtained with the EUROBALL array and its better efficiency at high energy allowed us to clearly identify two discrete decay paths from the lowest states of the highly-deformed band (see fig. 7): a cascade consisting of an E2 transition of 754 keV followed by an E1 transition of 1456 keV from the 21^- highly-deformed state to the 18^+ state of the S-band, and an E1 transition of 1492 keV directly from the

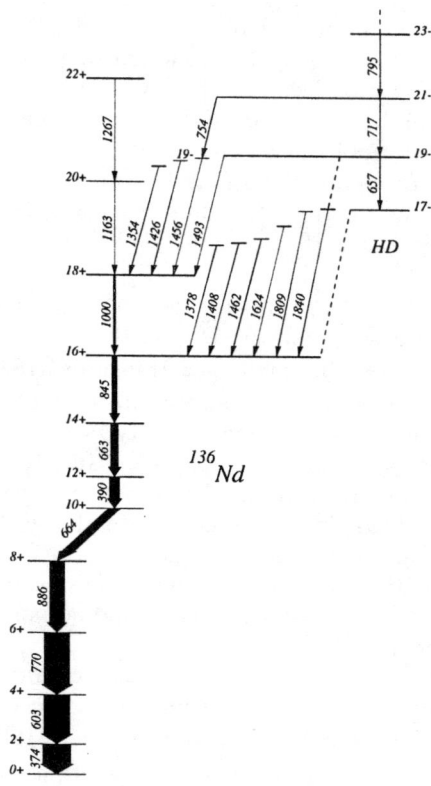

FIGURE 7. Decay out of the highly-deformed band of ^{136}Nd and of other transitions feeding the 16^+ and 18^+ states of the S-band.

19^- highly-deformed state to the same 18^+ state. The decay-out pattern in ^{136}Nd is very similar to the one observed for the yrast highly-deformed band of ^{134}Nd [15]. The lowest observed state of the highly-deformed band in ^{136}Nd is thus located at 7030 keV of excitation energy and has assigned spin and parity 17^-.

In addition to the observed high-energy linking transitions we could identify several new transitions feeding the 16^+ and 18^+ states of the S-band (see fig. 7). These transitions indicate the existence of several non-yrast levels with energies close to the energy of the lowest highly-deformed states. Unfortunately, until now we could not assign firm spins and parities to these levels. The observed non-yrast states could be similar to the 19^- state populated in the decay of the 21^- highly-deformed state. This non-yrast 19^- state decays by an E1 transition of 1456 keV, expected to be equally enhanced as the 1493 keV E1 transition from the 19^- state of the highly-deformed band ($\sim 10^{-3} - 10^{-4}$ W.u.). These states can be obtained from the coupling of the 3^- phonon with states of the S-band, which would give rise to multiplets of states with spins ranging from I-3 to I+3 for each member of the S-band. If the energy of the octupole phonon is of about 2.5 MeV (which is a reasonable value obtained from the extrapolation of the energy of the low-lying octupole strength of heavier Nd nuclei [48]), the maximum aligned $14^+ \otimes 3^-$ and $16^+ \otimes 3^-$ states with spins 17^- and 19^- respectively, would indeed be close to the states with the same spins of the highly-deformed band. The structure of the intermediate 19^- state would then be mainly $19^-_{ND} = |16^+ >_{ND} \otimes |3^- >$. The E1 transition of 1456 keV linking the $|16^+ >_{ND} \otimes |3^- >$ and $|18^+ >_{ND}$ states would be enhanced, since it connects states with one and zero 3^- phonons. In this scenario, the existence of the enhanced E1 transition of 1493 keV ($19^-_{HD} \rightarrow 18^-_{ND}$) can be explained as follows. Assuming the $(\pi h_{11/2})^2 \otimes \nu i_{13/2} f_{7/2}$ configuration for the highly-deformed band and a 16^+_{ND} state in which also the $(\nu f_{7/2})^2$ configuration has a certain contribution, the mixing between the $|19^- >_{HD} = |(\pi h_{11/2})^2; 10^+ > \otimes |\nu i_{13/2} f_{7/2}; 9^- >$ and the $|19^- >_{ND} = |16^+ >_{ND} \otimes |3^- >= |(\pi h_{11/2})^2; 10^+ > \otimes |\nu (f_{7/2})^2; 6^+ > \otimes |3^- >$ states would admix partial waves with octupole collectivity in the $|19^- >_{HD}$ state. The transition between the $|19^- >_{HD}$ and $|18^+ >_{ND}$ states would then be enhanced and easier to observe,

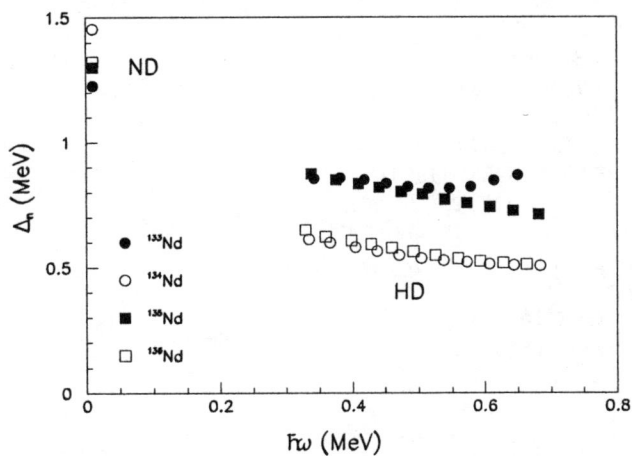

FIGURE 8. Experimental neutron pairing gaps for the normal-deformed ground states and for the highly-deformed configurations in Nd nuclei, extracted using the three point formula.

like the transitions between states with one and zero 3^- phonons.

This discussion is based on the essential assumption that the octupole collectivity plays a role at high spins in ^{136}Nd, and extrapolating, also in the neighboring ^{134}Nd nucleus, where a similar decay-out pattern of the negative-parity yrast highly-deformed band was observed (see fig. 2). The presence of octupole collectivity in nuclei of the A=130 mass region has been invoked when trying to explain the anomalous signature splitting of the negative-parity bands based on $\pi h_{11/2} d_{5/2}$ or $\nu s_{1/2} h_{11/2}$ configurations, which would apply also in the case of ^{134}Nd and ^{136}Nd. Anyhow, no experimental evidence exists on the 3^- state in these nuclei. Another open question about the proposed scenario is related to the existence and the collectivity of the octupole correlations based on two or four quasiparticle configurations present in the decay-out region. Hopefully, future experiments and calculations will answer these many exciting open questions.

The discovery of discrete linking transitions in ^{136}Nd completed the series of Nd nuclei, from ^{132}Nd to ^{137}Nd, with known excitation energies for the highly-deformed bands. This allows to estimate the pairing strength in the highly-deformed configuration, from the odd-even mass differences, like in the case of the neutron (proton) pairing gap Δ_n (Δ_p) for the normal-deformed matter. Using a first-order Taylor expansion [49] and the three-point formula [18], we extracted the experimental neutron pairing gap in the normal-deformed ground state and in the highly-deformed configurations for two odd-even and two even-even Nd nuclei. The pairing gap values for the highly-deformed configurations, shown in fig. 8, are reduced by approximately a factor of 2 with respect to the values for the normal-deformed ground states ($\Delta_n \sim 0.6 - 0.7$ MeV) and remain rather constant in a wide range of frequencies. In the normal-deformed case, the pairing gaps of the odd-even nuclei are smaller than those of the even-even ones, whereas in the highly-deformed configurations the opposite is true: the pairing gaps of the two-quasiparticle configurations present in even-even nuclei is smaller than the pairing gaps of the one-quasiparticle configurations present in odd-even nuclei. This peculiar feature is due to the blocking effects, which are stronger in the two-quasiparticle configurations based on intruder orbitals of different nature ($\nu i_{13/2}$ and $\nu f_{7/2}$) than in the one-quasiparticle $\nu i_{13/2}$ configurations.

REFERENCES

1. Vigezzi E., Broglia R.A., and Døssing T., *Phys. Lett.* **B249**, 163 (1990).
2. Ataç A., et al., *Phys. Rev. Lett.* **70**, 1069 (1993).
3. Henry R.G., et al., *Phys. Rev. Lett.* **73**, 777 (1994).
4. Døssing T., et al., *Phys. Rev. Lett.* **75**, 1276 (1995).
5. Khoo T.L., et al., *Phys. Rev. Lett.* **76**, 1583 (1996).
6. Hackman G., et al., *Phys. Rev. Lett.* **79**, 4100 (1997).

7. Lopez-Martens A., et al., *Phys. Lett.* **B380**, 18 (1996).
8. Hauschild K., et al., *Phys. Rev.* **C55**, 2819 (1997).
9. Wyss R., et al., *Phys. Lett.* **B215**, 211 (1988).
10. Dudek J., et al., *Phys. Rev.* **C38**, 940 (1988).
11. Petrache C.M., et al., *Phys. Lett.* **B335**, 307 (1994).
12. Petrache C.M., et al., *Phys. Lett.* **B415**, 223 (1997).
13. Bazzacco D., et al., *Phys. Lett.* **B309**, 235 (1993).
14. Bazzacco D., et al., *Phys. Rev.* **C49**, R2281 (1994).
15. Petrache C.M., et al., *Phys. Rev. Lett.* **77**, 239 (1996).
16. Deleplanque M.A., et al., *Phys. Rev.* **C52**, R2302 (1995).
17. Willsau P., et al., *Phys. Rev.* **C48**, R494 (1993).
18. Perriès S., et al., *to be published*.
19. Lunardi S., et al., *Phys. Rev.* **C52**, R6 (1995).
20. Petrache C.M., et al., *Nucl. Phys.* **A617**, 228 (1997).
21. Galindo-Uribarri A., et al., *Phys. Rev.* **C54**, R454 (1996).
22. Luo Y.X., et al., *Z. Phys.* **A329**, 125 (1988).
23. Semple A.T., et al., *Phys. Rev.* **C54**, 425 (1996).
24. Nolan P.J., et al., *J. Phys.* **G11**, L17 (1985).
25. Santos D., et al., *Phys. Rev. Lett.* **74**, 1708 (1995).
26. Hauschild K., et al., *Phys. Lett.* **B353**, 438 (1995).
27. Rossi Alvarez C., et al., *Nucl. Phys.* **A624**, 225 (1997).
28. Rossi Alvarez C., et al., *Phys. Rev.* **C54**, 57 (1996).
29. Galindo-Uribarri A., et al., *AECL-11132* (1994).
30. Galindo-Uribarri A., et al., *Phys. Rev.* **C50**, R2655 (1994).
31. Galindo-Uribarri A., et al., *Phys. Rev.* **C54**, 1057 (1996).
32. Riley M.A., et al., *Proc. Workshop on Gammasphere Physics*, Berkeley: World Scientific, 1995, pp. 241-252.
33. Wilson J.N., et al., *Phys. Rev. Lett.* **74**, 1950 (1995).
34. Petrache C.M., et al., *Nucl. Phys.* **A397**, 106 (1996).
35. Petrache C.M., et al., *Nucl. Phys.* **A603**, 50 (1996).
36. Petrache C.M., et al., *Nucl. Phys.* **A635**, 361 (1998).
37. Godfrey M.J., et al., *J. Phys.* **G15**, L163 (1989).
38. Hauschild K., et al., *Phys. Rev.* **C50**, 707 (1995).
39. Hartley D.J., et al., *Phys. Rev.* **C55**, R985 (1997).
40. Petrache C.M., et al., *Phys. Rev.* **C58**, R611 (1998).
41. Wadsworth R., et al., *Nucl. Phys.* **A526**, 188 (1991).
42. Riley M.A., et al., *Phys. Rev.* **C47**, R441 (1993).
43. Rao M.N., et al., *Phys. Rev.* **C58**, R1 (1998).
44. Petrache C.M., et al., *Phys. Lett.* **B387**, 31 (1996).
45. Petrache C.M., et al., *Phys. Rev.* **C57**, R10 (1998).
46. Klemme T., et al., *to be published*.
47. Petrache C.M., et al., *Phys. Lett.* **B373**, 275 (1996).
48. Cottle P.D., and Zamfir N.V., et al., *Phys. Rev.* **C54**, 176 (1996).
49. Madland D., and Nix J., *Nucl. Phys.* **A476**, 1 (1988).

Highly Deformed Structures in the Mass~130 Region and Their Relative Quadrupole Moments

F.G. Kondev[†], M.A. Riley[†], E.S. Paul[‡], T.B. Brown[†1], R.M. Clark[¶], M. Devlin[§], P. Fallon[¶],
D.J. Hartley[† 2], I.M. Hibbert[‖ 3], D.T. Joss[‡], D.R. LaFosse[§ 4], R.W. Laird[†], F. Lerma[§],
M. Lively[†], P.J. Nolan[‡], N.J. O'Brien[‖], J. Pfohl[†], D.G. Sarantites[§], R.K. Sheline[†],
S.L. Shepherd[‡], J. Simpson[+] and R. Wadsworth[‖]

[†] Department of Physics, Florida State University, Tallahassee, Florida 32306
[‡] Oliver Lodge Laboratory, University of Liverpool, Liverpool L69 7ZE, United Kingdom
[¶] Nuclear Science Division, Lawrence Berkeley Laboratory, Berkeley, California 94720
[§] Department of Chemistry, Washington University, St. Louis, Missouri 63130
[‖] Department of Physics, University of York, York Y01 5DD, United Kingdom
[+] CLRC, Daresbury Laboratory, Daresbury, Warrington, WA4 4AD, United Kingdom

Abstract. The quadrupole moments for variety of configurations, involving the $9/2^+[404]$ $(g_{9/2})$ proton, $1/2^+[660](i_{13/2})$ and $1/2^-[541]$ $(f_{7/2}, h_{9/2})$ neutron orbitals, were measured using the Doppler-shift attenuation method in a wide range of nuclei in the mass~130 region. While the involvement of the first two orbitals leads to quadrupole deformations that are comparable to those observed for the so-called superdeformed bands in this mass region, the β_2 values for structures that include the $1/2^-[541]$ neutron are found to lie intermediate between those for normally deformed and highly deformed bands.

INTRODUCTION

The recent studies of highly deformed ($\beta_2=0.3-0.4$), sometimes referred to as "superdeformed", structures in the region near mass~130 have revealed an important interplay between microscopic shell effects, such as the occurrence of large gaps in the nucleon single-particle energies, and the occupation of high-j low-Ω (intruder) orbitals in driving the nucleus towards higher deformation. Initially, it was thought that only the involvement of one or more $i_{13/2}$ neutrons could result in a strong polarization on the nuclear shape in this mass region [1]. However, it was recently shown that certain strongly coupled bands, built upon the $9/2^+[404]$ $(g_{9/2})$ proton orbital in the odd-Z ^{131}Pr [2] and ^{133}Pm [3] isotopes, exhibit quadrupole deformations comparable to the values found for highly deformed structures which include $i_{13/2}$ neutrons. Furthermore, for nuclei below $N = 73$ where the occupancy of the $\nu i_{13/2}$ orbital is energetically unfavored, there are indications that bands involving the $1/2^-[541]$ $(f_{7/2}, h_{9/2})$ neutron may also be highly deformed [4,5].

In order to provide further experimental evidence about the highly deformed character of these structures, as well as to elucidate the impact of the occupation of specific orbitals along with the "core stabilization" effect from the shell gaps on deformation, accurate lifetime measurements for a large number of bands have been measured. While the quadrupole moment, Q_0, for some of these bands were measured in the past using the Doppler-shift attenuation method (DSAM), conclusive comparisons were limited owing to systematic distinctions between experimental setups such as varying reactions and target retardation properties. Specifically, due to differences in the parameterization of the nuclear and electronic stopping powers, which act as an "internal clock" in the DSAM lifetime measurements, large variations in the measured Q_0 values are reported for the

---

[1]) Present address: Chemistry Department, University of Kentucky, Lexington, Kentucky 40506.
[2]) Present address: Department of Physics and Astronomy, University of Tennessee, Knoxville, Tennessee 37996.
[3]) Present address: Oliver Lodge Laboratory, University of Liverpool, Liverpool L69 7ZE, United Kingdom.
[4]) Present address: Department of Physics and Astronomy, SUNY at Stony Brook, New York 11794.

CP481, *Nuclear Structure 98*, edited by C. Baktash
© 1999 American Institute of Physics 1-56396-858-4/99/\$15.00

same band by different authors. The absence of adequate experimental information on the time structure of the quasicontinium sidefeeding contributions also results in an additional inaccuracy on the measured quadrupole moments. In the current work we have greatly reduced these systematic problems by measuring the lifetime decay properties of a large selection of bands in different nuclei under nearly identical experimental conditions in terms of angular momentum input, excitation energy and recoil velocity profile. Furthermore, by exploiting the high efficiency and resolving power of the modern γ-ray detector arrays it was possible, in cases of favorable statistics, to greatly minimize the effect of sidefeeding on the measured quadrupole deformations, by gating on shifted transitions at the top of the band of interest, thus gaining some insight into the nature and time scale of the sidefeeding.

EXPERIMENTAL CONDITIONS AND DATA REDUCTION

High-spin states in a wide range ($Z = 58-61$) of nuclei were populated after fusion of a ^{35}Cl beam with ^{105}Pd target nuclei. Thin and backed target experiments were performed at the 88-Inch Cyclotron at the Lawrence Berkeley National Laboratory with beam energies of 180 (thin target) and 173 MeV (backed target). The thin target consisted of an isotopically enriched ^{105}Pd foil with a thickness of 500 μg/cm^2. The backed target was a 1 mg/cm^2 thick ^{105}Pd foil mounted on a 17 mg/cm^2 Au backing. Emitted γ-rays were collected using the GAMMASPHERE spectrometer [6] consisting of 57 (thin target) and 97 (backed target) HPGe detectors arranged in seventeen rings located at $\theta = 17.3°$, 31.7°, 37.4°, 50.1°, 58.3°, 69.8°, 79.2°, 80.7°, 90.0°, 99.3°, 100.8°, 110.2°, 121.7°, 129.9°, 142.6°, 148.3° and 162.7°. The evaporated charged particles were identified with the MICROBALL detector system [7], whose selection capabilities allowed a clean separation of the different charged particle channels. The present work focuses on the properties of structures which involve the important $9/2^+[404]$ ($g_{9/2}$) proton, $1/2^+[660](i_{13/2})$ and $1/2^-[541]$ ($f_{7/2},h_{9/2}$) neutron orbitals, in the odd-N ($Z=60$) ^{133}Nd (populated in the αp2n channel) and ^{135}Nd (3p2n) isotopes, and the odd-Z ($Z=59$) ^{130}Pr (2α2n), ^{131}Pr (2α1n) and ^{132}Pr (1α2p2n) nuclei. Typically more than about 50×10^6 (thin target) and 20×10^6 (backed target) events (of a fold $\geq$3) per particle gated channel were collected.

In the off-line analysis, the thin target data were unfolded into γ^3 coincidences and incremented into three-dimensional cubes. Background subtracted, doubly-gated spectra were then produced for each particular cascade and were examined in order to construct the level schemes. Non-symmetrical (31.7° and 37.4° vs. 90°) matrices were also constructed which enabled γ-ray multipolarities to be deduced using the method of directional correlation from oriented states (DCO).

The backed target data were used to extract the quadrupole deformation using the centroid-shift technique in conjunction with the Doppler-shift attenuation method [8]. This was done in two ways. In the first method, the data were sorted into two-dimensional matrices in which one axis consisted of "forward" (31.7° and 37.4°) or "backward" (142.6° and 148.3°) group of detectors and the other axis was any coincident detector. Spectra were generated by summing gates on the cleanest, fully stopped transitions at the bottom of the band of interest and projecting the events onto the "forward" and "backward" axes. These spectra were then used to extract the fraction of the full Doppler shift, F(τ), for transitions within the band of interest. The use of this approach allowed the sidefeeding contributions to take full effect when the Q_0 values are deduced, since the measured F(τ) values depend not only on the lifetime of a particular state, but also on the time history of all levels within cascades which precede it. In the second method, the data were sorted into a number of double-gated spectra which contained counts registered by particular group of detectors. Specifically, for the relatively strongly populated $\nu i_{13/2}$ bands in ^{133}Nd and ^{135}Nd, gates were also set on in-band "moving" transitions (i.e., those which showed Doppler shift) in any ring of detectors and data were incremented into separate spectra for events detected at "forward", "90°" and "backward" angles. Since the gating transitions are Doppler-shifted, ring-by-ring energy-dependent Doppler corrections were applied in the sorting procedure using the formula:

$$C = C_0(1 + F(\tau)\beta_0 \cos\theta) \tag{1}$$

where C_0 and C are the gain matched raw and Doppler-corrected channel numbers, respectively, and β_0 refers to the initial velocity of the recoiling nucleus. The F(τ) values for transitions within the band of interest were calculated (see below in the text) iteratively by varying the in-band Q_0 until the Doppler-corrected peak centroids were perfectly matched in all detector rings. This allowed the correct coincidence relationships to be achieved in the gating procedure. Consequently, non-Doppler corrected spectra were written on disk and were used to extract the final F(τ) values. Examples of resulting spectra, after background subtraction, for

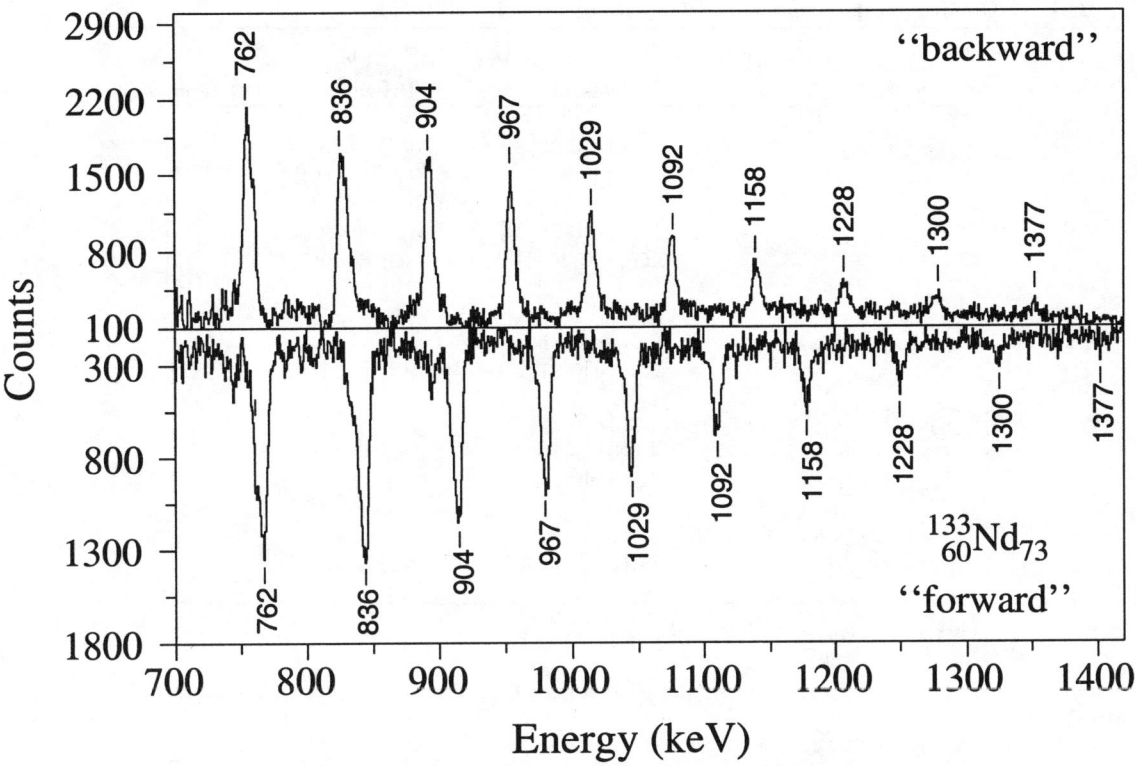

FIGURE 1. Coincidence γ-ray spectra for the $1/2^+[660]$ ($i_{13/2}$) band in ^{133}Nd formed from combinations of all double gates on in-band transitions from 345 keV up to 1228 keV. The peaks are labeled with the unshifted energies.

the $\nu 1/2^+[660]$ ($i_{13/2}$) band in ^{133}Nd are shown in fig. 1. It should be noted, that the implementation of this method made it possible to eliminate the effect of sidefeeding for states lower in the cascade, i.e. for those below the levels depopulated by the gating transitions.

In the off-line analysis various programs from the RADWARE interactive package [9] were employed.

EXPERIMENTAL RESULTS AND DISCUSSION

In order to extract the intrinsic quadrupole moments from the experimental $F(\tau)$ values, calculations using the code FITFTAU [10] were performed. The $F(\tau)$ curves were generated under the assumption that the band has a constant Q_0 value. In the modeling of the slowing process of the recoiling nuclei, the stopping powers were calculated using the 1995 version of the code TRIM [11] which uses the most recent evaluation of existing data. The corrections for multiple scattering were introduced using the prescription given by Blaugrund [12]. Where appropriate, the sidefeeding into each state was taken into account according to the experimental in-band intensity profile using a rotational cascade of three transitions with the same Q_0 as the in-band states. The impact of the number of sidefeeding cascades on Q_0 was investigated and found to be much smaller than the statistical uncertainties. Table 1 summarizes the measured Q_0 values and the corresponding quadrupole deformations, β_2. It should be emphasized, that although the uncertainties in the stopping powers and the modeling of the sidefeeding may contribute an additional systematic error of 15−20% in the absolute Q_0 values, the relative deformations are considered to be accurate to a level of 5−10% since the bands were studied under nearly identical conditions. Such precision allowed a clear differentiation in the Q_0 values to be made, which was used in turn as evidence for the involvement of specific orbitals within a band configuration.

TABLE 1. Quadrupole moments and deformations for selected bands.

| Nucleus | Configuration [a] | Q_0, eb $[\beta_2]$ [b] | | Reference |
		Present	Previous	
^{133}Nd	$\nu i_{13/2}$	5.8(4) [0.32(3)]	6.0(7) [0.33(3)]	[13]
		6.4(4) [0.36(3)] [c]	6.7(7)[0.37(4)] [d]	[13]
			7.4(7)[0.41(3)] [e]	[14]
^{135}Nd	$\nu i_{13/2}$	5.1(5) [0.28(3)]	5.4(10) [0.30(5)]	[15]
		5.5(6) [0.30(4)] [c]	7.4(10) [0.40(5)] [f]	[15]
			7.3(10) [0.37(5)] [g]	[16]
^{130}Pr	$\pi g_{9/2} \otimes \nu h_{11/2}$	6.1(4) [0.35(2)]		
^{131}Pr	$\pi g_{9/2}$	5.3(4) [0.31(2)]	5.5(8) [0.32(5)]	[2]
^{130}Pr	$\pi h_{11/2} \otimes \nu(f_{7/2}, h_{9/2})$ (band 1)	4.3(3) [0.25(2)]		
	$\pi h_{11/2} \otimes \nu(f_{7/2}, h_{9/2})$ (band 2)	4.5(3) [0.26(2)]		
^{132}Pr	$\pi h_{11/2} \otimes \nu(f_{7/2}, h_{9/2})$	4.1(3) [0.24(1)]		
^{130}Pr	$\pi h_{11/2} \otimes \nu d_{5/2}$	3.4(2) [0.20(1)]		
^{131}Pr	$\pi h_{11/2}$	3.3(2) [0.19(1)]	3.9(3) [0.23(2)]	[2]

[a] $\pi g_{9/2}$: $9/2^+[404]$, $\pi h_{11/2}$: $3/2^-[541]$, $\nu h_{11/2}$: $7/2^-[523]$, $\nu d_{5/2}$: $5/2^+[402]$, $\nu(f_{7/2}, h_{9/2})$: $1/2^-[541]$, $\nu i_{13/2}$: $1/2^+[660]$.
[b] Deduced using the centroid-shift technique with sidefeeding quadrupole moment, Q_{0sf} equal to Q_0.
[c] Deduced by gating above the level of interest (see the text for details).
[d] Deduced using the lineshape technique with $Q_{0sf}=5.3$ eb.
[e] Deduced using the lineshape technique with sidefeeding time distributions about 1.4 times slower than the in-band lifetimes.
[f] Deduced using the lineshape technique with $Q_{0sf}=3.5$ eb.
[g] Deduced using the centroid-shift technique with $Q_{0sf}=3.0(5)$ eb. Note that although nearly identical Q_0 values were measured in ref. [15,16] the quoted β_2 values differ by $\sim 7\%$.

Bands involving the $1/2^+[660]$ ($i_{13/2}$) neutron orbital

Collective structures built upon the $1/2^+[660]$ ($i_{13/2}$) intruder neuton orbital have been observed in the chain of odd-N (Z=60) Nd isotopes from ^{133}Nd up to ^{137}Nd [17,18]. These bands were connected to the normally deformed structures [18–23], so that their spin, parity and excitation energy are unambiguously determined. In addition, the g-factor experiment performed in the case of ^{133}Nd [14] independently confirms the $\nu 1/2^+[660]$ configuration assignment. Quadrupole moment measurements were carried out previously using both the centroid-shift and lineshape DSAM techniques [13–16,24]. Lifetimes of low-spin members of the band in ^{135}Nd were also measured via the Doppler-shift recoil-distance method [25]. The F(τ) values and the corresponding quadrupole deformations deduced in the current work when gates were set on the stopped 409, 440, 513 and 603 keV transitions in ^{133}Nd, and 546 and 676 keV γ-rays in ^{135}Nd are shown in Figs. 2c and 2d. Our observations are in agreement with the previously measured quadrupole deformations for the band in ^{133}Nd [13], as well as with the centroid-shift results for the structure in ^{135}Nd reported by Diamond *et al.* [15]. Importantly, they clearly indicate that the band in ^{133}Nd is more deformed than that in ^{135}Nd. The comparison of the intensity profiles for those two bands, shown in Figs. 2a and 2b, reveals that the sequence in ^{135}Nd is fed more from the side over a range of transitions for which the F(τ) values change very rapidly. Such a behavior led to speculations that the sidefeeding lifetimes could be much slower when compared to that for the in-band levels. In fact, the lineshape analysis of Diamond *et al.* [15], as well as recent centroid-shift observations by Petrache *et al.* [16], suggested that the sidefeeding lifetimes were about factor of four slower, thus leading to a quadrupole deformation that exceeded the value in ^{133}Nd (see Table 1). Figures 2e and 2f show our observations, when spectra gated on the Doppler-shifted in-band 1029, 1092, 1158, 1228 and 1300 keV γ-rays in ^{133}Nd, and 947, 1012, 1078, 1146, and 1216 keV γ-rays in ^{135}Nd, were used. We found a roughly 10% increase in the deformation of both these two bands, compared to values deduced when gates were set on stopped transitions. We estimated that the sidefeeding lifetimes are only about 1.3−1.4 times slower than those

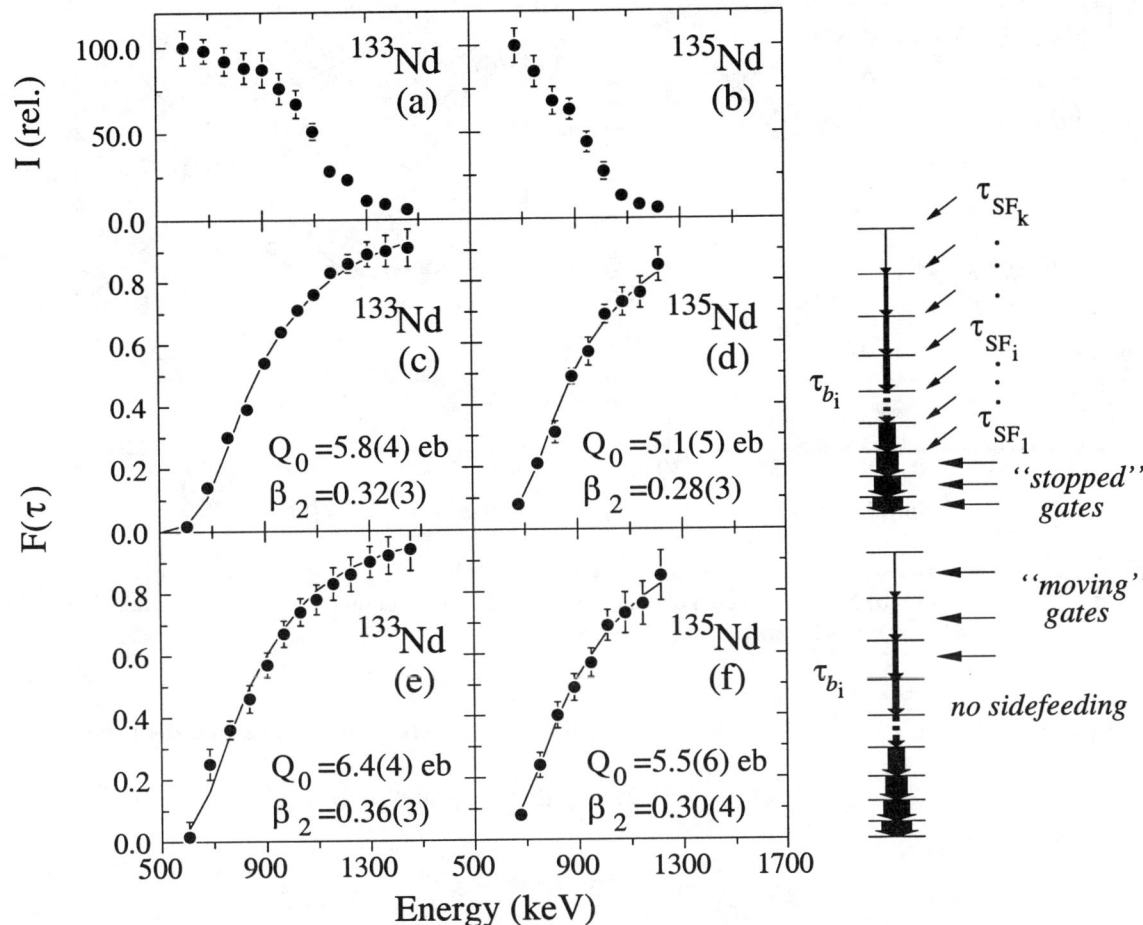

FIGURE 2. Intensity profiles (a) and (b), $F(\tau)$ values and corresponding quadrupole deformations deduced by gating on stopped transitions (c) and (d), and "moving" transitions (e) and (f) for the $1/2^+[660]$ ($i_{13/2}$) bands in ^{133}Nd and ^{135}Nd, respectively.

for the in-band levels. Similar results for the ratio of in-band to sidefeeding lifetimes were recently reported by Clark *et al.* [26] in studying the relative deformations of structures which include one or two $\nu i_{13/2}$ orbitals in the neighboring ^{131}Ce and ^{132}Ce isotopes.

The present observations, together with the values for the band in ^{137}Nd (Q_0=4.0(5) [β_2=0.22(3)]) [13], indicate that in the odd-N Nd nuclei there is a systematic decrease in the deformation of the $\nu i_{13/2}$ band as the neutron number increases. Similar behavior, for the same configuration, was also observed in the neighboring ^{135}Sm (N=73) and ^{137}Sm (N=75) isotopes [27]. Such a trend was predicted previously by Total Routhian Surface and Ultimate Cranker calculations with pairing [13,21], as well as by Cranked Nilsson-Strutinsky calculations [28] which do not include pairing. The excitation energy as a function of spin for the $\nu i_{13/2}$ bands in ^{133}Nd (N=73), ^{135}Nd (N=75) and ^{137}Nd (N=77) are shown in Fig. 3a. As can be seen, the band in ^{133}Nd is lower in energy compared to those in ^{135}Nd ($\Delta E\sim0.6$ MeV) and ^{137}Nd ($\Delta E\sim1.6$ MeV), thus implying that the neutron Fermi level is located closer to the $1/2^+[660]$ Nilsson configuration at $N=73$. As a consequence of the rapid downslope of this orbital with increasing the deformation, the higher excitation energy of the bands in ^{135}Nd and ^{137}Nd supports the experimental observations that they are less deformed, as illustrated in Fig. 3b.

Another interesting feature is the extra deformation for the $\nu i_{13/2}$ band in Z=58 ^{131}Ce (Q_0=7.4(3) eb [β_2=0.43(1)]) [26] compared to the values measured for the same configuration in the Z=60 ^{133}Nd and Z=62 ^{135}Sm (Q_0=7.0(7) eb [β_2=0.37(4)]) [27] isotones ($N=73$). The presence of two proton "holes" in the upsloping $g_{9/2}$ proton orbital at Z=58 (Ce) and their absence in the Nd and Sm isotopes is suggested [28] to result in an additional enhancement of the deformation in the Ce nuclei. By comparing the relative deformations within

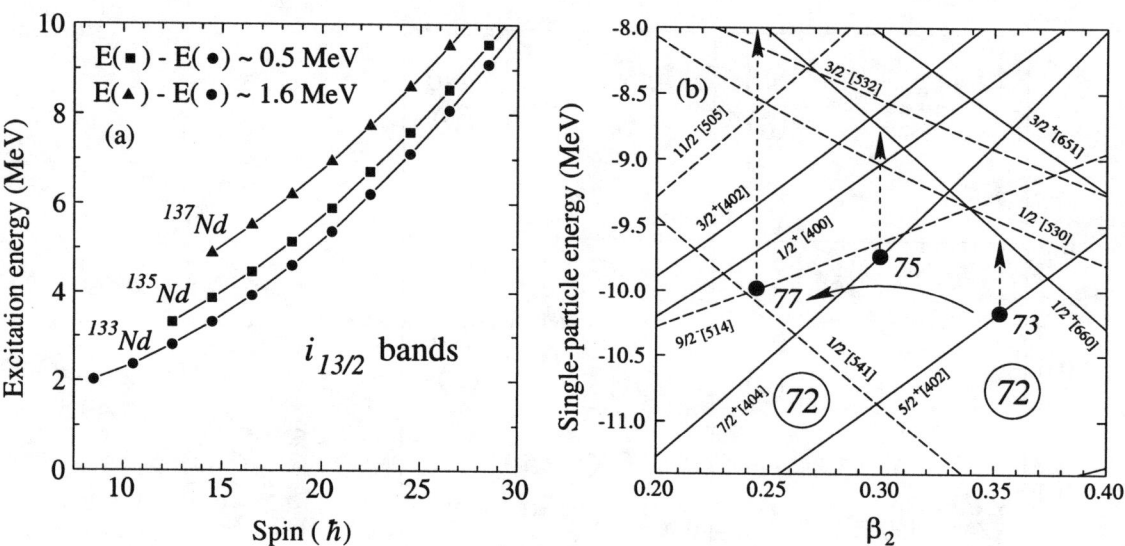

FIGURE 3. (a) Excitation energy as a function of spin for the $1/2^+[660]$ ($i_{13/2}$) bands in ^{133}Nd ($N = 73$), ^{135}Nd ($N = 75$) and ^{137}Nd ($N = 77$). (b) Neutron single-particle levels as a function of quadrupole deformation ($\gamma=0°$, $\beta_4=0$) calculated using the Woods-Saxon potential.

the $N = 73$ isotones the contribution of the partial occupation of the $\pi g_{9/2}$ orbital can be deduced as:

$$\delta\beta_2 = \beta_2(^{131}Ce) - \beta_2(^{133}Nd) = 0.43(1) - 0.36(3) = 0.7(4) \tag{2}$$

$$\delta\beta_2 = \beta_2(^{131}Ce) - \beta_2(^{135}Sm) = 0.43(1) - 0.37(4) = 0.6(5) \tag{3}$$

Thus, taking the measured quadrupole deformations for the bands in ^{135}Nd and ^{137}Nd as a base, then one may expect $Q_0 \sim 6.5$ and 5.1 eb for the $\nu i_{13/2}$ structures in ^{133}Ce and ^{135}Ce, respectively, which are not yet measured.

Bands involving the $9/2^+[404]$ ($g_{9/2}$) proton orbital

Initially, a band built upon the $9/2^+[404]$ ($g_{9/2}$) proton orbital was observed in ^{131}Pr by Galindo-Uribarri *et al.* [2]. This work has established a value of $Q_0=5.5(8)$ eb [$\beta_2=0.32(5)$] for this band which is much larger than $Q_0=3.9(3)$ eb [$\beta_2=0.23(2)$] deduced for the normally deformed $\pi 3/2^-[541]$ ($h_{11/2}$) structure in the same nucleus. Recently, Brown *et al.* [29] have observed a strongly coupled band in the neighboring odd-odd ^{130}Pr isotope which was suggested to include the $9/2^+[404]$ ($g_{9/2}$) proton orbital coupled to the $7/2^-[523]$ ($h_{11/2}$) neutron. The measured and calculated $F(\tau)$ values for this structure, as well as those for the normally deformed $\pi 3/2^-[541]$ ($h_{11/2}$)$\otimes\nu 5/2^+[402]$ ($d_{5/2}$) band are shown in Fig. 4a. The results for the $\pi g_{9/2}$ and $\pi h_{11/2}$ configurations in ^{131}Pr, deduced from the current work, are presented in Fig. 4b. We report $Q_0=6.1(5)$ eb [$\beta_2=0.35(3)$] for the $\pi g_{9/2}\otimes\nu h_{11/2}$ band [30], which is similar or perhaps slightly larger compared to the value for the $\pi g_{9/2}$ band in ^{131}Pr. Our observations for the quadrupole deformations of the $\pi g_{9/2}$ and $\pi h_{11/2}$ bands in ^{131}Pr are in agreement with the values reported by Galindo-Uribarri *et al.* [2]. The new results for the highly deformed band in ^{130}Pr is the first direct experimental confirmation of the deformation-driving character of the $\pi g_{9/2}$ orbital in an odd-odd nucleus in the mass~130 region. It is notable that the deformation of structures that involve the $\pi 9/2^+[404]$ ($g_{9/2}$) configuration is comparable to those for the $1/2^+[660]$ ($i_{13/2}$) bands in the neighboring nuclei ^{133}Nd and ^{135}Nd isotopes (see Table 1), thus confirming the important role played by the former orbital in building highly deformed structures in the region.

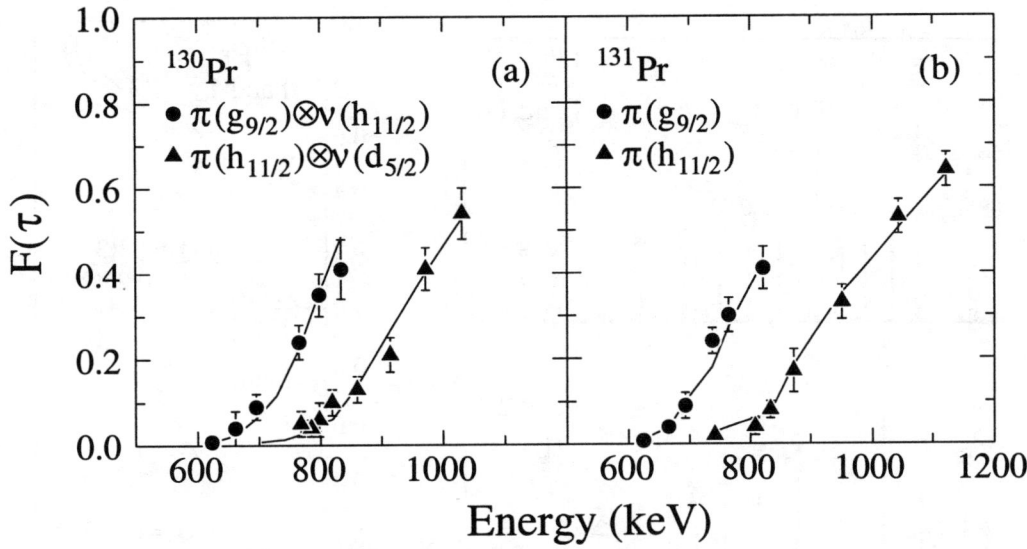

FIGURE 4. The experimental and calculated F(τ) values as a function of γ-ray energy for selected bands in ^{130}Pr and ^{131}Pr. Calculated curves as shown as solid lines and correspond to the best fit to the data.

Bands involving the $1/2^-[541]$ ($f_{7/2}, h_{9/2}$) neutron orbital

Two decoupled bands, referred to as band 1 and band 2 in the current work, were identified in ^{130}Pr in agreement with the parallel work of Smith *et al.* [31]. We confirm the previously reported decoupled band in ^{132}Pr [32,33], but we propose different spin values compared to the work of Shi *et al.* [33] (note that no spin assignments were given by Hauschild *et al.* [32]), and identify several additional in-band and inter-band transitions. Sample γ-ray coincidence spectra for all three bands from the thin target experiment are shown in Figs. 5a, 5b and 5c.

The band in ^{132}Pr decays via several branches to the yrast $\pi h_{11/2} \otimes \nu h_{11/2}$ structure [33,34] which enabled firm spin and parity (relative to the $\pi h_{11/2} \otimes \nu h_{11/2}$ structure) to be assigned. For example, the DCO ratio of 0.66(6), deduced for the 576.5 keV inter-band transition which feeds the 10^+ level of the $\pi h_{11/2} \otimes \nu h_{11/2}$ band [34], using the 541.3 keV stretched quadrupole transition as a gate, implies that the former γ-ray has a dipole character. Note, that a DCO ratio of approximately unity is expected for a $\Delta I = 2$ transition under the same gating condition. The unstretched I$\rightarrow$I assignment for the 576.5 keV γ-ray can be excluded given the relative population of these structures, thus suggesting odd spins for the decoupled band. The observation of the 707.5 keV γ-ray to the 9^+ state of the $\pi h_{11/2} \otimes \nu h_{11/2}$ band in parallel with the 576.5 keV transition and the measured intensity ratio of I$_\gamma$(708)/I$_\gamma$(577)=0.30(4) (see also Fig. 1) indicate that the former transition is a stretched quadrupole, thus establishing positive parity for the band. Band 1 in ^{130}Pr is also observed to decay to levels of the $\pi h_{11/2} \otimes \nu h_{11/2}$ band [34,35], in a way similar to that revealed for the decoupled structure in ^{132}Pr, and is also assigned odd spins and positive parity. The decay path of the second band in ^{130}Pr was difficult to establish, however, it was clear that it proceeded through level(s) at the bottom of the $\pi h_{11/2} \otimes \nu h_{11/2}$ band [34,35], as well (see Fig. 5b). The positive parity and even spins are tentatively assigned to this band. These observations, together with the measured rotational alignments, band crossing properties and the orbitals expected near both the proton and neutron Fermi surfaces, led us to conclude that the configuration of the decoupled structures includes the $1/2^-[541]$ ($f_{7/2}, h_{9/2}$) neutron orbital coupled to the $3/2^-[541]$ ($h_{11/2}$) proton. Such an interpretation is also supported by the measured quadrupole deformations, shown in Figs. 5d, 5e and 5f. It is noteworthy, that the observed $\beta_2 \sim 0.25$ for the decoupled bands in ^{130}Pr and ^{132}Pr lie between values of $\beta_2 \sim 0.20$, measured for the normally deformed structures, and those of $\beta_2 \sim 0.30 - 0.35$ reported for the highly deformed bands which involve the $\nu i_{13/2}$ or the $\pi g_{9/2}$ orbital (see Table 1). Thus, the occupancy of the $1/2^-[541]$ ($f_{7/2}, h_{9/2}$) neutron orbital results in the observation of enhanced deformed bands for nuclei below $N = 73$. The corresponding growth in quadrupole deformation values, however, is not as large as those observed when the $\nu i_{13/2}$ or $\pi g_{9/2}$ orbital is occupied.

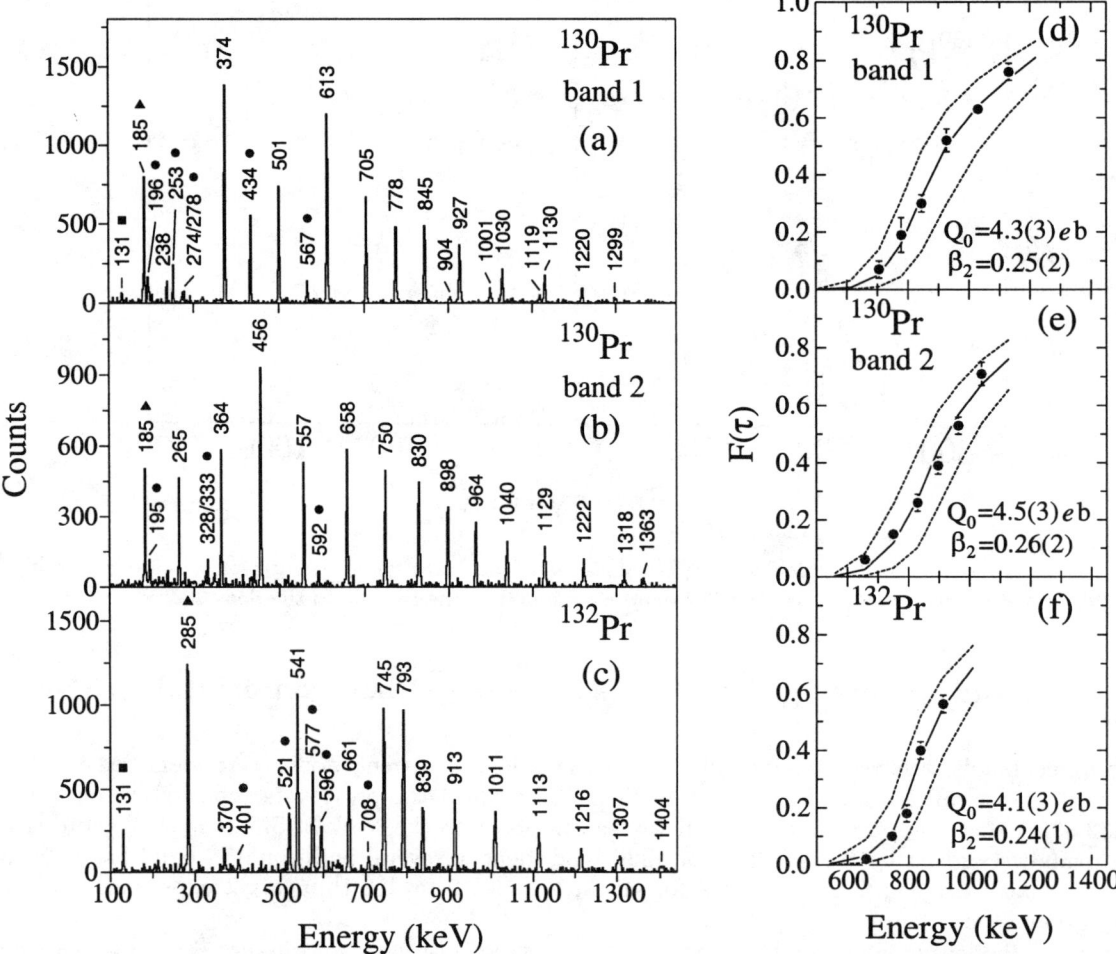

FIGURE 5. (a), (b) and (c) Doubly-gated coincidence γ-ray spectra from the thin target experiment for the decoupled bands in ^{130}Pr and ^{132}Pr. The peaks marked with circles correspond to inter-band transitions and those marked with triangles indicate transitions depopulating the $\pi h_{11/2} \otimes \nu h_{11/2}$ bands. The γ-rays labeled with squares are assigned as the $\pi h_{11/2} \otimes \nu h_{11/2}$ in-band transitions. (d), (e) and (f) The experimental (symbols) and calculated (solid and dashed lines) $F(\tau)$ values as a function of the γ-ray energy. The solid lines represent the best fit values and the deduced Q_0 and β_2 are indicated. The dashed curves correspond to values obtained with $Q_0 \pm 1$ eb.

SUMMARY AND OPEN PROBLEMS

The quadrupole moments of a number of bands, in several Nd and Pr nuclei, were measured using the Doppler-shift attenuation method as a part of our systematic study dedicated to understand the properties of highly deformed structures in mass~130 region. Differences in the observed deformations clearly demonstrate the important role played by the occupation of the $9/2^+[404]$ $(g_{9/2})$ proton, $1/2^+[660](i_{13/2})$ and $1/2^-[541]$ $(f_{7/2}, h_{9/2})$ neutron orbitals on the properties of the highly deformed bands.

These systematic data leave open several problems to be addressed by theory and further experiments.

(a) Due to differences in the experimental conditions, the parameterization of the stopping power data and the modeling of the sidefeeding lifetime contributions, the conclusive comparison between the quadrupole deformations reported by various authors is rather limited. The lack of experimental knowledge on the stopping powers is a difficult problem to overcome and direct measurements would be clearly valuable. Lifetime measurements, in the picosecond range, can be performed for in-band states using both the coincidence Recoil Distance Method, which is not sensitive to the stopping powers, and the DSAM lineshape technique. The comparison of the two sets of data could be used to extract useful information about the stopping powers. In addition, the improved efficiency of the new large detector arrays allows the reduction of some of the systematic pitfalls inherent in the DSAM techniques, by measuring the properties of a large number of structures simultaneously, in a single experiment. It has been shown here, that if the time structure of the sidefeeding is not taken into account correctly, the extracted quadrupole deformations for highly deformed bands will be lower (usually $\sim 10\%$) than the true values. One immediate approach that can be successfully applied in removing the effect of unknown sidefeeding is to gate on Doppler-shifted transitions at the top of the band of interest. It would also be valuable to perform quadrupole moment measurements as a function of spin using the DSAM lineshape analysis technique.

(b) The properties of the highly deformed bands are often judged only by the relative behavior of the dynamic moments of inertia, which depends not only on the variations in deformation, but also on the magnitude of the pairing field, which changes with the increase of both the seniority and rotational frequency, and the quasiparticle alignment effects. The deformation parameters of many structures in this mass region are successfully described by calculations which allow pairing correlations to take a full effect and also by those which completely ignore them, thus raising the question: "Is the pairing important and what role does it play?". The presence or absence of the pairing correlations will have also important consequences on the decay out mechanism of the highly deformed structures, as revealed for example in other mass regions.

(c) Some of the observed irregularities in collective bands at high spin are suggested to arise from a simple crossing between specific single-particle levels in a rotating frame. Such a scenario, however, may not be true in the case of odd-odd nuclei, as well as for high seniority bands in other nuclei, due to the additional effect of the residual nucleon-nucleon ($\pi - \nu$, $\pi - \pi$ and $\nu - \nu$) interactions. The role played by the residual interactions and their influence on the properties of the highly deformed structures clearly needs a further investigation. In the odd-odd nuclei there is a significant increase in the density of bands which results in greater fragmentation of the γ-ray intensity flow and consequently in lower population of a particular structure. In addition, there are difficulties and complications in linking together observed structures, particularly where low-energy transitions and/or long lifetimes are presented, which result in tentative spin and parity assignments, thus generating much speculations about the involved configurations. Many more highly deformed bands, formed by the coupling of different proton (neutron) orbitals to the deformation driving $\nu i_{13/2}$ or $\nu(f_{7/2}, h_{9/2})$ neutron ($g_{9/2}$ proton), are expected in odd-odd nuclei in this mass region. The observation of these structures and the elucidation of their detailed properties remain a challenge for further studies.

ACKNOWLEDGMENTS

The authors wish to extend their thanks to the staff of the LBNL GAMMASPHERE facility and the crew of the 88″ Cyclotron for their assistance during these experiments. The software support of D.C. Radford and H.Q. Jin is greatly appreciated. Discussions with L.L. Riedinger are gratefully acknowledged. Support for this work was provided by the U.S. Department of Energy under Contract No. DE–AC03–765F00098 and Grant No. DE–FG02–88ER40406, the National Science Foundation, the State of Florida and the U.K. Engineering and Physical Sciences Research Council. MAR and JS acknowledge the receipt of a NATO Collaborative Research Grant.

REFERENCES

1. Wyss, R., et al., *Phys. Lett.* **B215**, 211–217 (1988).
2. Galindo-Uribarri, A., et al., *Phys. Rev.* **C50**, R2655–R2659 (1994).
3. Galindo-Uribarri, A., et al., *Phys. Rev.* **C54**, 1057–1069 (1996).
4. Galindo-Uribarri, A., et al., *Phys. Rev.* **C54**, R454–R458 (1996).
5. Wadsworth, R., et al., *Nucl. Phys.* **A526**, 188–204 (1991).
6. Janssens, R., and Stephens, F., *Nucl. Phys. News* **6**, 9–17 (1996).
7. Sarantites, D.G., et al., *Nucl. Instrum. Methods Phys. Res.* **A381**, 418–432 (1996).
8. Alexander, T.K., and Forster, J.S., *Advances in Nuclear Physics*, New York: Plenum Press, 1978, vol.10, pp.197.
9. Radford, D.C., *Nucl. Instrum. Methods Phys. Res.* **A361**, 297–305 (1995).
10. Moore, E.F., et al., *Phys. Rev.* **C55**, R2150–R2154 (1997).
11. Ziegler, J.F., Biersack, J.P., and Littmark, U., *The Stopping and Range of Ions in Solids*, New York: Pergamon Press, 1985; Ziegler, J.F., (private communication).
12. Blaugrund, A.E., *Nucl. Phys.* **88**, 501–512 (1966).
13. Mullins, S.M., et al., *Phys. Rev* **C45**, 2683–2692 (1992).

14. Medina, N.H., *et al.*, *Nucl. Phys.* **A589**, 106–116 (1995).
15. Diamond, R.M., *et al.*, *Phys. Rev.* **C41**, R1327–R1331 (1990).
16. Petrache, C.M., *et al.*, *Phys. Rev.* **C57**, R10–R14 (1998).
17. Wadsworth, R., *et al.*, *J. Phys. G: Nucl. Phys.* **13**, L207–L212 (1987).
18. Beck, E.M., *et al.*, *Phys. Rev. Lett.* **58**, 2182–2185 (1987).
19. Bazzacco, D., *et al.*, *Phys. Lett.* **B309**, 235–240 (1993).
20. Bazzacco, D., *et al.*, *Phys. Rev.* **C49**, R2281–2284 (1994).
21. Deleplanque, M.A., *et al.*, *Phys. Rev.* **C52**, R2302–2305 (1995).
22. Lunardi, S., *et al.*, *Phys. Rev.* **C52**, R6–R10 (1995).
23. Petrache, C.M., *et al.*, *Nucl. Phys.* **A617**, 228–248 (1997).
24. Petrache, C.M., *et al.*, *Phys. Lett.* **B219**, 145–150 (1996).
25. Wilssau, P., *et al.*, *Phys. Rev.* **C48**, R494–497 (1993).
26. Clark, R.M., *et al.*, *Phys. Rev. Lett.* **76**, 3510–3513 (1996).
27. Regan, P.H., *et al.*, *J. Phys. G: Nucl. Phys.* **18**, 847–858 (1992).
28. Afanasjev, A.V., and Ragnarsson, I., *Nucl. Phys.* **A608**, 176–201 (1996).
29. Brown, T.B., *et al.*, *Phys. Rev.* **C56**, R1210–R1214 (1997).
30. Kondev, F.G., *et al.*, *Eur. Phys. J.* **A2**, 249–251 (1998).
31. Smith, B.H., *et al.* (to be published); Riedinger, L.L., *et al.* (private communication).
32. Hauschild, K., *et al.*, *Phys. Rev.* **C50**, 707–715 (1994).
33. Shi, S., *et al.*, *Phys. Rev.* **C37**, 1478–1484 (1988).
34. Petrache, C.M., *et al.*, *Nucl. Phys.* **A635**, 361–383 (1998).
35. Ma, R., *et al.*, *Phys. Rev.* **C37**, 1926–1931 (1988).

Superdeformation in the A=150 and A=190 Regions

M. P. Carpenter and R. V. F. Janssens

Argonne National Laboratory, Argonne, IL 60439

Abstract. Superdeformation has been established for over a decade in the mass 150 region and nearly as long in the A=190 region. The first measurements directed at nuclei in these regions concentrated on mapping out the superdeformed (SD) islands by identifying SD rotational bands in γ-ray coincidence data. These early studies provided new insights into the physics of superdeformation, but also raised unexpected issues. The new gamma-ray arrays (Gammasphere, Eurogam/Euroball and Gasp) have provided a wealth of new data on properties of SD states in these two mass regions. This paper highlights some of the more recent results from the large arrays which have addressed the outstanding issues in the field, namely, $\Delta I = 4$ staggering, identical bands, SD vibrational bands, and questions about the feeding into and the decay out of the SD well.

INTRODUCTION

Excited states associated with superdeformed (SD) shapes were first observed over thirty years ago in nuclei of the actinide region, around mass 240. These examples are the well known fission isomers, and a summary of both the experimental and theoretical work on this subject can be found in the review article by Bjornholm and Lynn [1]. While much information was obtained on the properties of these isomers in the 1960's and 1970's, little is known about the excited states built on these isomeric SD band heads.

This situation changed dramatically in 1986, with the identification of a SD rotational band in ^{152}Dy [2]. A subsequent lifetime measurement on the transitions in this SD band confirmed that the average quadrupole deformation of the band (β_2) was as large as that of the fission isomers, i.e., the deformation corresponds to a major to minor axis ratio of 2:1 for the prolate nucleus [3]. Superdeformation in the A=190 region was first reported in 1989 by Moore *et al.* [4] when a rotational SD band was identified in ^{191}Hg. The measured quadrupole moment of this band confirmed a large deformation corresponding to a major to minor axis ratio of 1.7:1. By now islands of superdeformation in these two mass regions have been well established and can be traced to large gaps in the calculated single-particle energies at large deformation. For the A=150 region, these gaps are calculated to occur at Z=66 and N=86 (^{152}Dy) while for A=190 they are calculated to be at Z=80 and N=112 (^{192}Hg). As a result, ^{152}Dy and ^{192}Hg are often referred to as doubly-magic SD nuclei. Figure 1 shows the extent of the experimentally established islands of superdeformation in both mass regions.

Many SD bands have been observed in both the A=150 and A=190 regions, and as a result, features common to the bands of a particular mass region have become apparent. For example, bands in the A=150 region have on average more coincident γ rays and the transition energies are twice as large as those observed in mass 190 SD bands. The latter observation is due primarily to the fact that the spin region over which the bands are observed is different. For the A=150 region, the SD bands cover a spin range from $I \sim 25 - 65\hbar$ while the SD bands in the A=190 region are observed from $I \sim 10 - 45\hbar$. The population pattern of the bands in the two regions exhibit the same general features: as the spin decreases the intensity in the SD band increases, reaches a maximum, and then drops rapidly over two or three transitions as the band depopulates and decays to the yrast levels. The transitions linking the SD band to the known yrast states are in general not observed and, thus, crucial information such as spins, parities and excitation energies of states in the SD well are not known experimentally. Isotopic identification comes from observing coincidences between known yrast transitions and members of the SD band.

Most of the studies on SD nuclei in these two mass regions carried out in the late 1980's and early 1990's focused on mapping out the two islands of superdeformation. A summary of this early work can be found in

CP481, *Nuclear Structure 98*, edited by C. Baktash
1999 American Institute of Physics 1-56396-858-4

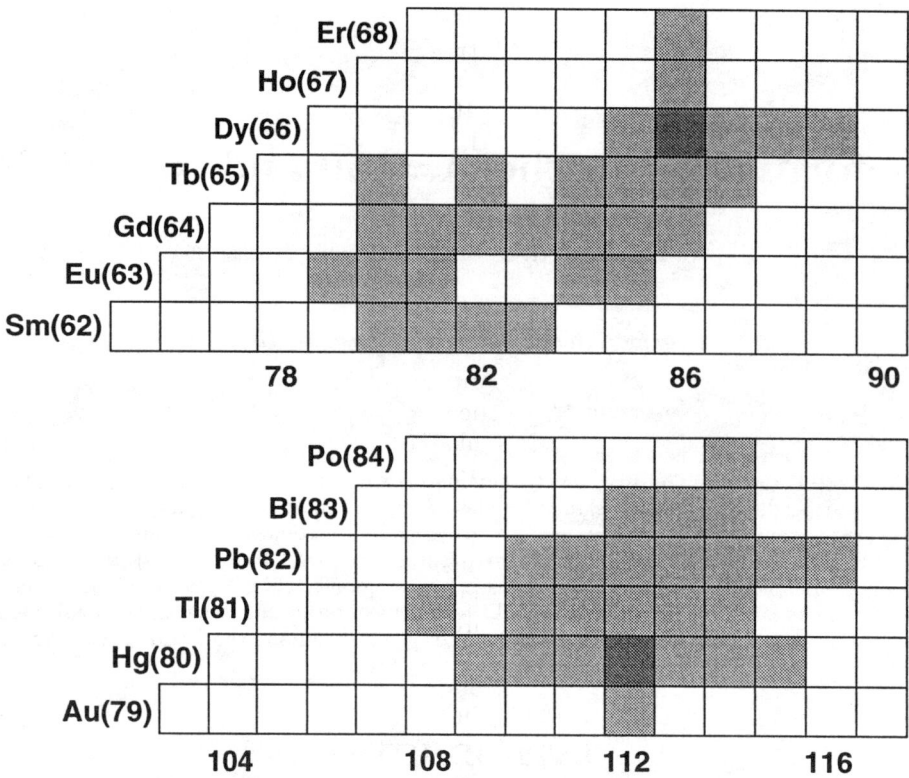

FIGURE 1. Island of superdeformation for the A=150 (upper) and A=190 (lower) regions. Nuclei in which SD bands have been identified are shaded. The darker shaded squares indicate the doubly-magic superdeformed nuclei ^{152}Dy and ^{192}Hg.

the reviews by Nolan and Twin [5] and Janssens and Khoo [6]. The measurements were performed with so-called second generation γ-ray arrays which consisted of 10-20 Compton-suppressed Ge (CSGe) spectrometers coupled to a sum-energy and multiplicity array. Since the spins of the SD bands were not known, it became a common practice to present the dynamical moment of inertia ($\mathcal{J}^{(2)}$) as a function of rotational frequency ($\hbar\omega$) in order to study the response of the nucleus under the stress of rotation. The quantity $\mathcal{J}^{(2)}$ is defined as the second derivative of the excitation energy with respect to spin. Experimentally, it is represented as $4/\Delta E_\gamma$ where ΔE_γ is the energy difference between successive transitions in a rotational band. Therefore, this quantity is independent of spin.

For the mass 150 region, the evolution of the $\mathcal{J}^{(2)}$ moment as a function of $\hbar\omega$ was found to be characterized by pronounced isotopic and isotonic variations. These variations have been attributed to differences in the occupation of specific high-N intruder orbitals, namely $j_{15/2}$ neutrons and $i_{13/2}$ protons [7]. This realization represents one of the early theoretical successes in understanding the underlying single-particle nature of SD bands. In contrast, the vast majority of SD bands in the mass 190 region were found to display the same smooth and rather pronounced increase in $\mathcal{J}^{(2)}$ with increasing $\hbar\omega$. It is generally accepted that this rise results from quasiparticle alignments of the same high-N, $j_{15/2}$ and $i_{13/2}$ intruder orbitals. This contrast between the $\mathcal{J}^{(2)}$ moments of inertia in the two mass regions highlights one of the more interesting differences between A=150 and A=190 SD bands: pair correlations have a major influence on the properties of the SD bands in the A=190 region while "single" particle features dominate near A=150.

With the construction of the present generation of powerful γ-ray spectrometers, Gammasphere, Eurogam/Euroball and Gasp, a new era in the study of nuclei at high angular momentum has begun. These new arrays are characterized by a gain in detection sensitivity of several orders of magnitude over earlier detection systems, and thus, are especially suitable for the study of SD bands which are weakly populated in heavy-ion induced fusion reactions. In the majority of nuclei where superdeformation had been established, multiple SD bands have now been observed. As examples of this fact, it is interesting to note that up to nine SD bands have been reported in a single nucleus in the A=190 region (^{192}Tl [8]), while in the A=150 region, thirteen SD bands have recently been observed in ^{149}Gd [9]. Configurations for all thirteen bands have been proposed in ref. [9], underscoring the fact that the intrinsic structure of SD bands is well understood in both mass regions. A review of the recent theoretical work which has led to this understanding can be found in the contribution of J. Dobaczewski to these proceedings [10].

In the remainder of this contribution, experimental results which address outstanding questions associated with superdeformation in the A=150 and 190 regions will be reviewed. Specifically, we will discuss the status of $\Delta I = 4$ staggering, of identical bands, of the presence of collective excitations in the SD well, and of feeding and decay of SD bands.

$\Delta I = 4$ BIFURCATION IN SUPERDEFORMED NUCLEI

Rotational states in nuclei are usually classified by two quantum numbers: parity and signature. The signature quantum number (α) is associated with the fact that the intrinsic Hamiltonian is invariant under a rotation by 180 degrees around an axis perpendicular to the symmetry axis (C_2 symmetry). As a result, rotational bands built on configurations with $K \neq 0$ are characterized by two $\Delta I = 2$ rotational sequences with each of these labeled by its signature quantum number (0 or 1 for even nuclei, 1/2 or -1/2 for odd nuclei). If these so-called signature partner bands do not undergo Coriolis mixing, they form a regular sequence of states which differ in spin by 1 $\hbar$. However, strong Coriolis mixing can cause the two sequences to decouple, and levels in one band are then pushed up in energy while levels in the other are pushed down relative to their respective unperturbed states. The splitting of the two signature partner bands is often referred to in the literature as $\Delta I = 1$ staggering or $\Delta I = 2$ bifurcation.

Recently, a new phenomenon, $\Delta I = 4$ bifurcation, has been observed in several A=150 SD nuclei. The first reported case was in ^{149}Gd from data taken with Eurogam [11]. In this experiment, the $\mathcal{J}^{(2)}$ moment for the yrast SD band in ^{149}Gd was observed to have an anomalous staggering for $\hbar\omega > 0.49$ MeV. In order to quantify this staggering, the differences in energy between two consecutive γ rays in the SD band were plotted after subtracting a smooth reference given by $\Delta E_\gamma^{ref}(I) = [\Delta E_\gamma(I + 2) + 2\Delta E_\gamma(I) + \Delta E_\gamma(I - 2)]/4$. When this quantity is plotted as a function of rotational frequency (see top panel in fig. 2), it was observed that the SD band splits into two sequences where successive members differ in spin by $4\hbar$. In analogy to $\Delta I = 2$ bifurcation, it was noted in ref. [11] that this bifurcation could be accounted for if levels in the SD band were alternatively pushed up and down by 58 ± 18 eV relative to the unperturbed positions defined by the reference band. This small perturbation of the levels can be compared with that observed in $\Delta I = 2$ bifurcation which can be on the order of 10 to 100 keV.

In ref. [11], it was suggested that this effect might reflect the fact that the intrinsic Hamiltonian of the system is invariant under a rotation by $\pi/2$ (C_4 symmetry), and this possibility was linked to the hexadecapole deformation. Several theoretical papers began to explore this possibility [12–14]. Hamamoto and Mottelson assumed that the C_4 symmetry was brought about by the Y_{44} deformation which produces hexadecapole distortions in the plane perpendicular to the symmetry axis of the prolate nucleus [12]. The Hamiltonian studied

$$H = AI_3^2 + B_1(I_1^2 - I_2^2)^2 + B_2(I_1^2 + I_2^2)^2, \tag{1}$$

is invariant with respect to the $C_{4\nu}$ point symmetry. By choosing appropriate values for the A, B_1 and B_2 coefficients ($B_2 = 0$, and $A/B_1 \sim 100$), a regular $\Delta I = 2$ staggering pattern occurs for the calculated transition energies. This pattern results from tunneling between the four equivalent minima present in the classical phase space.

Following the initial observation of $\Delta I = 4$ bifurcation in ^{149}Gd, more examples were reported in both the 150 and 190 SD regions [15–17]. However, the presence of bifurcation in all of the SD bands involved has been brought into question by subsequent measurements with larger data sets [8,18,19]. On the theoretical side, it has been demonstrated that the magnitude of hexadecapole deformation necessary to induce the staggering is much larger than is reasonable [20]. Other interpretations have been put forth as well. For example, several papers have suggested that the effect could be brought about by mixing between two or more bands [21,22], another discusses staggering in terms of intrinsic vortical motion [23], while yet another interpretation has suggested that $\Delta I = 4$ bifurcation is associated with the quadrupole deformation as defined within the interacting boson model [24].

A recent report by Haslip et al. [19] from data taken at Gammasphere identifies two more examples of $\Delta I = 4$ bifurcation in nuclei neighboring ^{149}Gd, i.e. ^{148}Eu and ^{148}Gd. Figure 2 presents the staggering plots

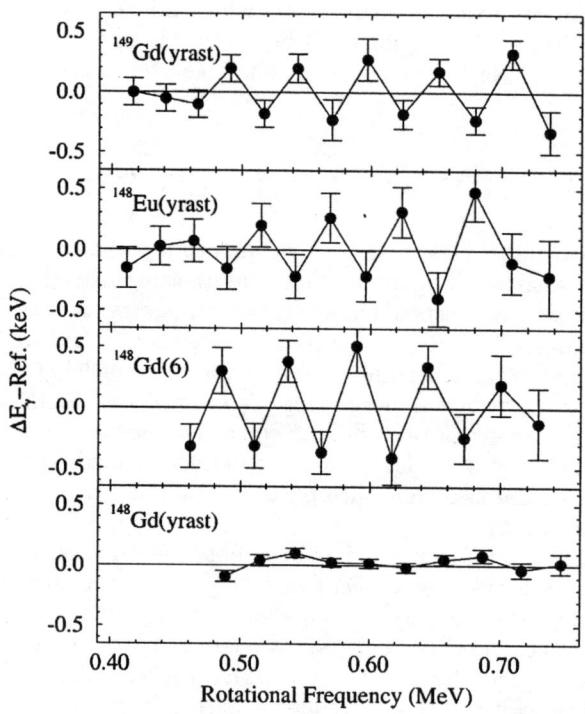

FIGURE 2. Staggering plots for ^{149}Gd(yrast), ^{148}Gd(yrast), ^{148}Gd(6), and ^{148}Gd(yrast) the SD bands. The figure is taken from ref. 19.

for SD bands ^{148}Gd(6) and ^{148}Eu(yrast), the two newly discovered examples. Also plotted in the figure are ^{149}Gd(yrast) (from ref. [11]) and ^{148}Gd(yrast) where no staggering occurs. The magnitude of the effect in the two new examples is comparable to that in ^{149}Gd(yrast). In addition, the ^{148}Gd(6) has identical γ-ray energies to those of ^{149}Gd(yrast) in the frequency range where staggering is observed. Interestingly, the phase of the oscillation is opposite in sign. The structure of ^{148}Gd(6) has been proposed to be the same as that of ^{149}Gd(yrast), except for a hole in the [411]1/2 neutron orbital. The ^{148}Eu yrast band is proposed to have the configuration ^{149}Gd$\otimes\pi$[301]1/2^{-1}, and its moment of inertia is identical to that of ^{149}Gd(yrast) in the staggering region.

One intriguing aspect of these new results is that they can be understood with the Hamiltonian given in (1). As pointed out in ref. [12], the tunneling amplitude can be expressed as a function of the angular momentum, $\cos[\frac{\pi}{2}(I - I_0')]$, where $I_0' = \sqrt{\frac{1}{4}(\frac{A}{B_1} - \frac{3}{2}\hbar^2)}$. By assuming the relative spins (for the 3 bands), ref. [19] has shown that the staggering as a function of spin can be reproduced by assuming a common value for $I_0' = (0.76 \pm 0.08)$. In addition, the model accounts for the fact that the staggering in the isospectral bands (^{149}Gd(yrast) and ^{148}Eu(yrast)) is opposite in phase. While this adds support to the Hamiltonian suggested by Hamamoto and Mottelson, it is fair to state that the mechanism responsible for the $\Delta I = 4$ bifurcation remains unclear at the moment. This is a topic where more experimental examples would be very useful or even required to help solve the puzzle.

IDENTICAL BANDS IN SUPERDEFORMED NUCLEI

One of the more fascinating properties associated with the study of SD nuclei is that of identical bands. This classification not only includes pairs of SD bands which share identical transition energies (isospectral), but also encompasses SD pairs sharing identical moments of inertia. When first reported, this phenomenon was very surprising since mass differences between neighboring nuclei were expected to result in differences between the transition energies of SD bands which are an order of magnitude larger than those observed in identical bands.

The first report of an isospectral SD pair came from an experiment which observed an excited SD band in ^{151}Tb whose transition energies were identical to the original ^{152}Dy SD band [25]. It was shown in ref. [26] that this phenomenon could be accounted for if the SD band in ^{151}Tb (i) was built on an intrinsic excitation with $\Omega = 1/2$, (ii) had a decoupling parameter of a=+1, and (iii) had an identical moment of inertia to that of the core. Requirements (i) and (ii) were shown to be satisfied in this case by attributing the SD band in ^{151}Tb to the configuration ^{152}Dy(yrast)$\otimes \pi[301]1/2^{-1}$ and assuming that the pseudo-SU(3) asymptotic limit was nearly reached at large deformations [26]. In this model, it is the decoupling and alignment of the spin of the valence nucleon which allows for isospectral bands to be observed in neighboring nuclei. However, no explanation was proposed in [26] as to why the bands were isospectral to such a high-degree of precision.

Shortly after this report, examples of identical bands were also reported in the A=190 region. The first examples in ^{191,193,194}Hg all showed an identical band relationship with the yrast SD band of ^{192}Hg. More surprising still was the suggestion that these bands exhibited an alignment of 1 $\hbar$ relative to ^{192}Hg [27]. The latter observation was based on the best estimates for the spins of these bands, since these quantities had not yet been measured experimentally. Stephens *et al.* [27] suggested that this phenomenon could be explained if the nucleonic spins from a quasiparticle pair decoupled and aligned with the axis of rotation while the orbital angular momentum remained strongly coupled to the symmetry axis. While this suggestion accounted for the relative spins of the bands, it offered no explanation for the fact that the bands were identical.

In the following years, many examples of identical bands have been uncovered, and there is at present no universally accepted explanation of the phenomenon. Theoretical suggestions have ranged from the possible manifestation of a new symmetry [28] to subtle cancellation effects between mass and pairing terms in the collective Hamiltonian [29], from the continuous readjustment of the self-consistent mean field with angular momentum [30] to a balance between residual p-n and pairing interactions [31]. A thorough review of both experimental data and theoretical work associated with identical bands is given in ref. [32].

With the current generation of γ-ray arrays much additional information has been obtained with regards to identical bands in the A=150 and A=190 regions. For example, many new pairs are now known, and the energies of these SD bands are determined with a high degree of accuracy. Several studies have tried to quantify the preponderance of identical bands in both normal deformed (ND) and SD nuclei [33,34]. In ref. [33], this was done by evaluating the fractional change (FC) in the $\mathcal{J}^{(2)}$ moment between two rotational bands where $FC(\omega) = [\mathcal{J}_X^{(2)}(\omega) - \mathcal{J}_Y^{(2)}(\omega)]/\mathcal{J}_X^{(2)}(\omega)$ and the mass of the nucleus corresponding to band X is always the largest. In addition, it can be shown that $FC = di_{eff}/dI_X$, where i_{eff} is the difference in spins between the two bands at fixed $\hbar\omega$. Thus, if i_{eff} is a linear function of I, the average slope of this function yields the average fractional change, $\overline{FC}$. This is the quantity presented in figure 3 for SD bands in the A=130, 150 and 190 regions. In panels a-c, all possible pairs are presented.

For A=150 SD nuclei (3-b), the distribution shows a number of sharp peaks. The one at $\overline{FC} = 0$ represent the identical bands. The other peaks in the distribution can be correlated with a change in the the number of occupied high-N intruder orbitals. For example, an analysis of the intruder content of the bands producing the large peak at 0.1 shows that 85% of the pairs tested differ in intruder content by one $i_{13/2}$ proton and one $j_{15/2}$ neutron, irrespective of their actual mass difference. Additionally, the $\overline{FC}$ values are predominantly positive which indicates that the $\mathcal{J}^{(2)}$ moment increases with mass. This again can be traced to the filling of the intruder orbitals as N and Z increase. In contrast, the A=190 (and A=130) SD nuclei show a broader distribution centered around $\overline{FC} = 0$, followed by secondary distributions on either side which are greatly reduced relative to the major distribution. The full distribution has been interpreted as resulting from the static pair correlations present in the A=190 SD nuclei. Support for this interpretation comes from the fact that ND nuclei show similar (although broader) distributions. This is illustrated in panels e and f of figure 3 where a subset of SD distributions are compared with normal deformed distributions. For SD nuclei, this subset corresponds to pairs of bands where the reference is the yrast band in a nucleus with mass A and the other member of the pair resides in the neighboring nucleus with A-1. For the normal deformed distribution odd-A

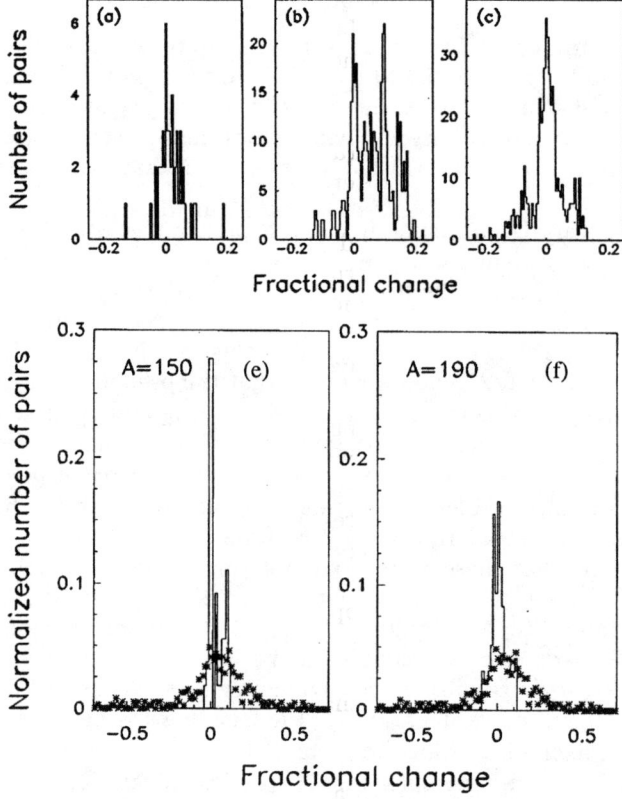

FIGURE 3. Panels a-c give the distributions for the average fractional change ($\overline{FC}$) in the $\mathcal{J}^{(2)}$ moment of SD bands in the mass 130, 150 and 190 regions. Panels e and f give the $\overline{FC}$ distributions for a subset of SD bands in the A=150 (e) and A=190 regions (f). In these cases, the comparison is performed for neighboring nuclei where the band in the heavier mass nucleus is always the yrast SD band. The stars show the corresponding distributions for normally deformed nuclei where odd-A nuclei are compared with yrast bands in even-even nuclei which are heavier in mass by one unit. The figure is taken from ref. 33.

rare-earth nuclei are referenced to the yrast band in the neighboring even-even nucleus. The distribution for ND nuclei is similar to that for A=190 SD bands, but is significantly wider as would be expected because of increased pair correlations.

While the work of ref. [33] does not offer an explanation for identical bands, it has shown that they are are more prevalent for SD nuclei than ND nuclei. This fact appears to be correlated with the decrease in the pairing strength in the SD well. In addition, the A=150 SD bands are observed to fall into distinct families defined by the number of high-N intruders occupied. A similar correlation cannot be made in the A=190 SD region.

Progress has also been made in characterizing the quadrupole moments of identical bands. This is accomplished by using the Doppler Shift Attenuation Method (DSAM) to measure the lifetimes of the SD states to high precision. The first such measurement using the new arrays was performed in the A=150 region by Savajols *et al.* [35] with the Eurogam spectrometer. Quadrupole moments were deduced for SD bands in ^{148,149}Gd nuclei produced in the reaction ^{30}Si + ^{124}Sn and for SD band 1 in ^{152}Dy produced in the ^{120}Sn(^{36}S,4n) reaction. These quadrupole moments were found to vary from 14.6 eb to 17.5 eb. Two interesting conclusions were drawn from the data: (i) a direct relationship was observed between the number of occupied high-N orbitals and the magnitude of the quadrupole moments, and (ii) identical bands were found to also be characterized by identical quadrupole moments, e.g. ^{152}Dy(1) and ^{149}Gd(4). A similar measurement by Nisius *et al.* [36] with Gammasphere on SD bands of ^{151,152}Dy and ^{151}Tb reached the same conclusions, and quantified the contributions of individual high-N orbitals to the quadrupole moments.

This link between identical SD bands in the A=150 and their high-N intruder content was examined in a recent theoretical investigation using the cranked Hartree-Fock (HF) Skyrme model [37]. In this study, an attempt was made to calculate relative quadrupole moments between SD bands by considering only contributions from the effective single-particle multipole moment $q_\lambda(i)$ relative to a common core (^{152}Dy,yrast). Once the individual $q_\lambda(i)$ values were determined, quadrupole moments in SD bands were constructed by directly adding the $q_\lambda(i)$'s to produce the desired SD configurations relative to ^{152}Dy(yrast). Both quadrupole ($\lambda = 2$) and hexadecapole ($\lambda = 4$) contributions were considered. The calculated relative quadrupole moments (δQ_2) using this extreme shell model approach were then compared to the experimental results of references [35] and [36], and excellent agreement between theory and experiment was found. The largest contributions to the quadrupole moments originate from the high-N intruder configurations. For identical bands, the difference in single-particle occupations come from orbitals which contribute very little to the quadrupole moment and apparently also to the moment of inertia.

More recent experimental lifetime measurements have attempted to test the limits of this concept of additivity of quadrupole moments as more and more particles are removed/added to the ^{152}Dy SD core. The work of Hackman et al. [38] reports on lifetime measurements in SD bands of ^{142}Sm, a nucleus lying at the lower corner in N and Z of the A=150 SD region, ten nucleons away from ^{152}Dy. In ^{142}Sm two SD bands are known and are found to have quadrupole moments of 11.7 eb and 13.2 eb. The two bands differ in intruder content by one i$_{13/2}$ proton and one j$_{15/2}$ neutron. The difference of 1.5 eb between the bands can be accounted for fully by this difference in intruder content. In addition, the quadrupole moments of these bands can be accounted for by starting from the ^{152}Dy core and subtracting the contributions from the single-particle orbitals. It was concluded that additivity of quadrupole moments applies when as many as 10 particle are removed from the core.

In contrast, lifetime measurements in ^{153}Dy appear to violate the additivity concept. Based on the assigned configuration for SD bands 1-3, additivity predicts that these bands have quadrupole moments between 17.3 and 17.9 eb. The experimental quadrupole moments are found to be $\sim$ 16.2 eb for all 3 bands [39]. Thus, the deformation of the core appears to shrink immediately above the N=86 single-particle gap. Interestingly, SD bands 2 and 3 are identical to ^{152}Dy(yrast), and as a result, this represents a case where identical bands do not have identical quadrupole moments.

The concept of additivity has also recently been extended to relative alignments between SD bands [40]. This approach builds on the pioneering work on effective alignments by Ragnarsson [41]. In this study, incremental alignments of specific orbitals were extracted from the data. These alignments were then used to calculate transitions energies for known SD bands by adding these effective alignments to a common core. The bands calculated in this way showed excellent agreement with the experimentally measured bands. This again stresses the concept of an extreme shell model picture for the SD bands of the A=150 region.

Precision lifetime measurements have also been made on SD bands in the A=190 region. One recent measurement has focused on the yrast SD band of ^{192}Hg and on the three SD bands of ^{194}Hg [42]. In this study, all SD bands were found to have the same quadrupole moment within errors ($Q_t \sim 17.5$ eb). The $\mathcal{J}^{(2)}$ moment of inertia of all 4 bands are similar, and the two excited bands in ^{194}Hg are identical to the yrast SD band in ^{192}Hg. On the other hand, a recent lifetime measurement in ^{190}Hg shows that the yrast and first excited SD band have quadrupole moments which are also identical to the heavier Hg isotopes, even though the $\mathcal{J}^{(2)}$ moments for these bands differ markedly from those measured in the heavier isotopes [43]. The fact that large deviations are not observed in the quadrupole moments in contrast to what is observed in the A=150 region can be explained by the presence of static pair correlations which minimize the polarizing effects of the high-N orbitals.

Somewhat puzzling are the results recently obtained for ^{193}Hg by Busse et al. [44] where the quadrupole moments measured for five SD bands in ^{193}Hg were found to be smaller than that for ^{192}Hg yrast by $\sim$ 15%. There is no apparent explanation for this observation, and it is unclear whether the effect is due to a real difference in quadrupole moments or to a difference in the properties of the sidefeeding. Another measurement of the relative quadrupole moments using a different reaction in ^{192,193}Hg is highly desirable.

With all these studies, have we moved any closer to understanding the identical band phenomenon? In the A=150 region, the characterization of the bands is well founded. For many cases quadrupole moments and transition energies can be described by the additivity of single-particle contributions. Current theoretical calculations using cranked-HFB or cranked relativistic mean field are able to reproduce the observables (moments of inertia, quadrupole moments) extremely well (see ref. [10] for a review), and one may be tempted to dismiss the phenomenon as a natural consequence of the mean field involving no new physics. However, there has been a recent theoretical realization that pseudo-spin symmetry can be linked to explicit symmetries in the relativistic mean field [45]. If so, it is interesting to wonder whether or not such symmetries could also be related to identical SD bands in the A=150 region. No such link have yet been established, but it is clearly an avenue which should be pursued.

In the A=190 region, the characterization of the bands is also well founded, and in a few instances, excitation energies and spins have now been determined (see below). However, pair correlations do not allow for a simple single-particle description of SD bands. Attempts to apply a so-called extreme quasiparticle picture have been made but found to work only in isolated cases [17]. Theoretical calculations are doing well in reproducing the moments of inertia of the bands, but not as well as in the A=150 region, due to the fact that one must deal with the consequences of pairing [46]. In addition, it is difficult to point to a specific common feature or structure for the identical bands in the A=190 region. For example, a grouping of identical bands is found to include excitations built on SD ground states, quasiparticle excitations and octupole vibrations (see following discussion). Thus, an understanding of identical bands based on some fundamental symmetry is less clear. The idea that this linking commonality could be pseudo-spin has been revived in a recent paper by Schuck et al., where a pair of newly observed SD bands in ^{191}Au is suggested to show pseudo-spin alignment for excitations with a pseudo orbital angular momentum ($\tilde{\Lambda}$) of one [47]. Evidence for such alignments in normally deformed nuclei with $\tilde{\Lambda} = 0, 1$ has recently been presented by Stephens et al. [48].

VIBRATIONAL EXCITATIONS IN THE SUPERDEFORMED WELL

The intrinsic structure of nearly all SD bands in the A=150 and A=190 regions can be understood in terms of either single-particle or quasiparticle excitations. In contrast, until recently, there had been no evidence for SD bands built on collective excitations. This came as somewhat of a surprise since rotational bands associated with quadrupole and octupole vibrations are commonplace features in nuclei at normal prolate deformation. Several recent theoretical investigations have suggested that collective octupole vibrations play a significant role in the SD wells of the A~150 and A~190 nuclei due to the presence near the Fermi surface of pairs of orbitals with opposite parity and $\Delta l = 3$ [49–51].

Strong evidence for an SD band built on an octupole vibrational state was found at Gammasphere in its early implementation phase. In this study, states in the SD well of ^{190}Hg were populated, and a newly identified SD band was found to decay over at least three levels to the yrast SD states rather than to the states of normal deformation [52]. Based on the branching to the yrast SD band, it was suggested that the newly observed band was built on an octupole vibration. A subsequent experiment with Eurogam measured the energies of the transitions linking the excited and yrast SD bands. The angular distributions of these connecting transitions were found to be consistent with $\Delta I = 1$ dipoles [53]. Assuming an E1 character results in transition matrix elements on the order of 10^{-3} Weisskopf units (W.u.) for these linking transitions. However, these rates were only estimates, since the quadrupole moment for either band was not known. In addition, the anomalously large $\mathcal{J}^{(2)}$ moment for the excited SD band left open the possibility that the deformation in the two SD bands was significantly different. Recently, the quadrupole moments of both bands have been measured [43] and found to be identical (~ 17.6 eb). With the quadrupole moments determined, the $B(E1)$ rates were extracted from the measured intensities. The B(E1) values are found to range between 1.5 and 2.5 $\times10^{-3}$ W.u. These numbers are orders of magnitude larger than those observed in heavy deformed nuclei, but are similar to those observed in actinide nuclei with strong octupole correlations. Further support for the octupole character of the band came from random-phase approximation (RPA) calculations based on the cranked Nilsson model (see figure 4).

Figure 4 presents both the partial SD level scheme for ^{190}Hg and a comparison of the RPA calculations with the data. In the upper panel, the $\mathcal{J}^{(2)}$ moment of the excited SD band is compared with the RPA calculation for the lowest lying octupole excitation. In the region where the band is observed, there is excellent agreement between the data and the calculations. In the lower panel, theoretical Routhians relative to the SD vacuum are plotted for the four lowest excited states of both signatures. Both vibrational and non-vibrational states are calculated in the RPA. However, the lowest calculated states are all members of the octupole multiplet. The experimental Routhian for the excited SD band relative to the yrast SD band (i.e. its energy and slope with $\hbar\omega$) match nicely with the lowest lying calculated Routhian. This lowest vibrational state at $\hbar\omega > 0.3$ MeV corresponds to the completely aligned octupole phonon. At low frequency, this state corresponds to the K=2 component of the multiplet and one would not expect decays to the K=0 yrast SD band. However, at the higher frequencies a substantial K=0 component is mixed into the wavefunction due to the Coriolis force,

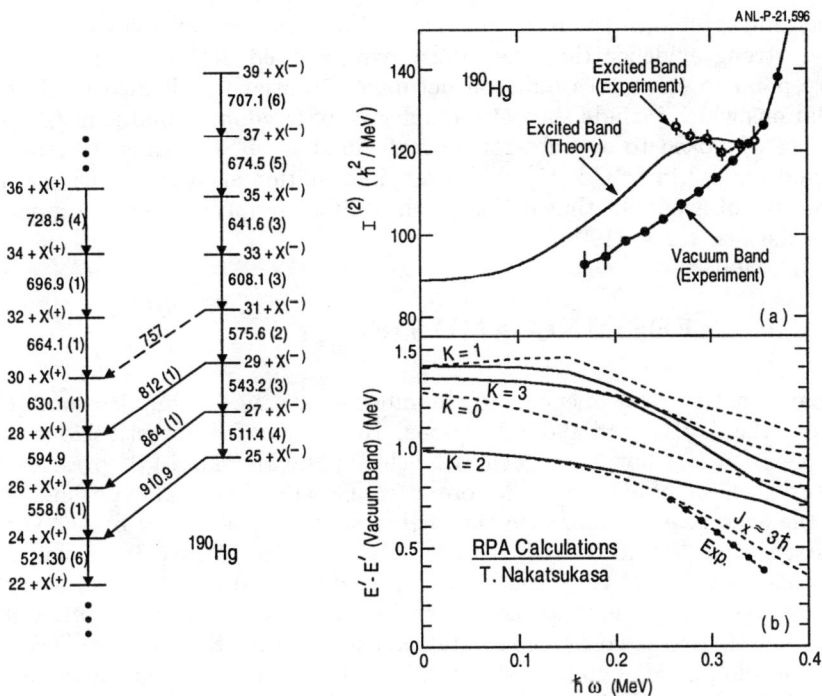

FIGURE 4. On the left, partial level structure for SD bands in ^{190}Hg. All members of the SD band proposed to be built on an octupole vibration are shown together with the transitions linking it with the yrast SD band. On the right, comparison of the data with results from RPA calculations based on the Nilsson potential. See text for details.

thus enhancing the E1 decays from the excited to the yrast SD band.

While the RPA calculations are able to account satisfactorily for the properties of the excited SD band in ^{190}Hg, the work of ref. [51] predicts that the lowest lying excited SD bands in ^{192,194}Hg should also be associated with octupole vibrations. In the case of ^{194}Hg, this was surprising since the two excited SD bands observed can be understood as signature partners of a two-quasiparticle excitation [54]. Of additional interest is the fact that these two excited bands exhibit an identical band relationship with the yrast SD band in ^{192}Hg. Direct transitions from the SD states to ND yrast states have recently been observed for both SD band 1 (yrast) and SD band 3 (see next section) [55,56]. In addition, transitions linking SD band 3 and SD band 1 have been measured. (see fig. 6). Therefore, ^{194}Hg represents the first case where the absolute spins, parities and excitation energies of the two bands have been established, i.e. band 1 has even spin, positive parity and band 3 has odd spins and negative parity. In addition, band 3 is estimated to lie only 0.8 MeV above the yrast SD band at $\hbar\omega = 0$, and the measured B(E1) values for the decay of band 3 to band 1 are $\sim 10^{-5}$ W.u., i.e. they are smaller than in ^{190}Hg. All these observations are in good agreement with the RPA calculations. In particular, the decrease in the B(E1) rate is predicted. A larger separation in energy between the low lying $K^\pi = 2^-$ component and the remaining members of the octupole multiplet results in reduced admixtures from the $K^\pi = 0^-$ and 1^- components, thus quenching the B(E1) rate. While the decay of band 2 is not observed to go through band 1 or band 3 nor are direct links to normal states observed, this band is assumed to be the signature partner to band 3. Assuming no signature splitting between bands 2 and 3 at low spins, it has been pointed out that the small increase in signature splitting with increasing spin is consistent with the octupole interpretation of these two bands [57]. Thus, there is strong experimental support for the theoretical interpretation of ref. [51] that bands 2 and 3 in ^{194}Hg are built on the K=2 octupole vibrational state.

Additional support for this interpretation comes from new data obtained with Eurogam II on ^{196}Pb, the isotone to ^{194}Hg. Three excited SD bands have been identified, all of which are reported to decay to the yrast SD band [58]. Excited SD bands 2 and 3 appear to be strongly coupled signature partners with extracted B(E1) rates on the order of $10^{-5} - 10^{-6}$ W.u., i.e. these bands appear to be analogues to SD bands 2 and 3 in ^{194}Hg. The interpretation of these structures as K=2 octupole vibrational bands is again consistent with the RPA calculations. Band 4 lies ~ 280 keV higher in energy than band 3 and its B(E1) decay rate to SD band

1 is $\sim 10^{-4}$ W.u. It has been interpreted as the K=0 member of the octupole multiplet.

In summary, there is now strong evidence that the lowest lying excited SD bands in even-even nuclei of the A=190 SD region correspond to structures built on octupole vibrations. All measured observables are consistent with RPA calculations which include the octupole degree of freedom. Finally, in the A=150 region, collective excitations have been invoked to account for a small number of SD bands, i.e., band 6 in ^{152}Dy [59], band 5 in ^{150}Gd [60] and band 3 in ^{148}Gd [15]. However, information such as branching ratios, relative excitation energies and spins are not available, thus making it difficult to perform detailed comparisons between experiment and theory as is the case for A=190.

FEEDING AND DECAY

The discovery of a SD band in ^{152}Dy [2] opened up a number of avenues of inquiry most of which were directed towards mapping out the A=150 and 190 SD regions. However, a few investigations did attempt to understand how SD bands were fed and how they decayed. Schiffer and Herskind [61] presented some of the first calculations describing the feeding of SD bands. Moore et al. [63] added critical experimental information to this topic by measuring the average entry points in the energy-spin (E,I) plane for both ND and SD states in ^{192}Hg and ^{152}Dy. Lauritsen et al. [62] measured, for the first time, the entry distribution of a SD band, and showed that this distribution originates from the higher spin components of the total entry distribution. Monte Carlo calculations were able to reproduce all the observables associated with the feeding of the SD band in the A=190 region and limits were extracted for the excitation energy of the SD band in ^{192}Hg. Recent work on ^{143}Eu has focused on the feeding of SD bands in the A=150 region. This work has attempted to account for all of the feeding into the SD well [64]. Indeed, it appears that a significant amount of intensity that passes through the SD well bypasses the SD yrast band. The transitions which account for this intensity have been interpreted as coming from rotationally damped states in the second well [65].

In the A=150 SD region, it was observed that mass-symmetric reactions enhance the population of SD bands relative to mass-asymmetric reactions forming the same compound system at the same excitation energy [66,67], and it was suggested that such effects could be explained by an increase in the fusion time for mass-symmetric reactions. The same effect could not be confirmed in the A=190 SD region [68]. This subject has recently been re-examined in the mass 150 region using Gammasphere [69]. In this study, the yrast SD band in ^{147}Gd was populated in both the ^{76}Ge + ^{76}Ge and ^{28}Si + ^{124}Sn reactions. The beam energy for each reaction was chosen such that the compound nucleus ^{152}Dy was produced with an excitation energy of 87 MeV. Relative to the ND states, a population enhancement by a factor of $\sim$ five was found in the mass-symmetric reaction for the ^{147}Gd yrast SD band. In addition, the feeding of the SD quasicontinuum in coincidence with the SD band was measured to be at least 12 times stronger in the mass-symmetric reaction. Interestingly enough, the relative population intensities as a function of γ-ray energy remained identical in the two reactions indicating no enhancement in the feeding of the highest-spin states. Rather, the additional feeding occurs over the entire band for the mass-symmetric reaction. While dissipative collision calculations suggest that the fusion time for the mass-symmetric reaction is about 6 times longer than that for the mass-asymmetric one, it is far from clear that this time difference is sufficient to account for the observed changes in the feeding.

The ability to link the SD states to known yrast levels of definite spin, parity and excitation energy has been a long outstanding challenge in the study of superdeformation. In the mass 130 region, the observed SD bands lie only $\sim$0.8 MeV above the ND yrast states, and it has been possible to identify many of the decay pathways between the SD and less deformed yrast states. In contrast, only two examples of direct links have been reported in the A=190 region and no examples have yet been observed conclusively in the A=150 region.

The first report that the excitation energy of a SD band in either region had been determined was in ^{143}Eu where 5-6 "sum peaks" were observed in coincidence with the yrast SD band [70]. These sum peaks were extracted from 3-fold coincidence events where one of the three detected γ rays was required to be a member of the SD band and the other two transitions were summed together. The sum peaks seen in this procedure were associated with the two-step decay of the SD band towards the yrast line, placing the lowest SD state at 3.635 MeV above the yrast $35/2\hbar$ level. However, two recent follow-up studies performed on ^{143}Eu at Eurogam II [71] and Gammasphere [72] are in contradiction over this result. Both works report discrete high-energy γ rays ($E_\gamma \geq 2.5$ MeV) in coincidence with the SD band, however, none of these lines can be established as a direct link between the SD band and the yrast structure. Furthermore, the Gammasphere measurement does not find evidence of the sum peaks discussed above while the Eurogam data reproduce only a subset of the sum-peaks proposed originally.

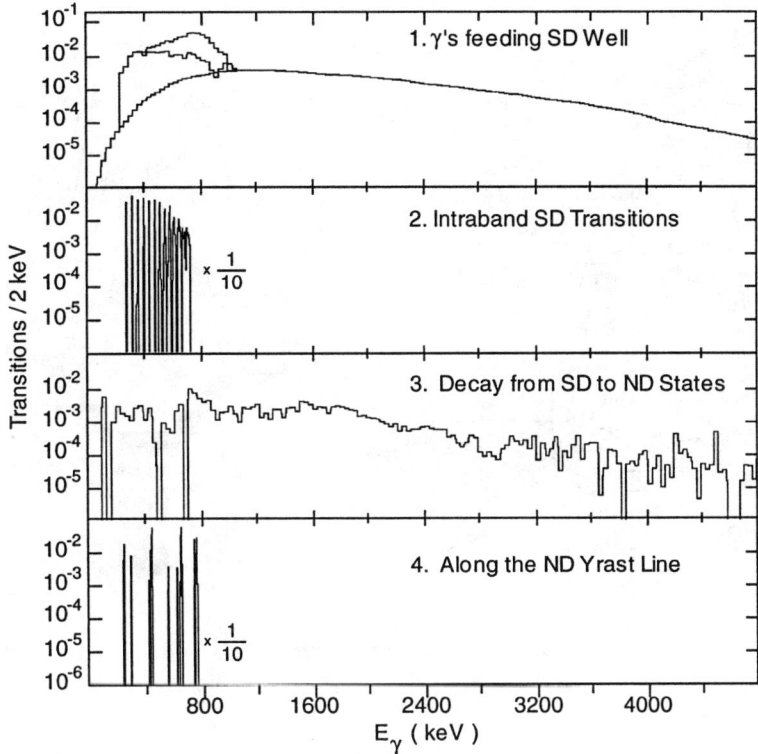

FIGURE 5. Deconvoluted components of the total spectrum in coincidence with the SD yrast band in ^{192}Hg. See text and ref. 73 for details.

In the mass 190 region, a different approach was adopted. Here, a total spectrum in coincidence with the known ^{192}Hg SD band was extracted from the data by placing pairwise coincidence gates on SD lines. This spectrum was then properly background subtracted and corrected for neutron interactions, coincidence summing, detector response and photo-peak efficiency [73]. In order to extract the total spectrum of γ rays connecting SD to ND states, transitions which feed the SD band, connect SD band members, and connect ND states at the end of the cascade were removed. The various components extracted from this deconvolution are shown in fig. 5, and panel 3 is the decay spectrum. This spectrum is characterized by a quasicontinuum of mostly dipole character with a statistical distribution which supports the suggestion of ref. [74] that an SD level decays to ND states when it acquires a small component of a hot compound state, and decays through the admixture of this component.

Other quantities extracted from the decay spectrum are: (i) the average number of transitions in the decay of the SD band to the yrast line (3.2 ± 0.6), and (ii) the excitation energy and spin of the SD band at the point of decay (4.3 ± 0.9 MeV and $10.1 \pm 0.7\hbar$). The decay spectrum also contains a noticeable bump between 1.3 and 2.2 MeV which sits on top of the statistical spectrum. A recent calculation by Dossing based on a model which treats self-consistently the weakening of pair correlations with increasing number of quasiparticle excitations reproduces the decay spectrum including this bump [75]. In the calculation, the bump arises from a combination of transitions from the sequential steps of the de-excitation cascade and from the last step of the decay where the statistically fed continuum states must cross the pair gap to the ground band.

While the extraction of the quasi-continuum decay spectrum in ^{192}Hg revealed much about the nature of the decay process, it could only provide limits on the excitation energy and spins of the SD band. In order to ascertain these physical quantities directly, it is necessary to observe so-called "one-step" linking transitions between SD states and states of known spin, parity and excitation energy. The first unambiguous observation of these one-step transitions reported three γ rays with energies of 3489, 4195, and 4485 keV which were shown to decay directly from SD band 1 to a member of the yrast negative parity band in ^{194}Hg [55] (see fig. 6). Angular distribution information confirmed that these transitions had dipole character and, based on the fact that the decay spectrum showed a statistical distribution [73], the γ rays were assumed to be electric dipoles

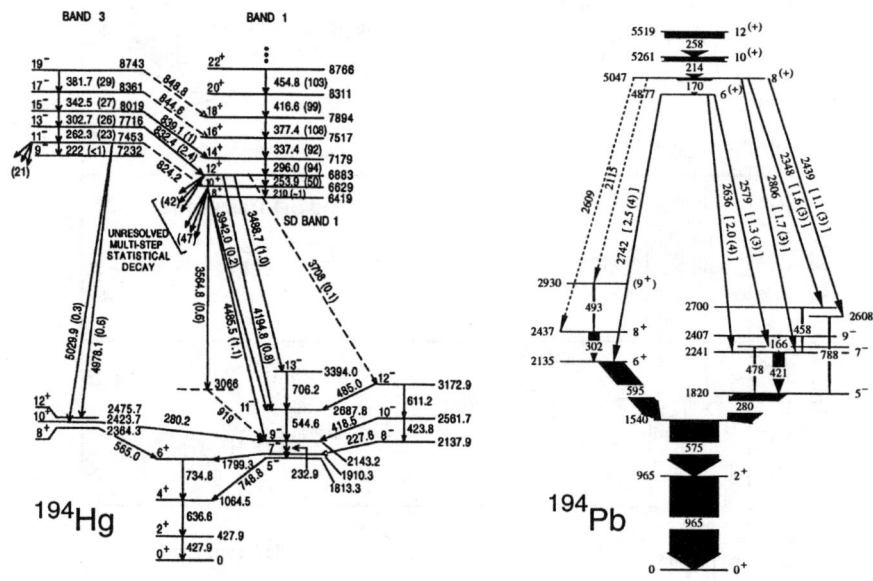

FIGURE 6. Linking transitions for ^{194}Hg and ^{194}Pb

and thus parity changing. With this information, the lowest observed member of the SD band was assigned a spin of 8, an excitation energy of 6419 keV, and positive parity. These "one-step" transitions only carried 3% of the decay strength. In a followup measurement, two one-step transitions of dipole character from SD band 3 to known positive parity states in ^{194}Hg were observed, thus establishing odd-spins and negative parities for the levels in this SD band [56] (see discussion above).

Immediately after the linking transitions were established for SD band 1 in ^{194}Hg, the observation of one-step decays in ^{194}Pb was reported from data taken at Eurogam [76] and subsequently confirmed by a study at Gammasphere [77]. The known placement of single-step decay transitions in ^{194}Hg and the single-step links deduced in ref. [76] for ^{194}Pb are summarized in the partial decay schemes of fig. 6. Interestingly, there are significant differences between the two cases:

- The yrast SD band in ^{194}Hg lies ~ 1.4 MeV higher in excitation energy than the yrast SD band in ^{194}Pb.

- The discrete one-step and two-step decay intensity in ^{194}Hg accounts for only $\sim$3% of the decay strength while for ^{194}Pb it is $\sim$21%.

- For ^{194}Hg, the one-step transitions are all of E1 multipolarity while in ^{194}Pb both E1 and M1/E2 one-step transitions are observed.

Based on these observations, lifetimes for the 8^+ and 10^+ levels of the yrast SD band in ^{194}Pb were measured using the recoil-distance Doppler shift method in order to ascertain whether the deduced electromagnetic properties are consistent with statistical decay [78]. These lifetime results were then used in a simple band mixing model in order to estimate the admixture of ND states into the SD band members at the points of decay. These admixtures are very small ($\sim$1%) and again consistent with the model suggested by Vigezzi *et al.* [74]. In addition, by correcting for this admixture, reduced transition probabilities were estimated for the ND-ND portion of the one-step decays. These are (i) B(E1) $\sim 10^{-6} - 10^{-5}$ W.u., (ii) B(M1) $< 5 \times 10^{-4}$ W.u., and (iii) B(E2) $< 3 \times 10^{-2}$ W.u.. These transition probabilities are all within the limits of statistical decays established, for example, from neutron resonance capture experiments. Therefore, even though variances in the SD decay properties are observed between ^{194}Hg and ^{194}Pb, they both appear to be statistical in nature.

While the one-step decay pathways between superdeformed and normally deformed states have been unambiguously established for ^{194}Hg and ^{194}Pb, the observation of more cases in the A=190 region has not been forth coming. For example, data sets of equivalent size have been taken on ^{192}Hg and approximately 50 weak transitions with energies ranging between 1 and 3.2 MeV have been observed in coincidence with SD band 1

[79]. However, none of these γ rays appears to represent a decay between the SD band and a known level, and as a result, the precise excitation energy of this SD band has not yet been established. One obvious question to be asked is why the one step decays are observed in only a few examples. Recent work has attempted to show experimentally that the observed linking transitions are due to fluctuations based on a Porter-Thomas distribution [80]. While the analysis is consistent with such a conclusion, the rather low probability represented by such decays (10^{-4}) cannot rule out the possibility that there are also special selection rules associated with the decay of SD bands.

SUMMARY

The study of SD nuclei in the A=150 and A=190 regions has made great strides over the last decade. Part of this success has come from the construction of very large and powerful γ-ray spectrometers. It is safe to say that a "very good" understanding of the characteristics of SD bands in both regions exists. However, several open problems still remain such as, (i) a fundamental explanation for $\Delta I = 4$ staggering, (ii) fundamental understanding of identical bands, and (iii) precise knowledge of excitation energies, spins and parities of SD bands beyond a few isolated examples.

ACKNOWLEDGMENTS

The authors would like to acknowledge the numerous contributions of current and past Argonne Physics Division members, T.L. Khoo, T. Lauritsen, I. Ahmad, E.F. Moore, R.R. Chasman, G. Hackman, D. Nisius, S.M. Fisher, I. Widenhoever, P. Reiter and B. Crowell and students H. Amro and S. Siem, who have collaborated with us on our own studies of superdeformed nuclei in these two mass regions. This work is supported by the U.S. Department of Energy, Nuclear Physics Division under contract W-31-109-ENG-38.

REFERENCES

1. S. Bjornholm and J.E. Lynn, Rev. Mod. Phys. **52**, (1980) 725.
2. P.J. Twin et al., Phys. Rev. Lett. **57** (1986) 811.
3. M.A. Bentley et al., Phys. Rev. Lett. **59** (1987) 2141.
4. E.F. Moore et al., Phys. Rev. Lett. **63** (1989) 360.
5. P.J. Nolan and P.J. Twin, Ann. Rev. Nucl. Part. Sci. **38** (1988) 533.
6. R.V.F. Janssens and T.L. Khoo, Ann. Rev. Nucl. Part. Sci. **41** (1991) 321.
7. T. Bengtsson, I. Ragnarsson, S. Aberg, Phys. Lett. **B208** (1988) 39.
8. S.M. Fischer et al., to be published.
9. T. Byrski et al., Phys. Rev. **C57** 1151.
10. J. Dobaczewski, proceedings to this conference.
11. S. Flibotte et al., Phys. Rev. Lett.**71** (1993) 4299.
12. I. Hamamoto and B. Mottelson, Phys. Lett. **B333** (1994) 294.
13. I.M. Pavlichenkov and S. Flibotte, Phys. Rev. **C51** (1995) R460.
14. A.O. Macchiavelli et al., Phys. Rev. **C51** (1995) R1.
15. G. de Angelis et al., Phys. Rev. **C53** (1996) 679.
16. B. Cederwall et al., Phys. Rev. Lett.**72** (1994) 3150.
17. S.M. Fisher et al., Phys. Rev. **C53** (1996) 2126.
18. R. Krucken et al., Phys. Rev. **C54** (1996) R2109.
19. D. Haslip et al., Phys. Rev. Lett.**78** (1997) 3447.
20. F. Donau, S. Frauendorf and J. Meng Phys. Lett. **B387** (1996) 667.
21. W. Reviol et al., Phys. Lett. **B371** (1996) 19.
22. Y. Sun, J.-Y. Zhang and M. Guidry, Phys. Rev. Lett.**75** (1995) 3398.
23. I.N. Mihailov and P. Quentin, Phys. Rev. Lett. **74** (1995) 3336.
24. H. Toki and L.-A. Wu, et al., Phys. Rev. Lett.**79** (1997) 2006.
25. T. Byrski et al., Phys. Rev. Lett. **64** (1990) 1650.
26. W. Nazarewicz et al., Phys. Rev. Lett. **64** (1990) 1654.
27. F.S. Stephens et al., Phys. Rev. Lett. **65** (1990) 301.
28. R.D. Amado et al., Phys. Rev. Lett. **67** (1991) 2777.

29. I. Ragnarsson Phys. Lett. **B264** (1991) 5.
30. B.Q. Chen *et al.*, Phys. Rev. **C46**, (1992) R1582.
31. J.-Y. Zhang *et al.*, Phys. Rev. Lett. **69** (1992) 1160.
32. C. Baktash, B. Haas and W. Nazarewicz, Annu. Rev. Nucl. Part. Sci. **45** (1995) 485.
33. G. de France, C. Baktash, B. Haas, W. Nazarewicz, Phys. Rev. **C53** (1996) R1070.
34. P. Fallon *et al.*, in *Proceedings of the Conference on Physics from Large γ-Ray Arrays*, Berkeley CA, Vol. II (1994) pp. 89-93.
35. H. Savajols *et al.*, Phys. Rev. Lett. **76** (1996) 4480.
36. D. Nisius *et al.*, Phys. Lett. **B392** (1997) 18.
37. W. Satula, J. Dobaczewski, J. Dudek and W. Nazarewicz, Phys. Rev. Lett. 77 (1996) 5182.
38. G. Hackman *et al.*, Phys. Lett. **B416** (1998) 268.
39. D. Nisius *et al.*, to be published.
40. B. Kharraja and U. Garg, Phys. Rev. Lett.**80** (1998) 1845.
41. I. Ragnarsson, Nucl. Phys. **A557** (1993) 167c.
42. E.F. Moore *et al.*, Phys. Rev. **C55** (1997) R2150.
43. H. Amro *et al.*, Phys. Lett. **B413** (1997) 15.
44. B. Busse *et al.*, Phys. Rev. **C57** (1998) R1017.
45. J.N. Ginocchio and A. Leviatan, Phys. Lett. **B425** (1998) 1.
46. P.-H. Heenen and R.V.F. Janssens, Phys. Rev. **C57** (1998) 159.
47. C. Schuck *et al.*, Phys. Rev. **C56** (1997) R1667.
48. F.S. Stephens *et al.*, Phys. Rev. **C57** (1998) R1565.
49. J. Skalski *et al.*, Nucl. Phys. **A551** (1993) 109.
50. S. Mizutori *et al.*, Nucl. Phys. **A557** (1993) 125c.
51. T. Nakatsukasa *et al.*, Phys. Rev. **C53** (1996) 2113.
52. B. Crowell *et al.*, Phys. Lett. **B333** (1994) 320.
53. B. Crowell *et al.*, Phys. Rev. **C51** (1995) R1599.
54. M. Riley *et al.*, Nucl. Phys. **A512** (1990) 178.
55. T.L. Khoo *et al.*, Phys. Rev. Lett. **76** (1996) 1583.
56. G. Hackman *et al.*, Phys. Rev. Lett. **79** (1997) 4900.
57. P. Fallon *et al.*, Phys. Rev. **C55** (1997) R999.
58. S. Bouneau *et al.*, Proceedings of the Conference on Nuclear Structure at the Limits (1997) 105.
59. P.J. Dagnall *et al.*, Phys. Lett. **B335** (1995) 313.
60. P. Fallon *et al.*, Phys. Rev. Lett. **73** (1994) 782.
61. K. Schiffer, B. Herskind and J. Gascon, Z. Phys. **A332**, 17 (1989).
62. T. Lauritsen *et al.*, Phys. Rev. Lett. **69**, 2479 (1992).
63. E.F. Moore *et al.*, Nuclear Structure and Heavy-ion Reaction Dynamics 1990, 171 (1991)
64. S. Leoni *et al.*, Phys. Lett. **B409** (1997) 71.
65. S. Leoni *et al.*, Phys. Rev. Lett. **76** (1996) 3281.
66. G. Smith *et al.*, Phys. Rev. Lett. **68** (1992) 158.
67. S. Flibotte *et al.*, Phys. Rev. **C45** (1992) R889.
68. F. Soramel *et al.*, Phys. Lett. **B350** (1995) 173
69. J.M. Nieminen *et al.*, Phys. Rev. **C58** (1998) R1.
70. A. Atac *et al.*, Phys. Rev. Lett. **70** (1993) 1069.
71. A. Atac *et al.*, Z. Phys. **A335** (1996) 343.
72. F. Lerma *et al.*, Phys. Rev. **CC56** (1997) R1671.
73. R.G. Henry *et al.*, Phys. Rev. Lett. **73** (1994) 777 .
74. E. Vigezzi *et al.*, Phys. Lett. **B249** (1990) 163.
75. T. Døssing *et al.*, Phys. Rev. Lett. **75**, 1276 (1995).
76. A. Lopez-Martens *et al.*, Phys. Lett. **B380** 18.
77. K. Hauschild *et al.*, Phys. Rev. **C55** (1997) 2819.
78. R. Krucken *et al.*, Phys. Rev. **C55** (1997) R1625.
79. A. Lopez-Martens *et al.*, Phys. Rev. Lett. **77** (1996) 1707.
80. A. Lopez-Martens *et al.*, submitted to Phys. Rev. C.

Unpaired Band Crossings in the A~160 Mass Region

J.C. Lisle*, J. Simpson†, A.P. Bagshaw*, M.A. Bentley‡,
D.M. Cullen§, P.J. Dagnall*, G.B. Hagemann¶, S.L. King§,
D. Napoli**, M. Riley††, S. Shepherd§, A.G. Smith*
and S. Tormanen¶.

* *Dept. of Physics and Astronomy, University of Manchester, Manchester M13 9PL, UK.*
† *CLRC, Daresbury Laboratory, Warrington WA4 4AD, UK.*
‡ *School of Sciences, University of Stafford, Stoke-on-Trent ST4 2DE, UK.*
§ *Dept. of Physics, University of Liverpool, Liverpool L69 3BX, UK.*
¶ *Niels Bohr Institute, Blegdamsvej 17, DK-2100 Copenhagen.*
** *INFN, Laboratori Nazionali di Legnaro, I-35020 Legnaro, Italy.*
†† *Dept of Physics, Florida State University, Tallahassee, USA.*

Abstract. Levels schemes for a number of low lying normally deformed bands in ^{161,162}Er have been extended to significantly higher spins than hitherto using the EUROBALL Ge detector array sited at the Legnaro National Laboratory. In the case of one band in ^{161}Er states have tentatively been observed to a spin in excess of $60\hbar$. Evidence is found for a band crossing at high spin in the yrast even parity, positive signature $(+,+\frac{1}{2})$ band in ^{161}Er. This can be explained in terms of a system in which the static pairing is quenched either involving a rearrangement of two neutrons into steeply aligning $i_{\frac{13}{2}}$ orbitals or possibly an interchange of state of one proton and one neutron. A similar type of band crossing is tentatively identified in the lowest $(-,-\frac{1}{2})$ band of ^{161}Er. Since a crossing involving the interchange of neutrons is blocked, this must involve protons. This, if confirmed, provides the first direct evidence for the quenching of static proton pairing at high spin in this mass region. Recent calculations [1] predict that such a crossing involving purely protons will occur in this band at $I \sim 60\hbar$.

INTRODUCTION

A band crossing in the lowest odd parity, positive signature band $(-,+\frac{1}{2})$ of ^{159}Er was observed at a rotational frequency of approximately 0.55 MeV by Simpson et al. [2]. Since this crossing was unique to this band in ^{159}Er and was not seen in neighbouring isotopes or isotones it was argued that this must be associated with a change in the occupied neutron orbitals in an unpaired regime. Specifically at high rotational frequencies it becomes energetically favourable for two N=5 neutrons to occupy $i_{\frac{13}{2}}$ intruder orbitals in such a way that parity and signature are conserved. Such a crossing is blocked in the yrast $(+,+\frac{1}{2})$ band in agreement with experimental observations. This provided the most direct evidence that neutron pairing in the rare earth region was largely quenched at high rotational frequencies.

The purpose of the present experiment was to extend the level scheme in ^{161}Er to higher spins to observe a similar crossing in the yrast $(+,+\frac{1}{2})$ band which is predicted to occur at a rotational frequency of ~ 0.75 MeV, corresponding to a spin in excess of $50\hbar$.

EXPERIMENTAL

The experiment was carried out using EUROBALL at the Tandem Laboratory in the Legnaro National Laboratory. ^{161,162}Er was excited using the fusion-evaporation reaction ^{130}Te(^{36}S,xn) at a bombarding energy of 170 MeV. The target consisted of two 0.5 mg cm^{-2} unbacked stacked foils of ^{130}Te. At the time of the experiment 14 seven-element Clover detectors, 26 four-element Cluster detectors and 30 single element tapered

CP481, *Nuclear Structure 98*, edited by C. Baktash

detectors were mounted in the EUROBALL array. In all 2.0×10^9 events with unsuppressed fold ≥ 5 were collected.

The data were sorted into a 3-dimensional matrix, the full data set leading to 2.0×10^{10} suppressed triples after unfolding. In the case of the Clover and Cluster detectors, when two neighbouring elements fired, this was regarded as being associated with Compton scattering and the two digitised outputs were added. However,

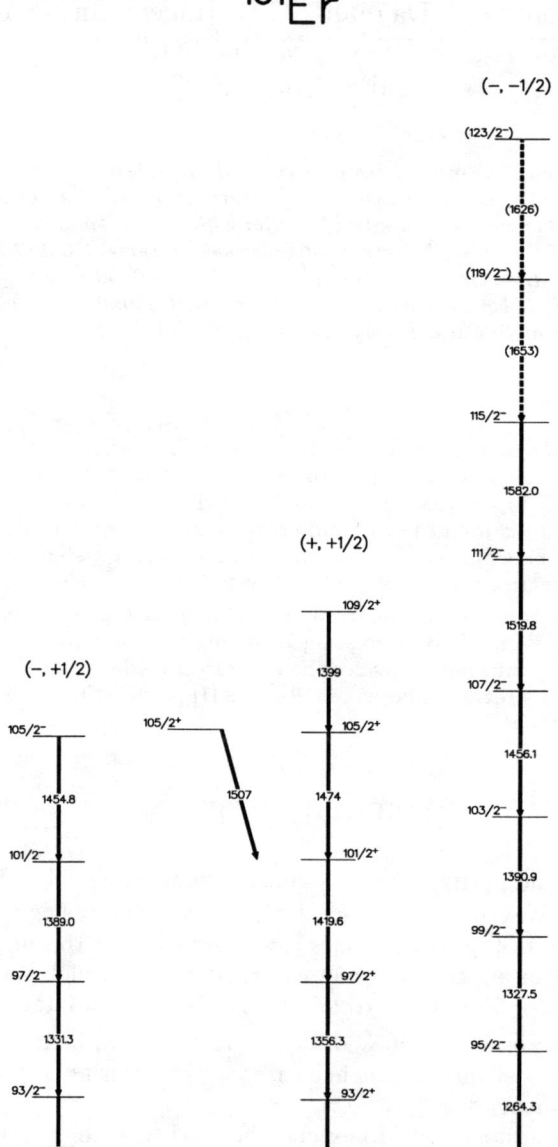

FIGURE 1. High spin level scheme for ^{161}Er.

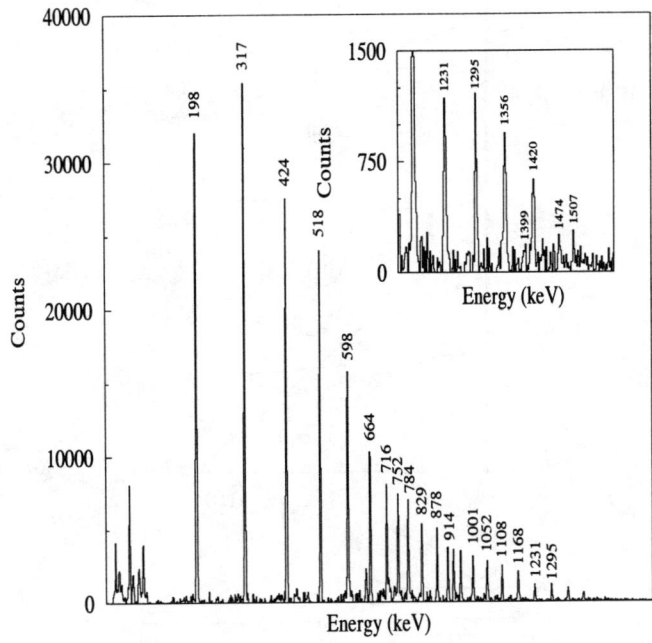

FIGURE 2. Multiple gated spectrum for the for $(+,+\frac{1}{2})$ in ^{161}Er.

when two non-neighbouring elements or more than two elements fired, the complete detector was eliminated from the sort.

RESULTS

The high spin states for the low lying rotationally aligned bands in ^{161}Er are shown in fig. 1. The ordering of the transitions in the bands is based mainly on relative intensities. All three bands have been extended to significantly higher spin compared with the level scheme presented by Riley et al. [3]. In particular the the $(-,-\frac{1}{2})$ band has tentatively been extended to $I^{\pi} = \frac{123}{2}^{-}$. The favoured ordering of the 1626 and 1653 keV transition is based on intensities but is not definitive. However, there is evidence for an anomaly, in which the smooth $I(I+1)$ behaviour is interrupted, whichever order is considered.

In the case of the $(+,+\frac{1}{2})$ band an anomalous sequence is also found. A double gated spectrum for this band, using multiple gates in both axes, is shown in figure 2 and indicates the existence of transitions at 1399, 1474 and 1507 keV. Spectra in which only one gate is set on the second axis indicate that the 1474 and 1507 keV transitions are not in coincidence, thus providing evidence for the existence of the band splitting into two branches as shown in figure 1.

The yrast band in ^{162}Er has been definitely identified up to the I=54 member with a tentative candidate for the $56 \rightarrow 54$ transition. No evidence for anomalous behaviour is found in this band up to I=56.

DISCUSSION

It has been noted in the previous section that anomalies are observed at high spin in the even parity and one of the negative parity bands of ^{161}Er.

In the case of the even parity $(+,+\frac{1}{2})$ band the anomaly can be explained using similar arguments to those used by Riley et al. [2] in terms of unpaired neutrons. This is illustrated in figure 3, which shows the neutron orbitals expected to be occupied for the even and two odd parity bands under discussion. In the case of the $(+,+\frac{1}{2})$ band the orbitals labelled $(+,-\frac{1}{2})$ [651] and $(+,\frac{1}{2})_2$ [642] are unoccupied at low frequencies. However, at high frequencies, because these orbitals are steeply aligning, it is energetically favourable for neutrons

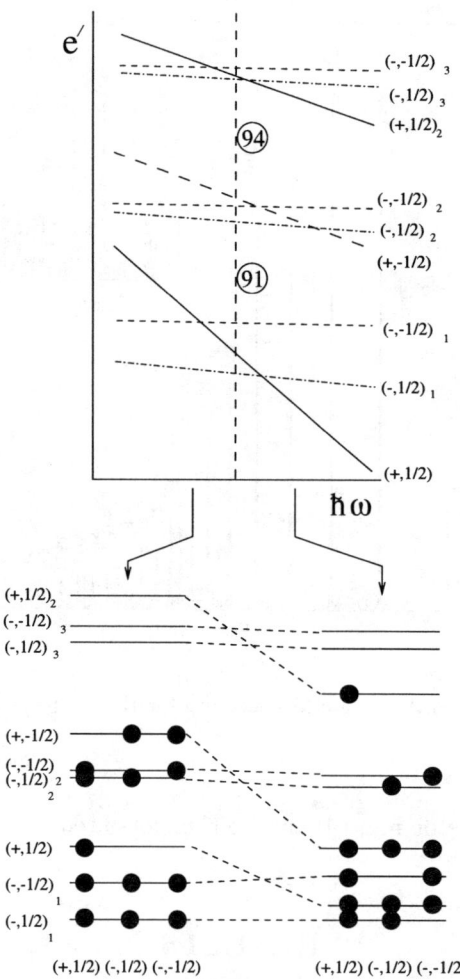

FIGURE 3. A schematic illustration of the expected occupation of unpaired neutron orbitals at low and high frequencies for the 3 lowest bands in ^{161}Er. The parity and signature assignments at the bottom relate to the bands and those at the sides to the orbitals.

previously in odd parity N=5 orbitals to occupy these even parity $i_{\frac{13}{2}}$ intruder orbitals. Such a crossing, in which there would be a considerable increase in alignment, would occur at a rotational frequency of ~ 0.75 MeV corresponding to a spin of $\sim 50\hbar$. It is interesting to note that recent calculations performed by Afansjev and Ragnarsson [1], using the cranked Nilsson-Strutinski approach without neutron or proton pairing, predict a crossing at about this spin which involves a change of configuration of both neutrons and protons. From the positions of the two $I = \frac{105}{2}$ states we can deduce that the interaction energy between the two crossing bands is ≤ 16 keV.

It can be seen from figure 3 that a purely neutron change of occupation would be blocked in the case of both odd parity bands. However, tentative evidence for an anomaly in the $(-,-\frac{1}{2})$ band, which apparently becomes yrast at high spin, is seen at $I = \frac{119}{2}$. The results of calculations, performed by Afansjev and Ragnarsson [1] and shown in figure 4, predict that a crossing involving a purely proton configuration change will occur at approximately this spin. In this figure each band is labelled with a set of parameters indicating the number of protons and neutrons outside the closed ^{146}Gd core. For a label [AB,CD], A is the number of $h_{\frac{11}{2}}$ and B the number of $i_{\frac{13}{2}}$ protons, while C is the combined number of $h_{\frac{9}{2}}$ and $f_{\frac{7}{2}}$ neutrons and D is the number of $i_{\frac{13}{2}}$ neutrons. Thus the predicted crossing involves the band [6,74], containing 6 $h_{\frac{11}{2}}$ protons being crossed by the [61,74] band containing 6 $h_{\frac{11}{2}}$ protons as before and 1 $i_{\frac{13}{2}}$ proton. The proton filling the $i_{\frac{13}{2}}$ orbital is obtained by particle-hole excitation from a N=4 orbital in the ^{146}Gd core. This observation, if confirmed by a more

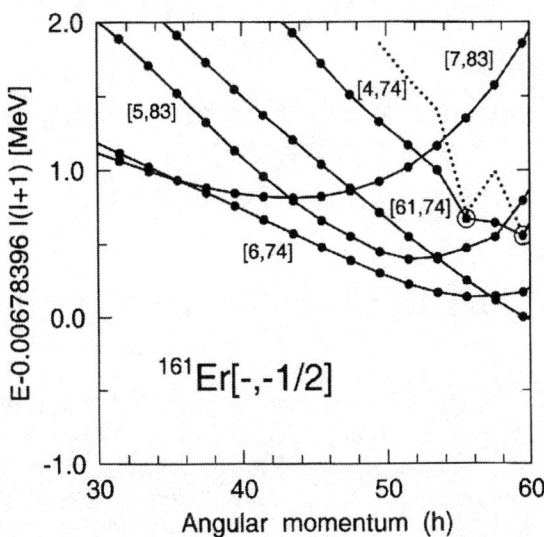

FIGURE 4. The result of a cranked Nilsson-Strutinski calculation for the $(-,-\frac{1}{2})$ band in ^{161}Er. The labelling notation is explained in the text.

detailed and thorough analysis of the experimental data, provides the most direct evidence for the quenching of proton pairing at high spin in normal deformed bands of rare earth nuclei.

The fact that no anomaly is observed in the even parity yrast band of ^{162}Er up to $I^{\pi} = 56^+$ is quite consistent with a picture involving purely neutron excitations, The negative signature [651] orbital, which was available for the even parity band of ^{161}Er, is blocked by the additional neutron.

We conclude that the discontinuity observed in the even parity, positive signature band of ^{161}Er can be explained by either an unpaired band crossing involving only neutrons or possibly by one involving both neutrons and protons. However, a discontinuity in the odd parity, odd signature band can only be explained as a particle-hole proton excitation in an unpaired regime. These Er nuclei and recent results for ^{156}Dy [4] constitute the first examples of spectroscopy at normal deformation above spin 50 in the unpaired regime. In the A~160 region the interesting possibility is raised for studying the competition between unpaired crossings and band termination. In the case of ^{158}Er [5] [6] it was demonstrated that the yrast band with an unpaired proton configuration, containing four $h_{\frac{11}{2}}$ protons, terminated at $I^{\pi} = 46^+$. However, with three additional neutrons in ^{161}Er the calculations of Afansjev and Ragnarsson [1] indicate that bands based on this terminating proton structure are about 0.5 MeV above the lowest bands of the same parity and signature (eg. figure 4). This appears to be consistent with the present experimental results for the $(+,+\frac{1}{2})$ and $(-,-\frac{1}{2})$ bands. The situation for the $(-,+\frac{1}{2})$ band is unclear since it seems to stop rather abruptly at a spin close to that expected for termination.

REFERENCES

1. Afansjev A. and Ragnarsson I., Private Communication (1998).
2. Riley M. et al., *Phys Rev. Lett.* **60** 553 (1988).
3. Riley M. et al., *J. Phys. G* **16** L67 (1990).
4. Kondev F.G. et al., *Phys. Lett.* In press.
5. Simpson J. et al., *Phys. Rev. Lett.* **53** 648 (1984).
6. Tjöm P. et al. *Phys Rev. Lett.* **55** 2405 (1985).

Multiple Triaxial SD Bands in ^{163,164}Lu Studied with EUROBALL

S. Törmänen[1], G.B. Hagemann[1], A. Harsmann[1], M. Bergström[1], R.A. Bark[1],
B. Herskind[1], G. Sletten[1], S. Ødegård[2], P.O. Tjøm[2], A. Görgen[3], H. Hübel[3],
J. Domscheit[3], B. Aengenvoort[3], U.J. van Severen[3], C. Fahlander[4,7],
D. Napoli[4], S. Lenzi[5], C.M. Petrache[5], C. Ur[5], H.J. Jensen[1,6], H. Ryde[7],
A. Bracco[8], S. Frattini[8], R. Chapman[9], D.M. Cullen[10] and S.L. King[10]

[1]NBI, Univ. of Copenhagen, DK-2100 Copenhagen, Denmark, [2] Univ. of Oslo, N-0316 Oslo, Norway,
[3] ISKP Univ. of Bonn, D-53115 Bonn, Germany, [4] LNL, INFN, I-35020 Legnaro, Italy,
[5] INFN, Univ. of Padova, I-35122 Padova, Italy, [6] KFA Jülich, D-52425 Jülich, Germany, [7] Univ. of Lund,
S-22362 Lund, Sweden, [8] Univ. of Milano, I-20133 Milano, Italy, [9] Univ. of Paisley, Paisley PA1 2BE, UK,
[10] Univ. of Liverpool, Liverpool L69 3BX, UK

Abstract. High-spin states in ^{163,164}Lu have been populated in the ^{139}La(^{29}Si,xn) reaction and their decay has been studied using the EUROBALL γ-ray spectrometer array. In ^{164}Lu eight new superdeformed (SD) bands have been found. In ^{163}Lu the previously known SD band is extended to lower and higher spins and its deacy to the normal-deformed states has been established. In addition, a new SD band is identified. Calculations with the Ultimate Cranker code predict large triaxiality ($\gamma \approx \pm 20°$) for all these bands.

Rotational bands with very large deformation were discovered in ^{163}Lu [1,2] and ^{165}Lu [3] some time ago. A similar observation has been reported recently for ^{167}Lu [4]. Large transition quadrupole moments, corresponding to $\beta_2 \approx 0.37$, were derived for the band in ^{163}Lu from lifetime measurements [2]. It was suggested that the band is built on the prolate deformation-driving [660]1/2$^+$proton orbital of i$_{13/2}$ origin. The band in ^{165}Lu has transition energies that are almost identical to the ones of the ^{163}Lu band. Therefore it was assumed that this band has also large deformation.

Potential energy surface calculations using different approaches [2–6] show that the nuclei around Z=72 and N=94 constitute a region of exotic shapes coexisting with normal prolate deformation. Large deformation minima exist for a γ deformation of approximately $\pm 20°$. In addition, a prolate hyperdeformed minimum is predicted. Calculations with the "Ultimate Cranker" (UC) code show that the large-deformation minima are expected for all symmetry groups (π, α) in ^{163,165}Lu [3]. That means the large deformation is not only due to the prolate-driving effect of the $\pi i_{13/2}$ intruder orbital, but also the result of a rearrangement of the core. In particular, for the neutron number 94 a large energy gap appears at $\gamma \approx 20°$. Calculations of the potential energy surfaces for the lowest expected configurations in ^{164}Lu with positive and negative parity are shown in fig. 1.

CP481, *Nuclear Structure 98*, edited by C. Baktash

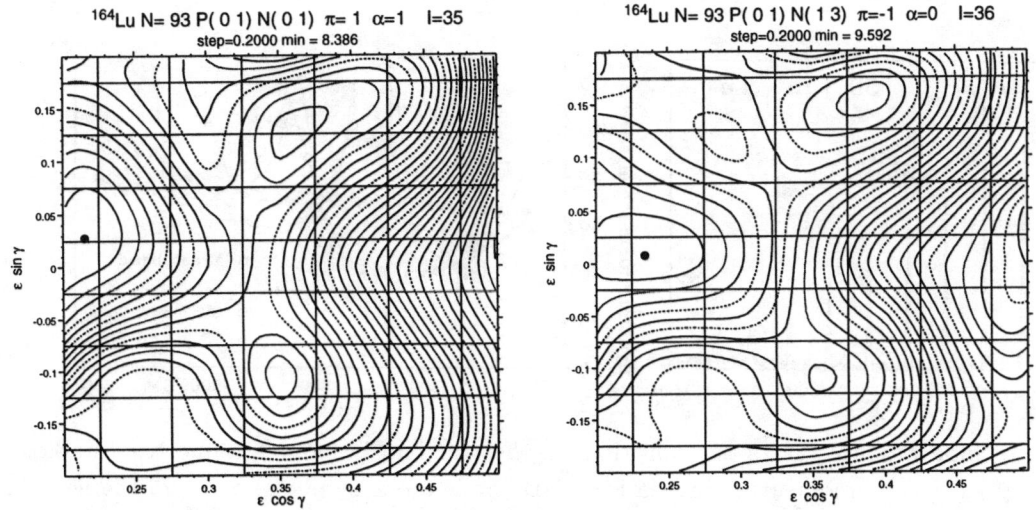

FIGURE 1. Calculated (UC) spin-adiabatic potential energy surfaces at I = 35 and 36 for the lowest configuration with positive and negative parity in ^{164}Lu.

In this contribution we report on an investigation of triaxial superdeformation in ^{163}Lu and ^{164}Lu using the EUROBALL spectrometer array. High-spin states in these isotopes were populated in the ^{139}La(^{29}Si,xn) reaction at a bombarding energy of 145 MeV. The beam was provided by the Legnaro XTU tandem accelerator. At the time of the experiment the EU-ROBALL spectrometer consisted of 13 Cluster Ge detectors, 25 Clover Ge detectors and 28 single-element tapered Ge detectors. A total of $3.8 \cdot 10^9$ events requiring 6 or more coincident Ge signals before Compton suppression was collected.

Since this experiment was one of the first measurements using EUROBALL, various problems were encountered. The stability and energy resolution had to be checked by dividing the data into small time segments. Most of the energy shifts could be recovered, but some of the detector elements had to be discarded, leaving 25 tapered detectors, 75 Cluster elements and 96 Clover elements for the analysis. This presorting resulted in $2.3 \cdot 10^9$ three- or higher-fold Compton-suppressed coincidence events.

The coincidence events were sorted into gated E_γ-E_γ matrices, a three-dimensional array (cube) and a four-dimensional array (hypercube) as well as into several coincidence spectra with various gating conditions [7,8]. For angular correlation information gated DCO matrixes were created with the forward and backward detectors on one axis and the near-90° detectors (Clovers) on the other axis.

The data contain information on several nuclei around ^{163,164}Lu, but here we will concentrate on these two isotopes. Our preliminary analysis has provided major extensions of their level schemes at normal deformation, in addition to many new, presumably triaxial, SD bands. As an example of the data, fig. 2 shows a double gated coincidence spectrum of one of the new SD bands in ^{164}Lu (band 1) together with the dynamic moments of inertia of several normal deformed (ND) and SD bands in that nucleus. The partial level scheme of ^{164}Lu is shown in fig. 3.

The partial level scheme of ^{163}Lu is displayed in fig. 4. The previously known SD band has been extended by two transitions at high spin and by one transition at the bottom. Its decay to the ND states is firmly established. In addition, the analysis revealed a second SD band which decays into SD 1 around spin 37/2. However, the connecting transitions were not found.

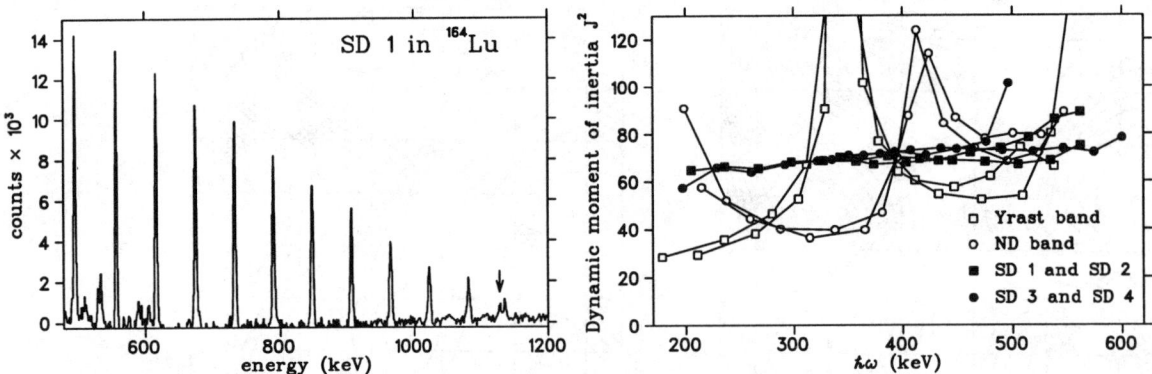

FIGURE 2. (a) Sum of double gates on transitions in the triaxial SD band 1 connected (marked) to ND states in ^{164}Lu. (b) Dynamic moments of inertia for 4 ND bands (open symbols) and 4 of the new SD bands (full symbols) in ^{164}Lu.

The previously known [1] strongly coupled bands which were not connected to the ground state are now all tied together in the level scheme presented in fig. 4.

The structure into which SD band 1 in ^{163}Lu decays via the 427, 529 and 697 keV γ-ray transitions was only partly known before our work [1,9]. The properties of this structure are very similar to the proton [411]$1/2^+$ bands occuring systematically in this mass region. In fact, the systematics of the excitation energies of these bands in neighbouring isotopes and isotones suggests that the [411]$1/2^+$ orbital might be the ground state of ^{163}Lu. Under this assumption and with the multipolarity assignments of the transitions determined from the DCO ratios, the band-head spin and parity of SD band 1 are $13/2^+$ as expected for a proton $i_{13/2}$ band. The transitions from the ND bands to the ground band confirm the spin and parity assignment of $1/2^+$ for the ground state.

The SD bands have quite similar dynamic moments of inertia $J^{(2)}$ at low frequency and a small variation in the moderate increase with increasing frequency (see fig.2). From the γ-ray energies SD 1 and SD 2 as well as SD 3 and SD 4 in ^{164}Lu could be signature partners, at least in a limited frequency range. The other bands, SD 5-8, are less populated and therefore probably located at higher excitation energy. The new bands may belong to any of the calculated minima with $\gamma \approx \pm 20°$ which have close to identical shapes but rotate around the small and intermediate axis, respectively. The calculated kinematic and dynamic moments of inertia $J^{(1)}$ and $J^{(2)}$ show a difference between the bands of positive and negative γ values. Furthermore, an appreciable difference is expected for the transition quadrupole moments for the two minima. However at present, without experimental knowledge of the transition quadrupole moments, only indirect evidence for the sign of γ can be given.

According to the calculation, the lowest excitation energies are expected for the states with positive γ values. In ^{164}Lu, the energies for the negative γ deformation are about 0.5 MeV larger than those for positive γ. The configuration with $(\pi, \alpha) = (-, 0)$ corresponds to the lowest excitation with $\gamma \approx +20°$ which is in agreement with the spin and parity suggested for the lowest SD band in ^{164}Lu (SD 1). The lowest positive-parity band is expected about 100 keV above this band. It should be noted, however, that the measured excitation relative to the ND yrast band shows that the triaxial SD minima in ^{164}Lu appear 0.5 - 1 MeV lower in energy than calculated.

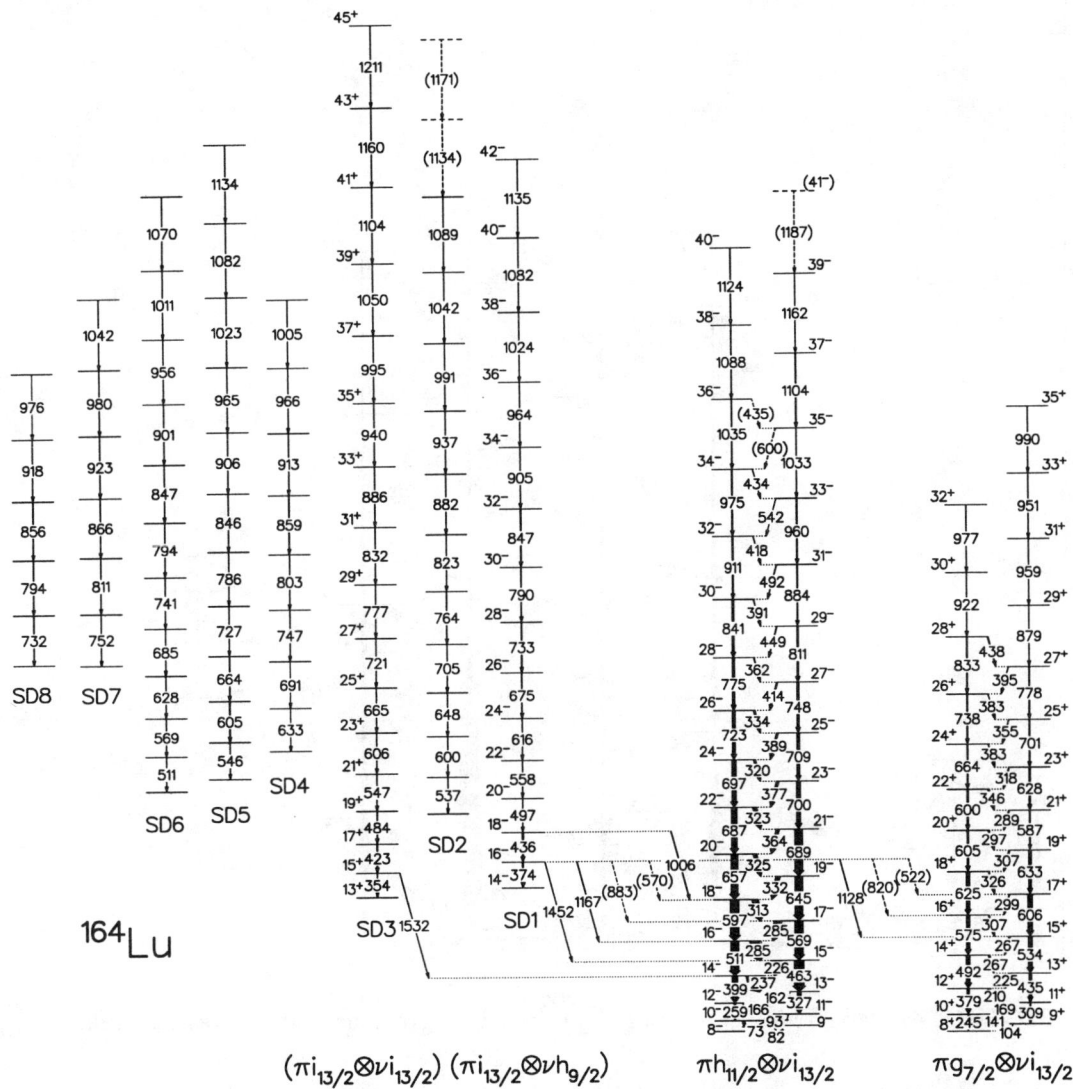

FIGURE 3. Partial level scheme of ^{164}Lu. Only the strongest populated ND bands are shown together with the new SD bands. For the 'hanging' bands the excitation energy is arbitrary.

In ^{163,165}Lu the identical triaxial SD bands were assigned to the lowest $i_{13/2}$ proton orbital with a gradual alignment of the first pair of $i_{13/2}$ quasineutrons (AB). In ^{164}Lu the lowest calculated configuration is obtained by coupling an $i_{13/2}$ proton to the lowest negative parity ($h_{9/2}$) quasineutron. Further support comes from the alignment of $\approx 2\hbar$ relative to the bands in ^{163,165}Lu which agrees with the expected alignment for the $h_{9/2}$ quasineutron in the observed frequency range. For SD 3 with $(\pi, \alpha) = (+, 1)$ the alignment relative to ^{163,165}Lu is $\approx 1.5\hbar$. This band could possibly be assigned to the configuration $\pi i_{13/2} \otimes \nu i_{13/2}$ with a gradual $i_{13/2}$ quasineutron alignment (BC). The less populated bands SD 5-8 could be candidates for the local SD minima with $\gamma \approx -20^\circ$.

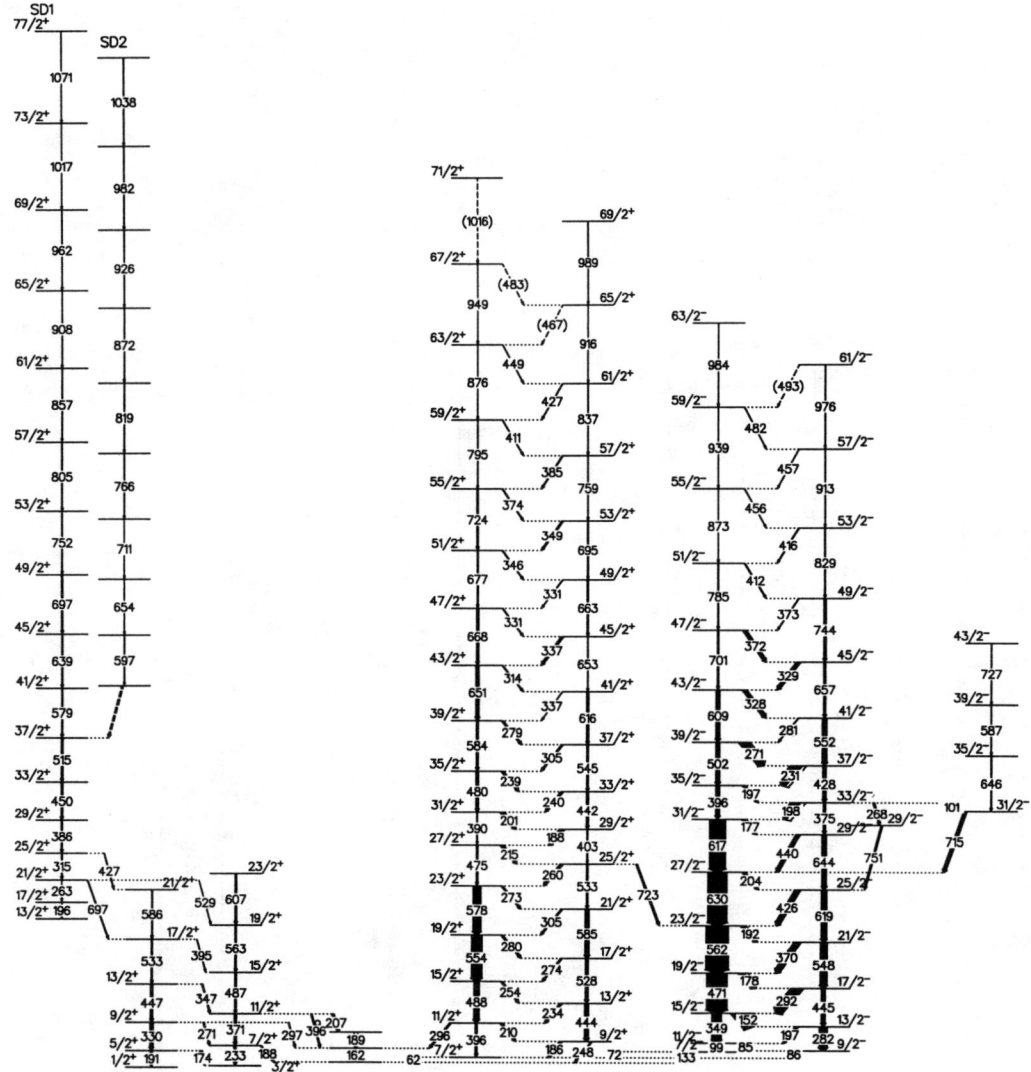

FIGURE 4. Partial level scheme of ^{163}Lu based on previous [1] and present work.

For the low-spin states of SD 1 in ^{163}Lu, $J^{(2)}$ shows large fluctuations which are due to mixing of the $21/2^+$ SD state with the state of the same spin and parity in the $[411]1/2^+$ band. The experimental energy difference of the two levels is only 111 keV. The strong decay branches from SD 1 to the $[411]1/2^+$ band via the 427 and 697 keV transitions which comprise about 40 % of the SD band intensity can be explained by that mixing. The situation is very similar to the decay of the SD bands in Nd nuclei [10,11] which has also been explained by mixing of SD and ND states lying close in energy. We assume that only the $21/2^+$ states at 2087 and 2198 keV are mixed with a constant interaction and that there is no interband transition strength. Under these assumptions we obtain from the in-band to out-of-band branching ratios a mixing amplitude $\alpha = 0.04$ and an interaction energy of 22 keV. This interaction results in a repulsion of the two $21/2^+$ levels of about 5 keV. If we correct the observed energy of the $21/2^+$ SD state by this amount we obtain a smooth dependence of the moment of inertia on rotational frequency. Such a large interaction energy and mixing amplitude are indicative of a very small or vanishing barrier between the SD and ND minima.

Acknowledgement: This project has been supported by the Danish Natural Science Foundation, the EU TMR project no ERBFMBICT961027, the Swedish Natural Science Research Council, the Research Council of Norway and the BMBF, Germany. The dedicated help from staff and EUROBALL support groups at the INFN laboratory in Legnaro is highly appreciated.

REFERENCES

1. W. Schmitz et al., Nucl. Phys. A 539, 112 (1992)
2. W. Schmitz et al., Phys. Lett. B 303, 230 (1993)
3. H. Schnack-Petersen et al., Nucl. Phys. A 594, 175 (1995)
4. C.X. Yang et al., Eur. Phys. J. A 1, 237 (1998)
5. S. Åberg, Nucl. Phys. A 520, 35c (1990) and refs. therein
6. I. Ragnarsson, Phys. Rev. Lett. 62, 2084 (1989)
7. S. Tormänen et al., Nuovo Cim., 111A, 685 (1998)
8. B. Aengenvoort, PhD thesis, Univ. Bonn (1998) unpublished
9. R.B. Firestone et al., Table of Isotopes, (Wiley 1996)
10. D. Bazzacco et al., Phys. Rev. C 49 , R2281 (1994)
11. C.M. Petrache, Contribution to this Conference

High-K spin-traps in the valley of stability

S. M. Mullins

Department of Nuclear Physics,
Research School of Physical Sciences and Engineering,
Australian National University,
Canberra, ACT 0200,
Australia

Abstract. Rotational bands based on very long-lived K-isomers in the stable hafnium isotopes ^{177,178,179}Hf have been studied with time-correlated particle-γ-γ techniques. Some physics highlights, and future possibiblities, will be discussed.

I INTRODUCTION

Nowhere better is the interplay between collective and intrinsic modes of nuclear excitation exemplified than in the deformed region near A = 180. At low excitation, angular momentum is generated from collective rotation of the prolate-deformed ground state, but the proximity of many high-Ω orbitals to the both the neutron and proton Fermi surfaces allows intrinsic multi-quasiparticle states to compete at higher spin. A multi-quasiparticle state of a particular configuration with seniority ν is formed from the coupling of ν individual high-Ω orbitals, so that $K = \sum \Omega$. Approximate conservation of the K quantum number leads to retarded decays between states when the change in K is greater than the angular momentum carried by the transition photon. These states are commonly referred to as "K-isomers".

The attraction of studying these multi-quasiparticle states is that they offer the chance to probe particular aspects of nuclear structure in well-defined configurations. These include the understanding of the K quantum number itself, the purity of which comes under individual or concerted attack from rotation, non-axiality and "non-yrastness". Also, the effect of blocking particular orbitals on pairing-dependent properties such as single-particle alignments and rotational parameters can be elucidated.

In some cases the multi-quasiparticle configuration is so favoured in energy that it cannot decay via a low-multipolarity photon. The state may then not only be K-isomeric, it also becomes a "spin-trap". Such states can have very long lifetimes, and some notable examples are found in the stable hafnium isotopes. These are listed in table 1. The best-known example is the $K^\pi = 16^+$ state in ^{178}Hf which has a half-life of 31 years [1,2]. Other notable cases are the $K^\pi = 37/2^-$ state in ^{177}Hf [3,4] and the $K^\pi = 25/2^-$ in ^{179}Hf [5–7], which have half-lives of 51 minutes and 25 days, respectively.

TABLE 1. Long-lived multi-quasiparticle K-isomers in the heaviest stable hafniums.

Nucleus	K^π	$T_{1/2}$	Rotational band?
^{177}Hf	$23/2^+$	1.1s	Yes (to $31/2^+$)
	$37/2^-$	51min	No
^{178}Hf	8^-	4.0s	Yes (to 14^-)
	16^+	31yr	No
^{179}Hf	$25/2^-$	25d	No
^{180}Hf	8^-	5.5h	No

When one is trying to characterize a K-isomeric state, the associated rotational band is usually a crucial source of information. The spectroscopy, however, of rotational bands based on these isomers in heavy hafniums

CP481, *Nuclear Structure 98*, edited by C. Baktash

is handicapped by the limited angular momentum of the ^{176}Yb(α,xn)$^{180-x}$Hf reaction, which is the only available fusion/neutron-evaporation reaction. This hurdle has been overcome at the ANU with the use of ^{9}Be-induced reactions to populate ^{177}Hf, ^{178}Hf and ^{179}Hf to sufficiently high spins to allow the study of states above the three aforementioned isomers, which include their respective rotational bands.

II ABOVE LONG-LIVED K-ISOMERS IN STABLE HAFNIUMS

A Experimental Details

The measurements were performed with the CAESAR γ-ray spectrometer, which consists of six Compton-suppressed HPGe detectors, in conjunction with a compact array of fourteen fast/slow phoswich plastic scintillators. This arrangement was used to collect time-correlated particle-γ-γ coincidences when a self-supporting metallic foil of ^{176}Yb was bombarded with beams of ^{9}Be ions supplied by the ANU 14UD accelerator at energies of 70, 60, 55, 45 and 38 MeV. States in hafnium nuclei were associated with emission of α particles, and the yields were found to be $\sim$20% of that from pure neutron-evaporation which led to tungsten isotopes. A large fraction of the α-particle yield came from the process termed "incomplete fusion" which arises due to moderately-bound Borromean structure ($\alpha + \alpha + n$) of the ^{9}Be beam. An over-simplified view of incomplete fusion is that of beam break-up in the entrance channel, where one fragment (an α particle in this case) continues with essentially the beam velocity, while the complementary fragment ("^{5}He") is transfered to the target, where it leads to fusion-evaporation. The yield of these pre-equilibrated α particles is expected to somewhat localized near to the grazing angle [8] for the collision, while evaporated α particles emitted after complete fusion are expected to be distributed isotropically in the centre-of-mass. These different distributions may be responsible for some features in the data which will be discussed below.

B Data Reduction and Results

Events were selected where an α particle was detected, and the associated pair of γ rays were incremented into an $E_{\gamma 1}$-$E_{\gamma 2}$ correlation matrix. At the higher bombarding energies (55, 60 and 70 MeV) separate matrices were formed according to whether the α-particle was detected at forward or middle/backward angles. In these cases the forward-gated matrix was expected to be dominated by events that arose from incomplete fusion, since grazing angles were $\sim$64°, $\sim$55° and $\sim$44°, respectively. For the 45 MeV ($\Theta_{gr} \simeq 95^\circ$) data, the forward- and middle-tagged data were combined into one matrix, while the back-gated coincidences were placed into a second matrix.

The matrices were analysed with the program ESCL8R [9], which allowed the coincidence intensities to be fitted and stored in a convenient fashion. Gamma decays above long-lived isomers were entirely decoupled from the remainder of the level scheme, which was consistent with the discovery of a number of "floating" bands that did not show coincidences with any known hafnium lines. Coincidences were observed, however, with hafnium X rays. The assignment of these bands was guided by an isotopic dependence of a ratio formed from the intensities measured in the forward- and middle/back-gated matrices. These are shown in figure 1 for $E_{Be} = $ (a) 55 MeV and (b) 70 MeV. There is a clear separation between the hafnium isotopes, where the case associated with fewest evaporated neutrons at each energy has the highest ratio, so that its yield is the most forward-peaked. At 55 MeV this corresponds to ^{179}Hf, which is populated via the $\alpha 2n$ exit channel. At 70 MeV there is no measurable yield into ^{179}Hf, and ^{178}Hf ($\alpha 3n$) is now the most forward-peaked product. This is probably due to the associated forward-peaked α particles having a higher average energy, so that the average excitation energy of the ^{181}Hf compound nucleus is lower, favouring the evaporation of fewer neutrons. The open symbols correspond to transitions that were previously known, and new transitions in coincidence with them. The filled symbols correspond to transitions in the "floating" bands mentioned above. Results for ^{178}Hf, ^{179}Hf and ^{177}Hf will be discussed below.

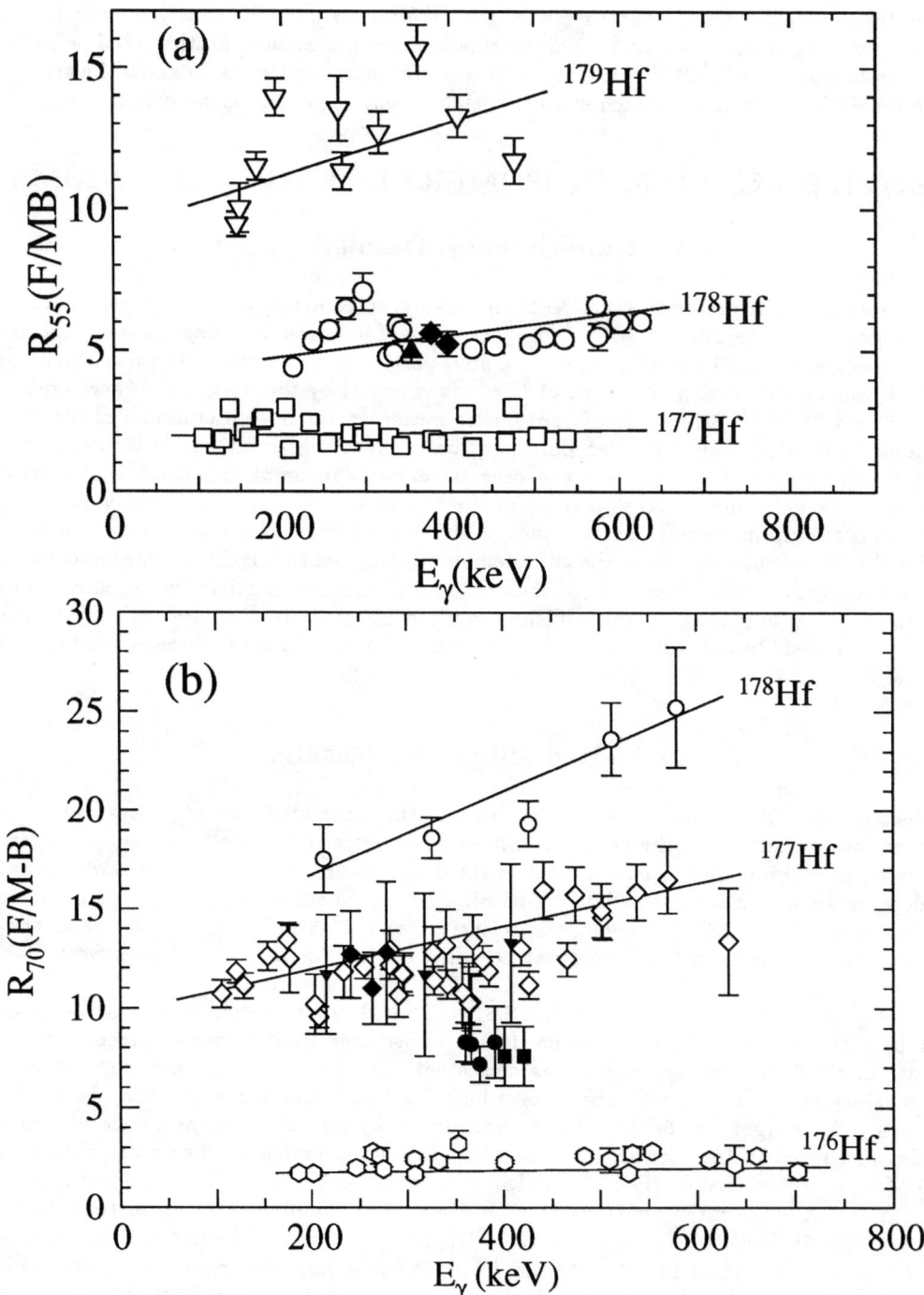

FIGURE 1. "Forward/(Middle-Back)" γ-ray intensity ratios for (a) the 55 MeV data, and (b) the 70 MeV data.

1 Rotation of the $K^{\pi} = 16^+$ isomer in ^{178}Hf

A sum-of-gates spectrum that shows the band associated with the $K^{\pi} = 16^+$, $T_{1/2} = 31$ year isomer in ^{178}Hf is presented in figure 2. The energy of the $17^+ \rightarrow 16^+$ transition (357keV) is consistent with the level-energy assignment made to the 17^+ state from (d,d') [10] and Coulomb excitation measurements [11] performed on a target in which micro-weight quantities of the isomer were present. In the ^{9}Be-induced measurement the band was clearly observed to its 22^+ state, with tentative evidence for the 23^+ level [12]. These spin values are unattainable with the $(\alpha,2n)$ reaction. It was now at last possible to extract some spectroscopic information for the band. The g_K values derived from the inband branching ratios agree with those expected for the previously suggested four-quasiparticle $\pi^2[7/2^+,9/2^-]\nu^2[7/2^-,9/2^+]$ configuration. No evidence was seen for $K^{\pi} = 19^+$, 20^- and 22^- states in the data, which are predicted to be the lowest six-quasiparticle configurations. This may indicate that they were too weakly populated, or that their lifetimes were prohibitively long.

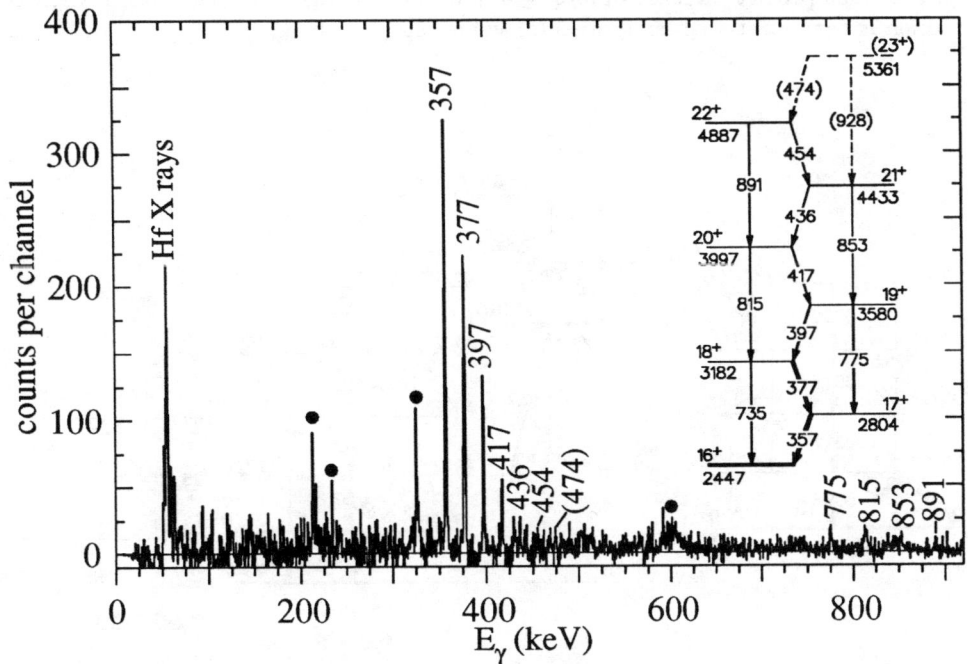

FIGURE 2. Sum-of-gates and partial level scheme for the rotational band assigned to the $K^{\pi} = 16^+$ isomer in ^{178}Hf.

2 Collective g-factor of the 25 day isomer in ^{179}Hf

The so-called collective g-factor (g_R) is sensitive to the partition of the total nuclear angular momentum between the proton and neutron components. A simple way of representing this is given in equation 1

$$g_R = \frac{\mathcal{J}^{\pi}}{\mathcal{J}^{\pi} + \mathcal{J}^{\nu}} \tag{1}$$

Hence, changes to the proton $(\mathcal{J}^{\pi})$ and/or neutron $(\mathcal{J}^{\nu})$ moments-of-inertia can effect g_R. The moments-of-inertia are expected to increase when pairing strengths are reduced due to the blocking of orbitals in multi-quasiparticle configurations.

In order to extract an experimental value for g_R, the magnetic and quadrupole moments of the bandhead need to be known, together with branching ratios of the associated rotational band, from which the quantity $(g_K - g_R)/Q_0$ can be evaluated. These can be placed into equation 2, and a value for g_R can be found.

$$g = \frac{\mu}{I} = g_R + (g_K - g_R)\frac{K^2}{I(I+1)} \tag{2}$$

In the case of the 25/2⁻ isomeric state [14] and the stable 9/2⁺ ground state [15] in ^{179}Hf, the magnetic moments had aleady been measured, but the associated rotational band was only known for the latter. The 25/2⁻ configuration arises from coupling the [624]9/2⁺ neutron orbital which forms the ground state to the 8⁻ [[404]7/2⁺, [514]9/2⁻] pair of quasiprotons. States above the 25/2⁻ isomer have now been found [16], as can be seen in figure 3. These include the band based on the 25/2⁻ isomer itself, into which a five-quasiparticle state with $K^\pi = (33/2^-)$ decays. A band has also been assigned to a three-quasiparticle $K^\pi = 23/2^+$ state, which arises from coupling the [514]7/2⁻ neutron orbital to the 8⁻ pair of quasiprotons. Analysis of the branching ratios from the 25/2⁻ and 23/2⁺ bands agree with the configurations mentioned above. Moreover, it is now possible to extract a value of g_R for the 25/2⁻ state and hence compare with that of the ground state. Values of $g_R(25/2^-) = 0.34(5)$ and $g_R(9/2^+) = 0.22(3)$ were found. Our blocked Lipkin-Nogami calculations suggest that the proton-pairing in the 25/2⁻ configuration is ~70% of that in the 9/2⁺ ground-state configuration. The g_R values are compared in table 2 with two calculated predictions based on (a) blocked Lipkin-Nogami and (b) HFBC pairing, which were combined with the Migdal two-fluid formula to calculate the moments-of-inertia. The two calculations predict increase of $\delta g_R \sim +0.07$ for the 25/2⁻ configuration when compared to the ground state, which is consistent with experimental value ($\delta g_R \sim +0.12(6)$).

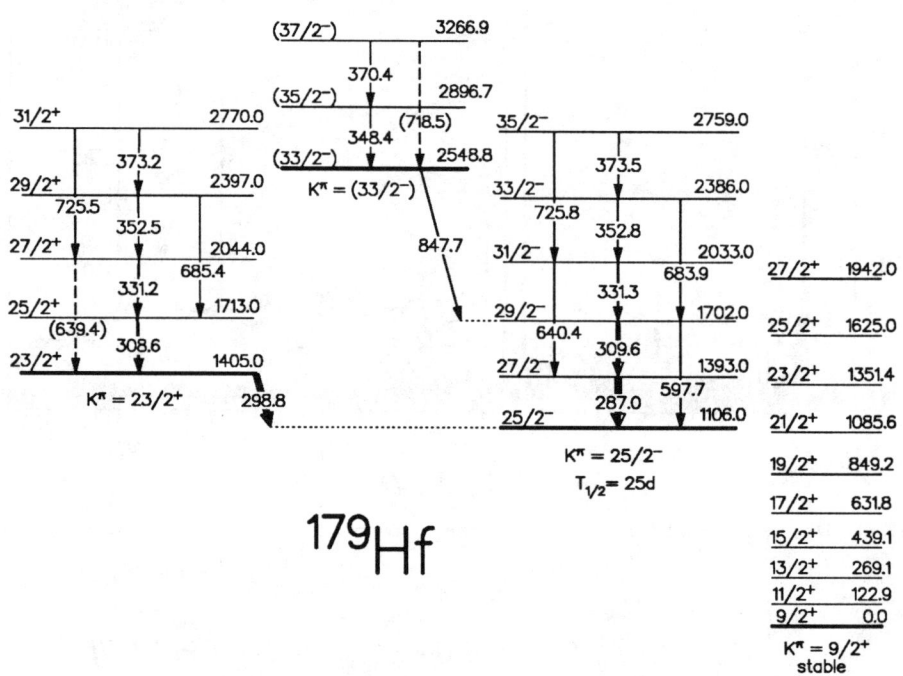

FIGURE 3. Partial level scheme for ^{179}Hf showing states above the $K^\pi = 25/2^-$, $T_{1/2} = 25$ day isomer.

TABLE 2. Experimental and calculated g_R values for the 9/2⁺ and 25/2⁻ states.

K^π	g_R (expt.)	g_R (calc.)	
		LN	HFBC
9/2⁺	0.22(3)	0.30	0.22
25/2⁻	0.34(5)	0.36	0.29

The highest bombarding energy used in these studies was 70 MeV, and this favoured the population of ^{177}Hf via the $\alpha 4n$ exit channel. A prime goal of this measurement was to establish the rotational band associated with the $K^\pi = 37/2^-$ $T_{1/2} = 51$min five-quasiparticle isomer. Not only was this achieved, but a new $K^\pi = 39/2^+$ five-quasiparticle state was discovered, together with the band with based on it, as is shown in figure 4. These two five-quasiparticle structures form part of a comprehensive level scheme obtained for ^{177}Hf [13], which includes considerable extensions to the previously known $K^\pi = 23/2^+$ and $25/2^-$ three-quasiparticle bands. Comparison with multi-quasiparticle calculations indicate that all of the most favoured three- and five-quasiparticle states were observed. Comparison of g_K values extracted from branching ratios for all the previously known and newly discovered bands agree well with simple Nilsson model predictions (table 3).

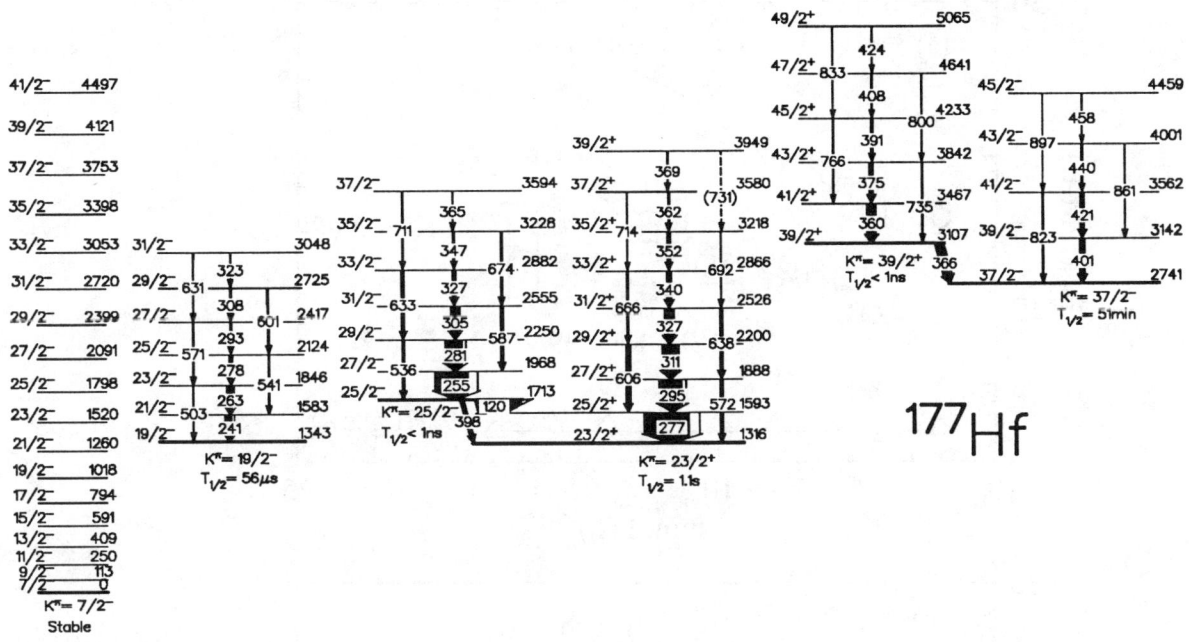

FIGURE 4. Three- and five-quasiparticle states in ^{177}Hf, along with their associated rotational bands.

TABLE 3. g_K factors for multi-quasiparticle states in ^{177}Hf.

| K^π | Configuration[a] | | Exp[b] | g_K Nilsson[c] | empirical[d] |
	ν	π			
$19/2^-$	$7/2^-$	$7/2^+, 5/2^+$	0.64(5) 0.71(5)†	0.74	0.72
$23/2^+$	$7/2^-$	$7/2^+, 9/2^-$	0.68(4) 0.75(4)†	0.79	0.77
$25/2^-$	$9/2^+$	$7/2^+, 9/2^-$	0.52(4) 0.59(4)†	0.55	0.59
$37/2^-$	$5/2^-, 7/2^-, 9/2^+$	$7/2^+, 9/2^-$	0.41(3)	0.38	0.40
$39/2^+$	$7/2^+, 7/2^-, 9/2^+$	$7/2^+, 9/2^-$	0.41(2)	0.36	0.40

[a] π: $7/2^+$: $7/2^+[404]$; $9/2^-$: $9/2^-[514]$; $5/2^+$: $5/2^+[402]$.
ν : $9/2^+$: $9/2^+[624]$; $7/2^-$: $7/2^-[514]$; $5/2^-$: $5/2^-[512]$; $7/2^+$: $7/2^+[633]$.
[b] $(g_K - g_R)/Q_0$ is positive, $Q_0 = 7.2(1)$ eb, $g_R = 0.23(2)$ (†, $g_R = 0.3$)
[c] Nilsson wave functions, $g_s = 0.7 g_{free}$ and $(\epsilon_2, \epsilon_4) = (0.245, 0.053)$.
[d] Empirical g_K values used wherever possible.

C Evidence for reaction-dependent spin-population?

Some previous investigations of incomplete fusion reactions have produced evidence that there may be a relatively narrow spin-distribution associated with this mechanism [17]. This has been linked to the limited range of impact parameters near to that for a grazing collision which are expected to lead to incomplete fusion, though it is far from clear that this is a general feature of these types of reactions. When the intensity ratios for rotational bands in ^{177}Hf are plotted as a function of spin, there are a number of features which may arise from different spin-distributions associated with complete and incomplete fusion. These ratios are shown in figure 5(a) for bands based on the one-quasiparticle $7/2^-$ and $9/2^+$ states, the three-quasiparticle $23/2^+$ state and the five-quasiparticle $39/2^+$ state. The ground-state bands (gsb) of ^{178}Hf and ^{176}Hf are also shown.

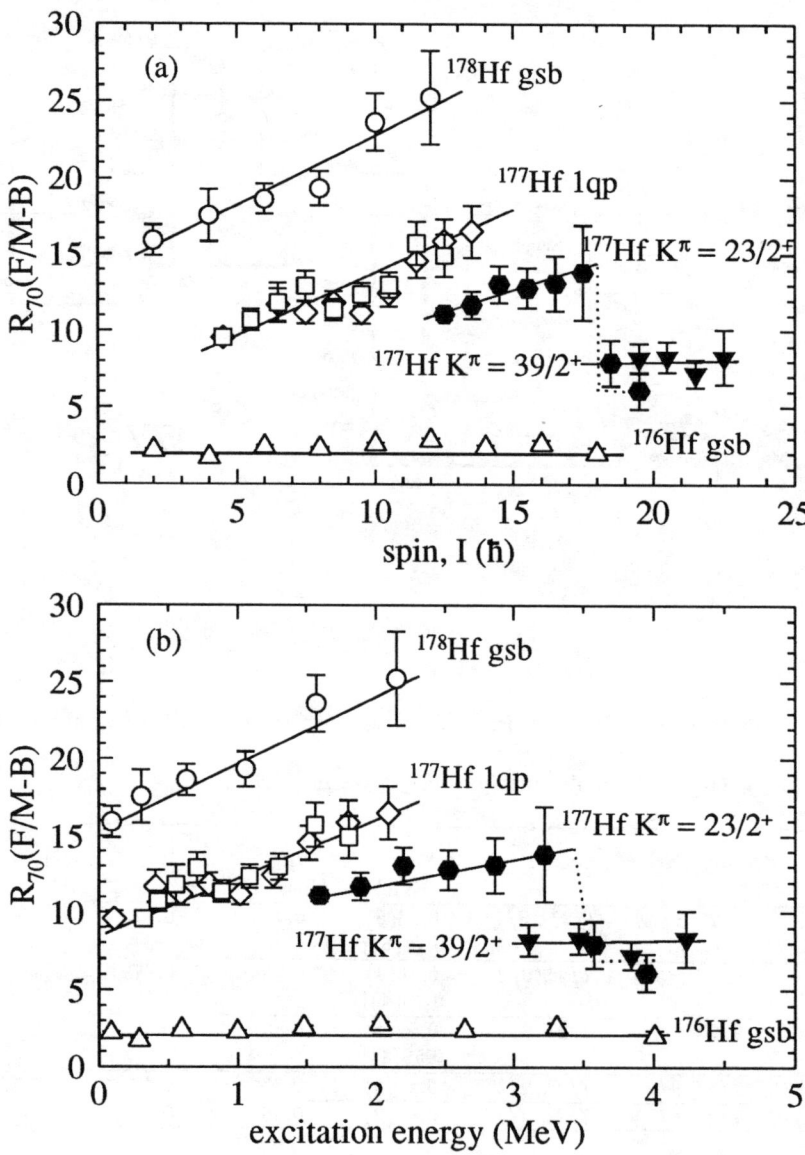

FIGURE 5. Intensity ratios for $E_{Be} = 70$ MeV data plotted versus (a) spin, and (b) excitation energy.

- The 7/2⁻ gsb and excited 9/2⁺ one-quasiparticle bands in ¹⁷⁷Hf show a gradual rise with spin similar to that of the gsb in ¹⁷⁸Hf.

- The ratios for the three-quasiparticle 23/2⁺ band in ¹⁷⁷Hf are lower than those of the one-quasiparticle bands at the same spin, and excitation energy (figure 5(b)).

- The ratios for the five-quasiparticle 39/2⁺ band in ¹⁷⁷Hf are markedly lower than those for the one-quasiparticle bands; the ratios are also lower than those for the three-quasiparticle 23/2⁺ band, until at I ≃ 18ℏ, they coincide. This seems to be due to a more isotropic distribution of α particles associated with the population of ¹⁷⁷Hf beyond ~18ℏ, which suggests that relatively more of the high-spin feeding comes from complete fusion.

- The ratios for the gsb of ¹⁷⁶Hf show no dependence on spin, which may indicate population via complete fusion only. If so, then these ratios should equal unity, which suggests that all the ratios should be empirically corrected by dividing by ~2.5.

III   CONCLUDING REMARKS; WHAT NEXT?

Incomplete fusion reactions are clearly a useful tool for nuclear structure studies to *moderate* spin for nuclei at or near the line of stability. The application of these reactions to populate states above very long-lived K-isomers in stable hafnium nuclei has produced a wealth of spectroscopic information. Similar measurements have been carried out on ¹⁸¹,¹⁸⁰Ta [18,19] with ¹¹B beams, and are in progress for ¹⁷⁷,¹⁷⁶,¹⁷⁵Lu [20] with ⁷Li beams. Higher seniority states in these nuclei are clearly beyond the reach of incomplete fusion reactions due to the limited spin-input. As an example, a comparision is shown in figure 6 between experiment and calculation for multi-quasiparticle states in ¹⁸⁰Ta [19]. There are a number of eight-quasiparticle states which come low in excitation energy due to favourable residual interactions in these configurations. It remains as a challenge to experiment to find a means to populate these states and then study their decay.

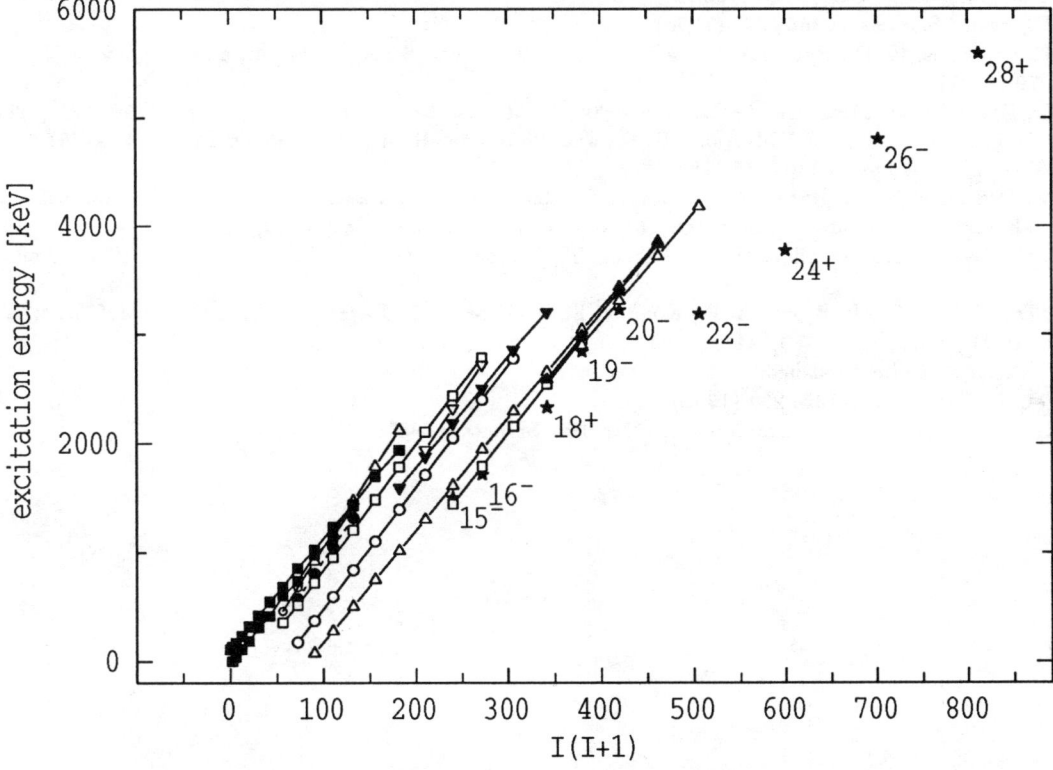

FIGURE 6. Comparison with calculated multi-quasiparticle states (stars) and experiment for ¹⁸⁰Ta.

425

Recent measurements with pulsed ^{238}U beams on thick rare-earth targets have produced some promising results [21,22]. Thus far the technique has been applied to studying the decay of new multi-quasiparticle isomers by measuring coincidences between beam pulses. It is not clear, however, how high in spin one can populate in these kind of reactions, and the assignment of rotational bands to these new states has yet to be achieved. A future direction could involve the use of a ^{132}Sn beam, produced by electro-fission of ^{238}U, in conventional fusion-evaporation reactions on targets such as ^{50}Ti ($\rightarrow$ ^{178}Hf + $4n$) and ^{51}V ($\rightarrow$ ^{180}Ta + $3n$).

ACKNOWLEDGEMENTS

The measurements on the hafnium isotopes were performed in collaboration with G. D. Dracoulis, A. P. Byrne, T. R. McGoram, S. Bayer, F. G. Kondev, W. A. Seale, R. T. Newman and R. A. Bark. The support of the technical staff at the ANU Heavy Ion Facility is gratefully acknowledged.

REFERENCES

1. Helmer, R. G. and Reich, C. W., *Nucl. Phys.* **A114**, 649 (1968).
2. Helmer, R. G. and Reich, C. W., *Nucl. Phys.* **A211**, 1 (1973).
3. Ward, T. E. and Haustein, P. E., *Phys. Rev. Lett.* **27**, 685 (1971).
4. Chu, Y. Y., Haustein, P. E., and Ward, T. E., *Phys. Rev.* **c66**, 2259 (1972).
5. Hübel, H., Naumann, R. A., Anderson, M. L., Larsen, J. S., Nielsen, O. B., and Roy Poulsen, N. O., *Phys. Rev.* **C1**, 1845 (1970).
6. Hübel, H., Naumann, R. A., and Hopke, P. K., *Phys. Rev.* **C2**, 1447 (1970).
7. Chu, Y. Y. and Ward, T. E., *Phys. Rev.* **C8**, 422 (1973).
8. For example, Bass, R., *Nuclear Reactions with Heavy Ions*, New York: Springer-Verlag, 1980, appendix C, pp. 403.
9. Radford, D. C., *Nucl. Instrum. Methods Phys. Rev.* **A306**, 297 (1995).
10. Deylitz, S., *et al.*, *Phys. Rev.* **C53**, 1266 (1996).
11. Lubkiewicz, E., *et al.*, *Z. Phys.* **A355**, 377 (1996).
12. Mullins, S. M., Dracoulis, G. D., Byrne, A. P., McGoram, T. R., Bayer, S., Seale, W. A., and Kondev, F. G., *Phys. Lett.* **B393**, 279 (1997).
13. Mullins, S. M., Byrne, A. P., Dracoulis, G. D., McGoram, T. R., and Seale, W. A., *Phys. Rev.* **C58**, 831 (1998).
14. Hübel, H., Freitag, K., Schoeters, E., Silverans, R. E., and Naumann, R. A., *Phys. Rev* **C12**, 2013 (1975).
15. Buttengach, S., *et al.*, *Z. Phys.* **A260**, 157 (1973).
16. Mullins, S. M., Dracoulis, G. D., Byrne, A. P., Bayer, S., McGoram, T. R., and Seale, W. A. to be published.
17. Inamura, T., Ishihara, M., Fukuda, T., Shimoda, T., and Hiruta, H., *Phys. Lett.* **68B**, 51 (1977).
18. Dracoulis, G. D., Byrne, A. P., Mullins, S. M., Kibédi, T., Kondev, F. G., and Davidson, P. M., *Phys. Rev.* **C58** 1837, (1998)
19. Dracoulis, G. D., Mullins, S. M., Byrne, A. P., Kondev, F. G., Kibédi, T., Bayer, S., Lane, G. J., McGoram, T. R., and Davidson, P. M., *Phys. Rev.* **C58** 1444, (1998).
20. McGoram, T. R., *et al.* to be published.
21. Wheldon C. *et al.*, *Phys. Lett* **B425**, 239 (1998).
22. D'Alarcao, R., *et al.*, abstract to this conference, ("Nuclear Structure '98" - Abstracts, p. 22).

COLLECTVE MODES OF EXCITATIONS

Nuclear Collective Motion: Strength and Limitations of Group Theoretical Methods

Caroline De Coster[a,c], Bruno Decroix[a] and Kris Heyde[a,b]

[a] Subatomic and Radiation Physics, INW, Proeftuinstraat 86, B-9000 Gent (Belgium)
[b] present address: EP-Isolde, CERN, CH-1211, Geneva 23
[c] postdoctoral fellow of the Fund for Scientific Research - Flanders (Belgium)

Abstract. We discuss both the strength of the Interacting Boson Model (IBM) as a unifying structure, emphasizing the robust elements as well as the limitations in an early U(6) group structure of interacting s and d bosons. We briefly address the rich structure of the IBM-2 approach, when treating proton and neutron degrees of freedom explicitly, which gives rise to non-symmetric spatial excitation modes of which a number have been detected experimentally. Finally, the very clear observation of particle-hole excitations near to closed-shell regions, leads to an extension of the IBM to include both particle and hole bosons. The basic concepts are presented as well as the compelling experimental evidence pointing out the existence of larger group-structures than the U(6) group structure.

INTRODUCTION

When studying the atomic nucleus within the shell model, one starts from the concept of interacting valence nucleons outside an inert core. As the number of active nucleons increases, coherent motion may give rise to collective properties. The latter are successfully studied using geometric models, which describe dynamic and static deformation of the nucleus using shape variables. A third complementary approach to study the nucleus starts from the symmetry concept, which is now proven to have successful applications in subatomic physics and quantum mechanics in general. Several fermionic models have been developed, such as the Elliott SU(3) model, the pseudo-SU(3) and the symplectic model, the fermion dynamical symmetry model [1]. Here we intend to address the bosonic approach, aiming at a description of low-lying collective phenomena observed in medium-heavy and heavy nuclei.

The building blocks of the interacting boson model (IBM) are nucleon pairs with angular momentum $L^\pi = 0^+$ and 2^+ which are mapped onto s and d bosons, respectively [2]. This severe truncation of the shell model space, which is reasonable because of the importance of pairing and low-lying quadrupole collectivity in medium-heavy and heavy nuclear systems, makes the many-body problem more tractable. Moreover, the finite boson number obtained explicitly appears in the IBM predictions of observables, hence implying a characteristic variation of the latter across a major shell mostly in accord with the data [3]. This is a very robust feature of the model in that, if the data were different, the model predictions couldn't be altered accordingly.

After briefly addressing the basic features of the IBM-1, we indicate how the introduction of proton and neutron bosons leads to additional degrees of freedom in the IBM-2 and therefore allows for non-symmetric modes of motion. We also briefly review the F-spin concept.

Alternatively, we enlarge the IBM-1 model space by introducing particle and hole bosons. In this way we break up the inert core to describe particle-hole excitations across a closed shell on equal footing with regular low-lying collective excitations. Starting from an analogous algebraic structure as in the IBM-2, we introduce I spin in the extended (E) IBM and illustrate its validity and use in mass regions where such intruder excitations have been found experimentally, e.g., the Z=50 and Z=82 region. We briefly indicate how mixing of regular and intruder excitations can be described in such a generalized scheme.

CP481, *Nuclear Structure 98*, edited by C. Baktash
© 1999 American Institute of Physics 1-56396-858-4/99/$15.00

Finally, we combine both proton-neutron and particle-hole degrees of freedom in the EIBM-2, indicating the still larger algebraic structure with possible applications.

DYNAMICAL SYMMETRIES IN THE IBM-1

The six single-boson states (s, d_μ), $\mu = -2, \ldots, +2$ are subject to a unitary U(6) group transformation. The associated algebra can be reduced into three algebraic "chains", all containing the O(3) algebra, corresponding with rotational invariance and good angular momentum L. These dynamical symmetry limits each have a definite geometrical interpretation : the U(5) or anharmonic vibrator limit, the SU(3) or axially symmetric rotor limit and the O(6) or γ-soft rotor limit [4,5],i.e.,

$$
\begin{aligned}
U(6) \supset U(5) &\supset O(5) \supset O_L(3) \\
&\supset O(6) \supset O(5) \supset O_L(3) \\
&\supset SU(3) \supset O_L(3) \quad .
\end{aligned}
$$ (1)

The dynamical symmetries can be used as benchmarks, that is, many actual nuclei can be easily and well described in terms of a relatively small amount of symmetry breaking.

The Hamiltonian of the IBM-1 is expressed in terms of creation and annihilation operators for s and d bosons, and contains one- and two-body terms. Rewritten in its multipole form [4], it reads

$$
\hat{H} = \epsilon \hat{n}_d + a_0 \hat{P}^+ \hat{P} + a_1 \hat{L} \cdot \hat{L} + a_2 \hat{Q} \cdot \hat{Q} + a_3 \hat{T}_3 \cdot \hat{T}_3 + a_4 \hat{T}_4 \cdot \hat{T}_4 \quad ,
$$ (2)

with $\hat{n}_d$ the d-boson number operator, $\hat{P}^+ \hat{P}$ the pairing term, $\hat{L}$ the angular momentum operator,

$$
\hat{Q} = (s^+ \tilde{d} + d^+ s)^{(2)} + \chi (d^+ \tilde{d})^{(2)}
$$ (3)

the quadrupole operator and

$$
\hat{T}_3 = (d^+ \tilde{d})^{(3)} \quad ,
$$ (4)

$$
\hat{T}_4 = (d^+ \tilde{d})^{(4)} \quad .
$$ (5)

If the d-boson is interpreted as a kind of quadrupole vibrational phonon, the first term would give a vibrational spectrum. In general anharmonicity terms can be included for a U(5) dynamical symmetry. The fourth quadrupole-quadrupole term gives SU(3); O(6) results when the second pairing term dominates. Thus single terms determine the structure of the symmetries and hence transition regions between two symmetries only depend on the ratio of the two relevant coefficients. One can therefore represent the basic structure of the IBM in terms of a symmetry triangle, now commonly denoted as "the Casten triangle" [6], where each vertex represents a symmetry and each leg is a transitional region which can be calculated as a function of a simple ratio of two parameters (see Figure 1). Since most nuclei turn out to be situated along the legs of the triangle, most calculations involve one to three parameters only.

Nowadays, the IBM has been subject to many critical tests [3] and is commonly acknowledged as a useful complementary approach to the nuclear many-body problem. In this talk I only want to emphasize the simplicity and use of the symmetry concept.

PROTON AND NEUTRON BOSONS: THE IBM-2

The interpretation of s and d bosons as pairs of valence nucleons originated from a mapping of the truncated shell-model space onto the boson space, based on a one-to-one correspondence of boson and fermion pair matrix elements respectively [7]. The need for the introduction of proton and neutron bosons was a very natural result of such a microscopic foundation of the IBM.

The charge character of the bosons can be introduced in two different ways: (i) one considers two distinct systems for neutron and proton bosons and treats them exactly as in the IBM-1; (ii) one considers the boson charge states as the components of a two-valued state vector and labels them with a quantum number called F spin [8]. For unification of both ways of explicitly taking into account the neutron and proton degrees of freedom an embedding symmetry can be constructed [9]:

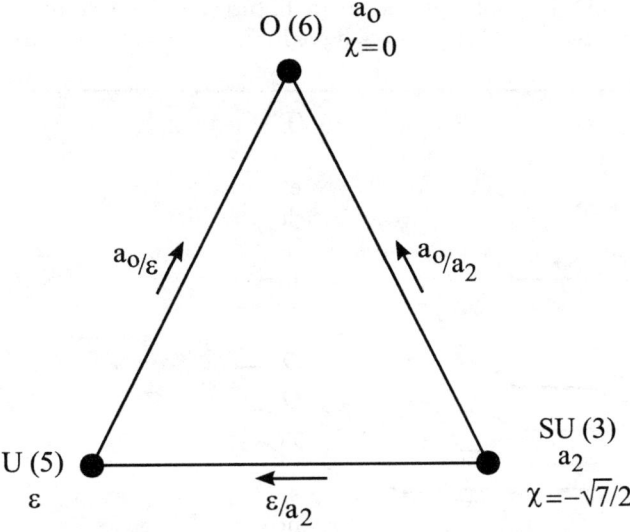

FIGURE 1. Symmetry triangle of the IBM. The parameters quoted refer to the Hamiltonian (2).

$$U(12) \supset U_\pi(6) \otimes U_\nu(6) \supset U_{\pi+\nu}(6) \supset G_\lambda \supset O_L(3)$$
$$\supset (U_{\pi+\nu}(6) \supset G_\lambda \supset O_L(3)) \otimes SU_F(2) \quad , \tag{6}$$

where G_λ stands for the different reduction schemes of U(6) in IBM-1. The former reduction of a product algebra leads to two-rowed representations of the resulting $U_{\pi+\nu}(6)$ algebra, which correspond to different symmetries of the proton-neutron boson wave function. These can alternatively be labelled by the quantum number F spin. When F spin is maximum, the wave function is totally symmetric: both its spatial and its charge part are symmetric. When $F \neq F_{max}$ both parts are non-symmetric, the total wave function still being symmetric. The corresponding states are commonly denoted as mixed-symmetry states and are in the geometrical picture interpreted as out-of-phase vibrational or rotational oscillations of valence protons and neutrons.

The prediction of such states is a remarkable feature of the IBM-2, especially since such excitations have been found experimentally, e.g., the non-symmetric vibrational 2^+ excitations in the A $\sim$ 130 region [10] and the so-called "scissors" 1^+ excitation, first discovered in Darmstadt in 1984 [11] in an (e,e') experiment on ^{156}Gd and subsequently studied over the whole rare-earth and actinide region as well as for lighter nuclei in (e,e'), (p,p') and (γ, γ') experiments [12,13].While the orbital character of the excitation supported the interpretation in terms of a mixed-symmetry rotational IBM-2 state, it was found to be much more fragmented due to microscopic structure effects. This illustrates nicely both the strength and limitations of the model. Moreover in the IBM-2 sum rules were developed [14–17] that allowed to describe in a quantitative way the remarkable relationships between low-lying summed magnetic dipole strength with deformation, isotopic shift and $B(E2; 0_1^+ \rightarrow 2_1^+)$, that have been found experimentally in the rare-earth and actinide region [18–20].

Another nice feature of the algebraic reduction (6) is the concept of F-spin multiplets. If the Hamiltonian is an F-spin scalar, its eigenstates have good F spin, or in other words, a definite symmetry character. Then F spin is a dynamical symmetry. If F spin is a true symmetry the energies are degenerate for the same value of F regardless of the number of proton and neutron bosons. Hence, one can construct multiplets involving different nuclei. The concept is supported by experimental data at least approximately. A good example is the multiplet containing ^{158}Dy [21].

PARTICLE AND HOLE BOSONS: THE EIBM

Motivation

At or near to closed shells, phenomena appear below the pair gap which cannot be explained in terms of an inert core and a number of interacting active valence nucleons outside the core. One of the best examples are the excited 0^+ states in the Pb isotopes shown in Figure 2, which come down in energy as mid-shell is approached. In $^{192-196}$Pb excited $2^+, 4^+, 6^+$ states have been observed which are decaying to the 0^+ by strong

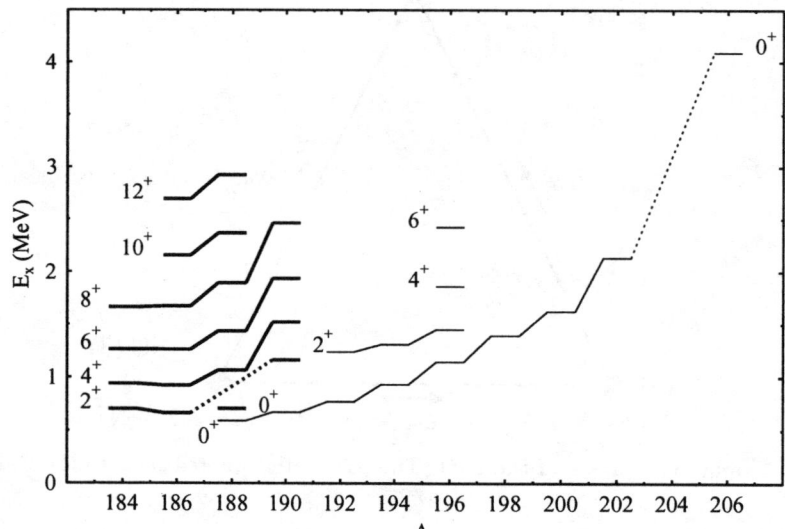

FIGURE 2. Systematics of intruder excitations in the Pb isotopes. The thick (thin) lines indicate the prolate 4p-4h (oblate 2p-2h) structures. Data are taken from refs. [23–26,40].

electric quadrupole (E2) transitions [22]. Moreover, for the light $^{184-190}$Pb isotopes, clear rotational band structures are observed [23–26]. From potential energy surface (PES) analysis, these phenomena have been interpreted as a manifestation of shape coexistence [27,28].

In the shell model these intruder excitations are described as particle-hole (p-h) excitations across the closed shells, which break up the "inert" core and gain a lot of energy through the pairing interaction and the quadrupole proton-neutron interaction, resulting in a parabolic energy dependence on valence neutron number with a minimum at midshell [29].

More generally, these intruder particle-hole excitations became already apparent both in odd-mass nuclei (e.g., In, Sb; Tl, Bi; ...) [30] and adjacent even-even nuclei (e.g., Sn, Cd; Pb, Hg, Pt, Po; ...) [22]. It is possible that they are present in the whole nuclear mass region. Our aim is to describe these excitations on an equal footing with regular excitations using symmetry concepts.

In this contribution we present an overview of the resulting algebraic reduction schemes, which have been worked out in a more elaborate way in ref. [31]. Two important classification schemes of particle-hole excitations result:

(i) A "horizontal" classification in multiplets using a new quantum number I spin (or intruder spin) to establish similarities between intruder and regular band structures in different nuclei belonging to the same multiplet. This is illustrated with data from the Z=82 mass region.

(ii) A "vertical" classification, creating multi-particle-multi-hole excitations through pair scattering within one nucleus using non-compact algebraic structures. In this way it is possible to treat mixing between regular and intruder excitations, which, in the framework of the IBM-1 and IBM-2, is shown to be important to describe the Cd isotopes [32–34] as well as the Hg and Pt isotopes [35–37].

Intruder Spin Classification of Excited States

In the extended IBM, the core is allowed to break up, and particle and hole pairs are now treated separately as particle and hole bosons, thus leading to a direct product algebra $U_p(6) \otimes U_h(6)$. Starting there, one can construct an embedding $U(12)$ algebra which can be reduced following two schemes:

$$(i) \quad U(12) \supset U_p(6) \otimes U_h(6) \supset U_{p+h}(6) \tag{7}$$

$$\begin{array}{ccc} | & | & | \\ [N_p] & [N_h] & [N_p + N_h - i, i] \end{array}$$

and

$$(ii) \quad U(12) \supset U_{p+h}(6) \otimes SU_I(2) \otimes U_+(1) \quad . \tag{8}$$

The former reduction is analogous to the IBM-2 which distinguishes between proton and neutron bosons. The two-rowed representations of the algebra $U_{p+h}(6)$ with label $i \neq 0$ are said to correspond to configurations with mixed particle-hole symmetry character. The latter reduction contains the invariant of the system, the total number of bosons N in $U_+(1)$, and a spin object corresponding with the $SU(2)$ algebra, which we denote by I spin or intruder spin. We can then develop an I-spin formalism in analogy to F spin in the IBM-2 [8].

If $U(12)$ is chosen as the dynamical algebra, the corresponding physical system contains states belonging to the representations $[N_p] \otimes [N_h]$ of $U_p(6) \otimes U_h(6)$ with $N_p + N_h = N$ constant. The "horizontal" classification of particle-hole excitations in terms of the resulting I-spin multiplet has been tested in the Z=50 and Z=82 regions [38,39]. Data clearly point to the validity of I-spin dynamical symmetry.

As an example we show the I=2 multiplet involving the 4p-4h excitations in ^{188}Pb and the 2p-6h excitations in ^{184}Pt in Figure 3. Note the local deviation of the 0^+ energy in ^{184}Pt, which points to mixing with the regular

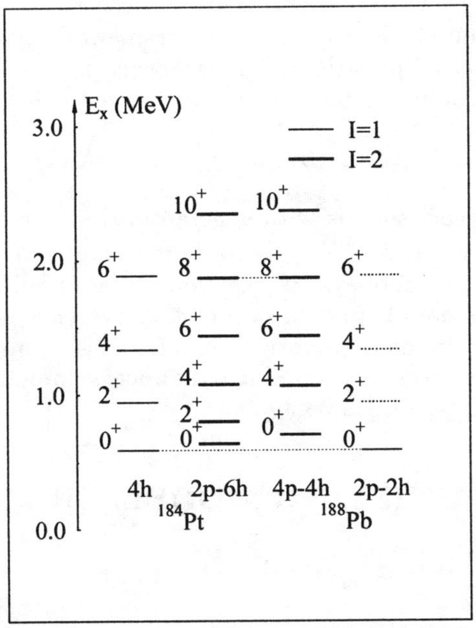

FIGURE 3. Bands in ^{188}Pb and ^{184}Pt belonging to the I=1 and I=2 multiplet, normalized resp. to the 0^+ and 8^+ energy. The dotted lines are predicted values assuming I-spin symmetry. Data are taken from ref. [23,40,42,43].

4h excitations (see following section). In the same figure, we show the 2p-2h 0^+ excitation in ^{188}Pb, recently measured independently by Bijnens [40,41] and Page [42]. By normalizing the 4h 0^+ energy in ^{184}Pt to this energy, one can predict from the 4h band structure in ^{184}Pt, the energy of the other 2p-2h band members in ^{188}Pb.

Algebraic Description of Multi-Particle-Multi-Hole Excitations

Besides the global behaviour of particle-hole intruder excitations in a certain mass region which can be well described in terms of I-spin multiplets, one needs to account for local perturbations in the energy systematics on top of the smooth behaviour, which can be attributed to the mixing between regular and intruder configurations. This has been demonstrated earlier by IBM-1 and IBM-2 mixing calculations [32,33,37], in which electromagnetic properties which are more wave function sensitive observables, were studied as well.

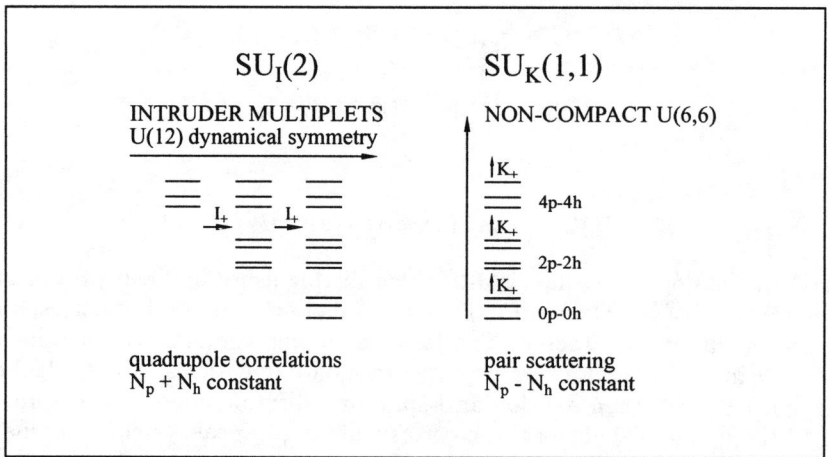

FIGURE 4. Horizontal and vertical classification of particle-hole excitations, respectively realized within a U(12) algebra, using I spin, and a non-compact U(6,6) algebra, using K spin. While multi-particle-multi-hole states within one nucleus are realized through pair scattering, quadrupole correlations are the driving force for the global behaviour of such excitations throughout a multiplet.

To describe such mixing in a consistent way using the framework of the extended IBM, one introduces operators that increase the number of particle and hole bosons by one, keeping $N_p - N_h$ constant. The resulting non-compact algebraic structure U(6,6) can be further reduced as

$$U(6,6) \supset U_{p-h}(6) \otimes SU_K(1,1) \otimes U_-(1) \quad . \tag{9}$$

The non-compact SU(1,1) algebra corresponds with generators that can be seen as three components of an object K spin, where the ladder operators $\hat{K}_+(\hat{K}_-)$ create (annihilate) a particle-hole pair. As a consequence of the non-compactness, however, this object obeys different commutation rules and therefore does not correspond to a general angular momentum, as does I spin. In Figure 4 an overview of both schemes is given.

One can show that an IBM-1 calculation as performed in ref. [37] is a limiting case in this general framework. More limiting cases have been studied recently, coupling dynamical symmetries U(5)-O(6) and O(6)-SU(3) with applications in the Z=50 and Z=82 region [34,38,39].

A UNIFYING MODEL: EIBM-2

Algebraic Structure

If one wants to combine the charge degree of freedom and the particle-hole degree of freedom, one can proceed by introducing particle and hole bosons in the IBM-2 [44]. One thus obtains four types of bosons as depictured in Figure 5, $(b_{\lambda\mu})_\alpha^{(\rho)}$, $\lambda = 0,2$; $\mu = -\lambda,\ldots,+\lambda$; $\rho = \pi,\nu$; $\alpha = p,h$. These 24 single boson states are now subject to a unitary U(24) group transformation. To obtain a classification of the many-boson states that can be constructed within the model space, one needs to determine the possible reductions of the associated U(24) algebra. Thereby, we aim at recovering the quantum numbers L, I and F, respectively containing information about angular momentum, charge and particle-hole symmetry of the wave function. The only possible reduction reads

$$U(24) \supset \otimes \begin{array}{c} U_L(6) \supset G_\lambda \supset O_L(3) \\ \\ SU_{IF}(4) \supset SU_I(2) \otimes SU_F(2) \end{array} \quad . \tag{10}$$

It is interesting to note that we can recover the EIBM by breaking the SU(4) symmetry in an $SU_{I(\rho)}(2)$ symmetry, $\rho = \pi,\nu$ when describing proton or neutron particle-hole excitations respectively. So, the EIBM

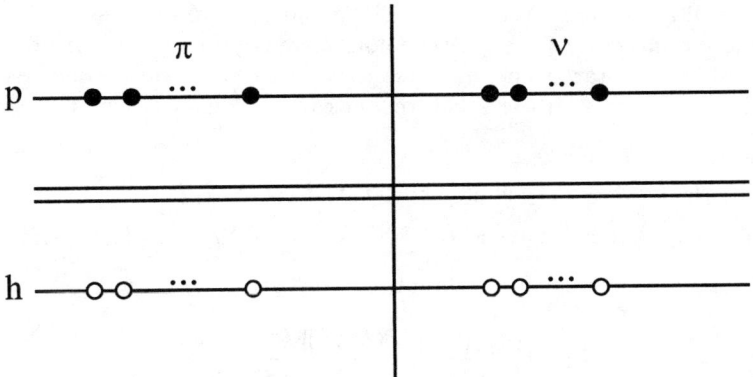

FIGURE 5. Schematic presentation of the four types of bosons included in EIBM-2.

will have applications for nuclei at or near a single-closed proton (neutron) shell, where the neutrons (protons) act as a reference state [38]. The corresponding reduction scheme reads

$$U(24) \supset U^{(\rho)}(12) \supset \otimes \quad \begin{array}{c} U_L(6) \supset G_\lambda \supset O_L(3) \\ \\ SU_{I^{(\rho)}}(2) \end{array} \quad . \tag{11}$$

As for the IBM-2, it can be recovered by selecting only those states which are symmetric in the particle-hole degree of freedom, or, in other words, have maximum F spin. It is however not straightforward to associate a reduction scheme of the U(24) algebra similar to (11) with it, since in usual IBM-2 calculations bosons are counted from the nearest closed shell, hence implicitly having a particle or hole character, which is changing when moving throughout a shell. A good example are F-spin multiplets of nuclei that differ by an α particle [5].

SU(4) multiplets

In previous sections we have indicated how the existence of F-spin and I-spin multiplets points to an underlying symmetry of the Hamiltonian describing the system. We can now similarly study the possible realization of the even more stringent SU(4) symmetry conditions on the Hamiltonian [45].

We have found a possible candidate for SU(4) symmetry in the Z=8 mass region. In the nucleus $^{16}_{8}O_8$ there is experimental evidence for a $K^\pi = 0^+$ band and a $K^\pi = 2^+$ band associated with 4p-4h excitations, featuring the properties of an asymmetric rotor [46,47]. The microscopic structure associated herewith is a $\pi(2p-2h)\nu(2p-2h)$ excitation [48]. In the EIBM-2 context, this means that these levels belong to the SU(4) multiplet with the total number of bosons $N = 4$. If the SU(4) symmetry is a real symmetry for this multiplet, the same structure should be found in the nucleus $^{24}_{12}Mg_{12}$. Experimentally one indeed finds a similar structure in the groundband and $K^\pi = 2^+$ band of the latter. The multiplet of course contains other members, such as $^{24}_{8}O_{16}$ or $^{24}_{16}S_8$ but these extremely proton and neutron deficient nuclei are at the drip lines and thus out of reach experimentally. Other members are the $\pi(4p-4h)$ and $\nu(4p-4h)$ states in $^{16}_{8}O_8$. We have tentatively assigned such configurations to experimentally detected higher-lying levels both for the $K^\pi = 0^+$ ground band and $K^\pi = 2^+$ triaxial band. The resulting multiplet is shown in Figures 6 and 7.

1^+ Mixed-Symmetry States in the EIBM-2

Intuitively it is clear that within the EIBM-2 two types of mixed-symmetry states can occur: the proton-neutron mixed-symmetry states, which were also encountered in the IBM-2 (see previous section) and the particle-hole mixed-symmetry states. Since nothing is known so far about this class of states, neither experimentally, nor theoretically, it is interesting to study their basic properties, such as excitation energy and magnetic dipole transition probability to the ground state. The former is still under study, starting from the seniority scheme in the shell model [50]. Within the EIBM-2 it is possible to estimate the relative strength of the M1 transition to the ground state for a particle-hole mixed-symmetry 1^+ excitation with respect to the scissors excitation. Therefore, we start from their respective wave functions in the O(6) dynamical symmetry limit, since the 2p-2h intruder states in the Z=50 region can be considered as featuring the O(6) symmetry [38]. Thus,

$$| 1^+ \rangle_a \equiv | (N_\alpha^{(\beta)})[N-1,1](N-1,1,0)(1,1)1; I = N/2, F = N/2 - 1 \rangle \quad , \tag{12}$$

$$| 1^+ \rangle_b \equiv | (N_\alpha^{(\beta)})[N-1,1](N-1,1,0)(1,1)1; I' = N/2 - 1, F' = N/2 \rangle \quad , \tag{13}$$

with $\alpha = p, h$ and $\beta = \pi, \nu$. We then calculate the ratio

$$R(a/b) \equiv \frac{\langle 0^+ \| \hat{T}(M1) \| 1^+ \rangle_a}{\langle 0^+ \| \hat{T}(M1) \| 1^+ \rangle_b} \quad . \tag{14}$$

For this more technical issue we refer to [49]. Finally we obtain the resulting simple expression

$$R(a/b) = \frac{g_p^\pi + g_h^\pi - g_p^\nu - g_h^\nu}{g_p^\pi - g_h^\pi + g_p^\nu - g_h^\nu} \quad . \tag{15}$$

Introducing some approximations, and starting from available data on magnetic moments in the Z=50 mass region, one can derive an estimate of this ratio for nuclei with N $\sim$ 66 and Z just below 50 [49]. It should lie between 1 and 1.5 approximately. Therefore the ratio of the magnetic dipole transition probabilities ranges between 1 and 2.3.

CONCLUSION

After briefly reviewing the basic algebraic structure of the IBM-1 and IBM-2, we have introduced an extension of the IBM that allows to describe regular and multi-particle-multi-hole excitations on an equal footing. In a first step, we restricted ourselves to an extension of IBM-1, which can to a large extent be developed in analogy with IBM-2. We then presented a unifying model, the EIBM-2, where both the proton-neutron and particle-hole degrees of freedom are included. This may be a starting point for a more complete description of low-energy nuclear structure phenomena.

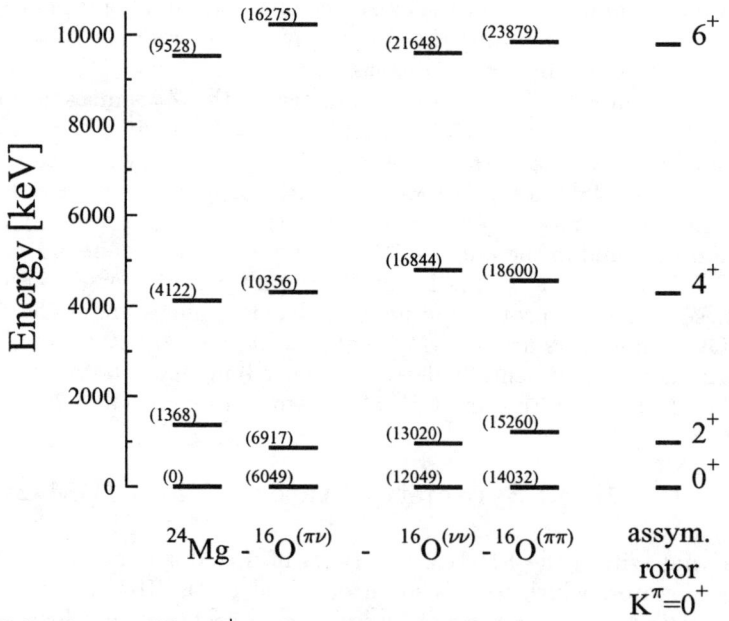

FIGURE 6. SU(4) multiplet for the $K^\pi = 0^+$ band as described in the text. The $\pi(2p - 2h)\nu(2p - 2h)$, $\pi(4p - 4h)$ and $\nu(4p - 4h)$ excitation levels are denoted respectively as $^{16}O^{(\pi\nu)}$, $^{16}O^{(\pi\pi)}$ and $^{16}O^{(\nu\nu)}$.

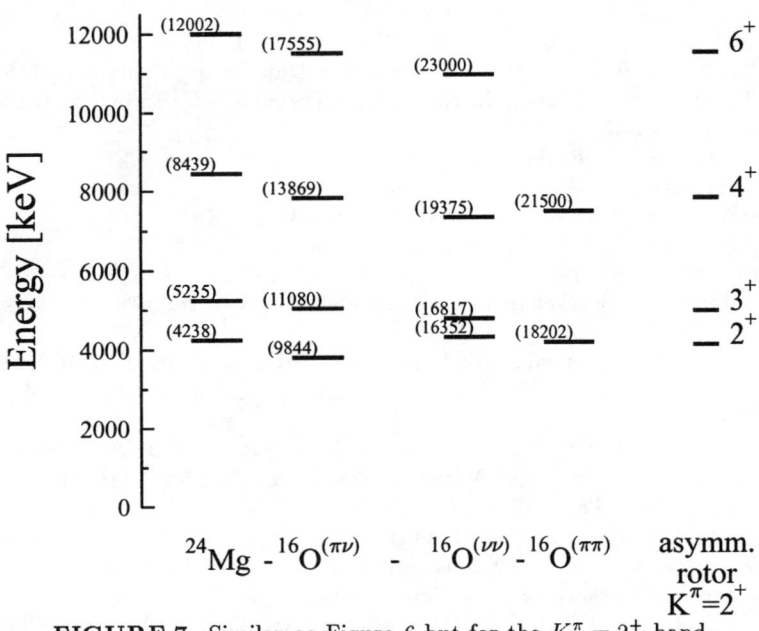

FIGURE 7. Similar as Figure 6 but for the $K^\pi = 2^+$ band.

ACKNOWLEDGEMENTS

The authors thank P. Van Isacker, F. Iachello, B.R. Barrett and G. Rosensteel for stimulating discussions. They are grateful to J. Jolie, H. Lehmann, A.M. Oros and J.L. Wood for active contribution to applications of the intruder spin concept. They acknowledge financial support from the F.W.O. and the I.W.T. One of them (KH) is grateful to CERN for financial support.

REFERENCES

1. Draayer, J.P., in *Algebraic Approaches to Nuclear Structure*, Harwood Academic Publishers, 1993, ch. 7.
2. Arima, A., and Iachello, F., *Phys. Rev. Lett.* **35**, 1069 (1975).
3. Casten, R.F., and Warner, D.D., in *Algebraic Approaches to Nuclear Structure*, Harwood Academic Publishers, 1993, ch. 3.
4. Iachello, F., and Arima, A., *The Interacting Boson Model*, Cambridge University Press, 1987.
5. Frank, A., and Van Isacker, P., *Algebraic Methods in Molecular and Nuclear Structure Physics*, John Wiley & Sons, Inc., 1994.
6. Casten, R.F., and Warner, D.D., *Rev. Mod. Phys.* **60**, 389 (1988).
7. Otsuka, T., in *Algebraic Approaches to Nuclear Structure*, Harwood Academic Publishers, 1993, ch. 4.
8. De Coster, C., and Heyde, K., *Int. J. Mod. Phys.* A **4**, 3665 (1989) and refs. therein.
9. Frank, A., and Van Isacker, P., *Phys. Rev.* C **832**, 1770 (1985).
10. Pietralla, N., et al., *Phys. Rev.* C **58**, 796 (1998).
11. Bohle, D., Richter A., Steffen, W., Dieperink, A.E.L., Lo Iudice, N., Palumbo, F., Scholten, O., *Phys. Lett.* B **137**, 27 (1984).
12. Richter, A., in *The Building Blocks of Nuclear Structure*, ed. Covello, A., World Scientific, 1993, p. 335 and refs. therein.
13. Kneissl, U., Margraf, J., Pitz, H.H., von Brentano, P., Herzberg, R.-D., and Zilges, A., *Prog. Part. Nucl. Phys.* **34**, 285 (1995) and refs. therein.
14. Ginocchio, J., *Phys. Lett.* B **265**, 6 (1991).
15. Heyde, K., De Coster, C., and Ooms, D., *Phys. Rev.* C **49**, 156 (1994).
16. von Neumann-Cosel, P., Ginocchio, J.N., Bauer, H., and Richter, A., *Phys. Rev. Lett.* **75**, 4178 (1995).
17. Lo Iudice, N., *Phys. Rev.* C **57**, 1246 (1998).
18. Ziegler, W., Rangacharyulu, C. Richter, A., and Spieler, C., *Phys. Rev. Lett.* **65**, 2515 (1990).
19. Rangacharyulu, C., Richter, A., Wörtche, H.-J., Ziegler, W., and Casten, R.F., *Phys. Rev.* C **43**, R949 (1991).

20. Margraf, J., et al., *Phys. Rev.* C **45**, R521 (1992) and **47**, 1474 (1993).
21. Lipas, P.O., von Brentano, P., and Gelberg, A., *Rep. Prog. Phys.* **53**, 1355 (1990).
22. Wood, J.L., Heyde, K., Nazarewicz, W., Huyse, M., and Van Duppen, P., *Phys. Rep.* **215**, 101 (1992).
23. Heese, J., Maier, K.H., Grawe, H., Grebosz, J., Kulge, H., Meczynski, W., Schramm, M., Schubart, R., Spohr, K., and Styczen, J., *Phys. Lett.* B **302**, 390 (1993).
24. Baxter, A.M., et al., *Phys. Rev.* C **48**, R2140 (1993).
25. Julin, R., private communication (1998).
26. Dracoulis, G., Byrne, A.P., and Baxter, A.M., *Phys. Lett.* B **432**, 37 (1998).
27. Bengtsson, R., and Nazarewicz, W., *Z. Phys.* A **334**, 269 (1989).
28. Nazarewicz, W., *Phys. Lett.* B **305**, 195 (1993).
29. Heyde, K., Jolie, J., Moreau, J., Ryckebusch, J., Waroquier, M., Van Duppen, P., Huyse, M., and Wood, J.L., *Nucl. Phys.* A **466**, 189 (1987).
30. Heyde, K., Van Isacker, P., Waroquier, M., Wood, J.L., and Meyer, R. *Phys. Rep.* **102**, 291 (1983).
31. De Coster, C., Heyde, K., Decroix, B., Van Isacker, P., Jolie, J., Lehmann, H., and Wood, J.L., *Nucl. Phys.* A **600**, 251 (1996).
32. Délèze, M., Drissi, S., Kern, J., Tercier, P. and Vorlet, J., *Nucl. Phys.* A **551**, 269 (1993).
33. Délèze, M., Drissi, S., Jolie, J., Kern, J., and Vorlet, J., *Nucl. Phys.* A **554**, 1 (1993).
34. Lehmann, H., and Jolie, J., *Nucl. Phys.* A **588**, 623 (1995).
35. Barfield, A., and Barrett, B.R., *Phys. Rev.* C **44**, 1454 (1991).
36. Vretenar, D., Paar, V., Bonsignori, G., and Savoia, M., *Phys. Rev.* C **47**, 2019 (1993).
37. Harder, M., Tang, K. and Van Isacker, P., *Phys. Lett.* B **405**, 25 (1997).
38. Lehmann, H., Jolie, J., De Coster, C., Decroix, B., Heyde, K., and Wood, J.L., *Nucl. Phys.* A **621**, 767 (1997).
39. De Coster, C., Decroix, B., Heyde, K., Wood, J.L., Jolie, J., and Lehmann, H., *Nucl. Phys.* A **621**, 802 (1997).
40. Bijnens, N., PhD thesis, Leuven (1998) unpublished and refs. therein.
41. Bijnens, N., et al., *Z. Phys.* A **356**,3 (1996).
42. Page, R., private communication (1998).
43. Carpenter, M.P., et al., *Nucl. Phys.* A **513**, 125 (1990).
44. Decroix, B., De Beule, J., De Coster, C., Heyde, K., Oros, A.M., and Van Isacker, P., *Phys. Rev.* C **57**, 2329 (1998).
45. Decroix, B., De Beule, J., De Coster, C., Heyde, K., *Phys. Lett.* B (1998) in press.
46. Larsson, S.E., Leander, G., Nilsson, S.G., Ragnarsson, I., and Sheline, R.K., *Phys. Lett.* B **47**, 422 (1973).
47. Aberg, S., Ragnarsson, I., Bengtsson, T. and Sheline, R.K., *Nucl. Phys.* A **391** 327.
48. Becchetti, F.D., Overway, D., Jänecke, J., and Jacobs, W.W., *Nucl. Phys.* A **344**, 336 (1980) and refs. therein.
49. Decroix, B., De Beule, J., De Coster, C., Heyde, K., Oros, A.M. and Van Isacker, P., *Phys. Rev.* C **58**,232 (1998).
50. J. De Beule, private communication.

Phase Transitions and Phase Coexistence in Finite Nuclei

R. F. Casten[1] and N. V. Zamfir[1,2]

[1] *Yale University, New Haven, CT 06520, USA*
[2] *Clark University, Worcester, MA 01610, USA*

Abstract. Recent experiments on ^{152}Sm have revealed a possibly unique type of phase transitional behavior and phase coexistence in nuclei and focus attention on a potential signature for such a phenomenon. The question of the existence of true phase transitional behavior and phase coexistence in finite nuclei is discussed in terms of the discrete nature of nucleon number and a possible control parameter for nuclear structural evolution.

INTRODUCTION

The question of the existence of true phase transitional behavior in nuclei, accompanied by the coexistence of different phases, has long been debated. We refer here to phase transitions as a function of N and Z, and not liquid-gas or confinement-deconfinement phase transitions as a function of temperature and density. Two issues arise. One is whether a finite system of nucleons (with an even smaller number of valence nucleons) can undergo such a phase transition. It is often argued that this is impossible since finite systems exhibit fluctuations and, often, discrete spectra. A second, related, issue is that a phase transition is usually defined in terms of properties of derivatives of various observables (1), but, in nuclei, the nucleon number changes from one nucleus to another by an integer number and it is difficult to even define a proper derivative.

We will first show evidence for possible phase coexistence near A~150, which is disclosed most clearly in ^{152}Sm. We will then discuss a possible way of avoiding the discrete nucleon number problem, of defining a useful control parameter for phase transitional behavior, and of identifying its critical value and an appropriate order parameter.

PHASE COEXISTENCE IN ^{152}Sm

The experimental data, recorded with the Osiris detector system at Köln (2) and the YRAST Ball array at Yale, are the essence of simplicity. They involve the use of high efficiency γ-ray spectroscopy of the decay of a standard ^{152}Eu source. Figure 1 shows a partial level scheme for ^{152}Sm and a portion of the singles γ-ray spectrum. The key γ line, previously unobserved, is at 401 keV and represents a transition from the 2^+_3 level (bandhead of a quasi-γ-band in the usual terminology - but see below) to the 0^+_2 state. The essential point, on which the interpretation rests, is the *weakness* of this transition, both absolutely and relatively. The lifetime of the 2^+_3 level is known, and the transition strength corresponding to these data gives a B(E2: $2^+_3 \rightarrow 0^+_2$) = 0.17 Wu. Relative to the $2^+_3 \rightarrow 0^+_1$ transition the data give a branching ratio

$$R^\gamma_{og} = \frac{B(E2:2^+_3 \rightarrow 0^+_2)}{B(E2:2^+_3 \rightarrow 0^+_1)} = 0.048\,(4) \tag{1}$$

We note the following two important points. The data of ref. 2 did not involve coincidence measurements and hence the issue of the correct placement of the 401 keV transition arises. However, the level scheme of ^{152}Sm is very well known up to the energy of the β-decaying ^{152}Eu parent. There is no other combination of level energies where the 401 keV transition fits. Secondly, since we are dealing with a weak transition, the possibility of contaminant lines (e.g., from ^{154}Eu in the source sample) at a similar energy arises. However, we recall that our interpretation is based on how

CP481, *Nuclear Structure 98*, edited by C. Baktash

weak -- not how strong -- the transition is. Any contamination from other γ-rays would render the $2^+_3 \rightarrow 0^+_2$ transition in ^{152}Sm even weaker, reinforcing the arguments below.

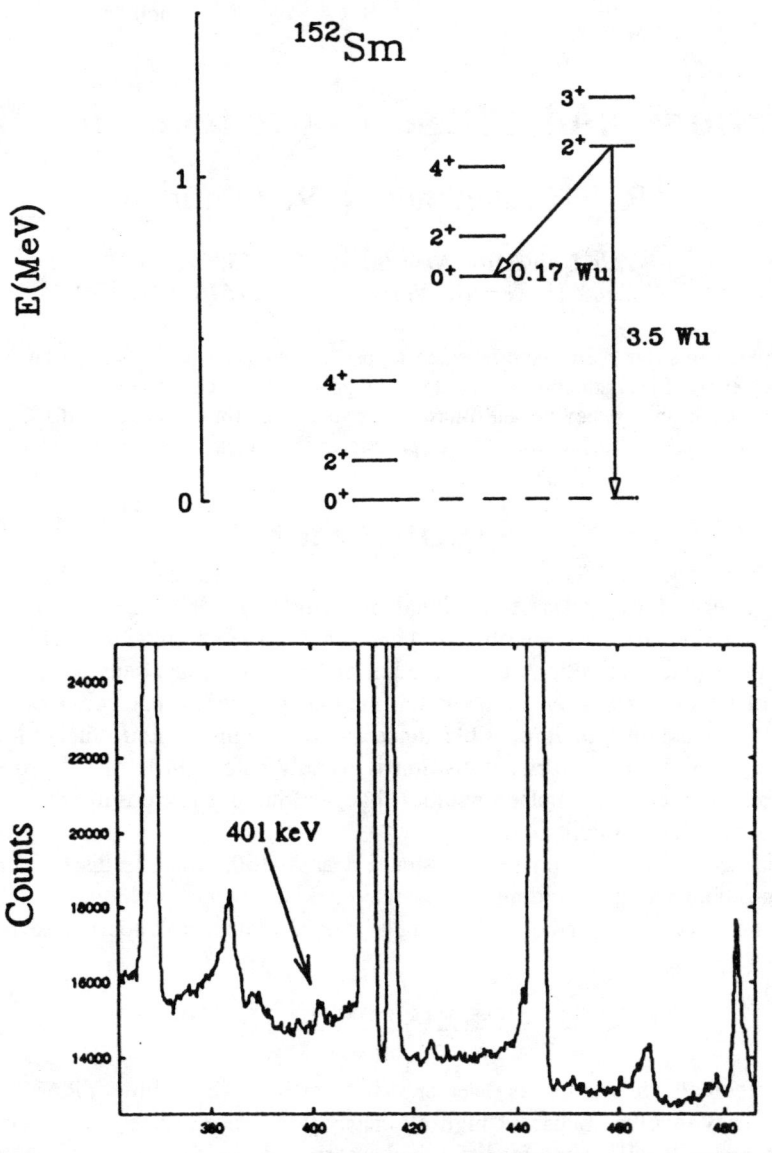

FIGURE 1. Partial level scheme of ^{152}Sm and the portion of the γ-ray singles spectrum showing the key 401 keV transition.

The interest in this transition is the following. In traditional paradigms of structure (vibrator and rotor models) the $2^+_3 \rightarrow 0^+_2$ transition is allowed and collective. This is illustrated in Fig. 2. In a vibrator (left) this transition is a collective 3-phonon → 2-phonon transition while in a rotor it is from the γ-bandhead to the bandhead of a K=0$^+$ band. This transition is also allowed and collective. The latter statement is well known in the context of the Interacting Boson Model (IBA) (3) where, in the SU(3) limit, the K=0 and 2 bands belong to the same representation and hence are connnected by collective E2 transitions. [Interestingly, given our introductory comments, this last statement is true only for *finite* nuclei: in the limit $N_B \rightarrow \infty$ (N_B is the boson number) transitions between these bands actually vanish in the IBA. It is interesting therefore that the weakness of the $2^+_3 \rightarrow 0^+_2$ transition as a signature of phase coexistence

440

is *only* valid in finite nuclei.] Even if one departs from SU(3), as is necessary in realistic IBA calculations of deformed nuclei, the collectivity of $\gamma \rightarrow K=0^+_2$ band E2 transitions persists.

In the collective model for deformed nuclei it is often stated, however, that such transitions, which would be γ vibration $\rightarrow \beta$ vibration transitions that destroy one vibrational excitation and create another, are forbidden. However, it is interesting to note that actual calculations in the Geometric Collective Model (GCM) (4) -- which embodies the collective model of deformed nuclei -- also invariably yield collective $\gamma \rightarrow K=0^+_2$ transitions. Whether this implies that the lowest $K=0^+$ modes in the geometric model for deformed nuclei are not, in fact, β vibrations is an important issue that is beyond the scope of this paper. The essential point here is that, even in the geometric model, the $2^+_3 \rightarrow 0^+_2$ transition is collective for deformed nuclei.

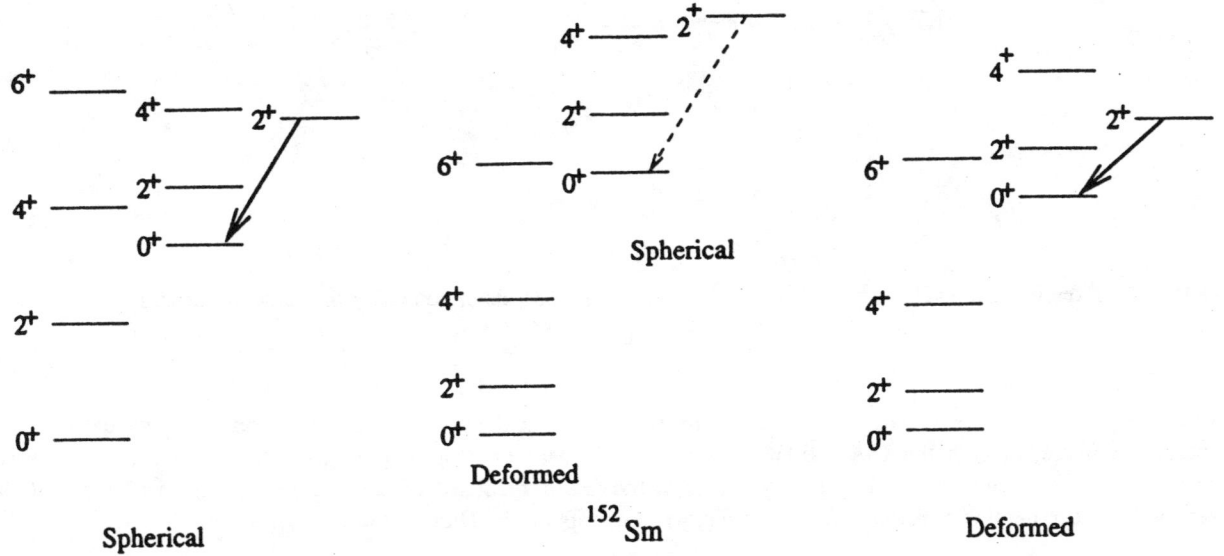

FIGURE 2. Schematic illustration of the expected allowed or forbidden character of the $2^+_3 \rightarrow 0^+_2$ transition for spherical and deformed nuclei and in a coexistence picture.

Thus, how can one understand the weak experimental value of this transition strength in ^{152}Sm? Detailed calculations in both the IBA and the GCM will be discussed below. However, the essential point is clear from the middle panel of Fig. 2, in terms of a coexistence picture (5). ^{152}Sm is in the middle of a shape transition region near N=90. ^{150}Sm is spherical in its ground state [E(2^+_1) = 334 keV, $R_{4/2}$ = 2.32] while ^{154}Sm is quite well deformed [E(2^+_1) = 82 keV, $R_{4/2}$ = 3.25]. The yrast states of ^{152}Sm form an intermediate situation of a quasi deformed rotational band with $R_{4/2}$ = 3.01. It was suggested in ref. 5 that the non-yrast excited states of ^{152}Sm starting with the 0^+_2 level form a vibrational sequence typical of an anharmonic vibrator [$R_{(4/2)2} \equiv$ (E(4^+_2) - E(0^+_2))/(E(2^+_2) - E(0^+_2)) = 2.68]. In this sequence, the 2^+_3 state is a 2-phonon vibrational excitation and hence its decay to the 0-phonon 0^+_2 level is forbidden. Figure 3 shows a fuller picture of the low lying levels of ^{152}Sm. One sees, in fact, candidates for a complete set of nearly degenerate 2-phonon levels -- the 0^+, 2^+, 4^+ triplet -- and the 6^+, 4^+, 3^+, 2^+, and (possibly) the 0^+ levels of the 3-phonon quintuplet.

We can immediately see four prerequisites for such a situation: a deformed or quasi-deformed ground state, the 0^+_2 level below the 2^+_2 and 2^+_3 levels, $R_{(4/2)2} < R_{(4/2)1}$ by a substantial amount and the weak B(E2: $2^+_3 \rightarrow 0^+_2$) value. (The inverse case of a vibrational yrast level sequence with deformed states lying higher in energy would be analogously defined.) Unfortunately, such a situation is rare. We will see later, from the theoretical side, why this rarity is expected. There may in fact be no other examples near the valley of stability although candidate nuclei could occur in the A~100 region. Of course, access to greatly expanded horizons in the nuclear chart with radioactive beams may well bring other examples to light: this is yet another new and very exciting motivation for studies of exotic nuclei.

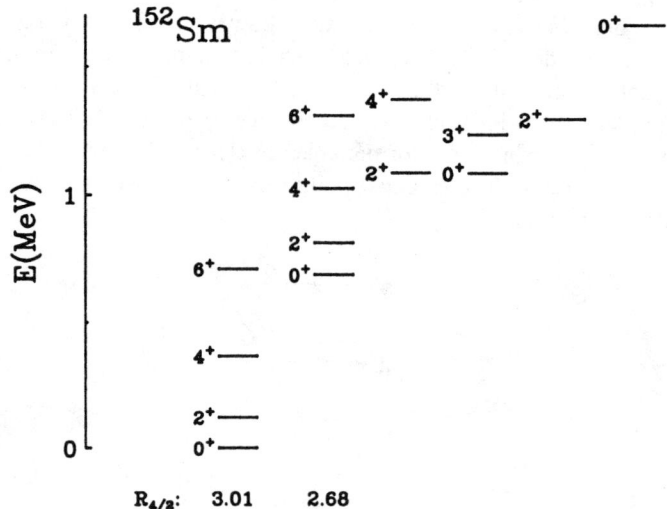

FIGURE 3. Experimental levels of ^{152}Sm. The $R_{4/2}$ ratios for the near-deformed and vibrational sequences are given at the bottom.

The interpretation above is suggestive but needs firm theoretical support. To this end we have calculated the properties of nuclei in both the IBA and GCM models. Similar conclusions emerge in both. We stress the IBA, where the analysis is complete, and will only illustrate the GCM results. We carried out extensive IBA calculations covering the entire parameter region of the symmetry triangle using the Hamiltonian

$$H = \varepsilon n_d + \kappa Q \cdot Q \qquad (2)$$

$$\text{with } Q = (s^\dagger d + d^\dagger s) + \chi (d^\dagger d)^{(2)} \qquad (3)$$

where $\kappa=0$ gives U(5) and $\varepsilon=0$, with $\chi = -7^{1/2}/2$ gives SU(3). The vibrator to rotor transition is obtained, therefore, by varying the ratio ε/κ.

For $\varepsilon/\kappa = 23.5$ and $\chi = -7^{1/2}/2$, the IBA calculations reproduce the ^{152}Sm data quite well, including energies, branching ratios and absolute B(E2) values, as shown in Fig. 4. Indeed, the model reproduces B(E2) values ranging over 3 orders of magnitude (144 Wu to 0.17 Wu) to within a factor of 2-3.

The most interesting aspect of the results is the minuscule range of IBA parameters for which the B(E2: $2^+_3 \rightarrow 0^+_2$) value is small and for which $R^\gamma_{og} \ll 1$. This is illustrated in Fig. 5 in terms of contour plots for this and several other observables as a function of ε/κ and χ. We see that $R^\gamma_{og} \ll 1$ only in a small range of both ε/κ and χ. It is significant that it is in *exactly* this same region of parameters that $R_{(4/2)1} > 3.0$, that $E(0^+_2) < E(2^+_\gamma)$, and that $R_{(4/2)2} < 3.0$ -- that is, that all the other key ranges of observables applicable to ^{152}Sm are also reproduced. This confluence is remarkable and clearly suggests that the physics embodied in the IBA calculations for this region of parameters reflects the actual situation in ^{152}Sm.

It is the sharp, 3 orders of magnitude or more, drop in the B(E2: $2^+_3 \rightarrow 0^+_2$) value, within a 2% range in ε/κ, that brings to mind the concept of phase transitions and the accompanying phase coexistence. The rapidity of this change in B(E2) value shows why the phenomenon observed in ^{152}Sm is so rare. As we have noted, adjacent nuclei (e.g., isotopes) differ by integer numbers of nucleons and hence change properties by discrete amounts. The probability that any given nucleus happens to land in the minimum would seem to be highly unlikely.

To pursue the coexistence interpretation further, we consider both the wave functions for the 0^+ states (and the 2^+ and 4^+ states built above them) and the potential energy surface corresponding to the IBA Hamiltonian (5). For our purposes it is useful to decompose the IBA wave functions into their n_d probabilities since n_d is a good quantum number for vibrational levels ($n_d = 0$, 1, 2 for the ground, one and 2-phonon levels) while the wave functions for deformed states are highly spread out in n_d.

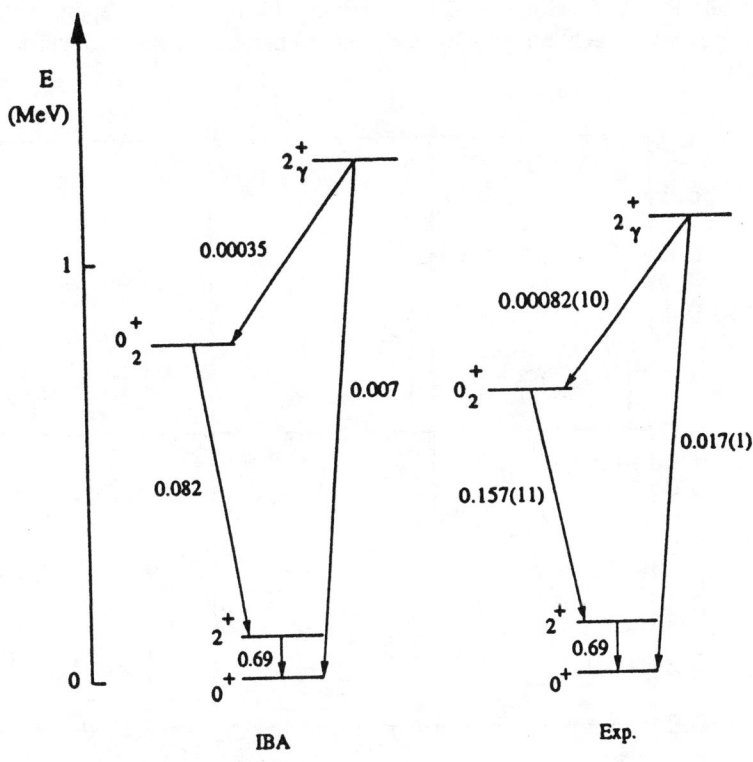

FIGURE 4. Comparison of the data for ^{152}Sm with IBA calculations. The B(E2) values are given in e^2b^2 and normalized for the $2^+_1 \rightarrow 0^+_1$ transition. The value 0.69 e^2b^2 corresponds to 144 Wu. The IBA parameters used were ε=470 keV, κ=-20 keV, and $\chi = -7^{1/2}/2$.

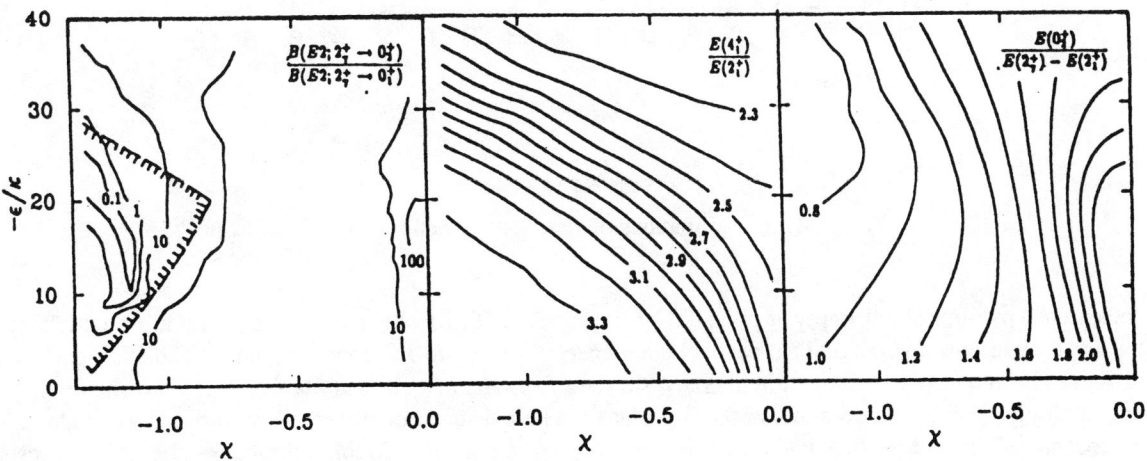

FIGURE 5. Contour plots of IBA calculations of three key observables against $-\varepsilon/\kappa$ and χ. In the left panel $R_{4/2} > 3.0$ and $E_{0+/2} < [E(2^+_\gamma) - E(2^+_1)]$ occur only to the left of the hatch marking.

The IBA wave functions are shown in Fig. 6 and immediately confirm the basic conclusions above (although, not surprisingly, they also show evidence for mixing of the phases). The 0^+_1 wave function is widely spread in n_d, as are the 2^+_1 and 4^+_1 levels, while the 0^+_2 level has a 60% probability for $n_d = 0$, as appropriate for the base level of a vibrator. The 2^+_2 level appears as largely a 1-phonon level and, in particular, the 2^+_3 level has a large amplitude for $n_d = 2$ (as does the 4^+_2 level).

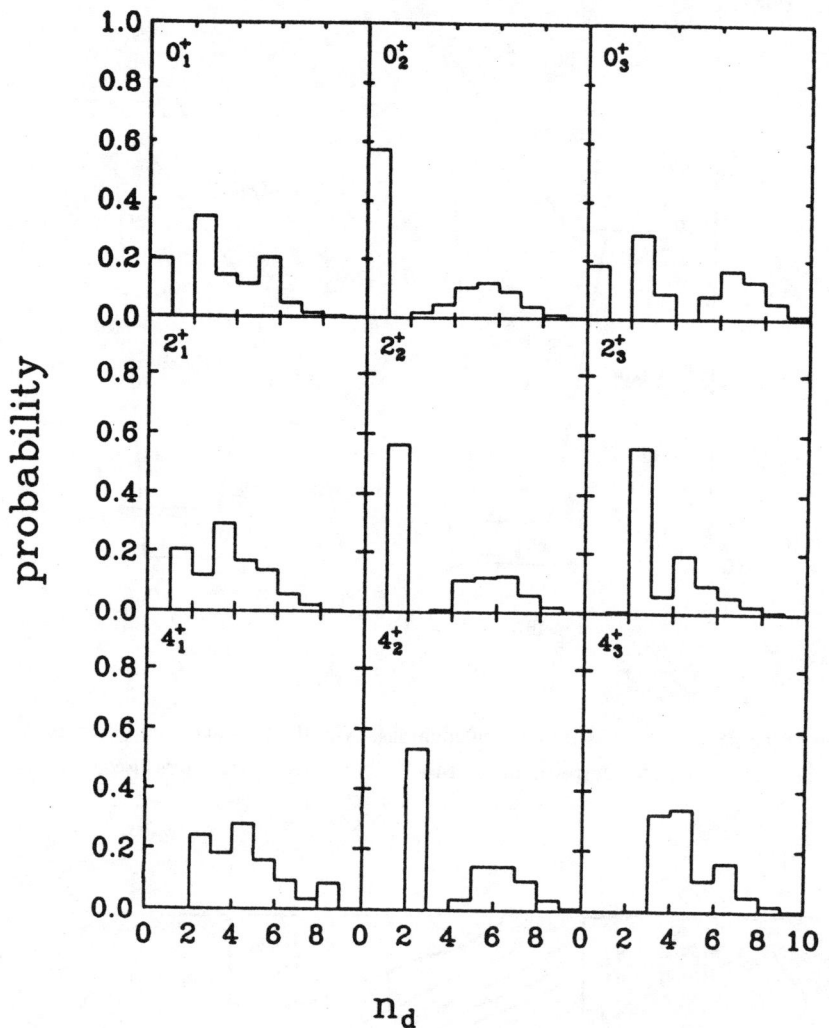

FIGURE 6. n_d probability distributions of IBA wave functions for key states in ^{152}Sm.

The coexistence picture should imply a potential energy surface (PES) either with two minima or with a structure where the wave functions of states of different excitation energy span quite different β values. The PES corresponding to any IBA Hamiltonian can be obtained in the intrinsic state formalism. The results for ^{152}Sm are shown in Fig. 7, both in a cut along $\gamma = 0$ and as contours in the β-γ plane. The existence of two shallow minima is evident. It has not been recognized heretofore that the IBA-1 was capable of producing double-minima in the PES, presumably because it does so only for such a minute, nearly singular, region of parameter space.

It is essential to recognize that this is true coexistence, produced within a single Hilbert space, and not the ad hoc variety encountered in the usual Duval-Barrett (particle-hole/intruder) formalism (6) that applies to other regions such as the Cd and Hg isotopes.

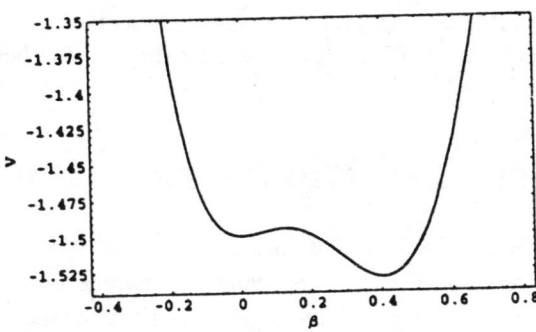

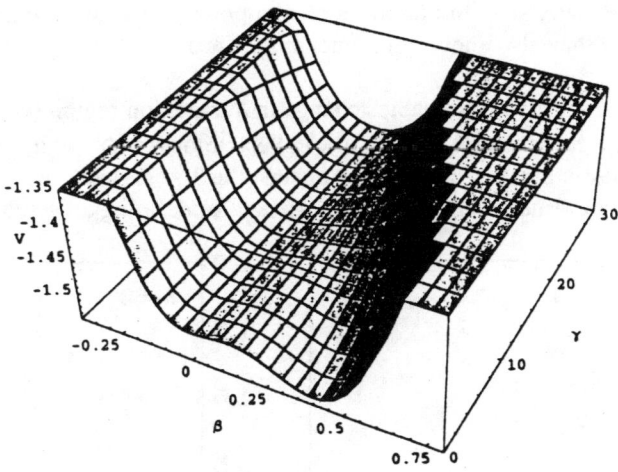

FIGURE 7. Potential energy surface for the IBA calculations in Fig. 4. Top: PES for $\gamma=0^0$. Bottom: PES as a function of β and γ.

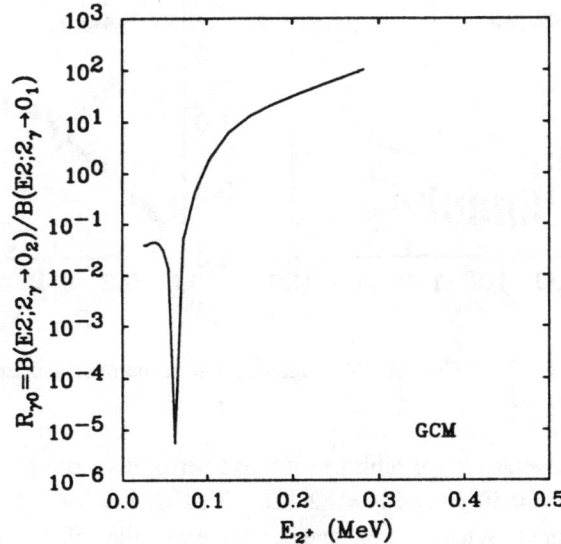

FIGURE 8. B(E2: $2^+_3 \rightarrow 0^+_1$)/B(E2: $2^+_3 \rightarrow 0^+_2$) in the GCM as a function of E(2^+_1).

The present interpretation does not seem to be particularly model dependent. In the GCM, there is also only a small trajectory of parameter values that gives $R^{\gamma}_{og} \ll 1$ and, for these same parameters, also reproduces the other properties of ^{152}Sm. This will be discussed elsewhere. Suffice it to show here (Fig. 8) the behavior of the $B(E2: 2^+_3 \rightarrow 0^+_2)$ value in the GCM.

PHASE TRANSITIONAL BEHAVIOR

Having identified phase *coexistence* in one nucleus, it is incumbent to see if we can understand, or at least, track, the phase *transition* itself across this region. Here we encounter the discrete nucleon number problem alluded to earlier. If we plot, for example, the energy of the 4^+_1 level against neutron number in Sm, as in Fig. 9, we see a gradual decrease as rotational structure sets in, but nothing striking appears that reflects a sharply changing observable. The problem is that the data points are separated, and can only legitimately be connected by straight line segments. We can now see the dichotomy between theory and experiment. Predictions of phenomenological models vary according to their parameter values, which are continuous, but real nuclei vary with discrete changes in nucleon number.

Plotting $E(4^+_1)$ for the entire major shell to obtain more data points does not help, as also shown in Fig. 9 (lower left). The data show trends but they also show large fluctuations and no obvious discontinuities. Similar results are obtained for other observables.

Since the relations *between* collective observables change in a transition region (e.g., $R_{4/2}$ goes from 2.0 to 3.33 for a vibrator to rotor region) it might be useful to plot one collective observable against another. We illustrate this in Fig. 9, for the same 4^+_1 energies in Sm, but plotted now against the 2^+_1 energies. Again, a discrete, downward bending set of points results, suggesting a structural evolution but useless for defining an actual phase transition.

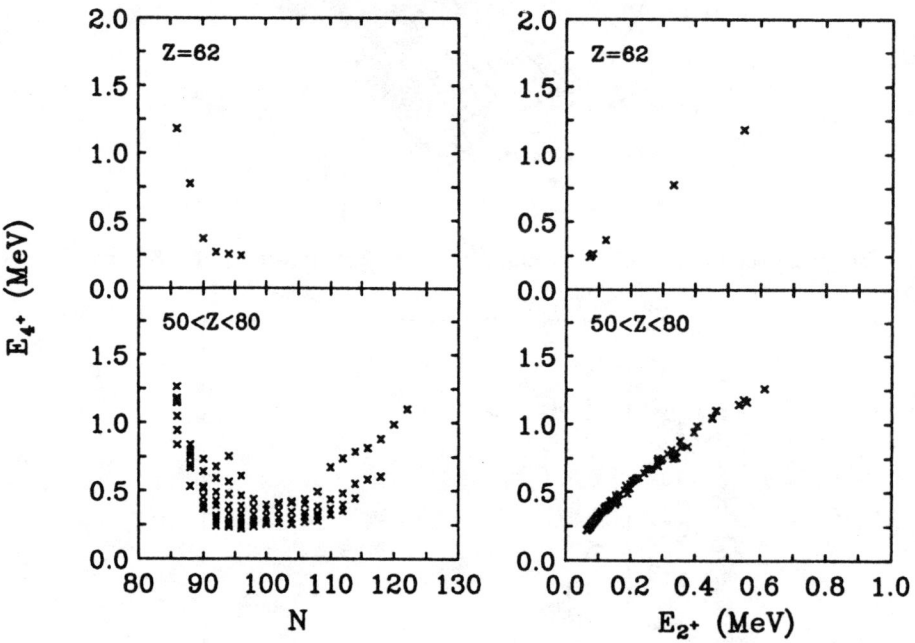

FIGURE 9. Plots of $E(4^+_1)$ against neutron number and $E(2^+_1)$.

However, a remarkable result emerges if we return to a broad perspective and plot $E(4^+_1)$ against $E(2^+_1)$ for *all the collective even-even nuclei* from $Z = 50$-80 and $N = 82$-126. This is done in Fig. 9 (lower right). Each elemental sequence individually looks the same. When plotted together, therefore, they all fall on the *same* trajectory but not at the same abscissa values. For further discussion and to show the universality of this plot, we show a similar correlation in Fig. 10, but now expanded to the entire region $Z = 38$-82.

These results are nothing short of remarkable and are not at all understood (although they are reproduced by both the IBA and GCM models). They are, however, very useful. Each element occupies a distinct set of points, shifted relative to each other: Because they all lie on the *same* trajectory there results a *filling-in* of the plot so that it becomes essentially *continuous* and therefore amenable to construction of a derivative. Moreover, the empirical trajectory observed is itself special -- most of the data lie along a straight line of slope 2.00 (an anharmonic vibrator) and then, at about $E(2^+_1) = 120$ keV, a sharp link occurs to a slope of 3.33 (rotor).

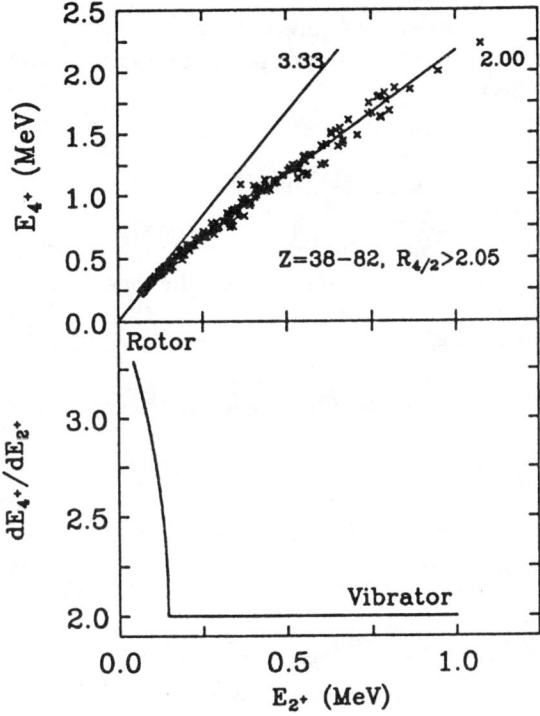

FIGURE 10. Top: $E(4^+_1)$ against $E(2^+_1)$ for the region Z=38-82. Bottom: The slope of $E(4^+_1)$ as a function of $E(2^+_1)$, plotted against $E(2^+_1)$, is obtained by differentiating an analytic function that fits the data in the upper panel. (See text).

Looked at in this way, it is the *slope* that becomes the critical "observable." But it is only "observable" because of the *compact* and *continuous* trajectory in Figs. 9 or 10. The slope is plotted in Fig. 10 (bottom). Its behavior is exactly that exhibited by critical phase transitions in condensed matter systems. The slope becomes the order parameter (technically, if S is the slope, the order parameter is S-2) and $E(2^+_1)$ becomes the control parameter with a critical value at about 120 keV. We note that some of these ideas have been expressed before (7), in the context of Fig. 10, but not the explicit use of the global systematics to circumvent the integer nucleon number problem by a construct that gives a continuous distribution.

We may try to understand this point as follows. Finite nuclear systems produce discrete spectra of bound levels. In order to search for and discuss phase transitional behavior one needs to be able to define a derivative of some relevant observable. This cannot normally be done for finite nuclei or (see Figs. 9) for ensembles of finite nuclei. However, by the approach of a) considering such ensembles *and*, b) plotting one collective observable against another (e.g., $E(4^+_1)$ against $E(2^+_1)$), we see that nuclear structure happens to evolve (for reasons we do not yet understand) in such a way that such correlation plots are both simple and follow compact, essentially continuous trajectories. This then allows us to *look for* phase transitional behavior (sharp kinks or discontinuous derivatives). It turns out that, in fact, we do observe such kinks. In this way, the 2^+_1 energy acts as a control parameter, the slope of $E(4^+_1)$ against $E(2^+_1)$ is essentially an order parameter and this process of correlating the data has circumvented the finite nucleon number problem.

To recapitulate, models are continuous because their parameters are. Now, one seems to be able to achieve a corresponding empirical continuity by exploiting the remarkable compactness shown in the evolution of structure over broad regions of the nuclear chart in Figs. 9 and 10. We therefore urge further study of the phase transitional behavior of nuclei and of the role of $E(2^+_1)$ as a control parameter.

We close by noting that it is now clear why ^{152}Sm acts as it does. Almost by chance, its 2^+_1 energy (122 keV) happens to occur nearly at the precise critical value denoting the phase transition. Other nuclei, shifted only slightly, "jump over" the phase transition point: they exhibit a structural change as a function of, say, neutron number but no individual nucleus happens to land at the critical point.

We are very grateful to our collaborators on much of this and related work, especially to P. von Brentano, M. Wilhelm, F. Iachello, Jing ye Zhang and M. Caprio. Work supported by the U.S. DOE under contract numbers DE-FG02-91ER40609 and DE-FG02-88ER40417.

REFERENCES

1. See for example Collins, M.F., *Magnetic Critical Scattering*, New York: Oxford University Press, 1989, pp. 3-15 and Landau, L.D. and Lifshitz, E.M., *Statistical Physics*, MA: Addison-Wesley Reading, 1958, pp. 430-445.

2. Casten, R.F., Wilhelm, M., Radermacher, E., Zamfir, N.V. and von Brentano, P., *Phys. Rev. C* **57,** R1553 (1998).

3. Iachello, F. and Arima, A., *The Interacting Boson Model*, Cambridge: Cambridge University Press, 1987.

4. Gneuss, G., Mosel, V. and Greiner, W., *Phys. Lett. B* **30**, 397 (1969).

5. Iachello, F., Zamfir, N.V., and Casten, R.F., *Phys. Rev. Lett.* **81**, 1191 (1998).

6. Duval, P.D. and Barrett, B.R., *Nucl. Phys. A* **376**, 213 (1982).

7. Casten, R.F., Zamfir, N.V. and Brenner, D.S., *Phys. Rev. Lett.* **71**, 227 (1993).

Low Energy Q-Phonon Excitations in Nuclei

P. von Brentano[1], N. Pietralla[1], C. Fransen[1], C. Friessner[1], A. Gade[1], A. Gelberg[1], R.-D. Herzberg[1,2], U. Kneissl[3], H. Meise[1], T. Otsuka[4], H.H. Pitz[3], V. Werner[1], I. Wiedenhöver[1,5]

[1] *Institut für Kernphysik, Universität zu Köln, 50937 Köln, Germany*
[2] *Oliver Lodge Laboratory, University of Liverpool, Liverpool L69 7ZE, UK*
[3] *Institut für Strahlenphysik, Universität Stuttgart, 70569 Stuttgart, Germany*
[4] *Department of Physics, University of Tokyo, Hongo, Bunkyo-ku, Tokyo 113-0033, Japan*
[5] *Argonne National Laboratory, Argonne, Illinois 60439*

Abstract. One-, two-, and multi-Q-phonon excitations in heavy nuclei are discussed. We describe the Q-phonon scheme for low-lying, isoscalar, positive parity states in γ-soft nuclei and compare the predictions of the Q-phonon scheme to new data on the nucleus ^{132}Ce. We report on the experimental proof for the quadrupole-octupole coupled two-phonon nature of the lowest-lying 1^- state in the semi-magic $N = 82$ nuclei ^{142}Nd and ^{144}Sm. Finally, low-lying proton-neutron asymmetric mixed-symmetry states are discussed in terms of the Q-phonon scheme. We report on recent lifetime measurements of the mixed-symmetry one-Q-phonon excitation, the 2^+_{ms} state, in the nuclei ^{126,128}Xe, ^{136}Ba, and ^{144}Nd.

I INTRODUCTION

Phonons are quantized collective excitation modes of many-body systems [1,2]. Simultaneous excitation of several phonons leads to multi-phonon states. In a pure bosonic picture the properties of multi-phonon states can be easily understood from the properties of the constituting phonons. This makes the phonon picture an attractive tool to understand the collective excitation spectrum of quantized many-body systems. The phonon concept has been successfully used in nuclear physics [3–5]. However, pure bosonic multi-phonon states are rare in nuclei. Due to anharmonicities and residual interactions the pure harmonic phonon picture is often strongly disturbed and the multi-phonon nature of collective excitations is difficult to identify experimentally. For comparison to experiment it is often more convenient to replace the bosonic formulation of phonons and to consider instead a new scheme – the Q-phonon scheme. The Q-phonon scheme was recently suggested by the Tokyo group [6] and was developed in the frame of the Tokyo–Köln collaboration [7–15]. In this scheme the collective excitations are discussed in terms of transition operators acting on the true ground state. The transition operators describe the experimentally observable transitions and, thus, they have no boson character, i.e., they do not fulfill the boson commutation relations [6].

The basic low-lying collective excitation modes in nuclei are electric quadrupole $E2$ and the electric octupole $E3$ excitations. We emphasize that we focus on low-lying bound states and not on unbound giant resonances. The electric quadrupole transition operator Q (and the electric octupole transition operator O) can be used to obtain the wave functions of low-lying collective states relative to the ground state. Here

$$|2^+\rangle = Q^{\mathrm{val}}|0^+_1\rangle \tag{1}$$

$$|3^-\rangle = O^{\mathrm{val}}|0^+_1\rangle \,, \tag{2}$$

where the electric multipole transition operators are restricted to the valence space in order to generate the low-lying, bound modes. In the following we will use the Interacting Boson Model as an effective model of the valence space and we will drop the superscripts on the transition operators (e.g., $Q^{\mathrm{val}} = Q^{\mathrm{IBM}} = Q$). This scheme is called the Q-phonon scheme. The Q-phonon scheme can be used in many nuclear structure models. It was introduced by the Tokyo group [6,7] in the Interacting Boson Model (IBM) in order to give an intuitive explanation for the τ-selection rules [16,17] and for the decay of low-lying states in nuclei close to

CP481, *Nuclear Structure 98*, edited by C. Baktash

the O(6) dynamical symmetry limit. In these nuclei the low-lying positive parity states are multiple Q-phonon excitations [8,9]. Later it was found that many excited nuclear states predicted by other collective models [10–12] and the low-lying excitation spectrum predicted by the IBM outside the O(6) dynamical symmetry limit [13–15,18] are similar to or can at least be well approximated by one or two Q-phonon configurations.

The low-lying isoscalar positive parity states are generated by one kind of Q-phonon, namely the isoscalar electric quadrupole operator in the valence shell Q. Two-Q-phonon excitations built up by different types of constituent phonons are expected if different collective excitation modes are involved in the formation of collective levels. The most prominent example of such two-phonon states formed by different types of phonons is the $(2^+ \otimes 3^-)$ coupling of the isoscalar quadrupole phonon Q and the isoscalar octupole phonon O in vibrational nuclei [19]. This coupling yields a quintuplet of two-phonon states with spin and parity quantum numbers $J^\pi = 1^- \ldots 5^-$. Experimental evidence for quadrupole-octupole coupled two-phonon states was reported by different groups, see e.g. Refs. [20–29]. Other examples of two-phonon states formed by different types of Q-phonons are the isoscalar-isovector mixed-symmetry states [30–33], where different proton-neutron degrees of freedom are involved. The best-known example of such type is the fragmented $J^\pi = 1^+$ scissors mode [34,35] discovered in the rare earth nuclei in Darmstadt [36].

In the next section we give the Q-phonon expressions [7] for the low-lying positive parity isoscalar excitations in γ-soft nuclei. We will present recent data for the nucleus ^{132}Ce from β-decay and for the nucleus ^{124}Xe from an $(^3$He,$2n)$ reaction. We compare the data to the simple estimates of the Q-phonon scheme and to detailed IBM-1 calculations. In the third section we discuss quadrupole-octupole coupled two-phonon $J^\pi = 1^-$ states in $N = 82$ isotones, which were discussed previously by Metzger [20,21]. We give a survey about the systematics of this two-phonon state and show the results of the observations of the defining one-phonon annihilation transitions to the remaining one-phonon states. The fourth section deals with the Q-phonon structure [33] of mixed-symmetry states in the IBM-2. The basic mixed-symmetry Q-phonon excitation is the 2^+_{ms} state. In nonrotational nuclei evidences for the 2^+_{ms} state were found by several groups [37–48]. In cases where lifetimes are unknown, the mixed-symmetry interpretation was questioned by other groups [49]. We discuss observations of the 2^+_{ms} state in nuclei of the $A = 130$ mass region, for which we measured the lifetimes using (α, n) and (γ, γ') reactions [47,48].

II Q-PHONON SCHEME FOR NUCLEI WITH $A = 130$

The low-lying positive parity excitation spectrum of nuclei in the mass region around $A = 130$ are well described in the frameworks of the Rot-Vib-Model by Faessler [50,51] and by the sd-IBM-1 [17]. Excitation

TABLE 1. Multi Q-phonon structure of the low-lying O(6) $[\sigma = N]$ states of the Interacting Boson Model (IBM–1). The left column gives the quantum numbers for spin and parity, d boson seniority τ, and the auxiliary O(6) quantum number ν_δ. The low-lying states are expressed by a single Q-phonon configuration, where the number of Q-phonons equals the value of τ. The square brackets denote angular momentum coupling. $Q = d^\dagger s + s^\dagger \tilde{d} = T(E2)/e$.

$\lvert I^\pi, \tau, \nu_\Delta \rangle$	$=$	$\lvert 0^+_1, 0, 0 \rangle$	$\propto$	$\lvert 0 \rangle$	
$\lvert I^\pi, \tau, \nu_\Delta \rangle$	$=$	$\lvert 2^+_1, 1, 0 \rangle$	$\propto$	$Q\lvert 0 \rangle$	
$\lvert I^\pi, \tau, \nu_\Delta \rangle$	$=$	$\lvert 4^+_1, 2, 0 \rangle$	$\propto$	$[QQ]^{(4)}\lvert 0 \rangle$	$\propto$ $[Q\lvert 2^+_1 \rangle]^{(4)}$
$\lvert I^\pi, \tau, \nu_\Delta \rangle$	$=$	$\lvert 6^+_1, 3, 0 \rangle$	$\propto$	$[QQQ]^{(6)}\lvert 0 \rangle$	$\propto$ $[Q\lvert 4^+_1 \rangle]^{(6)}$
$\lvert I^\pi, \tau, \nu_\Delta \rangle$	$=$	$\lvert 2^+_2, 2, 0 \rangle$	$\propto$	$[QQ]^{(2)}\lvert 0 \rangle$	$\propto$ $[Q\lvert 2^+_1 \rangle]^{(2)}$
$\lvert I^\pi, \tau, \nu_\Delta \rangle$	$=$	$\lvert 4^+_2, 3, 0 \rangle$	$\propto$	$[QQQ]^{(4)}\lvert 0 \rangle$	$\propto$ $[Q\lvert 2^+_2 \rangle]^{(4)}$
$\lvert I^\pi, \tau, \nu_\Delta \rangle$	$=$	$\lvert 3^+_1, 3, 0 \rangle$	$\propto$	$[QQQ]^{(3)}\lvert 0 \rangle$	$\propto$ $[Q\lvert 2^+_2 \rangle]^{(3)}$
$\lvert I^\pi, \tau, \nu_\Delta \rangle$	$=$	$\lvert 5^+_1, 4, 0 \rangle$	$\propto$	$[QQQQ]^{(5)}\lvert 0 \rangle$	$\propto$ $[Q\lvert 3^+_1 \rangle]^{(5)}$
$\lvert I^\pi, \tau, \nu_\Delta \rangle$	$=$	$\lvert 0^+_2, 3, 1 \rangle$	$\propto$	$[QQQ]^{(0)}\lvert 0 \rangle$	$\propto$ $[Q\lvert 2^+_2 \rangle]^{(0)}$
$\lvert I^\pi, \tau, \nu_\Delta \rangle$	$=$	$\lvert 2^+_3, 4, 1 \rangle$	$\propto$	$\alpha([QQ]^{(2)}[QQ]^{(2)})^{(2)}\lvert 0 \rangle + \beta[QQ]^{(2)}\lvert 0 \rangle$	
$\lvert I^\pi, \tau, \nu_\Delta \rangle$	$=$	$\lvert 4^+_3, 4, 0 \rangle$	$\propto$	$\alpha'([QQ]^{(2)}[QQ]^{(2)})^{(4)}\lvert 0 \rangle + \beta'[QQ]^{(4)}\lvert 0 \rangle$	

energies and electromagnetic decay branching ratios are close [52] to the predictions derived analytically for the O(6) dynamical symmetry. The relative wave functions in the O(6) limit can be easily pinned down in the Q-phonon scheme. Table 1 shows the Q-phonon configurations of the lowest-lying states with O(6) seniority quantum number $\sigma = N$, where N is the number of bosons, i.e., pairs of valence particles (or holes). $E2$ transitions are expected to be the dominant decay modes of the low-lying states. Furthermore, $E2$ transitions should be enhanced between states with Q-phonon configurations which differ by one unit. This fact corresponds to the τ–selection rule in the O(6) dynamical symmetry limit of the IBM [16,17].

In order to test the validity of the Q-phonon scheme it is necessary (i) to identify the low-lying excitations and to assign to them spin and parity quantum numbers, (ii) to verify that $E2$ transitions are the dominant decay modes and (iii) to measure the γ-decay branching ratios among these states. This aim can be reached in $\gamma\gamma$-coincidence spectroscopy experiments using "complete reactions", such as (α,n), $(\alpha,2n)$, $(^3He,2n)$ etc, or an extremely clean reaction, such as the β-decay.

For the study of the excitation spectrum of the nucleus ^{132}Ce we produced radioactive ^{132}Pr nuclei in the reaction ^{117}Sn$(^{19}$F,4n$)^{132}$Pr at a beam energy of $E(^{19}F) = 75$ MeV in the focus of the Cologne cube $\gamma\gamma$-coincidence spectrometer at the Cologne TANDEM accelerator facility. The beam current was pulsed: After 1 s irradiation time the beam was switched off and $\gamma\gamma$-coincidences of the γ transitions following the β-decay of the produced nuclides were taken off-beam for 1 s in each cycle.

Figure 1 shows a part of the total $\gamma\gamma$-coincidences (upper part) and a γ-spectrum observed in coincidence with the $2_1^+ \rightarrow 0_1^+$ transition in ^{132}Ce (lower part). Due to the off-beam data acquisition the γ spectra are extremely clean with low background. Even weak γ-lines can be observed. From the $\gamma\gamma$-coincidence data obtained with two different angular combinations, we could construct unambigously the low-lying level scheme, and the multipolarities of many γ-transitions could be determined (see Table 2).

Moreover, we investigated the low-lying excitation spectrum in the noble gas nucleus ^{124}Xe by means of the ^{123}Te$(^3$He,2n$)^{124}$Xe reaction using beam energies of 16 MeV $< E(^3$He$) <$ 24 MeV. We observed many γ-lines and $\gamma\gamma$-coincidences from decays of low-lying states. We determined decay intensity ratios and many $E2/M1$

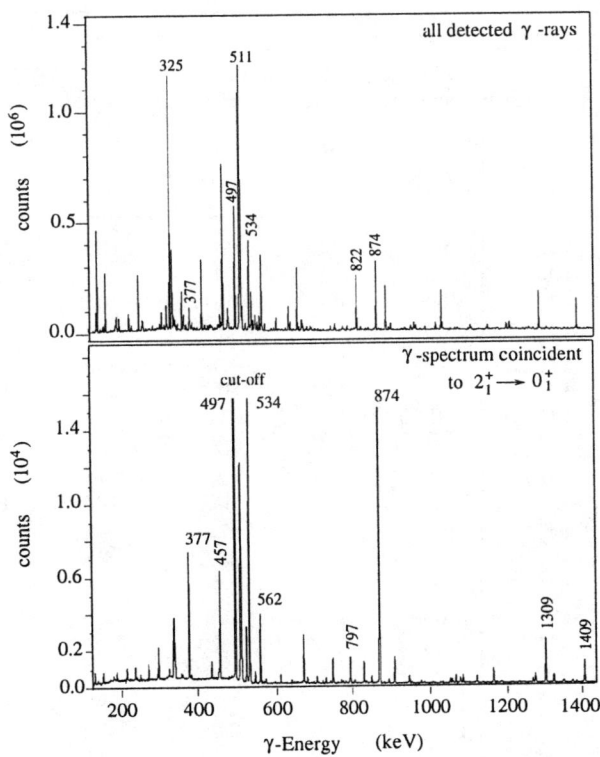

FIGURE 1. Measured γ-spectrum of ^{132}Ce following the β-decay of ^{132}Pr. Shown are the total $\gamma\gamma$-projections (upper part) and a γ-spectrum gated with the $2_1^+ \rightarrow 0_1^+$ transition in ^{132}Ce (lower part). As the γ-spectra are taken off-beam between the beam pulses, they are extremely clean with very low background.

mixing ratios. Some reduced intensity ratios are displayed in Table 2.

The level scheme of ^{124}Xe is compared to an sd-IBM-1 calculation in Figure 2. For the calculation we used the Hamiltonian

$$H = \epsilon n_d - \kappa Q^\chi \cdot Q^\chi + \lambda L \cdot L - \beta n_d^2, \qquad (3)$$

which contains a d boson energy term, a quadrupole-quadrupole interaction, a rotational energy term and a τ-compression term [53]. n_d is the d boson number operator and $Q^\chi = s^\dagger \tilde{d} + d^\dagger s + \chi[d^\dagger \tilde{d}]^{(2)}$ is the quadrupole operator with the structural parameter χ. L is the boson angular momentum operator. The term proportional to n_d^2 is used for τ-compression [53] and improves the moment of inertia. The five-parameter fit describes the level scheme very well.

In the consistent Q formalism (CQF) [54] the structure of the $E2$ transition operator is fixed to be the same as the quadrupole operator in the Hamiltonian, $T(E2) = e_B Q^\chi$, where e_B is the effective boson quadrupole charge. $B(E2)$ ratios are independent of the effective charge. Table 2 shows the reduced intensity ratios I_γ/E_γ^5, which were observed in ^{132}Ce and ^{124}Xe in comparison to the sd-IBM-1 predictions obtained in fits using the Hamiltonian from Eq. (3). In many cases the multipolarity of the γ transition could be determined. In particular the low-lying states decay predominantly by $E2$ transitions. This justifies the comparison to the IBM-1, which cannot describe $M1$ transitions. In case of pure $E2$ character the reduced intensity ratios I_γ/E_γ^5

TABLE 2. Experimental reduced intensity (I_γ/E_γ^5) ratios in ^{132}Ce [55] and in ^{124}Xe compared to IBM-1 calculations. For some transitions $E2/M1$ mixing ratios could be measured. The fraction of the γ intensity, which has $E2$ character, is given in the columns labeled $I_\gamma(E2)/I_\gamma$. If the M1–admixtures are negligible, the numbers given in the columns labeled "Expt" equal the B(E2) ratios. The errors of the reduced intensity ratios are 20% unless particular values are given. IBM-1 fits with τ-compression are included [see Eq. (3)]. The $E2$ transition operator is restricted according to the consistent Q formalism [54].

B(E2) ratios
Comparison: Experiment – IBM 1

$Q_i \to$	Q_f	$I_i \to I_f$	^{132}Ce $\frac{I_\gamma(E2)}{I_\gamma}$	Expt	IBM-1	^{124}Xe IBM-1	Expt	$\frac{I_\gamma(E2)}{I_\gamma}$
QQ→	Q	$2_2^+ \to 2_1^+$	0.99	100	100	100	100	0.98
→	1	$\to 0_1^+$	1.00	6.1	6.1	2.4	2.4	1.00
QQQ→	QQ	$3_1^+ \to 2_2^+$	1.00	100	100	100	100	1.00
→	QQ	$\to 4_1^+$	1.00	29.1(7)	30.7	32	32	0.04/0.94
→	Q	$\to 2_1^+$	0.96	4.0	5.3	2.9	3.0	0.92
QQQ→	QQ	$4_2^+ \to 2_2^+$		100	100	100	100	1.00
→	QQ	$\to 4_1^+$	> 0.30	59	65.5	69	58	0.84
→	Q	$\to 2_1^+$		0.42	0.21	0.5	0.1	1.00
→ QQQ		$\to 3_1^+$			29	19		
QQQQ→	QQQ	$5_1^+ \to 3_1^+$			100	100	100	1.00
→ QQQ		$\to 4_2^+$			47.0	46	98	0.96
→ QQQ		$\to 6_1^+$			33.9	37	71	
→	QQ	$\to 4_1^+$			2.1	1.6	2.6	(0.74)
QQQ→	QQ	$0_2^+ \to 2_2^+$	1.00	100	100	100	100	1.00
→	Q	$\to 2_1^+$	1.00	0.56	2.73	21	21	1.00
QQQQ→	QQQ	$2_3^+ \to 0_2^+$	1.00	100	100	100	100	1.00
→ QQQ		$\to 3_1^+$	[0.1, 0.9]	≤45	109.9	68		
→	QQ	$\to 2_2^+$	0.14	1.4(4)	0.28	4.2	2.7	0.0004
→	QQ	$\to 4_1^+$			0.6	11	4.9(29)	
→	Q	$\to 2_1^+$	0.66	0.10	0.005	0.001	0.4	
→	Q	$\to 0_1^+$	1.00	0.13	0.04	0.007	0.3	1.00

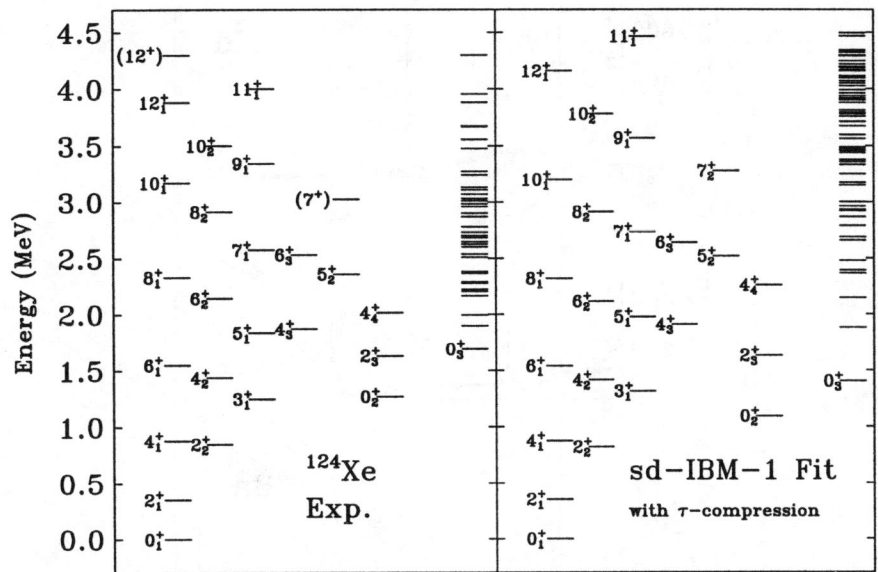

FIGURE 2. Comparison of the level scheme of ^{124}Xe with an sd-IBM-1 calculation using the Hamiltonian from Eq. (3). Additional levels are shown on the right.

from Table 2 coincide with the values of the $B(E2)$ ratios. The agreement with the IBM-1 calculation for the $B(E2)$ ratios, optimized for ^{132}Ce and ^{124}Xe respectively, is remarkable.

The quantitative agreement between the calculation and the experiment is very good not only for the lowest-lying levels. The $B(E2)$ ratios in ^{132}Ce are rather similar to the $B(E2)$ ratios in ^{124}Xe. They obey the selection rules of the Q-phonon scheme. "Q-phonon-allowed" $E2$ transitions, where the number of Q-phonons changes by one unit, are enhanced by an order of magnitude or more. The $E2$ decays of low-lying isoscalar positive parity states in the nuclei around $A = 130$ are, thus, qualitatively understood in the simple Q-phonon scheme.

III QUADRUPOLE-OCTUPOLE TWO-PHONON STATES IN $N = 82$ ISOTONES

In this section we discuss two-phonon states which are created by the coupling of two different types of phonons: The isoscalar electric quadrupole phonon Q and the isoscalar electric octupole phonon O. In particular we discuss the quadrupole-octupole coupled two phonon $J^\pi = 1^-$ dipole excitations, which were extensively investigated in the past in photon scattering experiments [20–23,26–29,56,57] and by other experimental approaches [24,25,58,59]. We will report on the experimental proof [58,59] for the quadrupole-octupole coupled two-phonon nature of the 1^- states below 4 MeV for $N = 82$ isotones.

These states were suggested earlier [21–23,27] to have two-phonon nature due to the observation of strong $E1$ excitations in photon scattering experiments. Figure 3 shows examples of the measured photon scattering spectra. The γ spectra are very clean. Above the nonresonant background one observes – besides the photon flux calibration lines and some decay lines from $J^\pi = 2^+$ excitations – one strong γ line, which dominates the (γ,γ')-spectrum below 4.1 MeV. This strong γ line stems from the decay of the lowest, photo-excited $J^\pi = 1^-$ state back to the ground state. Figure 4 shows the $E1$ strength and the excitation energy of this 1^- state in the $N = 82$ isotones. Both the excitation energy and the $E1$ excitation strength vary smoothly with the nuclear mass number A. This fact hints at the collective character of this 1^- state. Moreover, its excitation energy closely correlates to the sum energy of the 2_1^+ state and the 3_1^- state, the basic quadrupole and octupole one-phonon states. From this correlation it was believed that the 1^- state has a quadrupole-octupole coupled two-phonon structure.

Let us first pin down the phonon configurations of the relevant negative parity states

$$|3_1^-\rangle \propto O|0_1^+\rangle \qquad (4)$$

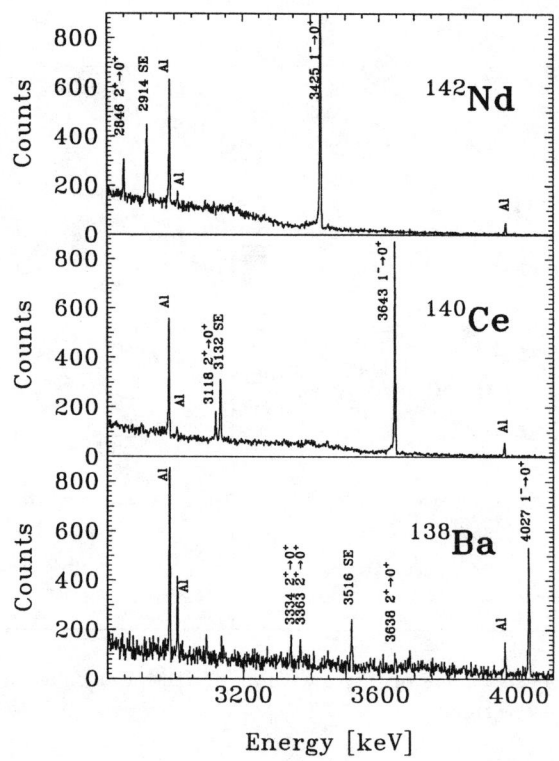

FIGURE 3. γ–spectra from photon scattering experiments on N=82 isotones performed at the bremsstrahlung facility of the Stuttgart Dynamitron accelerator (taken from Ref. [27]).

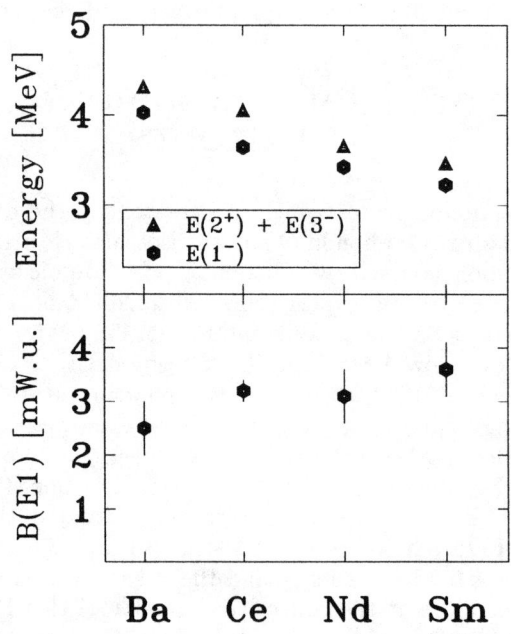

FIGURE 4. Properties of the lowest 1^- state in $N = 82$ isotones [27]. The upper part shows the sum energy of the quadrupole and the octupole phonons (triangles) and the excitation energy of the 1^- state (hexagons). Its $E1$ excitation strength is displayed in milli Weisskopf units in the lower part.

$$|J^-_{2^+\otimes 3^-}\rangle \propto [QO + OQ]^{(J)}|0^+_1\rangle. \tag{5}$$

In this general, symmetrized phonon formulation the ground state may contain correlations and the quadrupole and octupole operators may not be bosons and may not commute, and the phonon coupling does not have to be harmonic. But as the 3^-_1 state and the two-phonon $(2^+ \times 3^-)$ states differ by one quadrupole phonon, they are connected by strong $E2$ transitions. If one assumes a purely bosonic picture, where the transition operators are one-boson operators and where the ground state is the boson vacuum, one finds

$$B(E2; J^-_{2^+\otimes 3^-} \to 3^-_1) = B(E2; 2^+_1 \to 0^+_1) . \tag{6}$$

The collective one-phonon annihilation transition to the remaining one-phonon states is the defining signature of the quadrupole-octupole coupled two-phonon states.

In order to verify the two-phonon nature of the $J^\pi = 1^-$ two-phonon candidates we must measure its $E2$ transition strength to the 3^-_1 state. This collective $E2$ transition - the decay to the 3^-_1 state - is a very weak decay channel of the 1^- state. It must compete with the strong, higher energetic $E1$ transition to the ground state, which can carry more than 99% of the total decay intensity of the 1^- state. Since the total level width of the 1^- state in semi-magic nuclei is known from photon scattering experiments, it is sufficient to measure the decay intensity ratio $I_\gamma(1^- \to 3^-_1)/I_\gamma(1^- \to 0^+_1)$ in order to determine the absolute $B(E2; 1^- \to 3^-_1)$ value. This was done in $(p,p'\gamma)$-particle-γ-coincidence experiments [58,59] at the Cologne cube spectrometer.

The $(p,p'\gamma)$ experiments were performed for the nuclei ^{142}Nd and ^{144}Sm, for which the $J^\pi = 1^-$ two-phonon candidates lie at 3.425 MeV and 3.226 MeV, respectively. Their lifetimes are known from earlier photon scattering experiments [20,22,60]. Low-spin negative parity states were dominantly excited by choosing proton beam energies of $E_p = 10.25$ MeV for ^{142}Nd and $E_p = 10.20$ MeV for ^{144}Sm. These energies match the energies of the isobaric analog resonances. The energies of the scattered protons were measured with Si-detectors. Energy spectra of γ-rays were taken in coincidence to the scattered protons. Scattered protons with an energy loss, which matches the excitation energy of the 1^- state, can have directly excited these 1^- states. The coincident γ-spectrum is dominated by the $1^- \to 0^+_1$ $E1$ γ-transition. At γ-energies $E_\gamma = E(1^-) - E(3^-_1)$ we observe [58,59] the γ-transition searched for: the $1^- \to 3^-_1$ decay. From the branching ratios $I_\gamma(1^- \to 3^-_1)/I_\gamma(1^- \to 0^+_1)$ and the effective elastic photon scattering widths Γ^2_0/Γ [20,22,60] we determine the $E2$ transition strengths $B(E2; 1^- \to 3^-_1)$. According to Eq. (6) we compare these $B(E2)$ values to the

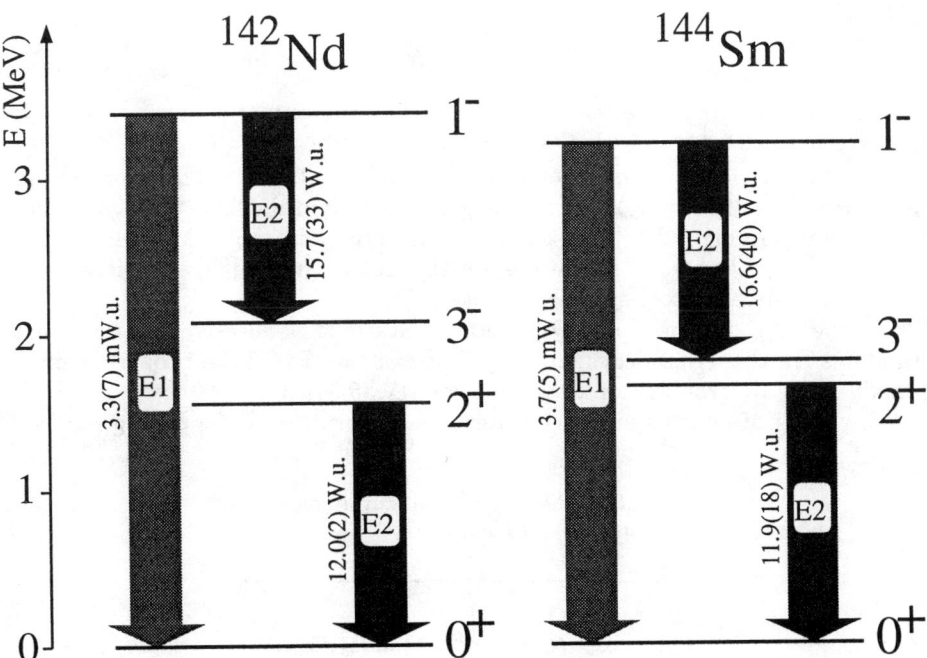

FIGURE 5. Two-phonon character of the strongly photo-excited 1^- states in $N = 82$ isotones. The $1^- \to 3^-$ $E2$ transition is as collective as the $2^+ \to 0^+$ transition (taken from Refs. [58,59]).

$B(E2; 2_1^+ \rightarrow 0_1^+)$ values in Fig. 5. The $1^- \rightarrow 3_1^-$ $E2$ transition is as collective as the $E2$ decay of the 2_1^+ to the ground state. Within the experimental errors the $B(E2; 1^- \rightarrow 3_1^-)$ values agree with the $B(E2)$ values of the corresponding $2_1^+ \rightarrow 0_1^+$ transition. We find Eq. (6) to be fulfilled. This proves the quadrupole-octupole coupled two-phonon nature of the 1^- states at 3.425 MeV in ^{142}Nd and at 3.226 MeV ^{144}Sm. Their excitation energies lie close to the sum energy of the corresponding one-phonon states $E(2_1^+) + E(3_1^-) = 3.659$ MeV for ^{142}Nd and $E(2_1^+) + E(3_1^-) = 3.470$ MeV for ^{144}Sm. From this we conclude a very harmonic coupling of the octupole and the quadrupole phonon.

Similar 1^- states with corresponding properties have been observed recently in other semi-magic and nearly semi-magic nuclei at the shell closures $N = 82, Z = 50, N = 50, N = 28$ [28,29,56,57,61,62].

IV   ISOVECTOR Q-PHONON EXCITATIONS

Above we have discussed isoscalar Q-phonon excitations. In the following we discuss isovector Q-phonon excitations [33]. In order to consider isospin degrees of freedom we must distinguish proton operators and neutron operators. The isoscalar (s) quadrupole operator used before has the form $Q = Q_s^{\mathrm{val}} = Q_\pi^{\mathrm{val}} + Q_\nu^{\mathrm{val}}$. As there are two types of quadrupole operators in the valence space, the proton quadrupole operator Q_π^{val} and the neutron quadrupole operator Q_ν^{val}, there exists – besides the isoscalar sum – a second linear combination $Q_{\mathrm{ms}}^{\mathrm{val}}$, which generates a one-$Q$-phonon state orthogonal to the 2_1^+ state. This linear combination is $Q_{\mathrm{ms}} \propto Q_\pi^{\mathrm{val}} - \alpha Q_\nu^{\mathrm{val}}$, where the coefficient α is obtained by the orthogonality condition $\langle 0_1^+ | Q_s^{\mathrm{val}} Q_{\mathrm{ms}}^{\mathrm{val}} | 0_1^+ \rangle = 0$. This coefficient α can be calculated in a nuclear structure model, which treats the valence space of a heavy nucleus and which distinguishes protons and neutrons. Such a model is, for instance, the proton-neutron version of the Interacting Boson Model, the IBM-2. We note, that the formalism of F-spin in the IBM-2 does not exactly coincide with the isospin formalism. However, Q-phonon boson operators can be defined in an analogous way and we use the isospin language in analogy. In the F-spin limit of the IBM-2 one finds $\alpha = N_\pi / N_\nu$ where N_π and N_ν are the numbers of proton bosons and neutron bosons. This was previously shown by the Tokyo group [33]. In the IBM-2 one obtains the basic F-scalar and mixed-symmetric Q-phonon operators [33]

$$Q_s = Q_\pi + Q_\nu \tag{7}$$

and

$$Q_{ms} = \frac{Q_\pi}{N_\pi} - \frac{Q_\nu}{N_\nu} . \tag{8}$$

In the dynamical symmetry limits of the IBM-2, the operators Q_s and Q_{ms} generate the symmetric and the mixed-symmetric one-phonon 2^+ states with F-spin quantum numbers $F = F_{\mathrm{max}} = \frac{N_\pi + N_\nu}{2}$ and $F = F_{\mathrm{max}} - 1$. The 2_1^+ state and the 2_{ms}^+ state are a natural doublet, which is split due to different proton-neutron symmetry. Other mixed-symmetry states of $F = F_{\mathrm{max}} - 1$ can be generated from the 2_{ms}^+ state by application of Q_s along with angular momentum coupling [33,48]. Mixed-symmetry states in the IBM-2 have been firstly discussed in terms of mixed-symmetry Q-phonon excitations by the Tokyo group [33]. Table 3 shows the Q-phonon configurations of the lowest-lying mixed-symmetry states.

Like the 2_1^+ state the 2_{ms}^+ state is a one-Q-phonon excitation. Collective $E2$ transitions must thus be expected for both states. In the dynamical symmetry limits of the IBM-2 the properties of the 2_{ms}^+ state have been calculated analytically by group-theoretical methods [17,35,31]. Outside the dynamical symmetry limits selection rules for the decay of mixed symmetry states can be formulated depending on the IBM-2 Hamilton

TABLE 3. Q-phonon configurations of the lowest-lying mixed-symmetry states.

$\lvert 2_{\mathrm{ms}}^+ \rangle$	$\propto$	$Q_{\mathrm{ms}} \lvert 0_1^+ \rangle$
$\lvert 1_{\mathrm{sc}}^+ \rangle$	$\propto$	$(Q_s Q_{\mathrm{ms}})^{(1)} \lvert 0_1^+ \rangle = (Q_s \lvert 2_{\mathrm{ms}}^+ \rangle)^{(1)}$
$\lvert 3_{\mathrm{ms}}^+ \rangle$	$\propto$	$(Q_s Q_{\mathrm{ms}})^{(3)} \lvert 0_1^+ \rangle = (Q_s \lvert 2_{\mathrm{ms}}^+ \rangle)^{(3)}$

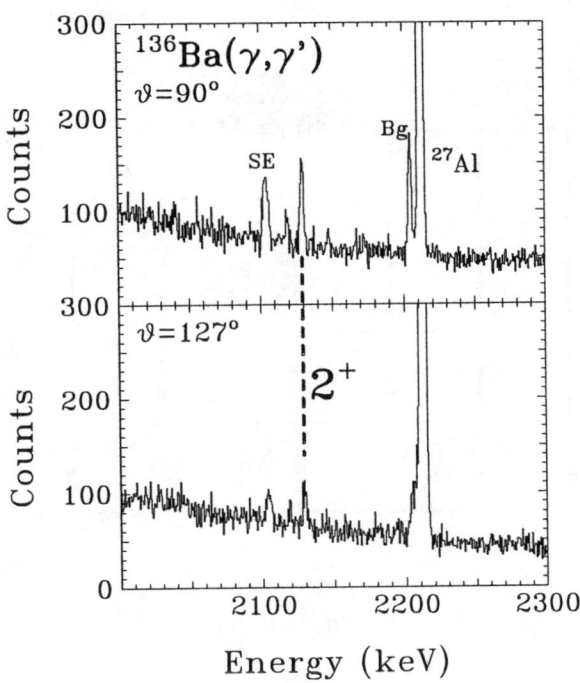

FIGURE 6. Photon scattering spectrum of ^{136}Ba measured at the photon scattering facility of the Stuttgart Dynamitron accelerator. Scattering angles of $\theta = 90°$ and $\theta = 127°$ were used. At 2129 keV a weakly collective, short-lived electric quadrupole excitation is observed [48].

operator [63,64]. Beside this weakly collective $E2$ transition to the ground state, the mixed-symmetry one-phonon excitation, i.e. the 2^+_{ms} state, decays by a strong $M1$ transition to the 2^+_1 state. Due to these strong decay modes the 2^+_{ms} state decays rapidly and is, thus, short-lived. In summary, the experimental signatures for the 2^+_{ms} are: the short lifetime τ, the large $B(M1; 2^+_{ms} \to 2^+_1)$ value, which leads to the small $E2/M1$ mixing

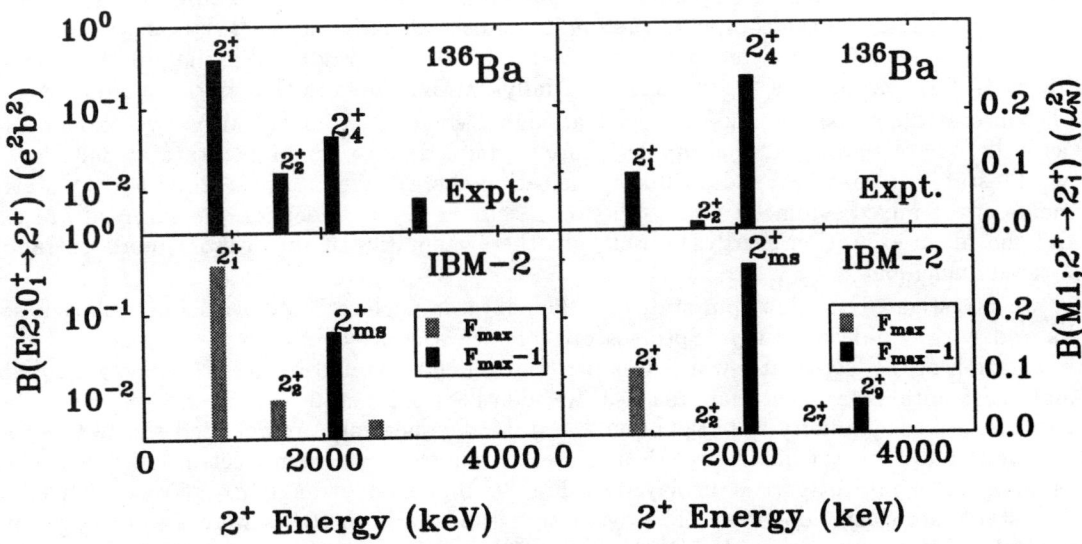

FIGURE 7. $E2$ and $M1$ transition strength distributions for 2^+ states in ^{136}Ba compared to an IBM-2 calculation using an F-scalar Hamiltonian [48]. The 2^+_4 state is the strongest $E2$ excitation above the 2^+_1 state. The 2^+_4 state is the only 2^+ state observed, which decays by a strong $M1$ transition to the 2^+_1 state.

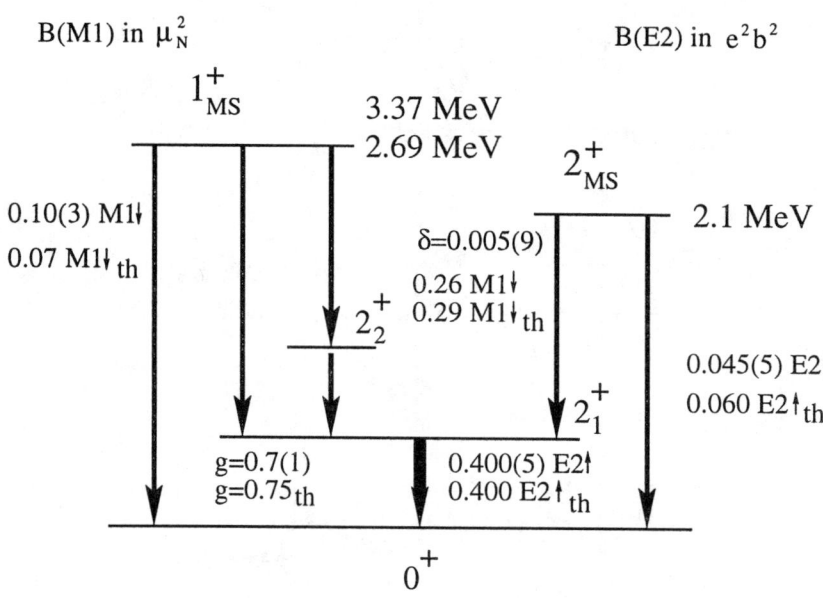

$B(M1)$ in μ_N^2 $\qquad\qquad\qquad\qquad\qquad$ $B(E2)$ in e^2b^2

FIGURE 8. Properties of the mixed-symmetry states in ^{136}Ba in comparison to the IBM-2 calculation from Ref. [48].

ratio δ for the transition to the 2_1^+ state, and the weakly collective (≈ 1 W.u.) $E2$ transition to the ground state.

Experiments on the 2_{ms}^+ state have been performed previously by several groups [37–48]. An efficient method for measuring the lifetimes of short-lived, low-multipolarity ground state excitations in stable nuclei is the scattering of real photons in a (γ, γ') reaction. We have performed a photon scattering experiment on the vibrational nucleus ^{136}Ba. Photon scattering spectra are shown in Fig. 6. At an excitation energy of 2129 keV we observed [48] a short-lived 2^+ state, which possesses all properties expected for a 2_{ms}^+ state: it has a short lifetime $\tau = 67(7)$ fs and a vanishing $E2/M1$ mixing ratio $\delta = +0.005(9)$ for the transition to the 2_1^+ state. This transition has a large $M1$ strength $B(M1; 2^+ \to 2_1^+) = 0.26(3)\mu_N^2$. Furthermore, the $E2$ ground state excitation is weakly collective with an $E2$ strength $B(E2; 0_1^+ \to 2^+) = 0.045(5)e^2b^2 = 2.1$ W.u. For this 2^+ state at 2129 keV we have gathered the complete set of signatures for the 2_{ms}^+ state. Moreover, from Fig. 7 one sees that the 2_4^+ state is the only one among the observed 2^+ states, which exhibits the properties of the 2_{ms}^+ state. The (γ, γ') reaction is a complete reaction if one uses a continuous energy bremsstrahlung beam in the entrance channel. Therefore, we can conclude, that the 2^+ state at 2129 keV is the only large fragment of the 2_{ms}^+ state in ^{136}Ba, i.e., that the 2_{ms}^+ state is essentially unfragmented in this vibrational nucleus.

In the photon scattering spectrum we observed at higher energies around 3 MeV two-phonon dipole excitations, as well. We observed the quadrupole-octupole coupled two-phonon 1^- state at 3436 keV and we detected two 1^+ states at 2694 keV and 3370 keV excitation energy. We interpret these two 1^+ states as the main fragments of the mixed-symmetry two-Q-phonon 1^+ state. Their properties, the summed $M1$ excitation strength and the mean excitation energy, fit well into the systematics of the scissors mode extrapolated to weakly deformed nuclei [65].

Recently, we investigated the low-spin states in the vibrator nucleus ^{94}Mo with both photon scattering experiments and with a β-decay study. Spectroscopy of γ-rays emitted after β decay can be a useful tool for the investigation of the 2_{ms}^+ state, which was previously demonstrated by the Florence group [45,66]. In the combination of both experimental approaches, we identified a 2^+ state, which shows a decay behaviour that points to the mixed-symmetry one-Q-phonon 2_{ms}^+ state. Furthermore, we detected the two-Q-phonon 1_{sc}^+ state and we measured the short lifetimes of these states. A photon scattering spectrum of ^{94}Mo taken at the Dynamitron accelerator in Stuttgart is displayed on Fig. 9. Signals of strongly excited, short-lived $J^\pi = 1^+$ and $J^\pi = 2^+$ states are clearly observed. The decay transitions of both states were also observed in our $\gamma\gamma$ coincidence study of the γ rays emitted after the ^{94}Tc$\to^{94}$Mo β-decay performed at the Cologne coincidence cube spectrometer. A γ-spectrum is shown in Fig. 10. In the very clean spectra decay transitions of proposed mixed-symmetry states are observed with high statistics. The data analysis is currently going on.

In radioactive, unstable nuclei it is more difficult to identify mixed-symmetry states. Due to the lack of a

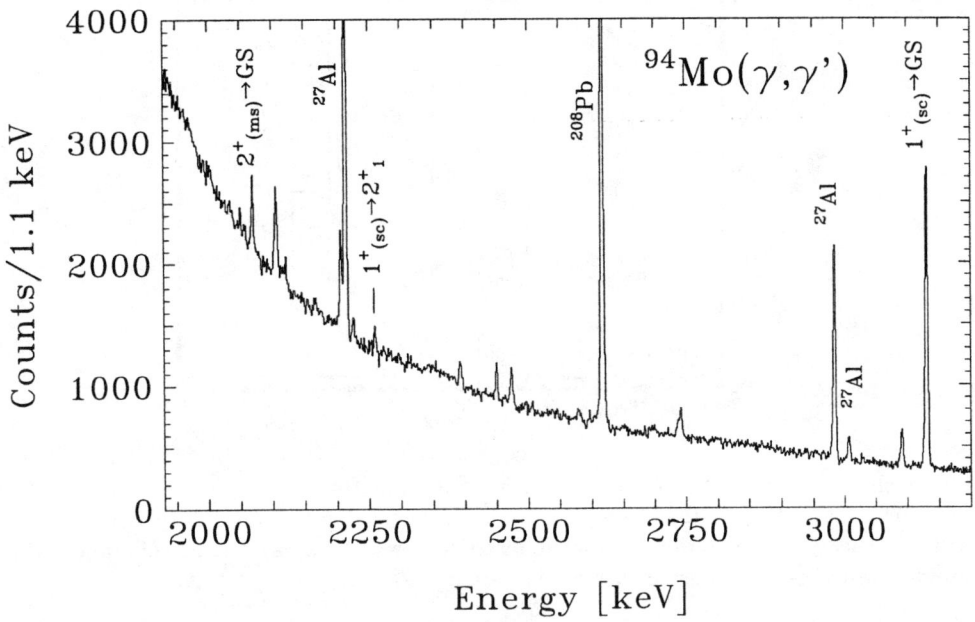

FIGURE 9. Photon scattering spectrum of ^{94}Mo in the energy range of the lowest lying mixed-symmetry states. Clear signals of dipole and quadrupole excitations are observed. The spectrum was taken at the bremsstrahlung facility of the Stuttgart Dynamitron accelerator.

stable target the typical reactions for lifetime measurements of mixed-symmetry states, namely (γ, γ'), (e, e'), Coulomb excitation and $(n, n'\gamma)$, are not practicable. Nevertheless, we observed the 2^+_{ms} state in the radioactive nucleus ^{128}Xe [47] at the coincidence cube spectrometer in Cologne. We measured the short lifetime of the 2^+_{ms} state by observing the angle-dependent Doppler shift of the $2^+_{ms} \rightarrow 2^+_1$ transition emitted during the stopping process in a backing material after the population in an (α,n)-reaction. The Doppler shift is shown in Fig. 11

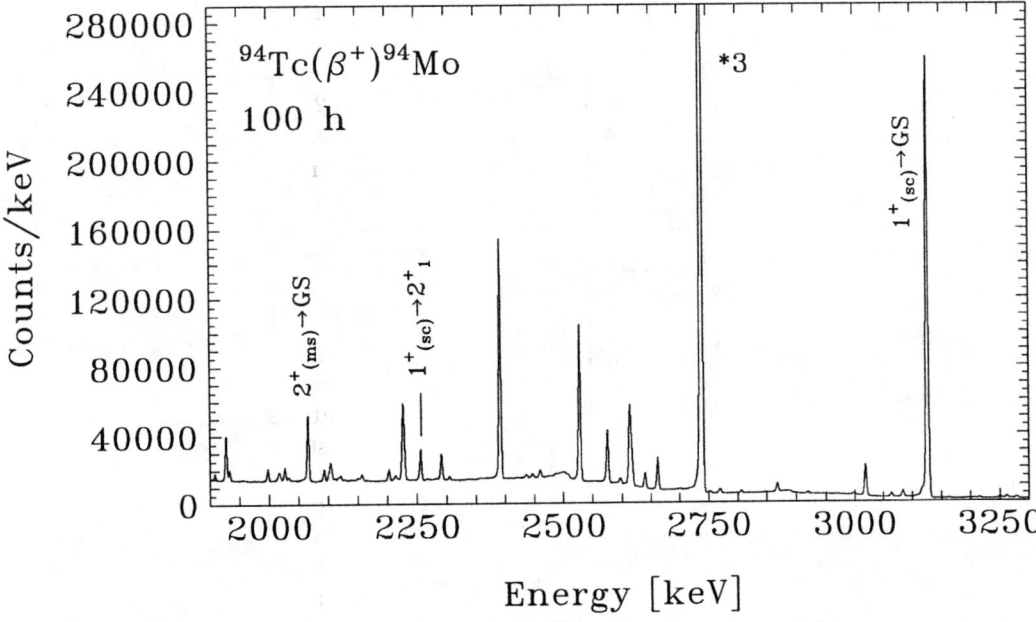

FIGURE 10. Spectrum of ^{94}Mo populated in the β–decay of ^{94}Tc. Low-spin states are populated with high statistics. The spectrum was taken at the coincidence-cube spectrometer at the Cologne Tandem accelerator.

459

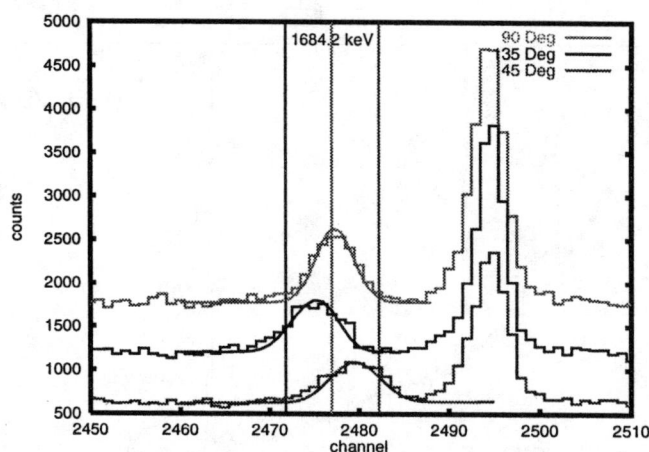

FIGURE 11. Line shape of the 1684.2 keV transition in ^{128}Xe gated with 442.8 keV $M1$ $2_1^+ \to 0_1^+$ transition. The spectra observed at 90°, 135°, 45° (top to bottom) are displaced vertically. The angle-dependent Doppler shift is clearly visible: the detected γ energy increases from backward angles to forward angles. The neighboring unshifted γ line from a long lived state demonstrates the accurate energy calibration.

From the known time scale of the stopping process and from the observed Doppler shift we measured [47] a lifetime of $\tau = 0.17(7)$ ps for the 2_{ms}^+ state in ^{128}Xe.

Recently, we performed the analog experiment on the neighboring nucleus ^{126}Xe [67]. Measured γ spectra are displayed in Fig. 12. From the observed angle-dependent Doppler shift of the $2_4^+ \to 2_1^+$ $M1$ transition in coincidence with the $2_1^+ \to 0_1^+$ ground state transition we determined the lifetime of the 2_4^+ state in ^{126}Xe. The 2_4^+ state in ^{126}Xe has a lifetime of $\tau = 0.35(7)$ ps. It decays to the 2_1^+ state by a pure $M1$ transition. The $M1$ strength amounts to $B(M1; 2_4^+ \to 2_1^+) = 0.025(6)\mu_N^2$.

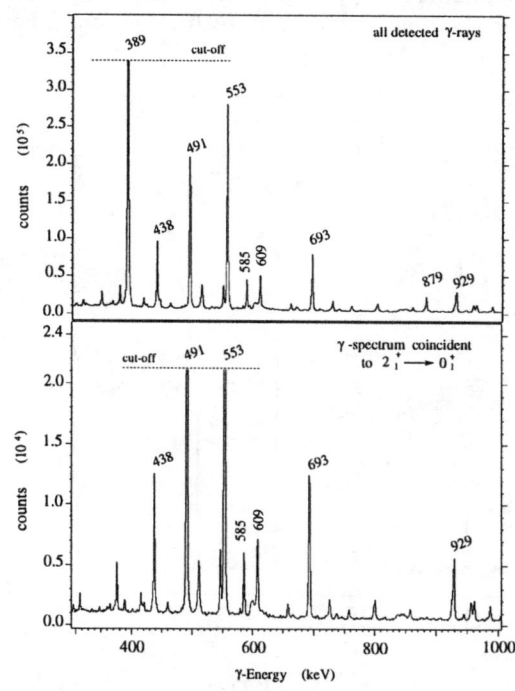

FIGURE 12. γ-spectrum of ^{126}Xe populated in the ^{123}Te(α, n) reaction. The upper part shows a part of the total γ rays observed. The lower part the γ spectrum observed in coincidence with the $2_1^+ \to 0_1^+$ transition.

460

TABLE 4. Summary of the relevant observables for nuclei in the $A = 130$ mass region, where 2^+_{ms} states or its fragments have been identified from measured absolute transition strengths: We give excitation energies, lifetimes, and $E2/M1$ mixing ratios δ^2 of 2^+_{ms} states, and different transition strengths. For more than one fragment the reduced transition strengths have been summed up. $M1$ transition strengths of the scissors mode are from the Stuttgart-Cologne photon scattering collaboration.

Nucleus	^{112}Cd	^{126}Xe	^{128}Xe	^{134}Ba	^{136}Ba	^{142}Ce	^{144}Nd
Reaction	$(n,n'\gamma)$	(α,n)	(α,n)	$(n,n'\gamma)$	(γ,γ')	Coulex $(n,n'\gamma)$	(γ,γ')
Ref.	[46]	[67]	[47]	[43]	[48]	[40,44]	[37,62]
E [keV]	2156 2231	2064 (2359)	2127	2029 2088	2129	2004	2072
τ [fs]	310(35) 220(20)	351(74) 74(30)	170(70)	230(23) 85(7)	67(7)	65(7)	66(20)
$\delta^2(2^+_{ms} \to 2^+_1)$	0.085(24) -0.02(3)	0.00(5) 0.8(8)	0.05(5)	-0.31(5) 0.02(5)	0.005(9)	-0.26(14)	0.08
$B(E2; 2^+_{ms} \to 0^+_1)$ [W.u.]	0.14(2)			2.0(2)	2.1(2)	2.5(2)	2.2(8)
$B(E2; 2^+_1 \to 0^+_1)$ [W.u.]	32	41	39	33	19	20	25
$B(M1; 2^+_{ms} \to 2^+_1)[\mu^2_N]$	0.10(1)	0.06(3)	0.07(2)	0.22(4)	0.26(3)	0.23(2)	0.23(9)
$B(M1; 1^+_{sc} \to 0^+_1)[\mu^2_N]$				0.19(3)	0.10(3)		0.17(1)

Table 4 summarizes the observables, which are relevant to the identification of mixed-symmetry 2^+_{ms} states as mentioned above. For nuclei in the $A = 130$ mass region we only consider 2^+ states for which absolute transition strengths were measured. The properties of this 2^+ state vary smoothly. These data begin to allow for a systematic analysis of the proton-neutron mixed-symmetric quadrupole excitation in the valence space of heavy nuclei.

V SUMMARY

We have discussed one-, two- and multi-Q-phonon excitations in heavy nuclei. The low-lying, isoscalar, positive parity excitation spectrum of even-even nuclei in the $A = 130$ mass region can be well understood in the Interacting Boson Model in terms of multiple excitations of the isoscalar quadrupole phonon from the ground state. Decay transitions between these states have predominant $E2$ character. The strongest $E2$ transitions connect states, which differ by one Q-phonon. Other Q-phonon forbidden $E2$ transitions are hindered by one to three orders of magnitude. This corresponds to the τ-selection rule of the IBM. Branching ratios and excitation energies can be well reproduced by few-parameter sd-IBM-1 fits.

A good example for a multi-phonon state, which is generated by different types of Q-phonons, is the $J^\pi = 1^-$ member of the quadrupole-octupole coupled $(2^+ \otimes 3^-)$ two-phonon quintuplet. From the experimental systematics it was known that the excitation strength and the excitation energy of the 1^- state vary smoothly as a function of the mass number A. This fact - together with the short lifetime - was the previously used argument for its collective two-phonon nature. We proved this quadrupole-octupole two-phonon character by measuring the collectivity of the $1^- \to 3^-_1$ $E2$ transition. It has about the same strength as the $2^+_1 \to 0^+_1$ transition, showing that the $1^- \to 3^-_1$ transition is a one-quadrupole-phonon annihilation.

Recently the Q-phonon scheme for mixed-symmetry states has been formulated by the Tokyo group. The fundamental mixed-symmetry one-phonon excitation is the mixed-symmetry 2^+_{ms} state, which can be characterized as the isovector quadrupole excitation in the valence shell. Safe identifications of mixed-symmetry states can be done by lifetime measurements. We measured lifetimes of 2^+_{ms} states in stable nuclei by (γ,γ') photon scattering experiments and in radioactive Xe nuclei by Doppler shift measurements after (α,n) reactions. Multi-Q-phonon mixed-symmetry states are currently under investigation.

461

ACKNOWLEDGMENTS

We thank Professor A. Zilges, Dr. K.-H. Kim, Dr. M. Wilhelm, Mrs. D. Belic, Mr. T. Diefenbach, Mr. A. Fitzler, Mr. A. Nord and Mr. I. Schneider for their help with the experiments and for discussions. We acknowledge discussions with Professors R.F. Casten, I. Hamamoto, F. Iachello, A. Richter, S.W. Yates and Dr. P. von Neumann-Cosel. This work is supported by the *Deutsche Forschungsgemeinschaft* under contracts Br 799/8-2/9-1 and Kn 154/30, and by the *Japan Society for the Promotion of Sciences* under Grant-in-Aid No. 0604 4249 and the DFG-JSPS cooperation.

REFERENCES

1. S. Flügge, Ann. Phys. **39**, 373 (1941).
2. L.D. Landau, J. Phys. (UdSSR) **5**, 71 (1941).
3. A. Bohr and B. Mottelson, *Nuclear Structure* I, (Benjamin, Reading, 1969).
4. A. Bohr and B. Mottelson, *Nuclear Structure* II, (Benjamin, Reading, 1978).
5. P. Ring and P.Schuck, *The Nuclear Many-Body Problem*, (Springer-Verlag, New York, 1980).
6. T. Otsuka and K.H. Kim, Phys. Rev. C **50**, R1768 (1994).
7. G. Siems, U. Neuneyer, I. Wiedenhöver, S. Albers, M. Eschenauer, R. Wirowski, A. Gelberg, P. von Brentano, T. Otsuka, Phys. Lett. **320 B**, 1 (1994).
8. P. von Brentano, A. Dewald, A. Gelberg, N. Pietralla, O. Vogel, I. Wiedenhöver, in *International Symposium on the Frontiers of Nuclear Structure Physics*, Tokyo, Japan, 1994, edt. by M. Ishihara *et al.* (World Scientific, Singapore, 1994).
9. P. von Brentano, O. Vogel, N. Pietralla, A. Gelberg, I. Wiedenhöver, in *International Conferenceon the Perspectives of the Interacting Boson Model*, Padua, Italy, 1994, edt. by R.F. Casten *et al.* (World Scientific, Singapore, 1994).
10. K.-H. Kim and T. Otsuka, Phys. Rev. C **52**, 2792 (1995).
11. R.V. Jolos, A. Gelberg, and P. von Brentano, Phys. Rev. C **53**, 168 (1996).
12. K.-H. Kim, T. Otsuka, A. Gelberg, P. von Brentano, and P. van Isacker, Phys. Rev. Lett. **76**, 3514 (1996).
13. N. Pietralla, P. von Brentano, R. F. Casten, T. Otsuka, N. V. Zamfir, Phys. Rev. Lett. **73**, 2962 (1994).
14. N. Pietralla, P. von Brentano, T. Otsuka, and R. F. Casten, Phys. Lett. **B 349**, 1 (1995).
15. N. Pietralla, T. Mizusaki, P. von Brentano, R.V. Jolos, T. Otsuka, and V. Werner, Phys. Rev. C **57**, 150 (1998).
16. A. Arima and F. Iachello, Ann. Phys. **123**, 201 (1979).
17. F. Iachello, A. Arima, *The Interacting Boson Model*, (Cambridge 1987).
18. Yu. V. Palchikov, P. von Brentano, R.V. Jolos, Phys. Rev. C **57**, 3026 (1998).
19. P.O. Lipas, Nucl. Phys. **82**, 91 (1966).
20. F.R. Metzger, Phys. Rev. C **17**, 939 (1978).
21. F.R. Metzger, Phys. Rev. C **18**, 1603 (1978).
22. H.H. Pitz, R.D. Heil, U. Kneissl, S. Lindenstruth, U. Seemann, R. Stock, C. Wesselborg, A. Zilges, P. von Brentano, S.D. Hoblit, and A.M. Nathan, Nucl. Phys. **A509**, 587 (1990).
23. P. von Brentano, A. Zilges, R.D. Heil, R.-D. Herzberg, U. Kneissl, H.H. Pitz, C. Wesselborg, Nucl. Phys. **A 557**, 593c (1993).
24. R.A. Gatenby, E.L. Johnson, E.M. Baum, S.W. Yates, D. Wang, J.R. Vanhoy, M.T. McEllistream, T. Belgya, B. Fazekas, G. Molnár, Nucl. Phys. **A560**, 633 (1993).
25. S.J. Robinson, J. Jolie, H.G. Börner, P. Schillebeeckx, S. Ulbig, Phys. Rev. Lett. **73**, 412 (1994).
26. K. Govaert, L. Govor, E. Jacobs, D. De Frenne, W. Mondelaers, K. Persyn, M.-L. Yoneama, U. Kneissl, J. Margraf, H.H. Pitz, K. Huber, S. Lindenstruth, R. Stock, K. Heyde, R. Vdovin, V. Yu., Ponomarev, Phys. Lett. **B335**, 113 (1994).
27. R.-D. Herzberg, I. Bauske, P. von Brentano, Th. Eckert, R. Fischer, W. Geiger, U. Kneissl, J. Margraf, H. Maser, N. Pietralla, H.H. Pitz, and A. Zilges, Nucl. Phys. **A592** 211 (1995).
28. R. Schwengner, G. Winter, W. Schauer, M. Grinberg, F. Becker, P. von Brentano, J. Eberth, J. Enders, T. von Egidy, R.-D. Herzberg, N. Huxel, L. Käubler, P. von Neumann-Cosel, N. Nicolay, J. Ott, N. Pietralla, H. Prade, S. Raman, J. Reif, A. Richter, C. Schlegel, H. Schnare, T. Servene, S. Skoda, T. Steinhardt, C. Stoyanov, H.G. Thomas, I. Wiedenhöver, A. Zilges, Nucl. Phys. **A620**, 277 (1997); Nucl. Phys. **A624**, 776 (1997).
29. J. Enders, P. von Brentano, J. Eberth, R.-D. Herzberg, N. Huxel, H. Lenske, P. von Neumann-Cosel, N. Nicolay, N. Pietralla, H. Prade, J. Reif, A. Richter, C. Schlegel, R. Schwengner, S. Skoda, H.G. Thomas, I. Wiedenhöver, G. Winter, A. Zilges, Nucl. Phys. **A636**, 139 (1998).
30. T. Otsuka, A. Arima, F. Iachello, Nucl. Phys. **A309**, 1 (1978).
31. P. Van Isacker, K. Heyde, J. Jolie, A. Sevrin, Ann. Phys. (NY) **171**, 253 (1986).

32. I. Talmi, Phys. Lett. **B 405**, 1 (1997).

33. K.-H. Kim, T. Otsuka, P. von Brentano, A. Gelberg, P. Van Isacker, R.F. Casten, in the proceedings of the 9th Int. Symp. on *Capture Gamma Ray Spectroscopy and Related Topics*, Budapest 1996, eds. G.L Molnár, T.Belgya, Zs.Réva, (Springer, Budapest, 1997), 195.

34. N. Lo Iudice, F. Palumbo, Phys. Rev. Lett. **41**, 1532 (1978).

35. F. Iachello, Phys. Rev. Lett. **53**, 1427 (1984).

36. D. Bohle, A. Richter, W. Steffen, A. E. L. Dieperink, N. Lo Iudice, F. Palumbo und O. Scholten, Phys. Lett. **B137**, 27 (1984).

37. W.D. Hamilton, A. Irbäck, J.P. Elliott, Phys. Rev. Lett. **53**, 2469 (1984).

38. P. Park, A.R.H. Subber, W.D. Hamilton, J.P. Elliott, and K. Kumar, J. Phys. G: Nucl. Phys. **11**, L251 (1985).

39. G. Molnár, R.A. Gatenby, S.W. Yates, Phys. Rev. C **37**, 898 (1988).

40. W.J. Vermeer, C.S. Lim, R.H. Spear, Phys. Rev. C **38**, 2982 (1988).

41. R. De Leo, L. Lagamba, N. Blasi, S. Micheletti, M. Pignanelli, M. Fujiwara, K. Hosono, I. Katayama, N. Matsuoka, S. Morinobu, T. Noro, S. Matsuki, H. Okamura, J.M. Schippers, S.Y. Van der Werf, M.N. Harakeh, Phys. Lett. **B226**, 5 (1989).

42. J. Takamatsu, T. Nakagawa, A. Terakawa, A. Narita, T. Tohei, M. Fujiwara, S. Morinobu, I. Katayama, H. Ikegami, K. Katori, S.I. Hayakawa, Y. Fujita, M. Tosaki, S. Hatori, Phys. (Paris) Colloque **51**, C6-423 (1990).

43. B. Fazekas, T. Belgya, G. Molnár, A. Veres, R.A. Gatenby, S.W. Yates, T. Otsuka, Nucl. Phys. **A 548**, 249 (1992).

44. J.R. Vanhoy, J.M Anthony, B.M. Haas, B.H. Benedict, B.T. Meehan, S.F. Hicks, C.M. Davoren, C.L. Lundstedt, Phys. Rev. C **52**, 2387 (1995).

45. A. Giannatiempo, A. Nannini, A. Perego, P. Sona, D. Cutoiu, Phys. Rev. C **53**, 2770 (1996).

46. P.E. Garrett, H. Lehmann, C.A. McGrath, Minfang Yeh, and S.W. Yates, Phys. Rev. C **54**, 2259 (1996).

47. I. Wiedenhöver, A. Gelberg, T. Otsuka, N. Pietralla, J. Gableske, A. Dewald and P. von Brentano, Phys. Rev. C **56**, R2354 (1997).

48. N. Pietralla, D. Belic, P. von Brentano, C. Fransen, R.-D. Herzberg, U. Kneissl, H. Maser, P. Matschinsky, A. Nord, T. Otsuka, H.H. Pitz, V. Werner, and I. Wiedenhöver, Phys. Rev. C **58**, 796 (1998).

49. J. Rikowska, N.J. Stone, P.M. Walker, W.B. Walters, Nucl. Phys. **A505**, 145 (1989).

50. A. Faessler, W. Greiner, Z. Phys. **168**, 425 (1962).

51. U. Meyer, A. Faessler, S.B. Khadkikar, Nucl. Phys. **A624**, 391 (1997).

52. R.F. Casten, P. von Brentano, Phys. Lett. **B 152**, 22 (1985).

53. Xing-Wang Pan, T. Otsuka, Jin-Quan Chen, and A. Arima, Phys. Lett. **B 287**, 1 (1992).

54. R.F. Casten, D.D. Warner, Rev. Mod. Phys. **60**, 389 (1988).

55. A. Gade, I. Wiedenhöver, M. Diefenbach, A. Gelberg, M. Luig, H. Meise, N. Pietralla, M. Wilhelm, T. Otsuka, and P. von Brentano, submitted.

56. K. Govaert, F. Bauwens, J. Bryssinck, D. De Frenne, E. Jacobs, W. Mondelaers, L. Govor, V. Yu. Ponomarev, Phys. Rev. C **57**, 2229 (1998).

57. C. Fransen, O. Beck, P. von Brentano, T. Eckert, R.-D. Herzberg, U. Kneissl, H. Maser, A. Nord, N. Pietralla, H.H. Pitz, and A. Zilges, Phys. Rev. C **57**, 129 (1998).

58. M. Wilhelm, E. Radermacher, A. Zilges, and P. von Brentano, Phys. Rev. C **54**, R449 (1996).

59. M. Wilhelm, S. Kasemann, G. Pascovici, E. Radermacher, P. von Brentano, and A. Zilges, Phys. Rev. C **57**, 577 (1998).

60. J. Margraf, R.D. Heil, U. Kneissl, U. Maier, H.H. Pitz, H. Friedrichs, S. Lindenstruth, B. Schlitt, C. Wesselborg, P. von Brentano, R.-D. Herzberg, and A. Zilges, Phys. Rev. C **47**, 1474 (1993).

61. R.-D. Herzberg, P. von Brentano, J. Eberth, J. Enders, R. Fischer, N. Huxel, T. Klemme, P. von Neumann-Cosel, N. Nicolay, N. Pietralla, V.Yu. Ponomarev, J. Reif, A. Richter, C. Schlegel, R. Schwengner, S. Skoda, H.G. Thomas, I. Wiedenhöver, G. Winter, and A. Zilges, Phys. Lett. **B 390**, 49 (1997).

62. T. Eckert, O. Beck, J. Besserer, P. von Brentano, R. Fischer, R.-D. Herzberg, U. Kneissl, J. Margraf, H. Maser, A. Nord, N. Pietralla, H.H. Pitz, S.W. Yates, and A. Zilges, Phys. Rev. C **56**, 1256 (1997); **57**, 1007 (1998).

63. P. Van Isacker, H. Harter, A. Gelberg, P. von Brentano, Phys. Rev. C **36**, 441 (1987).

64. N. Pietralla, P. von Brentano, A. Gelberg, T. Otsuka, A. Richter, N. Smirnova, I. Wiedenhöver, Phys. Rev. C **58**, 191 (1998).

65. N. Pietralla, P. von Brentano, R.-D. Herzberg, N Lo Iudice, U. Kneissl, H. Maser, H.H. Pitz, A. Zilges, Phys. Rev. C. **58**, 184 (1998).

66. T.F. Fazzini, A. Giannatiempo, A. Nannini, A. Perego, D. Cutoiu, Z. Phys. **A 346**, 21 (1993).

67. A. Gade, Diploma thesis, University of Cologne 1998 (unpublished).

Two-Phonon Excitations in ^{170}Er

W. Younes*, D.E. Archer*, J.A. Becker*, L.A. Bernstein*
P.E. Garrett**, N. Warr†, M. Kadi†, A. Martin†, S.W. Yates†
G.D. Johns††, R.O. Nelson††, W.S. Wilburn††

*Lawrence Livermore National Laboratory, Livermore, CA 94550-0808
**present address: Lawrence Livermore National Laboratory
† University of Kentucky, Lexington, KY 40506-0055 USA
†† Los Alamos National Laboratory, Los Alamos, NM 87545-1663

Abstract. Recent experiments at the GEANIE/WNR facility and the University of Kentucky accelerator have yielded strong evidence for a two-gamma excitation in ^{170}Er. This new case can be added to a handful of previously identified examples of two-gamma vibrations, all of them discovered in this decade. In this paper the experimental evidence for a two-phonon excitation ^{170}Er is presented and the current state of understanding of these structures is reviewed in the context of this and other recent findings.

INTRODUCTION

The existence of multiphonon excitations in deformed nuclei has been the subject of continuing debate among both theorists and experimentalist for the last 30 years. On the theoretical side, some models [1] maintain that two-gamma vibrations should only exist in a very few select cases, while others [2,3] present a less pessimistic picture. On the experimental side, the argument has been made [4] that single-phonon hexadecapole excitations can mimic the decay properties of two-phonon gamma states, fueling the controversy over the interpretation of experimental data.

The expected excitation of two-gamma states above the pairing gap where they can lose their character through mixing with a multitude of quasiparticle states is at the heart of both theoretical and experimental difficulties. This simple prediction has far-reaching consequences; to the theorist, it means that the very definition of a $\gamma\gamma$ state must depend on some arbitrary cutoff contribution in the wavefunction, and to the experimentalist, it means that such states will be difficult to observe and to identify unequivocally. However, by their very existence, multiphonon states in deformed nuclei, and their deviations from a simple harmonic oscillator description, provide a direct window into the microscopic basis of collective behavior, and are therefore of keen interest to theorists and experimentalists alike. In what follows, we present a brief review of the properties of recently proposed two-gamma phonon states.

Overview

The history of multiphonon states in deformed nuclei can be traced back to the late 1960s, when early hints [5–7] of such structures were found in ^{154}Gd, ^{168}Er and ^{240}Pu. Subsequent experiments could not confirm these structures, and in 1981 Soloviev and Shirikova [8] presented their case, using the Quasiparticle-Phonon Nuclear Model (QPNM), against the existence of multiphonon states in deformed nuclei. For a decade, their conclusion was borne out as no single irrefutable example of a two-phonon vibration in a deformed system could be found. In 1991, however, with the development of the Gamma-Ray-Induced Doppler-broadening (GRID) technique, the lifetimes of members of the candidate $K^{\pi} = 4^{+}$ band in ^{168}Er were finally measured [9], and found to be consistent with theoretical predictions for a $\gamma\gamma$ vibration. In the following years, more cases were discovered, and in 1993 the first example of a nearly harmonic two-phonon excitation was discovered in the well-deformed nucleus ^{232}Th [10]. Recently, this particular case has sparked some controversy when a remeasurement of the

CP481, *Nuclear Structure 98*, edited by C. Baktash
1999 American Institute of Physics 1-56396-858-4

angular distribution called into question the $K^\pi = 4^+$ assignment for the $\gamma\gamma$ candidate state [11]. However, an even more recent measurement [12] has confirmed the original assignment.

The $\gamma\gamma$-phonon multiplet results from the coupling of two ($K^\pi = 2^+$) γ phonons, and therefore has two components: one with $K^\pi = 0^+$ and the other with $K^\pi = 4^+$. In practice, high-lying $K^\pi = 0^+$ states frequently have complex structures which are difficult to interpret in a unique manner. The $K^\pi = 0^+$ two-gamma state has been clearly identified in only one nucleus to date: ^{166}Er [13,14]. Therefore this paper will focus on the $K^\pi = 4^+$ member of the doublet. The experimental evidence for $K^\pi = 4^+$ $\gamma\gamma$ states has come from different types of experiments relying on a variety of techniques to measure their decay properties. We have compiled the data for some key signatures from these different sources in Table 1.

Some broad features of $\gamma\gamma$ states can be extracted from this table. The measured energies of the two-phonon states seem to fall into two categories: some, like Os and ^{232}Th are near the harmonic limit of $2\times E_x(2_\gamma^+)$ while the others (^{164}Dy, ^{166,168}Er) display anharmonicities of $\approx 2.5\text{-}2.8\times E_x(2_\gamma^+)$. This dichotomy in the energy systematics seems to be correlated with the position of the $4_{\gamma\gamma}^+$ relative to the proton and neutron pairing gaps which are calculated using ref. [15]. and listed in Table 2. The energy ratio $E_x(4_{\gamma\gamma}^+)/E_x(2_\gamma^+)$ approaches the harmonic limit when the $4_{\gamma\gamma}^+$ energy is within the gaps and becomes anharmonic when the $4_{\gamma\gamma}^+$ energy lies above one or both of the gaps. The B(E2) are more closely related to the wavefunction of the $\gamma\gamma$ state and therefore do not follow an easily predictable pattern. Generally, the B(E2; $4_{\gamma\gamma}^+ \to 2_\gamma^+$) is comparable in order of magnitude to the B(E2; $2_\gamma^+ \to 0_{gs}^+$), although their actual ratio varies significantly.

TABLE 1. Experimental data on proposed two-gamma phonon states. Nuclei in parenthesis indicate that transfer reaction data disagree with two-phonon interpretation [4]. Identification of the $4_{\gamma\gamma}^+$ and experimental data are from the main references listed unless otherwise specified. All energies are in keV; all B(E2) are in W.u.[a]

Nucleus	main ref.	$E_x(2_\gamma^+)$	$E_x(4_{\gamma\gamma}^+)$	B(E2; $2_\gamma^+ \to 0_{gs}^+$)	B(E2; $4_{\gamma\gamma}^+ \to 2_\gamma^+$)	B(E2; $4_{\gamma\gamma}^+ \to 3_\gamma^+$)
^{106}Mo	[16]	710.4	1434.6			
(^{154}Gd)	[17]	996.3	1645.8	$5.91^{+0.28}_{-0.24}$[b]		
(^{156}Gd)	[18]	1154.1	1510.5	4.45 ± 0.24[c]	1.81 ± 0.25[d]	3.6 ± 0.4[d]
(^{160}Dy)	[18]	966.2	1694.4	3.40 ± 1.70[c]	0.18 ± 0.04[d]	0.17 ± 0.04[d]
^{164}Dy	[21]	761.8	2173.1	4.27 ± 0.22[c]	$4.1^{+3.0}_{-1.7}$	$1.5^{+1.8}_{-0.8}$
^{166}Er	[13] [14]	785.9	2028.2	$5.11^{+0.53}_{-0.52}$[e]	7.4 ± 2.5	9.5 ± 3.2
^{168}Er	[9] [23]	821.2	2055.9	4.73 ± 0.18[f]	7.1 ± 1.6	4.5 ± 1.6
^{186}Os	[18] [25]	767.5	1351.9	$9.42^{+0.46}_{-0.24}$	$24.9^{+5.9}_{-5.6}$	$40.7^{+5.0}_{-14.1}$
^{188}Os	[18] [25]	633.0	1279.1	$7.29^{+0.06}_{-0.27}$	$12.0^{+1.2}_{-0.8}$	$23.8^{+7.4}_{-2.0}$
(^{190}Os)	[18] [25]	558.0	1163.2	$6.07^{+0.25}_{-0.19}$	10.2 ± 1.3	$41.1^{+3.8}_{-18.5}$
(^{192}Os)	[18] [25]	489.1	1069.6	$5.62^{+0.21}_{-0.10}$	$10.4^{+1.0}_{-1.4}$	$44.8^{+6.2}_{-11.3}$
^{232}Th	[10] [26]	785.3	1414	3.47 ± 0.17	14 ± 4	12 ± 3

[a] $\text{B(E2)}_{W.u.} = \frac{1}{4\pi}\left(\frac{3}{5}\right)^2 (1.2A^{1/3})^4 \times 10^{-4}\ e^2 b^2$

[b] From ref. [19]

[c] From ref. [20]

[d] From ENSDF

[e] From ref. [22]

[f] From ref. [24]

A variety of models have been used to calculate properties of $\gamma\gamma$ states. These include the Quasiparticle-Phonon Nuclear Model (QPNM) [27], the Multi-Phonon Method (MPM) [3], the Self-consistent Collective Coordinate Method (SCCM) [2], the Dynamic Deformation Model (DDM) [28] and some very recent calculations in the intrinsic-state formalism of the Interacting Boson Approximation (IBA) [29]. Table 3 summarizes theoretical predictions from the harmonic limit, QPNM, MPM and SCCM. In all calculated cases, the models uniformly predict a sizeable anharmonicity in energy which, as was noted above, is not always the case experimentally. The predicted ratios B(E2; $4_{\gamma\gamma}^+ \to 2_\gamma^+$)/B(E2; $2_\gamma^+ \to 0_{gs}^+$) bracket the corresponding experimental value, with the MPM typically giving the lower bound and with all calculated values a factor of two or more less than the harmonic limit expectation.

TABLE 2. Average pairing gap, calculated using the formalism of ref. [15].

Nucleus	$2\overline{\Delta}_p$ (keV)	$2\overline{\Delta}_n$ (keV)
^{106}Mo	2356	2178
^{154}Gd	2330	1996
^{156}Gd	2256	1936
^{160}Dy	2264	1932
^{164}Dy	2122	1814
^{166}Er	2204	1872
^{168}Er	2134	1814
^{186}Os	2108	1758
^{188}Os	2050	1710
^{190}Os	1990	1662
^{192}Os	1928	1614
^{232}Th	1708	1396

The Erbium Isotopes

The erbium isotopes occupy a priviledged place in the history of $\gamma\gamma$ vibrations with ^{168}Er providing the first confirmed example and ^{166}Er the first–and to date only–case where both the $K^\pi = 0^+$ and $K^\pi = 4^+$ members of the $\gamma\gamma$ doublet have been identified and characterized [13,14]. Part of the reason for these landmark successes can be attributed to the position of these isotopes in the nuclear chart which makes them easily accessible by non-selective reactions such as (n,γ) and $(n,n'\gamma)$.

Soloviev et al. have proposed an explanation [33] for the existence of the $\gamma\gamma$ $K^\pi = 4^+$ state in ^{166}Er. The authors assert that a fortuitous dearth of $K^\pi = 4^+$ states near the energy of the unperturbed $\gamma \otimes \gamma$ level, along with a small value of the coupling between one- and two-phonon states combine to minimize the fragmentation of the $\gamma\gamma$ $K^\pi = 4^+$ level. In ^{168}Er the situation is not as clear: in the QPNM [1] only 30% of the wavefunction of the $K^\pi = 4^+$ state at 2055 keV is attributed to the coupling of two γ phonons. The largest contribution (60%) is due to the one-hexadecapole phonon structure. By comparison, the two-gamma-phonon component in ^{166}Er represents 73% of the $K^\pi = 4^+$ wavefunction. There are currently no published predictions for $\gamma\gamma$ states in ^{170}Er.

THE EXPERIMENTAL CASE FOR ^{170}ER

Based on the successful cases of ^{166}Er and ^{168}Er, a search for $\gamma\gamma$ vibrations was undertaken in the heaviest stable erbium isotope, ^{170}Er. This search was begun with the newly assembled GEANIE spectrometer [34], and a candidate for the $K^\pi = 4^+$ $\gamma\gamma$ state was identified. A follow-up experiment was carried out at the University of Kentucky accelerator to measure the lifetime of this candidate state.

The GEANIE Experiment

In March 1997, ^{170}Er was studied using the $(n,n'\gamma)$ reaction. A white neutron source with energies ranging from about 1 MeV to over 200 MeV is produced by the LANSCE/WNR linac at a 6% duty cycle through proton spallation of a ^{nat}W target. A flux of $\approx 10^6$ neutrons/cm^2/s spread over all energies > 1 MeV was delivered onto a sample of 8g of 99.9%-isotopically-enriched ^{170}Er in oxide powder form, located 20.34 m from the spallation target. Gamma rays from the reaction were observed using the GEANIE spectrometer which at the time of the experiment included 13 coaxial and 7 planar detectors. Both gamma-ray energy and corresponding neutron time-of-flight (TOF) were recorded for each event. A 55-hour run was carried out in this configuration.

The data from the coaxial detectors were sorted into a gamma-ray energy vs. neutron energy matrix, with the neutron energy obtained from the TOF information. The matrix was searched for a pair of gamma rays with energies $E_\gamma^{(1)}$ and $E_\gamma^{(2)}$ and a single corresponding neutron threshold energy at E_n consistent with the

TABLE 3. Theoretical predictions of the $4^+_{\gamma\gamma}$ state properties. QPNM calculations are taken from [31] for ^{156,158}Gd and from [1] for ^{162,164}Dy and ^{166,168}Er, MPM calculations are from ref. [3], and SCCM results from ref. [2]. B(E2) ratios in the harmonic limit are calculated using eq. (4-91) and (4-92) in ref. [30]

Nucleus	Model	$E_x(4^+_{\gamma\gamma})$ (keV)	$\dfrac{E_x(4^+_{\gamma\gamma})}{E_x(2^+_\gamma)}$	$B(E2;\,4^+_{\gamma\gamma}\to 2^+_\gamma)$ (W.u.)	$\dfrac{B(E2;4^+_{\gamma\gamma}\to2^+_\gamma)}{B(E2;2^+_\gamma\to0^+_{gs})}$
^{156}Gd	Harmonic		2		2.78
	QPNM[a]				
^{158}Gd	Harmonic		2		2.78
	QPNM[a]				
^{162}Dy	Harmonic		2		2.78
	QPNM[a]				
	SCCM	2248	2.50	10.5	1.86
	MPM	2302	2.64		0.59
^{164}Dy	Harmonic		2		2.78
	QPNM[b]	2150	2.80	7.65	1.7
		2230	2.80	4.09	1.0
	SCCM	2007	2.64	11.6	1.83
	MPM	2135	2.87		0.55
^{164}Er	Harmonic		2		2.78
	SCCM	2157	2.51	10.5	1.89
	MPM	2324	2.74		0.52
^{166}Er	Harmonic		2		2.78
	QPNM	2050	2.56	9.2	1.8
	SCCM	2021	2.57	10.6	1.89
	MPM	2152	2.82		0.57
^{168}Er	Harmonic		2		2.78
	QPNM	2000	2.40	3.18	0.73
	SCCM	2080	2.54	8.4	1.89
	MPM	2237	2.80		0.53
^{186}Os	Harmonic		2		2.78
	SCCM	1770	2.30	12.6	2.00
^{188}Os	Harmonic		2		2.78
	SCCM	1589	2.51	13.9	1.94
^{190}Os	Harmonic		2		2.78
	SCCM	1464	2.63	13.5	1.89
^{192}Os	Harmonic		2		2.78
	SCCM	1313	2.69	12.8	1.89

[a] QPNM predicts dominant hexadecapole structure for lowest calculated $K^\pi=4^+$ bandheads
[b] both parameter choices which predict $\gamma\gamma$ for the first $K^\pi = 4^+$ are listed

decay of a single state to the lowest two members of the γ band at $E_x(2^+_\gamma) = 934$ keV and $E_x(3^+_\gamma) = 1010$ keV. This translates into the conditions:

$$E_\gamma^{(1)} + 934\,keV = E_n$$
$$E_\gamma^{(2)} + 1010\,keV = E_n$$

The gamma-ray pair $(E_\gamma^{(1)} = 1518$ keV, $E_\gamma^{(2)} = 1441)$ occurring at a threshold energy $E_n \approx 2.45$ MeV was found to satisfy these criteria best. The excitation functions for these two gamma rays are shown in Figure 1.

An intensity ratio was calculated from these excitation functions and is plotted in Figure 2 as a function of incident neutron energy. Despite large experimental uncertainties, the ratio is consistent with the $K^\pi = 4^+$ assignment. The 1518 keV line had to be separated from another line less than a FWHM away, and therefore the ratio is contaminated. The quantity plotted in Figure 2 represents an upper bound to the actual intensity ratio.

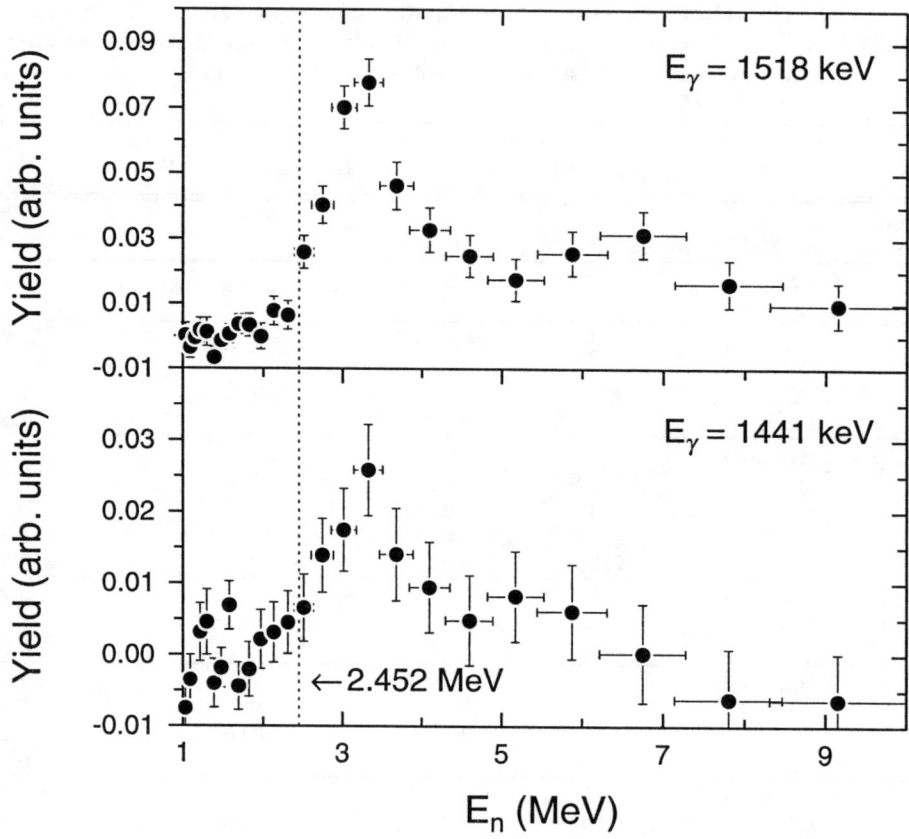

FIGURE 1. Excitation functions for the 1441 keV and 1518 keV gamma rays from GEANIE data.

The Kentucky Experiment

After the initial identification of a $\gamma\gamma$ candidate state, a follow-up experiment was carried out at the University of Kentucky to measure its lifetime using the Doppler-Shift Attenuation Method (DSAM) [35]. Monochromatic 2.7 MeV neutrons were produced by the reaction ^{3}He(p,n) with $E_p = 3.5$ MeV. A sample of 56 g of 99% isotopically enriched ^{170}Er in oxide powder form was suspended in the neutron flux. A single 52% germanium detector at ≈ 1 m from the sample was used to observe gamma rays from the reaction. The gamma-ray sources ^{24}Na and ^{137}Cs were placed near the detector for in-beam calibration. A seven-day run was carried out in March 1998 with the detector rotated through 11 distinct angles. This experiment was preceded by a 5-day coincidence measurement at $E_n = 3.4$ MeV using the KEGS array [36].

The coincidence data were used to confirm the placement of the 1441 and 1518 keV gamma rays. Gates on these lines are shown in Figure 3: the spectra reveal several contaminants for the 1518 keV line which are tentatively placed in the partial level scheme in Figure 4. Most of these contaminants do not pose serious obstructions to the analysis; they are either weak, originate from higher-spin off-yrast states which are not as strongly populated in the $(n, n'\gamma)$ reaction, or are sufficiently different in gamma-ray energy to be distinguished from the line of interest. There is one exception, however for the ≈ 1519 keV gamma ray feeding the 2^+ member at $E_x = 960$ keV of the $K^\pi = 0^+$ band. This line originates from a level less than 30 keV higher than the $\gamma\gamma$ candidate. The contaminant gamma-ray energy is only about 1 keV higher than the line of interest. In the DSAM analysis, the resolution of the germanium detector proved inadequate to separate the two lines. Consequently, our analysis has focused on the 1441 keV candidate $4^+_{\gamma\gamma} \rightarrow 3^+_\gamma$ line.

In addition to the standard sources ^{24}Na and ^{137}Cs, previously well-known lines in ^{170}Er were used for calibration. We also used some lines precisely measured during our experiment using an in-beam ^{226}Ra calibration source with an ^{170}Er spectrum taken at a lower incident neutron energy of 2.3 MeV. In all, 61 lines were selected for energy calibration, with most of them in the range of 200-1500 keV. A nonlinearity for the system

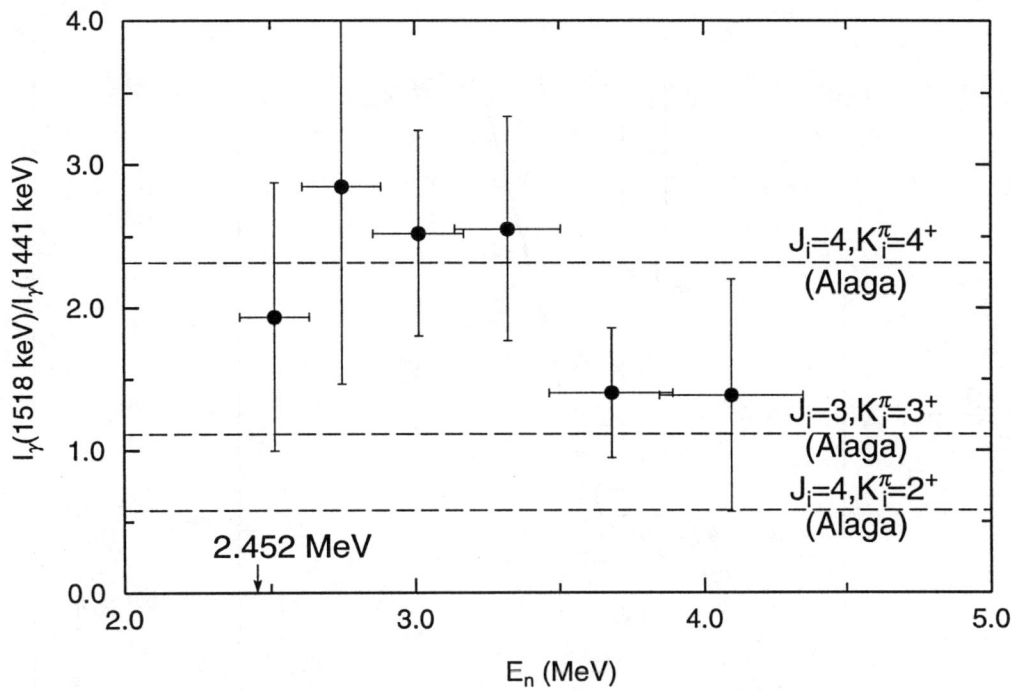

FIGURE 2. Intensity branching ratio from GEANIE data. Also shown are the expected intensity ratios from Alaga rule assuming different values of the K quantum number for the parent state.

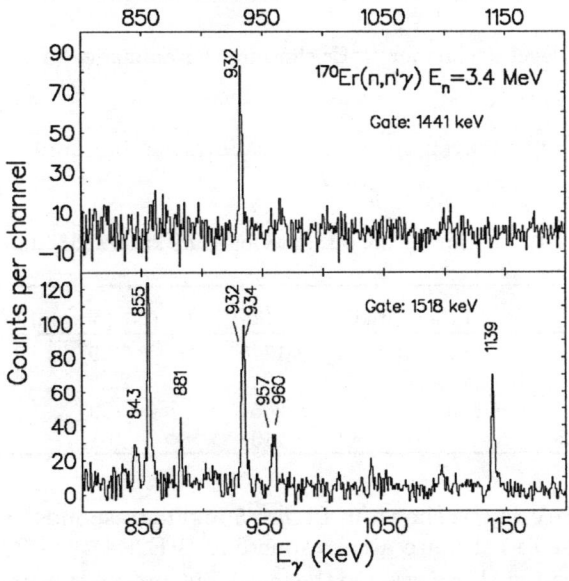

FIGURE 3. Gates on the 1441 and 1518 keV gamma rays from the KEGS coincidence run.

was extracted at each angle by fitting the residual from a linear calibration with a fifth-order polynomial. In this way, the Doppler-shifted energies of the 1441 keV gamma ray could be determined to better than one part in 10000.

Lifetimes were extracted for known states previously studied [37] using nuclear resonance fluorescence to test the calibration. The results are summarized in Table 4. The agreement is excellent for both gamma rays depopulating the level at $E_x = 1825$ keV, good for the 2054 keV gamma ray from the level at $E_x = 2133$ keV and poor for the 2133 keV gamma ray from that same level. It should be noted however that the 2133 keV

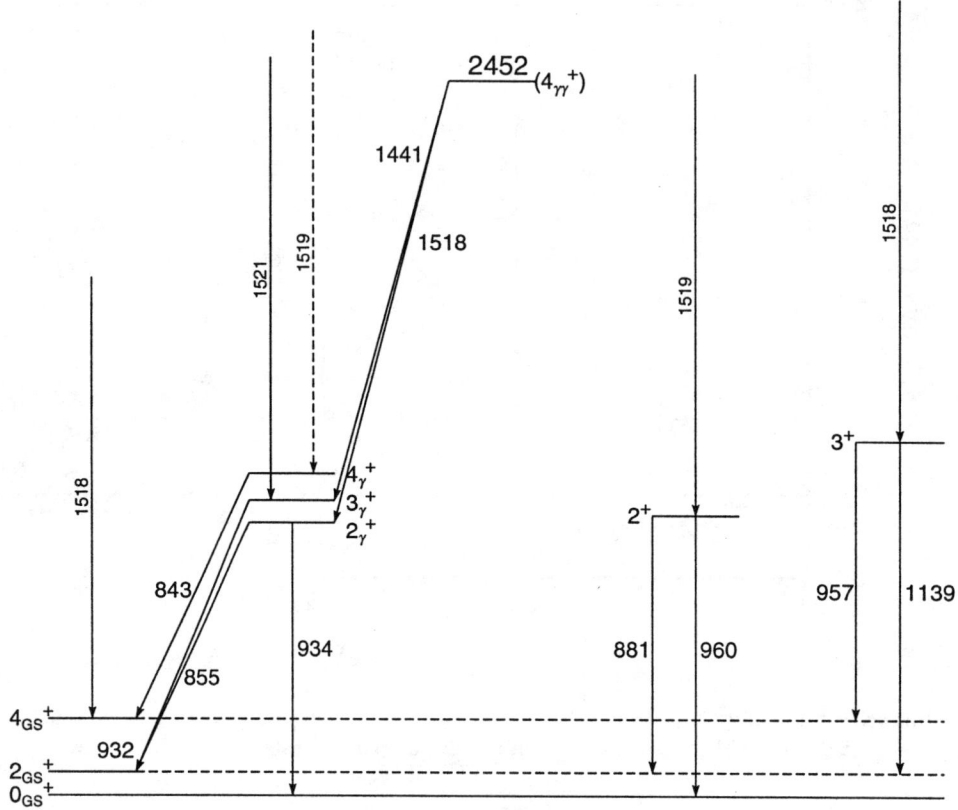

FIGURE 4. Partial level scheme for ^{170}Er showing contaminants for the 1518 keV line.

gamma ray lies within a 230 keV gap between the last two calibration lines and its Doppler shift can therefore not be reliably extracted.

TABLE 4. Lifetimes from our data for some known [37] dipole excitations.

E_x (keV)	τ (fs) (accepted)	E_γ (keV)	τ (fs) (our value)
1825	20.7 ± 3.4	1746	$20.9^{+2.4}_{-1.9}$
		1825	$21.8^{+3.4}_{-1.9}$
2133	173 ± 41	2054	129^{+35}_{-31}
		2133	61^{+8}_{-8}

The Doppler shift of the 1441 keV line is shown in Figure 5 and corresponds [35] in our experimental setup to a lifetime of 84^{+37}_{-24} fs. In order to calculate a corresponding B(E2; $4^+_{\gamma\gamma} \rightarrow 2^+_\gamma$), the relative intensities of decays out of the $4^+_{\gamma\gamma}$ level are required. If we assume the decay out proceeds primarily to the 2^+_γ and 3^+_γ states with relative intensities following the Alaga rule, as suggested by the GEANIE data, we find:

$$B(E2; 4^+_{\gamma\gamma} \rightarrow 2^+_\gamma) = 15.0^{+6.1}_{-4.6} \ W.u.$$

$$\frac{B(E2; 4^+_{\gamma\gamma} \rightarrow 2^+_\gamma)}{B(E2; 2^+_\gamma \rightarrow 0^+_{gs})} = 4.08^{+1.83}_{-1.34}$$

(1)

This B(E2) value is larger than would be expected in the harmonic limit (2.78 W.u.). However, as was noted above, the Alaga branching ratio, which is suggested by the GEANIE data, may be an overestimate. Few of the observed $\gamma\gamma$ states in other nuclei follow the Alaga rules in their decays. If we use instead the branching ratio found in ^{166}Er, the B(E2) values become:

$$B(E2; 4^+_{\gamma\gamma} \to 2^+_\gamma) = 10.9^{+4.3}_{-3.4} \; W.u.$$

$$\frac{B(E2; 4^+_{\gamma\gamma} \to 2^+_\gamma)}{B(E2; 2^+_\gamma \to 0^+_{gs})} = 2.96^{+1.30}_{-0.97}$$

(2)

which is significantly closer to the decay properties of observed $\gamma\gamma$ states in other nuclei. A detailed level scheme of ^{170}Er is currently being assembled from the KEGS coincidence data and could shed some light on the decay pattern of the candidate $\gamma\gamma$ state in ^{170}Er.

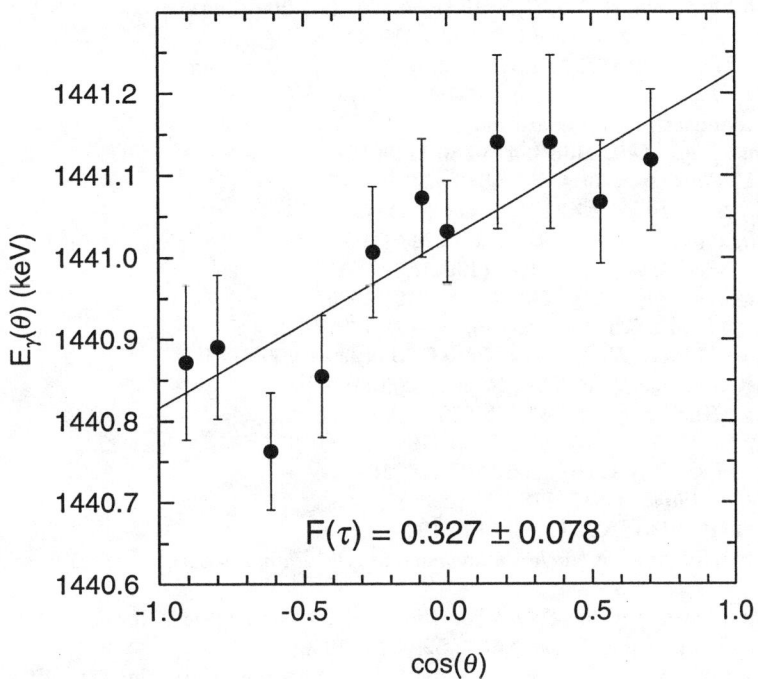

FIGURE 5. Measured Doppler shift for the 1441 keV line.

CONCLUSION

A search for $\gamma\gamma$ vibrations in ^{170}Er has yielded a strong candidate for the $K^\pi = 4^+$ member of the multiplet. Two gamma rays with energies of 1441 keV and 1518 keV have been observed depopulating the candidate state at 2452 keV and feeding the γ band's 3^+ and 2^+ states respectively. A lifetime of 84^{+23}_{-37} fs was measured for the state from the Doppler shift of the 1441 keV gamma ray which, assuming a branching ratio similar to ^{166}Er yields a $B(E2; 4^+_{\gamma\gamma} \to 2^+_\gamma)/B(E2; 2^+_\gamma \to 0^+_{gs}) = 2.96^{+1.30}_{-0.97}$, close in value to other observed $\gamma\gamma$ phonon decays. The value is still somewhat larger than in ^{166,168}Er; however, the calculation assumes no other decays out of the $\gamma\gamma$ state, which is known not to be the case in either ^{166}Er or ^{168}Er. Additional decay paths would further lower the ratio, bringing it more in line with experimental and theoretical values for ^{166,168}Er. The ongoing analysis of coincidence data from the Kentucky experiment might reveal these decay paths.

Encouraging though these results may be, a word of caution is warranted with regards to the role of hexadecapole modes in this structure. There is currently no transfer information available for ^{170}Er. Several β-decay experiments have been performed to study many levels in this nucleus although our candidate state has not been observed before. Transfer and β-decay data for this nucleus would be a welcome and necessary addition in identifying the structure of our candidate state and building a stronger case for a $K^\pi = 4^+$ $\gamma\gamma$ state.

This work is supported by the U.S. Department of Energy under contract numbers W-7405-ENG-48 (LLNL), W-4705-ENG-36 (LANL) and by the National Science Foundation (Kentucky)

REFERENCES

1. Soloviev, V.G., *et al.*, *Intl. J. Mod. Phys.* **E6**, 437-473 (1997).
2. Matsuo, M., and Matsuyanagi, K., *Prog. Theor. Phys.* **78**, 591-608 (1987).
3. Jammari, M.K., and Piepenbring, R., *Nucl. Phys.* **A487**, 77-91 (1988).
4. Burke, D.G., *Phys. Rev. Lett.* **73**, 1899-1902 (1994).
5. Meyer, R.A., *Phys. Rev.* **170**, 1089 (1968).
6. Michaelis, W., *et al.*, *Nucl. Phys.* **A150**, 161-186 (1970).
7. Schmorak, M.R., *et al.*, *Phys. Rev. Lett.* **24**, 1507-1511 (1970).
8. Soloviev, V.G., and Shirikova, N. Yu, *Zeit. Phys* **A301**, 263-269 (1981).
9. Börner, H.G., *et al.*, *Phys. Rev. Lett.* **66**, 691-694 (1991).
10. Korten, W., *et al.*, *Phys. Lett.* **B317**, 19-24 (1993).
11. Gerl, J., *et al.*, *Prog. Part. Nucl. Phys.* **38**, 79-85 (1997).
12. Martin, A., *et al.*, manuscript in preparation.
13. Fahlander, C., *et al.*, *Phys. Lett.* **B388**, 475-480 (1996).
14. Garrett, P.E., *et al.*, *Phys. Rev. Lett.* **78**, 4545-4548 (1997).
15. Jensen, A.S., *et al.*, *Nucl. Phys.* **A431**, 393-418 (1984).
16. Guessous, A., *et al.*, *Phys. Rev. Lett.* **75**, 2280-2283 (1995).
17. Wu, X., *et al.*, *Phys. Rev.* **C49**, 1837-1844 (1994).
18. Wu, C.Y., and Cline, D., *Phys. Lett.* **B382**, 214-219 (1996).
19. Wollersheim, H.J., and Elze, Th.W., *Zeit. Phys.* **A280**, 277-279 (1977).
20. McGowan, F.K., and Milner, W.T., *Phys. Rev.* **C23**, 1926-1937 (1981).
21. Corminboeuf, F., *et al.*, *Phys. Rev.* **C56**, R1201-1205 (1997).
22. Fahlander, C., *et al.*, *Nucl. Phys.* **A537**,183-206 (1992).
23. Oshima, M., *et al.*, *Phys. Rev.* **C52**, 3492-3495 (1995).
24. Baktash, C., *et al.*, *Phys. Rev.* **C10**, 2265-2267 (1974).
25. Wu, C.Y., *et al.*, *Nucl. Phys.* **A607**, 178-234 (1996).
26. Korten, W., *et al.*, *Zeit. Phys.* **A351**, 143-147 (1995).
27. Soloviev, V.G., *Theory of Atomic Nuclei. Quasiparticle and Phonons*, Bristol and Philadelphia, Institute of Physics Publishing, 1992.
28. Kumar, K., *Nuclear Models and the Search for Unity in Nuclear Physics*, Norway, Universitetforlaget, 1984.
29. Garcia-Ramos, J.E., *et al.*, *Nucl. Phys.* **A637**, 529-536 (1998).
30. Bohr, A., and Mottelson, B.R., *Nuclear Structure Vol. II*, Reading, Massachusetts, W.A. Benjamin Inc., 1975.
31. Soloviev, V.G., *et al.*, *Nucl. Phys.* **A568**, 244-264 (1994).
32. Soloviev, V.G., and Sushkov, A.V., *Zeit. Phys.* **A345**, 155-161 (1993).
33. Soloviev, V.G., *et al.*, *Phys. Rev.* **C51**, 551-558 (1995).
34. Becker, J.A., and Nelson, R.O., *Nucl. Phys. News Intl.* **7**, 11-14 (1997).
35. Belgya, T., *et al.*, *Nucl. Phys.* **A607**, 43-61 (1996).
36. McGrath, C.A., *et al.*, *Nucl. Instr. and Methods*, in press.
37. Maser, H., *et al.*, *Phys. Rev.* **C53**, 2749-2762 (1996).

Stable and Vibrational Octupole Modes in
Neutron Rich Nuclei

J.H. Hamilton[1], S.J. Zhu[1,2,3], A.V. Ramayya[1], J.K. Hwang[1],
L.K. Peker[1], E.F. Jones[1], P.M. Gore[1], X.Q. Zhang[1]

[1]*Physics Department, Vanderbilt University, Nashville, TN 37235, USA*
[2]*Physics Department, Tsinghua University, Beijing, P.R. China*
[3]*Joint Institute for Heavy Ion Research, Oak Ridge TN 37830, USA*

Abstract. Evidence for stable octupole deformation in neutron-rich nuclei bounded by $Z = 54-58$ and $N = 85-92$ is described. More vibrational type octupole modes are observed to either side of this region in ^{140}Ba and ^{152,154}Nd. The largest stable octupole deformation ($\beta_3 \sim 0.1$) is reported in ^{144}Ba$_{88}$. The theoretically predicted quenching ($\beta_3 \to 0$) of stable octupole deformation with increased rotation is observed in ^{146}Ba. There is good agreement between theory and experiment for the strongly varying electric dipole moments as a function of mass for $^{142-148}$Ba. In odd-A ^{143}Ba and odd-Z, ^{145}La, two pairs of positive and negative parity bands with opposite spins, parity doublets are observed. In ^{145}La the ground band with symmetric shape coexists with the asymmetric octupole shape which stabilizes above about spin $19/2^+$. The isotope ^{109}Mo was identified and a new region of stable octupole deformation is identified in ^{107,109}Mo centered around $N = 64-66$ as earlier predicted. This is the first case of stable octupole deformation involving only one pair of orbitals.

INTRODUCTION

Exciting new vistas of the varying structures of neutron-rich nuclei are coming out of studies of the spontaneous fission (SF) of heavy elements like ^{252}Cf, and ^{248}Cm, with large detector arrays like Gammasphere and Eurogam [1,2]. An important early discovery with a small array was that of stable octupole deformation centered around $Z = 56,58$, $N = 88$ in ^{144,146}Ba [3] and ^{146}Ce [4] associated with the $\nu f_{7/2}$-$\nu i_{13/2}$ and the $\pi d_{5/2}$-$\pi h_{11/2}$ orbitals. With much larger detector arrays, stable octupole deformation is reported to higher spins in $^{140-148}$Ba and in odd-A ^{143}Ba [1,5-11] and odd-Z ^{145}La [9,12,13]. In ^{145}La bands built on both symmetric and asymmetric shapes are observed for the first time. The first evidence for the theoretical prediction [14] of the vanishing of stable octupole deformation with increasing rotation is observed [1,5,8].

Static octupole deformation may be induced in nuclei where the Fermi level lies between single particle orbitals of the type $|N,L,j\rangle$ and $|N + 1, L + 3, j +3\rangle$ (e.g., $p_{3/2}$-$g_{9/2}$, $d_{5/2}$-$h_{11/2}$, $f_{7/2}$-$i_{13/2}$, etc.), because such orbitals couple strongly through the octupole potential term Y_{30} [15,16]. Calculations indicate that stable octupole effects (evidenced by enhanced E1 transitions between bands) should be observed in nuclei with Z and N around (34), 56, 88, 134. It was thought that octupole coupling for both protons and neutrons in a nucleus was needed to give sufficient enhancement to produce observable stable octupole effects. This is the case in the Ba-Ce and Rn regions. The occurrence of stable octupole deformation in $N = 64-66$ nuclei also has been predicted by Cottle [17]. Recently, evidence for stable octupole deformation was found in ^{107}Mo$_{65}$ and the newly discovered ^{109}Mo$_{67}$ [18] based only on one nucleon pair of orbitals for the first time. More vibrational-like octupole structures are observed in ^{140}Ba (6,8,10) and ^{152,154}Nd [19]. However, octupole deformation becomes stabilized with rotation in ^{140}Ba.

CP481, *Nuclear Structure 98*, edited by C. Baktash
© 1999 American Institute of Physics 1-56396-858-4/99/$15.00

STABLE OCTUPOLE DEFORMATION IN $^{140-148}$Ba NUCLEI

With Gammasphere (1993 ^{252}Cf SF data) new high spin states were observed in $^{142-146}$Ba, and the first evidence for octupole deformation in an odd-A nucleus in this region was reported in N = 87, ^{143}Ba [1,5]. Our extended $^{140-148}$Ba level schemes from our 1995 Gammasphere data [6,7,8] are given in Fig. 1. These level schemes are very similar to those simultaneously extracted from Eurogam II in SF of ^{248}Cm [10, 11]. The only significant difference in level assignments is in the 641-722 keV cascade into the 3153.2 keV 11$^-$ level in ^{142}Ba where we assign this cascade to the negative parity band [6] and Urban et al. [10] assign the 641.0 as an E1 crossing transition with the 722 placed in the yrast cascade with only a dashed yrast transition to connect the 3794.0 keV level to the 10$^+$ level. Important contributions of Urban et al. [10] and Jones et al. [11] were the establishment of the spins and parities of many levels assigned earlier on systematics in the even- and odd-A Ba isotopes from directional correlation and polarization measurements.

A striking feature seen in Fig. 1 is the intertwined E1 transitions between the positive and negative parity bands to quite high spin in ^{144}Ba but not continuing above 10$^+$ in ^{146}Ba. These E1 transitions are strongly enhanced in ^{142,144}Ba (see Table I) with B(E1) ~1 x 10^{-3} w.u. to give evidence for stable octupole deformation but are reduced by a factor of 10-100 in ^{146}Ba even when seen there. This drop in E1 strength in ^{146}Ba was first noted by Phillips et al. [3]. From Table I, B(E1) ~5 x 10^{-6} w.u., a factor of 200 lower than for ^{144}Ba. In Fig. 2 the intrinsic electric dipole moments, D_0, extracted from the B(E1) values are given averaged for high I > 8$\hbar$ and low spin states I < 7$\hbar$ data. D_0 are also reported in refs. [1,5,8-10] and here for ^{142,144,146}Ba, and in ref. [10] and here for ^{148}Ba. In ^{142,144}Ba, D_0 is significantly larger for the higher spin data to show an enhancement of the stable octupole deformation with rotation. However, in ^{146}Ba the reverse occurs with a sharp decrease and then undetectable E1 transition strength above spin 10$^+$ to indicate a vanishing of β_3 at higher spins [1,5,6]. The ^{148}Ba D_0 for high spins is similar to those of ^{142,144}Ba to indicate that stable octupole deformation is stabilized there.

The D_0 calculations of Butler and Nazarewicz [20] in the shell correction method based on the reflection asymmetric Woods-Saxon model nicely reproduce the data (Fig. 2) while the Interacting Boson Model cannot [21]. The D_0 in ^{146}Ba is dramatically reduced because of the very small shell correction term (cancellation between proton and neutron contribution to D_0^{shell}) and the negligible macroscopic contribution D_0^{mac} (cancellation between the reorientation term and the neutron-skin term).

In the moments of inertia for $^{142-148}$Ba (Fig. 3), one sees in ^{146}Ba a sudden sharp increase (backbend) in J_1 and a sharp rise and drop in J_2 above 10$^+$ that is not seen in ^{142,144,148}Ba where only smooth upbends are seen. Cranked shell model calculations for the Ba-Ce nuclei indicate that around N = 88 at intermediate spins the stable octupole deformation increases and the octupole minima are much better separated (see Fig. 4) [14]. However, at higher spin and rotational frequency the octupole deformation is predicted to disappear as seen in Fig. 4. A shape transition towards $\beta_3 = 0$ is expected in ^{146}Ba at frequencies above 0.3 MeV after the alignment of $\nu i_{13/2}$ and $\pi h_{11/2}$ pairs [14]. The new data for ^{146}Ba show the ground band is crossed above 10$^+$ (Fig. 3), and no intertwined connecting transitions are seen above 10$^+$. These data [1,5,8] provided the first evidence for the theoretical prediction of the vanishing of β_3 above $\hbar\omega = 0.3$ MeV [14]. However, in ^{144}Ba, where there is no backbend in J_1, the new data to 16$^+$ indicate no reduction in the E1 strength for these higher spins compared to the lower spin states. Indeed, the average D_0 values are larger for the higher spin states than for the lower spin states (Fig. 2), to indicate rotation enhances the stable octupole deformation in ^{144}Ba in agreement with theory.

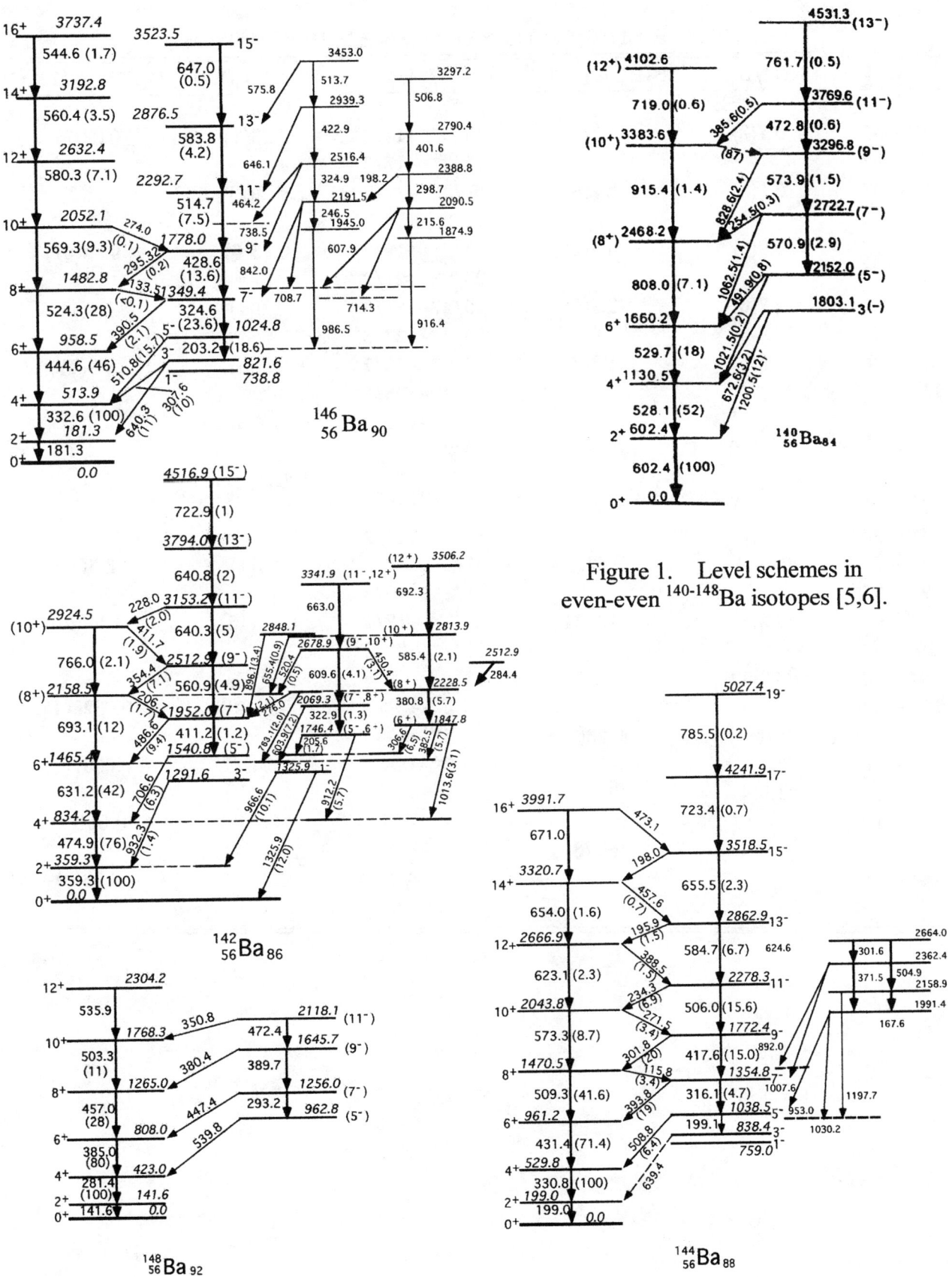

Figure 1. Level schemes in even-even $^{140-148}$Ba isotopes [5,6].

475

TABLE I. B(E1)/B(E2) (10^{-6}fm^{-2}) values in ^{142,143,144,146}Ba.

$I_i^\pi \rightarrow I_f^\pi$	B(E1)/B(E2)	B(E1)$_{W.U.}(10^{-3})$	$I_i^\pi \rightarrow I_f^\pi$	B(E1)/B(E2)	B(E1)$_{W.U.}(10^{-3})$
	^{142}Ba			^{143}Ba, s=-i	
$7^- \rightarrow 5^-$	0.7(2)	0.7(2)	$19/2^+ \rightarrow 15/2^+$	0.25(7)	0.28(7)
$7^- \rightarrow 6^+$			$19/2^+ \rightarrow 17/2^-$		
$8^+ \rightarrow 6^+$	2.1(8)	2.2(8)	$21/2^- \rightarrow 17/2^-$	1.2(2)	1.4(2)
$8^+ \rightarrow 7^-$			$21/2^- \rightarrow 19/2^+$		
$9^- \rightarrow 7^-$	0.9(2)	0.9(2)	$23/2^+ \rightarrow 19/2^+$	0.73(6)	0.83(6)
$9^- \rightarrow 8^+$			$23/2^+ \rightarrow 21/2^-$		
$10^+ \rightarrow 8^+$	2.7(6)	2.8(6)	$25/2^- \rightarrow 21/2^-$	1.0(1)	1.1(1)
$10^+ \rightarrow 9^-$			$25/2^- \rightarrow 23/2^+$		
$11^- \rightarrow 9^-$	1.3(3)	1.3(3)	$27/2^+ \rightarrow 23/2^+$	2.3(6)	3.0(7)
$11^- \rightarrow 10^+$			$27/2^+ \rightarrow 25/2^-$		
	^{144}Ba		$29/2^- \rightarrow 25/2^-$	0.7(3)	0.9(4)
$7^- \rightarrow 5^-$	0.16(2)	0.21(3)	$29/2^- \rightarrow 27/2^+$		
$7^- \rightarrow 6^+$				^{143}Ba, s=+i	
$8^+ \rightarrow 6^+$	1.4(1)	1.8(1)	$17/2^+ \rightarrow 13/2^+$	0.36(6)	0.37(6)
$8^+ \rightarrow 7^-$			$17/2^+ \rightarrow 15/2^-$		
$9^- \rightarrow 7^-$	0.48(4)	0.62(5)	$21/2^+ \rightarrow 17/2^+$	0.3(1)	0.4(1)
$9^- \rightarrow 8^+$			$21/2^+ \rightarrow 19/2^-$		
$10^+ \rightarrow 8^+$	0.9(2)	1.2(3)	$23/2^- \rightarrow 19/2^-$	0.20(15)	0.26(20)
$10^+ \rightarrow 9^-$			$23/2^- \rightarrow 21/2^+$		
$11^- \rightarrow 9^-$	0.88(12)	1.1(1)		^{146}Ba	
$11^- \rightarrow 10^+$			$5^- \rightarrow 3^-$	0.0016(2)	0.0028(4)
$12^+ \rightarrow 10^+$	0.75(4)	0.98(5)	$5^- \rightarrow 4^+$		
$12^+ \rightarrow 11^-$			$7^- \rightarrow 5^-$	0.0043(2)	0.0067(3)
$13^- \rightarrow 11^-$	1.5(4)	2.0(6)	$7^- \rightarrow 6^+$		
$13^- \rightarrow 12^+$			$9^- \rightarrow 7^-$	0.007(3)	0.011(4)
$14^+ \rightarrow 12^+$	0.4(2)	0.5(3)	$9^- \rightarrow 8^+$		
$14^+ \rightarrow 13^-$			$10^+ \rightarrow 8^+$	0.02(2)	0.03(3)
			$10^+ \rightarrow 9^-$		

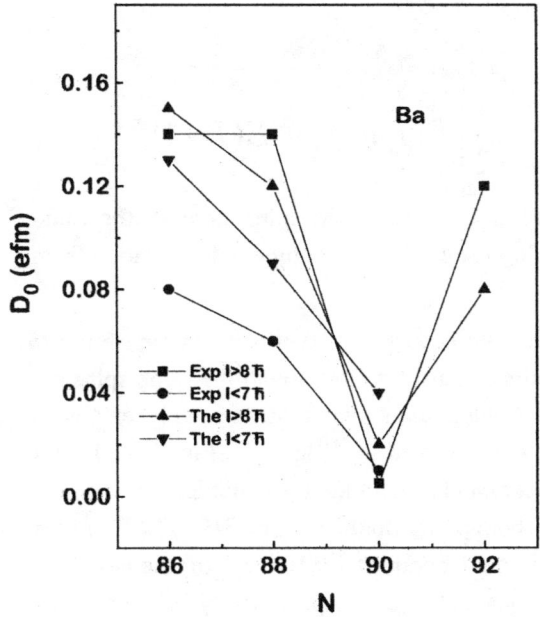

Figure 2. Experimental and calculated intrinsic dipole moments.

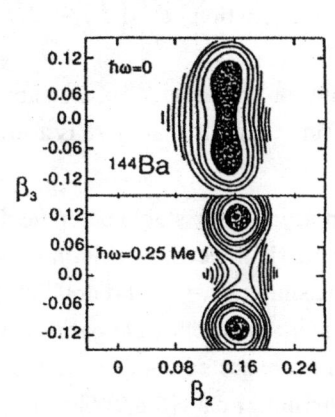

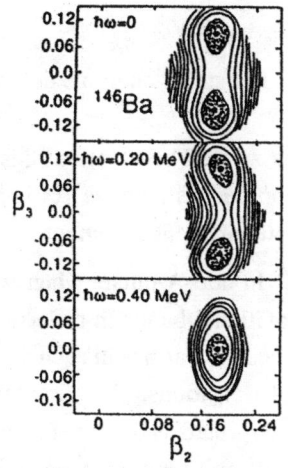

Figure 4. The influence of β_3 on the total Routhian surface in the (β_2, β_3) plane for ^{144}Ba at $\hbar\omega$=0, 0.2, and 0.4 MeV [14]

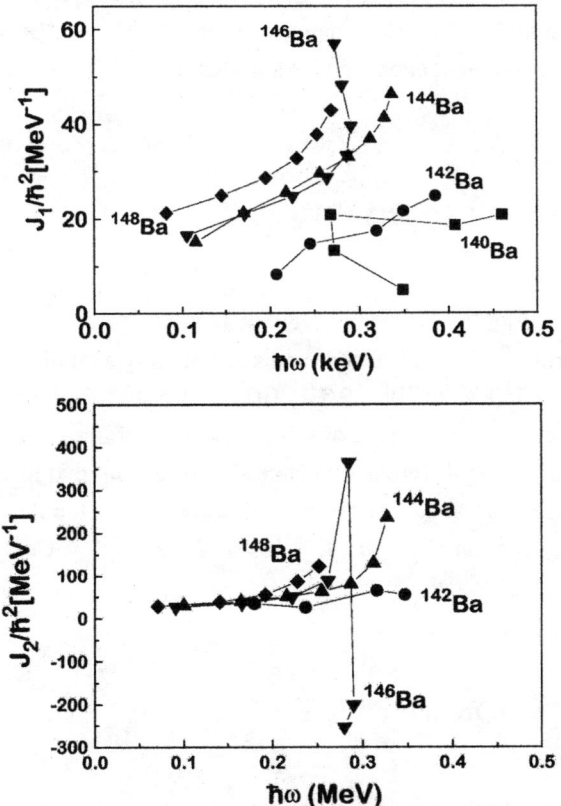

Figure 3. Yrast band moments of inertia (J_1 and J_2) as a function of rotational frequency ($\hbar\omega$) for even-even Ba nuclei.

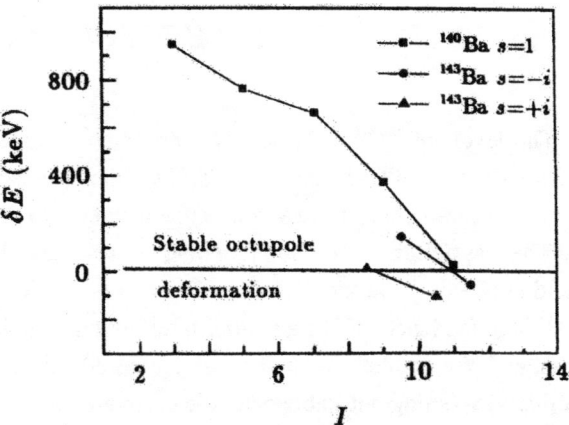

Figure 5. δE as a function of spin for the ^{140}Ba yrast band and for ^{143}Ba parity doublets.

For stable octupole deformation, the function $\delta E(I) = E(I)^{\mp} - \frac{1}{2}\left[E(I+1)^{\pm} + E(I-1)^{\pm}\right]$ should be zero. The values of this function for ^{140}Ba and ^{143}Ba are shown in Fig. 5. Note as the spins increase, the value approaches 0 above 10^+ in ^{140}Ba. This suggests that there is a shift at high spin to stable octupole deformation from rotational enhancement in ^{140}Ba.

In odd-A nuclei when reflection asymmetry stable octupole deformation occurs, the level patterns are similar to rotational bands in reflection-asymmetric molecules including two pairs of parity doublets with the same spins but opposite parities in each doublet, the simplex s = i and -i doublets [22]. Each doublet pair is intertwined by enhanced E1 transitions. The first two, E1 intertwined opposite parity bands were observed in ^{143}Ba [1,5] (Fig. 6) and were interpreted as the s = -i members of the expected doublet pairs. Jones et al. [11] found the s = i doublet pair with the same spins but opposite parities. Our newer data [6,8,9] likewise show both parity doublets. The B(E1)/B(E2) ratios are given in Table I for both bands. B(E1) ~1 x 10^{-3} w.u. for the s = -i doublet and ~0.4 x 10^{-3} w.u. for the new s = i doublet. In addition, in our work we find three new E1 crossing transitions and five new presumably M1 transitions between the s = i and -i bands which were not reported in ref. 11. These E1 transitions between the two doublets measure the degree of mixing of the s = i and -i doublets.

The levels of ^{145,147}Ba are strikingly different from those of ^{143}Ba [6,11] with the ground bands (Fig. 6) having quite different structures with no evidence for stable octupole deformation in ^{145,147}Ba. This change between ^{143}Ba and ^{145}Ba is surprising since the ^{145}Ba core, ^{144}Ba N = 88 has the strongest D_0's and most enhanced E1's in this region.

OCTUPOLE DEFORMATION IN ^{144,146}Ce AND ^{140}Xe

The levels of ^{144,146,148}Ce with N = 86,88 and 90 are shown in Fig. 7. The new levels of N = 86 ^{144}Ce show intertwined enhanced E1 transitions [1,5,8] similar to those seen earlier in N = 88 ^{146}Ce (4) and subsequently extended [1,6]. The E1 transitions have similar enhancements to those in N = 86,88 ^{142,144}Ba. B(E1)/B(E2) ~1 × 10^{-6} fm^{-2} in ^{146}Ce. The less deformed N = 86 ^{140}Xe in Fig. 7 also has a band structure characteristic of stable octupole deformation as found in N = 86 ^{142}Ba and ^{144}Ce. However, N = 90 ^{148}Ce has significantly different side band structure from that in ^{146}Ba with no evidence of E1 transitions intertwined with the ^{148}Ce yrast cascade. The absence of observable octupole strength in ^{148}Ce is likely related to its increased quadrupole deformation (β_2 ~ 0.14, 0.18, 0.22 for ^{144,146,148}Ce, respectively) washing out stable octupole correlations.

STABLE OCTUPOLE DEFORMATION IN ^{145,147}La

Neutron-rich $^{145-147}$La with Z=57 and N = 88,90 lie between the Ba(Z=56) and Ce(Z=58) nuclei where stable octupole deformation is reported as discussed above. Thus, they are expected to be good candidates for octupole deformation. As our extensive studies of these nuclei from ^{252}Cf SF was being completed [13], evidence for octupole correlations in ^{145,147}La up to intermediate spins were reported from SF studies of ^{248}Cm [12]. However, evidence for

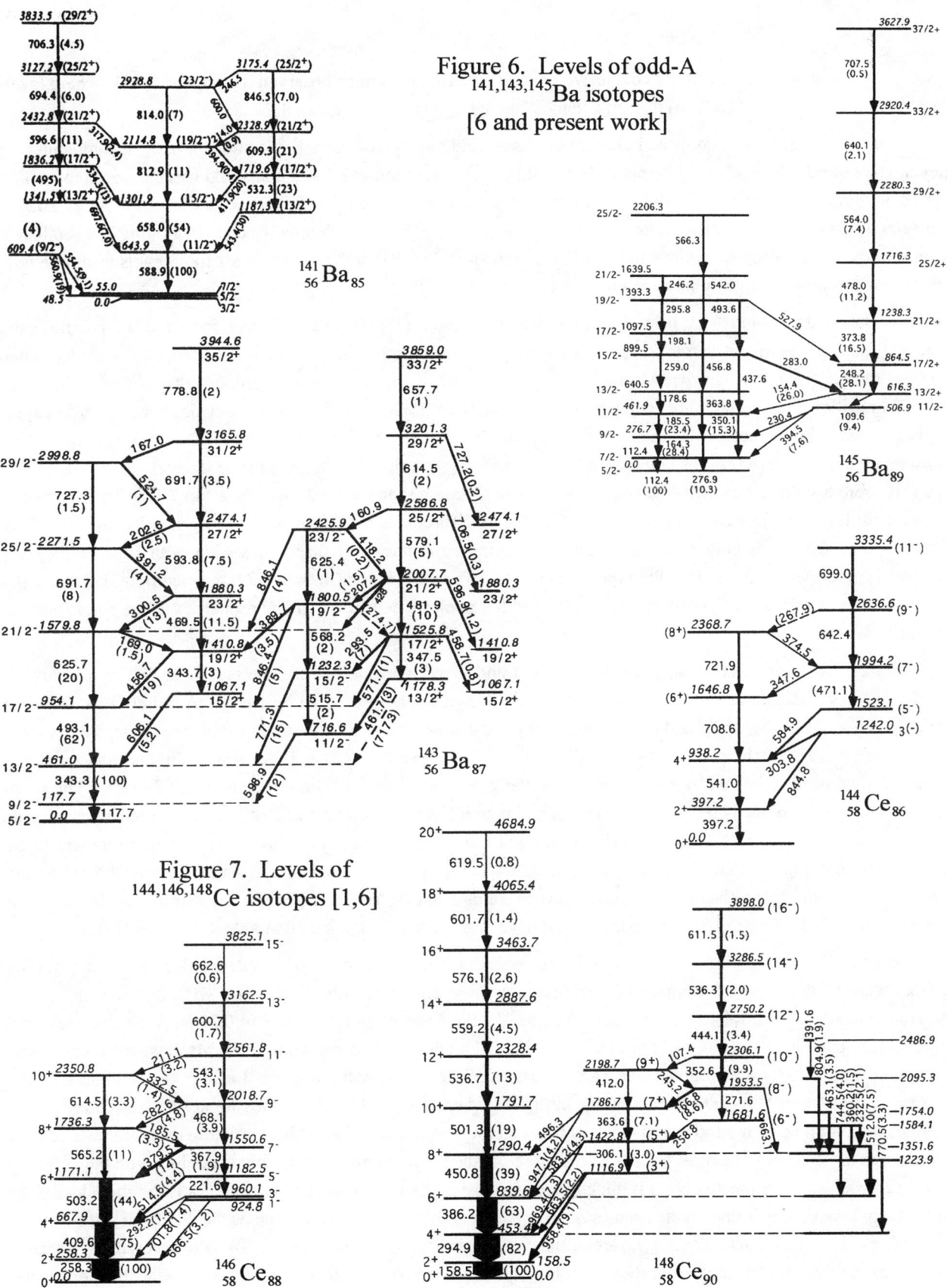

Figure 6. Levels of odd-A
^{141,143,145}Ba isotopes
[6 and present work]

Figure 7. Levels of
^{144,146,148}Ce isotopes [1,6]

479

stable octupole deformation was not conclusive up to the intermediate spins observed in that work [12], where levels to $31/2^-$, $17/2^+$, $23/2^+$, and $25/2^-$ were seen in bands labeled 1, 3, 4, 5, in our Fig. 8.

In ^{145}La, Fig. 8, we established eleven new levels and twenty-two new γ transitions mostly at high spin (13) beyond that of ref. 12. The total internal conversional coefficient of the 143.2 keV transition was measured to be 0.05 ± 0.05 in agreement with an E1 value of 0.08 but not with an M1 value of 0.40. This establishes the parity change between bands -4 and -5. Of particular interest is the discovery of a new high-spin band (2) beginning at 2186.0 keV with intertwined crossing transitions to band (1), as shown in Fig. 8. In ^{147}La, Fig. 8, six new levels and nine new transitions primarily at high spins were identified.

Bands -3 and -4 in ^{145}La, from the ground state up to the I^π (19/2$^+$) or (21/2$^+$) level, form a strong-coupled band with well-deformed symmetric rotor shape. This strong-coupled ground band in ^{145}La is assigned a $\pi g_{7/2}$ configuration coupled to the ^{144}Ba core. This band is not seen in ^{147}La. Band -4 shows a sharp upbend in J_1 beginning at the 23/2$^+$ → 19/2$^+$ transition, Fig. 9, that is not seen in band -5 to suggest the 19/2$^+$ (or 23/2$^+$) to 31/2$^+$ levels are not a continuation of band -4 even though observable E2 strength connects all states of what is called band -4. The large and remarkably constant eight B(E1)/B(E2) ratios, Fig. 10 for bands -4 and -5, from the 19/2$^+$ level to the 33/2$^-$ level indicate that there is a shift from a symmetric shape to a stable asymmetric octupole shape around spin (19/2$^+$) in ^{145}La. The average of these ratios, 1.3×10^{-6} fm^{-2}, is nearly twice those found in its ^{144}Ba core and in the octupole parity doublets in ^{143}Ba. This first evidence for bands built on a stable octupole asymmetric deformed shape coexisting with bands built on a symmetric deformed shape confirm the expectations of Nazarewicz [23] and Chasman [24] that octupole-deformed and reflection-symmetric shapes may occur in the same odd-A nucleus, depending on the properties of the single particle orbitals.

For bands -1 and -3 at low spin and for bands -1 and -2 at high spin in ^{145}La, one sees quite different behavior. The low B(E1)/B(E2) ratios for bands -1 and -3 indicate weaker octupole correlations, as noted earlier [12]. Our new band-2 seen at the higher spins and our extension of band-1 now give strong additional evidence for stable octupole deformation [9,13]. For bands -1 and -2 at higher spin in ^{145}La, the ratios start at a factor of six higher than for bands -1 and -3, then peak, followed by a sharp drop. Similar behavior is seen in ^{147}La for bands 1 and 2 there.. The sharp peaks in both cases are likely related to a strong reduction in E2 strength related to a change in structure between bands -2 and -3 in ^{145}La around 25/2$^+$ and a similar change at the same spin in ^{147}La rather than a sudden increase in E1 strength. Note in Fig. 9, the missing 25/2$^+$-21/2$^+$ transition is shown as a filled diamond, and a sharp backbend occurs there to likewise indicate a change in structure between bands 2 and 3. The B(E1)/B(E2) ratios of the two highest spin states in bands 1, 2 where the band crossing should have negligible effect, are comparable to those in ^{144}Ba.

The bands 1 and 2 in ^{145,147}la assigned s = +i are very similar in structure. These bands are also remarkably similar to those with stable octupole deformation in their neighboring even-even isotones ^{144}Ba and ^{146}Ba discussed above. For the ground band in ^{146}Ba, J_1 backbends at $\hbar\omega \sim 0.28$ MeV, which agrees with the value of 0.27 MeV for the backbend of the (11/2$^-$) band in ^{147}La (Fig. 8). The (11/2$^-$) band in ^{145}La does not backbend until 0.30 MeV, which is consistent with the upbend in ^{144}Ba at this same value. A comparison of the s = +i band levels in ^{145}La and ^{147}La with those of the N=88 and 90 isotones ^{144}Ba and ^{146}Ba is shown in Fig. 11 where the band head energies are normalized to zero. The energies of both the positive and negative bands between the isotones are remarkably similar to each other. Thus, the s = +i bands in ^{146}La and ^{147}La most probably originate from the single-particle $\pi h_{11/2}$ orbital coupling with the ^{144}Ba and ^{146}Ba even-even cores, respectively. It is quite interesting that at high spins, the B(E1)/B(E2) ratios in the odd-Z ^{145}La and ^{147}La nuclei are larger than in the even-even cores, ^{144}Ba and ^{146}Ba. Finally, one notes that the B(E1)/B(E2) ratios for the three highest spin state data in ^{147}La are similar to those in ^{145}La (in contrast to ref. 12). Moreover, the ^{147}La ratios are similar to those in ^{144}Ba and not to those in ^{146}Ba which are a factor of 200 smaller. This indicates the odd proton

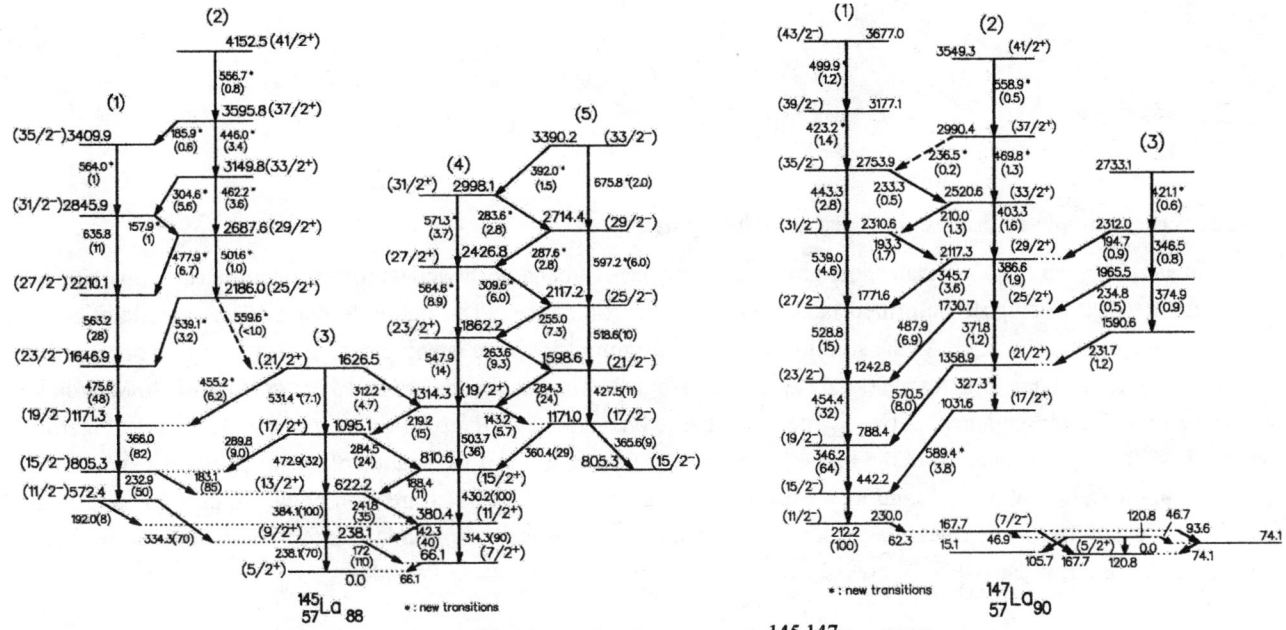

Figure 8. Level schemes of ^{145,147}La [13]

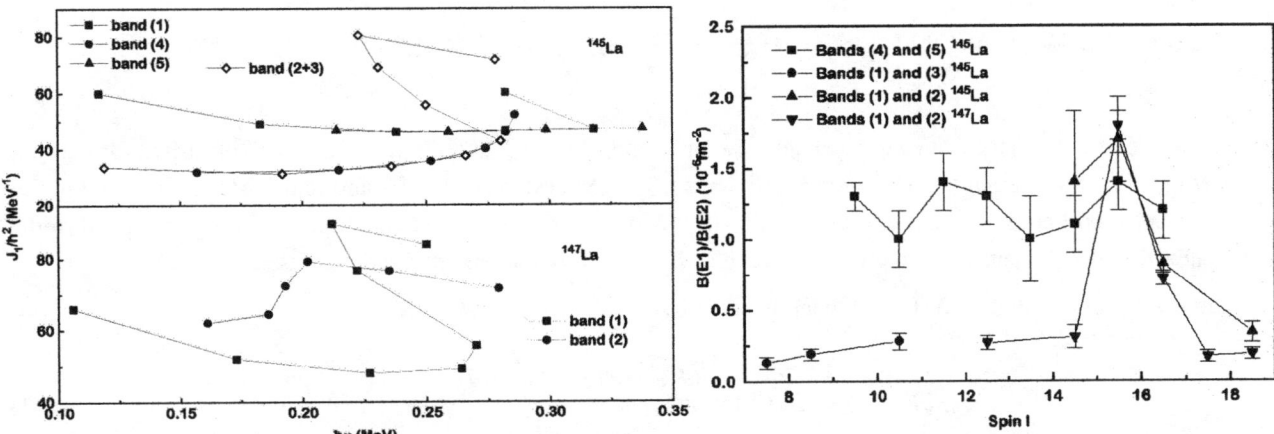

Figure 9. J_1 moments of inertia for levels in ^{145,147}La (as labeled in figure 8.)

Figure 10. B(E1)/B(E2) for ^{145,147}La [13]

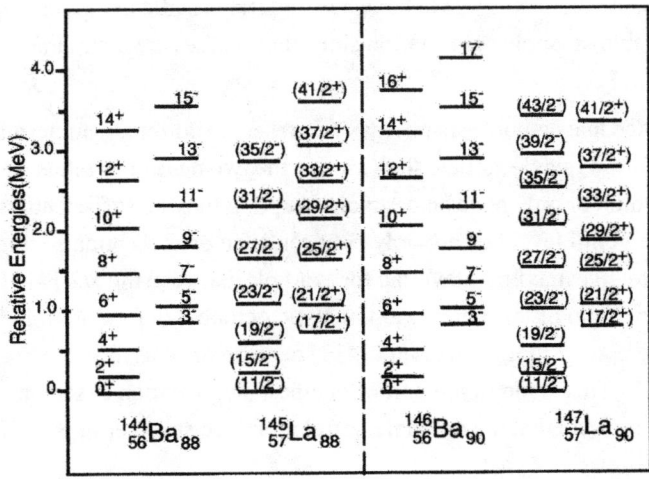

Figure 11. Comparison of the energies of the yrast and octupole bands in ^{144,146}Ba and those in bands 1,2 in ^{145,147}La [13] where the 11/2$^-$ energies are normalized to zero.

enhances the octupole deformation at least at higher spins.

In summary, in ^{145}La, a quadrupole ground state collective band with well-deformed symmetric shape coexists and competes with octupole deformed bands with asymmetric shape. The octupole correlations are stabilized at intermediate spins in ^{145,147}La. As the spin increases, the asymmetric shape is enhanced. This is the first observation for these structural characteristics in odd-Z nuclei in this region. The backbends observed at $\hbar\omega \sim 0.26$-0.30 MeV have been interpreted as the alignment of two $i_{13/2}$ neutrons based on cranked shell model calculations [9,13] Taken together, bands -1, -2, and -4, -5 (band -4 at the higher spins) likely form the set of two bands of parity doublets expected for stable octupole deformation as found in ^{143}Ba. At low spin, a 365.6 keV M1 transition between the 17/2$^-$ and 15/2$^-$ members of the two parity doublets was observed. Crossing transitions at the bottom of the two parity doublet structures are also observed in ^{143}Ba [6].

NEW REGION OF STABLE OCTUPOLE DEFORMATION IN ^{107,109}Mo

The levels of ^{107}Mo have been extended and new structures observed and the levels of ^{109}Mo identified for the first time (Fig. 12) [18]. Measured conversion coefficients in ^{107,109}Mo establish one E2 and four E1 transitions and parity changes between bands a, b and d, e in Fig. 12. In ^{107}Mo, parity doublets are reported with s = +i assigned to the sequence 7/2$^-$, 9/2$^+$, ... and s = -i to the sequence 5/2$^-$, 7/2$^+$, ... Only one intertwined band is seen in ^{109}Mo.

The level energy splitting, $\delta E(I)$ of a band is given by:

$$\delta E(I^{\pm}) = E(I^{\pm}) - \frac{(I + 1)E(I - 1)^{\mp} + IE(I + 1)^{\mp}}{2I + 1} \tag{1}$$

The superscripts $\pm$ indicate the parity of the levels and I_0 the ground state spin. For stable octupole deformation, $\delta E(I)$ should be close to zero. The parity doublet bands in ^{107,109}Mo show fluctuations along the $\delta E(I)=0$ line similar to those in ^{143}Ba (Fig. 13) but with smaller fluctuations to indicate that ^{107,109}Mo have a well defined static octupole shape. The B(E1)/B(E2) ratios in Table II likewise support stable octupole deformation since their values are comparable to those in ^{143,144}Ba and ^{145,147}La.

Nazarewicz [21] and Chasman [22] have noted that octupole shapes may be present in odd-mass nuclei and absent in the neighboring even- even systems. In the deformed single particle shell model, the two neutron orbitals $\nu d_{5/2}$-$\nu h_{11/2}$, lie near the Fermi surface for N = 64-67. Only proton or only neutron octupole coupling may be sufficiently strong to observe the effects of an octupole shape when the Fermi level lies not only between such orbitals but also the closest Nilsson levels near the Fermi level have $\Omega_1 = \Omega_2$, to give maximum overlap for example, between the 5/2$^+$[413] and 5/2$^-$ [532] neutron orbitals. It is proposed [18] that the origin of this new region of static octupole deformation in ^{107,109}Mo is related to the presence and strong overlap of the $h_{11/2}$5/2 [532] and $d_{5/2}$5/2 [413] neutron orbitals near the Fermi level with $\Omega_1 = \Omega_2$ for deformed nuclei with N = 64-67. This is the first report of octupole deformation based on only one pair of nucleon orbitals. The B(E1)/B(E2) ratios indicate that the octupole correlations are stronger in N = 67 than in N = 65.

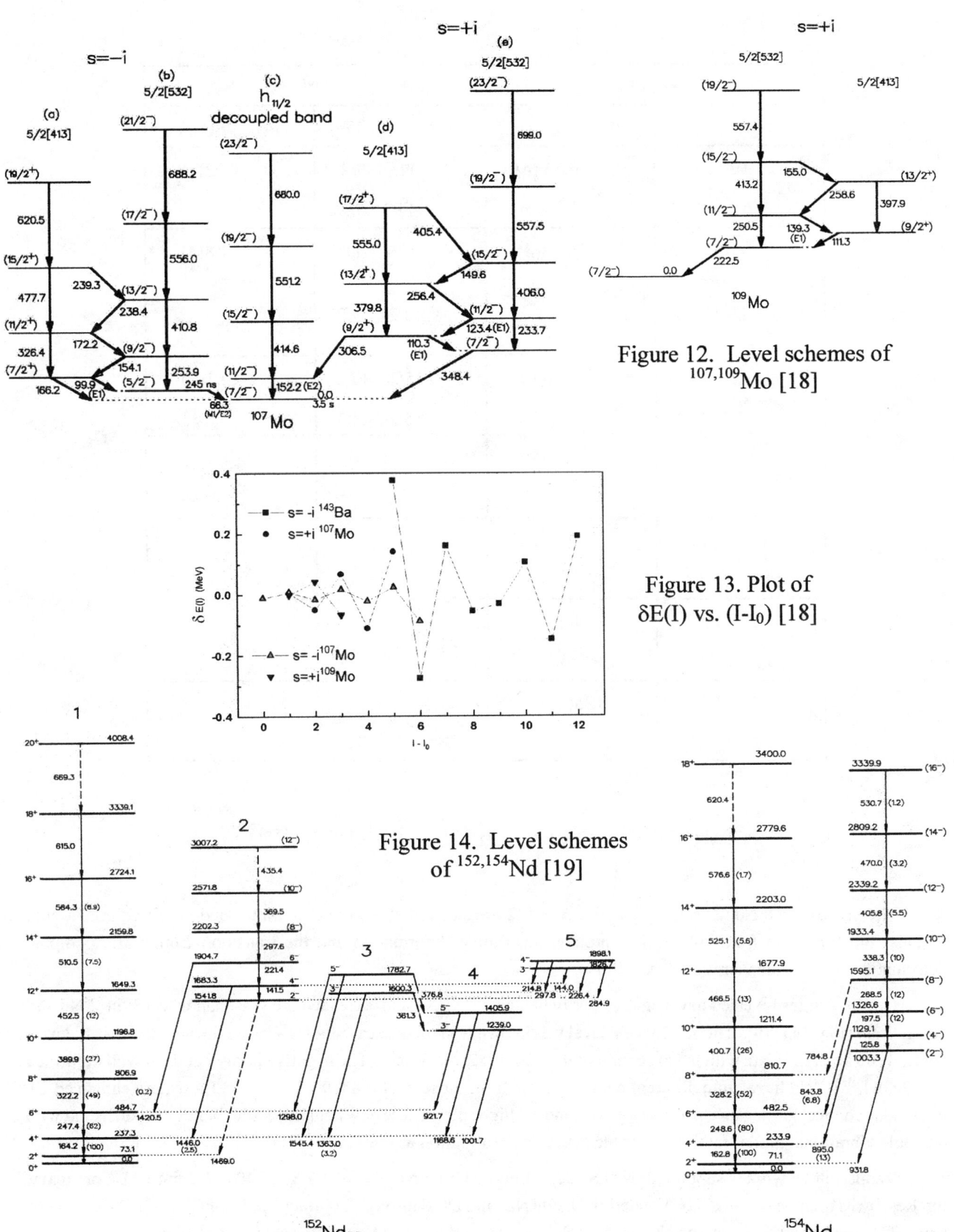

Figure 12. Level schemes of ^{107,109}Mo [18]

Figure 13. Plot of $\delta E(I)$ vs. $(I-I_0)$ [18]

Figure 14. Level schemes of ^{152,154}Nd [19]

483

Table II. B(E1)/B(E2) branching ratios in units 10^{-6}fm^{-2} ^{107,109}Mo.			
^{107}Mo $I^\pi \rightarrow I^\pi$	B(E1)/B(E2)	^{109}Mo $I^\pi \rightarrow I^\pi$	B(E1)/B(E2)
$9/2^-$ $7/2^+$ $9/2^-$ $5/2^-$	0.15 (4)	$11/2^-$ $9/2^+$ $11/2^-$ $7/2^-$	0.73(18)
$11/2^+$ $9/2^-$ $11/2^+$ $7/2^-$	0.28(8)	$13/2^+$ $11/2^-$ $13/2^+$ $9/2^+$	1.8(5)
$13/2^-$ $11/2^+$ $13/2^-$ $9/2^-$	0.4(1)	$15/2^-$ $13/2^+$ $15/2^-$ $11/2^-$	2.1(6)
$11/2^-$ $9/2^+$ $11/2^-$ $7/2^-$	0.29(9)		
$13/2^+$ $11/2^-$ $13/2^+$ $9/2^+$	0.14(4)		
$15/2^-$ $13/2^+$ $15/2^-$ $11/2^-$	0.4(1)		
$17/2^+$ $15/2^-$ $17/2^+$ $13/2^+$	0.35(10)		

NEW OCTUPOLE VIBRATIONAL BANDS IN ^{152,154}Nd

The neutron-rich Nd (Z=60) nuclei with A $\geq$ 142 are situated at the important intersection of two transitional regions: the transition from spherical to prolate quadrupole deformation and the transition from static octupole deformation to octupole vibrational excitations.

Our SF studies revealed new negative-parity bands in ^{152,154}Nd and extended the yrast bands to 18^+ in ^{154}Nd and 20^+ in ^{152}Nd (Fig. 14) [19]. The 1003.9 keV level was known from beta decay studies as were the negative parity levels below 1900 keV, in ^{152}Nd, with the exception of the new 1782.7 keV one. The negative parity bands in well deformed N = 92, 94, ^{152,154}Nd have quite different decay patterns from those in N = 88, 90, 92 ^{144,146,148}Ba where enhanced E1 transitions connect the negative parity and yrast bands to high spin. These new negative parity bands are consistent with octupole vibrational modes rather than stable octupole deformation as seen in $^{142-148}$Ba - ^{144,146}Ce.

Vanderbilt U. work is supported by U.S. Dept. Energy Grant No. DE-FG05-88ER40407, Joint Inst. For Heavy Ion Res. by its members, Univ. TN, Vanderbilt U., ORNL, and DOE through Contract No. DE-FG05-87ER40311 with Univ. TN, and Tsinghua U. by the Nat. Natural Sci. Found. and Science Found. for Nuc. Ind., China.

1. J.H. Hamilton *et al. Prog. Part. Nucl. Phys.* **35**, 635 (1995).

2. See papers in *Fission and Properties of Neutron Rich Nuclei*, J. H. Hamilton and A. V. Ramayya, eds., Singapore: World Scientific (1998).

3. W.R. Phillips *et al. Phys. Rev. Lett* **57**, 3257 (1986).

4. W.R. Phillips *et al. Phys. Lett* **B212**, 402 (1988).

5. S.J. Zhu *et al. Phys. Lett.* **B357**, 273 (1995).

6. J.H. Hamilton *et al. Prog. Part. Nucl. Phys.* **38**, 273 (1997).

7. S.J. Zhu *et al. Chin. Phys. Lett.* **14**, 569 (1997).

8. S. J. Zhu *et al.*, in *Fission and Properties of Neutron Rich Nuclei*, J. H. Hamilton and A. V. Ramayya, eds., Singapore: World Scientific (1998), p. 234.

9. J. H. Hamilton *et al.*, *Highlights of Modern Nuclear Structures*, A. Corvello, ed., Singapore: World Scientific (in press).

10. W. Urban *et al.*, *Nucl. Phys.* **A613**, 107 (1997).

11. M.A. Jones *et al.*, *Nucl. Phys.* **A605**, 133 (1996).

12. W. Urban *et al.*, *Phys. Rev.* **C54**, 945 (1997).

13. S.J. Zhu *et al.*, *Phys Rev.* (to be published).

14. W.Nazarewicz and S. Tabor, *Phys. Rev.* **C45**, 2226 (1992).

15. G.A. Leander *et al.*, *Nucl. Phys.* **A388**, 452 (1982).

16. W.Nazarewicz *et al.*, *Nucl. Phys.* **A429**, 269 (1984).

17. P.D. Cottle, *Phys. Rev.* **C42**, 1264 (1990).

18. J.K. Hwang *et al.*, *Phys. Rev.* **C56**, 1344 (1997).

19. X.Q. Zhang *et al.*, *Phys. Rev.* **57**, 2040 (1998).

20. P.A. Butler and W. Nazarewicz, *Nucl. Phys.* **A533**, 249 (1991).

21. H. Mach *et al. Phys. Lett.* **B230**, 21 (1990).

22. W. Nazarewicz and P. Olanders, *Nucl. Phys.* **A441**, 420 (1985).

23. W. Nazarewicz, *Nucl. Phys.* **A520**, 333c (1990).

24. R.R. Chasman, *Phys. Lett.* **96B**, 7 (1980).

Two-Phonon Octupole Excitations in ^{208}Pb: Status and Perspective

Steven W. Yates

Department of Chemistry, University of Kentucky, Lexington, Kentucky 40506-0055

Abstract: Progress toward the identification of the two-phonon octupole multiplet in ^{208}Pb is reviewed. Measurements with the ^{208}Pb(n,n′γ) reaction have permitted the characterization of a 0^+ state at 5241 keV, which decays by a cascade of E3 transitions. Additional measurements with this reaction have led to the identification of candidates for the 2^+ and 4^+ members of the quartet. Recent inelastic light-ion and neutron transfer reaction studies have led to further clarification of these candidates. In none of these studies or recent Coulomb excitation measurements has a reasonable candidate for the 6^+ excitation been identified. Quasiparticle-phonon model calculations of Ponomarev and von Neumann-Cosel seemingly account for all of the observed experimental data and add new information about the fragmentation of these excitations and the possible signatures for their decays.

INTRODUCTION

Vibrations are important in physical systems such as solids (phonons), plasmas (plasmons), and metallic clusters, but nuclei provide the smallest systems in which this fundamental mode can be examined in detail. One way in which vibrations in nuclei can be better understood is by the observation of multiphonon excitations. The question of the existence and collectivity of multiphonon excitations is important in understanding of the interplay of single-particle and collective degrees of freedom in nuclei and the effects of the Pauli principle.

Vibrations in Spherical Nuclei

While the role of vibrational excitations in nuclei has been a subject of study for many years, our knowledge of these modes remains incomplete. There are many examples in even-even nuclei near closed shells where the E_{4+}/E_{2+} ratio is near the harmonic value of two, but a closely spaced 0^+, 2^+, 4^+ triplet of two-phonon states is seldom observed. The Cd nuclei provide some of the best examples of vibrational nuclei in nature [1]. Evidence for three-phonon quadrupole states was sparse until the suggestion of a complete three-phonon vibrational quintuplet of levels in ^{118}Cd was made [2]. Recent studies [3-5] have focused on ^{112}Cd and ^{114}Cd, and the anharmonicities and ambiguities involved in describing the multiphonon structure in these nuclei. However, fundamental questions remain unanswered. How high in excitation do multiphonon states persist? To what degree do other excitations, such as intruder configurations (2p-4h excitations), affect the multiphonon structure?

Complete three-phonon multiplets have been suggested in several Cd nuclei, but until recently there existed no clear proof that the multiplet retains a collective structure. Measurements [3] on ^{112}Cd performed in our laboratory of the B(E2) values of the transitions decaying from the suggested 3^+ and 4^+ members of the 3-phonon quintuplet provided convincing evidence for collective enhancement, consistent with that expected for 3-phonon states. On the other hand, the proposed 2^+ member of the multiplet showed no enhancement in its decay. It was suggested that the interplay between the phonon, intruder, and mixed-symmetry degrees of freedom leads to such a mixed wave function that the decay pattern is severely perturbed. It was also found [6] that two-thirds of the expected B(M1) strength from the lowest 2^+ mixed-symmetry state was located in two states just above the 3-phonon 2^+ state, and that these states appeared to be mixed with intruder configurations. While an understanding of multiphonon quadrupole excitations is coming into focus, the role of higher-order vibrations remains obscure.

CP481, *Nuclear Structure 98*, edited by C. Baktash

Octupole Excitations in Spherical Nuclei

For about forty years, searches for a two-phonon octupole quartet of states (0^+, 2^+, 4^+, and 6^+) at approximately twice the energy of the octupole phonon have been conducted in spherical even-even nuclei, with no clear-cut identification of the members of this quartet emerging. On the other hand, individual stretch-coupled states of the two-phonon type have been established in ^{147}Gd [7] and in the $N = 84$ isotones ^{144}Nd [8], ^{146}Sm [8], and ^{148}Gd [9,10]. The identification of these states by the characteristic cascades of two enhanced E3 transitions was possible because they occur fortuitously as yrast states (or yrast traps) and lower multipolarity decays do not occur readily. However, because these states involve the coupling of particles to the two-phonon octupole excitation–*e.g.*, $\nu \otimes (3^- \otimes 3^-)_{6+}$ in ^{147}Gd and $\nu^2 \otimes (3^- \otimes 3^-)_{6+}$ in ^{148}Gd–their descriptions are not straightforward.

From an examination of the compilation on octupole transitions by Spear [11], three regions of nuclei emerge where multiphonon excitations might be expected. In the regions near ^{96}Zr (doubly closed subshell), ^{146}Gd (closed neutron shell), and ^{208}Pb (doubly closed shell) large E3 strengths, *i.e.*, B(E3;$3^- \rightarrow 0^+$) values greater than 30 W.u are observed. In addition, ^{146}Gd and ^{208}Pb are unique among even-even nuclei in that their first excited state is the 3^- octupole phonon. Detailed studies of ^{96}Zr [12] and ^{146}Gd [13,14] have yielded possible candidates for two-phonon excitations, but firm identifications have proven illusive. The octupole phonon in ^{146}Gd occurs more than an MeV lower than in ^{208}Pb and would seem to favor searches in this nucleus; however, the fact that ^{146}Gd is not stable limits the nuclear probes that can be used in its study. Nuclear structure arguments also appear to mitigate the possibility of observing the characteristic cascade of E3 transitions as lower multipolarity transitions are readily possible. On the other hand, the shell structure of ^{208}Pb leads to a paucity of low-spin states at low energy and, therefore, a seemingly greater probability that E3 transitions to the 3^- octupole phonon might be observed. For these reasons, ^{208}Pb has been the focus of many attempts to observe evidence of the two-phonon octupole quartet.

EXPERIMENTAL METHODS

A variety of probes and spectroscopic methods have been employed to search for two-phonon octupole vibrations. It would, of course, be desirable if these excitations could be populated in a selective reaction and that the detection process would exhibit great sensitivity. In a sense, the fusion-evaporation reactions used to populate the two-phonon octupole states in the region near ^{146}Gd are selective in that they populate these states that exist as yrast traps [7-10]. In general, however, the selectivity of reactions that might populate the quartet in an even-even nucleus is limited. Coulomb excitation has been employed recently using very heavy ions and particle and γ- ray detection arrays to search for these excitations in stable nuclei [15,16]. While this method shows great promise, it is currently limited in sensitivity to the highest-spin member(s) of the quartet.

At the University of Kentucky accelerator laboratory, we have employed the inelastic neutron scattering (INS) reaction, *i.e.*, (n,n'γ), in searching for two-phonon octupole excitations. This reaction offers significant advantages over other reactions.

- With monoenergetic, accelerator-produced neutrons (and no Coulomb barrier), levels can be examined close to the threshold for their excitation without the attendant complications associated with radiation from higher-lying levels.
- At low incident-neutron energies, this reaction is generally non-selective. It is limited by the amount of angular momentum that can be brought into the system by low-energy neutrons, but this is not a serious deficiency in low-spin studies.
- As γ rays are detected with HPGe detectors, the energy resolution is good.

On the other hand, the (n,n'γ) reaction is practically limited to stable nuclei and large amounts of material, frequently enriched isotopic samples, are required.

The accurate determination of the yield threshold for a particular γ ray is used to uniquely place the level from which the transition arises, and these excitation functions of γ-ray yields (cross sections) are also useful for inferring level spins and parities. The INS reaction leads to an alignment of the excited nuclei, so the γ-ray angular distributions from the decays of the excited levels exhibit anisotropies reflecting this alignment, the spins of the levels, and the multipolarities of the transitions.

A very exciting lifetime regime can be investigated in (n,n'γ) measurements by employing the Doppler-shift attenuation method (DSAM). While the recoil velocity imparted (v/c $\leq$ 0.001) in neutron scattering reactions on heavy nuclei is small, it is sufficient to produce measurable Doppler shifts, and lifetimes in the fs to ps range can be determined. The

Doppler-shifted γ-ray energy, $E_\gamma(\theta)$, measured at a detector angle of θ with respect to the incident neutrons can be related to E_γ, the energy of the γ ray emitted by a nucleus at rest, by the expression,

$$E_\gamma(\theta) = E_\gamma [1 + F_{exp}(\tau) v_{cm} \cos \theta /c] \tag{1}$$

where v_{cm} is the velocity of the center of mass in the inelastic neutron scattering collision with the nucleus, and c is the speed of light. $F_{exp}(\tau)$ is the experimental attenuation factor determined from the measured Doppler shift and is compared with calculated attenuation factors to determine the lifetime. The techniques for measuring nuclear lifetimes using DSAM methods with the $(n,n'\gamma)$ reaction have been developed in recent years [17]. Relative uncertainties of γ-ray energies can be determined to < 10 eV, so the largest source of uncertainty now resides in our lack of knowledge of the stopping powers of the recoiling residual nuclei.

For several years, we have performed γ-γ coincidence measurements using "beams" of fast neutrons produced with a forced-reflection collimator developed in our laboratories. These measurements were initiated to permit the isotopic identification of coincident γ rays in experiments with samples of natural isotopic abundance; however, they proved so successful in providing new, unambiguous information that we now apply this method with isotopically enriched samples as well. The current four-detector "array" consists of the HPGe detectors in a close geometrical arrangement located as near as possible to the sample which is irradiated with collimated neutrons; the scattering sample and detector array are less than one meter from the source of neutrons. Much of this new methodology has been described in a manuscript providing details of our experimental facility [18].

SEARCH FOR TWO-PHONON OCTUPOLE EXCITATIONS IN ^{208}PB

The first excited state of ^{208}Pb at 2614 keV has spin-parity 3^- and has been interpreted as a surface vibration of octupole character; its collectivity is confirmed by the observed $B(E3;3^- \rightarrow 0^+)$ value of 34 W.u. [11]. The quartet of two-phonon octupole states (0^+, 2^+, 4^+, and 6^+) is expected to occur with small anharmonicities at about 5.2 MeV [19]. A number of searches for these excitations in ^{208}Pb have been conducted in recent years using neutron capture reactions [20,21], inelastic light-ion scattering and transfer reactions [22,23], and Coulomb excitation [15,16,24-27], but none of the suggested candidates for these states have survived detailed scrutiny. Only recently have we provided firm evidence [28], in the form of an observed cascade of two E3 transitions from a 0^+ state at 5241 keV, for the lowest-spin member of the two-phonon octupole quartet in ^{208}Pb. This 0^+ (2-phonon) $\rightarrow 3^-$ (1-phonon) $\rightarrow 0^+$ (ground state) cascade exhibits the characteristic signature of a two-phonon octupole excitation, and the absence of other decay branches is suggestive of the collective nature of the 2626-keV $0^+ \rightarrow 3^-$ transition.

While the identification of the E3 transition from the 5241-keV 0^+ state in ^{208}Pb is the best evidence to date for a two-phonon octupole excitation built on the ground state of an even-even nucleus (and the first for such an excitation in a nucleus outside the N = 82 region), crucial transition rate data are lacking as the lifetime of this state is only known to be greater than 1 ps [28]. To address this shortcoming, we have proposed to measure the lifetime of the 5241-keV 0^+ state, to determine directly the multipolarity of the 2626-keV transition, and to measure the E3/E0 branching ratio in experiments at the Accelerator Laboratory at the University of Jyväskylä, Finland.

The location of the other members (2^+, 4^+, and 6^+) of the two-phonon octupole quartet in ^{208}Pb remains an important question in the description of these vibrational excitations. A unique signature for identifying members of the two-phonon octupole quartet would be the observation of enhanced E3 decays to the 3^- one-phonon state, with B(E3) values similar in magnitude to that of the $3^- \rightarrow 0^+$ decay. However, other lower-multipolarity decays are possible, thus making the observation of a two-phonon to one-phonon E3 transition, as observed [28] for the $0^+ \rightarrow 3^-$ E3 decay, unlikely, and other identifying signatures of two-phonon octupole states must be sought.

One possibility is that these states may decay by fast E1 transitions, as has been observed for quadrupole-octupole coupled states in the N = 82 region [29-34]. For example, the 2^+ member of the two-phonon octupole quartet of ^{208}Pb is expected to decay by an E1 transition to the first 3^- excited state, the octupole phonon at 2614 keV, by what will be referred to as an "octupole E1 transition"–$i.e.$, an E1 transition accompanying the destruction of an octupole phonon. Similarly, the 4^+ multiplet member could decay by E1 transitions to either the 3^- octupole excitation at 2614 keV or the 5^- state at 3197 keV; although, the $4^+ \rightarrow 3^-$ transition would seem more favorable, particularly in view of the energy factor and the "octupole E1" nature of the decay transition. For the spin 6^+ state, an E1 transition to the 5^- state at 3197 keV seems most likely. It is evident that knowledge of the relevant E1 transition rates could play an important role in characterizing the two-phonon octupole states.

E1 Transitions in the [208]Pb Region

The decay of two-phonon excitations in [208]Pb might be evinced by comparing the observed E1 transitions rates with those of similar "octupole E1 transitions" in neighboring odd-A nuclei. We have taken this approach in suggesting "candidates" for the 2^+ and 4^+ members of the two-phonon octupole quartet in [208]Pb [35]. Low-lying collective octupole excitations in odd-A nuclei near [208]Pb arise from particle-vibration coupling of single-particle (or hole) configurations to the 3^- octupole vibration of the [208]Pb core. Certainly, the most spectacular example of weak coupling in this region is the well-known $3^- \otimes h_{9/2}$ septuplet of [209]Bi. The members of the septuplet decay by the expected enhanced E3 transitions, with strengths comparable to the E3 decay of the octupole vibration of [208]Pb, and E1 transitions to lower-lying single-particle states. The absolute transition rates of these E1 decays are known, but because these E1 transitions are not strong, they are very sensitive to small admixtures in their wave functions.

To obtain additional information about "octupole E1 transition" rates, we initiated a study of [207]Pb with the $(n,n'\gamma)$ reaction [36]. The lifetimes of the $5/2^+$ and $7/2^+$ components of the $3^- \otimes p_{1/2}^{-1}$ doublet of [207]Pb are known, but E1 decay to the $p_{1/2}^{-1}$ state is forbidden. We, therefore, sought to identify the $3^- \otimes f_{5/2}^{-1}$ sextuplet of [207]Pb and examine the E1 decays from this multiplet to the $f_{5/2}^{-1}$ excitation, the first excited state. Because these E1 decays in [207]Pb and the aforementioned decays in [209]Bi represent the destruction of an octupole phonon, their transition rates might be expected to be similar to octupole phonon E1 transitions in [208]Pb.

Based on energy and E1 transition rate arguments, we suggested candidates for the 2^+ and 4^+ members of the two-phonon quartet in [208]Pb. Recently, however, Ponomarev and von Neumann-Cosel [37] have performed quasiparticle-phonon model (QPM) calculations of E1 transition rates in [208]Pb. They find that, while the calculated E1 transition strengths are of the same order of magnitude as those that we have measured for the proposed two-phonon to one-phonon transitions, other comparable E1 transition rates are quite common. Therefore, the process of identifying multiphonon octupole excitations based on E1 transition rates is called into question. Clearly, our knowledge of E1 transition rates in this region should be expanded if E1 transitions are to be used to deduce information about two-phonon octupole excitations.

Fragmentation of the Two-Phonon Octupole Strength

The purity of the two-phonon octupole states is a question of considerable importance, as fragmentation of their strength through mixing with other excitations would dilute the distinct qualities of these excitations and make their identification more complicated, if not impossible. Such fragmentation of the 6^+ member of the quartet has been suggested as an explanation of the absence of population of candidates through Coulomb excitation [16]. Credence to this argument has been provided by the results of recent calculations with the quasiparticle-phonon model which indicate that the 0^+ and 4^+ members of the $3^- \otimes 3^-$ quartet remain practically pure while the 2^+ and 6^+ states are more fragmented [37].

According to Ponomarev and von Neumann-Cosel [37], the purity of the 0^+ state is understood easily because many other 0^+ excitations with which it might mix are greatly removed in energy; the neutron and proton pairing vibrations are possible exceptions. The fragmentation of the 2^+ and 4^+ states depends on the magnitude of the interaction matrix elements between these and other nearby states–*i.e.*, they are small in the case of the 4^+ state but larger for the 2^+ excitation. Pauli corrections are predicted to move the 6^+ state to higher energies in a region of larger density of other configurations. As the interaction matrix elements are larger, it mixes with a large number of other configurations and its strength is fragmented over an energy region of about 2 MeV.

Status and Perspective

While candidates for the members of the two-phonon octupole states is [208]Pb have been nominated in various studies, none of these has yet been successfully elected–*i.e.*, the characteristic cascade of collective E3 transitions observed in nuclei near [146]Gd has not been observed–and they are not anticipated in the decays of the higher-spin members of the multiplet. Moreover, it has been suggested that the expected E1 decays will not provide distinctive signatures of these states [37]. It is now clear that additional measurements will be required to answer the long-standing questions about the existence of two-phonon octupole excitations in [208]Pb. Indeed, a complete characterization of the states below 6 MeV in [208]Pb may be necessary. On the other hand, considerable progress has been made in attempts to characterize these excitations, to confirm the picture of octupole vibrations in spherical nuclei, and to understand the fragmentation of these excitations. A brief summary of the status of the possible members of the two-phonon octupole quartet in ^{208}Pb is presented below and in Fig. 1.

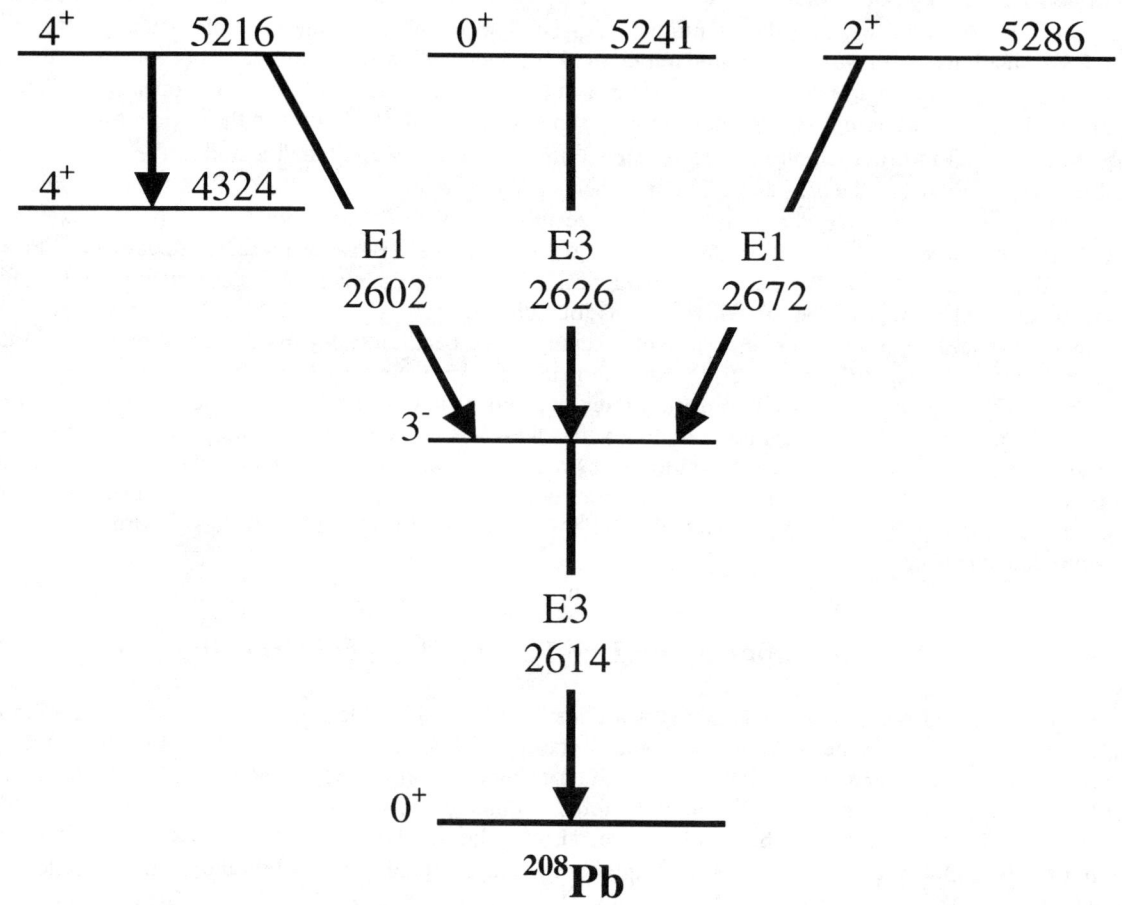

FIGURE 1. Simplified diagram of decays of two-phonon octupole candidates in ^{208}Pb.

The 0$^+$ Member of the Two-Phonon Octupole Quartet of ^{208}Pb

The evidence for the 5241-keV state as the lowest-spin member of the quartet is the strongest for any of the candidates [28]. The crucial information, the B(E3) of the 0$^+$ → 3$^-$ transition, remains to be measured; however, experiments designed to determine the lifetime of the 5241-keV state are in progress. A finding that this transition is not collective would, of course, raise serious questions about the existence of two-phonon octupole states in ^{208}Pb.

The 2$^+$ Member

While the state at 5286-keV has been suggested as a likely candidate for this excitation, the case is hardly compelling. Following the suggestion by Yeh *et al.* [35] that a state at this energy with spin-parity of (2,4)$^+$ might be the sought excitation, Graw *et al.* [38] determined the spin to be 2$^+$ in support of this conjecture. The 2$^+$ state at 5561-

keV [35] could represent the anticipated fragmentation of the 2^+ octupole strength, but no additional 2^+ candidates have been identified.

The 4^+ Member

The state at 5216-keV, very close to twice the one-phonon energy, emerges as the strongest candidate for the 4^+ member of the quartet [35]. According to the calculations of Ponomarev and von Neumann-Cosel [37], little fragmentation of this state is anticipated. Additional 4^+ states in this energy region occur at 5194 and 5690 keV, but their properties do not suggest that they are of two-phonon origin.

The 6^+ Member

While no clear candidate for the highest-spin member of the quartet is currently available, it may be that the prospects for locating the 6^+ two-phonon octupole strength is greatest. In the Coulomb excitation measurements of Vetter *et al.* [16], only the 4424-keV state was populated with approximately 20% of the total expected E3 strength for the two-phonon octupole 6^+ excitation, and only limits could be placed on the population of higher-lying candidates. New Coulomb excitation measurements should provide significantly greater sensitivity than previous studies and hold the promise of observing this strength, even if it is as fragmented as predicted [37].

ACKNOWLEDGMENTS

The author wishes to acknowledge the contributions of Minfang Yeh, M. Kadi, P. E. Garrett, C. A. McGrath, and T. Belgya and his other colleagues at the University of Kentucky. Of particular note are valuable discussions with H. Amro, D. Cline, G. Graw, K. Heyde, R. Janssens, R. Julin, K. H. Maier, E. F. Moore, W. Nazarewicz, V. Ponomarev, B. D. Valnion, K. Vetter, and P. von Neumann-Cosel. This work was supported by the U. S. National Science Foundation under Grant No. PHY-9803784.

REFERENCES

1. Kern, J., Garrett, P. E., Jolie, J., and Lehmann, H., Nucl. Phys. **A593**, 21 (1995).
2. Aprahamian, A., Brenner, D. S., Casten, R. F., Gill, R. L., and Piotrowski, A., Phys. Rev. Lett. **59**, 535 (1987).
3. Lehmann, H., Garrett, P. E., Jolie, J., McGrath, C. A., Yeh, M., and Yates, S. W., Phys. Lett. B **387**, 259 (1996).
4. Casten, R. F., Jolie, J., Börner, H. G., Brenner, D. S., Zamfir, N. V., Chou, W. T., and Aprahamian, A., Phys. Lett. B **297**, 19 (1992).
5. Délèze, M., Drissi, S., Jolie, J., Kern, J., and Vorlet, J. P., Nucl. Phys. **A554**, 1 (1993).
6. Garrett, P. E., Lehmann, H., McGrath, C. A., Yeh, M., and Yates, S. W., Phys. Rev. C **54**, 2259 (1996).
7. Kleinheinz, P., Styczen, J., Piiparinen, M., Blomqvist, J., and Kortelahti, M., Phys. Rev. Lett. **48**, 1457 (1982).
8. Bargioni, L., Bizzetti, P. G., Bizzetti-Sona, A. M., Bazzacco, D., Lunardi, S., Pavan, P., Rossi-Alvarez, C., de Angelis, G., Maron, G., and Rico, J., Phys. Rev. C **51**, R1057 (1995).
9. Lunardi, S., Kleinheinz, P., Piiparinen, M., Ogawa, M., Lach, M., and Blomqvist, J., Phys. Rev. Lett. **53**, 1531 (1984).
10. Piiparinen, M., Kleinheinz, P., Blomqvist, J., Virtanen, A., Ataç, A., Müller, D., Nyberg, J., Ramsøy, T., and Sletten, G., Phys. Rev. Lett. **70**, 150 (1993).
11. Spear, R. H., At. Data Nucl. Tables **42**, 55 (1989).
12. Molnár, G., Belgya, T., Fazekas, B., Veres, A., Yates, S. W., Kleppinger, E. W., Gatenby, R. A., Julin, R., Kumpulainen, J., Passoja, A., and Verho, E., Nucl. Phys. **A500**, 43 (1989).
13. Yates, S. W., Julin, R., Kleinheinz, P., Rubio, B., Mann, L. G., Henry, E. A., Stoeffl, W., Decman, D. J., and Blomqvist, J., Z. Phys. **A234**, 417 (1986).
14. Yates, S. W., Mann, L. G., Henry, E. A., Decman, D. J., Meyer, R. A., Estep, R. J., Julin, R., Passoja, A., Kantele, J., and Trzaska, W., Phys. Rev. C **36**, 2143 (1987).
15. Vetter, K., *et al.*, Phys. Rev. C **56**, 2316 (1997).
16. Vetter, K., *et al.*, (to be published).
17. Belgya, T., Molnár, G., and Yates, S. W., Nucl. Phys. **A607**, 43 (1996).
18. McGrath, C. A., Villani, M. F., Garrett, P. E., and Yates, S. W., Nucl. Instrum. Methods, (to be published).
19. Blomqvist, J., Phys. Lett. **33B**, 541 (1970).

20. Mariscotti, M. A. J., Bes, D. R., Reich, S. L., Sofia, H. M., Hungerford, P., Kerr, S. A., Schreckenbach, K., Warner, D. D., Davidson, W. F., and Gelletly, W., Nucl. Phys. **A407**, 98 (1983).

21. Belgya, T., Kasztovszky, Zs., Revay, Zs., Molnár, G., Yeh, M., Garrett, P. E., and Yates, S. W., Phys. Rev. C **57**, 2740 (1998).

22. Julin, R., Kantele, J., Kumpulainen, J., Luontama, M., Passoja, A., Trzaska, W., Verho, E., and Blomqvist, J., Phys. Rev. C **36**, 1129 (1987).

23. Valnion, B. D., Oelmaier, W., Hofer, D., Zanotti-Müller, E., and Graw, G., Z. Phys. **A350**, 11 (1994).

24. Wollersheim, H. J., *et al.*, Z. Phys. **A341**, 137 (1992).

25. Schramm, M., Grawe, H., Heese, J., Kluge, H., Maier, K. H., Schubart, R., Broda, R., Grebosz, J., Krolas, W., Maj, A., and Blomqvist, J., Z. Phys. **A344**, 121 (1992).

26. Fahlander, C., *et. al.*, Phys. Scr. **T56**, 243 (1995).

27. Moore, E. F., *et al.*, Nucl. Instrum. Methods Phys. Res., Sect. B **99**, 308 (1995).

28. Yeh, M., Garrett, P. E., McGrath, C. A., Yates, S. W., and Belgya, T., Phys. Rev. Lett. **76**, 1208 (1996).

29. Belgya, T., Gatenby, R. A., Baum, E. M., Johnson, E. L., DiPrete, D. P., Yates, S. W., Fazekas, B., and Molnár, G., Phys. Rev. C **52**, R2314 (1995).

30. Gatenby, R. A., Vanhoy, J. R., Baum, E. M., Johnson, E. L., Yates, S. W., Belgya, T., Fazekas, B., Veres, Á., and Molnár, G., Phys. Rev. C **41**, R414 (1990).

31. Gatenby, R. A., Johnson, E. L., Baum, E. M., Yates, S. W., Wang, D. W., Vanhoy, J. R., McEllistrem, M. T., Belgya, T., Fazekas, B., and Molnár, G., Nucl. Phys. **A560**, 633 (1993).

32. Wilhelm, M., Radermacher, E., Zilges, A., and von Brentano, P., Phys. Rev. C **54**, R449 (1996).

33. Wilhelm, M., Kasemann, S., Pascovici, G., Radermacher, E., von Brentano, P., and Zilges, A., Phys. Rev. C **54**, R449 (1996).

34. Fransen, C., Beck. O., von Brentano, P., Eckert, T., Herzberg, R.-D., Kneissl, U., Maser, H., Nord, A., Pietralla, N., Pitz, H. H., and Zilges, A., Phys. Rev. C **57**, 129 (1998).

35. Yeh, M., Kadi, M., Garrett, P. E., McGrath, C. A., Yates, S. W., and Belgya, T., Phys. Rev. C **57**, R2085 (1998).

36. Kadi, M., doctoral dissertation, University of Kentucky, 1998.

37. Ponomarev, V. Yu., and von Neumann-Cosel, P., (to be published).

38. Graw, G., Valnion, B. D., Eisermann, Y., Gollwitzer, A., Hertenberger, R., Metz, A., and Schiemenz, P., "Light ion induced excitations in ^{208}Pb and double octupole multiplett states," presented at the 6th International Spring Seminar on Nuclear Physics; Highlights of Modern Nuclear Structure, S. Agatha sui due Golfi, May 18-22, 1998.

Double and triple octupole excitations in the $A \approx 150$ region

P.G. Bizzeti[1], A.M. Bizzeti–Sona[1], L. Bargioni[1], S. Lunardi[2], Zs. Podolyák[3],
F. Banci Buonamici[1], D. Bazzacco[2], G. de Angelis[3], M. De Poli[3], A. Dewald[4],
D.R. Kasemann[4], T. Klemme[4], T. Klug[4], R. Krücken[4], G. Maron[3], D.R. Napoli[3],
P. Pavan[2], C.M. Petrache[2], R. Peusquens[4], C. Rossi Alvarez[2], H. Tiesler[4], and L.H. Zhu[3]

[1] *Università di Firenze and I.N.F.N., Sezione di Firenze – Largo E.Fermi 2, Firenze, Italy*
[2] *Università di Padova and I.N.F.N., Sezione di Padova – Via Marzolo 8, Padova, Italy*
[3] *Laboratori Nazionali di Legnaro, I.N.F.N. – Via Romea 4, Legnaro (Pd), Italy*
[4] *Institut für Kernphysik, Universität zu Köln – Zülpiker Strasse 77, Köln, Germany*

Abstract. The results of recent experiments performed at L.N.L. on multiple octupole excitations in the $N = 84$ isotones are reviewed. In particular, the results on $B(E3)$ strengths in ^{146}Sm are discussed and expectations for higher-order octupole excitations are compared with the present experimental situation.

I INTRODUCTION

Collective nuclear excitations of "vibrational" nature are known to result from the coherent contribution of several particle–hole excitations of equal tensor rank. Such coherence persists, at least to certain extent, when the collective excitation of the core is coupled to single–particle degrees of freedom or to a different kind of collective excitation. It is therefore interesting to test how far the vibrational spectrum would extend when several identical phonons are coupled to each other. While many examples of multi–phonon excitations of the nuclear quadrupole vibrator type have been found, very little is known in the case of collective octupole excitations.

In a simple geometrical model [1], quadrupole and octupole excitations naturally arise as the lower modes of surface vibration. In this case, the characteristic frequency of the octupole mode (or, if we prefer, the energy of the octupole phonon) is almost a factor of two larger than that of the quadrupole mode – what is not necessarily true in real nuclei, and is certainly not true in our region (fig. 1). It can be useful, however, to examine first an example of nuclear *quadrupole* vibrations, which are by far more common and easier to study than the octupole ones.

Fig. 2 shows the level scheme of ^{98}Ru, that can be considered to be a "good" quadrupole oscillator. One can observe the complete 2-phonon triplet and at least the higher–spin members of the 3-phonon quintuplet, while for the highest–spin state of the 4-phonon multiplet, the collective strength seems to be split among at least two levels. There are, however, several discrepancies with respect to the spectrum of an ideal harmonic oscillator. Energies and strengths do not follow exactly the harmonic-oscillator rules, and a few "forbidden" transitions are observed. In this particular case, most of the apparent anharmonicities can be accounted for in the frame of the more comprehensive IBA2 model [3,4] (already discussed in this conference by C. de Coster and by P. von Brentano).

In the case of an octupole oscillator, one should expect, therefore, similar – or even larger – anharmonicities (to be interpreted, as far as possible, in the frame of a more comprehensive model).

A glance to the systematics of $N = 82$ and $N = 84$ nuclei, shown in fig. 1 is sufficient to realize that particle–hole excitations of the $Z = 64$ proton core must be very effective in producing collective octupole excitations, as shown by the low excitation energy (1579 keV) of the lowest 3^- state of ^{146}Gd and by its large transition strength towards the ground state ($B(E3, 3^- \rightarrow 0^+) \approx 52$ W.u.) [6].

CP481, *Nuclear Structure 98*, edited by C. Baktash

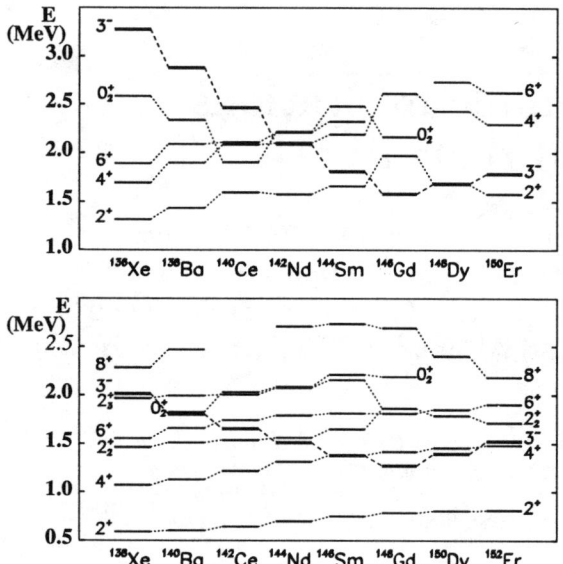

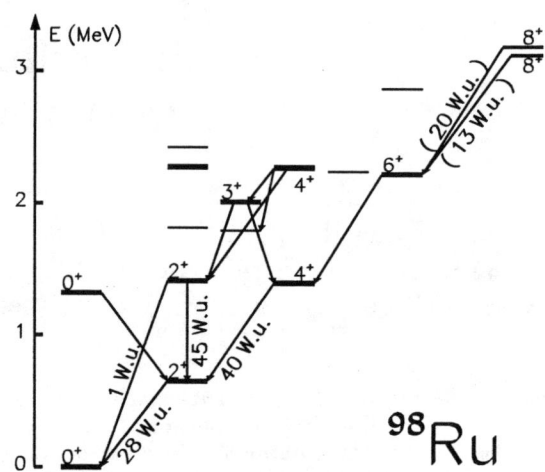

FIGURE 1. Systematics of low-lying excitations in the $N = 82$ and $N = 84$ isotones. Note the minima in the excitation energy of the lowest 3^- state in correspondence to the sub-shell closure at $Z = 64$ (Gd). A more extensive discussion of this systematics can be found in ref. [2].

FIGURE 2. Low lying levels of ^{98}Ru. Thick lines correspond to levels interpreted as having "full symmetry" in IBA2, thin lines to "mixed-symmetry" levels. Whenever the E2 strengths are known, their value (in W.u.) is indicated. Mixing of two 8^+ states with F-spin equal to F_{max} and $F_{max} - 2$ results in the two observed 8^+ levels, whose *calculated* decay strengths are shown. (From ref. [5], by courtesy of Società Italiana di Fisica).

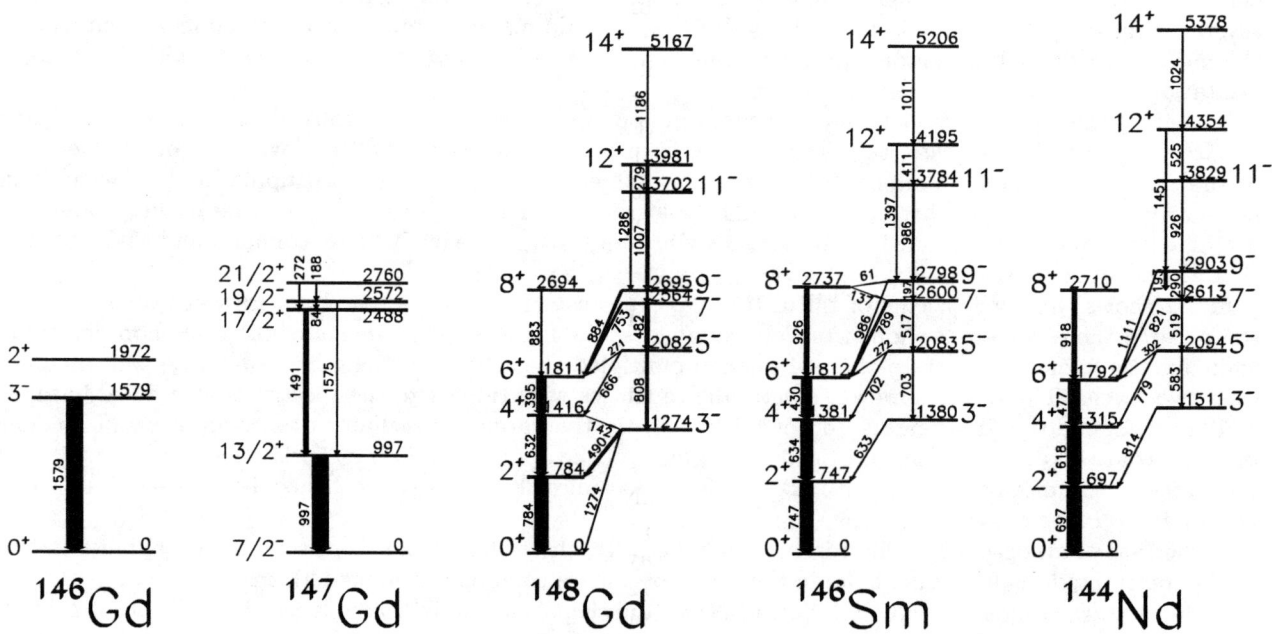

FIGURE 3. Enhanced E3 transitions ($3^- \to 0^+$, $\frac{19}{2}^- \to \frac{13}{2}^+ \to \frac{7}{2}^-$, $12^+ \to 9^- \to 6^+$) in ^{146,147,148}Gd (ref. [6–11]), ^{146}Sm, and ^{146}Nd [12,13]. Note the strict similarity of the $12^+ \to 9^- \to 6^+$ cascades in the three $N = 84$ isotones.

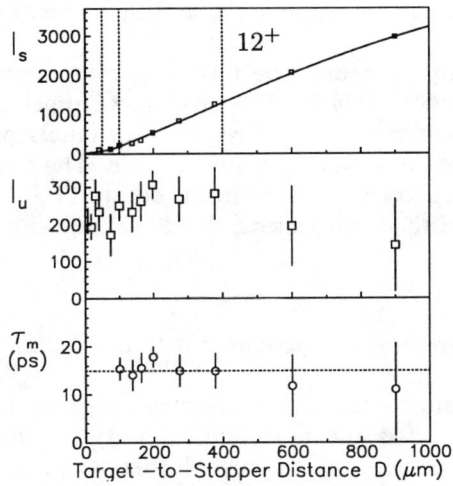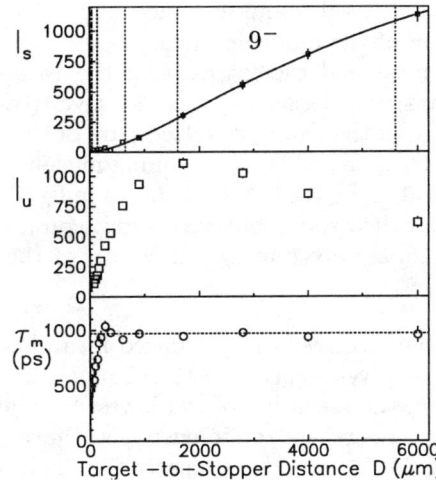

FIGURE 4. Intensity corresponding to the decay in flight (I_s) and at rest (I_u) of the 12^+ level and of the 9^- level of ^{146}Sm and deduced values of the mean life, $\tau_m = I_u/(v\,dI_s/dD)$, as a function of the target–to–stopper distance D. The horizontal lines correspond to the adopted value of the mean life, i.e. 15 ± 2 ps for the 12^+ level and 0.97 ± 0.05 ns for the 9^- level. (From ref. [5], by courtesy of Società Italiana di Fisica).

When a valence $2f_{7/2}$ neutron is coupled to the ^{146}Gd core, one obtains a weak–coupling multiplet, whose highest spin member ($J^\pi = \frac{13}{2}^+$) decays to the ground state with an enhanced E3 transition (44 W.u.) [7]. Already in 1982, a further E3 transition (of 57 W.u.) was found [8] between a higher-lying $J^\pi = \frac{19}{2}^-$ state and the $J^\pi = \frac{13}{2}^+$ member of the multiplet. The $\frac{19}{2}^-$ state was interpreted as resulting from the stretched coupling of two octupole phonons with the valence $f_{7/2}$ neutron – and this was the first example of a double octupole excitation.

Soon after, a similar structure was identified in ^{148}Gd [9], where, however, the cascade of two E3 transitions is *not* built upon the ground state, but upon the highest–spin member ($J^\pi = 6^+$) of the $(\nu f_{7/2})^2$ configuration (fig. 3). A more recent measurement of the E3 strength for the first transition in the $12^+ \rightarrow 9^- \rightarrow 6^+$ cascade definitely confirmed its collective nature and the *double–octupole* character of the 12^+ state [10].

In a series of measurements performed at Laboratori Nazionali di Legnaro, we have tried to extend the investigation of the collective octupole excitations observed in ^{148}Gd, *vertically* to higher–lying states and *horizontally* to neighboring nuclei. In the two lighter isotones, ^{146}Sm and ^{144}Nd, we have identified E3 cascades with patterns very similar to that of ^{148}Gd, and found that at least the $9^- \rightarrow 6^+$ transitions have very similar strengths in the three nuclei. These results are published [12,13] and we shall not discuss them in detail. Instead, we will spend some more words on the measurements concerning the strength of the $12^+ \rightarrow 9^-$ transition in ^{146}Sm, which is the most recent part of our work.

II THE 12^+ LEVEL OF ^{146}Sm

A Measurement of the mean life

We have measured the mean life of the lowest 12^+ state (and also of the lowest 9^- state) of ^{146}Sm by the Recoil-Distance Method, using the GASP array of L.N.L. [14] (in configuration II, i.e overall peak efficiency $\approx 5.8 \cdot 10^{-2}$, Ge counters at about 22 cm from the target) and the Plunger device of the group of Köln [15,16]. As for the analysis, the Ge counters of the GASP array have been grouped in seven "rings" according to their angle to the beam direction ($31.7^0 - 36^0$, $58.3^0 - 60^0$, 72^0, 90^0, 108^0, $120^0 - 121.7^0$, $144^0 - 148.3^0$, respectively). A total of 49 coincidence matrices of 6000×6000 channels, 0.25 keV/channel, have been constructed, one for each combination of detector rings. Obviously, only the most forward or backward rings, i.e. rings 1, 2, 6 and 7, show a Doppler shift large enough to observe separate peaks for the unshifted and the Doppler shifted line (in the relevant region of γ ray energies) and are therefore useful for the lifetime measurement. All data, however, can be used – and have actually been used – for auxiliary measurements, and in particular for the determination of the branching ratios.

We have measured the intensity of the Doppler shifted component and the unshifted component of the 411 keV transition deexciting the 12^+ level (corresponding to decays in flight and at rest, respectively) in coincidence with the Doppler shifted part of the feeding transition of 1011 keV (see fig. 3), to make sure that the 12^+ level was fed while the recoiling nucleus was still in flight between target and stopper. The presence of another transition of 411 keV (not shown in fig 3), fed at the level of a few percents in the decay of the 12^+ level, was somewhat disturbing, but we could account for it. Results of the analysis with the differential-decay-curve method [15,16] are given in fig. 4). Values of the lifetime

$$\tau_m = I_u/(v \ dI_s/dD)$$

(where v is the recoil velocity) deduced for different distances, are in good agreement with one another. Their "best average" gives $\tau_m(12^+) = 15 \pm 2$ ps.

In addition, the mean life of the lowest 9^- state has been remeasured. As the direct γ rays from this level have too low energy and/or intensity, we have measured the shifted and unshifted intensity for the second transition of the $9^- \to 8^+ \to 6^+$ cascade (926 keV), in coincidence with the Doppler shifted component of the 986 keV γ ray feeding the 9^- level, and derived the corresponding intensities (shown in fig. 4) for the first transition of the cascade by means of the relations [15]

$$I_s(9^-) = I_s(8^+) + v \ \tau_m(8^+) \ dI_s(8^+)/dD$$
$$I_u(9^-) = I_u(8^+) - v \ \tau_m(8^+) \ dI_s(8^+)/dD \ ,$$

using the mean life $\tau_m(8^+) = 16 \pm 6$ ps given in ref. [22]. Errors shown in the figure include the uncertainty on $\tau_m(8^+)$. From a great average, we obtain $\tau_m(9^-) = 0.97 \pm 0.05$ ns, to be compared with the previous value of 1.2 ± 0.2 ns given in ref [17].

B Branching ratio for 12^+ level and the sum-up effect

Once obtained, with reasonable precision, the mean life of the 12^+ level, one needs a *similar precision* in the measurement of the branching ratio, in order to deduce the $B(E3)$ strength. Our most recent measurements have enough statistics for this purpose, but the precision is limited by the possible contribution of the "sum-up" to the observed peak of the E3 transition. The sum-up effect results from the simultaneous detection of two coincident γ rays in the same counter, which cannot be distinguished from the detection of the true cross-over γ ray.

A method to evaluate the sum-up effect when composite detectors – like EUROBALL "Clusters" – are used, has been proposed recently by Radermacher, Wilhelm, and von Brentano [18], and was used to investigate weak transitions in ^{205,207}Pb. We use instead a new method, exploiting the high symmetry of the GASP array[1] and the fact that (if polarization is not detected) the angular correlation among the reaction γ-rays is invariant when the direction of any one of the γ rays is reversed. As a consequence, the detection of two coincident γ rays in two given *diametrically opposite* counters, or of both γ rays in any one of the two counters is equally probable. The latter events produce sum-up, and we can evaluate their contribution to the γ spectrum by taking coincidences between opposite counters and adding their signals together. This procedure remains valid also if one requires additional coincidences with one or more γ rays, provided they are detected by any one of the *other* counters (*i.e.*, coincidence spectra in a given counter should exclude coincidences with the diametrically opposite one).

Obviously, this is an ideal scheme, and in practice the situation might be somewhat more complex than expected. First, our conclusions would be strictly true only if the efficiencies of all counters were exactly equal. If, however, the sum-up is added over all counters, the first-order effects of efficiency asymmetries cancel each other and second order terms can usually be neglected. In fact, if one considers one pair of opposite counters (a and b), the overall number of sum–up events in the two counters is

$$I_s = I(\gamma_1)I(\gamma_2) \left[\epsilon_a(\gamma_1)\epsilon_a(\gamma_2) + \epsilon_b(\gamma_1)\epsilon_b(\gamma_2)\right]$$

while the number of coincidences with either γ_1 or γ_2 detected in counter a and the other γ detected in b is

$$I_c = I(\gamma_1)I(\gamma_2) \left[\epsilon_a(\gamma_1)\epsilon_b(\gamma_2) + \epsilon_b(\gamma_1)\epsilon_a(\gamma_2)\right] \ ,$$

---

[1] Actually, the 40 Ge counters of GASP are arranged in 20 pairs of diametrically opposite counters.

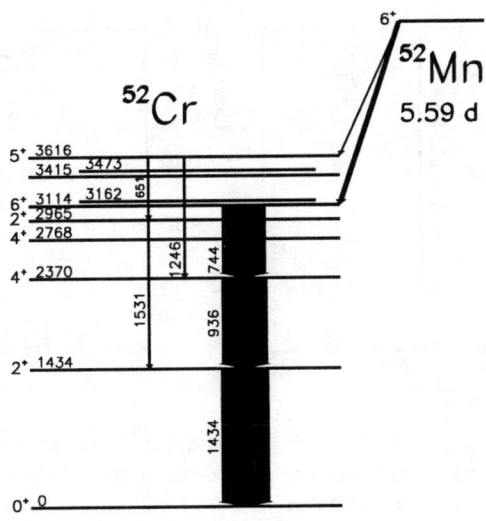

FIGURE 5. Simplified decay scheme of ^{52}Mn (5.6 d) to ^{52}Cr.

and, in terms of the efficiency asymmetry $\eta(\gamma) = ((\epsilon_a(\gamma) - \epsilon_b(\gamma)) / ((\epsilon_a(\gamma) + \epsilon_b(\gamma))$, one obtains

$$I_c/I_s = \frac{1 - \eta(\gamma_1)\eta(\gamma_2)/4}{1 - \eta(\gamma_1)\eta(\gamma_2)/4} \approx 1 - \eta(\gamma_1)\eta(\gamma_2)/2 .$$

For $\eta(\gamma_1) \approx \eta(\gamma_2)$ as large as 20%, the correction would amount to only 2%.

Other more subtle effects could bias our evaluation of the sum–up. "Simulated" sum–up spectra could be affected by a loss in coincidence counting due to – e.g. – electronic dead time or too narrow time gates, while the "real" sum–up would not. Before using this method for sum–up subtraction in critical cases of very weak transitions, an independent check of its reliability is therefore necessary.

C   Correcting for the sum–up: Test of the method and results

In order to test our method of sum-up evaluation, in conditions as close as possible to those of the real experiment, an auxiliary measurement has been performed at GASP (in configuration I), with a radioactive source of ^{52}Mn. A simplified decay scheme of ^{52}Mn is shown in fig 5. The most intense branch of the 5.6d decay feeds the lowest 6^+ level of ^{52}Cr and is followed by three γ rays of 744 keV, 936 keV, and 1434 keV, respectively. It is therefore possible to observe the sum–up of two of these three γ transitions in coincidence with the third one (fig. 6). This measurement has been performed very recently and data analysis is still under way. Results reported here refer to only a small part of the available statistics.

In Table 1, the integrals of the *observed* sum peaks are compared with those evaluated from coincidences at 180°, for each combination of two γ rays in coincidence with the third one. The agreement is good, within

TABLE 1. Test of the method of evaluation of Sum-up in the $\gamma - \gamma$ coincidence spectra of ^{52}Cr, populated by the decay of ^{52}Mn (5.6d: Comparison of measured integrals of sum-up peaks in the γ coincidence spectra (coincidences at 180° excluded) with those obtained from the sum of energies of γ rays detected at 180°.

COINCIDENCE GATE	SUM-UP PEAK	INTEGRAL THE OF SUM-UP PEAK	
E_γ (keV)	$E_{\gamma_1} + E_{\gamma_2}$ (keV)	Estimated	Observed
1434	744 + 936	3860 ± 128	4194 ± 140
936	744 + 1434	3891 ± 128	3904 ± 136
744	936 + 1434	4009 ± 130	3882 ± 134

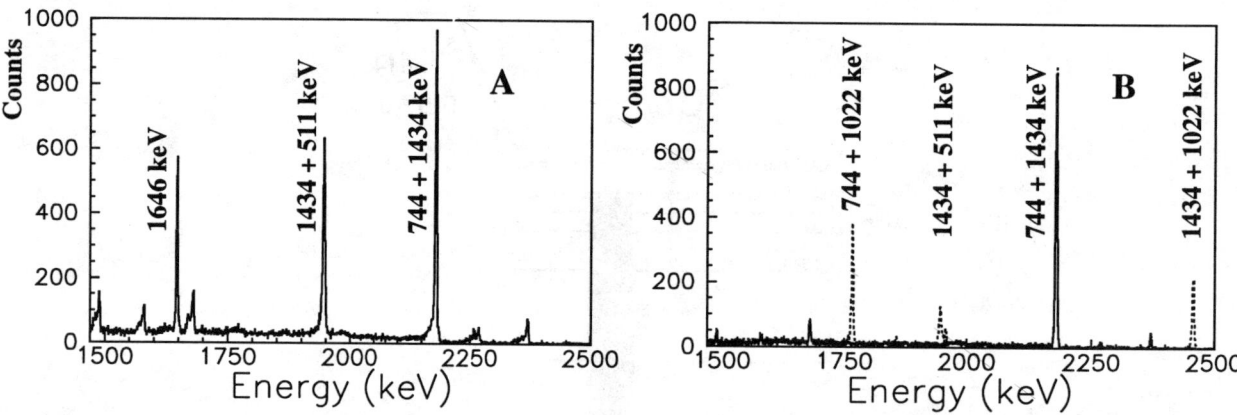

FIGURE 6. Part of the γ spectrum of ^{52}Cr from ^{52}Mn (5.6d) decay, in coincidence with the 936 keV transition: (A) Coincidence spectrum (γ at 180° excluded) showing the sum peak of the 744 keV and 1434 keV transitions, and (B) corresponding part of the "simulated sum–up spectrum" (sum of energies of γ rays detected at 180°). In the latter, the complete coincidence spectrum is shown as a dashed line, while the full line corresponds to coincidences excluding a small energy interval around 511 keV. For the γ rays from $e^+ - e^-$ annihilation, in fact, the assumed symmetry in the angular distribution does not hold and, as a consequence, sum peaks involving one or two 511 keV photons are not correctly reproduced in the "simulated sum-up" spectrum.

the experimental errors. Just to estimate the level of the effect (and of the corresponding errors), it can be sufficient to give a look to the part A of fig. 6, and compare the sum-up peak at 744 + 1434 keV with that at 1646 keV. The latter corresponds to a true γ transition in ^{52}Cr (not shown in fig. 5), which is fed by a weak branch of the ^{52}Mn decay and has a relative intensity of only $5 \cdot 10^{-4}$.

Now, as the sum-up estimation seems to work, we have applied the method to ^{146}Sm, and in particular to the E3 branch from the 12^+ level. Part of the coincidence spectrum of ^{146}Sm, including the 1397 keV transition corresponding to the $12^+ \to 9^-$ cross-over, is shown in fig. 7, together with the corresponding spectrum of simulated sum-up (obtained from coincidences at 180°). A sum-up peak appears around 1416 keV, and is well reproduced. The contribution of sum-up to the 1397 keV peak is small, and has been evaluated to be about 13% of the total. After correction for the sum-up, the branching ratio for the $12^+ \to 9^-$ decay results to be $(5.7 \pm 1.3) \cdot 10^{-3}$.

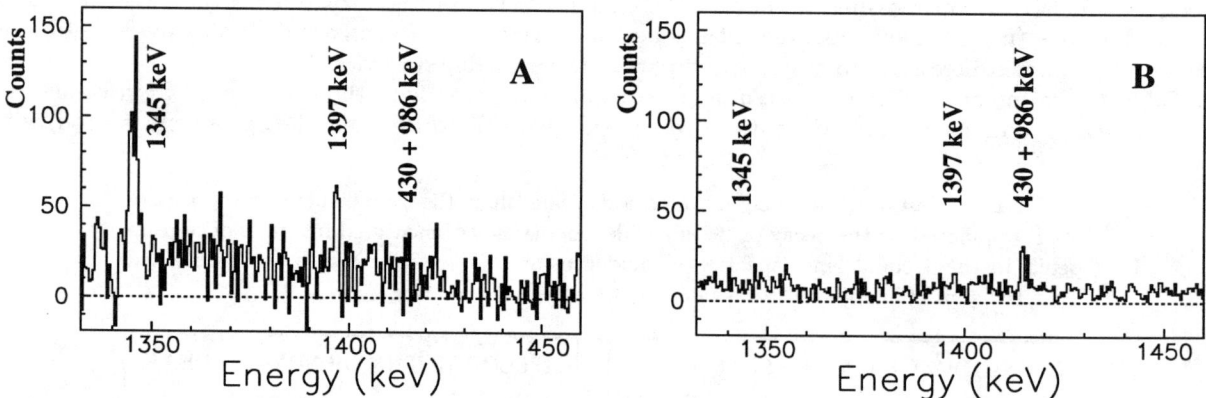

FIGURE 7. Estimate of sum-up contribution to the 1397 keV line in the gamma spectrum of ^{146}Sm, in coincidence with the gating signal (1011 keV, unshifted). Part A: Coincidence spectrum, excluding coincidences between diametrically opposite counters. Part B: Sum of pulses from diametrically opposite counters, in coincidence with the gating signal from a third counter.

TABLE 2. Comparison of E3 strengths in $N = 84$ Isotones

| E3 transition | EXPERIMENTAL | | | | | | THEOR. | |
| | ^{144}Nd | | ^{146}Sm | | ^{148}Gd | | ^{148}Gd | |
	E_γ (keV)	B(E3) (W.u.)	E_γ (keV)	B(E3) (W.u.)	E_γ (keV)	B(E3) (W.u.)	E_γ (keV)	B(E3) (W.u.)
$9^- \rightarrow 6^+$	1111	55 ± 10	986	46 ± 9	884	52 ± 4	781	48
$12^+ \rightarrow 9^-$	1451	?	1397	50 ± 12	1286	77 ± 11	1116	78
$15^- \rightarrow 12^+$					?	?	1573	98

III   DISCUSSION OF THE RESULTS

From the mean life of the 12^+ and 9^- levels of ^{146}Sm and the value of the branching ratio for E3 decay, properly corrected for the sum-up contribution, one obtains the reduced E3 strengths (50 ± 12 W.u. and 46 ± 9 W.u., respectively), to be compared with those of neighboring nuclei. The available information on $N = 84$ isotones is summarized in Table 2. As for the $9^- \rightarrow 6^+$ transition, our new measurement of $\tau_m(9^-)$ brings the value of ^{146}Sm very close to those of ^{144}Nd and of ^{148}Gd. The apparent decrease of B(E3,$12^+ \rightarrow 9^-$) from ^{148}Gd to ^{146}Sm is still at the limit of experimental errors: it would be very interesting, at this point, to check what happens in ^{144}Nd, but this concerns (possibly) our future work.

The ratio of reduced strengths for the $12^+ \rightarrow 9^-$ and the $9^- \rightarrow 6^+$ transition is smaller than 2 – the factor one would expect for harmonic oscillators – but, as we have seen, this happens also in the case of quadrupole vibrations. A more comprehensive model is necessary, also in the present case, to provide a more satisfactory description of the experimental results. In particular, the excitation of valence $2f_{7/2}$ neutrons into the higher lying $1i_{13/2}$ orbit must certainly be taken into account [19,20]. For ^{148}Gd, such a calculation has been performed some years ago by Blomqvist, Kleinheinz and coworkers [21,10]. The last column of Table 2 shows the results of their calculation, which take into account the effect of Pauli blocking in the multi-phonon states as well

TABLE 3. Collective E3 transitions of ^{148}Gd and interference between core excitation and valence neutrons excitation from the $2f_{7/2}$ orbit (f) and the $1i_{13/2}$ orbit (i).

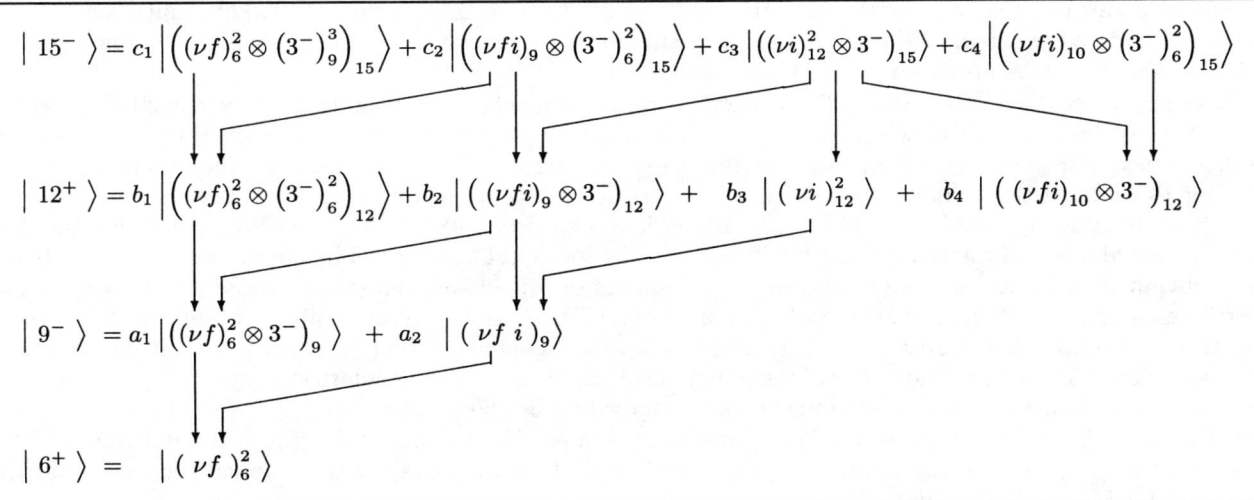

as the excitation of valence neutrons from the $2f_{7/2}$ orbit to $1i_{13/2}$ orbit, as shown in Table 3. The latter contribution turns out to combine coherently with that of the core, entering as an additional component in the structure of the octupole phonon (although with obvious limitations for what concerns the multi-phonon states). In this way, the model obtains a remarkable agreement with the experimental values of E3 strengths in the $12^+ \rightarrow 9^- \rightarrow 6^+$ cascade and, in addition, it predicts a new collective level $J^\pi = 15^-$, corresponding to a *triple* octupole excitation built over the lowest 6^+ state. The search for this new level has been the object of several measurements we have recently performed at L.N.L..

IV SEARCH FOR A TRIPLE OCTUPOLE EXCITATION IN ^{148}Gd

A Measurements with the ^{124}Sn(^{28}Si,4n)^{148}Gd reaction

Already in the first paper [9] on double octupole excitations in ^{148}Gd, a level with $J = 15$ and unknown parity was reported at an excitation energy of 5389 keV. Owing to its selective decay, this state could have been identified with the collective 15^- state. However, the level scheme proposed in a later paper by Drigert *et al.* [22] gives a different order of transitions for a few γ–ray cascades in this region of excitation energy, and – as a consequence – different level positions.

In order to solve this discrepancy and to identify – if possible – the E3 decay from the collective 15^- state, a new measurement has been performed with the GASP array of L.N.L., which provided a significant improvement in efficiency and in resolving power with respect to previous experiments [9,22]. We used, for this purpose, the reaction ^{124}Sn(^{28}Si,4n)^{148}Gd at $E_{beam} = 135$ MeV, *i.e.* the same reaction used by Drigert *et al* [22]. The order of relevant γ rays (fig. 3) resulted to be in agreement with that of ref. [9]. Detection of a (weak) 1409 keV branch from the 5389 keV level to the collective 12^+ state at 3981 keV would have assigned negative parity to the parent state, and would have identified it as the predicted three-octupole-phonon state. Unfortunately, the search for this cross–over transition resulted to be very difficult, because γ transitions involved in the analysis have energies very close (within 1 keV or less) to those of other transitions situated higher in the level scheme, and it has only been possible to assign an upper limit of about 20% to the possible cross-over transition.

B New measurements with the reaction ^{11}B + ^{141}Pr

Therefore, a further measurement has been performed at GASP (in configuration I [14]) with the reaction ^{141}Pr(^{11}B,4n)^{148}Gd at $E_{beam} = 50$ MeV, which directly populates a region of angular momenta and excitation energies immediately above the region of interest. In this way we could identify a new transition of 363 keV (fig. 8) from the 5389 keV level to the yrast 14^+ state, which confirmed the position of the parent state but also resulted to be peaked at small angles, and therefore not consistent with a stretched dipole character. As a sizeable M2 mixing with E1 appears to be unreasonable in this case, the assignment of negative parity to the parent level can be excluded. By the way, the observed intensity of the crossover 1409 keV transition (if any) turns out to be very small, at the level of the expected sum-up. Therefore, the three-octupole-phonon state – if it exists – cannot be identified with the 5389 keV level.

However, with the ^{141}Pr(^{11}B,4n)^{148}Gd reaction we have identified a number of new transitions feeding directly the 12^+ level, out of which several have an angular distribution consistent with that of a stretched octupole or quadrupole (the experimental errors being too large to distinguish between the two possibilities). Among these transitions, at least one case results to be particularly promising. It concerns the 1316 keV transition from a new level at 5296 keV, which is found to decay also to the two lowest 14^+ levels. The angular distribution of the transition (of 270 keV) to the lowest 14^+ results to be consistent with a stretched E1. Obviously, nobody can exclude that a proper mixing of dipole and quadrupole could mimic the angular distribution of a stretched E3 transition (and non-stretched transitions can certainly be observed in our case), but if both the 1316 keV transition and the 270 keV one are assumed to be pure multipole, our results would uniquely fix to 15^- the spin and parity of the parent level and to E3 the multipolarity of the 1316 keV transition. In order to confirm this possible assignment, next step would be the measurement of the lifetime. If the 1316 keV transition is really an E3 of ≈ 100 W.u., as predicted in ref. [10] (or less), taking into account the observed branching ratio of $\approx 50\%$ the mean life would be close to 1 ns (or more), while a much shorter lifetime is expected for M1 or E2 multipolarities.

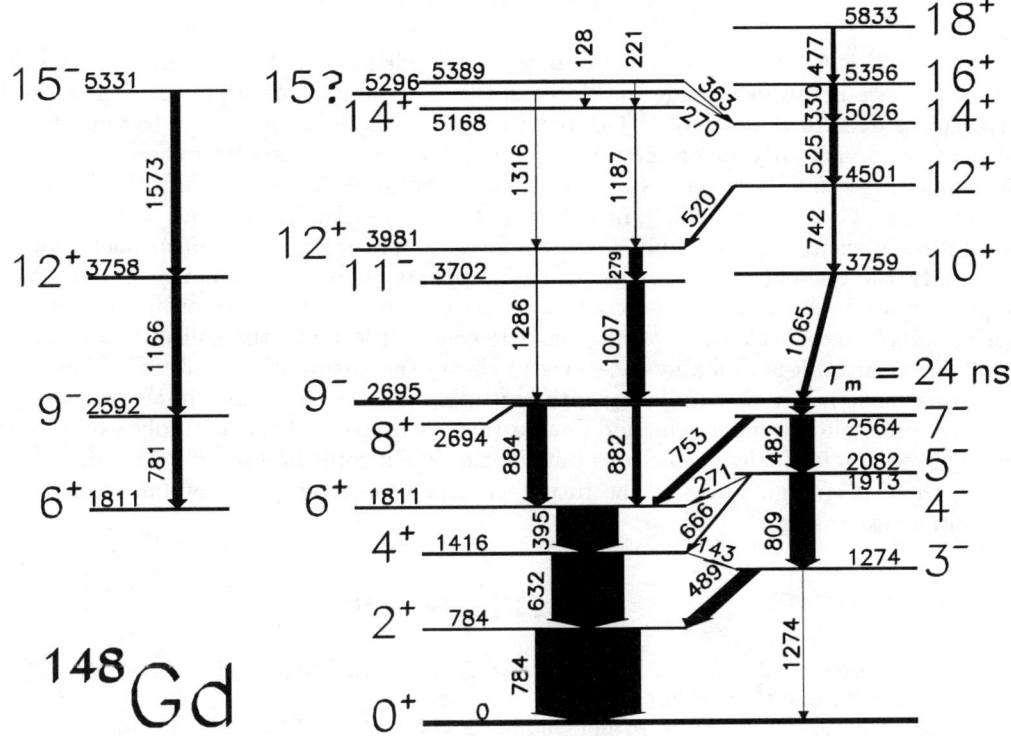

FIGURE 8. Partial level scheme of ^{148}Gd. The calculated sequence of octupole excitations (built upon the 6^+ state of the basic two–neutron configuration), given in ref. [10], is shown for comparison in the left part of the figure.

C New measurement at EUROBALL

Since, in the meantime, the EUROBALL array started its first period of operation at the L.N.L., we have submitted a proposal for a Recoil-Distance measurement with the new array. The proposal was approved, and the measurement has been performed in April 1998, with the reaction ^{11}B + ^{141}Pr and with the same Plunger device of the Köln group which had already been used in our previous measurements with the GASP array.

The Euroball array includes 15 "cluster" detectors (grouped in two rings, covering the angular ranges $149° - 163°$ and $129° - 145°$, respectively), 26 "clover" detectors (at $108°$ and $72°$), and three rings of individual "tapered" detectors in the forward hemisphere (at $52°$, $35°$ and $15°$ to the beam), for a total of 239 Ge crystals.

We have measured at 17 distances, ranging from almost 0 to 2000 μm, and started to analyse the results with the standard procedure already used at GASP. The analysis took a lot of time, also due to the large number of counters. However, we soon realized that our standard procedure of analysis is not sufficient in the present case. Actually, no sizeable γ transitions feeding the 5296 keV level – to be used for coincidence – have been observed, and coincidence spectra gated with γ transitions deexciting the 12^+ level, turn out not to be clean enough in the region of interest.

We have therefore to restart our analysis from the very beginning. We plan to exploit the presence of the relatively long mean life (24 ns) of the 9^- state at 2695 keV to clean-up the spectrum of γ rays lying above this level, by requiring a *delayed* coincidence with one of the γ rays belonging to the isomer decay. If this system will work, we hope to obtain at least an approximate value of the mean life of the 5296 keV level, in order to confirm the possible E3 character of the 1316 keV transition. Moreover , thanks to the large efficiency of EUROBALL, we can hope to identify other new levels decaying to the collective 12^+ state, (or perhaps weak branches from known levels), which could possess a fraction of the collective three-octupole-phonon strength. At the moment, the identification of the expected 15^- state cannot be confirmed.

V FINAL REMARKS

Summarizing, in the $A \approx 150$ region there exists clear evidence of double octupole excitations in several nuclei with one or two neutrons outside the closed shell. The question of triple octupole excitation, for which definite predictions exist in the case of ^{148}Gd, remains still open. It is interesting to remark that the presence of valence neutrons apparently favors octupole collectivity (or, at least, its experimental observation), not only in the region $A \approx 150$. New results about enhanced octupole transitions in other regions of nuclei have been reported in this Conference [23,24], in the case of ^{104}Sn (by H. Grawe) and of ^{211}Po (by G.J. Lane). In both cases, octupole excitations are built upon the highest-spin state of the basic shell-model configuration. In ^{104}Sn, already the one-octupole-phonon strength appears to be split among at least two levels, but in ^{211}Po one observe a strong cascade of two enhanced E3 transitions that – according to the authors – involve a constructive interference between a valence-particle contribution and the collective excitation of the ^{208}Pb core. More theoretical work is probably necessary to clarify the nature of the E3 transitions observed in ^{211}Po (and neighboring nuclei), but the similarity with multiple octupole excitation in the region of ^{146}Gd is really impressive, and one could wonder whether double (and perhaps higher) octupole excitations are easier to observe in nuclei with a few valence nucleons outside the ^{208}Pb core than in ^{208}Pb itself.

As you see, there is enough work for the future, in order to investigate how far extends the coherence of collective octupole excitations.

REFERENCES

1. Bohr, Aa., and Mottelson, B., *Nuclear Structure, Vol. II, Ch. 6 and 6A*. Benjamin, London (1975).
2. Oros, A.M., Ph.D. Thesis, Köln 1996.
3. Giannatiempo, A., Nannini, A., Sona, P, and Cutoiu, D., *Phys. Rev. C* **52**, 2969 (1995).
4. Arima, A., *et al.*, *Phys. Lett. B* **66**, 205 (1977); Otsuka, T., Arima, A., Iachello, F., and Talmi, I., *Phys. Lett. B* **76**,139 (1978).
5. Bizzeti, P.G., *et al.*, *Nuovo Cimento A*, in the press.
6. Kleinheinz, P., *et al.*, *Z. Phys. A, Atoms and Nuclei* **286**, 27 (1978).
7. Kleinheinz, P., *et al.*, *Z. Phys. A, Atoms and Nuclei* **290**, 279 (1979).
8. Kleinheinz, P., *et al.*, *Phys. Rev. Letters* **48**, 1457 (1982).
9. Lunardi, S., *et al.*, *Phys. Rev. Letters* **53**, 1531 (1984).
10. Piiparinen, M., *et al.*, *Phys. Rev. Letters* **70**, 150 (1993).
11. Podolyák, Zs., *et al.*, in *LNL–INFN (Rep) 105/96*, p.65.
12. Bargioni, L., *et al.*, *Phys. Rev. C* **51**, R1057 (1995).
13. Bizzeti, P.G., *et al.*, in *New Perspectives in Nuclear Physics* (Proc. Ravello 1995, edited by Covello, A.) p.267.
14. Rossi Alvarez, C., *Nuclear Physics News* **3**, n.3 (1993).
15. Dewald, A., *et al.*, *Z. Phys. A* **334**, 163 (1989).
16. Böhm, G., *et al.*, *Nucl. Instr. Meth. A* **329**, 248 (1993).
17. Rozak, S., Funk, E.G., and Mihelich, J.W., *Phys. Rev. C* **25**, 3000 (1982).
18. Radermacher, E., Wilhelm, M., and von Brentano, P., *Nucl. Instr. Methods A* **383**, 480 (1996); Radermacher, E., Wilhelm, M., von Brentano, P., and Jolos, R.V., *Nuclear Phys. A*, **620**, 151, (1997).
19. Cottle, P.D., and Kempler, K.W., *Phys. Rev. C* **53**, 2017 (1996).
20. Bargioni, L., *et al.*, *Phys. Rev. C* **53**, 2020 (1996).
21. Piiparinen, M., *et al.*, *Z. Phys. A, Atoms and Nuclei* **337**, 387 (1990).
22. Drigert, M.W., *et al.*, *Nuclear Phys. A* **515**, 466 (1990).
23. Grawe, H., *these Proceedings*.
24. Lane, G.J., *these Proceedings*.

Magnetic Rotation in Nuclei in the A~110 and 200 Mass regions

R. Wadsworth[*] and R.M. Clark[†]

[*] *Dept. of Physics, University of York, York YO10 5DD, UK*
[†] *Lawrence Berkeley National Laboratory, Berkeley, CA 94720*

Abstract. This paper will briefly review the properties of M1 bands in the lead nuclei in the A~200 mass region and discuss how these structures can be interpreted within the tilted axis cranking (TAC) model. Following this, the objectives are to compare the results of new Doppler shift attenuation method lifetimes for M1 (shears) bands in the light lead nuclei with the predictions of the TAC and Donau and Frauendorf models. Recent results for M1 bands in the A~110 region will be shown and discussed in terms of the TAC model. In addition, the possibility of the existence of a new type of shears band based on E2 transitions will be presented. Lifetime measurements for such a structure in ^{110}Cd will be discussed together with a comparison of the predictions of the TAC model.

INTRODUCTION

Many rotational-like bands consisting of magnetic dipole transitions have been observed in lead nuclei with mass numbers ranging from 191 to 202 (see for example, [1,2]). These structures are also called *shears bands* or *magnetic rotational bands*, this terminology will be discussed. In some nuclei multiple bands are observed. The intensities of these structures varies from 1–10% of the reaction channel and, furthermore, the B(M1) values are large, typically a few Wu [3]. In addition, the cross-over E2 transitions are very weak or not observed at all, leading to large B(M1)/B(E2) ratios of 20–40 $(\mu_N/eb)^2$ or greater. This paper will discuss the interpretation of these structures in terms of the tilted axis cranking (TAC) model [4] and present new results for the lifetimes of states in M1 bands in the lead isotopes. Several M1 bands are also known to exist in the Z~50, A~110 mass region. We have identified new bands of this type in light Cd,Sn,Sb isotopes. The structure of these bands will be discussed in terms of the TAC model. Finally, the latest evidence for the existence of a new type of shears band, based on a $\Delta I=2$ sequence of energy levels, will be presented. The results discussed in this paper are from data which have been analysed at LBNL and York.

INTERPRETATION OF M1 BANDS IN LEAD NUCLEI - TAC MODEL

The interpretation of the M1 bands in the lead isotopes involves protons in high-Ω ($h_{9/2}$, $i_{13/2}$) orbitals coupled to a spin of j_π and low-Ω $i_{13/2}$ neutrons coupled to j_ν in nuclei with small oblate deformations ($\beta_2 \sim -0.1$). At the band-head the proton particles and neutron holes combine to maximise the overlap of their wavefunctions and consequently j_π and j_ν are essentially perpendicular to each other. This coupling scheme has recently been confirmed in an experiment which measured the g-factor of the band-head of an M1 band in ^{193}Pb [5]. The total angular momentum vector **J**, which is the vector sum of j_π and j_ν, lies between the two component vectors.

The tilted axis cranking model (TAC) has been used to interpret the structure of the M1 bands in the lead region [4]. Figure 1 shows a schematic diagram of the angular momentum coupling at two different frequencies. As noted above at the band-head j_π and j_ν are essentially perpendicular. Higher angular momentum states are then generated by aligning the two component vectors along the direction of the total angular momentum vector **J**. (Since this action resembles the closing of the blades of a pair of garden or sheep shears these bands have been called *shears bands*.) The majority of the spin is generated in this way, however, there is a small

CP481, *Nuclear Structure 98*, edited by C. Baktash

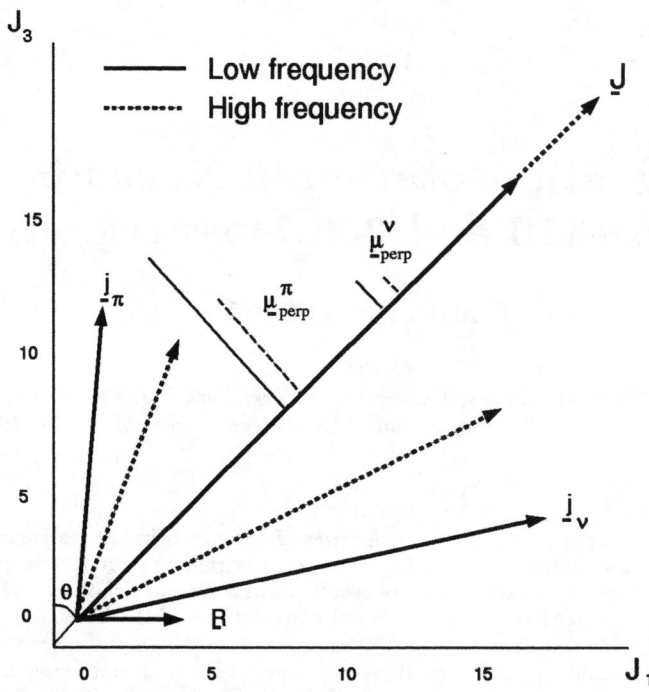

FIGURE 1. The angular momentum coupling scheme and the mode of generating angular momentum (the 'shears mechanism') as suggested by the TAC model. The neutron (proton) angular momentum vector is represented by $\underline{j}_\nu$ ($\underline{j}_\pi$) while the perpendicular component of the magnetic moment is labelled by $\underline{\mu}^\nu_{perp}$ ($\underline{\mu}^\pi_{perp}$). The collective component of the angular momentum is represented by $\underline{R}$, the total angular momentum by $\underline{J}$, and the tilting angle by θ. The solid (dashed) arrows show the approximate coupling at low (high) frequency.

contribution from the collective rotation of the weakly deformed core and this is represented by the vector **R** in fig. 1.

Both the TAC and standard cranking models are able to reproduce the experimental routhians, angular momenta and moments of inertia for the M1 bands in the lead nuclei (e.g. see [1,2]). However, the models predict different behaviour for the B(M1) values, thus it should be possible to distinguish between them. For example, the standard cranking approaches such as the Donau and Frauendorf formalism [6], which assumes a fixed K value and an alignment which is perpendicular to the symmetry axis, predict a very gradual decrease of the B(M1)'s with increasing spin. On the other hand as can be seen from fig. 1, a consequence of the near perpendicular coupling of the proton and neutron configurations at the band-head is the presence of a large perpendicular component of the magnetic moment vector, μ_{perp}, which is the sum of the perpendicular components for the protons and neutrons. As the angular momentum component vectors gradually align along the total angular momentum vector, **J**, μ_{perp} rapidly decreases. Since the B(M1) values are proportional to μ^2_{perp} then the prediction is for rapidly decreasing B(M1) values as the spin increases up the band. This prediction can be tested through the measurement of lifetimes of states within these bands.

The states in the lead shears bands follow a rotational-like pattern with the excitation energies obeying the expression $\Delta E(I) = E(I) - E(I_b) \sim A(I-I_b)^2$, where I is the spin of the state and I_b is the spin of the band-head. These bands are however based on weakly deformed oblate shapes ($\beta_2 \sim \leq 0.1$) and it is therefore intriguing as to why such structures should have rotational-like properties. A new form of quantal rotation known as *magnetic rotation* has been suggested by Stefan Frauendorf as a possible explanation [7]. In this case it is the anisotropic arrangement of nucleon currents, from which the blades of the shears arise, that is responsible for the symmetry breaking (see fig. 2). This results in the rotation of a large magnetic dipole vector about the total angular momentum vector **J**. This is in contrast to the more conventional view of nuclear (electric) rotation which occurs when an intrinsic deformation breaks the spherical symmetry.

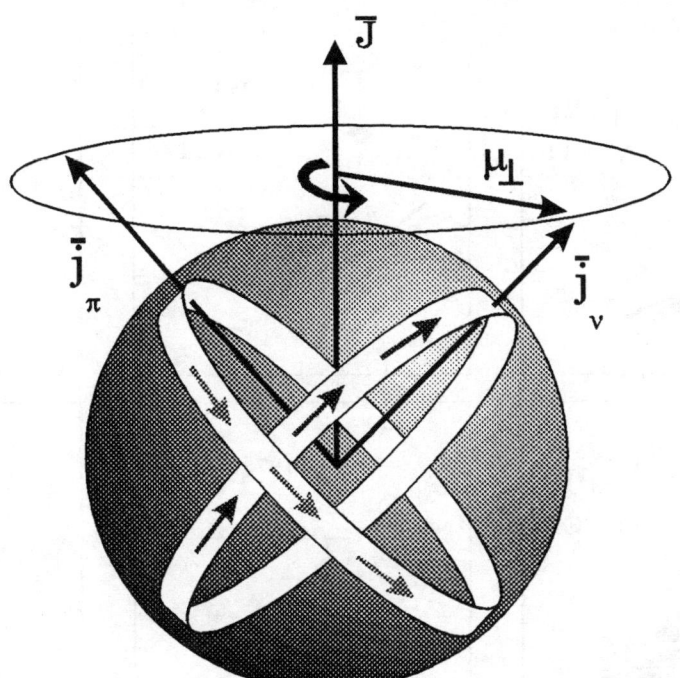

FIGURE 2. The few protons and neutrons which take part in the shears mechanism form current loops embedded in the spherical mass distribution of the nucleus. These current loops enable us to define angular momentum vectors, j_π and j_ν, for the active protons and neutrons. The total angular momentum **J** is along the axis about which the near-spherical nucleus can rotate. $\mu_\perp$ is the perpendicular component of the magnetic moment.

NEW RESULTS IN THE A~200 MASS REGION

Experimental details and Data Analysis

In order to try and determine the structure of the M1 bands in the lead nuclei and to investigate the applicability of the TAC and standard cranking models to these structures, we have measured lifetimes for several states within these bands in a range of lead nuclei. The ^{26}Mg + ^{176}Yb(^{174}Yb, ^{172}Yb) reactions have been used at 139 (137,135) MeV respectively, together with GAMMASPHERE containing 97 large volume HPGe detectors and ~1mg/cm² Yb targets on 12mg/cm² lead backings to measure lifetimes using the Doppler shift attenuation method. Approximately 9×10^8 quadruple and higher fold events were collected per reaction. The data were subsequently unfolded and sorted in to (a) γ-gated angle dependent spectra and (b) γ- gated E_γ–E_γ correlation matrices. At least one M1 band was analysed in $^{193-196}$Pb and two bands were analysed in ^{197}Pb.

Lifetimes were extracted from Doppler broadened lineshapes using the codes of Wells and Johnson [8]. The complete stopping of the recoiling nuclei was modelled using the prescription of Gascon et al [9]. The tabulations of Northcliffe and Schilling [10] with shell corrections were used for the electronic stopping powers. The detailed slowing down history of the recoils in the target and backing material was simulated using a Monte carlo technique (5000 histories with a time step of 0.002ps) and then sorted according to detector geometry. Calculated lineshapes for each transition were obtained assuming (a) feeding into the top of the band through a cascade of five transitions with the same moment of inertia as the in-band states. The lineshape for the highest transition observed was fitted and used as an input parameter to extract the lifetimes of states lower down in the cascade, (b) sidefeeding into each state made up of a rotational cascade of five transitions - the intensity of the sidefeeding being constrained to reproduce that observed experimentally from thin target experiments. The sidefeeding lifetimes were always found to be shorter than the in-band lifetimes (typically up to 2 times shorter).

Simultaneous fits were made to forward/backward and transverse spectra and the final results were obtained from a global fit of the cascade with independent variable lifetimes for each state and the associated sidefeeding.

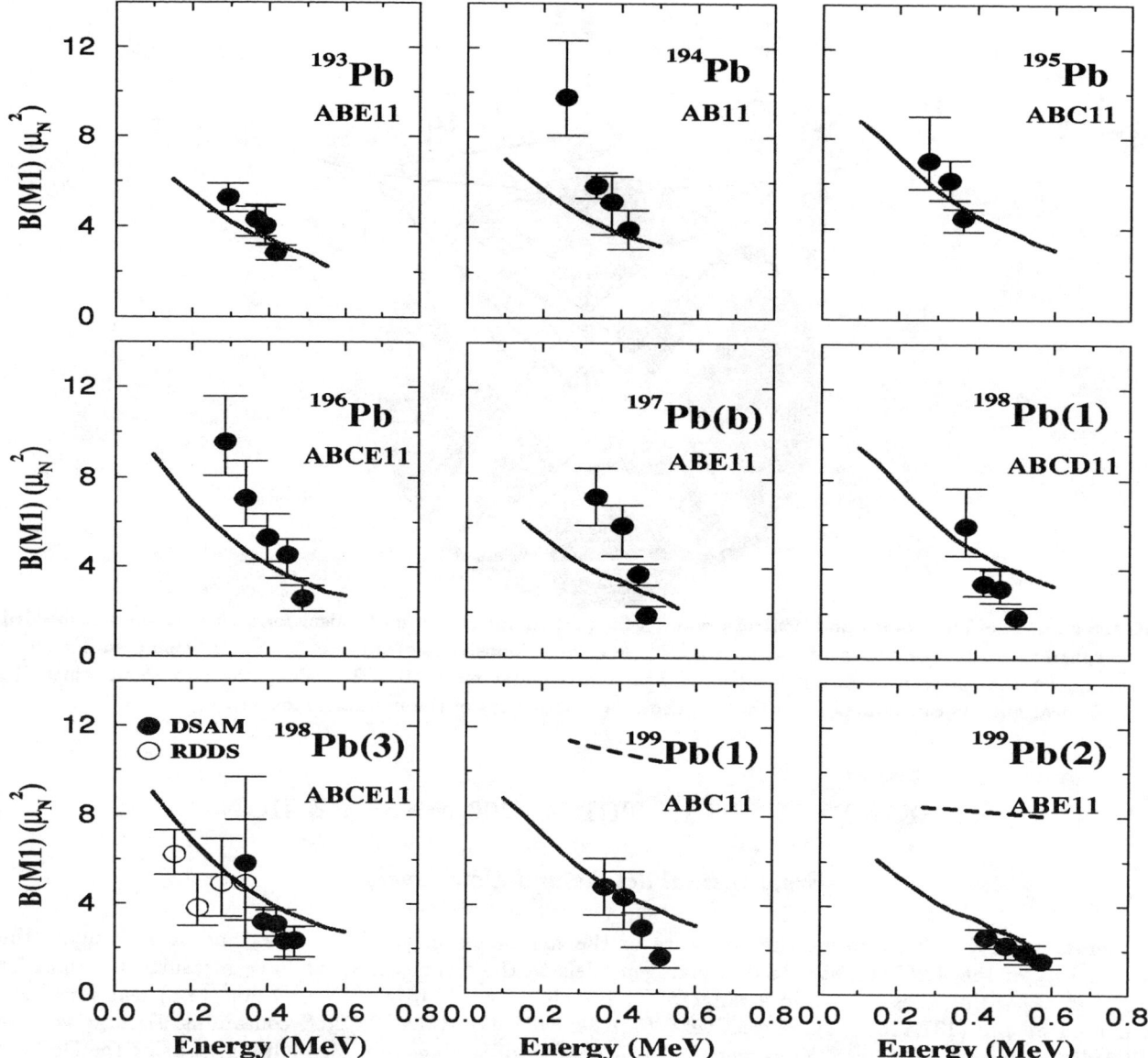

FIGURE 3. Plots of B(M1) versus transition energy, E_γ, for bands in $^{193-199}$Pb for which lifetimes of states have been determined. The solid lines represent the results of TAC calculations for the configurations shown. A, B, C, D denote $i_{13/2}$ quasineutrons, E, F the natural parity quasineutrons, and the proton configuration is denoted by its aligned spin. The RDDS data for ^{198}Pb (band 3) were taken from Krucken et al [12]. The dashed lines for the two bands in ^{199}Pb are the results of Donau and Frauendorf calculations using the same configurations as for the TAC calculations.

(It should be noted that both the number of transitions in the sidefeeding cascade and the moment of inertia of this cascade were varied in order to ascertain the affect on the fits. The best χ^2 fits were obtained with the parameters given above.) Further details of this work, including the actual lifetimes, can be found in ref [11]. In deducing the B(M1) values the transitions were assumed to be pure dipoles and the branching ratios were taken as 1.0 for cases where no cross-over E2 transitions were observed. The B(M1) values for a range of shears bands in the lead isotopes are shown in fig. 3 together with the results of TAC calculations for the configurations which have been previously assigned to these bands. This figure also contains the results from earlier work on ^{198,199}Pb [3] for comparison. The RDDS data for band 3 in ^{198}Pb were taken from [12]. The notation for the configurations is given in the figure caption. Clearly the agreement between the TAC and the experimental data is quite good and certainly much better than the agreement with the Donau and Frauendorf

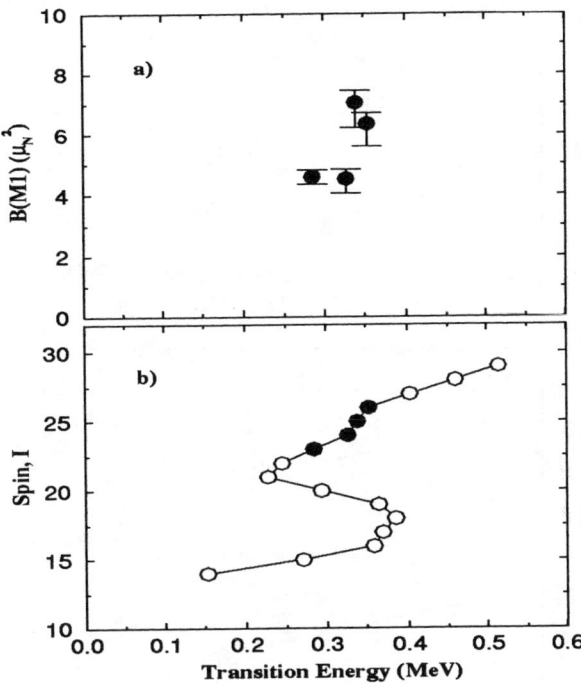

FIGURE 4. Plots of Angular momentum and B(M1) values versus transition energy for band 2 in ^{197}Pb. The filled circles show the points for which lifetimes have been measured. At low spin the band has an A11 configuration. The large backbend results from the alignment of a pair of $i_{13/2}$ neutrons (BC), whilst the short up-bend at an energy of 0.35 MeV results from the alignment of a pair of natural parity neutrons (EF). See caption to fig. 3 for the notation used.

semi-classical model results shown for ^{199}Pb. A similar situation is observed for all the Pb isotopes where M1 bands have been identified to date. From these results it would appear that the experimental B(M1) values fall off faster than the TAC model predicts. It has been found that a reasonable adjustment of TAC parameters, such as the strength of the quadrupole–quadrupole (Q.Q) coupling constant, can yield a sharper decrease in the calculated B(M1) values [13]. These results put the TAC model on a firm experimental footing and support the existence of the shears mechanism and the phenomenon of magnetic rotation.

Figure 4a shows the deduced B(M1) values for the second band in ^{197}Pb. Instead of decreasing smoothly with increasing spin/ γ-ray energy there is a sudden rise at an energy of 0.35 MeV. This can be understood qualitatively within the TAC model by looking at fig. 4b. For the data points where the lifetimes exist there is clearly an upbend in spin. This results from the alignment of a pair of natural parity neutrons (EF, $f_{5/2}$). At this point the neutron vector $\mathbf{j}_\nu$ will become longer and consequently the angle between $\mathbf{j}_\pi$ and $\mathbf{j}_\nu$ will increase. This should result in a larger perpendicular component of the magnetic moment and hence an increase in the B(M1) value. It would clearly be very interesting to test this hypothesis further by measuring the lifetimes of states lower in this band around the $i_{(13/2)}^2$ neutron alignment using the recoil distance method.

A~110 MASS REGION: A NEW REGION OF SHEARS BANDS?

Introduction

Many M1 bands are known to exist in Cd,In,Sn,Sb nuclei in this mass region. These have generally been assigned structures which are based on one or two holes in the $g_{9/2}$ proton orbital and particles in the $h_{11/2}, g_{7/2}, d_{5/2}$ neutron orbitals (e.g., see [14,15]) and they are consequently expected to have a small prolate deformation $\beta_2 \sim 0.1$ The important ingredients as far as the TAC model is concerned are the proton $g_{9/2}$ holes and the low-Ω $h_{11/2}$ neutrons plus the small intrinsic deformation.

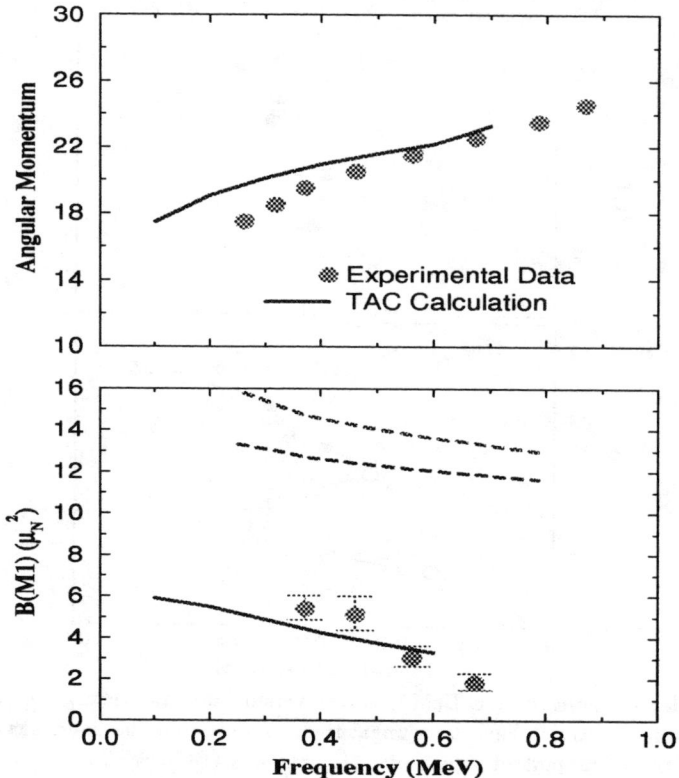

FIGURE 5. Angular momentum and B(M1) values as a function of frequency (γ-ray energy) for the M1 band in ^{110}Cd. The solid line represents the results of a TAC calculation for the configuration discussed in the text. The dashed lines show the results of Donau and Frauendorf calculations for the same configuration. The two lines represent the limits obtained when g-factors for $g_{7/2}$ or $d_{5/2}$ neutrons are used in the calculation.

The TAC model predicts that magnetic rotation should exist around other proton or neutron closed shell regions [7,16,17], one of the more promising of which lies around Z=50 and A$\simeq$110. In this region the dipole bands might be expected to result from the near perpendicular coupling at the band–head of the angular momentum vectors from a proton configuration involving $g_{9/2}$ holes and neutrons in the $h_{11/2}$ orbital. The remaining active protons and neutrons occupying the near degenerate $g_{7/2}$,$d_{5/2}$ shells.

Experiments have been performed which have enabled us to observe new dipole bands in ^{106,108}Sn, ^{108}Cd and ^{108}Sb. Lifetimes have also been measured for states within some of these bands using the DSAM technique discussed earlier. The data for the bands in ^{106,108}Sn and ^{108}Sb were obtained by using the ^{58}Ni + ^{54}Fe reaction (thin and backed target data) at a beam energy of 243 MeV , whilst that for ^{108,110}Cd were obtained from the ^{16}O + ^{96}Zr (backed target data) and ^{18}O + ^{96}Zr (thin and backed target data) reactions at 72 and 70 MeV respectively. Gammasphere (with 95-99) detectors was used for all of these experiments.

Cd Nuclei

In ^{110}Cd lifetimes have been measured (see [18] for further details) for states in the M1 band which had previously been assigned a $\pi g_{9/2}^{-2} \otimes \nu [h_{11/2}^2 (g_{7/2}d_{5/2})^2]$ configuration [19]. In the present work no E2 cross-over transitions could be observed for this structure. Fig. 5 shows the angular momentum and deduced B(M1) values as a function of frequency (γ-ray energy). The B(M1) values were again deduced assuming that the transitions are pure dipoles and a branching ratio of 1.0 was also used. The solid lines represent the TAC calculations for the configuration given above. These show good agreement with the experimental results. The dashed lines show the results of calculations using the Donau and Frauendorf standard cranking approach for the same configuration, the two lines corresponding to the limits obtained when g-factors for $g_{7/2}$ or $d_{5/2}$ neutrons are used in the calculation.

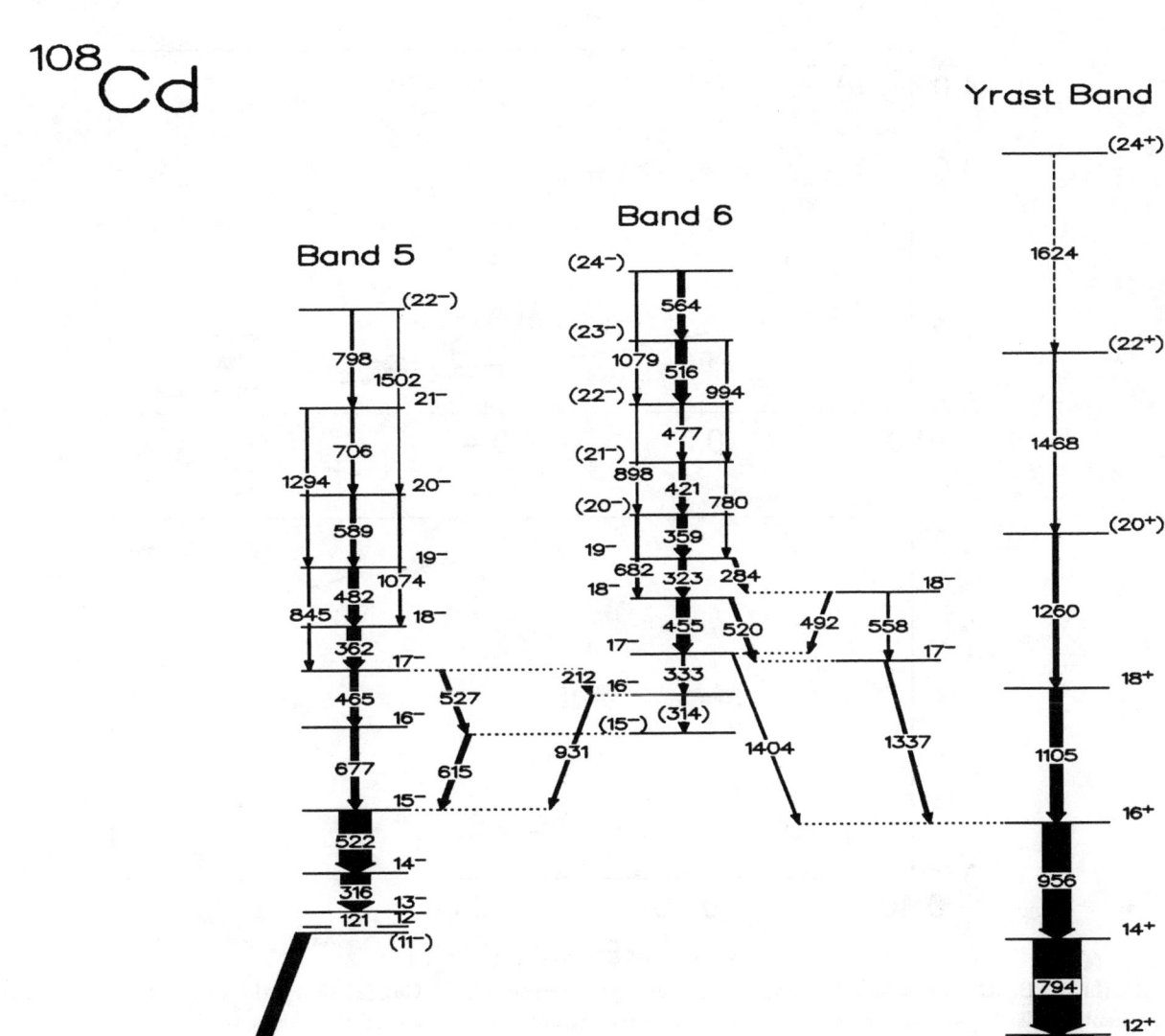

FIGURE 6. Partial decay scheme for ^{108}Cd deduced from the present work.

Figure 6 shows a partial decay scheme for ^{108}Cd deduced from the present work. Here two M1 bands are observed and in both cases weak cross-over E2 transitions are seen. Both structures are assigned negative parity based on an analysis of the decay properties of the bands and the DCO ratios for the feedout transitions. Preliminary lifetimes have been extracted for the states in band 5. This band has previously been tentatively assigned a $\pi g_{9/2}^{-2} \otimes \nu h_{11/2}^1 (g_{7/2}, d_{5/2})^3$ configuration at low spin [20], however, as can be seen in figure 7 such a configuration does not reproduce the experimental B(M1) values within the TAC model. Band 5 shows evidence for the alignment of a pair of $h_{11/2}$ neutrons at $\hbar\omega \sim 0.45$ MeV. Evidence for this change in structure can be seen in fig. 6 through the appearance of E2 transitions above the backbend. The low experimental B(M1) values strongly suggests that there are fewer than two $g_{9/2}$ holes involved in the configuration. In ^{106}Cd a similar negative parity M1 band has been observed and assigned a $\pi[g_{9/2}^{-3}g_{7/2}^1] \otimes \nu h_{11/2}^1 (g_{7/2}, d_{5/2})^1$ configuration [21]. It is clear from fig. 7 that the experimental B(M1) values for band 5 are in much better agreement with the TAC calculations for this latter configuration, both below and above the $\nu(h_{11/2})^2$ neutron crossing. We therefore tentatively assign band 5 to have the same configuration as the M1 band in ^{106}Cd.

Since the E2 transitions are also observed for this band we have been able to extract an estimate of the

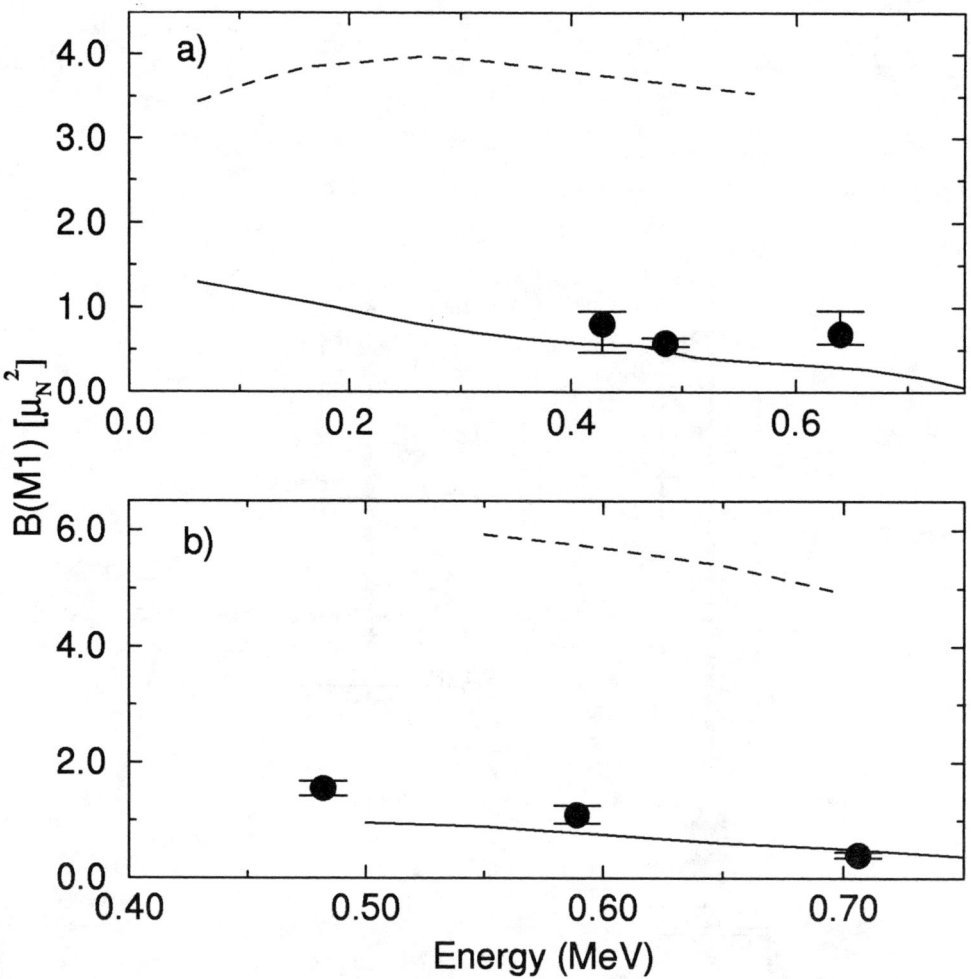

FIGURE 7. B(M1) values as a function of γ-ray energy for band 5 in ^{108}Cd. (a) shows the values below the $\nu(h_{11/2})^2$ alignment, whilst (b) shows the values after the alignment has occured. The solid lines show the TAC predictions for the $\pi g_{[9/2}^{-3} g_{7/2}^1]\otimes\nu h_{11/2}^1(g_{7/2},d_{5/2})^1$ and the $\pi g_{[9/2}^{-3} g_{7/2}^1]\otimes\nu h_{11/2}^3(g_{7/2},d_{5/2})^1$ configurations in (a) and (b) respectively. The dashed lines are the results for similar calculations for the (a) $\pi g_{9/2}^{-2}\otimes\nu h_{11/2}^1(g_{7/2},d_{5/2})^3$ and (b) $\pi g_{9/2}^{-2}\otimes\nu h_{11/2}^3(g_{7/2},d_{5/2})^3$ configurations.

quadrupole deformation. Total routhian surface calculations indicate that ^{108}Cd is essentially prolate for all configurations. With this assumption the average quadrupole deformation, after the $(h_{11/2})^2$ neutron alignment is found to be β_2=0.11±0.01. Near the top of the band this reduces to β_2=0.06. These results are in good agreement with the deformations obtained for the shears bands in the lead nuclei [3].

Sn Nuclei

Four new bands, consisting of magnetic dipole transitions, have been observed in ^{106}Sn (2 bands) and ^{108}Sn (2 bands) (see [15] for further details). For the bands in ^{108}Sn E2 crossover transitions are observed and the angular momentum and B(M1)/B(E2) values can be nicely reproduced by the two lowest energy configurations in the TAC model (see fig. 8). However, in ^{106}Sn the dipole bands have no associated E2 crossover transitions, which leads to very high B(M1)/B(E2) ratios of >100 $(\mu_N/\text{eb})^2$ in both cases. Moreover, there are no configurations within the TAC model which can reproduce these very high values. A similar situation was also observed for an M1 band in ^{105}Sn [22]. We have obtained some preliminary lifetime results for band 1 in ^{106}Sn which has recently been observed to branch into two bands at the top (see fig. 9). One branch appears to be the continuation of band 1 whilst the other shows evidence for an alignment at a frequency of $\sim$0.6 MeV/$\hbar$. Figure

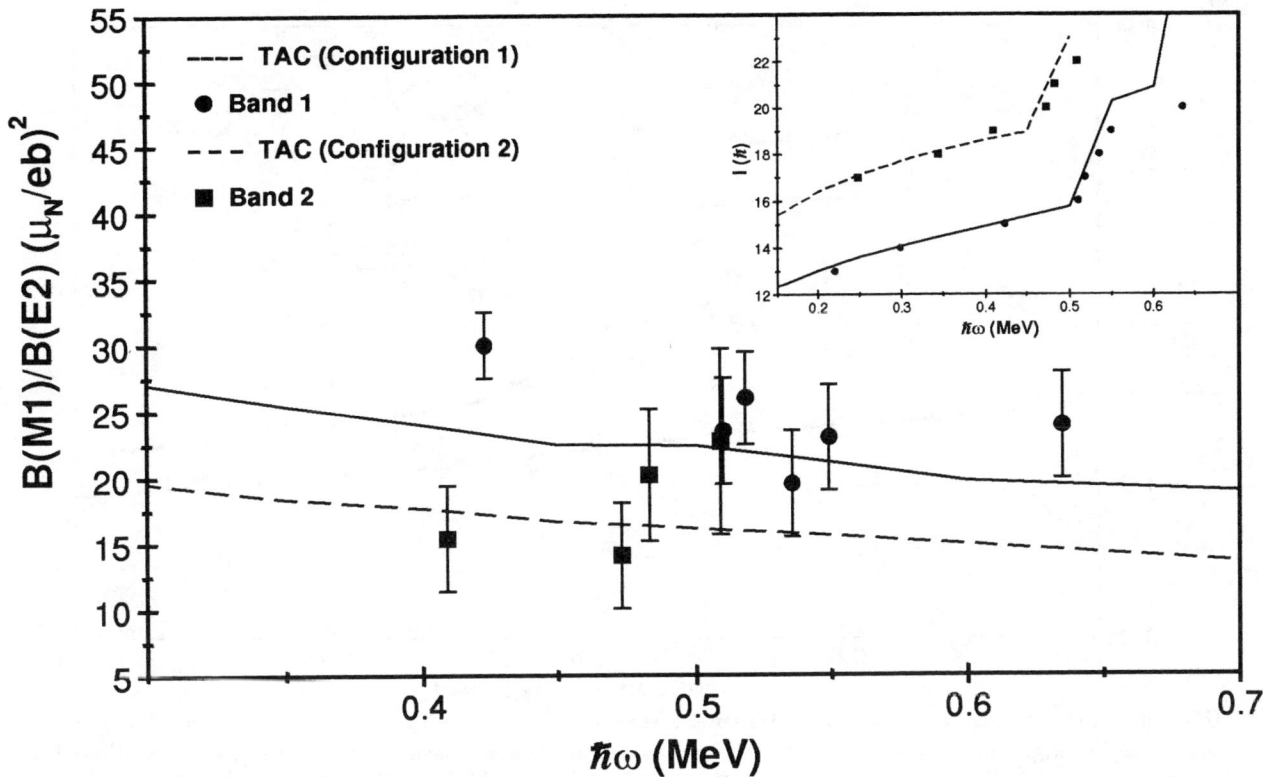

FIGURE 8. B(M1)/B(E2) values as a function of transition energy ($\hbar\omega$) for the two M1 bands in ^{108}Sn. The solid lines show the TAC calculations for the $\pi[g_{9/2}^{-1}g_{7/2}^1]\otimes\nu h_{11/2}^1(g_{7/2},d_{5/2})^1$ (band 1) and $\pi g[_{9/2}^{-1}g_{7/2}^1]\otimes\nu h_{11/2}^1(g_{7/2},d_{5/2})^3$ (band 2) configurations. The inset shows the calculated and experimental spins as a function of frequency using the same configurations as specified above.

9 shows the preliminary B(M1) values as a function of γ-ray energy (frequency). The solid line shows the results of a TAC calculation for the $\pi[g_{9/2}^{-1}g_{7/2}^1]\otimes\nu[h_{11/2}^1(g_{7/2},d_{5/2})^3]$ configuration for this band. The rise in the theoretical B(M1) values at ~0.6 MeV results from the alignment of a pair of $h_{11/2}$ neutrons. The dot-dashed line represents the continuation of the initial configuration (i.e., no $h_{11/2}$ neutron alignment). This agrees reasonably well with the experimental results (filled squares).

At the present time we only have one data point (open square) from the branch which shows the $(h_{11/2})^2$ neutron alignment. This shows an increase in B(M1) strength as predicted. Fig. 9 also shows the results of calculations (dashed lines) performed with the Donau and Frauendorf formalism for the same configuration as used in the the TAC calculations. The two lines correspond to calculations where either the $d_{5/2}$ or $g_{7/2}$ g-factors have been used and therefore represent the limits. The TAC calculations show much better agreement with the data. Analysis of these data is continuing.

In order to understand why the TAC model fails to reproduce the B(M1)/B(E2) values for the bands in the lightest tin isotopes whilst being able to provide a reasonable description of the B(M1) values it is necessary to look at the ingredients of the model. For example, higher deformation multipoles, induced by the high-j particles involved in the configurations (e.g., hexadecapole deformation), are not fully treated within the TAC model. It is clear however that such multipoles exist from the density distributions of the particles and holes. Since the minimisation that is carried out for the deformation parameters did not involve higher order multipoles it is likely that the quadrupole deformation increased to compensate for their absence. Thus in ^{105,106}Sn the TAC predicts too large a deformation for all configurations, and, as a consequence, yields B(M1)/B(E2) ratios which are much smaller than the observed values. In ^{108}Sn the quadrupole deformation will be larger than it is in ^{105,106}Sn, hence the effect of the absence of the higher order multipoles in the calculations might be expected to be less in this case.

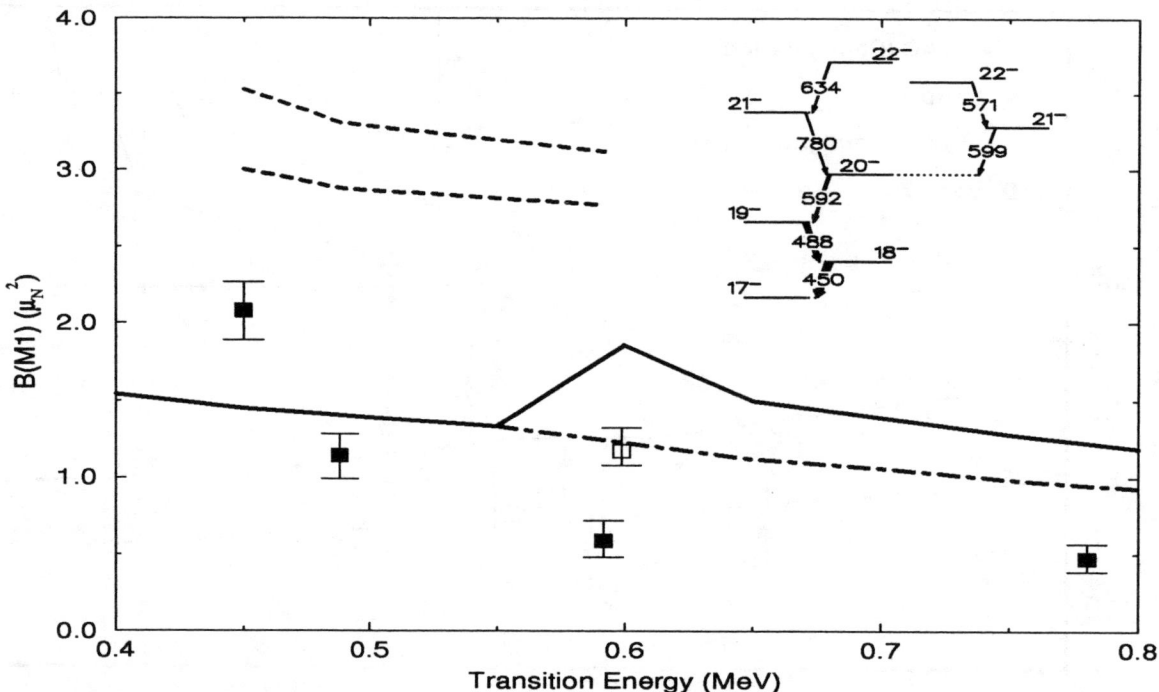

FIGURE 9. B(M1) values as a function of transition energy ($\hbar\omega$) for band 1 in ^{106}Sn. The solid shows the TAC calculation for the $\pi[g_{9/2}^{-1}g_{7/2}^{1}]\otimes\nu[h_{11/2}^{1}(g_{7/2},d_{5/2})^{3}]$ configuration. The dot-dashed line represents the continuation of the initial configuration (i.e., no $h_{11/2}$ neutron alignment). The dashed lines show the results of the Donau and Frauendorf standard cranking calculations for the same configuration (see text). The insert at the top right shows the new decay scheme for this band.

Sb Nuclei

Two strongly coupled negative parity bands have been observed in ^{108}Sb, one of these (band 1) was known from previous work [23] but has been extended in the present work. The second band is new. Further details of the extended level scheme can be found in [24]. The configurations assigned to the two bands are $\pi[g_{7/2}^{2}\otimes g_{9/2}^{-1}]\otimes\nu h_{11/2}$ (band 1) and $\pi[h_{11/2}g_{7/2}\otimes g_{9/2}^{-1}]\otimes\nu g_{7/2}$ (band 2). The B(M1)/B(E2) ratios for both bands are shown in fig. 10 together with the TAC calculations for the lowest energy negative parity configurations. The results show good agreement with the model predictions. The rise in the B(M1)/B(E2) ratios for band 1 at a frequency of ~ 0.55 MeV can be understood in terms of the alignment of the second and third pair of $h_{11/2}$ neutrons. As discussed previously for band 2 in ^{197}Pb, this results in an opening of the blades of the shears and consequently an increase in μ_{perp} and hence the B(M1) values. It is debateable however whether band 2 can really be thought of as a shears type structure at low spin/frequency since it does not involve any $h_{11/2}$ neutrons. The rise in the B(M1)/B(E2) values at ~ 0.4 MeV results from the alignment of a pair of $h_{11/2}$ neutrons. This could be sufficient to set the shears mechanism in action. Unfortunately, the experimental data are not good enough to test the predictions at high frequencies. This is primarily because of the near degeneracy of some of the γ-rays and poor statistics for the highest transitions. Lifetime data are currently being analysed for these bands.

Conclusion

The data presented above suggest that the shears mechanism (and the phenomenon of magnetic rotation) is present in the weakly deformed prolate nuclei in the Z$\sim$50, A$\sim$110 mass region. It should be noted however that for nuclei which are not too far removed from this region there are strongly coupled bands which can be adequately explained by the standard cranking models such as the Donau and Frauendorf formalism. For example, the dipole bands in ^{110,112}Te (Z=52) are found to agree well with the standard calculations [25]. One of the important features which appears to distinguish between the applicability of standard and tilted axis

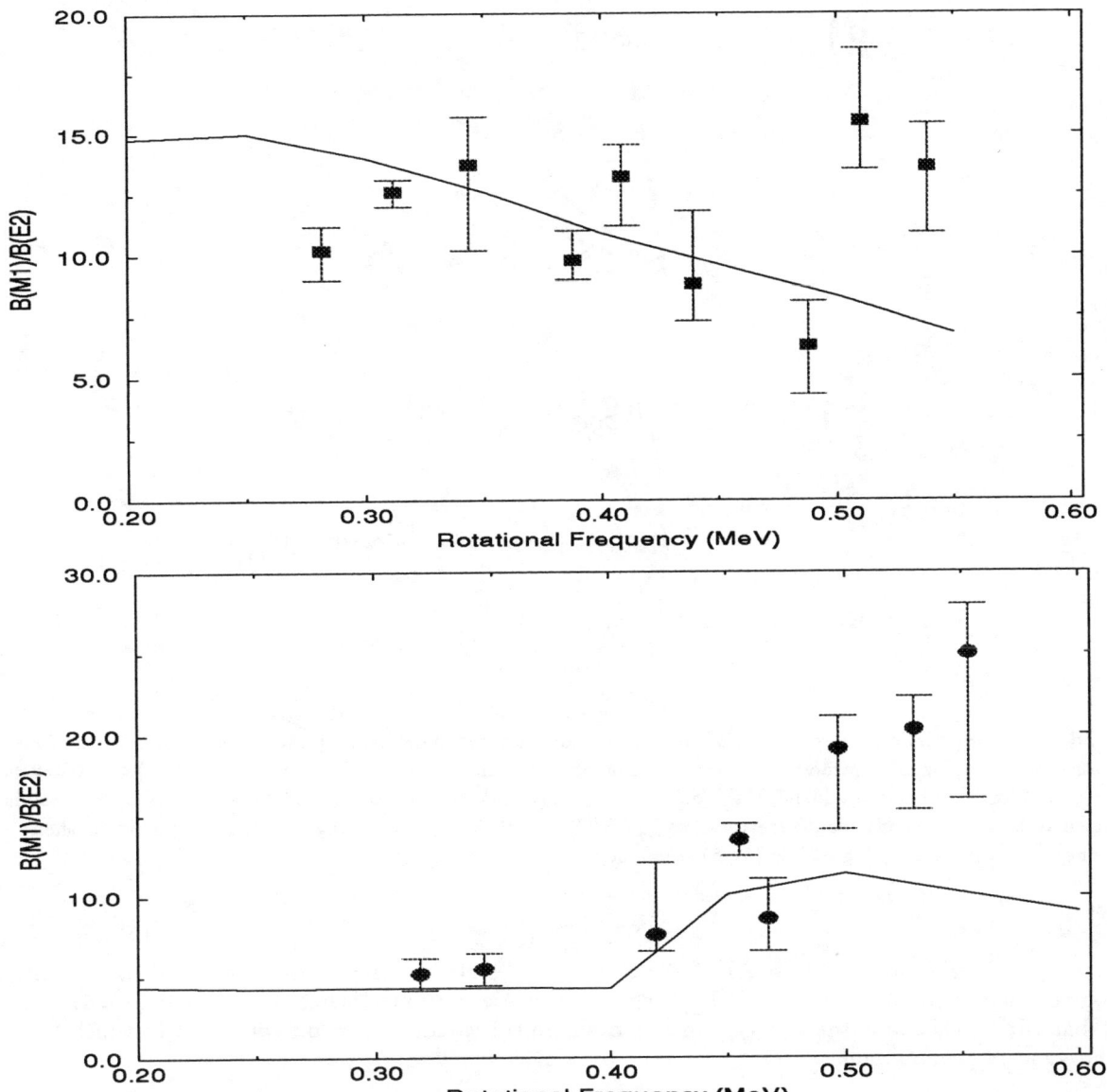

FIGURE 10. Experimental B(M1)/B(E2) ratios as a function of frequency (γ-ray energy) for dipole bands 1 (squares) and 2 (circles) in ^{108}Sb. The solid lines show the TAC calculations for the $\pi[g_{7/2}^2 \otimes g_{9/2}^{-1}] \otimes \nu h_{11/2}$ (band 1) and $\pi[h_{11/2} g_{7/2} \otimes g_{9/2}^{-1}] \otimes \nu g_{7/2}$ (band 2) configurations.

cranking to these structures is the deformation of the nucleus for the particular configuration. If the nucleus is weakly deformed ($\beta_2 \sim 0.1$), the spin is mainly generated by the shears mechanism, however, if the nucleus is more strongly deformed ($\beta_2 \sim 0.2$), then the collective contribution plays an important role and effectively dominates any shears type mechanism effects.

$\Delta I = 2$ SHEARS BANDS

So far the discussion has focused on shears bands with vectors as shown in fig. 11a, i.e., the blades of the shears result from different types of particles. However, another type of shears mechanism has been proposed by Stefan Frauendorf where the blades are composed of vectors resulting from the same type of particle. This is illustrated in fig. 11b where the two $g_{9/2}$ holes are effectively coupled to K=0 at the band-head. As the spin increases these vectors gradually align along the $\nu(h_{11/2})^2$ vector. In this case the shears mechanism produces states of increasing angular momentum but now there is no large magnetic dipole moment. The reflection

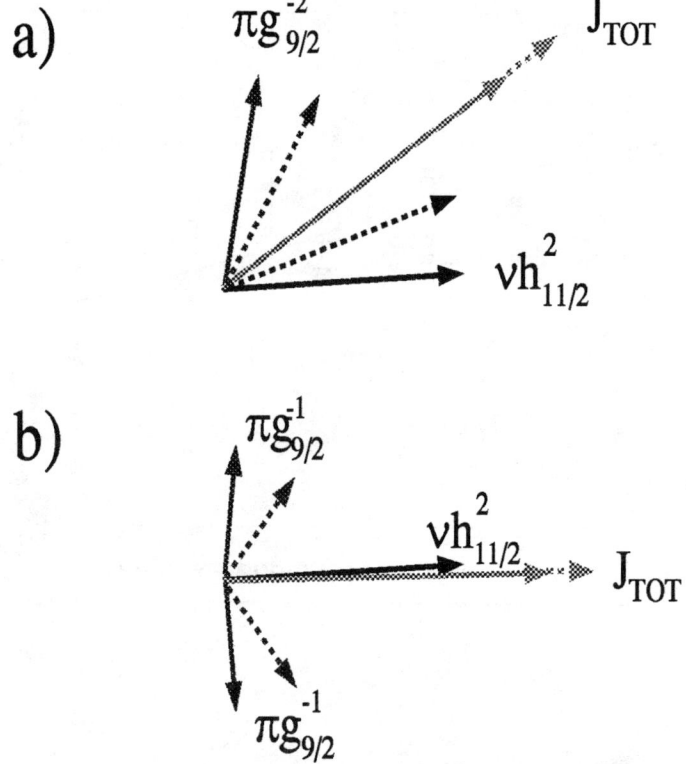

FIGURE 11. Schematic representation of the 'shears mechanism' and the coupling schemes for: a) the M1–shears type structures (giving a large perpendicular component of the magnetic dipole moment, $\mu_\perp$,); b) the $\Delta I=2$ type of shears structures (which result in no M1 strength and only weak E2's – see text). Note, in ^{110}Cd, two bands were predicted to exist with similar configurations ($\pi g_{9/2}^{-2} \otimes \nu h_{11/2}^{2}$) but different coupling schemes. The solid (dashed) arrows show the approximate coupling at low (high) frequency.

symmetry of the system leads to a $\Delta I=2$ structure and the states are connected by weak E2 transitions in near-spherical nuclei. Moreover, the B(E2)'s should decrease with increasing spin. It was proposed that the yrast band in ^{110}Cd above the $\nu(h_{11/2})^2$ neutron alignment would be a good candidate in which to observe this new shears phenomenon.

Results for ^{110}Cd

The ^{18}O + ^{96}Zr reaction at a beam energy of 70 MeV was used to determine the lifetimes of states in the yrast band above the $\nu(h_{11/2})^2$ band crossing in ^{110}Cd. Further details of the experiment and data analysis can be found in [18]. Previous work on this nucleus had assigned the first band crossing in the yrast band to a pair of $h_{11/2}$ neutrons [19]. Fig. 12 shows the angular momentum and B(E2) values as a function of frequency, together with results of a TAC calculation for the $\pi g_{9/2}^{-2} \otimes \nu h_{11/2}^2$ configuration. Clearly the agreement with the calculations is good, however, standard cranking calculations can also reproduce the experimental data just as well. The trend of the experimental B(E2) values suggests that in this case a significant fraction of the total spin is carried by the core (i.e., if the shears mechanism had been solely responsible a faster drop off in B(E2) values with increasing spin would be expected). The TAC calculations indicate that ~50% of the angular momentum along the band comes from the shears mechanism, the rest being generated by the collective rotation of the core. Unfortunately, the data in this instance do not provide a clear signature of the $\Delta I=2$ shears mechanism. This phenomenon should be investigated further, for example, in lighter Cd nuclei where the core deformation may be smaller, or possibly in Pd nuclei where the hole vectors are larger and hence should produce a greater effect.

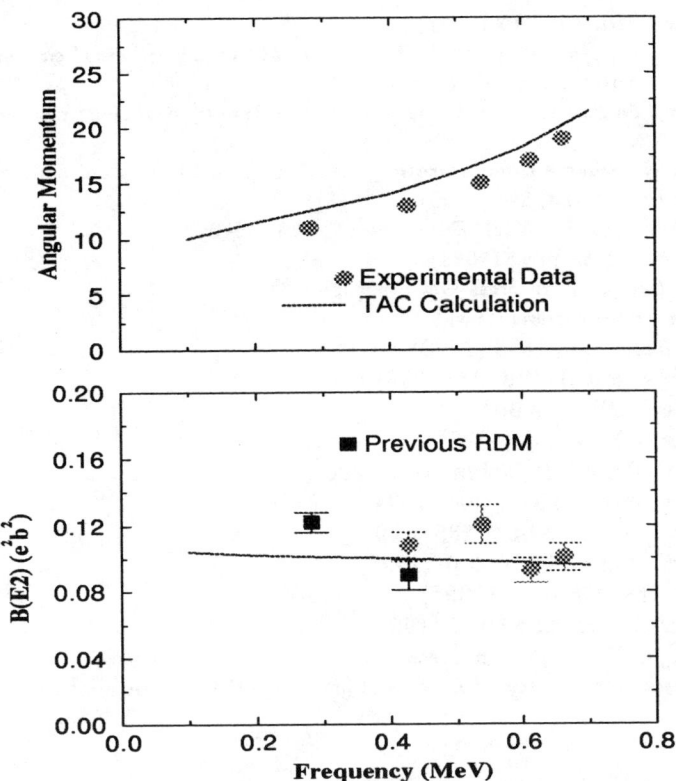

FIGURE 12. Plots as functions of frequency (γ-ray energy) of: (top) angular momentum for the yrast band in ^{110}Cd above the $\nu h(_{11/2})^2$ alignment (the proposed $\Delta I=2$ shears band) and (bottom) the experimentally deduced B(E2) values for this band (in the lowere figure the circles are results from the current work whilst the squares show the results of previous RDM measurements.). In both cases the solid line shows the results of TAC calulations for the $\pi g_{9/2}^{-2} \otimes \nu h(_{11/2})^2$ configuration.

SUMMARY

There is now a large body of data which exists in both the A~110 and 200 mass regions which supports the existence of the phenomenon of magnetic rotation in nuclei which have either a small prolate or oblate deformation. It is clear however that the TAC model has some limitations, particularly in nuclei where the quadrupole deformation is quite small (e.g., ^{105,106}Sn). The model clearly needs a certain level of deformation to work successfully. In this regard it would be interesting to investigate the detailed predictions of other approaches, for example, see the contribution to this conference by A. Macchiavelli, where the M1 structures are described in terms of a residual intereaction between the proton and neutron spin vectors arising from a particle-vibration coupling. This approach does not require the existence of deformation and consequently may provide a better interpretation of these structures in the Sn region as the nuclei become more spherical. This latter approach may also be interesting in order to study the competition between the shears type structures and collective rotation in nuclei (i.e., bands which can be explained by the standard cranking type approach such as the Donau and Frauendorf formalism). Finally, it would also be interesting to compare the current experimental results with the latest shell model calculations.

REFERENCES

1. Clark, R.M. et al, *Nucl. Phys* **A562**, 121 (1993).
2. Baldsiefen, G. et al, *Nucl. Phys* **A574**, 521 (1994).
3. Clark, R.M. et al, *Phys. Rev. lett.* **78**, 1868 (1997).
4. Frauendorf, S., *Nucl. Phys.* **A557**, 259c (1993).

5. Chmel, S. et al, *Phys. Rev. lett.* **79**, 2002 (1997).

6. Donau, F. and Frauendorf, S., *proceedings of the International Conference on High Angular Momentum properties of Nuclei, Oak Ridge*, New York: Harwood, 1983, p. 143.

7. Frauendorf, S., *Proc. Conf. on Physics from Large gamma Ray Arrays, Berkeley* LBL-35687, CONF-940888, UC-413 Vol. 2, 1994, p. 52.

8. Wells, J.C. and Johnson, N., *private communication*, modified from the original code by Bacelar, J.

9. Gascon, J. et al, *Nucl. Phys.* **A513**, 344 (1990).

10. Northcliffe, L.C. and Schilling, R.F., *Nucl. Data tables* **7**, 233 (1970).

11. Clark, R.M. et al Phys. lett. **B**, in press (1998).

12. Krucken, R. et al, *Phys. Rev.* **C58**, in press Oct (1998).

13. Frauendorf, S., *private communication* (1998).

14. Schnare, H. et al, *Phys. Rev.* [**C54**, 1598 (1996).

15. Jenkins, D.G. et al, *Phys. lett.* **B428**, 23 (1998).

16. Frauendorf, S. et al, *Nucl. Phys.* **A601**, 41 (1996).

17. Frauendorf, S., *Z. Phys.* **A356**, 263 (1996).

18. Clark, R.M. et al, *Phys. Rev. Lett.*, submitted (1998).

19. Juutinen, S. et al, *Nucl. Phys.* **A573**, 306 (1994).

20. Thorslund, I. et al, *Nucl. Phys.* **A564**, 285 (1993).

21. Regan, P.R., et al, *Nucl. Phys.* **A586**, 351 (1995).

22. Gadea, A. et al, *Phys. Rev.* **C55**, R1 (1997).

23. Cederkall, J. et al, *Nucl. Phys.* **A581** 189 (1995).

24. Jenkins, D.G. et al, *Phys. Rev.* **C58**, in press Nov (1998).

25. Boston, A., *Ph.D. Thesis, University of Liverpool* (1998) and to be published.

Semi-classical Shears and Effective Forces [1]

A.O. Macchiavelli, R.M. Clark, M.A. Deleplanque, R.M. Diamond, P. Fallon
I.Y. Lee, F.S. Stephens and K. Vetter

Nuclear Science Division
Lawrence Berkeley National Laboratory, Berkeley, California 94720

Abstract. We present a semi-classical analysis of the $B(M1)$ and $B(E2)$ transition probabilities in the shears bands in $^{193-199}Pb$ as a function of the shears angle. This provides a *semi-empirical* confirmation of the shears mechanism proposed by S. Frauendorf using the Tilted-Axis-Cranking model. We interpret this as a consequence of a residual interaction between the proton and neutron blades, and it is shown that the main ingredient of this effective force can be described by a $P_2(\theta)$ term with a strength of $400 - 600\ keV$. Such an interaction can be mediated through the core by particle-vibration coupling. The competition between the shears and the rotation of the core is investigated within the framework of a classical solution of two particles-plus-rotor model.

INTRODUCTION

Using the Tilted-Axis-Cranking (TAC) model, S.Frauendorf [1] interpreted the rotational behavior of the so-called M1 bands in neutron-deficient Pb isotopes [2] as a coupling of $h_{9/2}$ and $i_{13/2}$ protons and $i_{13/2}$ neutron-holes to a slightly deformed core. The total angular momentum is generated by aligning the proton and neutron spin vectors, $\vec{j}_\pi$ and $\vec{j}_\nu$, in a way that resembles the closing of a pair of shears, hence the name usually given to these structures: *shears bands*. Recent Gammasphere data on the lifetimes of states in the M1 bands in $^{193-199}$Pb isotopes [3] provide the most sensitive test of the shears mechanism. This new mode of "rotational motion" has also been observed in other mass regions, for example, in neutron-deficient Cd, Sn, In and Sm nuclei [4].

A comprehensive review of the experimental data obtained over the last few years and an overview of the TAC results were presented by R. Wadsworth [5]. In this work we discuss a semi-classical analysis [6] of the $B(M1)$ and $B(E2)$ values recently measured for the shears bands in neutron-deficient Pb isotopes [3].

The good agreement between the data and both the TAC calculations and the present analysis suggests to us that the important degree of freedom is indeed the shears angle. Therefore, we go a step further and obtain information on the nature of the effective interaction between the proton and the neutron constituents of the blades from the correlation between the shears angle and the level energies above the bandhead. We show evidence that a main ingredient of this interaction can be described by a $V_2 P_2(\theta)$ term and interpret its origin from particle-vibration coupling acting in second order [7].

Finally, we will discuss the competition between the shears mechanism and the rotation of the core based on the results of a classical particles-plus-rotor model. In this simple framework, the competition between these two modes is governed by the relative strength of the residual force between the blades and the rotational energy of the core.

ANALYSIS OF ELECTROMAGNETIC PROPERTIES

Following the nomenclature introduced in Fig. 1, we start by defining θ_π and θ_ν as the angles of the proton and neutron spin vectors, $\vec{j}_\nu$ and $\vec{j}_\pi$, with respect to the total angular momentum, $\vec{I} = \vec{j}_\nu + \vec{j}_\pi$. The shears

---

[1] Work supported by the US Department of Energy under grant DE-AC03-76SF00098.

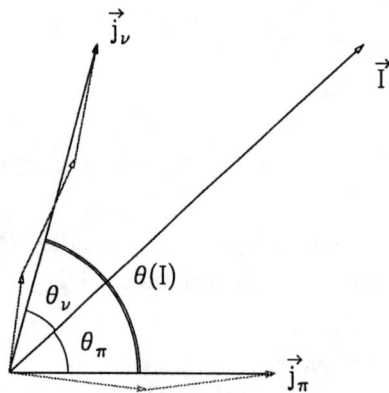

FIGURE 1. Schematic representation of the angular momentum coupling between protons and neutrons in a shears band.

angle θ that corresponds to a given state in the band can be derived using the semi-classical expression

$$cos\theta = \frac{\vec{j}_\nu \cdot \vec{j}_\pi}{j_\nu j_\pi} = \frac{\vec{I}^2 - \vec{j}_\nu^2 - \vec{j}_\pi^2}{2j_\nu j_\pi} \qquad (1)$$

Since the $B(M1)$ values are proportional to the square of the component of the magnetic moment perpendicular to the spin vector [8,9] they should show a characteristic drop as the shears close (i.e. $\theta \approx 90^0 \to \theta \approx 0^0$). From the geometry specified in Fig. 1, this dependence is given by

$$B(M1, I \to I-1) = \frac{3}{4\pi}\frac{1}{2}\vec{\mu}_\perp^2 = \frac{3}{4\pi}g_{eff}^2 j_\pi^2 \frac{1}{2}sin^2\theta_\pi \qquad [\mu_N^2] \qquad (2)$$

as a function of the proton angle $\theta_\pi = \theta_\pi(\theta)$ and where we have introduced an effective gyromagnetic factor, $g_{eff} = g_\pi - g_\nu$. The relation between θ and θ_π is given by the formula $tan\theta_\pi = j_\nu sin\theta/(j_\pi + j_\nu cos\theta)$. In a similar way, from the parallel component of the magnetic moment we can obtain an expression for the gyromagnetic factor of a state with angular momentum I as

$$g(I) = g_{eff}\left(\frac{1}{1 + tan\theta_\pi/tan\theta_\nu}\right) + g_\nu \qquad (3)$$

The rather large values of $B(M1)$ suggest that a proton configuration is playing an important role. In our analysis we fix that configuration to $j_\pi = 11$ (corresponding to the 11^- K-isomer seen in several of the Pb nuclei) and the neutron spin j_ν is determined to reproduce the spin of the bandhead state in each band, under the assumption of perpendicular coupling [10]. The angle of the blade, θ, for higher spins is then obtained from Eq. (1). From the measured $B(M1)$'s we derive g_{eff} for the different bands in $^{193-199}Pb$ using Eq. (2) and these values are shown in Fig. 2. We can use the measured values [11] for the configurations $(\pi h_{9/2} \otimes \pi i_{13/2})_{11^-}$ (g_π) and $(\nu i_{13/2})_{12+}^{-2}$ (g_ν) to estimate $g_{eff} \approx 1.12$ as a representative value for the Pb isotopes. In ^{110}Cd, a similar estimate gives $g_{eff} \approx 1.45$ from the measured g-factors [11] of the $(\pi g_{9/2})_{8+}$ and $(\nu h_{11/2})$ configurations. As observed in the figure, the results seem consistent with these estimates (dotted line) and while most of the isotopes considered show a g-factor somewhat smaller than expected, ^{194}Pb seems peculiar in that respect and may invite a closer look.

Overall, the approximate constancy of g_{eff} indicates that expression (2) provides a reasonable description of the experimental data. Indeed, this is shown in a different way in Fig. 3a) by plotting $B(M1)$'s for $^{198,199}Pb$ as a function of the shears angle together with the results of Eq. (2) using an average $g_{eff} = 0.92$. Moreover, if we use the configurations above, Eq.(3) gives $g \approx 0.6$ for the bandhead of ^{193}Pb, in agreement with the result $g = 0.68 \pm 0.03$, reported in Ref. [10].

Following Refs. [8,9] we can also derive a similar expression for $B(E2, I \to I-2)$ values, which are proportional to the square of the $\mathcal{M}(E2, \mu = 2)$ component of the electric quadrupole tensor. Here we have

$$B(E2, I \to I-2) = \frac{5}{16\pi}(eQ)_{eff}^2 \frac{3}{8}sin^4\theta_\pi \qquad [e^2b^2] \qquad (4)$$

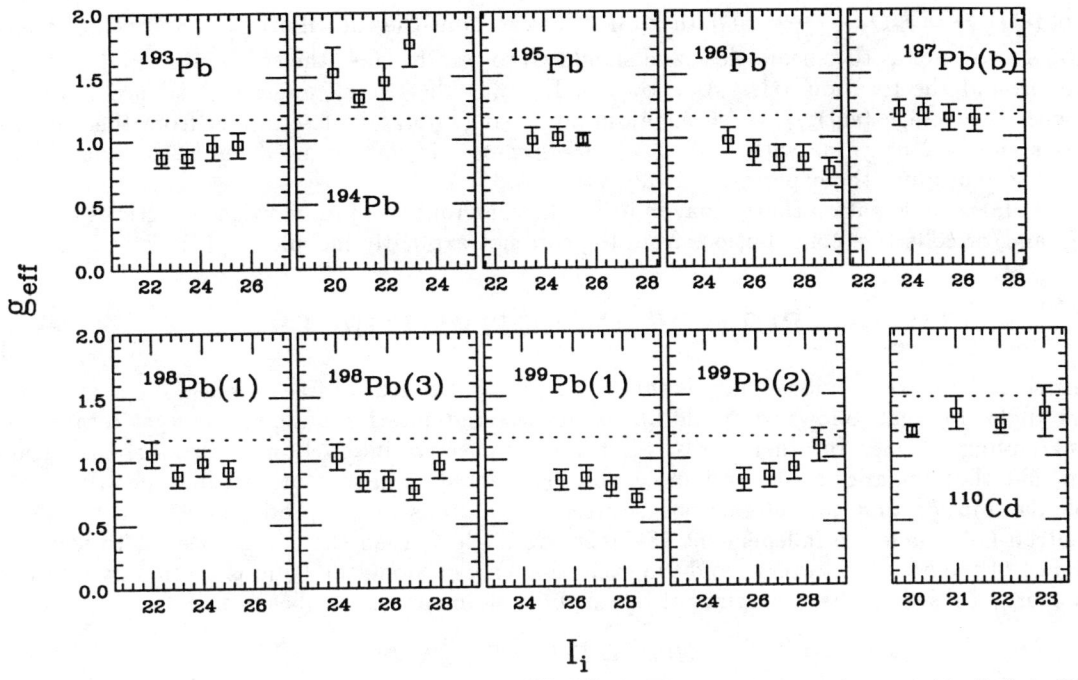

FIGURE 2. Effective g-factors obtained for the bands in $^{193-199}Pb$. The labels follow those used in Ref. 3. The dotted line is the value expected from the configurations $(\pi h_{9/2} \otimes \pi i_{13/2})_{11^-}$ and $(\nu i_{13/2})^{-2}_{12^+}$ in the Pb nuclei and $(\pi g_{9/2})_{8^+}$ and $(\nu h_{11/2})$ for ^{110}Cd

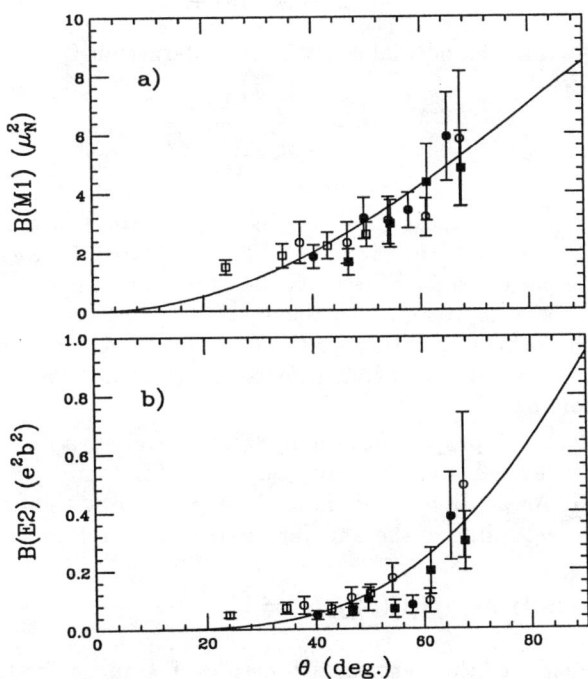

FIGURE 3. a) B(M1) values as a function of the shears angle. The solid line is the result of Eq. (2) with $g_{eff} = 0.92$. b) B(E2) values as a function of the shears angle, the solid line is from Eq. (4) using $(eQ)_{eff} = 6.5eb$. Symbols are as follows, solid-circle: $^{198}Pb(1)$; circle: $^{198}Pb(3)$; solid-square: $^{199}Pb(1)$; square: $^{199}Pb(2)$.

in terms of $(eQ)_{eff} = e_\pi Q_\pi + (\frac{j_\pi}{j_\nu})^2 e_\nu Q_\nu$ that takes into account the contributions from protons and neutrons. The $B(E2)$'s also drop as the shears close and should go to zero because the charge distribution then becomes symmetric around the rotation axis. As shown in Fig. 3b) the overall angle dependence is reproduced by Eq. (4) with an average $(eQ)_{eff} \approx 6.5eb$. Assuming that $(eQ)_{eff}$ is determined from the contributions of ~ 2 protons and ~ 2 neutrons with radii $< r_\pi^2 > \sim < r_\nu^2 > \sim 7^2 fm^2$ we need an $E2$ polarization charge [8] $(e_{pol})_{E2} \approx 3$ to reproduce the experimental value.

This rather large polarization charge may indicate the contribution of more valence particles to the estimates of $(eQ)_{eff}$ and/or collective contributions from the core not explicitly included in (4).

ROLE OF EFFECTIVE FORCES

It seems clear from the previous analysis that the important degree of freedom in these structures is indeed the shears angle. In what follows we would like to address the question of how do we get rotational behavior from this coupling scheme. In other words, what kind of effective interaction in terms of the shears angle is needed so that the dynamics of the system can be represented with a rotational-like spectrum. Making the reasonable assumption that the interactions between the protons in the proton blade and the neutron-holes in the neutron-hole blade are independent of the shears angle θ, then the energy needed to form each of the blades is part of the bandhead energy and the excitation energy along the band is given only by the change in potential energy caused by the recoupling of the angular momenta in the shears, i.e.

$$V(I(\theta)) = E(I) - E_{bandhead} \qquad (5)$$

Therefore, knowing the angle θ between $\vec{j}_\pi$ and $\vec{j}_\nu$ and the level energies $E(I)$ it is possible to obtain information about the nature of this effective interaction, V, between the protons and the neutrons [12,13]. If we restrict ourselves to spatial forces we can expand this interaction in even multipoles as [14]:

$$V(\theta) = V_0 + V_2 P_2(\theta) + ... \qquad (6)$$

Following Schiffer [12], we look at the normalized effective interaction :

$$\frac{V(I(\theta))}{\bar{V}} = \frac{V(I(\theta))}{\frac{\sum_I (2I+1)V(I)}{\sum_I (2I+1)}} \qquad (7)$$

This is shown in Fig. 4a) for the bands in ^{198,199}Pb as a function of θ and for comparison, the result expected for a P_2 term for $j_\nu = j_\pi = 15$ in Fig. 4b). The similarity between the two suggests to us that the main ingredient of this residual interaction between the protons and neutrons in the shears arises from the long-range part of the nuclear force, represented by the P_2 term.

Experimentally, the strength of this interaction for the Pb region is found to be $\sim 2.3\ MeV$, and taking into account ~ 2 protons and $\sim 2 - 3$ neutrons as constituents of the blades, we derive a value of $400 - 600\ keV$ for a single proton, neutron-hole pair.

We will now show that indeed the properties of a P_2 term can give rise to a rotational spectrum. Let us assume for simplicity that we have a neutron- and a proton-blade of the same j coupled to spin I and interacting via a term of the form $V_2 P_2(\theta)$. As we discussed before, the energy along the band is given only by the change in potential energy due to the recoupling of the angular momenta, therefore from Eqs. (5) and (6):

$$E(I) \propto < jjI|V(\theta)|jjI > = V_2 \frac{(3cos^2\theta(I) - 1)}{2} \qquad (8)$$

In Fig. 5 we show the dependence of this term as a function of I and θ for the particle-particle (hole-hole) and particle-hole cases. As can be seen, the minimum of the potential energy for the particle-particle case (V_2 negative) occurs at $\theta = 180^0$, $I = 0$ where the overlap between the particles' wave functions is maximum. Since for $I < j$ we have $sin\theta \approx (I/j)$, it follows from Eq. (8) that the low-spin members of the $(2j + 1)$-multiplet are split approximately by I^2 as shown with the dashed curve. This result leads to the interesting prediction for the existence of excited *shears bands* in near spherical $N = Z$ nuclei where protons and neutrons occupy the same high-j particle-particle (hole-hole) orbits. It is however uncertain whether these bands will be low enough in excitation energy to be seen in (HI,xn) reactions.

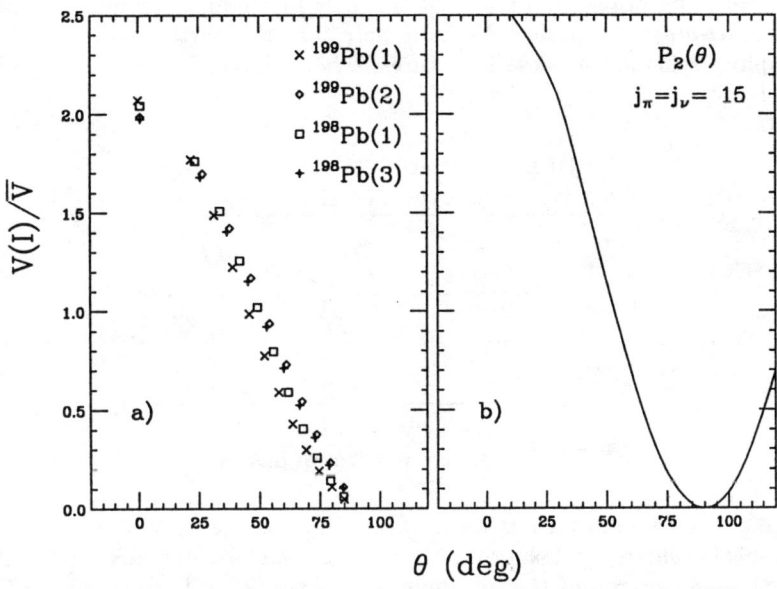

FIGURE 4. Effective interaction between blades as a function of the shears angle. The results derived from the excitation energies in the M1 bands in ^{198,199}Pb (a)) are compared with the expected dependence of a P_2 term for the case of $j_\pi = j_\nu = 15$ (b)). Labels for the different bands follow those used in Ref. [3]

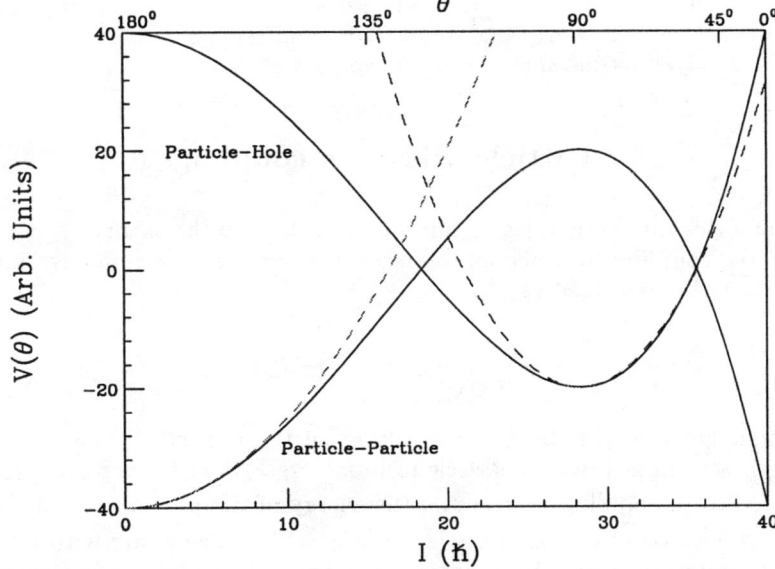

FIGURE 5. Particle-particle(hole) potential as a function of angular momentum and θ for an interaction of the form $V_2 P_2(\theta)$. The dashed lines correspond to a rotational approximation. The example is for $j_\pi = j_\nu = j = 20$.

If we now turn our attention to the particle-hole channel, which is the situation in the Pb *shears bands*, the spatial interaction we are considering changes sign, i.e. $V \to -V$ [14] (V_2 positive), and the minimum now occurs at $\theta = 90^0$, $I_{90} = \sqrt{2}j$. We then obtain $E \propto (I - I_{90})^2$, as observed in experiment with I_{90} representing the spin of the bandhead. It is clear that one can also couple $\vec{j}_\pi$ and $\vec{j}_\nu$ to angles greater than 90^0. An inspection of Fig. 5 shows that these members of the multiplet will lie on the unfavoured side of the parabola. The observation of these states would provide an important signature of the shears mechanism.

It has also been suggested by Frauendorf [15] that this new mechanism should give rise to $E2$ rotational-like bands in the so-called *Anti-magnetic Rotor*. We note here that the particle-particle or hole-hole channels can provide the basic coupling scheme for these structures, where, having identical particles(holes), the allowed spins differ now by $\Delta I = 2$.

TABLE 1. Moments of inertia of shears bands in different mass-regions.

A	$\mathcal{J}$ [a] $(\hbar^2/MeV)$	$\mathcal{J}_0$ [b] $(\hbar^2/MeV)$	$\mathcal{J}/\mathcal{J}_0$
200	21.2	95	0.23
140	13.2	52.4	0.25
110	9.4	35	0.26
105	8.6	32.5	0.26

[a] From I vs. ω plots

[b] The rigid-sphere moment of inertia $\frac{A^{5/3}}{72}$

Moments of Inertia.- A question that readily comes to mind concerns the mass dependence of the moment of inertia , $\mathcal{J}$, of these M1 bands, and *a priori* there is not an obvious answer for the shears mechanism. However, from Eq. (8) we have around the minimum $E \approx (3V_2/2j^2)I^2$ which gives $\mathcal{J}=j^2/3V_2$, and because we expect the overall dependence $j \sim A^{1/3}$ and $V_2 \sim A^{-1}$ then $\mathcal{J} \sim A^{5/3}$ as in the case of normal rotational bands. Although the available information is still limited to a few examples that span a broad range of masses, they seem to confirm this prediction as shown in Table 1.

As a final point in this discussion, let us consider a spin-dependent force, represented for example by a $\vec{j}_\pi \cdot \vec{j}_\nu$ interaction. A P_1 term proportional to $cos\theta$ can now appear in Eq. (6) and naturally give rise to a rotational spectrum. However, since this term favors 0^0 or 180^0 coupling depending on its sign, one should realize that its contribution must be small based on the fact that the bandhead configuration is known to have perpendicular coupling [10]. Besides, the angles shown in Fig. 2 are derived under this assumption and explain the $B(M1)$ behavior rather well. Likewise, a large contribution from a short-range component can be ruled out since, for example, a delta force cannot give rise to a rotational-like spectrum. While it remains as an intriguing question whether a spin-dependent term or a short-range term may manifest themselves in the shears mechanism, it is clear that a major role is played by the spatial long-range part of the force.

Particle-vibration coupling

In order to understand this interaction, let us turn our attention to the theory of particle-vibration coupling [16]. To second order the coupling to a phonon of order λ gives rise to an effective interaction between two particles, 1 and 2, that can be expressed as:

$$V_\lambda = \frac{(2\lambda + 1)}{4\pi C_\lambda} k_\lambda(r_1)k_\lambda(r_2)P_\lambda(\vartheta_{12}) \tag{9}$$

with ϑ_{12} being the angle between the the position vectors of the particles. This effective interaction is of a P_λ type with a strength determined by the particle form factors k_λ and the restoring force parameter C_λ, the latter of which is related to the amplitude (α_λ) and the energy of the phonon ($\hbar\omega$) by $C_\lambda = \frac{(2\lambda+1)\hbar\omega_\lambda}{|<1||\alpha_\lambda||0>|^2}$. We now apply this result to the case of the quadrupole mode($\lambda = 2$). To compare with the experimental data we have to evaluate the expectation value of this interaction potential in the angular momentum coupled state $|j_1j_2I>$, which for large values of j gives [17]

$$< j_1 j_2 I | V_2 | j_1 j_2 I > = (\frac{1}{4}) \frac{5}{4\pi C_2} < k_2(1) >< k_2(2) > P_2(\theta) \tag{10}$$

Note that this is expressed in terms of the angle θ, between the angular momentum vectors (i.e. the shears angle) which is the variable used in the experimental analysis; this transformation introduces the factor $\frac{1}{4}$ explicitly indicated in Eq. (10) (See Ref. [13]). We estimate $C_2 \approx 500\ MeV$ for the light Pb region from a extrapolation of the systematics of the restoring force parameters given in Ref. [8] (Vol. II, p. 529) for both Hg and Pb nuclei. Using $< k_2 >\approx 50\ MeV$ [18] we obtain $V_2 \approx 500\ keV$, to compare with the experimental estimate of $400 - 600\ keV$. Due to the uncertainty in the extrapolation of the C_2 coefficients, it is important to stress here that we should view this result as a qualitative agreement.

The particle-vibration coupling also gives rise to a renormalization of the electric charge, usually expressed in terms of a polarization charge that, with the parameters above, can be estimated to be $(e_{pol})_2 \approx 2$, not quite as large as the value we estimated from the analysis of $B(E2)$ values; as we argued before this may be an indication of a permanent deformation of the core.

From these results, it seems that a particle-vibration coupling scenario may provide a complementary approach to the TAC model, that requires a deformed mean field for nuclei that are almost spherical. In our picture the particle-phonon coupling splits the otherwise degenerate multiplet generated by the coupling of the blades, and can give rise to the shears mechanism in spherical nuclei.

COMPETITION BETWEEN CORE ROTATION AND SHEARS

We now consider the competition between the shears mechanism and the rotation of the core. To provide an answer to this interesting question we add to Eq. (8) a term representing the rotational energy of the core

$$E(I) = \frac{\vec{R}^2}{2\Im} + V_2 P_2(\theta) \tag{11}$$

This expression for the total energy now looks like the familiar Hamiltonian of the Particle-plus-Rotor Model [8]. Using the standard substitution, $\vec{R} = \vec{I} - \vec{j}_\pi - \vec{j}_\nu$, it goes into

$$E(I) = \frac{(\vec{I} - \vec{j}_\pi - \vec{j}_\nu)^2}{2\Im} + V_2 P_2(\theta) \tag{12}$$

which we solve in the classical limit by the requirement that at each spin the shears angle minimizes the energy. We assume once again for simplicity, that $j_\pi = j_\nu = j$. It is convenient to introduce the parameter $\chi = \Im/(j^2/3V_2)$ that relates the core moment of inertia to that of the shears, or in other words, $\chi = E_{2^+}(j^2)/E_{2^+}(core)$ the ratio of the 2^+ energies of the shears and the core. Measuring the angular momentum in units of the maximun available from the shears, i.e. $\hat{I} = I/2j$, we rewrite Eq. (12) in the form

$$\hat{E}(\hat{I}) = E(I)/(2j^2/\Im) = \hat{I}^2 - 2\hat{I}\cos\frac{\theta}{2} + \frac{1}{2}\cos\theta + \frac{\chi}{4}\cos^2\theta + \frac{1}{2} - \frac{\chi}{12}$$

The minimization condition $(\frac{\partial \hat{E}}{\partial \theta})_{\hat{I}} = 0$ determines the function $\theta_{min}(\hat{I}, \chi)$ that gives the shears angle at each $\hat{I}$ for a given χ. Intuitively we expect that the competition between these two modes should be governed by the parameter χ that measures the relative strength of the residual interaction to the rotational energy of the core. The sign of χ describes the particle-particle (negative) and particle-hole (positive) cases. In the following we study only the latter case, applicable to all the M1 bands. Fig. 6 shows the shears angle as a function of angular momentum for different values of χ. On one hand, if $\chi \to 0$, the energy needed to recouple the blades to a given angular momentum is small compared to the energy associated with the core rotation and the solution gives exactly the angle between two angular momentum vectors coupled to spin I, i.e. $\cos\theta \to 2\hat{I}^2 - 1$. On the other hand, if $\chi \to \infty$, it is more favorable for the core to rotate and for the shears to remain static at $\theta \approx 90^0$. The gradual transition beween the two limiting cases discussed above is clearly seen. It is interesting to note in the plot the spin at which the shears close ($\theta = 0$); it can be shown analytically that this occurs at $\hat{I} = 1 + \chi$. Knowing the shears angle it is now possible to derive for example the core contribution to the total angular momentum, as shown in Fig 7. Its behavior follows from Fig. 6; when $\chi \to 0$ it is so expensive for the core to rotate that the contribution is almost zero until, of course, the shears close at $\hat{I} = 1$, from then on

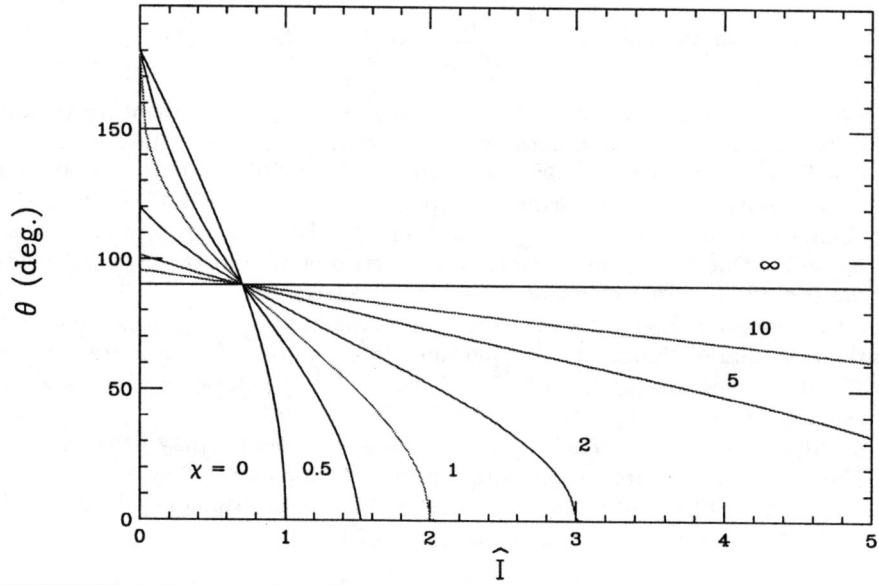

FIGURE 6. Shears angle as function of I for different values of χ from 0 to ∞.

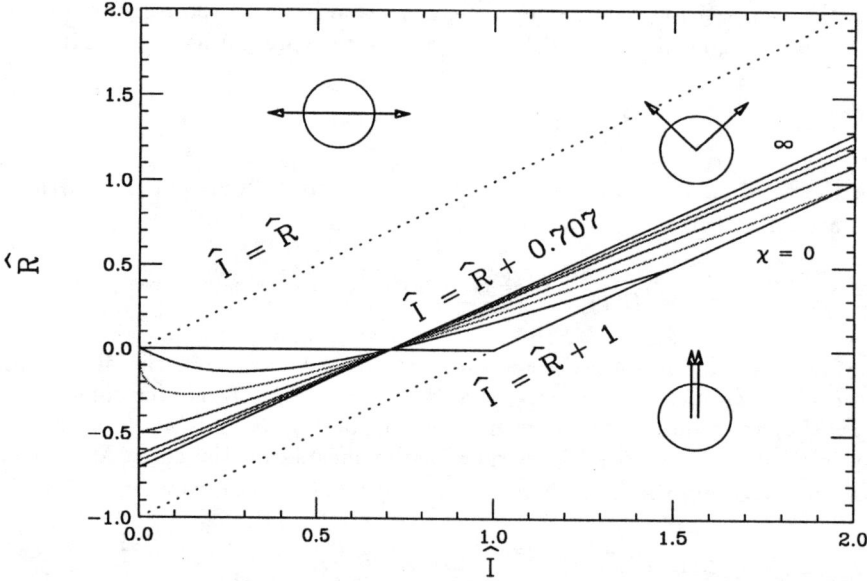

FIGURE 7. Contribution of the core to the total angular momentum as function of $\hat{I}$. Values of χ are as in Fig. 6

the angular momentum is generated only by the core but with an alignment of $2j$ as indicated by the equation and the inset. As discussed before, when $\chi \to \infty$ the shears are at $\theta \approx 90^0$ and contribute an alignment of $\frac{\sqrt{2}}{2}2j$, thus $\hat{I} = \hat{R} + 0.707$.

In an effort to present our results in a concise way we would like to introduce a "phase diagram" between shears and core. *A priori*, one may want to define the transition at the point where the shears close. It then follows from the result above that even for large values of χ there is always shears, a conclusion which seems intuitively wrong. We know that the dynamic response of a rotating system is related to changes in angular momentum, ΔI. In terms of inertias, it is the dynamic moment that governs those changes. If we characterize these two phases in terms of their contributions to the total change in angular momentum, $\Delta I = \Delta R(core) + \Delta j(shears)$, we can now properly define the transition point (for a given $\hat{I}$ and χ values) when each of the contributions is 50%. This "phase diagram" is shown in Fig. 8, where the shaded region corresponds

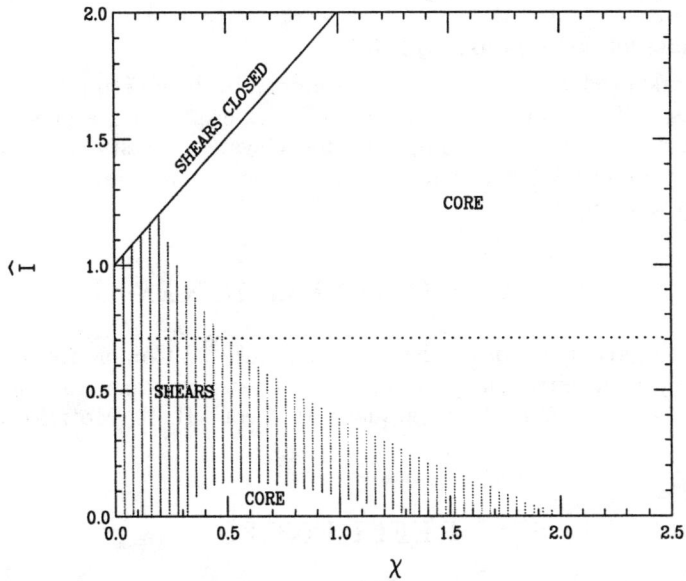

FIGURE 8. A "phase transition" diagram between the shears and the core rotation in the $\hat{I}$ *vs.* χ space (See text for details).

to a dominant shears mechanism. When χ is small the system prefers to generate angular momentum by the shears and the transition to core rotation occurs once the spins from the blades is exhausted. As the value of χ increases it is more economic to follow the rotation of the core while the shears contribute less and less to the generation of angular momentum. The transition occurs much earlier than the spin at which the blades close: they are still closing, but "very slowly". Indicated by the the dotted line is the bandhead spin $(\sqrt{2}j)$; the region above this line is that accessible experimentally since the states below it, having a lower spin and higher excitation energy, are likely non-yrast. Qualitatively, we propose to associate a quenching of the shears mechanism when, already at the bandhead, the contribution from the core is more than 50% . As seen in Fig. 8 this happens at $\chi \sim 0.5$. From the arguments presented in the preceding section we expect that this value of χ will be common to all mass-regions. Using estimates from the Pb region, where we know that the shears mechanism dominates, we have $E_{2+}(j^2) \sim 150\ keV$ requiring $E_{2+}(core) \gtrsim 300\ keV$ and a deformation of the core $\epsilon_2 \lesssim 0.12$, in agreement with the experimental observations and the deformations used in the TAC calculations.

SUMMARY AND CONCLUSIONS

To summarize, we have presented a semi-classical analysis of the $B(M1)$ and $B(E2)$ values in the shears bands in $^{193-199}Pb$. By deriving the angle between $\vec{j}_\pi$ and $\vec{j}_\nu$ we calculate the dependence of the reduced transition probabilities and compare with the experimental results. This procedure seems to give a consistent picture for both $B(M1)$ and $B(E2)$ values and provides an additional confirmation of the shears picture.

We have presented evidence that this mechanism can also be interpreted as an effective P_2 force between the neutron holes and the protons in the shears. Such an interaction can be mediated through the core by particle-vibration coupling acting in second order and estimates based on such a picture are in qualitative agreement with experiment. This may provide a complementary approach to the TAC model, which has the problem of requiring a deformed mean field for nuclei that are almost spherical. In our scenario we overcome this difficulty because the particle-phonon coupling is responsible for the splitting of the otherwise degenerate multiplet generated by the coupling of the blades, and can give rise to the shears mechanism in spherical nuclei. Detailed particle-phonon coupling calculations will be neccesary to quantitatively confirm this idea.

We have addressed the question of the competition between the shears and the core within the framework of a classical particles-plus-rotor model. The competition between these two modes depends on the relative strength of the residual interaction and the rotational energy of the core. Our results indicate that the shears mechanism should dominate for deformations $\epsilon_2 \lesssim 0.12$.

In conclusion, the shears mechanism is on solid grounds, both experimentally and theoretically. However, some open questions still remain to be answered. For example: i) the experimental characterization of antimagnetic-rotation (particle-particle coupling) ii) a more systematic study on the role of the deformation and iii) whether relistic two-body forces can indeed explain the rather simple picture that seems to emerge from our analysis of the data.

ACKNOWLEDGMENTS

It has been a pleasure to participate in the Nuclear Structure 98 Conference. We would like to thank the members of the Organizing Committee, in particular Dr. Cyrus Baktash, for the invitation to present this work. Enlightening discussions with Prof. S. Frauendorf, Prof. B. R. Mottelson and Dr. M. Redlich are gratefully acknowledged.

REFERENCES

1. S.Frauendorf, *Nucl. Phys.* **A557**, 259c (1993).
2. R.M.Clark et al., *Phys. Lett.* **B275** 247 (1992); G.Baldsiefen et al., *Phys. Lett.* **B275** 252 (1992) and A.Kuhnert et al., *Phys. Rev.* **C46** 133 (1992).
3. R.M.Clark et al., *Phys. Rev. Lett* **78** 1868 (1997), and *Phys. Lett.* **B**, in press.
4. See *BAPS* **43** (1998), pp. 1098-1100.
5. R.Wadsworth, These Proceedings
6. A.O.Macchiavelli et al. *Phys. Rev.* **C57** R1073 (1998).
7. A.O.Macchiavelli et al. *Phys. Rev.* **C58** R621 (1998).
8. A.Bohr and B.R.Mottelson, *Nuclear Structure* Vol. I and II, W.A. Benjamin, New York, 1969 and 1974.
9. F. Dönau and S.Frauendorf, *Proceedings of the International Conference on High Angular Momentum Properties of Nuclei*, Oak Ridge, 1982, Nucl. Sci. Res. Conf. Series Vol. 4, p. 143.
10. S.Chmel et al. , *Phys. Rev. Lett* **79** 2002 (1997).
11. R.B.Firestone, V.S.Shirley , et al., *Table of Isotopes 8th Edition.*, Appendix E., John Wiley & sons Inc., 1996.
12. J.P.Schiffer and W.W.True, *Rev. Mod. Physics.* **48**, 191, (1976), and references therein.
13. A.Molinari, M.B.Johnson, H.A.Bethe and W.M.Alberico *Nucl. Phys.* **A239**, 45, (1975).
14. A. de Shalit and I. Talmi, *Nuclear Shell Theory*, Academic Press, New York, 1963.
15. S. Frauendorf, *Proceedings of the Conference on Physics from Large γ- Detector Arrays*, Berkeley 1994, Report LBL-35687, Vol. 2, p. 52.
16. A.Bohr and B.R.Mottelson, *op. cit.* Vol. II, Chapter 6, section 6-5.
17. A.Bohr and B.R.Mottelson, *Mat. Fys. Medd. Dan. Vid. Selsk.*, **27**, no. 16 (1953).
18. See Ref. [8] Vol. II, pag. 420. If we use for example the harmonic oscillator we get $k = M\omega_0^2 R_0^2 \approx 55\ MeV$

Unsafe Coulomb Excitation of $^{240-244}$Pu

I.Wiedenhöver[1], R.V.F. Janssens[1], G. Hackman[1], I. Ahmad[1], J.P. Greene[1], H. Amro[1],
M.P. Carpenter[1], D.T. Nisius[1], P. Reiter[1], T. Lauritsen[1], C.J. Lister[1], T.L. Khoo[1],
S. Siem[1,2], J. Cizewski[1,3], D. Seweryniak[1], J. Uusitalo[1], A.O. Macchiavelli[4]
P. Chowdhury[5], E.H. Seabury[5], D. Cline[6], C.Y. Wu[6]

[1]Argonne National Laboratory, Argonne, Il 60439
[2]Lawrence Berkeley National Laboratory, Berkeley, CA 94720
[3]University of Oslo, Oslo, Norway
[4]Rutgers University, New Brunswick, NJ 08903
[5] University of Massachusetts, Lowell, MA 01854
[6]WSRL, University of Rochester, Rochester, NY 14627

Abstract. The high spin states of ^{240}Pu and ^{244}Pu have been investigated with GAMMASPHERE at ATLAS, using Coulomb excitation with a ^{208}Pb beam at energies above the Coulomb barrier. Data on a transfer channel leading to ^{242}Pu were obtained as well. In the case of ^{244}Pu, the yrast band was extended to $34\hbar$, revealing the completed $\pi i_{13/2}$ alignment, a "first" for actinide nuclei. The yrast sequence of ^{242}Pu was also extended to higher spin and a similar backbend was delineated. In contrast, while the ground state band of ^{240}Pu was measured up to the highest rotational frequencies ever reported in the actinide region ($\sim$300 keV), no sign of particle alignment was observed. In this case, several observables such as the large B(E1)/B(E2) branching ratios in the negative parity band, and the vanishing energy staggering between the negative and positive parity bands suggest that the strength of octupole correlations increases with rotational frequency. These stronger correlations may well be responsible for delaying or suppressing the $\pi i_{13/2}$ particle alignment.

INTRODUCTION

The stable actinide nuclei remain an interesting source of nuclear structure information as, on the one hand, they offer the possibility to investigate the interplay between collective rotation and octupole degrees of freedom, and on the other provide a unique opportunity to investigate the behavior with frequency of the proton $i_{13/2}$ and neutron $j_{15/2}$ high-j orbitals. Recent progress in the study of these nuclei has come from the work of Ward et al. [1] and Hackman et al. [2] where the power of inelastic scattering at beam energies slightly above the Coulomb barrier on thick targets was used to study collective excitations with high sensitivity. This technique is often referred to as "Unsafe" Coulomb Excitation. Besides the higher than usual beam energy, which optimizes the population of the highest spin levels, the technique is also associated with the use of thick targets and with the selection of the cascades of the highest multiplicity as a means to enhance the γ rays emitted after the excited nucleus has come to rest. In this way most of the transitions in a collective cascade are measured with the intrinsic resolution of the Ge detectors.

EXPERIMENT

The experiments described here represent the first "production" runs with GAMMASPHERE at ATLAS. Targets of isotopically enriched ^{240}Pu and ^{244}Pu were bombarded with a 1300 MeV ^{208}Pb beam. The Pu material was electroplated onto a thick (50 mg/cm^2) Au foil, which served as a backing to stop both the recoiling Pu nuclei and the beam. The targets had a thickness of $\approx 300 \mu g/cm^2$. In addition, a second, thin Au foil ($\approx 50 \mu g/cm^2$) was mounted in front of each target to minimize the danger of contaminating the experimental equipment by the release of radioactive target material. It should be pointed out that the relatively low target thickness achievable with electroplating constitutes the main experimental limitation in these measurements.

CP481, *Nuclear Structure 98*, edited by C. Baktash
1999 American Institute of Physics 1-56396-858-4

During three days of beam time, coincidence events of fold three or higher were collected. Most of the subsequent data analysis was performed on the quadruple γ coincidence events, where the data were sorted into a coincidence cube gated on known transitions from the nuclei of interest. Proper subtraction of random coincidence events proved essential in order to remove the contamination of the spectra with the intense ^{197}Au Coulomb excitation γ rays. The data analysis was performed with the programs of the RADWARE package [3].

Fig. 1 presents sample spectra of the yrast bands of ^{244}Pu and ^{242}Pu obtained as sums of the cleanest coincidence gates. These spectra were measured with the ^{244}Pu target. Here ^{242}Pu is produced by a direct two-neutron transfer reaction, as demonstrated by the presence of the ^{210}Pb ground-state transition in the coincidence spectra.

RESULTS

$\pi i_{13/2}$ Quasiparticle Alignment

In ^{244}Pu, the yrast rotational band could be traced up to the 34^+ level. A pronounced backbending is present at a rotational frequency of $\hbar\omega \approx 240$ keV (see figs. 1 and 2). This result represents the first instance in the actinide region where a backbending is now fully delineated. The resulting gain in alignment ($\sim 10\hbar$) is consistent with a $i_{13/2}$ quasi-proton alignment, but is too low for a $j_{15/2}$ quasi-neutron alignment. Thus, the present result settles a long standing debate about the nature of quasiparticle alignments in the actinide region [1,4]. The observed behavior can be reproduced both by cranked shell model and by HFB calculations which find that the quasi-proton alignment occurs at lower frequency than the quasi-neutron one. The behavior with frequency of the ^{242}Pu yrast band is very similar to that seen in ^{244}Pu: a strong backbending is present here as well, starting at a frequency of ~ 250 keV.

Because of the success of microscopic calculations in understanding the yrast sequences of ^{242}Pu and ^{244}Pu, the behavior of the ground-state band of ^{240}Pu appears to be rather puzzling. Being a member of the same isotopic chain, with a deformation essentially identical to that of the heavier even-even Pu nuclei, ^{240}Pu would be expected to exhibit a similar backbend at roughly the same frequency. Yet its yrast sequence, which is extended in the present work up to 32^+ and a rotational frequency of $\hbar\omega \approx 300$ keV, shows no indication of any irregularity. In fact, this ground-state band exhibits all the characteristics of a single, "well-behaved" rotational band, without any sign of a band crossing in either the alignment plot (fig. 2) or the evolution of the dynamic moment of inertia with frequency (not shown). Thus, the alignment process of the $i_{13/2}$ protons and/or of the $j_{15/2}$ neutrons appears to be either suppressed or at least delayed towards higher rotational frequencies. Possible explanations for this observation can be found in the behavior of the negative parity bands discussed below.

Negative-Parity Bands

The presence of low lying rotational bands of negative parity is a characteristic feature of the level structure of all actinide nuclei. These bands have been associated with various components of the octupole phonon [1,2]. In both ^{240}Pu and ^{244}Pu, these octupole bands have now been followed up to high spin (29^- and $\hbar\omega \sim 270$ keV in ^{240}Pu, 27^- and $\hbar\omega \sim 230$ keV in ^{244}Pu). The measured alignments are again quite different for the two isotopes (fig. 2). In the lighter ^{240}Pu isotope, an initial gain in alignment of $3\hbar$ is followed by a value which becomes constant with frequency as would be expected for the $K=0$ component of the octupole phonon [2]. While a similar low frequency rise of the alignment towards the same $3\hbar$ value occurs in ^{244}Pu, it is followed by a further increase at higher frequencies which mirrors that seen in the yrast sequence. Thus, the behavior discussed previously for the yrast sequences is also present in the negative parity bands.

The behavior of the negative parity bands in the two isotopes is also found to be markedly different when the excitation energy of the levels is considered. In fig.3, the so-called energy staggering S(I) between the odd-spin, negative parity and even-spin, positive parity bands is presented. This staggering is defined as:

$$S(I) = E(I) - \frac{E(I+1)*(I+1) + E(I-1)*I}{2I+1} \tag{1}$$

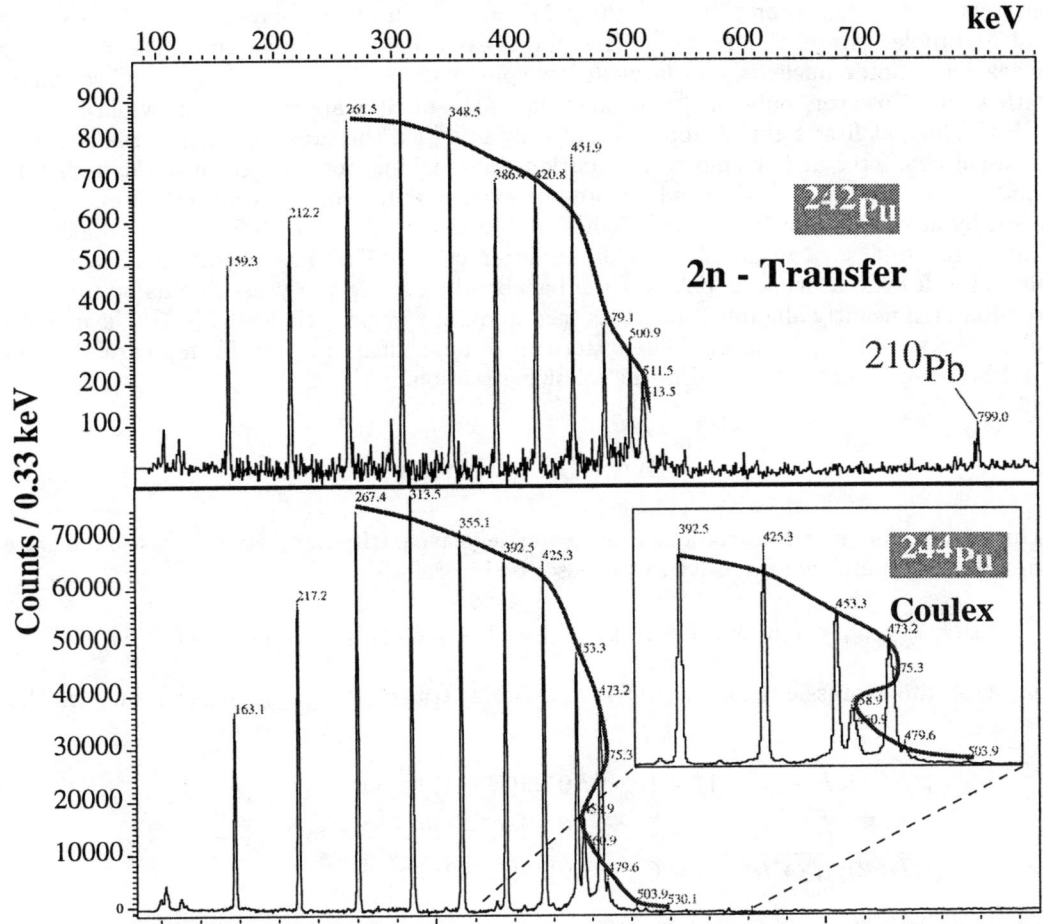

FIGURE 1. Sample coincidence spectra for the ^{242,244}Pu yrast bands. ^{242}Pu was obtained as a direct, two-neutron transfer reaction product from the ^{244}Pu target, as illustrated by the presence of the ^{210}Pb ground-state transition.

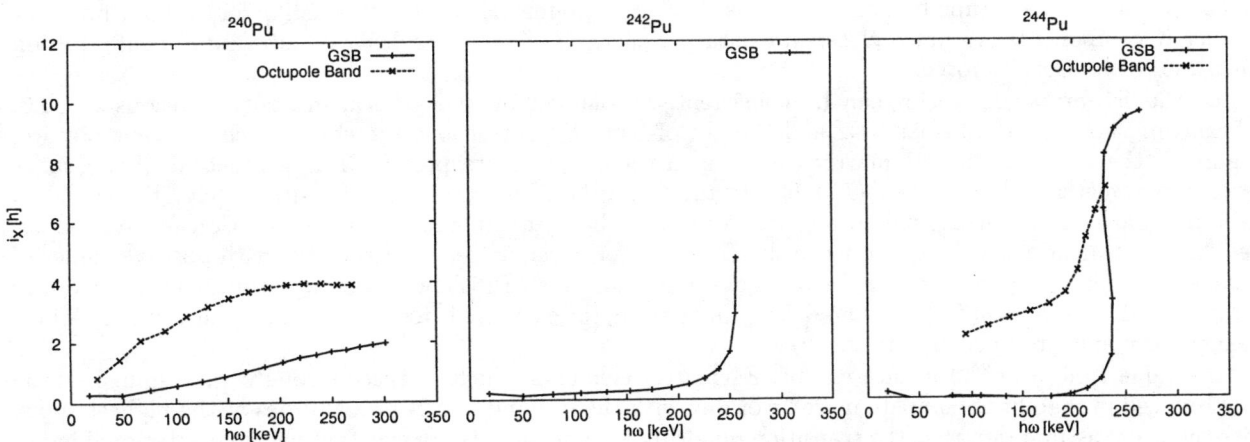

FIGURE 2. Alignment of the ground state and octupole rotational bands in ^{240,242,244}Pu. In all cases the same reference is subtracted, with the Harris parameters $J_0 = 65\hbar^{-2}\text{MeV}^{-1}$, $J_1 = 369\text{MeV}$.

and is a measure of the extent to which the two bands of opposite parity can actually be regarded as a single, rotational octupole excitation, i.e. the degree to which the odd-spin level of spin I has an excitation energy located in between those of the two neighboring even spin states with respective spins $I-1$ and $I+1$.

The experimental S(I) values for ^{240}Pu, ^{244}Pu, ^{238}U and ^{222}Th are displayed in fig. 3. As expected, the quintessential "octupole-deformed" nucleus ^{222}Th exhibits essentially zero staggering. The staggering observed in the three heavier actinide nuclei is very large at low spin as expected for an octupole vibrational band, but decreases with spin. However, only in ^{240}Pu does the value of S(I) approach zero, while S(I) levels off in ^{238}U and ^{244}Pu. Thus, at first sight, it appears that only in ^{240}Pu the yrast and near-yrast structure evolves from a vibrational character at low spin to a situation approaching octupole rotation at higher frequencies. If this interpretation is correct, ^{240}Pu would represent experimental confirmation of the model for octupole bands proposed by Jolos and von Brentano [5] where an octupole vibration evolves with rotational frequency into an octupole rotation as a result of the Coriolis interaction. This interpretation also would provide a natural explanation for the absence or delay of the backbending along the yrast line as octupole correlations are known to alter significantly alignment patterns seen in axially symmetric nuclei [6,7]. Nevertheless, further experimental evidence is highly desirable. The latter can come, at least in part, from a careful examination of the E1 and E2 branching ratios discussed in the following section.

Branching Ratios

The branching ratios for the E1 transitions linking the negative parity band to the yrast band were analyzed with the formalism of generalized intensity ratios described in ref. [9].

$$\sqrt{B(E\lambda)} = Q_t < I_i K_i \lambda(K_f - K_i)|I_f K_f > (1 + q(E\lambda)\,[I_f(I_f + 1) - I_i(I_i + 1)]) \tag{2}$$

In the present case, the expressions for the B(E1) and B(E2) transitions probabilities reduce to the following formulae:

$$\sqrt{B(E1, I_i \to I_i - 1)} = D_0 < I_i 0\,10|(I_i - 1)0 > (1 + q(E1)\,[-2I_i])$$
$$\sqrt{B(E1, I_i \to I_i - 1)} = D_0 < I_i 0\,10|(I_i + 1)0 > (1 + q(E1)\,[2(I_i + 1)])$$
$$\sqrt{B(E2), I_i \to I_i - 2)} = Q_0 < I_i 0\,20|(I_i - 2)0 > \tag{3}$$

In these expressions, the effects of Coriolis mixing can be found in the so-called decoupling coefficient $q(E1)$, which can be calculated in models of the intrinsic motion, such as the cranked RPA model [10], for example. In fig. 4, the $\sqrt{B(E1)}$-branching ratios, corrected with the appropriate Clebsch-Gordan coefficient of equation 3, are displayed for the cases of ^{238}U [1], ^{240}Pu and ^{244}Pu. The data can be described with a very small decoupling coefficient in the ^{240}Pu and ^{238}U cases. In contrast, the data for ^{244}Pu require a much larger value of $q(E1) = 0.065\hbar^{-1}$. This situation agrees nicely with cranked RPA calculations [2,11], which predict that the K=0 component of the octupole phonon is lowest in ^{240}Pu (giving rise to a very small $q(E1)$ value), but that the other K-components are much closer in excitation energy in ^{244}Pu, resulting in substantial configuration mixing due to the Coriolis force.

While the E1–branching ratios can be consistently explained by the general intensity relations of eq.3, the branching between the in-band E2 and the out-of-band E1 transitions present a much different picture. In figure 5, the ratios of the E1 matrix elements to the E2 ones are presented as a function of the spin–difference parameter $[I_f(I_f + 1) - I_i(I_i + 1)]$. In this plot, the $I_i^- \to (I_i - 1)^+$ transitions (e.g. the so-called "downhill" transitions) correspond to negative values of the spin–difference parameter. Conversely, the so-called "uphill" transitions, corresponding to the $I_i^- \to (I_i + 1)^+$ sequence, are associated with positive values of the same parameter. Again, the data for the measured ratios in ^{244}Pu show a good agreement with the values calculated with the same $q(E1) = 0.065$ decoupling parameter extracted from the E1-branching ratios. Thus, a clear picture appears to emerge in this case.

At the same time, the ^{240}Pu data exhibit E1 matrix elements which increase strongly with spin, i.e. the B(E1) reduced probabilities compete more favorably with the in-band B(E2) probabilities at the highest spins. Under the usual assumption that the transition quadrupole moments Q_0 are constant within a rotational band, this result implies an increase in E1 (i.e octupole) collectivity with angular momentum. This behavior cannot be reproduced with a single value of the $q(E1)$ parameter.

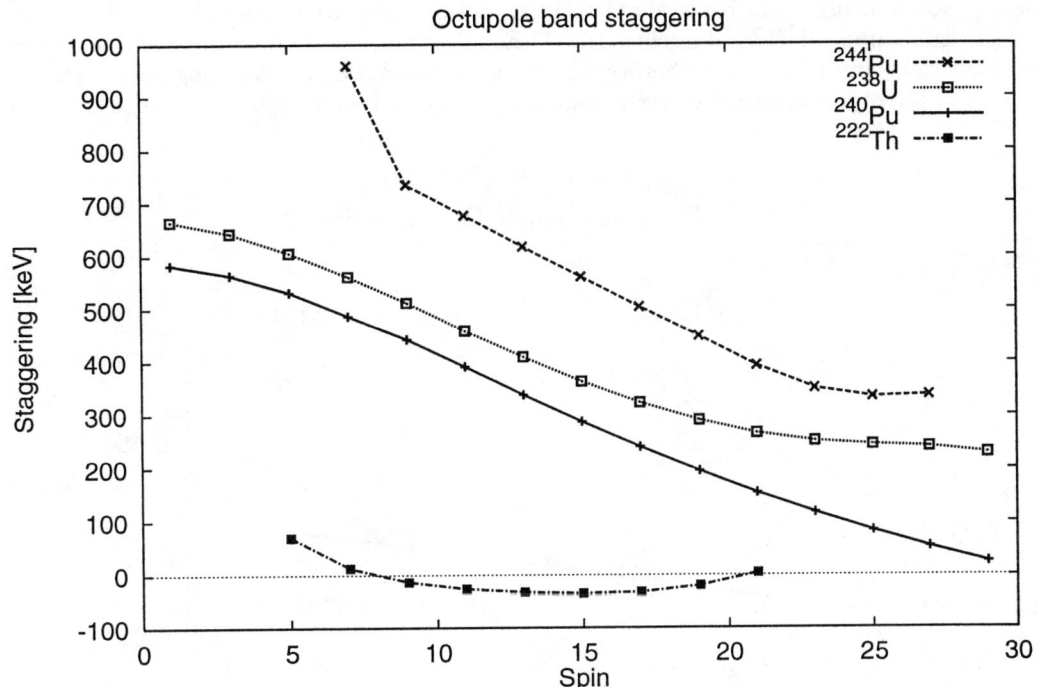

FIGURE 3. staggering parameter (see eq. 1) for negative and positive parity bands in ^{244}Pu, ^{238}U, ^{240}Pu and ^{222}Th.

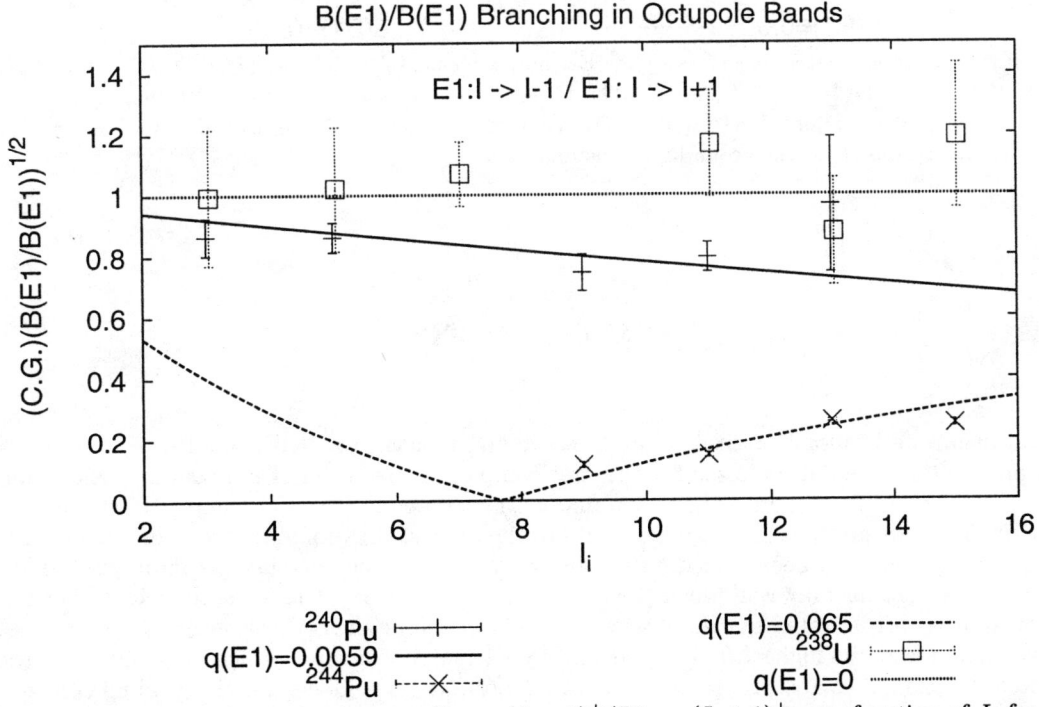

FIGURE 4. Ratio of the E1-matrix elements $I_i^- \rightarrow (I_i - 1)^+ / I_i^- \rightarrow (I_i + 1)^+$ as a function of I_i for the octupole bands of ^{240}Pu, ^{244}Pu and ^{238}U [1]

The right y-axis of fig. 5 is labeled with the transition dipole moment values D_0 derived from the measured transition probability ratios assuming a constant $Q_0 = 25.88eb$ quadrupole moment. The latter value of Q_0 is adopted from the measured B(E2) value of the $2^+ \rightarrow 0^+$ ground state transition [8]. It can be seen from fig. 5 that the D_0 values at high spin become quite large. In fact, they are comparable to the dipole moments observed in ^{222}Th, an octupole deformed nucleus [7]. Although experimental data on absolute lifetimes and/or B(E2) reduced transitions probabilities are highly desirable, the present analysis strongly suggests a substantial increase of the transition dipole moment D_0 with spin, until it reaches values consistent with a sizable octupole deformation.

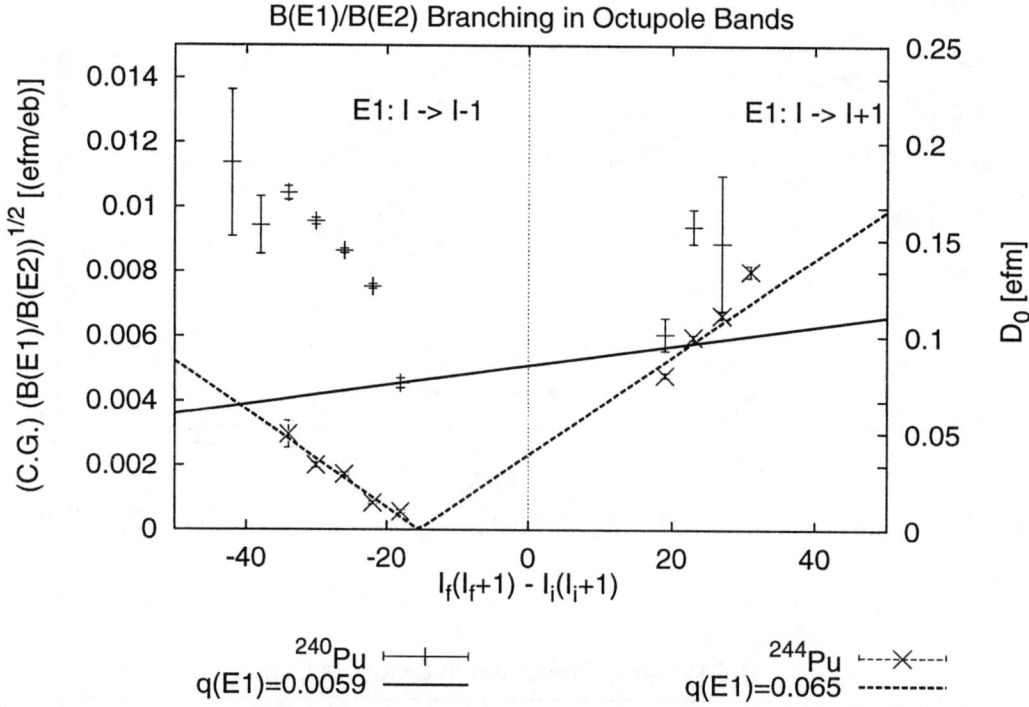

FIGURE 5. Ratio of the E1-matrix elements to the E2-matrix elements $E1 : I_i^- \rightarrow (I_i \pm 1)^+ / E2 : I_i^- \rightarrow (I_i - 2)^-$ as a function of the parameter $I_f(I_f + 1) - I_i(I_i + 1)$. The $I_i^- \rightarrow (I_i - 1)^+$ decays correspond to the negative x-values, the $I_i^- \rightarrow (I_i + 1)^+$ to the positive values. The values of D_0 given on the right hand side are calculated assuming rotational E2-matrix elements with the Q_0 of the ground state band.

CONCLUSIONS

The method of unsafe Coulomb excitation at GAMMASPHERE has provided a wealth of information on the nuclei ^{244}Pu and ^{240}Pu. The data allowed for the first complete delineation of a quasiparticle alignment in an actinide yrast sequence. This alignment has been understood to be of proton character. In sharp contrast, the yrast band of ^{240}Pu shows no signs of backbending up to the highest rotational frequencies. In the present work this surprising observation has been linked to the presence of octupole correlations increasing in strength at high spin. These correlations may well block the expected alignment or, at least, shift it to higher frequencies. This interpretation is corroborated by the observation of a vanishing energy–staggering between the octupole and ground state bands at the highest frequencies and by E1 matrix elements which also increase strongly with spin. This interpretation, if correct, represents the first experimental evidence for the gradual change of a band associated with an octupole vibration into one of collective, rotational octupole character for a well deformed nucleus.

ACKNOWLEDGMENT

The authors are indebted for the use of ^{240}Pu and ^{244}Pu to the office of Basic Energy Sciences, U.S. Dept. of Energy through the transplutonium element production facilities at the Oak Ridge National Laboratory. This work is supported by the U.S. Department of Energy under Contract Nos. W–31–109–ENG–38, DE-FG02-94ER40848 and DE-AC0376SF00098 and by the National Science Foundation.

REFERENCES

1. D. Ward et al., Nucl.Phys. **A600** (1996) 88
2. G. Hackman et al., Phys.Rev. **C57** (1998) R1059
3. D.C. Radford: Nucl. Instr. Meth. **A361**, 306 (1995)
4. W. Spreng et al., Phys. Rev. Lett. **51** (1983) 1522, J. Dudek et al., Phys. Rev. **C26** (1998) 1708
5. R.V. Jolos and P. von Brentano, Nucl.Phys. **A587** 377 (1995)
6. S. Frauendorf and V.V. Pashkevich, Phys.Lett. **141B** (1984) 23
7. P.A. Butler and W. Nazarewicz, Reviews of Modern Physics, **68** (1996) 349
8. C.E.Bemis Jr., F.K.McGowan, J.L.C.Ford Jr., W.T.Milner, P.H.Stelson, R.L.Robinson, Phys.Rev. **C8**, 1466 (1973)
9. A.Bohr and B. Mottelson, in *Nuclear Structure* II, Benjamin, Reading, Massachusetts, 1975
10. T. Nakatsukasa, K. Matsuyanagi, S. Mizutori and Y.R. Shimizu, *Conference on Nuclear Structure at the Limits*: Proceedings, Argonne, ANL–PHY97/1, p. 111 (1997)
11. T. Nakatsukasa, K. Matsuyanagi, S. Mizutori and Y.R. Shimizu, Phys. Rev. **C 53**, 2213 (1996).

Observation of the GT Resonance in the β^+-Decay of ^{150}Ho 2^-

J. Agramunt[a], A. Algora[a], L. Batist[d], R. Borcea[c], D. Cano-Ott[a], R. Collatz[c], A. Gadea[a], J. Gerl[c], M. Gierlik[b], M. Górska[c], O. Guilbaud[c], H. Grawe[c], M. Hellström[c], Z. Hu[c] Z. Janas[b], M. Karny[b], R. Kirchner[c], P. Kleinheinz[a], W. Liu[c], T. Martinez[a], F. Moroz[d], A. Płochocki[b], M. Rejmund[c], E. Roeckl[c], B. Rubio[a], K. Rykaczewski[b], M. Shibata[c], J. Szerypo[b], J.L. Tain[a], V. Wittmann[d] and the German Euroball Col.

[a]Instituto de Física Corpuscular, Dr. Moliner 50, E-46100 Burjassot-Valencia, Spain
[b]Institute of Experimental Physics, Warsaw University, PL-00-681 Warsaw, Poland
[c]Gesellschaft für Schwerionenforschung, D-64220 Darmstadt, Germany
[d]St. Petersburg Nuclear Physics Institute, 188-350 Gatchina, Russia

Abstract. The Gamow-Teller beta decay of ^{150}Ho 2^- isomer has been studied with a Total Absorption Spectrometer (TAS), with an array of 6 Euroball CLUSTER Germanium detectors (CLUSTER CUBE), and with an alpha detector. The three techniques complement each other. The results provide the first observation of an extremely sharp resonance in GT beta decay. The analysis of the data provides a precise value of the β-strength within the decay window.

INTRODUCTION

Although the problem of the missing strength in Gamow-Teller (GT) decay is one of long-standing it remains unresolved. In light nuclei there is a great deal of experimental information, from both beta decay and charge-exchange reactions, which systematically indicates that $\sim 40\%$ of the strength is missing [1]. For heavier nuclei the beta decay information is sparse because of the difficulty of accessing nuclei with allowed decays. In this work we will report studies of GT beta-decay in heavier nuclei. In comparison with charge-exchange reactions they are reaction-model independent and free of background uncertainties. In addition they permit the study of nuclei far from stability. However, due to the selection rules, very few GT decays are allowed above the heaviest N~Z particle stable nuclei. This is because in general the required orbitals for allowed decays lie outside the beta-window. There are only two regions where the $\sigma\tau$ resonance is accessible in β-decay. In both of these regions a proton in a high J orbital decays into its spin-orbit partner neutron orbital with J-1, which is in general less bound than the J proton orbital and therefore empty. The nuclei we refer to lie in the vicinity of ^{100}Sn and ^{146}Gd.

The quantity to be measured is the β-strength S_β as a function of the final level excitation energy E_x which can be experimentally determined from the β intensity or feeding probability (per unit energy interval) I_β:

$$S_\beta(E_x) = \frac{I_\beta(E_x)}{f(Q_\beta - E_x)T_{1/2}} \equiv \frac{1}{ft} \tag{1}$$

where f stands for the statistical rate Fermi function which depends on the energy $Q_\beta - E_x$ available to the decay, and $T_{1/2}$ is the β-decay half-life. This quantity can be compared with theoretical estimates obtained as the average over an energy interval ΔE_x of the β-decay reduced transition probabilities B(GT)$_{i \to f}$ to the levels in that interval [2]:

$$S_\beta(E_x) = \frac{1}{D'}\frac{1}{\Delta E_x}\sum B(GT)_{i \to f} \tag{2}$$

CP481, *Nuclear Structure 98*, edited by C. Baktash

The numerical constant D' takes the value 3862 s when the B(GT) is expressed in units of $g_A^2/4\pi$, being g_A the axial coupling constant of the weak interaction.

The task of making precise β-strength measurements is far from trivial. Traditional high resolution detection techniques, based on a few Ge detectors to determine the $I_\beta(E_x)$ by measuring the balance of gamma-rays feeding and de-exciting the levels, often fail to detect significant strength. The β-decay can proceed to levels at high excitation energy in the daughter nucleus, where the level density is high and the β-strength fragmented into many individual levels. A further fragmentation in the emitted gammas will arise from the different paths through which every level will de-excite, including highly energetic gamma rays which are difficult to detect. As a result there will be a displacement to levels lying at low excitation energy of the deduced β-intensity, and furthermore, there will be an underestimation of the integrated strength, due to the renormalization effect of the Fermi function f (Eq. 1).

There are two possible alternatives to improve the experimental situation. a) To use a highly efficient detector, hereafter called a Total Absorption Spectrometer (TAS), which is sensitive to the β-population of the nuclear levels rather than to the individual gamma rays. This can be achieved using a large scintillation crystal covering the entire solid angle. In relation to high resolution techniques this method lacks energy resolution and all the advantages inherent in γ-γ coincidence measurements. b) A second possibility is to use an array of closely packed Ge detectors, with greatly enhanced efficiency, of the kind developed in recent years for in-beam gamma studies. We have installed these two alternative experimental set-ups at the GSI On-line Mass-separator and have performed a series of experiments in the rare earth region as well as on nuclei in the neighbourhood of ^{100}Sn. In the present paper we will describe both methods and present the results for the case of the decay of ^{150}Ho 2^- isomer. This nucleus is ideal to test these methods because the strength is expected to be concentrated at high excitation energy but still lie inside the Q_β-window. One difficulty arises however in this case. The ground state of the daughter nucleus ^{150}Dy is a known alpha emitter. It might happen then, that the process of β-delayed alpha emission competes with gamma de-excitation for sufficiently high excitation energy in the daughter nucleus. Even proton emission will be energetically allowed. In order to check for this possibility a third experiment was performed with a Si particle telescope.

The present experiments should provide the following outcomes: a) The β-strength distribution over a large range of excitation energy, b) a comparison of the total absorption technique and the high resolution technique, c) fragmentation of the GT strength in a region of high level density.

EXPERIMENTAL TECHNIQUES

The experiments were carried out at the GSI on-line mass-separator. A 1.8 mg/cm^2 ^{96}Ru target 96.5% enriched, located in front of a thermal ionization source, was bombarded with ^{58}Ni beams delivered by the UNILAC with energies up to ~ 5 MeV/u and typical intensities of 40 pnA. After ion extraction and mass separation, the secondary beam with A=150 was deflected to the adequate experimental station. The secondary beam was collected on tape and transported to the measuring position. The measurement cycle (collection time = measuring time = 120 s) was chosen to optimize the activity of ^{150}Ho 2^- ($T_{1/2} = 72$ s).

The ^{150}Ho activity can only be produced in the chosen reaction, as the decay product of ^{150}Er. The 0^+ ground state of ^{150}Er ($T_{1/2} = 18.5$ s) decays 95.4% [3] to the 1^+ state which de-excites through the 476 keV γ-ray to the 2^- isomer in ^{150}Ho and will never populate the ^{150}Ho 9^+ isomer [4]. However, a small contribution ($\sim 5\%$) from the decay of the 9^+ isomer was visible in the recorded spectra due to target contaminations.

TAS Measurements

The working principle of a TAS is to absorb all the energy of the γ-rays produced in the de-excitation of a level fed in β-decay. A practical TAS will not have the ideal 100% efficiency, but must come close to it to be useful. In the present case this is achieved with the arrangement shown in Fig. 1 (left). A more detailed description of the spectrometer can be found in Ref. [5].

The radioactive sources are transported into the central cavity of a large NaI crystal ($\oslash 35$ cm$\times 35$ cm). However, this is not sufficient to perform measurements on proton rich nuclei. Here β^+-decay and the EC-process normally compete, and the response of the apparatus to these two types of event will be quite different. The positron emitted in the β^+-process can penetrate the crystal, and even when it does not, it will produce two 511 keV quanta which will sum up with the rest of the γ-rays detected. We need to differentiate between

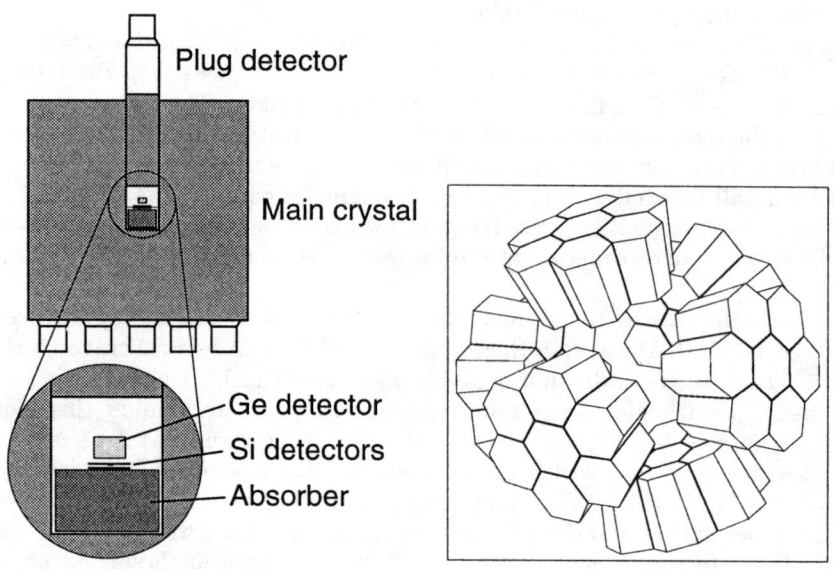

FIGURE 1. The two different gamma-ray spectrometers. Left: NaI(Tl) Total Absorption Spectrometer (TAS). Right: Array of 6 Euroball CLUSTER Ge detectors (CLUSTER CUBE)

the two processes. In the present set-up this is done by gating on the X-rays emitted in the EC process and detected in a Ge X-ray detector, and by gating on the positrons detected in two ΔE-Silicon detectors 1 mm thick and 18 mm diameter (see insert in Fig. 1 left). The threshold set in the Si detectors (≈ 300keV) minimizes the contribution of γ-ray and conversion electron interactions. It should be noted that the introduction of these detectors inevitably worsens the peak efficiency of the spectrometer, especially at low energy. Moreover, in order to reduce the effect of the positron penetration we have introduced a 2.3 cm thick polyethylene absorber between the lowest Si detector and the NaI. The peak efficiency of the spectrometer in the present arrangement is 62% at 1.3 MeV and 52% at 4.5 MeV, while the total efficiencies are 95% and 88% respectively.

Si Particle Telescope Measurements

A set-up consisting of a Si ΔE-E telescope and a germanium detector was employed to investigate the delayed particle emission in the ^{150}Ho decay. The closely placed Si detectors with thickness 22 μm and 530 μm subtended a solid angle of 20% of 4π. The particle detection efficiency of the telescope relative to the Ge γ-ray detection efficiency was determined from the known ratio [6] of α- to β-decay of ^{150}Dy g.s.

CLUSTER CUBE Measurements

We have used 6 Euroball CLUSTER detectors [7] arranged forming a cube, see Fig. 1 (right), to measure the decay of ^{150}Ho. Each CLUSTER consists of seven hexagonal tapered Ge detectors which are individually encapsulated and mounted closely packed in a common cryostat. The mass-separated sources are placed in the middle of the cube at a distance of ~11 cm from the central capsule of the CLUSTERS. All type of events were recorded in list mode, including those where only one hit was detected in one capsule. This allowed the reconstruction during the off-line sort of singles spectra as well as coincidence matrices. With this arrangement the peak detection efficiency for a γ-ray of 1.33 MeV was 17.5% (6.9% for 4.5 MeV), when the energy deposited by the γ-ray in neighbouring capsules was added up (add-back mode). A careful gain matching of all individual Ge capsules allowed to reach an energy resolution of 5.1 keV at 4.5 MeV photon energy. Sources of ^{152}Eu and ^{56}Co as well as the three peaks from the 6130 keV γ-ray in ^{16}O were used for this purpose. The energy calibration gave the literature values of the background 7632 and 7645 keV peaks in ^{57}Fe to less than 1 keV.

RESULTS

Total Absorption Spectroscopy

Two independent spectra were obtained from the sort of the list-mode data: a) a TAS spectrum (from now on EC spectrum) in coincidence with the K_{α_1} and K_{α_2} Dy X-ray lines seen in the Ge detector. b) a TAS spectrum (from now on β^+ spectrum) in coincidence with the signals from the Si detectors. In this way we select the EC and the β^+ part of the decay free from background contamination. Additionally, the EC spectrum selected in this way is free from isobaric contaminations. The β^+ spectrum was contaminated with the decays of ^{150}Tm ($Q_{EC} = 11.2$ MeV, $T_{1/2} = 2.2$ s), ^{150}Er ($Q_{EC} = 4.1$ MeV, $T_{1/2} = 18.5$ s), and ^{150}Dy $Q_{EC} = 1.8$ MeV, $T_{1/2} = 7.2$ min). Selecting in the off-line sort the last 30 s of the measuring cycle (120 s) eliminates the contribution from ^{150}Tm and reduces to a non-disturbing level the contribution from ^{150}Er. The upper panels of Fig. 2 show the EC (left) and β^+ (right) spectra.

The only remaining spectral distortion is electronic pulse pile-up. Although this effect was minimized, it remains important because it shows up close to the end-point of the spectrum. Our method [8] of correction is based on the digitized amplifier pulse shape, allowing a precise reproduction of the pulse pile-up spectrum shape, which is then normalized to the part of the experimental spectrum extending beyond the end-point.

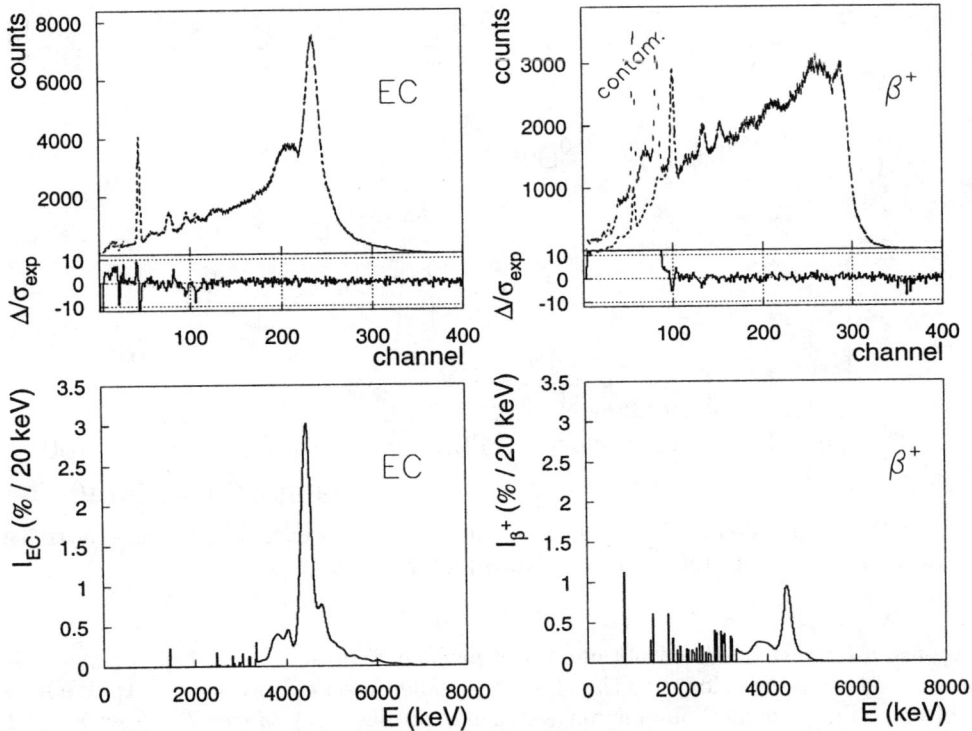

FIGURE 2. Analysis of TAS spectra obtained in the ^{150}Ho 2^- EC (left) and β^+ (right) decay. Upper panels: Comparison of measured and reconstructed (dashed line) TAS spectra. Lower panels: Feeding probability distribution obtained by the Bayesian Iterative Method (see text).

Finally, in order to combine the results from the analysis of both EC and β^+ spectra, the normalization factors are obtained by least-squares fit of a linear combination of the EC and β^+ spectrum to the singles spectrum.

The important remaining task is to analyze the EC and β^+ spectra in order to extract the feeding probabilities. The relation between a measured TAS spectrum $\mathbf{d}$ and the level feeding distribution $\mathbf{f}$ ($\equiv NI_\beta$, $N=$ total number of decays) is best represented by the equation:

$$d_i = \sum_{j=0}^{j_{max}} R_{ij} f_j, \qquad i = 1, i_{\max}$$

$$\text{or} \quad \mathbf{d} = \mathbf{R} \cdot \mathbf{f} \tag{3}$$

Each column of the response matrix $\mathbf{R}$ represents the probability that the decay into a given level produces a count in a certain channel. The response $\mathbf{R}$ can be obtained from the response to individual quanta emitted in the decay and the subsequent electromagnetic cascade. This requires the knowledge of the cascade branching ratios (b.r.). In general this knowledge is not available for the decay into highly excited levels, and it will be necessary to assume some decay pattern for those levels. The assumption can be improved iteratively based on the comparison between the generated spectra and the measured one. For a good TAS (*i.e.* one with large efficiency) the dependence of $\mathbf{R}$ on the b.r. will be small, and it will be possible to achieve a good reproduction of the data in a limited number of steps.

The response to the individual quanta (γ-rays, positrons,...) can only be obtained by Monte Carlo simulation. We have chosen the GEANT3 code [9], with a powerful geometry package which allows the implementation of the apparatus description with the required detail. With the inclusion of the non-linearity of the light yield in NaI(Tl) scintillation crystals we are able to reproduce with great accuracy the response for several calibration sources [10].

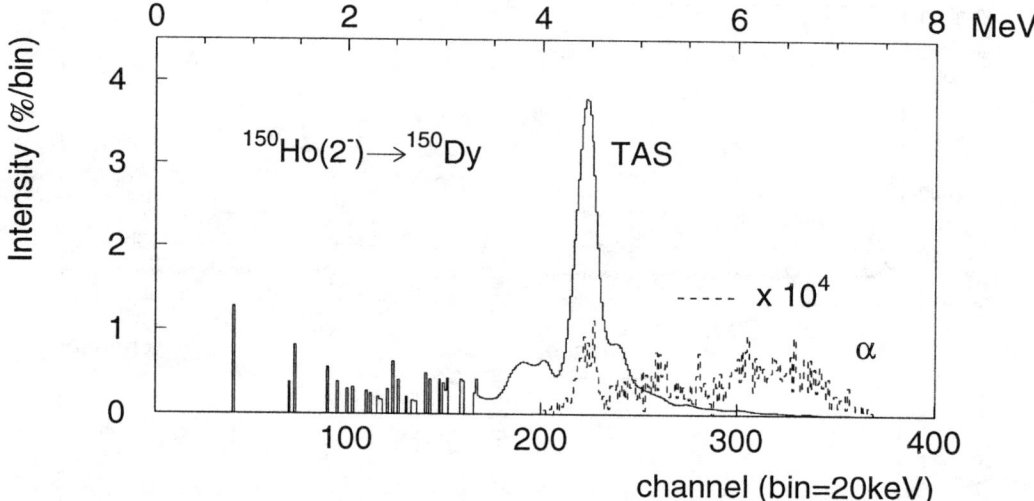

FIGURE 3. The ^{150}Ho 2^- beta-decay intensity I_β followed by γ-ray de-excitation, measured with TAS (solid line), and by alpha emission, measured with the particle telescope (dashed line).

Once the response matrix $\mathbf{R}$ has been obtained, the problem represented by Eq. (3) must be solved. This task is far from trivial because it belongs to the class of so-called *ill-posed* problems [11]. We have investigated the suitability of three different methods for our particular problem: 1) *Linear (or Tykhonov) Regularization Method* [12], 2) *Maximum Entropy Method* [13] and 3) *Bayesian Iterative Method* [14]. The comparison of the results from these three completely different methods reveals that the feedings obtained are essentially the same. Details will be given elsewhere.

We present here the results obtained using the Bayesian Iterative Method for the ^{150}Ho 2^- decay. A preliminary analysis of the CLUSTER CUBE data for the ^{150}Ho 2^- decay gave us the low energy discrete level scheme up to 3.3 MeV excitation energy. Above this energy we assumed that all levels de-excited in the same way to the discrete levels. Level energies were grouped in 20 keV bins. As an initial guess for the b.r. we took the apparent feeding to these levels observed in the high resolution experiment. We only needed to change the b.r. to the three lowest excited levels to obtain an excellent agreement with the data, simultaneously for the EC and β^+ decays as is shown in the upper panels of Fig. 2. The feeding intensities I_β which reproduce the data are displayed in the lower panels of Fig. 2.

The combined EC and β^+ result for the β-intensity is shown as a solid line in Fig. 3. A sharp resonance of $\sim$240 keV width emerges at 4.5 MeV excitation energy. The EC to β^+ ratio as a function of the energy is very sensitive to the end-point energy Q_{EC} and can actually be used to determine it [15]. From a preliminary analysis we obtain a value of $Q_{EC} = 7.4$ MeV. This result is in accord with the recently determined [16] using a Penning trap mass spectrometer: $Q_{EC} = 7372(27)$ keV. Using this value, the measured half-life [6] $T_{1/2} = 72(4)$ s, and Eq. (1), we calculate the strength S_β displayed as a dashed line in Fig. 4. The integrated $\log ft$ up to 7.0 MeV excitation energy amounts to 3.85(3), which is equivalent to a total B(GT)= 0.55(4) in units of $g_A^2/4\pi$.

β-Delayed Particle Spectroscopy

The result of our measurement with the particle telescope confirmed for the first time the existence of β-delayed alpha emission in the decay of ^{150}Ho 2^-. No proton emission was detected associated with this decay, although proton decay with a half-life compatible with ^{150}Tm ($T_{1/2} = 2.2$ s) was observed. The intensity of the alpha emission was extremely small. The total α-branching intensity relative to the total γ-branching intensity was determined to be:

$$\frac{I_\alpha}{I_\gamma} = 7 \times 10^{-5} \tag{4}$$

The normalized α spectrum is shown as a dashed line in Fig. 3. In the figure the α-intensity has been enhanced by a factor 10^4. It is remarkable that the resonance at $E_x \sim 4.5$ MeV shows-up in these data also.

High Resolution Spectroscopy

A total of 5.4×10^8 γ-γ coincidences involving one, two or three neighbouring capsules were registered with the CLUSTER CUBE for the decay of ^{150}Ho 2^-. The analysis of these coincidences gave a very rich level scheme. We have located 1124 γ-transitions de-exciting 334 levels up to $E_x = 5450$ keV. The levels populated around 4.5 MeV excitation energy are clearly evident in the spectra. In general the spin and parity could be determined unambiguously for the levels populated with enough intensity. The β feeding probability to the levels $I_\beta(E_x)$ was determined from γ-ray intensity balance. This was transformed into the β-strength distribution S_β which is shown as a continuous line in Fig. 4. The strength distribution is rebinned with the bin size (20 keV) of the TAS results. The strength around 4.5 MeV is concentrated in few levels: only 19 levels have intensities larger than 1%. The total B(GT) up to the highest observed level (5.45 MeV) is 0.26 ($\log ft = 4.18$), i.e. 47% of the total TAS result. In the restricted energy interval the TAS result is B(GT)= 0.44 ($\log ft = 3.94$), still significantly larger.

The CLUSTER CUBE data clearly demonstrate the correctness of the analysis method for the TAS experiments: Both experiments give a similar shape, but the high resolution experiment loses sensitivity towards higher energies.

DISCUSSION

In a very simple-minded approach, the 2^- isomer in ^{150}Ho can be viewed as $(\pi d_{3/2} \nu d_{7/2})_{2^-} (\pi^2)_{0^+}$. Thus, the only allowed decay is when the pair of protons is in the $\pi h_{11/2}$ orbital, and the GT transition $(\pi h_{11/2}^2)_{0^+} \to (\pi h_{11/2} \nu h_{9/2})_{1^+}$ can occur. The resulting decay will populate 4 quasiparticle (qp) states at $\sim$4 times the pairing gap energy (4-5 MeV) with spin-parity $J^\pi = 1, 2, 3^-$. In an extreme single particle (ESP) picture, where there are n protons in a $\pi j_>$ orbital ($j_> = l + 1/2$, l: orbital angular momentum) and the $\nu j_<$ ($j_< = l - 1/2$) is empty, the total B(GT) (in units of $g_A^2/4\pi$) for the transition $\pi j_> \to \nu j_<$ is given by [17]:

$$B(GT)^{ESP} = n \frac{4l}{2l+1} \tag{5}$$

Assuming that two protons occupy the $\pi h_{11/2}$ orbital gives B(GT)$^{ESP} = 3.64$, a value 6.6 times larger than the TAS result. First order corrections to the ESP picture were investigated by Towner [18] as a function of the number of protons n in the orbital. He considered the effect of partial occupancy of the orbits involved (pairing

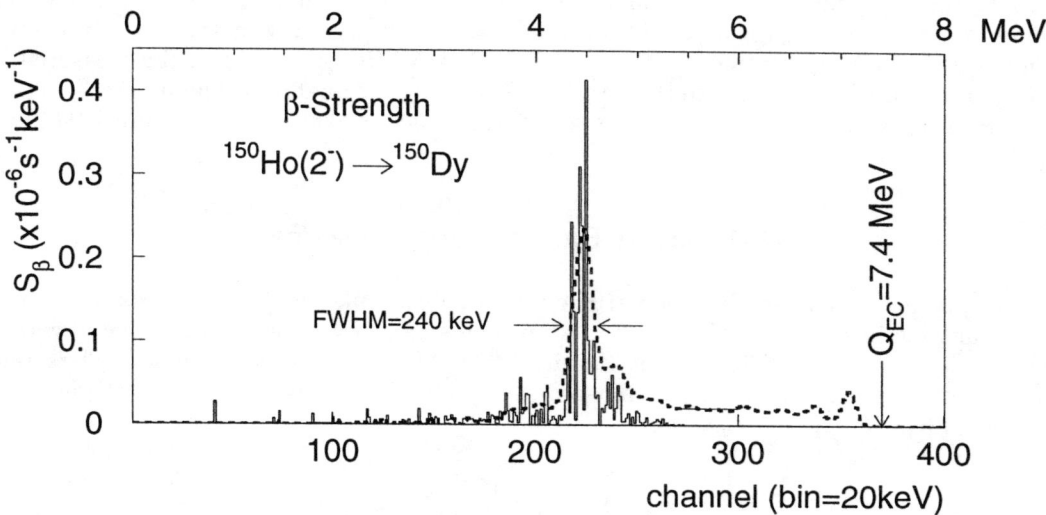

FIGURE 4. The TAS (dashed line) and CLUSTER CUBE (solid line) results for the ^{150}Ho 2^- β strength .

correlations) and of ground state correlations. The latter were estimated in first order perturbation theory (core polarization) for the most important contributions. According to these estimations the B(GT)^{ESP} for $n = 2$ should be reduced by a factor 2.4-3.0 depending on the interaction used. The decay of ^{150}Ho 2^- is very similar to the decay of ^{148}Dy (Q_{EC} =2678(10) MeV) where the strength is expected to lie at low excitation energy. Indeed the B(GT) measured in a conventional Ge spectroscopy experiment [19] is 0.44(3), close to our value. It appears therefore that an additional reduction factor of ~2 is needed to match the experiment.

If the absolute value of the strength remains difficult to understand then the distribution of the strength should not be so complicated. The shape observed, with an extremely narrow resonance at 4.5 MeV excitation, tells us that the 4 qp states primarily populated in the decay mix very little with other states. Now that the shape has been clearly measured, realistic calculations should also aim to reproduce the distribution of the strength. This could be investigated with QRPA calculations similar to those performed in the past [20,21].

Other experiments in this region have been done or are planned, which will allow a systematic investigation of the GT strength magnitude and distribution as a function of the number of protons occupying the $\pi h_{11/2}$ orbital. Such study will be an essential contribution to the understanding of the mechanisms responsible for the hindrance of GT transition probabilities in heavy nuclei.

CONCLUSIONS

We have taken a big step forward in terms of observing the GT resonance in β-decay. We have used three experimental approaches, only one of which is really sensitive to the β-feeding over the whole energy range, the total absorption spectroscopy. The high resolution data fails to detect the full strength but shows that the distribution obtained with the total absorption spectroscopy is correct. However this technique is very important in revealing the fine structure of the resonance.

The application of the two techniques presented here revealed a sharp GT resonance in the decay of ^{150}Ho, hidden until now by the experimental difficulties. These techniques which are applicable to other regions of the nuclide table might help us to understand transition probabilities governed by the $\sigma\tau$ operator.

ACKNOWLEDGMENTS

Work supported by C.I.C.Y.T. (Spain) under project AEN96-1662, by E.C.C. contract ERBCIPD-CT-950083, by the Russian Fund for Basic Research and Deutsche Forschungsgemeinschaft contract 436-RUS-113/201/0(R) and by Polish Committee of Scientific Research grant KBN-2-P03B-039-13.

REFERENCES

1. B.A. Brown and B.H. Wildenthal, *At. Dat. Nucl. Dat. Tab.* **33**, 348 (1985).
2. P.G. Hansen, *Adv. Nucl. Phys.* **7**, 159 (1973).
3. P. Kleinheinz *et al*, *IKP-KFA 1988 Ann. Rep.*, 34 (1989).
4. H. Salewski *et al*, *Z. Phys.* **A337**, 161 (1990).
5. M. Karny *et al*, *Nucl. Instrum. Methods* **B126**, 411 (1997).
6. R.B. Firestone, *Table of Isotopes*, Wiley, New York (1996).
7. M. Wilhelm *et al*, *Nucl. Instrum. Methods* **A381**, 462 (1996).
8. D. Cano *et al*, submitted to *Nucl. Instrum. Methods* **A**.
9. *GEANT: Detector Description and Simulation Tool*, CERN Program Library W5013, Geneve (1994).
10. D. Cano *et al*, submitted to *Nucl. Instrum. Methods* **A**.
11. W.H. Press, S.A. Teukolsky, W.T. Vetterling and B.P. Flannery, *Numerical Recipes*, Cambridge University Press, Cambridge (1992).
12. A.N. Tykhonov and V.Y. Arsenin, *Solutions of Ill-Posed Problems*, Wiley, New York, (1977).
13. D.M. Collins, *Nature* **298**, 49 (1982).
14. G. D'Agostini, *Nucl. Instrum. Methods* **A362**, 487 (1995).
15. A. Gadea *et al* in *The Future of Nuclear Spectroscopy*, ed. W. Gelletly *et al*, NCSR Demokritos, Athens (1994).
16. D. Beck *et al*, *Nucl. Phys.* **A626**, 343c (1997).
17. W.F. Hornyak, *Nuclear Structure*, Academic Press, New York (1975).
18. I.S. Towner, *Nucl. Phys.* **A444**, 402 (1985).
19. P. Kleinheinz *et al*, *Phys. Rev. Lett.* **55**, 2664 (1985)
20. J. Engel *et al*, *Phys. Rev.* **C37**, 731 (1988)
21. J. Suhonen *et al*, *Phys. Lett.* **B202**, 174 (1988)

INTERFACE OF NUCLEAR STRUCTURE
AND ASTROPHYSICS

THE rp-PROCESS IN X-RAY BURSTS

Michael Wiescher, Ani Aprahamian, Joachim Döring, Joachim Görres

Department of Physics, University of Notre Dame, Notre Dame, IN 46556, USA

Abstract. The rp-process was first suggested by Wallace and Woosley (1981) as the dominant nucleosynthesis process in explosive hydrogen burning at high temperature and density conditions. The process is characterized by a sequence of fast proton capture reactions and subsequent β-decays. The reaction path of the rp-process runs along the drip line up to $Z\approx50$. Within a sufficiently long time scale for the thermonuclear runaway mainly N=Z even-even nuclei are produced in the associated nucleosynthesis. Extended model calculations indicate a large production of mass A=80 isotopes. The resulting abundances are extremely sensitive to the structure, lifetimes and decay pattern of the relevant isotopes. Experimental results are presented here for ^{80}Zr along with a discussion of the associated nucleosynthesis.

I INTRODUCTION

Presently the interpretation of type-I X-ray bursts [1] is based on accretion processes in a close binary system. Accretion takes place from the filled Roche-Lobe of an extended companion star onto the surface of a neutron star. Typical predictions for the accretion rate vary from 10^{-10} to 10^{-9} $M_\odot$ y^{-1}. The accreted matter is continuously compressed by the freshly accreted material until it reaches sufficiently high pressure and temperature conditions which allow the thermonuclear ignition [2–5].

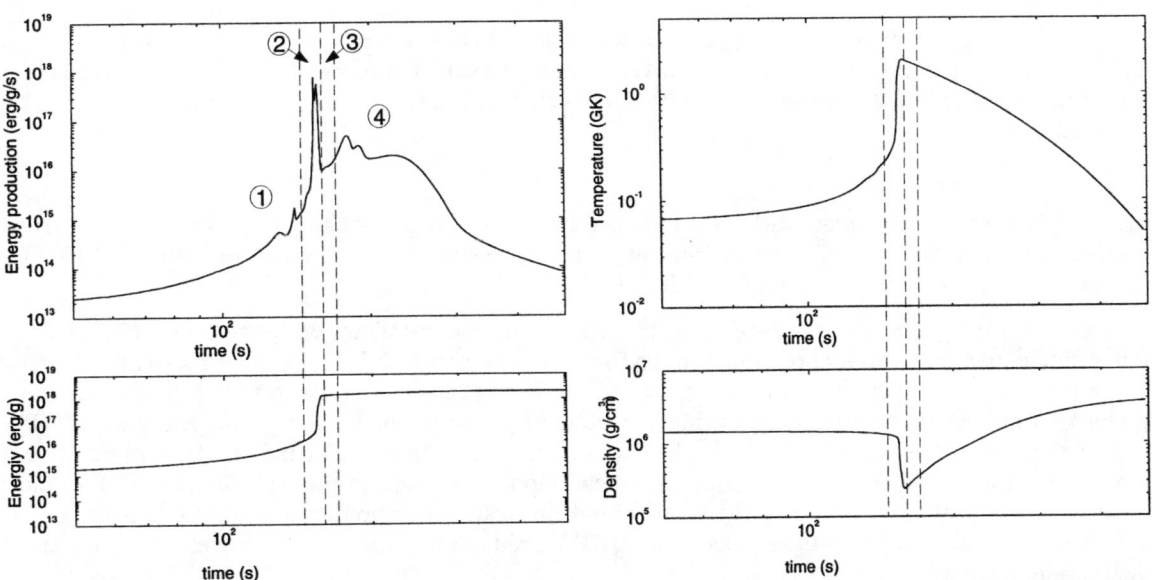

FIGURE 1. The figure shows the energy production rate (luminosity), the total energy production, the temperature, and the density as a function of time for the thermonuclear runaway in a simplified X-ray-burst model. Indicated are different phases in the thermonuclear runaway. Phases 1, 2, and 4 correspond to periods of rapid nucleosynthesis, and phase 3 corresponds to a dormant period at the maximum of the temperature curve. For details see text.

CP481, *Nuclear Structure 98*, edited by C. Baktash

Nuclear burning is ignited at high density, $\rho \geq 10^6$ g/cm^3, in the accreted envelope, via the pp-chains and the hot CNO-cycles. Yet the released energy is not sufficient for triggering the thermonuclear runaway at electron degenerate conditions near the bottom of the accreted layer. This requires the ignition of the triple α process and the break-out from the hot CNO cycles via α capture on the CNO waiting points ^{14}O, ^{15}O, and ^{18}Ne.

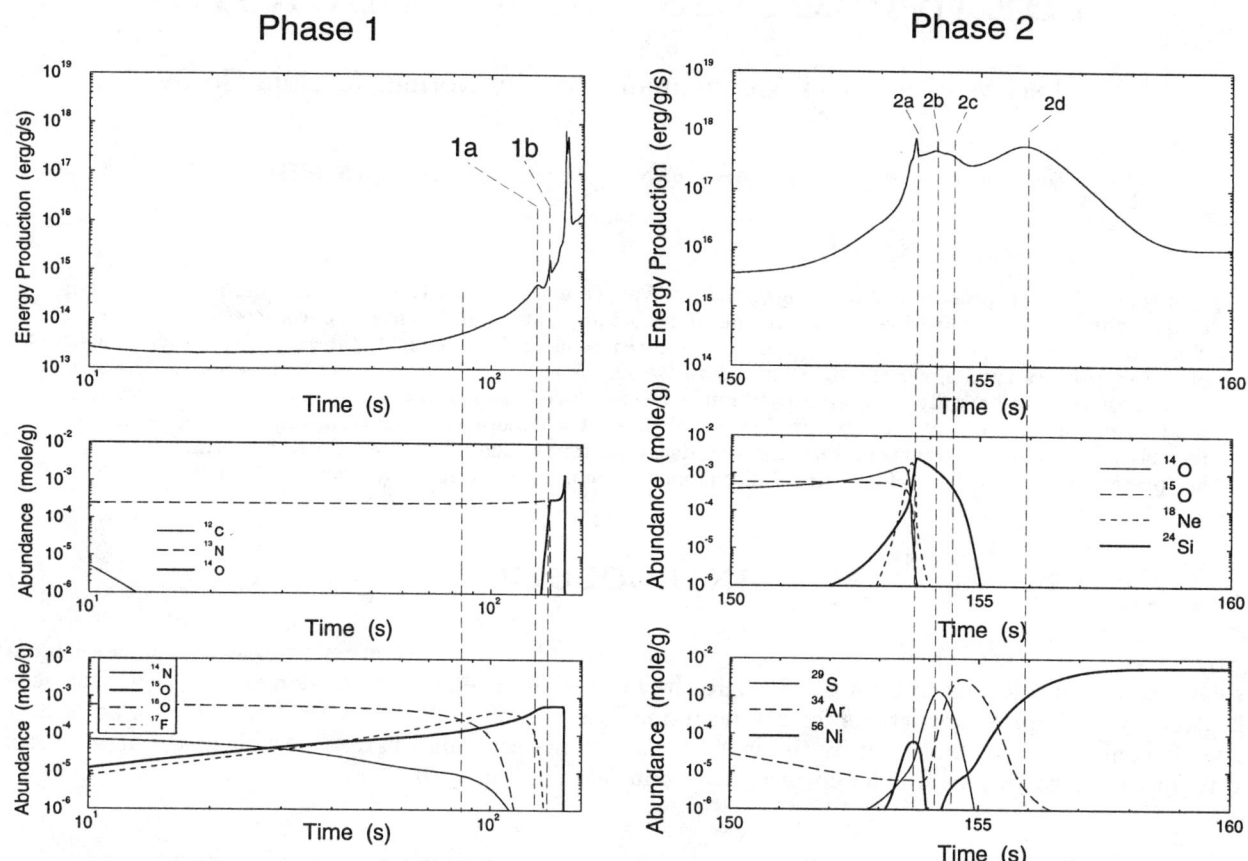

FIGURE 2. Energy production (luminosity) and nucleosynthesis in the on-set (phase 1) and during the thermonuclear runaway (Phase 2) of an X-ray burst. Peak 1a and 1b correspond to the burning of ^{17}F and ^{13}N. The structure of the luminosity peak during phase 2 is associated with the destruction of the rp-process waiting point nuclei ^{29}S, ^{34}Ar, and the final enrichment in ^{56}Ni.

While the triple α reaction delivers the CNO-fuel, the break-out triggers the rp-process which in turn releases the required energy for initiating a thermonuclear runaway [6–10]. Peak temperatures up to $2 \cdot 10^9$ K can be reached before the degeneracy is completely lifted [11].

The time-scale for the thermal runaway and the subsequent cooling phase varies between 10 s and 100 s [10] depending on the particular model parameters for the accretion process. Within this time-scale, the rp-process proceeds well up to ^{56}Ni [6] or even further up to ^{96}Cd [7–10,12]. The details of the rp-process reaction path during the thermonuclear runaway and during the cooling phase is not very well understood yet. The main handicap for the interpretation is the lack of experimental and theoretical information about the masses, life-times, decay rates, and reaction rates for the nuclei along the reaction path at the limits of stability. The model calculations presented here are therefore based on systematic predictions for masses on the basis of the FRDM-mass model [13], life-times based on QRPA-predictions, and capture rates based on the Hauser Feshbach model [8].

In the present contribution, we will discuss the detailed characteristics of the rp-process in the framework of a simple one-zone model to investigate its influence on the time structure of the luminosity burst and its possible endpoint in the range of the highly deformed N=Z nuclei near A=80.

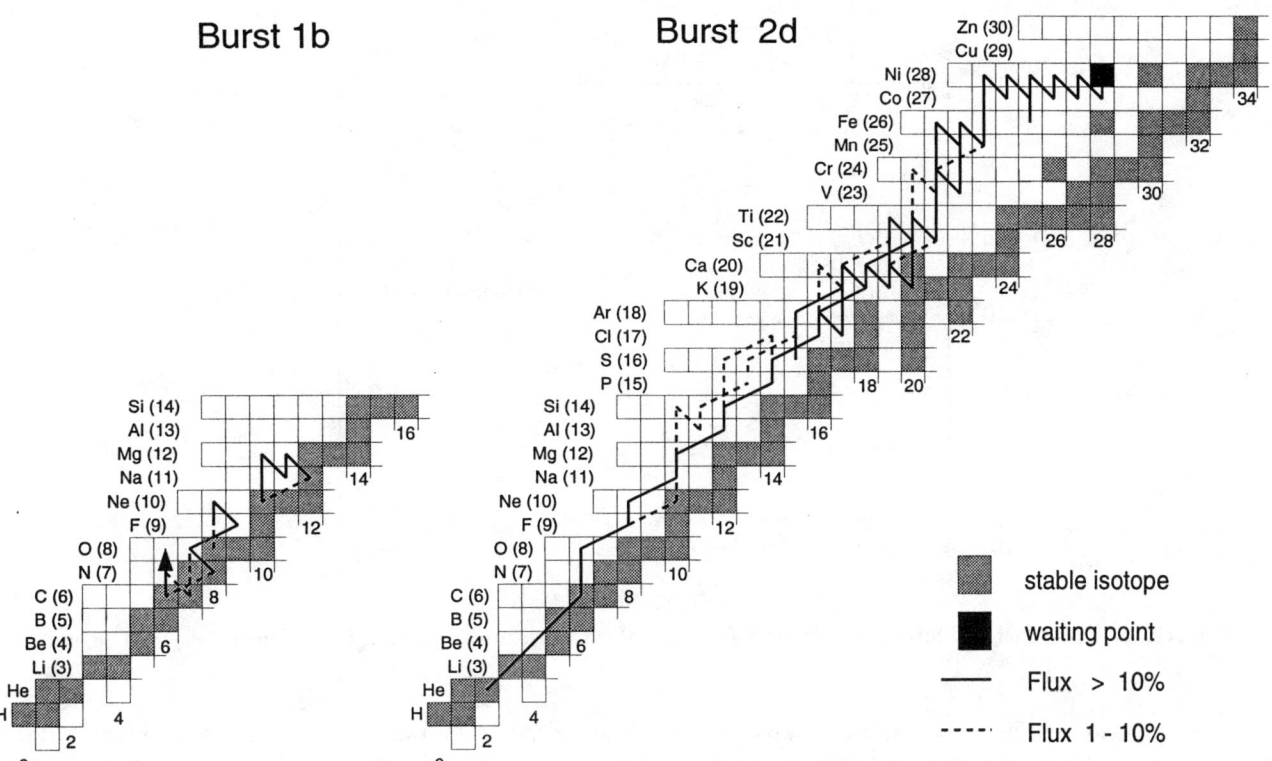

FIGURE 3. The αp- and rp-process reaction path during the on-set (1) and the thermonuclear runaway phase (2) of the X-ray burst.

A   The rp-process in the thermal runaway of the X-ray burst

The following section will focus on the rp-process characteristics calculated selfconstistently for temperature conditions during the thermonuclear runaway in an one mass zone X-ray burst model. For this model, hydrostatic equilibrium is maintained. Figure 1 shows the energy production, the total released energy, the temperature, and the density in the accreted envelope (burning-zone) calculated as a function of time over the duration of the X-ray burst. The ignition is correlated with a rapid increase in energy production which is responsible for most of the released energy. This energy release in turn causes a rapid increase of temperature and a parallel drop in density while maintaining constant pressure from the degenerate electron gas [8,9].

The energy production curve exhibits a pronounced structure which is directly correlated with the waiting point concept of the rp-process [8,9]. The energy production in the burst can be characterized by four periods, the ignition phase of the burst (1), the peak of the burst (2), the dormant phase of the burst (3), and the phase of the after-burst (4). These four different phases are characterized by different temperature and density conditions for the nucleosynthesis and energy generation. Phase (1) is characterized by the hot CNO cycles triggered by proton capture reactions of the accreted hydrogen on the carbon, nitrogen, and oxygen isotopes which have not been destroyed by spallation processes in the outer atmosphere of the accreting neutron star [14]. The energy production and the nucleosynthesis are shown in figure 2. The two peaks in the energy production are caused by the conversion of the initial abundance of ^{12}C into ^{14}O and of ^{16}O into ^{15}O by two subsequent proton capture reactions. Because of the slow decay of these two isotopes the CNO process is halted and the energy production drops. The main reaction flow is confined to the CNO cycles, but, as seen in figure 2 the temperature has already increased sufficiently to trigger phase (2) of the burst, the ignition of the triple-α-reaction.

Phase (2) is initiated at a temperature of $T \approx 2.4 \cdot 10^8$ K via the triple α-process. Parallel to these processes the ^{14}O and ^{15}O waiting point nuclei are rapidly depleted by α-capture processes. The fine structure of the energy burst is characterized by the various waiting points along the process path. The burst is initiated by the conversion of the hot CNO waiting point isotopes ^{14}O, ^{15}O, and ^{18}Ne by the αp-process to ^{24}Si. After the decay of ^{24}Si the process is reignited at higher temperatures by proton and α capture reactions leading

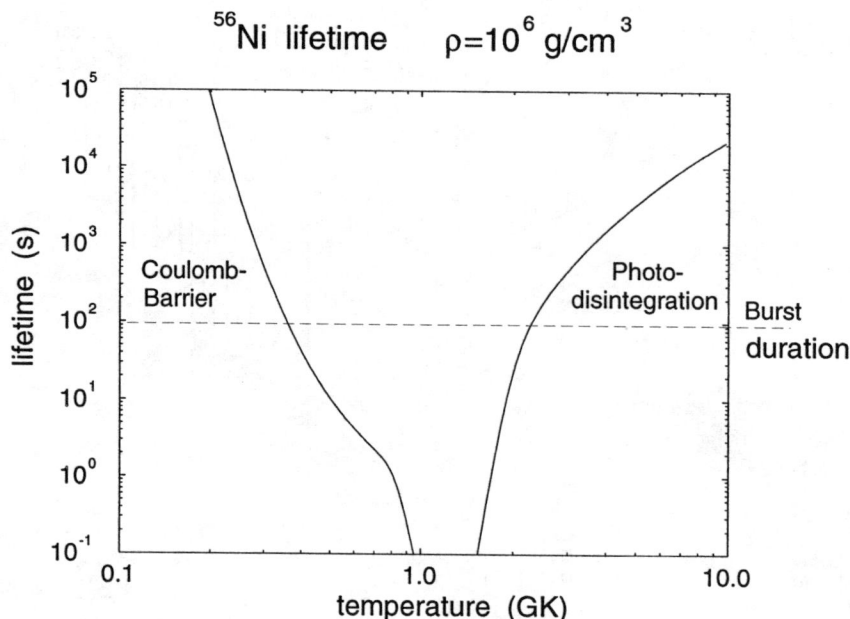

FIGURE 4. The effective lifetime of ^{56}Ni as a function of temperature calculated for a density of $\rho=10^6$ g/cm^3.

to the production of the next waiting point isotopes ^{29}S and ^{34}Ar. The rapid increase in temperature allows subsequent α-capture to bridge these waiting points and leads to a rapid conversion of the waiting point isotopes to ^{56}Ni. At this time a continuous reaction flow converts the accreted He-abundance via the αp-process and the rp-process to ^{56}Ni. The reaction flow, integrated over the duration of the last phase (2d) of the burst is shown in figure 3. Since most of the reaction path is characterized by (α,p) reactions considerably more helium is burned than hydrogen. At this point, peak temperatures of $T \approx 1.5 \cdot 10^9$ K have been reached and further processing is halted by the ^{56}Ni(p,γ)-(γ,p) equilibrium [8]. The energy production drops rapidly while most of the initial heavy isotope abundances as well as a large fraction of the initial helium remains stored in the waiting point nucleus ^{56}Ni. The drop in energy production causes a slow down in the temperature increase just before the peak temperature is reached.

At these peak temperature conditions ^{56}Ni has a lifetime of approximately 100 s because the proton capture is largely balanced by the inverse photodisintegration of ^{57}Cu (see figure 4). However, with the decrease in temperature and the parallel increase in density the ^{56}Ni(p,γ)-(γ,p) falls out of equilibrium and the effective lifetime of ^{56}Ni decreases down to a fraction of a second versus proton capture at temperatures $T \approx 1.5 \cdot 10^9$ - $1 \cdot 10^9$ K as shown in figure 4. For temperatures below $1 \cdot 10^9$ K the lifetime increases again rapidly because of the temperature dependence in the reaction rate for ^{56}Ni(p,γ)^{57}Cu. The small lifetime of ^{56}Ni within this temperature window allows the ignition of phase (4) of the energy burst by proton capture on ^{56}Ni. This phase is characterized by nucleosynthesis via the rp-process beyond ^{56}Ni which operates during the cooling period until the Coulomb barriers prevent any further proton capture processes. Figure 5 shows the reaction flow integrated over the duration of the cooling phase (4c). Notice that due to the lower temperature the αp-process is only dominant below sulfur, at higher masses the reaction path is characterized by the rp-process pattern leading up to ^{100}Sn. Figure 5 also shows the details of the burst structure during this cooling phase of the burst. Several peaks in the energy production are due to the depletion of ^{56}Ni and the further processing towards the N=Z waiting point nuclei ^{64}Ge and the subsequent nucleosynthesis towards ^{68}Se and ^{72}Kr. These waiting point nuclei originally have been suggested as possible termination points for the rp-process because of their rather long β-decay lifetime [15–17]. Yet, it has been shown [8,9] that the effective lifetime of these nuclei is actually shortened by two sequential proton capture reaction which bridge the barely bound isotope ^{65}As [18] as well as the unbound isotopes ^{69}Br [16,17] and ^{73}Rb [17,19]. High density and temperature conditions in the accreted layer during this phase allow two-proton capture processes to become prevalent. It should be pointed out that the two-proton capture rates used in this study depend heavily on the reliability of predicted masses for the relevant isotopes [8,13]. In the final phase ^{68}Se and ^{72}Kr are eventually converted to heavier isotopes up to the mass 100 region. In the final phase most ($\geq$90%) of the initial helium as well as most of the other isotopes has been converted to heavy isotopes with masses A$\geq$72. The most abundant emerging isotopes

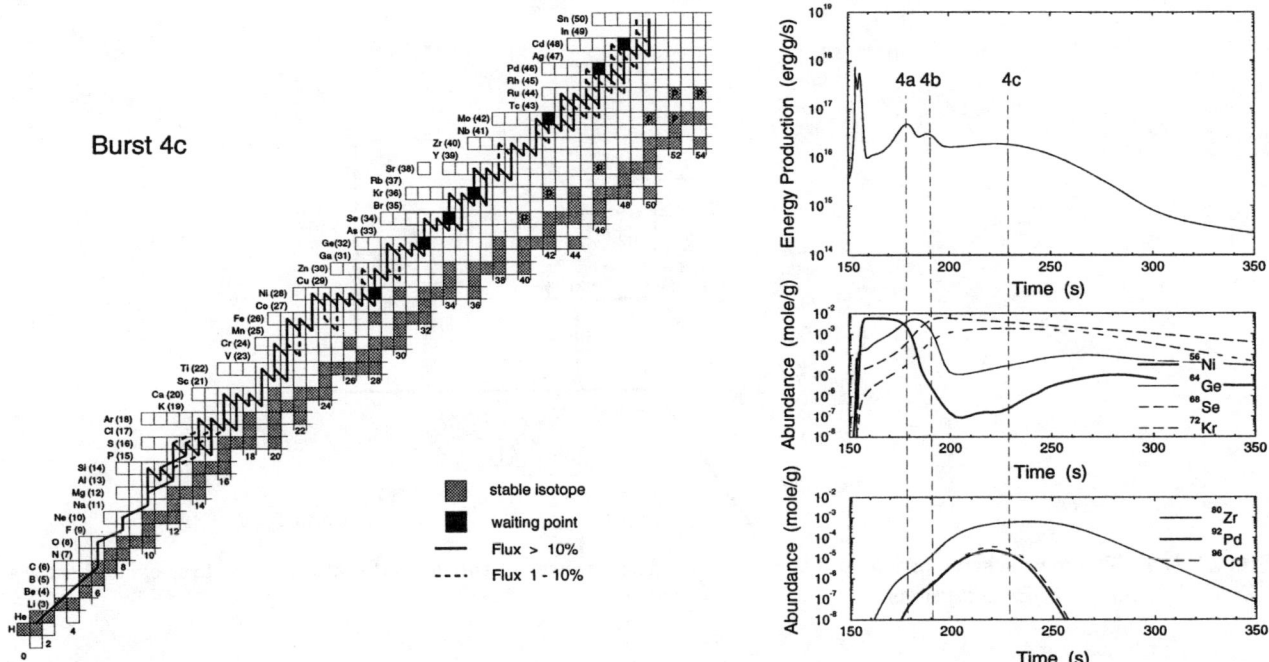

FIGURE 5. The right hand side of the figure shows the reaction flux in the late cooling phase (4d) of the X-ray burst. The overall flux is dominated by the rp-process. Only for very light Z nuclei (Z≤14) is the Coulomb barrier low enough for α capture reactions to compete with the proton capture and the β-decay. Also indicated are the light p-nuclei which are possibly produced in the X-ray burst. The left hand side of the figures shows the energy production (luminosity) and nucleosynthesis in the late phase (4) of an X-ray burst. The peaks in the luminosity correspond to the destruction of waiting point nuclei.

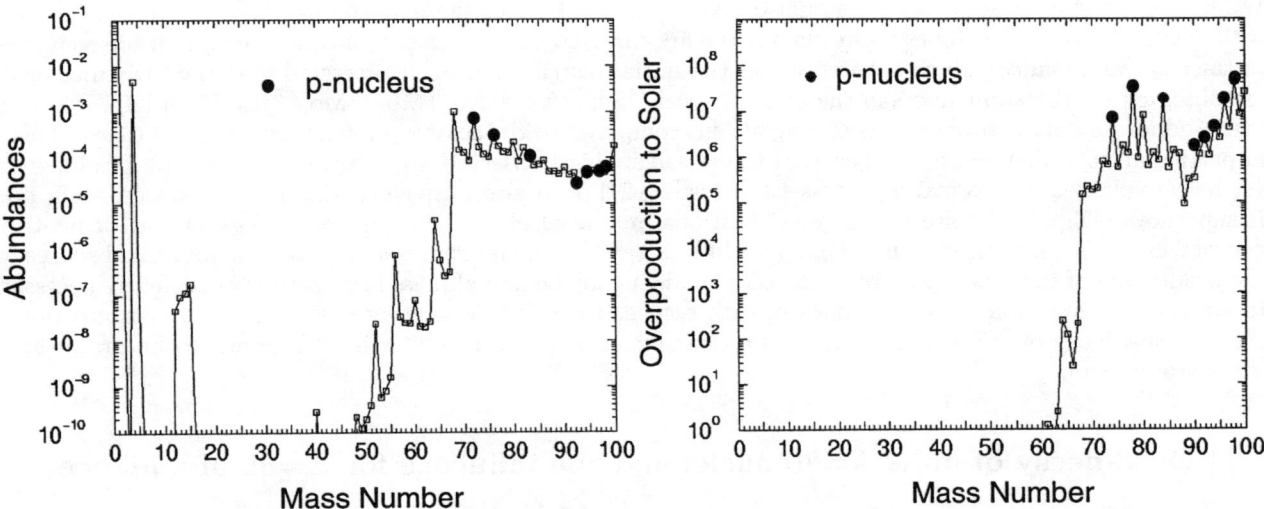

FIGURE 6. The abundance distribution after the thermonuclear runaway of a single X-ray burst. The figre on the right demonstrates the relative overabundance to observed solar values.

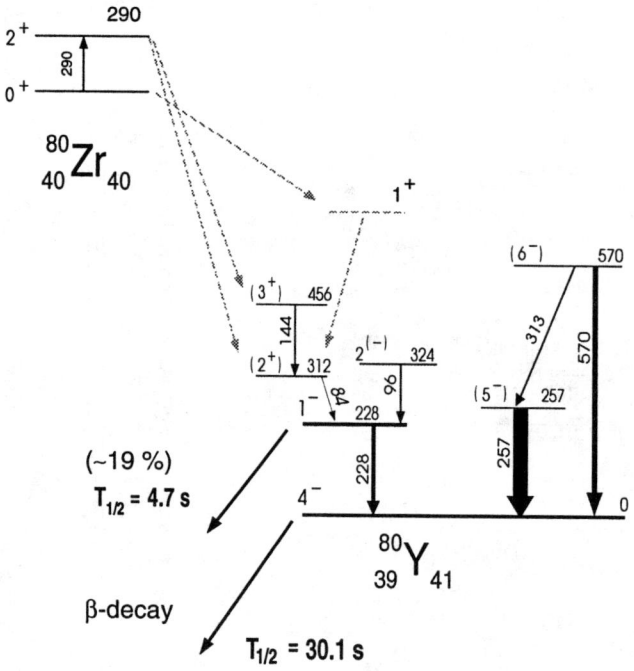

FIGURE 7. The level scheme of ^{80}Y and the proposed decay routes of the ground state and the first excited state in ^{80}Zr at high temperature conditions.

are the N=Z nuclei ^{80}Zr and ^{84}Mo as indicated in figure 5.

B Nucleosynthesis in the mass 80 range

Very limited experimental information is available about nuclear reaction and nuclear decay processes for nuclei along the rp-process path in the mass 76 to 100 range. The calculations are entirely based on model predictions for masses, lifetimes, and cross sections [8]. The nucleosynthesis predictions may therefore carry considerable uncertainties. The overall abundance conditions after the freeze out of the thermonuclear runaway are demonstrated in figure 6. The left part of the figure shows the abundance distribution of the material and the right shows the overabundances which are calculated in comparison with the solar abundance distribution. While there is still an appreciable amount of hydrogen, the bulk of the material has been converted to nuclei with masses A≥70. All isotopes above mass A=68 are enriched by more than four orders of magnitude compared to the solar abundances which have served as the initial distribution of the accreted material. Distinct peaks are observed for the abundances of the light p-nuclei ^{74}Se, ^{78}Kr, ^{84}Sr, ^{92}Mo, ^{94}Mo, ^{96}Ru, ^{98}Ru [20]. They are overproduced by up to seven orders of magnitudes compared to the solar abundances of these isotopes. This is in particular noticable due to the fact that the relatively high observed abundances of these isotopes have not yet been explained in classical p-process scenarios [21–24] or in alternative neutrino-induced processes in type II supernovae [25]. If, despite the large gravitational potential of the neutron star, radiation and/or neutrino driven winds [26–29] cause suffiently high mass loss, X-ray bursts may represent a potential additional source for the production of these isotopes. More detailed modeling of the actual mass loss conditions is clearly necessary to address this question. Another concern, with respect to the nucleosynthesis is the large overproduction of ^{80}Kr by the decay of ^{80}Zr. This represents a clear inconsistency since ^{80}Kr is well known to be produced in the s-process [30].

C β-decay of mass A=80 nuclei and the influence for A=80 abundance

The high production of the progenitor ^{80}Zr in the rp-process model calculations is mainly due to the long lifetime of $T_{1/2}$=6.85 s which has been predicted by QRPA model calculations for a prolate ground state deformation of ϵ_2=0.383 [8]. Recent attempts to verify the lifetime predictions experimentally failed [1], but led to the identification of an isomeric J^π=1$^-$ state in ^{80}Y at 228 keV excitation energy. This isomeric state β/EC decays with a 19 % branching to low excited states in ^{80}Sr with a half life of 4.7 s [31]. The half life of the ^{80}Y ground state was determined to be $T_{1/2}$=30.1 s. In the aforementioned calculations the β-decay lifetimes of ^{80}Zr and ^{80}Y was based on the ground state decay only. However, β-decay processes in high temperature scenarios also involve β-decay of thermally excited states [8]. Their lifetime can be significantly different from the lifetime of the ground state due to the significantly higher Q_β-values. For the highly deformed nucleus ^{80}Zr, the excitation energy of the first 2$^+$ state is rather low, $E_x \approx$290 keV [32], therefore the decay through the thermally excited state is initiated at temperatures above $\approx$ 1 GK. The allowed β-decay will populate 1$^+$ and/or 2$^+$ states which are predicted at an excitation range $E_x \approx$300-600 keV [31], possibly to the 2$^+$ (6 μs) isomeric state at 312 keV which decays by 84 keV γ-emission to the 1$^-$ isomeric state at 228 keV [31,33]. The decay pattern of ^{80}Zr is demonstrated in figure 7. A QRPA calculation for such a decay suggests an effective half life of ^{80}Zr to $\approx$3 s for the temperature conditions in the cooling phase of the X-ray burst. Figure 8 shows the development of the abundances of the N=Z isotopes ^{56}Ni, ^{64}Ge, ^{68}Se, and ^{80}Zr calculated on the assumption of a pure ground state decay of ^{80}Zr. In comparison the figure also shows the abundances calculated with the addition of a fast decay sequence through thermal excitation of ^{80}Zr with the subsequent β^+ decay to the isomeric states in ^{80}Y. The figure clearly shows a factor four reduction of the ^{80}Zr abundances while the abundances of the other isotopes remain unaffected. For the formation of ^{80}Kr in the freeze out phase the decay of all A=80 neutron deficient isotopes have to be considered. Assuming only ground state decay,

---

[1] A recent measurement at the Oak Ridge RMS separator has been sucessfull, the data are still being analyzed (A. Piechaczek, private communication).

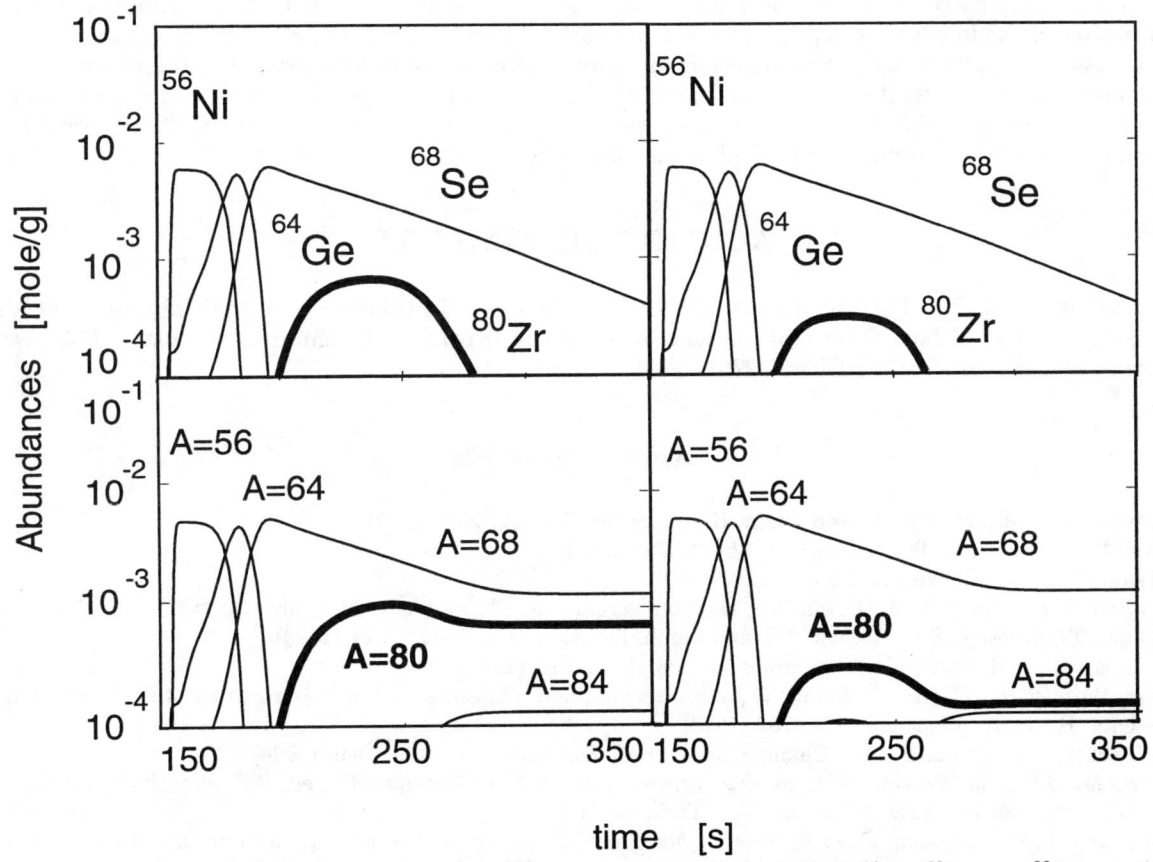

FIGURE 8. The two top figures show the development of the abundances for ^{56}Ni, ^{64}Ge, ^{68}Se, and ^{80}Zr as a function of time. The figure on the top left shows abundances based on pure ground state decay of ^{80}Zr while the one on the right side includes the decay via the excited states. The lower part of the figure shows the total abundances for mass A=56, 64, 68, 80, and 84 nuclei calculated at the same conditions.

significant enrichment can be observed for ^{80}Y along the rp-process path due to its long half-life of 30.1 s. The feeding of the isomeric states in ^{80}Y by the decay of ^{80}Zr reduces the effective half life of ^{80}Y significantly and subsequently reduces its abundance during the cooling phase. This is demonstrated in figure 8 comparing the development of the total A=80 abundance for the two discussed cases. Including thermal excitation of ^{80}Zr with subsequent decay through the isomeric states of ^{80}Y reduces the A=80 abundance by about one order of magnitude compared to ground state decay only.

II   CONCLUSION

It has been shown within the framework of a simplified hydrostatic X-ray burst model that the rp-process is sufficiently fast to process light mass material from the CNO range up to the mass A=100 region. The time structure of the nucleosynthesis process and the associated energy production depends on the effective lifetimes of the waiting point nuclei along the rp-process path. These are in turn determined by the temperature and density conditions in the burning zone. The main delay for the ignition of the thermonuclear runaway are the break-out reactions of the hot CNO cycles. The main impedance for the rp-process nucleosynthesis is the ^{56}Ni(p,γ)-(γ,p) equilibrium which can only be bridged in the cooling phase of the burst. For the subsequent rp-process nucleosynthesis up to the mass A=100 range the dominant impedances are the long effective lifetimes of ^{64}Ge, ^{68}Se, and ^{72}Kr. These lifetimes are determined by the temperature and density dependent two-proton capture reaction rates and by the β-decay of their ground states and possibly their thermally excited states. The last waiting point is the largely deformed N=Z nucleus ^{80}Zr which cannot be depleted by two-proton capture since ^{81}Nb is highly proton unbound [8]. It was shown in this study that the effective lifetime of ^{80}Zr can be sufficiently reduced by thermal population of its first excited state and its subsequent β-decay feeding the short-lived isomeric 1^{-} level in ^{80}Y [31]. This reduces the effective lifetime of ^{80}Zr by more than 50% which leads to a significant reduction of the final abundance of A=80 nuclei. Reliable experimental information about masses of the involved isotopes, their β-decay patterns, and the β-feeding and decay of isomeric states in the mass A=80 range are necessary to perform a more reliable and consistent study of the rp-nucleosynthesis in this mass range. To further investigate possible X-ray burst nucleosynthesis contributions to the galactic abundances, more detailed theoretical studies are clearly necessary in order to evaluate the various mass loss effects out of the gravitational potential of the neutron star.

ACKNOWLEDGMENTS

We want to thank F.K Thielemann from the Universität Basel, L. Bildsten from the University of California at Berkeley, and H. Schatz from GSI Darmstadt for many fruitful and helpful discussions. This work was supported by the NSF-Grant PHY94-02761.

REFERENCES

1. Lewin, W., van Paradijs, J., and Taam, R., *Space Sci.Rev.* **62**, 233 (1993)
2. Ayasli, A., and Joos, P., *Astrophys.J.* **256**, 637 (1982)
3. Taam, R., *Ann.Rev.Nucl.Sci.* **35**, 1 (1985)
4. Fujimoto, M., Sztajno, M., Lewin, W., and van Paradijs, J., *Astrophys.J.* **319**, 902 (1987)
5. Taam, T., Woosley, S.E., Weaver, T., and Lamb, D., *Astrophys.J.* **413**, 324 (1993)
6. Wallace R., and Woosley, S.E., *Astrophys.J.Suppl.* **45**, 389 (1981)
7. Van Wormer, L., Görres, J., Iliadis, C., Wiescher, M., and Thielemann, F.K., *Astrophys.J.* **432**, 326 (1994)
8. Schatz, H., *et al.*, *Phys.Rep.* **294**, 167 (1998)
9. Wiescher, M., Schatz, H., and Champagne A., *Phil.Trans.Roy.Soc.* **356**, 1949 (1998)
10. Woosley S.E., and Weaver, T.A. in *High Energy Transients in Astrophysics*, ed. S.E. Woosley, AIP Conf. Proc. **115**, (American Institute of Physics, New York, 1984)
11. Bildsten, L. in *The many Faces of Neutron Stars*, eds. A. Alpar, L. Buccheri, J. van Paradijs (Dordrecht, 1997)
12. Wallace, R., and Woosley, S.E. in *High Energy Transients in Astrophysics*, ed. S.E. Woosley, AIP Conf. Proc. **115**, (American Institute of Physics, New York, 1984)
13. Möller, P., Nix, J.R., Myers, D., and Swiatecki, W.J., *At.Data.Nucl.Data.Tab.* **59**, 185 (1995)
14. Bildsten, L., Salpeter, E.E., and Wassermann, I., *Astrophys.J.* **384**, 143 (1992)

15. Wallace, R.K., and Woosley, S.E. in *Proceedings of Accelerated Radioactive Beam Workshop, Parksville, Canada*, eds. L. Buchmann, J. D'Auria, (TRIUMF, Vancouver, 1985)
16. Blank, B., *et al.*, *Phys.Rev.Lett* **74**, 4611, (1995)
17. Pfaff, R., *et al.*, *Phys.Rev.C* **53**, 1753 (1996)
18. Winger, J., *et al.*, *Phys.Rev.C* **48**, 3097 (1993)
19. Mohar, M.F., *et al.*, *Phys.Rev.Lett* **66**, 1571 (1991)
20. Lambert, D.L., *Astr.Astrophys.Rev.* **3**, 201 (1992)
21. Woosley, S.E., and Howard, W.M., *Astrophys.J.* **354**, L21 (1990)
22. Rayet, M., Prantzos, N., and Arnould, M., *Astr.& Astrophys.* **227**, 271 (1990)
23. Howard, W.M., Meyer, B.S., and Woosley, S.E., *Astrophys.J.* **373**, L5 (1991)
24. Rayet, M., *et al.*, *Astr.& Astrophys.* **298**, 517 (1995)
25. Hoffman, R.D., Woosley, S.E., Fuller, G.M., and Meyer, B.S., *Astrophys.J.* **460**, 478 (1996)
26. Taam, R.E., Woosley, S.E., and Lamb, D.Q., *Astrophys.J.* **459**, 271 (1996)
27. Duncan, R.C., Shapiro, S.L., and Wasserman, I., *Ap.J.* **309**, 141 (1986)
28. Chevalier, R.A., *Ap.J.* **411**, L33 (1993)
29. Chevalier, R.A., *Ap.J.* **459**, 322 (1996)
30. Raiteri, C., *et al.*, *Astrophys.J.* **367**, 228 (1991)
31. Döring, J., *et al.*, *Phys.Rev.* C **57**, 1159 (1998)
32. Lister, C.J., *et al.*, *Phys.Rev.Lett.* **59**, 1270 (1987)
33. Regan, P.H., *et al.*, *Acta Phys.Pol. B* **28** , 431(1997)

Beta-decay from r-process waiting-point nuclei

J. Engel and M. Bender

Department of Physics and Astronomy, University of North Carolina at Chapel Hill, U.S.A.

Abstract. Beta-decay rates in neutron-rich nuclei with closed neutron shells govern the relative abundances of heavy elements made by the astrophysical r process. We calculate the beta-decay half-lives of these important nuclei in a fully microscopic self–consistent HFB+QRPA model, using the same Skyrme force for the particle-hole channel in both HFB and QRPA and a two-Gaussian neutron-proton (np) particle–particle interaction. In most cases we find smaller half-lives than do other groups; the reason is our inclusion of the residual particle-particle interaction. Our results are closest to those of others for doubly magic nuclei, in which the particle-particle interaction has little effect.

INTRODUCTION

The astrophysical r-process, consisting of rapid neutron capture punctuated frequently by beta decay, is responsible for half of all nuclei heavier than $A = 70$ [1]. The distribution of elements that it produces has peaks corresponding to neutron closed shells at $N = 50$, 82, and 126. The peaks form because these nuclei capture neutrons only reluctantly and are slow to beta decay; they therefore are not only built up themselves but also restrict the amount of heavier material that is synthesized. The neutron-separation energies and beta-decay lifetimes of these select nuclei are the nuclear properties with the largest effect on the r-process.

For this reason, simulations of the astrophysical r-process require accurate predictions of beta-decay lifetimes in closed-neutron shell nuclei far from stability. Unfortunately beta decay is hard to calculate. Most of the strength associated with the beta-decay operator $\sigma\tau_+$ gets shoved into a high-lying resonance (the Gamow-Teller resonance) above decay threshold. The strength that actually contributes to beta decay is the small low-energy tail of the Gamow-Teller distribution; tails are not easy to calculate accurately. In addition, beta-decay lifetimes depend sensitively on the nuclear binding energy released by the process, and small errors in the calculated Q values can cause much larger errors in decay rates[1]. These problems are complicated enough to demand a coherent and systematic approach to beta decay, which we present here.

Most prior work on beta decay far from stability [3–5] has used schematic forces and blended macroscopic and microscopic methods in nonself-consistent ways, so that deficiencies in the treatment are not always easy to correct. Here we develop a fully microscopic self-consistent mean-field model to describe the beta decay of even-even nuclei; we calculate masses and single–particle energies of the initial nuclei in the Hartree–Fock–Bogolyubov model (HFB), and properties of states in the final nuclei in the neutron-proton quasiparticle random phase approximation (QRPA), which can be viewed as part of a lowest-order correction to the mean field. We restrict our study to semi–magic nuclei with closed neutron shells, which are crucial for the r-process and conveniently have spherical shapes. Our focus on a limited region of the nuclide chart and the self-consistency of our methods gives us confidence that we can make good predictions as we move towards the drip line.

FRAMEWORK

The starting point for our calculations is the coordinate-space Skyrme HFB model described in Ref. [6] (the question of what interaction to use is discussed below). By working in coordinate space we are able to

---

[1] For a review of the problem of calculating masses far from stability, see Ref. [2].

CP481, *Nuclear Structure 98*, edited by C. Baktash

include admixtures of the continuum that become important near the neutron drip line. The resulting ground-state energies, the quasiparticle energies, and the quasiparticle wave functions all play an important role in determining beta lifetimes. The other important ingredient is the residual np interaction, which we treat in the QRPA.

Although it is possible to formulate a coordinate-space QRPA, here we stick with the more conventional matrix version. The only problem with this approach is that if we want to include the entire single-quasiparticle spectrum that emerges from the HFB calculation we are faced with large numbers of two-quasiparticle configurations. We can reduce these numbers, however, by working in the "canonical" HFB basis, which diagonalizes the density matrix. Within this basis the model space can be truncated by restricting the single-quasiparticle states to those with an energy-expectation value less than some cutoff value. Canonical states are all localized and the low-lying canonical spectrum simulates a set of bound states and resonances, with the background continuum pushed to higher energies. The QRPA matrices can be written in this basis with the only change coming in the form of non-diagonal single-quasiparticle contributions. We discuss truncation in the canonical basis and its influence on actual calculations further in the next section.

Together the HFB and QRPA allow predictions of both the decay Q-value and the transition strengths to final states. Instead of dealing with the Q-value itself, we directly compute the binding energy released in the transition from the ground state of the initial nucleus to the many 1^+ QRPA states through the following approximation:

$$E(Z, N) - E_{1^+}(Z + 1, N - 1) \approx \lambda_n - \lambda_p + \delta - E_{\text{QRPA}} \quad , \tag{1}$$

where the λ's are the proton and neutron Fermi energies, δ is the neutron-proton mass difference, and E_{QRPA}, the eigenvalue of the QRPA matrix, can be interpreted as the excitation energy relative to a fictitious 0^+ state in the daughter nucleus. The approximation is good to the extent that the binding energies change linearly with N and Z. The transition matrix elements come directly from the QRPA in the usual way through the quasiboson approximation.

THE EFFECTIVE INTERACTION

Our self-consistent model requires the same effective interaction in the HFB and QRPA parts of the calculation. It must be able to describe not only the mean-field ground-state properties of the initial nucleus, which for $N \gg Z$ depend primarily on the particle-hole and like-particle pairing interactions, but also the excited states in the daughter nucleus and transitions to them. The transitions are most sensitive to the np particle-hole force, which pushes most of the β^- strength into the Gamow-Teller resonance at high energies, and the np particle-particle force, which shifts some strength to very low energies. The beta-decay lifetimes depend on the balance between these two channels. The mean field in neutron-rich nuclei depends almost not at all on the np particle-particle interaction.

For the particle-hole channel we use the Skyrme energy functional, which is flexible enough to give a good description of nuclear ground-state properties throughout the chart of nuclides [2]. Unfortunately, almost none of the existing Skyrme interactions is able to describe beta decay. This is not too surprising because the parameters of the forces have seldom been fit to charge-changing reactions. One can see that most forces are unsuitable without actually carrying out the HFB+QRPA calculation by simply looking at properties of symmetric nuclear matter, where a single number, the Landau–Migdal parameter g'_0, governs the interaction in the spin- and isospin-changing channel. Most of the Skyrme interactions in the literature give a value of g'_0 much smaller than the value 0.7 (in "pionic" units) determined by experiment [7]. Only one, developed by P.–G. Reinhard, is even close; it gives $g'_0 = 0.62$. In honor of this welcome value the interaction is called [8] SkG0 (the parameters are in Table 1). Although symmetric nuclear matter in its ground state is different from finite neutron-rich nuclei, SkG0 complemented with a simple delta interaction in the like-particle (HFB) pairing channel does a reasonable job of reproducing beta spectra measured in (p, n) experiments [9]. The Gamow-Teller resonances usually contain the correct fraction of the strength and come out at about the right energy. The total experimental strength, of course, is always less than that implied by the Ikeda sum rule [10], a fact we account for in beta decay by taking the axial vector coupling constant g_A to be 1.0 instead of 1.26.

A good description of the Gamow–Teller resonance, however, is only the first step towards the successful description of β^- decay. While the resonance is built almost entirely of particle-hole excitations the low-lying strength responsible for beta decay involves np particle-particle correlations and is therefore sensitive to a completely different channel of the force. The residual attractive np particle-particle force is neglected in most

TABLE 1. Parameters of the Skyrme interaction SkG0.

t_0	t_1	t_2	t_3
-2103.65 MeV fm^3	303.352 MeV fm^5	791.674 MeV fm^5	13553.252 MeV fm$^{3+3\alpha}$

x_0	x_1	x_2	x_3
-0.21070	-2.81075	-1.46160	-0.42988

W_0	b_4' [a]	α	
353.156 MeV fm^4	-198.74899 MeV fm^4	0.25	

[a] see [11] for definition

existing global calculations of beta decay. We find it absolutely necessary, however; without it beta-decay lifetimes in the neutron-rich semi-magic nuclei are always too long. The particle-particle interaction pulls strength down in energy, producing shorter lifetimes.

To keep matters relatively simple in constructing the np particle-particle interaction, we restrict ourselves to a density-independent force acting only in the $T = 0$, $S = 1$ and $T = 1$, $S = 0$ channels. QRPA calculations of double beta decay [12] have shown that the a good value for the ratio of the $T = 1$ to $T = 0$ particle-particle force is 0.6, a value we use to reduce the number of free parameters in our calculation. While a simple delta force can be used successfully for like-particle pairing when describing nuclear ground states [6,2], it turns out to be inadequate here. The half-lives fail to converge as single-particle states are added to the model space used in the QRPA. Any purely attractive interaction apparently suffers from the same problem. The situation is better, however, when we use a combination of a short-range repulsive Gaussian and a weaker longer-range attractive Gaussian:

$$V_{12} = V_1 \sum_{j=1}^{2} g_j \, e^{-r_{12}^2/\mu_j^2} \left[\hat{\Pi}_{S=1,T=0} + 0.6 \, \hat{\Pi}_{S=0,T=1} \right] \quad , \qquad (2)$$

where the $\hat{\Pi}_{S,T}$'s are projectors onto spin and isospin. We take the ranges $\mu_1 = 1.2$ fm, $\mu_2 = 0.7$ fm of the two Gaussians from the Gogny interaction [13], and choose the strengths $g_1 = 1$, $g_2 = -2$ so that the force is repulsive at small distances. The only remaining free parameter left to be adjusted to experimental data is V_1, the overall (negative) strength.

Though we use this two-Gaussian interaction for the results presented below, we are actually also able to get convergence from a simplified Skyrme-like interaction with the same spin–isospin structure as the finite-range force. Its parameters: $t_0 = 6.1697$ MeV fm^3, $t_1 = -2.5849$ MeV fm^5, $x_0 = x_1 = 0.25$, and all others equal to zero. Figure 1 depicts the rate of convergence of a predicted half-live as a function of the particle-particle strength V_1 for this simplified Skyrme force, the two-Gaussian interaction, and a δ interaction. The latter gives results that diverge badly with the size of the model space.

FIXING THE PARTICLE-PARTICLE STRENGTH

The calculations as outlined above contain only a single free parameter: the overall strength V_1 multiplying the two-Gaussian force in the $T = 0$, $S = 1$ np particle-particle interaction. In recent years progress has been made in measuring properties of neutron–rich nuclei where the r-process path comes closest to stability (see, e.g., [14,15]). We use the experimental half-lives of such nuclei to fit V_1. Near the $N = 82$ closed shell, for example, we include the nuclei ^{124}Cd, ^{126}Cd, ^{128}Cd, and ^{130}Cd in our fit. We omit nuclei such as ^{122}Cd and ^{134}Sn that have such small Q-values (and long lifetimes) that a slight error in the predicted Q-value has a large effect. Figure 2 displays the ratio of calculated to experimental lifetimes of the nuclei in our fit near $N = 50$ as a function of the strength of the particle-particle interaction. The lines intersect at a point very near 1, showing that we can reproduce all the lifetimes with a particle-particle strength of 250 MeV. We can therefore use this value to predict the lifetimes of $N = 50$ nuclei farther from stability. We repeat the procedure near $N = 82$. The fit is of the same quality, with the only drawback that we must use a very different value of the particle-particle strength: 150 MeV instead of 250 MeV. The reason may be connected with differences in the quality of the underlying mean-field (single-particle spectra and Q-values) in the two regions. The fact that we have to change V_1 by such a large amount to compensate demonstrates the sensitivity of beta-decay to these properties.

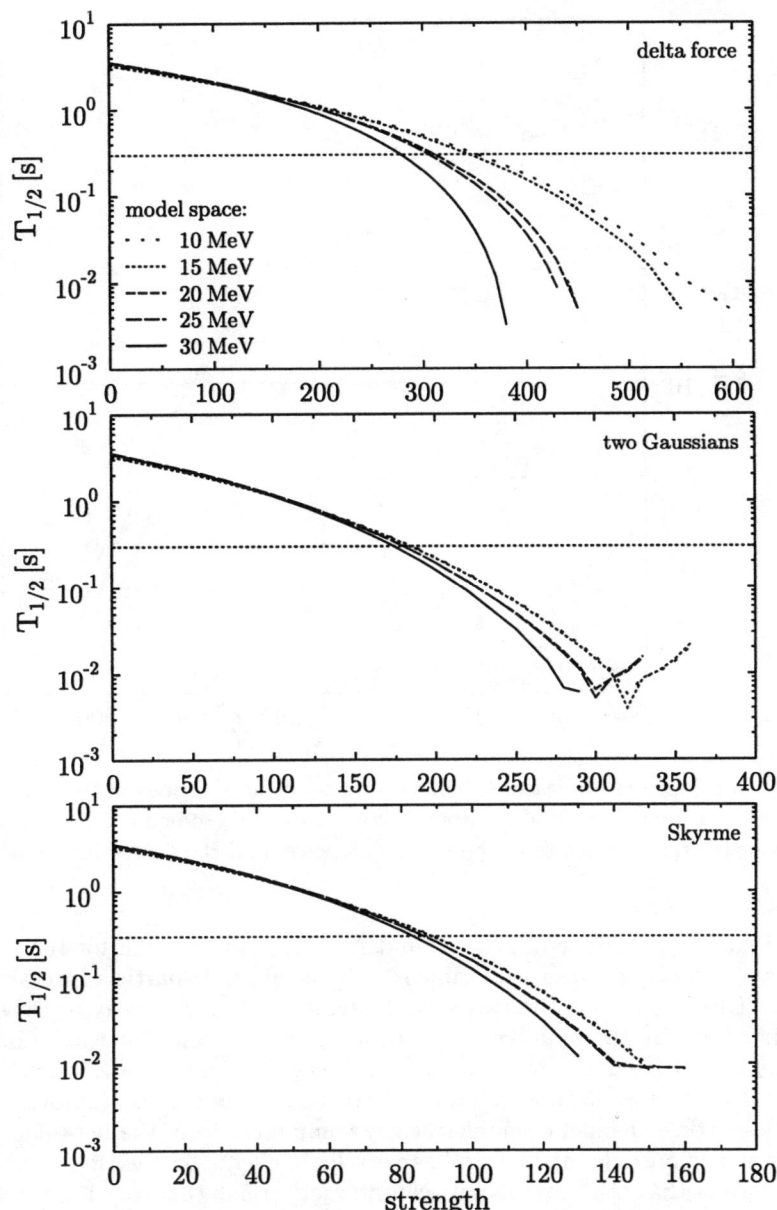

FIGURE 1. The dependence of the half-life of ^{128}Cd on the size of the single-quasiparticle model space above the Fermi surface, plotted versus the strength of the particle-particle interaction for three different forms. The combination of two Gaussians is the interaction in Eq. 2 we use to obtain our results. The Skyrme interaction is the simplified density-independent version discussed in the text. All single-quasiparticle orbits below the Fermi surfaces are included in the calculations; the different lines in each panel correspond to different limits on the energies of the quasiparticle orbits above the Fermi surfaces.

In the $N = 126$ region there are not enough data to support a fit and so we use the same value of the particle-particle strength as in the $N = 82$ region. Our predictions for the highest-mass nuclei are therefore less certain than the rest.

RESULTS FOR R-PROCESS NUCLEI

Figure 3 shows our predictions and those of other groups for the half-lives of the crucial closed-neutron-shell nuclei along the r-process path. Interestingly, our results agree fairly well with those of Möller *et al.* [3] for

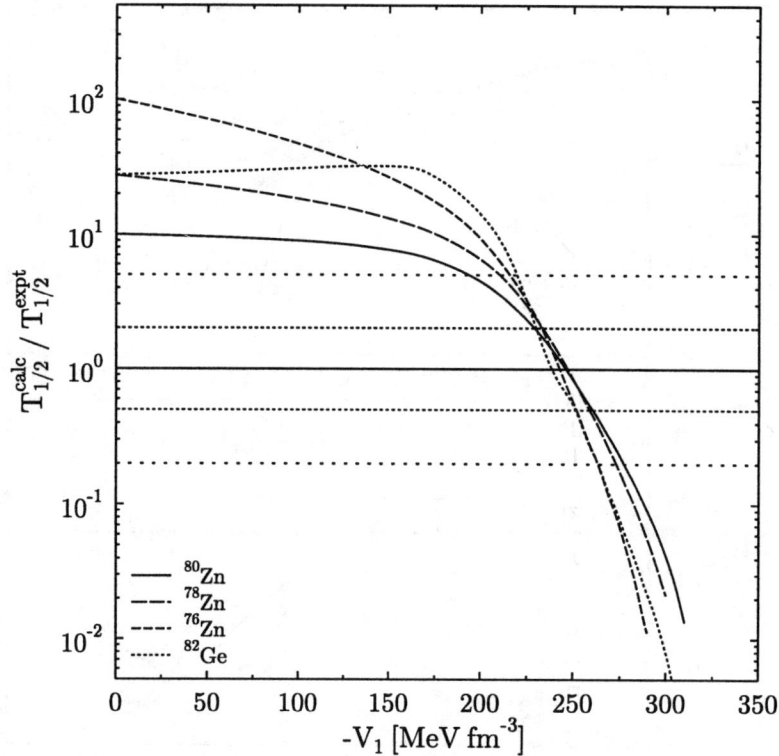

FIGURE 2. The ratio of calculated to measured half-lives for four nuclei near $N = 50$ as a function of the strength of the np particle-particle interaction. The solid horizontal line corresponds to equal values for measured and calculated half-lives, the short-dashed horizontal line to a factor of 2 difference, and the dotted line to a factor of 5 difference.

the very neutron-rich nuclei (at least with $N = 50$ and $N = 82$) but less well for the nuclei closer to stability. To understand this we also display results obtained with the particle-particle interaction switched off, as it is in the calculations of Ref. [3]. Now the two sets of predictions agree nearly everywhere. The influence of the np particle–particle force on the half-lives depends strongly on shell structure. For the semi-magic nuclei investigated here, it is largest for nuclei with two proton holes (i.e. ^{76}Fe with $Z = 26$, $N = 50$ and ^{130}Cd with $Z = 48$, $N = 82$) and decreases as Z approaches the next magic number from above. Such nuclei are doubly magic and the particle-particle channel decouples nearly completely from the beta-decay particle-hole channel with the result that particle-particle interaction has very little effect. The existence of proton closed shells at $Z = 20$ and $Z = 40$ means that the particle-particle interaction has the least influence in the most unstable nuclei we examine. It also has little influence in a few nuclei closer to stability, e.g. ^{78}Ni, for the same reason. For $N = 126$ there are no obivous closed proton shells far from stability. Our results there are in closer agreement with those of Ref. [4] than those of Ref. [3].

SUMMARY AND OUTLOOK

We have presented results for beta-decay half-lives of neutron-rich neutron-closed-shell nuclei important in the astrophysical r process. The residual np particle–particle interaction has a significant effect on the half-lives of most of these nuclei, the exceptions being those at or just above closed proton shells.

How can these calculations be improved? The most important step is a better effective interaction. Although the Skyrme force used here gives a good overall description of both nuclear ground-states and beta-decay spectra, it has some problems (which it shares with other effective interactions). It fails, for example, to accurately describe single-particle spectra in the region around ^{132}Sn. We suspect that this difficulty is the source of the significant difference between the best values for the strength of the residual np particle-particle interaction in the $N = 50$ and $N = 82$ regions. In addition, the strength we use for the $S = 0$, $T = 1$ channel

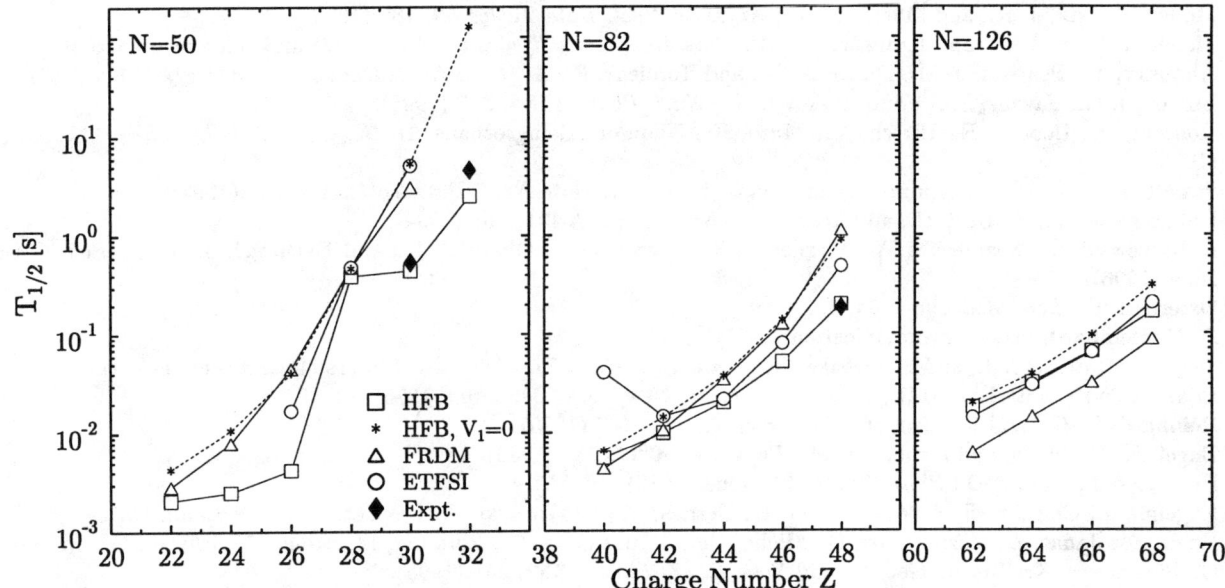

FIGURE 3. Predictions for the half-lives of closed neutron-shell nuclei along the r-process path. Our results appear with ("HFB") and without ("HFB, $V_1 = 0$") the particle-particle interaction. Also plotted are the results of Refs. [3] ("FRDM") and [4] ("ETFSI"), and experimental data where available.

of the residual interaction might differ somewhat from the best value[2]. Finally, it would be nice to go beyond the QRPA approximation for charge-changing excitations. A more refined many-body treatment will be easier in these semi-magic nuclei than elsewhere.

What are the implications of our results for the r-process? They remain to be seen (we are exploring them now) but are probably not as large as we expected when we began this research. Several factors conspire to make the residual np particle-particle interaction less important for beta decay in the neutron-rich closed neutron-shell nuclei than elsewhere. Although pairing correlations play an important role in closed shell nuclei, they are absent in the HFB treatment of pairing unless projection onto states with good particle number is performed. Furthermore, as already mentioned, at or just above proton closed proton shells the lifetimes are relatively insensitive to the particle-particle interaction. In spite of these facts, however, our model reliably describes the beta decay of nuclei near neutron closed shells with only one free parameter. The quality with which we are able to fit that parameter gives us confidence in our predictions for the half-lives of unexplored r-process nuclei.

Acknowledgments

We thank our collaborators J. Dobaczewski, W. Nazarewicz, and P.–G. Reinhard for their contributions to this research. This work was supported in parts by the U.S. Department of Energy under Contract No. DE–FG02–97ER41019 with the University of North Carolina and and Contract No. DE–FG02–96ER40963 with the University of Tennessee.

REFERENCES

1. For reviews see Cowan, J.J., Thielemann, F.-K., and Truran, J. W., *Phys. Rep.* **208**, 257 (1991); Meyer, B.S., *Ann. Rev. Astron. Astrophys.* **32**, 153 (1994).
2. Bender, M., Rutz, K., Bürvenich, T., Reinhard, P.–G., Maruhn, J. A., and Greiner, W., contribution to this volume and references therein.

---

[2] We can view these difficulties as providing constraints to be incorporated into future Skyrme-forces fits.

3. Möller, P., Nix, J. R., and Myers, W. D., *At. Data Nucl. Data Tables* **59**, 185 (1995); Möller, P., Nix, J. R., and Kratz, K.-L., *At. Data Nucl. Data Tables* **66**, 131 (1997) and references therein.

4. Aboussir, Y., Pearson, J. M., Dutta, A. K., and Tondeur, F., *At. Data Nucl. Data Tables* **61**, 127 (1995); Borzov, I. N., Fayans, S. A., and Trykov E. L., *Nucl. Phys.* **A584**, 335 (1995).

5. Homma, H., Bender, E., Hirsch, M., Muto, K., Klapdor-Kleingrothaus, H. V., Oda. T., *Phys. Rev.* **C54**, 2972 (1996); Staudt, A., Muto, K., Klapdor-Kleingrothaus, H. V., *At. Data Nucl. Data Tables* **53**, 165 (1993).

6. Dobaczewski, J., Flocard, H., and Treiner, J., *Nucl. Phys.* **A422**, 103 (1984), Dobaczewski, J., Nazarewicz W., Werner T. R., Berger J.-F., Chinn C. R., and Decharge, J., *Phys. Rev.* *C* **53**, 2809 (1996).

7. Osterfeld, F., *Rev. Mod. Phys.* **64**, 491 (1992).

8. P.-G. Reinhard, private communication (1998).

9. See, e.g., Rapaport, J., and Sugarbaker, E., *Ann. Rev. Nucl. Part. Sci.* **44**, 109 (1994) and references therein.

10. Bohr, A., and Mottelson, B.R., *Nuclear Structure*, New York, Benjamin, 1975, vol. II.

11. Reinhard, P.-G., and Flocard, H., *Nucl. Phys.* **A584**, 467 (1995).

12. Engel, J., Vogel, P., and Zirnbauer, M., *Phys. Rev.* **C37**, 731 (1988).

13. Decharge, J., Gogny, D., *Phys. Rev.* *C* **21**, 1568 (1980).

14. Engelmann, Ch., Ameil, F., Armbruster, P., Bernas, M., Czajkowski, S., Dessagne, Ph., Dinzaud, C., Geissel, H., Heinz, A., Janas, Z., Kozhuharov, C., Miehé, Ch., Münzenberg, G., Pfützner, M., Röhl, C., Schwab, W., Stéphan, C., Sümmerer, K., Tassan-Got, L., and Voss, B., *Z. Phys.* **A352**, 251 (1995).

15. Ameil, F., Bernas, M., Armbruster, P., Czajkowski, S., Dessagne, Ph., Geissel, H., Hanelt, E., Kozhuharov, C., Miehé, C., Donzaud, C., Grewe, A., Heinz, A., Janas, Z., de Jong, M., Schwab, W., and Steinhäuser, S., *Eur. Phys. J.* **A1**, 275 (1998).

FUTURE DETECTOR SYSTEMS

GRETA: Status and physics potentials [1]

I.Y. Lee

Nuclear Science Division
Lawrence Berkeley National Laboratory, Berkeley, California 94720

Abstract. A gamma-ray tracking detector is a new concept for a detector array composed of highly segmented Ge detector elements. The detector would give the energy and the position of all the interaction points and by using the angle-energy relation of the Compton scattering, the scattering sequence of the gamma rays can be reconstructed. Such a detector will have a high efficiency and a good peak-to-background ratio. Research and development are being carried out in the production of highly segmented Ge detectors, understanding the signal, and development of tracking algorithms. Recent progress indicated that this type of detector system appears to be feasible and would have a large impact on a wide variety of physics.

INTRODUCTION

Gamma-ray detector systems are important in a broad range of science and new capabilities are continuously being developed. The most recent step in the evolution of detectors for nuclear spectroscopy is the construction of large arrays such as GAMMASPHERE and Euroball. These arrays, consisting of approximately 100 modules of Compton suppressed Ge detectors, have a total peak efficiency of about 0.1 (for a 1.3 MeV gamma ray) and a peak-to-total ratio (P/T) of about 0.6. The power that Gammasphere brought was large; it provided a factor of 100 improvement over previous detector systems in the ability to resolve weak features in a complex spectrum and many examples of new physics have been presented in this conference.

To improve this performance, the efficiency and/or the P/T have to be increased. However, the highest efficiency that can be reasonably achieved for a Compton suppressed Ge detector array is limited to about 0.15. This is partly due to the scattered gamma rays escaping from the Ge detector (a 1.3 MeV gamma ray deposits full energy in a 7cm-by-7cm detector only 20% of the time) and partly due to the solid angle lost to the Compton shield (about 50% for GAMMASPHERE).

It has been suggested that a shell of closely packed Ge detectors could provide a solution to the efficiency limitation. In this case, the entire solid angle is covered by Ge detectors, and by adding the signal from neighboring detectors, the escaped energy is recovered and much higher efficiency can be achieved. For example, a 9 cm thick shell of Ge will have a peak efficiency of about 0.6 for a 1.3 MeV gamma ray. However, for events with many coincident gamma rays, such as long cascades in the decay of nuclear high-spin states, the summing of two gamma rays hitting neighboring detectors reduces the efficiency and increases the background. In order to reduce this summing, a large number of detectors, of the order of 1000, is required. The cost of such a detector array will be prohibitive.

GRETA CONCEPT

We have developed a new concept, the Gamma-Ray Energy Tracking Array (GRETA), which uses highly segmented Ge detector elements to determine the location and energy of every interaction of each gamma ray. The interaction points belonging to a particular gamma ray would then be identified from the position and energy using a procedure called "tracking". The full gamma-ray energy is obtained by summing only the interactions belonging to that gamma-ray and the problem of summing two gamma rays is avoided. Tracking

---

[1] Work supported by the US Department of Energy under grant DE-AC03-76SF00098.

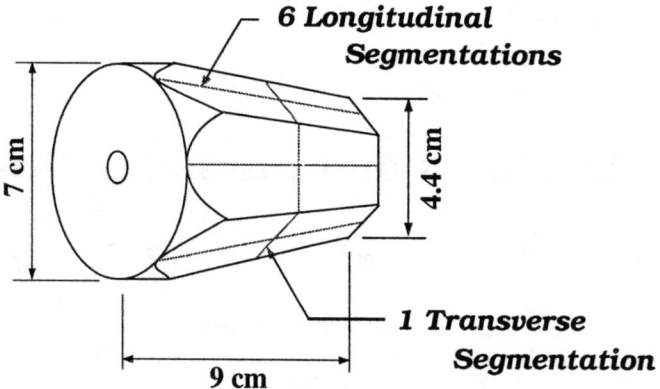

FIGURE 1. The first prototype of GRETA detector has a tapered hexagonal shape with six longitudinal segmentations and one transverse segmentation.

makes use of all the information deposited in the detector and the energy-angle relation given by the Compton scattering formula. Such an array could be constructed from 100-200 highly segmented Ge detectors and with a cost comparable to that of GAMMASPHERE.

The technology needed to realize GRETA can be divided in to three areas. They are: 1) manufacture of segmented detectors which can provide signals for resolving and locating individual interaction points; 2) electronics and signal processing methods for determining energy, time and position based on pulse shape digitization and digital processing of signals; and 3) tracking algorithms, using the energy and position information, to identify interaction points belonging to a particular gamma ray.

STATUS OF DEVELOPMENT

Most of the technology needed to realize the GRETA concept exists today. Some of it is available commercially and we have developed some of the other important aspects. In the last three years, we have followed a plan, proceeding with R&D in the three key areas. The goal of the R&D effort at this stage is to achieve the "proof of principle" of this concept. Detailed study for an optimal design will be done in the next stage. In the following sections, we will discuss the status of the R&D efforts.

Segmented Detector

The basic element of GRETA is a Ge detector which can give three dimensional position information on the gamma-ray interaction points. This can be achieved by segmenting one of the electrodes in two dimensions to provide two of the coordinates. The third coordinate will be derived from the drift time of the charges. There are several possible segmentation and packing schemes. One possibility is spherical packing, as is used in Gammasphere, using tapered hexagonal and pentagonal detectors with a segmented outside surface. A prototype detector of this type has been designed, produced and extensively tested. As shown in Figure 1, this detector has a tapered hexagonal shape with six longitudinal segmentations and one transverse segmentation which divide the outer surface into 12 segments. Signals from each of the segments as well as from the center electrode are read out. This detector has an energy resolution of 1.8 keV for all the segments which is better than the unsegmented detector due to the low capacitance of each of the segments.

Currently, we are in the process of having this detector further segmented into six transverse regions, giving a total of 36 segments. According to our simulation studies, this segmentation will be sufficient for achieving a position resolution of 1-2 mm under ideal conditions, which will be a significant improvement over the performance of the 12-segment detector. This second prototype will be available in fall of 1998. The purpose of this prototype is to push the segmentation technology and to test the position and energy resolution achievable. These results will be available in early 1999 and will be used to determine the segmentation scheme that will be used in the next prototype. In addition, we will determine the performance of the detector electronics. New miniaturized preamplifiers will be tested for their wider bandwidth and their noise and cross talk properties.

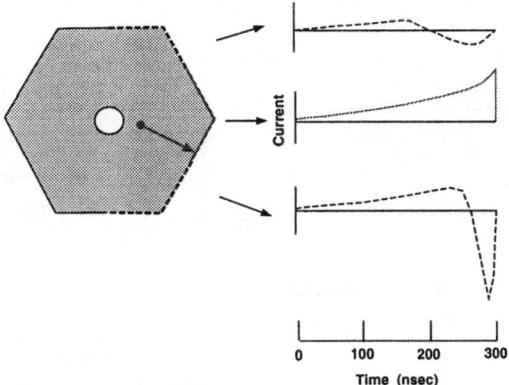

FIGURE 2. Current signal on three electrodes from a charge drifting toward the outside electrodes.

Energy resolution of gamma rays from the summing of their individual interaction points, will be measured to determine the requirements on the electronics. With this prototype, the specification and performance of a single element of GRETA can be fully characterized.

The next step of detector development is the construction and testing a small multi-element array. This prototype will address the question of packaging scheme and is necessary to study the problems associated with tracking across detector boundaries. The number of crystals in a single crystat is one of the important questions we will study before the packaging scheme of the prototype is decided. Currently, we are considering our module to be a cluster of 7 crystals in a single cryostat or 3 clusters of 3 crystals each.

Signal Processing

To determine the position in three dimensions accurately, a detailed understanding of the signal shape is necessary. A signal is produced when electrons and holes formed by the slowing down of the photo- or Compton-electrons induce an image charge of opposite sign on the electrodes. As the charge drifts toward the electrodes, the amount of the image charge changes and currents flow into or out of the electrodes. Figure 2 shows a schematic diagram of the current signal expected from a segmented detector. When the charge is at a large distance from the electrodes, the induced charge is distributed over several electrodes and their current signals are similar. As the charge moves closer to the destination electrode, the induced charge on this electrode increases and charges on the other electrodes decrease. The current continues to increase on the destination electrode until the charge finally reaches the electrode and neutralizes the image charge and the current stops. The duration of the current signal (drift time) can be used to determine the drift distance. The integral of the current gives the charge which provides the energy measurement. The signal from the neighboring electrode has a bipolar shape with a zero net charge. This fact can be used to identify the electrode where the charge is collected, which gives a position resolution on the order of the size of the electrode. However, a better position resolution can be obtained by analyzing the shape of the transient charge signal from the neighboring electrodes. As shown in Figure 2, the maximum amplitude and time of zero crossing of these signals are sensitive to the transverse position and they can be used as input information for the localization.

In order to use the pulse shape information, the preamplifier should have a high bandwidth and the pulse shape has to be digitized in small time steps and with sufficient resolution in amplitude for further processing. Two different designs of preamplifier are being tested on the first prototype. The first one gives a rise time of the differentiated charge signal of approximately 20 nsec (10% to 90%). We will choose one of them for the second prototype based on their bandwidth, noise and stability performance. Several commercial digital pulse shape acquisition systems have been used in the testing of the first prototype. In addition, a specially designed module for signal processing based on digital signal processing is being developed by X-ray Instrumentation Associates. A prototype electronic module is presently being tested. For the multi-detector module we will use one of these commercial systems. Another benefit of the digital processing will be the ability to resolve pulses close in time, and therefore an increase of the acceptable counting rate. We expect a factor of at least five increase in count rate per segment (and a factor of 30 overall due to the many segments).

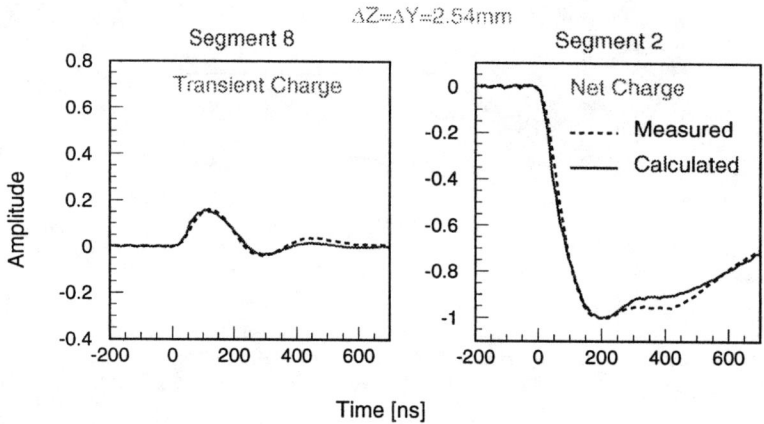

FIGURE 3. Measured and calculated charge signals from the 12-fold segmented GRETA prototype detector. The right panel shows the net charge signals from the electrode where the charge is collected. The left panel shows the induced signals at the neighboring electrode.

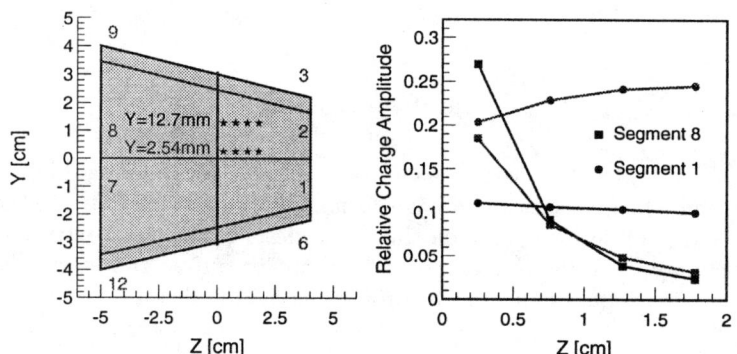

FIGURE 4. Measured amplitudes of transient charge signals. The left drawing indicates the locations of the collimated ^{242}Am (E = 60 keV) source illuminating segment 2 of the GRETA prototype detector. The right panel shows the maximum amplitides of the transient charge signals in segment 1(circles) and segment 8 (squares) as function of the longitudinal position.

We are developing algorithms to analyze the digitized signal to obtain the position of the interaction points. The goal of this on-going study is to determine the segmentation needed to achieve a given position resolution. As a first step we have compared the measured signal from the first prototype (see Fig. 3) with calculated signals and confirmed the accuracy of our calculations. Analysis of the charge drift time from the measured signals using a simple algorithm gave a radial position resolution of about 3-4 mm.

Figure 4 shows a measure of the sensitivity of the transient charge signal to the distance of an interaction from the boundary of the transverse segmentation. The collimated 60 keV gamma rays from a ^{241}Am source were used to ensure that only one interaction occurred in the crystal. However, with such a low energy, the noise is large and approximately 100 signals have been averaged to produce the curves in Fig. 4. The sensitivity is maximum near one boundary (as shown in the figure, segment 8, Y=12.7 mm, Z=2.54 mm) where the transient charge signal occurs mostly in segment 8. It can be represented by the change in the signal amplitude per change in distance from the boundary, in our case 3% per mm. Thus, with a measured noise of 10 keV, the

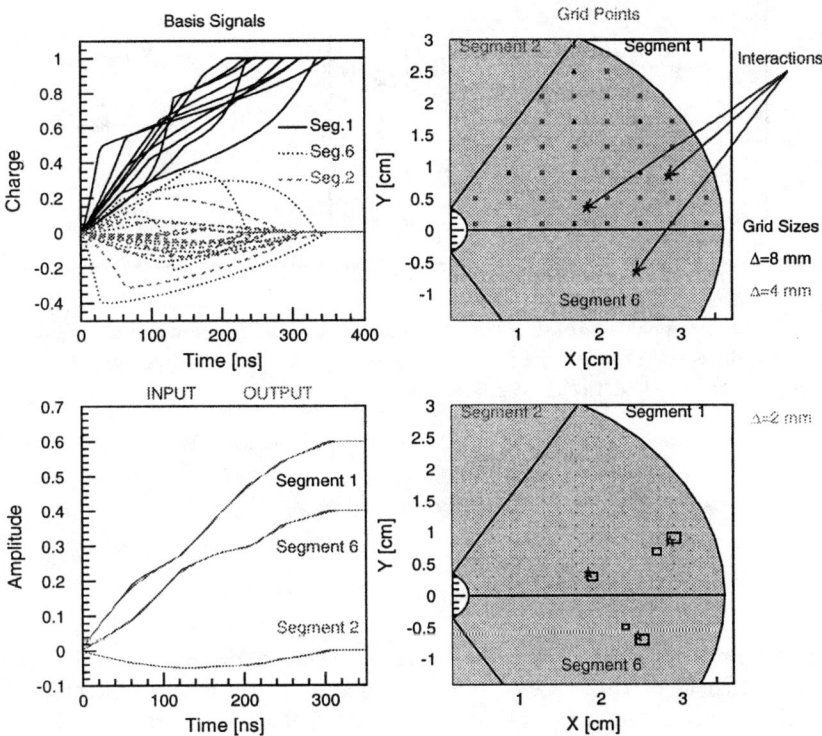

FIGURE 5. Demostration of the decomposition of signals due to multiple interactions into their individual components. The upper right panel indicates grid positions of interactions and the signals they produce are shown on the upper left panel. Assuming three interactions indicated by the three stars on the upper right figure, which result in the signal in the lower left figure, the decomposition procedure is able to find the closest grid points to the interactions, as shown in the lower right figure.

transverse position can be obtained with a resolution of about 5 mm for a 60 keV gamma ray if the interaction point is within 1 cm from the neighboring segment. For a 600 keV gamma ray, the signal to noise can be scaled and a position resolution of 0.5 mm can be anticipated.

To resolve multiple interactions within a segment the measured signals need to be decomposed into their individual components. Algorithms are being developed to separate multiple interaction points in one segment, and one example is summarized in Fig. 5. All signals from an interaction at each point of a rectangular grid (as shown in the upper right plot) are calculated and stored in a lookup table as "basis" functions. The upper left portion of the figure shows 10 signals (for an 8 mm grid size) from the segment containing the interaction and 10 signals from the two nearest neighbors which have a transient charge signal. The deviation from the measured composite signal is minimized by varying the number of interactions, the positions of the interactions and the corresponding amplitudes. By an iterative method, which improves both the position resolution (from 8 mm to 1 mm) and amplitude resolution (from about 10% to only a few percent) we were able to show that one can resolve the composite signal into the original interactions. This is demonstrated in the lower part of Fig. 5. The left hand side shows the input (simulated measured) signals and the output (reconstructed) signals are indistinguishable. The input signals are generated from three interactions at positions indicated by stars on the upper right plot. The lower right hand panel shows the amplitude distribution of interactions found on the grid. Interactions found in neighboring grid points are assumed to correspond to a single "real" interaction. The energy-weighted average between these points then gives the position of this interaction. The deviation in all three positions is less than 1 mm. The next step is to consider realistic signals which include the electronic response of the processing system and noise, and an algorithm to treat signals in a three-dimensional

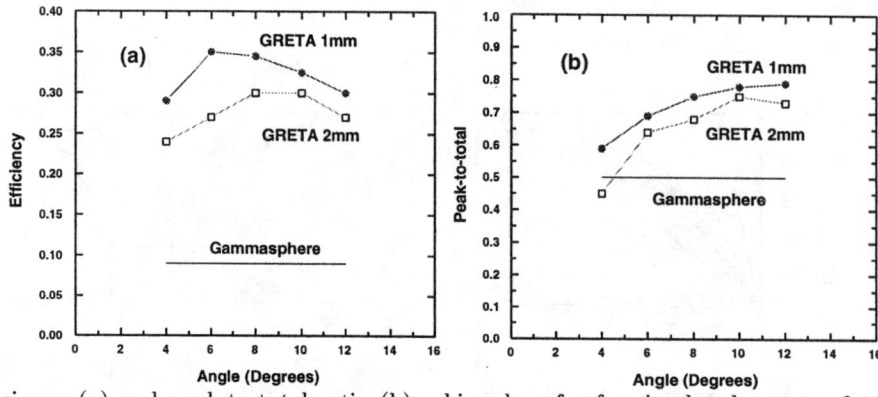

FIGURE 6. Efficiency (a) and peak-to-total ratio (b) achieved so far for simulated events of 25 gamma rays with energy of 1.33 MeV and a detector position resolution of 1 mm and 2 mm respectively. The curves are plotted as a function of the angle parameter used in the cluster identification.

segmented crystal, as well as to optimize the computing time.

Tracking Algorithm

Algorithms are also being developed for identifying interaction points belonging to a given gamma ray. The current algorithm consists of three steps. Cluster identification is the first step of the algorithm. The interaction points within a given angular separation as viewed from the target are grouped into a cluster. In the second step, each cluster was evaluated by tracking to determine whether it contains all the interaction points belonging to a single gamma ray. The tracking algorithm uses the angle-energy relation of the Compton scattering to determine the most likely scattering sequence from the position and energy of the interaction points. If the interaction points have infinite position and energy resolution, the tracking would be exact and the properly identified full-energy clusters will show no deviation from the scattering formula ($\chi^2 = 0$). Wrongly identified clusters or partial-energy clusters will deviate from the formula and the separation of the good and bad clusters would be easy. However, in reality, with finite position and energy resolution, the good clusters will also have a non-zero χ^2 and they cannot be separated cleanly from the bad clusters. This causes a lower efficiency and poorer P/T ratio. In the third step, we try to recover some of the wrongly identified gamma rays. For example, one type of incorrectly identified cluster comes from a single gamma ray being separated into two clusters. This gamma ray can be correctly identified by tracking together all pairs of two bad clusters. When the result gives a small χ^2, the gamma ray is recovered by adding the two clusters. Similarly, the case of two gamma rays wrongly identified as one cluster can also be separated by tracking. The clusters which do not satisfy any of the above criteria are rejected.

These simulations were carried out for a number of different conditions such as the multiplicity and energies of the gamma rays as well as position resolution of the detector. Figure 6 shows the efficiency and P/T achieved so far for simulated events with 25 gamma rays each with an energy of 1.33 MeV and detector position resolutions of 1 and 2 mm. Several improvements of this algorithm are being developed. Information such as the energy dependence of the angular distribution of the Compton scattering cross section and the distribution of the distance between two interaction points will be used in the tracking. Other methods such as a neural network are also being considered. It should be noted that the numbers in Fig. 6 represent the capabilities of the initial tracking algorithm and are not estimates of GRETA's final capabilities.

EXPECTED PERFORMANCE AND PHYSICS OPPORTUNITIES

The estimates of GRETA's final capabilities come from simulations using a spherical-shell geometry with gaps and absorbers appropriate to packaging in clusters of three crystals. The efficiencies and peak-to-total ratios assume signal processing and tracking algorithms that recover all the information deposited in the crystals. These efficiencies are: 0.82, 0.61, 0.47, 0.39, 0.25, 0.17, and 0.11, for gamma-ray energies: 0.1, 0.6, 1.3, 2.0,

5.0, 10, and 15 MeV, respectively, with a peak-to-total ratio of 1.0 in all cases. A variety of physics can be addressed with the new capabilities of GRETA. Tracking can determine the scattering sequence and the linear polarization of a gamma ray and thereby define its electric or magnetic character, important information in most nuclear structure studies. The localization of the first interaction point with GRETA can define the angle of emission of a gamma ray to better than one degree, and thus eliminate Doppler broadening for nuclei having v/c less than 10% and greatly reduce it for higher velocities. As an example consider the study of neutron-rich light nuclei, where the production of the interesting near-drip-line nuclei is by fragmentation reactions resulting in product recoil velocities of $v/c = 30\%$, or higher. For a 1 MeV gamma ray emitted at 90° to the recoil direction, the contribution to the FWHM of the gamma-ray peak due to the Doppler broadening would be 39 keV with a standard Gammasphere detector, but only 3.7 keV with GRETA. Clearly this improvement can have a large effect on the extracted physics, both in detecting weak gamma rays and in separating close-lying peaks. Localization is important for many experiments, especially ones using inverse-reaction kinematics and heavy-ion Coulomb-excitation studies.

The very high efficiency for high-energy gamma rays, together with excellent energy resolution, will open up new studies of giant resonances in nuclei. For example, it will be possible to tag the giant-resonance gamma rays with known low-lying gamma rays to define the process being studied. The efficiency, response function and background suppression of GRETA will be important in many experiments where yields are low, e.g., in experiments with ISOL-type RIB facilities studying the most neutron- and proton-rich nuclei and in determining some important cross sections and level schemes involved in astrophysical processes. The nucleus provides an excellent laboratory for many types of studies, and tests of the standard model, as well as of basic quantum mechanics can be made using GRETA. In many types of high-spin nuclear-structure studies, GRETA will improve the sensitivity by a factor of 1000 or more. This will be essential to discover the predicted very elongated hyperdeformed shapes and the expected pre-fission Jacobi shapes. It will also be necessary to elucidate the order-to-chaos transition in nuclei and to understand the superfluidity of this finite quantal system.

CONCLUSIONS

A gamma ray energy tracking array which is based on highly segmented Ge detector elements would provide far better efficiency, peak-to-back ground, counting rate and localization than any existing gamma ray array. These new capabilities will have a major impact in many areas of physics. Recent progress in detector segmentation, signal processing and gamma ray tracking indicate that this new concept is feasible and a cost effective array can be constructed.

ACKNOWLEDGMENTS

This work was carried out by the nuclear structure group at the 88-Inch Cyclotron of Lawrence Berkeley Laboratory. The gamma-ray tracking work was carried out mainly by Greg Schmid. Kai Vetter did the measurements and calculations of the segmented detector signals. Other members of this effort include Marie-Agnes Deleplanque, Paul Fallon, Augusto Macchiavelli and Frank Stephens.

Analysis of High-Fold Gamma Data

D.C. Radford[1], M. Cromaz[2] and C.J. Beyer[3]

[1]*Physics Division, Oak Ridge National Laboratory,*
Oak Ridge, Tennessee 37831
[2]*Nuclear Science Division, Lawrence Berkeley National Laboratory,*
Berkeley, California 94720
[3]*Department of Physics and Astronomy, Vanderbilt University,*
Nashville, Tennessee 37235

Abstract. Historically, γ-γ and γ-γ-γ coincidence spectra were utilized to build nuclear level schemes. With the development of large detector arrays, it has became possible to analyze higher fold coincidence data sets. This paper briefly reports on software to analyze 4-fold coincidence data sets that allows creation of 4-fold histograms (hypercubes) of at least 1024 channels per side (corresponding to a 43 gigachannel data space) that will fit onto a few gigabytes of disk space, and extraction of triple-gated spectra in a few seconds. Future detector arrays may have even much higher efficiencies, and detect as many as 15 or 20 γ rays simultaneously; such data will require very different algorithms for storage and analysis. Difficulties inherent in the analysis of such data are discussed, and two possible new solutions are presented, namely adaptive list-mode systems and "list-list-mode" storage.

1. INTRODUCTION

Large γ-ray detector arrays such as GAMMASPHERE and EUROBALL now generate three-, four-, and even five-fold HPGe coincidence data in useful quantities. The unsurpassed sensitivity of these arrays has allowed observation of γ-ray cascades that are well beyond the detection limits of previous-generation arrays. However, in order to realize this sensitivity, the data must be analyzed in high fold, and the size of data sets generated by these instruments can make it very difficult to extract all of the useful but detailed information contained in them. Furthermore, limited available computer resources have meant that researchers have typically only been able to make simple histograms of data sets of up to 3-fold; for folds higher than three, list-mode storage has had to be used, or the events reduced to fold less than four by applying gates during the sorting process.

General analysis of γ-γ and γ-γ-γ coincidence data to construct complete and consistent nuclear level schemes is facilitated by sophisticated computer programs such as "escl8r" and "levit8r" [1], that can extract the physically interesting numbers from the raw data and present them to the physicist in an easily assimilated manner, keep track of all γ-ray assignments and expected coincidence intensities, and to quickly find and report major discrepancies between a proposed level scheme and the data so that they can be understood and corrected. These programs include an interactive display of the user's proposed level scheme as an integral part of the Graphical User Interface. An equivalent tool for 4-fold analysis ("4dg8r") has recently been developed. The techniques that allow creation and storage of 4-fold histograms (hypercubes) that will fit onto a few gigabytes of disk space, and yet allow access to triple-gated spectra in only a few seconds, are briefly described below in section 3.

The next generation of γ-ray detector arrays is already being discussed; if constructed, an array such as GRETA could produce copious quantities of very-high fold ($F \sim 15$) data that would require analysis in 6 or 7 dimensions. Analysis of these data would present very significant new problems, many of which are discussed below in section 2. The most significant problems have less to do with actual storage of the data than with the time required to generate gated spectra; in a simple-minded list-mode scheme, all the data need to be read and processed when gates are applied. Techniques which overcome these problems will need to be developed; two attempts to do so are described in sections 4 and 5.

CP481, *Nuclear Structure 98*, edited by C. Baktash
© 1999 American Institute of Physics 1-56396-858-4/99/$15.00

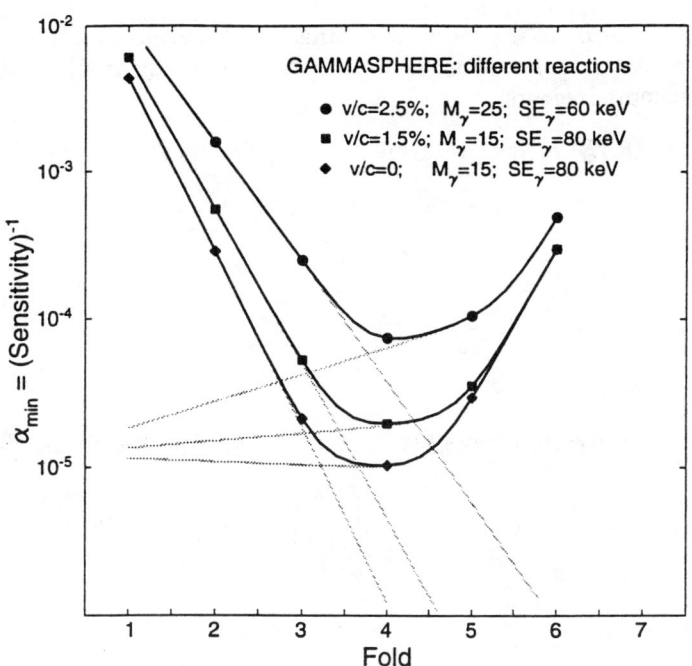

FIGURE 1. Calculated sensitivity limits for GAMMASPHERE used in conjunction with three different values of γ-ray multiplicity, recoil velocity, and average γ-ray energy separation. The calculations are performed as described in ref. [2].

2. DIFFICULTIES INHERENT IN HIGH-FOLD ANALYSIS

As new detector arrays have improved in efficiency and resolving power, their sensitivity has improved tremendously. Higher efficiency allows collection of higher-fold coincidences, and each additional fold allows an increase in the peak-to-background ratio for weak coincidences, typically by a factor of about 4 to 10. This is illustrated in Figure 1, which shows calculated sensitivity limits for GAMMASPHERE used in conjunction with three different values of γ-ray multiplicity and recoil velocity. Note that in all cases, the optimum calculated sensitivity occurs for fold 4, so that to realize the full sensitivity, the analysis must somehow be performed in a four-dimensional space.

Unfortunately, the large size of a four-dimensional data-set gives problems with data storage and data access speeds. Until recently, this meant that researchers were generally restricted to 2D and 3D histograms (matrices and cubes), so that a four-dimensional analysis required some form of data selection (gating) during the replay (histogram construction) phase. Furthermore, the complexity of the analysis of these data makes computer-assistance in *examining* the data, for example as in the programs escl8r and levit8r mentioned above, even more imperative than for 2- and 3-fold analyses.

The next generation of detector arrays, such as GRETA, will greatly compound these problems. For example, the sensitivity plot corresponding to Figure 1 but calculated for GRETA would have optimum sensitivity for fold 7, requiring a 7-dimensional analysis.

In order to illustrate some of the problems inherent in 4- and higher-fold analysis, we first define a few useful terms:

- The **Gamma-Ray Multiplicity** M_γ is the number of γ rays produced in a single event, and varies with the reaction used in the experiment. For rare-earth nuclei produced in heavy-ion fusion-evaporation reactions, for example, M_γ is typically $\sim$ 25 to 30.

- The **Detected Fold** F_D in an event is $\sim M_\gamma \epsilon$, where ϵ is the *total* efficiency of the detector array, *i.e.* the efficiency including both photopeak and residual unsuppressed Compton-scattered gamma rays.

- The **Analysis Fold** F_A is the fold or dimensionality in which the data are analyzed. Obviously we require that $F_D \geq F_A$.

- The **Number of Reactions** N_R produced during the experiment depends on beam current, target thickness and total reaction cross-section, and possibly also on any auxiliary detectors that perform some sort of selection on accepted events.

- The **Number of Events** N_E recorded during the experiment varies with the detected fold F_D as

$$N_E[F_D] \approx N_R \, ^{M_\gamma}C_{F_D} \, \epsilon^{F_D} \, (1 - \epsilon)^{M_\gamma - F_D}. \tag{1}$$

where the number of combinations

$$^{M_\gamma}C_{F_D} = \frac{M_\gamma!}{F_D! \, (M_\gamma - F_D)!}. \tag{2}$$

- The **Number of Counts** N_C in a (possibly imaginary) F_A-fold histogram depends on the analysis fold F_A as

$$N_C[F_A] = \sum_{F_D} N_E[F_D] \, ^{F_D}C_{F_A}, \tag{3}$$

where the number of combinations

$$^{F_D}C_{F_A} = \frac{F_D!}{F_A! \, (F_D - F_A)!}. \tag{4}$$

With these definitions, we can now consider examples of typical values for these quantities for three different detector arrays. In a typical high-spin experiment, using a fusion-evaporation reaction to produce rare-earth nuclei, with several pnA of beam current, one can expect a reaction rate of the order of 10^5 per second. A three-day experiment thus produces $N_R \sim 3 \times 10^{10}$ reactions, with a typical γ-ray multiplicity of $M_\gamma \sim 25$.

The second-generation 8π-spectrometer array has a total efficiency of $\epsilon \sim 0.01$, resulting in $N_E[F_D \geq 2] \approx 7 \times 10^8$ events, and $N_C[F_A = 2] \approx 9 \times 10^8$ counts in a two-dimensional histogram (matrix). Due to a small fraction (10%) of 3- and higher-fold events, there are an average of about 1.3 counts generated by each event.

GAMMASPHERE and EUROBALL are third-generation arrays, with efficiencies of $\epsilon \sim 0.15$, resulting in $N_E[F_D \geq 4] \approx 1.6 \times 10^{10}$ events, and $N_C[F_A = 4] \approx 1.9 \times 10^{11}$ counts in a four-dimensional histogram (hypercube). Due to a significant fraction of 5- and higher-fold events, there are now an average of about 12 counts generated by each event.

The next generation of detector arrays, such as GRETA, might have efficiencies approaching $\epsilon \sim 0.6$, resulting in an average detected fold of about 15. Thus we get a useful event for essentially every reaction, $N_E[F_D \geq 7] \approx 3 \times 10^{10}$ (neglecting dead-time losses), and $N_C[F_A = 7] \approx 4 \times 10^{14}$ counts in an imaginary 7-dimensional histogram. The large detected fold means that there are now about 13000 "counts" generated by each event!

Incidentally, it is these large values of $^{F_D}C_{F_A}$ that produce "spikes" [3] in incorrectly-produced spectra from the sum of many gates. If each event is unpacked into F_A-tuples before the gates are applied, and many resulting gated spectra added together, then it is possible for many F_A-tuples from one individual event to meet the gating conditions. Thus a single γ-ray energy from a single event may be added into the resulting spectrum many times, producing "spiking". In order to avoid this problem with high-fold data, one should either avoid summing many gated spectra (since an individual $(F_A - 1)$-fold gate on F_A-fold data contains no spikes), or produce the summed spectrum directly from the raw data, without first unpacking into F_A-tuples, and make sure that each energy for each event is added to the spectrum at most once.

One conclusion that follows from the above considerations is that for analysis folds above four, storage of the data in histograms is no longer feasible. Firstly, the storage space requirements quickly become prohibitive, since the number of channels grows by several orders of magnitude with each fold. Secondly, as the detected fold becomes very large, many counts are produced by each event, so that compression algorithms are not able to make use of a small count density to greatly reduce the histogram size. Lastly, the problem of spiking, together with the need to sum many combinations of gates to obtain useful statistics, requires preserving the data in its raw form, rather than unpacking it into F_A-tuples as would be done in producing a histogram. Thus for $F_A \geq 5$, list-mode storage, where each event is essentially saved as a simple list of γ-ray energies, becomes the preferred storage mode even for very large numbers of events.

List-mode storage for quintuple γ-ray coincidence events has been used at several laboratories for some time [4]. This type of storage typically requires several bytes per event, and the main limitation is that in order to generate gated spectra, one needs to read through entire data-set and process every event. For large quantities of higher-fold data, this quickly becomes unrealistic, and in order to achieve reasonable access times we will require better schemes than those of simple list-mode storage.

3. FOUR-DIMENSIONAL HISTOGRAMS

The "RadWare" programs "escl8r" and "levit8r" [1] are designed to facilitate analysis of γ-γ and γ-γ-γ coincidence data, respectively. One unique aspect of these programs is that they include an interactive display of the current version of the user's proposed level scheme as an integral part of the Graphical User Interface, in order to assist the user in examining the data and extracting γ-ray intensities *etc.* Information about the detector efficiency and resolution, as a function of energy, allows the software to calculate *expected* coincidence spectra, on the basis of the proposed level scheme, for comparison with the corresponding *observed* spectra. It also enables least-squares fitting of the 2- or 3-dimensional data, so that γ-ray energies and intensities can be extracted from the data quickly and easily. In this section, we briefly describe an equivalent tool for 4-fold analysis ("4dg8r") that has recently been developed.

In developing this software for 4-fold histograms (hypercubes), the following goals were considered to be of primary importance:

- It should be able to run on standard workstations, without special requirements such as especially large amounts of RAM or disk storage.

- It should be able to handle a large number of counts (the current version uses up to two bytes per channel).

- The user should be able to extract triple-gated spectra interactively, *i.e.* in a few seconds.

- A typical histogram should occupy less than 9 GB of space on disk.

- Since some file systems have limits on the size of disk partitions and files, and since users may need to spread their hypercube over a number of different disks, it is important that the hypercube be able to be broken into several sub-files.

In order to cover a large energy range with as few channels as possible, the programs allow the use of a non-linear energy dispersion, as for "levit8r" [1]. The current version by default allows up to 1024 channels per side of the hypercube; this can be extended without major modifications. By symmetrizing the hypercube (*i.e.* ordering the four channel indices, $w \leq x \leq y \leq z$) we can reduce the corresponding number of hypercube channels to $(1024 \times 1025 \times 1026 \times 1027)/(4!)$. It should be noted, however, that this number is still in excess of 40 billion channels.

Since the number of channels in the hypercube is much greater than the the number of bytes in which we would like to store it, the only way to do this is to use data compression. The average count density in the hypercube is usually very low (about one count per channel or less), so that good compression ratios are obtainable. Since we require fast random access to the hypercube, the compression/decompression algorithm needs to be very fast, and we also need to compress small pieces of the hypercube individually. For this reason, the hypercube is subdivided into "minicubes", each 4 channels on a side, or $4^4 = 256$ channels total. These minicubes are individually compressed, and occupy a space of between 1 and 512 bytes on disk.

The compression algorithm used is explicitly designed for such small histogram segments, and provides an excellent trade-off between performance and high speed. It consists essentially of storing an optimum number of "bitplanes" of the hypercube, together with addresses of channels which have overflowed this number of bits. Each bitplane doubles the range of counts that can be stored without overflows, at the cost of $256/8 = 32$ bytes (since there are 256 channels in the minicube). Also, each overflowed channel within the minicube can be addressed with a single byte. Thus, once 32 or more overflows have occurred, the overflows occupy as much space as a bitplane, and the algorithm simply allocates that additional bitplane. A one-byte header for each minicube stores enough information for the decompression routine to know how many bitplanes and overflows are stored.

Individual compression of the minicube histogram segments, however, has the drawback that they now have different lengths on disk, so that locating one individual segment for decompression and analysis is no longer simply a matter of skipping over a calculable number of bytes. Rather, tables of pointers, or indices that map minicube numbers to locations on disk, are required, and there will in general be a trade-off between the size of these tables and the average number of interpolations (and hence time) required to find a given minicube. Furthermore, during the histogram construction phase of the analysis, the software needs to read and decompress portions of the hypercube, increment the channels, and recompress/rewrite. Since the count density increases during this process, the size of the hypercube portions will in general grow, and they cannot be written back into the space from which they were read. Thus the data need to be moved and space reallocated, giving rise to additional complexity in the file structure, and in the book-keeping aspects of the program.

Using this design, a package of programs for four-dimensional analysis has been written and included in the "RadWare" software package. In addition to "4dg8r" (the 4D version of "levit8r"), there are programs "4play" for data replay (histogram construction), and "pro4d" for histogram projection to 3D, 2D and 1D. Replay of a typical high-statistics GAMMASPHERE/EUROGAM experiment, on a standard workstation, can be expected to take of the order of 50 hours. Extraction of triple-gated spectra from the hypercube typically takes 1 to 5 seconds per gate, depending on the size of the cube and disk speeds, *etc.*

4. ADAPTIVE LIST-MODE STORAGE

As discussed in section 2, at folds 5 or above histogramming events is no longer practical. The alternative is list-mode based systems, which grow linearly rather than geometrically with fold. It is important to note that in list-mode systems, one does not unfold coincidences as this will cause the size of the database to grow geometrically. Realistically, the number of coincidences in a given experiment with current (or future arrays) will be less than 10^{10} as it is limited by detector efficiency and count rate considerations for Ge detectors. Given this and the capacity of current mass storage systems, space constraints on list-mode systems is no longer a major issue.

The challenge in list-mode systems is to quickly access events which meet a specific gating condition. Exhaustively testing each coincidence in a multi-gigabyte data set is not acceptable if one wishes to set gates interactively. To accomplish this, effective list-mode systems require two elements. First, events must be sorted so that coincidences likely to be accessed in a given query are "close" together in the database. For the purposes of high-fold gamma-ray spectroscopy, "close" refers to the Euclidean distance between coincidences in the hypercube. As data is fundamentally represented in computer memory (or on disk) as a one-dimensional array, implicit in this sorting is an approximate method to map the channel numbers $(x_1, x_2, x_3, \ldots, x_{F_D})$ to an integer key which may then be ordered. The second requirement for a list-mode system is an indexing mechanism for the sorted list of coincidences, so that groups of interesting coincidences can be accessed quickly. Unlike histograms, where the location of an interesting element can be calculated directly from its indices, list-mode systems require a set of indices so that entry points into the database can be specified.

These two goals of effectively sorting and indexing multi-fold coincidences can be met with an adaptive list-mode system. This is based on a data structure called a kd-tree, introduced by Bentley [5] for multi-key databases. The index for such a database is given by a binary tree where each node of the tree represents a volume of the hypercube. Using this data structure, one is able to partition into volumes whose size depends on the density of counts in that part of the hypercube. This adaptive list-mode system allows for greater granularity in densely populated areas of the cube, thus yielding greater performance.

Such a system has been implemented by Mario Cromaz at the Lawrence Berkeley National Laboratory, in the software package *blue*. It provides facilities for sorting into and taking gates on multi-fold list-mode data sets. Adaptive list-mode systems have good performance characteristics enabling one to take $(F_D - 1)$-fold gates on F_D-fold data sets in a few seconds. Gating time tends to be independent of fold but strongly dependent on the total counts in the list. This is a consequence of the ordering and indexing mechanism chosen.

5. LIST-LIST-MODE DATA STORAGE AND ANALYSIS

As discussed above in section 2, GRETA data would require $F_A \approx 7$, and should produce $N_E[F_D \geq 7] \sim 10^{10}$ events in a typical high-spin experiment. The large detected fold $< F_D > \sim 15$, together with the (relatively!) low analysis fold, means that there are now an average of about 13000 septuples generated by each event. This presents a real problem; since we need to be able to apply $F_A - 1$ gates on the data set, and each event can meet those gating conditions in $^{F_D}C_{F_A-1} \sim 10^4$ possible ways, it seems that we need to read and inspect each and every event to see if it meets the gating conditions. Even the adaptive list-mode system of the previous section will suffer from this; it is primarily designed for setting $(F_D - 1)$-fold gates on F_D-fold data, and for the very-high-fold data discussed here almost all of the indices would have to be checked for matching events. In this section, a new solution to this problem is presented, a solution which has been entitled "list-list-mode".

Firstly, it is worth pointing out that a relatively small subset of the full data set produces most of the counts in a gated spectrum. The distribution of detected fold $N_E[F_D]$ is given by equation (1) above. For $\epsilon = 0.6$ and $M_\gamma = 25$, the resulting distribution has a mean at $F_D \approx 15$, and is rather wide. If we apply a random 6-fold gate to to that distribution, 75% of the counts in the resulting spectrum come from events with $F_D \geq 17$, which comprise only 27% of the total number of events. Indeed, 55% of the counts come from events with $F_D \geq 19$, or only 3% of the events. Thus if we save only events with $F_D \geq 17$, we lose very few statistics, but save a factor of 3 or 4 in storage requirements. The number of events remaining would be $N_E[F_D \geq 17] \sim 3 \times 10^9$, with an average fold of $< F_D > \sim 19$, and corresponding to $N_C[F_A = 7] \sim 8 \times 10^{13}$ 7-fold combinations.

We now address the question of how such events are to be stored on disk. We first look at one simple scheme that illustrates the principles involved, but does not perhaps give the optimum storage. More complicated compression schemes, that further reduce the space required to store the data set, could be added later, perhaps at the cost of a minor increase in processing time when the data are read back.

We assume that in each event there are, on average, about 18 energies, each distributed over 2048 channels. Let us order the energies in ascending channel number, so that an event consists of the set of channel numbers $(x_1, x_2, x_3, \ldots, x_{F_D})$, with $x_1 \leq x_2 \leq \ldots \leq x_{F_D-1} \leq x_{F_D}$. Then the average *difference* between the consecutive energies, $x_i - x_{i-1}$, is about 2048/18, or 114 channels. Since one byte (8 bits) can store numbers between zero and 255, we can use a byte to store each of the differences between the consecutive energies in the event, with some extra conventions to handle the special cases where a difference is 255 or more. In this way, an 18-fold event can usually be stored in 18 to 20 bytes. Different events will have different lengths, depending on the detected fold, and on the number of extra-large energy differences that require additional bytes.

In this list-list-mode scheme, we allocate different event-storage files for each of the different possible *event lengths* (in bytes). For example, for the data set discussed above, with $N_E[F_D \geq 17] \sim 3 \times 10^9$, we would have 14 or so different event lengths, from 17 up to about 30 bytes. The average length would be about 19 bytes, so the full data set could be stored in less than 60 GB.

The problem then becomes how to quickly find those very few events that meet the conditions of a 6-fold gate. That is, how do we find the needle of interesting events in the haystack of $\sim 3 \times 10^9$ events which fail to meet our gating conditions, without reading through all 60 GB to check each event in turn?

Imagine that for each channel x we could pick out the subset $\{e_x\}$ of all events that involve that channel x, that is, those events for which at least one of the detected energies is $E_\gamma = x$. Then, for a quadruple gate (w, x, y, z), for example, we want to specify four channels of the interesting events, so we want the subset

$$\{e_{wxyz}\} = \{e_w\} \bigcap \{e_x\} \bigcap \{e_y\} \bigcap \{e_z\} \tag{5}$$

One simple way to achieve this is to give each event a unique identifying "event-ID", that lets us find it on disk very quickly, so that once an event has been determined to be of interest, it can be accessed with a single read operation. The most obvious scheme to do this is to give consecutive IDs sequentially to each of the events in each event file, since every event in one file has the same length, and we can therefore easily compute the number of bytes that we should seek over in order to access one particular event of interest. Then, for each of the 2048 channel numbers, we create a *list of event-IDs*, containing the IDs of all events that involve that particular channel number. For example, the one-hundredth such list would then correspond to the set of event-identifiers for all events $\{e_{100}\}$ that involve channel 100. The event-ID of each 18-fold event would be put into 18 such event-ID-lists.

These event-ID-lists are the second list in "list-list-mode", and are stored on disk using the same idea of storing the *difference* between consecutive IDs, rather than the IDs themselves. In this way, we again require about one byte per number, or about 50 - 60 GB total. The average length of the 2048 different event-ID-lists is about 25 MB.

If the average single gate is about 3 channels wide, out of the total of 2048 channels, then on average a single gate accepts about $18 \times 3 \div 2048 = 2.6\%$ of all events. Here the 18 comes from the average event fold, which is the number of chances each event has to meet the gate. A random five-fold gate would be expected to accept about

$$18!/(18 - 5)! \times (3 \div 2048)^5 = 6.9 \times 10^{-9} \tag{6}$$

of all events. In this example, that corresponds to only about 20 events, for a total of about $(18-5) \times 20 = 260$ counts. Gates on strong coincidences would of course produce many more events; this random-gate estimate corresponds essentially to a "background" gate.

Once these event-ID-lists have been generated, then it is simple to see how to use them to create gated spectra. Consider, for instance, the above example of a five-fold gate that is three channels wide in each of five dimensions. By reading 15 event-ID-lists, we can easily construct five ordered lists of IDs of all events that match the gate on the five different dimensions. By looking for IDs that appear in all five of these lists, we can find those few events that meet the five-fold gate. Then it takes very little time to locate the events in the energy-list files, read them from disk, and create the gated spectrum. When the sum of many five-fold gates is required, it is a simple matter to ensure that each energy in each event is used only once, thus avoiding the problem of spiking.

Sums of gates are usually constructed by taking all combinations of multi-dimensional gates from one or more sets of one-dimensional gates. For example, five-fold gates for a superdeformed band of 12 transitions may be created by taking the $^{12}C_5 = 792$ five-fold combinations of the twelve individual gates. In such a case, again assuming that each one-dimensional gate is three channels wide, we can find the interesting events simply by reading a total of 36 different event-ID-lists, and searching for IDs common to at least five of the twelve one-dimensional gates.

The total space required, for this example data set, would be about 120 GB, one-half for the event energy-lists themselves, and the rest for the event-ID-lists for each channel. Even using today's technology, this would be quite easy to do by using fast disk arrays, and/or by spreading the data over about 10 workstations and programming them for parallel processing over a local-area network. Since the time required to construct multi-dimensionally-gated spectra is essentially determined by the time spent seeking and reading from disk, and the event-ID-lists can be made contiguous on disk so that the required number of disk seeks is minimized, access to gated spectra would be very fast. We can estimate that a single five-fold gated spectrum could require as little as 5 seconds to create, and the sum of 792 five-fold gates on a (weak) cascade of twelve transitions would take about 30 seconds total. This time will decrease as the speed and capacity of available disks increases. Within five years from now, this should be technologically trivial, or nearly so.

One potential difficulty, however, has been glossed over in the above discussion, namely that of background subtraction. To perform a complete and accurate background subtraction, all folds of gate from one up to $F_A - 1$ need to be accessed [6]. Clearly, folds 1 through 3 can be dealt with by using histogrammed projections of the data set, but higher folds may require reading a greater subset of the events, which could significantly slow down the construction of gated spectra. The CPU time for calculation of the background could also be significant. Clever, sophisticated background-subtraction algorithms may need to be investigated to determine the full extent of this problem.

6. TEST OF LIST-LIST-MODE

The concept of list-list-mode storage and analysis has been tested with an artificial pseudo-data set, generated by Monte-Carlo methods, and resembling an extremely weak (0.005%) band added to a flat background.

The data had fold F_D distributed between 10 and 20, with a mean at $F_D \approx 14$. Twenty million events were generated. Over 99.995% of the energies had a flat, random distribution over channels 0 to 1023; about 1000 of the events, however, contained a regularly-spaced cascade of up to thirteen transitions. Members of this cascade were "detected" with a probability of 0.7, except that at higher energies the band was cut off over some distribution of "spins". Events containing the band had the same fold distribution as background events, with the remaining energies having the same flat distribution as the background events.

Twenty million events is not a large data set; nevertheless, if these data were to be unpacked into five-folds, they would generate 74 billion quintuples. The events were stored in 15 different files, depending on the event length in bytes, and required 280 MB for the energy-lists and another 280 MB for the event-ID-lists.

The list-list-mode gating scheme was tested with up to 8-fold gates, three channels wide in all dimensions. Background was not subtracted for any of the spectra. Spectra were generated for single F_{A-1}-fold gates, and also by taking all combinations of F_{A-1}-fold gates from a list of 8 one-dimensional gates on different peaks of the band. Examples of the gated spectra produced are shown in Figures 2 and 3.

The time required to construct the gated spectra was measured as a function of analysis fold and the number of summed gates. For low analysis fold, a large fraction of the events met the gating conditions, so that many of the event-list records had to be read from disk, resulting in very long gating times. As the analysis fold was increased, fewer events were accessed, and the time required dropped to about 5 seconds for a single five-fold gate ($F_A = 6$). Above this, for $F_A \geq 7$, the larger number of event-ID-lists that needed to be read dominated the gating time, so that it increased linearly with analysis fold.

The time required for setting gates is, as expected, dominated by disk access times. If the data have been previously accessed and cached in RAM, the gating times decrease greatly. For this reason, in the above timing tests, care was taken to ensure that any previously-accessed data had been flushed from cache memory. Summaries of the time taken and the number of events accessed, as a function of analysis fold, are shown in Figure 4. These tests were performed on a Pentium-II 300 MHz computer, running linux, with the data stored on a local SCSI disk.

6. CONCLUSIONS

High-fold γ-ray coincidence data such as that provided by GAMMASPHERE and EUROGAM often require analysis in four or even five dimensions, so that four-dimensional histograms (hypercubes) can be a powerful tool in extracting as much of the physics information as possible. Software to create and analyse such histograms using standard computer resources available to most physicists has been developed, and the design of the programs and data structures has been briefly presented here. They allow creation of hypercubes of at least 1024 channels per side (corresponding to a 43 gigachannel data space) that will fit onto a few gigabytes of disk space, and extraction of triple-gated spectra in a few seconds.

A new generation of detector arrays, such as GRETA, could produce copious quantities of very-high fold data that would require analysis in 6 or 7 dimensions. Some of the new problems presented by such data have been discussed. The most significant problems have less to do with actual storage of the data than with the time required to generate gated spectra. Two new approaches which attempt to overcome these problems have been presented, namely adaptive list-mode systems and "list-list-mode" storage.

Even with well-organised data structures and very fast gating techniques, the analysis of high-fold coincidence data can be an overwhelming task due to the tremendous quantity of detailed information that they represent. Computer-assistance, in the form of clever programs to keep track of level schemes and help in examining gated spectra, is vitally important. Such programs are now available for 4-fold. By using techniques such as list-list-mode and adaptive list-mode systems, software for higher-fold analysis should not be too difficult to create, even for the very high folds generated by GRETA.

Research at Oak Ridge National Laboratory is sponsored by the U.S. Department of Energy under contract number DE-AC05-96OR22464 with Lockheed Martin Energy Research Corporation.

REFERENCES

1. Radford, D.C., *Nucl. Instr. and Meth. in Phys. Res.* **A361**, 297 (1995).
2. Radford, D.C., *Proceedings of the International Seminar on the Frontier of Nuclear Spectroscopy, Kyoto, October 1992*, ed. Y. Yoshizawa, H. Kusakari and T. Otsuka, World Scientific, 1993, p. 229.
3. Beausang, C.W. *et al.*, *Nucl. Instr. and Meth. in Phys. Res.* **A364**, 560 (1995).
4. See, for example, Flibotte, S. *et al.*, *Nucl. Instr. and Meth. in Phys. Res.* **A320**, 325 (1992).
5. Bentley, J.L., *Commun. ACM* **18**, 509 (1975)
6. Radford, D.C., *Nucl. Instr. and Meth. in Phys. Res.* **A361**, 306 (1995).

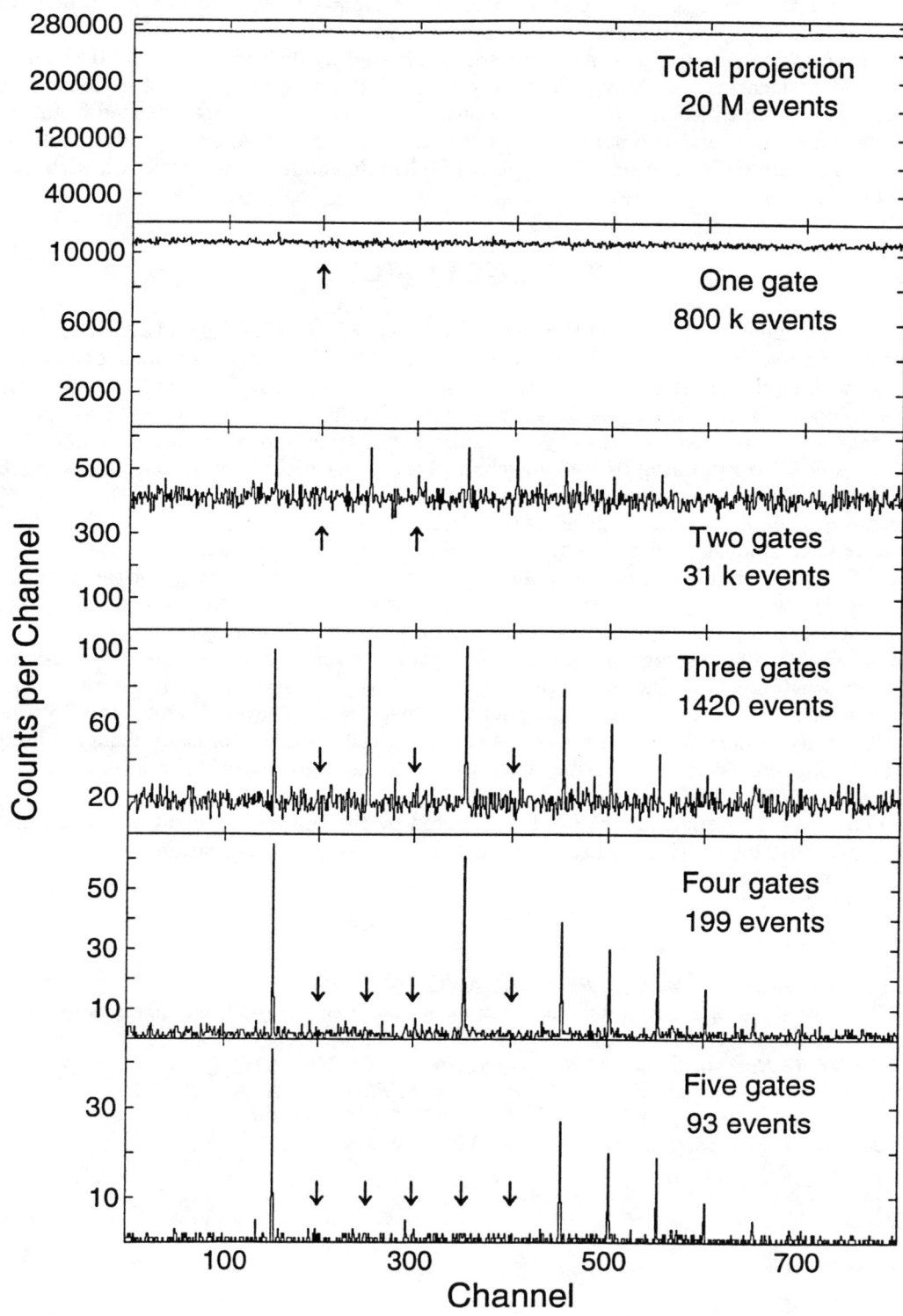

FIGURE 2. Spectra from the set of 2×10^7 events of pseudo-data stored in list-list-mode. Shown are spectra generated by placing a single multi-fold gate, of dimension 1 to 5, on the regularly-spaced "band". The number of events that met the gating conditions, and therefore needed to be read from the energy-lists, is also given for each spectrum. The arrows show the locations of the one-dimensional gates.

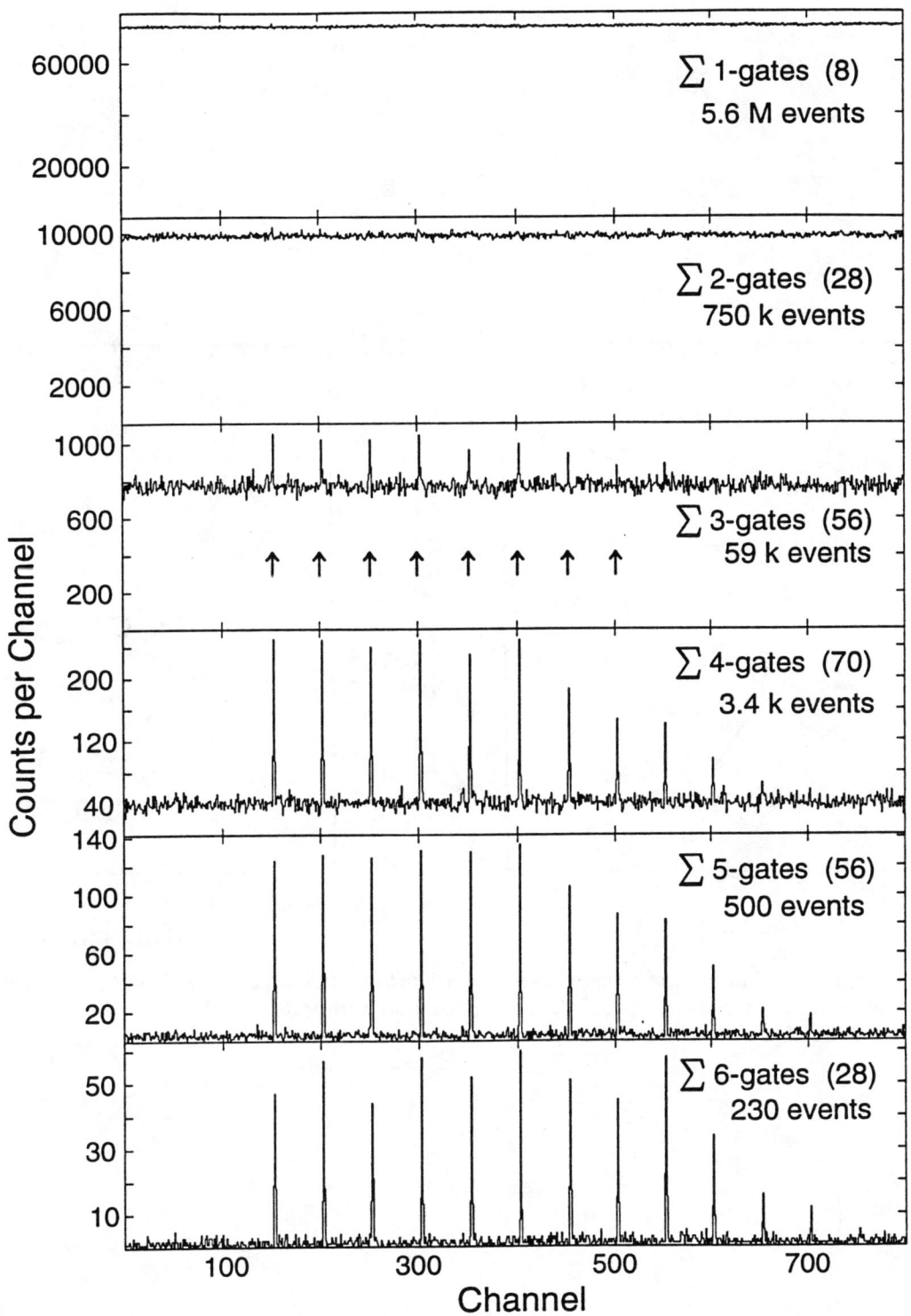

FIGURE 3. Pseudo-data spectra generated by placing all possible combinations of F-fold gates, where F ranges from 1 to 6, from a list of eight one-dimensional gates, on the regularly-spaced "band". The number of events that met the gating conditions, and therefore needed to be read from the energy-lists, is also given for each spectrum, as is the number of F-fold gate combinations, ${}^{8}C_{F}$ (shown in parentheses.) The arrows show the locations of the one-dimensional gates.

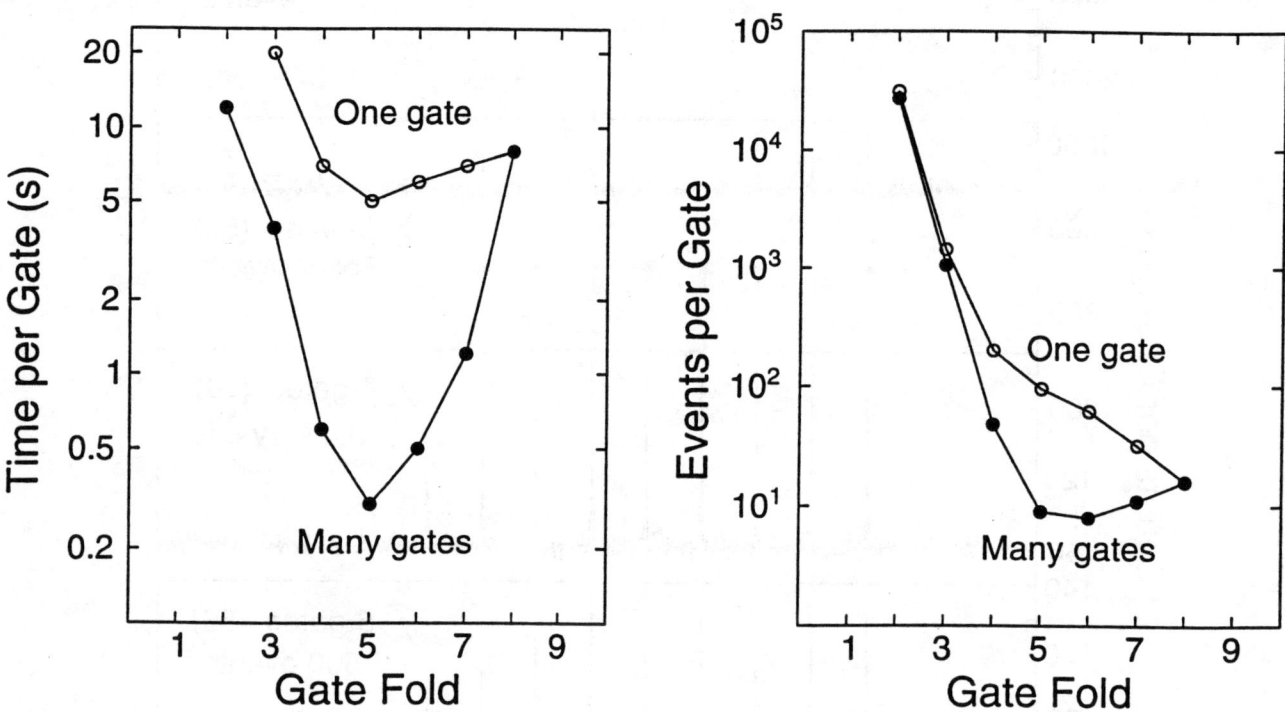

FIGURE 4. Time taken and number of accepted events as a function of gating fold, for tests of list-list-mode storage. Open symbols are for single multi-fold gates, while solid symbols correspond to the case of summing all possible combinations from a list of eight one-dimensional gates.

Program

Monday, August 10, 1998

8:45 a.m.   Welcome/General Announcements

SESSION I-A   **CHAIR:**
M. R. STRAYER (OAK RIDGE NATIONAL LABORATORY)

9:00 a.m.   W. Nazarewicz (University of Tennessee)
Nuclear Structure Near the Drip Lines

9:35 a.m.   A. Covello (Universita di Napoli Federico II)
Realistic Effective Interactions and Nuclear Structure Calculations

10:10 a.m.   H. Grawe (GSI)
Nuclear Structure Near Doubly-Magic Nucleus ^{100}Sn

10:45 a.m.   Break

SESSION I-B   **CHAIR:**
J. MARUHN (UNIVERSITY OF FRANKFURT)

11:15 a.m.   H. Mach (Uppsala University)
Spectroscopy Near ^{132}Sn

11:50 a.m.   D. J. Dean (Oak Ridge National Laboratory)
Shell Model Monte Carlo

12:25 p.m.   Lunch Break, Tennessee Ballroom 2

SESSION I-C   **CHAIR:**
P. B. SEMMES (TENNESSEE TECHNOLOGICAL UNIVERSITY)

2:00 p.m.   A. L. Goodman (Tulane University)
Neutron-Proton Pairing in $N = Z$ Nuclei

2:25 p.m.   R. Wyss (Royal Institute of Technology)
Competition Between $T = 0$ and $T = 1$ Pairing

2:50 p.m.   S. D. Paul (Oak Ridge National Laboratory)
Band Structure in ^{79}Y and the Question of $T = 0$ Pairing

3:15 p.m.   Break

SESSION I-D   CHAIR:
                     P. H. HEENEN (UNIVERSITE LIBRE DE BRUSSELS)

3:45 p.m.   D. Vretenar (University of Zagreb)
                     Relativistic Mean Field Description of Nuclei at the
                     Drip Lines

4:20 p.m.   M. Bender (University North Carolina, Chapel Hill)
                     Exotic Nuclei in Self-Consistent Mean Field Models

4:45 p.m.   W. Satula (Warsaw University)
                     Wigner Energy, Odd-Even Mass Staggering, and the Time-Odd Mean Field

5:20 p.m.   H. Wollnik (University of Giessen)
                     Mass Measurements with Storage Rings

5:45 p.m.   Close of Session

Tuesday, August 11, 1998

SESSION II-A   CHAIR:
                     L. L. RIEDINGER (UNIVERSITY OF TENNESSEE)

9:00 a.m.   J. Dobaczewski (Warsaw University)
                     Superdeformation: Perspectives and Prospects

9:35 a.m.   C. Svensson (McMaster University)
                     Superdeformation and Band Termination in A ~ 60 Nuclei

10:10 a.m.   D. Rudolph (Lund University)
                     Particle Decays of Deformed and Superdeformed Bands Near ^{56}Ni

10:45 a.m.   Break

SESSION II-B   CHAIR:
                     J. C. WADDINGTON (MCMASTER UNIVERSITY)

11:15 a.m.   T. Otsuka (University of Tokyo)
                     Monte Carlo Shell Model Calculations of Medium-Mass Nuclei

11:50 a.m.   M. Matsuo (Kyoto University)
                     Tetrahedral and Triangular Deformations of N = Z Nuclei in A ~ 60 - 80 Mass
                     Region

12:15 p.m.	F. Lerma (Washington University) *Comparative Transitional Quadrupole Moments of Superdeformed Bands in A ~ 80 Region*
12:40 p.m.	Lunch Break, Tennessee Ballroom 2
SESSION II-C	**CHAIR:** **F. S. STEPHENS (LAWRENCE BERKELEY NATIONAL LABORATORY)**
2:00 p.m.	M. P. Carpenter (Argonne National Laboratory) *Superdeformation in the Dy and Hg Regions*
2:35 p.m.	M. Hass (Weizmann Institute of Science) *Magnetic Moments of High-Spin Normally-Deformed and Superdeformed States in $^{193,194}Hg$*
3:00 p.m.	H. Hübel (Bonn University) *Triaxial Superdeformed Bands in $^{163,164}Lu$*
3:25 p.m.	Break
Session II-D	**Chair:** **J. A. Cizewski (Rutgers University)**
3:55 p.m.	K. Matsuyanagi (Kyoto University) *Exotic Shapes in ^{32}S Predicted by Symmetry-Unrestricted Cranked HF Calculations*
4:20 p.m.	C. De Coster (University of Gent) *Nuclear Collective Motion: Strengths and Limitations of Group-Theoretical Methods*
4:55 p.m.	R. F. Casten (Yale University) *Phase Transitions, Phase Co-existence and Control Parameters in Finite Nuclei*
5:30 p.m.	Close of Session
6:00 p.m.	HRIBF Users Meeting, Tennessee Ballrooms 3 and 4

Session III-A   **Chair:**
                        R. F. Casten (Yale University)

9:00 a.m.   P. J. Woods (Edinburgh University)
                       Proton Radioactivity

9:40 a.m.   J. C. Batchelder (ORISE)
                       Proton Decay of the Light Lu, Tm, and Ho Isotopes

10:05 a.m.   M. Thoennessen (Michigan State University)
                       Two Proton Decay of the First Excited State in ^{17}Ne

10:30 a.m.   Break

Session III-B   **Chair:**
                        P. Fallon (Lawrence Berkeley National Laboratory)

11:00 a.m.   J. H. Hamilton (Vanderbilt University)
                       Octupole Degree of Freedom in n-Rich Rare-Earth Nuclei

11:35 a.m.   A. G. Smith (University of Manchester)
                       Spectroscopy of n-rich Nuclei Populated by SF of ^{252}Cf

12:10 p.m.   G. J. Lane (Lawrence Berkeley National Laboratory)
                       Population of High-Spin States in ^{234}U by Incomplete-Fusion Reaction

12:35 p.m.   Lunch, Tennessee Ballroom 2

2:00 p.m.   Departure for tour of HRIBF

Following the tour, JIHIR will sponsor a reception 5:00-6:00 p.m.
Buses return to Park Vista Hotel before 8:00 p.m.

Thursday, August 13, 1998

Session IV-A	**Chair:** **K. Rykaczewski (Oak Ridge National Laboratory)**
9:00 a.m.	J. C. Hardy (Texas A&M University) *Superallowed Fermi Beta Decay and Coulomb Mixing in Nuclei*
9:35 a.m.	W. F. Mueller (University of Leuven) *Beta Decay of Neutron-Rich Ni and Co: Probing Single-Particle States Above N=40 Subshell Closure*
10:00 a.m.	J. L. Tain (Instituto de Fisica Corpuscular, Valencia) *The Gamow-Teller Resonance Revealed in Beta-Decay Using New Experimental Techniques*
10:25 a.m.	Break
Session IV-B	**Chair:** **T. L. Khoo (Argonne National Laboratory)**
10:50 a.m.	S. Juutinen (University of Jyväskylä) *RDT Studies of Onset of Collectivity Near Z = 82 Shell Gap*
11:25 a.m.	P. Reiter (Argonne National Laboratory) *Structure and Formation Mechanisms of the Transfermium Isotope ^{254}No*
11:50 a.m.	W. Reviol (University of Tennessee) *Shape Competition in the Hg-Pb Region: The Case of ^{183}Tl and ^{183}Pb*
12:15 p.m.	A. R. Junghans (GSI) *Fission Studies of Secondary Beams from Relativistic U Projectiles*
12:40 p.m.	Lunch, Tennessee Ballroom 2

Session IV-C   Chair:
                          M. A. Riley (Florida State University)

2:00 p.m.   R. Wadsworth (University of York)
                          Magnetic Rotation in the A ~ 110 and 200 Mass Regions

2:35 p.m.   A. O. Macchiavelli (Lawrence Berkeley National Laboratory)
                          Semi-Classical Shears and Effective Forces

3:10 p.m.   S. M. Mullins (Australian National University)
                          High-K Spin Traps in the Valley of Stability

3:35 p.m.   Break

Session IV-D   Chair:
                          D. B. Fossan (SUNY, Stony Brook)

4:00 p.m.   C. M. Petrache (INFN, Padova)
                          Decay Out in the A = 130 Mass Region: Experimental Δ_υ for the Highly Deformed Nd Nuclei

4:35 p.m.   F. G. Kondev (Florida State University)
                          Second Minimum Structures in the A ~ 135 Region and Their Relative Quadrupole Moments

5:00 p.m.   D. R. LaFosse (SUNY, Stony Brook)
                          Band Termination in the A ~ 110 Mass Region

5:35 p.m.   J. C. Lisle (University of Manchester)
                          Competition Between Collective and Terminating Structures in the A ~ 160 Mass Region

6:00 p.m.   Close of Session

7:00 p.m.   Banquet, Tennessee Ballroom 2

8:00 p.m.   Jazz Music and Cash Bar
                          Music by Donald Brown and Friends is sponsored by SCIONIX

Friday, August 14, 1998

Session V-A	**Chair:** **V. E. Oberacker (Vanderbilt University)**

9:00 a.m.   R. Machleidt (University of Idaho)
                   Nuclear Forces and Nuclear Structure

9:35 a.m.   S. C. Pieper (Argonne National Laboratory)
                   Quantum Monte Carlo Calculations of Light Nuclei

10:10 a.m.   B. R. Barrett (University of Arizona)
                   Large-Scale No-Core Shell-Model Calculations for p-Shell

10:35 a.m.   Break

Session V-B   **Chair:**
                   M. S. Smith (Oak Ridge National Laboratory)

11:00 a.m.   D. G. Ravenhall (University of Illinois)
                   Neutron Stars and Equation of State

11:35 a.m.   M. Wiescher (University of Notre Dame)
                   Nuclear Structure Issues Relevant to Nucleosynthesis

12:10 p.m.   J. Engel (University of North Carolina, Chapel Hill)
                   Beta Decay and Neutrino Capture in r-Process Waiting-Point Nuclei

12:35 p.m.   Lunch, Tennessee Ballroom 2

Session V-C   **Chair:**
J.D. Garrett (Oak Ridge National Laboratory)

2:00 p.m.   J. R. Beene (Oak Ridge National Laboratory)
Recent Developments in Studies of Giant Resonances

2:35 p.m.   H. Sagawa (University of Aizu)
Isoscalar and Isovector Dipole Mode in Drip Line Nuclei

3:00 p.m.   P. Von Brentano (University of Köln)
Low-Lying Double-Phonon Excitations in Heavy Nuclei

3:35 p.m.   Break

Session V-D   **Chair:**
F. Hannachi (Orsay)

4:00 p.m.   R. Grzywacz (University of Tennessee)
Microsecond Isomers Beyond ^{68}Ni

4:25 p.m.   P. G. Bizzeti (University of Florence)
Two and Three Octupole Phonon Excitations

4:50 p.m.   S. W. Yates (University of Kentucky)
Double-Phonon Octupole Excitations in ^{208}Pb

5:20 p.m.   W. Younes (Lawrence Livermore National Laboratory)
Double-Phonon Excitations in Deformed ^{170}Er

5:45 p.m.   Close of Session

Saturday, August 15, 1998

Session VI-A   **Chair:**
D. G. Sarantites (Washington University)

9:00 a.m.   I. Y. Lee (Lawrence Berkeley National Laboratory)
GRETA: Status and Physics Potentials

9:35 a.m.   D. C. Radford (Oak Ridge National Laboratory)
Analysis of High-Fold Gamma Data

10:10 a.m.   T. Dössing (Niels Bohr Institute)
Quasicontinuum Gamma Rays: A Theoretical Perspective

10:45 a.m.   Break

Session VI-B   **Chair:**
B. M. Sherrill (Michigan State University)

11:15 a.m.   I. Wiedenhoever (Argonne National Laboratory)
Unsafe Coulomb Excitation of ^{240}Pu and ^{244}Pu

11:40 a.m.   F. Azaiez (Institut de Physique Nucleaire, Orsay)
Gamma Spectroscopy of Neutron-Rich Nuclei at GANIL

12:10 p.m.   T. Glasmacher (Michigan State University)
Proton Scattering on Radioactive Proton sd-shell Nuclei and Comparison to Coulomb Excitation

12:35 p.m.   Close of Meeting

Poster Presentations

"Lifetime Measurements and Dipole Transition Rates for Superdeformed States in ^{190}Hg"
H. Amro, R. V. F. Janssens, G. Hackman, S. Fischer, M. P. Carpenter,
I. Ahmad, T. L. Khoo, T. Lauritsen, D. Nisius, E. F. Moore, B. Crowell,
A. N. Wilson, and J. Timar

"High-Spin Spectroscopy of the Neutron-Deficient ^{110}Te Nucleus"
A. J. Boston, E. S. Paul, C. J. Chiara, M. Devlin, D. B. Fossan, D. R. LaFosse,
G. J. Lane, I.-Y. Lee, A. O. Macchiavelli, P. J. Nolan, D. G. Sarantites,
J. M. Sears, A. T. Semple, J. F. Smith, A. V. Afanasjev, and I. Ragnarsson

"Exploring the Isotopes"
S. Y. (Frank) Chu

"High-Spin States in Stable Nuclei Populated as Fission Fragments in Heavy-Ion Reactions"
J. A. Cizewski, N. Fotiades, D. P. McNabb, K. Y. Ding, D. E. Archer,
J. A. Becker, L. A. Bernstein, K. Hauschild, W. Younes, R. M. Clark,
P. Fallon, I.-Y. Lee, A. O. Machiavelli, and R. W. MacLeod

"Investigation of the Reaction ^{12}C(^{16}O,^{16}O*)^{12}C* $\rightarrow$ ^{8}Be + α Using the New Megha Strip Detector Array"
G. K. Dillon and S. P. G. Chappell

"Excited States in ^{155,157}Lu Using Recoil-Decay Tagging of Gammasphere"
K. Y. Ding, D. Seweryniak, J. A. Cizewski, H. Amro, L. T. Brown,
M. P. Carpenter, C. N. Davids, N. Fotiades, G. Hackman, R. V. F. Janssens, T. Lauritsen, C. J. Lister, A. O. Macchiavelli, D. Nisius, P. Reiter, S. Siem,
J. Uusitalo, and I. Wiedenhöver

"The Oak Ridge HRIBF Recoil Mass Spectrometer"
T. N. Ginter, C. J. Gross, J. H. Hamilton, A. V. Ramayya, D. Shapira,
J. W. Johnson, J. W., McConnell, W. T. Milner, J. F. Mas, Y. A. Akovali,
C. Baktash, J. C. Batchelder, C. R. Bingham, A. Galindo-Uribarri,
B. D. MacDonald, S. D. Paul, A. Piechaczek, D. C. Radford, W. Reviol,
K. Rykaczewski, K. S. Toth, C.-H. Yu, and E. F. Zganjar

"Identification of Gamma Rays in the Ground State Proton Emitter ^{113}Cs"
C. J. Gross, Y. A. Akovali, C. Baktash, J. C. Batchelder, C. R. Bingham,
M. P. Carpenter, C. N. Davids, T. Davinson, D. Ellis, A. Galindo-Uribarri,
T. N. Ginter, R. Grzywacz, R. V. F. Janssens, J. W. Johnson, F. Liang,
C. J. Lister, J. Mas, B. D. MacDonald, S. D. Paul, A. Piechaczek,
D. C. Radford, W. Reviol, K. Rykaczewski, W. Satula, D. Seweryniak,
D. Shapira, K. S. Toth, W. Weintraub, P. J. Woods, C.-H. Yu, E. F. Zganjar, and J. Uusitalo

"Investigation of ^{123}Xe in the High Spin Regime"
S. Naguleswaran, B. R. S. Babu, P. J. Binns, S. V. Förtsch, J. J. Lawrie,
C. Rigollet, J. F. Sharpey-Schafer, B. R. S. Simpson, F. D. Smit, and
G. F. Steyn

"Gas Filled Split-Pole Spectrograph for Nuclear Reaction Studies at HRIBF"
J. F. Liang, J. R. Beene, A. Galindo-Uribarri, J. Gomez del Campo,
C. J. Gross, M. L. Halbert, A. I. D. MacNab, D. Shapira, R. L. Varner, and
K. Zhao

"Relativistic Mean Field Investigation on the Superdeformed Bands in the A = 60 Mass Region"
H. Madokoro and M. Matsuzaki

"Nuclear Mean Field Calculations in the Spline-Galerkin Representation"
V. E. Oberacker and A. S. Umar

"High-Spin Study of ^{119}Xe and Terminating Bands in ^{117,119}Xe"
E. S. Paul, Scraggs, A. J. Boston, D. M. Cullen, J. F. C. Cocks, K. Helariutta,
P. M. Jones, R. Julin, S. Juutinen, H. Kankaanpää, M. Muikku, A. Savelius,
C. M. Parry, R. Wadsworth, A. V. Afanasjev, and I. Ragnarsson

"Collectivity and Pairing Strength in the Second Minimum of ^{136}Nd"
S. Perries, A. Astier, L. Ducroux, M. Meyer, N. Redon, C. M. Petrache,
D. Bazzacco, G. Falconi, S. Lunardi, M. Lunardon, C. Rossi-Alvarez,
C. A. Ur, R. Venturelli, G. Viesti, I. Deloncle, M. G. Porquet, G. de Angelis,
M. De Poli, C. Fahlander, E. Farnea, D. Foltescu, A. Gadea, D. R. Napoli,
Zs. Podolyak, A. Bracco, S. Frattini, S. Leonie, B. Cederwall, A. Johnson,
and R. Wyss

"Magnetic Dipole Bands in ^{82}Rb, ^{83}Rb and ^{84}Rb"
H. Schnare, R. Schwengner, S. Frauendorf, F. Donau, L. Käubler, H. Prade,
E. Grosse, A. Jungclaus, K. P. Lieb, C. Lingk, S. Skoda, J. Eberth,
G. de Angelis, A. Gadea, E. Farnea, C. A. Ur, and G. Lo Bianco

"Coulomb Excitation of a ^{78}Rb Radioactive Beam"
J. Schwartz, C. J. Lister, D. H. Henderson, S. M. Fischer, P. Reiter,
A. Aprahimian, J. A. Cizewski, C. N. Davids, R. de Haan, R. V. F. Janssens,
D. Nisius, D. Seweryniak, and S. M. Vincent

"Exploring the N = Z Line at A ~ 60-70"
S. Skoda, B. Fiedler, F. Becker, J. Eberth, S. Freund, T. Steinhardt, O. Stuch,
O. Thelen, H. G. Thomas, L. Käubler, J. Reif, H. Schnare, R. Schwengner,
T. Servene, G. Winter, V. Fischer, A. Jungclaus, D. Kast, K. P. Lieb, C. Teich,
C. Ender, T. Härtlein, F. Köck, D. Schwalm, and P. Baumann

"Identification of Excited States in ^{112}Xe"
J. F. Smith, C. J. Chiara, D. B. Fossan, G. J. Lane, J. M. Sears, M. Devlin,
D. R. LaFosse, D. G. Sarantites, A. J. Boston, E. S. Paul, A. T. Semple,
I.-Y. Lee, and A. O. Macchiavelli

"Smooth Band Termination in ^{112}Te and ^{113}I N = 60 Isotones"
K. Starosta, C. J. Chiara, D. B. Fossan, D. R. LaFosse, G. J. Lane, J. M. Sears,
J. F. Smith, A. J. Boston, E. S. Paul, M. Devlin, D. G. Sarantites, I.-Y. Lee, and A. O.
Macchiavelli

"Structure of ^{112}I and Proton $h_{11/2}$ Neutron $h_{11/2}$ Systematics vs. N"
K. Starosta, C. .J. Chiara, D. B. Fossan, D. R. LaFosse, G. J. Lane, J. M. Sears,
J. F. Smith, A. J. Boston, E. S. Paul, M. Devlin, D. G. Sarantites, I.-Y. Lee,
A. O. Macchiavelli, and S. G. Rohizinski

"Shell Corrections in the Drip-Line Region"
T. Vertse, A. T. Kruppa, R. J. Liotta, W. Nazarewicz, N. Sandulescu, and
T. R. Werner

"Shell Model Monte Carlo Results for Nuclear Structure in Rare Earth Nuclei"
J. White

NUCLEAR STRUCTURE '98 ATTENDEES

AKOVALI, Y.
Oak Ridge National Laboratory
Physics Division
P. O. Box 2008, Bldg 6000, MS-6371
Oak Ridge, TN 37831-6371
USA
Tele: 423 574 4695
Fax: 423 574 8902
akovali@mail.phy.ornl.gov

AZAIEZ, F.
Inst de Physique Nucleaire
B.P. No. 1
Orsay F-91406 Cedex
FRANCE
Tele: 33 169 15 7105
Fax: 33 169 15 7196
azaiez@ipno.in2p3.fr

BALABANSKI, D. L.
University of Leuven
Instituut voor Kern- en Stralingsfysica
Celestijnenlaan 200D
Leuven B-3001
BELGIUM
Tele: 32 16 32 72 62
Fax: 32 16 32 79 85
dimiter.balabanski@fys.kuleuven.ac.be

BARRETT, B. R.
Department of Physics
P O Box 210081
University of Arizona
Tucson, AZ 85721
USA
Tele: 520 621 2979
Fax: 520 621 4721
bbarrett@physics.arizona.edu

AMRO, H. M.
Argonne National Laboratory
Physics Division
PHY 203
Argonne, IL 60439-4843
Tele: 630 252 4730
Fax: 630 252 6210
amro@sun0.phy.anl.gov

BAKTASH, C.
Oak Ridge National Laboratory
Physics Division
P. O. Box 2008, Bldg. 6000, MS-6371
Oak Ridge, TN 37831-6371
USA
Tele: 423 576 7949
Fax: 423 574 1268
baktash@mail.phy.ornl.gov

BARMORE, B.
Joint Institute for Heavy Ion Research/ORNL
Physics Division
P. O. Box 2008, Bldg 6008, MS6374
Oak Ridge, TN 37831-6374
USA
Tele: 423 576-8763
Fax: 423 576-5780
barmore@mail.phy.ornl.gov

BATCHELDER, J. C.
Oak Ridge Institute for Science and Education/ORNL
Physics Division
P O Box 2008, Bldg. 6008, MS-6374
Oak Ridge, TN 37831-6374
USA
Tele: 423 576 7656
Fax: 423 576 1268
batcheld@mail.phy.ornl.gov

BEENE, J. R.
Oak Ridge National Laboratory
Physics Division
P O Box 2008, Bldg 6000, MS6368
Oak Ridge, TN 37831-6368
USA
Tele: 423 574 4622
Fax: 423 574 1268
beene@orph14.phy.ornl.gov

BENDER, M.
University of North Carolina at Chapel Hill
Department of Physics and Astronomy
CB 3255 Phillips Hall
Chapel Hill, NC 27599
Tele: 919 962 1571
Fax: 919 962 0480
bender@physics.unc.edu

BINGHAM, C. R.
University of Tennessee
Department of Physics
401 Nielsen Physics Building
Knoxville, TN 37996-1200
USA
Tele: (423) 974-7802
Fax: (423) 974-7843
cbingham@utk.edu

BOSTON, A. J.
The University of Liverpool
Department of Physics
Oliver Lodge Laboratory
Liverpool L69 7ZE
UNITED KINGDOM
Tele: 44 1517943381
Fax: 44 1517943348
ajb@ns.ph.liv.ac.uk

BELLEGUIC, M.
IPN Orsay
Rue Clemenceau
Orsay Cedex 91406
FRANCE
Tele: 33 1 69 15 4459
Fax: 33 1 69 15 7196
bellegui@ipno.in2p3.fr

BERTRAND, F. E.
Oak Ridge National Laboratory
Physics Division
P O Box 2008, Bldg 6000, MS-6369
Oak Ridge, TN 37831-6369
USA
Tele: 423 574 4737
Fax: 423 574 1268
feb@ornl.gov

BIZZETI, P. G.
University of Florence
Dipartimento di Fisica
Largo E. Fermi 2
I-50125 Firenze
ITALY
Tele: 39 055 230 7641
Fax: 39 055 229 330
bizzeti@fi.infn.it

BROWN, T.
University of Kentucky
Department of Physics
Lexington, KY 40506
USA
Tele: 606-257-6704
brown@nucmar.physics.fsu.edu

CARPENTER, M. P.
Argonne National Laboratory
Physics Division D-203
9700 South Cass Avenue
Argonne, IL 60439
USA
Tele: (630) 252-5365
Fax: (630) 252-2864
carpenter@anlphy.phy.anl.gov

CASTEN, R. F.
Yale University
Physics Department
Wright Nuclear Structure Laboratory
New Haven, CT 06520-8124
USA
Tele: 203-432-6174
Fax: 203-432-3522
rick@riviera.physics.yale.edu

CIZEWSKI, J. A.
Physics Department
Rutgers University
136 Frelinghuysen Road
Piscataway, NJ 08854-8019
USA
Tele: 732-445-3884
Fax: 732-445-4343
cizewski@physics.rutgers.edu

CULLEN, D. M.
University of Liverpool
Dept. of Physics, Oliver Lodge Laboratory
P. O. Box 147
Liverpool L69 7ZE
UNITED KINGDOM
Tele: 011-44-151-794-3713
Fax: 011-44-151-794-3348
dmc@ns.ph.liv.ac.uk

CARTER, H. K.
Oak Ridge Institute for Science and Education/ORNL
Physics Division
P. O. Box 2008, Bldg. 6008, MS-6374
Oak Ridge, TN 37831-6374
USA
Tele: 423 576 2642
Fax: 423 576 5780
carter@mail.phy.ornl.gov

CHU, S. Y. (FRANK)
Lawrence Berkeley National Laboratory
Nuclear Science Division
1 Cyclotron Road
Berkeley, CA 94720
USA
Tele: 510 486 7648
Fax: 510 486 5757
syfchu@lbl.gov

COVELLO, A.
Dipartimento di Scienze Fisiche
Universita' di Napoli
Federico II, Complesso Universitario di Monte
S. Angelo, Via Cintia, Napoli 80126
ITALY
Tele: 39 81 7253402
Fax: 39 81 676346
covello@na.infn.it

DE COSTER, C.
University of Gent
Inst of Theoretical Physics
Vakgroep Subatomaire en Stralingsfysica
Proeftuinstraat 86, Gent B-9000
BELGIUM
Tele: 32-9-264-65-29
Fax: 32-9-264-66-99
caroline@inwfaxp1.rug.ac.be

DEAN, D. J.
Oak Ridge National Laboratory
Physics Division
P. O. Box 2008, Building 6003, MS-6373
Oak Ridge, TN 37831-6373
USA
Tele: 423 576 5229
Fax: 423 574 4745
deandj@ornl.gov

DIAMOND, R. M.
Lawrence Berekeley National Laboratory
Nuclear Science Division
Bldg 88, 1 Cyclotron Road
Berkeley, CA 94720
USA
Tele: 510 486 5720
Fax: 510 486 7983
rmdiamond@lbl.gov

DING, K. Y.
Rutgers University
Department of Physics & Astronomy
Frelinghuysen and Allison Road
Piscataway, NJ 08855-0849
USA
Tele: 732 445 2405
Fax: 732 445 4343
dingding@physics.rutgers.edu

DOBACZEWSKI, J.
Institute of Theoretical Physics
Warsaw University
Hoza 69
Warsaw PL-00-681
POLAND
Tele: 48 22 628 3396
Fax: 48 22 621 9475
dobaczew@fuw.edu.pl

DELEPLANQUE-STEPHENS, M.-A.
Lawrence Berkeley National Laboratory
Nuclear Science Division
Bldg 88-250, 1 Cyclotron Road
Berkeley, CA 94720
USA
Tele: 510 486-5384
Fax: 510 486-7983
mads@lbl.gov

DILLON, G. K.
Nuclear and Astrophysics Laboratory
Department of Physics
University of Oxford, Keble Road
Oxford OX1 3RH
UNITED KINGDOM
Tele: 01904 432244
Fax: 01904 432214
dillon@s23.physics.ox.ac.uk

DJERROUD, B.
McMaster University
Department of Physics & Astronomy
1280 Main Street West
Hamilton Ontario L8S 4M1
CANADA
Tele: 905 525 9140/27458
Fax: 905 546 1252
djerroud@physics.mcmaster.ca

DOERING, H. J.
University of Notre Dame
Physics Division
Notre Dame, IN 46556
USA
Tele: 219 631 7716
Fax: 219 631 5952
jdoering@sirius.phys.nd.edu

DÖSSING, T.
Niels Bohr Institute
Blegdamsvej 17
Copenhagen 0, DK 2100
DENMARK
Tele: 45 35 32 52 57
Fax: 45 35 32 54 00
dossing@nbi.dk

FALLON, P.
Lawrence Berkeley National Laboratory
Nuclear Science Division
1 Cyclotron Road, Bldg 88-253
Berkeley, CA 94720
USA
Tele: (510) 486-7018
Fax: (510) 486-7983
pfallon@lbl.gov

FOSSAN, D. B.
State University of New York-Stony Brook
Nuclear Structure Laboratory
Physics Department
Stony Brook, NY 11794-3800
USA
Tele: 516- 632-8113
Fax: 516-632-8573
fossan@nuclear.physics.sunysb.edu

GALINDO-URIBARRI, A.
Oak Ridge National Lab
Physics Division
P. O. Box 2008, Bldg. 6000, MS-6371
Oak Ridge, TN37831-6371
USA
Tele: 423-574-6124
Fax: 423-574-1268
uribarri@mail.phy.ornl.gov

ENGEL, J.
University of North Carolina at Chapel Hill
Department of Physics and Astronomy
CB3255 Phillips Hall
Chapel Hill, NC 27599-3255
Tele: 919 962 2619
Fax: 919 962 0480
engelj@physics.unc.edu

FLIBOTTE, S.
McMaster University
Dept of Physics & Astronomy
1280 Main Street, West
Hamilton L8S 4M1 Ontario
CANADA
Tele: 905-525-9140 Ext 23632
Fax: 905-546-1252
flibotte@zinnia.physics.mcmaster.ca

FOX, J.
Florida State University/ORNL
Physics Division
P. O. Box 2008, Bldg 6000, MS6368
Oak Ridge, TN 37831-6368
USA
Tele: 423-576-6915
Fax: 423-574-1268
jfox@fsulcd.physics.fsu.edu

GARRETT, J. D.
Oak Ridge National Laboratory
Physics Division
P. O. Box 2008, Bldg 6000 , MS-6368
Oak Ridge, TN 37831-6368
USA
Tele: 423 576 5489
Fax: 423 574 1268
garrettjd@ornl.gov

GINTER, T. N.
Department of Physics & Astronomy
Vanderbilt University
Station 1807-B
Nashville, TN 37235
USA
Tele: 423-574-4105
Fax: 423-574-1268
gintertn@ctrvax.vanderbilt.edu

GOODMAN, A. L.
Tulane University
Physics Department
New Orleans, LA 70118
USA
Tele: 504 862 3172
Fax: 504 862 8702
alan.goodman@tulane.edu

GROSS, C. J.
Oak Ridge Institute for Science and Education/ORNL
Physics Division
P. O. Box 2008, Bldg 6000, MS-6371
Oak Ridge, TN 37831-6371
USA
Tele: 423 576 7698
Fax: 423 574 1268
cgross@mail.phy.ornl.gov

HAMILTON, J. H.
Vanderbilt University
Department of Physics & Astronomy
Station 1807-B
Nashville, TN 37235
USA
Tele: 615 322-2828
Fax: 615-322-4936
hamiltj1@ctrvax.vanderbilt.edu

GLASMACHER, T.
Michigan State University
National Superconducting Cyclotron Lab
East Lansing, MI 48824-1321
USA
Tele: 517-355-9672 Ext: 418
Fax: 517-353-5967
glasmacher@nscl.msu.edu

GRAWE, H.
GSI Darmstadt
Planckstr
KPII
Darmstadt D-64291
GERMANY
Tele: 49 6159 71 2430
Fax: 49 6159 71 2902
h.grawe@gsi.de

GRZYWACZ, R.
University of Tennessee
Department of Physics
401 Nielsen Physics Bldg
Knoxville, TN 37996-1200
USA
Tele: 423 574-4498
Fax: 423 576-5780
grzywacz@mail.phy.ornl.gov

HANNACHI, F.
CSNSM
Batiments 104-108
Orsay, Campus 91405
FRANCE
Tele: 33-1-6941-5081
Fax: 33-1-6941-5008
hannachi@csnsm.in2p3.fr

HARDY, J. C.
Cyclotron Institute
Texas A&M University
College Station, TX 77843
USA
Tele: 409-845-1411
Fax: 409-845-1899
hardy@comp.tamu.edu

HASHIMOTO, Y.
University of Tsukuba
Institute of Phsyics
Tennodai 1-1-1
Tsukuba Ibaraki 305-8571
JAPAN
Tele: 0298 53 4281
Fax: 298 53 4492
hashi@nucl.ph.tsukuba.ac.jp

HEENEN, P.-H.
Universite Libre de Bruxelles
Physique Nucleaire Theorique
PNTPM-CP 229, Ave. Roosevelt 50
Brussels B-1050
BELGIUM
Tele: 32 2 650 5558
Fax: 32 2 650 5045
phheenen@ulb.ac.be

JAMES, A. N.
Univ. of Liverpool
Dept. of Physics
P. O. Box 147
Liverpool L69 3BX
UNITED KINGDOM
Tele: 44 151 794 3385
Fax: 44 151 794 3444
anjames@ns.ph.liv.ac.uk

HARTLEY, D. J.
University of Tennessee
Department of Physics
401 Nielsen Physics Bldg
Knoxville, TN 37996-1200
USA
Tele: 423 974 7803
Fax: 423 974 7843
daryl@spinno.phys.utk.edu

HASS, M.
Department of Particle Physics
Weizmann Institute
Rehovot 76100
ISRAEL
Tele: 972 8 934 3997
Fax: 972 8 934 4166
fnhass@wis.weizmann.ac.il

HÜBEL, H. C.
Institute für Strahlen- und Kernphysik
University of Bonn
Nussallee 14-16
D-53115 Bonn
GERMANY
Tele: 49-228-73-3277
Fax: 49-228-73-2505
hubel@iskp.uni-bonn.de

JOHNSON, N. R.
Oak Ridge National Laboratory
Physics Division
P. O. Box 2008, Bldg 6007, MS 6371
Oak Ridge, TN 37831-6371
USA
Tele: 423-574-4739
Fax: 423-574 1268
johnson@orph01.phy.ornl.gov

JUNGHANS, A. R.
GSI Darmstadt
Planckstr. 1
Darmstadt D-64291
GERMANY
Tele: 49 6159 71 2738
Fax: 49 6159 71 2785
a.junghans@gsi.de

KHOO, T. L.
Argonne National Laboratory
Physics Division 203
9700 South Cass Avenue
Argonne, IL 60439
USA
Tele: 630 252-4034
Fax: 630 252-6210
khoo@anl.gov

KORMICKI, J.
Vanderbilt University
Department of Physics & Astronomy
Station 1807-B
Nashville, TN 37235
Tele: 615-322-0656
Fax: 615-343-7263
kormicj0@ctrvax.vanderbilt.edu

LAIRD, R. W.
Florida State University
Department of Physics
120 Nuclear Research Bldg
Tallahassee, FL 32306-4350
USA
Tele: 850 644 2878
Fax: 850 644 9848
laird@nucmar.physics.fsu.edu

JUUTINEN, S.
University of Jyväskylä
Dept. of Physics
P. O. Box 35
Jyväskylä SF-40351
FINLAND
Tele: 358 14 602 368
Fax: 358 14 602 351
juutinen@jyfl.jyu.fi

KONDEV, F. G.
Florida State University
Department of Physics
Tallahassee, FL 32306-4350
Tele: 850-644-6226
Fax: 850-644-9848
kondev@nucott.physics.fsu.edu

LAFOSSE, D. R.
SUNY at Stony Brook
Department of Physics & Astronomy
Stony Brook, NY 11794-3800
USA
Tele: 516-632-7023
Fax: 516 -632 -8573
lafosse@nuclear.physics.sunysb.edu

LANE, G. J.
Lawrence Berkeley National Laboratory
Nuclear Science Division
1 Cyclotron Road
Berkeley, CA 94720
USA
Tele: 510-486-5702
Fax: 510-486-7983
gjlane@lbl.gov

LAWRIE, J. J.
National Accelorator Centre
P. O. Box 72
Faure
SOUTH AFRICA
Tele: 021 843 3820
Fax: 021 843 3525
lawrie@nac.ac.za

LERMA, F.
Washington University
Chemistry Department
Campus Box 1134, One Brookings Drive
St. Louis, MO 63130-4889
USA
Tele: 314-935-6570
Fax: 314-935-6184
lerma@proton.wustl.edu

LISLE, J. C.
Schuster Laboratory
University of Manchester
Department of Physics
Manchester M13 9PL
UNITED KINGDOM
Tele: 44 161 275 4407
Fax: 44 161 275 5509
hfl@mags.ph.man.ac.uk

MACCHIAVELLI, A. O.
Lawrence Berkeley National Laboratory
Nuclear Science Division
1 Cyclotron Road, Bldg 88-253
Berkeley, CA 94720
USA
Tele: 510-486-4428
Fax: 510-486-7983
aom@lbl.gov

LEE, I.-Y.
Lawrence Berkeley National Laboratory
Nuclear Sciences Division
1 Cyclotron Road, Bldg 88
Berkeley, CA 94720
USA
Tele: 510-486-5727
Fax: 510-486-7983
iylee@lbl.gov

LIANG, F.
Oak Ridge National Laboratory
Physics Division
P. O. Box 2008, Bldg 6000, MS-6368
Oak Ridge, TN 37831-6368
Tele: 423-574-4109
Fax: 423-574-1268
liang@mail.phy.ornl.gov

LIVELY, M. L.
Florida State University
Department of Physics
Tallahassee, FL 32306-4473
USA
Tele: 850-574-0655
Fax: 850-644-9848
lively@nucmar.physics.fsu.edu

MACH, H.
Uppsala University
Department of Neutron Research
S 61182, Nykoping
SWEDEN
Tele: 46 155 221 839
Fax: 46 155 263 001
henryk@studsvik.uu.se

MACHLEIDT, R.
University of Idaho
Department of Physics
Moscow, ID 83844
USA
Tele: 208-885-8951
Fax: 208-885-4055
machleid@uidaho.edu

MAIER, M.
Lawrence Berkeley National Laboratory
Department of Engineering
1 Cyclotron Road
Berkeley, CA 94720
USA
Tele: 510-486-5599
Fax: 510-486-7557
mrmaier@lbl.gov

MATSUYANAGI, K.
Kyoto University
Faculty of Science
Department of Physics
Kitashirakawa Kyoto 606 8502
JAPAN
Tele: 81-75-753-3841
Fax: 81-75-753-3886
ken@ruby.scphys.kyoto-u.ac.jp

MULLINS, S. M.
Australian National University
Department of Nuclear physics
RS Phys SE,
Canberra, ACT 0200
AUSTRALIA
Tele: 61 2 6249 0375
Fax: 61 2 6249 0748
smm103@nuc.anu.edu.au

MADOKORO, H.
Kyushu University
Theoretical Nucl Phys Lab, Dept of Phys
Faculty of Sci, 6-10-1 Hakozaki, Higashi-ku
Fukuoka 812-81
JAPAN
Tele: 011-81-92-642-2558
Fax: 011-81-92-642-2553
madokoro@nthp1.phys.kyushu-u.ac.jp

MATSUO, M.
Yukawa Inst for Theoretical Physics
Kyoto University
Kitashirakawa
Kyoto 606 8502
JAPAN
Tele: 81 757 537 021
Fax: 81 757 537 010
matsuo@yukawa.kyoto-u.ac.jp

MUELLER, W. F.
Instituut voor Kern- en Stralingsfysica
Celestijnenlaan 200D
B-3001 Leuven
BELGIUM
Tele: 32 16 32 72 71
wilhelm.mueller@fys.kuleuven.ac.be

NAZAREWICZ, W.
University of Tennessee/ORNL
Physics Division
P. O. Box 2008, Bldg 6003, MS 6373
Oak Ridge, TN 37831-6373
USA
Tele: 423-574-4580
Fax: 423-574-4745
witek-nazarewicz@utk.edu

OBERACKER, V.E.
Vanderbilt University
Station 1807-B
Dept of Physics and Astronomy
Nashville, TN 37235
Tele: 615-322-2828
Fax: 615-343-7263
volker@compsci.cas.vanderbilt.edu

OZEN, C.
University of Tennessee
Department of Physics
401 Nielsen Physics Bldg
Knoxville, TN 37996-1200
USA
Tele: 423-974-3342
Fax: 423-974-7843
cozen@utk.edu

PAUL, S. D.
Oak Ridge National Laboratory
Physics Division
P. O. Box 2008, Bldg 6000, MS-6371
Oak Ridge, TN 37831-6371
Tele: 423-574-4708
Fax: 423-574-1268
sdpaul@mail.phy.ornl.gov

PETRACHE, C.
INFN, Sezione di Padova
Dipartimento di Fisica, Universita di Padova
8 via Marzolo
Padova I-35131
ITALY
Tele: 39 49 8277176
Fax: 39 49 8277102
petrache@pd.infn.it

OTSUKA, T.
University of Tokyo
Department of Physics
7-3-1 Hongo Bunkyo-ku
Tokyo 113
JAPAN
Tele: 81 3 3818 7490
Fax: 81 3 5684 9642
otsuka@phys.s.u-tokyo.ac.jp

PAUL, E. S.
University of Liverpool
Department of Physics
Oliver Lodge Laboratory
Liverpool L69 7ZE
UNITED KINGDOM
Tele: 011 44 151 794 3382
Fax: 011 44 151 794 3348
esp@ns.ph.liv.ac.uk

PERRIES, S.
INPL
43, Bd du 11/11/1918
Villeurbanne G-69622
FRANCE
Tele: 33 4 72 43 1058
Fax: 33 4 72 44 8004
perries@ipnl.in2p3.fr

PIECHACZEK, A.
LSU/ORNL
Physics Division
P. O. Box 2008, Bldg 6000, MS6371
Oak Ridge, TN 37831-6371
USA
(423) 574-4723
(423) 574-8902
andreas@mail.phy.ornl.gov

PIEPER, S. C.
Argonne National Laboratory
Physics Division
9700 S. Cass Avenue
Argonne, IL 60439
USA
Tele: 630 252 4232
Fax: 630 252 3903
pieper@theory.phy.anl.gov

RADFORD, D. C.
Oak Ridge National Laboratory
Physics Division
P. O. Box 2008, Bldg 6000, MS6371
Oak Ridge, TN 37831-6371
Tele: 423-241-5332
Fax: 423-574-8902
radfordd@mail.phy.ornl.gov

REITER, P.
Argonne National Laboratory
Physics Division
9700 S. Cass Avenue
Argonne, IL 60439-4843
USA
Tele: 630-252-4048
Fax: 630-252-2864
reiter@sun0.phy.anl.gov

RIEDINGER, L. L.
University of Tennessee
Physics Department
401 Nielsen Physics Bldg.
Knoxville, TN 37996-1200
USA
Tele: 423-974-7805
Fax: 423-974-7843
lrieding@utk.edu

PLASIL, F.
Oak Ridge National Laboratory
Physics Division
P. O. Box 2008, Bldg 6003, MS6372
Oak Ridge, TN 37831-6372
USA
Tele: 423-574-4711
Fax: 423-576-2822
plasil@orph01.phy.ornl.gov

RAVENHALL, D. G.
University of Illinois
Department of Physics
1110 West Green Street
Urbana, IL 61801
USA
Tele: 217-333-3865
Fax: 217-333-9819
ravenhal@uiuc.edu

REVIOL, W.
University of Tennessee
Physics Department
401 Nielsen Physics Bldg.
Knoxville, TN 37996-1200
USA
Tele: 423-974-7802
Fax: 423-974-7843
wreviol@utk.edu

RILEY, M. A.
Florida State University
Department of Physics
Tallahassee, Fl 32306
USA
Tele: 850-644-2066
Fax: 850-644-8630
mriley@nucott.physics.fsu.edu

RUDOLPH, D.
Lund University
Division of Cosmic and Subatomic Physics
Professorsgatan 1
S-223 63 Lund
SWEDEN
Tele: 46 46 222 7633
Fax: 46 46 222 4015
dirkr@alpha.kosufy.lu.se

SAGAWA, H.
University of Aizu
Center for Mathematical Sciences
Ikki-machi Aitsu-Wakamatsu
Fukushima 965 -80
JAPAN
Tele: 81 242 37 2725
Fax: 81 242 37 2752
sagawa@u-aizu.ac.jp

SATULA, W.
Inst of Theoretical Physics
University of Warsaw
ul. Hoza 69
00-681 Warsaw
POLAND
Tele: 48 22 628 3396
Fax: 48 22 621 9475
satula@fuw.edu.pl

SCHNARE, H.
Forschungszentrum Rossendorf
Postfach 51 01 19
Dresden D-01314
GERMANY
Tele: 49 351 260 3352
Fax: 49 351 260 3700
schnare@fz-rossendorf.de

RYKACZEWSKI, K. P.
Oak Ridge National Laboratory
Physics Division
P. O. Box 2008, Bldg. 6000, MS6371
Oak Ridge, TN 37831-6371
Tele: 423 576-2636
Fax: 423 574-8902
rykaczew@phy.ornl.gov

SARANTITES, D. G.
Washington University
Department of Chemistry
University City
St. Louis. MO 63130
USA
Tele: 314-935-6504
Fax: 314-935-6184
dgs@proton.wustl.edu

SCHALY, B.
McMaster University
Department of Physics & Astronomy
1280 Main Street, West
Hamilton L8S 4M1 Ontario
CANADA
Tele: 905-525-9140 Ext 27458
Fax: 905-546-1252
schalyb@physics. mcmaster.ca

SCHWARTZ, J.
Yale University/Argonne National Laboratory
Department of Physics
9700 S. Cass Avenue
Argonne, IL 60439-4803
USA
Tele: 630-252-9114
Fax: 630-252-2864
schwartz@anph04.phy.anl.gov

SEMMES, P. B.
Tennessee Technological University
Department of Physics
Box 5051
Cookeville, TN 38505
USA
Tele: 931 372-3145
Fax: 931 372-6351
psemmes@tntech.edu

SHERRILL, B.
National Superconducting Cyclotron Lab
Michigan State University
South Shaw Lane
East Lansing, MI 48824-1321
USA
Tele: 517-633-6322
517-353-5967
sherrill@nscl.msu.edu

SMITH, A. G.
University of Manchester
Department of Physics and Astronomy
Brunswick Street
Manchester M13 9PL
UNITED KINGDOM
Tele: 44 161 275 4156
Fax: 44 161 272 5509
gavin.smith@man.ac.uk

SMITH, M. S.
Oak Ridge National Laboratory
Physics Division
P. O. Box 2008, Bldg 6010, MS6354
Oak Ridge, TN 37831-6354
USA
Tele: 423-576-5037
Fax: 423-576-8746
msmith@orph01.phy.ornl.gov

SHAPIRA, D.
Oak Ridge National Laboratory
Physics Division
P. O. BOX 2008, Bldg 6000, MS6368
Oak Ridge, TN 37831-6368
USA
Tele: 423-576-2548
Fax: 423-574-1268
shapira@orph01.phy.ornl.gov

SHIMIZU, Y. R.
Kyushu University
Department of Physics
6-10-1 Hakozaki, Higashi-ku
Fukuoka 812
JAPAN
Tele: 81-92-642-2558
Fax: 81-92-642-2553
yrsh2scp@mbox.nc.kyushu-u.ac.jp

SMITH, J. F.
Nuclear Physics Group
Schuster Laboratory
The University of Manchester
Manchester M13 9PL
UNITED KINGDOM
Tele: 44 161 275 4155
Fax: 44 161 275 5509
jfs@mags.ph.man.ac.uk

SONA-BIZZETI, A. M.
University of Florence
Dipartimento di Fisica
Largo E. Fermi 2
I-50125 Firenze
ITALY
Tele: 39 055 230 7641
Fax: 39 055 229 330
annamaria@fi.infn.it

STAROSTA, K.
SUNY at Stony Brook
Department of Physics
Stony Brook, NY 11794-3800
USA
Tele: 516-632-7023
Fax: 516-632-8573
starosta@nuclear.physics.sunysb.edu

STEVENSON, P. D.
Oak Ridge National Laboratory
Physics Division
P. O. Box 2008, Bldg 6003, MS6373
Oak Ridge, TN 37831-6373
USA
Tele: 423-576-4295
Fax: 423-574-4745
paul.stevenson@balliol.ox.ac.uk

STRAYER, M. R.
Oak Ridge National Laboratory
Physics Division
P. O. Box 2008, Bldg 6003, MS6373
Oak Ridge, TN 37831-6373
USA
Tele: 423-574-4590
Fax: 423-574-4745
strayermr@ornl.gov

TAIN, J. L.
CSIC
Instituto de Fisica Corpuscular
Avda Dr. Moliner 50
Burjassot Valencia E-46111
SPAIN
Tele: 34 96 386 4581
Fax: 34 96 386 4583
tain@evala4.ific.uv.es

STEPHENS, F. S.
Lawrence Berkeley National Laboratory
Nuclear Science Division
Bldg 88-252, 1 Cyclotron Road
Berkeley, CA 94720
USA
Tele: 510-486-5724
Fax: 510-486-7983
fss@lbl.gov

STRACENER, D.
Oak Ridge National Laboratory
Physics Division
P. O. Box 2008, Bldg 6000, MS6368
Oak Ridge, TN 37831-6368
USA
Tele: 423-574-4725
Fax: 423-574-1268
stracener@orph01.phy.ornl.gov

SVENSSON, C. E.
McMaster University
Department of Physics & Astronomy
1280 Main Street West
Hamilton L8S 4M1 Ontario
CANADA
Tele: 905 525 9140 Ext 27457
Fax: 905 546 1252
sven@poppy.physics.mcmaster.ca

TANAKA, T.
IPNS, HEARO
Theory Group
Midori-Machi 3-2-1
Tanashi Tokyo 188-8501
JAPAN
Tele: 81 424 69 9543
Fax: 81 424 62 0763
tanakat@thaxp1.tanashi.kek.jp

THOENNESSEN, M.
Michigan State University
Cyclotron Laboratory
S. Shaw Lane, Bldg. 164
East Lansing, MI 48824
USA
Tele: 517-333-6323
Fax: 517-353-5967
thoennessen@nscl.msu.edu

VARMETTE, P. G.
The Niels Bohr Institute
Physics Department
Risφ
Roskilde DK-4000
DENMARK
Tele: 45 46 77 53 66
Fax: 45 46 37 35 16
pgv@talws10.nbi.dk

VON BRENTANO, P.
University of Koln
Institut fur Kernphysik
Zulpicher Str. 77
Koln 41 D-50937
GERMANY
Tele: 49-221-470-3456
Fax: 49-221-470-5168
brentano@ikp.uni-koeln.de

WADDINGTON, J. C.
McMaster University ABB241
Dept of Phys & Astronomy
1280 Main Street West
Hamilton L8S 4M1 Ontario
CANADA
Tele: 905-525-9140 Ext.23635
Fax: 905-546-1252
jcw@physics.mcmaster.ca

TOTH, K. S.
Oak Ridge National Laboratory
Physics Division
P. O. Box 2008, Bldg 6000, MS6371
Oak Ridge, TN 37831-6371
USA
Tele: 423 574 4732
Fax: 423 574 1268
toth@mail.phy.ornl.gov

VERTSE, T.
Institute of Nuclear Research
Hungarian Academy of Sciences
P. O. Box 51
Debrecen H-4001
HUNGARY
Tele: 36 52 417266 Ext 1327
Fax: 36 52 416181
vertse@tigris.klte.hu

VRETENAR, D.
University of Zagreb
Physics Department
Bijenicka 32
Zagreb 10 000
CROATIA
dario_vretenar@physik.tu-muenchen.de

WADSWORTH, R.
University of York
Department of Physics
York Y010 5DD
UNITED KINGDOM
Tele: 44 1904 432242
Fax: 44 1904 432214
oew@yksc.york.ac.uk

WHITE, J. A.
California Institute of Technology
Physics Department
1200 E. California Blvd, MC-106-38
Pasadena, CA 91125
USA
Tele: 626-395-4666
Fax: 626-564-8708
jody@krl.caltech.edu

WIESCHER, M.
Department of Physics
University of Notre Dame
225 Nieuwland Science Hall
Notre Dame, IN 46556
USA
Tele: 219-631-6788
Fax: 219-631-5952
wiescher.1@nd.edu

WOLLNIK, H.
Institut fur Kernphysik
Strahlenzentrum
Leihgsterner Wef 217
D-6300 Giessen
GERMANY
Tele: 49 641 99 33440
Fax: 49 64 17 022 672
wollnik@piggy.physik.uni-giessen.de

WU, C.-L.
Chung Yuan University
Physics Department
Chung-Li Taiwan
PRC
Tele: 886 3 456 3171 Ext 3223

WIEDENHOEVER, I.
Argonne National Laboratory
Department of Physics
9700 S. Cass Avenue
Argonne, IL 60439
USA
Tele: 630-252-6870
Fax: 630-252-6210
wiedenhoever@anlphy.phy.anl.gov

WINCHELL, D.
Brookhaven National Laboratory
National Nuclear Data Center
Bldg 197D
Upton, NY 11973-5000
USA
Tele: 516-344-5081
Fax: 516-344-2806
winchell@bnl.gov

WOODS, P. J.
University of Edinburgh
Physics Dept.
James Clerk Maxwell Bldg. Kings Bldgs.
Edinburgh EH9 3JZ
UNITED KINGDOM
Tele: 131 650 5287
Fax: 131 650 5212
pjw@np.ph.ed.ac.uk

WYSS, R. A.
Royal Institute of Technology (KTH)
Physics Department
Frescativagen 24
Stockholm S-10405
SWEDEN
Tele: 46 8 161107
Fax: 46 8 158674
wyss@msi.se

YATES, S. W.
University of Kentucky
Department of Chemistry
Lexington, KY 40506
USA
Tele: 606-257-7085
Fax: 606-323-1069
yates@pop.uky.edu

YU, C.-H.
Oak Ridge National Laboratory
Physics Division
P. O. Box 2008, Bldg 6000, MS6371
Oak Ridge, TN 37831-6371
USA
Tele: 423-574-4493
Fax: 423-574-1268
chy@mail.phy.ornl.gov

YOUNES, W.
Lawrence Livermore National Laboratory
North Division
7000 E. Avenue
Livermore, CA 94550
USA
Tele: 925-423-1873
Fax: 925-422-0883
younes@llnl.gov

ZHANG, J.-Y.
University of Tennessee
Department of Physics
401 Physics Building
Knoxville, TN 37996-1200
USA
Tele: 423-974-7803
Fax: 423-974-7843
jingye@utk.edu

A

Achouri, N. L., 243, 257
Aengenvoort, B., 412
Afanasjev, A., 358
Agramunt, J., 534
Ahmad, I., 121, 283, 527
Akovali, Y., 216
Alcorta, M., 121
Algora, A., 283, 534
Amro, H., 527
Amzal, N., 121
Angelique, J. C., 243, 257
Aprahamian, A., 545
Archer, D. E., 464
Astier, A., 368
Azaiez, F., 243

B

Bagshaw, A. P., 407
Baiborodin, D., 257
Baktash, C., 168, 353
Banci Buonamici, F., 493
Bargioni, L., 493
Bark, R. A., 412
Barrett, B. R., 18
Batchelder, J. C., 216, 232
Batist, L., 534
Bazzacco, D., 368, 493
Becker, J. A., 464
Belleguic, M., 243
Bender, M., 81, 554
Benlliure, J., 294
Bentida, R., 257
Bentley, M. A., 407
Béraud, R., 257
Bergström, M., 412
Bernstein, L. A., 464
Beyer, C. J., 570
Billowes, J., 283
Bingham, C. R., 216, 232, 257
Birriel, I., 168
Bizzeti, P. G., 493
Bizzeti-Sona, A. M., 493
Blumenfeld, Y., 253
Böckstiegel, C., 294
Borcea, C., 243, 257
Borcea, R., 534
Boston, A., 358
Bourgeois, C., 243
Bracco, A., 368, 412

C

Cano-Ott, D., 534
Carpenter, M. P., 121, 232, 393, 527
Casten, R. F., 439
Catford, W. N., 257
Cederwall, B., 368
Chapman, R., 412
Chewter, A. J., 121
Chiara, C. J., 358
Chowdhury, P., 527
Chromik, M. J., 187
Cizewski, J. A., 121, 232, 527
Clark, R. M., 168, 353, 358, 383, 503, 517
Clerc, H.-G., 294
Cline, D., 527
Cocks, J. F. C., 226
Collatz, R., 534
Coraggio, L., 56
Cottle, P. D., 253
Covello, A., 56
Cromaz, M., 570
Cullen, D. M., 407, 412
Cunningham, R. A., 168

D

Dagnall, P. J., 283, 407
Daugas, J. M., 243, 257
Davids, C. N., 121
Davinson, T., 216
Dean, D. J., 28
de Angelis, G., 177, 368, 493
De Coster, C., 429
Decroix, B., 429
de France, G., 257
de Jong, M., 294
Deleplanque, M. A., 517
Deloncle, I., 243, 368
De Oliveira, F., 243, 257
De Poli, M., 368, 493
Devlin, M., 168, 353, 358, 383
Dewald, A., 493
Diamond, R. M., 517
Ding, K. Y., 121
Dlouhy, Z., 243

Brown, T. B., 383
Bruyneel, B., 264
Bürvenich, T., 81
Butler, P. A., 121
Byrne, A. P., 304